A Level Physics for OCR

A

Series Editor
Gurinder Chadha

Authors
Graham Bone
Gurinder Chadha
Nigel Saunders

OXFORD
UNIVERSITY PRESS

OXFORD
UNIVERSITY PRESS

Great Clarendon Street, Oxford, OX2 6DP, United Kingdom

Oxford University Press is a department of the University of Oxford. It furthers the University's objective of excellence in research, scholarship, and education by publishing worldwide. Oxford is a registered trade mark of Oxford University Press in the UK and in certain other countries

First published in 2015

British Library Cataloguing in Publication Data
Data available

978-0-19-835218-1

11

Paper used in the production of this book is a natural, recyclable product made from wood grown in sustainable forests. The manufacturing process conforms to the environmental regulations of the country of origin.

Printed in Great Britain by Bell and Bain Ltd, Glasgow

This resource is endorsed by OCR for use with specification H156 AS Level GCE Physics A and H556 A Level GCE Physics A. In order to gain endorsement this resource has undergone an independent quality check. OCR has not paid for the production of this resource, nor does OCR receive any royalties from its sale. For more information about the endorsement process please visit the OCR website www.ocr.org.uk

MIX
Paper from responsible sources
FSC
www.fsc.org
FSC® C007785

AS/A Level course structure

This book has been written to support students studying for OCR A Level Physics A. It covers all A Level modules from the specification. These are shown in the contents list, which also shows you the page numbers for the main topics within each module. There is also an index at the back to help you find what you are looking for. If you are studying for OCR AS Physics A, you will only need to know the content in the blue box.

AS exam

Year 1 content

1 Development of practical skills in physics
2 Foundations in physics
3 Forces and motion
4 Electrons, waves, and photons

Year 2 content

5 Newtonian world and astrophysics
6 Particles and medical physics

A level exam

A Level exams will cover content from Year 1 and Year 2 and will be at a higher demand. You will also carry out practical activities throughout your course.

Learning outcomes

- → At the beginning of each topic, there is a list of learning outcomes.
- → These are matched to the specification and allow you to monitor your progress.
- → A specification reference is also included.
  Specification reference: 2.1.3

Study Tips

Study tips contain prompts to help you with your understanding and revision.

Synoptic link

These highlight the key areas where topics relate to each other. As you go through your course, knowing how to link different areas of physics together becomes increasingly important. Many exam questions will require you to bring together your knowledge from different areas.

This book contains many different features. Each feature is designed to support and develop the skills you will need for your examinations, as well as foster and stimulate your interest in physics.

Terms that you will need to be able to define and understand are highlighted by **bold text**.

Application features

These features contain important and interesting applications of physics in order to emphasise how scientists and engineers have used their scientific knowledge and understanding to develop new applications and technologies. There are also practical application features, with the icon, to support further development of your practical skills.

1 All application features have a question to link to material covered with the concept from the specification.

Extension features

These features contain material that is beyond the specification. They are designed to stretch and provide you with a broader knowledge and understanding and lead the way into the types of thinking and areas you might study in further education. As such, neither the detail nor the depth of questioning will be required for the examinations. But this book is about more than getting through the examinations.

1 Extension features also contain questions that link the off-specification material back to your course.

Summary Questions

1 These are short questions at the end of each topic.

2 They test your understanding of the topic and allow you to apply the knowledge and skills you have acquired.

3 The questions are ramped in order of difficulty. Lower-demand questions have a paler background, with the higher-demand questions having a darker background. Try to attempt every question you can, to help you achieve your best in the exams.

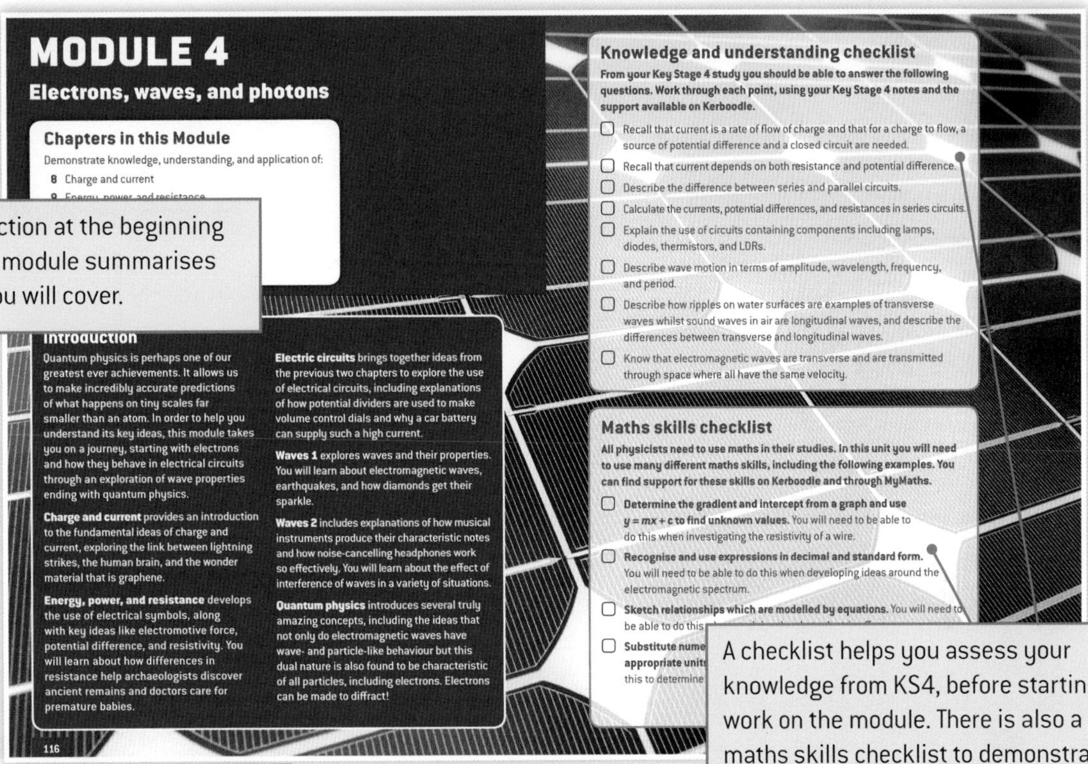

MODULE 4
Electrons, waves, and photons
Chapters in this Module
Demonstrate knowledge, understanding, and application of:
8 Charge and current
Introduction
Quantum physics is perhaps one of our greatest ever achievements. It allows us to make incredibly accurate predictions of what happens on tiny scales far smaller than an atom. In order to help you understand its key ideas, this module takes you on a journey, starting with electrons and how they behave in electrical circuits through an exploration of wave properties ending with quantum physics.
Charge and current provides an introduction to the fundamental ideas of charge and current, exploring the link between lightning strikes, the human brain, and the wonder material that is graphene.
Energy, power, and resistance develops the use of electrical symbols, along with key ideas like electromotive force, potential difference, and resistivity. You will learn about how differences in resistance help archaeologists discover ancient remains and doctors care for premature babies.
Electric circuits brings together ideas from the previous two chapters to explore the use of electrical circuits, including explanations of how potential dividers are used to make volume control dials and why a car battery can supply such a high current.
Waves 1 explores waves and their properties. You will learn about electromagnetic waves, earthquakes, and how diamonds get their sparkle.
Waves 2 includes explanations of how musical instruments produce their characteristic notes and how noise-cancelling headphones work so effectively. You will learn about the effect of interference of waves in a variety of situations.
Quantum physics introduces several truly amazing concepts, including the ideas that not only do electromagnetic waves have wave- and particle-like behaviour but this dual nature is also found to be characteristic of all particles, including electrons. Electrons can be made to diffract!
116
Knowledge and understanding checklist
From your Key Stage 4 study you should be able to answer the following questions. Work through each point, using your Key Stage 4 notes and the support available on Kerboodle.
Recall that current is a rate of flow of charge and that for a charge to flow, a source of potential difference and a closed circuit are needed.
Recall that current depends on both resistance and potential difference.
Describe the difference between series and parallel circuits.
Calculate the currents, potential differences, and resistances in series circuits.
Explain the use of circuits containing components including lamps, diodes, thermistors, and LDRs.
Describe wave motion in terms of amplitude, wavelength, frequency, and period.
Describe how ripples on water surfaces are examples of transverse waves whilst sound waves in air are longitudinal waves, and describe the differences between transverse and longitudinal waves.
Know that electromagnetic waves are transverse and are transmitted through space where all have the same velocity.
Maths skills checklist
All physicists need to use maths in their studies. In this unit you will need to use many different maths skills, including the following examples. You can find support for these skills on Kerboodle and through MyMaths.
Determine the gradient and intercept from a graph and use y = mx + c to find unknown values. You will need to be able to do this when investigating the resistivity of a wire.
Recognise and use expressions in decimal and standard form. You will need to be able to do this when developing ideas around the electromagnetic spectrum.
Sketch relationships which are modelled by equations.

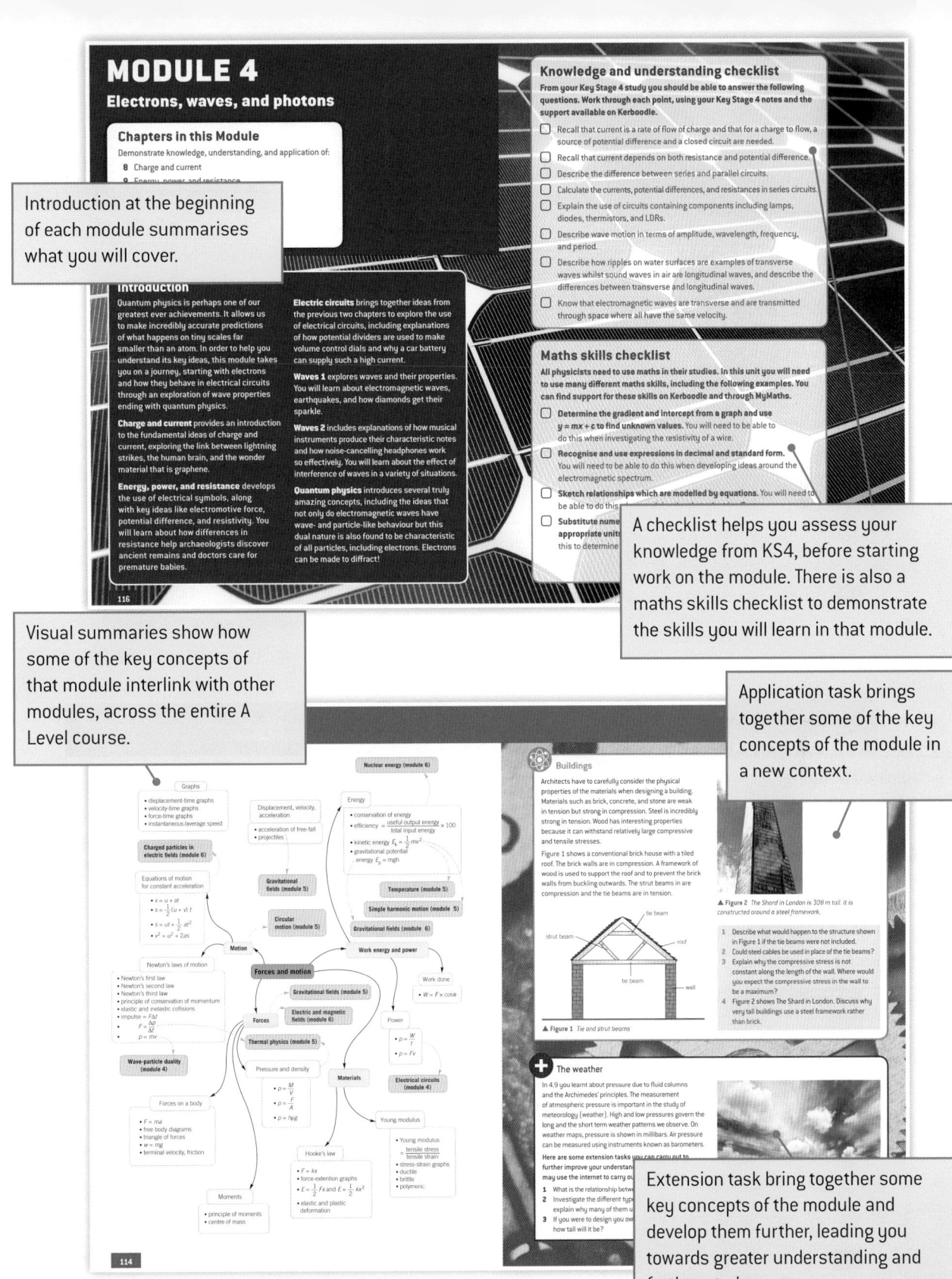

Introduction at the beginning of each module summarises what you will cover.

A checklist helps you assess your knowledge from KS4, before starting work on the module. There is also a maths skills checklist to demonstrate the skills you will learn in that module.

Visual summaries show how some of the key concepts of that module interlink with other modules, across the entire A Level course.

Application task brings together some of the key concepts of the module in a new context.

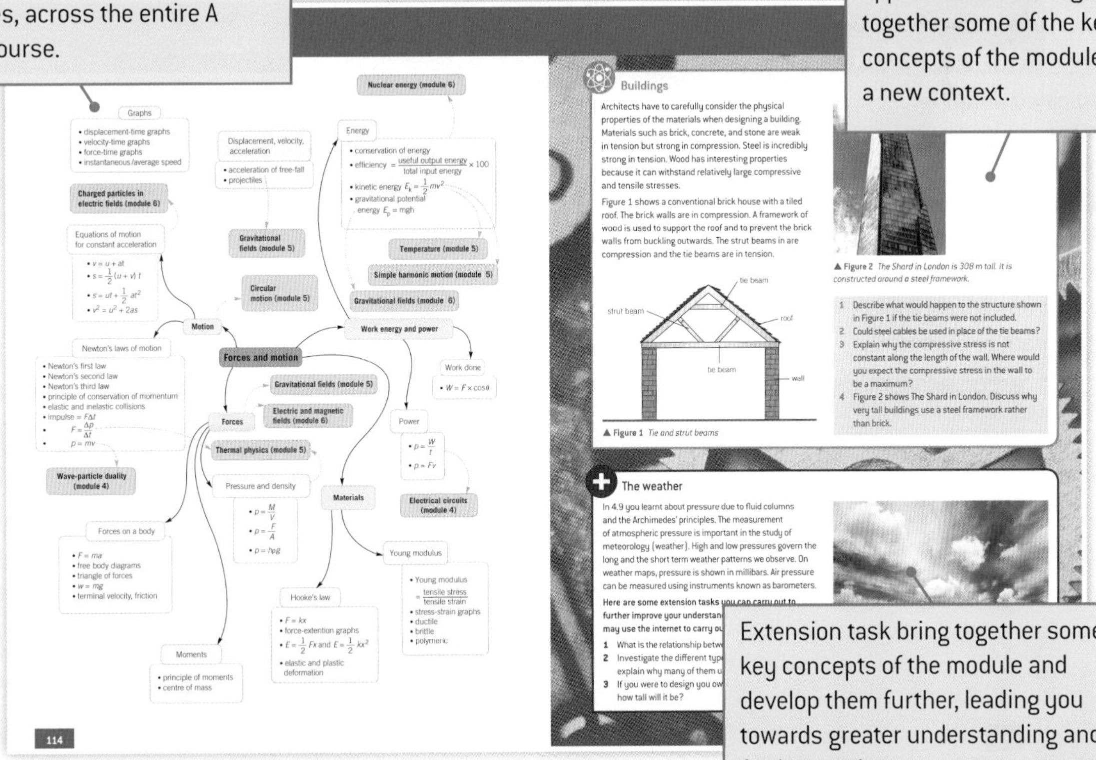

Extension task bring together some key concepts of the module and develop them further, leading you towards greater understanding and further study.

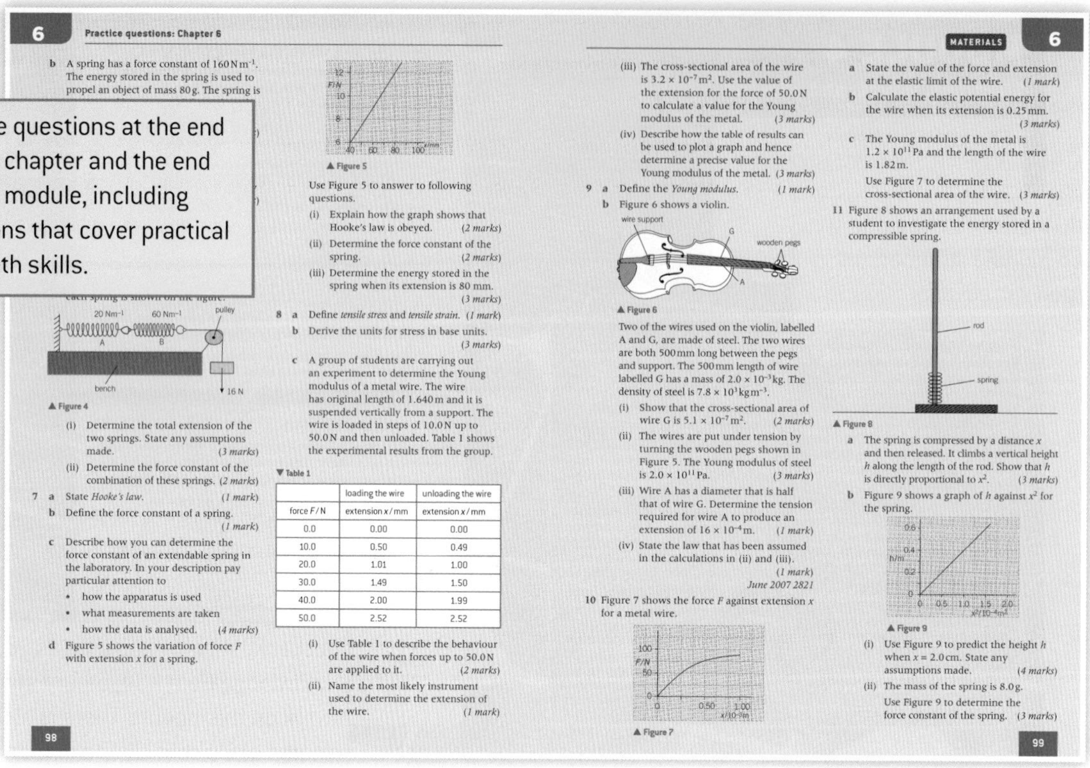

Practice questions at the end of each chapter and the end of each module, including questions that cover practical and math skills.

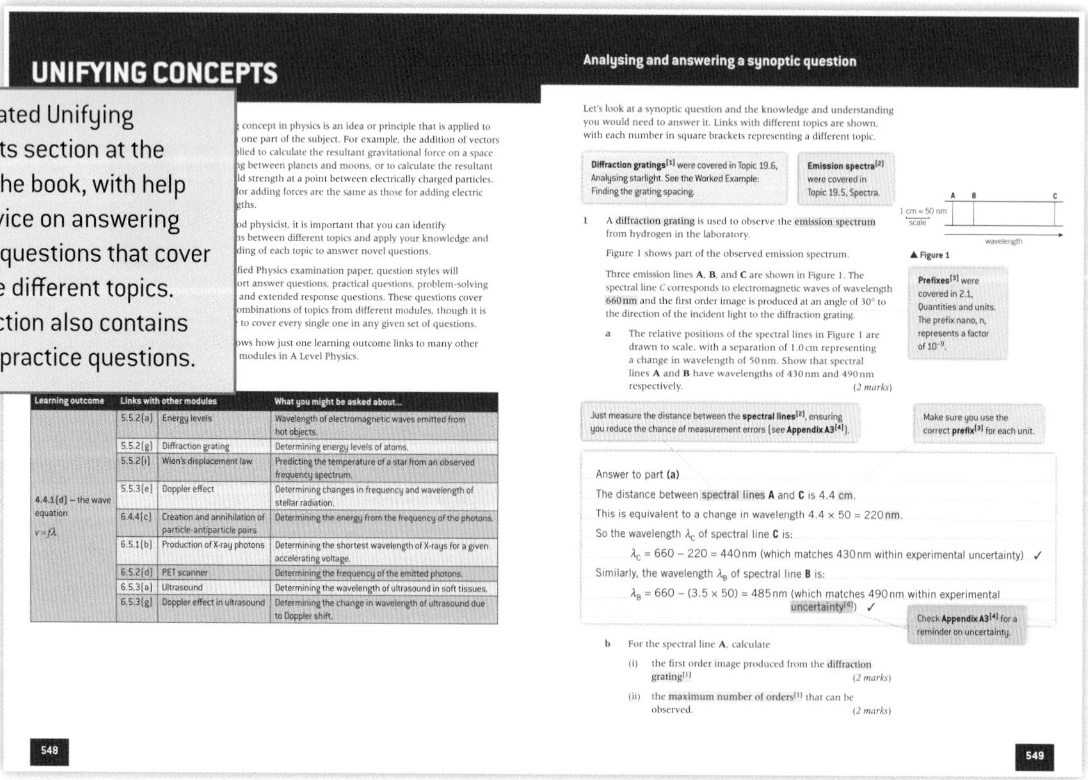

A dedicated Unifying Concepts section at the end of the book, with help and advice on answering unified questions that cover multiple different topics. This section also contains further practice questions.

This book is supported by next generation Kerboodle, offering unrivalled digital support for independent study, differentiation, assessment, and the new practical endorsement.

If your school subscribes to Kerboodle, you will also find a wealth of additional resources to help you with your studies and with revision.

- Study guides
- Maths skills boosters and calculation worksheets
- On your marks activities to help you achieve your best
- Practicals and follow up activities to support the practical endorsement
- Interactive objective tests that give question-by-question feedback
- Animations and revision podcasts
- Self-assessment checklists

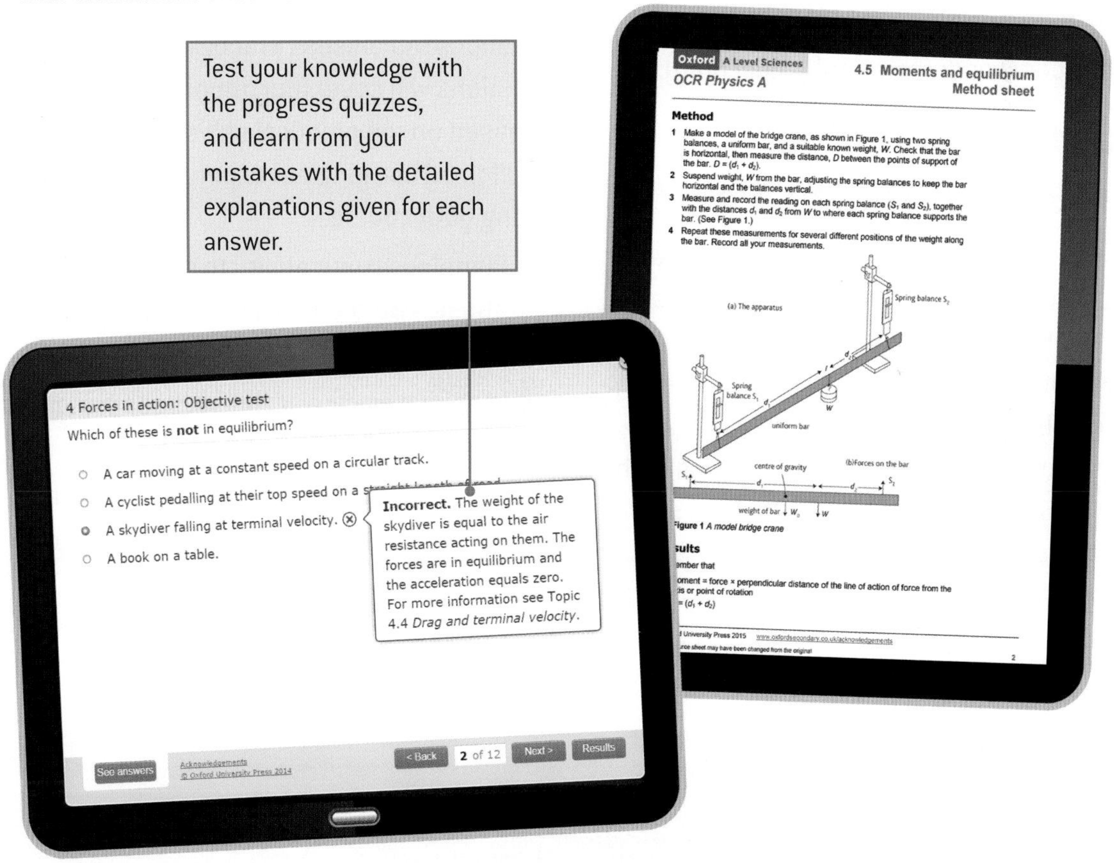

For teachers, Kerboodle also has plenty of further assessment resources, answers to the questions in the book, and a digital markbook along with full teacher support for practicals and the worksheets, which include suggestions on how to support and stretch students. All of the resources are pulled together into teacher guides that suggest a route through each chapter.

MODULE 1

Development of practical skills in physics

Physics is a practical subject and experimental work provides you with important practical skills, as well as enhancing your understanding of physical theory. You will be developing practical skills by carrying out practical and investigative work in the laboratory throughout both the AS and the A level Physics course. You will be assessed on your practical skills in two different ways:

- written examinations (AS and A level)
- practical endorsement (A level only)

Practical coverage throughout this book

Practical skills are a fundamental part of a complete education in science, and you are advised to keep a record of your practical work from the start of your A level course that you can later use as part of your practical endorsement. You can find more details of the practical endorsement from your teacher or from the specification.

In this book and its supporting materials practical skills are covered in a number of ways. By studying Application boxes and Exam-style questions in this student book, and by using the Practical activities and Skills sheets in Kerboodle you will have many opportunities to learn about the scientific method and carry-out practical activities.

1.1 Practical skills assessed in written examinations

In the written examination papers for AS and A level, at least 15% of the marks will be from questions that assess practical skills. The questions will cover four important skill areas, all based on the practical skills that you will develop by carrying out experimental work during your course.

- Planning – your ability to solve a physics problem in a practical context.
- Implementing – your understanding of important practical techniques and processes.
- Analysing – your interpretation of experimental results set in a practical context and related to the experiments that you would have carried out.
- Evaluating – your ability to develop a plan that is fit for the intended purpose.

1.1.1 Planning

- Designing experiments
- Identifying variables to be controlled
- Evaluating the experimental method

Skills checklist

- ☐ Selecting apparatus and equipment
- ☐ Selecting appropriate techniques
- ☐ Selecting appropriate quantities of materials and substances and scale of working
- ☐ Solving physical problems in a practical context
- ☐ Applying physics concepts to practical problems

1.1.2 Implementing

- Using a range of practical apparatus
- Carrying out a range of techniques
- Using appropriate units for measurements
- Recording data and observations in an appropriate format

Skills checklist

- [ ] Understanding practical techniques and processes
- [ ] Identifying hazards and safe procedures
- [ ] Using SI units
- [ ] Recording qualitative observations accurately
- [ ] Recording a range of quantitative measurements
- [ ] Using the appropriate precision for apparatus

1.1.3 Analysis

- Processing, analysing, and interpreting results
- Analysing data using appropriate mathematical skills
- Using significant figures appropriately
- Plotting and interpreting graphs

Skills checklist

- [ ] Analysing qualitative observations
- [ ] Analysing quantitative experimental data, including
  - calculation of means
  - amount of substance and equations
- [ ] For graphs,
  - selecting and labelling axes with appropriate scales, quantities, and units
  - drawing tangents and measuring gradients

1.1.4 Evaluation

- Evaluating results to draw conclusions
- Identify anomalies
- Explain limitations in method
- Identifying uncertainties and errors
- Suggesting improvements

Skills checklist

- [ ] Reaching conclusions from qualitative observations
- [ ] Identifying uncertainties and calculating percentage errors
- [ ] Identifying procedural and measurement errors
- [ ] Refining procedures and measurements to suggest improvements

1.2 Practical skills assessed in practical endorsement

You will also be assessed on how well you carry out a wide range of practical work and how to record the results of this work. These hands-on skills are divided into 12 categories and form the practical endorsement. This is assessed for A level Physics qualification only.

The endorsement requires a range of practical skills from both years of your course. If you are taking only AS Physics, you will not be assessed through the practical endorsement but the written AS examinations will include questions that relate to the skills that naturally form part of the AS common content to the A level course.

The practicals you do as part of the endorsement will not contribute to your final grade awarded to you. However, these practicals must be covered and your teacher will go through how this is to be done in class. It is important that you are actively involved in practical work because it will help you with understanding the theory and also how to effectively answer some of the questions in the written papers.

The practical activities you will carry out in class are divided into Practical Activity Group (PAGs). PAG1 to PAG 6 will be undertaken in Year 1, PAG7 to PAG 10 in Year 2, and PAG11 to PAG 12 throughout the two-year course.

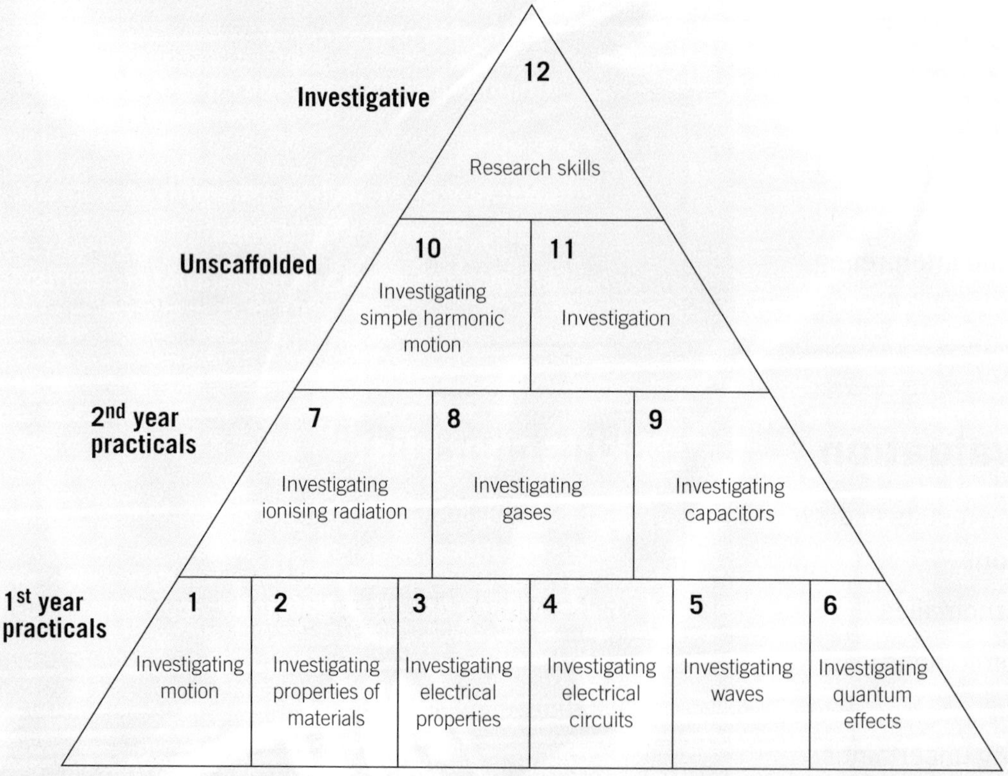

The PAGs are summarised below, together with the topic reference in the book that relates to the specific PAG.

PAG	Topic reference
1 Investigating motion	3.7, 4.4
2 Investigating properties of materials	6.1, 6.4
3 Investigating electrical properties	9.5, 9.7
4 Investigating electrical circuits	10.4, 10.6
5 Investigating waves	11.2, 11.7, 12.2–12.6
6 Investigating quantum effects	13.1
7 Investigating ionising radiation	25.4
8 Investigating gases	15.2
9 Investigating capacitors	21.4
10 Investigating simple harmonic motion	17.1
11 Investigation	Throughout
12 Research skills	Throughout

Maths skills and How Science Works across Module 1

In order to develop your knowledge and understanding in A Level Physics, it is important to have specific skills in mathematics. All the mathematical skills you will need during your physics course have been embedded into the individual topics for you to learn as you meet them. An overview is available in each of the module openers and these skills are further supported by the worked examples, summary questions, and examination-style questions.

How Science Works (HSW) is another area required for success in A Level Physics, and helps you to put science in a wider context, helping you to develop your critical and creative thinking skills in order to solve problems in a variety of contexts. Once again, this has been embedded into the individual topics covered in the books, particularly in application boxes and examination-style questions. The application and extension boxes cover some of the HSW elements.

You can find further support for maths skills and HSW on Kerboodle.

Topics in this module

2.1 Quantities and units

2.2 Derived units

2.3 Scalar and vector quantities

2.4 Adding vectors

2.5 Resolving vectors

2.6 More on vectors

Appendices

A1 Physical quantities and units

A2 Recording results

A3 Measurements and uncertainties

Introduction

Physics is not only a collection of concepts about everything from subatomic particles to the whole Universe. It is a set of different ways of thinking that have led to countless successful descriptions and explanations of the way the Universe works. Physicists have learned new ways of thinking as they look for deeper and deeper explanations of physical phenomena, searching for the most fundamental answer that can be applied across the widest range of disciplines.

Foundations of physics introduces the important ideas and conventions that permeate the fabric of physics. You will develop your skills in critical thinking, reasoning and logic, and mathematics. With these you will be able to build models to describe a wide variety of systems and to make predictions about different circumstances.

Through an exploration of **units** you will learn about the well-defined and universally understood methods used by physicists to measure physical phenomena, and methods that help physicists across the globe effectively communicate their ideas within the scientific community.

By developing your understanding of **vectors** you will build a powerful mathematical toolkit that you will use throughout your studies.

You will hone your ability to make approximations and **estimations** in order to gain a sense of magnitudes and to know what sort of answers to expect. A study of **errors, uncertainty, precision, and accuracy** develops your understanding of the limitations of experimentation. You will learn how to present your data appropriately and express numerically a level of confidence in your findings.

Some of these key ideas and skills relate to and will be developed by experimental work that you will meet across the different modules – you will find these topics covered in detail in the appendices.

Knowledge and understanding checklist

From your Key Stage 4 study you should be able to do the following. Work through each point, using your Key Stage 4 notes and the support available on Kerboodle.

- ☐ Use vector diagrams to illustrate forces, a net force, and equilibrium situations.
- ☐ Explain the vector–scalar distinction as it applies to displacement, distance, velocity, and speed.
- ☐ Make calculations using ratios and proportional reasoning to convert units.

Maths skills checklist

All physicists use maths. In this unit you will need to use many different maths skills, including the following examples. You can find support for these skills on Kerboodle and through MyMaths.

- ☐ **Recognise and make use of appropriate units in calculations.** You will need to be able to do this when identifying the correct units for physical properties and converting between units with different prefixes, for example, km and m.
- ☐ **Use calculators to find and use power functions.** You will need to be able to do this to calculate resultant vectors.
- ☐ **Use sin, cos, and tan in physical problems.** You will need to be able to do this when resolving vectors into components.
- ☐ **Visualise and represent two-dimensional and three-dimensional forms, including 2D representations of 3D objects.** You will need to be able to do this to solve problems involving the addition of vectors through the use of scale diagrams.
- ☐ **Use Pythagoras' theorem and the angle sum of a triangle.** You will need to be able to do this to solve various problems involving vectors.
- ☐ **Estimate results.** You will need to be able to do this when estimating the effect of changing a value during an experiment.
- ☐ **Find arithmetic means.** You will need to be able to do this when presenting data in tables from various pieces of experimental work.
- ☐ **Identify uncertainties in measurements and use simple techniques to determine uncertainty when data is combined.** You will need to be able to do this when analysing experimental data.

MyMaths.co.uk Bringing Maths Alive

2 FOUNDATIONS OF PHYSICS

2.1 Quantities and units

Specification reference: 2.1.1, 2.1.2

Learning outcomes

Demonstrate knowledge, understanding, and application of:

→ units for physical quantities

→ SI base quantities and units, their symbols and prefixes.

▲ Figure 1 *The correct use of units would have prevented the destruction of the Mars Climate Orbiter*

Measurements

Measurements are very important in physics. Not only must they be recorded accurately, they must also be communicated clearly. In 1998 NASA launched the Mars Climate Orbiter, a mission costing almost £195 million. When the probe arrived at Mars a few months later, it disintegrated in the planet's upper atmosphere instead of going into orbit. The disaster had a simple cause: one of NASA's teams worked in feet and pounds, whilst the other team worked in metres and kilograms. Each team assumed that the other was using the same units.

In A Level Physics, failure to use units correctly may not cost millions of pounds but it will cost you valuable marks in the examination.

Quantities

A physical **quantity** is a property of an object or of a phenomenon that can be measured. Some quantities are just numbers. For example, proton number, efficiency, and magnification are numbers. They have a numerical magnitude or size, but no units. Many other quantities consist of numbers *and* units. For example, length is a quantity that has units. It has many different units, including metres, inches, and miles. To avoid problems like the one NASA experienced with the Mars Climate Orbiter, scientists use a standard system of units called the *Système International d'Unités* (International System of Units), abbreviated to **SI**.

SI base units

SI is built around seven **base units**, six of which are shown in Table 1. The seventh unit, the unit for luminous intensity (the candela, cd), is not assessed in the A Level Physics course.

▼ Table 1 *SI base units*

Quantity	Base unit	Unit symbol
length	metre	m
mass	kilogram	kg
time	second	s
electric current	ampère	A
temperature	kelvin	K
amount of substance	mole	mol

Symbols

A unit symbol is written in lower case, for example, m rather than M for metres, unless the unit is named after a person. In that situation, its name still begins with a lower-case letter but its symbol has a capital letter. The unit of electric current is named after André-Marie Ampère, so its name is the ampère (often just amp) and its symbol is A.

Prefixes

SI uses prefixes to show multiples and fractions of units (Table 2). For example, km stands for kilometre. The **prefix** is the 'kilo', and the unit is the 'metre'.

Notice that, apart from k for kilo, the prefixes for multiples all have initial capitals. Similarly, the prefixes for fractions are all lower case (µ is the lower-case Greek letter mu).

▼ Table 2 *Prefixes for SI units*

Prefix name	Prefix symbol	Factor
peta	P	10^{15}
tera	T	10^{12}
giga	G	10^{9}
mega	M	10^{6}
kilo	k	10^{3}
deci	d	10^{-1}
centi	c	10^{-2}
milli	m	10^{-3}
micro	µ	10^{-6}
nano	n	10^{-9}
pico	p	10^{-12}
femto	f	10^{-15}

Worked example: Using prefixes

a Convert 1.25 kA into A.

1.25 kA = 1.25×10^{3} A (or 1250 A)

b Convert 234 µm into m.

234 µm = 234×10^{-6} m = 2.34×10^{-4} m

c Convert 0.567 s into ms.

There are 10^{3} ms in 1 s. To change from seconds to milliseconds, you have to *multiply* by a factor of 10^{3}.

Therefore, 0.567 s = 0.567×10^{3} = 567 ms

Study tip

Standard form is used to display very small or very large numbers in a scientific way. For scientific notation it is ideally expressed in the form $n \times 10^{m}$, where $1 < n < 10$, and m is an integer.

Summary questions

1 A student records the following figures in his notes: 60 cm and 40 ms.
   - **a** Name the two quantities being measured. *(2 marks)*
   - **b** Change these measurements into their base units. *(2 marks)*

2 **a** A collision between two molecules lasts for about 100 picoseconds. Write this time in seconds. *(1 mark)*
   - **b** A chemical bond is approximately 0.15 nanometres long. Write this length in metres. *(1 mark)*
   - **c** The Sun's core has a temperature of approximately 16 megakelvin. Write this temperature in kelvin. *(1 mark)*

3 Convert the following measurements to their base units. Write your answers in standard form.
   a 200 pm; **b** 0.40 Mm; **c** 35 µs; **d** 0.25 mA; **e** 756 ns. *(5 marks)*

4 There are 86 400 s in a day. Alternatively you could say there are 86.4 ks in a day.
   - **a** The distance by train from London to Edinburgh is 5.34×10^{5} m. What is this distance in km?
   - **b** The diameter of the Earth is 1.274×10^{7} m. What is this diameter in Mm?
   - **c** The thickness of a human hair is about 7.5×10^{-5} m. What is this thickness in µm?
   - **d** The electric current in a nerve cell is about 1.4×10^{-7} A. What is this current in nA? *(4 marks)*

Standard form

You can show small and large numbers in **standard form.**

For example, instead of writing 230 km or 230×10^{3} m, we could express this distance as 2.3×10^{5} m.

Write 45 ns (45×10^{-9} s) in standard form.

Study tip

Take care when you are writing prefixes and units. For example, ms means milliseconds, but Ms means megaseconds.

2.2 Derived units

Specification reference: 2.1.2

Learning outcomes

Demonstrate knowledge, understanding, and application of:

→ derived units of SI base units and the quantities that use them.

▼ Table 1 *Some derived units*

Derived quantity	Derived unit
area	m^2
volume	m^3
acceleration	$m\,s^{-2}$
density	$kg\,m^{-3}$

Study tip

You can determine derived units from the equation for the derived quantity. For example, for density, you need the equation that links density, mass, and length:

$$\text{density} = \frac{\text{mass}}{\text{volume}}$$

(where volume = length^3)

The derived unit for density is therefore the unit for mass (kg) divided by the unit for volume (m^3): $kg\,m^{-3}$.

▲ Figure 1 *Speed is measured in $m\,s^{-1}$, a derived unit in SI*

Beyond base units

The seven base units are used to measure the base quantities that they represent. However, there are many more quantities to measure than just mass, length, electric current, time, and the other three base quantities. For example, what are the units for speed and force? Quantities like these are called **derived quantities**. They use **derived units**, which can be worked out from the base units and the equations relating derived quantities to the base quantities. With derived units any quantity can be communicated.

Names and symbols

Derived units without special names

You already know some derived units. For example, the unit for speed is $m\,s^{-1}$. It comes from the equation that links average speed with two base quantities – distance and time.

$$\text{average speed} = \frac{\text{distance travelled}}{\text{time taken}}$$

Since m is the unit for distance, s is the unit for time, and we are *dividing* m by s, the derived unit for speed is m/s, written $m\,s^{-1}$ at A Level ($s^{-1} = \frac{1}{s}$). We write derived units like this because it is better for more complex units, such as the unit for specific heat capacity, $J\,kg^{-1}\,K^{-1}$, which is much clearer than J/(kg K).

Table 1 shows some derived units without any special names.

Derived units with special names

Some derived quantities are used so often that they have special names. SI has 22 derived units with special names and symbols, but you will not need to know them all for your physics course. Table 2 shows a small selection of these units.

▼ Table 2 *Some named derived units*

Derived quantity	Unit name	Unit symbol	Unit expressed in other SI units
force	newton	N	$kg\,m\,s^{-2}$
pressure	pascal	Pa	$N\,m^{-2}$
energy or work done	joule	J	N m
power	watt	W	$J\,s^{-1}$
electric potential difference	volt	V	$J\,C^{-1}$
electric resistance	ohm	Ω	$V\,A^{-1}$
electric charge	coulomb	C	A s
frequency	hertz	Hz	s^{-1}

SI units can be combined to form a huge range of other derived units. You may be familiar with some of these already. For example, the moment of a force is measured in newton metres, N m.

Temperature

The SI base unit for temperature is the kelvin, K. In everyday life you are likely to use a different unit for temperature, a derived unit called the degree Celsius, °C. To convert from °C to K you add 273, so 20°C is 293 K and 100°C is 373 K.

A difference of 1°C is the same as a difference of 1 K, so temperature *differences* do not need conversion. For example, if you warm some water from 20°C to 100°C its temperature increases by 80°C, which is also 80 K.

1. Converting from K to °C is equally simple. Convert 298 K to °C.
2. The degree Fahrenheit, °F, is a non-SI unit for temperature. To convert from °F to °C you subtract 32, multiply by 5 then divide by 9. For example, 68°F = $(68 - 32) \times \frac{5}{9}$ = 20°C. Deduce the temperature that has the same value, whether given in °F or in °C.

Summary questions

1. The unit of mass is the kg. Acceleration has the derived unit m s^{-2}. The force acting on an object can be determined using the equation force = mass × acceleration. Determine the derived unit for force in base units. *(2 marks)*

2. Use the equations given to determine the derived unit of each quantity in base units.
   a. $\text{force constant} = \dfrac{\text{force}}{\text{extension}}$

      Extension is the change in length. Determine the derived unit for force constant. *(2 marks)*
   b. work done = force × distance moved in direction of force

      Determine the derived unit for work done. *(2 marks)*
   c. $\text{pressure} = \dfrac{\text{force}}{\text{cross-sectional area}}$

      Determine the derived unit for pressure. *(2 marks)*

3. State the difference between 1 N m, 1 nm, 1 mN and 1 MN. *(3 marks)*

4. In electrical work, it is useful to define a quantity known as *number density* of free electrons. Number density of free electrons is the number of electrons per unit volume. What is the unit for number density in base units? *(2 marks)*

2.3 Scalar and vector quantities

Specification reference: 2.3.1

Learning outcomes

Demonstrate knowledge, understanding, and application of:

→ scalar and vector quantities.

Going up

Flyboarding is a sport in which the rider stands on a board with a long hose attached that hangs into a lake. Water from the lake is forced through the hose and into jets under the board. The water rushes out of the jet nozzles, pushing the rider into the air. Skilled flyboarders can perform all sorts of aerial acrobatics, thanks to practice in judging scalar and vector quantities.

▲ Figure 1 *Flyboarders can hover up to 15 m above the water*

Scalar quantities

A **scalar quantity** has magnitude (size) but no direction. For example, the *distance* between a flyboarder and the surface of the water is a scalar quantity, and so is his *mass* and the *time* he can stay in the air. Table 1 shows some examples of scalar quantities with their SI units.

▼ Table 1 *Some scalar quantities and units*

Scalar quantity	SI unit
length	m
mass	kg
time	s
speed	$\mathrm{m\,s^{-1}}$
temperature	K, °C
volume	$\mathrm{m^3}$
energy	J
potential difference	V
power	W

Adding and subtracting scalar quantities

Scalar quantities can be added together or subtracted from one another in the usual way. For example, if your mass is 55 kg and you pick up a 5 kg bag, your new total mass is (55 + 5) = 60 kg. If you sharpen a 16 cm pencil and remove 1 cm as you do so, the new length of the pencil is (16 – 1) = 15 cm.

Scalar quantities must have the same units when you add or subtract them. If you time something in an experiment you cannot add together 1 *minute* and 30 *seconds* as (1 + 30). Instead, you would convert the time from minutes into seconds and then add the times: (60 + 30) = 90 s. Alternatively, you could work in minutes to get a time of (1 + 0.5) = 1.5 minutes.

Multiplying and dividing scalar quantities

Scalar quantities can also be multiplied together or divided by one another. However, in this case the units can be the same or different, unlike adding and subtracting. It is important that you work out the final units correctly.

Worked example: Lighter than air

A balloon is inflated with $6.1 \times 10^{-3}\,\mathrm{m^3}$ of helium. Its mass increases by 0.98 g. Calculate the density of helium.

Step 1: The equation for density is

$$\text{density} = \frac{\text{mass}}{\text{volume}}$$

Step 2: Consider the units of the equation.

You are dividing together two scalar quantities. The SI base unit for mass is the kg. Volume has the unit $\mathrm{m^3}$. The mass must be converted into kg before substitution; mass = 9.8×10^{-4} kg. →

Step 3: Substitute the values into the equation and calculate the density.

$$\text{density} = \frac{9.8 \times 10^{-4}}{6.1 \times 10^{-3}} = 0.16\,\text{kg}\,\text{m}^{-3}$$

Vector quantities

A **vector quantity** has magnitude *and* direction. For example, the weight of a flyboarder is a vector quantity, and so is the force from the rushing water from the jet nozzles. Table 2 shows some examples of vector quantities and their SI units.

▼ **Table 2** *Some vector quantities and units*

Vector quantity	SI unit
displacement	m
velocity	$\text{m}\,\text{s}^{-1}$
acceleration	$\text{m}\,\text{s}^{-2}$
force	N ($\text{kg}\,\text{m}\,\text{s}^{-2}$)
momentum	$\text{kg}\,\text{m}\,\text{s}^{-1}$

Synoptic link

You find out more about vector quantities when studying motion, forces, and momentum in Chapters 3, 4, and 7 of this book.

Distance and displacement

Distance and displacement are both measured in m, but distance is a scalar quantity and displacement is a vector quantity. This is illustrated in Figure 2.

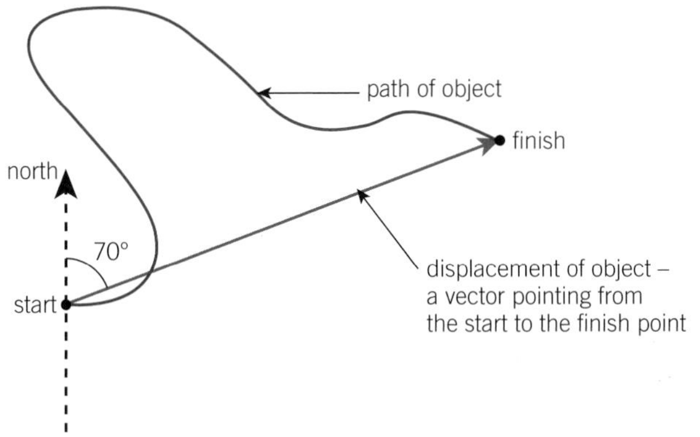

▲ **Figure 2** *Distance travelled is the length of the red path, whereas the magnitude of the displacement is the length of the blue arrow and the direction of the displacement is 70° off due north*

Synoptic link

In Chapter 3, you will come across two important vector quantities – velocity and acceleration.

Summary questions

1. Explain what is wrong with the following calculation:
   $\text{mass}_1 = 150\,\text{g}$, $\text{mass}_2 = 0.500\,\text{kg}$; total mass = 150 + 0.500 = 150.5 g *(2 marks)*
2. Compare and contrast distance and displacement. *(2 marks)*
3. You can calculate power by dividing energy by time. Explain whether power is a scalar or a vector quantity. *(2 marks)*
4. Figure 2 shows the path of a beetle that takes 20 s to travel from the start to the finish. The diagram is drawn to 1:1 scale. Determine:
   a. the distance travelled, using a length of string; *(1 mark)*
   b. the magnitude of the displacement; *(1 mark)*
   c. the average speed of the beetle. *(2 marks)*
5. Explain why the magnitude of the displacement of an object can never be greater than the distance travelled by the object. *(1 mark)*

2.4 Adding vectors

Specification reference: 2.3.1

Learning outcomes

Demonstrate knowledge, understanding, and application of:

→ addition of two vectors with scale drawings and with calculations.

▲ Figure 1 *What effect will the flowing water have on the dog's progress across the river?*

Going against the flow

Many dogs love to jump into rivers to fetch sticks thrown for them. When a dog swims back to a point on the river bank, it has to swim against the current. The velocity of the flowing water and the velocity of the dog's paddling are vector quantities, so it is possible to work out the overall or **resultant** velocity of the dog by adding the two vectors together.

Vectors in one dimension

As you have already seen with displacement in Topic 2.3, a vector quantity is represented by a line with a single arrowhead:

- the length of the line represents the magnitude of the vector, drawn to scale
- the direction in which the arrowhead points represents the direction of the vector.

For example, Figure 2 shows a line representing a single vector. It is drawn to a scale of 1.0 cm ≡ 1.0 N, so a line 5.0 cm long represents a force of 5.0 N.

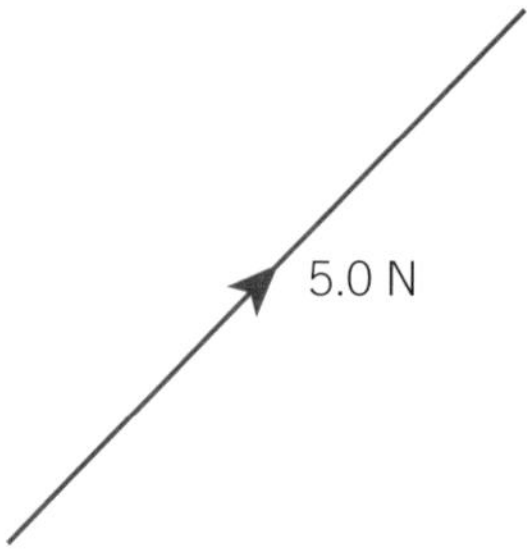

▲ Figure 2 *Representing a vector quantity, in this example a force of 5.0 N*

Parallel vectors

Where two vectors are **parallel** (they act in the same line and direction), you just add them together to find the **resultant vector**. The direction of the resultant is the same as the individual vectors but its magnitude is greater. For example, if two forces of 3.0 N and 4.0 N act in the same direction on an object, the resultant force is 7.0 N.

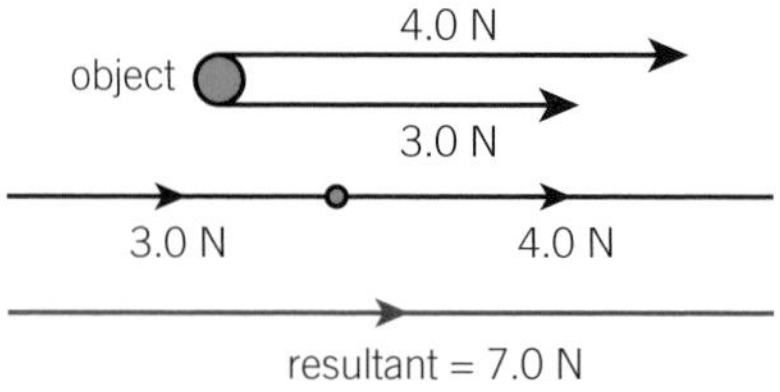

▲ Figure 3 *Two parallel forces acting on an object are shown at the top, with the corresponding vector diagrams below*

Antiparallel vectors

Where two vectors are **antiparallel** (they act in the same line but in opposite directions), you call one direction positive and the opposite direction negative (it does not matter which), and then add the vectors together to find the resultant. The magnitude and direction of the resultant will depend on the magnitude of the two vectors.

Worked example: Vectors in opposite directions

Two forces act in opposite directions on an object, as shown in Figure 4. Calculate the magnitude and direction of the resultant force.

Step 1: Assign positive and negative values to the vectors.

Assume that the positive direction is towards the right, so the two forces are –3.0 N and +4.0 N.

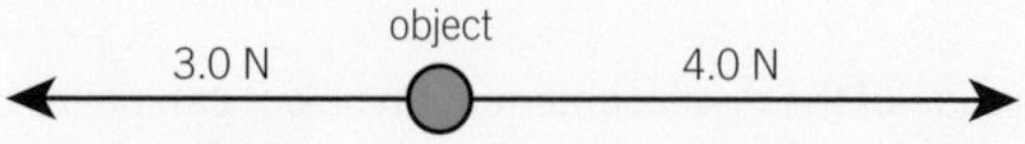

▲ Figure 4 *Two forces acting in opposite directions*

Step 2: Calculate the resultant force.

resultant = –3.0 + 4.0 = +1.0 N towards the right

Two perpendicular vectors

Perpendicular vectors act at right angles to each other. Figure 5a represents two perpendicular forces of magnitudes 4.0 N and 3.0 N acting on an object.

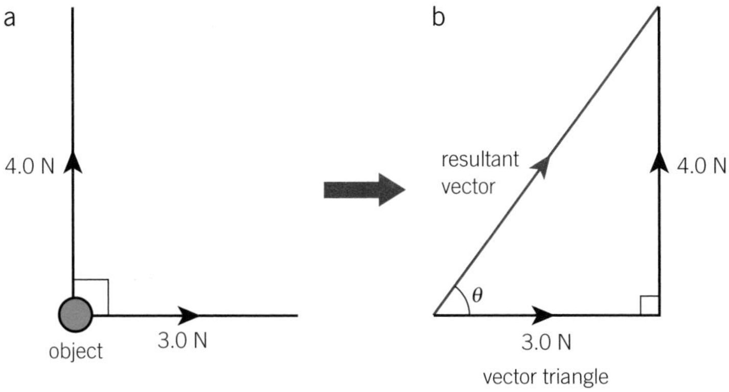

▲ **Figure 5** *Two perpendicular forces: (a) the two forces acting on the object; (b) the vector triangle used to determine the resultant vector*

The resultant vector can be found either by calculation or by a scale drawing of a **vector triangle**. Follow the rules below when adding any two vectors.

1. Draw a line to represent the first vector.
2. Draw a line to represent the second vector, starting from the *end* of the first vector.
3. To find the resultant vector, join the start to the finish. You have created a vector triangle (Figure 5b).

The method can be used to determine the resultant vector for any two vectors – displacements, velocities, accelerations, and so on. The angle between the vectors need not be 90°; any triangle works.

In this case, since the angle is 90°, you can also determine the magnitude of the resultant force F using **Pythagoras' theorem**.

$$F^2 = 4.0^2 + 3.0^2$$

$$F = \sqrt{4.0^2 + 3.0^2} = \sqrt{25}$$

$$F = 5.0\,\text{N}$$

To find the direction of the resultant force, you can calculate the angle θ made with the 3.0 N force.

$$\tan\theta = \frac{\text{opp}}{\text{adj}} = \frac{4.0}{3.0} = 1.333$$

$$\theta = 53°$$

Summary questions

1 The steps on an escalator move upwards at $0.5\,\text{m}\,\text{s}^{-1}$. Calculate the resultant vertical velocity of a person:

a standing still on the escalator; *(1 mark)*

b walking upwards at $2.0\,\text{m}\,\text{s}^{-1}$; *(1 mark)*

c walking downwards at $1.0\,\text{m}\,\text{s}^{-1}$. *(1 mark)*

2 The diagrams in Figure 6 represent forces acting on an object. For each one, draw a vector triangle and therefore determine the magnitude and direction of the resultant force. *(10 marks)*

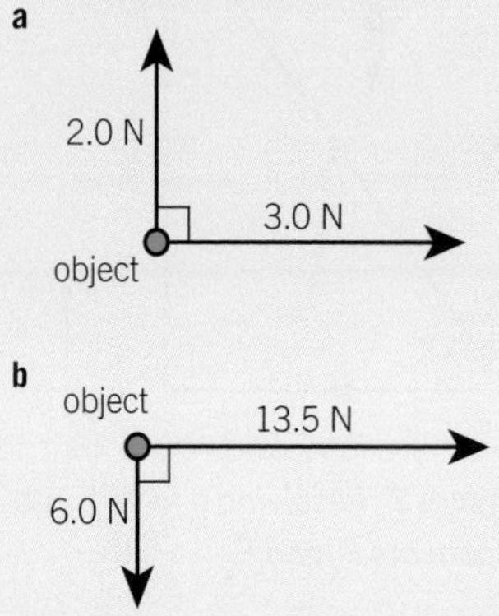

▲ **Figure 6**

3 A river flows due north at $0.90\,\text{m}\,\text{s}^{-1}$. A dog swims at $0.30\,\text{m}\,\text{s}^{-1}$. Calculate the magnitude and direction of the resultant velocity when the dog swims:

a due north; *(2 marks)*

b due south; *(2 marks)*

c due east. *(3 marks)*

2.5 Resolving vectors

Specification reference: 2.3.1

Learning outcomes

Demonstrate knowledge, understanding, and application of:

→ resolution of a vector into two perpendicular component vectors.

▲ **Figure 1** *Pilots must compensate for the effect of crosswinds during take-off and landing*

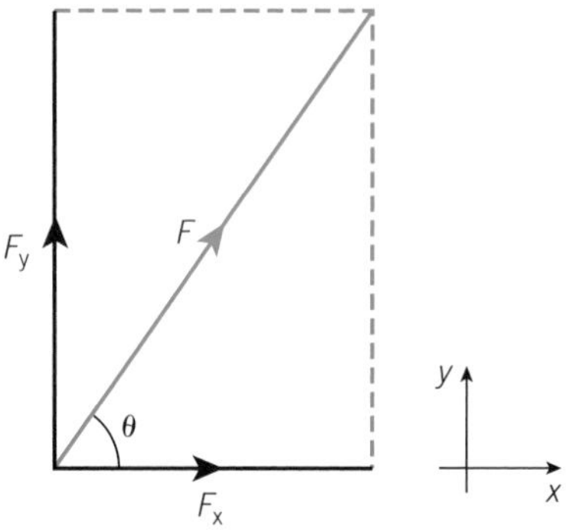

▲ **Figure 2** *Resolving a force F into components F_x and F_y*

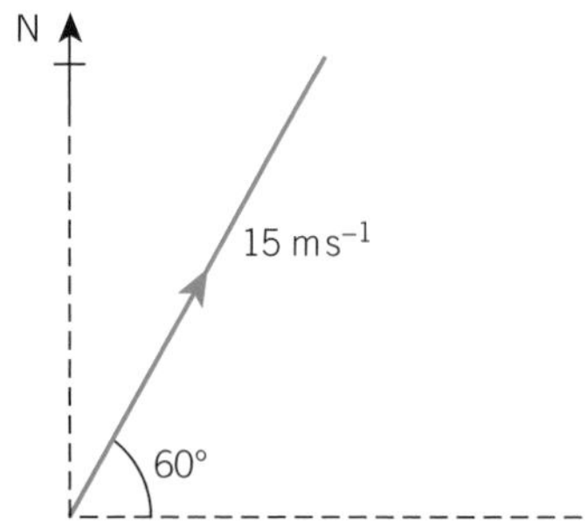

▲ **Figure 3**

Crosswinds

The wind can be helpful to aircraft. A tailwind, blowing in the same direction as the aircraft is travelling, reduces the journey time and saves fuel. On the other hand, a headwind can increase the journey time and waste fuel. Crosswinds can blow an aircraft off course unless the pilot takes them into account. An understanding of vectors is helpful in situations like these.

Resolving a vector into two components

You already know how to add together two perpendicular vectors to find a resultant vector. You can reverse this procedure to split a vector into two perpendicular components. This is called **resolving the vector**. It can be done using a scale drawing, but more often vectors are resolved by calculation.

To resolve a force F into the x and y directions, the two **components** of the force are

- $F_x = F\cos\theta$
- $F_y = F\sin\theta$

where θ is the angle made with the x direction. These equations can be used with any vector in the place of x.

Worked example: A crosswind

At an airport, a horizontal wind is blowing at 15 m s^{-1} at an angle of 60° north of east (Figure 3). Calculate the components of the wind velocity in the north and east directions.

Step 1: Select the equations for resolving vectors.

- $v_x = v\cos\theta$
- $v_y = v\sin\theta$

Step 2: Substitute the values into the equations and calculate the components.

velocity component due east = $v_x = 15 \times \cos 60° = 7.5\,\text{m s}^{-1}$

velocity component due north = $v_y = 15 \times \sin 60° = 13\,\text{m s}^{-1}$

You can quickly check your answer using Pythagoras' theorem.

$$v^2 = v_x^2 + v_y^2 = 7.5^2 + 13^2 = 56.25 + 169$$

$$v = 15\,\text{m s}^{-1}$$

Worked example: Going down

A freely falling object has a vertical acceleration of $9.81\,m\,s^{-2}$. The object is placed on a smooth ramp that makes an angle of 30° to the horizontal (Figure 4). Calculate the component of the acceleration a down the ramp.

Step 1: Select the equation.

acceleration component down the ramp = $a\cos\theta$ where θ is the angle a makes to the slope.

Step 2: Substitute the values into the equations and calculate the component.

component = $9.81 \times \cos 60° = 4.91\,m\,s^{-2}$

You could have used $9.81 \times \sin 30°$ instead. The answer will be the same because sin 30° is the same as cos 60°.

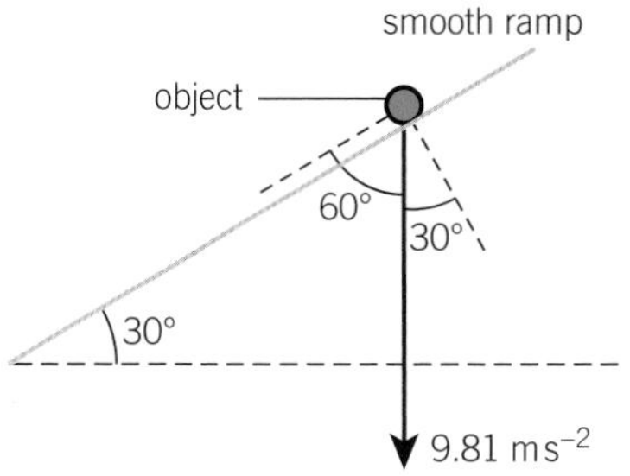

▲ Figure 4

Study tip

Always check that your calculator is in the correct mode – in this case degrees – when you resolve vectors.

Summary questions

1 A force of 10 N acts on an object at an angle θ to the horizontal. Calculate the horizontal component of the force when $\theta = 0$, $\theta = 45°$, and $\theta = 90°$. Comment on your answers. *(4 marks)*

2 A parascender is attached by a rope to a boat travelling at a constant velocity (Figure 5a). The rope is angled at 35° to the surface of the sea, and the tension in the rope is 1650 N. Calculate the horizontal and vertical components of the tension in the rope. *(2 marks)*

3 A sailing boat is travelling north. It is moving because of a force due to the wind, which is 350 N blowing towards 40° east of north (Figure 5b). Calculate the components of the force from the wind:

a towards the north (the direction in which the boat is moving); *(1 mark)*

b towards the east (perpendicular to the direction in which the boat is moving). *(1 mark)*

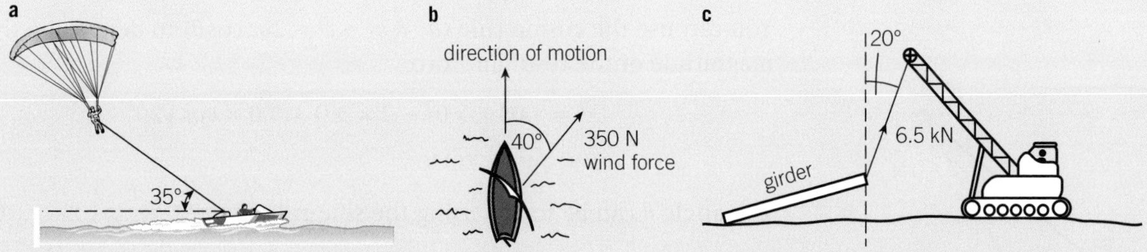

▲ Figure 5

4 One end of a steel girder is lifted off the ground by a crane. The cable is at 20° from the vertical and the tension in the cable is 6.5 kN (Figure 5c). Calculate the vertical and horizontal components of this force. *(2 marks)*

2.6 More on vectors

Specification reference: 2.3.1

Learning outcomes

Demonstrate knowledge, understanding, and application of:

→ calculations involving vectors.

Tugboats

A tugboat is a small but powerful boat that pushes or pulls larger vessels such as barges and tankers. Tugboats manoeuvre these large ships through crowded waterways and harbours. Larger, ocean-going tugboats can tow damaged ships to safety. Sometimes even the most powerful tugboats need to work in pairs or groups. Tugboat captains must understand the vectors involved so that the towed vessel travels in the right direction.

▲ **Figure 1** *Tugboats towing an oil platform*

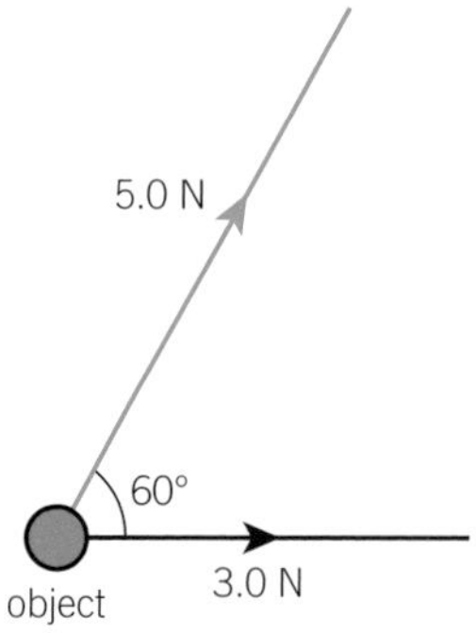

▲ **Figure 2** *Two non-perpendicular forces acting on an object*

Adding non-perpendicular vectors

There are several techniques you can use to add together two non-perpendicular vectors. They all rely on constructing a clear vector triangle. We will apply each of the techniques in turn to the following problem in order to demonstrate how to use them.

Two forces, of 5.0 N and 3.0 N, act on a single point at 60° to each other (Figure 2). What is the magnitude and direction of the resultant force?

Technique 1 – Scale diagram

Choose an appropriate scale for the drawing of your vector triangle. Use the rules outlined in Topic 2.4 to construct your vector triangle (Figure 3).

Carefully measure the length of the resultant vector: it is 7.0 cm. With 1.0 cm representing 1.0 N in the diagram, the resultant force must equal 7.0 N. The angle made by the resultant and the 4.0 N force is 38°.

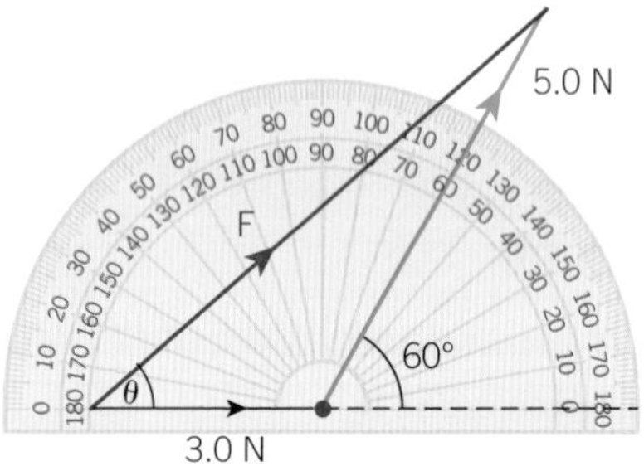

▲ **Figure 3** *A vector triangle drawn to scale*

Technique 2 – Calculations using cosine and sine rules

Figure 4 shows a sketch of the vector triangle. The angles and magnitudes of the vectors are all shown. The resultant force is F.

You can use the cosine rule ($a^2 = b^2 + c^2 - 2bc\cos\theta$) to determine the magnitude of the resultant force.

$$F^2 = 3.0^2 + 5.0^2 - 2 \times 3.0 \times 5.0 \times \cos 120°$$

$$F = \sqrt{49} = 7.0\,\text{N}$$

The angle θ can be found using the sine rule $\left(\frac{a}{\sin A} = \frac{b}{\sin B}\right)$.

$$\frac{5.0}{\sin\theta} = \frac{7.0}{\sin 120}$$

$$\sin\theta = \frac{5.0 \times \sin 120}{7.0} = 0.6186$$

$$\theta = 38°$$

The magnitude of the resultant force is 7.0 N at an angle of 38° relative to the 3.0 N force.

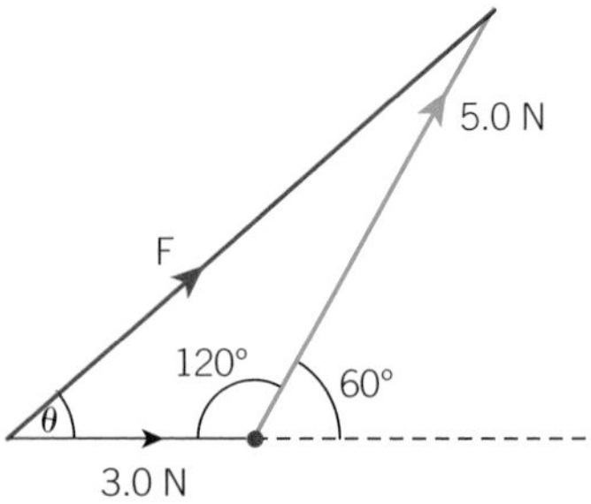

▲ **Figure 4** *A vector triangle with angles and forces shown*

Technique 3 – Calculations using vector resolution

This technique relies on choosing convenient perpendicular axes. One of the vectors is resolved along each axis so that the magnitude of the resultant vector can be determined using Pythagoras' theorem (Figure 5).

total force in x direction $= 3.0 + 5.0\cos 60° = 5.5\,\text{N}$

total force in y direction $= 5.0\sin 60° = 4.33\,\text{N}$

resultant force $F = \sqrt{5.5^2 + 4.33^2} = 7.0\,\text{N}$

$$\theta = \tan^{-1}\left(\frac{4.33}{5.5}\right) = 38°$$

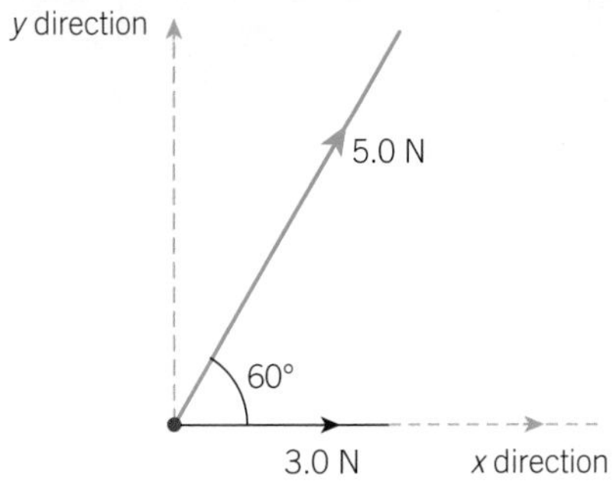

▲ **Figure 5** *Two non-perpendicular vectors shown as part of a right-angled triangle*

Subtracting vectors

Two vectors are represented by **X** and **Y**. To subtract **Y** from **X**, you simply reverse the direction of **Y** and then add this new vector to **X** (Figure 6).

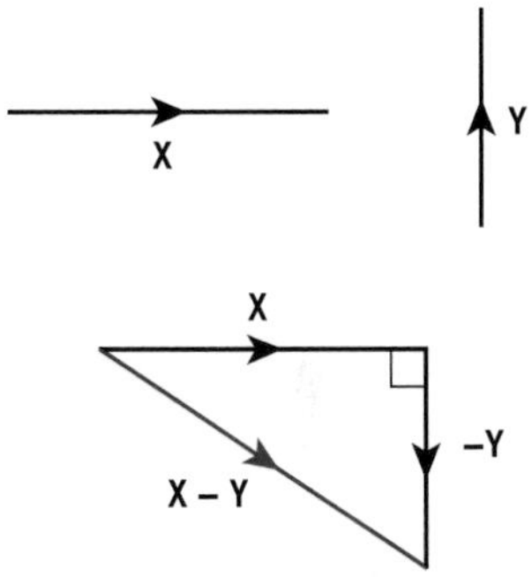

▲ **Figure 6** *Subtracting vectors*

Summary questions

1 Three forces act on an object (Figure 7). Calculate the magnitude and direction of the resultant force. *(4 marks)*

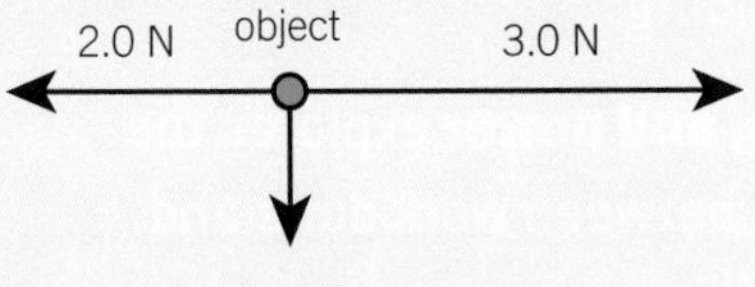

◀ **Figure 7**

2 Two tugboats are pulling a ship, each with a force of 8.0 kN, and with an angle of 40° between the cables (Figure 8). Calculate the magnitude and direction of the resultant force. *(4 marks)*

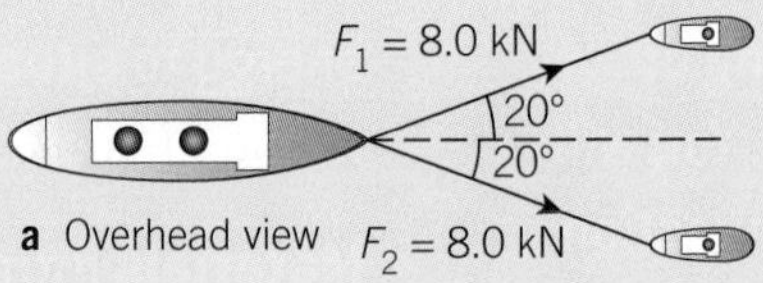

▲ **Figure 8**

3 Three tugboats are towing an object at sea. The forces and angles between the cables are shown in Figure 9. Calculate the magnitude and direction of the resultant force on the object. *(6 marks)*

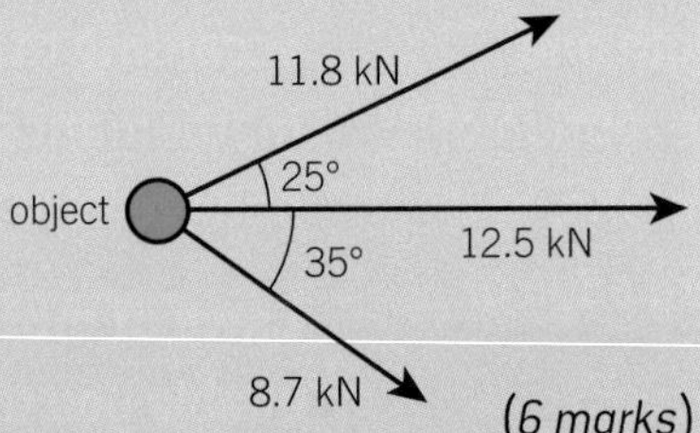

▲ **Figure 9**

4 Figure 10 shows two vectors, **A** and **B**. Determine the magnitude and the direction of the resultant vector **A** – **B**. *(4 marks)*

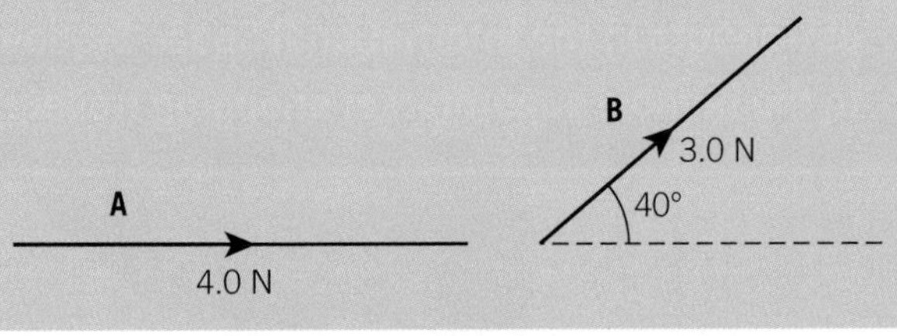

◀ **Figure 10**

MODULE 3

Forces and motion

Chapters in this module

3 Motion

4 Forces in action

5 Work, energy, and power

6 Materials

7 Laws of motion and momentum

Introduction

Force and motion are tightly knitted together. They form a central part of every physicist's understanding of the Universe around us.

In this module you will learn how to mathematically model the motion of objects and will develop your understanding of the effects forces have on objects. You will also learn about the important connection between force and energy.

Motion explores the key ideas used to describe and analyse motion in both one and two dimensions, including the motion of Olympic swimmers, sprinting cheetahs, and parachutists jumping from the very edge of space.

Forces in action develops ideas about the effect of forces on objects. In this chapter you will learn how the motion of an object changes when it experiences a resultant force, and how several balanced forces are essential in contexts including rock climbing and bridge building.

Work, energy, and power explores the important link between work done and energy. You will learn how to apply the important principle of conservation of energy to situations from wind turbines to roller coasters.

Materials introduces several ideas that are essential in engineering. In this chapter you will learn how to classify different materials according to their properties, and the mathematics of the differences between a bungee cord and the latest aluminium alloy.

Laws of motion and momentum will enable you to combine the ideas developed in the previous chapters. You will learn how Newton's laws are used to predict the motion of all colliding or interacting objects, from astronauts in the International Space Station to the humble supermarket shopping trolley.

Knowledge and understanding checklist

From your Key Stage 4 study you should be able to do the following. Work through each point, using your Key Stage 4 notes and the support available on Kerboodle.

- ☐ Relate changes and differences in motion to appropriate distance–time and velocity–time graphs.
- ☐ Apply formulae relating distance, time, and speed for uniform motion, and for motion with uniform acceleration.
- ☐ Recall examples of ways in which objects interact and describe how such examples involve interactions between pairs of objects that produce a force on each object.
- ☐ Apply Newton's first law to explain the motion of objects and apply Newton's second law in calculations relating forces, masses, and accelerations.
- ☐ Describe and calculate changes in energy and explain the definition of power as the rate at which energy is transferred.
- ☐ Calculate energy efficiency for any energy transfer, and describe ways to increase efficiency.

Maths skills checklist

All physicists use maths. In this unit you will need to use many different maths skills, including the following examples. You can find support for these skills on Kerboodle and through MyMaths.

- ☐ **Change the subject of an equation, including nonlinear equations.** You will need to be able to do this to solve mathematical problems when dealing with energies.
- ☐ **Use an appropriate number of significant figures.** You will need to be able to do this in solving a variety of problems in the motion topic, including projectiles.
- ☐ **Plot two variables from experimental or other data and use $y = mx + c$.** You will need to be able to do this when studying Hooke's law.
- ☐ **Calculate the gradient from a graph (including tangents).** You will need to be able to do this in experiments to determine g by free fall.
- ☐ **Understand the possible physical significance of the area between a curve and the x-axis and be able to calculate it or estimate it by graphical methods.** You will need to be able to do this when studying motion graphs.

3 MOTION
3.1 Distance and speed
Specification reference: 3.1.1

Learning outcomes
Demonstrate knowledge, understanding, and application of:
- → average speed
- → instantaneous speed
- → distance–time graphs to determine speed.

Average speed
Laws limit the speed at which vehicles can travel on roads. Speed cameras help the police enforce these limits. Average speed check areas use cameras along the journey (Figure 1). Automatic numberplate recognition technology identifies individual vehicles, so the time they take to travel between cameras can be measured. If the distance between the cameras is also known, the average speed can be calculated.

Calculating average speed
The **average speed** v of an object can be calculated from the distance x travelled and the time t taken using the equation

$$\text{average speed} = \frac{\text{distance travelled}}{\text{time taken}}$$

You can write this in algebraic form as

$$v = \frac{\Delta x}{\Delta t}$$

The Greek capital letter Δ (delta) means 'change in'. From the SI base units for distance and time, the unit of speed is m s^{-1}.

▲ **Figure 1** *To protect the people repairing the road, your average speed on this section of the road must not be more than 50 miles per hour*

Worked example: Fine or not?
A car travels 2.5 km in 1 minute 22 seconds. The average speed limit for the road is 50 mph ($22\,\text{m s}^{-1}$). Has the driver exceeded this average speed limit?

Step 1: Identify the equation and list the known values.

$$v = \frac{\Delta x}{\Delta t}$$

$$\Delta x = 2500\,\text{m},\ \Delta t = 1\text{ minute }22\text{ seconds} = 60 + 22 = 82\,\text{s}$$

Step 2: Substitute the values into the equation and calculate the answer.

$$v = \frac{2500}{82} = 30.49\,\text{m s}^{-1}$$

$$v = 30\,\text{m s}^{-1}\ (2\text{ s.f.})$$

Yes, the driver has exceeded the average speed limit.

Distance–time graphs

Graphs of distance against time are used to represent the motion of objects.

- Distance is plotted on the y-axis (vertical axis).
- Time is plotted on the x-axis (horizontal axis).

In a distance–time graph, a stationary object is represented by a horizontal straight line. An object moving at a **constant speed** is represented by a straight, sloping line. The **gradient** of that line is equal to the distance travelled divided by the time taken, $\Delta x/\Delta t$, in other words, to the speed of the object.

- speed = gradient of a distance–time graph

In Figure 2

change in distance $\Delta x = 1400 - 400 = 1000\,\text{m}$

change in time $\Delta t = 70 - 20 = 50\,\text{s}$

speed $v = \frac{\Delta x}{\Delta t} = \frac{1\,000}{50} = 20\,\text{m}\,\text{s}^{-1}$

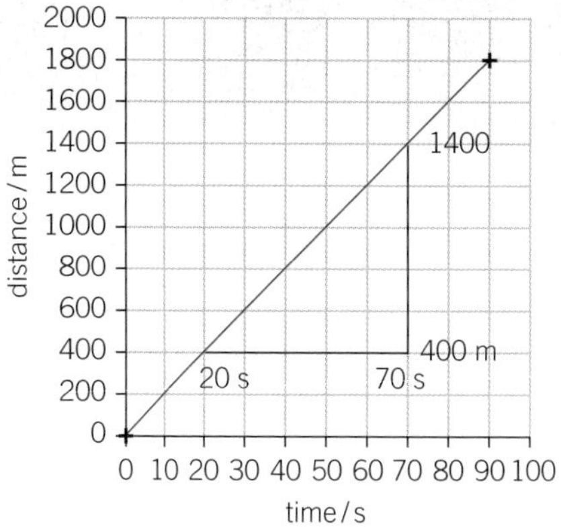

▲ **Figure 2** *A distance–time graph for an object moving at constant speed – the gradient of the graph represents the speed*

Synoptic link

You will find more information on graphs in Appendix A2, Recording results.

Instantaneous speed

A criticism of checks on average speed is that a vehicle can travel faster than the average speed allowed for part of the journey, then travel slowly to increase the total journey time sufficiently to avoid a fine.

Instantaneous speed is the speed of the car over a very short interval of time. The instantaneous speed at a particular time is found by drawing the tangent to the distance–time graph at that time, then determining the gradient of this tangent (Figure 3). The greater the gradient, the greater the instantaneous speed.

Study tip

When you calculate a gradient, make sure the triangle that you draw on the graph is large enough to provide an accurate answer. It is advisable to use more than half the length of the line to determine your gradient.

Summary questions

1 Calculate the average speed in m s^{-1} for the following bicycle journeys:
  a a distance of 180 m covered in a time of 9.0 s; *(2 marks)*
  b a distance of 2.0 km covered in 6.5 minutes. *(2 marks)*

2 A snail travels 19.2 m in 1 day (24 hours). Calculate its average speed in m s^{-1}. *(2 marks)*

3 A car travels for 19 s at an average speed of 31 m s^{-1}. How far does it travel? *(2 marks)*

4 An aircraft travels at an average speed of 240 m s^{-1} for 12 000 km. Calculate the time taken in seconds and in hours. *(3 marks)*

5 A lorry travels on a motorway for 2.0 minutes at a constant speed of 25 m s^{-1}. It then struggles on a hill and travels 800 m in 50 s. Calculate:
  a the total distance travelled; *(2 marks)*
  b the average speed of the lorry. *(2 marks)*

6 Use Figure 3 to determine the instantaneous speed of the object at time t = 80 s. *(3 marks)*

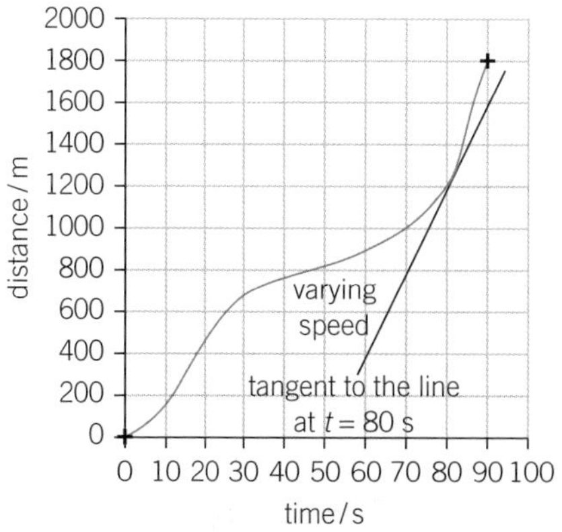

▲ **Figure 3** *A distance–time graph for an object moving at varying speed – you can determine the instantaneous speed from the gradient of the tangent to the graph*

3.2 Displacement and velocity

Specification reference: 3.1.1

Learning outcomes

Demonstrate knowledge, understanding, and application of:

- → displacement
- → velocity
- → displacement–time graphs to determine velocity.

▲ **Figure 1** *Each swimmer has an average speed, but if they do two laps of the pool they all have an average velocity of zero*

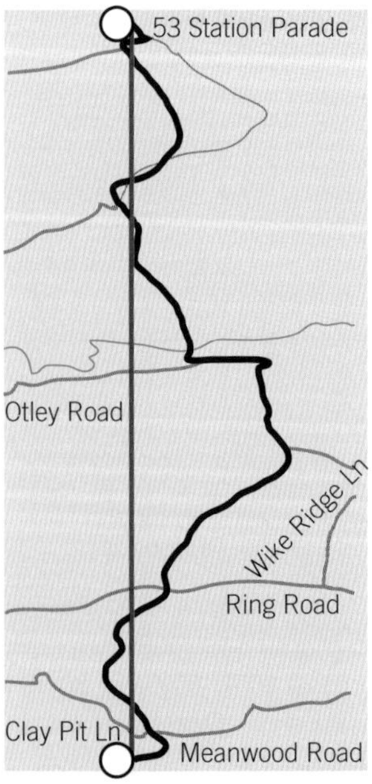

▲ **Figure 2** *The distance travelled by road on a journey from Harrogate to Leeds is different from the displacement*

Scalar and vector quantities

A 100 m swimming race in an Olympic-sized pool takes two laps. During the London Olympic Games in 2012, the American swimmer Missy Franklin beat a previous world record for the backstroke with her time of 58.33 s, giving her an average speed of 1.71 m s^{-1}. However, she finished where she started, which means that, despite all the hard work she did in racing 100 m, her total displacement and average velocity were zero.

The vector nature of velocity

Displacement s is a vector quantity, unlike distance, which is scalar. Displacement has both magnitude and direction. Speed is a scalar quantity calculated from distance, but velocity is a vector quantity calculated from displacement. The **average velocity** v of an object can be calculated from the change in displacement and the time taken.

$$\text{average velocity} = \frac{\text{change in displacement}}{\text{time taken}}$$

You can write this in algebraic form as

$$v = \frac{\Delta s}{\Delta t}$$

where Δs is the change in displacement and Δt is the time taken. The SI unit for velocity is m s^{-1}.

The worked example below shows that average speed and average velocity are very different quantities.

Worked example: Speed and velocity

a Leeds is about 21 km south of Harrogate, but the distance by road is about 24 km (Figure 2). It takes 37 minutes to travel from Harrogate to Leeds by road. Calculate the average speed and the average velocity.

Step 1: Identify the equations needed.

$$\text{average speed } v = \frac{\Delta x}{\Delta t}, \text{ average velocity } v = \frac{\Delta s}{\Delta t}$$

Step 2: Substitute the values in SI units into the equations and calculate the answers.

$$\text{time taken } \Delta t = 60 \times 37 = 2220\,\text{s}$$

$$\text{average speed } v = \frac{\Delta x}{\Delta t} = \frac{24\,000}{2220} = 10.8\,\text{m s}^{-1} = 11\,\text{m s}^{-1} \text{ (2 s.f.)}$$

$$\text{average velocity } v = \frac{\Delta s}{\Delta t} = \frac{21\,000}{2220} = 9.5\,\text{m s}^{-1} \text{ (2 s.f.)}$$

The magnitude of the average velocity is 9.5 m s^{-1} and its direction is due south from Harrogate.

b What would happen to the magnitude of the average velocity if the journey was from Harrogate to Leeds and then back to Harrogate?

The overall change in displacement would be zero and therefore the average velocity would be zero.

Displacement–time graphs

Graphs of displacement against time are used to represent the motion of objects.

- Displacement is plotted on the *y*-axis (vertical axis).
- Time is plotted on the *x*-axis (horizontal axis).

Figure 3 shows the displacement–time graph for a car travelling along a straight road. The car is travelling at a constant velocity between $t = 0$ and $t = 20\,\text{s}$, as can be seen from the first straight-line section of the graph. The horizontal section of the graph between $t = 20\,\text{s}$ and $t = 30\,\text{s}$ shows that the displacement of the car remains constant. Therefore, the car must be stationary. After $t = 30\,\text{s}$, the graph is still a straight line but has a negative slope. The displacement of the car is getting smaller with time. The car must therefore be returning at a constant velocity.

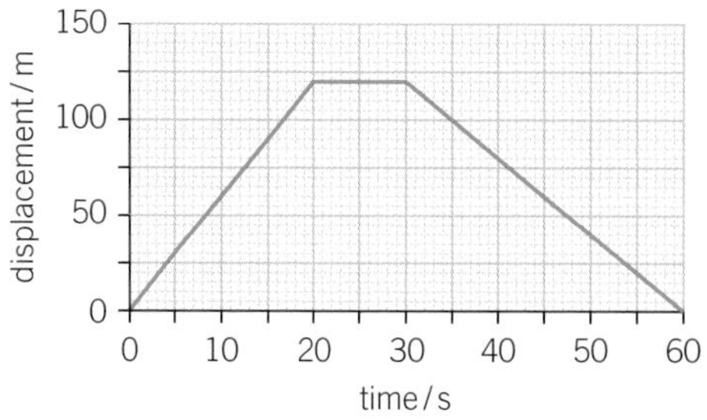

▲ **Figure 3** *A displacement–time graph for a car journey*

You can determine the velocity of an object from the gradient of its displacement–time (s–t) graph. If the graph is not a straight line, draw a tangent to the graph, then calculate the gradient of this tangent for the instantaneous velocity, as illustrated in Figure 4.

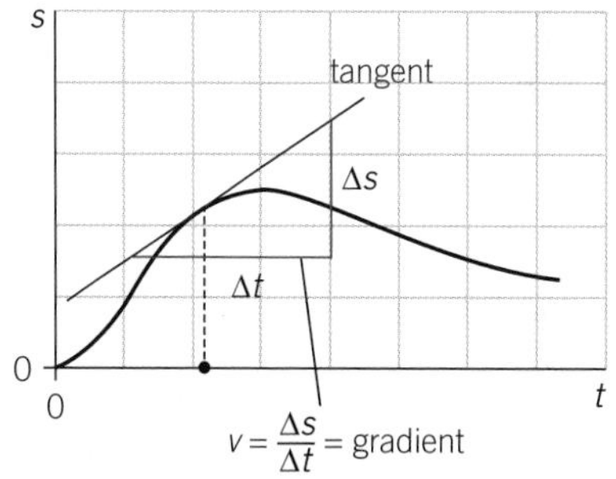

▲ **Figure 4** *Velocity can be determined from the gradient of the displacement–time graph*

Worked example: Forwards and backwards

Use Figure 3 to determine the velocity of the car at $t = 10\,\text{s}$ and $t = 40\,\text{s}$.

Step 1: Identify the equation needed and how to obtain the values from the graph.

$$\text{velocity } v = \frac{\Delta s}{\Delta t}$$

The right-hand terms are equivalent to the change on the *y*-axis (s) divided by the change on the *x*-axis (t).

Step 2: Substitute the values into the equation and calculate the answer.

The velocity at $t = 10\,\text{s}$ can be determined from the gradient of the straight-line graph between $t = 0$ and $t = 20\,\text{s}$.

$$\text{velocity } v = \frac{\Delta s}{\Delta t} = \frac{120-0}{20-0} = 6.0\,\text{m}\,\text{s}^{-1}$$

Synoptic link

You will find more information about gradients and tangents in Appendix A2, Recording results.

The velocity at $t = 40\,s$ can be determined from the gradient of the straight-line graph between $t = 30\,s$ and $t = 60\,s$.

$$\text{velocity } v = \frac{\Delta s}{\Delta t} = \frac{0-120}{60-30} = -4.0\,m\,s^{-1}$$

The negative sign for the velocity shows that the car is travelling in the opposite direction to its motion between 0 and 20 s.

Summary questions

1 Describe the journey of the object with the displacement–time (*s*–*t*) graph shown in Figure 5. *(3 marks)*

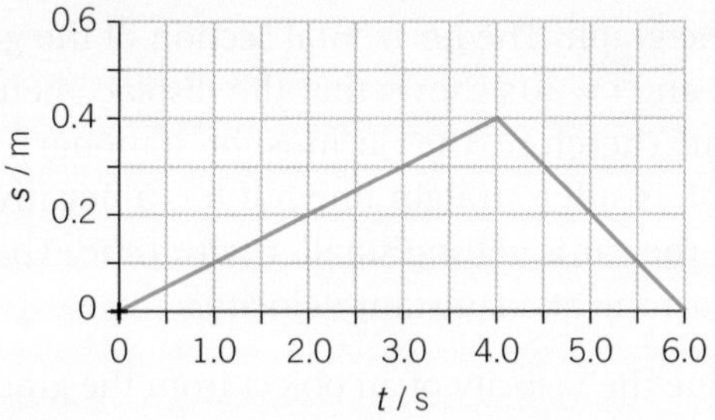

▲ **Figure 5**

2 Determine the velocity of the object in Figure 5 at time $t = 2.0\,s$ and $t = 5.0\,s$. *(5 marks)*

3 Determine the average speed of the object in Figure 5 between time $t = 0.0\,s$ and $t = 6.0\,s$. *(5 marks)*

4 A particle travels in a circular path of radius 80 cm. It starts from point **A** and takes 8.0 s to travel once round the circle (Figure 6).

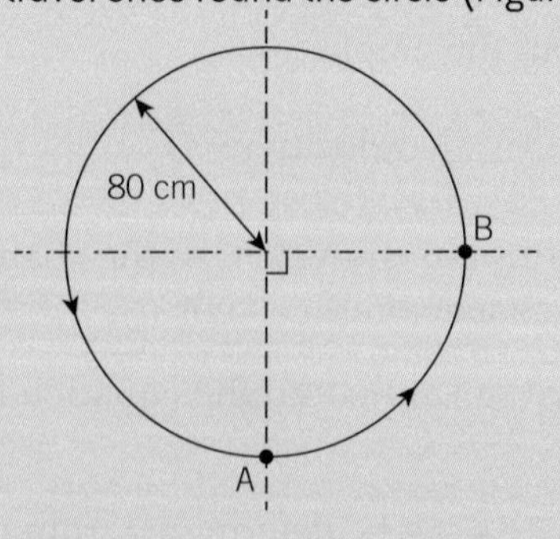

▲ **Figure 6**

Calculate the:

a average speed of the particle; *(3 marks)*

b average velocity of the particle from **A** to **B**. *(3 marks)*

3.3 Acceleration

Specification reference: 3.1.1

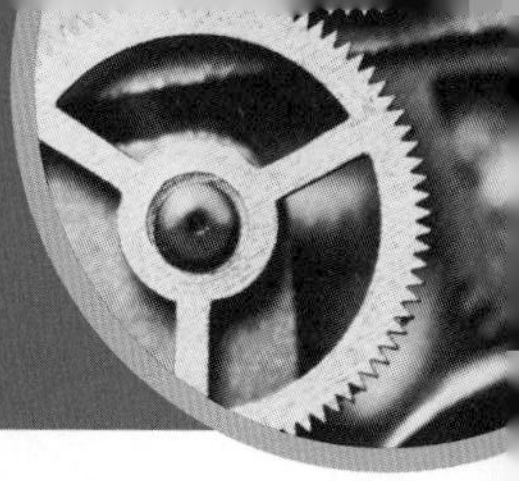

Bursts of acceleration

In 2013, scientists fitted wild cheetahs with collars containing a global positioning system (GPS) module and electronic motion sensors, so the animals' velocity and acceleration could be studied. The cheetahs ran at speeds of up to $26\,\mathrm{m\,s^{-1}}$ and were able to turn very quickly, with an acceleration of up to $13\,\mathrm{m\,s^{-2}}$, more than that of the fastest production car that year, the Bugatti Veyron.

Learning outcomes

Demonstrate knowledge, understanding, and application of:

→ acceleration

→ velocity–time graphs to determine acceleration.

Determining acceleration

The **acceleration** of an object is defined as the rate of change of velocity. In mathematical form, the acceleration a is

$$a = \frac{\Delta v}{\Delta t}$$

where Δv is the change in velocity and Δt is the time taken for the change. The unit of acceleration is $\mathrm{m\,s^{-2}}$. Since acceleration is determined from velocity, it too is a vector quantity – it has magnitude and direction. A negative acceleration is often called deceleration.

Acceleration can be determined by calculation, or from a velocity–time (v–t) graph.

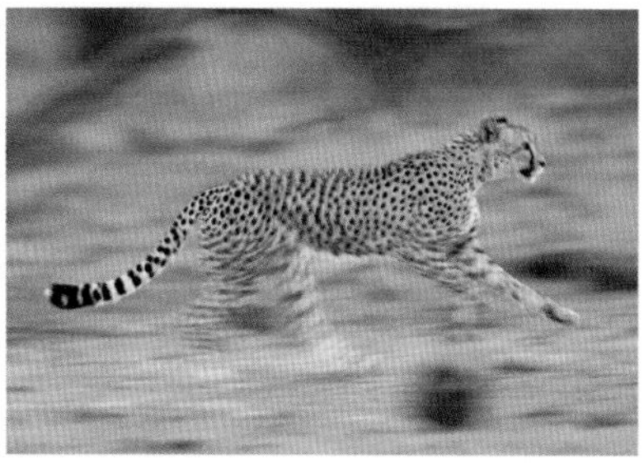

▲ **Figure 1** *Cheetahs need to be able to make rapid changes in speed and direction in order to hunt*

Calculating acceleration

You can calculate acceleration if you know the change in velocity of an object and the time taken for this change.

Worked example: 0 to 62, then slam the brakes on

a A Bugatti Veyron can accelerate from 0 to 100 km/h ($27.8\,\mathrm{m\,s^{-1}}$) in 2.46 s. Calculate its average acceleration.

Step 1: Identify the equation needed.

$$a = \frac{\Delta v}{\Delta t}$$

Step 2: Substitute the values into the equation and calculate the answer.

$$a = \frac{\Delta v}{\Delta t} = \frac{27.8-0}{2.46} = 11.3\,\mathrm{m\,s^{-2}}$$

b The car takes 2.34 s to stop from 100 km/h under braking. Calculate its acceleration, assuming that this is constant during braking.

$$a = \frac{\Delta v}{\Delta t} = \frac{0-27.8}{2.34} = -11.9\,\mathrm{m\,s^{-2}}$$

The negative sign means that the velocity of the car is decreasing over time – it is decelerating.

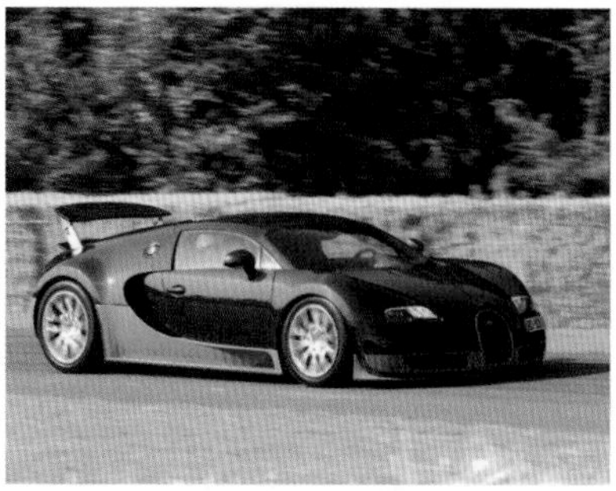

▲ **Figure 2** *The Bugatti Veyron has similar acceleration to that of a cheetah*

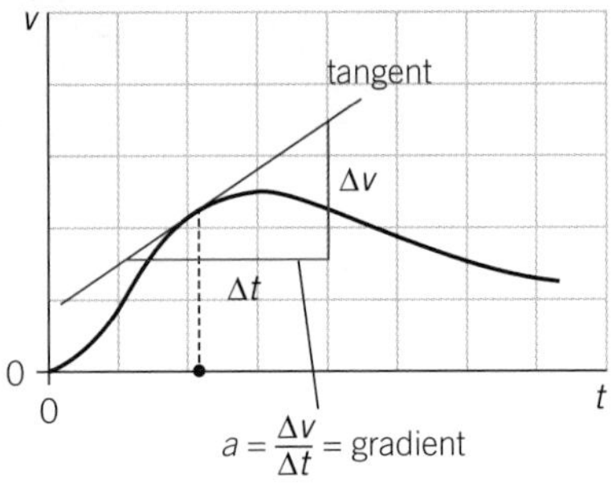

▲ **Figure 3** *Acceleration can be determined from the gradient of the velocity–time graph*

Study tip

Many candidates confuse displacement–time and velocity–time graphs. Look carefully at the labels on the vertical axis.

Velocity–time graphs

Since $a = \frac{\Delta v}{\Delta t}$, it follows that the acceleration of an object can be determined from the gradient of a velocity–time graph. You have seen how to determine gradients for straight-line graphs and for non-linear graphs in the previous topics in this chapter; the only difference in this case is that the y-axis of the graph represents velocity v rather than displacement s.

- acceleration = gradient of velocity–time graph

Figure 4 shows how the motion of an object can be deduced from the velocity–time graph.

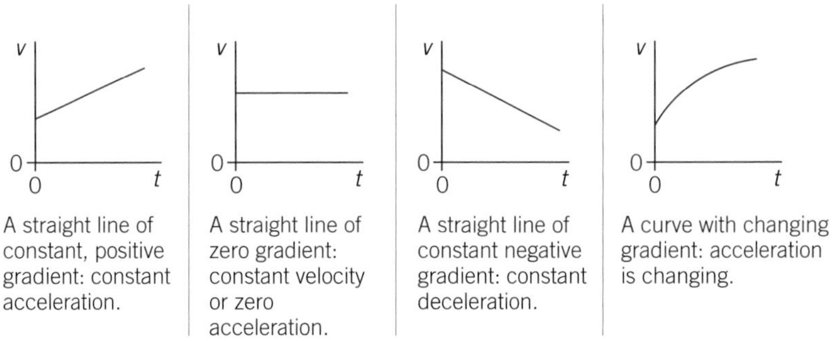

v
0
0
t
A straight line of constant, positive gradient: constant acceleration.
A straight line of zero gradient: constant velocity or zero acceleration.
A straight line of constant negative gradient: constant deceleration.
A curve with changing gradient: acceleration is changing.

▲ **Figure 4** *Interpreting velocity–time graphs*

Summary questions

1 A racing cyclist starts from rest and reaches a velocity of 8.0 m s^{-1} after 12 s. Calculate the acceleration of the cyclist. *(2 marks)*

2 A train slows down from 40 m s^{-1} to 10 m s^{-1} over 60 s. Calculate the magnitude of the deceleration of the train. *(2 marks)*

3 Describe the journey of the object represented in the velocity–time graph shown in Figure 5. *(2 marks)*

4 **a** Use Figure 5 to determine the maximum acceleration of the object. Explain your answer. *(2 marks)*
b Sketch an acceleration–time graph from $t = 0$ to $t = 3.0$ s. *(2 marks)*

5 A ball is held above the surface of water and released. It falls through the air and then through the water. Figure 6 shows the velocity–time graph for this ball.
a Calculate the acceleration of the ball as it falls through the air. *(2 marks)*
b Determine the magnitude of the deceleration at time $t = 1.0$ s. *(3 marks)*

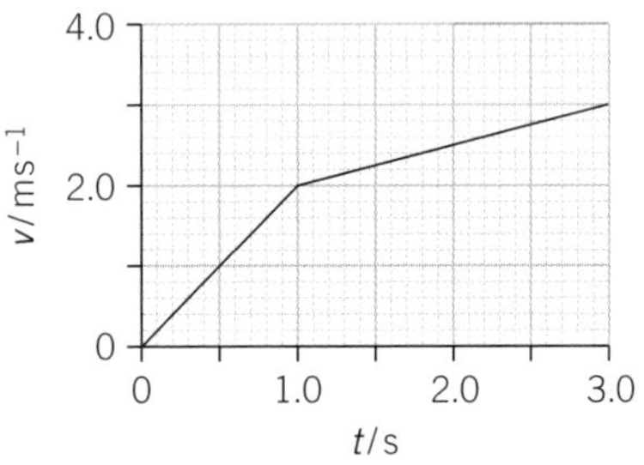

▲ **Figure 5**

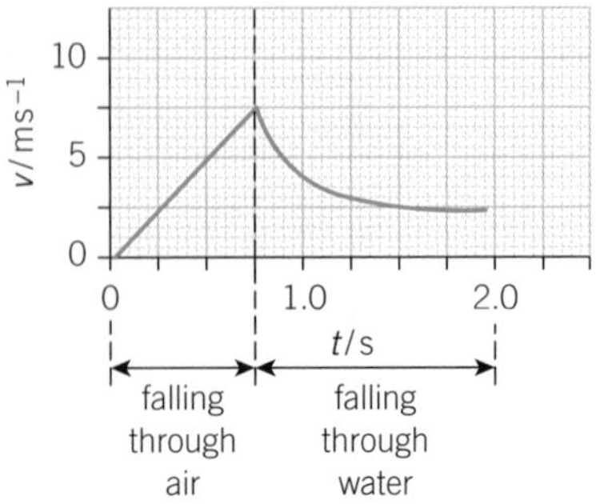

▲ **Figure 6**

3.4 More on velocity–time graphs

Specification reference: 3.1.1

The spy in the cab

Tiredness can affect the ability to drive safely as much as alcohol can. The law sets time limits to ensure that drivers of heavy vehicles take proper rest breaks and do not drive whilst tired. Tachographs record the speed and distance travelled by a vehicle. Modern tachographs are digital, but older ones use a stylus to record the information on a circular chart. This chart rotates once in 24 hours, providing a permanent record of the journey so that the authorities can check that the driver has taken regular breaks.

Learning outcomes

Demonstrate knowledge, understanding, and application of:

→ velocity–time graphs to determine displacement.

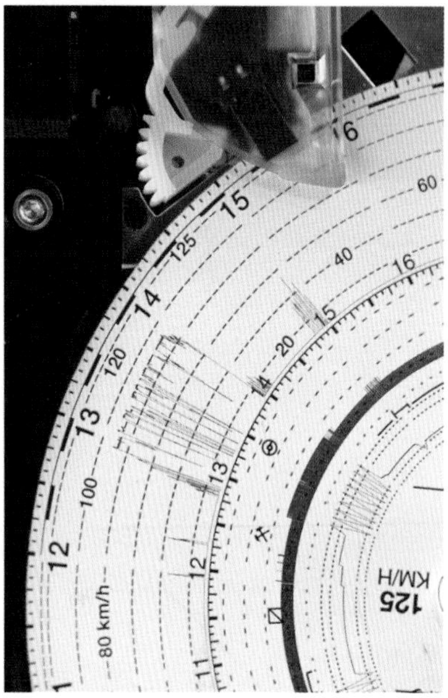

▲ **Figure 1** *An analogue tachograph record of road speed against time*

Area under the graph

In the previous topic you saw that acceleration can be determined from the gradient of a velocity–time graph. In addition, we can read the displacement of the object from the area under the graph. Figure 2 shows why this is so.

You will recall that the average velocity v is given by the equation

$$v = \frac{\Delta s}{\Delta t}$$

where Δs is the displacement in the time interval Δt. For the instantaneous velocity, assume that Δt is very small indeed – the velocity of the object is not going to change much. The change in the displacement $\Delta s \approx v\Delta t$. If you look at the graph in Figure 2, this is the area of the very thin rectangular strip marked under the graph. Therefore, the change in displacement is equal to the area of this strip. If you add similar strips for a longer interval of time, then clearly the area under the velocity–time graph is the total displacement of the object.

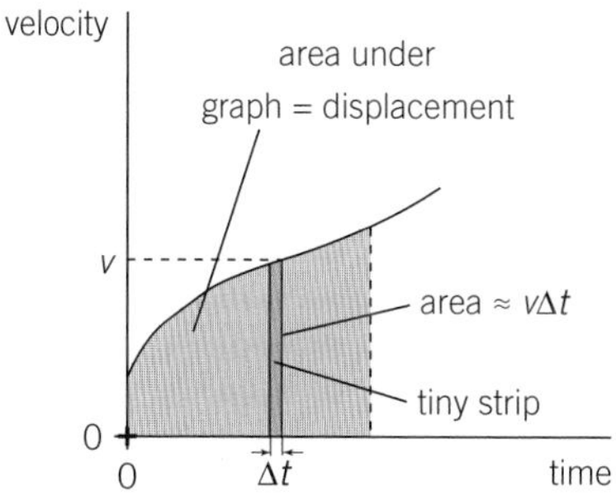

▲ **Figure 2** *The area under the velocity–time graph is equal to displacement*

Calculating displacement for constant accelerations

Displacement is easy to calculate when the acceleration is constant, because the areas can be broken down into rectangles and right-angled triangles. This is illustrated in the worked example below for the short journey of a cyclist.

Worked example: Displacement of a cyclist

The velocity–time graph for a cyclist travelling along a straight road is shown in Figure 3.

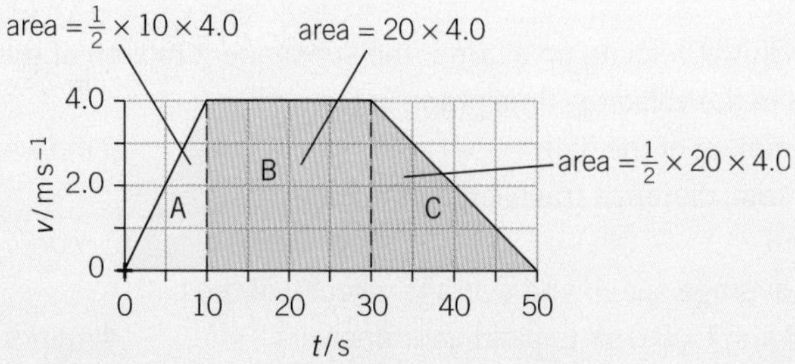

▲ **Figure 3**

→

Calculate the total distance travelled by the cyclist (the cyclist's displacement) in the period of 50 s.

Step 1: Identify the method needed.

distance travelled = area between the graph and the time axis

Step 2: Calculate the answer.

distance travelled = area of triangle **A** + area of rectangle **B** + area of triangle **C**

$$= \left(\tfrac{1}{2} \times 10 \times 4.0\right) + \left(20 \times 4.0\right) + \left(\tfrac{1}{2} \times 20 \times 4.0\right)$$

$$= 20 + 80 + 40$$

$$= 140\,\text{m}$$

Step 3: There is an alternative method to determine the area under the graph. Calculate the area of the trapezium using the formula

$$\text{area} = \tfrac{1}{2} \times (\text{sum of the parallel sides}) \times \text{vertical height}$$

$$\text{distance} = \tfrac{1}{2} \times (50 + 20) \times 4.0 = 140\,\text{m}$$

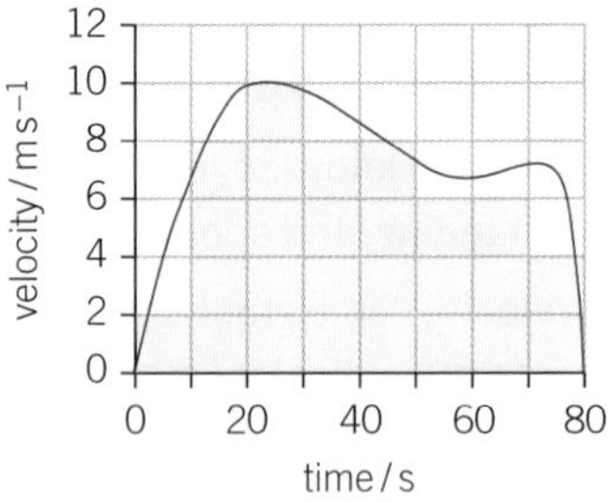

▲ **Figure 4** *Calculating the area under a non-linear velocity–time graph*

Calculating displacement for changing accelerations

For non-linear velocity–time graphs, you can determine the area under the graph by counting squares. Taking Figure 4 as an example, you would start by counting the squares that are complete or nearly complete (yellow). Then count the remaining squares that lie mostly beneath the graph. Omit squares that are mostly above the graph.

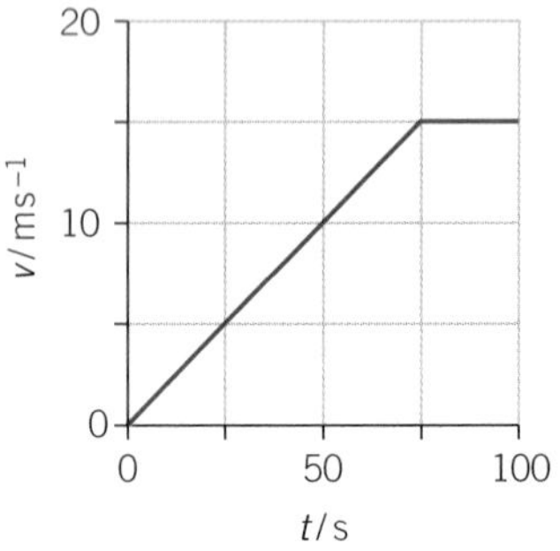

▲ **Figure 5**

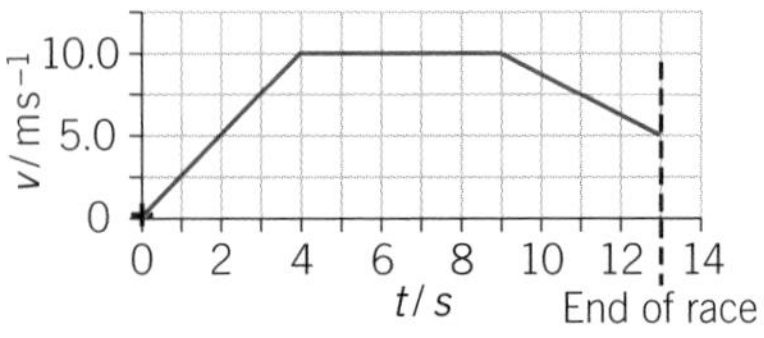

▲ **Figure 6**

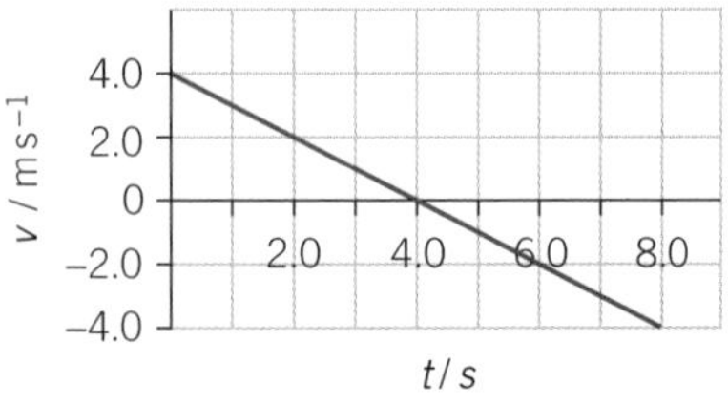

▲ **Figure 7**

Summary questions

1. The velocity–time graph for a car travelling on a straight road is shown in Figure 5. Without carrying out calculations, describe how the acceleration of the car varies between $t = 0$ s and $t = 100$ s. *(2 marks)*

2. Use the velocity–time graph shown in Figure 5 to calculate:
   a. the total distance travelled by the car in 100 s; *(3 marks)*
   b. the average velocity of the car. *(2 marks)*

3. The velocity–time graph for a sprinter running along a straight track is shown in Figure 6.
   a. Calculate the total distance travelled by the runner in 13 s. *(4 marks)*
   b. Calculate the average velocity of the runner. *(2 marks)*

4. A ball is given an initial velocity up a ramp. The subsequent motion of the ball is illustrated in the velocity–time graph in Figure 7.
   a. Describe the motion of the ball. *(3 marks)*
   b. Calculate the total distance travelled by the ball up and down the ramp. *(2 marks)*
   c. What are the average speed and average velocity of the ball between $t = 0$ and $t = 8.0$ s? Explain your answers. *(4 marks)*

3.5 Equations of motion

Specification reference: 3.1.2

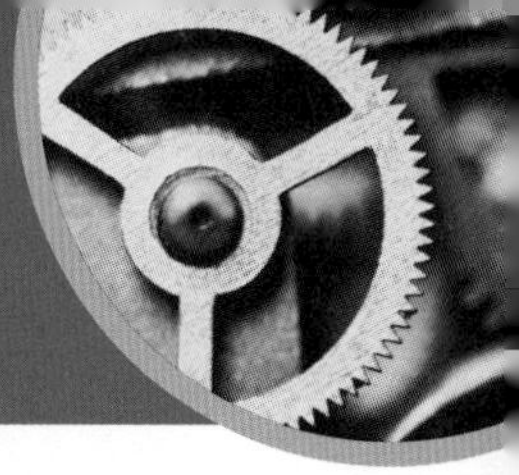

Predicting motion

We can use a knowledge of physics to predict the motion of accelerating or decelerating objects: for example, the impact speed of a small meteor about to hit the Earth, the initial speed of a car from skid marks left on the road, and the final speed of a space probe landing on a distant planet. It is amazing that all this is possible with just four equations.

Learning outcomes

Demonstrate knowledge, understanding, and application of:

→ the equations of motion for constant acceleration in a straight line.

Equations of motion: the *suvat* equations

You need four equations to calculate quantities involving motion in a *straight line* at a constant acceleration. These equations of motion are often informally referred to as the '*suvat* equations' after the symbols for the quantities involved.

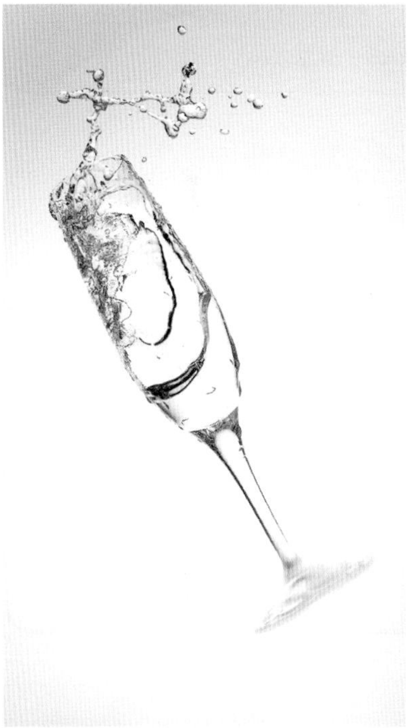

▲ **Figure 1** *Can you predict the time of fall for this glass of water?*

Deriving the equations of motion

Figure 2 shows the velocity–time graph for an accelerating object. The initial velocity of the object is u. After a time t the final velocity of the object is v. The object has a constant acceleration a, as you can see from the straight-line graph.

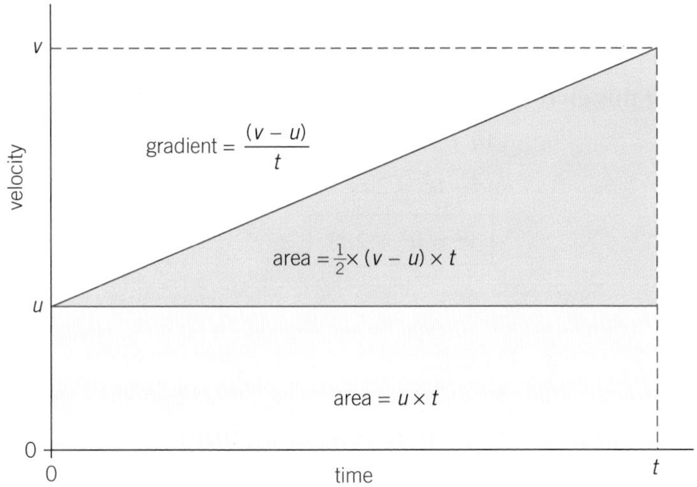

▲ **Figure 2** *Velocity–time graph showing how the* suvat *equations are derived*

Equation without *s*

From the graph in Figure 2

$$a = \frac{\Delta v}{\Delta t} = \frac{v-u}{t}$$

This can be rearranged to give

$$\boxed{v = u + at} \qquad \textbf{(1)}$$

Equation without *v*

You will recall that the area under the velocity–time graph is equal to the displacement s.

- the rectangular area = ut
- the triangular area = $\frac{1}{2} \times (v-u) \times t$

▼ **Table 1** *The* suvat *quantities*

Symbol	Quantity
s	displacement (or distance travelled)
u	initial velocity
v	final velocity
a	acceleration
t	time taken for the change in velocity

From Equation **1** above, $(v - u) = at$. If you substitute this into the expression for the area of the triangle, you get $\frac{1}{2} \times at \times t$. With ut for the area of the rectangle, this gives the total area s.

$$\boxed{s = ut + \frac{1}{2}at^2} \qquad \textbf{(2)}$$

Equation without *a*

If you treat the area under the graph as the area of a trapezium (with u and v as the parallel sides of the trapezium, and t as the perpendicular separation between them), this becomes

$$\boxed{s = \frac{1}{2}(u + v)t} \qquad \textbf{(3)}$$

In other words, the displacement s is the average velocity, $\left(\frac{u+v}{2}\right)$, multiplied by the time t.

Equation without *t*

You need Equations **1** and **3** to derive the last useful equation of motion.

According to Equation **1** the time t is given by the equation

$$t = \frac{(v - u)}{a}$$

This equation for t can be substituted into Equation **3** to give

$$s = \frac{1}{2}(u + v) \times \frac{(v - u)}{a}$$

Rearranging this gives

$$(u + v)(v - u) = 2as$$

$$v^2 - u^2 = 2as$$

$$\boxed{v^2 = u^2 + 2as} \qquad \textbf{(4)}$$

Study tip

The four *suvat* equations are

$v = u + at$

$s = ut + \frac{1}{2}at^2$

$s = \frac{1}{2}(u + v)t$

$v^2 = u^2 + 2as$

Equation without *u*

There is one more equation, $s = vt - \frac{1}{2}at^2$.

See if you can derive it from the velocity–time graph.

Study tip

You can always make a good start by jotting down the *suvat* values to see what you know.

▲ **Figure 3** *Accelerating from rest*

Worked example: Car accelerating from a standing start

A car on a straight road accelerates from rest to a velocity of $12\,\mathrm{m\,s^{-1}}$ in a time of 9.0 s. Calculate the acceleration of the car and the distance travelled in this time.

Step 1: Write down the quantities given in the order *suvat* and identify the equations needed.

$$s = ?,\ u = 0,\ v = 12\,\mathrm{m\,s^{-1}},\ a = ?,\ t = 9.0\,\mathrm{s}$$

Use Equation **1** to calculate the acceleration a and Equation **3** to calculate s.

Step 2: Substitute the values into the equations and calculate the answers.

Acceleration can be calculated from Equation **1**.

$$v = u + at$$

$$12 = 0 + a \times 9.0$$

$$a = \frac{12 - 0}{9.0} = 1.33\,\mathrm{m\,s^{-2}}$$

The distance travelled is the displacement *s* along the straight road. Equation **3** can be used to calculate *s*.

$$s = \frac{1}{2}(v + u)t$$

$$s = \frac{1}{2} \times (12 + 0) \times 9.0$$

$$s = 54\,\mathrm{m}$$

Worked example: Particle accelerating

A particle travels a distance of 16 m as it accelerates from $4.0\,\mathrm{m\,s^{-1}}$ to $12\,\mathrm{m\,s^{-1}}$. Calculate its acceleration.

Step 1: Again, start with the *suvat* values and identify the equation needed.

$s = 16\,\mathrm{m}$, $u = 4.0\,\mathrm{m\,s^{-1}}$, $v = 12\,\mathrm{m\,s^{-1}}$, $a = ?$, $t = ?$

The equation for *v*, *u*, and *s* is Equation **4.** We can use this to calculate the acceleration *a*.

Step 2: Substitute the values into the equation and calculate the answer.

$$v^2 = u^2 + 2as$$

$$12^2 = 4.0^2 + 2 \times a \times 16$$

$$a = \frac{12^2 - 4.0^2}{2 \times 16} = 4.0\,\mathrm{m\,s^{-2}}$$

Study tip

Remember that acceleration is the change in velocity with time, so although this first sentence refers to acceleration the values given are two different velocities. If you are ever confused, simply look at the units.

Synoptic link

You can find more information about units in Appendix A1, Physical quantities and units.

Worked example: Falling to Earth

An apple falls from rest in a tree towards soft ground from a height of 1.50 m. Objects falling to Earth have an acceleration (free fall) of $9.81\,\mathrm{m\,s^{-2}}$. Calculate the time taken for the apple to reach the ground. Assume air resistance has negligible effect on the motion.

Step 1: List the *suvat* values and identify the equation needed.

$s = 1.50\,\mathrm{m}$, $u = 0\,\mathrm{m\,s^{-1}}$, $a = 9.81\,\mathrm{m\,s^{-2}}$, $t = ?$

Use Equation **2** to calculate the time *t*.

Study tip

Note that $0 \times t = 0$ and not t.

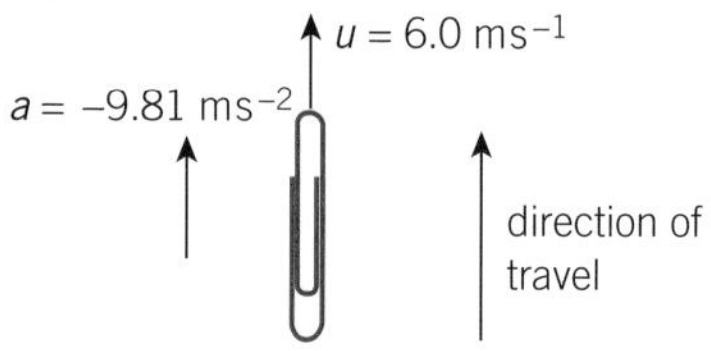

▲ Figure 4

Step 2: Substitute the values into the equation and calculate the answer.

$$s = ut + \frac{1}{2}at^2$$

$$1.50 = (0 \times t) + \frac{1}{2} \times 9.81 \times t^2$$

$$t^2 = \frac{2 \times 1.50}{9.81} = 0.306 \text{ (3 s.f.)}$$

$$t = 0.553\,\text{s}$$

In this calculation, the equation is effectively $s = \frac{1}{2}at^2$, because $u = 0$.

Worked example: What goes up

A paper clip is flicked vertically up in the air at 6.0 m s^{-1} (Figure 4). Calculate its maximum height. Assume air resistance has negligible effect on the motion.

Step 1: List the known *suvat* values and identify the equation needed.

The paper clip will decelerate as it moves vertically, therefore, a must be *negative*. At maximum height it will stop momentarily, therefore, $v = 0$.

$$s = ?,\ u = 6.0\,\text{m s}^{-1},\ v = 0\,\text{m s}^{-1},\ a = -9.81\,\text{m s}^{-2}$$

The equation for v, u, and s is Equation **4.** We can use this to calculate the height s.

Step 2: Substitute the values into the equation and calculate the answer.

$$v^2 = u^2 + 2as$$

$$0 = u^2 + 2as$$

$$s = -\frac{u^2}{2a} = -\frac{6.0^2}{2 \times -9.81}$$

$$s = 1.83\,\text{m} = 1.8\,\text{m (2 s.f.)}$$

Summary questions

1. A vehicle accelerated from 13.4 m s^{-1} to 22.3 m s^{-1} in 8.70 s. Calculate the distance travelled. *(3 marks)*
2. A runner changes her velocity from 3.2 m s^{-1} to 4.2 m s^{-1}. In this time she travels 200 m. Calculate her acceleration. *(3 marks)*
3. In 2009 Usain Bolt sprinted 100 m in 9.58 s, setting a record. Assuming that he travelled at a constant acceleration from start to the finish, calculate his acceleration. *(3 marks)*
4. A dragster travels 0.25 miles in a time of 4.6 s along a straight track from a standing start. 1 mile ≈ 1600 m. Calculate:
   a. its acceleration; *(3 marks)*
   b. its final velocity. *(3 marks)*
5. An apple is dropped from a tall building. Calculate the distance it travels between time t = 3.0 s and t = 5.0 s. Assume the acceleration of the falling apple is 9.81 m s^{-2}. *(4 marks)*
6. A car is travelling at 28 m s^{-1}. The driver applies the brakes. The car skids for a distance of 30 m before stopping. Calculate the magnitude of the deceleration of the car. *(3 marks)*

3.6 Car stopping distances

Specification reference: 3.1.2

Stopping distances

Modern road vehicles transport people in comfort at speeds that would have astonished our great-grandparents. This speed is not always a good thing. An alert driver who has left enough room can stop in good time in good driving conditions on seeing a hazard. However, in an emergency or in poor conditions, the vehicle may still be moving when it meets an obstacle, even if the driver is braking hard.

Components of stopping distances

The **stopping distance** is the total distance travelled from when the driver first sees a reason to stop, to when the vehicle stops. It has two components:

- **thinking distance**, the distance travelled between the moment when you first see a reason to stop, to the moment when you use the brake
- **braking distance**, the distance travelled from the time the brake is applied until the vehicle stops.

Many factors influence these distances, including the speed of the vehicle, the condition of the brakes, tyres, and road, the weather conditions, and the alertness of the driver.

Thinking distance

It takes time for a driver to react to a need to stop. For a vehicle moving at constant speed

$$\text{thinking distance} = \text{speed} \times \text{reaction time}$$

Worked example: Reaction time

In the UK Highway Code, the thinking distance at 30 mph ($13.4\,\text{m s}^{-1}$) is shown as 9.0 m. Calculate the corresponding reaction time.

Step 1: Identify the equation needed.

$$\text{reaction time} = \frac{\text{thinking distance}}{\text{speed}}$$

Step 2: Substitute the values into the equation and calculate the answer.

$$\text{reaction time} = \frac{9.0}{13.4} = 0.67\,\text{s (2 s.f.)}$$

The greater the speed or the reaction time, the further a vehicle will travel before its driver applies the brakes. Assuming a constant reaction time of 0.67 s, the thinking distance will be about 21 m at the UK national speed limit of 70 mph ($31.1\,\text{m s}^{-1}$). This is equivalent to the total length of five average cars lined up.

Learning outcomes

Demonstrate knowledge, understanding, and application of:

→ thinking distance and braking distance

→ the effect of reaction time on total stopping distance of a vehicle.

▲ **Figure 1** *Some motorways have chevron markings to help drivers judge the safe distance from the vehicle in front*

Study tip

Do not confuse distance and time. If you write that 'it takes longer to stop' you must make it clear whether you mean the distance or the time.

▲ **Figure 2** *Vehicles following each other too closely may not be able to stop in a short enough distance in an emergency*

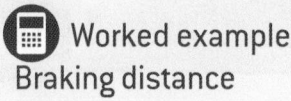

Worked example: Braking distance

Step 1: Once again, start with the *suvat* quantities and identify the equation you need.

$s = 14.0\,\text{m}$

$u = 13.4\,\text{m s}^{-1}$

$v = 0$

$a = ?$

Use the equation $v^2 = u^2 + 2as$.

Step 2: Substitute the values into the equation and calculate the answer.

$$a = \frac{v^2 - u^2}{2s}$$

$v = 0$

Therefore

$$a = -\frac{u^2}{2s}$$

$$= -\frac{13.4^2}{2 \times 14.0}$$

$$= -6.4\,\text{m s}^{-2}\ \text{(2 s.f.)}$$

The magnitude of the deceleration is about $6.4\,\text{m s}^{-2}$.

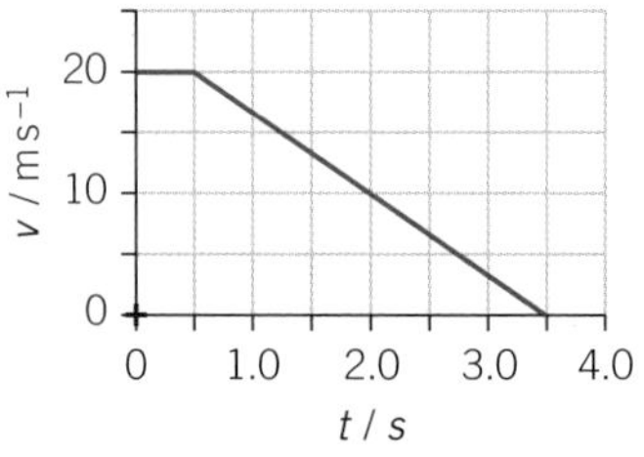

▲ Figure 3

Braking distance

In the UK Highway Code, the braking distance at 30 mph ($13.4\,\text{m s}^{-1}$) is shown as 14.0 m. If you assume constant deceleration from $13.4\,\text{m s}^{-1}$ to $0\,\text{m s}^{-1}$, you can use one of the equations of motion to determine the magnitude of the deceleration.

▼ **Table 1** *Thinking, braking, and overall stopping distances according to the Highway Code*

Speed / mph	20	30	40	50	60	70
Speed / m s^{-1}	8.9	13.4	17.8	22.2	26.7	31.1
Thinking distance / m	6	9	12	15	18	21
Braking distance / m	6	14	24	38	55	75
Stopping distance / m	12	23	36	53	73	96

Summary questions

1 The reaction time of a tired driver is 1.5 s. The speed of the car is $22\,\text{m s}^{-1}$. The braking distance of the car is 38 m. Calculate the stopping distance of the car. *(3 marks)*

2 According to a student, thinking distance is directly proportional to the speed of the car. Show that this is the case. *(2 marks)*

3 Use Table 1 to answer this question. A car is travelling at 70 mph ($31.1\,\text{m s}^{-1}$) on the motorway when it has to stop for an emergency. Calculate:

a the deceleration of the car when travelling at this speed; *(4 marks)*

b the time taken for the car to stop when the brakes are applied. *(3 marks)*

4 The velocity–time graph in Figure 3 shows the motion of a car from the instant the driver sees a hazard on the road.

Calculate the thinking, braking, and stopping distances. Explain your answer. *(3 marks)*

5 According to a student, braking distance is directly proportional to the $(\text{speed})^2$. Show that this is the case. *(3 marks)*

3.7 Free fall and *g*

Specification reference: 3.1.2

From the edge of space

Felix Baumgartner made a record-breaking leap from the edge of space on 14th October 2012. A giant helium-filled balloon lifted his capsule 39.0 km above the surface of the Earth. Then he stepped off. Baumgartner accelerated as he fell, reaching a maximum speed of $380\,m\,s^{-1}$ – greater than the speed of sound – after just 50 seconds. He fell 36.4 km in 4 minutes 20 seconds before deploying his parachute and landing safely just under 5 minutes later.

Acceleration due to gravity

Objects with mass exert a gravitational force on each other. The Earth is so massive that its gravitational pull is enough to keep us on its surface. An object released on the Earth will accelerate vertically downwards towards the centre of the Earth. When an object is accelerating under gravity, with no other force acting on it, it is said to be in **free fall**. The **acceleration of free fall** is denoted by the label g (not g, which means grams). Since g is an acceleration, it has the unit $m\,s^{-2}$.

Value for *g* close to Earth's surface

The value for g varies depending upon factors including altitude, latitude, and the geology of an area. For example, g is $9.825\,m\,s^{-2}$ in Helsinki, $9.816\,m\,s^{-2}$ in London, but only $9.776\,m\,s^{-2}$ in Singapore. A value of $9.81\,m\,s^{-2}$ is generally used.

Determining *g*

The basic idea behind determining g in the laboratory is to drop a heavy ball over a known distance and time its descent. The problem is that it all happens very quickly, about 0.45 s for a 1.0 m fall. Methods for measuring g are described here.

Electromagnet and trapdoor

An electromagnet holds a small steel ball above a trapdoor (Figure 2). When the current is switched off, a timer is triggered, the electromagnet demagnetises, and the ball falls. When it hits the trapdoor, the electrical contact is broken and the timer stops. The value for g is calculated from the height of the fall and the time taken.

Learning outcomes

Demonstrate knowledge, understanding, and application of:

- → the equations of motion for falling objects in a uniform gravitational field
- → the acceleration due to free fall g
- → an experiment to determine g.

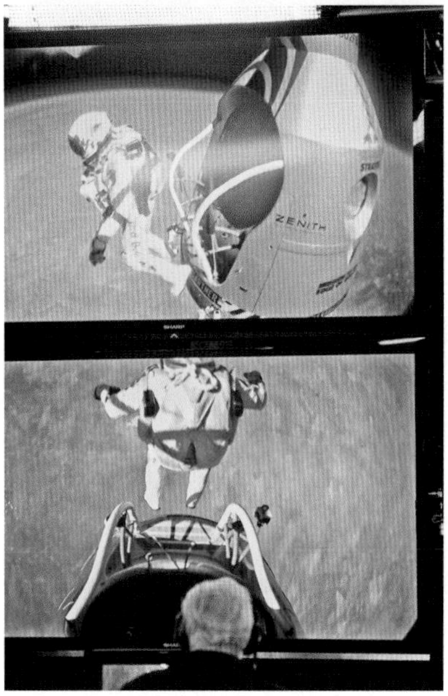

▲ **Figure 1** *Staff at the flight control centre monitoring Felix Baumgartner leaving his capsule*

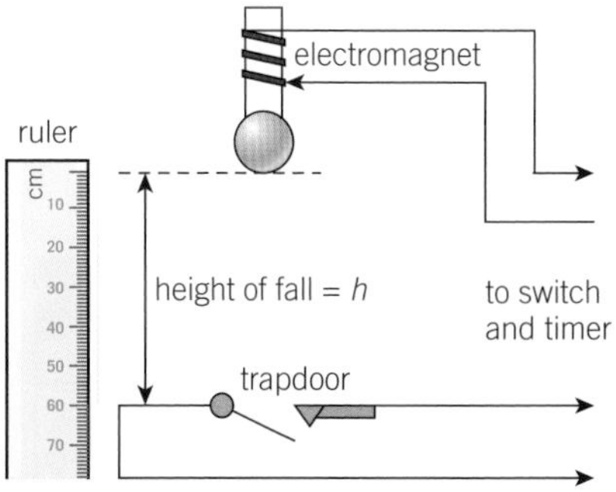

▲ **Figure 2** *Determining g using an electromagnet and timer*

Worked example: An experimental value for g

A ball drops 85.6 cm from an electromagnet to a trapdoor in 0.421 s. Use this information to determine a value for g.

Step 1: List the *suvat* values and identify the equation needed.

$s = 0.856\,\text{m}$, $u = 0\,\text{m}\,\text{s}^{-1}$, $a = g = ?$, $t = 0.421\,\text{s}$ (the distance must be converted into the SI unit m.)

Using $s = ut + \frac{1}{2}at^2$, we have

$$s = \frac{1}{2}at^2 \text{ because } u = 0$$

Step 2: Substitute the values into the equation and calculate the answer.

$$a = \frac{2s}{t^2} = \frac{2 \times 0.856}{0.421^2}$$

$$a = 9.66\,\text{m}\,\text{s}^{-2} \text{ (3 s.f.)}$$

The inaccuracy in this experiment is caused by the presence of air resistance and the slight delay in the release of the steel ball because of the finite time taken for the magnet to demagnetise. The accuracy may be improved by using a heavier ball and a much longer drop.

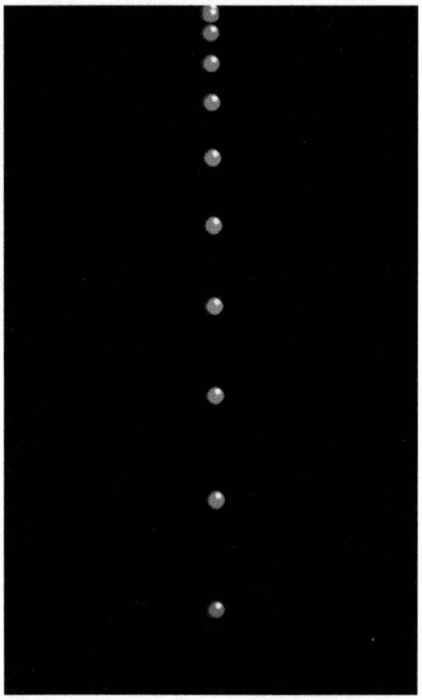

▲ **Figure 3** *The photos are taken at regular intervals and the distance between each image of the ball increases as it falls vertically towards the ground, showing that it is accelerating*

Light gates

The electromagnet and trapdoor introduce tiny delays into the timing. Instead we can use 'light gates', two light beams, one above the other, with detectors connected to a timer. When the ball falls through the first beam, it interrupts the light and the timer starts. When the ball falls through the second beam a known distance further down, the timer stops.

Taking pictures

A small metal ball is dropped from rest next to a metre rule, and its fall is recorded on video or with a camera in rapid-fire repeating mode. Alternatively, a stroboscope illuminates the scene with rapid flashes. The camera shutter is held open, producing a photograph with multiple images of the falling ball. The position of the ball at regular intervals is then determined by examining the recording.

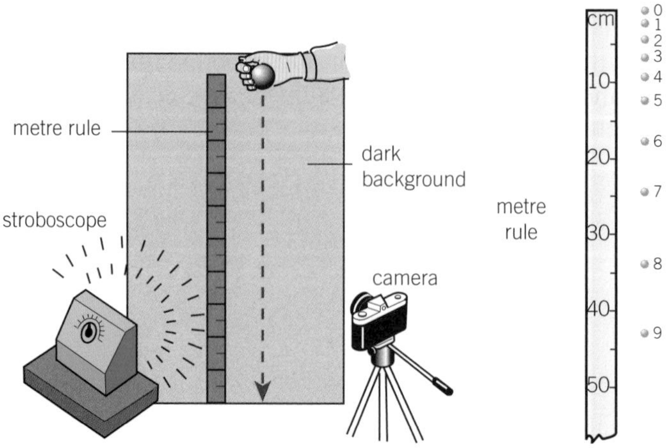

▲ **Figure 4** *Determining g using a camera and stroboscope*

Synoptic link

You will find more information about lines of best fit in Appendix A3, Measurements and uncertainties.

Determining g by plotting a graph

The table shows data obtained from images of a ball in free fall.

The acceleration g of free fall can be determined using the equation $s = ut + \frac{1}{2}at^2$.

Since the object is dropped from rest and $a = g$

$$s = \frac{1}{2}gt^2$$

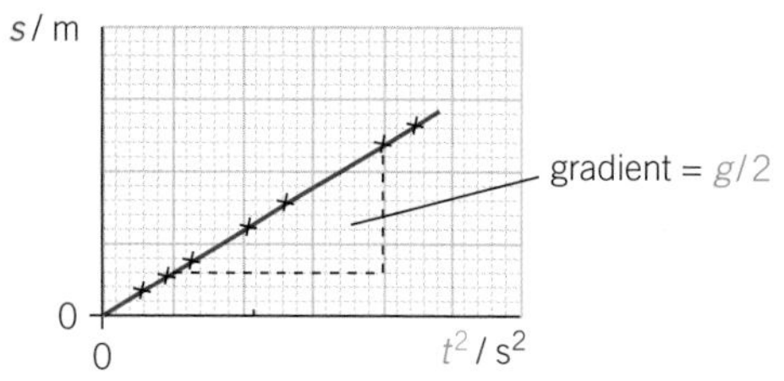

▲ **Figure 5** *Graph of the experimental results plotted as s against t^2*

▼ **Table 1** *Results from the experiment in Figure 4*

Time of fall t / s	0.132	0.165	0.198	0.231	0.264	0.297	0.330	0.363	0.396	0.429
Distance fallen s / m	0.085	0.134	0.192	0.262	0.368	0.433	0.534	0.646	0.769	0.903

From the general equation for a straight-line graph, $y = mx + c$, you should get a straight line if you plot a graph of s against t^2 ($y = s$, $m = \frac{1}{2}g$, $x = t^2$, $c = 0$). The gradient, m, will be equal to $\frac{g}{2}$.

For each value of t, calculate t^2. Plot a graph of s (y-axis) against t^2 (x-axis). Draw a straight line of best fit, ignoring any anomalous points.

1. Determine the gradient of the line using a large triangle.
2. Calculate the experimental value for g.
3. What is the percentage difference between the experimental and accepted values for g?

Summary questions

1. Two different heavy objects are dropped from the same height. State the acceleration of free fall of each object. State any assumptions made. *(2 marks)*
2. A marble is dropped from a height H. It lands on the ground below after a time of 2.3 s. Calculate H. Assume the acceleration of free fall is 9.81 m s^{-2}. *(3 marks)*
3. Plan a simple experiment to estimate the acceleration g of free fall using a stopwatch and a tape measure. Explain how the experiment can be made precise. *(4 marks)*
4. A coin is dropped from the top of a bridge towards the water 9.5 m below. The coin is in free fall for a time of 1.5 s.
   a. Estimate the acceleration of free fall. *(3 marks)*
   b. Explain why the answer is not 9.81 m s^{-2}. *(1 mark)*
5. Figure 6 shows water dripping from a tap. The time between successive water drops is 0.040 s.
   a. Use the position of water drop **A** to determine the acceleration of free fall. *(3 marks)*
   b. Repeat (a) for another drop and therefore determine an average experimental value for g. *(4 marks)*

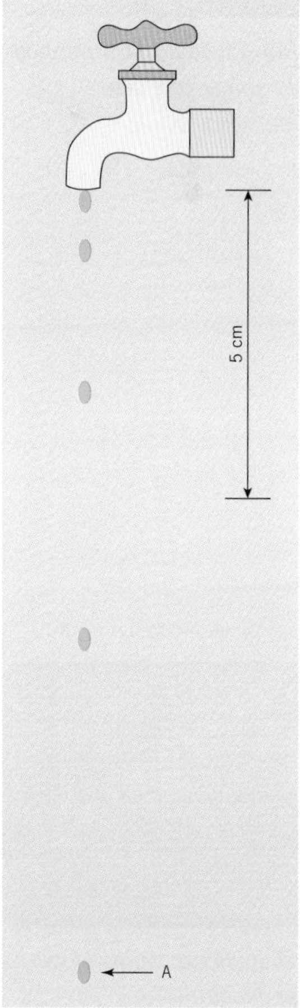

◀ **Figure 6**

3.8 Projectile motion

Specification reference: 3.1.3

Learning outcomes

Demonstrate knowledge, understanding, and application of:

→ the two-dimensional motion of projectiles

→ the equations of motion for projectiles.

▲ **Figure 1** *How far could a cannonball travel before landing in the sea?*

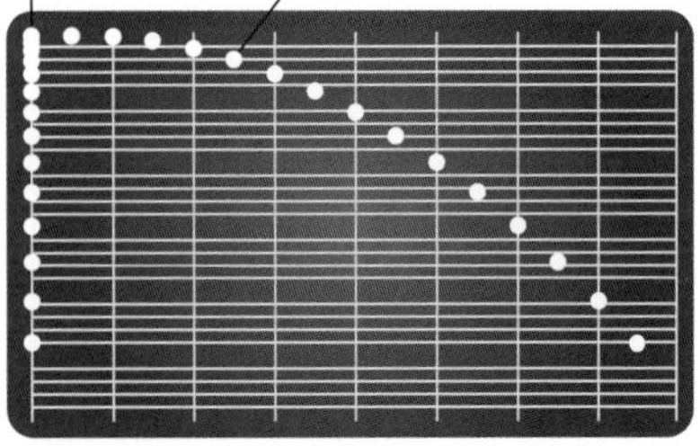

▲ **Figure 2** *Multiple exposures of two objects released together*

Hitting the target

In the past, cannons were sited on clifftops to defend against attack from the sea. How far **projectiles** like cannonballs can travel depends on several factors. Ignoring the effect of air resistance, once a cannon has fired, the range depends on the height of the cannon above the sea and the initial velocity of the ball.

Independent motion

Figure 2 shows multiple images of two balls. The time interval between successive images is the same. One ball was dropped *vertically*, whilst the other was thrown *horizontally*. You will notice that both balls fall at the same rate – they are both at the same height at the same time. It does not matter whether the ball is moving horizontally. The vertical and horizontal motions of the ball are independent of each other.

Assuming no air resistance

- the vertical velocity changes due to acceleration of free fall
- the vertical displacement and **time of flight** can be calculated using equations of motion
- horizontal velocity remains constant.

Why does the horizontal velocity of the projectile remain constant? Remember that acceleration and velocity are vectors. The acceleration of free fall is vertically downwards. The component of this acceleration in the horizontal direction is zero.

$$\text{horizontal acceleration} = g\cos 90° = 0$$

The horizontal velocity is therefore unaffected by the fall.

Worked example: Cannonball

A cannonball is fired horizontally from a clifftop 44.1 m above the sea. The initial horizontal velocity of the cannonball is $304\,\text{m s}^{-1}$. Calculate:

a the time of flight;

b the horizontal distance it travels.

Remember that the vertical motion and the horizontal motion are independent of each other.

a There is acceleration in the vertical direction.

Step 1: Identify the equation needed and list the known values.

We can use the equation $s = ut + \frac{1}{2}at^2$ to calculate the time t of flight.

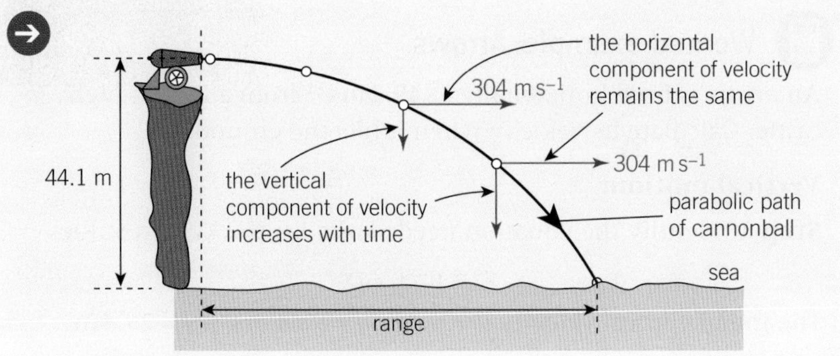

▲ **Figure 3** *Trajectory of a cannonball*

The initial vertical velocity $u = 0$ (initial vertical velocity $u = 304 \times \cos 90° = 0$).

$$s = 44.1\ \text{m},\ u = 0,\ a = g = 9.81\ \text{m s}^{-2},\ t = ?$$

Step 2: Substitute the values into the equation and calculate the answer.

$$s = \frac{1}{2}at^2$$

$$t^2 = \frac{2s}{a} = \frac{2 \times 44.1}{9.81} = 8.991\ \text{s}^2$$

$$t = 3.00\ \text{s}$$

b There is no acceleration in the horizontal direction.

Step 1: Identify the equation needed.

No acceleration, therefore
horizontal distance = horizontal velocity × time

Step 2: Substitute the values into the equation and calculate the answer.

$$\text{horizontal distance} = 304 \times 3.00 = 912\ \text{m}$$

The horizontal range of the cannonball is 912 m, almost 1 km.

Vector calculations

The path described by the cannonball in the worked example is curved because the vertical component of its velocity increases with time whilst the horizontal component is unaffected.

The magnitude of the actual velocity v of the cannonball, or any other projectile, can be calculated from the vertical and horizontal components v_x and v_y of this velocity. You just use Pythagoras' theorem (Figure 4).

$$\text{Actual velocity } v = \sqrt{v_x^2 + v_y^2}$$

The angle θ made by the velocity to the horizontal is given by

$$\theta = \tan^{-1}\left(\frac{v_y}{v_x}\right)$$

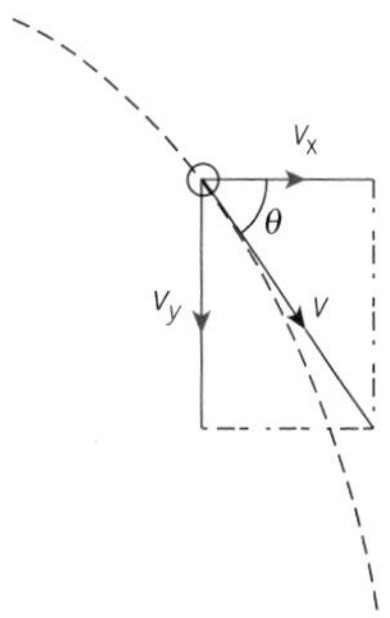

▲ **Figure 4** *The velocity v of a projectile has vertical and horizontal components*

Worked example: Arrows

An arrow is fired horizontally at $49.2\,m\,s^{-1}$ from a 29.4 m high castle. Calculate its velocity when it hits the ground.

Vertical motion:

Step 1: Identify the equation needed and list the known values.

$$v^2 = u^2 + 2as$$

The initial vertical velocity $u = 0$, $a = g = 9.81\,m\,s^{-2}$, $s = 29.4\,m$

Step 2: Substitute the values into the equation and calculate the answer.

$$v^2 = 0^2 + (2 \times 9.81 \times 29.4) = 576.8\,m^2\,s^{-2}$$

final vertical velocity $v_y = 24.0\,m\,s^{-1}$

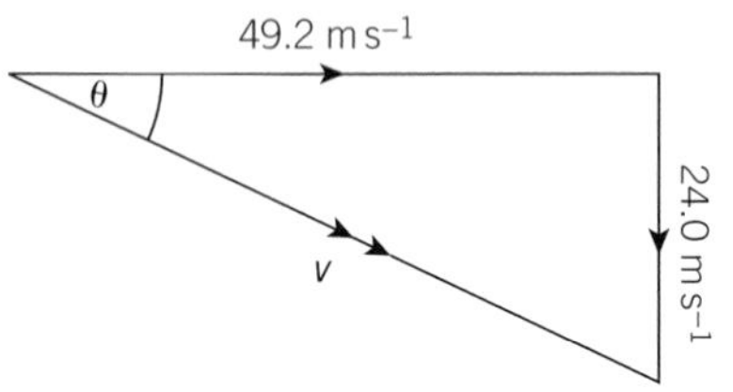

▲ Figure 5

Horizontal motion:

Step 1: Identify the method needed and list the known values.

Use Pythagoras' theorem (Figure 5).

The horizontal component of the velocity $v_x = 49.2\,m\,s^{-1}$.

Step 2: Substitute the values into the equation and calculate the answer.

$$\text{velocity of arrow} = \sqrt{24.0^2 + 49.2^2} = 54.7\,m\,s^{-1}$$

$$= \tan^{-1}\frac{24.0}{49.2} = 26.0^\circ \text{ to the horizontal.}$$

Firing upwards

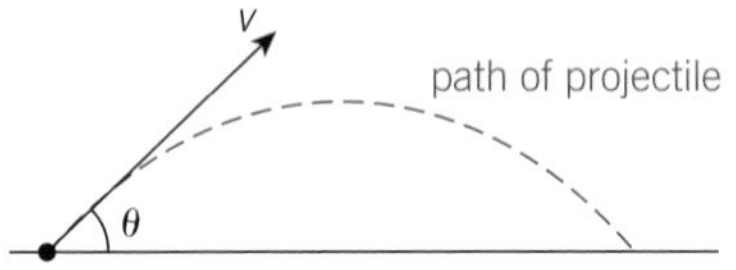

▲ **Figure 6** *The curved path described by a projectile fired upwards*

Figure 6 shows the path of a projectile fired at an angle θ to the horizontal. The initial velocity of the projectile is v. The motion of this projectile can still be analysed in terms of the independence of motion in the horizontal and vertical directions.

The horizontal component of the velocity is $v\cos\theta$ and the initial vertical upwards component of the velocity is $v\sin\theta$.

Worked example: Kicking a ball

A ball is kicked with an initial velocity of $15\,m\,s^{-1}$ at an angle of 30° to the horizontal. Calculate the range R of this ball.

Step 1: Identify the method needed.

Work out the time of flight from the vertical motion, then the horizontal range from the time (Figure 7).

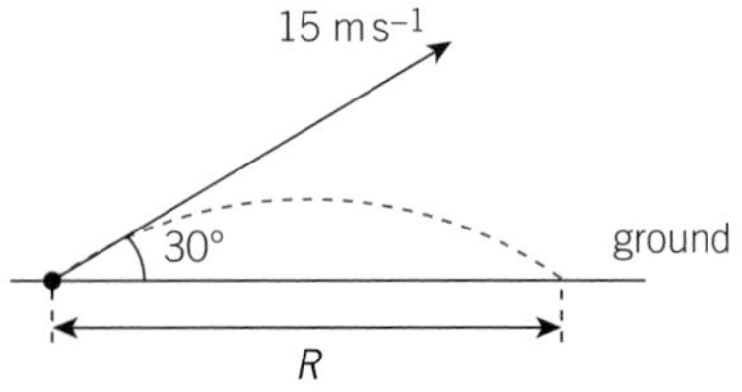

▲ **Figure 7** *How far will this ball travel?*

Step 2: List the values and the equation for the first part.

Vertical motion:

The time of the ball in flight can be calculated from its vertical motion. Note that the velocity of the ball when it hits the ground has the same magnitude as, but the opposite sign to, its upward velocity when it left the ground.

$$u = 15\sin 30° = 7.5\,\mathrm{m\,s^{-1}},\ v = -7.5\,\mathrm{m\,s^{-1}},\ a = -9.81\,\mathrm{m\,s^{-2}},\ t = ?$$

$$v = u + at$$

Step 2: Substitute the values into the equation and calculate the answer.

$$v = u + at$$

$$t = \frac{v-u}{a} = \frac{-7.5-7.5}{-9.81} = 1.53\ \mathrm{s}$$

Step 3: Use the answer for vertical motion to calculate the answer for the range.

Horizontal motion:

The horizontal component of the velocity is $15\cos 30° = 13.0\,\mathrm{m\,s^{-1}}$. This remains constant throughout the flight. Therefore

$$\text{range } R = (15\cos 30°) \times 1.53 = 20\ \mathrm{m}\ (2\ \text{s.f.})$$

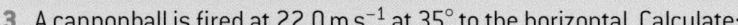

Summary questions

1 A ball is kicked into the air. Figure 8 shows the velocity components of the ball at a particular instant. Calculate the velocity *v* of the ball. *(2 marks)*

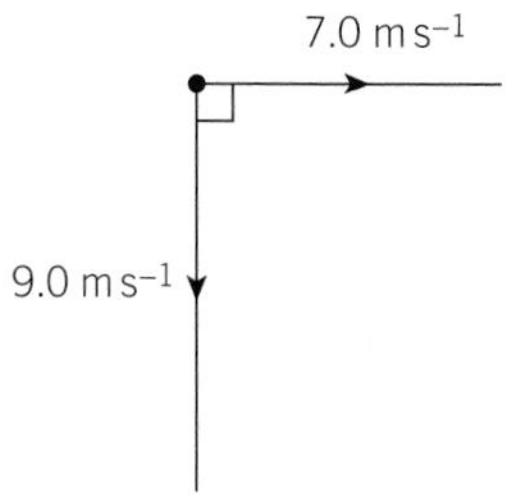

▲ **Figure 8**

2 A cannonball is fired horizontally from a cliff 29 m above the sea. The initial horizontal velocity of the cannonball is $320\ \mathrm{m\,s^{-1}}$. Calculate:

- **a** the time of flight; *(3 marks)*
- **b** the horizontal distance it travels; *(3 marks)*
- **c** the speed at which it hits the sea. *(3 marks)*

3 A cannonball is fired at $22.0\ \mathrm{m\,s^{-1}}$ at 35° to the horizontal. Calculate:

- **a** the maximum vertical height of the ball; *(4 marks)*
- **b** the horizontal distance travelled by the ball. *(5 marks)*

4 Sketch the vertical velocity–time graph for the cannonball in **3**. *(3 marks)*

Practice questions

1 a Copy and complete by stating the value or name of each of the remaining three prefixes.

▼ Table 1

prefix	value
micro (μ)	10^{-6}
mega (M)	
	10^{-9}
tera (T)	

(3 marks)

b Write down all the scalar quantities in the list below.

density **weight** **velocity**
volume **acceleration** *(1 mark)*

c The distance between the Sun and the Earth is 1.5×10^{11} m. Calculate the time in minutes for light to travel from the Sun to the Earth. The speed of light is $3.0 \times 10^{8}\,\mathrm{m\,s^{-1}}$. *(2 marks)*

May 2010 G481 Mechanics

2 a State the difference between a *scalar* quantity and a *vector* quantity. *(2 marks)*

b Define *velocity* and derive its base units. *(2 marks)*

c Figure 1 show the displacement *s* against time *t* for an object.

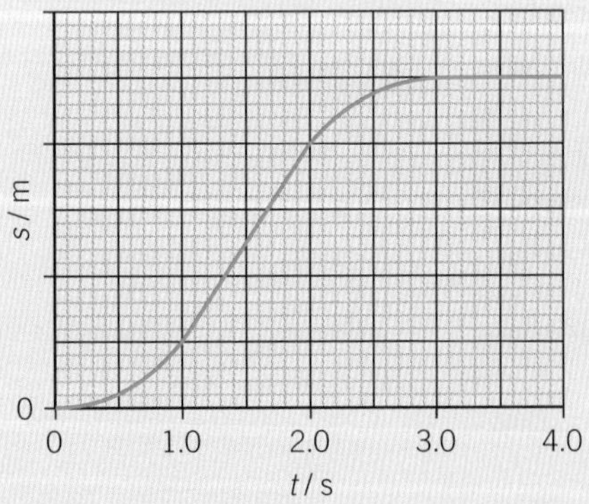

▲ Figure 1

Use the graph to describe and explain how the velocity for this object changes with time from $t = 0$ to $t = 4.0$ s. *(5 marks)*

3 A student uses the apparatus shown in Figure 2 to determine the acceleration of free fall *g*.

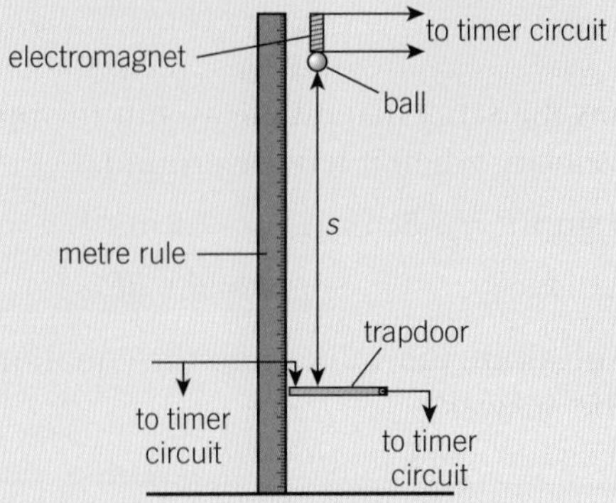

▲ Figure 2

The metal ball drops when the electromagnet is switched off. This also starts the timer. The timer stops when the ball opens the trapdoor below. The time of fall of the ball is *t*. The distance *s* between the trapdoor and the bottom of the metal ball is changed. Figure 3 shows the graph plotted by the student.

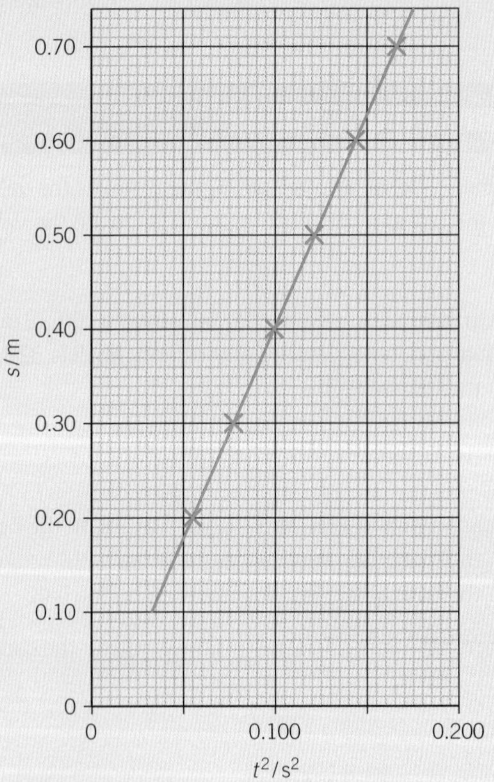

▲ Figure 3

a Use an equation of motion to explain why the graph of *s* against t^2 is a straight line. *(2 marks)*

b Use the graph to determine the acceleration of free fall of the ball. *(3 marks)* c Calculate the percentage difference between your value in **(b)** and the accepted value for g. *(1 mark)*

d Explain how you can deduce that there is a systematic error in this experiment. Suggest what this error is likely to be. *(2 marks)*

4 a Define *acceleration*. *(1 mark)*

b A super-tanker cruising at an initial velocity of $6.0\,\mathrm{m\,s^{-1}}$ takes 40 minutes (2400 s) to come to a stop. The super-tanker has a constant deceleration.

(i) Calculate the magnitude of the deceleration. *(3 marks)*

(ii) Calculate the distance travelled in the 40 minutes it takes the tanker to stop. *(2 marks)*

(iii) On a copy of Figure 4, sketch a graph to show the variation of distance x travelled by the super-tanker with time t as it decelerates to a stop.

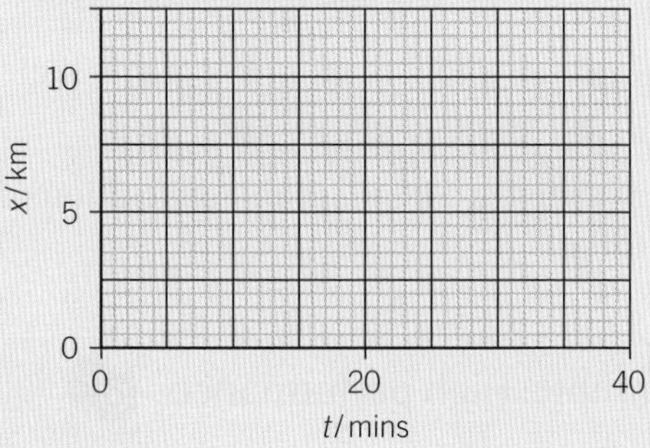

▲ Figure 4

(2 marks)

Jan 2012 G481

5 Figure 5 shows the variation of velocity v with time t for a small rocket.

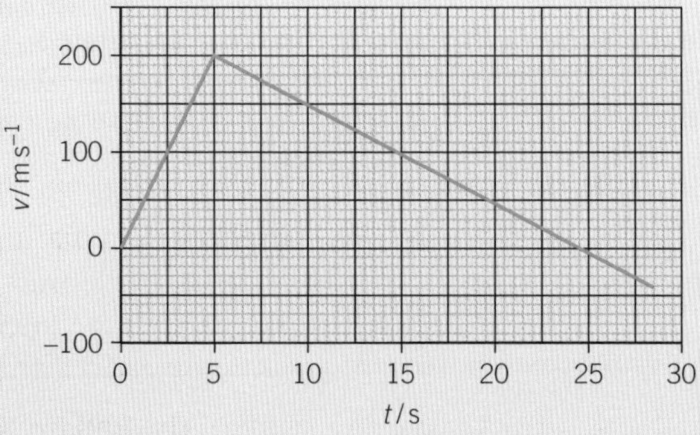

▲ Figure 5

The rocket is initially at rest and is fired vertically upwards from the ground. All the rocket fuel is burnt after a time of 5.0 s when the rocket has a vertical velocity of $200\,\mathrm{m\,s^{-1}}$. Assume that air resistance has a negligible effect on the motion of the rocket.

(i) Without doing any calculations, describe the motion of the rocket

1 from $t = 0$ to $t = 5.0\,\mathrm{s}$

2 from $t = 5.0\,\mathrm{s}$ to $t = 25\,\mathrm{s}$. *(3 marks)*

(ii) Calculate the maximum height reached by the rocket. *(3 marks)*

(iii) Explain why the rocket has a speed greater than $200\,\mathrm{m\,s^{-1}}$ as it hits the ground. *(1 mark)*

Jan 2013 G481

6 Figure 6 shows the path of a ball falling from the top of a table. The initial velocity of the ball is in the horizontal direction and has magnitude $2.0\,\mathrm{m\,s^{-1}}$.

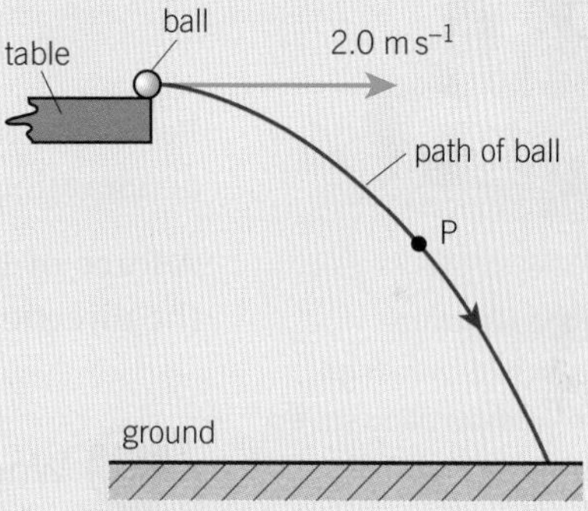

▲ Figure 6

a State the direction of the acceleration of the ball when at point **P**. *(1 mark)*

b Describe and explain the variation of the vertical component of the velocity of the ball as it travels towards the ground. *(3 marks)*

c The vertical component of the velocity at **P** is $2.9\,\mathrm{m\,s^{-1}}$. Calculate

(i) the velocity of the ball at **P**, *(3 marks)*

(ii) the angle made by the velocity of the ball with the horizontal when at **P**. *(2 marks)*

4 FORCES IN ACTION

4.1 Force, mass, and weight

Specification reference: 3.2.1

Learning outcomes

Demonstrate knowledge, understanding, and application of:

- → the net force on an accelerating object
- → the newton
- → weight.

▲ **Figure 1** *One of the UK's three standard kilograms, a 39.17 mm high cylinder of platinum–iridium alloy stored in a bell jar at the National Physical Laboratory*

Study tip

Remember that $F \propto a$ *when* m *is constant.*

$a \propto \frac{1}{m}$ when F is constant.

Synoptic link

The equation $F = ma$ is often referred to as Newton's second law. As you will see in Topic 7.3, Newton's second law of motion, this law is defined in terms of rate of change of momentum. $F = ma$ is just a special case for constant mass.

The last artefact

The SI **base units** for length, time, current, and light intensity are directly or indirectly defined in terms of the speed of light. The base unit of mass is still defined by an artefact (an actual object). The international prototype kilogram or IPK was made in 1889. It is stored in a vault near Paris. By definition, the IPK has a mass of exactly 1 kg. Copies are used as standards around the world.

Force, mass, and acceleration

The **mass** of an object is one of its physical properties, and depends on the amount of matter it contains. A *net (resultant)* **force** acting on the object will make the object accelerate in the direction of the net force. The net force F, mass m of the object, and acceleration a of the object are related by the equation

$$F = ma$$

Force is measured in newtons (N), mass in kilograms (kg), and acceleration in metres per second squared (m s^{-2}).

A force of 1 newton will give a 1 kg mass an acceleration of 1 m s^{-2} in the direction of the force.

Worked example: High performance

An electric car has a mass of 2.1×10^3 kg. It accelerates from rest to 27 m s^{-1} in 5.4 s. Calculate the net force acting on the car.

Step 1: Identify the equation needed and list the known values.

$$m = 2.1 \times 10^3\,\text{kg},\ u = 0,\ v = 27\,\text{m s}^{-1},\ t = 5.4\,\text{s}$$

$$F = ma$$

Step 2: Substitute the values into the equation and calculate the answer.

$$F = 2.1 \times 10^3 \times \left(\frac{27-0}{5.4}\right) = 1.05 \times 10^4\,\text{N}$$

The net force acting on the car = 1.1×10^4 N (2 s.f.)

Mass and weight

In physics it is important to distinguish between mass and weight. You cannot afford to confuse these two quantities (Table 1).

▼ **Table 1** *Mass and weight*

Quantity	Unit	Comment
mass	kg	constant for a specific object or particle
weight	N (kg m s^{-2})	magnitude is variable – it depends on location

The **weight** of an object on the surface of the Earth is the gravitational force acting on the object. An object in free fall has an acceleration g of 9.81 m s^{-2}. The only force acting on the object is its weight W. Since $F = ma$, it follows that

$$W = mg$$

You can calculate the weight of an object on the Moon or other planets. Remember that the value of g will be different, but the mass m will be the same.

You can determine the weight of an object using a newtonmeter, which is calibrated to show the gravitational force acting on an object in newtons. A 1.0 kg object hanging from the newtonmeter will show a weight reading of about 9.8 N. The same object on the Moon would give a smaller reading of 1.6 N – the acceleration of free fall on the Moon is only 1.6 m s^{-2}. The mass remains constant, but weight is a variable.

▲ **Figure 2** *This astronaut weighs less than on the Earth and can walk in a heavy suit with little effort because the value of g on the surface of the Moon is much less than 9.81 m s^{-2}*

Study tip

The equations $F = ma$ and $W = mg$ are not provided in the examinations so you will have to remember them.

Understanding mass

The definition of mass goes back to the 17th century when Isaac Newton related mass, acceleration, and force. Mass was regarded as constant for a given object or particle (and we will treat it as constant in this course).

In 1905 Albert Einstein came up with the model of relativistic mass in his special theory of relativity. He found that the mass m of a particle depends on its speed v, according to the equation

$$m = \frac{m_0}{\sqrt{1 - \left(\frac{v}{c}\right)^2}}$$

where m_0 is the rest mass of the object and c is the speed of light in a vacuum ($c = 3.00 \times 10^8$ m s^{-1}).

Your mass will not alter much at the speeds at which we move around. You would have to travel close to the speed of light for your mass to change significantly.

1. Sketch a graph of relativistic mass m against the speed v of a particle.
2. The rest mass of the electron is 9.11×10^{-31} kg. Calculate its mass at:
   a 0.10 c (10% speed of light); **b** 0.999 c.
3. Explain whether or not an electron can travel at the speed of light c.

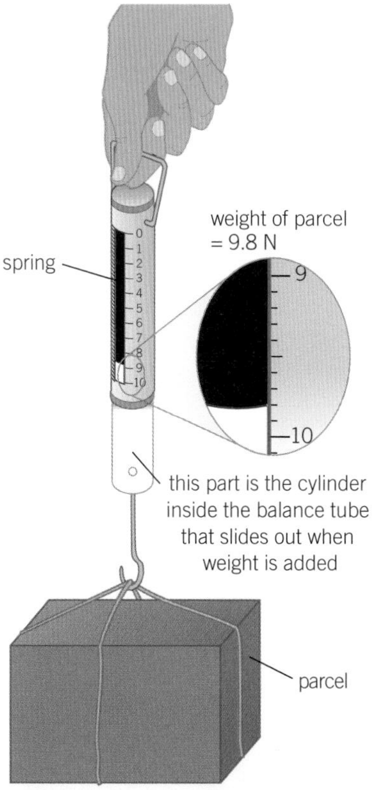

▲ **Figure 3** *You can use a newtonmeter to determine weight*

Summary questions

1 A resultant force of 500 N acts on a stationary car of mass 1200 kg. Calculate:

a the car's acceleration; *(2 marks)*

b its velocity after 6.0 s. *(2 marks)*

2 Calculate the mass in grams of a mobile phone of weight 1.1 N. *(2 marks)*

3 A golf ball has a mass of 46 g. It is hit with a force of 5.8 kN. Calculate the initial acceleration of the ball. *(2 marks)*

4 The weight of a car is 1.8×10^4 N. It accelerates from rest to a velocity of 28 m s^{-1} in a time of 9.6 s. Calculate the acceleration of the car. *(5 marks)*

5 The forces acting on a proton (mass 1.7×10^{-27} kg) are shown in Figure 4. Calculate the magnitude and direction of the acceleration of the proton. *(6 marks)*

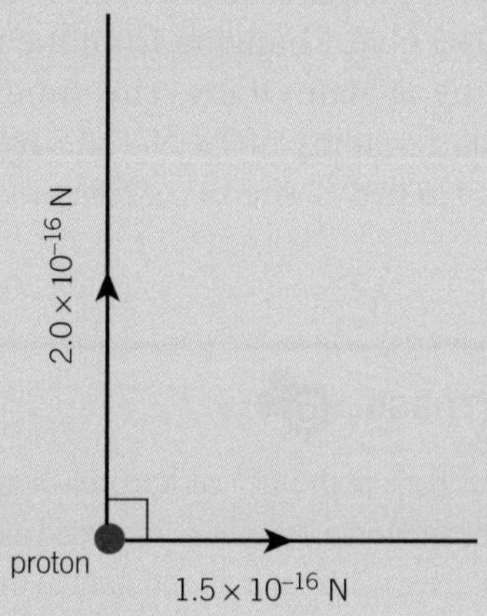

▲ **Figure 4**

6 A 8.0 g pellet travelling at 420 m s^{-1} hits a wooden crate. The pellet penetrates 98 mm into the crate. Calculate the average magnitude of the force acting on the pellet. *(3 marks)*

4.2 Centre of mass

Specification reference: 3.2.3

A balancing act

You need a sense of balance to stay upright. If you lean too far to one side you may fall over. Idol Rock in North Yorkshire (Figure 1) is an example of a natural balancing act. Over many years, the weaker layers of millstone grit have been eroded, leaving 200 tonnes of rock perched perilously on a small pyramid. As long as its **centre of mass** does not move to one side, Idol Rock will stay upright.

Learning outcomes

Demonstrate knowledge, understanding, and application of:

- → centre of gravity
- → an experiment to determine the centre of gravity of an object.

Where weight acts

Imagine pushing a spanner floating freely in space with your finger. The spanner will rotate, or move in a straight line, or both. The spanner can be made to move in a straight line if the force is applied along a line of action that coincides with its centre of mass. The centre of mass of an object is a point through which any externally applied force produces straight-line motion but no rotation.

Figure 2 shows an irregular object made up of identical atoms, each having a tiny weight w. The resultant gravitational force on the object, its total weight W, will act through a point often called the object's **centre of gravity**, which coincides with its centre of mass. The centre of mass of an object is an imaginary point where the entire weight of an object appears to act. The centre of mass of a uniform metre rule will be at its 50.0 cm mark. If you stand upright, your centre of mass is just behind your navel.

It is much easier to analyse and solve physics problems if we think about the weight of an object acting through its centre of mass rather than as a collection of many tiny forces on each part of the object. In Figure 3, for example, the complicated movement of the stunt rider can be represented by the smooth parabolic path described by his centre of mass.

▲ **Figure 1** *Idol Rock at Brimham Rocks near Harrogate in North Yorkshire*

▲ **Figure 3** *It is easier to analyse the motion of the stunt rider by looking at the motion of his centre of mass instead of his whole body*

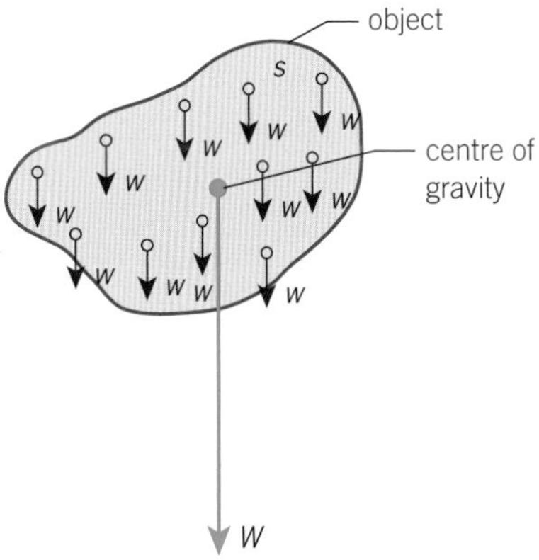

▲ **Figure 2** *The centre of gravity, through which the object's weight acts, coincides with its centre of mass*

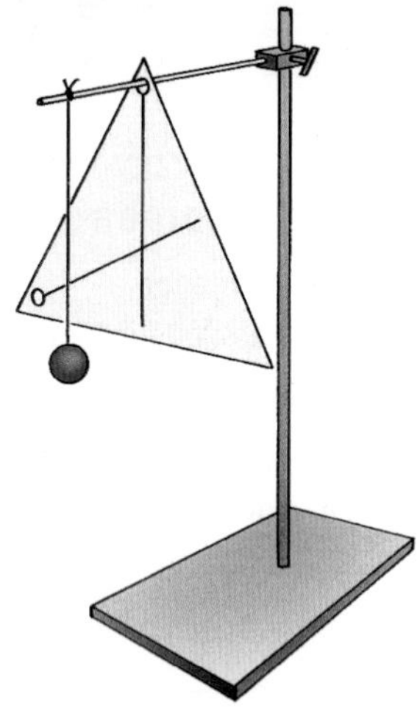

▲ **Figure 4** *Determining centre of gravity*

Study tip

If an object's centre of gravity falls outside of the base of the object, it will topple as a moment (turning force) is introduced.

Study tip

On the Earth, the centre of mass and the centre of gravity are at the same point.

Finding the centre of gravity

A freely suspended object will come to rest with its centre of gravity vertically below the point of suspension. This is the idea behind a **plumb-line**, used in construction. A heavy object, the plumb-bob, is suspended from a piece of string. When the plumb-bob comes to rest, the string is vertical.

You can use a plumb-line to find the centre of gravity of an object. This can be difficult in practice with complex three dimensional objects, but it is easy to do with objects made from card.

Make small holes along the edges of the object made from card. Insert a pin through one of the holes and hold the pin firmly in a clamp. Allow the object to swing freely. It will come to rest with its centre of gravity vertically below the pin. Hang a plumb-line from the pin and draw a line along the vertical string of the plumb-line. Repeat this process for other holes. The centre of gravity will be the point of intersection of the lines.

You can check that the position of the centre of gravity is correct by removing the card and seeing whether it will balance on a pin, or your finger, at this point.

Summary questions

1 The centre of gravity of a metre rule is found to be at its 48.3 cm mark. Suggest why it is not at the 50.0 cm mark. *(1 mark)*

2 Use diagrams to show the centre of mass of:
  a a flat circular plate; *(1 mark)*
  b a rectangular table; *(1 mark)*
  c a triangular card. *(1 mark)*

3 Describe how you could determine the centre of gravity of an irregularly shaped piece of card using the edge of the ruler instead of the plumb-line method above. *(4 marks)*

4 Explain why the centre of gravity for a table-tennis ball is in the empty space inside the ball, rather than in the plastic of the ball itself. *(2 marks)*

5 Suggest how you can locate the centre of mass of an object in space where there is no detectable gravitational field. *(4 marks)*

4.3 Free-body diagrams

Specification reference: 3.2.1

Hanging around

How can we analyse the forces acting on an object? The easiest way to do this is to draw a **free-body diagram**, which isolates all the forces acting on a particular object. The photograph of the cliff climber in Figure 1 reveals many forces in operation: there are forces within the rope, forces acting on the climber, and forces acting on the cliff face too. Fortunately, we can isolate three key forces when analysing the stability of this climber – her *weight*, the *tension* in the rope, and the **normal contact force** between her shoe and the cliff face.

Learning outcomes

Demonstrate knowledge, understanding, and application of:

→ free-body diagrams.

▲ **Figure 1** *What are the major forces acting on this climber?*

Some important forces

Table 1 summarises some of the forces that you will meet in your study of mechanics.

▼ **Table 1** *A summary of key forces*

Force	Comment	Force diagram
weight	the gravitational force acting on an object through its centre of mass	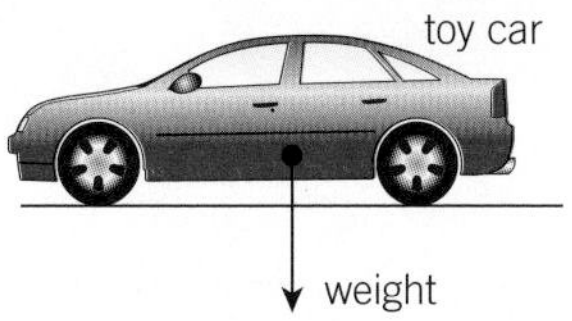
friction	the force that arises when two surfaces rub against each other	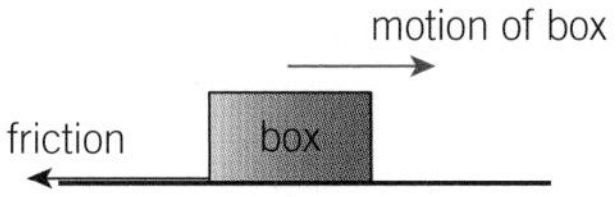
drag	the resistive force on an object travelling through a fluid (e.g., air and water); the same as friction	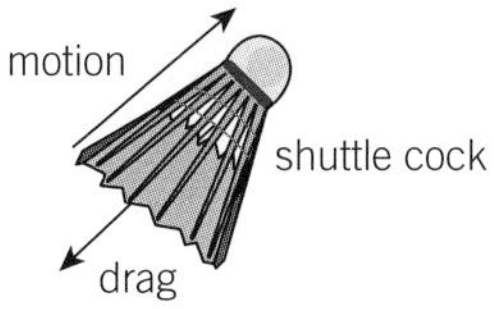
tension	the force within a stretched cable or rope	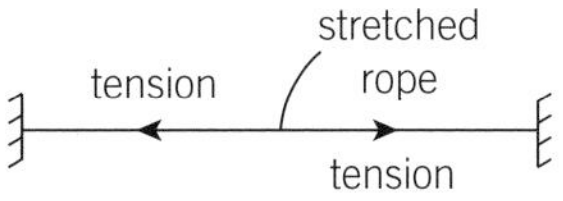
upthrust	an upward buoyancy force acting on an object when it is in a fluid	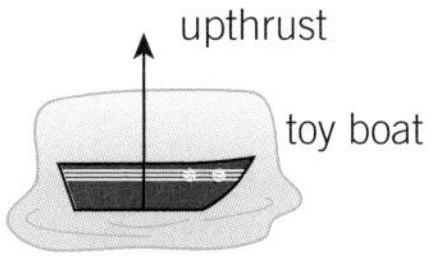
normal contact force	a force arising when one object rests against another object	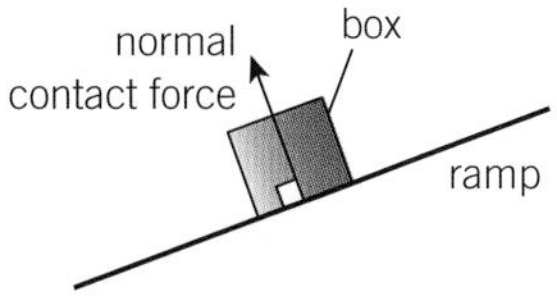

Study tip

'Normal' means 'at right angles to'.

Representing forces

In a free-body diagram

- each force vector is represented by an arrow labelled with the force it represents
- each arrow is drawn to the same scale (the longer the arrow, the greater the force).

Figure 2 shows the free-body diagram of the climber from the start of this topic.

▲ **Figure 2** *The free-body diagram for a climber*

On a slope

Figure 3 shows an object on a smooth inclined slope.

Assume that there is no friction – the only force acting on the object is its weight. This weight can be resolved into two components, parallel and perpendicular to the slope.

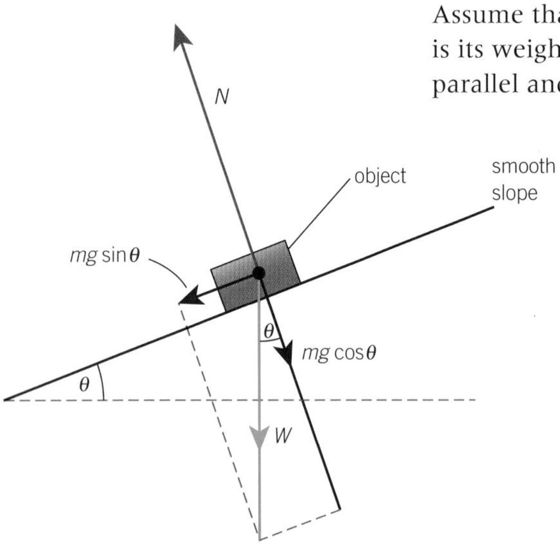

▲ **Figure 3** *Object on a slope*

Force parallel to the slope = $W\sin\theta$ or $F_x = mg\sin\theta$

Force perpendicular to the slope = $W\cos\theta$ or $F_y = mg\cos\theta$

The component of the weight down the slope is responsible for the acceleration of the object down the slope. There is no acceleration of the object perpendicular to the slope. Therefore, this component of the weight must be equal to the **normal contact force** N acting on the object, that is

$$F_y = N = mg\cos\theta$$

The worked example shows how you can analyse the motion of an object down the slope.

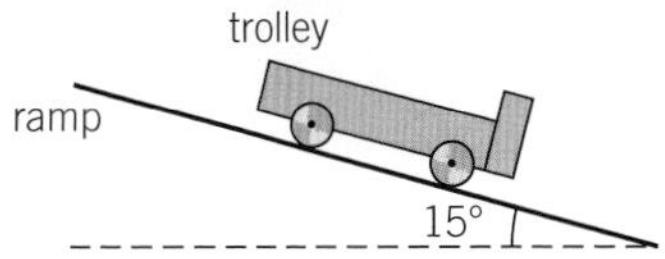

▲ **Figure 4** *Trolley on a slope*

Worked example: Down the slope

An 850 g trolley is held at the top of a 1.2 m long ramp (Figure 4). The ramp makes an angle of 15° to the horizontal. The trolley is released from rest. Calculate the acceleration a of the trolley as it travels down the ramp and the time *t* it takes to reach the bottom of the ramp.

Step 1: Identify the equations needed.

$$\text{force on trolley down the ramp} = mg\sin\theta$$

$$F = ma$$

Step 2: Substitute the values into the equation and calculate the answer.

acceleration of trolley $a = \frac{F}{m} = \frac{mg\sin\theta}{m} = g\sin\theta$ (note that the acceleration is independent of the mass)

$$a = 9.81 \times \sin 15° = 2.54\,\text{m}\,\text{s}^{-2}$$

You can now use the equation of motion $s = ut + \frac{1}{2}at^2$ to calculate the time *t*.

$$1.2 = \frac{1}{2} \times 2.54 \times t^2 \ (u = 0)$$

$$t = \sqrt{\frac{2 \times 1.2}{2.54}} = 0.97\text{s (2 s.f.)}$$

Summary questions

1 Draw a labelled free-body diagram for:

a a ball falling vertically through the air; *(1 mark)*

b a toy boat resting on the surface of water. *(1 mark)*

2 Figure 5 shows the free-body diagram for a bag on the floor of a lift. The mass of the bag is 8.0 kg. The normal contact force is *N*. The lift travels vertically upwards with an acceleration of 1.5 m s^{-2}. Calculate the resultant force on the bag and therefore the magnitude of the force *N*. Explain your answer. *(4 marks)*

3 A 20 g wooden block is placed on a smooth ramp that makes an angle of 30° to the horizontal. The block is released from rest. Calculate its acceleration down the ramp. *(3 marks)*

4 Calculate the acceleration of the block in question 3 assuming a constant friction of 0.10 N acts against the motion of the block. *(4 marks)*

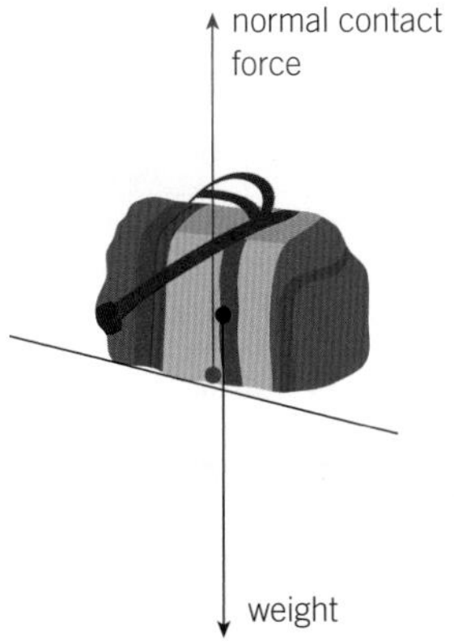

▲ **Figure 5** *A bag on the floor of a lift*

4.4 Drag and terminal velocity

Specification reference: 3.2.2

Learning outcomes

Demonstrate knowledge, understanding, and application of:

→ drag

→ the motion of objects in the presence of drag

→ terminal velocity.

▲ **Figure 1** *A peregrine falcon*

Stooping to prey

The peregrine falcon is one of the fastest birds in the world. In level flight it can reach $30\,\mathrm{m\,s^{-1}}$, but when it goes into a controlled dive, or 'stoop', to catch its prey, it folds its wings back to minimise air resistance and hurtles vertically downwards at up to $108\,\mathrm{m\,s^{-1}}$. At this speed the bird has reached its terminal velocity, the velocity at which the drag force on it balances its weight.

Moving through a fluid

An object moving through a fluid, for example, air or water, experiences a **drag force** from the fluid. Drag is a frictional force that opposes the motion of the object. Its magnitude depends on several factors, including the speed of the object, the shape of the object, the roughness or texture of the object, and the density of the fluid through which it travels. The two most important factors that affect the magnitude of the drag force are *speed* of the object and its *cross-sectional area.*

Large cross-sectional areas result in greater drag force. For most objects, including those falling through air, the drag force is directly proportional to speed^2. This means that, for example, when the speed of an object is doubled, the drag force increases by a factor of four (Figure 2).

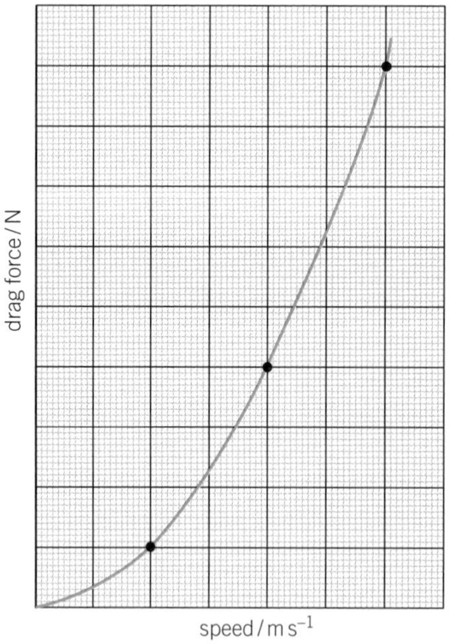

▲ **Figure 2** *The drag force–speed graph for many objects: drag force ∝ speed^2*

The drag force experienced by objects moving in air is often called **air resistance**. Modern vehicles have smooth, streamlined shapes to reduce the air resistance exerted against them whilst travelling (Figure 3). This increases the top speed of the vehicle and also reduces the amount of fuel used on a journey.

▲ **Figure 3** *Streamlining reduces drag force*

Terminal velocity

During a vertical fall through air or another fluid, the weight of the object remains constant but the drag force increases as the speed increases.

At the instant an object starts to fall, there is no drag force on the object. The total force is equal to the weight. The acceleration of the object is g, the acceleration of free fall.

As the object falls, its speed increases and this in turn increases the magnitude of the opposing drag force. The resultant (net) force on the object decreases and the instantaneous acceleration of the object becomes less than g.

Eventually the object reaches **terminal velocity**, when the drag force on the object is equal and opposite to its weight. At terminal velocity, the object has zero acceleration and its speed is a constant.

Figure 4 shows the velocity–time graph for an object falling through air and the corresponding free-body diagrams at three different times. The weight of the object is mg, the variable drag force is D, and the instantaneous acceleration is a. The gradient of the graph gives the instantaneous acceleration of the object.

Synoptic link

You will find more information about gradients of graphs in Appendix A2, Recording results.

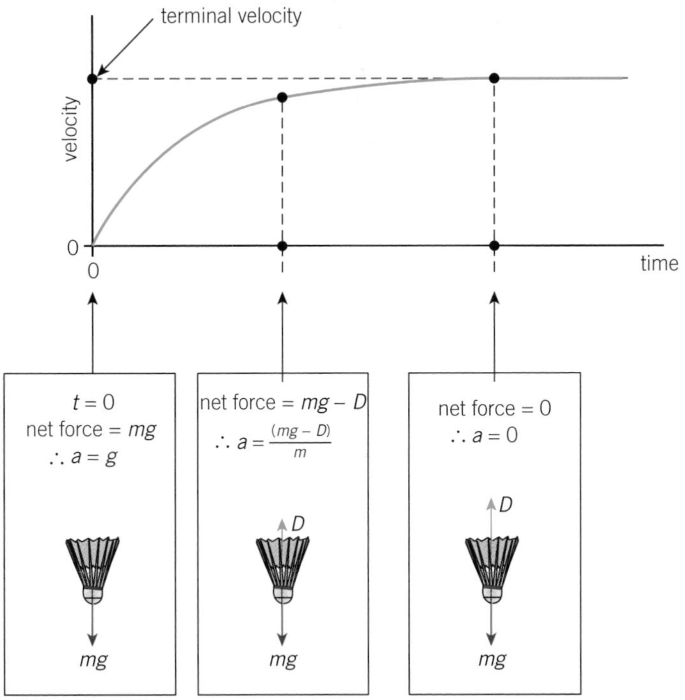

▲ **Figure 4** *Velocity–time graph for an object falling through air: the object has an acceleration of g at the start of the fall and zero acceleration when terminal velocity is reached*

Investigating motion in a fluid

You can easily investigate the motion of an object falling affected by a drag force by using a motion sensor connected to a data-logger or a laptop. The falling object is attached to a light polystyrene ball by a thin thread passed over a pulley. The object is then dropped through a cylinder of liquid such as water or glycerol, pulling the polystyrene ball vertically upwards. The motion of this ball is identical to that of the object falling through the fluid. You can generate and analyse velocity–time and acceleration–time graphs with this arrangement.

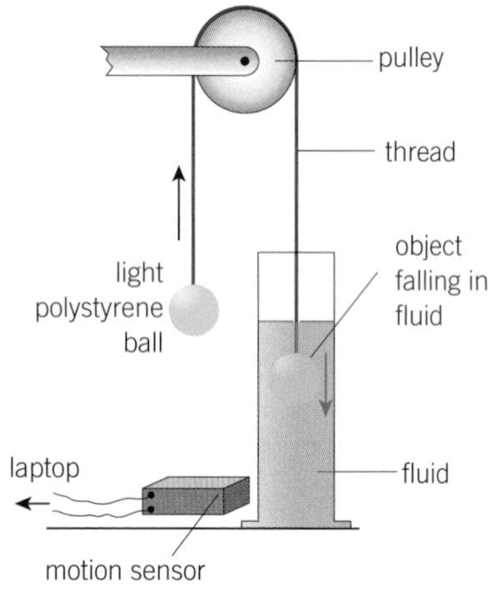

▲ **Figure 5** *Investigating an object falling through a fluid*

1 A student suggests that the motion sensor can be pointed directly towards the object falling in the fluid. Explain why this will not produce any useful data.

2 Describe how you could *estimate* the terminal velocity of the falling object without using the motion sensor.

Summary questions

1 A skydiver is falling towards the Earth at a terminal velocity of 45 m s^{-1}. Describe what she could do to change her terminal velocity. *(1 mark)*

2 A rubber ball of mass 0.120 kg is dropped from a tall building. Calculate the magnitude of the drag force as it falls through the air at its terminal velocity. Explain your answer. *(3 marks)*

3 At 10 m s^{-1} the drag force acting on a car is 1.0 kN. What is the drag force on the same car travelling at 30 m s^{-1}? Explain your answer. *(3 marks)*

4 Determine the instantaneous acceleration of each object in Figure 6. *(10 marks)*

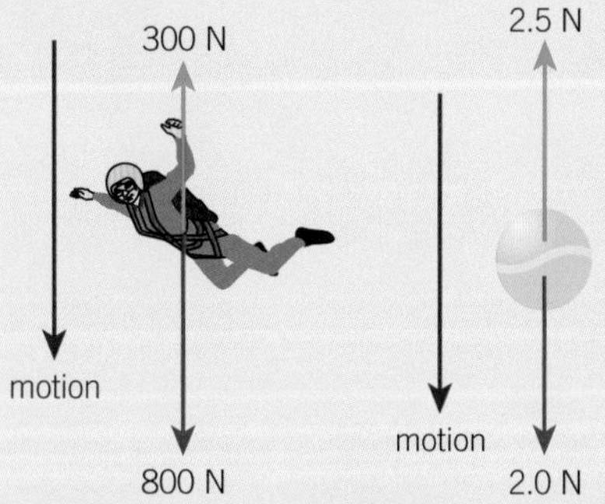

▲ **Figure 6**

5 The magnitude of the drag force D in newtons (N) acting on a 0.30 kg object falling through air is given by the expression

$D = 0.20v^2$

where v is the speed of the object in m s^{-1}. Calculate:

a the instantaneous acceleration of the object at 1.5 m s^{-1}; *(5 marks)*

b the terminal velocity of this object. *(4 marks)*

4.5 Moments and equilibrium

Specification reference: 3.2.3

Crossing rivers

The Tees Transporter Bridge crosses the River Tees at Middlesbrough. It is an unusual solution to building a river crossing that allows ships to pass underneath. Instead of going over the bridge, pedestrians and vehicles travel in a gondola suspended by cables from a system of wheels and rails. The engineers who designed it had to take into account the changing forces as the gondola moves across the river.

Learning outcomes

Demonstrate knowledge, understanding, and application of:

→ moment of force

→ the principle of moments.

The moment of a force

The weight of the gondola shown in Figure 1 creates a turning or a twisting effect about its two supports. The engineers took this into account when thinking about the stability of the structure.

The **moment** of a force is the turning effect of a force about some axis or point. It is defined as follows (Figure 2):

moment = force × perpendicular distance of the line of action of force from the axis or point of rotation

moment = Fx

The SI unit for the moment of a force is N m.

▲ **Figure 1** *The Tees Transporter Bridge, opened in 1911, is one of only twenty such bridges ever built*

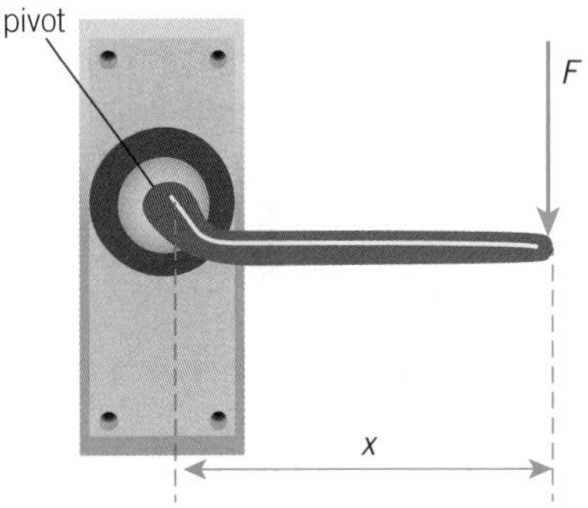

▲ **Figure 2** *Moment of a force*

Perpendicular distance

It is important that you use the perpendicular distance in calculations involving moments, not just the distance from force to **pivot**. Figure 3 illustrates what is meant by the perpendicular distance. You can calculate the perpendicular distance x from the pivot using trigonometry.

$$x = 0.20 \cos\theta$$

The clockwise moment of the force must therefore be

$$\text{moment} = F \times 0.20 \cos\theta = 0.20 F \cos\theta$$

There is another method, in which you simply resolve the force F into two perpendicular directions. The perpendicular component of the force, $F\cos\theta$, has a perpendicular distance of 0.20 m from the pivot in Figure 3. Its clockwise moment about the pivot is

$$\text{moment} = F\cos\theta \times 0.20 = 0.20 F \cos\theta$$

This is exactly the same as in the first method. The other component of the force, $F \sin\theta$, has zero perpendicular distance from the pivot, so its contribution towards the moment is zero.

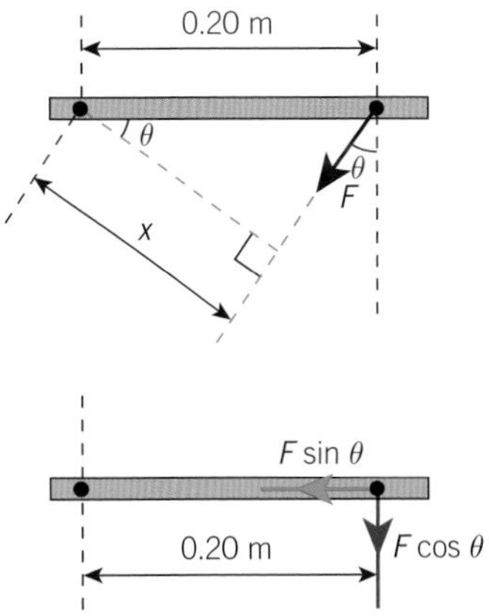

▲ **Figure 3** *Understanding perpendicular distance*

The principle of moments

When a body is in **equilibrium**, the net force acting on it is zero and its net moment is zero. You can use the **principle of moments** to solve problems where an object is in rotational equilibrium.

Study tip

When you define the moment of a force, make sure you state 'perpendicular' and not just 'distance'.

Principle of moments: For a body in rotational equilibrium, the sum of the anticlockwise moments about any point is equal to the sum of the clockwise moments about that same point.

The principle of moments was recognised almost 2000 years ago by Archimedes. Engineers still use this important principle when designing buildings and other complex structures. The worked examples below show how you can analyse problems using this principle. As you can see, it is important to draw clearly labelled free-body diagrams.

Worked example: A simple see-saw

Figure 4 shows a metre rule pivoted at its 50 cm mark. Two objects of weights 0.70 N and W are placed on the ruler as shown to balance it. Assume the weight of the ruler is negligible.

Calculate the size of W and the force acting at the pivot.

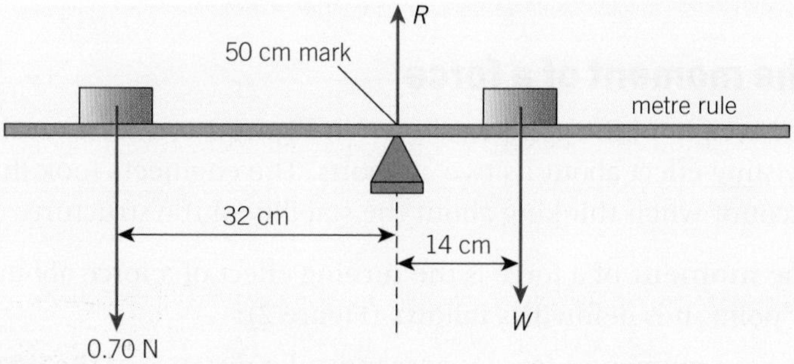

▲ Figure 4

Step 1: Identify the equation needed.

According to the principle of moments

sum of clockwise moments = sum of anticlockwise moments

Step 2: Substitute the values into the equation and calculate the answer.

$0.14 \times W = 0.70 \times 0.32$ (note that the line of action of the normal contact force R passes through the pivot, so its moment is zero)

Now rearrange this equation to determine W.

$$W = \frac{0.70 \times 0.32}{0.14} = 1.6\,\text{N}$$

The size of W is 1.6 N.

Step 3: To calculate the force R at the pivot, you can now use the idea that the net force on the ruler is zero. There are two downwards vertical forces, 0.70 N and 1.6 N. This implies that there must be an upwards vertical force R at the pivot equal to sum of these two forces. Therefore

$$R = 0.70 + 1.6 = 2.3\,\text{N}$$

Worked example: A loaded bridge

Figure 5 shows a model of a loaded section of a bridge made using a uniform wooden beam.

The wooden beam is 120 cm (1.2 m) long and has weight 15 N. One end of the beam rests on a support and the other end is fixed to a vertical string. Calculate the vertical force R at the support.

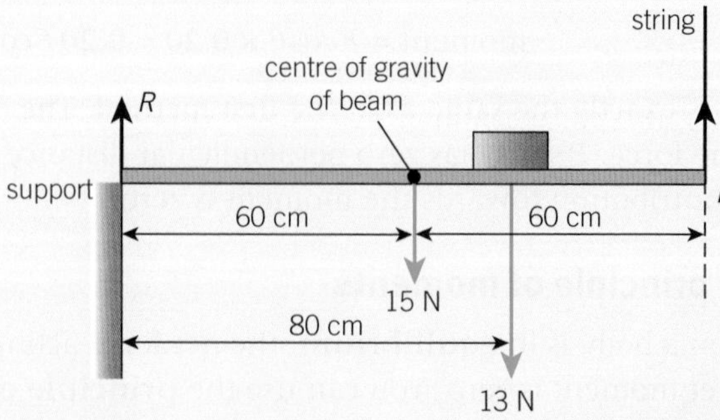

▲ Figure 5

Step 1: To solve this problem you need to examine where to take moments. Taking moments about the support will not be helpful. The most convenient point to take moments would be the end **A**, where the string is tied. The tension T in the string will have zero moment about **A**.

The weight of the beam acts through its centre of gravity.

Taking moments about **A**, we have

sum of clockwise moments = sum of anticlockwise moments

Step 2: Substitute the values into the equation and calculate the answer.

$R \times 1.20 = (15 \times 0.60) + (13 \times 0.40)$ (note that all perpendicular distances are from **A**)

$$R = \frac{142}{1.20} = 11.83\,\text{N} = 12\,\text{N (2 s.f.)}$$

You could now calculate the tension T from the fact that the net force on the beam is zero.

$$T = 16\text{ N (2 s.f.)}$$

Summary questions

1 Calculate the moment from each force about the pivot in Figure 6. *(3 marks)*

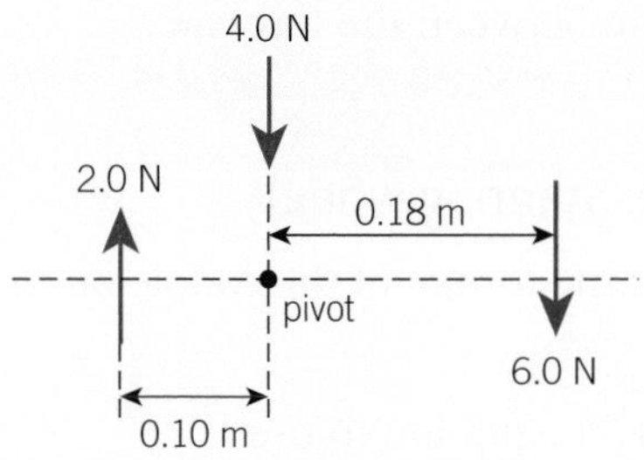

▲ **Figure 6**

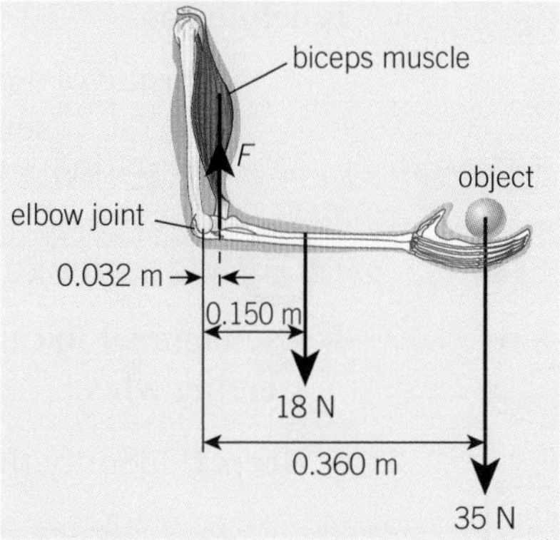

◀ **Figure 7**

2 Figure 7 shows a human forearm held horizontally and still. Calculate:

a the clockwise moment about the elbow joint; *(3 marks)*

b the force F in the muscle. *(3 marks)*

3 A uniform cylinder has height 10.0 cm and diameter 3.0 cm. The cylinder is placed with its circular base resting on a horizontal table. Calculate the maximum angle through which it can tip before it continues to fall by itself. *(3 marks)*

4 Calculate the magnitude of the force F in Figure 8. *(6 marks)*

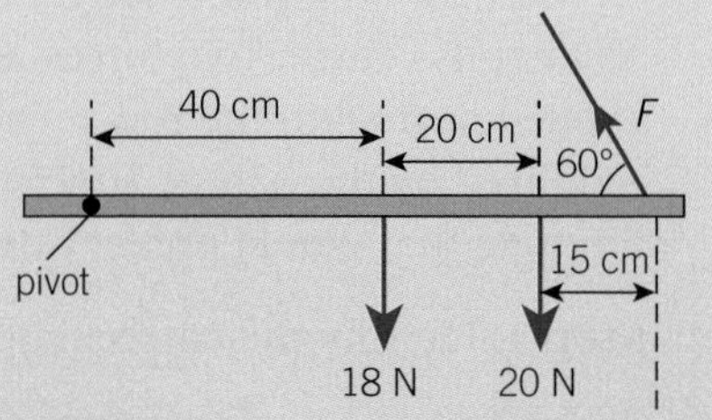

▲ **Figure 8**

4.6 Couples and torques

Specification reference: 3.2.3

Learning outcomes

Demonstrate knowledge, understanding, and application of:

- → couples and torques
- → equilibrium of objects under the action of forces and torques.

▲ **Figure 1** *An example of a couple: a pair of equal but opposite forces are applied to the pedals*

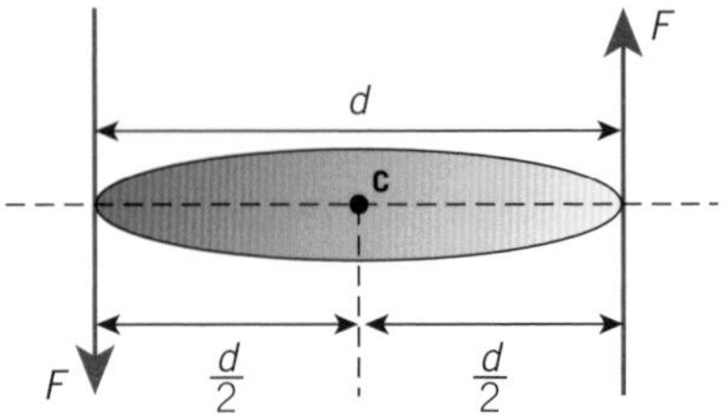

▲ **Figure 2** *The moment of this couple is called a torque*

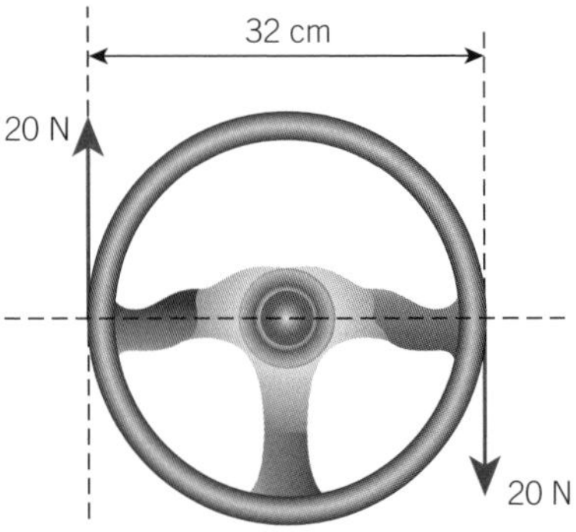

▲ **Figure 3** *The couple on this steering wheel will make it turn*

Couples

Imagine pushing the side of your calculator on the table with your finger. If the line of action of this force does not coincide with the centre of mass of the calculator, then it will both slide along the table (translate) and rotate. How can you make an object spin without any translational motion? The trick is to apply a pair of equal but opposite forces to the object. These two forces must be parallel and along different lines. Such a pair of forces is referred to as a **couple**.

Torque of a couple

Figure 2 shows a couple applied to an object. The magnitude of each force is F and the perpendicular distance between them is d. The moment of this pair of equal but opposite forces about the centre point **C** is

$$\text{moment} = \left(F \times \frac{d}{2}\right) + \left(F \times \frac{d}{2}\right) = Fd$$

The moment of a couple is known as a **torque**. The torque of a couple is defined as

$$\text{torque of a couple} = \text{one of the forces} \times \text{perpendicular separation between the forces} = Fd$$

Worked example: Steering wheel

Use Figure 3 to calculate the torque of the couple on this car steering wheel.

Step 1: Identify the correct equation to use.

$$\text{torque} = Fd$$

Step 2: Substitute the values in and calculate the torque.

$$\text{torque} = 20 \times 0.32$$

$$= 6.4\,\text{N m}$$

The torque applied is 6.4 N m in a clockwise direction.

Worked example: Preventing rotation

Figure 4 shows a rod of length 50 cm with a disc of radius 10 cm fixed at its centre. Two forces, each of magnitude 30 N, are applied normal to the rod at each end. Calculate the torque produced by the pair of 30 N forces and the minimum tension in the rope that would prevent the disc from rotating.

Step 1: Identify the correct equation to use.

$$\text{torque} = Fd$$

Step 2: Substitute the values in and calculate the torque.

$$\text{torque} = 30 \times 0.25 = 7.5\,\text{N m}$$

Step 3: To prevent rotation, the moment of the tension T in the rope must be equal but opposite to this torque.

$$\text{moment} = T \times \text{distance} = T \times 0.10 = 7.5$$

$$T = \frac{7.5}{0.10} = 75\,\text{N}$$

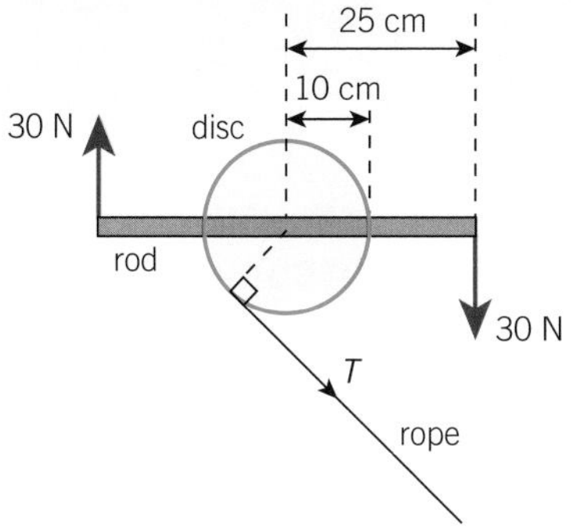

▲ **Figure 4**

Summary questions

1 A snooker ball is resting on a table. A single off-centre force is applied to its surface with a cue. Describe the subsequent motion of the ball. *(2 marks)*

2 The top of a kitchen tap has diameter 4.0 cm. Estimate the torque required to open such a tap using your thumb and one of the other fingers. *(3 marks)*

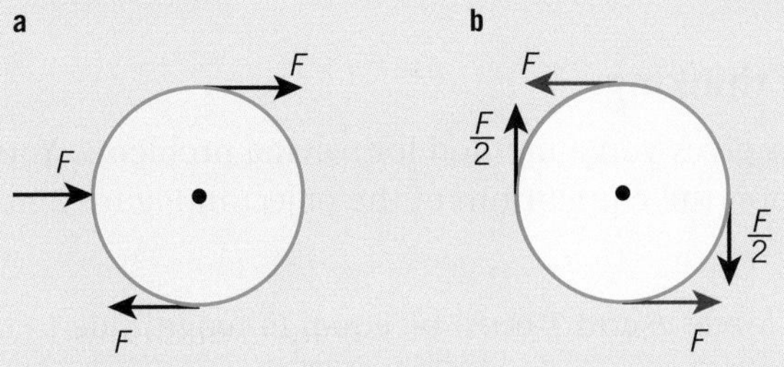

▲ **Figure 5**

3 Figure 5 shows two discs placed on a smooth horizontal surface.

Describe qualitatively the type of motion each disc will perform. *(4 marks)*

4 Figure 6 shows a couple acting on an object.

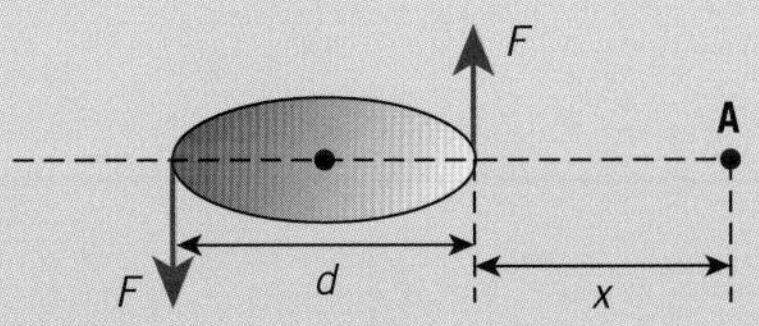

▲ **Figure 6**

a Determine the total moment of the couple about the point A. *(2 marks)*

b What can you deduce from this calculation? *(1 mark)*

Synoptic link

You will find more information on estimations in Appendix A3, Measurements and uncertainties.

4.7 Triangle of forces

Specification reference: 3.2.3

Learning outcomes

Demonstrate knowledge, understanding, and application of:

→ the equilibrium of three coplanar forces

→ the rule of the triangle of forces.

▲ **Figure 1** *A baby having fun with forces*

Synoptic link

You know how to add two vector forces from Topic 2.4, Adding vectors.

Bouncing babies

Baby bouncers help babies to develop their leg muscles so that they are ready to learn to walk. The bouncer is securely suspended from a door frame and the baby is supported in a padded seat. The weight acts vertically downwards and is balanced by the **tension** in the straps.

Forces in equilibrium

A heavy ball is held at rest by two ropes (Figure 2, left). The tensions in the ropes have magnitudes F and T. The weight of the ball is W. The resultant of these three coplanar forces must be zero.

To add three vector forces you simply extend the procedure for adding two vectors. Figure 2 below shows the vectors in the free-body diagram for the ball (centre) and in a **triangle of forces** (Figure 3).

- Arrows are drawn to represent each of the three forces end-to-end.
- The triangle is closed because the net force is zero and so the object is in equilibrium.

Different ways of thinking

The triangle of forces gives you a method for solving problems. You can, however, interpret the equilibrium of the object in Figure 2 in two other ways.

- The resultant of forces F and T must be equal in magnitude to the third force W but in the opposite direction. The same is true for any pair of forces in Figure 2.
- The resultant force vertically must be zero and the resultant horizontal force must also be zero. Therefore, the force T can be resolved into its vertical and horizontal components, with $T\cos\theta = F$ and $T\sin\theta = W$.

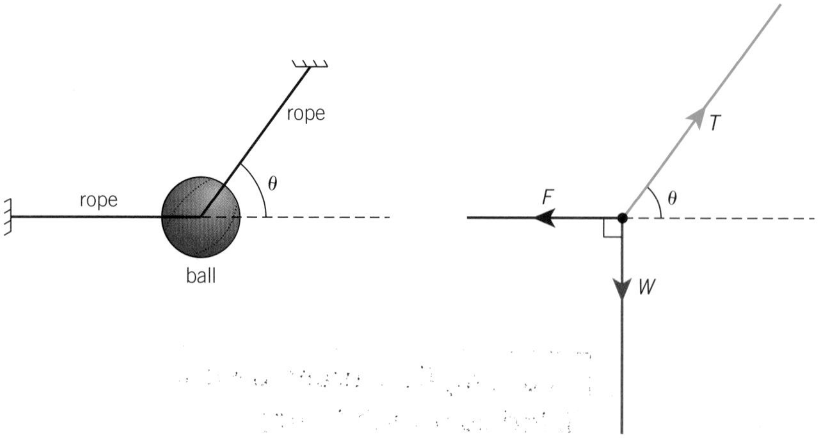

▲ **Figure 2** *All three forces acting on the ball act through the same point, as shown in the free-body diagram in the centre*

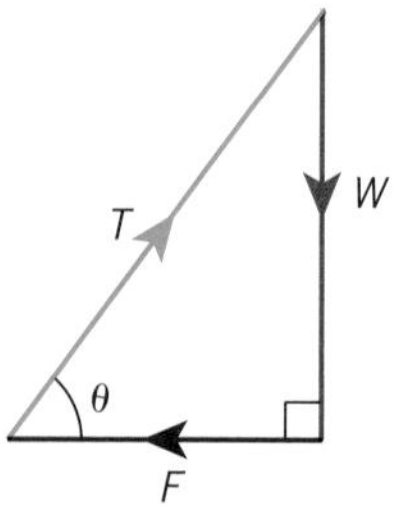

▲ **Figure 3** *Triangle of forces showing the situation in Figure 2 – all arrows follow one another*

Worked example: Resting on a slope

Figure 4 shows a block of wood weighing 1.20 N resting on a rough ramp that makes an angle of 30° to the horizontal. The normal contact force between the block and the ramp is N and the frictional force on the block is F. Draw a triangle of forces and determine the magnitude of the forces N and F.

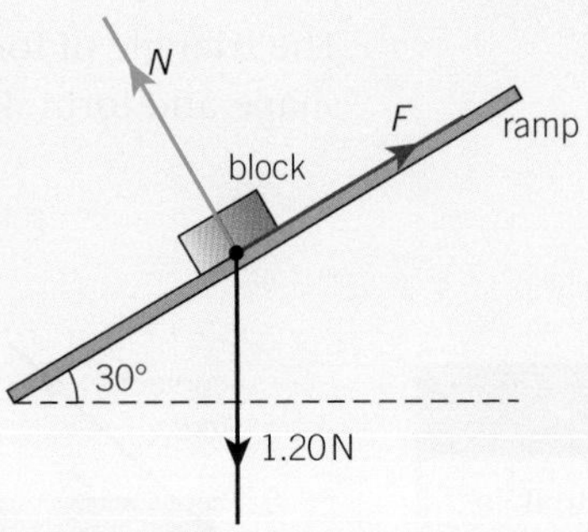

▲ **Figure 4** *A block resting on a ramp*

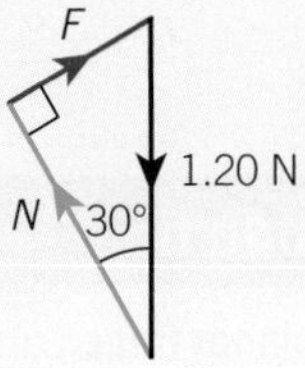

▲ **Figure 5**

Step 1: Draw a triangle of forces (Figure 5).

Step 2: Calculate the unknown forces.

$\cos 30° = \frac{N}{1.20}$ therefore $N = 1.20 \times \cos 30° = 1.04\,\text{N}$

$\sin 30° = \frac{F}{1.20}$ therefore $F = 1.20 \times \sin 30° = 0.60\,\text{N}$

You can check the answers using Pythagoras' theorem.

$$0.60^2 + 1.04^2 = 1.20^2$$

Worked example: Baby bouncer

A baby and its bouncer seat weigh 120 N (Figure 6). Two straps holding the seat are suspended from a plastic bar. The tension in each strap is the same, and they meet at a point on the seat at an angle of 50°. Figure 6 shows the corresponding free-body diagram.

Step 1 (one possible method): Resolve the diagonal forces by inspection.

There is no net force in any direction as the forces balance. Therefore

$$T\cos 25° + T\cos 25° = 120$$

$$2T\cos 25° = 120$$

$$T = \frac{120}{2\cos 25°} = 66\,\text{N (2 s.f.)}$$

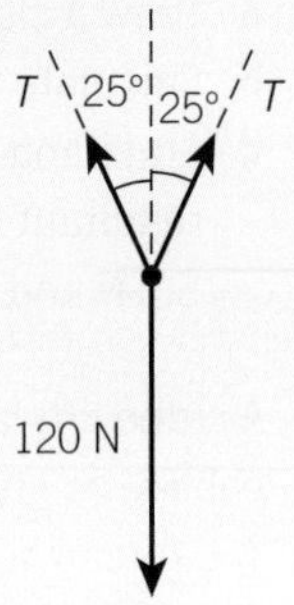

▲ **Figure 6**

You can of course draw a triangle of forces to determine the magnitude of T, but the method above is quick and neat.

The horizontal components of T are equal and opposite and therefore balance each other.

Three coplanar forces acting on an extended object

The triangle of forces method can also be applied to objects that have shape and form, from bridges to bicycles.

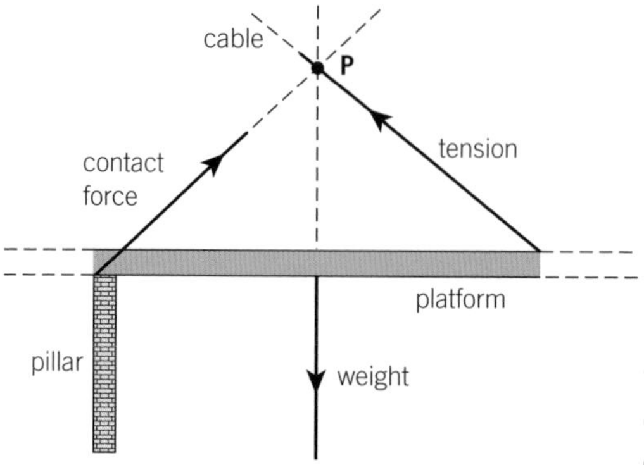

◀ **Figure 7** *The lines of action of all three forces on the platform act through point P*

> **Study tip**
>
> Coplanar forces are forces that lie in the same plane.

Figure 7 shows the free-body diagram of a section of a bridge platform. All three coplanar forces pass through a point **P** in space, so you can draw a triangle of forces for the forces passing through **P**. This method simplifies a complex problem to the equilibrium of this imaginary point in space.

Summary questions

1 Three coplanar forces act on an object. The vectors representing these three forces form a closed triangle (triangle of forces). State the resultant force acting on the object. *(1 mark)*

2 Figure 8 shows an object in equilibrium.
 a Draw a clearly labelled triangle of forces. *(2 marks)*
 b Calculate the magnitude of the force T. *(2 marks)*
 c State and explain the magnitude of the resultant of the forces 5.0 N and 12 N. *(2 marks)*

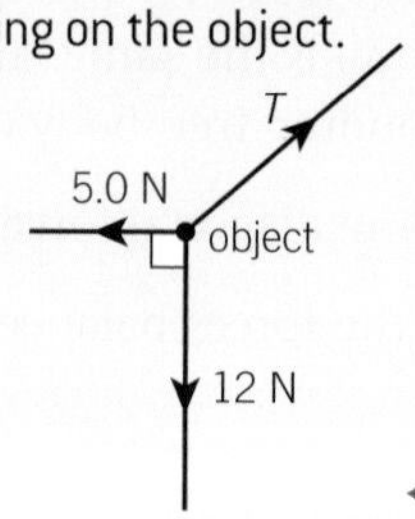

◀ **Figure 8**

3 A crane lifts a girder using a hook and two cables, as shown in Figure 9. The forces acting on the hook are shown. Calculate the magnitudes of the forces T_1 and T_2. *(4 marks)*

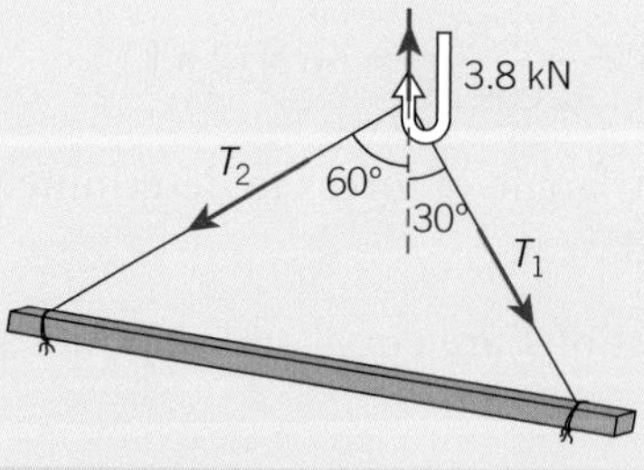

◀ **Figure 9**

4 Two newtonmeters are used to lift 500 g of slotted masses (Figure 10). The masses are at rest.
 a The angles θ_1 and θ_2 are 60° and 50°, respectively. Calculate the readings T_1 and T_2. *(4 marks)*
 b Explain why it would be impossible for both angles to be zero. *(2 marks)*

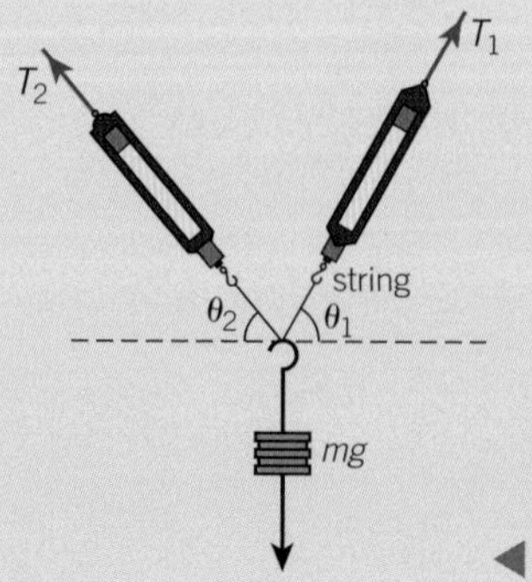

◀ **Figure 10**

4.8 Density and pressure

Specification reference: 3.2.4

Light for its size

Aerogel is one of the least dense solids known. It is so lightweight and wispy that it is sometimes called frozen smoke. Made from silica, aerogel has a density of just 1.9 kg m^{-3}. By comparison, the density of air is about 1.3 kg m^{-3} and the density of water is 1000 kg m^{-3}. Its low density means that aerogel has very little mass for its volume.

Learning outcomes

Demonstrate knowledge, understanding, and application of:

→ density $\rho = \frac{m}{V}$

→ pressure $p = \frac{F}{A}$.

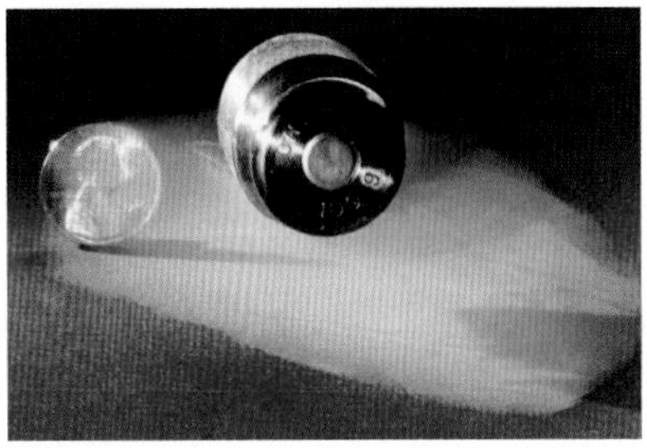

▲ **Figure 1** *A 0.70 g piece of aerogel supporting a 100 g mass – you can see a coin through it*

Density

The **density** of a substance is defined as its mass per unit volume. You can use the following equation to calculate density ρ.

$$\rho = \frac{m}{V}$$

where m is mass and V is volume. The SI unit of density is kg m^{-3}.

Worked example: Dense osmium

A 50.0 cm^3 sample of osmium has a mass of 1.13 kg. Calculate its density.

Step 1: Select the equation for density.

$$\rho = \frac{m}{V}$$

Step 2: Substitute in the known values in SI units and calculate the density.

$m = 1.13$ kg $\quad V = 50 \times 10^{-6}$ m^3 (note that 1 cm^3 = $(10^{-2}\,\text{m})^3 = 10^{-6}\,\text{m}^3$)

$$\rho = \frac{m}{V} = \frac{1.13}{50.0 \times 10^{-6}} = 2.26 \times 10^4\,\text{kg m}^{-3}$$

Osmium is 22.6 times denser than water.

Determining density

You need to know mass and volume to determine the density of a substance. The mass can be measured directly using a digital balance. For liquids, you can use a measuring cylinder to determine the volume. The volume of a regular-shaped solid can be calculated from measurements taken with a ruler, digital callipers, or a micrometer. The volume of irregular solids can be determined by displacement (Figure 2).

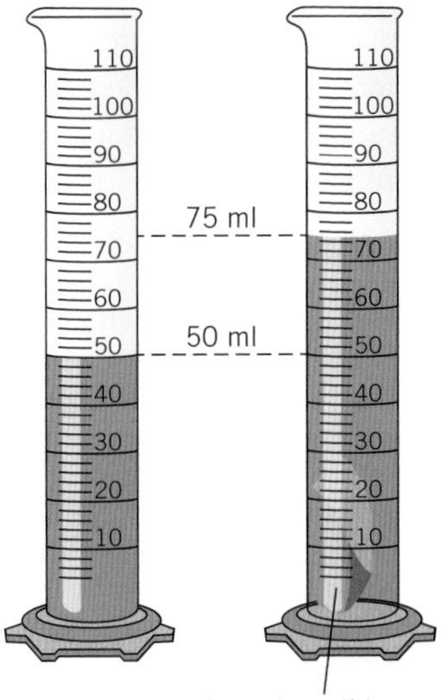

▲ **Figure 2** *Determining the volume of an irregular solid by displacement of a liquid: the volume is the difference between the two water levels*

Synoptic link

You will find more information on estimations in Appendix A3, Measurements and uncertainties.

Pressure

Drawing pins are designed to be pushed into noticeboards without hurting you. The pin's head has a much larger surface area than its point. This means that, when you push it into a noticeboard, the pressure exerted on your thumb is much less than the pressure exerted on the board. If you used the drawing pin the other way round it would hurt.

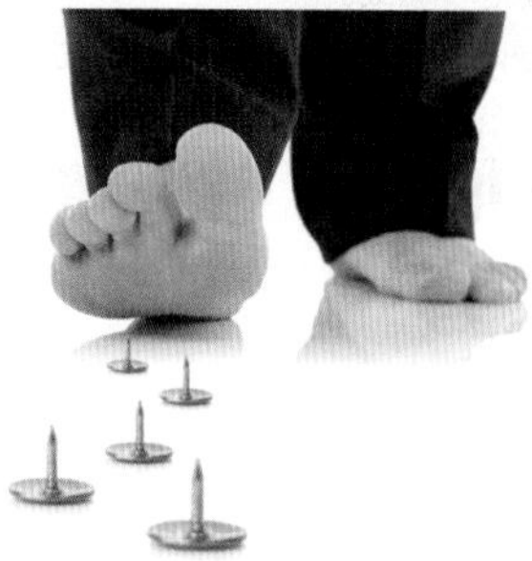

▲ **Figure 3** *Spreading weight over a smaller area increases the pressure exerted*

Pressure is the normal force exerted per unit cross-sectional area. You can use the following equation to calculate pressure p:

$$p = \frac{F}{A}$$

where F is normal force and A is cross-sectional area. The SI unit of pressure is N m^{-2} or pascal (Pa), where $1\text{ Pa} = 1\text{ N m}^{-2}$.

We are under pressure all the time. The Earth's atmosphere exerts about 1.0×10^5 Pa on everything on its surface, including us. Tiny fluctuations in this pressure are responsible for the variety of weather patterns we see on the Earth.

Worked example: Standing still

Estimate the pressure you exert on the floor when standing up.

An estimate is a rough calculation, not a guess. The numbers you use should be realistic.

Step 1: Select the equation for pressure and estimate the numbers you need for the calculation.

$$p = \frac{F}{A}$$

You can estimate the area of one shoe by measuring its width and length.

cross-sectional area of each shoe $A = 0.25\text{ m} \times 0.10\text{ m}$
$= 2.5 \times 10^{-2}\text{ m}^2$ (2 s.f.)

The force exerted on the floor is your weight, mg. In this calculation m is estimated to be 65 kg.

Step 2: Substitute your estimates into the equation and calculate the estimated pressure.

$$p = \frac{F}{A} = \frac{65 \times 9.81}{2 \times 2.5 \times 10^{-2}} = 1.3 \times 10^4\text{ Pa (2 s.f.)}$$

Summary questions

1. Calculate the mass of air in a room of volume 140 m^3. *(2 marks)*
2. A circular head of a drawing pin has a diameter of 7.5 mm. A thumb presses against the head with a force of 8.0 N. Calculate the pressure exerted on the head. *(2 marks)*
3. A measuring cylinder is filled with water up to its 70 cm^3 mark. An irregular rock of mass 0.080 kg is gently placed in the water. The new water level mark is 85 cm^3. Calculate the density of the rock. *(2 marks)*
4. Calculate the vertical force exerted by the atmosphere on the surface of a calculator of length 15 cm and width 7.5 cm. *(3 marks)*
5. Calculate the average density of a neutron star of mass 3.0×10^{24} kg and radius 12 km. *(2 marks)*
6. An 18 carat 'gold' bar of volume $3.34 \times 10^{-6}\text{ m}^3$ comprises 58.2% gold and 41.8% copper by volume. Calculate the density of the bar. State any assumptions made.
$\rho_{gold} = 1.93 \times 10^4\text{ kg m}^{-3}$ and $\rho_{copper} = 8.96 \times 10^3\text{ kg m}^{-3}$. *(6 marks)*

4.9 $p = h\rho g$ and Archimedes' principle

Specification reference: 3.2.4

Deep water

You can safely swim underwater in the sea at depths of a few tens of metres. However, humans need a pressurised submersible to do any work at greater depth, because of the enormous pressure due to the weight of the water. Figure 1 shows a research submarine with a clear acrylic bubble housing that is 9.5 cm thick. At a depth of 610 m, the pressure acting on this submarine is about 6 000 000 Pa.

Learning outcomes

Demonstrate knowledge, understanding, and application of:

→ $p = h\rho g$

→ upthrust and Archimedes' principle.

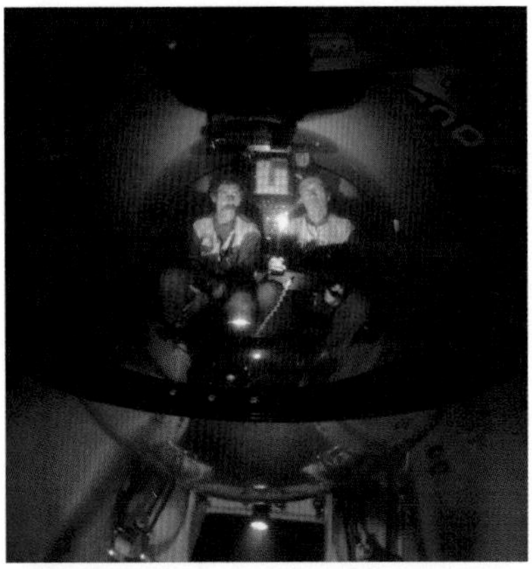

▲ **Figure 1** *This submarine has a spherical shape to resist the pressure of the surrounding water, which increases with its depth*

Pressure in fluids

Gases and liquids are **fluids** – substances that can flow. Gases, such as air, exert pressure on surfaces because of the constant bombardment by their molecules. Liquids also exert pressure for the same reason.

The pressure exerted by the atmosphere of the Earth varies with altitude. At sea level, atmospheric pressure is about 101 kPa. At the top of Ben Nevis (Britain's highest mountain) the pressure is only 87 kPa.

Liquids

You can calculate the pressure p exerted by a vertical column of any liquid from its weight and the cross-sectional area of the base.

$$p = h\rho g$$

where h is the height of the liquid column, ρ is the density of the liquid, and g is the acceleration of free fall ($9.81\,\text{m}\,\text{s}^{-2}$).

It is important to understand how this equation is derived. Figure 2 shows a cylindrical column of liquid with height h and base of cross-sectional area A.

The pressure at the base is equal to the weight W of the column divided by A.

$$W = \text{mass of column} \times g$$

The mass of the column is the density × the volume.

$$W = (\rho V) \times g$$

The volume V of the column is Ah.

$$W = \rho \times Ah \times g$$

Finally, the pressure p is given by

$$p = \frac{\rho \times A \times h \times g}{A} = h\rho g$$

This equation shows that pressure does not depend on the cross-sectional area. It also clearly shows that $p \propto h$, so water pressure increases with depth. The term ρ shows that denser liquids will exert greater pressure.

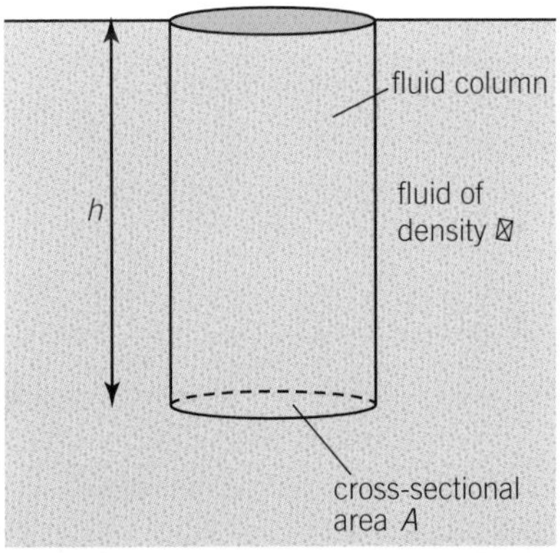

▲ **Figure 2** *Pressure in a column of liquid*

The pressure in a fluid at any particular depth has an unusual feature – it is the same in all directions.

Worked example: Feeling the pressure

Calculate the total pressure acting on a submarine at a depth of 20 m.

atmospheric pressure = 101 kPa

density of seawater = $1.03 \times 10^3\,\text{kg}\,\text{m}^{-3}$

Step 1: Select the equation needed to calculate pressure in a fluid.

Take care! The pressure on the submarine will be the sum of the atmospheric pressure and the pressure due to the water. Therefore

pressure = atmospheric pressure + $h\rho g$

pressure = $1.01 \times 10^5 + (20 \times 1.03 \times 10^3 \times 9.81)$

pressure = 3.03×10^5 Pa

The pressure exerted by the seawater is twice that of the atmosphere.

Upthrust

Try pushing a small piece of wood into water and then letting it go. The wood will immediately pop out of the water, bob up and down on the surface, and remain afloat. The buoyant force on the submerged wood can be explained in terms of the pressure differences at its upper and lower surfaces.

Figure 3 shows the position of a submerged rectangular block of wood of cross-sectional area A. The density of the fluid is ρ. This block of wood will displace the fluid.

force at the top surface = $h\rho gA$

force at the bottom surface = $(h + x)\rho gA$

resultant upward force = $(h + x)\rho gA - h\rho gA = x\rho gA$

This resultant force is called **upthrust**.

upthrust = $Ax\rho g$

The volume of the block of fluid displaced is Ax and it has mass $(Ax)\rho$. So the upthrust is equal to the weight $(Ax\rho)$g of the fluid displaced by the block of wood. This idea was known to Archimedes almost 2000 years ago and is still important to engineers, who design ships and structures under water. It applies to fully or partially submerged objects.

Archimedes' principle: The upthrust exerted on a body immersed in a fluid, whether fully or partially submerged, is equal to the weight of the fluid that the body displaces.

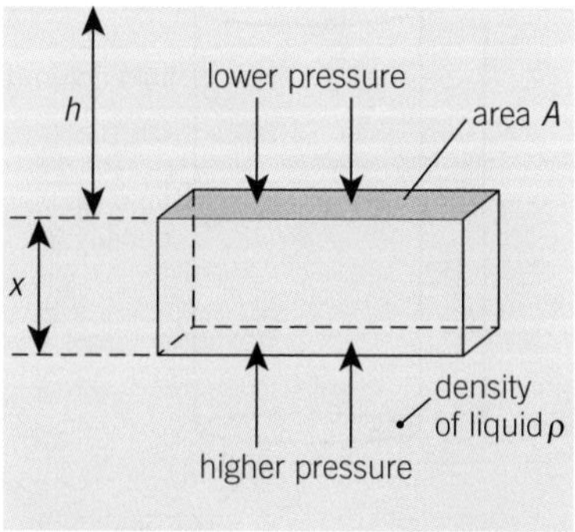

▲ **Figure 3** *Upthrust is due to pressure differences*

An object will sink if the upthrust is less than the weight of the object. For a floating object, such as a ship or a person in water, the upthrust must equal the weight of the object. This in turn means that the weight of a floating object must be equal to the weight of the fluid it displaces.

Icebergs

It is said that nine-tenths of an iceberg lie hidden underwater. At 0°C the density of ice is about $900\,\text{kg}\,\text{m}^{-3}$, whereas the density of water is about $1000\,\text{kg}\,\text{m}^{-3}$. A cube of ice with sides 1.0 m will therefore have a weight of $(900 \times 9.81) = 8.83 \times 10^3$ N. When it floats, it will displace water of weight 8.83 × 103 N. This has a volume of $(8.83 \times 10^3)/(1000 \times 9.81) = 0.90\,\text{m}^3$. So the $1.0\,\text{m}^3$ cube of ice will sink until it has displaced $0.90\,\text{m}^3$ of water, that is, nine-tenths of the cube is below the surface of the water.

▲ **Figure 4** *Most of an iceberg is underwater*

Deep-water divers

Barotrauma is the physical damage to body tissues caused by a difference in the pressure within the body and that in the surrounding fluid. Divers swimming in deep water can sometimes suffer ear or lung damage.

A descent of 10 m in water doubles the external pressure on a diver. The record for deep-water scuba diving stands at about 330 m. Divers attempting such records are highly trained.

1. Explain, with the help of a calculation, how the external pressure on a diver at a depth of 10 m is doubled.
2. Estimate the vertical force acting on a diver's hand at a depth of 330 m. Assume the cross-sectional area of the hand is $1.2 \times 10^{-2}\,\text{m}^2$.
3. Suggest why a scuba diver who breathes air from the tank at 10 m depth and then ascends without exhaling can cause lung damage.

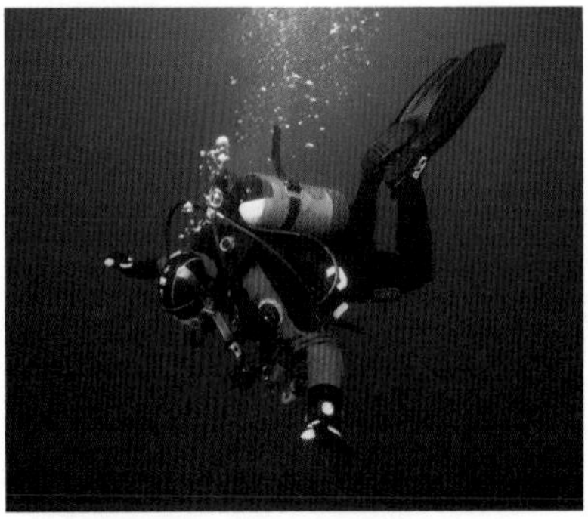

▲ **Figure 5** *The differences in pressure at different depths pose a danger to divers*

Summary questions

Assume the density of water = $1.0 \times 10^3\,\text{kg}\,\text{m}^{-3}$.

1. The density of mercury is $1.35 \times 10^4\,\text{kg}\,\text{m}^{-3}$. Calculate the pressure exerted by a column of mercury at a depth of 0.765 m. *(2 marks)*
2. Show that the pressure exerted at a depth of 610 m in water is about 6 million pascals. *(2 marks)*
3. A table-tennis ball is held under water and then released. Describe and explain the subsequent motion of the ball. *(5 marks)*
4. A metal bar is suspended from a newtonmeter. The reading is 1.54 N in air but only 1.34 N when the bar is fully submerged in water. Calculate: **a** the upthrust on the bar; **b** the density of the bar. *(4 marks)*
5. A cube of a substance of density ρ_s floats when placed in water. Show that the fraction of the cube under water is $\frac{\rho_s}{\rho}$, where ρ is the density of water. *(4 marks)*

Practice questions

1 a Define the *newton* and derive its base units. (*3 marks*)

b Figure 1 shows a rocket on the surface of the Earth.

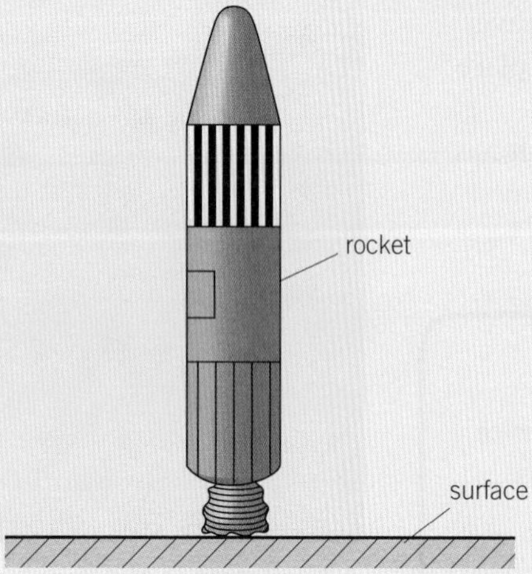

▲ Figure 1

The mass of the rocket is 3.0×10^6 kg. At lift off, the vertical upward thrust on the rocket is 34 MN.

(i) Calculate the initial vertical acceleration of the rocket. (*3 marks*)

(ii) The upward thrust on the rocket remains the same. Explain why after some time, the vertical acceleration is much larger than the value calculated in **(i)**. (*2 marks*)

2 Figure 2 shows the vertical forces acting on a helium-filled weather balloon just before lift off.

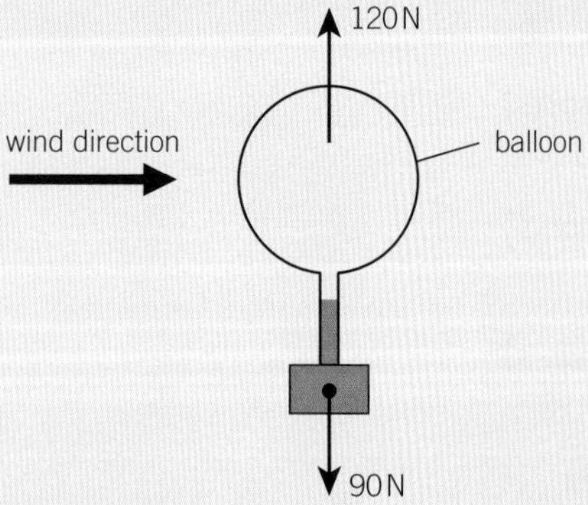

▲ Figure 2

The balloon experiences an upward vertical force (upthrust) equal to 120 N. The weight of the balloon and its contents is 90 N. The magnitude of the horizontal force provided by the wind is 18 N.

a Determine the magnitude of the resultant force acting on the balloon and the angle this resultant force makes with the horizontal. (*4 marks*)

b As the balloon rises through the air, it experiences a drag force. State *two* factors that affect the magnitude of the drag force on this balloon. (*2 marks*)

May 2012 G481

3 Figure 3 shows a lamp supported by two cables. The weight of the lamp is 24 N.

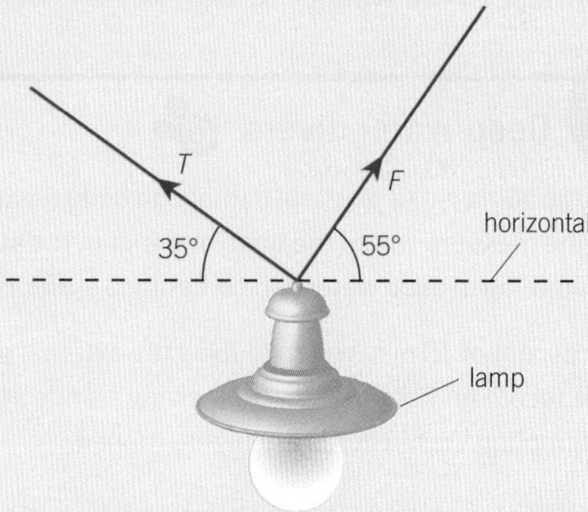

▲ Figure 3

The lamp is in equilibrium. The tensions in the cables are T and F.

a Without any calculations, explain the value of the resultant force due to T and F. (*2 marks*)

b Calculate the magnitude of the forces T and F. (*4 marks*)

c The angle made by the force T with the horizontal is decreased. Explain the effect this has on the tension T. (*2 marks*)

4 a Define *density*. (*1 mark*)

b Figure 4 shows the variation of density of the Earth with **depth** from the surface.

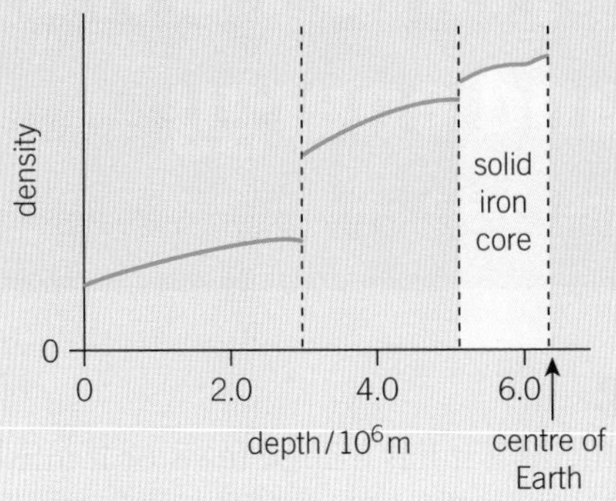

▲ Figure 4

(i) Suggest how Figure 4 shows that the Earth consists of a number of distinct layers. *(1 mark)*

(ii) Geophysicists believe that the central core of the Earth is solid iron and nickel. This central core is surrounded by a layer of molten metal. The central core starts at a **depth** of 5.1×10^6 m. The solid iron core accounts for 18% of the mass of the Earth. The mass of the Earth is 6.0×10^{24} kg and its radius is 6.4×10^6 m. Calculate the mean density of the central core of the Earth. Volume of a sphere $= \frac{4}{3}\pi r^3$ *(3 marks)*

May 2011 G481

5 a The atmosphere of the Earth exerts pressure on all objects on its surface. At a depth *d* in water, the total pressure is *P*. On a copy of the axes below, sketch a graph to show the variation of *P* with *d*. *(2 marks)*

b Figure 5 shows an object held under water.

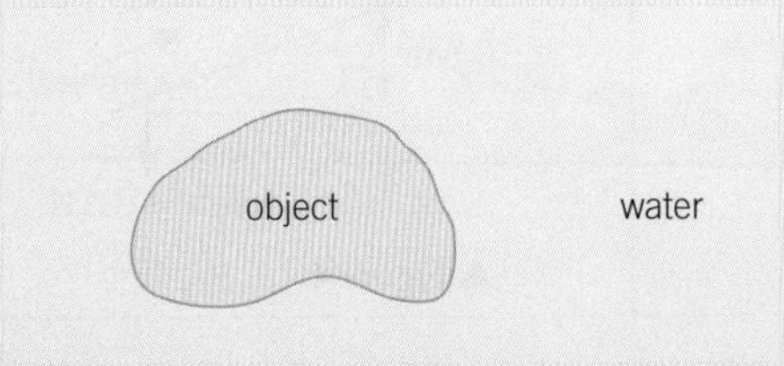

▲ Figure 5

The density of the object is 400 kg m^{-3} and it has a volume of 6.0 cm^3. The density of water is 1000 kg m^{-3}.

(i) Explain why the object experiences an upthrust. *(3 marks)*

(ii) Calculate initial upwards acceleration of the object when it is released. *(5 marks)*

6 a State what is meant by the *centre of gravity* of an object. *(1 mark)*

b Define *moment of a force*. *(1 mark)*

c Figure 6 shows a baby's mobile toy.

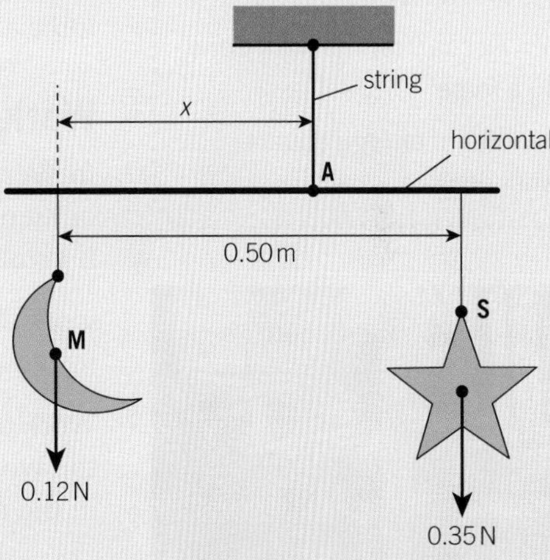

▲ Figure 6

The toy consists of a horizontal rod from which two objects shaped as a crescent moon **M** and a star **S** are suspended from lengths of string. The weight of the rod is negligible and it is pivoted about an axis passing through point **A** perpendicular to the plane of the diagram. The weights of **M** and **S** and the separation between the strings are shown in Figure 6. The distance between the string on the left and point **A** is *x*. The arrangement shown in Figure 6 is in equilibrium.

(i) State **two** conditions necessary for the rod to be in equilibrium. *(2 marks)*

(ii) By taking moments about **A** determine the distance *x*. *(3 marks)*

(iii) Determine the magnitude of the tension in the string attached to **A**. *(1 mark)*

May 2013 G481

5 WORK, ENERGY, AND POWER
5.1 Work done and energy

Specification reference: 3.3.1

Learning outcomes

Demonstrate knowledge, understanding, and application of:

→ work done by a force

→ the unit joule

→ $W = Fx \cos\theta$ for the work done by a force

→ transfer of energy equal to work done.

▲ **Figure 1** *How much work is done raising the boxes to this height?*

Moving up and along

Moving an object can be difficult if it has a large mass or it must be moved a long distance. Fork-lift trucks are used in warehouses and other places where a load may need to be stacked or moved. A system of hydraulics powered by a motor in the base of the machine provides the force to move the platform. The energy needed depends upon the force and the distance over which this force acts.

Work done and energy

In everyday conversation, the word **work** can mean any kind of physical or mental activity. Work done in physics has a very precise definition. It involves a force F and distance x moved in the direction of the force.

work done = force × distance moved in the direction of the force

$$W = Fx$$

Work done therefore has the unit N m, or joule (J). 1 J = 1 N m. 1 J is the work done when a force of 1 N moves its point of application 1 m in the direction of the force.

You do no work when holding a heavy book in your hand. You exert a force on the book, but there is no movement. You do work when lifting the book vertically. You apply a force on the book and it moves through a certain distance. Energy is transferred when you do work. The book has gained gravitational potential energy. If you let the book fall, then its weight will do work, which is transferred into kinetic energy. In fact

work done = energy transferred

This is not surprising, because energy is defined as the capacity to do work.

Worked example: Work done

How much work is done by:

pushing a box on a rough surface at a constant speed (Figure 2)?	a phone falling freely to Earth (Figure 3)?	an object being moved up a smooth slope (Figure 4)?
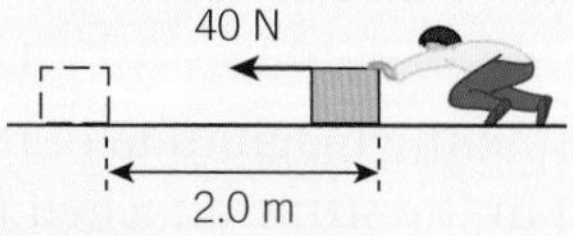 ▲ **Figure 2**	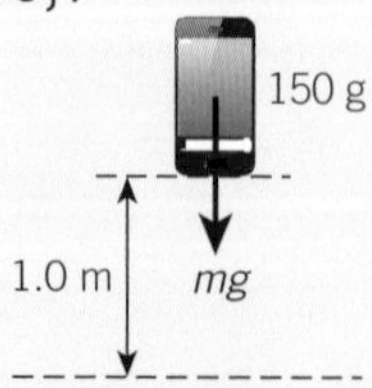 ▲ **Figure 3**	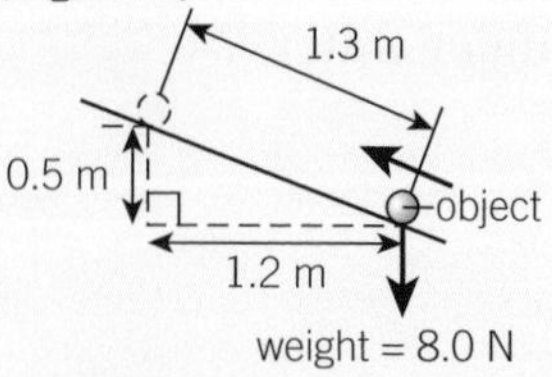 ▲ **Figure 4**

$W = Fx$ $W = 40 \times 2.0 = 80$ J	The force acting on the object is its weight *mg*. $W = Fx$ $W = (0.150 \times 9.81) \times 1.0$ $= 1.5$ J (2 s.f.)	The distance travelled in the direction of the force (the weight) is 0.5 m, and not 1.3 m nor 1.2 m. $W = Fx$ $W = 8.0 \times 0.5 = 4.0$ J
The work done **on** the box is transferred into thermal energy of the box and the surface below.	The work done **by** the force of gravity on the object is transferred to kinetic energy.	The work done **against** the force of gravity is transferred into gravitational potential energy.

Work done at an angle to motion

Quite often, a force is applied at an angle to the direction in which an object can move, as in Figure 5. How do you calculate the work done by the force?

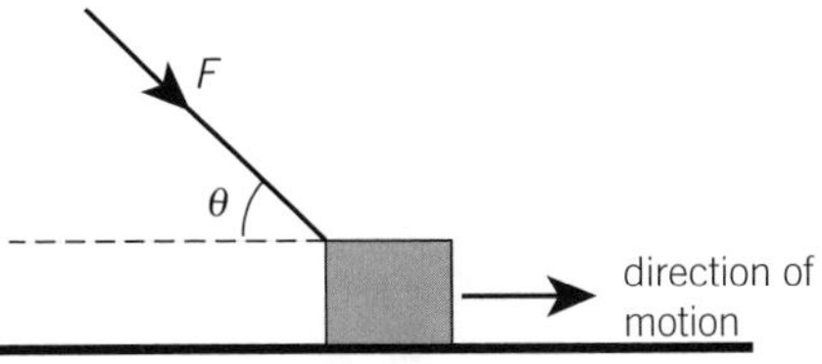

▲ **Figure 5** *Force exerted on an object at an angle to the direction of motion*

The component of the force F in the direction of motion is $F\cos\theta$. Therefore

$$\text{work done } W = (F\cos\theta) \times x$$

Or simply

$$W = Fx\cos\theta$$

The work done for the object moving up the slope in Figure 4 can be calculated using this equation. You just need to know the angle θ between the force and direction of motion. Try it for yourself with this alternative equation – you will get the same answer.

Study tip

When a force acts in the direction of travel, $\theta = 0°$ and $\cos 0° = 1$. This means you can use a simpler expression, $W = Fx$.

Summary questions

1 Calculate the work done when a force of 24 N moves an object a distance of 0.50 m in the direction of the force. *(2 marks)*

2 The thrust from a hovercraft fan is 430 N. Calculate the work done when the thrust moves the hovercraft by 1.0 km. *(2 marks)*

3 Calculate the work done by a person of mass 60 kg to climb to the top of a 5.8 m wall. *(2 marks)*

4 A shopper pushes a 38 kg shopping trolley at a constant speed up a car park ramp. The ramp is at 10° to the horizontal and is 3.1 m long. Calculate the work done to push the trolley to the top of the ramp. *(3 marks)*

5 A person exerts a force of 65 N at 52° to the horizontal floor to push a box 5.0 m across a floor at constant speed. Calculate the work done on the box. Explain, in terms of energy, what happens to the work done on the box. *(3 marks)*

6 A bullet travels straight through a piece of wood of thickness 30 mm. The change in the kinetic energy of the bullet is 1.4 kJ. Calculate the average force exerted by the wood on the bullet. *(4 marks)*

5.2 Conservation of energy

Specification reference: 3.3.1

Learning outcomes

Demonstrate knowledge, understanding, and application of:

→ the principle of conservation of energy

→ energy in different forms, transfer and conservation.

▲ **Figure 1** *In this design for a perpetual motion machine, weights falling on one side of the wheel make weights on the other side move upwards, supposedly keeping the wheel turning forever. In reality this will not happen – why not?*

▲ **Figure 2** *The spinning blades of a wind turbine have kinetic energy, and they gained gravitational potential energy when they were lifted into place by a crane*

Perpetual motion

There have been many attempts to build a perpetual motion machine – a device which, once started, will continue to move and do work without any further input of energy. None of them work and all eventually stop. In fact, perpetual motion is impossible according to the **principle of conservation of energy**. Would-be inventors should know that a patent will never be granted because their designs violate this well-established physical law.

Energy

Energy is the capacity for doing work. It is a scalar quantity, with magnitude but not direction. The SI unit for energy is the joule (J), the same unit as for work done.

Forms of energy

The energy of systems with mass can be classified as **kinetic energy** and **gravitational potential energy**.

- Kinetic energy is the energy due to the movement of an object.
- Gravitational potential energy is the energy due to the position of an object in the Earth's gravitational field.

The term *potential* is often used in physics to mean 'hidden' or 'stored'. Table 1 summarises some other forms of energy you have come across in your physics lessons.

▼ **Table 1** *Forms of energy*

Form of energy	Description	Examples
kinetic energy	energy due to motion of an object with mass	moving car moving atoms
gravitational potential energy	energy of an object due to its position in a gravitational field	child at the top of a slide water held in clouds
chemical energy	energy contained within the chemical bonds between atoms – it can be released when the atoms are rearranged	energy stored within a chemical cell energy stored in petrol and released when it is burnt
elastic potential energy	energy stored in an object as a result of reversible change in its shape	a stretched guitar string a squashed spring
electrical potential energy	energy of electrical charges due to their position in an electric field	electrical charges on a thundercloud static charge on a charged balloon

▼ **Table 1** *Continued*

Form of energy	Description	Examples
nuclear energy	energy within the nuclei of atoms – it can be released when the particles within the nucleus are rearranged	energy from fusion processes in the Sun energy from nuclear fission reactors
radiant (or electromagnetic) energy	energy associated with all electromagnetic waves, stored within the oscillating electric and magnetic fields	energy from the hot Sun energy from an LED
sound energy	energy of mechanical waves due to the movement of atoms	energy emitted when you clap output energy from your headphones
internal (heat or thermal) energy	the sum of the random potential and kinetic energies of atoms in a system	a hot cup of tea has more thermal energy than a cold one

▲ **Figure 3** *In a fire, chemical energy stored in the substances that make up the building and its contents is converted to thermal energy and radiant energy*

The principle of conservation of energy

Energy can be converted from one form to another. For example, an archer's bow usefully converts elastic potential energy into kinetic energy in an arrow. The bow also converts some of the elastic potential energy into thermal energy and sound energy, but the *total* final energy is always equal to the *total* initial energy.

The **principle of conservation of energy** states that the total energy of a closed system remains constant: energy can never be created or destroyed, but it can be transferred from one form to another.

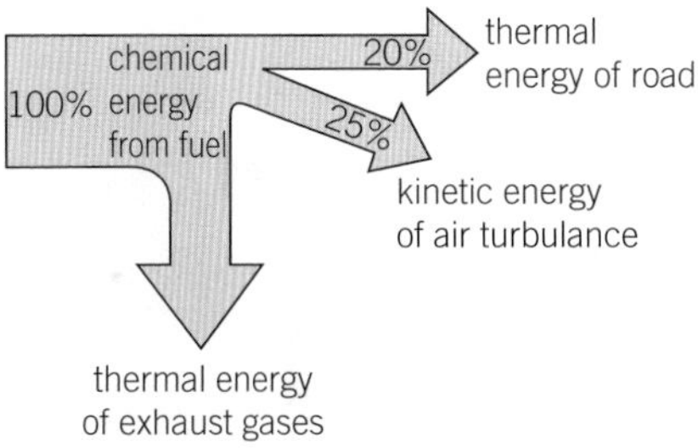

▲ **Figure 4** *Energy conversions from fuel burnt*

Summary questions

1 a State what is meant by the term *potential* in physics. *(1 mark)*
b State the energy changes taking place when you rub your hands together. *(1 mark)*

2 A lamp converts 20 J of electrical energy into 5 J of light energy and one other form of energy. Suggest what this other form of energy is, and calculate its quantity. *(2 marks)*

3 Describe the useful energy conversions that happen in:
a a filament lamp; *(1 mark)*
b the headphones connected to a mobile phone. *(1 mark)*

4 A car is travelling on a level road at constant speed. Figure 4 is a visual representation of the energy conversions of the chemical energy in the fuel – the diagram is called a Sankey diagram.
a Explain why the diagram does not show the kinetic energy of the car. *(1 mark)*
b Calculate the percentage of total energy wasted as thermal energy. *(1 mark)*

5.3 Kinetic energy and gravitational potential energy

Specification reference: 3.3.2

Learning outcomes

Demonstrate knowledge, understanding, and application of:

- → kinetic energy of an object; $E_k = \frac{1}{2}mv^2$
- → gravitational potential energy of an object in a uniform gravitational field; $E_p = mgh$
- → the exchange between gravitational potential energy and kinetic energy.

▲ **Figure 1** *The loop in this rollercoaster plunges thrill-seekers vertically downwards, converting GPE into KE*

Don't look down

Roller coasters have been thrilling people for at least two hundred years. In a typical rollercoaster, the cars are hauled up a steep slope by a motor, and then the force of gravity takes over. As the cars hurtle along the track, gravitational potential energy (GPE) and kinetic energy (KE) are interchanged repeatedly, enough to excite or terrify the passengers without causing injury. It is possible to analyse the motion of the rollercoaster from its GPE and KE.

Kinetic energy

Kinetic energy is energy associated with an object as a result of its motion. You can calculate the KE E_k of an object in linear motion from its mass m and speed v using the equation

$$E_k = \frac{1}{2}mv^2$$

Notice that for objects travelling at the same speed, the KE is directly proportional to the mass. For a given object, the KE is directly proportional to the square of its speed.

Worked example: A meteor hitting the Earth

The Chelyabinsk meteor was a near-Earth asteroid that crashed into Russia in February 2013. It had a mass of 1.2×10^7 kg and a speed on impact of $19\,\text{km s}^{-1}$. Calculate its KE, and compare this energy with the KE of a 1.2×10^4 kg truck travelling at $20\,\text{m s}^{-1}$.

Step 1: Identify the equation needed and list the known values.

$$E_k = \frac{1}{2}mv^2$$

Meteor: $m = 1.2 \times 10^7$ kg, $v = 19 \times 10^3\,\text{m s}^{-1}$ ($1.9 \times 10^4\,\text{m s}^{-1}$)

Truck: $m = 1.2 \times 10^4$ kg, $v = 20\,\text{m s}^{-1}$

Step 2: Substitute the values into the equation and calculate the answer.

Meteor: $E_k = \frac{1}{2}mv^2 = \frac{1}{2} \times 1.2 \times 10^7 \times (1.9 \times 10^4)^2$ (note that you must convert the speed into m s^{-1})

$E_k = 2.17 \times 10^{15}\,\text{J} \approx 2.2 \times 10^{15}\,\text{J}$

Truck: $E_k = \frac{1}{2}mv^2 = \frac{1}{2} \times 1.2 \times 10^4 \times 20^2$

$E_k = 2.4 \times 10^6\,\text{J}$

The ratio of the kinetic energies is $\frac{2.17 \times 10^{15}}{2.4 \times 10^6} = 9.0 \times 10^8$ (2 s.f.)

So, the KE of the meteor on impact was equivalent to the KE of 900 million trucks!

Study tip

The equations $E_k = \frac{1}{2}mv^2$ and $E_p = mgh$ are not provided in the examinations so you will have to remember them.

Deriving $E_k = \frac{1}{2}mv^2$

You can derive the equation for KE by using ideas developed in this book already.

Figure 2 shows a constant force F acting on an object of mass m. The object is initially at rest. The acceleration of the object is a. After a distance s it has a speed v.

The distance s travelled by the object can be determined from the equation of motion $v^2 = u^2 + 2as$.

$$s = \frac{v^2 - u^2}{2a} = \frac{v^2}{2a} \qquad (u = 0)$$

The work done by the force is entirely transferred to the KE of the object. Therefore

$$\text{work done} = E_k = Fs$$

The force F is given by $F = ma$. Therefore

$$E_k = ma \times s = ma \times \frac{v^2}{2a}$$

$$\therefore E_k = \frac{1}{2}mv^2$$

(where ∴ denotes *therefore*).

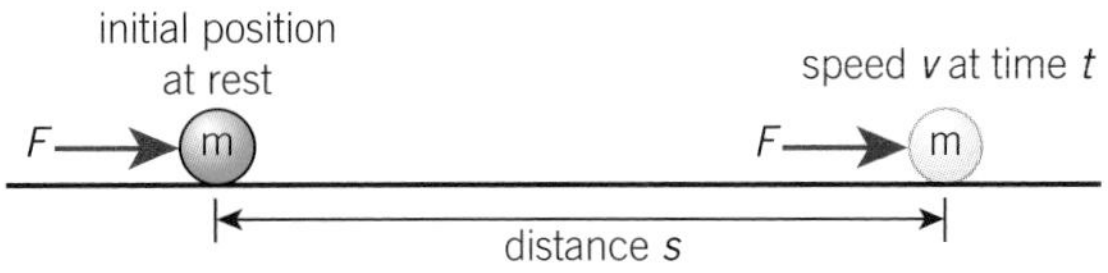

▲ **Figure 2** *Gaining kinetic energy*

1. Use the equation for E_k to derive the SI base units for kinetic energy.
2. Derive an equation for E_k in terms of speed v, weight W of an object, and acceleration of free fall g.

Gravitational potential energy

Gravitational potential energy is the capacity for doing work as a result of an object's position in a gravitational field. You can calculate the change in GPE E_p of an object in a **uniform gravitational field** from its mass m and the change in height h using the equation $E_p = mgh$

where g is acceleration of free fall.

You can use this equation for all objects close to the Earth's surface, where g may be assumed to be constant and to have a value of $9.81\,\text{m s}^{-2}$.

GPE is *gained* when an object gets higher, and is *lost* when an object gets lower.

You can explain the origin of the equation for GPE using the idea of work done by a force. Imagine lifting a mass vertically upwards through a height h at constant speed (so its KE does not change). You have to apply a force equal to its weight mg. The work done W by this force is transferred into GPE. Therefore

$$E_p = W = \text{force} \times \text{distance moved in the direction of force}$$

$$E_p = (mg) \times h \qquad E_p = mgh$$

Synoptic link

You can use work done = change in KE or $Fs = \frac{1}{2}mv^2$ to explain why the braking distance (s) of a particular car is directly proportional to (initial speed)^2 – see Topic 3.6, Car stopping distances.

Study tip

Remember that things fall without any extra energy being supplied, so GPE is dissipated.

▲ **Figure 3** *Parachutists gain gravitational potential energy on the way up in an aircraft, and lose it as they fall to the ground*

Worked example: Skydiver

A skydiver with a mass of 80 kg falls through a distance of 2.0 km at a terminal velocity of $45.0\,\text{m s}^{-1}$. Calculate the loss of GPE and explain what happens to the energy lost.

Step 1: Identify the equation needed.

$$\text{loss in GPE} = E_p = mgh$$

Step 2: Substitute the values into the equation and calculate the answer.

$E_p = 80 \times 9.81 \times 2000 = 1.6 \times 10^6$ J (remember h must be in metres)

There is no change in the KE of the skydiver, so the GPE is transferred to the thermal energy of the surrounding air.

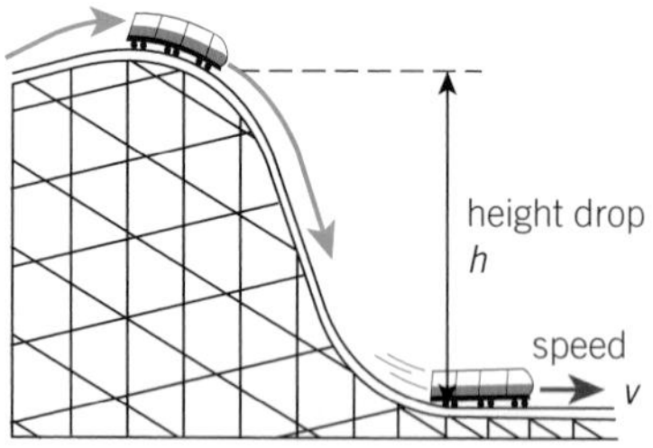

▲ **Figure 4** *The speed of the roller coaster at the bottom is given by $v = \sqrt{2gh}$, as long as there are no frictional losses*

Energy exchange

There are many situations where KE and GPE are exchanged. For example, as an object falls, its GPE decreases and its KE increases. You can see this happening for a waterfall or when you drop a pen. An object falling from rest will lose GPE. From the principle of conservation of energy, the object will gain an equal amount of KE. Therefore

$$mgh = \frac{1}{2}mv^2$$

where v is the final speed of the object.

The mass m on both sides of the expression cancels out. Therefore

$$gh = \frac{1}{2}v^2 \qquad v^2 = 2gh \qquad v = \sqrt{2gh}$$

The equation above is only valid if there are no resistive forces involved. The mass of the object has no bearing on its final speed. You should not be too surprised by this because you already know that the acceleration of free fall is the same for all objects. An object dropped from a height of only 7.0 m will hit the ground at a speed of 31 m s^{-1} (70 mph) – you can use the equation above to check this.

Summary questions

1 Calculate the kinetic energy of a 1500 kg object travelling at 10 m s^{-1}. *(2 marks)*

2 Calculate the gain in the gravitational potential energy of a plane of mass = 9.4×10^4 kg climbing a vertical distance of 1500 m. *(2 marks)*

3 A 120 g ball is dropped from rest from a height of 90 cm. Its rebound height is 70 cm. Calculate the energy lost to the ground. *(3 marks)*

4 A roller coaster of mass 400 kg begins at rest, then drops 55 m into a tunnel. Assume that air resistance has negligible effect on its motion.

a Calculate the initial GPE of the roller coaster. *(2 marks)*

b Calculate its KE as it enters the tunnel. *(1 mark)*

c Use your answer to (b) to calculate its speed. *(3 marks)*

5 Victoria Falls (Mosi-oa-Tunya) is a waterfall on the Zambezi River at the border of Zambia and Zimbabwe. It has a height of about 110 m. Calculate the maximum speed of the water at the bottom. State any assumptions made. *(3 marks)*

6 A book of mass 0.80 kg falls from a height of 1.2 m. Its speed is 3.2 m s^{-1} when it hits the ground. Calculate the work done against the drag force. *(3 marks)*

7 A bullet of mass 30 g is fired at a block of wood at a speed of 240 m s^{-1}. The bullet penetrates a distance of 8.5 cm into the wood. Calculate the average resistive force exerted on the bullet by the wood block. *(3 marks)*

5.4 Power and efficiency

Specification reference: 3.3.3

Electric mountain

The Dinorwig Power Station in north Wales is an unusual hydroelectric scheme. Built into a mountain in Snowdonia National Park, it has two reservoirs, one more than 500 m above the other. Several kilometres of water-filled tunnels pass through the mountain, carrying water between the reservoirs. A huge cavern excavated deep inside the mountain contains six reversible turbines.

When there is a high demand for electricity from the National Grid, water flows from the upper reservoir, driving the turbines and generators. During off-peak times, the turbines pump water back up from the lower reservoir. Dinorwig is the largest pumped storage facility in Europe, but although it is a brilliant solution to providing electricity at short notice, it is not 100% efficient.

Learning outcomes

Demonstrate knowledge, understanding, and application of:

→ power; the unit watt; $P = \frac{W}{t}$

→ $P = Fv$

→ efficiency of mechanical systems; *efficiency* = $\frac{\textit{useful output energy}}{\textit{total input energy}} \times 100\%$.

Energy and power

For a sprinter or a racing car, we are interested not only in the amount of energy but also in the *rate* at which energy is transferred. A powerful car is the one with the largest value for the rate of energy transfer.

Power is the rate of work done.

As an equation, this is written as

$$P = \frac{W}{t}$$

where P is the power and W is the work done in a time t. Since work done is equal to energy transfer, we can also define power as the rate of energy transfer.

Power is measured in joules per second (J s^{-1}) or in watts (W). 1 W is equal to one joule per second.

▲ **Figure 1** *The upper reservoir for the Dinorwig Power Station has a low water level in the daytime*

Worked example: Body power

A 60 kg person runs up a flight of steps in a time of 7.2 s. The gain in vertical height in this time interval is 5.0 m. Calculate the rate of work done against the force of gravity.

Step 1: Calculating the rate of work done against gravity is the same as calculating the power P. Also, the work done against the force of gravity is the same as the gain in the gravitational potential energy E_p of the person.

$$\text{power} = \frac{\text{work done}}{\text{time}}$$

$$P = \frac{E_p}{t} = \frac{mgh}{t}$$

→

Step 2: Substitute the values into the equation and calculate the answer.

$$P = \frac{60 \times 9.81 \times 5.0}{7.2} = 410\,\text{W (2 s.f.)}$$

The rate of work done against the force of gravity is about 410 W. This is the same as 410 J per second.

The actual power developed by the person is much greater than this value. Our muscles are incapable of transferring all available energy into movement – some is wasted as thermal energy within the muscles. This is why you get hot when you exercise. Muscles are not very efficient at transferring energy (see Table 1).

Power and motion

There are situation in physics where constant force has to be exerted on an object to maintain its constant speed. A good example of this is a car travelling on a level road at a constant speed (Figure 2). The net force on the car is zero. The rate of work done by the forward force provided by the car engine is equal to the rate of work done against the frictional forces acting on the car. It is possible to calculate the power P developed by the force provided by the car.

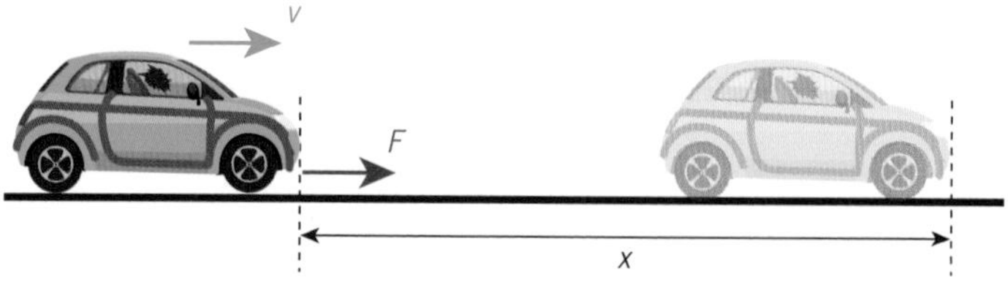

▲ **Figure 2** *The power developed by the car depends on the force F and the speed v*

A constant force F moves the car a distance x in a time t.

work done by the force $W = F \times x$

$$P = \frac{W}{t} = \frac{Fx}{t}$$

The speed v of the car is $\left(\frac{x}{t}\right)$, therefore

$$P = Fv$$

This equation is not just for cars. You can use it in a variety of situations where a constant force is necessary to maintain a constant speed. Examples include a swimmer travelling though water at constant speed (Figure 3) and a skydiver falling through the air at terminal velocity.

▲ **Figure 3** *Greater power is required to swim faster through the water*

Efficiency

You will remember the principle of conservation of energy: the total energy of a closed system remains constant. However, this does not mean that processes and machines convert all their energy into *useful* work. In particular, thermal losses can mean that some of the input energy is not converted into useful output energy.

You can calculate **efficiency** using the expression

$$efficiency = \frac{useful\ output\ energy}{total\ input\ energy} \times 100\%$$

The greater the efficiency, the greater the percentage of input energy converted. Some examples of efficiency are shown in Table 1.

▼ **Table 1** *Typical efficiencies – can you suggest why the electric heater is almost 100% efficient?*

System	filament lamp	muscles	petrol engine	solar cell	LED	diesel engine	wind-generator	electric heater
Typical efficiency /%	5	20	20	25	35	35	40	~ 100

Summary questions

1 Calculate the power of a lamp that transfers 240 J in 30 s. *(2 marks)*

2 Calculate the energy transferred by a 2.0 kW motor in a time of 60 s. *(2 marks)*

3 A lamp is about 5.0% efficient at producing light energy from electrical energy. Calculate the light energy produced from a 60 W lamp in a time of 1.0 hour. *(3 marks)*

4 A car of mass 1200 kg starting from rest reaches a speed of 18 m s^{-1} in 20 s. Calculate the average rate of work done on the car. *(3 marks)*

5 An aircraft has four jet engines, each producing 210 kN of thrust. Calculate the total power output of the engines when the aircraft is in level flight and travelling at a constant speed of 250 m s^{-1}. *(2 marks)*

6 A small 3.5 W electric motor raises a 15 N load through a vertical height of 1.4 m in 30 s (Figure 4). Calculate the efficiency of the motor. *(3 marks)*

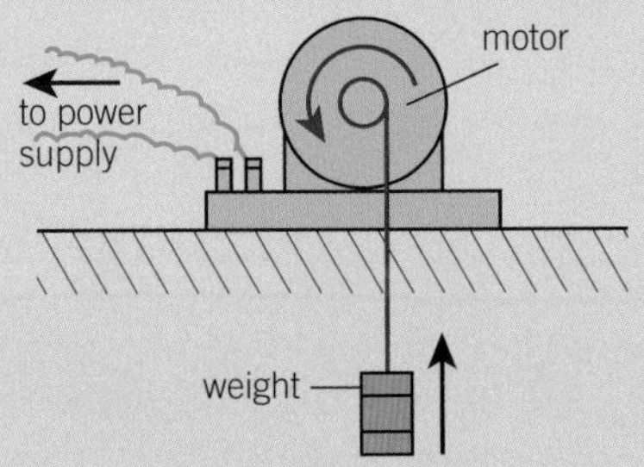

▲ **Figure 4**

7 A hydroelectric power station produces 600 MW of electrical power. Water falls 50 m before passing through its turbines. The transfer of gravitational potential energy to electrical energy is 40% efficient. Calculate the rate (volume/time) at which water passes through the turbines. The density of water is 1000 kg m^{-3}. *(5 marks)*

Practice questions

1 a Define power. *(1 mark)*

b An electric motor on a mechanical crane is used to lift heavy objects, see Figure 1.

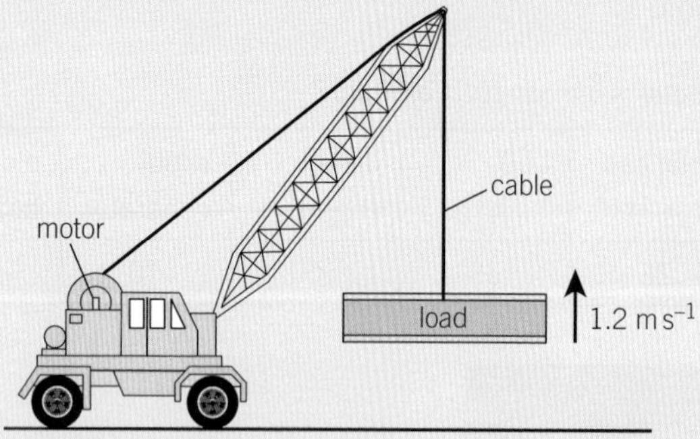

▲ **Figure 1**

The cable attached to the motor is used to lift a load of total mass of 1500 kg at a constant vertical velocity of 1.2 m s^{-1}.

(i) Calculate the tension in the cable. Explain your answer. *(3 marks)*

(ii) Calculate the minimum output power of the motor needed to raise the 1500 kg mass. *(3 marks)*

2 a Write a word equation for *kinetic energy*. *(1 mark)*

b A bullet of mass 3.0×10^{-2} kg is fired at a sheet of plastic of thickness 0.015 m. The bullet enters the plastic with a speed of 200 m s^{-1} and emerges from the other side with a speed of 50 m s^{-1}.

Calculate

(i) The loss of kinetic energy of the bullet as it passes through the plastic *(3 marks)*

(ii) The average frictional force exerted by the plastic on the bullet. *(2 marks)*

c Plan a simple experiment to determine the kinetic energy of a student running round a track. *(4 marks)*

3 a State the *principle of conservation of energy*. *(1 mark)*

b Define *work done* by a force and state its unit. *(1 mark)*

c Figure 2 shows a crater on the surface of the Earth.

▲ **Figure 2**

The crater was formed by a meteorite impact 50,000 years ago. The meteorite was estimated to have a mass of 3.0×10^{8} kg with an initial kinetic energy of 8.4×10^{16} J just before impact.

(i) State one major energy transformation that took place during the impact of the meteorite with the Earth. *(1 mark)*

(ii) Show that the initial impact speed of the meteorite was about 2.0×10^{4} m s^{-1}. *(2 marks)*

(iii) The crater is about 200 m deep. Estimate the average force acting on the meteorite during the impact. *(3 marks)*

Jan 2012 G481

4 a Derive the base units for work done. *(3 marks)*

b Figure 3 shows a person at the base of an escalator.

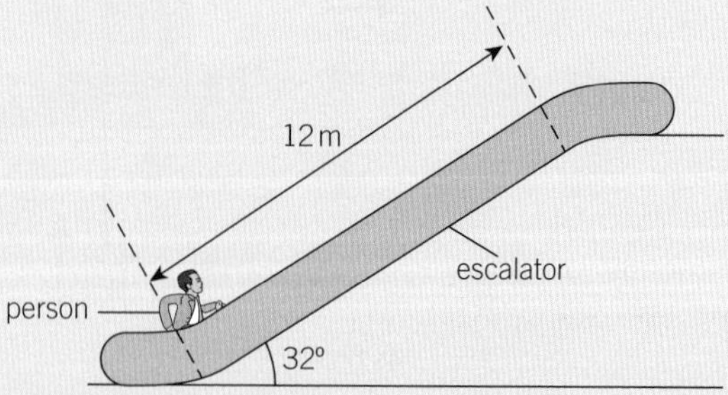

▲ **Figure 3**

The mass of the person is 70 kg. The escalator travels at an angle of 32° to the horizontal. The person travels a total distance of 12 m from the bottom to the top of the escalator in a time of 8.0 s.

(i) Calculate the kinetic energy of the person on the moving escalator. (*3 marks*)

(ii) Calculate the gravitational potential gained by the person. (*3 marks*)

(iii) Calculate the power required to lift the person from the bottom to the top of the escalator. (*2 marks*)

5 A skydiver jumps off a helicopter and falls vertically towards the ground. The skydiver reaches a steady velocity before he opens his parachute and lands safely on the ground.

a Describe the energy changes taking place from the time the skydiver jumps off the helicopter to the instant before he opens his parachute. (*4 marks*)

b The total mass of the skydiver is 80 kg and his terminal velocity is $45\,\mathrm{m\,s^{-1}}$. Calculate the rate of work done against drag. (*3 marks*)

6 a Explain what is meant by *energy* and relate it to *power*. (*2 marks*)

b Define the *watt*. (*1 mark*)

c A lift is used to carry people up a building. The mass of the lift is 1500 kg and it can carry a maximum of 8 people of average mass 70 kg. The vertical height travelled by the lift is 120 m and it takes 55 s.

(i) Calculate the gain in gravitational potential energy of the 8 people in the lift. (*2 marks*)

(ii) Calculate the minimum output power of the electrical motor used to operate the lift. (*3 marks*)

7 a In a downhill race, the total distance between start and finish is 5.00 km, and the total vertical drop is 520 m. The weight of a runner is 70 kg and the time taken for the descent is 15 minutes.

(i) Calculate the average kinetic energy of the runner during this race. (*3 marks*)

(ii) Calculate the total loss in the gravitational potential energy of the runner. (*2 marks*)

(iii) Explain why your answers to **(a)**(i) and **(a)**(ii) are not the same. (*1 mark*)

b (i) State the principle of conservation of energy. (*1 mark*)

(ii) Use the principle of conservation of energy to show that an object dropped through a vertical distance of 520 m can have a speed of more than $100\,\mathrm{m\,s^{-1}}$. (*3 marks*)

8 A stunt person, initially at rest, slides down a cable attached between a tall building and the ground.

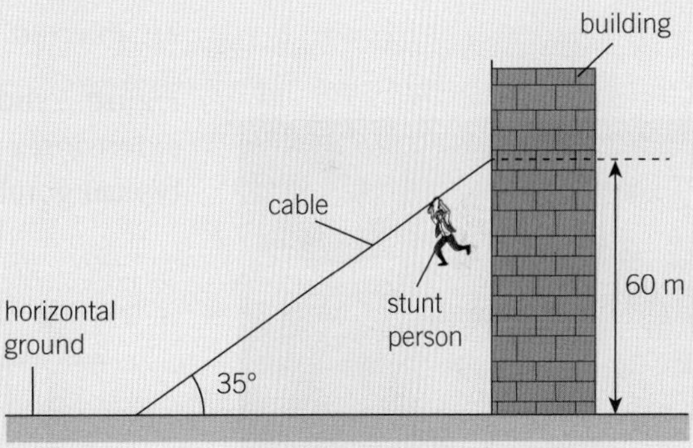

▲ **Figure 4**

The person has mass 72 kg. The speed of the person at the bottom of the cable is $20\,\mathrm{m\,s^{-1}}$.

a Calculate the loss in gravitational potential energy of the person. (*2 marks*)

b Calculate the work done against resistive forces. (*3 marks*)

c Calculate the average frictional force acting on the person. (*3 marks*)

6 MATERIALS

6.1 Springs and Hooke's law

Specification reference: 3.4.1, 3.4.2

Learning outcomes

Demonstrate knowledge, understanding, and application of:

- → tensile and compressive deformation
- → Hooke's law
- → force constant k of a spring or wire and $F = kx$
- → force–extension (or compression) graphs for springs and wires
- → elastic and plastic deformation of springs.

Tensile and compressive forces

You need a pair of equal and opposite forces to alter the shape of an object. For example, you can extend the length of a rubber band by pulling its ends. Forces that produce **extension** are known as **tensile forces**, and those that shorten an object (**compression**) are **compressive forces**. Imagine sitting on a stool. Your weight downwards and the contact force upwards from the ground provide the pair of compressive forces on the stool; these will shorten the length of the stool by a tiny amount.

In the suspension bridge shown in Figure 1, the cables and the huge vertical supports hold together the road structure below. Can you identify where the tensile and compressive forces act?

▲ **Figure 1** *The Forth Road Bridge in Scotland*

Hooke's law

A helical spring undergoes **tensile deformation** when tensile forces are exerted and **compressive deformation** when compressive forces are exerted.

Figure 2 shows a simple apparatus used to investigate how the extension x of a helical spring is affected by the applied force F, and a typical graph of force against extension obtained.

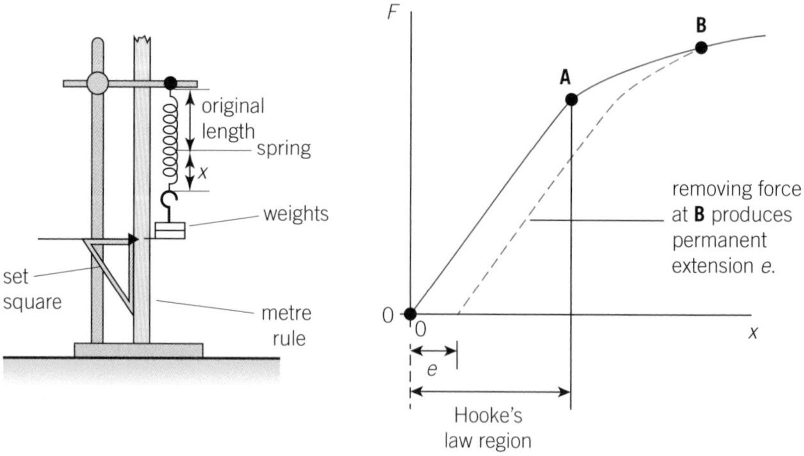

▲ **Figure 2** *Recording the force–extension graph for a spring*

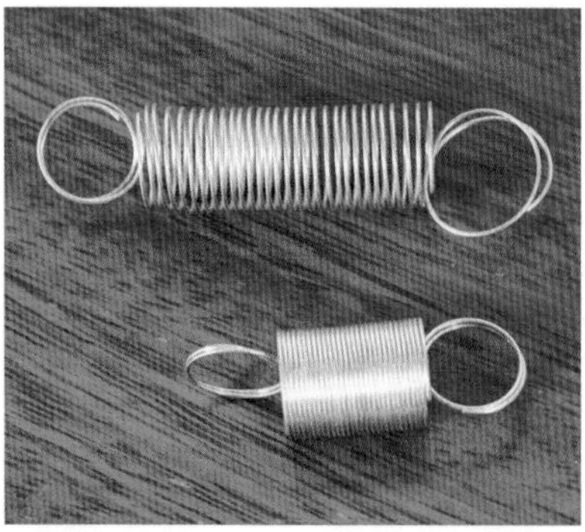

▲ **Figure 3** *You can easily identify the spring that has been stretched beyond its elastic limit*

The **force–extension graph** is a straight line from the origin up to the **elastic limit** (point **A**) of the spring. In this linear region, the spring undergoes **elastic deformation**. This means that the spring will return to its original length when the force is removed. Beyond point **A**, the spring undergoes **plastic deformation**: permanent structural changes to the spring occur and it does not return to its original length when the force is removed (Figure 3).

For forces less than the elastic limit of the spring, the spring obeys **Hooke's law**: The extension of the spring is directly proportional to the force applied. This is true as long as the elastic limit of the spring is not exceeded.

Force constant *k*

For a spring obeying Hooke's law, the applied force F is directly proportional to the extension x. Therefore

$$F \propto x$$

or

$$F = kx$$

where k is called the **force constant** of the spring (SI unit newton per metre, $N\,m^{-1}$). This is a measure of the **stiffness** of a spring. A spring with a large force constant is difficult to extend and you would refer to it as a stiff spring. You can also use the equation $F = kx$ for a compressible spring: x then represents the compression of the spring.

You can determine the force constant k from the gradient of the linear region of the force–extension graph.

Hooke's law also applies to wires under tension and concrete columns under compression. You can even model the behaviour of atoms in a solid using Hooke's law. In fact the equation $F = kx$ can be applied to almost any object that can be elastically squashed or extended.

Synoptic link

You will find more information about gradients of straight-line graphs in Appendix A2, Recording results.

Worked example: Force constant of a wire

A shelf of mass 14.00 kg is supported by four identical wires. The original length of each wire was 1.800 m. When attached to the shelf, the length of each wire is 1.804 m. Calculate the force constant of each wire.

Step 1: Select the correct equation to calculate the force F acting on each wire.

The weight W of the shelf can be calculate using $W = mg$; assume that this weight is shared equally amongst the four wires.

$$F = \frac{\text{weight}}{4} = \frac{14.00 \times 9.81}{4} = 34.3\,N$$

Step 2: Determine the extension x of each wire.

$$x = \text{new length} - \text{original length}$$
$$= 1.804 - 1.800 = 0.004\,m$$

Step 3: Select the correct equation to calculate k.

$$F = kx$$

$$34.3 = k \times 0.004$$

$$k = \frac{34.3}{0.004} = 9 \times 10^4\,N\,m^{-1}\ (1\ \text{s.f.})$$

The force constant of each wire is $9 \times 10^4\,N\,m^{-1}$ (1 s.f.).

Investigating Hooke's law

You can investigate Hooke's law using a spring and some standard masses (see Figure 2).

Attach the spring at one end using a clamp, boss, and clamp stand secured to the bench using a G-clamp or a large mass. Set up a metre rule with a resolution of 1 mm close to the spring. Suspend slotted masses from the spring and, as you add each one, record the total mass added and the new length of the spring.

You can improve the accuracy of the length measurements using a set square, and by taking readings at eye level to reduce parallax errors. You might also measure the mass of each slotted mass using a digital balance. To obtain reliable results, aim to take at least six different readings and to repeat each one at least once.

Synoptic link

You will find more information about accuracy of readings in Appendix A3, Measurements and uncertainties.

▲ **Figure 4** *Every year in summer the swans born on a section of the river Thames are caught, weighed and measured, and ringed before being released to grow to maturity*

Force meters

Hooke's law is used in the design and calibration of simple force meters or newtonmeters, often used for weighing. Such spring-loaded scales are useful in situations where scales must be mobile or robust and easy to repair. They are used to monitor babies' growth in clinics in developing countries, for example. In Figure 4 a cygnet – a young swan – is weighed in a wildlife survey.

1. The extension of the spring in the force meter is 12 mm when a cygnet with a mass of 4.0 kg is weighed. Predict the extension of the spring when a cygnet of 6.0 kg is weighed. State any assumptions made.
2. Determine the force constant of the spring in N m^{-1}.

Summary questions

1. Figure 5 shows the force *F* against compression *x* graphs for two metal rods **A** and **B**. Compare the behaviour of the rods. *(2 marks)*

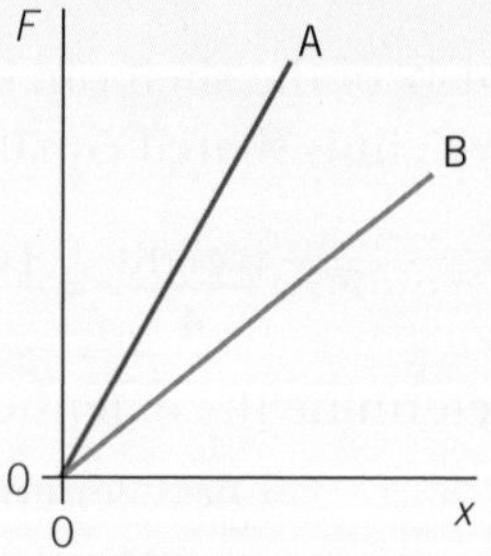

▶ **Figure 5**

2. The elastic limit of a wire is at 5.0 N. When a force of 2.5 N is applied the wire has an extension of 4 mm. Sketch a force–extension graph for this wire. *(3 marks)*
3. A spring is compressed by 5 mm by a force of 4.0 N.
   a. Calculate the force constant of the spring. *(2 marks)*
   b. Calculate the force applied when its compression is 32 mm. State any assumptions made. *(3 marks)*

4 Table 1 shows the results obtained for a spring in the experiment described above. Repeat readings could not be taken in this experiment because the spring was stretched beyond its elastic limit.

▼ **Table 1** *Results from an investigation into the extension of a spring*

Mass attached to the spring / g	Length of spring, L / 10^{-2} m
100	4.3
200	8.6
300	13.0
400	17.1
500	21.6
600	28.1
700	37.0

a Copy the table of results. Add a column for the force F/N acting on the spring by calculating $F = mg$, where m is the mass in kg and $g = 9.81\ \mathrm{m\,s^{-2}}$. *(1 mark)*

b Plot a graph of force F against length L of the spring. *(3 marks)*

c Explain how the graph shows that the spring obeys Hooke's law and state the value of the force at the elastic limit. *(3 marks)*

d Determine the force constant of the spring in $\mathrm{N\,m^{-1}}$. *(2 marks)*

5 A 200 mm long spring is suspended vertically. The length of the spring increases to 294 mm when a mass of 280 g is attached to it.

a Calculate the force constant of the spring. *(3 marks)*

b A second, identical spring is suspended alongside the first spring and both are then attached to a rod of negligible mass (Figure 6).

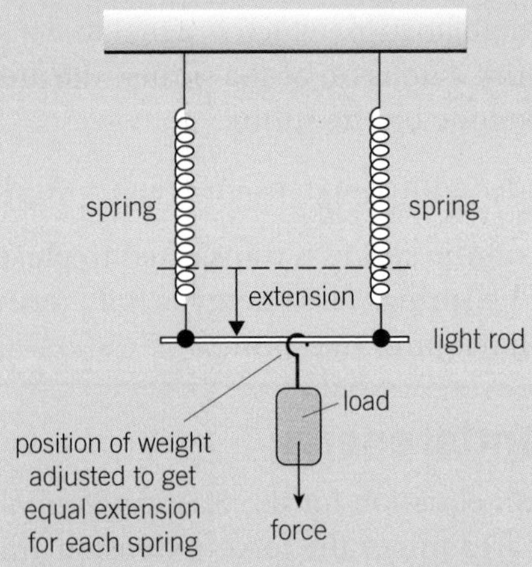

▲ **Figure 6**

i Calculate the combined force constant of this parallel arrangement. Explain your answer. *(3 marks)*

ii Calculate the combined force constant when the same springs are joined end-to-end in a series arrangement. Explain your answer. *(3 marks)*

6.2 Elastic potential energy

Specification reference: 3.4.2

Learning outcomes

Demonstrate knowledge, understanding and application of:

- → force–extension graph and work done
- → elastic potential energy
- → $E = \frac{1}{2}Fx$ and $E = \frac{1}{2}kx^2$.

▲ **Figure 1** *The stretched elastic of a catapult stores energy and transfers it to kinetic energy when released*

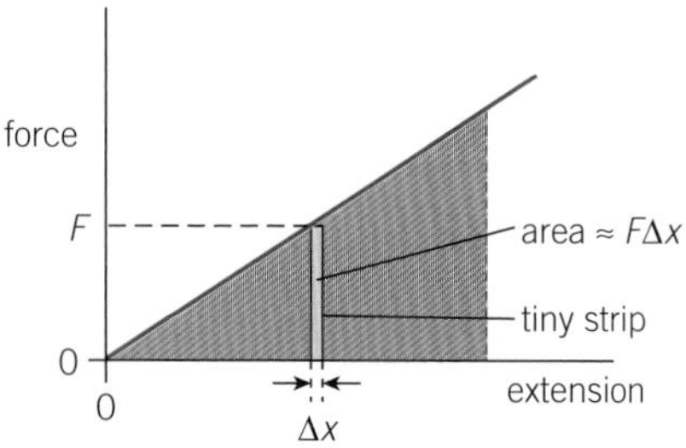

▲ **Figure 2** *Force–extension graph and work done*

Synoptic link

In Topic 5.1, Work done and energy, you learnt the definition for work done by a force.

Stored energy

Imagine holding a rubber band between your fingers and applying a force by stretching it. The rubber band will extend in the direction of the force. The work done on the rubber band is transferred into stored energy in the band. You can show that a stretched rubber band has stored energy when you suddenly let it go and it flies across the room.

Clockwork toys use coiled springs to store and then release energy. The stretched strings of a guitar store energy – when a string snaps, this energy is transferred to kinetic energy and the broken string flies away at $15\,\text{m s}^{-1}$. Figure 1 shows another store of energy.

Work done and springs

When a material is compressed or extended without going beyond its elastic limit, the work done on the material can be fully recovered. If the material has gone through plastic deformation, then some of the work done on the material has gone into moving its atoms to new permanent positions. This energy is not recoverable.

How can you determine the energy stored in an elastic material? A good start is the force–extension graph for a spring (Figure 2).

The small amount of work done ΔW by a force F in extending the spring by a small length Δx is given by the equation

$$\Delta W \approx F \times \Delta x$$

If Δx is very small indeed, the force F acting on the spring will change very little over the range Δx. In the graph in Figure 2, $F\Delta x$ is the area of the thin rectangular strip, which is equal to ΔW. If you add similar strips for the entire extension of the spring, the area under the graph is the total work done on the spring.

area under a force–extension graph = work done

The work done on the spring is transferred to **elastic potential energy** within the spring. This energy is fully recoverable because of the elastic behaviour of the spring.

Elastic potential energy

You can derive an equation for the elastic potential energy E for a spring from the area under the force–extension graph (Figure 3).

E = area under graph = area of shaded triangle

$$E = \frac{1}{2}Fx$$

where F is the force producing an extension x.

You can also interpret the equation above as 'work done = average force × final extension'.

A spring obeys Hooke's law, $F = kx$. Substituting this equation into $E = \frac{1}{2}Fx$ gives us another useful equation for elastic potential energy.

$$E = \frac{1}{2}Fx = \frac{1}{2}(kx) \times x$$

$$E = \frac{1}{2}kx^2$$

For a given spring, E is directly proportional to extension^2, so doubling the extension quadruples the energy stored.

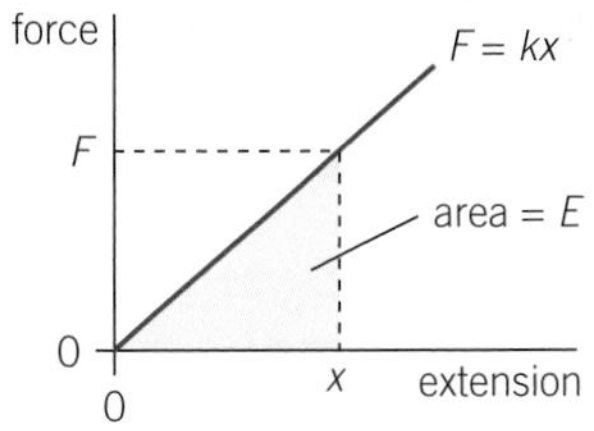

▲ **Figure 3** *Force–extension graph for a spring or wire*

Worked example: Firing a spring

A compressible spring of force constant $k = 50\,\text{N m}^{-1}$ and mass 4.0 g is placed around a short horizontal rod (Figure 4). The spring is compressed by 8.0 cm and then released.

a Calculate the elastic potential energy in the spring when compressed.

b Calculate the speed of the spring immediately after it has fully extended. State any assumptions made.

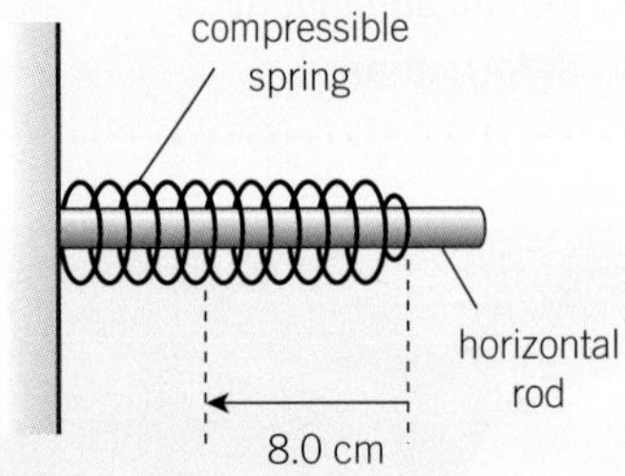

▲ **Figure 4**

a **Step 1:** Write down all the quantities given in this question in SI units.

$$k = 50\,\text{N m}^{-1},\ m = 4.0 \times 10^{-3}\,\text{kg},\ x = 0.08\,\text{m}$$

Step 2: Select the equation for the energy stored in the spring and calculate it.

$$E = \frac{1}{2}kx^2 = \frac{1}{2} \times 50 \times 0.08^2 = 0.16\,\text{J}$$

b **Step 3:** Assume all the elastic potential energy in the spring is transferred to its kinetic energy. Therefore

$$\text{kinetic energy} = \frac{1}{2}mv^2 = 0.16\,\text{J}$$

Step 4: Rearrange this equation for v and then substitute the mass of the spring to calculate the speed.

$$v = \sqrt{\frac{2 \times \text{kinetic energy}}{m}} = \sqrt{\frac{2 \times 0.16}{4.0 \times 10^{-3}}} = 8.9\,\text{m s}^{-1}\ (2\ \text{s.f.})$$

The speed of the spring is $8.9\,\text{m s}^{-1}$ (2 s.f.).

Summary questions

1 A rubber band is extended by 20 cm by a force of 18 N. Estimate the energy stored in this stretched rubber band. *(2 marks)*

2 The energy stored in a stretched cable is 1.5 J when it is extended by 2.0 mm. Calculate the force constant of the wire. *(3 marks)*

3 Figure 5 shows a force–extension graph for a spring. Calculate the work done on the spring when its extension changes from 5.0 cm to 15.0 cm. *(3 marks)*

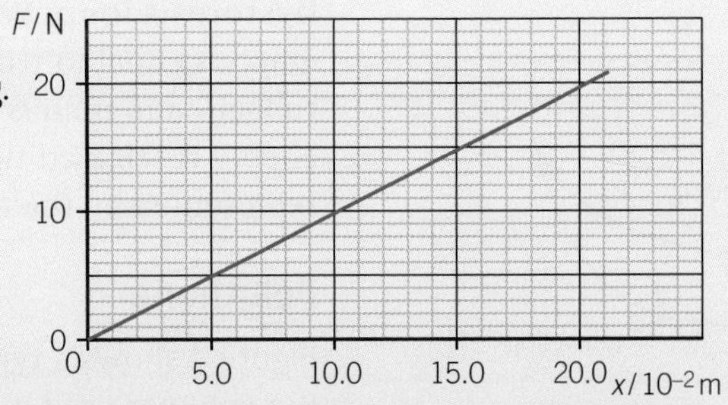

▶ **Figure 5**

4 A spring of mass 8.0 g has a force constant of $120\,\text{N m}^{-1}$. It is placed upright on a horizontal table and compressed by 4.0 cm. It is then released and it jumps vertically above the table. Calculate the maximum height gained by the spring above the table. State any assumptions made. *(5 marks)*

6.3 Deforming materials

Specification reference: 3.4.1, 3.4.2

Learning outcomes

Demonstrate knowledge, understanding, and application of:

- → force–extension graphs for metal, rubber, and polythene
- → elastic and plastic deformation.

▲ **Figure 1** *Bungee cord must be strong and elastic*

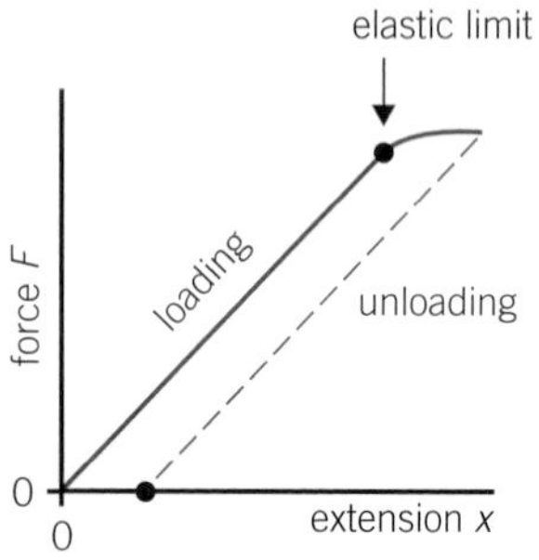

▲ **Figure 2** *Loading and unloading curves for a metal wire*

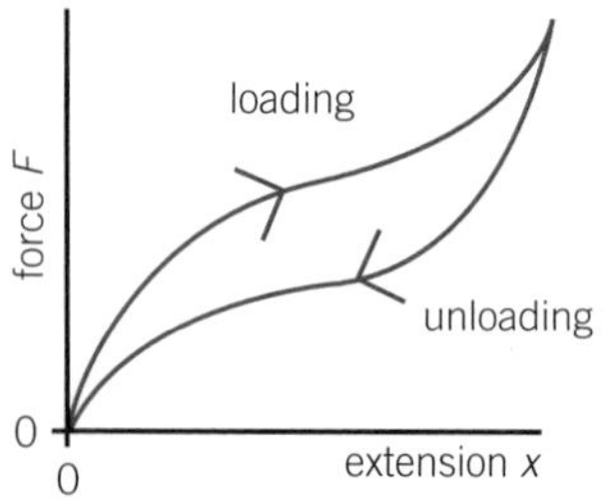

▲ **Figure 3** *Loading and unloading curve for a rubber band*

Bungee cords

Bungee jumpers leap from heights whilst securely attached to a long bungee cord. At first the jumper free falls, but then the slack is taken up, the cord stretches, and the jumper slows to a halt. As elastic potential energy in the cord is converted into kinetic energy and gravitational potential energy, the jumper accelerates upwards again. For safety, operators of bungee jumps need to take into account factors such as the weight of the jumper, the height of the jump, and the force–extension properties of the cord.

Loading and unloading

Different materials respond differently to tensile forces. Their extension increases as the force increases, then decreases again as the force is reduced. However, the **loading curve** and the corresponding unloading curve may not be the same.

Metal wire

Figure 2 shows a typical force–extension graph for a metal wire.

The loading graph in Figure 2 follows Hooke's law until the elastic limit of the wire. The unloading graph will be identical for forces less than the elastic limit. However, beyond the elastic limit it is parallel to the loading graph but not identical to it. The wire is permanently extended after the force is removed – it is longer than it was at the start. Like the springs in Topic 6.1, Springs and Hooke's law, the wire has suffered plastic deformation.

Rubber

Figure 3 shows a typical force–extension graph for a rubber band.

Rubber bands do not obey Hooke's law. The rubber band will return to its original length after the force is removed – elastic deformation – but the loading and unloading graphs are both curved and are different.

The 'loop' formed by the loading and unloading curves is called a **hysteresis loop**. You will recall that the area under a force–extension graph is equal to work done. More work is done when stretching a rubber band than is done when its extension decreases again. Thermal energy is released when the material is loaded then unloaded, represented by the area inside the hysteresis loop.

Polythene

Figure 4 shows a typical force–extension graph for a strip of polythene, the polymer used in plastic carrier bags.

A polythene strip does not obey Hooke's law. Thin strips of polythene are very easy to stretch and they suffer plastic deformation under relatively little force. As you know, shopping bags made from polythene do not return to their original size after being stretched.

Plastic deformation is not necessarily a bad thing. For example, steel sheet is pressed into car body parts, which must retain their new shape after manufacture.

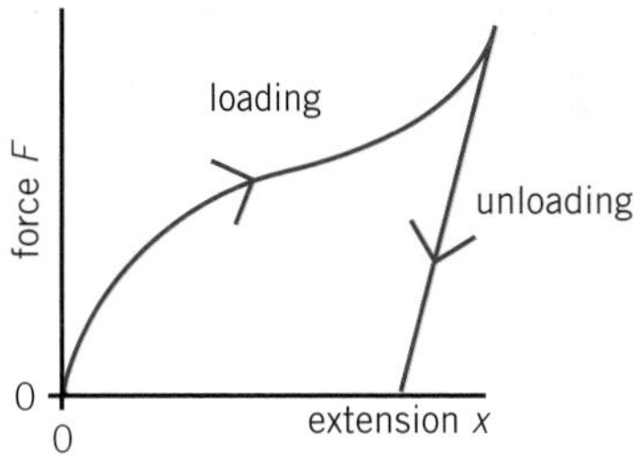

▲ **Figure 4** *Loading and unloading curve for a polythene strip*

Warming up

Rubber consists of squashed and tangled long-chain molecules. These can be untangled easily with small forces, but once straightened they require large forces to extend any further (Figure 5).

Rubber is an elastic material, but it is poor at storing energy. This makes it an ideal material for aeroplane tyres. Aeroplane tyres suffer sudden impact forces during landing. Their material makes landings smooth. The temperature of an aeroplane tyre can increase by as much as 100°C during landing.

1. Suggest how the shape of the loading curve for rubber shown in Figure 3 can be explained by the molecular structure of rubber.
2. Explain what is meant by the statement 'Rubber is an elastic material, but it is poor at storing energy'.
3. Use the force–extension graph in Figure 3 to explain why aeroplane tyres:
   a. reduce the bumpiness of landings;
   b. warm up during landing.

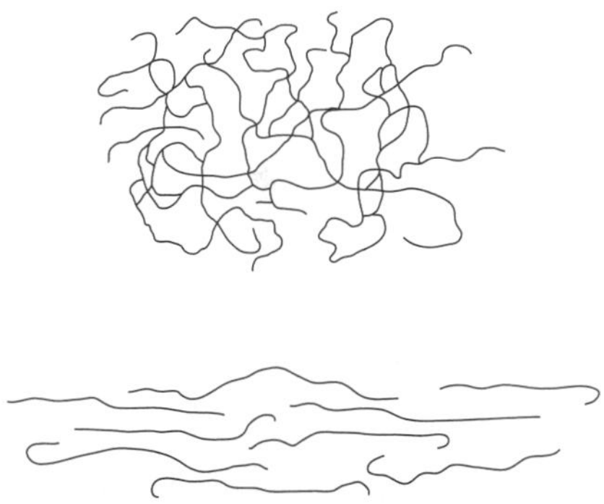

▲ **Figure 5** *The long chains of rubber molecules can easily be extended up to a certain point*

Summary questions

1. Rubber is an elastic material. Explain what this means. *(1 mark)*
2. Explain why a rubber band does not have a force constant. *(2 marks)*
3. Compare and contrast the behaviour of a metal wire and a strip of polythene. *(2 marks)*
4. Figure 6 shows the force–extension graph for a length of bungee cord. Use the graph to estimate the thermal energy released when the cord is stretched by 0.4 m and returns to its original length. *(3 marks)*

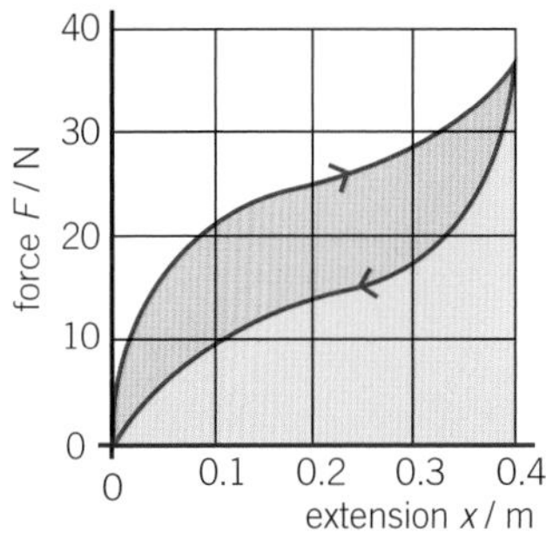

▲ **Figure 6**

6.4 Stress, strain, and the Young modulus

Specification reference: 3.4.2

Learning outcomes

Demonstrate knowledge, understanding, and application of:

→ stress, strain, and ultimate tensile strength

→ Young modulus

→ stress–strain graphs for ductile, brittle, and polymeric materials

→ elastic and plastic deformation.

▲ **Figure 1** *Spider web*

As strong as a cobweb

Spiders make their webs from a natural protein polymer called fibroin. A single strand of spider silk can be stretched up to 40% of its length before it breaks. As well as being very elastic, it is as strong as steel but has one-sixth the density.

Stretching materials

Imagine extending a wire by pulling at its ends. The extension will depend on the original length of the wire, its diameter, the tension in the wire, and of course the material of the wire. There are two helpful terms for the behaviour of materials under tensile forces: **tensile stress** and **tensile strain**.

Tensile stress

Tensile stress is defined as the force applied per unit cross-sectional area of the wire.

$$\text{tensile stress} = \frac{\text{force}}{\text{cross-sectional area}}$$

You can write this as

$$\sigma = \frac{F}{A}$$

where σ (Greek letter sigma) is the tensile stress (units pascals), F is the applied force, and A is the cross-sectional area.

Tensile strain

Tensile strain is defined as the fractional change in the original length of the wire.

$$\text{tensile strain} = \frac{\text{extension}}{\text{original length}}$$

You can write this as

$$\varepsilon = \frac{x}{L}$$

where ε (Greek letter epsilon) is the tensile strain, x is the extension, and L is the original length. Tensile strain is the ratio of two lengths, so has no units. Sometimes strain is written as a percentage, for example, 6.4% instead of 0.064.

Synoptic link

Tensile stress is similar to pressure, which you read about in Topic 4.8, Density and pressure, and therefore has the same units, $N\,m^{-2}$ or pascals (Pa).

Stress–strain graph for a metal

Figure 2 shows a typical stress against strain graph for mild (low-carbon) steel, a **ductile** material. A ductile material can easily be drawn into a wire or hammered into thin sheets.

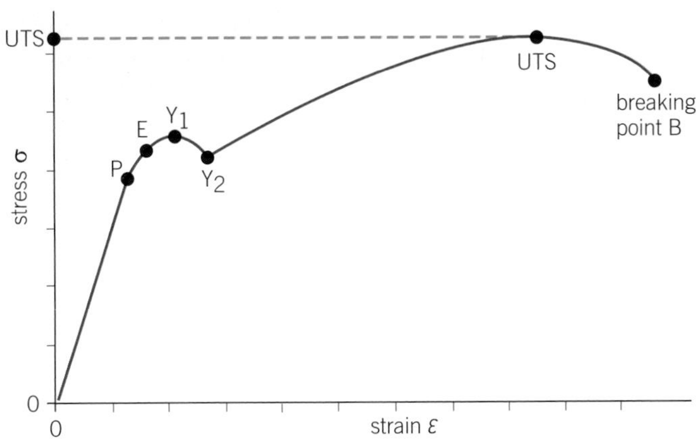

▲ **Figure 2** *A stress–strain graph for mild steel wire*

In this graph, the stress is directly proportional to the strain from the origin to P, the **limit of proportionality**. The material obeys Hooke's law in this linear region. E represents the elastic limit. Materials may obey Hooke's law up to this limit but not always. Elastic deformation occurs up to the elastic limit, and plastic deformation beyond it. Y1 and Y2 are upper and lower **yield points**, where the material extends rapidly. This part of the curve is typical of mild steel but may be absent from other ductile materials.

The stress at the point labelled UTS represents the material's **ultimate tensile strength**. This is the maximum stress that a material can withstand when being stretched before it breaks. Beyond this point, the material may become longer and thinner at its weakest point, a process called necking. The material eventually snaps at its breaking point, labelled B. The stress value at the point of fracture is known as the **breaking strength** of the material.

A **strong material** is one with a high ultimate tensile strength. For example, copper is stronger than lead, but mild steel is stronger than copper.

The Young modulus *E*

Within the limit of proportionality, stress is directly proportional to strain. The ratio of stress to strain for a particular material is a constant and is known as its **Young modulus**, *E*. That is

$$\text{Young modulus} = \frac{\text{tensile stress}}{\text{tensile strain}}$$

or

$$E = \frac{\sigma}{\varepsilon}$$

The unit of the Young modulus is the same as that for stress, N m^{-2} or Pa.

The Young modulus *E* of a material is the gradient of the linear region of the stress–strain graph ($y = mx + c$ is the same as $\sigma = E\varepsilon$). It depends only on the material, not its shape and size. An experiment carried out on a block of copper and a thin copper wire will give the same value for the Young modulus.

Synoptic link

For more information on gradients of straight-line graphs, see Appendix A2, Recording results.

You can compare the stiffness of materials by comparing their Young modulus values. A material with a large Young modulus is stiffer than one with a smaller Young modulus (Table 1).

▼ **Table 1** *Young modulus of some materials*

Material	E / Pa
polystyrene	$\sim 3 \times 10^9$
lead	1.8×10^{10}
aluminium	7.0×10^{10}
mild steel	2.1×10^{11}
graphene	1.1×10^{12}
diamond	1.2×10^{12}

Worked example: Crane cable

An object of weight 790 N is suspended vertically from a crane on a steel cable 5.0 m long and 6.0 mm in diameter. The Young modulus of the material of the cable is 2.0×10^{11} Pa. Calculate the extension of the cable.

Step 1: Select the equation for the stress in the cable and calculate it. Remember to convert the diameter into metres when calculating the cross-sectional area.

$$\sigma = \frac{F}{A} = \frac{F}{\pi r^2} = \frac{790}{\pi \times (3.0 \times 10^{-3})^2} = 2.794 \times 10^7\,\text{Pa}$$

Note: Although the data is given to two significant figures, the value for σ above is an intermediate value in the calculation, so it is best to retain as many significant figures as you can.

Step 2: Rearrange the equation for Young modulus to calculate the strain of the cable.

$$\varepsilon = \frac{\sigma}{E} = \frac{2.794 \times 10^7}{2.0 \times 10^{11}} = 1.397 \times 10^{-4}\,\text{Pa}$$

Step 3: Calculate the extension x. Rearrange the equation first to make x the subject.

$$\varepsilon = \frac{x}{L}$$

$$x = \varepsilon L = 1.397 \times 10^{-4} \times 5.0 = 7.0 \times 10^{-4}\,\text{m (2 s.f.)}$$

The extension of the cable is about 0.7 mm.

Determining the Young modulus of a wire

You can determine the Young modulus of a material in the form of wire by measuring its diameter, applying various loads to it, and measuring its length each time.

Figure 3 shows a simple method in which a wire of starting length greater than 1.00 m is clamped securely at one end, passed over a pulley, and loaded with slotted masses at the other end. You should wear eye protection in case the wire breaks.

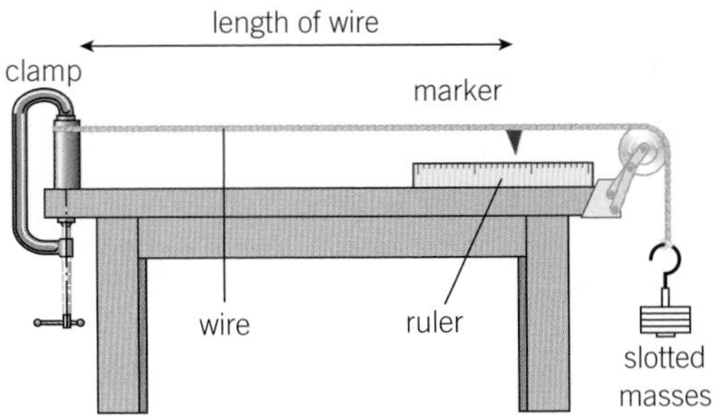

◀ **Figure 3** *Stretching a wire in the lab – the marker (tape) helps you to determine the length accurately*

The diameter d of the wire can be measured using a micrometer. The cross-sectional area A of the wire can be calculated from $A = \frac{\pi d^2}{4}$. You can obtain a more accurate diameter by averaging measurements from several places along the wire, and also by monitoring any changes during the experiment. The tensile force F acting on the wire can be calculated from the hanging mass m using $F = mg$, where g is the acceleration of free fall.

After applying each additional mass, the extension is calculated (x = extended length – original length L). You can improve accuracy by taking readings for at least six different masses, and repeating them.

The stress and strain values for each load are calculated and used to plot a stress–strain graph. The Young modulus of the material can be determined from the gradient of the linear section of the graph.

More stress–strain graphs

Not all materials behave in the same way. You can infer a great deal about the properties of a material from its stress–strain graph.

You have already seen the stress–strain graph for a ductile material (Figure 2). The stress–strain graphs for glass and cast iron, both **brittle** materials, are shown in Figure 4. A brittle material shows elastic behaviour up to its breaking point, without plastic deformation.

Polymeric materials are materials that consist of long molecular chains. (You have already met an example of a polymer, rubber, in Topic 6.3, Deforming materials). These behave differently depending on their molecular structure and temperature. Both rubber and polythene can stretch a great deal before breaking, but rubber shows elastic behaviour and polythene shows plastic behaviour (Figure 5).

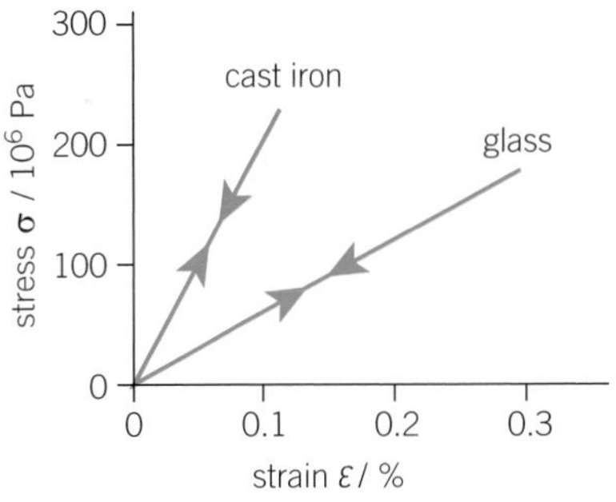

▲ **Figure 4** *Stress–strain graphs for two brittle materials – ultimate tensile strength is the same as breaking strength for a brittle material*

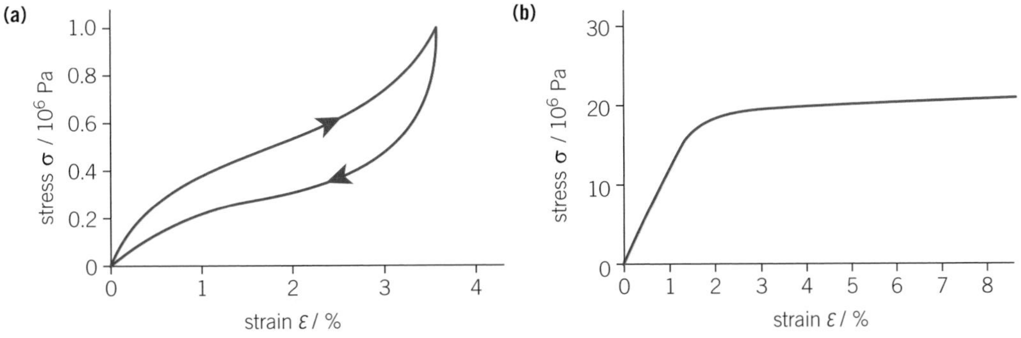

▲ **Figure 5** *Stress–strain graphs for polymers: (a) rubber and (b) polythene*

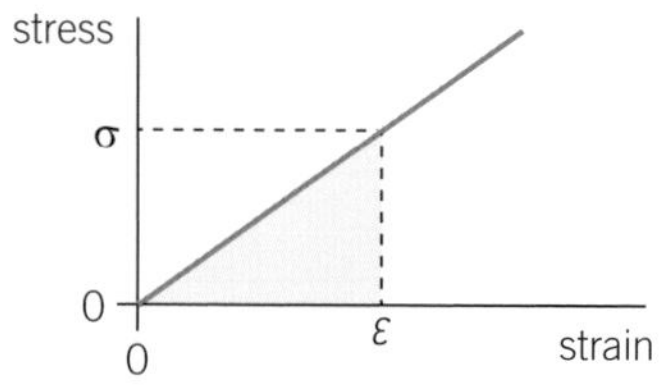

▲ **Figure 6** *Stress–strain graph for an elastic material*

Storing energy per unit volume

For a wire or a spring, you can use the area under a force–extension graph to determine work done and therefore, as you saw in Topic 6.2, Elastic potential energy, to determine the elastic potential energy. Figure 6 shows the stress–strain graph for an elastic material.

1. Show that the area under the graph is the energy stored per unit volume by the material.
2. Show that the base units for *energy per unit volume* and *tensile stress* are the same.
3. A single strand of spider silk can store 120 MJ of energy per unit volume. Use this and the information given in the introduction to estimate the typical stress for a spider's silk. State any assumptions made.

▲ **Figure 7** *A modern aeroplane*

Aeroplanes are engineered to exploit the useful properties of these materials. Their wings are made of an aluminium alloy that is both strong and stiff (high Young modulus). The rotor blades in its jet engines are made from ceramics that can withstand high temperatures and are very strong. Ceramics are brittle and show no plastic deformation. The tyres of planes are made from rubber, which is a polymeric material. Rubber has elastic properties and is an excellent shock absorber (Figure 7).

Summary questions

1. Explain what is meant by ultimate tensile strength of a material. *(1 mark)*
2. Use Figure 4 to describe the properties of cast iron. *(1 mark)*
3. A metal wire has diameter 0.20 mm. A force of 6.3 N changes its length from 1.035 m to 1.048 m. Calculate the Young modulus of the metal. *(3 marks)*
4. The ultimate tensile strength of a metal is 220 MPa. Calculate the maximum force that can be applied on a wire of diameter 1.2 mm made from this metal. *(3 marks)*
5. **a** Show that the Young modulus E of a material can be calculated using the equation

$$E = \frac{FL}{Ax}$$

where F is the force applied, L is the original length of the wire, x is the extension of the wire, and A is the cross-sectional area of the wire. *(2 marks)*

b Use the equation in (a) to determine E given the data: diameter of wire = 0.84 mm, original length of wire = 2.500 m, force applied = 300 N, extension = 1.4 cm. *(2 marks)*

c State one significant assumption made in your calculation in (b). *(1 mark)*

Practice questions

1 a Describe how you can determine the force constant of an extendable helical spring in the laboratory. *(4 marks)*

b (i) Sketch a graph to show the variation of elastic potential energy E for a spring with its extension x. *(2 marks)*

(ii) The energy stored in a spring is 0.10 J when it has an extension of 6.0 cm. Calculate the energy stored when the extension is 9.0 cm. Explain your answer. *(3 marks)*

2 A glider of mass 0.180 kg is placed on a horizontal frictionless air track. One end of the glider is attached to a compressible spring of force constant 50 N m^{-1}. The glider is pushed against a fixed support so that the spring compresses by 0.070 m, see Figure 1. The glider is then released.

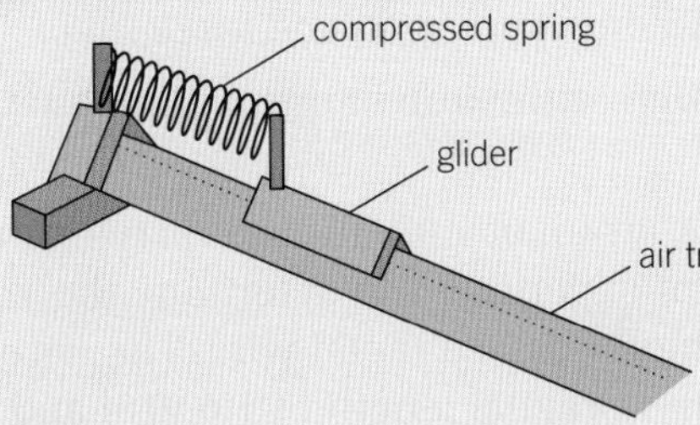

▲ **Figure 1**

(i) Calculate the horizontal acceleration of the glider **immediately** after release. *(3 marks)*

(ii) After release, the spring exerts a force on the glider for a time of 0.094 s. Calculate the average rate of work done by the spring on the glider. *(2 marks)*

May 2012 G481

3 A metal wire is suspended from a tall ceiling. The wire has diameter of (0.90 ± 0.01) mm and length 2.500 m. A mass of 8.00 kg hung from its lower end produces an extension of 4.0 mm.

a Calculate the absolute uncertainty in the value of the cross-sectional area. *(3 marks)*

b Calculate the Young modulus of the metal. State any assumptions made. *(4 marks)*

c (i) On a copy of the axes below, sketch the stress against strain graphs for a ductile metal and a brittle metal. *(2 marks)*

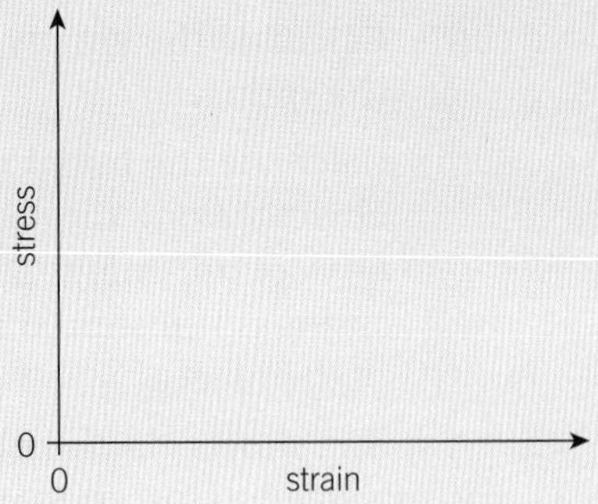

▲ **Figure 2**

(ii) Describe the behaviour of the ductile material as it is stretched. *(2 marks)*

4 a Define tensile stress and tensile strain. *(2 marks)*

b A metal wire of length 1.80 m and cross-sectional area 1.92×10^{-7} m^2 is extended by a force of 12.0 N. The metal has a Young modulus of 2.00 GPa.

(i) Calculate the extension of the wire. *(3 marks)*

(ii) A second wire made from the same metal has the same length but twice the diameter. State and explain whether the extension of this second wire under the same force is greater, the same or less than the first wire. *(3 marks)*

5 a Figure 3 shows the force F against extension x graph for a spring.

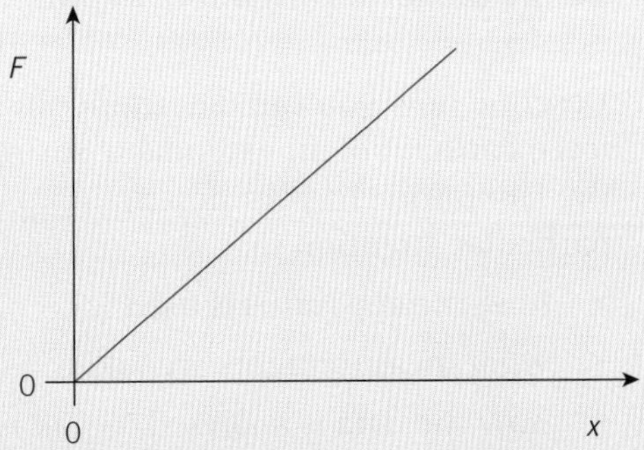

▲ **Figure 3**

State what is represented by

(i) the gradient of the graph *(1 mark)*

(ii) the area under the graph. *(1 mark)*

b A spring has a force constant of $160\,\text{N}\,\text{m}^{-1}$. The energy stored in the spring is used to propel an object of mass 80 g. The spring is compressed by 7.2 cm and then released.

(i) Calculate the energy stored in the spring. *(2 marks)*

(ii) Calculate the initial speed of the object leaving the spring, assuming 60% of the energy stored in the spring is transferred as kinetic energy of the object. *(3 marks)*

6 a A spring shows elastic behaviour when it is subjected to forces. State what is meant by the term *elastic*. *(1 mark)*

b Figure 4 shows two springs A and B, connected in series and supporting a weight of 16 N. The force constant of each spring is shown on the figure.

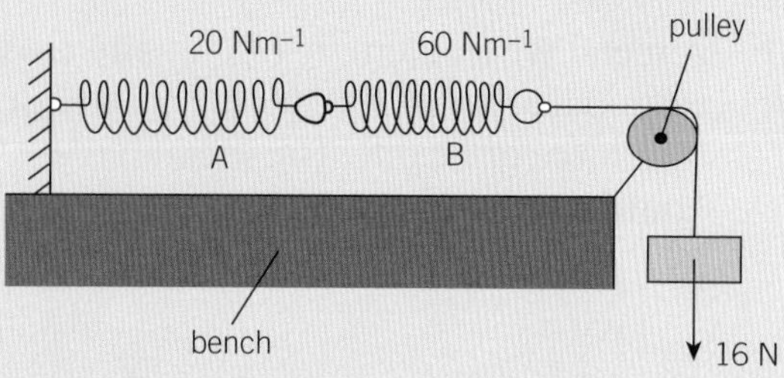

▲ **Figure 4**

(i) Determine the total extension of the two springs. State any assumptions made. *(3 marks)*

(ii) Determine the force constant of the combination of these springs. *(2 marks)*

7 a State *Hooke's law*. *(1 mark)*

b Define the force constant of a spring. *(1 mark)*

c Describe how you can determine the force constant of an extendable spring in the laboratory. In your description pay particular attention to

- how the apparatus is used
- what measurements are taken
- how the data is analysed. *(4 marks)*

d Figure 5 shows the variation of force *F* with extension *x* for a spring.

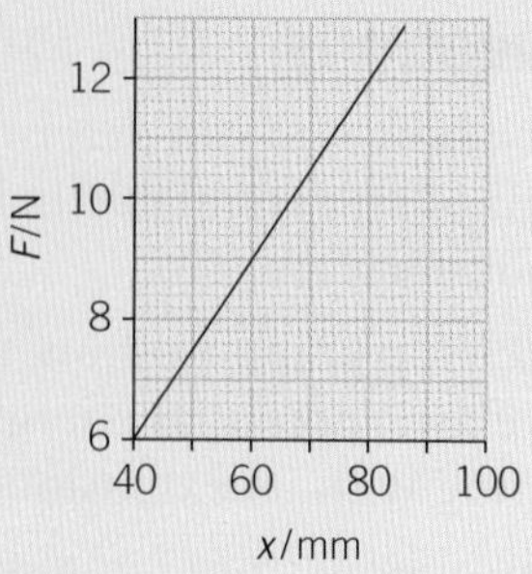

▲ **Figure 5**

Use Figure 5 to answer to following questions.

(i) Explain how the graph shows that Hooke's law is obeyed. *(2 marks)*

(ii) Determine the force constant of the spring. *(2 marks)*

(iii) Determine the energy stored in the spring when its extension is 80 mm. *(3 marks)*

8 a Define *tensile stress* and *tensile strain*. *(1 mark)*

b Derive the units for stress in base units. *(3 marks)*

c A group of students are carrying out an experiment to determine the Young modulus of a metal wire. The wire has original length of 1.640 m and it is suspended vertically from a support. The wire is loaded in steps of 10.0 N up to 50.0 N and then unloaded. Table 1 shows the experimental results from the group.

▼ **Table 1**

	loading the wire	unloading the wire
force *F* / N	extension *x* / mm	extension *x* / mm
0.0	0.00	0.00
10.0	0.50	0.49
20.0	1.01	1.00
30.0	1.49	1.50
40.0	2.00	1.99
50.0	2.52	2.52

(i) Use Table 1 to describe the behaviour of the wire when forces up to 50.0 N are applied to it. *(2 marks)*

(ii) Name the most likely instrument used to determine the extension of the wire. *(1 mark)*

(iii) The cross-sectional area of the wire is $3.2 \times 10^{-7}\,m^2$. Use the value of the extension for the force of 50.0 N to calculate a value for the Young modulus of the metal. *(3 marks)*

(iv) Describe how the table of results can be used to plot a graph and hence determine a precise value for the Young modulus of the metal. *(3 marks)*

9 **a** Define the *Young modulus*. *(1 mark)*

b Figure 6 shows a violin.

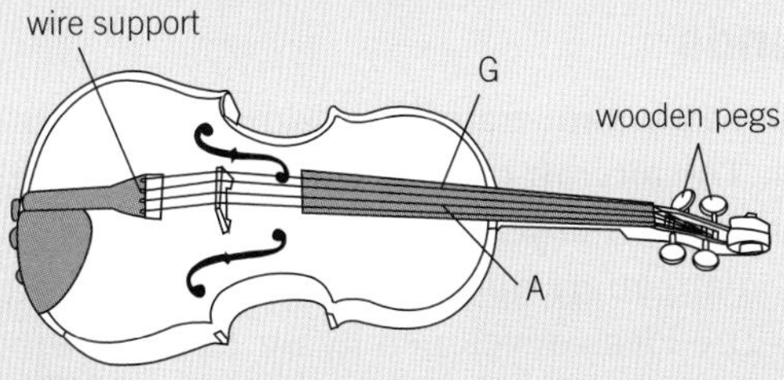

▲ Figure 6

Two of the wires used on the violin, labelled A and G, are made of steel. The two wires are both 500 mm long between the pegs and support. The 500 mm length of wire labelled G has a mass of 2.0×10^{-3} kg. The density of steel is $7.8 \times 10^{3}\,kg\,m^{-3}$.

(i) Show that the cross-sectional area of wire G is $5.1 \times 10^{-7}\,m^2$. *(2 marks)*

(ii) The wires are put under tension by turning the wooden pegs shown in Figure 6. Calculate the force on the wire when the extension is 0.4 mm. The Young modulus of steel is 2.0×10^{11} Pa. *(3 marks)*

(iii) Wire A has a diameter that is half that of wire G. Determine the tension required for wire A to produce an extension of 16×10^{-4} m. *(1 mark)*

(iv) State the law that has been assumed in the calculations in (ii) and (iii). *(1 mark)*

June 2007 2821

10 Figure 7 shows the force F against extension x for a metal wire.

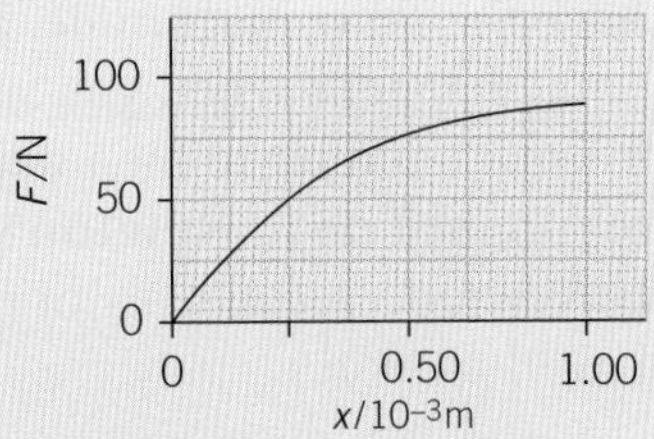

▲ Figure 7

a State the value of the force and extension at the elastic limit of the wire. *(1 mark)*

b Calculate the elastic potential energy for the wire when its extension is 0.25 mm. *(3 marks)*

c The Young modulus of the metal is 1.2×10^{11} Pa and the length of the wire is 1.82 m.

Use Figure 7 to determine the cross-sectional area of the wire. *(3 marks)*

11 Figure 8 shows an arrangement used by a student to investigate the energy stored in a compressible spring.

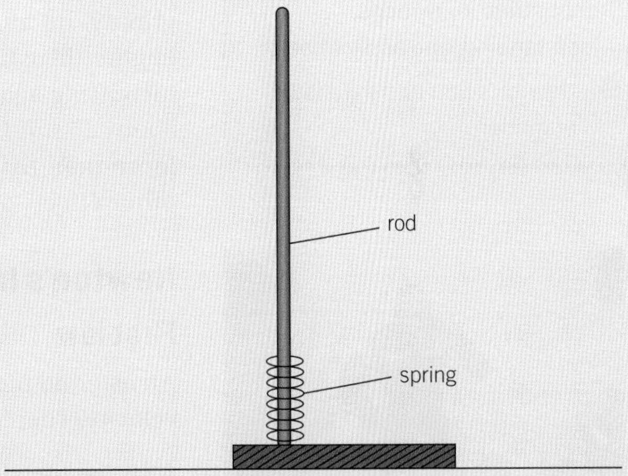

▲ Figure 8

a The spring is compressed by a distance x and then released. It climbs a vertical height h along the length of the rod. Show that h is directly proportional to x^2. *(3 marks)*

b Figure 9 shows a graph of h against x^2 for the spring.

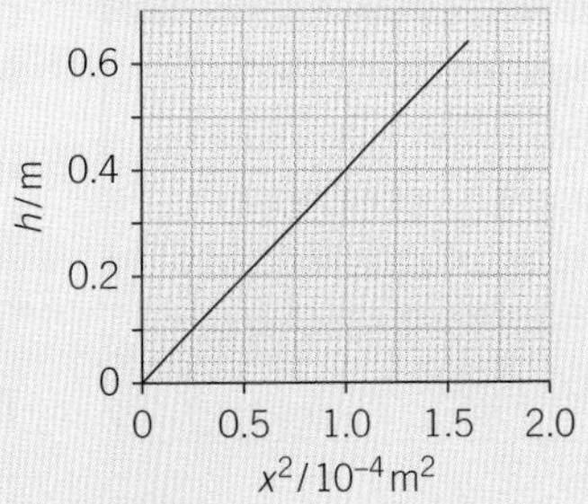

▲ Figure 9

(i) Use Figure 9 to predict the height h when $x = 2.0$ cm. State any assumptions made. *(4 marks)*

(ii) The mass of the spring is 8.0 g.

Use Figure 9 to determine the force constant of the spring. *(3 marks)*

7 LAWS OF MOTION AND MOMENTUM

7.1 Newton's first and third laws of motion

Specification reference: 3.5.1

Learning outcomes

Demonstrate knowledge, understanding, and application of:

→ Newton's first and third laws of motion.

▲ **Figure 1** *A trainee astronaut in 'zero gravity' – the padded interior reduces the risk of injury*

Synoptic link

You already know from Topic 4.1, Force, mass, and weight, that a resultant force is necessary for an object to accelerate. Without a resultant force it will not change direction or speed.

Study tip

When defining Newton's first law of motion, take care to refer to constant velocity rather than to constant speed.

Weightless seconds

Figure 1 shows a NASA astronaut training in simulated 'zero gravity' on an aircraft. The aircraft climbs, gradually reducing engine thrust, and then points downwards with the thrust increasing again. This manoeuvre allows trainees to experience about 25 s of free fall. The ideas developed by Sir Isaac Newton more than three centuries ago can predict and explain this and other effects of motion and forces.

Newton's laws of motion

First law

An asteroid moving in deep space will keep moving at constant velocity. There is no force acting on it to alter its motion.

A cyclist travelling on a straight road at constant velocity experiences several forces, including contact force, weight, and air resistance. Again, no resultant force acts on the cyclist and the acceleration is zero.

You have already met **Newton's first law of motion**. A formal statement of this law is given below.

Newton's first law of motion: An object will remain at rest or continue to move with constant velocity unless acted upon by a resultant force.

If an object's velocity changes, then you know a resultant force must be acting on the object. Remember that velocity is a vector quantity, so an object's velocity changes if its speed and/or direction changes.

Third law

Tap the screen of your mobile phone. Your finger will feel a force. You also exert a force on the screen. In fact, the forces acting on your finger and on the screen have the same magnitude, but they are in opposite directions. The same happens when you stand on the ground, clap your hands, or use a hammer to hit a nail. These interactions between objects are summed up in **Newton's third law of motion**. (You will study Newton's second law of motion in Topic 7.3, Newton's second law of motion.)

Newton's third law of motion: When two objects interact, they exert equal and opposite forces on each other.

When two objects interact, the pair of forces produced will always be equal and opposite. The forces acting on the interacting objects are always of the same type.

Electrons have a negative charge; they exert an *electrostatic* repulsive force on each other (Figure 2). The two electrons experience the same magnitude of force but in opposite directions. Similarly, the unlike poles of the two magnets exert a pair of *magnetic* forces. Once again the forces are equal and opposite. When you jump off a wall and fall towards the Earth, you and the Earth are interacting. The *gravitational* force exerted by the Earth, your weight, is equal and opposite to the gravitational force that you exert on the Earth. Notice how each interaction in Figure 2 involves a pair of forces of the same type each acting on a different object. In Newton's Third Law you never see both forces acting on the same object.

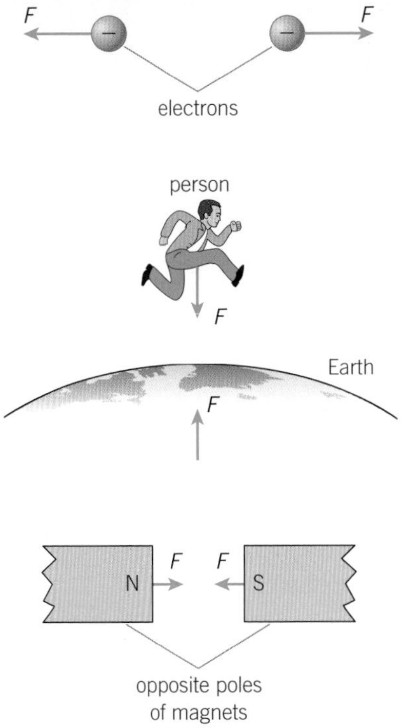

▲ **Figure 2** *Interacting objects exert forces of the same type on each other*

Types of interaction

All interactions can be explained in terms of four fundamental forces – gravitational, electromagnetic, strong nuclear, and weak nuclear. The two nuclear forces have an extremely short range and so very little impact on the things we observe in daily life. You are familiar with gravitational force. When you push your hands together, the contact force you feel is due to the electrostatic repulsive forces between the electron clouds around the atomic nuclei in your hands. Next time you walk, remember that you are exerting a backward force on the ground and the ground is exerting a forward force on you, both forces of electrical origin.

Summary questions

1 A car is travelling at constant velocity. State the resultant force on the car. *(1 mark)*

2 A person of weight 600 N is standing still on the ground. State and explain the force exerted by the person on the Earth. *(2 marks)*

3 Two magnets repel each other. State and explain the resultant force acting on the magnets. *(2 marks)*

4 Explain how Newton's laws can be used to explain how a person runs on a road. *(2 marks)*

5 Dalmatian pelicans are massive birds (Figure 3). One such bird weighs 150 N.

a State the acceleration of free fall of this bird. *(1 mark)*

b Calculate the acceleration of the Earth towards this bird. (The mass of the Earth is 5.97×10^{24} kg.) Explain your answer. *(3 marks)*

▲ **Figure 3** *This flying Dalmatian pelican is interacting with the Earth*

7.2 Linear momentum

Specification reference: 3.1.2, 3.5.1, 3.5.2

Learning objectives

Demonstrate knowledge, understanding, and application of:

- → linear momentum; $p = mv$; vector nature of momentum
- → principle of conservation of momentum
- → collisions and interactions of bodies in one dimension
- → perfectly elastic and inelastic collisions.

▲ **Figure 1** *An Atlas V rocket burning the fuel it carries*

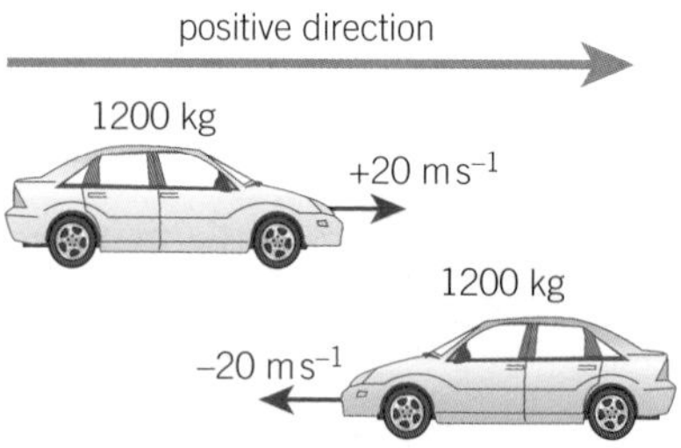

▲ **Figure 2** *Both cars have the same magnitude of momentum, 2.4×10^4 kg m s^{-1}, but the green car has a negative momentum*

Mission to Mars

The Mars Science Laboratory spacecraft was launched on top of an Atlas V rocket in November 2011. It landed on Mars on August 2012 on a mission to determine if Mars is, or was ever, able to support microbial life.

Because the mass of the rocket carrying it decreased as its fuel was consumed and expelled, the equation $F = ma$ cannot be used to predict the motion of the rocket. We need a new quantity – **linear momentum** – to analyse motion of objects such as rockets.

Momentum

The linear momentum (or simply momentum) p of an object depends on its mass m and its velocity v.

$$\text{momentum} = \text{mass} \times \text{velocity}$$

or

$$p = mv$$

The SI unit of momentum is kg m s^{-1}. Momentum is a vector quantity because it is a product of a scalar (mass) and a vector (velocity).

Figure 2 shows two identical cars, each of mass 1200 kg, travelling at the same speed of 20 m s^{-1} in opposite directions. The car moving to the right has a velocity of +20 m s^{-1} and the car moving left has a velocity of −20 m s^{-1}. Therefore

- momentum of car moving to right
  $= mv = 1200 \times (+20) = +2.4 \times 10^4\,\text{kg m s}^{-1}$
- momentum of car moving to left
  $= mv = 1200 \times (-20) = -2.4 \times 10^4\,\text{kg m s}^{-1}$

It does not matter which direction is taken as positive. The important idea is that one of the values for momentum is negative.

Conservation of momentum

What happens when two or more objects collide or interact? The objects transfer both momentum and kinetic energy between themselves, but the total momentum does not change, provided that no external forces act on the interacting objects. The group of interacting objects is referred to as a **closed system**. The **principle of conservation of momentum** is expressed as follows:

For a system of interacting objects, the total momentum in a specified direction remains constant, as long as no external forces act on the system.

This means that when objects collide, the total momentum before and after the collision is the same. You can use this principle to predict the motion of interacting objects, which could be atoms bouncing off each other, cars crashing, and even colliding distant galaxies.

Worked example: Air track collision

Two gliders are on a linear air track. Glider **A** is travelling at $0.20\,\mathrm{m\,s^{-1}}$ and has mass 0.10 kg. It is hit by glider **B** of mass 0.15 kg travelling at $0.40\,\mathrm{m\,s^{-1}}$ in the opposite direction. They stick together. Calculate their new velocity v.

Step 1: Write down the information given for each glider before and after the collision. Alternatively, you can do a quick sketch to help you to visualise the problem, as shown in Figure 3.

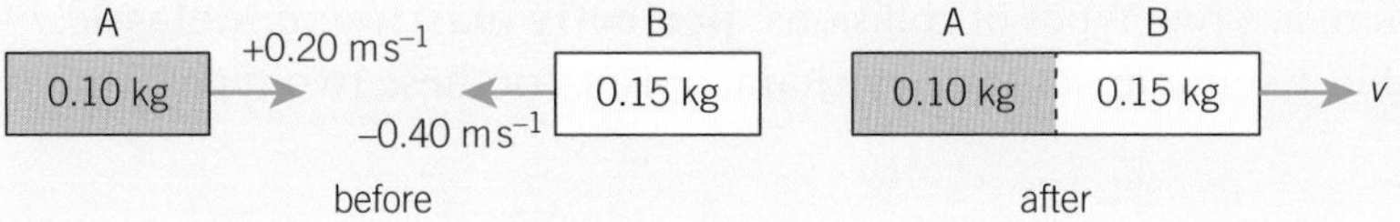

▲ **Figure 3** *Before and after sketches*

The velocity, and therefore momentum, of one glider must be negative because of its direction of travel.

Step 2: Write an equation for this collision using the principle of conservation of momentum.

total momentum before = total momentum after

$$(0.10 \times 0.20) + (0.15 \times -0.40) = (0.10 + 0.15)v$$

Step 3: Solve this equation to calculate v.

$$0.020 - 0.060 = 0.25v$$

$$-0.040 = 0.25v$$

$$v = \frac{-0.040}{0.25} = -0.16\,\mathrm{m\,s^{-1}}$$

The velocity of the joined gliders is $-0.16\,\mathrm{m\,s^{-1}}$.

Note: The negative sign shows that the gliders move in the direction in which glider **B** was originally travelling.

Zero momentum?

A gun recoils when a bullet is fired. The total momentum of this system remains the same and is equal to zero. The momentum of the gun and the momentum of the bullet have the same magnitude but act in opposite directions. The same physics can be used to explain a recoiling radioactive nucleus when it emits an alpha-particle, and an exploding firework (Figure 4).

▲ **Figure 4** *The total momentum of this exploding firework is zero*

Investigating momentum

There are several ways to investigate momentum in the laboratory. A linear air track is ideal because a cushion of air minimises the friction between the gliders and track, but trolleys and a horizontal runway also work (Figure 5). The velocity of each object is determined with a motion sensor and a laptop, light gates and a digital timer, ticker timers, or simply a stopwatch to measure the time taken to cover a known distance.

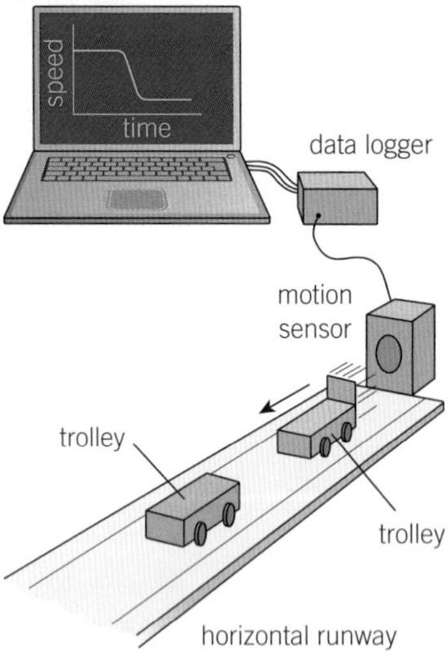

▲ **Figure 5** *Investigating linear momentum using trolleys and a motion sensor*

Elastic and inelastic collisions

Momentum and total energy are both conserved when two objects collide. All the kinetic energy could be retained by the objects, or the kinetic energy could be transformed into other forms such as heat and sound. In a collision of two cars, the kinetic energy could be completely transformed as the cars crumple and deform (Figure 6).

There are two types of collisions: **perfectly elastic** and **inelastic**. Table 1 summarises the key characteristics of these two types.

▲ **Figure 6** *An example of an inelastic collision*

▼ **Table 1** *Summary of the two types of collisions*

Type of collision	Momentum	Total energy	Total kinetic energy
perfectly elastic	conserved	conserved	conserved
inelastic	conserved	conserved	not conserved

Summary questions

1 State two quantities conserved in all collisions. *(2 marks)*

2 Two identical objects, each of mass 2.0 kg, collide. After the collision, the loss of momentum for one of the objects is 120 kg m s^{-1}. Calculate the change in momentum of the other object and the change in its velocity. *(3 marks)*

3 A stationary cannon of mass 1200 kg fires a 20 kg shell at a velocity of 300 m s^{-1}. Calculate the recoil speed of the cannon. Explain your answer. *(3 marks)*

4 A bumper car of mass 300 kg moving at 2.5 m s^{-1} collides head-on with another bumper car of mass 400 kg moving at 4.0 m s^{-1} in the opposite direction. The 400 kg bumper car stops after the collision.

- **a** Calculate the final velocity of the 300 kg bumper car. *(3 marks)*
- **b** Calculate the change in the kinetic energy of the bumper cars. *(3 marks)*
- **c** State and explain the type of collision between these bumper cars. *(2 marks)*

7.3 Newton's second law of motion

Specification reference: 3.5.1

Air bags

Air bags in modern cars are designed to reduce the chance of injury to occupants in an accident. Sensors detect the rapid changes in acceleration associated with a crash and trigger an explosive chemical reaction, which fills the air bag with nitrogen gas in just 30 ms to provide a cushion for the occupant's head and upper body. This reduces the momentum of the occupant to zero more gradually than sudden deceleration by contact with the dashboard or windscreen, and therefore reduces the force and minimises injuries.

Learning outcomes

Demonstrate knowledge, understanding, and application of:

- → Newton's second law of motion
- → net force = rate of change of momentum; $F = \frac{\Delta p}{\Delta T}$.

Newton's second law

Newton's second law of motion links the idea of net force acting on an object and rate of change of its momentum. A formal statement of the law is given below.

Newton's second law: The net (resultant) force acting on an object is directly proportional to the rate of change of its momentum, and is in the same direction.

Therefore

$$\text{net force} \propto \text{rate of change of momentum}$$

or

$$F \propto \frac{\Delta p}{\Delta t}$$

where F is the net force acting on the object and Δp is the change in momentum over a time interval Δt. The expression above can be rewritten as an equation with a constant of proportionality k.

$$F = \frac{k\Delta p}{\Delta t}$$

The value of k is made equal to 1 by defining the unit of force, the newton, as the force required to give a 1 kg mass an acceleration of $1\,\text{m}\,\text{s}^{-2}$.

Newton's second law of motion can therefore be written mathematically as

$$F = \frac{\Delta p}{\Delta t}$$

This astonishingly useful equation can predict the motion of any object subjected to forces, even when its mass changes with time, like a rocket.

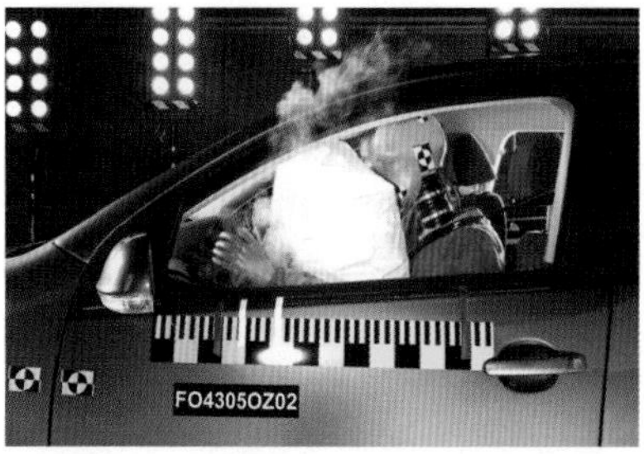

▲ **Figure 1** *Crash-test dummies are used to test air bags*

Study tip

Remember that the letter delta Δ is a shorthand for *change in*.

Worked example: Crash-test dummy

In a crash test at $14\,\text{m}\,\text{s}^{-1}$, a 4.5 kg dummy head hits the steering wheel and comes to rest in 9.1 ms. Calculate the net force acting on the head in the impact.

Synoptic link

The equation $F = ma$ was introduced in Topic 4.1, Force, mass, and weight.

Summary questions

1 In a period of 5.0 s the change in momentum of a car is $1.2 \times 10^4\,kg\,m\,s^{-1}$. Calculate the net force acting on the car. *(2 marks)*

2 A force of 150 N acts on a ball for a time of 0.025 s. Calculate the change in momentum of the ball. *(2 marks)*

3 The mass of a rocket is 2.3×10^6 kg and its velocity is $1.2 \times 10^3\,m\,s^{-1}$. After burning fuel for 200 s, its mass has decreased to 1.0×10^6 kg and its velocity has increased to $3.4 \times 10^3\,m\,s^{-1}$. Calculate the force acting on the rocket. *(2 marks)*

4 A 150 g ball travelling at $15\,m\,s^{-1}$ hits a wall at right angles and rebounds at the same speed. The ball is in contact with the wall for 0.025 s. Calculate the force exerted by the wall on the ball. *(3 marks)*

5 A hosepipe squirts water at a rate of $2.5\,kg\,s^{-1}$. The speed of the water in the hosepipe is $4.0\,m\,s^{-1}$. Calculate the force needed to push the water out of the hosepipe. *(2 marks)*

Step 1: Write down all the quantities given in SI units, and select the equation you need.

initial velocity $u = 14\,m\,s^{-1}$, final velocity $v = 0\,m\,s^{-1}$, $\Delta t = 9.1 \times 10^{-3}$ s, $m = 4.5$ kg

$$F = \frac{\Delta p}{\Delta t}$$

Step 2: Determine the change in momentum Δp.

$$\Delta p = \text{final momentum} - \text{initial momentum}$$
$$\Delta p = (4.5 \times 0) - (4.5 \times 14) = -(4.5 \times 14)$$

Step 3: Substitute the values above into the equation to calculate F.

$$F = \frac{\Delta p}{\Delta t} = \frac{-(4.5 \times 14)}{9.1 \times 10^{-3}} = -6.9 \times 10^3\,\text{N}$$

Note: The magnitude of the force is 6.9 kN, roughly 10 times your weight. The minus means that the net force is opposite to the initial velocity, so the net force causes a deceleration.

$F = ma$ – a special case

Figure 2 shows a constant force F acting on an object of constant mass m. The initial velocity of the object is u and after a time t it has a final velocity v. According to Newton's second law

$$F = \frac{\Delta p}{\Delta t} = \frac{mv - mu}{t} = m\left(\frac{v - u}{t}\right)$$

The term in brackets is the acceleration a of the object. Therefore

$$F = ma$$

This equation is just a special case of Newton's second law when the mass m of the object remains constant during the period of acceleration. In the worked example above, you could have used $F = ma$ to determine the force on the dummy's head.

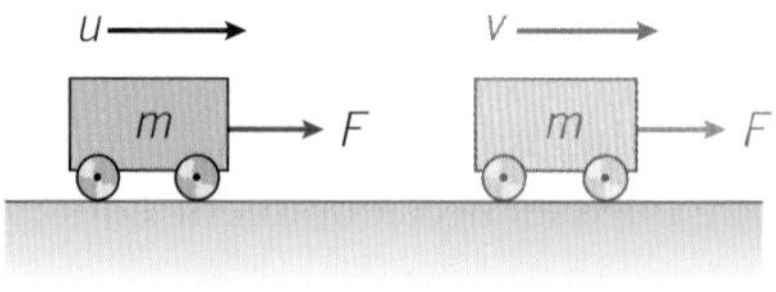

▲ **Figure 2** *The constant force F applied for a time t increases the momentum of the object*

▲ **Figure 3** *Two interacting objects*

Why is momentum conserved in collisions?

The principle of conservation of momentum is a natural consequence of Newton's laws. Figure 3 shows two interacting objects. According to Newton's third law, each experiences an equal but opposite force F.

The net force acting on the objects in this closed system is zero. According to Newton's second law $\frac{\Delta p}{\Delta t} = 0$. The change in momentum Δp of both objects must be zero; therefore, the total momentum of the objects does not change. Momentum is always conserved.

7.4 Impulse

Specification reference: 3.5.1

Squash

Squash is a fast-paced racket game played in a walled court. The 40 mm diameter rubber balls are hit at speeds of up to $76\,\mathrm{m\,s^{-1}}$. A cold squash ball does not bounce well, so it must be warmed up before play by hitting it around the court. The forces exerted on a squash ball change during impact with a racket (Figure 1), and can be analysed using **force–time graphs**.

Impulse of a force

Forces accelerating or decelerating an object usually change over time, for example, kicking a ball or crashing a car into a barrier. This type of motion can be analysed using the idea of **impulse**.

According to Newton's second law of motion

net force = rate of change of momentum

$$F = \frac{\Delta p}{\Delta t}$$

Rearranging this equation gives

$$F \times \Delta t = \Delta p$$

The product of force and time is equal to the change in momentum Δp.

Impulse of a force is defined as the product of force and the time for which this force acts on an object.

Therefore

impulse of a force = change in momentum

The unit of impulse is N s or $\mathrm{kg\,m\,s^{-1}}$.

Force–time graphs

Figure 2(a) is the force–time graph for an object experiencing a *constant* force F for a time t.

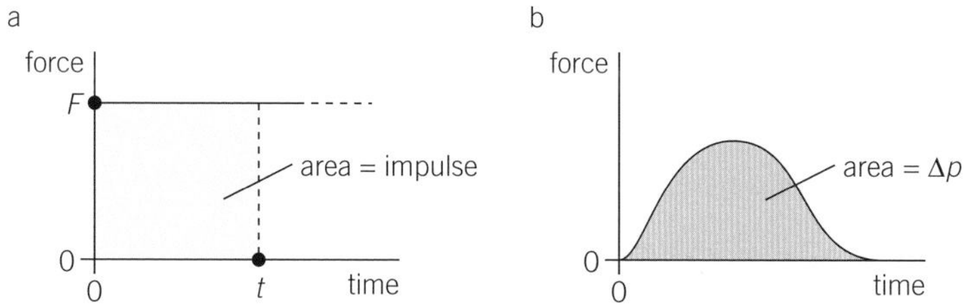

▲ **Figure 2** *Force–time graphs for (a) constant force; (b) changing force*

As you can see from Figure 2(a), the area under the graph is equal to Ft, which is the impulse of the force or the change in momentum of the object. In fact, the area under a force–time graph is always equal to the change in momentum, even when the force is changing (Figure 2(b)).

Learning outcomes

Demonstrate knowledge, understanding, and application of:

→ impulse of a force; impulse = $F\Delta t$

→ impulse as the area under a force–time graph.

▲ **Figure 1** *A ball stretches the racket's strings and changes shape itself when it is hit*

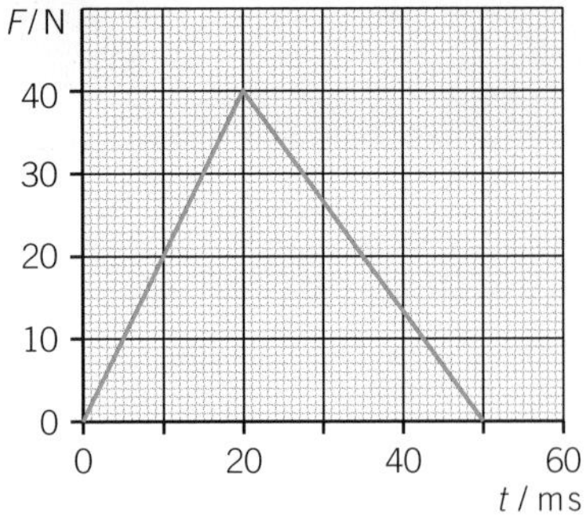

▲ Figure 3

Worked example: Hitting a squash ball

A stationary squash ball of mass 0.025 kg is hit with a racket. Use the force–time graph for the ball in Figure 3 to determine the final velocity v of the ball.

Step 1: The area under the graph is equal to the impulse of the force. Calculate the area of the triangle.

impulse = area = $\frac{1}{2} \times 40 \times 50 \times 10^{-3}$ (remember: $1\,\text{ms} = 10^{-3}\,\text{s}$)

$\Delta p = 1.0\,\text{N s}$ or kg m s^{-1}

Step 2: The impulse of the force is equal to the change in momentum. Use the mass of the ball to calculate the final velocity.

$\Delta p = mv - mu = 0.025v - 0$ (because $u = 0$)

$0.025v = 1.0$

$v = \frac{1.0}{0.025} = 40\,\text{m s}^{-1}$

The final velocity of the ball is $40\,\text{m s}^{-1}$.

Photons have momentum

You may have heard of photons, subatomic particles that travel at the speed of light and have no mass. According to quantum physics, as well as having energy, photons have momentum. The momentum p of a photon is given by the equation $p = \frac{h}{\lambda}$, where h is the Planck constant and λ is the wavelength of the photon.

Photons exert a tiny force when they collide with objects. Our Sun is a source of photons. These photons exert a radiation pressure of about 9.1 μPa at the Earth.

1 State and explain how the momentum of a photon depends on its wavelength.
2 Calculate the momentum of a visible light photon of wavelength 500 nm. The Planck constant $h = 6.63 \times 10^{-34}$ J s.
3 Estimate the area required to produce a force of 1 N at the Earth from the radiation pressure from the Sun.

Summary questions

1 State the two SI units for impulse and momentum. *(2 marks)*

2 A constant resultant force of 200 N acts on a car for 5.0 s. Calculate the impulse of the force. *(2 marks)*

3 A stationary 0.050 kg ball experiences an impulse of 1.1 N s. Calculate its final velocity. *(2 marks)*

4 The force F against time t graph for a proton is shown in Figure 4.
  a Calculate the change in momentum of the proton. *(2 marks)*
  b The initial velocity of the proton is $5.0 \times 10^{4}\,\text{m s}^{-1}$. Calculate its final velocity. The mass of a proton is 1.7×10^{-27} kg. *(3 marks)*

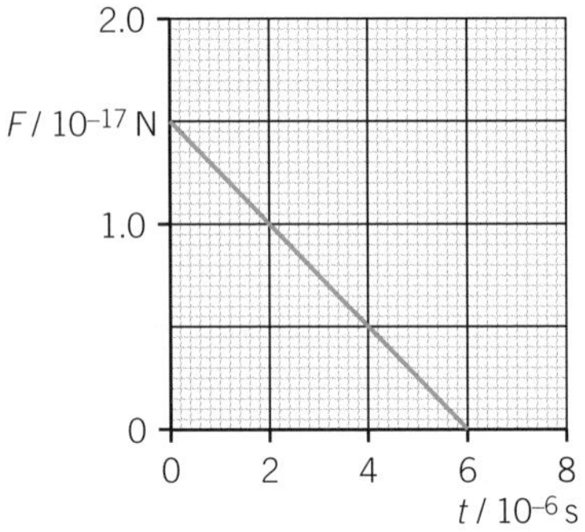

▲ Figure 4

7.5 Collisions in two dimensions

Specification reference: 3.5.2

Collisions big and small

When a snooker ball collides obliquely with an identical ball, they move off on paths at an angle of 90° to each other. This is an example of a collision in two dimensions. As with all collisions, both total energy and linear momentum are conserved.

Similar events also occur at microscopic levels. Figure 1 shows a collision between a helium nucleus and a proton. The angle after the collision is not 90° because the particles have different masses. Physicists can use particle tracks to determine the momentum of particles.

Conservation of momentum

In collisions and interactions, linear momentum is conserved in all directions. You can use your knowledge of vector triangles and of resolving vectors to analyse a variety of problems.

Adding momentum

Figure 2 shows an object **A** moving to the right with momentum p. It collides with a stationary object **B**. After the collision, **A** and **B** move off in different directions with momenta p_1 and p_2, respectively, as shown. Since linear momentum must be conserved, the vector sum p_1 and p_2 (total final momentum) must be equal to p (initial momentum). You can draw a vector triangle to add the vectors p_1 and p_2 together (Figure 2).

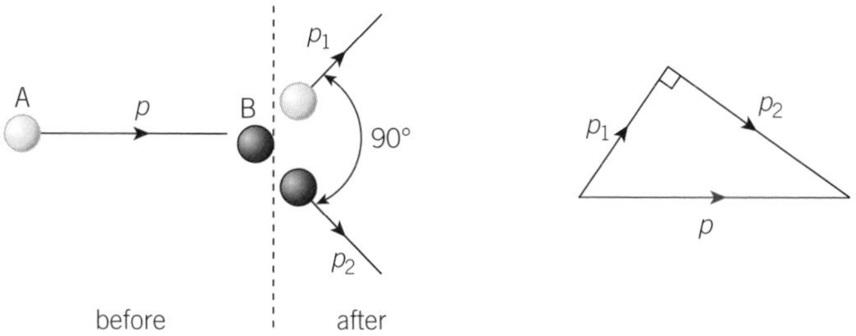

▲ **Figure 2** *Objects A and B before (left) and after the collision, and the vector triangle (right)*

Worked example: Snooker balls

A 160 g white ball travelling at $4.0\,\mathrm{m\,s^{-1}}$ hits a stationary 170 g black ball (Figure 3). After the impact, the balls move apart at approximately 90° to each other, with the white ball travelling at $2.5\,\mathrm{m\,s^{-1}}$. Calculate the magnitude of the final velocity of the black ball.

Step 1: Select the equation you need and calculate the momentum of each ball.

$$p = mv$$

Learning outcomes

Demonstrate knowledge, understanding, and application of:

→ collisions and interactions of bodies in two dimensions.

Study tip

You will not be assessed on two-dimensional collisions until A Level.

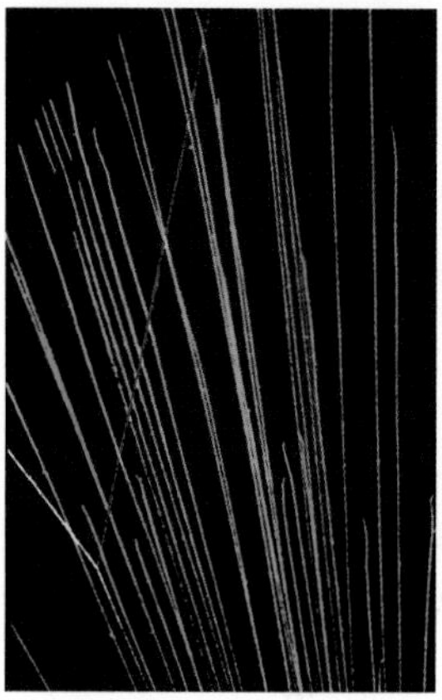

▲ **Figure 1** *After a collision between a helium nucleus and a proton, the proton shoots off to the right (red track) and the helium nucleus moves off to the left (yellow track)*

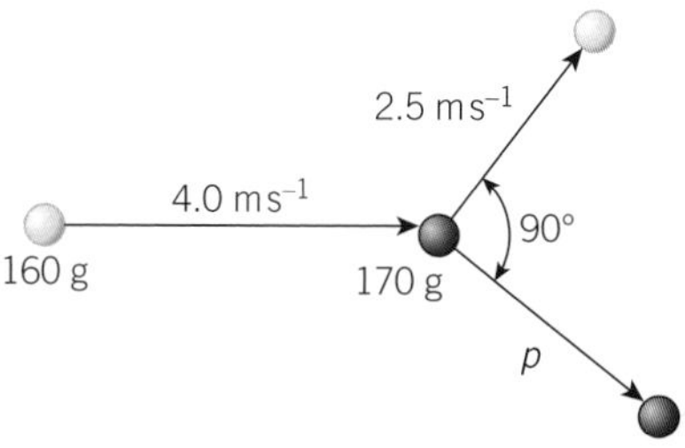

▲ **Figure 3**

initial momentum of white ball $= 160 \times 10^{-3} \times 4.0 = 0.64\,\text{kg m s}^{-1}$

initial momentum of black ball $= 0$

final momentum of white ball after impact $= 160 \times 10^{-3} \times 2.5 = 0.40\,\text{kg m s}^{-1}$

Step 2: Draw a vector triangle for the momenta after the impact (Figure 4). These must add up to the initial momentum of 0.64 kg m s^{-1}.

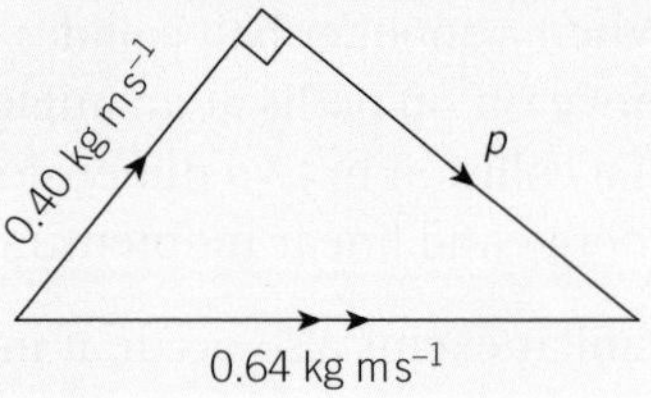

▲ **Figure 4**

Step 3: Use Pythagoras' theorem to determine the final momentum p of the black ball.

$$p^2 = 0.64^2 - 0.40^2 = 0.25 \qquad p = 0.50\,\text{kg m s}^{-1}$$

Step 4: Use the equation $p = mv$ to calculate the velocity of the black ball.

$$0.50 = 0.170 \times v \qquad v = \frac{0.50}{0.170} = 2.9\,\text{m s}^{-1}$$

The magnitude of the final velocity of the black ball is $2.9\,\text{m s}^{-1}$.

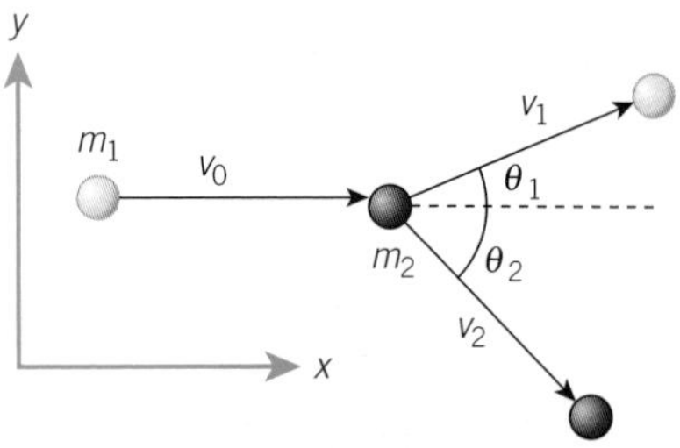

▲ **Figure 5** *The white object is originally moving in the x direction, so after the collision, the total momentum in the y direction must be zero*

Resolving momentum

Figure 5 shows a general collision involving two objects of mass m_1 and m_2. The white object travels at velocity v_0 and collides with the stationary black object. After the collision, the white object travels at angle θ_1 to its original direction with velocity v_1, and the black object travels at angle θ_2 with velocity v_2.

The momentum in any direction must be conserved. In this case, the momentum must remain the same in the x direction and y direction.

***x* direction:** total initial momentum = total final momentum

$$m_1 v_0 = m_1 v_1 \cos\theta_1 + m_2 v_2 \cos\theta_2$$

***y* direction:** total initial momentum = total final momentum

$$0 = m_1 v_1 \sin\theta_1 + m_2 v_2 \sin\theta_2$$

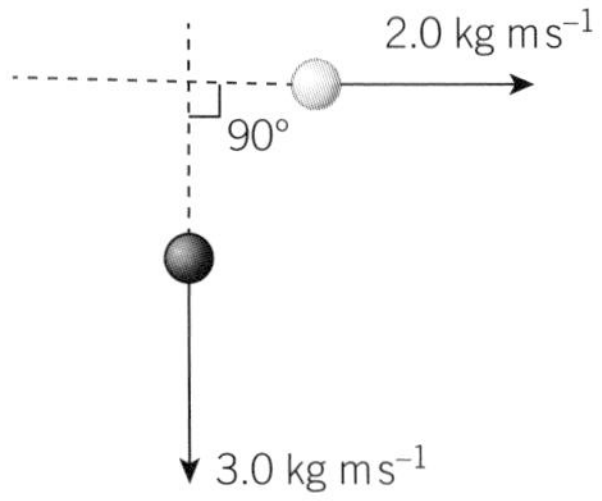

▲ **Figure 7**

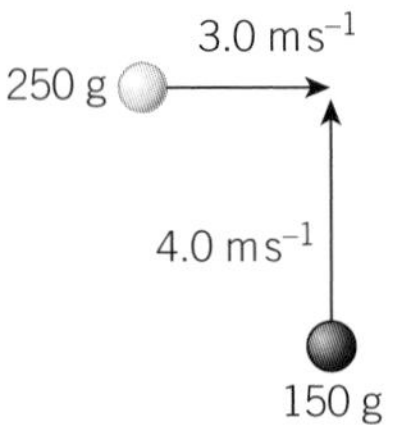

▲ **Figure 8**

Summary questions

1. Two objects of the same mass have an oblique collision. State the angle between the directions in which the objects travel after the collision. *(1 mark)*

2. Figure 6 shows the momentum of two objects **X** and **Y** before and after a collision. Explain how you can tell that the collision is incorrectly represented. *(2 marks)*

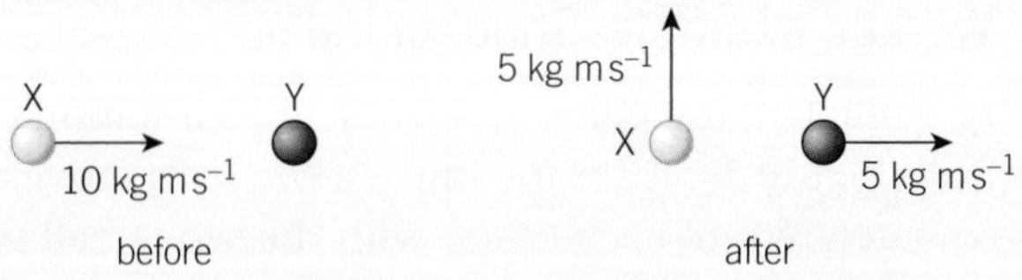

▶ **Figure 6**

3. Figure 7 shows the final momenta of two particles after a collision. Calculate the magnitude of the initial momentum of the particles. Explain your answer. *(3 marks)*

4. Two objects collide at 90° to each other and stick together (Figure 8). Calculate the magnitude of their final velocity. *(4 marks)*

Practice questions

1 **a** Derive the base units of momentum. *(3 marks)*

b A stationary tennis ball of mass 60 g is hit with a racquet. Figure 1 shows a graph of force *F* on the ball against time *t* of impact between the racquet and the ball.

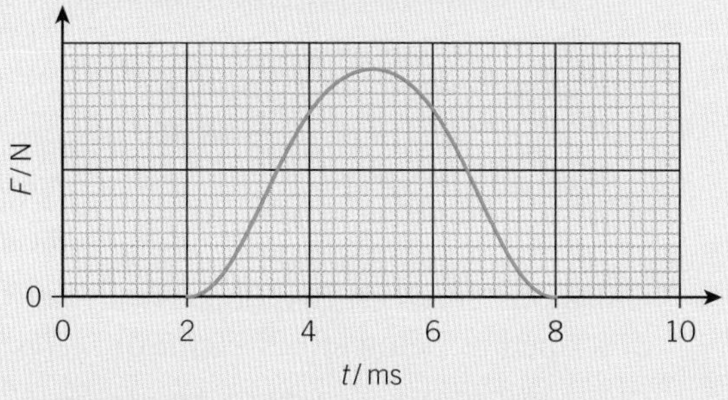

▲ **Figure 1**

The area under the graph is 3.2 N s.

(i) Calculate the final velocity of the ball. *(3 marks)*

(ii) Show that the maximum force acting on the ball is about 1 kN. *(3 marks)*

(iii) Describe and explain the motion of the racquet after it has hit the ball. *(3 marks)*

2 **a** Define impulse of a force. *(1 mark)*

b A driving force acts on a stationary car on a level road. Figure 2(a) shows the variation of the net force *F* acting on the car with time *t*.

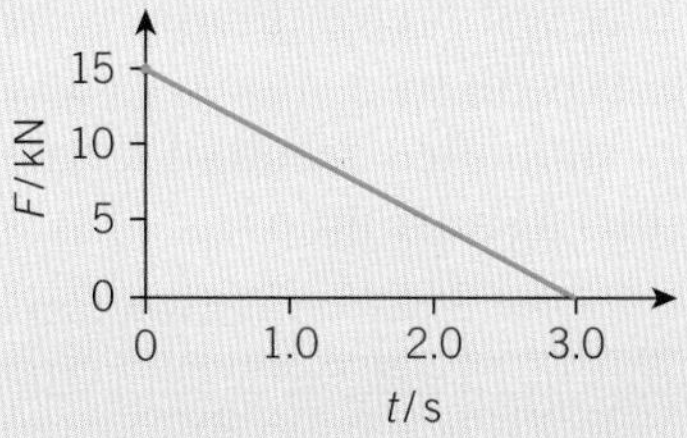

▲ **Figure 2a**

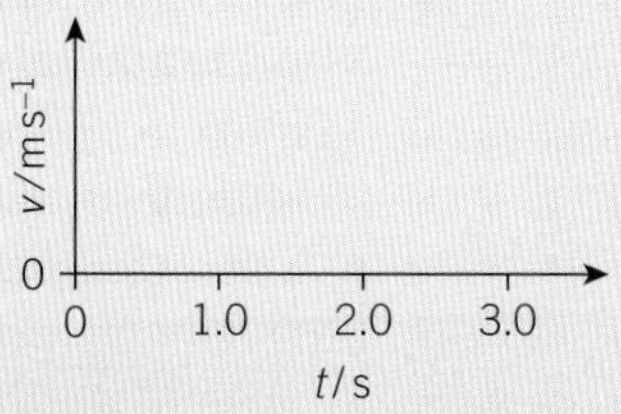

▲ **Figure 2b**

(i) Describe how the acceleration of the car changes time $t = 0$ to $t = 3.0$ s. *(2 marks)*

(ii) On a copy of Figure 2(b), sketch a graph to show the variation of the velocity of the car with time *t*. You are not expected to show the values on the *v*-axis. *(2 marks)*

(iii) The mass of the car is 1200 kg. Calculate the final kinetic energy of the car. *(5 marks)*

3 **a** State Newton's second law of motion. *(1 mark)*

b Use Newton's second law of motion to show how the net force *F* acting on an object is related to its mass *m* and acceleration *a*. *(3 marks)*

c An 80 g ball is dropped from a height *h* above the ground. It has a speed of 5.4 m s^{-1} just before it hits the ground. After hitting the ground, it has a rebound speed of 3.0 m s^{-1}.

(i) Calculate the height *h* of the ball. State any assumption made. *(4 marks)*

(ii) Figure 3 shows a graph of force *F* exerted by the ground on the ball against the time *t* of contact between the ball and the ground.

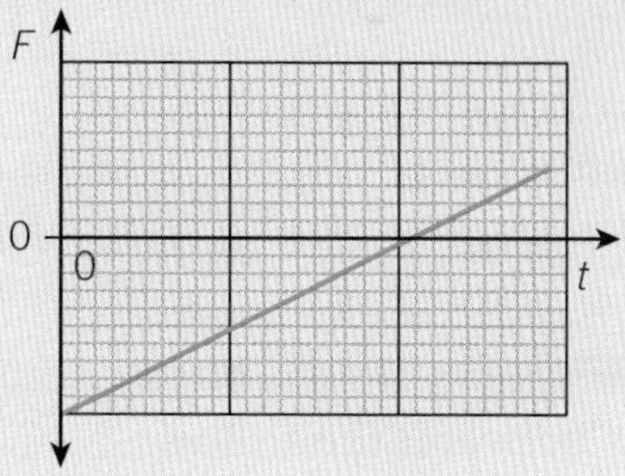

▲ **Figure 3**

On a copy of Figure 3 sketch a graph to show the variation of the force exerted by the ball on the ground. Explain your answer. *(2 marks)*

(iii) Calculate the magnitude of the change in momentum of the ball. *(2 marks)*

Jan 2013 G484

4 This question is about pressing a red hot bar of steel into a sheet in a rolling mill.

a A bar of steel of mass 500 kg is moved on a conveyor belt at $0.60\,\mathrm{m\,s^{-1}}$.

Calculate the momentum of the bar, giving a suitable unit for your answer. *(2 marks)*

b From the conveyor belt, the bar is passed between two rollers, shown in Figure 4. The bar enters the rollers at $0.60\,\mathrm{m\,s^{-1}}$. The rollers flatten the bar into a sheet with the result that the sheet leaves the rollers at $1.8\,\mathrm{m\,s^{-1}}$.

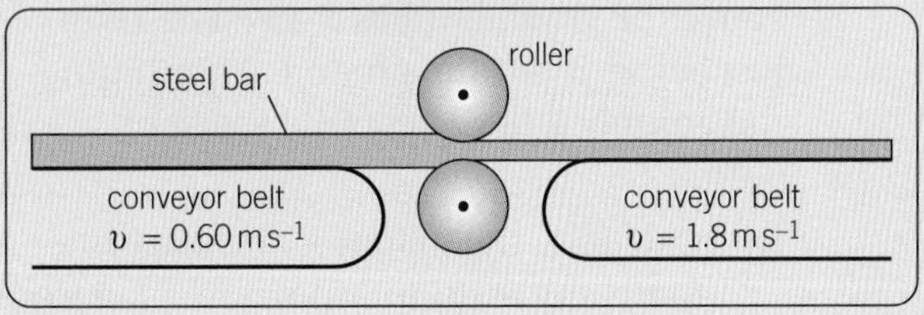

▲ **Figure 4**

(i) Explain why there is a resultant horizontal force on the bar at the point immediately between the rollers. *(2 marks)*

(ii) In which direction does this force act? *(1 mark)*

(iii) The original length of the bar is 3.0 m. Calculate the time it takes for the bar to pass between the rollers. *(1 mark)*

(iv) Calculate the magnitude of the resultant force on the bar during the pressing process. *(3 marks)*

5 Figure 5 shows the masses and velocities of two objects **A** and **B** moving directly towards each other. **A** and **B** stick together on impact and move with a common velocity *v*.

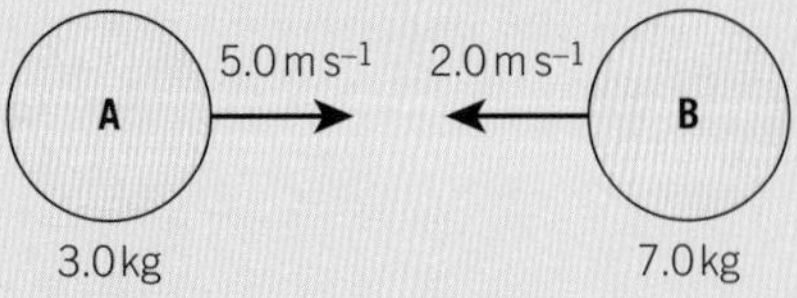

▲ **Figure 5**

(i) Determine the velocity *v*, stating its magnitude and direction. *(3 marks)*

(ii) Determine the impulse of the force experienced by the object **A** and state its direction. *(2 marks)*

(iii) Explain, using Newton's third law of motion, the relationship between the impulse experienced by **A** and the impulse experienced by **B** during the impact. *(2 marks)*

Jan 2013 G484

6 a Collisions between two objects can be described as being either *elastic* or *inelastic.*

Copy and complete Table 1 by placing a tick (✓) in the relevant row(s) for each statement that is true for each type of collision. *(4 marks)*

▼ **Table 1**

Statement	Elastic collision	Inelastic collision
Total momentum for the objects is conserved.		
Total kinetic energy of the objects is convserved.		
Total energy is conserved.		
Magnitude of the impulse on each object is the same.		

b A snooker ball is at rest on a smooth horizontal table. It is hit by a snooker cue. Figure 6 shows a simplified graph of force *F* acting on the ball against time *t*.

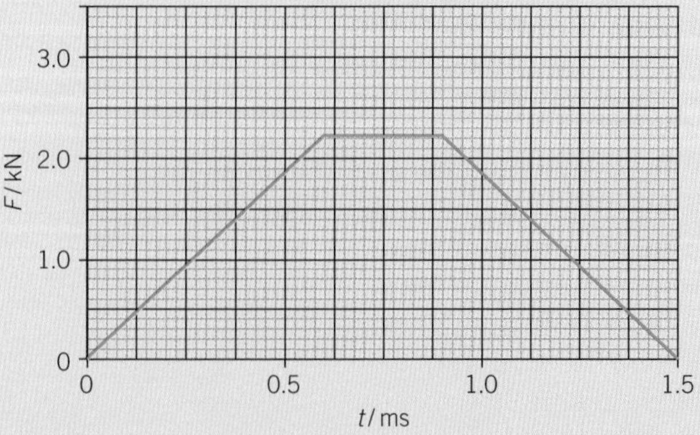

▲ **Figure 6**

(i) Describe how the velocity of the ball varies between $t = 0.6\,\text{ms}$ and $t = 0.9\,\text{ms}$. *(1 mark)*

(ii) Use Figure 6 to calculate the impulse acting on the ball. *(2 marks)*

(iii) The mass of the snooker ball is 140 g. Calculate the final speed of the snooker ball as it leaves the cue. *(2 marks)*

7 a Compare and contrast elastic and inelastic collisions. *(3 marks)*

b Figure 7 shows an object. It has two sections A and B. The mass of section A is 25% of the total mass of the object. An explosion within the object ejects A and B in opposite directions.

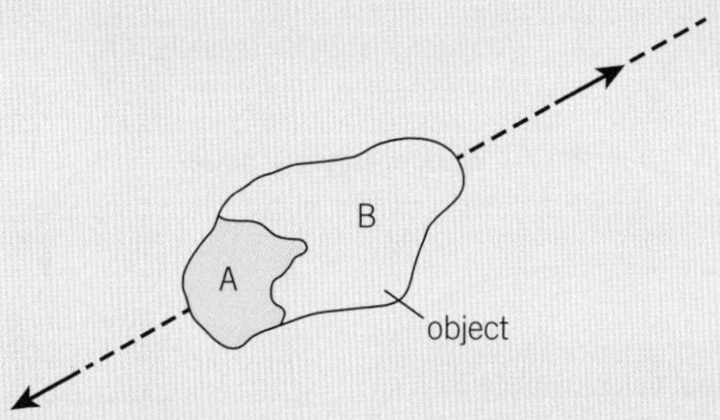

▲ **Figure 7**

(i) Use Newton's third law of motion to explain why A and B are ejected in opposite directions. *(2 marks)*

(ii) Calculate the ratio $\frac{\text{kinetic energy of A}}{\text{kinetic energy of B}}$. *(5 marks)*

8 a State what is meant by an *inelastic collision*. *(1 mark)*

b Two objects A and B collide. Figure 8 shows the variation momentum p of A with time t before, during, and after the collision.

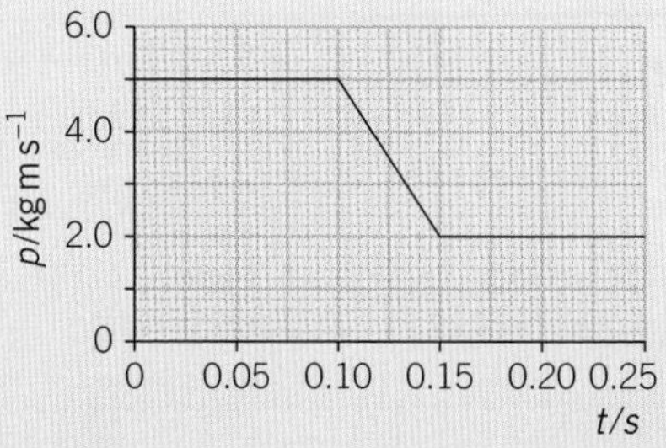

▲ **Figure 8**

Use Figure 8 to determine

(i) the magnitude of the change in momentum of B *(2 marks)*

(ii) the magnitude of the force acting on B during the collision. Explain your answer. *(3 marks)*

c Figure 9 shows the initial and final states of two trolleys involved in a collision.

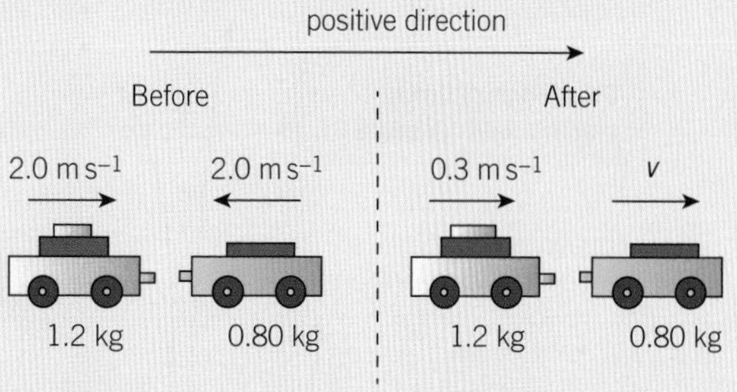

▲ **Figure 9**

The mass and the velocities are shown in Figure 9.

Calculate the velocity v of the 0.80 kg trolley after the collision. *(3 marks)*

d Figure 10 shows the initial and final states of two identical balls, X and Y, involved in a collision.

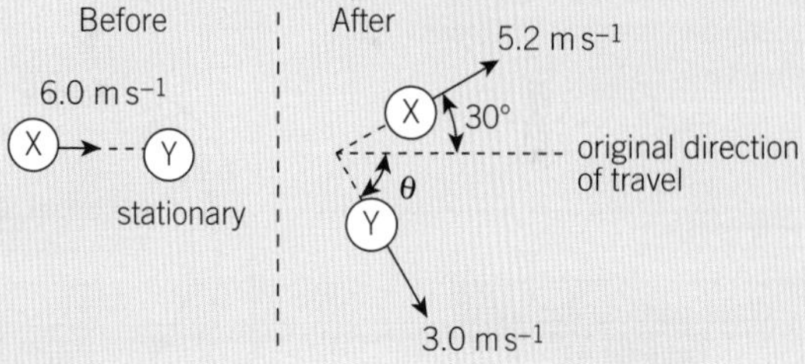

▲ **Figure 10**

Before the collision, X has a velocity of $6.0\,\text{m s}^{-1}$ and Y is stationary. After the collision, X has a velocity of $5.2\,\text{m s}^{-1}$ at an angle of 30° to its original direction of travel. Y is deflected at an angle θ and has a velocity of $3.0\,\text{m s}^{-1}$.

(i) Explain why the total momentum in the direction at right angles to the initial velocity of X is zero. *(1 mark)*

(ii) Calculate the angle θ. *(3 marks)*

Module 3 Summary

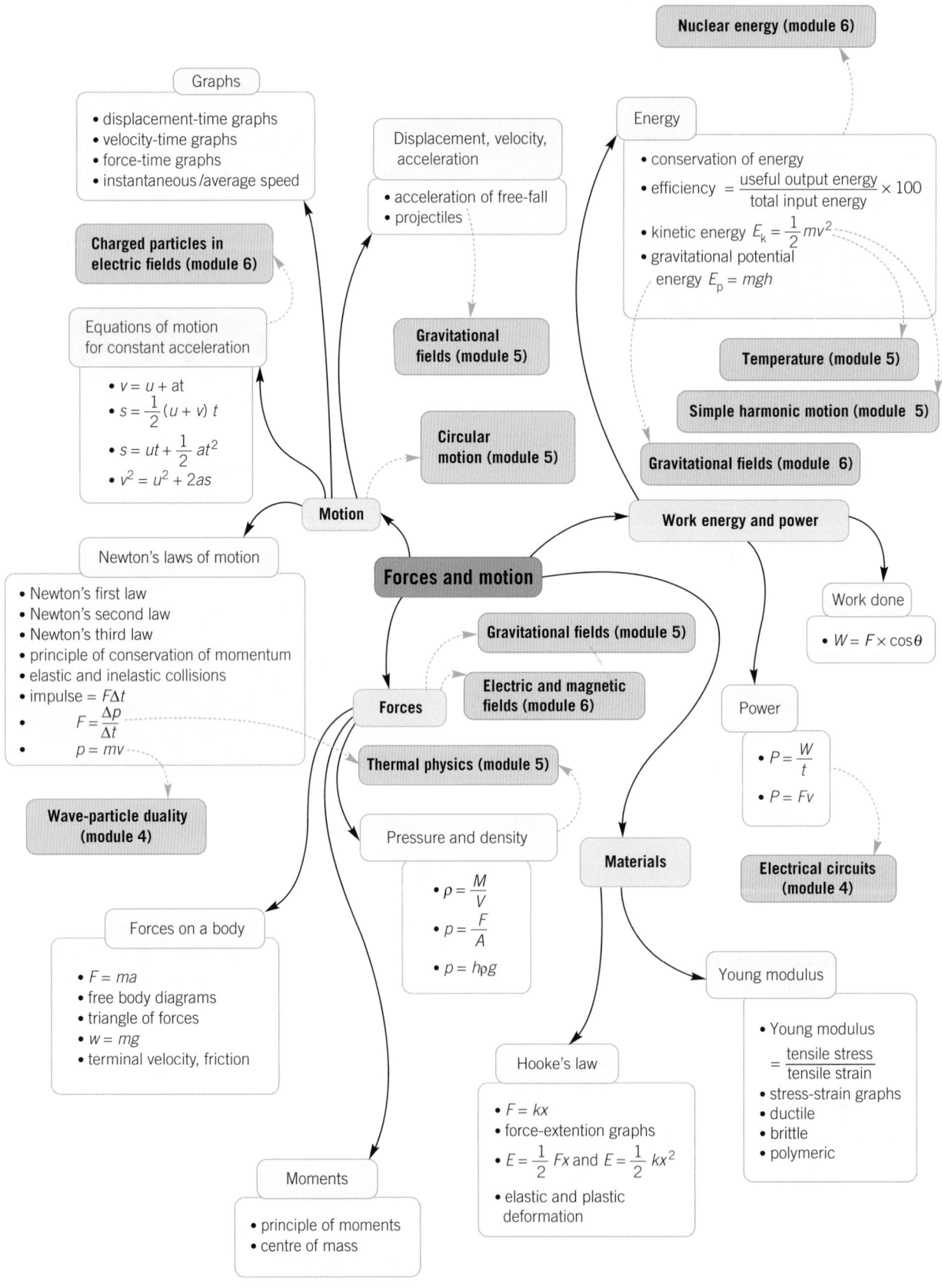

Forces and motion

Motion

Graphs
- displacement-time graphs
- velocity-time graphs
- force-time graphs
- instantaneous /average speed

Displacement, velocity, acceleration
- acceleration of free-fall
- projectiles

Gravitational fields (module 5)

Equations of motion for constant acceleration
- $v = u + at$
- $s = \frac{1}{2}(u + v)\,t$
- $s = ut + \frac{1}{2}at^2$
- $v^2 = u^2 + 2as$

Charged particles in electric fields (module 6)

Circular motion (module 5)

Newton's laws of motion
- Newton's first law
- Newton's second law
- Newton's third law
- principle of conservation of momentum
- elastic and inelastic collisions
- impulse = $F\Delta t$
- $F = \frac{\Delta p}{\Delta t}$
- $p = mv$

Wave-particle duality (module 4)

Work energy and power

Energy
- conservation of energy
- efficiency $= \frac{\text{useful output energy}}{\text{total input energy}} \times 100$
- kinetic energy $E_k = \frac{1}{2}mv^2$
- gravitational potential energy $E_p = mgh$

Nuclear energy (module 6)

Temperature (module 5)

Simple harmonic motion (module 5)

Gravitational fields (module 6)

Work done
- $W = F \times \cos\theta$

Power
- $P = \frac{W}{t}$
- $P = Fv$

Electrical circuits (module 4)

Forces

Gravitational fields (module 5)

Electric and magnetic fields (module 6)

Thermal physics (module 5)

Pressure and density
- $\rho = \frac{M}{V}$
- $p = \frac{F}{A}$
- $p = h\rho g$

Forces on a body
- $F = ma$
- free body diagrams
- triangle of forces
- $w = mg$
- terminal velocity, friction

Moments
- principle of moments
- centre of mass

Materials

Hooke's law
- $F = kx$
- force-extention graphs
- $E = \frac{1}{2}Fx$ and $E = \frac{1}{2}kx^2$
- elastic and plastic deformation

Young modulus
- Young modulus $= \frac{\text{tensile stress}}{\text{tensile strain}}$
- stress-strain graphs
- ductile
- brittle
- polymeric

Buildings

Architects have to carefully consider the physical properties of the materials when designing a building. Materials such as brick, concrete, and stone are weak in tension but strong in compression. Steel is incredibly strong in tension. Wood has interesting properties because it can withstand relatively large compressive and tensile stresses.

Figure 1 shows a conventional brick house with a tiled roof. The brick walls are in compression. A framework of wood is used to support the roof and to prevent the brick walls from buckling outwards. The strut beams in are compression and the tie beams are in tension.

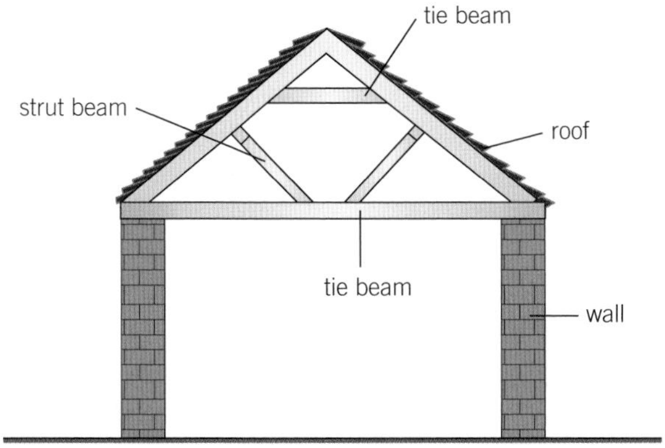

▲ **Figure 1** *Tie and strut beams*

▲ **Figure 2** *The Shard in London is 308 m tall. It is constructed around a steel framework.*

1 Describe what would happen to the structure shown in Figure 1 if the tie beams were not included.
2 Could steel cables be used in place of the tie beams?
3 Explain why the compressive stress is not constant along the length of the wall. Where would you expect the compressive stress in the wall to be a maximum?
4 Figure 2 shows The Shard in London. Discuss why very tall buildings use a steel framework rather than brick.

The weather

In 4.9 you learnt about pressure due to fluid columns and the Archimedes' principles. The measurement of atmospheric pressure is important in the study of meteorology (weather). High and low pressures govern the long and the short term weather patterns we observe. On weather maps, pressure is shown in millibars. Air pressure can be measured using instruments known as barometers.

Here are some extension tasks you can carry out to further improve your understanding of this topic. You may use the internet to carry out some of the research.

1. What is the relationship between millibars and pascals?
2. Investigate the different types of barometers and explain why many of them use mercury.
3. If you were to design you own barometer using water, how tall will it be?

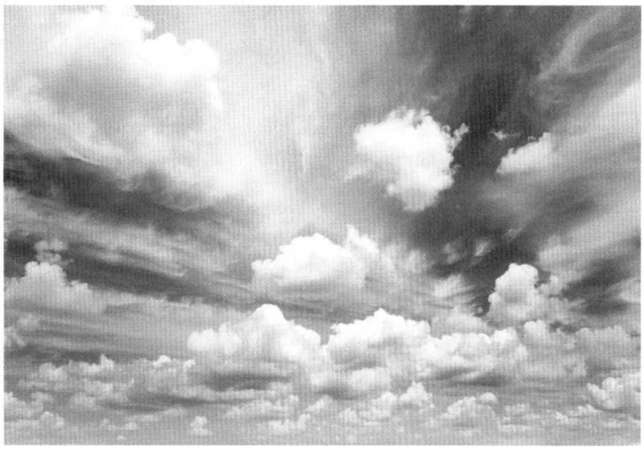

4. Design an aneroid barometer using everyday items around you.
5. How does air pressure vary with altitude?

Module 3 practice questions

Section A

1 Students A and B use micrometer screw gauges to measure the diameter of a copper wire in three different places along its length. The diameter of the wire according to the manufacturer is 0.278 mm. The results recorded by students A and B are shown in Figure 1.

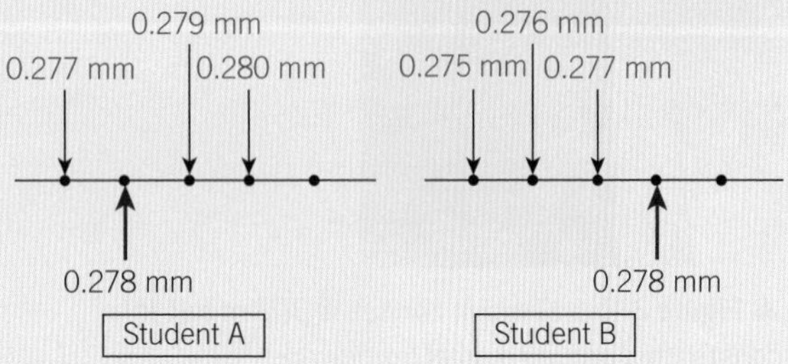

▲ **Figure 1**

Which statement is correct about the measurements made by the student B compared with those of student A?

A The measurements are more accurate.

B The measurements are not as precise.

C The measurements are both more accurate and more precise.

D The measurements are not accurate but are more precise. *(1 mark)*

From the AS Paper 1 practice questions

2 A spring of original length 3.0 cm and force constant $100\,\mathrm{N\,m^{-1}}$ is placed on a smooth horizontal surface. Its length is changed from 6.0 cm to 8.0 cm.

What is the change in the energy stored by the spring?

A 0.020 J

B 0.080 J

C 0.140 J

D 1.00 J *(1 mark)*

From the AS Paper 1 practice questions

3 A wooden block is held under water and then released, as shown in Figure 2.

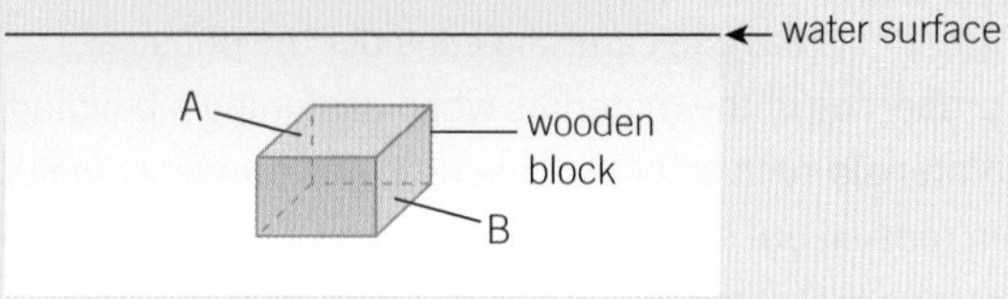

▲ **Figure 2**

The wooden block moves towards the surface of the water.

Which of the following statements is/are true about the block as soon as it is released?

1 The force experienced by the face B due to water is greater than the force experienced by the face A.

2 The upthrust on the block is equal to its weight.

3 The mass of the water displaced is equal to the weight of the block.

A 1, 2 and 3 are correct

B Only 1 and 2 are correct

C Only 2 and 3 are correct

D Only 1 is correct *(1 mark)*

From the AS Paper 1 practice questions

4 What are the correct base units for kinetic energy?

A $\mathrm{kg\,m}$

B $\mathrm{kg\,s^{-2}}$

C $\mathrm{kg\,m^2\,s^{-1}}$

D $\mathrm{kg\,m^2\,s^{-2}}$

5 Which statement is correct about impulse?

A Impulse is equal to the area under a force against distance graph.

B Impulse has the same unit as momentum.

C Impulse is equal to the rate of change of momentum.

D Impulse is not conserved in an inelastic collision.

6 Figure 3 shows the forces acting on an object of mass 3.0 kg.

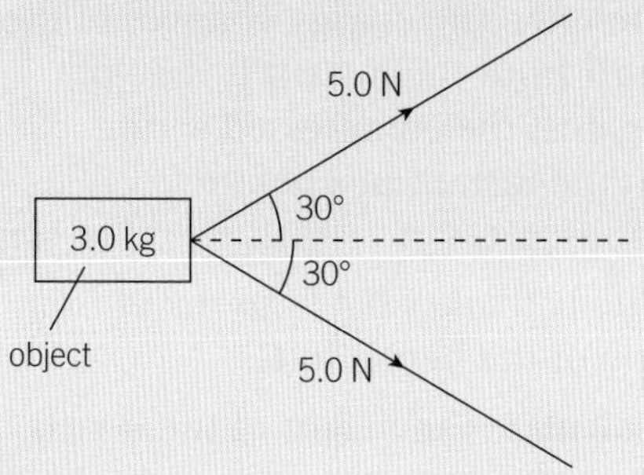

▲ **Figure 3**

What is the acceleration of the object?

A $0.83\,m\,s^{-2}$

B $1.4\,m\,s^{-2}$

C $1.7\,m\,s^{-2}$

D $2.9\,m\,s^{-2}$ *(1 mark)*

7 The point A shown in Figure 4 is a distance *x* below the surface of the water.

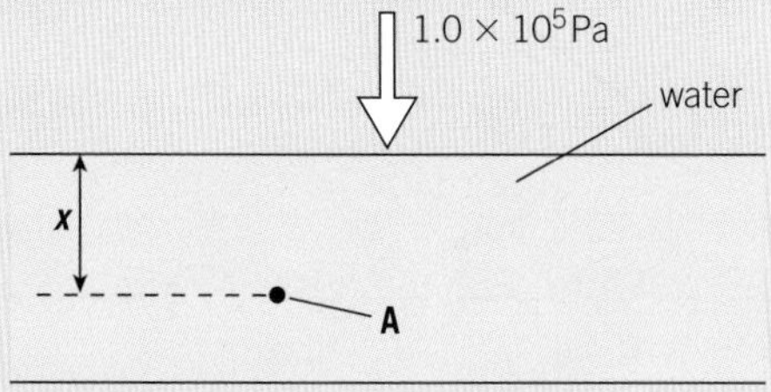

▲ **Figure 4**

The density of water is $1000\,kg\,m^{-3}$ and the atmospheric pressure is 1.0×10^5 Pa.
The pressure measured by A by a scuba diver is 1.5×10^5 Pa.
What is the value of *x*?

A 0.20 m
B 5.1 m
C 10 m
D 15 m *(1 mark)*

Section B

8 **a** Define *velocity*. *(1 mark)*

b The mass of an ostrich is 130 kg. It can run at a maximum speed of 70 kilometers per hour.

(i) Calculate the maximum kinetic energy of the ostrich when it is running. *(3 marks)*

(ii) Scientists have recently found fossils of a prehistoric bird known as Mononykus. Figure 5 shows what the Mononykus would have looked like.

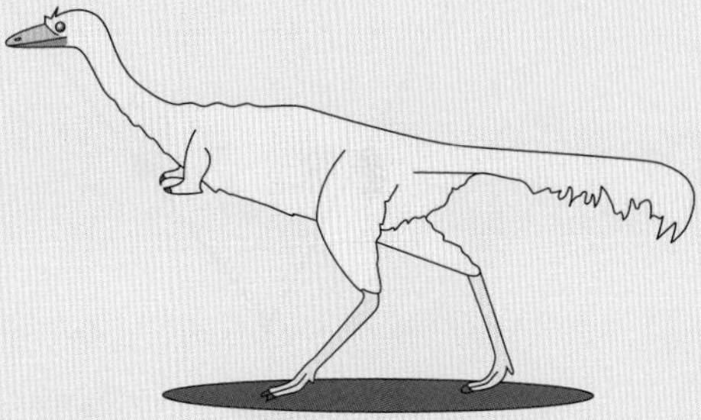

▲ **Figure 5**

According to a student, the Mononykus looks similar to our modern day ostrich. The length, height and width of the Mononykus were all **half** that of an ostrich. Estimate the mass of the Mononykus. Explain your reasoning. *(2 marks)*

G481 June 2014

From the AS Paper 1 practice questions

9 Figure 6 shows a block of wood held at rest at the top of a smooth ramp.

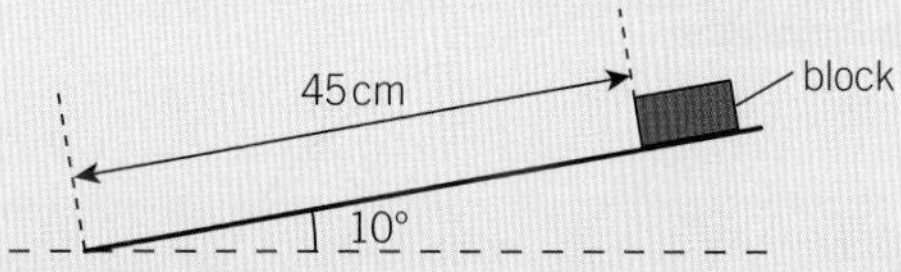

▲ **Figure 6**

The ramp makes an angle of 10° to the horizontal. The block is released and it slides down the ramp.

a Calculate the acceleration of the block along the length of the ramp. *(2 marks)*

b The block travels a total distance of 45 cm down the ramp. Calculate the time it takes to reach the bottom of the ramp. *(3 marks)*

c The speed of the block at the bottom of the ramp is v. Describe a simple experiment a student can carry out to determine an approximate value of the speed v. The student only has a metre rule and a stopwatch. *(3 marks)*

From the AS Paper 1 practice questions

10 a Figure 7a shows a 500 g mass suspended from two strings. The mass hangs vertically and is in equilibrium.

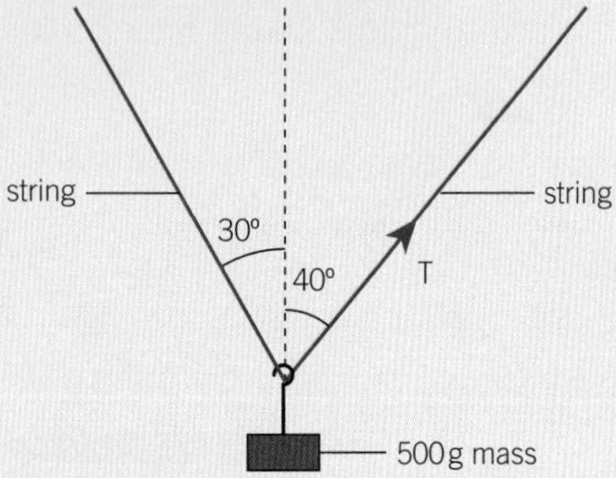

▲ **Figure 7a**

(i) Determine the tension T in one of the strings. *(4 marks)*

(ii) Describe how a student could determine the value of T experimentally in the laboratory. State one possible limitation of the experiment. *(2 marks)*

b Figure 7b shows an experiment designed by a student.

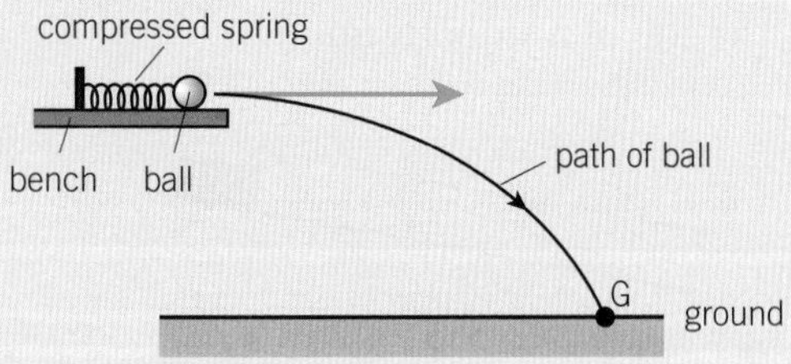

▲ **Figure 7b**

A metal ball is pushed against a compressible spring and then released. The ball has a horizontal velocity of $1.5\,\text{m s}^{-1}$. The ball leaves the horizontal bench and lands on the ground below at point **G**.

Assume friction has negligible effect on the motion of the ball.

(i) Describe the energy changes of the **ball** from the instant it is held against the compressed spring to the instant just before it lands at **G**. *(4 marks)*

(ii) The ball takes 0.42 s to travel from top of the bench to **G**.

Calculate the height of the bench from the ground. *(3 marks)*

From the AS Paper 2 practice questions

11 Figure 8 shows a simple pendulum. It consists of a metal ball of diameter 2.00 cm and a thin string.

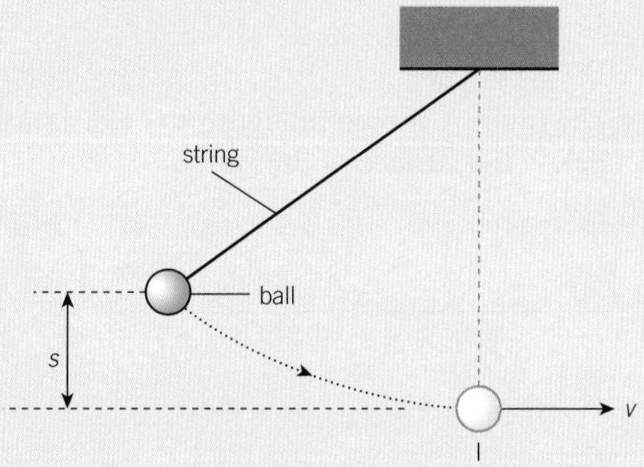

▲ **Figure 8**

The ball is raised to a vertical height s and then released.

a Show that its speed v at the bottom of its swing is given by the equation $v^2 = 2gs$, where g is the acceleration of free fall. *(3 marks)*

b Describe how a student could determine the speed v at the bottom of the pendulum's swing in the laboratory. State one possible limitation of your method. *(4 marks)* c The table below shows **some** of the results obtained by a student.

s/m	v/m s^{-1}	v^2/m^2 s^{-2}
0.210	2.03 ± 0.15	4.12 ± 0.61
0.330	2.54 ± 0.18	
0.410	2.83 ± 0.25	8.01 ± 1.42
0.490	3.10 ± 0.30	9.61 ± 1.86

(i) Copy and complete the table by determining the missing value for v^2 and the absolute uncertainty in this value. *(3 marks)*

(ii) The student plots a graph of v^2 against s. Explain how the graph may be used to determine the acceleration of free fall g. *(2 marks)*

From the AS Paper 2 practice questions

12 a Figure 9 shows a ball of mass 0.050 kg resting on the strings of a tennis racket held horizontally.

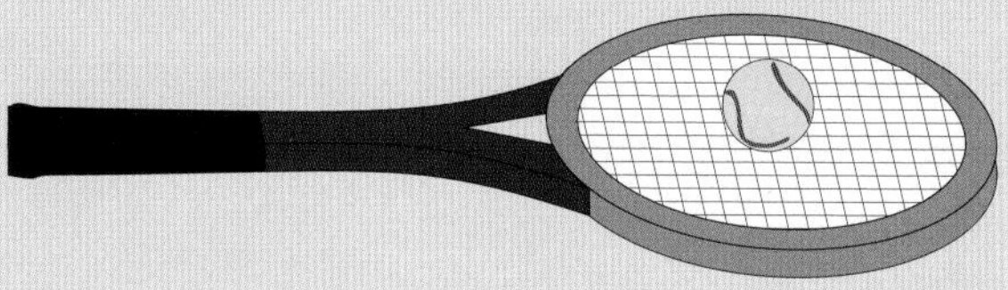

▲ **Figure 9**

(i) On a copy of Figure 5, draw and label arrows to represent the **two** forces acting on the ball. *(2 marks)*

(ii) Calculate the difference in magnitude between the two forces on the ball when the racket is accelerated upwards at 2.0 m s^{-2}. *(2 marks)*

b The ball is dropped from rest at a point 0.80 m above the racket head. The racket is fixed rigidly. Assume that the ball makes an elastic collision with the strings and that any effects of air resistance are negligible.

Calculate

(i) the speed of the ball just before impact, *(2 marks)*

(ii) the momentum of the ball just before impact, *(1 mark)*

(iii) the change in momentum of the ball during the impact, *(1 mark)*

(iv) the average force during the impact for a contact time of 0.050 s. *(1 mark)*

c The two forces you have drawn in **(a)(i)** are not a pair of forces as required by Newton's third law of motion. However each of these forces does have a corresponding equal and opposite force to satisfy Newton's third law. Describe these equal and opposite forces and state the objects on which they act. *(4 marks)*

Q1 2824 Jan 2010 paper

From the AS Paper 2 practice questions

MODULE 4

Electrons, waves, and photons

Chapters in this Module

Demonstrate knowledge, understanding, and application of:

8 Charge and current

9 Energy, power, and resistance

10 Electrical circuits

11 Waves 1

12 Waves 2

13 Quantum physics

Introduction

Quantum physics is perhaps one of our greatest ever achievements. It allows us to make incredibly accurate predictions of what happens on tiny scales far smaller than an atom. In order to help you understand its key ideas, this module takes you on a journey, starting with electrons and how they behave in electrical circuits through an exploration of wave properties ending with quantum physics.

Charge and current provides an introduction to the fundamental ideas of charge and current, exploring the link between lightning strikes, the human brain, and the wonder material that is graphene.

Energy, power, and resistance develops the use of electrical symbols, along with key ideas like electromotive force, potential difference, and resistivity. You will learn about how differences in resistance help archaeologists discover ancient remains and doctors care for premature babies.

Electric circuits brings together ideas from the previous two chapters to explore the use of electrical circuits, including explanations of how potential dividers are used to make volume control dials and why a car battery can supply such a high current.

Waves 1 explores waves and their properties. You will learn about electromagnetic waves, earthquakes, and how diamonds get their sparkle.

Waves 2 includes explanations of how musical instruments produce their characteristic notes and how noise-cancelling headphones work so effectively. You will learn about the effect of interference of waves in a variety of situations.

Quantum physics introduces several truly amazing concepts, including the ideas that not only do electromagnetic waves have wave- and particle-like behaviour but this dual nature is also found to be characteristic of all particles, including electrons. Electrons can be made to diffract!

Knowledge and understanding checklist

From your Key Stage 4 study you should be able to answer the following questions. Work through each point, using your Key Stage 4 notes and the support available on Kerboodle.

- ☐ Recall that current is a rate of flow of charge and that for a charge to flow, a source of potential difference and a closed circuit are needed.
- ☐ Recall that current depends on both resistance and potential difference.
- ☐ Describe the difference between series and parallel circuits.
- ☐ Calculate the currents, potential differences, and resistances in series circuits.
- ☐ Explain the use of circuits containing components including lamps, diodes, thermistors, and LDRs.
- ☐ Describe wave motion in terms of amplitude, wavelength, frequency, and period.
- ☐ Describe how ripples on water surfaces are examples of transverse waves whilst sound waves in air are longitudinal waves, and describe the differences between transverse and longitudinal waves.
- ☐ Know that electromagnetic waves are transverse and are transmitted through space where all have the same velocity.

Maths skills checklist

All physicists need to use maths in their studies. In this unit you will need to use many different maths skills, including the following examples. You can find support for these skills on Kerboodle and through MyMaths.

- ☐ **Determine the gradient and intercept from a graph and use $y = mx + c$ to find unknown values.** You will need to be able to do this when investigating the resistivity of a wire.
- ☐ **Recognise and use expressions in decimal and standard form.** You will need to be able to do this when developing ideas around the electromagnetic spectrum.
- ☐ **Sketch relationships which are modelled by equations.** You will need to be able to do this when studying the photoelectric effect.
- ☐ **Substitute numerical values into algebraic equations using appropriate units for physical quantities.** You will need to be able to do this to determine the wavelength of light from a double slit experiment.

MyMaths.co.uk Bringing Maths Alive

8 CHARGE AND CURRENT

8.1 Current and charge

Specification reference: 4.1.1

Learning outcomes

Demonstrate knowledge, understanding, and application of:

- → electric current as rate of flow of charge $I = \frac{\Delta Q}{\Delta t}$
- → the coulomb as the unit of charge
- → the elementary charge $e = 1.60 \times 10^{-19}$ C
- → net charge on a particle or an object is quantised and a multiple of *e*.

▲ **Figure 1** *The major path of ionised air through which billions of electrons have travelled to the ground is clearly visible here*

Synoptic link

You will recall from Topic 2.1, Quantities and units, that the ampere is one of the seven base units. It is used to define many other derived units in electricity, for example, the coulomb (A s) and the ohm ($kg\,m^2\,s^{-1}\,A^{-2}$).

Study tip

Be sure to convert time into SI units (seconds) when using the equation for current.

A bolt from the blue

A lightning storm is one of nature's most awesome spectacles. Each bolt of lightning shows the path taken by billions and billions of electrons travelling from the cloud to the ground (or occasionally the other way around). This flow of *charged* particles produces a massive **electric current**.

Whereas current of 1.2 A is enough to charge a typical smartphone, the current in a lightning strike is often in excess of 30 000 A and heats the surrounding air to over five times the temperature of the surface of the Sun.

Defining electric current

Electric current is measured in **amperes** (or just amps for short). The ampere is one of the seven SI base units.

Electric current is defined as the rate of flow of charge, and can be calculated using the equation

$$I = \frac{\Delta Q}{\Delta t}$$

where I is the electric current in amperes, ΔQ is the charge transferred in coulombs, and Δt is the time in seconds.

In simple terms this is the amount of charge passing a given point in a circuit per unit time. One ampere (1 A) is the same as one coulomb of charge passing a given point per second ($1\,C\,s^{-1}$), 12 kA is the same as 12 000 coulombs per second, and so on.

Worked example: Current in a heater

A charge of 0.26 MC passes through a heater in 6.0 hours. Calculate the average current in the heater in that time, giving your answer to an appropriate number of significant figures.

Step 1: Identify the equation needed.

$$\text{Use } I = \frac{\Delta Q}{\Delta t}$$

Step 2: Substitute known values in SI units into the equation and calculate the answer.

$$I = \frac{0.26 \times 10^6}{6.0 \times 3600}$$

Express the answer to the correct number of significant figures.

$I = 12$ A (2 s.f.)

What is electric charge?

Electric charge is a physical property, much like mass, volume, or temperature. You can think of it as a measure of 'chargedness'. Some particles are charged, like protons and electrons, and others are not, like neutrons. Any object that is not charged is called neutral.

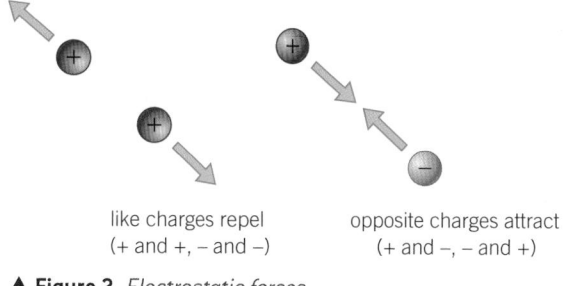

There are two types of charge, **positive** and **negative**. Objects with charge interact and exert forces on each other. Like charges repel each other, and opposite charges attract.

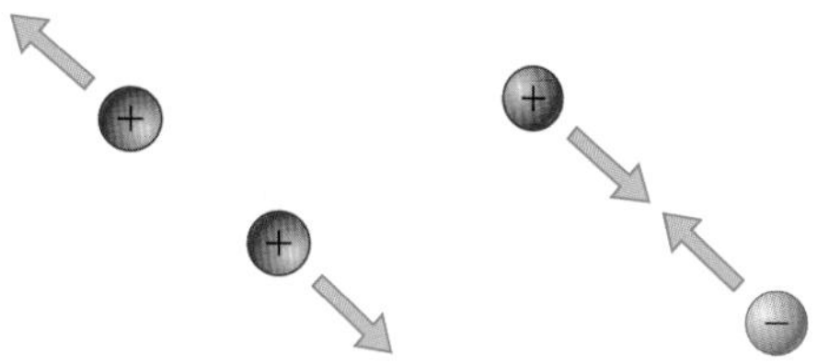
like charges repel
(+ and +, – and –)
opposite charges attract
(+ and –, – and +)

▲ **Figure 2** *Electrostatic forces*

Measuring electric charge

Electric charge is measured in **coulombs** (C). The coulomb is a derived unit, named after the French physicist Charles-Augustin de Coulomb. It is defined as the electric charge flowing past a point in one second when there is an electric current of one ampere.

From $\Delta Q = I\Delta t$, in base units, one coulomb (1 C) is equivalent to one ampere second (1 A s).

Any particle that has an **electric charge** is a charge carrier. Table 1 lists some examples of charge carriers and their charges.

You will be familiar with measuring charges as simply +1 or −2, for example, on **ions** in chemistry, but these are **relative charges**, that is, measured against the constant e. This is the **elementary charge** and is equal to 1.60×10^{-19} C. It is the same as the charge on one proton. A proton has a relative charge of $+1\,e$ and an electron $-1\,e$, not just +1 or −1.

If we know the value of the current in a metal wire, for example, we can also calculate the charge passing through it in a given time and even the number of electrons.

Study tip

Electric charges should not be confused with magnetic poles. There is a deep link between electricity and magnetism, but north and south poles are not the same as positive and negative charges.

▼ **Table 1** *Charges*

Charge carrier	Charge / C
proton	1.60×10^{-19}
electron	-1.60×10^{-19}
copper^{2+} ion	3.20×10^{-19}
sodium^{+} ion	1.60×10^{-19}
chloride^{-} ion	-1.60×10^{-19}

Worked example: Electrons passing through a lamp

The current in a lamp is 6.2 A. Calculate the number of electrons passing through one point in the lamp in 2.0 minutes.

Step 1: Identify the correct equation to use.

$$I = \frac{\Delta Q}{\Delta t}$$

Rearrange to make the charge the subject. $\Delta Q = I\Delta t$

Step 2: Substitute all the values in SI units and calculate the answer.

$$\Delta Q = 6.2 \times 120 = 744\,\text{C}$$

Step 3: The number of electrons responsible for the charge of 744 C can be determined by dividing this charge by the charge e on each electron.

$$\text{number of electrons} = \frac{744}{1.60 \times 10^{-19}}$$
$$= 4.65 \times 10^{21} \approx 4.7 \times 10^{21}$$

The number of electrons is about 4 700 000 000 000 000 000 000.

Net charge

The charge on most objects results from either a gain or a loss of electrons by the object. If electrons have been added to the object it will be negatively charged, if electrons have been removed it will have a positive charge. The size of the charge on a particular object can be expressed as a multiple of e. The net charge on an object is given by

$$Q = \pm ne$$

where Q is the net charge on the object in coulombs, n is the number of electrons (either added or removed), and e is the elementary charge.

We describe the charge on an object as being **quantised**. This is because charge can only have certain values. These values must be integer multiples of e. For example, an object with a charge of 1.92×10^{-18} C has a charge of $+12e$.

Millikan's experiment – the discovery of the quantisation of charge

In 1909 Robert Millikan, helped by Harvey Fletcher, carried out one of the most important physics experiments of the 20th century, now simply called 'the oil-drop experiment'. He analysed the motion of electrically charged oil droplets between two oppositely charged parallel plates. Oil droplets falling through the air experienced gravitational force, air resistance and upthrust.

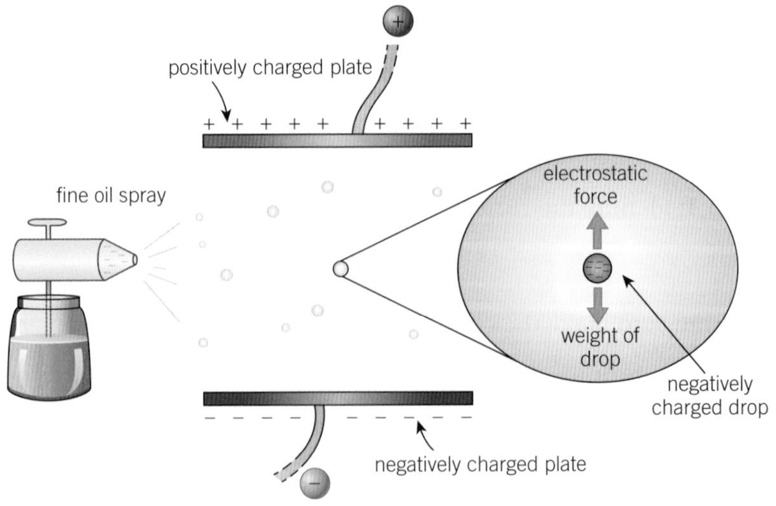

▲ **Figure 3** *A stationary charged oil droplet*

The occasional oil droplet was held stationary as the downward gravitational force was balanced by the upward attractive electrostatic force, whilst others drifted slowly through the electric field. Millikan was able to determine the charge on many droplets. He found that the charge on the droplets was quantised: it did not take just any value, but only values that were multiples of elementary charge.

Using his data Millikan calculated e to be -1.59×10^{-19} C. He was within 1% of the currently accepted value.

1 Using our current understanding of charge, explain why the charge on each oil drop was quantised.
2 Millikan was not able to directly determine the mass of each oil drop. Instead he took careful measurements of the diameter of each drop and calculated its weight using the density of oil. Suggest the steps he would have taken and the formulae he would have used.
3 It is now alleged that Millikan may have ignored some of his data in order to provide a smaller range of readings. Discuss the importance of including all raw data in scientific papers.

Summary questions

1 Calculate the charge in coulombs for the following relative charges:
a $+2.0\,e$ *(1 mark)*
b $-5.0\,e$ *(1 mark)*
c $-12\,e$ *(1 mark)*
d $+41\,e$. *(1 mark)*

2 An object acquires a charge when electrons are either removed from it or deposited on it. Determine the number of electrons deposited on an object to give a net charge of:
a $-10\,e$ *(1 mark)* b -15 C. *(3 marks)*

3 A mobile phone charger draws 500 mA and takes 4.0 hours to charge a phone. Calculate the charge transferred in that time. *(3 marks)*

4 Determine the value of the electric current if 5.0×10^{14} electrons pass through a wire in 1.0 s. *(2 marks)*

5 A negatively charged plate gradually loses electrons to the surrounding air. If the plate loses a charge of 9000 C in 2.0 hours, calculate the average number of electrons per second that leave the plate. *(3 marks)*

6 Calculate the number of electrons passing through the lamp in the second worked example if it were to be left switched on for two weeks. *(2 marks)*

7 A rechargeable battery pack is labelled 5000 mA h. Calculate how much charge the pack can deliver when fully charged. *(3 marks)*

Learning outcomes

Demonstrate knowledge, understanding, and application of:

- → current as the movement of charged particles
- → conventional current and electron flow.

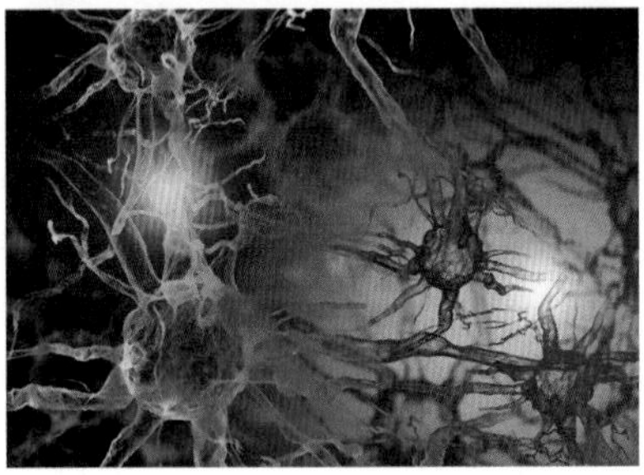

▲ **Figure 1** *Neurons within the brain conduct an electric current, but it is not just a simple flow of electrons*

A flow of charge, not just electrons

Electric currents are everywhere: not just confined to wires, but in the Earth's core, within the layers of the atmosphere, and even inside the cells in your body. We have already described an electric current as a flow of charge, a movement of charge carriers. Remember that this does not necessarily mean a flow of electrons: an electric current is a flow of any type of charge carrier – an electron is just one possibility. In metals the charge carriers are electrons, but in liquids the charge carriers tend to be **ions**.

The electric current in the nerve cells in your brain involves sodium and potassium ions passing through different parts of the membrane surrounding each cell.

Modelling electric current in metals

To help us better understand electric current, we use models to describe the movement of charge carriers through different materials. In metals an electric current is usually a flow of electrons. Because of the way atoms in metals are bonded, most electrons in metal atoms remain fixed to their atom. However, a small number of electrons from each atom are free to move (Figure 2).

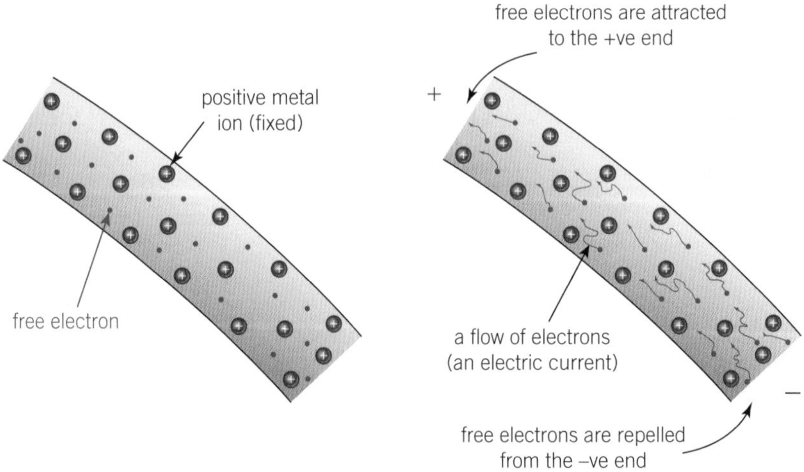

▲ **Figure 2** *A model of electric current in metals: free electrons move through the wire, randomly colliding with the positive ions as they drift past them*

The structure of a metal can be thought of as a regular crystal structure or lattice of positive ions, surrounded by a number of **free electrons** (sometimes called conduction or delocalised electrons). The positive ions are not free to move, but they do vibrate around fixed points, and they vibrate more vigorously as the temperature of the metal increases.

One way to make electrons move is to make one end of a wire positive and the other negative. The electrons in the metal of the wire will be attracted towards the positive end, and so move through the wire as an electric current. The greater the rate of charge flow, the greater the electric current in the wire. A larger current may be due to:

- a greater number of electrons moving past a given point each second (for example, a wire with a greater cross-sectional area)
- the same number of electrons moving faster through the metal.

Conventional current and electron flow

Electric current has been studied for centuries. **Conventional current** was defined long before the discovery of the electron as a current from a positive terminal towards a negative one. The direction of all electric currents is still treated as from positive to negative, regardless of the direction of movement of the charge carriers. In metals, for example, electrons travel from the negative terminal towards the positive terminal, but this flow of electrons is in the opposite direction to the conventional current (Figure 4).

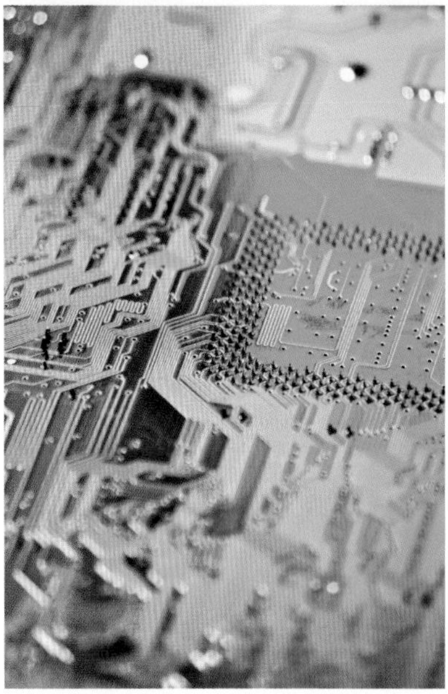

▲ **Figure 3** *Modern processors contains billions of connections, each one allowing a tiny flow of charge to pass through it. In 1985 the conductors inside chips were 1500 nm across, whereas in 2014 this size has fallen to just 14 nm (only a few tens of atoms across)*

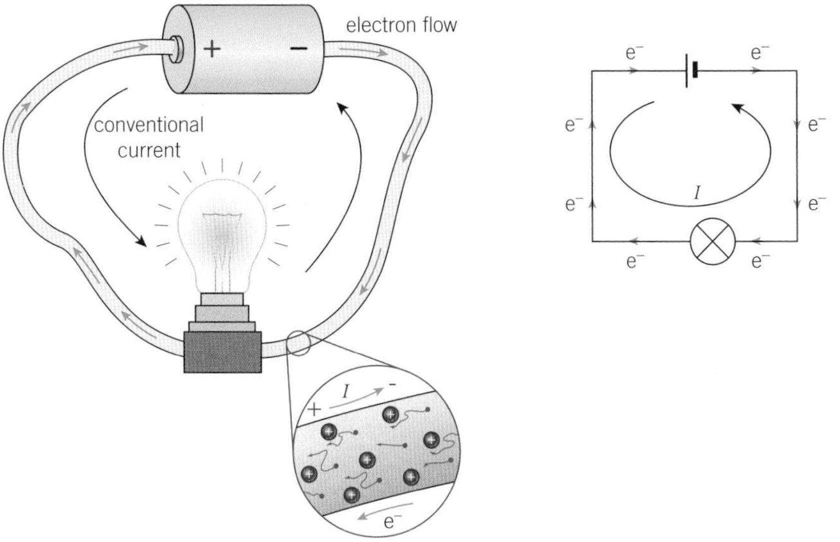

▲ **Figure 4** *Conventional current is always written from positive to negative, but in metals the electrons move the other way*

Electric current in electrolytes

It is not only metals which can conduct electricity. Several liquids can also conduct. Liquids that can carry an electric current are called **electrolytes**. In these cases the electric current is not a flow of electrons but a flow of ions. All electrolytes are either molten ionic compounds or, more commonly, **ionic solutions**. Pure water is an excellent insulator, but water from taps and rivers is an ionic solution – it contains dissolved ions, so can carry an electric current.

A common example of an ionic solution is salt (sodium chloride, NaCl) dissolved in water. The salt separates into positively charged sodium ions (**cations**), Na^+, and negatively charged chlorine ions (**anions**), Cl^- (Figure 5).

If a positive electrode (**anode**) and a negative electrode (**cathode**) are placed in the solution, ions are attracted to the electrodes. The Na^+ ions move towards the cathode and the Cl^- ions move towards the anode. This movement of ions is a flow of charge, an electric current.

Synoptic link

You will learn more about circuit diagrams in Topic 9.1, Circuit symbols.

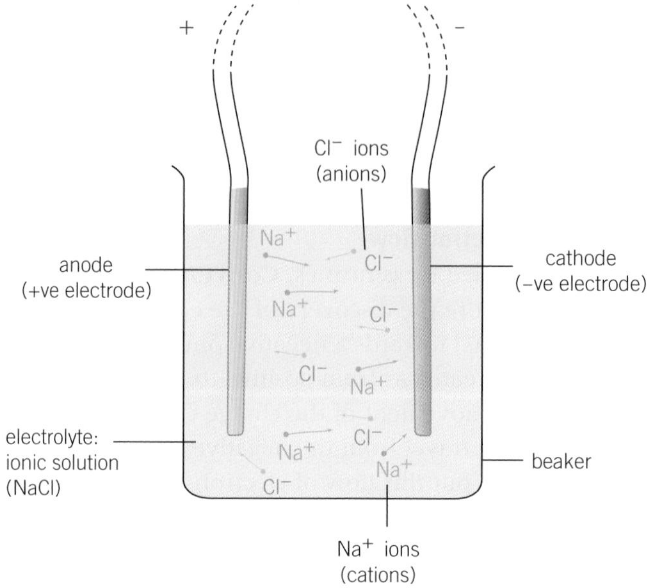

▶ **Figure 5** *In an electrolyte an electric current is a flow of ions*

When the Na^+ ions reach the cathode, they accept an electron, and when the Cl^- ions reach the anode they donate an electron, so electrons can flow through the metal part of the circuit.

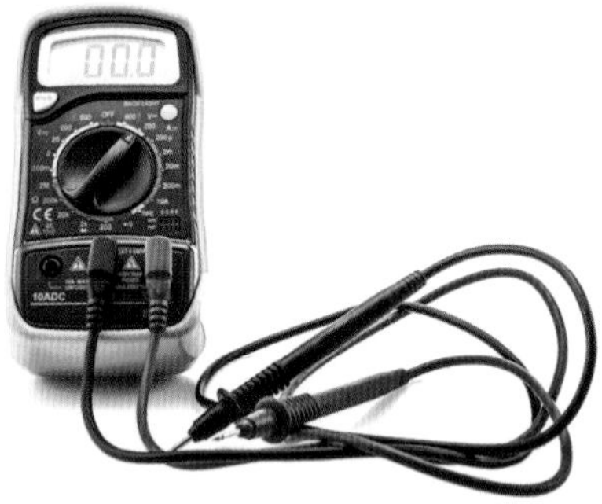

▲ **Figure 6** *Ammeters come in many different forms – modern ammeters are often one function of a digital multimeter*

Measuring electric current

An **ammeter** is used to measure the electric current at any point in a circuit. It is always placed directly in series in the circuit at the point where you want to measure the current.

As ammeters are placed in series they should have the lowest possible **resistance** in order to reduce the effect they have on the current – an ammeter with high resistance would decrease the current it should be measuring. The ideal (perfect) ammeter has zero resistance, and so has no effect on the current it measures.

Summary questions

1 Outline the difference between conventional current and electron flow in a metal wire. *(2 marks)*

2 Describe the similarities and differences between electric current in a metal wire and in an ionic solution. *(4 marks)*

3 A solution of copper sulfate contains Cu^{2+} and SO_4^{2-} ions. Sketch a diagram to show how an electric current is carried in a solution of copper sulfate. *(3 marks)*

4 Compare the direction of conventional current with the movement of the anions and cations in an electrolyte. *(2 marks)*

5 A solution of magnesium chloride contains Mg^{2+} and Cl^- ions. Calculate the current in the cathode in μA when 6.0×10^{14} cations come into contact with the cathode in 3.0 minutes. *(3 marks)*

Synoptic link

You will learn more about resistance in Topic 9.4, Resistance.

Conservation – it's the law

The large circle in Figure 1 is the outline of the Large Hadron Collider (LHC) near Geneva in Switzerland. The LHC is the most powerful particle accelerator ever built and forms part of the world's largest laboratory. In it, beams of subatomic particles are made to collide in order to replicate the conditions just after the Big Bang.

Physicists at the LHC analyse the data from each collision, applying several conservation laws. Each conservation law states that a particular, measurable physical quantity does not change. These laws are a cornerstone of physics, and include the conservation of energy, of linear momentum, and of charge. Using these laws, physicists are able to demonstrate the existence of exotic particles that vanish after a fraction of a second, including the Higgs boson.

Learning outcomes

Demonstrate knowledge, understanding, and application of:

→ Kirchhoff's first law and the conservation of charge.

▲ **Figure 1** *The Large Hadron Collider is a circular particle accelerator with a circumference of 27 km buried 100 m below the French–Swiss border*

Conservation of charge

In Topic 8.1, Current and charge, we described charge as a physical property (a measure of 'chargedness'). We don't know why some particles have charge and others do not, but if they didn't our universe could not exist in anything like its current form.

Charge is a fundamental physical property and one of only a handful of properties that must be conserved. That is to say in any interaction the total charge before and after must be the same. **Conservation of charge** states that electric charge can neither be created nor destroyed. The total amount of electric charge in the universe is constant.

Kirchhoff's first law

Gustav Kirchhoff was a German physicist born in 1824. He made many valuable contributions to science, including working with Robert Bunsen, who invented the burner, to discovering caesium and rubidium. He also studied electricity.

Kirchhoff's first law deals with electric current. It states that, for any point in an electrical circuit, the sum of currents into that point is equal to the sum of currents out of that point.

Synoptic link

You will learn more about Kirchhoff's second law in Topic 10.1, Kirchhoff's laws and circuits.

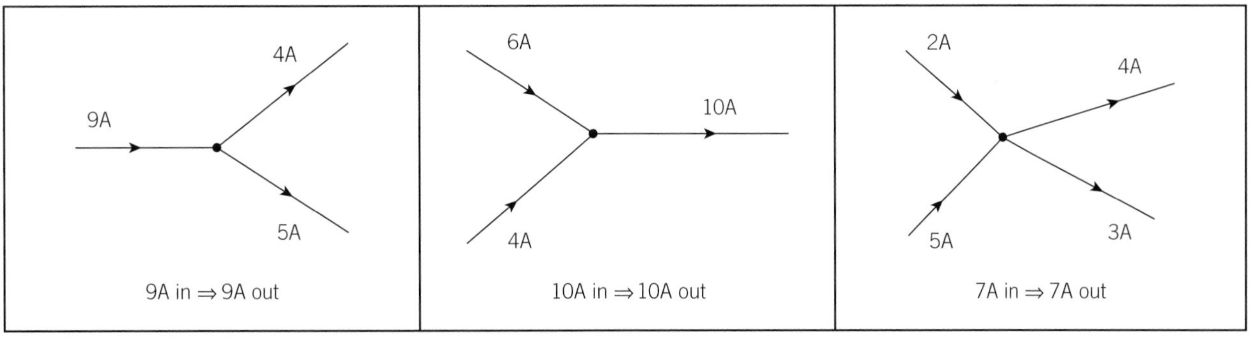

▲ **Figure 2** *Charge must be conserved at all points in a circuit*

The law can be written as

$$\Sigma I_{in} = \Sigma I_{out}$$

where Σ (Greek sigma) denotes 'sum of'. ΣI_{in} is the sum of the current into a point and ΣI_{out} is the sum of the current out of that point.

The law is based on conservation of charge, where the charge (measured in coulombs) is the product of the current (in amperes) and the time (in seconds). Charge cannot be destroyed, so the charge carriers entering a point in a given time must equal the total number of charge carriers leaving that same point during that time.

Summary questions

1 Explain the meaning of conservation of charge. *(2 marks)*

2 Draw a diagram to explain and illustrate an example of Kirchhoff's first law. *(2 marks)*

3 Copy and complete the diagrams below, stating both the magnitude and the direction of the missing electric currents. *(6 marks)*

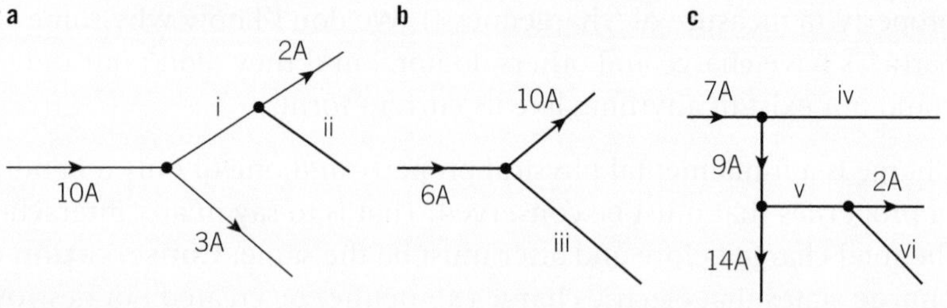

▲ **Figure 3**

4 A single wire is connected to two other wires, A and B. If the current in the single wire is 15 A, and 1.9×10^{21} electrons pass along wire A in 1.0 minutes, calculate the current in wire B to the nearest ampere. *(3 marks)*

5 Discuss how the idea of conservation of charge relates to the quantisation of charge studied in Topic 8.1, Current and charge. *(3 marks)*

6 Suggest how conservation of charge might be used to determine the existence of a short-lived, negatively charged particle created inside a particle accelerator when two protons collide together. *(2 marks)*

A miracle material?

Graphene is pure carbon, arranged in thin sheets just one atom thick. It is incredibly strong and at the same time extremely flexible.

Its unusual composition gives graphene a huge number of free electrons. The number of free charge carriers per unit volume (the **number density**) of graphene is even greater than that of copper, making it an outstanding electrical conductor.

Classification of materials

The number density is the number of free electrons per cubic metre of material. The higher the number density, the greater the number of free electrons per m^3 and so the better the electrical conductor.

We can classify materials into three groups according to their number density. Conductors have a very high number density (of the order of $10^{28}\,m^{-3}$), insulators have a much lower value with **semiconductors** in between the two, with number densities around $10^{17}\,m^{-3}$.

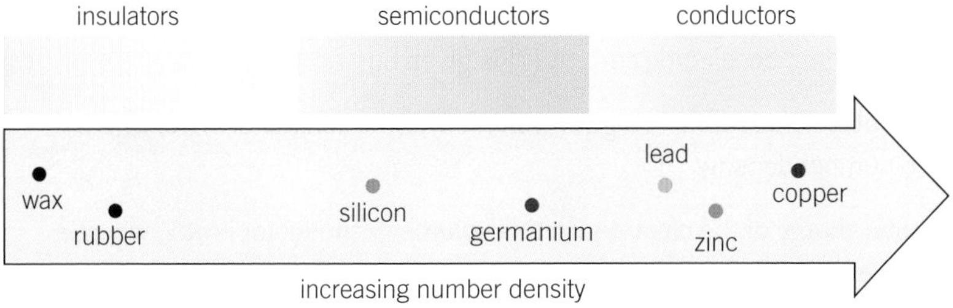

▲ **Figure 2** *The number density of a material determines how easily the material will conduct an electric current*

▼ **Table 1** *Conduction and number density*

Material	Type	n /m^{-3} (at 300 K)
copper	conductor	8.5×10^{28}
zinc	conductor	6.6×10^{28}
germanium	semiconductor	2.0×10^{18}
silicon	semiconductor	8.7×10^{15}

Semiconductors have a much lower number density than metals, so in order to carry the same current the electrons in semiconductors need to move much faster. This increases the temperature of the semiconductor, a fact relevant in the design of computers, which use processors made of silicon (Figure 3).

Learning outcomes

Demonstrate knowledge, understanding, and application of:

→ mean drift velocity of charge carriers

→ the equation $I = Anev$

→ distinction between conductors, semiconductors, and insulators in terms of n.

▲ **Figure 1** *Andre Geim and Konstantin Novoselov from the University of Manchester won the 2010 Nobel Prize in Physics for their work on graphene*

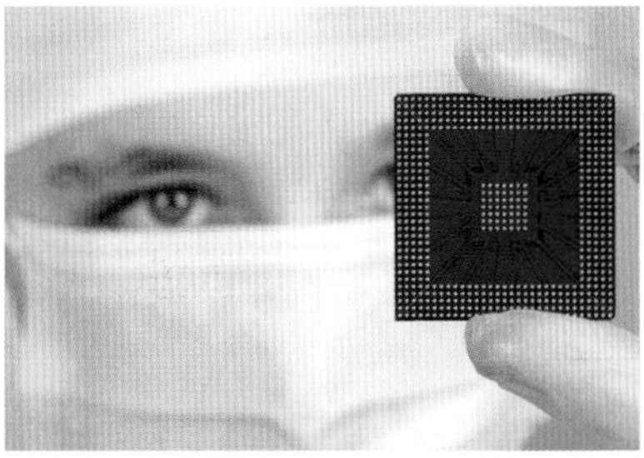

▲ **Figure 3** *Microprocessors are made largely of silicon, and can get very hot when a current passes through them, so computers need carefully designed cooling systems*

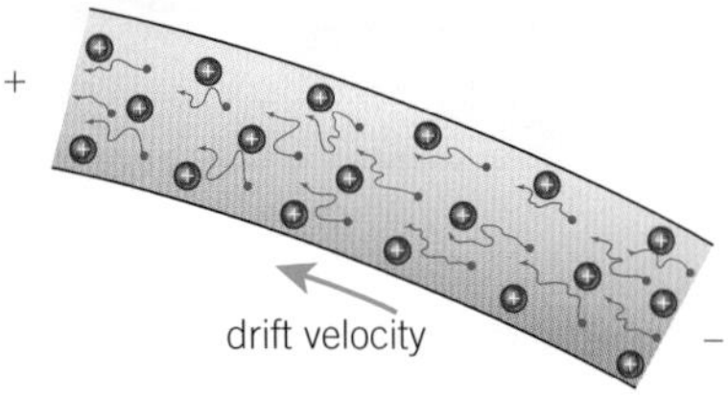

▲ **Figure 4** *Repeated collisions between positive ions and free electrons cause the random motion of the electrons*

How fast do charge carriers move?

When you flick a switch most filament lamps turn on almost instantly. You might imagine that electrons have rushed from the switch to the lamp, but this is not true. In fact, most charge carriers, like electrons, move slowly. Free electrons repeatedly collide with the positive metal ions as they drift through the wire towards the positive terminal. The reason that lights turn on so quickly is that all the free electrons in the wire start moving almost at once.

A new equation for electric current

There is an additional equation for electric current.

$$I = Anev$$

where I is the electric current in the conductor in amperes, A is the cross-sectional area of the conductor in m^2, e is the elementary charge (1.60×10^{-19} C), n is the number density, and v is the **mean drift velocity** of the charge carriers in $m\,s^{-1}$.

Derivation of $I = Anev$

We can derive the equation by considering the number density and dimensions of a conductor.

From its definition, electric current (I) is given by $I = \frac{\Delta Q}{\Delta t}$

The number of electrons in a given volume V of the conductor is nV, where n is the number density.

The total charge of the electrons in this volume of conductor is neV, where e is the elementary charge. This gives

$$I = \frac{neV}{\Delta t}$$

When there is an electric current in the conductor, a certain volume of charge carriers passes a given point each second. This volume depends on the cross-sectional area A of the conductor and the mean drift velocity v of the charge carriers.

$$\frac{V}{\Delta t} = Av$$

Substituting this into our previous equation for electric current gives

$I = \frac{neV}{\Delta t} = neAv$, more commonly written as

$$I = Anev$$

1. Show that the equation is homogenous with respect to base units.
2. Describe the effect on the mean drift velocity if all other factors are constant and:
   - **a** current increases;
   - **b** cross-sectional area decreases;
   - **c** number density doubles.

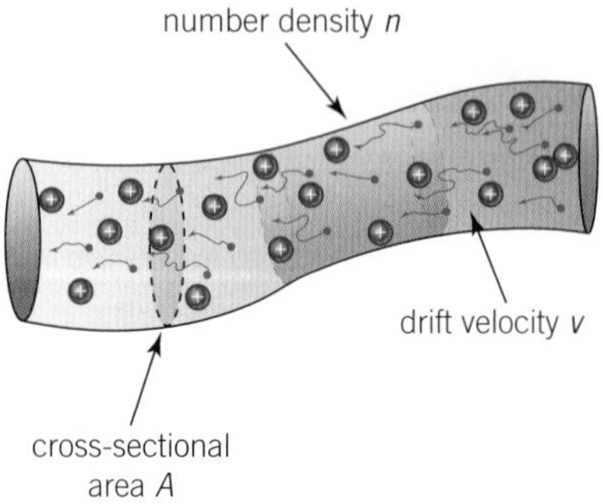

▲ **Figure 5** *$I = Anev$ can be derived by considering the properties of the wire and the free electrons*

Worked example: Calculating mean drift velocity

A copper wire has a cross-sectional area of $7.85 \times 10^{-7}\,\mathrm{m}^2$. The number density of copper is $8.50 \times 10^{28}\,\mathrm{m}^{-3}$. Calculate the mean drift velocity of the electrons through the wire when the current is 1.40 A.

Step 1: Identify the equation needed.

$$I = Anev$$

Rearrange to make v the subject. $v = \frac{I}{Ane}$

Step 2: Substitute the known values in SI units (including the cross-sectional area in m^2) into the equation and calculate the answer.

$$v = \frac{1.40}{7.85 \times 10^{-7} \times 8.50 \times 10^{28} \times 1.60 \times 10^{-19}}$$

$$v = 1.31 \times 10^{-4}\,\mathrm{m\,s^{-1}}\ \text{(3 s.f.)}$$

This is very slow: only $0.13\,\mathrm{mm\,s^{-1}}$. The electrons have such a low speed because of the large number of random collisions with the fixed positive ions.

The effect of changing cross-sectional area

If the cross-sectional area of a wire changes, so must the drift velocity. The narrower the wire, the greater the drift velocity must be in order for the current to be the same.

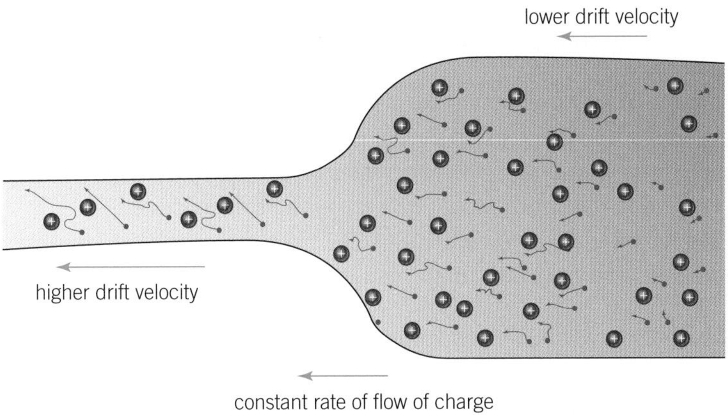

▲ **Figure 6** *In order to maintain the same current (rate of flow of charge), electrons must move faster through narrower wires – the mean drift velocity is inversely proportional to the cross-sectional area of the wire*

Worked example: Getting thinner

A piece of wire carrying a current narrows such that its radius halves. What effect does this have on the mean drift velocity of the electrons in the wire?

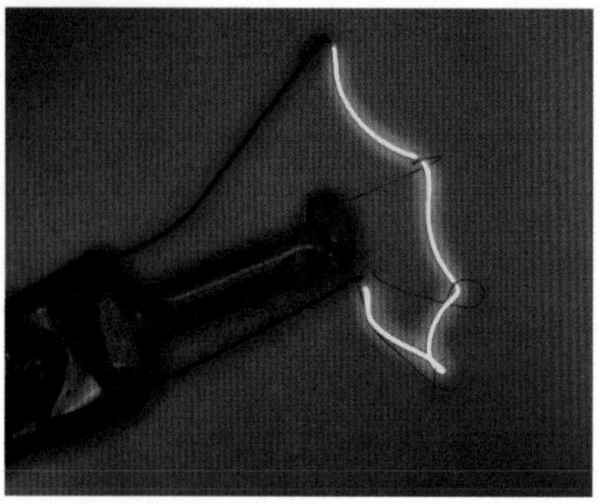

▲ **Figure 7** *The filament of an incandescent bulb is much narrower than the surrounding wire, so electrons move faster through this part of the wire, heating it up so much that it glows white*

Step 1: Identify the equation that relates v and A.

$$I = Anev$$

Rearrange to make v the subject. $v = \frac{I}{Ane}$

Step 2: Examine the effect of changing A.

According to Kirchhoff's first law, the current in the wire is the same. The elementary charge e and the number density n are also constants, and n does not change because it is the same material.

Therefore, $v \propto \frac{1}{A}$

If the radius halves the cross-sectional area will decrease by a factor of 4.

As a result the mean drift velocity must increase by a factor of 4.

Summary questions

1 Describe the meaning of the terms 'conductor', 'semiconductor', and 'insulator' in terms of their relative number densities. *(2 marks)*

2 Calculate the current through a copper wire with a cross-sectional area of $5.50 \times 10^{-8}\,m^2$, when the mean drift velocity of the charge carriers is $2.0 \times 10^{-3}\,m\,s^{-1}$. Use the values for n and e in the first worked example above. *(3 marks)*

3 A silver wire has a cross-sectional area of $7.10 \times 10^{-6}\,m^2$ and a number density of $5.86 \times 10^{28}\,m^{-3}$. Calculate the mean drift velocity of the electrons through the wire when the current is 500 mA. *(3 marks)*

4 State and explain the effect on the mean drift velocity of electrons if a constant current travels:

a into the same wire with a larger cross-sectional area; *(2 marks)*

b from a zinc wire into a copper wire of the same dimensions (see Table 1); *(2 marks)*

c into the same wire with $\frac{1}{3}$ of the radius. *(3 marks)*

5 A zinc wire has a diameter of 1.0 mm. Calculate the mean drift velocity of the electrons through the wire when the current through it is 3.0 mA. *(3 marks)*

6 A small piece of semiconducting material with a cross-sectional area of $8.2 \times 10^{-6}\,m^2$ is used in an experiment. When the current in the material is 12 mA, the mean drift velocity of the charge carriers is $72\,m\,s^{-1}$. Determine the ratio of the number density in the semiconductor and the number density of copper. *(4 marks)*

Practice questions

1 a State the base units for electrical charge. (*2 marks*)

b Figure 1 shows a simple electrical circuit constructed by a student.

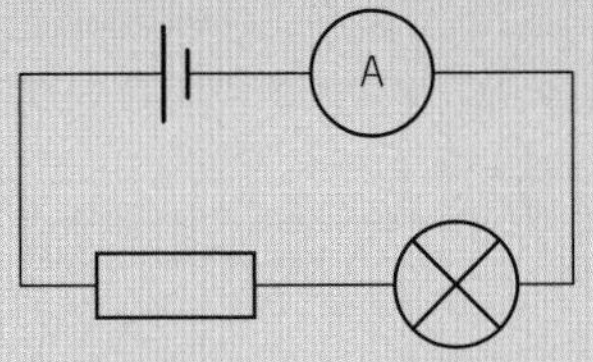

▲ Figure 1

The ammeter reading is 20 mA.

(i) State the direction of conventional current. (*1 mark*)

(ii) Calculate the number of electrons entering the resistor per second. (*3 marks*)

(iii) State and explain the value of the current in the lamp. (*2 marks*)

Jan 2011 G482

2 a A 12 V car battery contains an electrolyte. The battery is connected to an electric motor **M**. There is a current in the motor and the battery. See Figure 2.

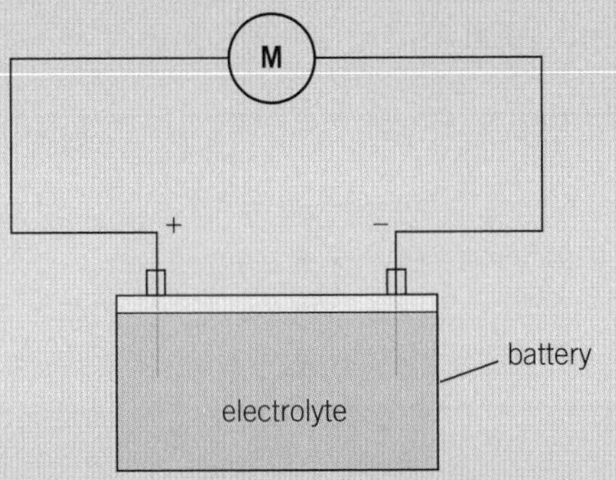

▲ Figure 2

State

(i) the charge carriers in the electrolyte, (*1 mark*)

(ii) the charge carriers moving through the electrolyte to the positive terminal of the battery, (*1 mark*)

(iii) the charge carriers moving through the wires to the positive terminal of the battery. (*1 mark*)

Jan 2011 G482

3 A resistor of length l is connected to a cell, see Figure 3.

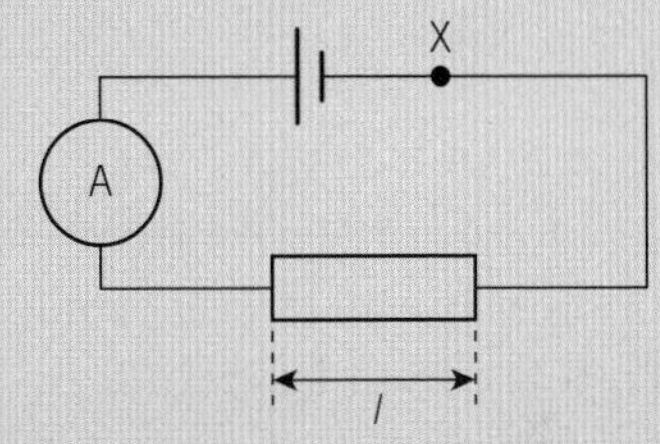

▲ Figure 3

a State the direction of the flow of electrons in the resistor. (*1 mark*)

b In a time of 2.0 minutes, the number of electrons passing through point **X** in the wire is 1.88×10^{20}. Calculate

(i) the charge flow through point **X**, (*2 marks*)

(ii) the current in the wire. (*2 marks*)

c On a copy of the axes of Figure 4, sketch a graph to show the variation of current I with the distance from one end of the resistor. (*2 marks*)

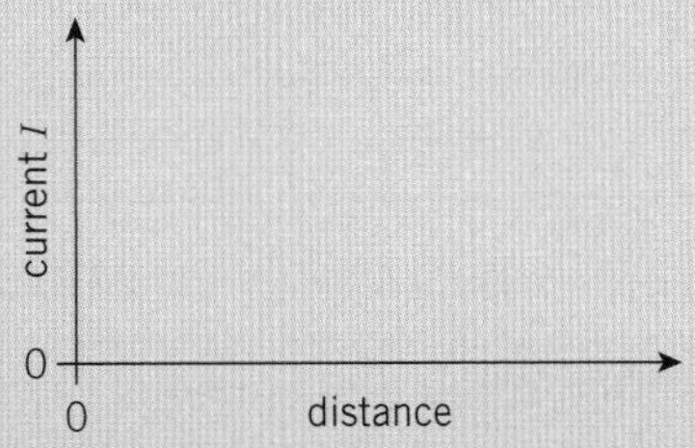

▲ Figure 4

4 Figure 5 shows an electrical circuit with three resistors **A**, **B** and **C**.

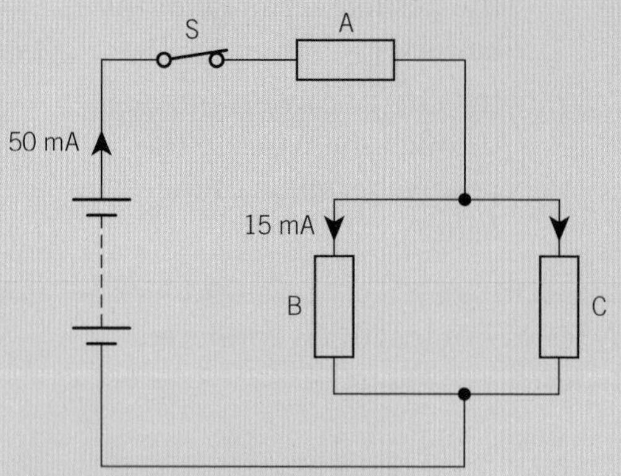

▲ **Figure 5**

The circuit shows the currents at different points in the circuit when the switch **S** is closed for a time of 30 s. Calculate

a the total charge flow into the resistor **B**, *(2 marks)*

b the number of electrons responsible for the charge in (a), *(2 marks)*

c the current in the resistor **C**. *(1 mark)*

5 a State Kirchhoff's first law and state the quantity conserved according to this law. *(2 marks)*

b Figure 6 shows part of an electrical circuit.

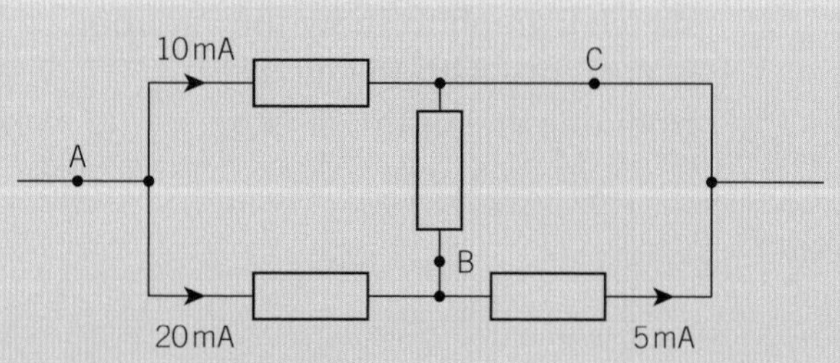

▲ **Figure 6**

(i) Determine the currents at points **A**, **B** and **C**. *(3 marks)*

(ii) State the direction of the current at **B**. *(1 mark)*

6 a Explain the term *mean drift velocity* of electrons in a metal wire. *(3 marks)*

b Figure 7 shows a resistor made from a conducting material deposited onto a plastic base.

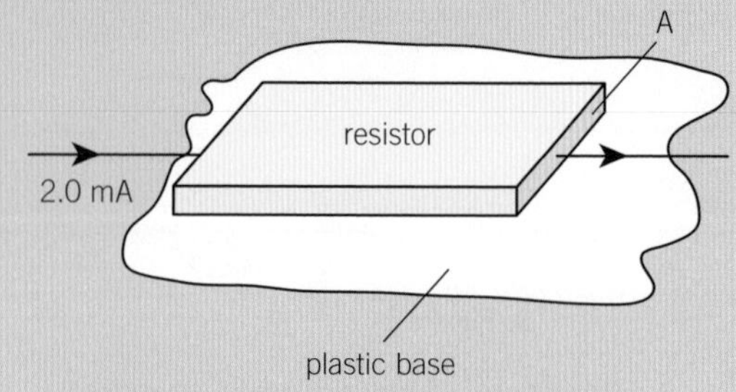

▲ **Figure 7**

The cross-sectional area *A* of the resistor is $4.2 \times 10^{-10}\,m^2$. The material used for the resistor has 8.0×10^{27} free electrons per unit volume. The current in the resistor is 2.0 mA.

(i) Calculate the mean drift velocity of the electrons inside the resistor. *(3 marks)*

(ii) The length of the resistor is 8.0 mm. Calculate the time taken for an electron to drift along the length of this resistor. *(1 mark)*

7 Figure 8 shows a lightning strike between a storm cloud and a tall building.

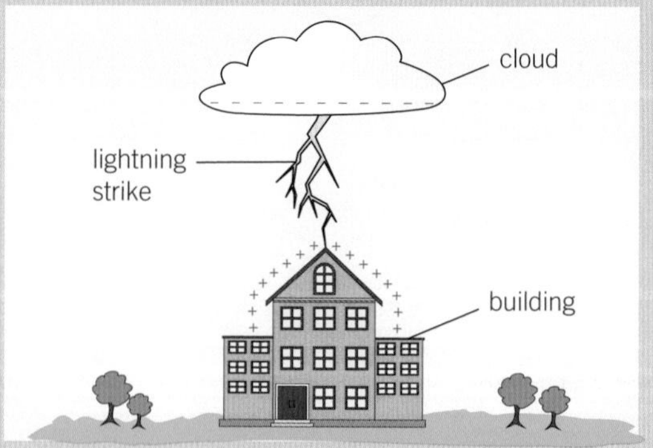

▲ **Figure 8**

a The bottom of the cloud has a negative charge and the building has a positive charge.

Indicate the direction of the conventional current in the lightning strike in Figure 8. (*1 mark*)

b The current in the lightning strike is 8200 A and the strike lasts for 120 ms.

(i) Calculate the total number of electrons transferred between the cloud and the building. (*3 marks*)

(ii) On a copy of the axes below, sketch a graph of charge Q transferred against time t. Explain the shape of the graph. (*3 marks*)

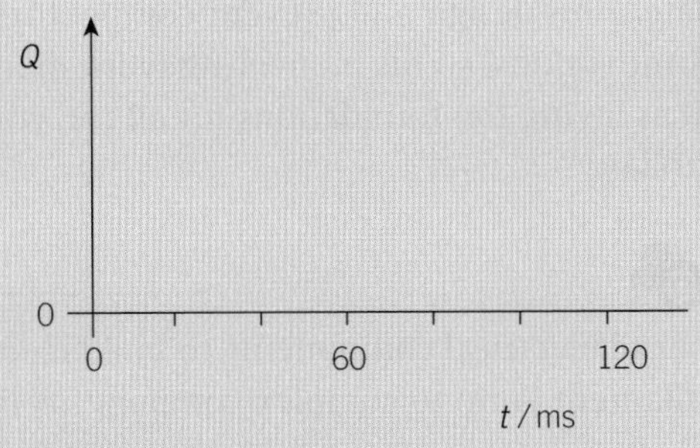

▲ **Figure 9**

8 a Explain what is meant by *electric current.* (*1 mark*)

b Determine the unit of charge in SI base units. (*1 mark*)

c Figure 10 shows a negatively charged metal sphere attached to a plastic rod.

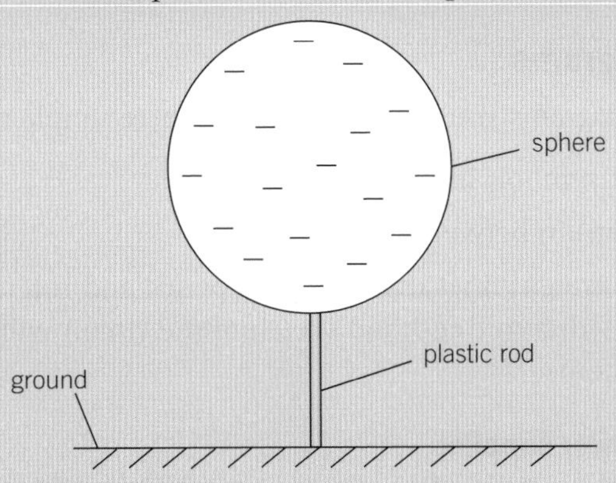

▲ **Figure 10**

The charge on the sphere is -5.4×10^{-9} C. The charge slowly leaks away through the rod.

This leakage current is 0.20 pA. Estimate the time in hours it would take for the sphere to be completely discharged. (*3 marks*)

9 a Explain what is meant by the *mean drift velocity* of electrons in a current-carrying wire. (*2 marks*)

b Figure 11 shows two wires, A and B, connected in series to a supply.

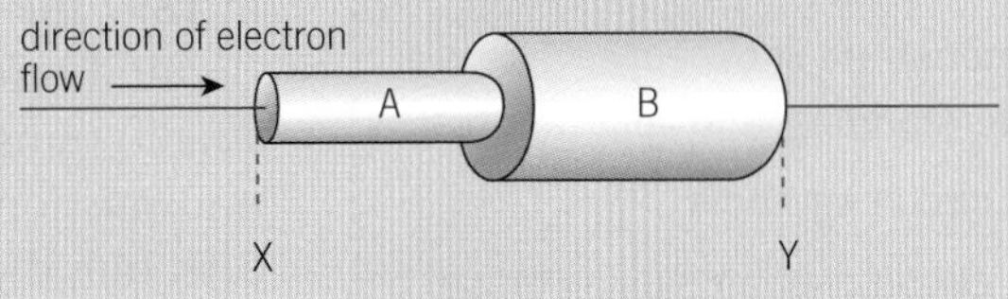

▲ **Figure 11**

Describe and explain the variation of the mean drift velocity of the electrons from X to Y. (*3 marks*)

c The current in a filament lamp is 0.30 A. The filament has a diameter of 5.0×10^{-5} m. The material of the filament has 3.4×10^{28} free electrons per unit metre.

Calculate the mean drift velocity of the electrons in the filament. (*3 marks*)

9 ENERGY, POWER, AND RESISTANCE

9.1 Circuit symbols

Specification reference: 4.2.1

Learning outcomes

Demonstrate knowledge, understanding, and application of:

→ circuit symbols

→ circuit diagrams using these symbols.

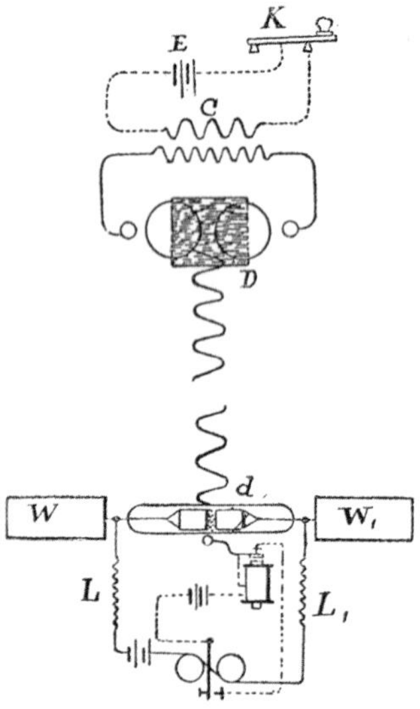

▲ **Figure 1** *This is the sketch Marconi submitted to the Patent Office, with some symbols we still use today*

Cornwall calling

In 1897 the Italian physicist Guglielmo Marconi submitted this circuit diagram to the UK Patent Office (Figure 1). It is a design for a radio transmitter that, a few years later, would send the first radio transmission across the Atlantic Ocean from Cornwall to Canada, and earn him the 1909 Nobel Prize in Physics. Using clearly defined circuit symbols he was able to patent his design and prevent competitors from simply copying his innovations, so his pioneering work made him very wealthy, as well as laying the foundations for all the wireless communications we use today.

Circuit diagrams

We can use a huge number of electrical components to build electrical circuits, from the simple filament lamp to complex components like the **capacitor** or diode.

When drawing circuits it is important to use an internationally recognised set of symbols to represent components, so that scientists and engineers around the world can construct similar circuits to test and verify claims made by anyone working in this field. Every type of component has its own unique circuit symbol – for example, all capacitors have the same symbol, no matter what size or colour they are. Figure 2 contains all the symbols we will require in A Level Physics.

Rules for circuit diagrams

When drawing circuit diagrams you should remember three simple rules.

1 Only use the circuit symbols in Figure 2.

2 Do not leave any gaps in between the wires.

3 When possible use straight lines drawn with a pencil and ruler. However, a carefully drawn free-hand sketch of the circuit will be adequate in assessments.

Cells, batteries, and power supplies

Most circuits will contain a single **cell**, a number of cells, or a mains power supply. A '**battery**' in physics means two or more cells connected end-to-end, or in **series**. In the case of a single cell or a battery, the longer terminal represents the positive terminal. When using a power supply, a small plus sign is often placed next to the positive terminal. Take care to ensure that this **polarity** is represented correctly when you use these symbols, because polarity is very important when using components such as diodes and light-emitting diodes.

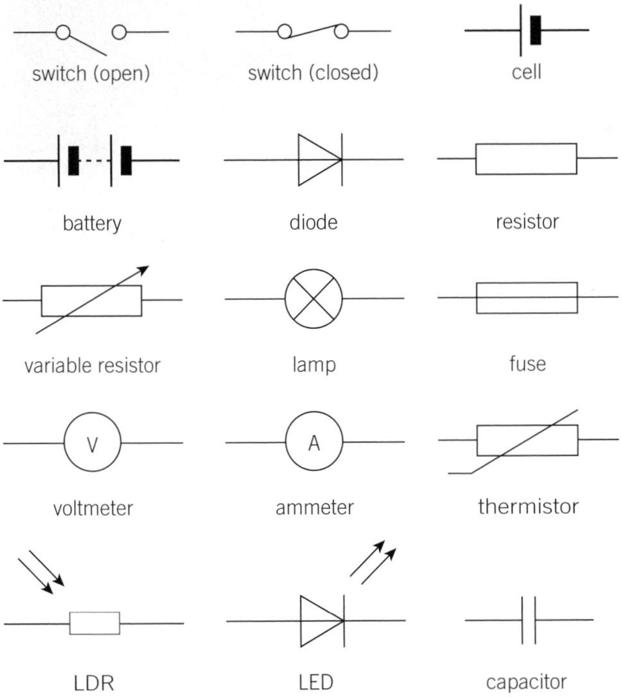

▲ **Figure 2** *Circuit symbols from ASE publication: Signs, symbols, and systematics (The ASE Companion to 16–19 Science, 2000)*

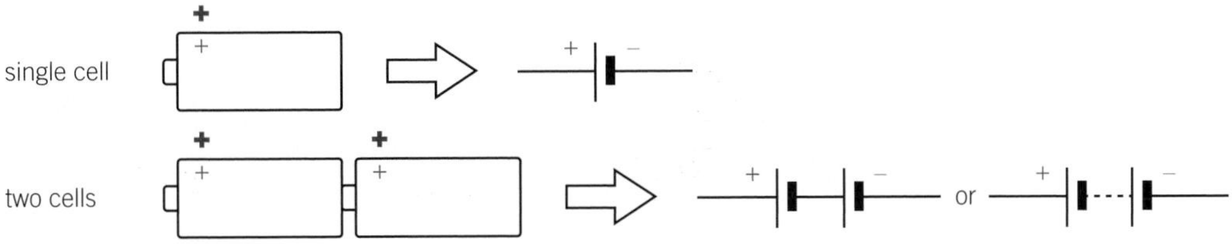

▲ **Figure 3** *You must carefully consider the polarity of a cell when building circuits or drawing circuit diagrams*

Summary questions

1 Name the components represented by the symbols in Figure 4. *(3 marks)*

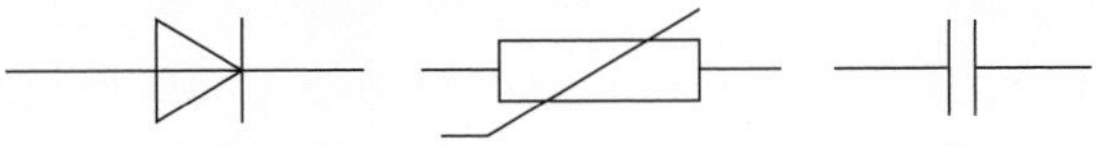

▲ **Figure 4**

2 In each case, draw a circuit diagram using the correct symbols.

a A single cell connected in series with a filament lamp. *(2 marks)*

b A battery connected to a resistor and an ammeter in series. *(3 marks)*

c A power supply connected to two resistors connected in series. *(2 marks)*

3 Draw a circuit containing a cell, an open switch, and a filament lamp. Show a voltmeter connected across the lamp (in parallel). *(2 marks)*

4 Identify two errors in the circuit diagram in Figure 5. *(2 marks)*

5 Draw the circuit diagram for the circuit shown in Figure 6. *(3 marks)*

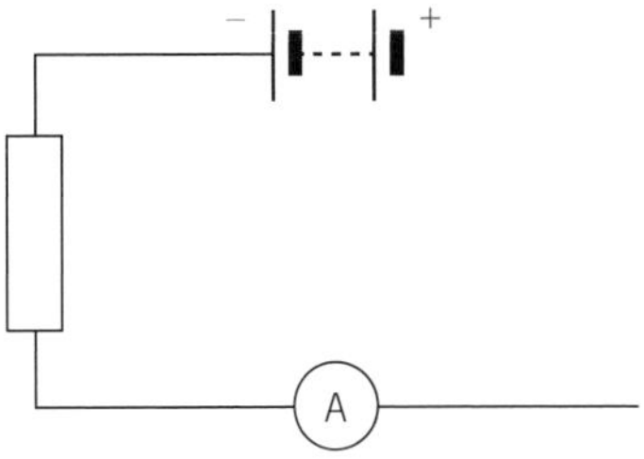

▲ **Figure 5**

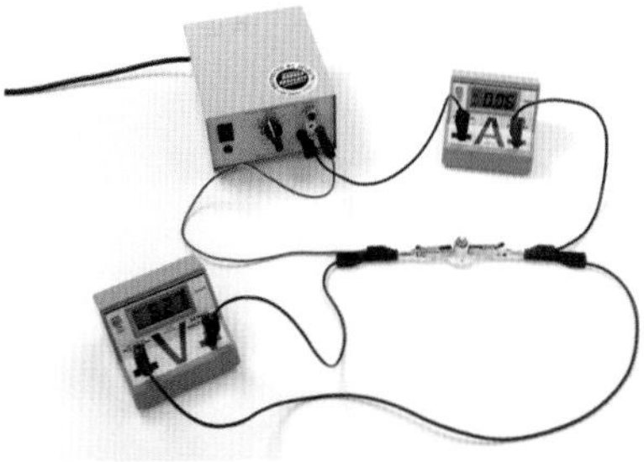

▲ **Figure 6**

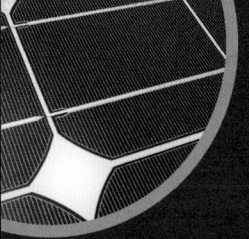

9.2 Potential difference and electromotive force

Specification reference: 4.2.2

Learning outcomes

Demonstrate knowledge, understanding, and application of:

- → potential difference and the volt
- → electromotive force
- → distinction between e.m.f. and p.d.
- → energy transfer $W = VQ$ and $W = \mathcal{E}Q$.

Potential difference

A Tesla coil in action is a truly impressive sight (Figure 1). It produces artificial lightning in the laboratory by generating a huge **potential difference** (p.d.), also called voltage. Potential difference is a measure of the transfer of energy by charge carriers. The p.d. across a component like a filament lamp is a result of electrical energy being transferred into heat and light as charge carriers move through the lamp. High voltages can be very dangerous because charge carriers can transfer enormous amounts of energy through conductors and, if the voltage is high enough, through insulators like air, or people.

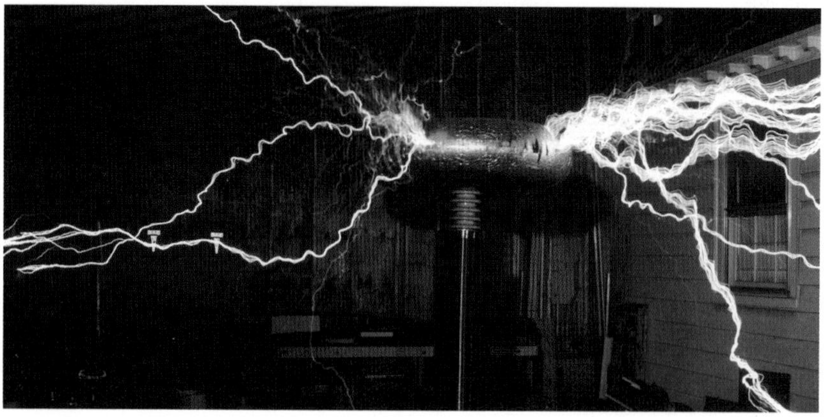

▲ **Figure 1** *A Tesla coil, like this one, can generate p.d.s in excess of 500 000 V. The coil is named after its inventor, the physics genius Nikola Tesla*

Study tip

Think of one volt as one joule per coulomb of charge.

The volt

Potential difference is measured in volts, named after Alessandro Volta, an Italian who invented the first battery.

One **volt** is the p.d. across a component when 1 J of energy is transferred per unit charge passing through the component.

$$1\,\text{V} = 1\,\text{J}\,\text{C}^{-1}$$

A p.d. of 1000 V means that 1000 J of energy is transferred per coulomb of charge. The term potential difference is used when charged particles lose energy in a component. Potential difference is defined as the energy transferred from electrical energy to other forms (heat, light, etc.) per unit charge. The equation for potential difference V is.

$$V = \frac{W}{Q}$$

where V is p.d. measured in volts, Q is charge in coulombs, and W is the energy transferred by charge Q.

The voltmeter

A **voltmeter** is used to measure p.d. Like ammeters, voltmeters come in all different shapes and sizes, but they are always connected in parallel across a particular component.

An ideal voltmeter should have an infinite resistance (see Topic 9.4, Resistance), so that when connected, no current passes through the voltmeter itself. Whilst this is not possible in reality, most voltmeters have a resistance of several million ohms.

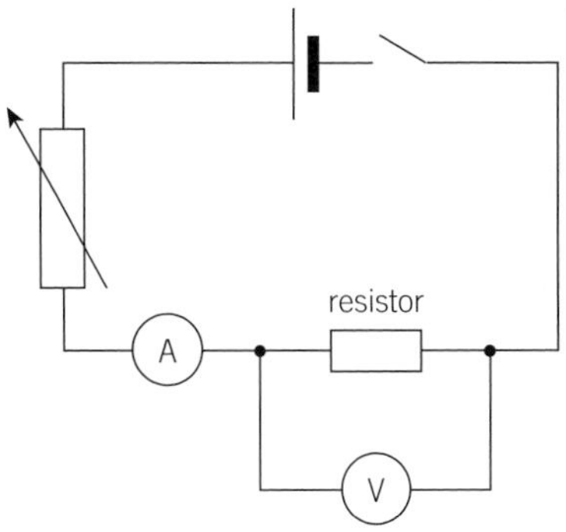

▲ **Figure 2** *Voltmeters are used to measure p.d. and are always connected in parallel*

Electromotive force

At times it is necessary to describe whether the charges in a circuit are losing or gaining energy. Two different terms are used depending on whether the charge carriers are doing work or work is being done on them.

Potential difference is used to describe when work is done *by* the charge carriers. Essentially the charges are losing energy as they pass through the component. The greater the p.d., the more energy per coulomb is transferred from electrical energy into other forms (like light or heat) as the charges move through the component.

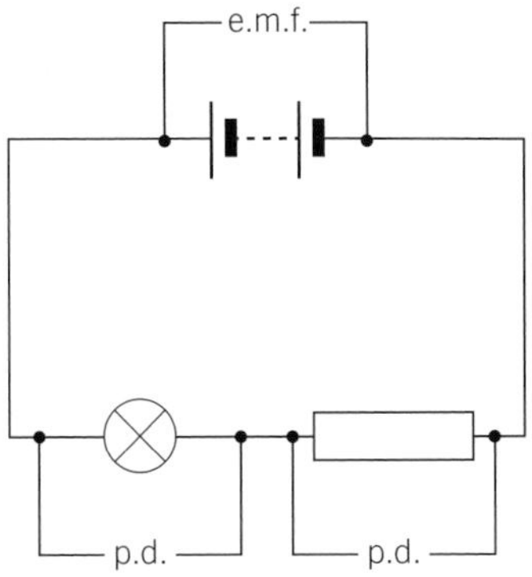

▲ **Figure 3** *There is an e.m.f. across the cell, and p.d.s across the other components*

Electromotive force (e.m.f.) is used to describe when work is done *on* the charge carriers. Essentially the charges are gaining energy as they pass through a component like a cell, battery, or power pack. The greater the e.m.f., the more energy per coulomb has been transferred (often from the form of chemical energy in a cell) into electrical energy. Other sources of e.m.f. include solar cells (from light), dynamos (movement), and thermocouples (heat).

The term electromotive force is used when charged particles gain energy from a source. Electromotive force is defined as the energy transferred from chemical energy (or another form) to electrical energy per unit charge. The equation for electromotive force $\mathcal{E}$ is

$$\mathcal{E} = \frac{W}{Q}$$

where $\mathcal{E}$ is e.m.f. measured in volts, Q is charge in coulombs, and W is the energy transferred by charge Q.

Study tip

Despite its name, e.m.f. is not a force, which is measured in newtons, whereas e.m.f. is measured in volts.

Calculating energy transfer

Whether it is a p.d. or an e.m.f., the energy transferred from or to the charges can be calculated from the defining equations. The energy transferred depends on the size of the p.d. and the charge passing through the component.

$$W = VQ \quad \text{or} \quad W = \varepsilon Q$$

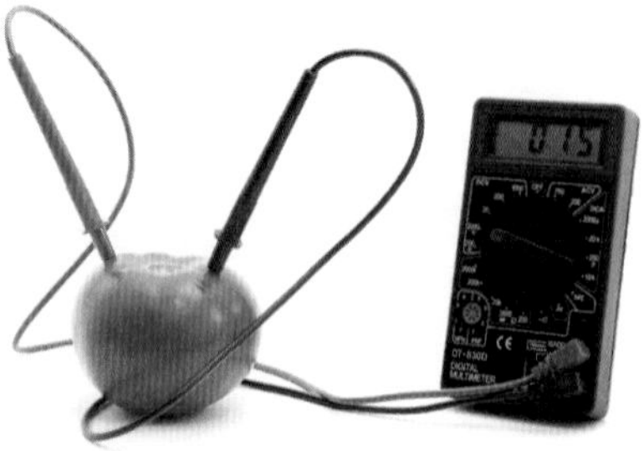

▲ **Figure 4** *A chemical reaction between the acid in the tomato and the different metal electrodes produces an e.m.f.*

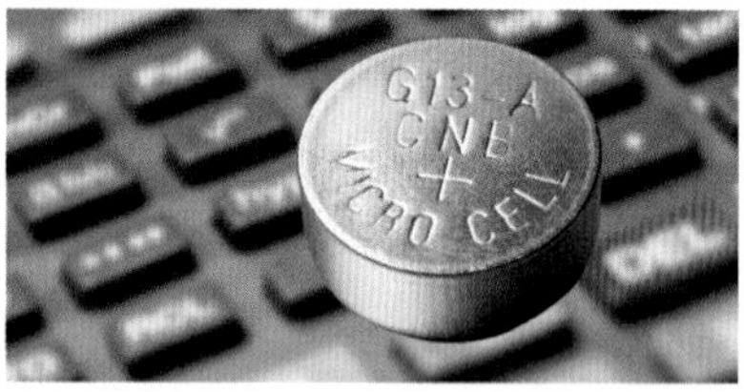

▲ **Figure 5** *A small button cell like the ones found in calculators contains a mixture of chemicals that react together to produce an e.m.f., transferring chemical energy into electrical energy*

Worked example: Energy transferred by an electrical heater

Calculate the energy transferred in 4.0 hours by an electrical heater that draws a current of 15 A when the p.d. across it is 90 V.

Step 1: Identify the correct equation to calculate the charge through the heater in that time.

$$\Delta Q = I\Delta t$$

Step 2: Substitute in the known values in SI units (time in seconds) and calculate the charge in coulombs.

$$\Delta Q = 15 \times 60 \times 60 \times 4.0$$

$$\Delta Q = 216\,000\,\text{C}$$

Step 3: Identify the correct equation to calculate the energy transferred.

$$W = VQ$$

Step 4: Substitute in known values in SI units and calculate the energy transferred in joules.

$$W = 90 \times 216\,000$$

$$W = 19\,\text{MJ (2 s.f.)}$$

Summary questions

1 Sketch a circuit diagram to show how a voltmeter may be used to measure the p.d. across a filament lamp. *(2 marks)*

2 Outline the difference between p.d. and e.m.f. *(2 marks)*

3 Calculate the energy transferred to 4.0 C of charge by a cell with an electromotive force of 80 V. *(2 marks)*

4 Calculate the p.d. across a filament lamp when 168 J of energy is transferred by the lamp by 14 C of charge. *(2 marks)*

5 Define the volt and express it in base units. *(4 marks)*

6 A resistor has a current through it of 500 mA and a p.d. across it of 1.0 kV. Calculate the energy transferred to the resistor in 6.0 hours. *(4 marks)*

Phasers set to stun?

Despite its name, an **electron gun** is not a sci-fi weapon, but an electrical device used to produce a narrow beam of electrons. These electrons can be used to ionise particles by adding or removing electrons from atoms, and they can have very precisely determined kinetic energies. Electron guns are used in scientific instruments such as electron microscopes, mass spectrometers, and oscilloscopes.

Learning outcomes

Demonstrate knowledge, understanding, and application of:

→ energy transfer $eV = \frac{1}{2}mv^2$ for electrons and other charged particles.

How does it work?

All electron guns need a source of electrons. In most cases a small metal filament is heated by an electric current. The electrons in this piece of wire gain kinetic energy. Some of them gain enough kinetic energy to escape from the surface of the metal. This process is called **thermionic emission** – the emission of electrons through the action of heat.

If the heated filament is placed in a vacuum and a high p.d. applied between the filament and an anode, the filament acts as a cathode, and the freed electrons accelerate towards the anode, gaining kinetic energy. If the anode has a small hole in it, then electrons in line with this hole can pass through it, creating a beam of electrons with a specific kinetic energy (Figure 2).

▲ **Figure 1** *Old-style computer monitors and TV screens – cathode-ray tubes – use electron guns to produce images on the screen*

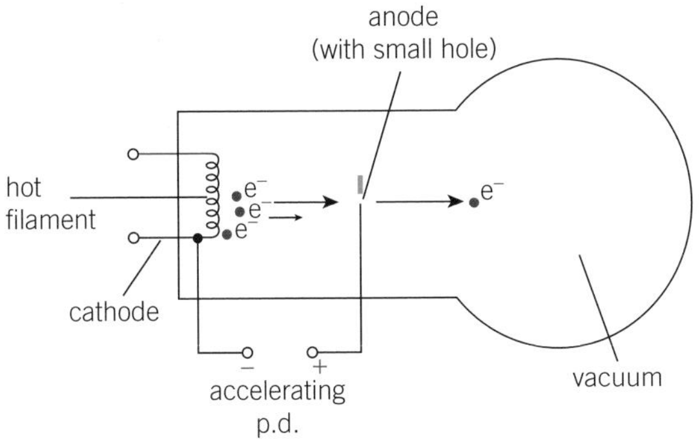

▲ **Figure 2** *An electron gun produces a narrow beam of electrons with a specific kinetic energy depending on the p.d. between the cathode and the anode*

Energy transfers

As the electrons accelerate towards the anode they gain kinetic energy. From the definition for p.d., the work done on a single electron travelling from the cathode to the anode is equal to eV, where e is the elementary charge, the charge on each electron, and V is the accelerating p.d.

By considering the law of conservation of energy we can derive an expression relating the work done on the electron to its increase in kinetic energy.

work done on electron = gain in kinetic energy

$$eV = \frac{1}{2}mv^2$$

This assumes the electrons have negligible kinetic energy at the cathode.

Changing the accelerating p.d. changes the kinetic energy of the electrons within the beam. The greater the p.d. the more energy is transferred to the electrons and so the faster they move.

Worked example: Calculating the velocity of an electron

Calculate the velocity of an electron from an electron gun with an accelerating p.d. of 4.0 kV.

Step 1: Select the equation relating the velocity of an electron and the p.d.

$$eV = \frac{1}{2}mv^2$$

Rearrange to make v the subject. $\sqrt{\frac{2eV}{m}} = v$

Step 2: Substitute in known values (including e and mass of an electron m) in SI units. $v = \sqrt{\frac{2 \times 1.60 \times 10^{-19} \times 4000}{9.11 \times 10^{-31}}}$

Calculate the velocity of the electron. $v = 3.7 \times 10^7\,\mathrm{m\,s^{-1}}$

Summary questions

1 Describe how an electron gun produces a beam of high-speed electrons. *(4 marks)*

2 Calculate the kinetic energy of an electron accelerated through a p.d. of 12 kV. *(3 marks)*

3 Calculate the velocity of an electron that has a kinetic energy of 1.8×10^{-15} J. *(2 marks)*

4 Calculate the magnitude of the accelerating p.d. that produces electrons with a velocity of 9% of the speed of light. (Speed of light $c = 3.00 \times 10^8\,\mathrm{m\,s^{-1}}$.) *(4 marks)*

5 State and explain how the velocity of an electron and a proton would compare when accelerated through the same p.d. *(3 marks)*

Particle accelerators

A linear particle accelerator (often shortened to LINAC) uses a series of cylindrical electrodes (drift tubes) to accelerate subatomic particles such as electrons. The polarity of the drift tubes is alternated between positive and negative with precise timing so that each time the electrons leave one of the tubes, the polarity changes in order to attract them to the next one. Figure 3 shows a simplified cross-section of a LINAC.

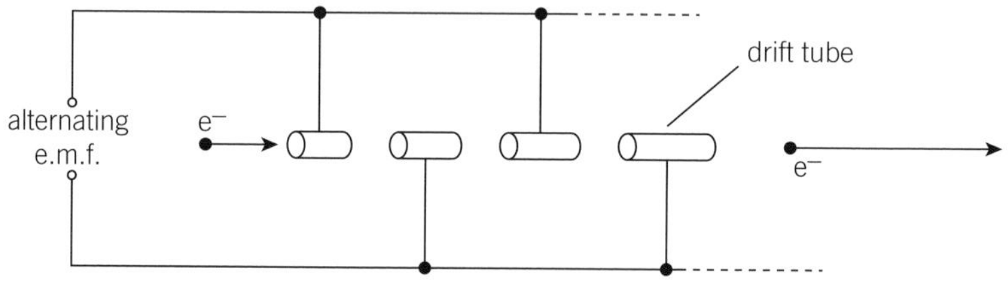

▲ **Figure 3** *A LINAC in cross-section*

Every time an electron moves from one tube to another it gains energy equal to eV, where V is the maximum e.m.f. of the alternating source connected to the tubes. With a large number of electrodes, electrons can be accelerated to extremely high velocities.

LINACs have many applications. They are often used as injectors for even higher energy particle accelerators like the LHC at CERN and they are commonly used in hospitals to produce the X-rays required for radiotherapy.

1 A linear accelerator uses 100 drift tubes, each with a maximum alternating e.m.f. of 50 V. Calculate the maximum speed of the electrons from the LINAC.

2 Suggest why it is necessary for the length of the drift tubes inside a linear accelerator to increase as they get further along the accelerator (Figure 3).

X marks the spot

Archaeologists use a number of different techniques to survey the ground at a dig site. An electrical resistance survey uses specialist detectors to measure the resistance of the Earth. Different features under the soil have different resistances. This property enables archaeologists to build up a map of the features beneath the surface so that they can start digging in a likely place.

What is resistance?

The electrical components called **resistors** have a known resistance, but in fact all components – including filament lamps, diodes, connecting wires, and even cells – have their own resistances. Each component resists the flow of charge carriers through it. It takes energy to push electrons through a component, and the higher the resistance of that component the more energy it takes.

Determining resistance

The term resistance has a very precise meaning in physics. The resistance of a component in a circuit can be determined by measuring the current I in the component and the p.d. V across the component. The resistance of the component R is defined as the ratio between V and I.

$$\text{resistance of component } R = \frac{\text{p.d. across component}}{\text{current in component}}$$

$$R = \frac{V}{I}$$

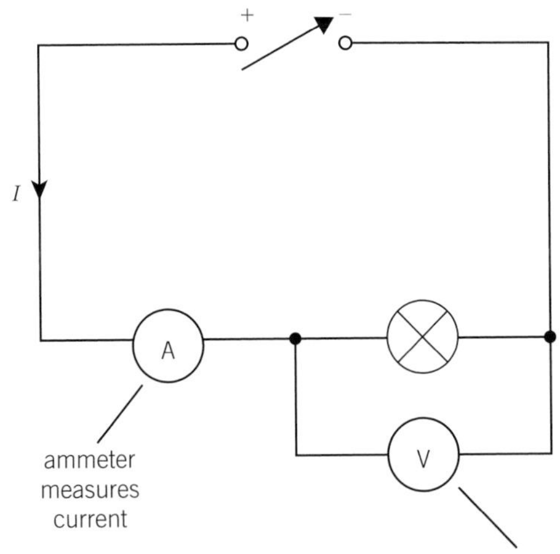

▲ **Figure 2** *The current in the component and the potential difference across it are used to determine its resistance*

Learning outcomes

Demonstrate knowledge, understanding, and application of:

→ resistance; $R = \frac{V}{I}$; the unit ohm

→ Ohm's law.

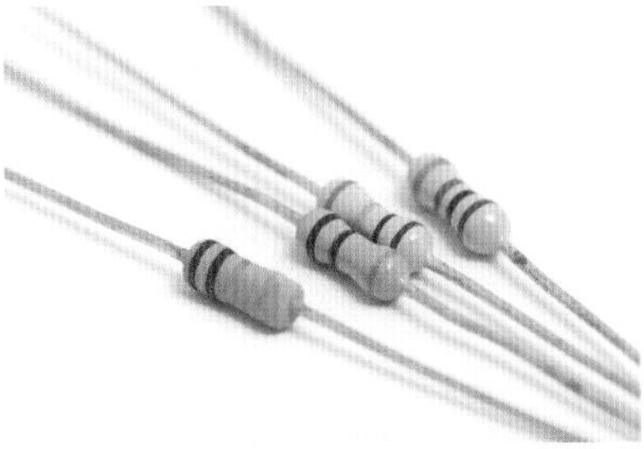

▲ **Figure 1** *Resistors come in all shapes and sizes, like these carbon film resistors – the coloured bands represent the resistance of each one*

The unit of resistance is the **ohm** (Ω). The ohm was named after the German Georg Ohm, who carried out pioneering work on electrical resistance in the 1800s.

The ohm is defined as the resistance of a component when a p.d. of 1 V is produced per ampere of current.

$$1\,\Omega = 1\,\text{V}\,\text{A}^{-1}$$

A component with a resistance of 1 Ω will have a p.d. across it of 1 V per ampere of current in it. A component with a resistance of 500 Ω would have a p.d. of 500 V when there is a current of 1 A in it.

Ohm's law

Ohm's investigations into the resistances of metallic conductors led him to derive what is now referred to as **Ohm's law**.

For a metallic conductor kept at a constant temperature, the current in the wire is directly proportional to the p.d. across its ends.

In other words, he found that when the p.d. across the wire (kept at constant temperature) doubled, the current in the wire also doubled.

Temperature and resistance

Figure 3 shows a length of insulated metallic wire in the form of a tight bundle connected in a circuit, alongside a graph of the variation of the current I in the circuit with time t. The switch **S** is closed at time $t = 0$. The p.d. across the wire remains constant at 1.5 V, but the current in the wire decreases with time. The shape of the graph can be explained as the resistance of the wire increasing with time. You can confirm this by calculating the resistance at $t = 0$ and at $t = 4.0$ s.

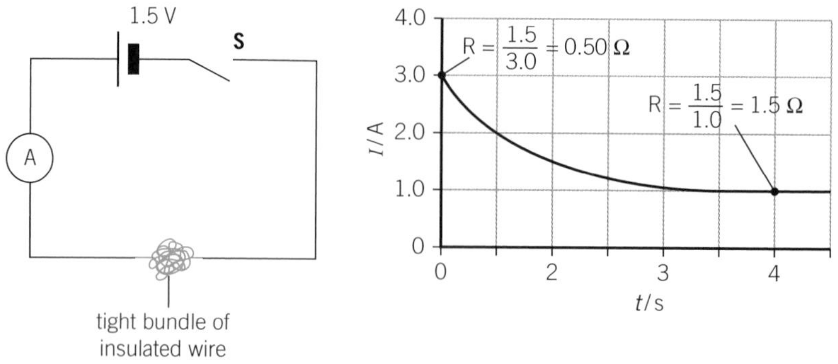

▲ **Figure 3** *A circuit including a bundle of thin wire carries less current as the wire gets hotter*

The current in the circuit changes because the temperature of the wire increases over time as a result of heating caused by the current. As the wire gets hotter, its resistance increases.

A microscopic explanation

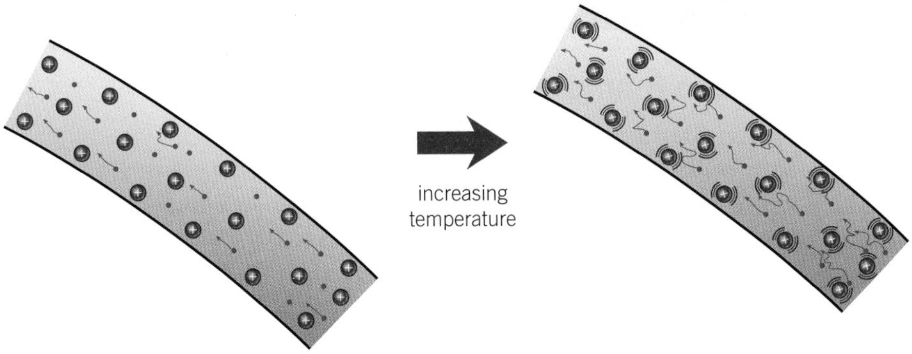

▲ **Figure 4** *As the wire gets hotter the positive ions vibrate more, increasing the resistance of the wire*

When the temperature of the wire increases the positive ions inside the wire have more internal energy and vibrate with greater amplitude about their mean positions. The frequency of the collisions between the charge carriers (free electrons in the metal) and the positive ions increases, and so the charge carriers do more work, in other words, transfer more energy as they travel through the wire.

Other factors as well as temperature affect the resistance of a length of wire. These are discussed in Topic 9.7, Resistance and resistivity.

Summary questions

1 Sketch a graph of I against V for a metallic conductor at constant temperature and use it to describe Ohm's law. *(4 marks)*

2 The current in a resistor is 0.50 A and the p.d. across it is 5.2 V. Calculate its resistance. *(1 mark)*

3 Define the ohm, and express it in base units. *(2 marks)*

4 A length of wire is connected to a cell. The current in the wire is 80 mA and the p.d. across the wire is 2.4 V. Calculate the resistance of the wire. *(3 marks)*

5 A resistor has a resistance of 1.2 kΩ. In 3.0 minutes a charge of 54 C flows through the resistor. Calculate the p.d. across the resistor. *(2 marks)*

6 The current in a filament lamp is 1.5 A. It is operated for a time of 1.0 minutes. The charges flowing through the lamp transfer 500 J of energy to the lamp. Calculate the resistance of the lamp. *(3 marks)*

Learning outcomes

Demonstrate knowledge, understanding, and application of:

→ I–V characteristics of resistors and filament lamps.

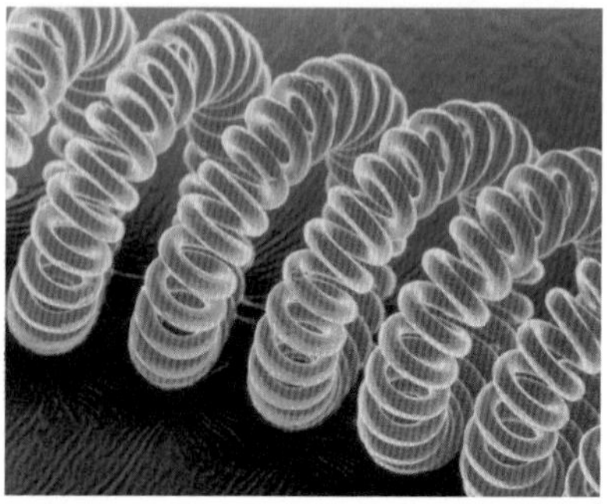

▲ **Figure 1** *The filament is designed so that, when there is a current in it, it gets so hot that it glows brightly*

Tungsten filaments

Tungsten **filament lamps** were invented in Hungary in 1904, and rapidly replaced the carbon filaments used up to that point. Tungsten is a remarkable metal. It is also used in radiation shielding and to make penetrating tips for high-speed military projectiles, because it is incredibly dense (around 1.7 times denser than lead) and very robust. However, it is tungsten's melting point that makes it so useful for filament lamps. It has the highest melting point of all the elements, at over 3000°C.

Filament lamps behave in a complex way when the current in them increases. In order to understand what is going on we need to look carefully at how the current and potential difference are related.

Graphs of I against V

The current–potential difference characteristic (or simply ***I*–*V* characteristic**) for any electrical component shows the relationship between the electric current I in a component and the potential difference V across it.

Collecting data for an I–V characteristic

Two methods are commonly used to collect the data needed to plot an I–V characteristic (Figure 2). The methods are very similar and both involve a simple way to vary either the current in a component or the potential difference across it.

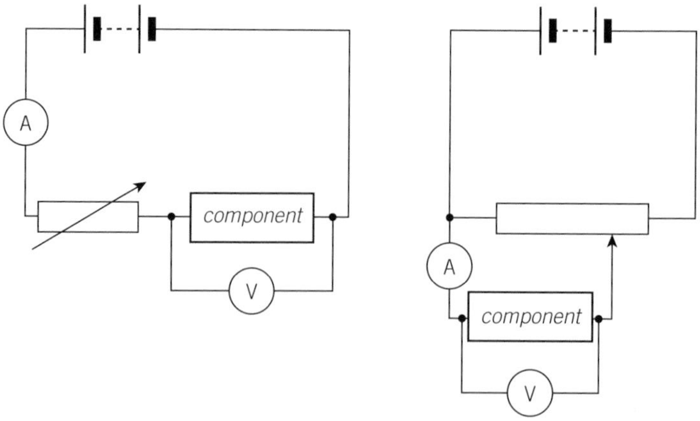

▲ **Figure 2** *Different techniques used to collect data to produce an I–V characteristic: the circuit on the left uses a variable resistor, the one on the right a potentiometer to provide different values of I and V respectively*

When investigating the I–V characteristic for a component it is important to look at whether or not the component behaves the same way if the current through it is in the opposite direction. This is normally achieved by reversing the polarity of the battery or the power supply.

Synoptic link

You will learn more about the potentiometer in Topic 10.6, Sensing circuits.

1 Identify the components in the circuit diagrams.
2 Describe how changing the resistance of the variable resistor in the first circuit affects the current in the component being tested.
3 Outline a method for producing an I–V characteristic for a filament lamp. You should include a circuit diagram, details of the measurements you would take, and the procedures you would use.

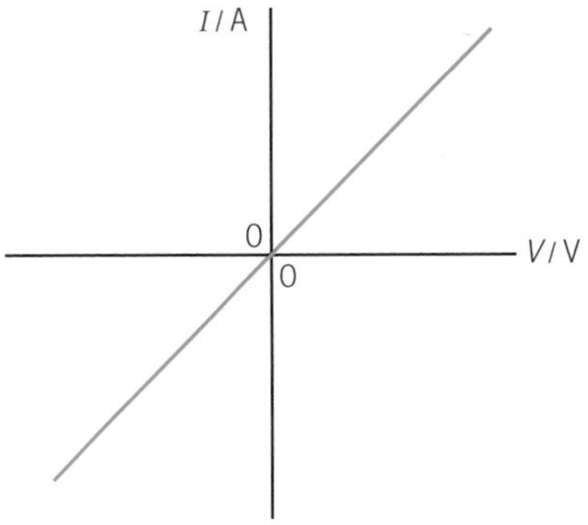

▲ **Figure 3** *The I–V characteristic for a fixed resistor is a straight line through the origin*

Resistors

Using either circuit in Figure 2 we can produce the I–V characteristic for a resistor. Fixed resistors are designed to ensure their resistance is constant, regardless of typical changes in temperature as the current in them varies.

Looking at the graph (Figure 3) we can draw a number of conclusions.

- The potential difference across the resistor is directly proportional to the current in the resistor ($V \propto I$). As a result
  - a resistor obeys Ohm's law, and so can be described as an **ohmic conductor**
  - the resistance of the resistor is constant.
- The resistor behaves in the same way regardless of the polarity.

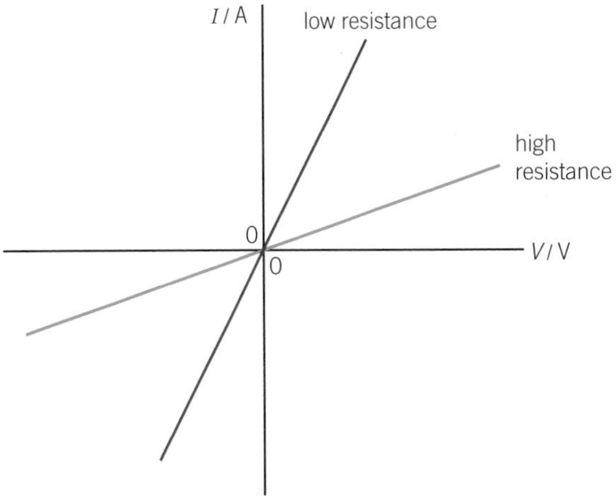

▲ **Figure 4** *Different resistors produce straight lines through the origin with different gradients – the shallower the line, the greater the resistance, which can be determined for each line using* $R = \frac{V}{I}$ *(the inverse of the gradient)*

Most wires and other metallic conductors behave in the same way as a resistor; they can be thought of as resistors with very low resistance.

Filament lamps

Repeating the same experiment for filament lamps produces very different results – see Figure 5.

Looking at the graph we can draw a number of conclusions.

- The potential difference across a filament lamp is not directly proportional to the current through the resistor. In other words
  - a filament lamp does not obey Ohm's law, and so can be described as a **non-ohmic component**
  - the resistance of the filament lamp is not constant.
- The filament lamp behaves in the same way regardless of the polarity.

The resistance of the filament increases as the p.d. across it increases. You can confirm this by determining $\frac{V}{I}$ at different points on the graph.

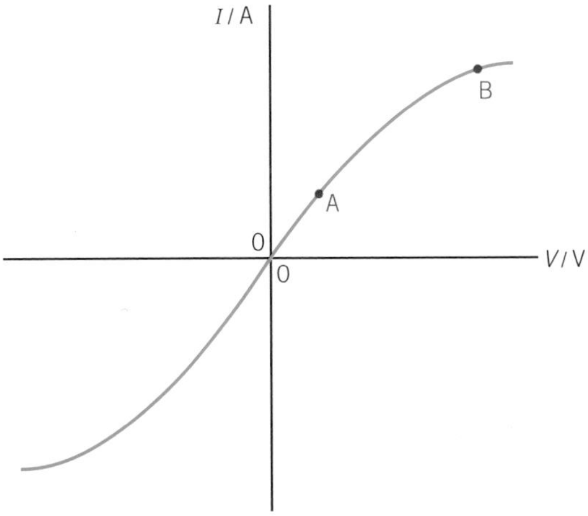

▲ **Figure 5** *The I–V characteristic for a filament lamp is most definitely not a straight line through the origin*

This increase in resistance is caused by the wire getting so hot that it glows. As the current increases so does the rate of flow of charge through the filament – more electrons per second pass through it, so more collisions occur between the electrons and the positive metal ions per second. When the electrons collide with the ions they transfer energy to the ions, causing the ions to vibrate more, or in other words to increase in temperature, and to collide with still more electrons.

Summary questions

1 The graph in Figure 6 shows the *I–V* characteristic for two different components.

a State and explain which component obeys Ohm's law. *(2 marks)*

b Calculate the resistance of each component at 4.0 V. *(2 marks)*

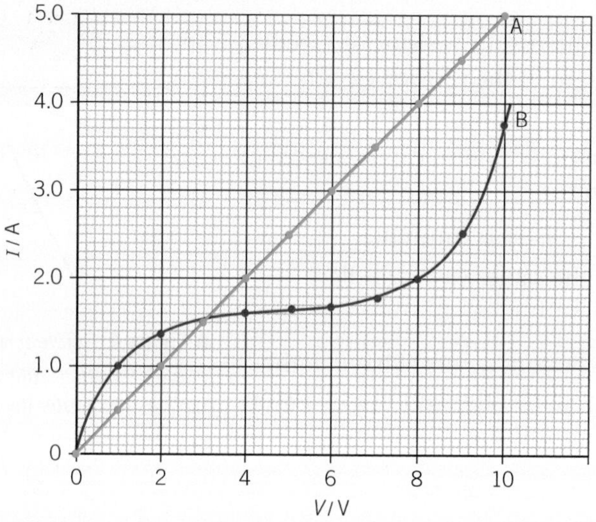

▲ **Figure 6**

2 Plot an *I–V* characteristic from the data in the table below. *(3 marks)*

I / A	−1.65	−1.63	−1.52	−1.34	−1.14	−0.82	0.00	0.81	1.12	1.36	1.53	1.62	1.64
V / V	−9.0	−7.5	−6.0	−4.5	−3.0	−1.5	0.0	1.5	3.0	4.5	6.0	7.5	9.0

3 Use your graph to calculate the resistance of the component at:

a 0.50 A; **b** 1.00 A; **c** 1.60 A. *(3 marks)*

4 Identify the component used to produce the *I–V* characteristic in question 2 and explain the shape of the graph. *(4 marks)*

5 On the same set of axes, sketch a labelled *I–V* characteristic for a metallic conductor:

a at room temperature; *(2 marks)*

b at a much higher temperature. Explain the shape of the graphs. *(2 marks)*

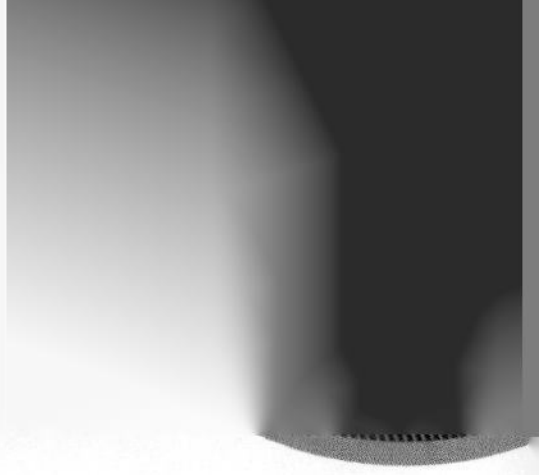

Lighting the way

If you look closely at rear lights on modern cars you will see they are very different from the lights on older cars (Figure 1). Instead of a single filament lamp, the lights are made up of arrays of **light-emitting diodes** (LEDs). They emit light by a process very different from that in a filament lamp – electrical energy is transferred directly into light, and LEDs do not get hot, so they are much more efficient and draw much less power.

Learning outcomes

Demonstrate knowledge, understanding, and application of:

→ I–V characteristics of diodes and light-emitting diodes.

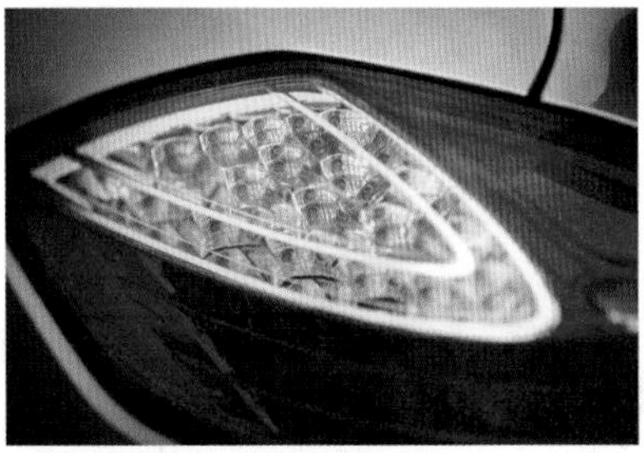

▲ **Figure 1** *Some modern cars have arrays of LEDs in place of filament lamps – LEDs are much more efficient and last much longer*

The diode

Diodes are everywhere, but you very rarely see them (Figure 2). They are a vital part of nearly every modern electronic circuit, from mobile phones to washing machines.

A diode only allows a current in one particular direction. This is the unique property that makes diodes so important for modern electronics. All the components you have seen so far are unaffected by the direction of the current – a filament lamp works equally well regardless of which way round you connect its terminals. Not so for a diode.

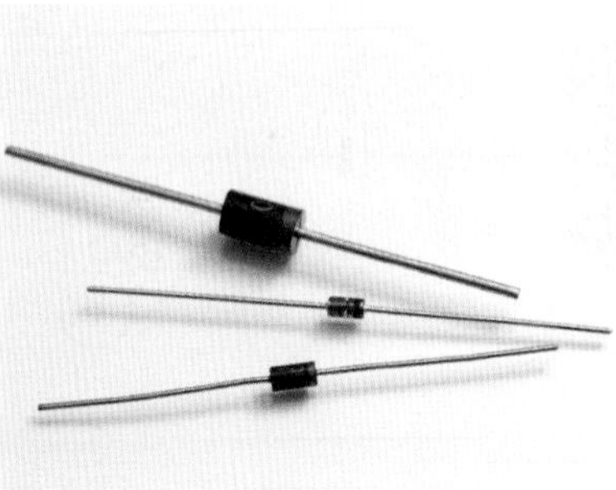

▲ **Figure 2** *Diodes don't appear to be anything special, but they are: unlike the components you have seen so far, a diode is made from a semiconductor*

The light-emitting diode

Some diodes are made of a material that emits light when they conduct. However, unlike most light sources, these light-emitting diodes (LEDs) emit light of a single specific wavelength (see Topic 11.2, Wave properties).

As LEDs are so efficient and take very little energy to run, they are sometimes used as simple indicators to show the direction of current through a particular part of a circuit. They light up when there is current in them, showing that that part of the circuit is live (Figure 5).

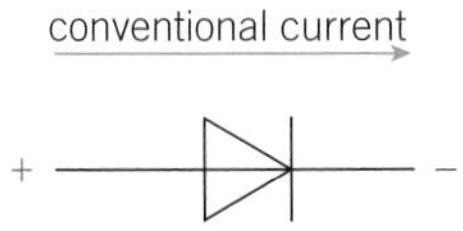

▲ **Figure 3** *The circuit symbol for a diode hints at its key feature: it allows current in one direction only*

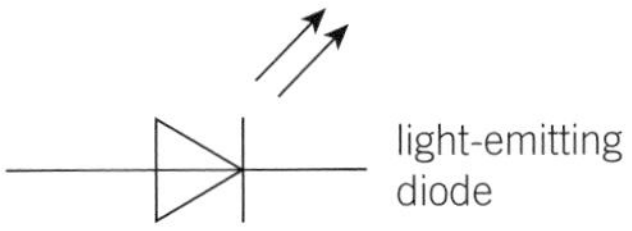

▲ **Figure 4** *Circuit symbol for a light-emitting diode*

I–V characteristic for a diode

Repeating the experiment described in Topic 9.5 to collect data for a diode produces a very different I–V characteristic from those for a resistor or a filament lamp (Figure 6).

Looking at the graph for a semiconducting diode we can draw a number of conclusions.

Synoptic link

Wavelengths of light are covered in Topic 11.2, Wave properties.

- The potential difference across a diode (or LED) is not directly proportional to the current through it. This means
  - a diode does not obey Ohm's law, and so can be described as a non-Ohmic component
  - the resistance of the diode is not constant.
- The diode's behaviour depends on the polarity.

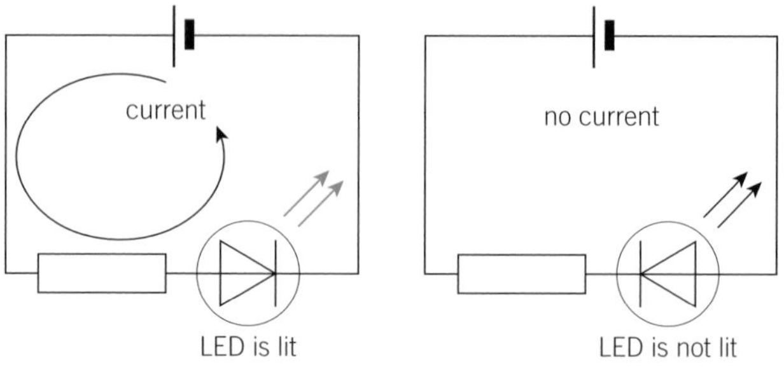

▲ **Figure 5** *An LED will only light up if the polarity allows a current to pass through it*

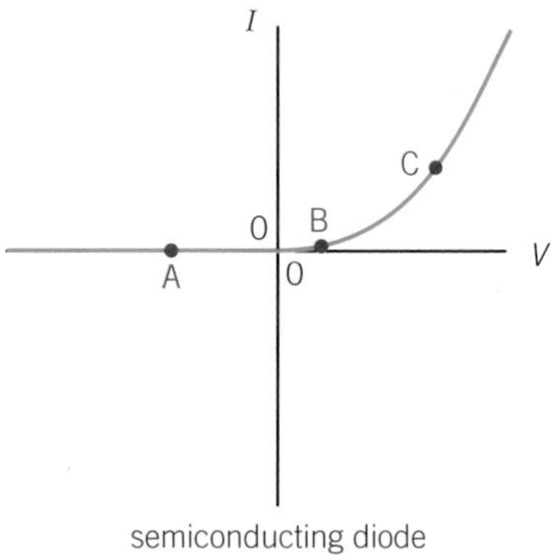

▲ **Figure 6** *The I–V characteristic for a diode*

At **A** in Figure 6 the resistance of the diode is very high – infinite for practical purposes. With the p.d. in this reverse direction, the diode does not conduct. At **B**, as the p.d. increases, the resistance gradually starts to drop. For a silicon diode this happens at around 0.7 V (the **threshold p.d.**). Above this value, the resistance drops sharply for every small increase in p.d. (at **C**). Above this point the diode has very little resistance.

Different LEDs have different values for their threshold p.d., related to the colour of the light they emit.

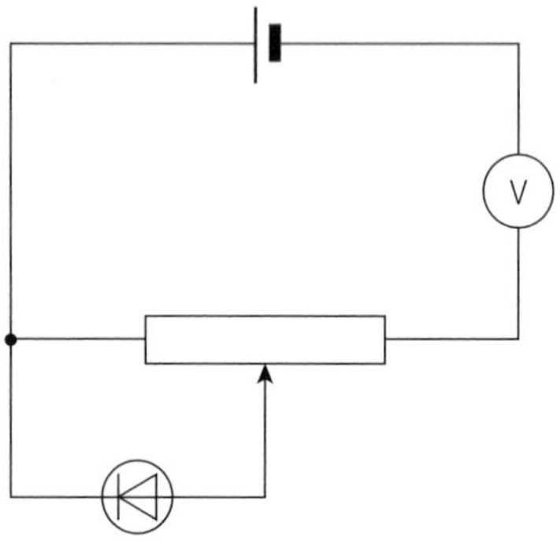

▲ **Figure 7**

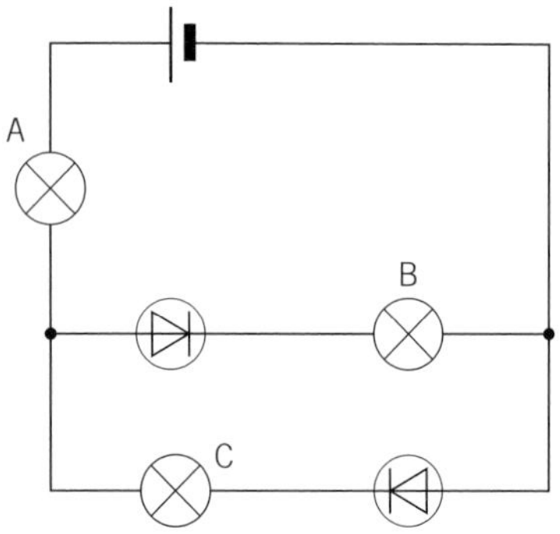

▲ **Figure 8**

Summary questions

1 The circuit in Figure 7 was used to collect data in order to produce an I–V characteristic for a diode. Identify two errors in the circuit. *(2 marks)*

2 State and explain which of the lamps in the circuit in Figure 8 would be lit. *(3 marks)*

3 Plot the I–V characteristic for a diode using the data in the table below. *(4 marks)*

I/mA	0.00	0.00	0.00	0.00	0.00	0.02	0.35	1.50	29.20	124.10
V/mV	−750	−500	−250	0	200	400	600	650	700	750

4 Use your graph to determine the resistance of the diode at:
a −300 mV; **b** 625 mV; **c** 0.72 V. *(3 marks)*

5 A diode and a fixed resistor are connected together in a circuit. Sketch the I–V characteristic when the diode and resistor are connected:
a in series; **b** in parallel. *(5 marks)*

Not all wires are the same

Gold is an excellent conductor. It has a number density almost as high as copper and is much less reactive, so it corrodes very little over time. It is used in high-energy electrical applications, and for the fine wires that connect the pins on computer chips to integrated circuit boards. Some expensive electrical cables use gold contacts, which have a much lower resistance than the tin in cheaper cables, but many audiovisual experts consider this simply a marketing ploy and a pointless expense.

A gold contact and a similar tin contact have different resistances. You already know that the temperature of a wire affects its resistance (Topic 9.4). Material is one of three factors beside temperature that affects the resistance.

- The material of the wire.
- The length of the wire L.
- The cross-sectional area of the wire A.

Resistance and material: resistivity

The term 'resistance' refers only to a specific component. The term '**resistivity**' is used to describe the electrical property of a material. For example, different components made from copper may have different resistances: copper wires may have different resistances as their lengths or cross-sectional areas differ, but copper has a unique resistivity.

Resistance and length

For any given current, increasing the length of the wire will increase the p.d. across it. Doubling the length doubles the p.d., so the resistance must have doubled.

The resistance R of a wire is directly proportional to its length L.

$$R \propto L$$

Resistance and cross-sectional area

When the cross-sectional area of the wire increases, the opposite happens – the resistance drops. Wires with a greater cross-sectional area have a lower resistance.

For any given p.d., doubling the cross-sectional area will double the current in the wire, so the resistance must have halved.

The resistance R of a wire is inversely proportional to its cross-sectional area A.

$$R \propto \frac{1}{A}$$

Learning outcomes

Demonstrate knowledge, understanding, and application of:

→ resistivity of a material; the equation $R = \frac{\rho L}{A}$

→ the variation of resistivity of metals and semiconductors with temperature.

▲ **Figure 1** *The resistance of a material depends on a number of factors*

Study tip

Great care should be taken in using this equation. In particular, make sure you have the cross-sectional area in m^2, not the diameter or radius of the wire.

Calculating resistance from resistivity

We can combine the two relationships above to give

$$R \propto \frac{L}{A}$$

The resistivity ρ of a particular material at a given temperature is the constant of proportionality in this equation.

$$R = \frac{\rho L}{A}$$

Resistivity has the unit ohm metre (Ω m). This equation may be used for a known constant temperature. As you will see later, the resistivity of a material is affected by its temperature too.

Worked example: Resistance of a nichrome wire

Nichrome has a resistivity of $1.50 \times 10^{-6}\,\Omega\,m$. Calculate the resistance of a wire made from nichrome with a length of 80 cm and a radius of 2.0×10^{-4} m.

Step 1: Identify the correct equation to calculate the resistance.

$$R = \frac{\rho L}{A}$$

Step 2: Determine the cross-sectional area of the wire in m^2 using $A = \pi r^2$.

$$A = \pi \times (2.0 \times 10^{-4})^2 = 1.26 \times 10^{-7}\,m^2$$

Step 3: Substitute known values in SI units into the equation for resistance.

$$R = \frac{\rho L}{A} = \frac{1.50 \times 10^{-6} \times 0.80}{1.26 \times 10^{-7}}$$

Calculating the answer to an appropriate number of significant figures gives

$$R = 9.5\,\Omega \text{ (2 s.f.)}$$

Defining resistivity

We can define the resistivity of a material by rearranging the equation above to make the resistivity the subject.

$$\rho = \frac{RA}{L}$$

The resistivity of a material at a given temperature is the product of the resistance of a component made of the material and its cross-sectional area, divided by its length.

The resistivity of a material varies with temperature in the same way as the resistance of most components varies with temperature. As the material gets hotter, its resistivity increases (see Topic 9.4).

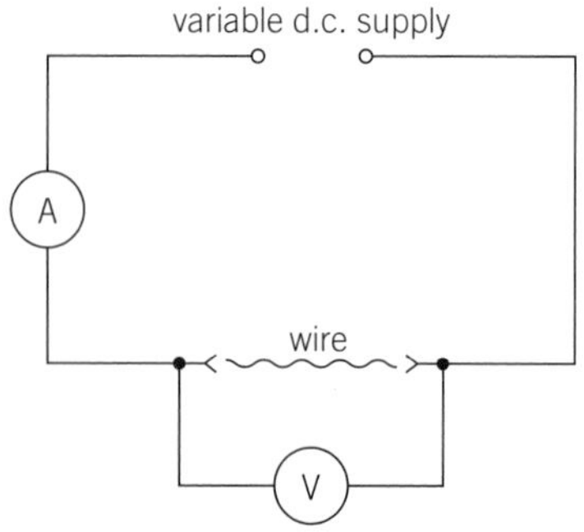

▲ **Figure 2** *Circuit for determining resistivity*

Determining ρ

A simple experiment can be carried out to determine the resistivity of a material, in this case a piece of wire. Amongst the different wires that might be used are copper and some common alloys including constantan and nichrome.

A typical way to determine the resistivity is to investigate how the resistance of a wire varies with its length. Using the circuit illustrated in Figure 2 we can obtain values for the p.d. across different lengths of wire. If the current in each wire is measured too, we can use $R = \frac{V}{I}$ to calculate the resistance for each length.

As $R \propto L$, a graph of R against L is a straight line through the origin. By considering the general equation for a straight line ($y = mx + c$) and the equation $R = \frac{\rho L}{A}$ we see that the gradient of this graph is $\frac{\rho}{A}$.

The resistivity can be determined by multiplying the gradient of the graph by the cross-sectional area of the wire.

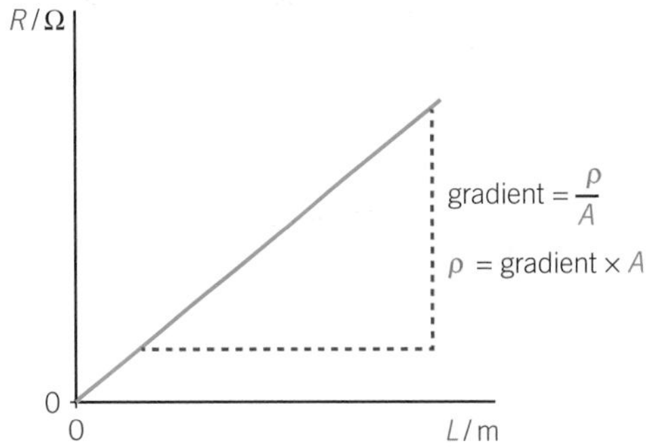

▲ **Figure 3** *A graph of the resistance of a wire against its length produces a straight line through the origin, with a gradient equal to* $\frac{\rho}{A}$

Conductors, semiconductors, and insulators

Different materials have widely different values for resistivity. Good conductors like metals have a resistivity of the order of $10^{-8}\,\Omega\,m$, insulators have a value of the order of $10^{16}\,\Omega\,m$, and semiconductors have values in between these extremes. Table 1 shows the resistivity values for some materials at 20 °C.

Superconductivity

A strange quantum effect occurs when some materials are cooled. As they get colder their resistivity drops, as expected, but then at a critical temperature the resistivity suddenly falls to zero. Not just very low, but truly zero (Figure 4).

This is **superconductivity**. Any components made from a superconducting material have no electrical resistance. No energy is lost when there is a current in the material, and so huge amounts of charge can pass through a superconductor without it even getting warm. The LHC at CERN uses superconducting wires carrying up to 20 000 A to produce exceptionally strong magnetic fields. The only problem is that the wires become superconducting at around 4 K (−269°C, even colder than deep space). This is a typical temperature for superconductors. A few high-temperature superconductors have been developed that become superconducting around 80 K, but this is still around −190°C. The race is on to develop room-temperature superconductors.

1. Compare how the resistance of a normal metallic wire and a superconductor changes with temperature.
2. Outline some potential advantages of room-temperature superconductors.

Synoptic link

You can find more information about the gradient of straight-line graphs in Appendix A2, Recording results.

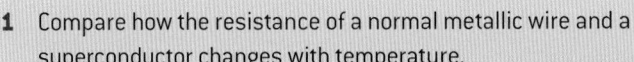

▼ **Table 1** *Some materials and their resistivity values*

Material	Type	ρ at 20°C / Ω m
silver	conductor	1.6×10^{-8}
copper	conductor	1.7×10^{-8}
iron	conductor	1.0×10^{-7}
carbon	semiconductor	3.5×10^{-5}
silicon	semiconductor	640
glass	insulator	$10^{10} - 10^{14}$
quartz	insulator	7.0×10^{17}

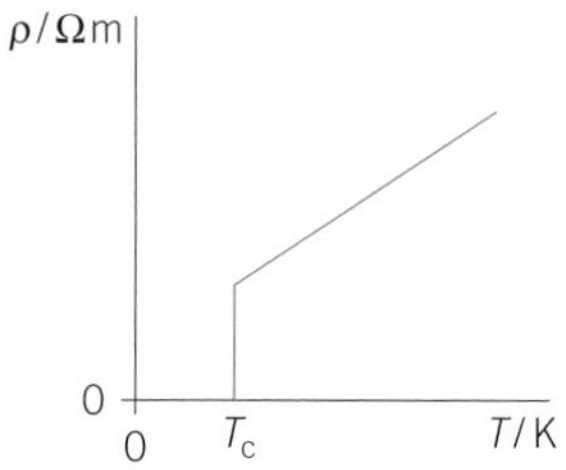

▲ **Figure 4** *Below a critical temperature the resistivity of some materials falls to zero*

▲ **Figure 5** *Superconductivity has another strange effect: a superconducting material does not allow magnetic fields to pass through it (the Meissner effect), which results in the permanent magnet hovering perfectly over the superconductor*

▼ **Table 2** *I–V data for a nichrome wire*

Length / m	p.d. / V	Current / A
0.100	0.31	0.50
0.200	0.61	0.50
0.300	0.92	0.50
0.400	1.23	0.50
0.500	1.53	0.50
0.600	1.84	0.50
0.700	2.14	0.50
0.800	2.45	0.50
0.900	2.76	0.50
1.000	3.06	0.50

Summary questions

1 Describe the difference between resistance and resistivity. *(2 marks)*

2 Calculate the resistance of a copper wire of length 1.0 m and cross-sectional area of $3.32 \times 10^{-6}\,\text{m}^2$. *(3 marks)*

3 Explain the effect of changing the temperature on the resistivity of a metal. *(3 marks)*

4 Calculate the resistivity of a piece of wire that has a resistance of 170 Ω when it has a length of 12 m and a radius of 3.0 mm. *(5 marks)*

5 A wire has resistance of 8.0 W. State and explain the effect on the resistance of a piece of wire if:

a the length of the wire doubles (all other factors remain the same); *(1 mark)*

b the cross-sectional area of the wire doubles (all other factors remain the same); *(2 marks)*

c the radius of the wire halves (the length remains constant); *(2 marks)*

d the volume remains constant but the length of the wire doubles. *(3 marks)*

6 Table 2 lists data collected for a nichrome wire. The diameter of the wire was measured in three places using a micrometer and an average value of 0.46 mm was calculated from these results.

a Explain why the diameter of the wire should be measured in several places along the wire. *(2 marks)*

b i Use the data in the table to plot a graph of the resistance of the wire against length. *(4 marks)*

ii Use the graph to determine the resistivity of the wire. *(4 marks)*

c Describe how the current is kept constant and why it is important to do so in this experiment. *(3 marks)*

7 Outline an experiment you could do using several different thicknesses of wire of set length to determine the resistivity of the wire. *(4 marks)*

Not too hot, not too cold

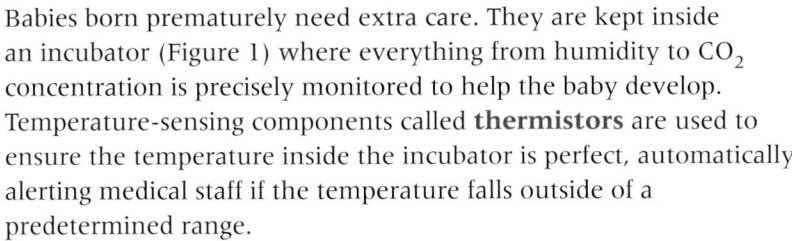

Babies born prematurely need extra care. They are kept inside an incubator (Figure 1) where everything from humidity to CO_2 concentration is precisely monitored to help the baby develop. Temperature-sensing components called **thermistors** are used to ensure the temperature inside the incubator is perfect, automatically alerting medical staff if the temperature falls outside of a predetermined range.

Learning outcomes

Demonstrate knowledge, understanding, and application of:

→ negative temperature coefficient thermistor

→ *I–V* characteristics of the thermistor.

Temperature and number density

We have already seen the effect of increasing temperature on the resistance of a metal wire. The hotter the wire becomes, the more the positive ions vibrate, and so the greater the resistance becomes. But a few materials behave differently. Some semiconductor components have a **negative temperature coefficient**, meaning that their resistance drops as the temperature increases. This sounds odd, but it is nothing new: it was first observed by Michael Faraday in 1833.

The effect can be explained in terms of the number density of the charge carriers within the material from which the component is made. In some semiconductors, as the temperature increases, the number density of the charge carriers also increases.

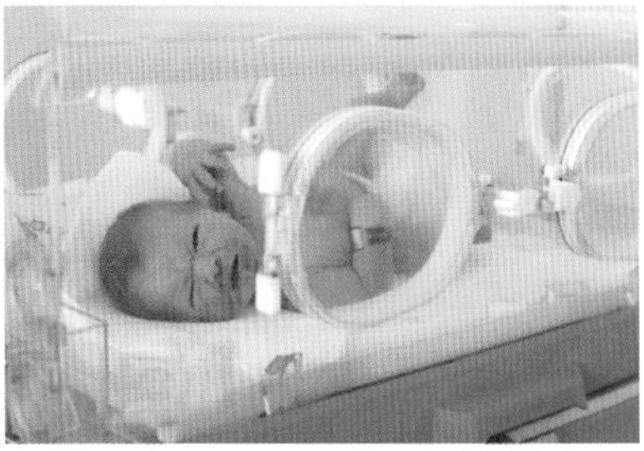

▲ **Figure 1** *Thermistors are sensitive enough to precisely monitor the temperature of incubators for premature babies*

The thermistor

A thermistor (Figures 2 and 3) is an electrical component made from a semiconductor with a negative temperature coefficient. As the temperature of the thermistor increases, its resistance drops.

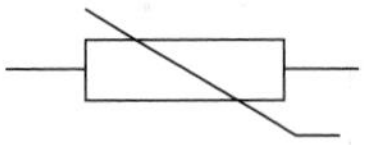

▲ **Figure 2** *The circuit symbol for a thermistor (the term 'thermistor' comes from combining 'thermal' and 'resistor')*

The change in resistance is often dramatic (Figure 4). This makes thermistors particularly useful in temperature-sensing circuits. A small change in temperature can be detected by monitoring the resistance of the thermistor.

Thermistors are used:

- in simple thermometers
- in thermostats to control heating and air-conditioning units
- to monitor the temperature of components inside electrical devices like computers and smartphones so that they can power down before overheating damages them
- to measure temperature in a wide variety of electrical devices like toasters, kettles, fridges, freezers, and hair dryers
- to monitor engine temperatures to ensure the engine does not overheat.

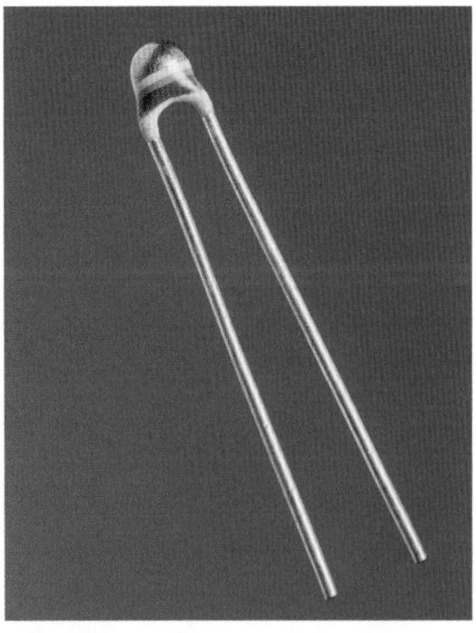

▲ **Figure 3** *Thermistors come in different shapes and sizes, but can be small enough to fit inside most electronic devices*

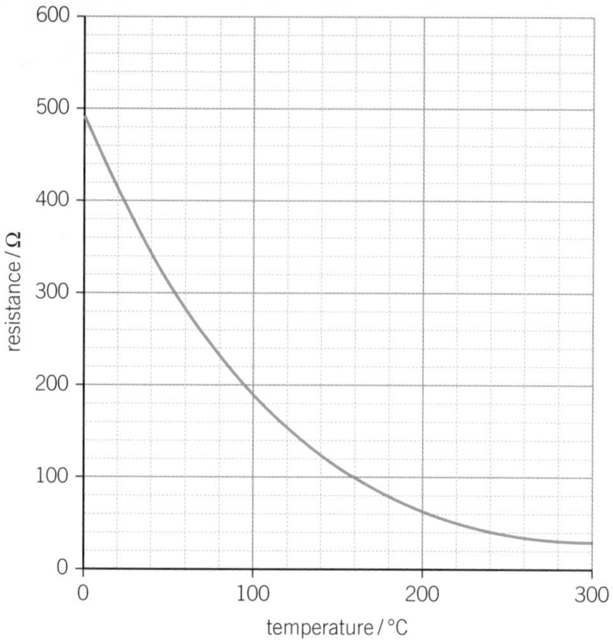

▲ **Figure 4** *The drop in resistance of a thermistor as the temperature increases is not linear – there is often a significant decrease, followed by a more gradual drop as it gets hotter and hotter*

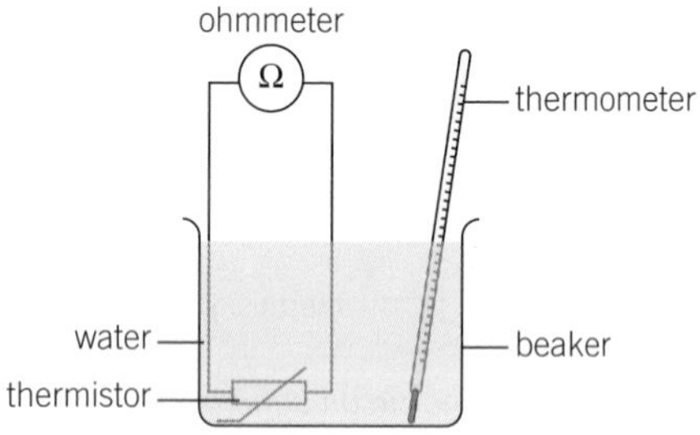

▲ **Figure 5** *Use a water bath to control the temperature of a thermistor and an ohmmeter for a quick and simple recording of the resistance in an experiment to investigate the effect of temperature on resistance*

Thermistor experiment

A simple investigation into how the resistance of a thermistor changes with temperature change be carried out using an ohmmeter and a water bath (Figure 5). Alternatively, an ammeter and a voltmeter can be used to measure current in the thermistor and p.d. across it at different temperatures. The resistance can then be calculated with $R = \frac{V}{I}$.

The results from this experiment may be used in the choice of a thermistor for a particular application. A thermistor is selected to ensure it provides the greatest change in resistance over the range of temperatures in which it will operate. In Figure 4, this change takes place at 20–50°C, making that particular thermistor perfect for incubators, but less useful for monitoring the temperature inside a car engine, typically around 100°C.

I–V characteristics of thermistors

Like most semiconducting components, thermistors are non-ohmic. The *I–V* characteristic has some features similar to that of a filament lamp, and one crucial difference (Figure 6). With a filament lamp, as the current increases, electrons transfer energy to the positive ions, which raises the temperature. This causes an increase in resistance (see Topic 9.4).

With a thermistor, like a lamp, as the current increases the temperature increases. But unlike the lamp, this temperature increase leads to a drop in resistance because the number density of charge carriers increases. This may be confirmed by comparing $R = \frac{V}{I}$ at various points on the graph.

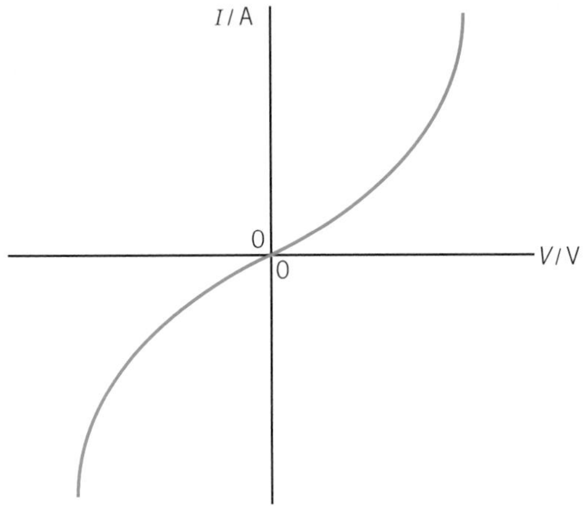

▲ **Figure 6** *The I–V characteristic for a thermistor is not a straight line, so its resistance must be changing*

An increase in temperature leads to an increase in the number density of the free electrons. This means that the resistance of the thermistor decreases as its temperature increases.

Summary questions

1. A thermistor is used in a thermostat in order to maintain a constant temperature inside a lorry delivering frozen goods.
   - **a** Describe how any changes in the temperature inside the lorry would affect the resistance of the thermistor. *(1 mark)*
   - **b** Suggest how the thermistor might be used to ensure the produce transported by the lorry is kept at a constant temperature. *(2 marks)*

2. Sketch a labelled graph to show how the resistance of a typical thermistor used inside an oven varies with temperature. You should include typical values for the temperature and an explanation of why you selected these values. *(3 marks)*

3. Table 1 contains data collected using a thermistor in a water bath. The thermistor was connected to a 1.5 V cell. Plot a graph of resistance against temperature for the thermistor. *(5 marks)*

▼ **Table 1** *Variation of current with changing temperature*

Temperature /°C	−20.0	−10.0	0.0	10.0	20.0	30.0
Current in the thermistor /mA	10.7	17.3	32.5	50.2	82.5	121.2

4. Use your graph from question 3 to determine the resistance of the thermistor at: **a** −5.0°C; **b** 25°C. *(2 marks)*

5. The thermistor used in question 3 is now used to measure the temperature in a water heater. The current in the thermistor was 45.2 mA and the p.d. across it was 1.03 V. Determine the temperature of the water heater. *(3 marks)*

Learning outcomes

Demonstrate knowledge, understanding, and application of:

→ light-dependent resistor

→ variation of resistance with light intensity.

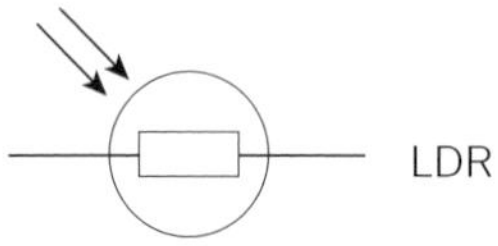

▲ **Figure 1** *The circuit symbol for a light-dependent resistor*

▲ **Figure 2** *LDRs are often made of cadmium sulfide formed into a flat disc on the surface of the LDR*

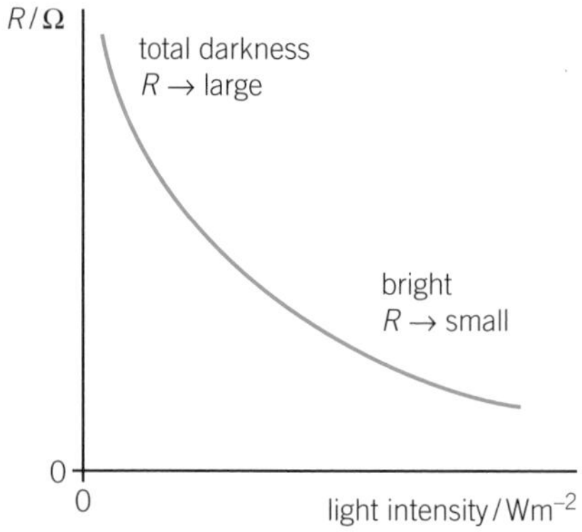

▲ **Figure 3** *The resistance of an LDR can vary from millions of ohms to just a few, depending on the light intensity*

Definitely not, just, cricket

A **light-dependent resistor** (LDR) is an essential part of many modern sports, including cricket and tennis. LDRs are small electrical components which change their resistance depending on the light intensity. If it gets too dark, play is postponed until the light intensity increases, maybe even to the following day.

As well as sports, LDRs are used in automatic street lights, the brightness meters in smartphones and laptops, and even in some space telescopes.

The light-dependent resistor

With the thermistor we saw how the resistance of some semiconductors varies in an unusual way. An LDR is another component that makes use of the unusual properties of a different type of semiconductor.

A typical LDR is made from a semiconductor in which the number density of charge carriers changes depending on the intensity of the incident light. In dark conditions the LDR has a very high resistance. The number density of the free electrons inside the semiconductor is very low, so the resistance is very high (often into MΩ). When light shines onto an LDR, the number density of the charge carriers increases dramatically, leading to a rapid decrease in the resistance of this component.

Investigating LDRs

Figure 4 shows an experiment to investigate how the resistance of an LDR varies with distance from a constant light source (like a simple filament lamp). A narrow tube made of black cardboard placed around the LDR will greatly reduce the effect of other background sources of light.

The results give a calibration graph that relates the resistance of the LDR to the light intensity. This graph can then be used to determine light intensity from different sources.

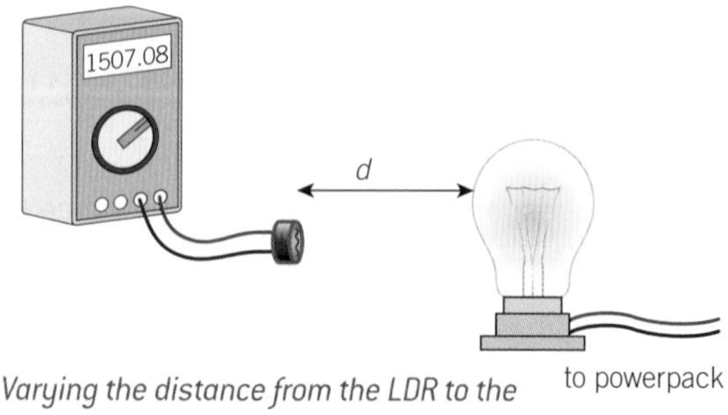

▲ **Figure 4** *Varying the distance from the LDR to the lamp has the effect of changing the light intensity received by the LDR. The further it is away from the filament lamp, the higher its resistance becomes.*

Infrared astronomy and LDRs

Some LDRs are particularly sensitive to infrared radiation (see Topic 11.6, Electromagnetic waves), so are useful as sensors for the very dim infrared received from space.

Our eyes cannot detect infrared wavelengths, so using infrared telescopes enables us to discover more about the universe than we ever could with our naked eyes. Astrophysicists can even see *inside* dense clouds of gas and dust with infrared, which passes through these clouds, whereas visible light does not. This allows us to see objects like new stars forming in stellar nurseries and to peer into the very centre of our galaxy.

But infrared astronomy has limitations. Infrared is absorbed by water in the atmosphere, so the European Space Agency has recently launched an infrared telescope into space, the Herschel Space Observatory.

1 Explain why most infrared telescopes are positioned on mountain tops.
2 Suggest how the changing resistance of an LDR might be used to detect extrasolar planets as they pass in front of distant stars.

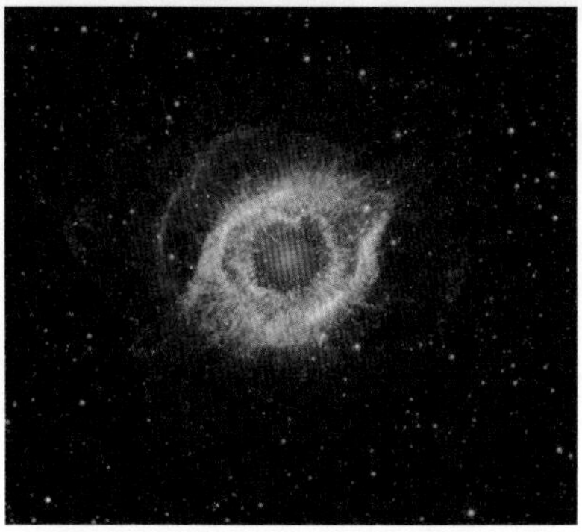

▲ **Figure 5** *This image of the Helix Nebula, made with data collected from infrared telescopes using LDRs, shows the fine detail in the clouds of hydrogen and helium around the centre*

Synoptic link

You can read more about intensity in Topic 11.5, Intensity.

Summary questions

1 LDRs are used in smartphones to measure the ambient light level and automatically adjust the screen brightness when necessary. Describe how the resistance of the LDR changes in different conditions. *(2 marks)*

2 The data in Table 1 was collected using an LDR and an ohmmeter in different conditions. Identify two likely errors in the table. *(2 marks)*

▼ **Table 1**

Conditions	complete darkness	dim light	normal daylight	very bright
Resistance of LED	400	2000	18 000	1 000 000

3 Compare a resistor, a thermistor, and an LDR. *(4 marks)*

4 The graph in Figure 6 is a logarithmic plot of the resistance of an LDR against light intensity. Use the graph to estimate the relative light intensity when the resistance is: **a** 1000 Ω; **b** 50 kΩ. *(2 marks)*

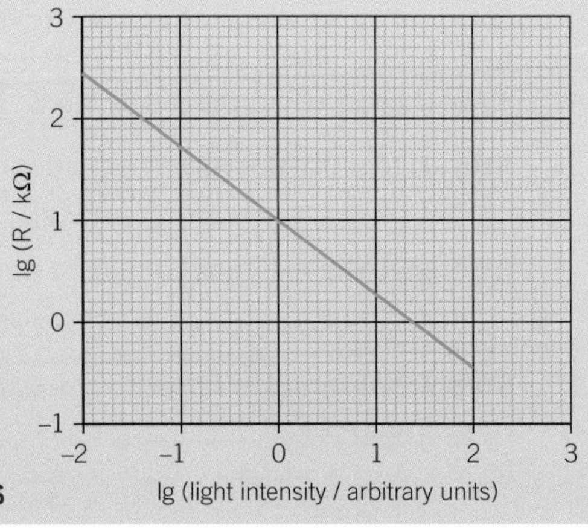

▶ **Figure 6**

9.10 Electrical energy and power

Specification reference: 4.2.5

Learning outcomes

Demonstrate knowledge, understanding, and application of:

→ the equations $P = VI$, $P = I^2R$, $P = \frac{V^2}{R}$

→ the equation $W = VIt$.

▲ **Figure 1** *The Three Gorges Dam*

Ten million kettles

The Three Gorges Dam, the world's largest power plant, is a hydroelectric power station that spans the Yangtze river in eastern China. It has an output of 22.5 GW, in other words 22 500 000 000 J s^{-1}, or enough to boil around 10 million average kettles at once.

The electricity generated by the dam is distributed to consumers through the Chinese grid. Electrical energy is relatively easy to transfer from one place to another and from one form to another, and this is one of the reasons it has become essential to us.

Transferring energy

Electric circuits are often used to transfer energy from one place to another, often from a power source like a cell or power supply to a component that transfers this energy into another form. A filament lamp transfers electrical energy to heat and light, whereas a loudspeaker transfers electrical energy to sound and heat.

Whenever there is a current in a component, energy is transferred from the power source to that component. Open a switch, or disconnect the power source, and the current falls to zero. No energy is transferred and the circuit is off.

Electrical power

The rate of energy transfer by each electrical component is called the electrical power (or just power). This depends on the current I in the component, measured in amperes (A), and the p.d. V across it, in volts (V).

The equation for electrical power P (in watts, W) is

$$\text{electrical power} = \text{p.d.} \times \text{current}$$

$$P = VI$$

Worked example: Charging a tablet

A typical tablet charger has a power of 12 W compared with 6.0 W for a mobile phone charger. Calculate the current drawn by a 12 W charger when the p.d. is 5.0 V.

Step 1: Identify the correct equation.

$$P = VI$$

Rearrange the equation to make I the subject.

$$I = \frac{P}{V}$$

Step 2: Substitute in known values and calculate the value for the current.

$$I = \frac{12}{5.0}$$

$$I = 2.4\,\text{A}$$

Additional equations for power

We can combine the equation above with $V = IR$ to give two additional equations for power.

$$P = VI$$
$$= (IR) \times I$$
$$= I^2R$$

Rearranging $V = IR$ to make I the subject gives $I = \frac{V}{R}$

Substituting this into $P = VI$ gives

$$P = V \times \frac{V}{R}$$
$$= \frac{V^2}{R}$$

Each version of the power equation is missing one of V, I, or R.

$$P = VI \qquad P = I^2R \qquad P = \frac{V^2}{R}$$

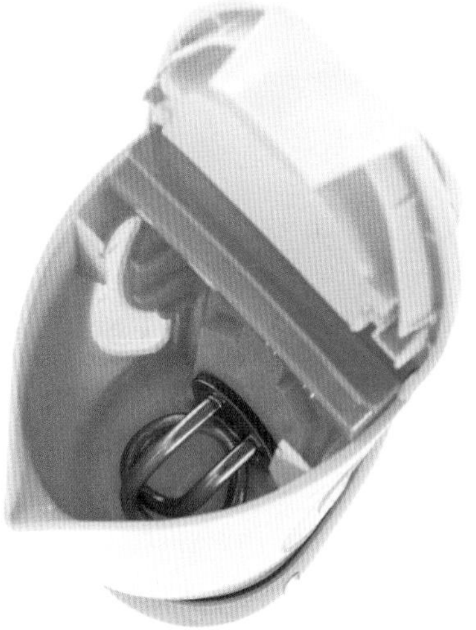

▲ **Figure 2** *Heating element of a kettle*

Worked example: Kettle power I

The heating element in a kettle has a small resistance, typically 20.0 Ω, in order to draw the large current needed to heat the element. For a current of 12 A, calculate the rate of energy transfer.

Step 1: Identify the correct equation to calculate the rate of energy transfer (the power).

$$P = I^2R$$

Step 2: Substitute the known values in SI units into the equation and calculate the power.

$$P = 12^2 \times 20.0 = 2880$$

Express the answer to an appropriate number of significant figures.

$$P = 2900 \text{ W (2 s.f.)}$$

Two significant figures are appropriate in this case as the value of the current (12 A) is to two significant figures.

Deriving $P = VI$

Our equation for electrical power can be derived from the generic equation

$$P = \frac{W}{t}$$

Using our defining equation for p.d., rearrange to make W the subject.

$$V = \frac{W}{Q}$$

$$W = VQ$$

Substituting work done into our generic equation for power gives

$$P = \frac{VQ}{t}$$

Synoptic link

If you need to review the equation for power, look at Topic 5.4, Power and efficiency.

However, $\frac{Q}{t}$ is equal to the current I. Therefore

$$P = VI$$

Calculating energy transferred

The energy transferred in a given time can be determined by combining our general equation for power and our equation for electrical power.

$$P = \frac{W}{t} \text{ is rearranged to give } W = Pt$$

Substituting in $P = VI$ gives

$$W = VIt$$

where W is the energy transferred in joules, V is the p.d. in volts, I is the current in amperes and t is the time in seconds.

Worked example: Kettle power II

Calculate the energy transferred into heat in 2.0 minutes by the kettle in the first worked example.

Step 1: Identify the correct equation to calculate energy transfer.

$P = \frac{W}{t}$ Rearrange to make the energy transferred the subject. $W = Pt$

Step 2: Substitute the values in SI units into the equation and calculate the value for the energy transferred.

$$W = 2900 \times 120 = 3.48 \times 10^5 \text{ J}$$

$$W = 3.5 \times 10^5 \text{ J (2 s.f.)}$$

Summary questions

1 Using the appropriate circuit symbols outline a simple experiment that could be used to determine the power of a filament lamp (you should include a circuit diagram in your answer). *(2 marks)*

2 Calculate the rate of energy transfer from a resistor with a current of 5.0 A and a p.d. of 8.0 V. *(2 marks)*

3 A 1.2 kW heater has a p.d. of 20 V when working normally. Calculate:

a the current in the heater; *(2 marks)*

b the energy transferred in one hour. *(2 marks)*

4 State and explain the effect on the rate of energy transfer for a component when:

a its resistance doubles (the current in the component remains unchanged);

b the current in the component doubles (the resistance is unchanged). *(4 marks)*

5 Using base units, show that one watt is equivalent to one volt amp (1 W = 1 V A). *(4 marks)*

Electricity in the home

What would our homes be like without electricity? Most of the devices we use at home require electricity. They transfer electrical energy into other useful forms for everything from heating and cooking to entertaining ourselves. Think about all the electrical devices used in your home in the last week, and what you would have to do without them.

However, there is a financial cost, and we need to pay to keep all of our devices running.

> **Learning outcomes**
>
> Demonstrate knowledge, understanding, and application of:
>
> → the kilowatt-hour (kW h) as a unit of energy
>
> → calculating the cost of energy.

Paying for energy

We pay electricity companies for the total amount of electrical energy they transfer to our homes. By law, each home contains an **electricity meter** that accurately records the transfer of energy from the National Grid to the house (Figure 1). All the electricity supplied to the house passes through the meter.

▲ **Figure 1** *A typical electricity meter, which is calibrated to ensure it accurately measures the electrical energy transferred to the home*

The energy transferred to each individual electrical device, and so how much it costs to run, depends on two factors (Figure 1):

- the power of the device
- how long the device is used for.

These two factors must be considered together when calculating the energy transferred by the device. For example, a powerful microwave oven may be cheaper to run than a less powerful one, despite transferring more energy per second, as it may take less time to cook food.

The defining equation for power is

$$\text{power} = \frac{\text{energy transferred}}{\text{time taken}} \qquad P = \frac{W}{t}$$

Rearranging this equation to determine the energy transferred by an electrical device gives

$$W = Pt$$

▲ **Figure 2** *A label on the base normally indicates the power rating of a device, meaning the energy transferred to the device each second*

The kilowatt-hour

The SI unit for energy is the joule, but one joule is a tiny amount on the scale of the energy transferred to our homes. It takes at least 100 kJ to boil just one cup of water. Electricity bills therefore use a derived unit, the **kilowatt-hour** (kW h), defined as the energy transferred by a device with a power of 1 kW operating for a time of 1 hour.

From its definition 1 kW h is equivalent to 3.6 MJ.

Our equation for energy transferred therefore has two versions, depending on the units used.

> **Synoptic link**
>
> To remind yourself of the meaning of power and its defining equation, look at Topic 5.4, Power and efficiency.

Worked example: Calculating energy transferred in kW h

A 1450 W dishwasher is used for 15.0 minutes. Calculate the energy transferred in kW h.

Step 1: Identify the correct equation to calculate energy transferred.

$$W = Pt$$

Step 2: Substitute in the values, paying close attention to the units.

$$W = 1.450 \times 0.250$$
$$= 0.363\,\text{kW h (3 s.f.)}$$

Study tip

Make sure that you are confident about the meaning of 'power'. Phrases like 'the rate of energy transfer', 'the energy transferred per second', 'the rate of work done' are all the same thing, just written differently.

Study tip

It's always important to look carefully at the units used in calculations. It's even more important where there might be different units used depending on the context, as with joules and kilowatt-hours.

SI units: energy transferred (J) = power of device (W) × time for which the device is used (s)

kW h: energy transferred (kW h) = power of device (kW) × time for which the device is used (h)

Electricity Statement

Electricity Readings					
Meter Serial no.	**Read Date**	***Read Type***	**Read**	**Last Read**	**Units Used**
		Removal	28619	28170	449
		Smart	1749	0	1749
Total units					**2198 kWh**

Electricity Charges 05 Jan 2010 - 30 Jun 2010				
Electricity supply standing charge	177 days	22.0p per day	£	38.94
Electricity total unit charge	2198.0 kWh	8.085p per kWh	£	177.71
Total supply charges			**£**	**216.65**
VAT @5.00%			£	10.83
Total cost of electricity			**£**	**227.48**

▲ **Figure 3** *An electricity bill shows how much energy has been transferred to each home in a certain time (usually each quarter) in kWh*

The cost of each kW h of electrical energy (sometimes simply referred to as a 'unit') varies between operating companies. It depends on the particular tariff from the electricity company, and even in some cases on the time of day the energy is transferred. Typical costs are around 6–15 p per kW h.

Summary questions

1 Calculate the energy transferred in joules by a 60 W television in 2.0 hours. *(2 marks)*

2 Two readings are taken from an electricity meter. Feb: 0034512 – Jun: 0035387. If each unit costs 12 p calculate the cost of the electrical energy transferred to the home during this time. *(2 marks)*

3 Show that one kilowatt-hour is equal to 3 600 000 J. *(3 marks)*

4 Calculate the energy transferred in kW h by a 9000 W power shower used for 15 minutes:
a in J; **b** in kW h. *(4 marks)*

5 If one kWh costs 11.2 p, calculate the cost of leaving a 60 W lamp on continuously for 5.0 weeks. *(3 marks)*

6 A household decides to swap all their filament lamps for energy-efficient LEDs. They have a total of 18 lamps each with a power of 100 W. The replacement LEDs have a power of 15 W.
If each lamp is used for an average of 2.0 hours during the day, calculate the difference in energy transferred over one year. *(5 marks)*

Practice questions

1 a With the help of a sketch graph, describe and explain how the resistance of a negative temperature coefficient (NTC) thermistor is affected by its temperature. *(3 marks)*

b Figure 1 shows a circuit for a simple light-meter designed by a student.

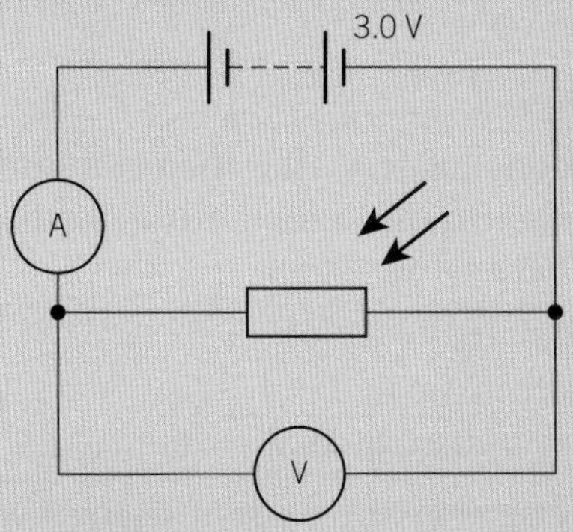

▲ **Figure 1**

The battery and the ammeter both have negligible internal resistances. Discuss how each meter responds as the intensity of light incident on the LDR is increased. *(4 marks)*

2 a State one similarity and one difference between potential difference and electromotive force (e.m.f.). *(2 marks)*

b The e.m.f of a cell is 1.5 V. The energy transformed by the cell is 100 J. Calculate the total charge flowing through the cell. *(2 marks)*

c According to a student, an electron accelerated from rest can be made to travel at a speed greater than 10 km s^{-1} by using the cell from (b) connected across two electrodes. With the help of a calculation, show whether or not this suggestion is correct. *(4 marks)*

3 a Show that the SI unit for resistivity is Ω m. *(2 marks)*

b Figure 2 shows a cube made from a material of resistivity 5.0×10^{-2} Ω m.

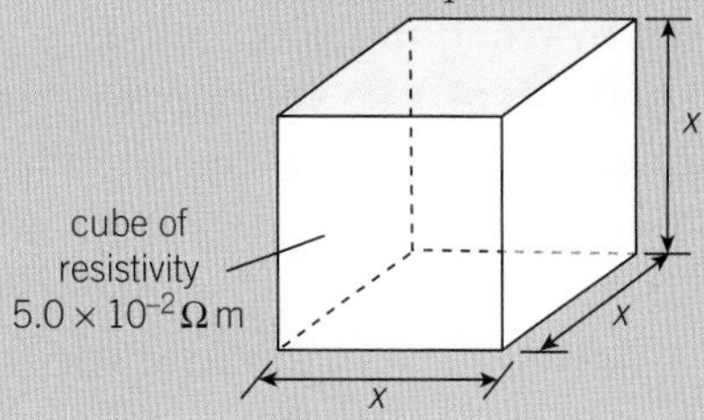

▲ **Figure 2**

The resistance of the cube across any of its two opposite faces is 5.0 Ω. The length of each edge of the cube is *x*. Calculate the length *x*. *(3 marks)*

c A student conducts an experiment to determine the resistivity of a metal in the form of a wire. Figure 2 shows a graph of resistance *R* against its length *L*.

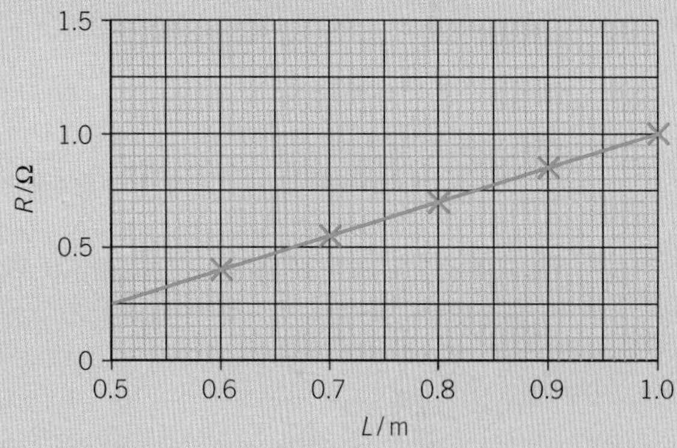

▲ **Figure 3**

(i) Explain how the student may have determined the resistance of each length of the wire using meters. Assume an ohmmeter is not available. *(3 marks)*

(ii) The wire has cross-sectional area of 7.8×10^{-7} m^2. Use Figure 3 to determine the resistivity of the metal. *(4 marks)*

4 The power of a 230 V mains filament lamp is 40 W.

a Define *power*. *(1 mark)*

b The lamp is connected to the 230 V supply. Calculate

(i) the current *I* in the filament, *(2 marks)*

(ii) the resistance *R* of the filament. *(1 mark)*

c The cross-sectional area of the filament is $3.0 \times 10^{-8}\,m^2$. The resistivity of the filament when the lamp is lit is $7.0 \times 10^{-5}\,\Omega\,m$. Use your answer to **(b)(ii)** to calculate the length L of the filament wire. *(3 marks)*

May 2012 G482

5 Figure 4 shows the I–V characteristic of a slice of semiconducting material.

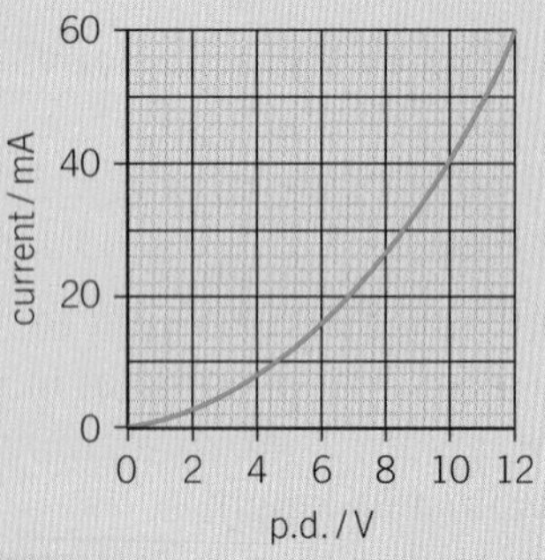

▲ **Figure 4**

a (i) Define *resistance*. *(1 mark)*

(ii) Show that the resistance of the slice is about 250 Ω when there is a current of 40 mA in it. *(2 marks)*

b The dimensions of the slice are shown in Figure 5.

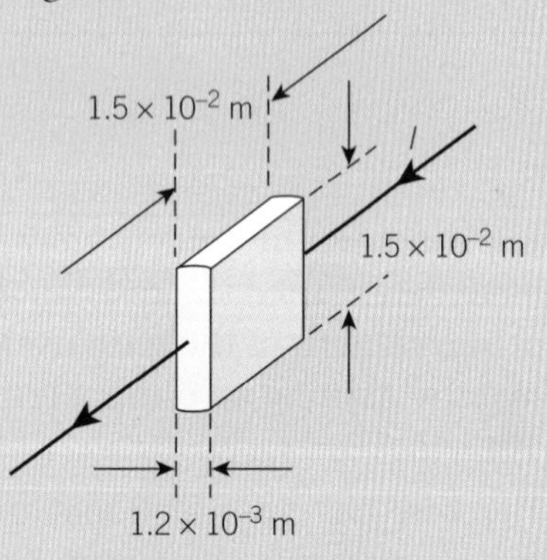

▲ **Figure 5**

Calculate the resistivity ρ of the semiconducting material when there is a current I of 40 mA in the slice. *(3 marks)*

c Explain how the I–V characteristic shows that the resistivity of the semiconducting material decreases with increasing temperature. *(4 marks)*

June 2013 G482

6 A student connects a component across a battery of negligible internal resistance. Figure 6 shows the variation of the current I in this component with time t from the moment the component is connected to the battery.

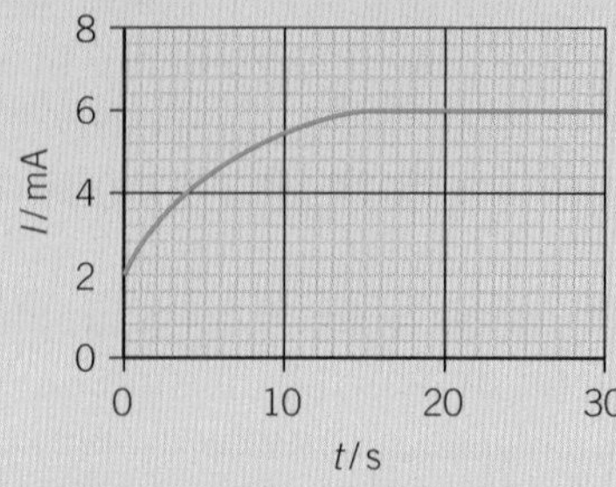

▲ **Figure 6**

The student suggests that the component must be a negative temperature coefficient (NTC) thermistor.

a Calculate the ratio $\dfrac{\text{power dissipated at } t = 30\,s}{\text{power dissipated at } t = 0}$. *(3 marks)*

b Explain why the current changes as shown in the graph of Figure 6 *(4 marks)*

7 This question is about the rigid copper bars which carry the very large currents generated in a power station to the transformers. Figure 7 shows such a copper bar.

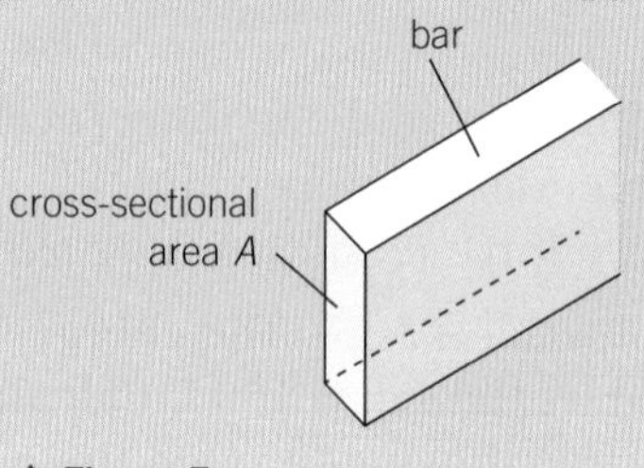

▲ **Figure 7**

a Write down a suitable word equation to define the *resistivity* of a material. *(1 mark)*

b (i) The cross-sectional area A of the bar is $6.4 \times 10^{-3}\,m^2$. Calculate the resistance of a 1.0 m length of the bar. The resistivity of copper is $1.7 \times 10^{-8}\,\Omega\,m$. *(2 marks)*

(ii) The bar carries a constant current of 8000 A. Calculate the power dissipated as heat along a 1.0 m length of it. *(3 marks)*

(iii) The bar is 9.0 m long. Estimate the total energy in kW h lost from the bar in one day. *(2 marks)*

(iv) Calculate the cost per day of operating the copper bar. The cost of 1 kW h is 15 p. *(1 mark)*

Jan 2012 G482

8 a Define electrical resistivity. *(1 mark)*

b Figure 8 shows the 'lead' of a pencil.

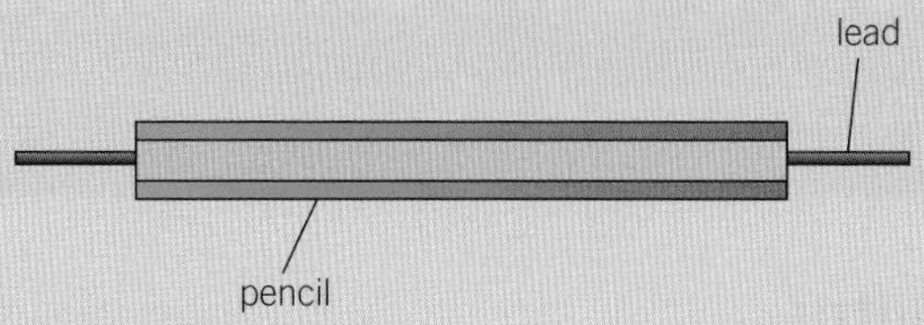

▲ **Figure 8**

Describe how you can determine the resistivity of the material of the 'lead' in the laboratory.

In your description pay particular attention to

- how the circuit is connected
- what measurements are taken
- how the data is analysed. *(5 marks)*

c Figure 9 shows a glass tube with some conducting paint.

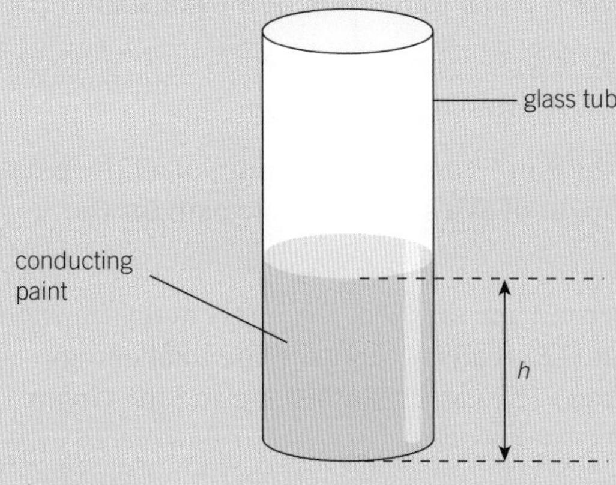

▲ **Figure 9**

The volume of the paint is $8.0 \times 10^{-6}\,m^3$. The internal cross-sectional area of the base of the tube is $1.2\,cm^2$.

(i) Calculate the height h of the paint column. *(2 marks)*

(ii) The resistivity of the paint is $5.2 \times 10^{-4}\,\Omega\,m$.

Calculate the resistance of the paint column. *(3 marks)*

(iii) State and explain how your answer to **(c)(ii)** would change if the same volume of paint is poured into another glass tube with double the internal diameter. *(3 marks)*

9 Figure 10 shows an electrical cable consisting of bare copper wires encased in plastic insulation.

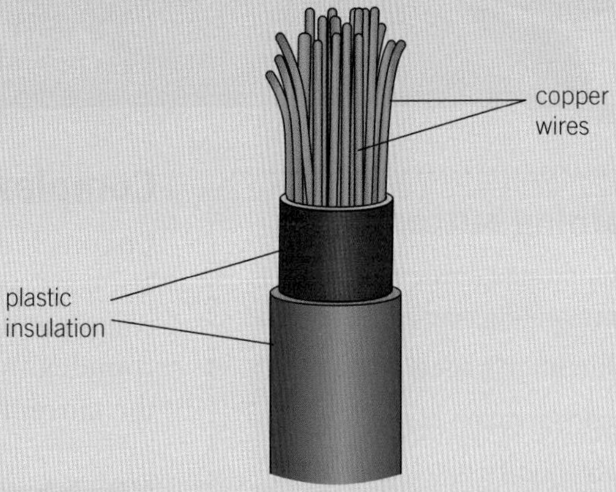

▲ **Figure 10**

a A particular cable contains 26 copper wires and is 12.0 m long. The radius of each copper wire is $3.50 \times 10^{-5}\,m$. The resistivity of copper is $1.70 \times 10^{-8}\,\Omega\,m$.

(i) Show that the resistance of a single copper wire is about $53\,\Omega$. *(3 marks)*

(ii) Explain why the resistance of the electrical cable is about $2\,\Omega$. *(1 mark)*

b Figure 11 shows two electrical cables used to connect a power supply to a lamp. Each cable has length 12.0 m and is identical to that described in **(a)**.

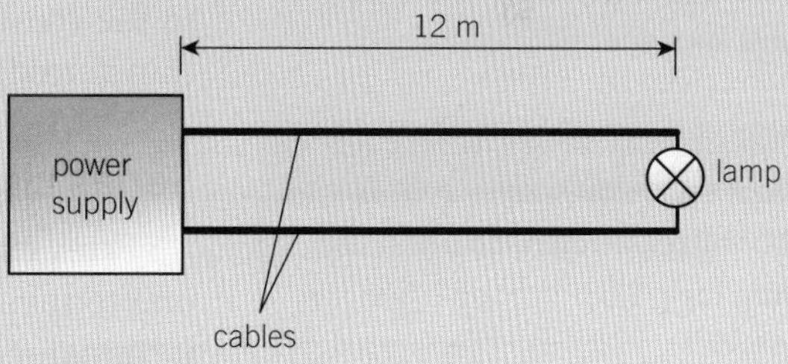

▲ **Figure 11**

The lamp is rated at 24 W, 6.0 V. The power supply has negligible internal resistance and its output is adjusted so that the potential difference across the lamp is 6.0 V.

(i) Calculate the resistance of the lamp when operating at 6.0 V *(2 marks)*

(ii) Explain why the e.m.f of the power supply is greater than 6.0 V. *(1 mark)*

(iii) Calculate the e.m.f. of the power supply. *(2 marks)*

May 2008 2822

10 ELECTRICA CIRCUITS

10.1 Kirchhoff's laws and circuits

Specification reference: 4.3.1

Learning outcomes

Demonstrate knowledge, understanding, and application of:

- → Kirchhoff's second law
- → conservation of energy
- → Kirchhoff's first and second laws applied to electrical circuits.

▲ **Figure 1** *Circuits found in modern electronic devices can be incredibly complex, but they rely on just a few simple laws and relationships*

Synoptic link

You will recall Kirchhoff's first law from Topic 8.3, Kirchhoff's first law.

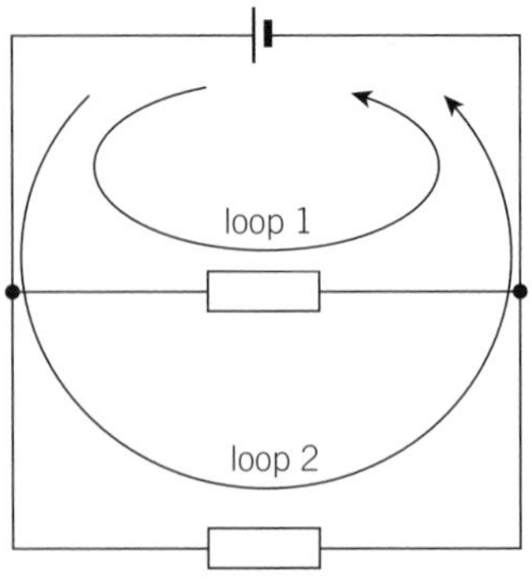

▲ **Figure 2** *A closed loop is one possible path for the current – in a series circuit there is only one loop, but in parallel there are often several possible loops*

Complex circuitry

Circuits found in modern electronic devices can look very daunting. They are rarely simple series or parallel circuits. Instead hundreds of components are connected together in complex ways. No matter how complex the circuit, though, it follows a few simple rules governing the current and p.d. through the circuit.

Kirchhoff's laws

We have already met Kirchhoff's first law, $\Sigma I_{in} = \Sigma I_{out}$. It is essentially the law of conservation of charge applied to electric circuits.

Kirchhoff's second law takes the law of conservation of energy and applies it to electrical circuits. It states: In any circuit, the sum of the electromotive forces is equal to the sum of the p.d.s around a closed loop.

$$\Sigma \mathcal{E} = \Sigma V \text{ around a closed loop}$$

where $\Sigma \mathcal{E}$ is the sum of the e.m.f.s and ΣV is the sum of all the p.d.s. A closed loop can be thought of as a single possible path for the current (Figure 2).

Essentially Kirchhoff's second law says that the total energy transferred to the charges in a circuit ($\Sigma \mathcal{E}$) is always equal to the total energy transferred from the charges (ΣV) as they move around the circuit.

Series circuits

A **series circuit** has only one path for the current, a single loop from one terminal of the source of e.m.f. (e.g., a cell) back to the other terminal (Figure 3).

In a series circuit the current is the same in every position.

From Kirchhoff's first law you know that the rate of flow of charge is the same at all points in the circuit, the charge is not used up; it just flows around the circuit.

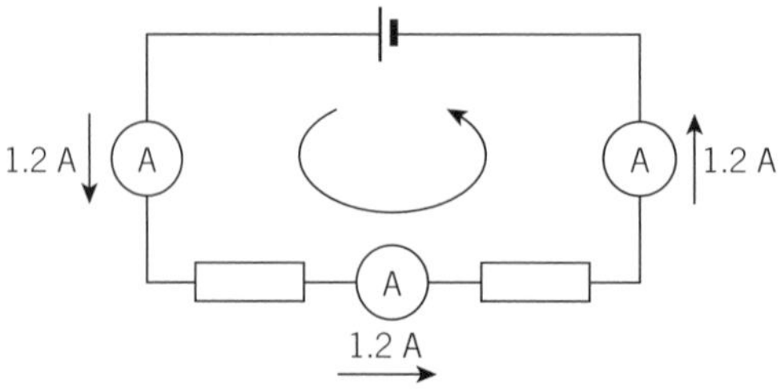

▲ **Figure 3** *In a series circuit there is one possible path, or one loop for the current, and the current is the same at all places*

Since a series circuit has only one closed loop, then from Kirchhoff's second law the e.m.f. is shared between the components. The sum of the p.d.s across the components is always equal to the e.m.f. (Figure 4). If the circuit contains two components with the same resistance, the e.m.f. is shared equally between them. If the components have different resistances the component with the greater resistance will take a greater proportion of the e.m.f.

In circuits with more than one source of e.m.f., the same rule applies, but we need to add the e.m.f. from each source, before sharing it between the components (Figure 5). The sources of e.m.f. are connected with opposing polarities. The sum of the e.m.f.s is equal to 9.0 V − 6.0 V = 3.0 V, and not 15 V.

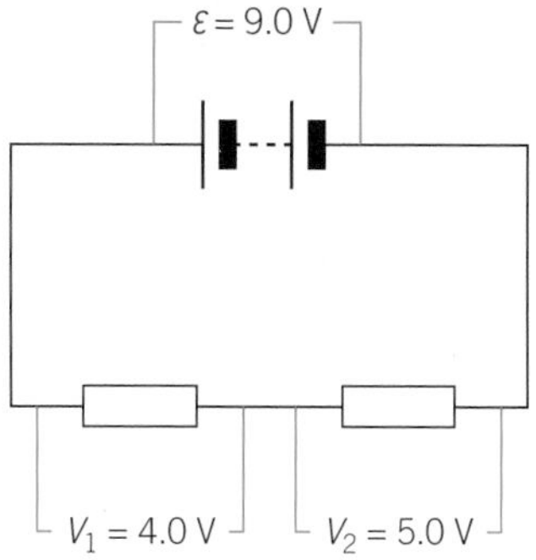

▲ **Figure 4** *In a series circuit the e.m.f. from the power source is shared between components: the total p.d. always adds up to the e.m.f.*

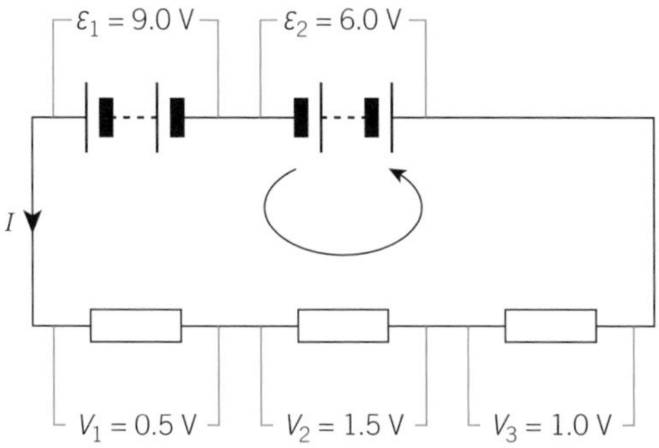

▲ **Figure 5** *Where there are multiple sources of e.m.f., the p.d. across each component must add up to the total e.m.f. (in this case 3.0 V)*

Parallel circuits

A **parallel circuit** provides more than one possible path for the charges. How much charge flows down each path depends on the resistance of the path. Kirchhoff's first law tells us that the current into each junction must be equal to the current out of that junction (Figure 6).

The greater the resistance of the branch, the lower the current that passes through it. If one branch has half the resistance of the other, it will have twice the current through it, so two-thirds of the total current will go through the branch with the lower resistance.

Each parallel branch can be thought of as a separate circuit. If changes are made to one branch, the other branches are not affected (Figure 7).

In a parallel circuit there are several different loops. Each branch forms its own loop. Kirchhoff's second law tells us that around each loop the e.m.f. must equal the p.d., so in other words the total p.d. across each branch is equal to the e.m.f. from the power supply (Figure 8). If one branch contains several components then the sum of the p.d.s across these components must equal the e.m.f. (Figure 9).

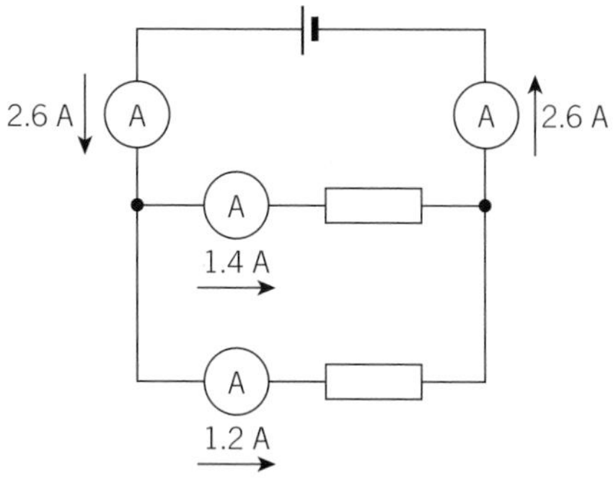

▲ **Figure 6** *In a parallel circuit the current in each branch might be different, but the current into each junction must always equal the total current leaving it (in this case 2.6 A = 1.4 A + 1.2 A)*

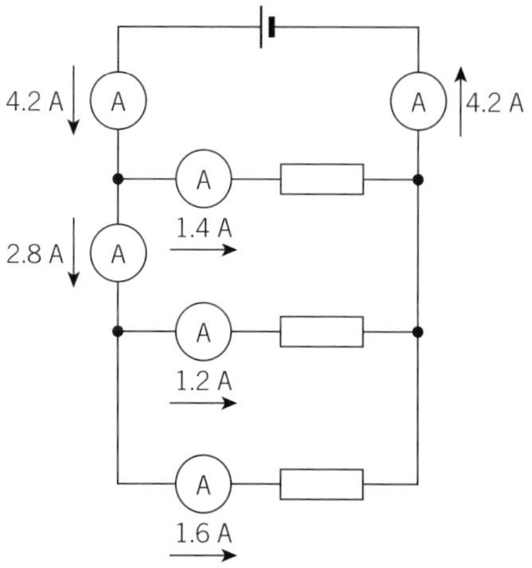

▲ **Figure 7** *Adding another branch has no effect on the current through the first two branches, but an additional 1.6 A is drawn from the cell for the final branch, increasing the current through the cell to 4.2 A*

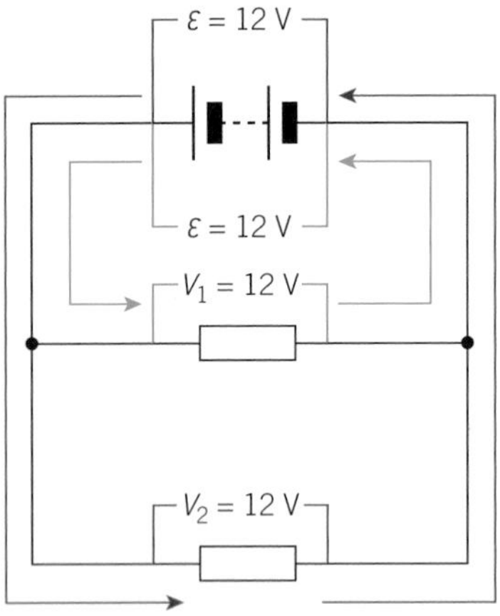

▲ **Figure 8** *In parallel circuits, the p.d. across each branch is the same as the e.m.f., no matter how many branches there are*

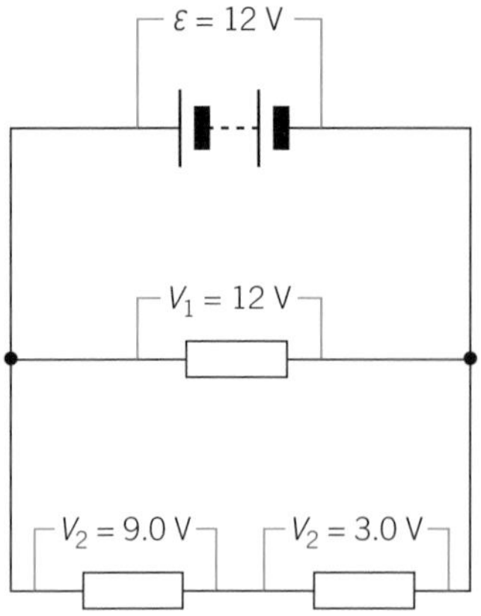

▲ **Figure 9** *If there is more than one component on a branch, then within that branch the sum of the p.d.s across the components must equal the e.m.f., as in a series circuit*

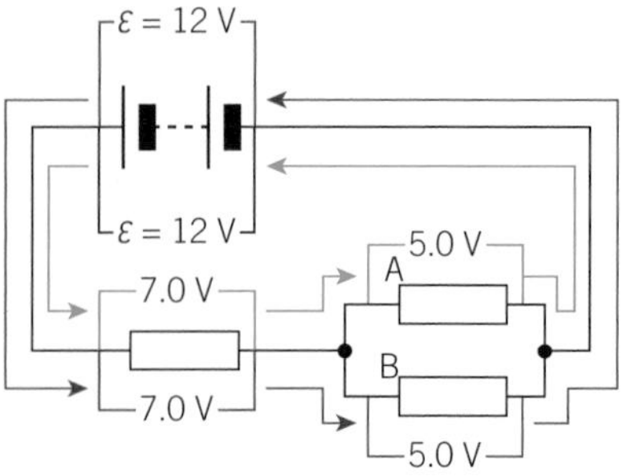

▲ **Figure 10** *It's often helpful to draw separate loops in more complex circuits to determine any unknown values: here resistors A and B are in parallel, so the p.d. across them must be the same; they don't 'share' the 5.0 V between them*

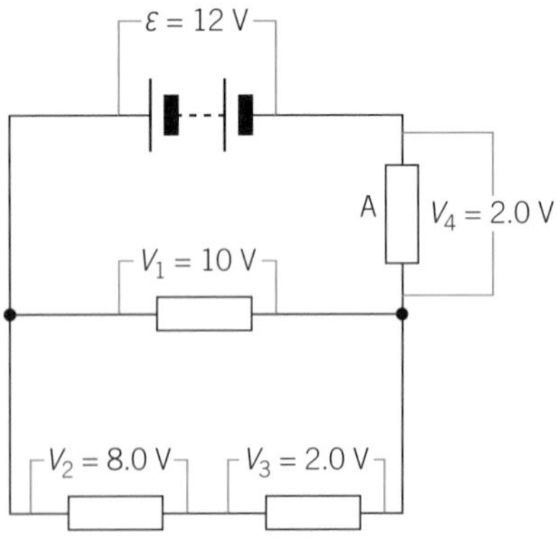

▲ **Figure 11** *In this example resistor A has a p.d. 2.0 V across it and so 10 V remains from the supplied 12 V for each branch – the p.d.s around each loop must add up to 12 V*

Multiple loops and adding components

In more complex circuits it is useful to consider each loop separately, paying particular attention to components in parallel (Figure 10).

Adding additional components in series reduces the e.m.f. that is shared between the original components (Figure 11). In this case the addition of a resistor means there is a lower p.d. across the original components as 2.0 V of the 12 V supplied has been taken up by the new resistor.

Study tip

When solving circuit problems, it is really helpful to label all the values you know for current, e.m.f., and p.d. on the circuit diagram. This often makes it easier to determine any missing values.

Summary questions

1 Fill in the six missing values A–F for the current in the circuits in Figure 12. *(6 marks)*

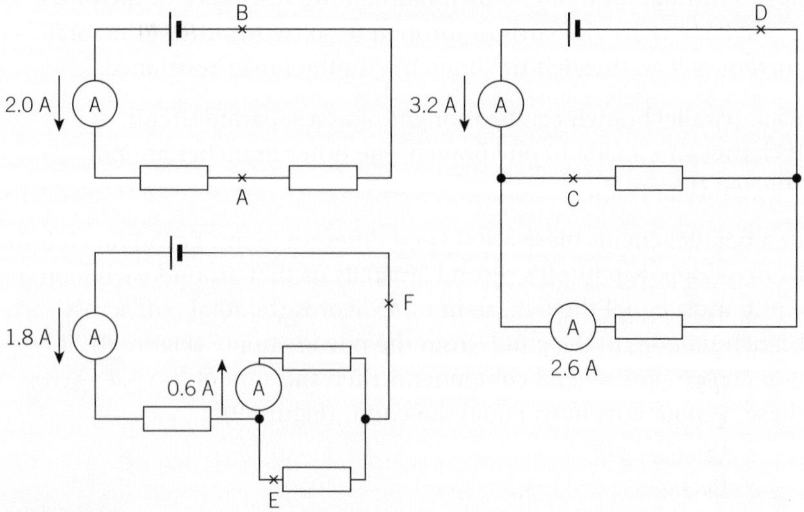

▲ **Figure 12**

2 Fill in the seven missing values **A–G** for the p.d.s in the circuits in Figure 13. *(7 marks)*

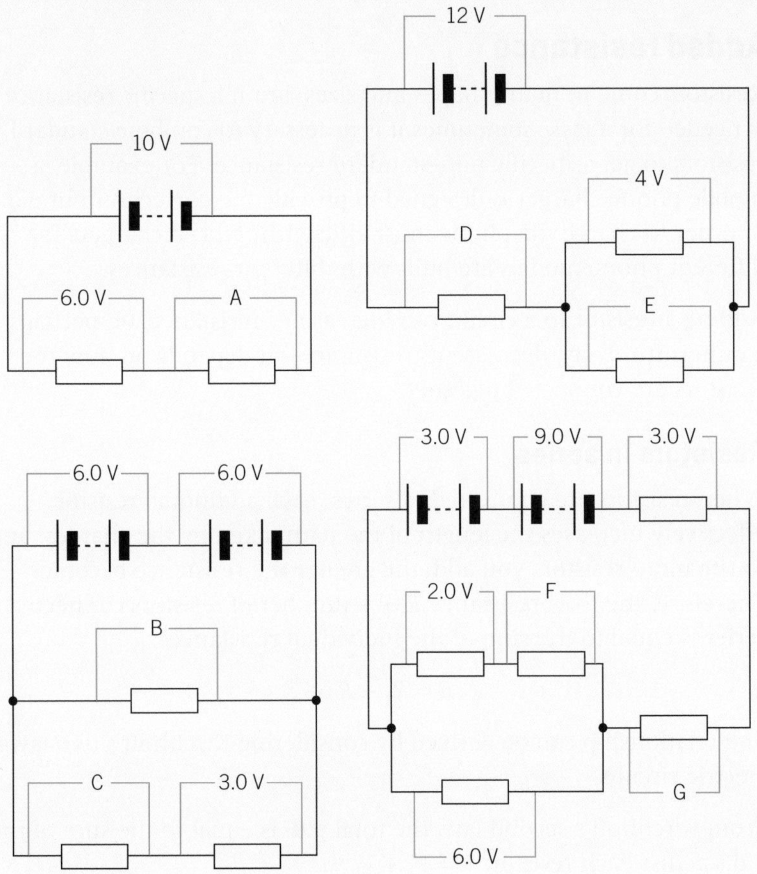

▲ **Figure 13**

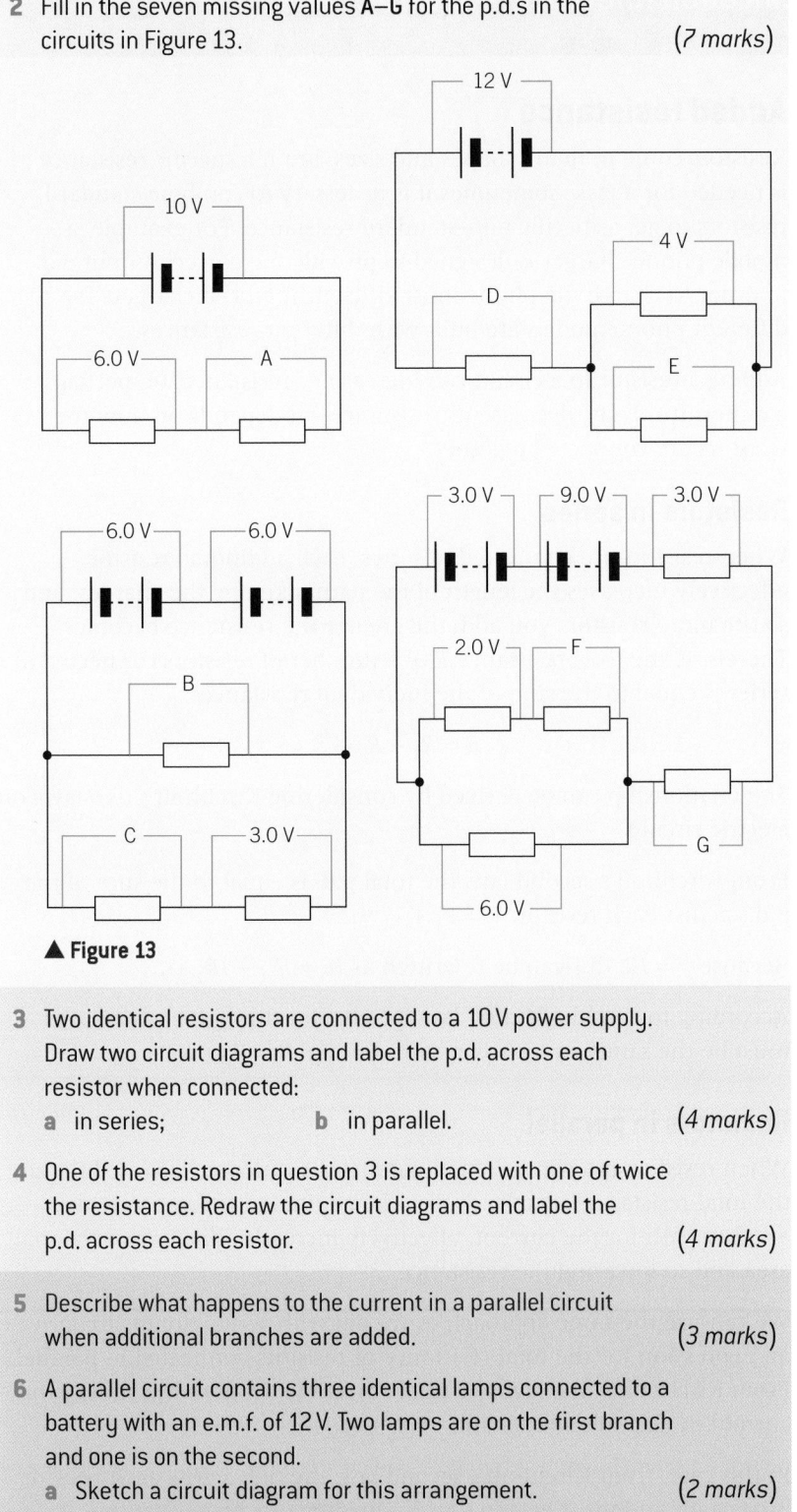

3 Two identical resistors are connected to a 10 V power supply. Draw two circuit diagrams and label the p.d. across each resistor when connected:

a in series; **b** in parallel. *(4 marks)*

4 One of the resistors in question 3 is replaced with one of twice the resistance. Redraw the circuit diagrams and label the p.d. across each resistor. *(4 marks)*

5 Describe what happens to the current in a parallel circuit when additional branches are added. *(3 marks)*

6 A parallel circuit contains three identical lamps connected to a battery with an e.m.f. of 12 V. Two lamps are on the first branch and one is on the second.

a Sketch a circuit diagram for this arrangement. *(2 marks)*

b If a current of 6.0 A is drawn from the cell, determine the current through each lamp in the circuit. *(4 marks)*

c Calculate the power dissipated by each lamp. *(4 marks)*

d Over time, the current drawn from the battery decreases as the battery begins to go flat. If a current of 5.0 A is drawn from the battery determine the current through each lamp. *(3 marks)*

Learning outcomes

Demonstrate knowledge, understanding, and application of:

→ total resistance of resistors in series $R = R_1 + R_2 + \ldots$

→ total resistance of resistors in parallel $\frac{1}{R} = \frac{1}{R_1} + \frac{1}{R_2} + \ldots$

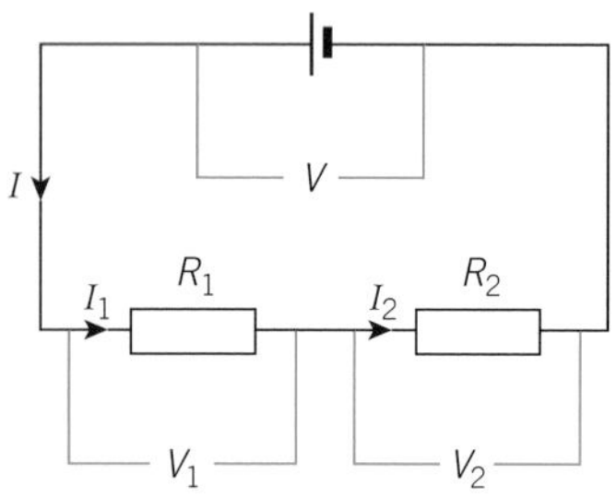

▲ **Figure 1** *In series, the total resistance is just the sum of the resistances of the resistors*

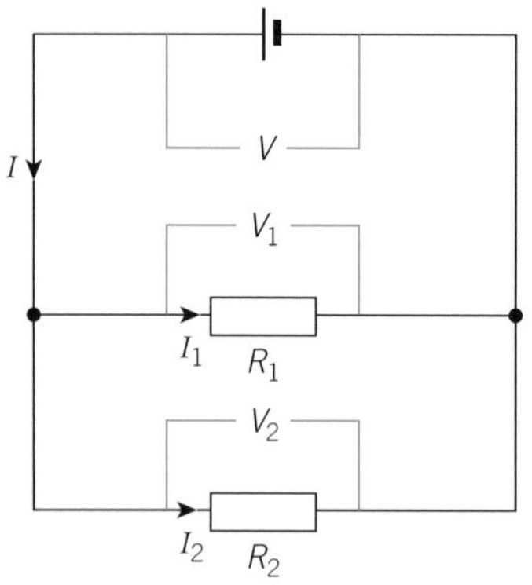

▲ **Figure 2** *The total resistance decreases as more resistors are added in parallel*

Added resistance

Resistors come in many shapes and sizes, but if a specific resistance is needed for a task, sometimes it is necessary to combine standard resistors to get a specific non-standard resistance. For example, a mobile phone charger is designed to provide the correct output p.d. in order to charge the phone most efficiently, and so chargers for different phone models are built with different resistances.

Adding a resistor to a circuit can increase its resistance, or, perhaps counterintuitively, decrease its resistance – it depends on how the resistors are connected together.

Resistors in series

When resistors are connected in series, each additional resistor effectively increases the length of the path taken by the charges, and so the more resistors you add, the greater the resistance becomes. Therefore, the total resistance R of a number of resistors connected in series is equal to the sum of the individual resistances.

$$R = R_1 + R_2 + \ldots$$

The relationship can be derived by considering Kirchhoff's two laws on electric circuits.

From Kirchhoff's second law, the total p.d. is equal to the sum of the p.d.s across each resistor: $V = V_1 + V_2 + \ldots$

Because $V = IR$, this can be rewritten as $IR = IR_1 + IR_2 + \ldots$

According to Kirchhoff's first law the current through each resistor must be the same, so I is a constant. Giving $R = R_1 + R_2 + \ldots$

Resistors in parallel

When resistors are connected in parallel the outcome is very different: the total resistance actually drops. The additional resistor provides another path for the current, effectively increasing the cross-sectional area and so lowering the resistance.

We can use the same approach as we did with series circuits to derive an expression for the total resistance of resistors connected in parallel. From Kirchhoff's first law, the total current is equal to the sum of the current in each resistor, giving $I = I_1 + I_2 + \ldots$

In this case, from Kirchhoff's second law, the p.d. across each resistor is constant and must be equal to V. Dividing our first equation by V gives

$$\frac{I}{V} = \frac{I_1}{V} + \frac{I_2}{V} + \ldots$$

$V = IR$, so $\frac{I}{V} = \frac{1}{R}$, so the equation for the total resistance R of resistors connected in parallel is

$$\frac{1}{R} = \frac{1}{R_1} + \frac{1}{R_2} + \ldots$$

Worked example: Three resistors in parallel

Resistors of 6.0 Ω, 4.0 Ω, and 3.0 Ω are connected in parallel. Calculate the total resistance of this combination.

Step 1: Identify the correct equation to calculate the resistance for resistors connected in parallel.

$$\frac{1}{R} = \frac{1}{R_1} + \frac{1}{R_2} + \frac{1}{R_3}$$

Step 2: Substitute in known values and calculate a value for $\frac{1}{R}$.

$$\frac{1}{R} = \frac{1}{6.0} + \frac{1}{4.0} + \frac{1}{3.0} = 0.75$$

Step 3: Invert $\frac{1}{R}$ to calculate a value for R.

$$\frac{1}{R} = 0.75 \text{ therefore } R = \frac{1}{0.75}$$

Calculate R and express it to an appropriate number of significant figures.

$$R = 1.3\,\Omega \text{ (2 s.f.)}$$

The total resistance is always lower than the lowest resistance of any resistor in the combination. This is a quick way to check whether your answer is correct.

Study tip

When using the equation for the total resistance of resistors connected in parallel, don't forget that your initial calculation gives you $\frac{1}{R}$. To find R you must find the reciprocal of your value (use the x^{-1} button on your calculator).

Resistor circuits

Different combinations of resistors in series and parallel can be used to build more complex **resistor circuits**. Using the relationships above we can determine the total resistance of each circuit (Figure 3).

Study tip

The equations for total resistance of series and parallel circuits can also be applied to other circuit components.

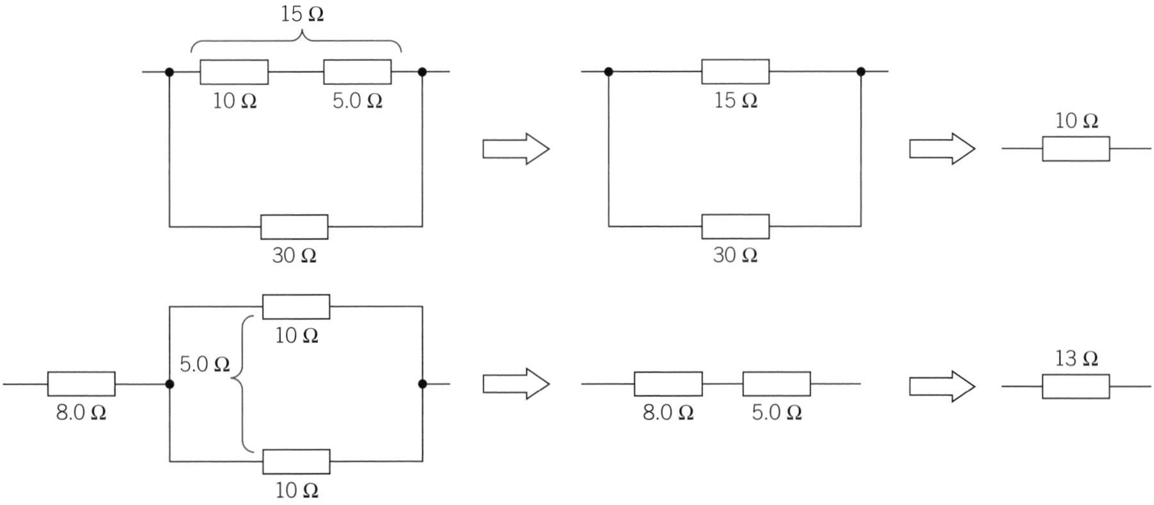

▲ **Figure 3** *It is possible to simplify the parts of a complex resistor circuit before calculating the total resistance*

Summary questions

1 State what happens to the resistance when resistors are connected:

a in series; *(1 mark)*

b in parallel. *(1 mark)*

2 Calculate the total resistance of the resistor circuits in Figure 4. *(5 marks)*

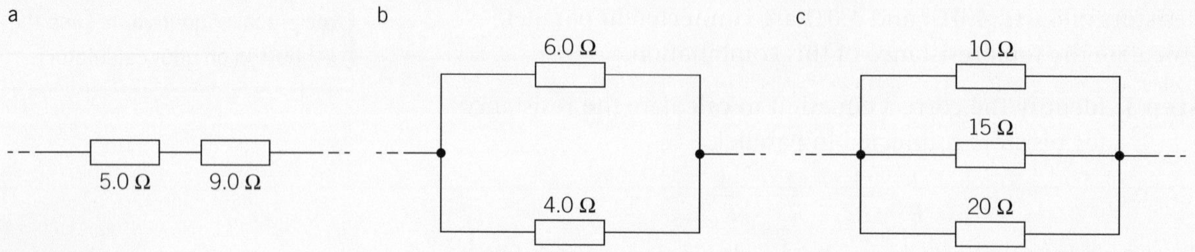

▲ **Figure 4**

3 Calculate the total resistance of the resistor circuits in Figure 5. *(6 marks)*

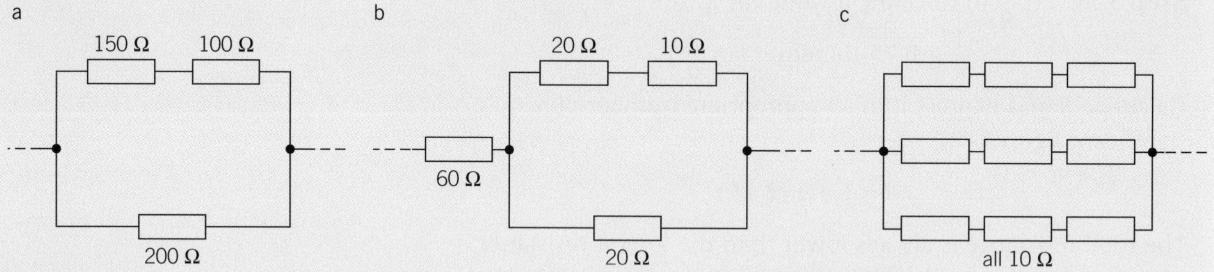

▲ **Figure 5**

4 Determine all the possible combinations of one or more resistors from a selection of three resistors with resistance 100 Ω, 50 Ω, and 200 Ω. For each combination draw a diagram of the resistor circuit and calculate the total resistance. *(15 marks)*

5 The total resistance of three resistors in parallel is 1030 Ω. Two of the resistors are known to have resistances 2.2 kΩ and 4.7 kΩ. Calculate the resistance of the third resistor. *(3 marks)*

Call an electrician

The Gaia space probe (Figure 1) launched in 2013 is now around 1.6 million kilometres from the Earth. Its five-year mission is to search for extrasolar planets, asteroids within the solar system, and to collect further evidence to test Einstein's theory of general relativity. All the complex electrical circuitry in Gaia had to be carefully designed and thoroughly tested using the rules you have been studying. If scientists got their calculations wrong, there would be no way to repair the probe so far from the Earth.

Learning outcomes

Demonstrate knowledge, understanding, and application of:

→ analysis of circuits with components in both series and parallel

→ analysis of circuits with more than one source of e.m.f.

▲ **Figure 1** *The circuits found in cutting-edge technologies still rely on Kirchhoff's laws and the other relationships you have studied*

An electrical mathematical toolkit

When tackling any complex circuit problem we should start with Kirchhoff's circuit laws and these four key electrical relationships.

$$I = \frac{\Delta Q}{\Delta t} \qquad V = \frac{W}{Q} \qquad P = VI \qquad \text{and most importantly} \qquad V = IR$$

In the following three examples we shall determine any missing values for p.d. or current. Resistors are used for simplicity, but circuits could contain a variety of components, including diodes, filament lamps, and so on.

Worked example: Circuit 1

For the circuit in Figure 2, calculate the current in, p.d. across, and power rating for each resistor and in total.

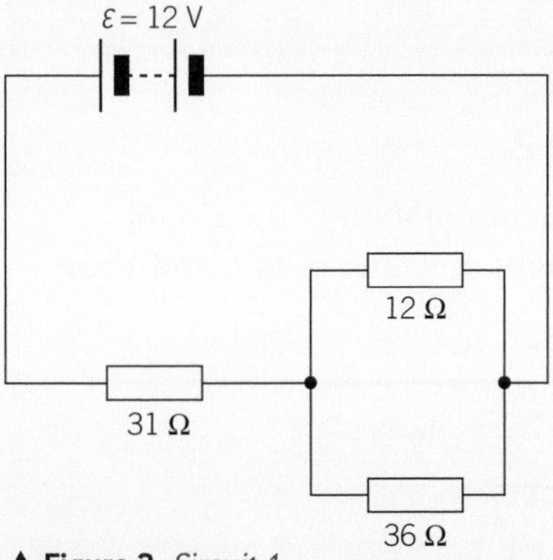

▲ **Figure 2** *Circuit 1*

Step 1: We can use our formulae for resistor networks to determine the total resistance of the network.

Using $\frac{1}{R} = \frac{1}{R_1} + \frac{1}{R_2} + \ldots$ we first find the resistance of the two resistors in parallel (9.0 Ω) and add this to the single resistor, giving a total resistance of $R = 40\,\Omega$.

Step 2: We can then use $V = IR$ to determine the current in the circuit, $I = 0.30\,A$. This must also be the current in the $31\,\Omega$ resistor, as it is in series.

Step 3: We can use $V = IR$ to determine the p.d. across this resistor, $V_{31\Omega} = 9.3\,V$.

Step 4: Using Kirchhoff's second law we now know that the p.d. across both the $12\,\Omega$ and $36\,\Omega$ resistors must be 2.7 V. Using $V = IR$ we can then calculate the current in each resistor and using $P = VI$ the power of each component.

The answers are summarised in Table 1. All values are expressed to two significant figures.

▼ **Table 1**

Component	Current / A	p.d. / V	Resistance / Ω	Power / W
31 Ω resistor	0.30	9.3	*31*	2.8
12 Ω resistor	0.23	2.7	*12*	0.62
36 Ω resistor	0.075	2.7	*36*	0.20
total circuit	0.30	12	40	3.6

The total values can be used to check for any calculation errors (using $V = IR$ and $P = VI$).

If we are given a known time we could also determine the energy transferred by each component and the charge passing through each component in that time using $I = \frac{\Delta Q}{\Delta t}$ and $V = \frac{W}{Q}$.

Worked example: Circuit 2

In our second example circuit the orientation of the circuit appears different and an assortment of values of R, I, and V are available to us.

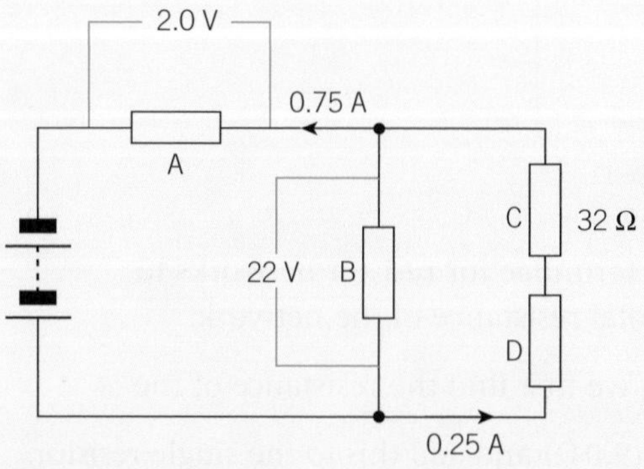

▲ **Figure 3** *Circuit 2*

Step 1: By considering the loop including the battery, resistor B, and resistor A, we can determine that the e.m.f. must be 24 V (2 s.f.). We can also use Kirchhoff's first law to find the current in resistor B, $I_B = 0.50$ A.

Step 2: Using $V = IR$ we can calculate the p.d. across resistor C ($R_c = 8.0$ V) and from Kirchhoff's second law the p.d. across resistor D must be 14 V.

Step 3: From this point we can use $V = IR$ and $P = VI$ to determine all the other values for V, I, R, and P, summarised in Table 2. All values are expressed to two significant figures.

▼ **Table 2**

Component	Current / A	p.d. / V	Resistance / Ω	Power / W
resistor A	*0.75*	*2.0*	*2.7*	1.5
resistor B	0.50	*22*	44	11
resistor C	*0.25*	8.0	*32*	2.0
resistor D	*0.25*	14	56	3.5
total circuit	0.75	24	32	18

Worked example: Circuit 3

In our final example circuit we have two sources of e.m.f., several resistors of unknown resistance (although we know the power of resistor A) and a known p.d. across resistor C.

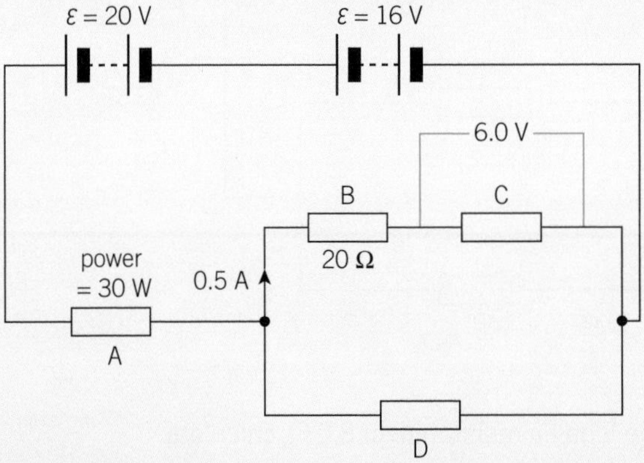

▲ **Figure 4** *Circuit 3*

Step 1: We know that the total e.m.f. in this circuit must be 36 V. We can use $V = IR$ to determine the p.d. across resistor B (10 V), giving 16 V for the branch. From Kirchhoff's second law it follows that the p.d. across resistor A must be 20 V, and the p.d. across resistor D must be 16 V.

Step 2: We can also determine the resistance of resistor C using $V = IR$ ($R_C = 12\,\Omega$).

Step 3: Using $P = VI$ we can determine the current through resistor A as 1.5 A. From Kirchhoff's first law it follows that the current in resistor D must be 1.0 A.

Step 4: We can then determine the resistance of resistors A and D using $V = IR$, making sure to use the appropriate values for the p.d. across each resistor and the current in the resistor.

$$R_A = \frac{V_A}{I_A} = \frac{20}{1.5} = 13\,\Omega$$

$$R_D = \frac{V_D}{I_D} = \frac{16}{1.0} = 16\,\Omega$$

Step 5: We now have all values for current, p.d., and resistance, and can calculate the power of each component.

The total power in this case is 54 W, with 30 W from the 20 V battery and 24 W from the 16 V battery. These values should be checked carefully using $P = IV$.

Component	Current / A	p.d. / V	Resistance / Ω	Power / W
resistor A	1.5	20	13	*30*
resistor B	0.50	10	*20*	5.0
resistor C	0.50	*6.0*	12	3.0
resistor D	1.0	16	16	16
total circuit	1.5	36	24	54

Summary questions

1 Name all the quantities represented in the following equations:

$$I = \frac{\Delta Q}{\Delta t} \qquad V = \frac{W}{Q} \qquad P = VI \qquad V = IR$$ *(4 marks)*

2 Using circuit 1 calculate:

a the charge passing through the 31 Ω resistor in 45 seconds; *(2 marks)*

b the energy transferred by the 36 Ω resistor in 2.0 minutes. *(2 marks)*

3 For the circuit in Figure 5:

a determine the current through resistor B; *(1 mark)*

b calculate the p.d. across resistors A and D; *(4 marks)*

c show that the resistance of resistor D = 2.0 Ω. *(2 marks)*

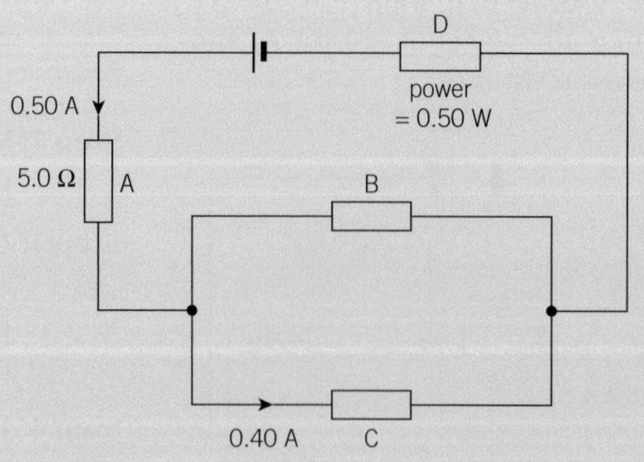

▲ **Figure 5**

4 If resistor B in Figure 5 has a resistance of 8.0 Ω, calculate the power of each resistor and the e.m.f. of the cell. *(6 marks)*

Not all batteries are created equal

Different power sources have different internal resistances, often by design, depending on the job they are made for. The key consideration is the current in the circuits they power. If a large current is needed, a power source with a small **internal resistance** is required.

Car batteries have a very low internal resistance so that they can provide the large current needed (often hundreds of amperes) to turn the starter motor in the car. Even if you connected enough AA batteries together to give an e.m.f. of 12.6 V, the same as a car battery, they would not provide the necessary current because of their internal resistance.

Internal resistance and lost volts

Whenever there is a current in a power source, work has to be done by the charges as they move through the power source. In a chemical cell this work is due to reactions between the chemicals. In a solar cell it is due to the resistance of the materials of the cell.

As a result some energy is 'lost' (transferred into heat) when there is a current in the power source, and not all the energy transferred to the charge is available for the circuit. The p.d. measured at the terminals of the power source (the **terminal p.d.**) is less than the actual e.m.f. We call this difference **lost volts**.

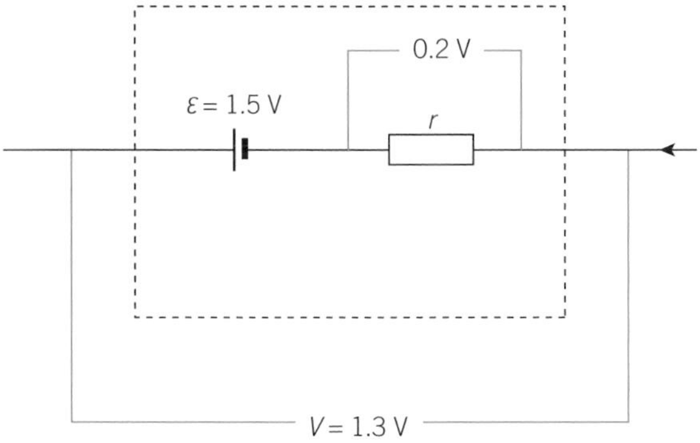

▲ **Figure 2** *This cell has an e.m.f. ε of 1.5 V, but the terminal p.d. V across it is only 1.3 V, because 0.2 V has been 'lost' to the internal resistance of the cell* r

From Kirchhoff's second law, the relationship between the e.m.f., the terminal p.d., and the lost volts is

electromotive force = terminal p.d. + lost volts

In normal use the e.m.f. does not change. However, changing the current affects the lost volts and the terminal p.d. Increasing the current means that more charges travel through the cell each second and so more work is done by the charges, increasing the lost volts. This lowers the terminal p.d.

Learning outcomes

Demonstrate knowledge, understanding, and application of:

→ the source of e.m.f.

→ internal resistance

→ terminal p.d. and 'lost volts'

→ the equations $\varepsilon = I(R + r)$ and $\varepsilon = V + Ir$

→ determining r.

▲ **Figure 1** *Car batteries have a negligible internal resistance and so they can supply very large currents*

If we apply the equation $V = IR$ to the internal resistance we can see that

$$\text{lost volts} = I \times r$$

where r is the internal resistance in ohms. If r remains fixed then the current in the power source is directly proportional to the lost volts.

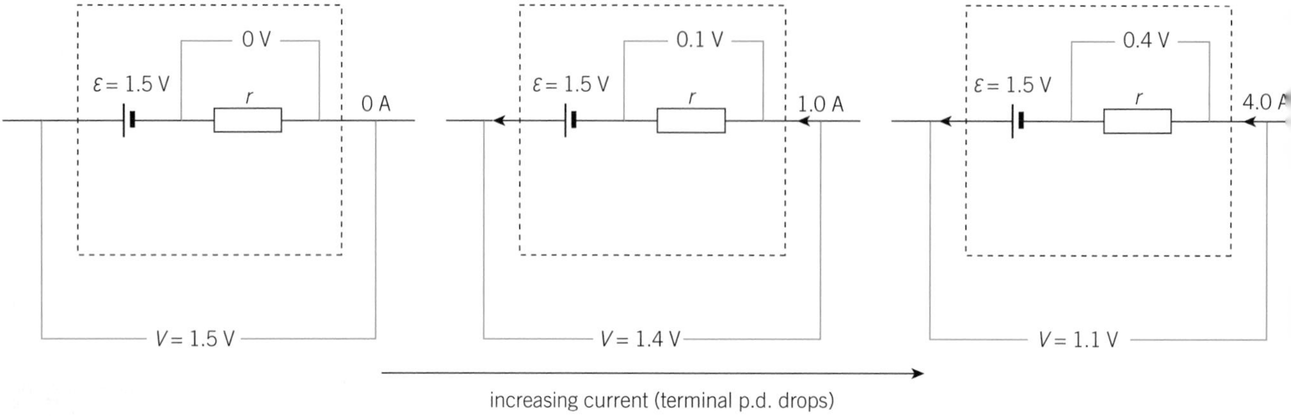

▲ **Figure 3** *As the current through the cell increases, the e.m.f. does not change, but the terminal p.d. drops as the lost volts increase*

The e.m.f. is always more than the terminal p.d. (unless there is no current). When the current is very small $\mathcal{E} \approx V$. This is why a high-resistance digital voltmeter connected directly across a cell gives a reading that approximates the e.m.f.

Combining the relationship electromotive force = terminal p.d. + lost volts and the equation above, we can derive an equation for the e.m.f. from a power source.

$$\mathcal{E} = V + \text{lost volts}$$

$$\mathcal{E} = V + Ir$$

where V is the terminal p.d. in volts, I is the current though the power supply in amperes, and r is the internal resistance of the power supply in ohms.

The terminal p.d. V is also equal to IR, where R is the resistance of the circuit, and so

$$\mathcal{E} = IR + Ir$$

As the current through the circuit and through the power supply must be the same, I is a common factor.

$$\mathcal{E} = I(R + r)$$

This relationship is essentially a version of $V = IR$ that takes into account the internal resistance of the power source.

Connecting cells

We can connect cells together to produce either a higher e.m.f. or a higher current. Depending on the desired effect we can connect them in series or in parallel.

Connecting cells in series increases the available e.m.f., but also increases the internal resistance. This limits the current that the combination can produce (Figure 4).

The same two cells connected in parallel produce the same e.m.f. as one cell, but have a much smaller internal resistance, so provide a greater current.

1 Two identical 1.5 V cells, each with an internal resistance of 0.75 Ω, are connected in series. Calculate the terminal p.d. when:
   a the cells supply a current of 0.80 A to an external circuit;
   b the cells are connected to a resistor of 10 Ω.
2 Repeat the above calculations for the cells connected in parallel.

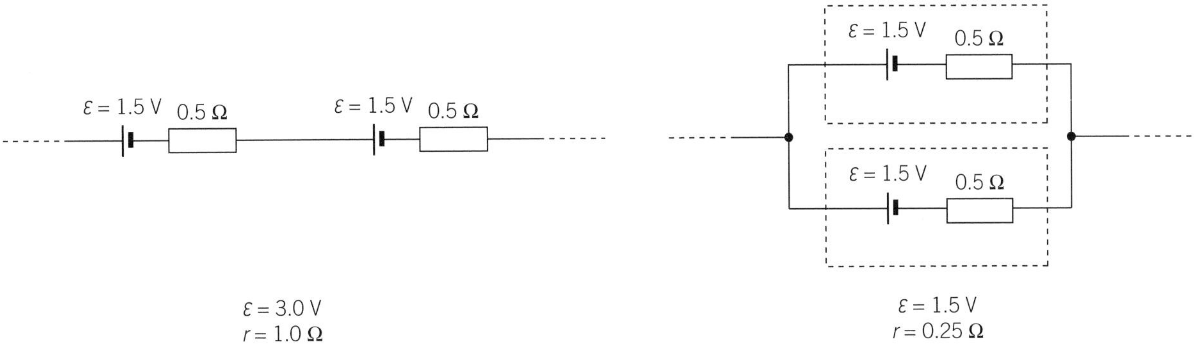

▲ **Figure 4** *The same two 1.5 V cells produce very different outcomes when connected in series and when connected in parallel*

Investigating internal resistance

Using the circuit in Figure 5, we can record values for terminal p.d. for different values of current. We use the variable resistor to change the resistance of the circuit, drawing different currents from the power source. Alternatively, several different resistors can be connected in various combinations.

By considering the general equation for a straight line ($y = mx + c$) and rearranging the equation $\mathcal{E} = V + Ir$ to give $V = -rI + \mathcal{E}$, we can see that if we plot a graph of terminal p.d. (V) against current (I) we can find both the e.m.f. $\mathcal{E}$ (constant) and the internal resistance of a power source r. The graph of V against I will have a gradient equal to $-r$ and a y-intercept equal to the e.m.f. $\mathcal{E}$ of the power source.

Sketch the graph and you will see that, as the current through the cell increases, the terminal p.d. drops and the lost volts increase. When the current is zero the terminal p.d. is equal to the e.m.f. As the current increases so do the lost volts. The lost volts and the terminal p.d. always add up to the e.m.f. The current should not be allowed to get too high or it will raise the temperature of the cell, increasing its internal resistance.

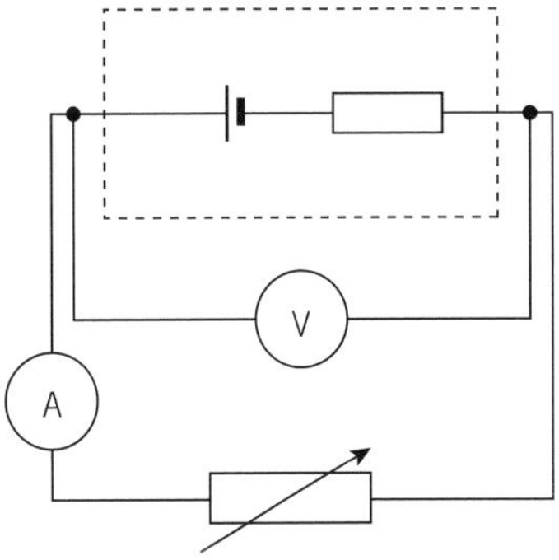

▲ **Figure 5** *Using a variable resistor we can take measurements of the terminal p.d. for various values of current in the cell*

Synoptic link

You will find more information about gradients of straight-line graphs in Appendix A2, Recording results.

▼ **Table 1**

Terminal p.d. / V	Current / A
1.52	0.15
1.50	0.20
1.47	0.25
1.44	0.30
1.43	0.35
1.40	0.40
1.37	0.45
1.34	0.50
1.32	0.55

Table 1 contains data collected using a typical AA cell.

1. Use the data in Table 1 to plot a graph of the terminal p.d. against current.
2. Explain how the graph suggests that the internal resistance is not affected by increasing the current through the cell.
3. Use the graph to determine the e.m.f. and the internal resistance of the cell.
4. Explain why the cell should be disconnected between each reading.

Graphs of *V* against *I*

Different power sources (including different cells) have different e.m.f.s and internal resistances. The graphs for different power sources will follow the same general trend but have different values for the e.m.f. and *r*. Three examples can be seen in Figure 6.

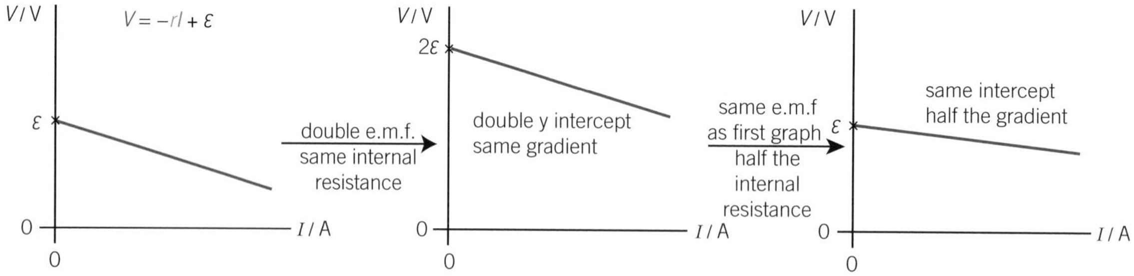

▲ **Figure 6** *Changing the e.m.f. ε or the internal resistance r of the power source changes the graph*

High or low internal resistance?

Like car batteries, many rechargeable batteries, including those in mobile phones and laptops as well as the more traditional types, have a small internal resistance. This allows them to be recharged using higher currents without overheating or wasting a lot of energy, so that recharging is fast. Some batteries in mobile phones charge to 80% capacity from flat in under an hour.

In contrast, the high-voltage power supplies used in classrooms have a very high internal resistance (often millions of ohms). This acts as a safety feature, preventing the power supply from delivering a fatal electric current.

Summary questions

1 With examples, outline why it is necessary for different power sources to have different internal resistances. *(3 marks)*

2 Describe what happens to the terminal p.d. from a power source as the current through the source increases. *(2 marks)*

3 A 9.0 V battery with an internal resistance of 2.0 Ω is connected in series with a filament lamp. The lamp draws a current of 1.5 A. Calculate:

  a the lost volts; *(1 mark)*
  b the terminal p.d.; *(1 mark)*
  c the energy lost per second by the battery; *(1 mark)*
  d the resistance of the lamp. *(1 mark)*

4 A 12 V battery with an unknown internal resistance is connected with three resistors as shown in Figure 7. If the current through the cell is 0.10 A, calculate:

  a the total resistance of the resistor circuit; *(2 marks)*
  b the terminal p.d. and the lost volts; *(2 marks)*
  c the internal resistance of the battery. *(1 mark)*

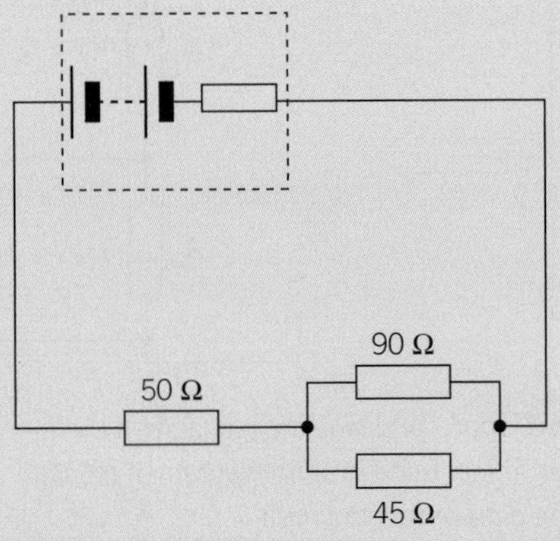

▲ **Figure 7**

5 Sketch a graph of V against I and label the internal resistance r and the e.m.f. $\mathcal{E}$. Sketch a second graph to show how the graph changes if two identical cells are connected in series. *(4 marks)*

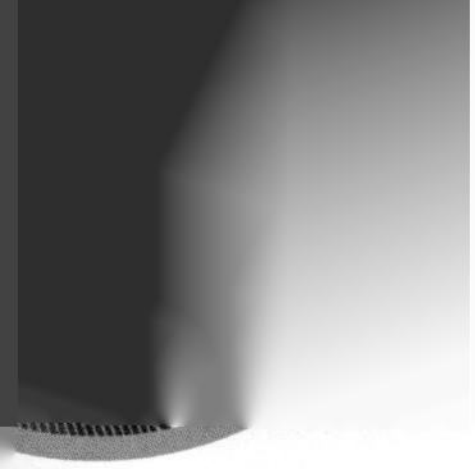

Learning outcomes

Demonstrate knowledge, understanding, and application of:

- → potential divider circuit with components
- → potential divider equations, for example, $V_{out} = \left(\frac{R_2}{R_1 + R_2}\right) \times V_{in}$ and $\frac{V_1}{V_2} = \frac{R_1}{R_2}$

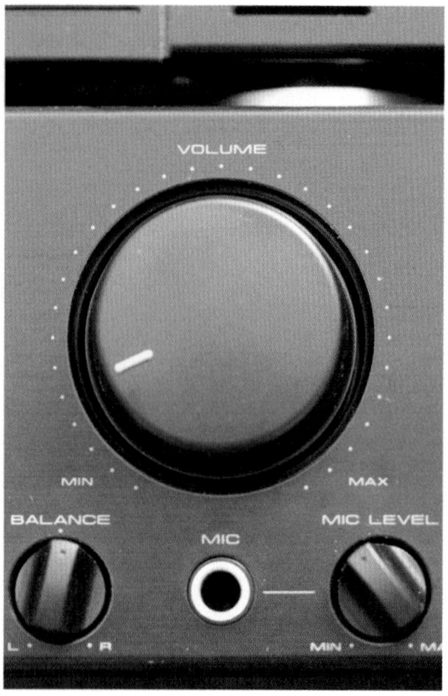

▲ **Figure 1** *A mechanical volume control uses a type of potential divider to change the p.d. across a speaker, increasing or decreasing the intensity of sound*

Turn it up to 11

Mains-powered speakers are always connected to the same power source with a fixed p.d. So why do they not all always produce the same volume?

Speakers are amongst hundreds of electrical devices that make use of **potential divider** circuits. These circuits can vary the p.d. across an output (like a speaker) when connected to a fixed input.

Potential dividers

You may find you need a p.d. of 4.0 V for a specific task, but the power source is a 9.0 V battery. You can use potential dividers to divide the p.d. to give any value you require up to the maximum supplied from the power source (Figure 2).

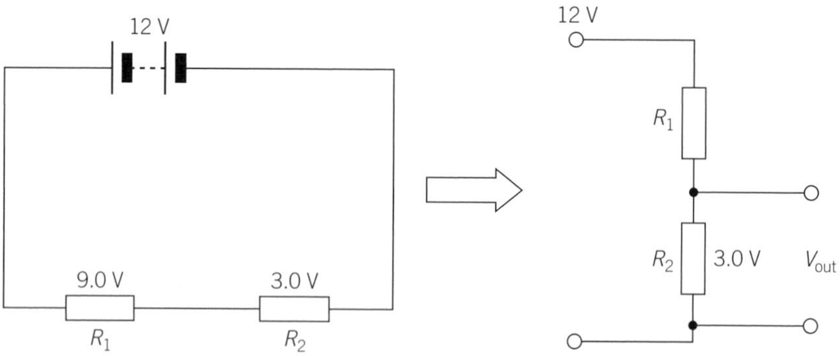

▲ **Figure 2** *A pair of resistors in series will 'share' or divide the p.d. across them, depending on the ratio of their resistances, a fact used in simple potential divider circuits, which are normally drawn like the diagram on the right*

A circuit can be connected across one of the resistors in parallel. The p.d. supplied to this circuit (V_{out}) can be varied to any value from zero to the maximum supplied from the power source, depending on the resistances of R_1 and R_2. From Kirchhoff's second law the p.d. across each resistor must always add up to the p.d. from the power source.

Ratio of resistances

The p.d. across each resistor in a potential divider depends on their resistances. If they have the same resistance then the p.d. is shared equally between them. If one has twice the resistance of the other, then this one will receive two-thirds of the total p.d.

Mathematically, this can be expressed as

$$\frac{V_1}{V_2} = \frac{R_1}{R_2}$$

where V_1 is the p.d. (in volts) across the resistor with resistance R_1 (in ohms) and V_2 is the p.d. (in volts) across the resistor with resistance R_2 (in ohms).

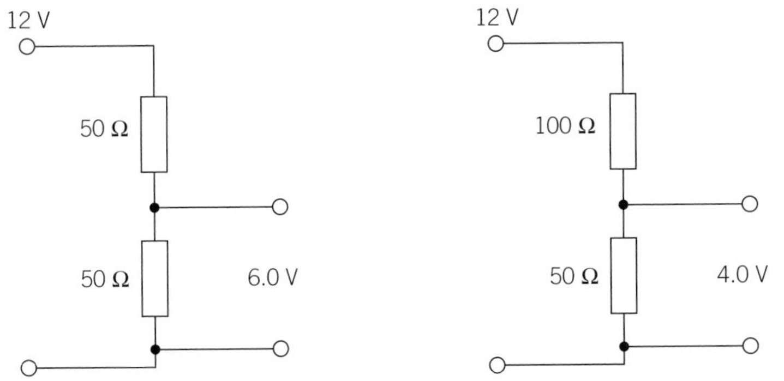

▲ **Figure 3** *When the two resistors have the same resistance the p.d. is split evenly between them (left), but when they are different the p.d. is shared according to the ratio of the resistances, so here (right) the upper resistor makes up two-thirds of the total resistance and so receives two-thirds of the p.d., leaving one-third or 4.0 V for the second resistor*

The potential divider equation

By considering the total p.d. V_{in} and the fraction of the total resistance provided by R_2 we can determine the value of V_{out}.

$$V_{out} = \left(\frac{R_2}{R_1 + R_2}\right) \times V_{in}$$

This relationship is often simply called the **potential divider equation**.

Worked example: Calculating V_{out}

A 270 Ω resistor and a 170 Ω resistor are connected as part of a potential divider circuit to a 36 V supply. The output is connected across the 270 Ω resistor. Calculate V_{out}.

Step 1: Identify the correct equation to calculate V_{out}.

$$V_{out} = \left(\frac{R_2}{R_1 + R_2}\right) \times V_{in}$$

Step 2: Substitute in known values (taking care to ensure that R_2 is the correct resistor – in this case 270 Ω) and calculate V_{out}.

$$V_{out} = \left(\frac{270}{170 + 270}\right) \times 36 = 22\text{ V}$$

Study tip

For a potential divider, the current I is the same in each resistor. Since $V = IR$, this means that $V \propto R$. This means that the ratio of potential differences is the same as the ratio of the resistances.

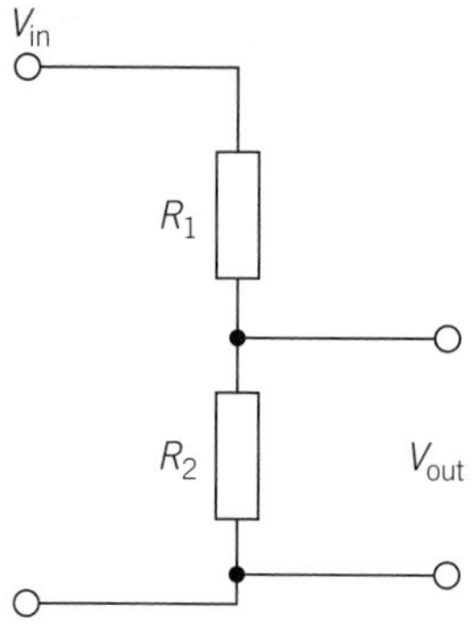

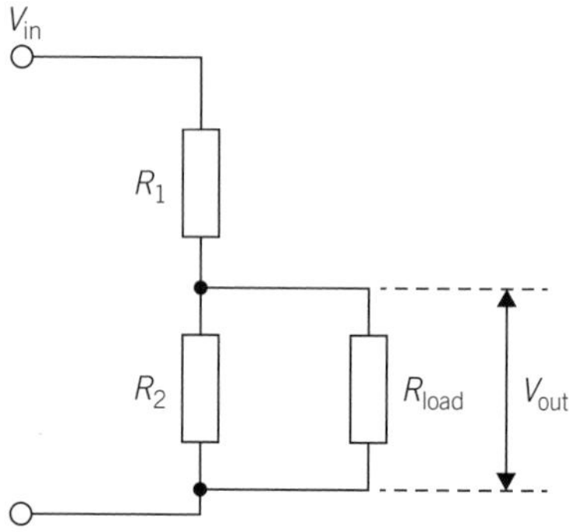

▲ **Figure 4** *When loaded, the resistance of the loaded part of the circuit is less than R_2 and so V_{out} drops*

Loading a potential divider

Loading refers to connecting a component or circuit to V_{out}, that is, placing a component in parallel with R_2. This lowers the resistance of this part of the potential divider circuit, which lowers the fraction of the total p.d. across this part of the circuit, and so lowers V_{out}.

Adding a large load (high resistance) to the circuit has little effect on V_{out}, but if the load has a small resistance, V_{out} is significantly reduced.

1 Outline why V_{out} drops when a component is connected across it.

2 A 2.2 kΩ resistor and a 4.7 kΩ resistor are connected as part of a potential divider circuit to a 12 V supply. V_{out} is connected across the 4.7 kΩ resistor. Calculate V_{out} when

- **a** the potential divider is not loaded
- **b** the potential divider is loaded with a resistor of resistance 10 kΩ
- **c** the potential divider is loaded with a resistor of resistance 100 Ω.

Summary questions

1 Outline how a pair of identical resistors can be used in a potential divider to produce an output of 10 V from a 20 V supply. Include a labelled diagram in your answer. *(4 marks)*

2 Calculate V_{out} from the two potential dividers in Figure 5. *(4 marks)*

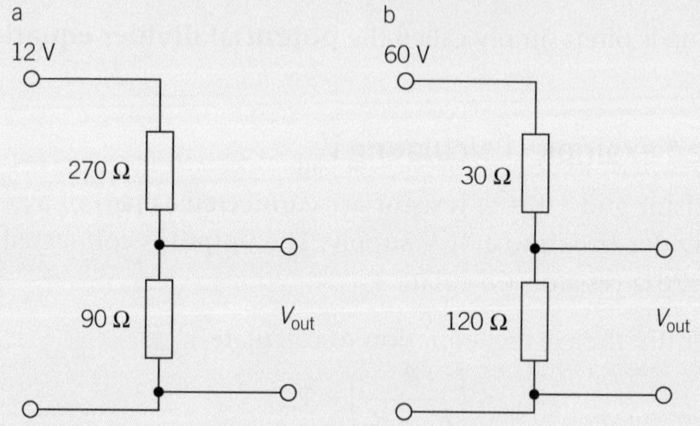

▲ **Figure 5**

3 Using a 24 V battery with negligible internal resistance, a 30 Ω resistor, and a 90 Ω resistor, draw a labelled diagram of a potential divider that would produce a V_{out} of:

a 6.0 V; **b** 18 V. *(2 marks)*

4 A potential divider is connected to a 360 V power source with negligible internal resistance. The resistance of R_1 is 110 Ω. Calculate the resistance of R_2 if V_{out} is 3.0 V. *(3 marks)*

Stay in lane

A line-follower robot is a mobile robot that can follow a coloured line painted on the ground. Early designs used simple sensor circuits made up of potential dividers with an LDR in place of one of the resistors. The robot can be programmed to respond to a changing V_{out} caused by a change in the resistance of the LDR as it moves, ensuring that it always follows the line.

This technology has come a long way. Some cars now include lane departure warning systems. These warn the driver if the car begins to move out of its lane unless it is indicating to do so. The system is designed to reduce accidents caused by drivers drifting out of their lane when tired or distracted.

Learning outcomes

Demonstrate knowledge, understanding, and application of:

→ potential divider circuits with variable components.

Producing a varying V_{out}

Using a pair of fixed resistors in series in a potential divider has the effect of splitting the p.d., but what if you need V_{out} to vary?

The simplest way to vary V_{out} is to replace one of the fixed resistors with a variable resistor (Figure 1). In the configuration shown, increasing the resistance of the variable resistor will increase V_{out} and vice versa.

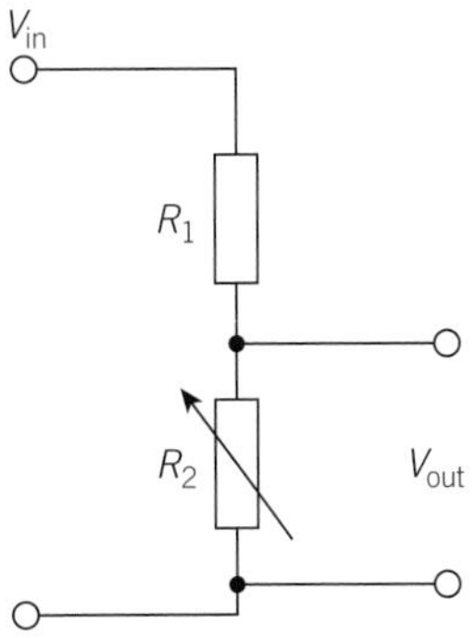

▲ **Figure 1** *A variable resistor can be used to change the output voltage*

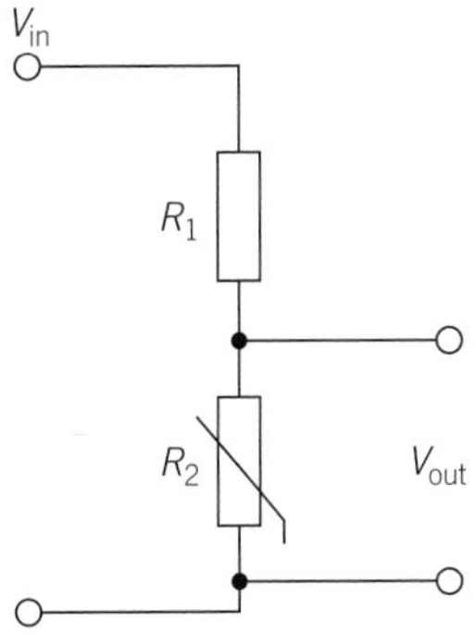

▲ **Figure 2** *The output voltage will depend on the temperature of the thermistor*

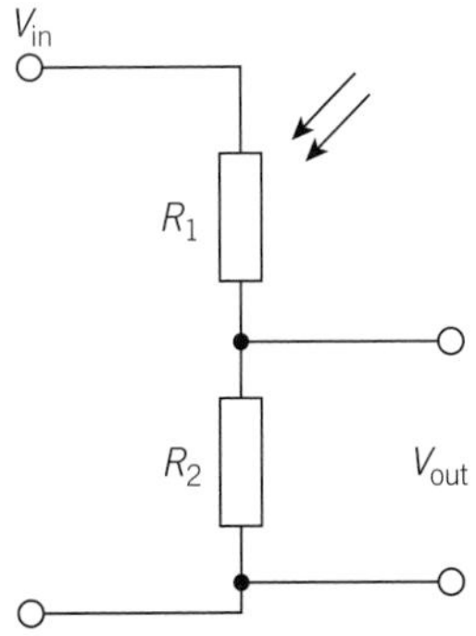

▲ **Figure 3** *The output voltage will depend on the intensity of light*

Temperature-sensing circuits – using a thermistor

Replacing the variable resistor with a thermistor allows V_{out} to vary automatically depending on the temperature of the surroundings.

As the temperature increases, the resistance of the thermistor decreases and so V_{out} drops (Figure 2).

Light-sensing circuits – using an LDR

An LDR can be used in the same way as a thermistor, producing a potential divider that gives an output that depends on the light intensity.

In this configuration (Figure 3), as the light intensity increases, the resistance of the LDR falls and so the p.d. across it decreases. R_2 receives a greater proportion of the p.d. and so V_{out} increases.

1. Design a potential divider which increases V_{out} as the temperature increases.
2. An LDR and a resistor with a resistance of 1000 Ω are connected to a 9.0 V battery with negligible internal resistance as part of a potential divider. The resistance of the LDR varies from 500 Ω in bright light to 50 MΩ in darkness. Calculate the minimum and maximum values for V_{out}.

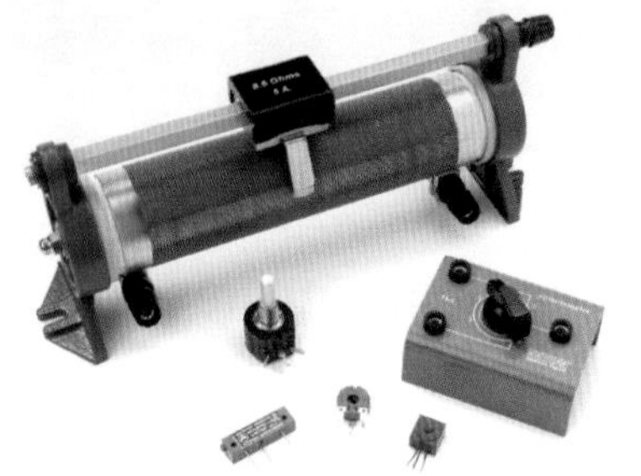

▲ **Figure 4** *Potentiometers with sliding contacts*

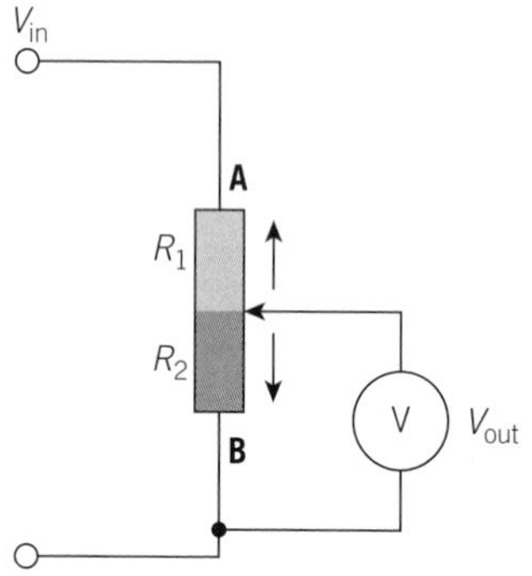

▲ **Figure 5** *The sliding contact found inside every potentiometer is a simple, cheap, and often compact way to produce a variable V_{out}*

The potentiometer

Many low-voltage electrical circuits that need a varying p.d. use a **potentiometer** rather than a potential divider.

A potentiometer is a variable resistor with three terminals and a sliding contact. Adjusting this contact varies the p.d. between two of the terminals, giving a variable V_{out} (Figures 4 and 5). Potentiometers can be made very compact, making them useful for portable electronic devices.

When the contact is moved towards **A**, V_{out} increases, until at **A** it is equal to V_{in}. When the contact is moved towards **B**, V_{out} decreases until at **B** it is zero.

Sometimes a dial is used rather than a slider, making the potentiometer even more compact. The potentiometer can also be constructed so that the change in resistance is either linear or logarithmic.

Summary questions

1 State two advantages of using a potentiometer over a potential divider. *(2 marks)*

2 Design a potential divider in which V_{out} increases as it gets darker. *(3 marks)*

3 A thermistor and a resistor with a resistance of 2.2 kΩ are connected to a 12 V battery with negligible internal resistance as part of a potential divider. V_{out} is connected across the thermistor. Calculate the resistance of the thermistor when V_{out} is:
a 6.0 V; **b** 10 V; **c** 1.0 V. *(6 marks)*

4 The potential divider used in question 3 is modified so that V_{out} is connected across the 2.2 kΩ resistor. Describe how changes in temperature would affect the value of V_{out} from this new potential divider circuit. *(2 marks)*

5 A thermistor and a resistor of resistance 220 Ω are connected in series as part of a potential divider to a 12 V battery with negligible internal resistance. The output is connected across the resistor. The thermistor has a resistance of 200 Ω at 0°C and 50 Ω at 100°C.
a Sketch a graph of resistance against temperature for the thermistor. *(2 marks)*
b Use your graph to estimate the temperature that would give an output of 4.0 V. *(3 marks)*

6 Suggest a reason for connecting a variable resistor, rather than a fixed resistor, with a thermistor as part of a potential divider. *(2 marks)*

Practice questions

1 a A student is given three resistors of resistances 10 Ω, 20 Ω and 40 Ω. These resistors are connected in different combinations. Calculate

(i) the minimum possible resistance, *(3 marks)*

(ii) the maximum possible resistance. *(2 marks)*

b A filament lamp connected to a power supply glows at its maximum brightness.

The output voltage from the power supply is halved. Explain why the current in the lamp is not halved. *(2 marks)*

2 a Figure 1 shows combination of resistors connected to a power supply of e.m.f. ε.

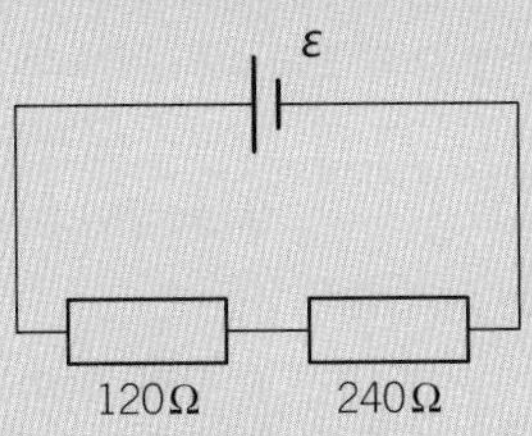

▲ **Figure 1a**

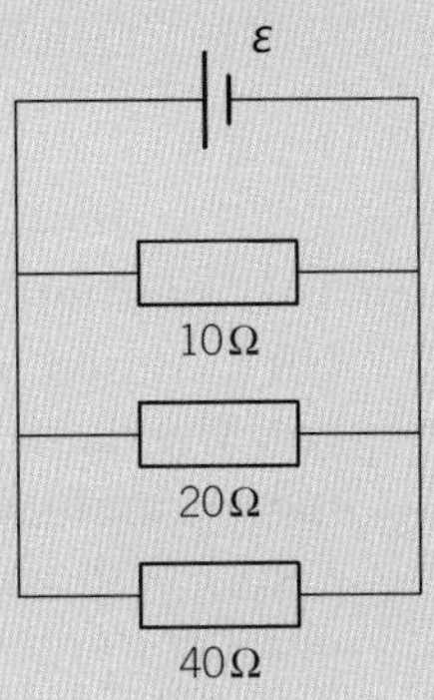

▲ **Figure 1b**

(i) For the circuit of Figure 1a

1 calculate the total resistance R_s, *(1 mark)*

2 state one electrical quantity which is the same for both resistors. *(1 mark)*

(ii) For the circuit of Figure 1b

1 calculate the total resistance R_p, *(2 marks)*

2 state one electrical quantity which is the same for all the resistors. *(1 mark)*

Jan 2012 G482

3 Figure 2 shows a circuit connected to a d.c. supply.

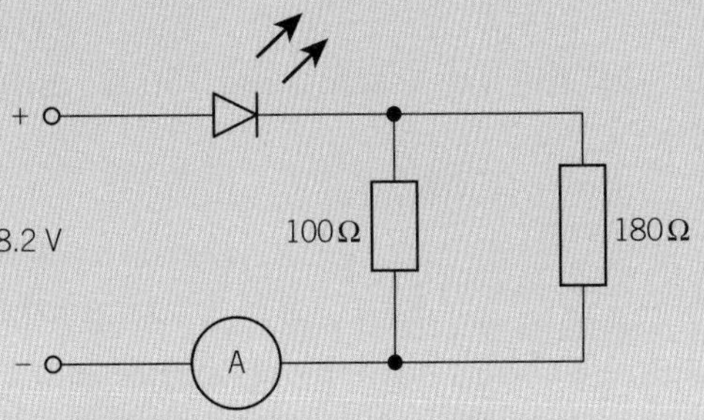

▲ **Figure 2**

The supply has e.m.f. 8.2 V and negligible internal resistance.

a Calculate the total resistance of the resistors. *(2 marks)*

b The current measured by the ammeter is 100 mA. Calculate

(i) the potential difference across the LED, *(3 marks)*

(ii) the total power dissipated in the resistors. *(2 marks)*

4 a Kirchhoff's laws can be used to analyse any electrical circuit. State each of Kirchhoff's laws and the physical quantity associated with each law that is conserved in the circuit.

(i) Kirchhoff's first law *(2 marks)*

(ii) Kirchhoff's second law *(2 marks)*

b The circuit in Figure 3 consists of a battery of e.m.f. 45 V and negligible internal resistance, and three resistors.

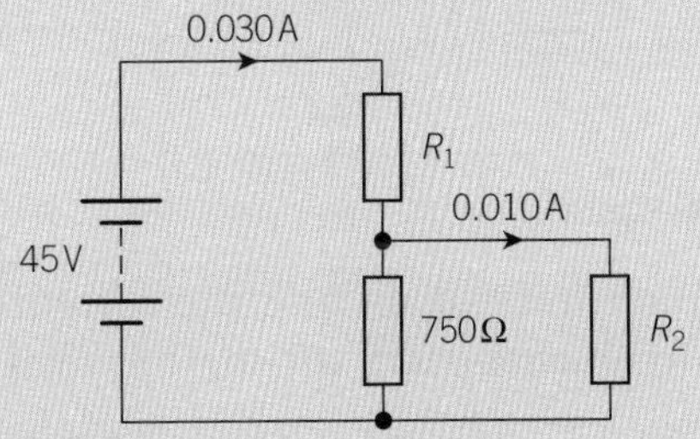

▲ **Figure 3**

The resistors have resistances R_1, R_2 and 750 Ω. The current in the resistor of resistance R_1 is 0.030 A. The current in the resistor of resistance R_2 is 0.010 A.

Calculate

(i) the current I in the 750 Ω resistor, *(1 mark)*

(ii) the p.d. V across the 750 Ω resistor, *(1 mark)*

(iii) the resistances R_1 and R_2. *(2 marks)*

May 2013 G482

5 Figure 4 shows how the resistance of a thermistor varies with temperature.

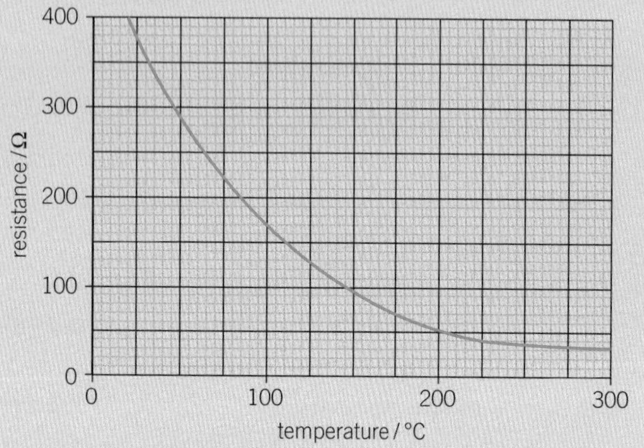

▲ **Figure 4**

The thermistor is used in the potential divider circuit of Figure 5 to monitor the temperature of an oven. The 6.0 V d.c. supply has zero internal resistance and the voltmeter has infinite resistance.

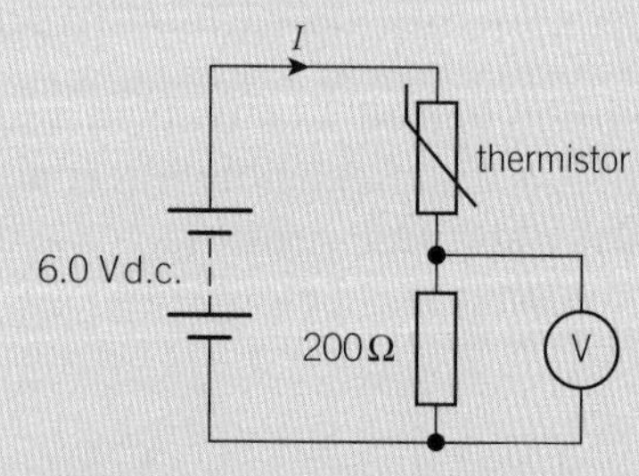

▲ **Figure 5**

a State and explain how the current I in the circuit changes as the thermistor is heated. *(3 marks)*

b Use Figure 4 to calculate the voltmeter reading when the temperature of the oven is 240 °C. *(4 marks)*

c A light-dependent resistor (LDR) is another component used in sensing circuits.

(i) Copy and complete Figure 6 with an LDR between **X** and **Y**.

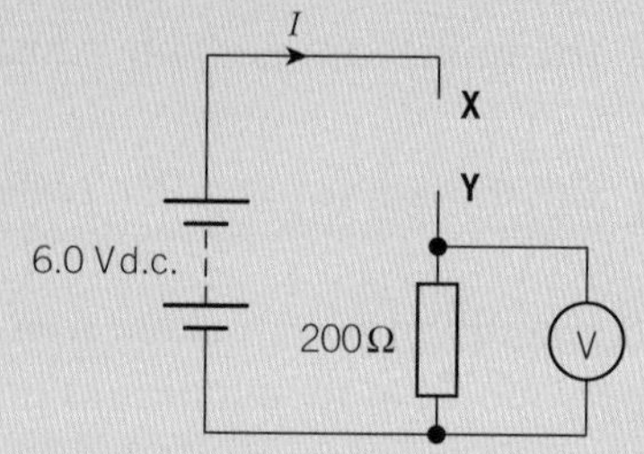

▲ **Figure 6**

(1 mark)

(ii) State with a reason how the voltmeter reading varies as the intensity of the light incident on the LDR increases. *(2 marks)*

May 2012 G482

6 This question is about possible heating circuits used to demist the rear window of a car. The heater is made of 8 thin strips of a metal conductor fused onto the glass surface. Figure 7 shows the 8 strips connected in parallel to the car battery of e.m.f. $\mathcal{E}$ and internal resistance r.

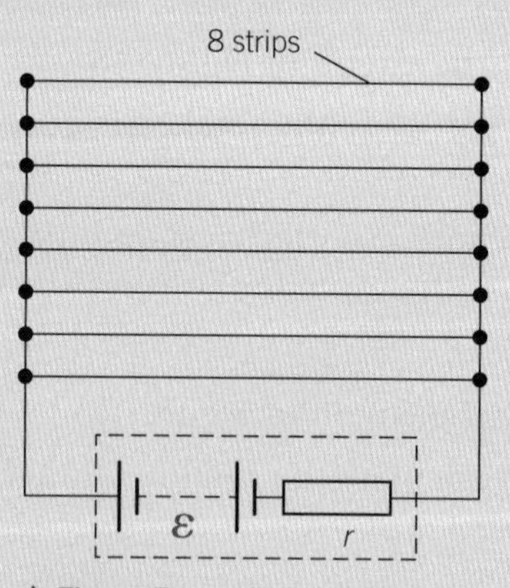

▲ **Figure 7**

▲ **Figure 7**

a The potential difference across each strip is 12 V when a current of 2.0 A passes through it.

(i) Calculate the resistance r_P of one strip of the heater. *(1 mark)*

(ii) Calculate the total resistance R_P of the heater. (*3 marks*)

(ii) Show that the power P dissipated by the heater is about 200 W. (*2 marks*)

b Each strip is 0.90 m long, 2.4×10^{-4} m thick and 2.0×10^{-3} m wide.

Calculate the resistivity ρ of the metal of the strip. Give the unit with your answer.

(*4 marks*)

June 2011 G482

7 Figure 8 shows an electrical circuit.

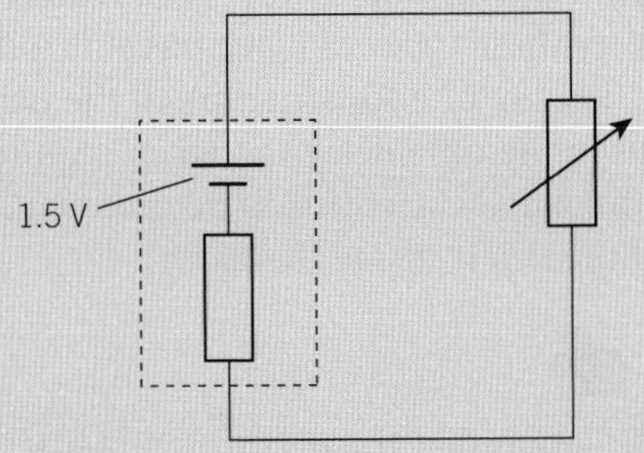

▲ **Figure 8**

The cell has e.m.f. 1.5 V. The resistance of the variable resistor is set to 1.0 Ω. The current in the cell is 0.50 A.

a Calculate the internal resistance of the cell. (*3 marks*)

b The resistance R of the variable resistor is changed from 1.0 Ω to 4.0 Ω.

(i) Copy and complete Table 1 to show the current I in the circuit and the power P dissipated in the variable resistor. (*2 marks*)

▼ **Table 1**

R/Ω	I/A	P/W
1.0		
2.0		
4.0		

(ii) Use your answer to (i) to suggest how the power P dissipated in the variable resistor is linked to the value of the internal resistance of the cell. (*1 mark*)

8 Figure 9 shows a circuit used to monitor the level of water in a container.

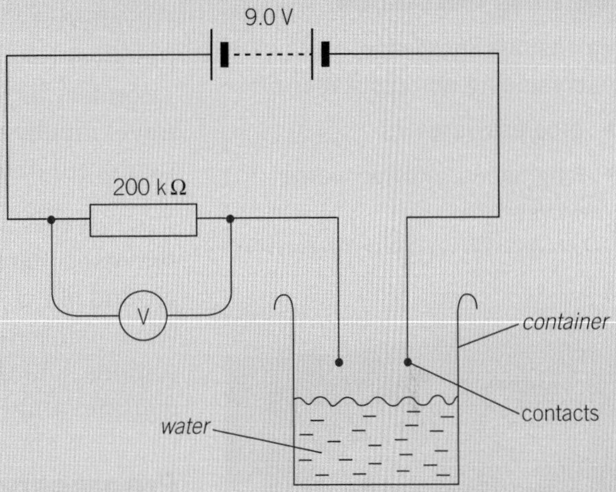

▲ **Figure 9**

The battery has electromotive force (e.m.f.) of 9.0 V and negligible internal resistance. The digital voltmeter shows a reading of 0.0 V when the contacts are dry and 6.5 V when the contacts are in water.

a Explain why the voltmeter reading is 0.0 V when the contacts are dry. (*2 marks*)

b Calculate the resistance of the water between the contacts when they are placed in the water. (*3 marks*)

c Without doing any calculations, explain how the voltmeter reading would change when the contacts are in water and the resistance of the resistor is much smaller than 200 kΩ. (*3 marks*)

11 WAVES 1

11.1 Progressive waves

Specification reference: 4.4.1

Learning outcomes

Demonstrate knowledge, understanding, and application of:

→ progressive waves

→ longitudinal and transverse waves.

▲ **Figure 1** *By studying how seismic waves travel through the Earth, scientists may one day be able to predict earthquakes, potentially saving lives*

▲ **Figure 2** *A ripple on water is another example of a progressive wave: it transfers energy from the centre of the ripple to the edges, but the water molecules vibrate perpendicular to the direction of energy transfer without moving out from the centre*

Earth-shaking

The effects of an earthquake can be devastating. An earthquake produces two main types of seismic wave, primary or **P-waves** and secondary or **S-waves**. Both are types of **progressive wave** that travel rapidly through parts of the Earth's interior. They can cause significant damage to structures on the surface.

The P-waves are **longitudinal waves** and S-waves are **transverse waves**. Differences in the way these types of wave travel through the ground allow scientists to study the interior structure of the Earth. The wave paths are calculated from the time delays between monitoring stations, and this information enables scientists to determine the densities and thicknesses of the layers inside the Earth.

Progressive waves

A progressive wave is an oscillation that travels through matter (or in some cases, through a vacuum). All progressive waves transfer energy from one place to another, but not matter. In other words, although the particles in the matter vibrate, they do not move along the wave.

Sound is an example of a progressive wave. When you hear someone talking to you, vibrations travel to your ears, but the air particles do not. Instead they vibrate in a plane parallel to the direction of energy transfer as the wave passes through the air.

When a progressive wave travels through a medium, like air or water, the particles in the medium move from their original **equilibrium position** to a new position. The particles in the medium exert forces on each other. A displaced particle experiences a **restoring force** from its neighbours and it is pulled back to its original position.

For example, waves on the surface of water propagate via water molecules interacting with their neighbours. As one molecule moves up it attracts its neighbours, which in turn pull the original molecule back down towards its equilibrium position, whilst at the same pulling up the neighbouring particles. No single water molecule moves along the wave. Instead they oscillate at right angles to the energy transfer.

Transverse waves

When you think of waves, the first image in your mind is likely to be a transverse wave like a water wave. In a transverse wave the oscillations or vibrations are perpendicular to the direction of energy transfer (Figure 3).

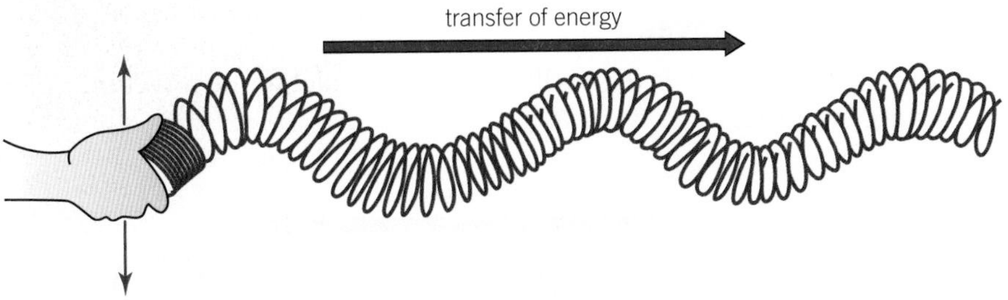

▲ **Figure 3** *A transverse wave travelling along a horizontal slinky spring*

As the wave moves from left to right, the oscillations are at 90° to the direction of the wave's movement – up and down, or side to side, or at any orientation as long as it is in a plane that is perpendicular to the direction of energy transfer.

Transverse waves have **peaks** and **troughs** where the oscillating particles are at a maximum displacement from their equilibrium position (Figure 4).

Examples of transverse waves include:

- waves on the surface of water
- any electromagnetic wave – radio waves, microwaves, infrared, visible light, ultraviolet, X-rays, and gamma rays
- waves on stretched strings
- S-waves produced in earthquakes.

You will learn more about electromagnetic waves in Topic 11.6, Electromagnetic waves.

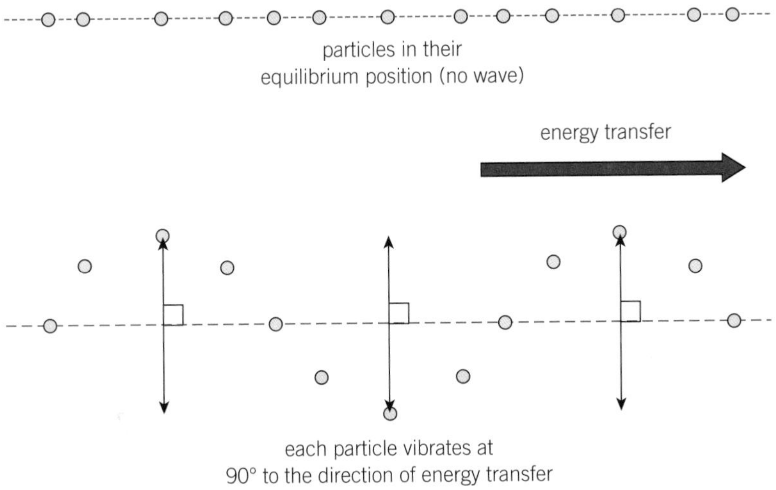

▲ **Figure 4** *For a transverse wave, the particles of the medium vibrate in a plane at 90° to the direction of energy transfer*

Longitudinal waves

In longitudinal waves the oscillations are parallel to the direction of energy transfer.

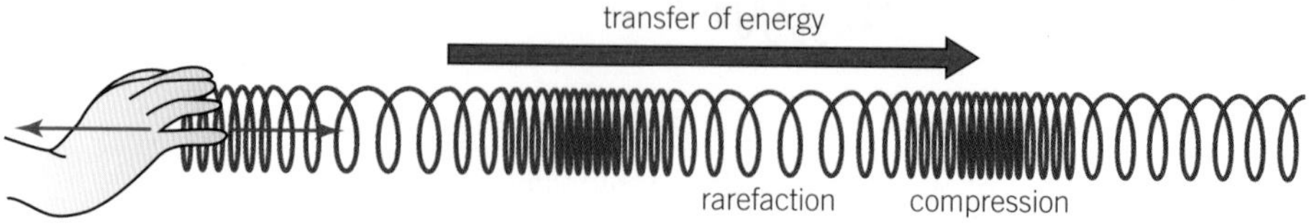

▲ **Figure 5** *A longitudinal wave looks very different to a transverse wave*

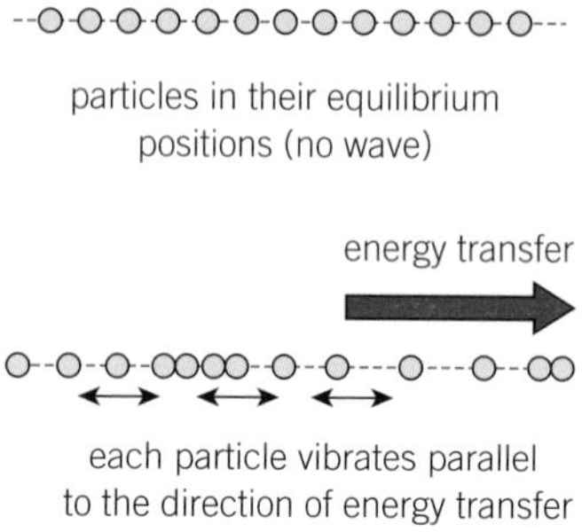

▲ **Figure 6** *In a longitudinal wave the particles vibrate parallel to the direction of energy transfer*

Examples of longitudinal waves include:

- sound waves
- P-waves produced in earthquakes.

Longitudinal waves are often called compression waves. When they travel through a medium they create a series of **compressions** and **rarefactions**.

When sound waves travel through air, air particles are displaced and bounce off their neighbours. These collisions provide the restoring force. As the wave moves, regions of higher pressure (where the particles are close together) and regions of lower pressure (where the particles are more spread out) travel through the air, but no single air particle travels along the wave. Instead they oscillate about their equilibrium positions.

Study tip

Do not be tempted to use language like 'side to side' and 'up and down' to describe progressive waves. Instead use 'perpendicular' or 'parallel' to describe the oscillations.

Summary questions

1 Describe the similarities and the differences between transverse and longitudinal waves. *(3 marks)*

2 Describe how you can produce the two types of wave using a slinky spring. *(2 marks)*

3 Copy Figure 7 of a transverse wave on a rope and label the direction of motion of the particles labelled A, B, and C. *(3 marks)*

4 Suggest why the speed of sound is faster through a medium with a higher density. *(2 marks)*

5 Sketch a series of diagrams to show how the particles in a sound wave vibrate about their equilibrium positions during one complete oscillation. *(4 marks)*

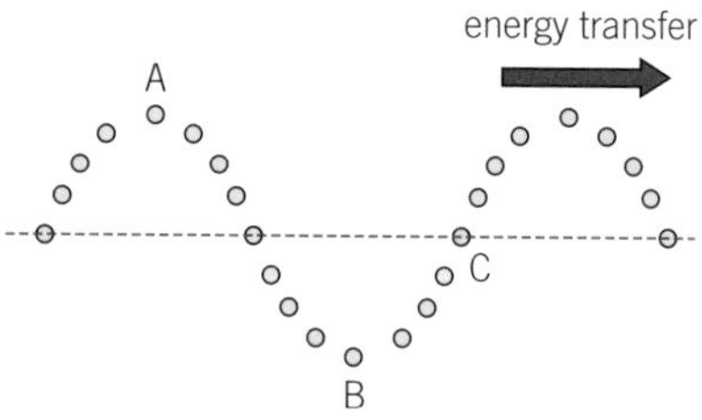

▲ **Figure 7**

Notes on terminology

Humans have played flutes and similar instruments for 40 000 years. A modern flute can produce notes with frequencies ranging from around 250 Hz to over 2 kHz. The characteristics of these notes can be described with musical terms like pitch and volume, but they can also be described with scientific terms like frequency and wavelength.

Table 1 contains a list of key terms used to describe waves.

▼ **Table 1** *Wave terminology*

Term	Symbol	Unit	Definition
displacement	s	m	distance from the equilibrium position in a particular direction; a vector, so it can have either a positive or a negative value
amplitude	A	m	maximum displacement from the equilibrium position (can be positive or negative)
wavelength	λ	m	minimum distance between two points in phase on adjacent waves, for example, the distance from one peak to the next or from one compression to the next
period of oscillation	T	s	the time taken for one oscillation or time taken for wave to move one whole wavelength past a given point
frequency	f	Hz	the number of wavelengths passing a given point per unit time
wave speed	v (or c)	m s^{-1}	the distance travelled by the wave per unit time

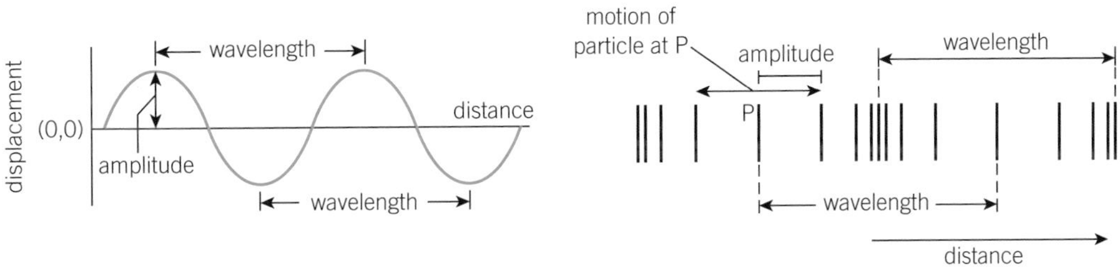

▲ **Figure 2** *Wavelength and amplitude of a transverse and a longitudinal wave*

The wave equation

From the definitions above we can derive the **wave equation**. This relates the frequency f in hertz, the wavelength λ in metres, and the wave speed v in m s^{-1}.

$$v = f\lambda$$

Learning outcomes

Demonstrate knowledge, understanding, and application of:

- → displacement, amplitude, wavelength, period, phase difference, frequency, and speed of a wave
- → the equation $f = \frac{1}{T}$
- → the wave equation $v = f\lambda$
- → graphical representations of transverse and longitudinal waves.

▲ **Figure 1** *Playing the flute*

If a wave has a frequency of 5.0 Hz, each second there are 5 complete oscillations. If each wave has a wavelength of 2.0 m, then the wave has travelled 10 m from the source in that time. Therefore, its speed must be $10\,\text{m s}^{-1}$. So, for a frequency f, the wave would have travelled a distance of $f \times \lambda$ per second, that is, the wave speed v.

We can also see from the definitions that the period of oscillation and the frequency of a wave are reciprocals of each other. If a wave has a frequency of 2.0 Hz, there are two complete wave cycles each second; therefore, the period for each wave must be 0.50 s or $\frac{1}{2.0}$. Therefore, we have a second important equation relating the frequency f of a wave in Hz to its period T in s.

$$f = \frac{1}{T}$$

Worked example: Finding the wavelength of a musical note

A flute produces a high-pitched note that has a time period of 0.45 ms. The speed of sound through air is $330\,\text{m s}^{-1}$. Calculate the wavelength of the note produced.

Step 1: Identify the correct equation to calculate the speed of the wave.

$$v = f\lambda$$

As $f = \frac{1}{T}$ substituting this into the first equation gives $v = \frac{\lambda}{T}$

Rearranging gives $\lambda = vT$.

Step 2: Substitute in known values in SI units (including the time in seconds) and calculate the wavelength of the note.

$$\lambda = 330 \times 0.45 \times 10^{-3} = 0.15\,\text{m (2 s.f.)}$$

Wave profile: displacement–distance graphs

A graph showing the displacement of the particles in the wave against the distance along the wave is sometimes called the **wave profile** (Figure 3). It may be helpful to think of such a graph as a 'snapshot' of the wave.

The wave profile can be used to determine the wavelength and amplitude of both types of wave. As the displacement of the particles in the wave is continuously changing, the wave profile changes shape over time.

Figure 4 shows how the wave profile for a progressive wave changes shape for four consecutive quarters of the period T, starting at $t = 0$ and increasing by $\frac{T}{4}$ each time. After one complete period the particles are back in their original positions.

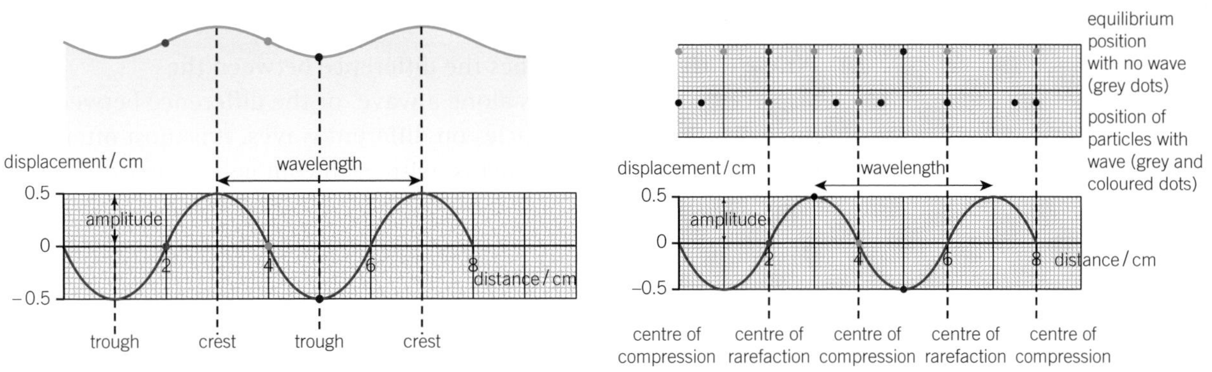

▲ **Figure 3** *Displacement–distance graphs for transverse and longitudinal waves are identical*

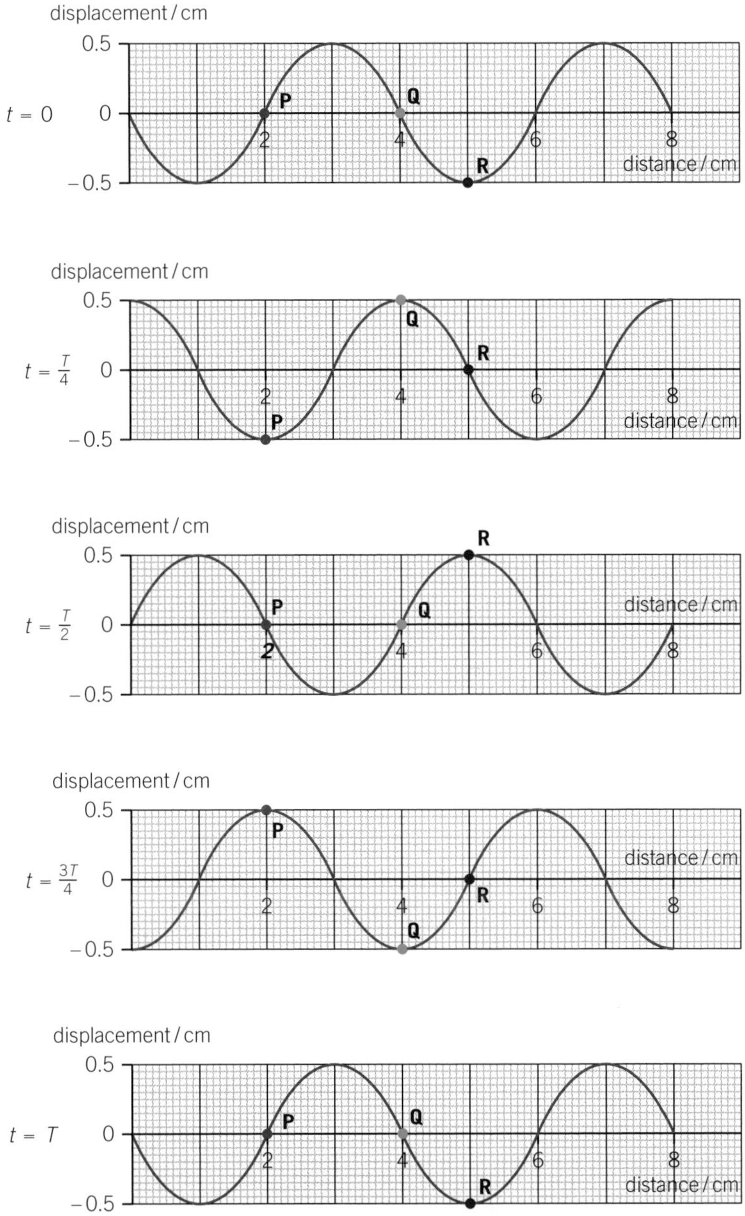

▲ **Figure 4** *Particles at P, Q, and R oscillate from their equilibrium position to a maximum positive displacement, back through their equilibrium position to a maximum negative displacement, and back again*

Phase difference

Phase difference describes the difference between the displacements of particles along a wave, or the difference between the displacements of particles on different waves. It is most often measured in degrees or radians, with each complete cycle or wave representing 360° or 2π radians.

If particles are oscillating perfectly in step with each other (they both reach their maximum positive displacement at the same time) then they are described as **in phase**. They have a phase difference of zero.

If two particles are separated by a distance of one whole wavelength (Figure 5), we say their phase difference is 360°, or 2π radians (angles can also be measured in radians). If they are two complete cycles out of step their phase difference is 720° or 4π radians, and so on.

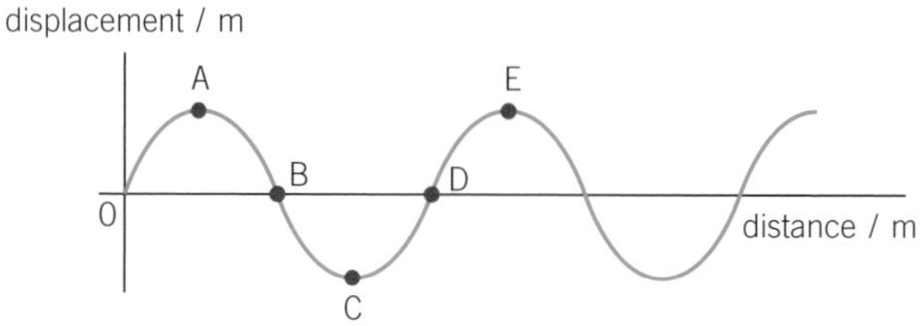

▲ **Figure 5** *Wave profiles are helpful when determining phase differences: particles at A and E have a phase difference of 360°, particles at C and D have a phase difference of 90°, whereas those at C and E are in antiphase and have a have a phase difference of 180°*

If particles are oscillating completely out of step with each other (one reaches its maximum positive displacement at the same time as the other reaches its maximum negative displacement) then they are described as being in **antiphase**. They have a phase difference of 180°, or π radians.

Two particles can have any phase difference as phase difference depends on the separation of particles in terms of the wavelength.

Relating position and phase difference

The phase difference ϕ between two points on a wave of wavelength λ separated by a distance x is given by

$$\phi = \frac{x}{\lambda} \times 360°$$

From this relationship we can see that if the distance between two points is equal to one wavelength the phase difference will be 360°.

1 Calculate the phase difference in degrees between two points on a wave of wavelength 40 cm separated by:
a 20 cm; **b** 40 cm; **c** 80 cm.

2 Calculate the distance between two points on a wave of wavelength 1.60 m when they have a phase difference of:
a 90°; **b** 540°; **c** 5π radians.

Displacement–time graphs

A second type of graph can be used to show how the displacement of a given particle of the medium varies with time as the wave passes through the medium. This graph looks the same for both transverse and longitudinal waves (Figure 6).

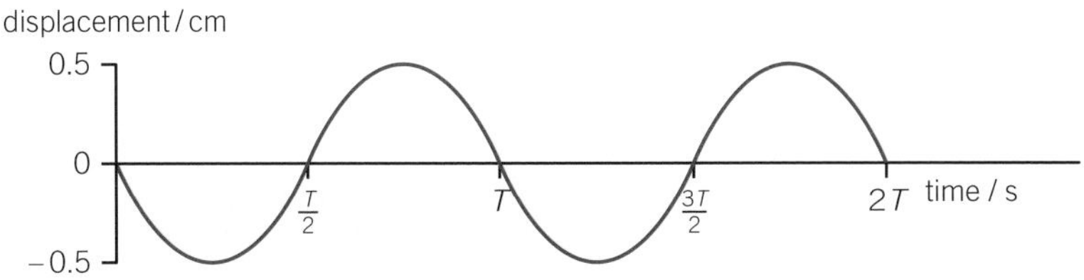

▲ **Figure 6** *A graph of displacement against time for a wave looks similar to a wave profile*

A graph of displacement against time can easily be used to determine the period T and amplitude of both types of wave. From the graph we can see that at time $t = 0$ the displacement of the particle is zero. After one-quarter of the period ($t = \frac{T}{4}$) the particle is at its maximum possible negative displacement −0.5 m. At time $t = \frac{T}{2}$ the particle is back in its equilibrium position (displacement = 0), before moving to its maximum positive displacement at $t = \frac{3T}{4}$ and returning again to its equilibrium position after one complete cycle ($t = T$). The amplitude of the progressive wave is 0.5 cm.

Using an oscilloscope to determine wave frequency

An **oscilloscope** can be used to determine the frequency of a wave. For example, using a microphone we can produce a trace on the screen (Figure 7). The oscilloscope screen shows a graph of p.d. against time for any signal fed into it.

Each horizontal square on the oscilloscope screen represents a certain time interval. This is called the **timebase**. If this is set to 1.0 ms cm^{-1}, then each square represents a time interval of 1.0 ms (the squares are normally 1 cm across). The height of the trace on the screen can be changed by adjusting the y sensitivity, measured in V cm^{-1}. For example, a setting of 10 V cm^{-1} would result in each square representing a p.d. of 10 V.

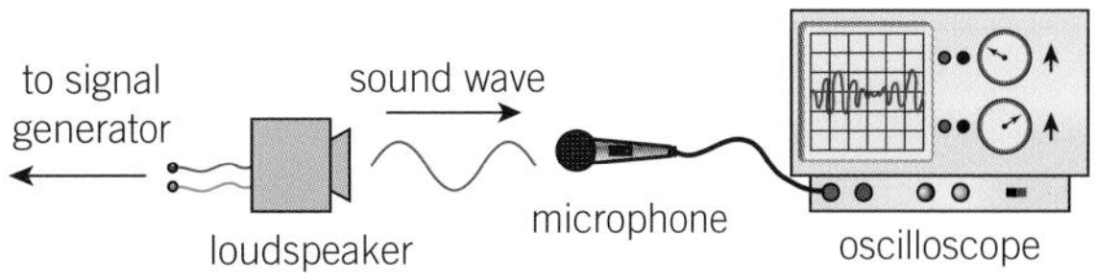

▲ **Figure 7**

From the timebase of the oscilloscope we can determine the time period T of the wave. Using this we can calculate the frequency with

$$f = \frac{1}{T}$$

In Figure 8 the timebase is set to 0.50 ms cm^{-1}. The horizontal distance from one peak to the next is exactly two squares on the screen, giving a period of 1.0 ms and therefore a frequency of 1.0 kHz.

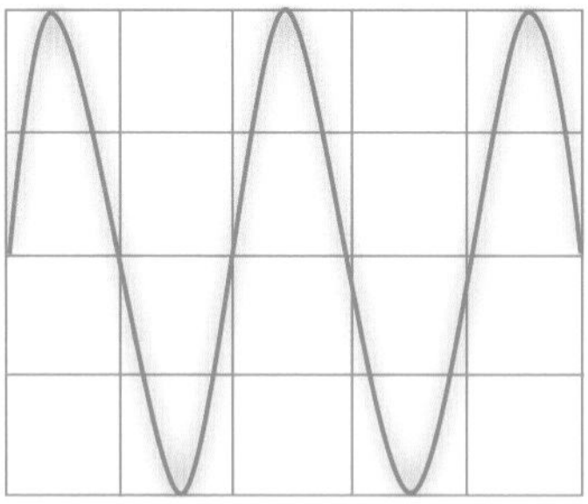

▲ **Figure 8**

As the frequency of the varying p.d. from the microphone is the same as the sound frequency, we can use the oscilloscope to determine the sound frequency.

1 Describe and explain the effect on the trace produced on the oscilloscope screen from a source of constant frequency if the timebase is changed.
2 Sketch the trace on an oscilloscope screen with a timebase set to 0.02 s cm^{-1} when a loudspeaker producing a sound at 50 Hz is directed towards a microphone connected to the oscilloscope.

Summary questions

1 Calculate the speed of a water wave with a wavelength of 50 cm and a frequency of 2.0 Hz. *(2 marks)*

2 Copy Figure 5 and label the direction of the velocity of particles at A and D. *(2 marks)*

3 Plan an experiment to determine the frequency of sound emitted from a whistle. *(4 marks)*

4 A flute produces a note with a time period of 2.0 ms. The speed of sound through air is 340 m s^{-1}. Calculate the frequency of the note and wavelength of the sound produced. *(4 marks)*

5 The wave profile in Figure 8 shows several particles at $t = 0$. The wave has a period of 4.0 seconds.

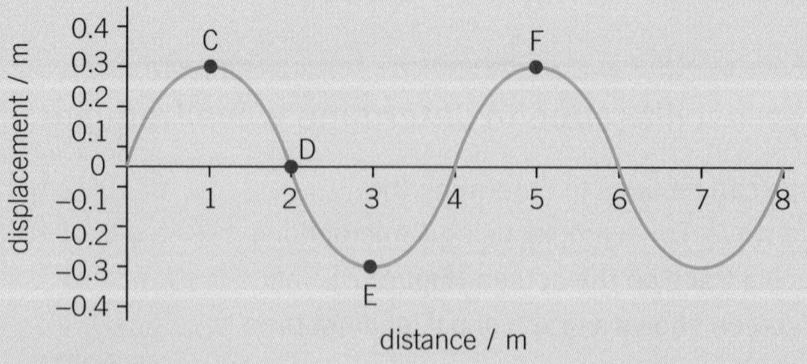

▲ **Figure 9**

a Sketch a wave profile showing the positions of the particles after:
i 1.0 s; ii 2.0 s. *(4 marks)*

b Determine the displacement of particle C after:
i 4.0 s; ii 11 s; iii 1 minute. *(4 marks)*

6 Determine the phase difference in degrees and radians between the following particles in Figure 8 ($360° = 2\pi$ radians)
a CD; b CE; c DF. *(3 marks)*

Getting the right shot

There are certain features of some photographs that make them enthralling. One key aspect of a good shot is what the light is doing in the photo: how it reflects off and highlights different objects, and sometimes unusual effects caused by the light changing direction as it passes from one medium to another.

All waves can be reflected and refracted. Not just light but sound, radio waves, and even X-rays can be reflected and refracted under the right conditions.

Learning outcomes

Demonstrate knowledge, understanding, and application of:

→ reflection and refraction of all waves.

Reflection

Reflection occurs when a wave changes direction at a boundary between two different media, remaining in the original medium.

A simple example is light reflecting off a mirrored surface (Figure 1). The light waves remain in the original medium (the air). We often represent the direction taken by a wave as a **ray** (like those in Figure 1). The ray shows the direction of energy transfer and so the path taken by the wave.

The **law of reflection** applies whenever waves are reflected. It states that the **angle of incidence** is equal to the **angle of reflection**.

When waves are reflected their wavelength and frequency do not change. This can be seen by reflecting water waves using a ripple tank. In Figure 2 we have represented the wave as a series of **wavefronts**. Each wavefront is a line joining points of the wave which are in phase. They can be thought of as the peak of each ripple. By definition (see Topic 11.2, Wave properties) the distance between wavefronts is equal to the wavelength of the wave.

Like plane (straight) waves, circular waves – like ripples from dropping a stone into a pond – can be reflected too (Figure 3).

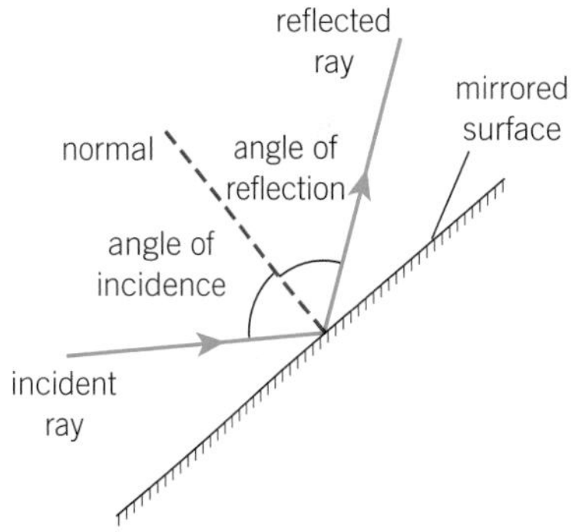

▲ **Figure 1** *When waves are reflected they always obey the law of reflection (note that all angles are measured to the normal)*

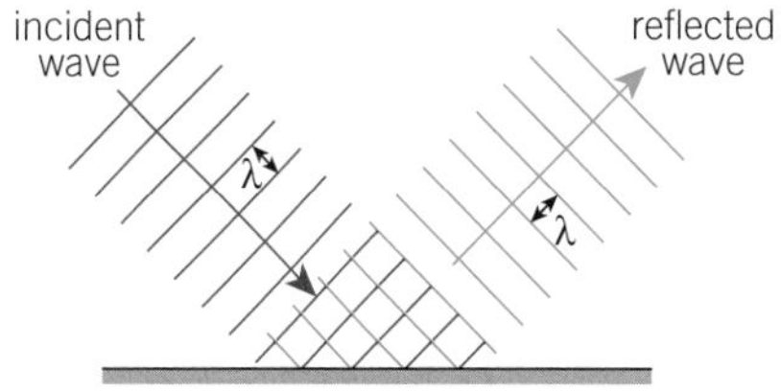

▲ **Figure 2** *Plane waves can be made by bobbing a ruler up and down in a ripple tank – when they reflect off a surface their frequency and wavelength remain unchanged*

Refraction

Refraction occurs when a wave changes direction as it changes speed when it passes from one medium to another. You will look at refraction of light in more detail in Topic 11.8, Refractive index.

Whenever a wave refracts there is always some reflection off the surface (partial reflection).

If the wave slows down it will refract towards the normal, if it speeds up it refracts away from the normal. Sound waves normally speed up when they enter a denser medium, whereas electromagnetic waves, like light, normally slow down. This results in the waves refracting in different directions.

Study tip

When drawing ray diagrams, always measure angles to the normal.

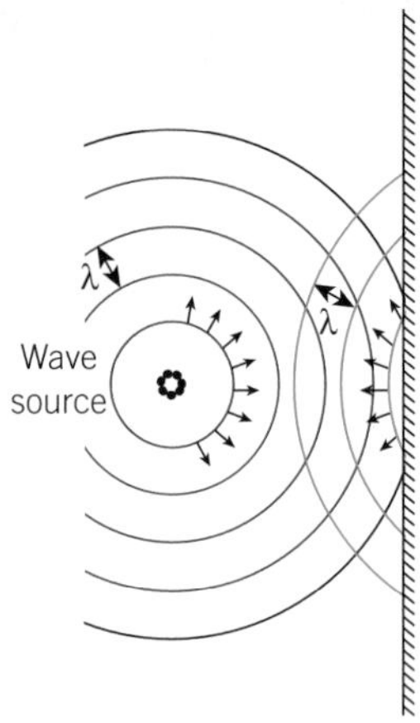

▲ **Figure 3** *When circular waves reflect off surfaces their wavelength and frequency remain the same, just like plane waves*

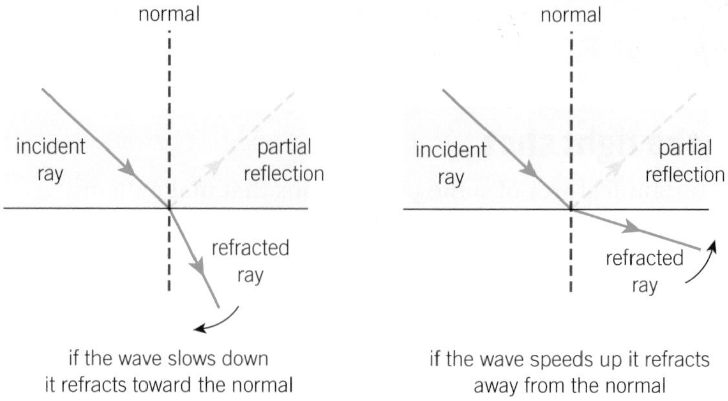

▲ **Figure 4** *A change in wave speed causes the wave to refract towards or away from the normal*

Unlike reflection, refraction does have an effect on the wavelength of the wave, but not its frequency. If the wave slows down its wavelength decreases and the frequency remains unchanged, and vice versa.

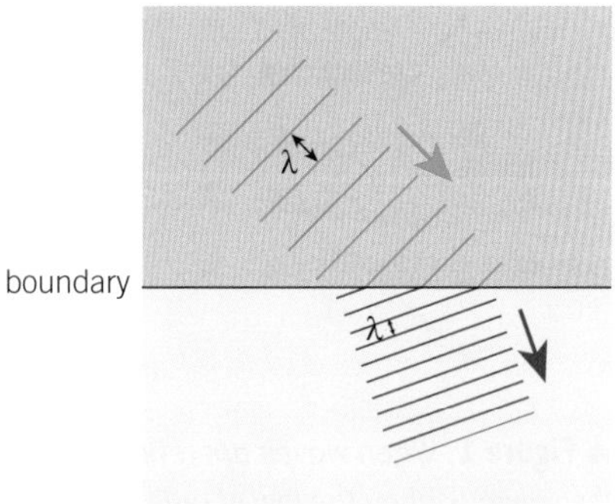

▲ **Figure 5** *When waves refract their wavelength changes: if the wave slows down, as shown here, the wavelength decreases so the wavefronts get closer together*

Refraction of water waves

The speed of water waves is affected by changes in the depth of the water, which gives us an easy way to investigate refraction of water waves. When a water wave enters shallower water, it slows down and the wavelength gets shorter.

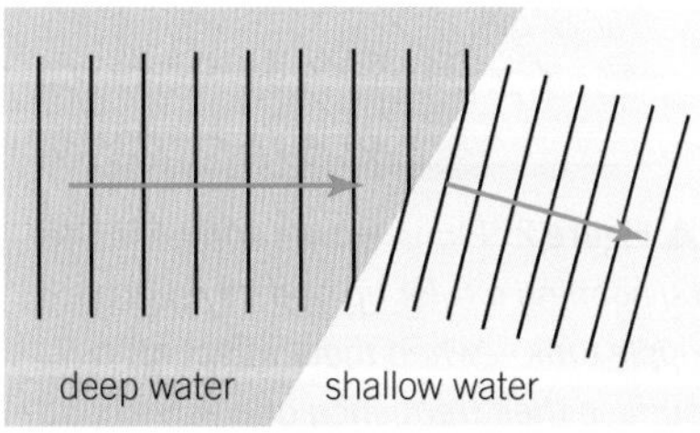

▲ **Figure 6** *Water waves are refracted when there is a change in depth*

Summary questions

1 Outline the similarities and differences between reflection and refraction. *(3 marks)*

2 Complete the diagrams in Figure 9 to show the reflection of light off various mirrored surfaces. *(4 marks)*

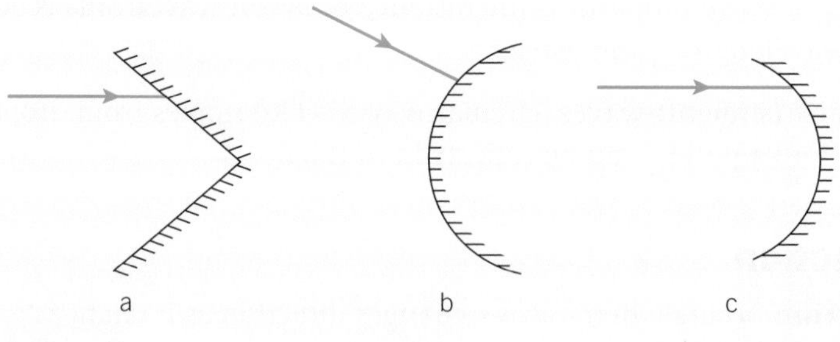

▲ **Figure 7**

3 Sketch a diagram to show a ray of light refracting when it travels from water into air. *(2 marks)*

4 Use the wave equation to explain why the wavelength of a refracted wave changes when it enters a different medium. *(2 marks)*

5 A swimming pool changes in depth from the shallow end to the deep end. Draw a wavefront diagram to show how circular ripples from a source in the centre of the pool travel to the edges. *(3 marks)*

Seeing atoms

Optical microscopes have enabled us to make huge advances in the fields of science and medicine. However, they have a limitation. We cannot keep on magnifying an object, seeing ever more detail with light. There is a limit.

This limit is around a few hundred nanometres. It results from a particular property of all waves. At high magnifications the image gets blurry because of the spreading of light, diffraction, as it passes through the apertures in the microscope. Diffraction cannot be avoided, and so if we want further magnification we must use different microscopes that do not rely on light. A scanning tunnelling microscope (STM) uses electrons to form images with much greater magnification. Objects down to individual atoms can be detected using this technology.

Learning outcomes

Demonstrate knowledge, understanding, and application of:

→ diffraction

→ polarisation.

▲ **Figure 1** *This image of a small number of gold atoms (around 5 nm across at their base) on top of a graphite layer (with individual carbon atoms seen as green dots) could not possibly be seen using a light microscope – it was recorded with a scanning tunnelling microscope*

Diffraction

Diffraction is a property unique to waves. When waves pass through a gap or travel around an obstacle, they spread out.

All waves can be diffracted. The speed, wavelength, and frequency of a wave do not change when diffraction occurs.

How much a wave diffracts depends on the relative sizes of the wavelength and the gap or obstacle. Diffraction effects are most significant when the size of the gap or obstacle is about the same as the wavelength of the wave. This is why sound diffracts when it passes through a doorway, allowing you to hear conversations around the corner. The wavelength of the sound is similar to the size of the gap. However, light has a much smaller wavelength, so it does not diffract through such a large gap. In order to observe the diffraction of light we need a much smaller gap.

Polarisation

It is also possible to polarise some waves. **Polarisation** means that the particles oscillate along one direction only (e.g., up and down in the vertical direction), which means that the wave is confined to a single plane. This 'plane of oscillation' contains the oscillation of the particles and the direction of travel of the wave. The wave is said to be **plane polarised**. You will learn more about the polarisation of electromagnetic waves in Topic 11.7.

Light from an **unpolarised** source, like a filament lamp, is made up of oscillations in many possible planes. As light is a transverse wave, these oscillations are always at 90° to the direction of energy transfer. If you could observe the wave travelling towards you, you might see these oscillations as up and down, side to side, or at any angle.

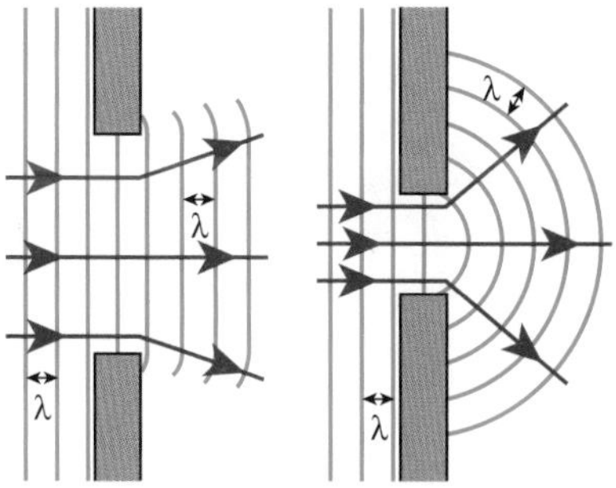

▲ **Figure 2** *The size of the gap compared with the wavelength of the wave affects how much diffraction takes place, and the spacing between the wavefronts shows that there is no change in wavelength*

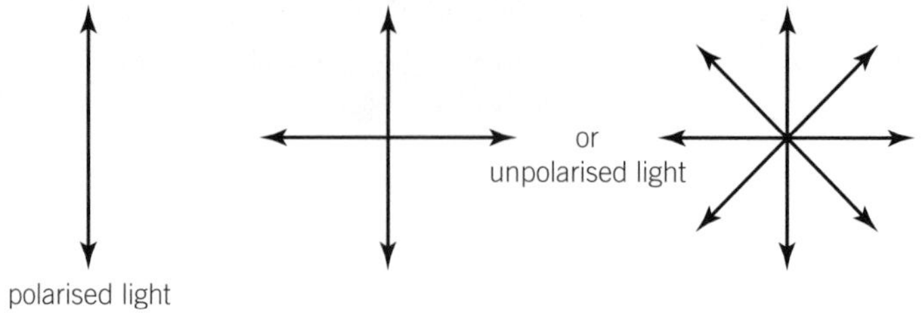

▲ **Figure 3** *A plane polarised wave is a wave in which the oscillations are in one direction only (left), whereas an unpolarised wave has oscillations in several directions*

In longitudinal waves, the oscillations are always parallel to the direction of energy transfer, so longitudinal waves cannot be plane polarised. Their oscillations are already limited to only one plane (the direction of energy transfer).

Partial polarisation

When transverse waves reflect off a surface they become **partially polarised**. This means there are more waves oscillating in one particular plane, but the wave is not completely plane polarised. For example, light reflected off the surface of water is partially polarised. Most of the light reflected off the surface becomes horizontally polarised. Some sunglasses contain polarising filters. These only allow light oscillating in one plane to pass through them, reducing the glare reflected off flat surfaces like lakes.

Study tip

When drawing diagrams to show diffraction, make sure the wavelength does not change.

Summary questions

1 Explain why it is not possible to polarise a sound wave. *(2 marks)*

2 Give two examples of a wave that can be plane polarised. *(2 marks)*

3 Explain why the diffraction of sound is regularly observed, but the diffraction of light is observed less frequently. *(2 marks)*

4 Two different waves pass through a 3.0 m gap. The first wave has a wavelength of 3.0 cm, the second wave 3.0 m. Describe the effect of the gap on each wave. *(3 marks)*

5 Explain why it is possible to receive long-wavelength radio signals at the bottom of some valleys in which the higher-frequency TV signal cannot be received. *(3 marks)*

6 Sound waves are directed towards a slit of width 0.30 m. The speed of sound in air is $340\ \text{m s}^{-1}$. State and explain whether or not each of the following frequency sound waves will be diffracted significantly at this slit:

a 1200 Hz; *(2 marks)*

b 1.0 MHz. *(2 marks)*

▲ **Figure 4** *This composite image shows the effect of a polarising filter (right) on the reflections off the surface of a rock pool – when the polarised reflections are screened out by the filter it is much easier to see the seaweed growing in the pool*

Can you hear me at the back?

The Andromeda galaxy (Figure 1) is one of the most distant objects you can detect with the naked eye. It is around 24 000 000 000 000 000 000 000 m (2.4×10^{22} m) from Earth. Like any progressive wave, the light from the galaxy spreads out as it travels further away from the source. The energy and power transferred becomes less concentrated. In the case of light, this means the further it travels from a source, the dimmer it becomes.

If the power of the wave source is known, this drop in brightness can be used to calculate how far the source is from the receiver. Certain types of supernovae always reach the same maximum brightness. Astronomers can use this information to determine our distance from the Andromeda galaxy and other astronomical objects.

This decrease in intensity with distance from the source occurs for all waves. Sound gets quieter and water waves decrease in amplitude.

Learning outcomes

Demonstrate knowledge, understanding, and application of:

→ intensity of a progressive wave $I = \frac{P}{A}$

→ intensity $\propto$ (amplitude)^2.

▲ **Figure 1** *The Andromeda galaxy is around two million light years away, so its light has taken two million years to reach us*

Intensity

The **intensity** of a progressive wave is defined as the radiant power passing through a surface per unit area. Intensity has units watts per square metre (W m^{-2}) and can be calculated using the equation

$$I = \frac{P}{A}$$

where I is the intensity of the wave at a surface, P is the radiant power passing through the surface, and A is the cross-sectional area of the surface.

Intensity and distance – an inverse square relationship

When the wave travels out from a source the radiant power spreads out, reducing the intensity. For a point source of a wave, the energy and power spread uniformly in all directions, that is, over the surface of a sphere (Figure 2).

The total radiant power P at a distance r from the source is spread out over an area equal to the surface area of the sphere ($A = 4\pi r^2$). Substituting this area into our equation for intensity gives

$$I = \frac{P}{A} = \frac{P}{4\pi r^2}$$

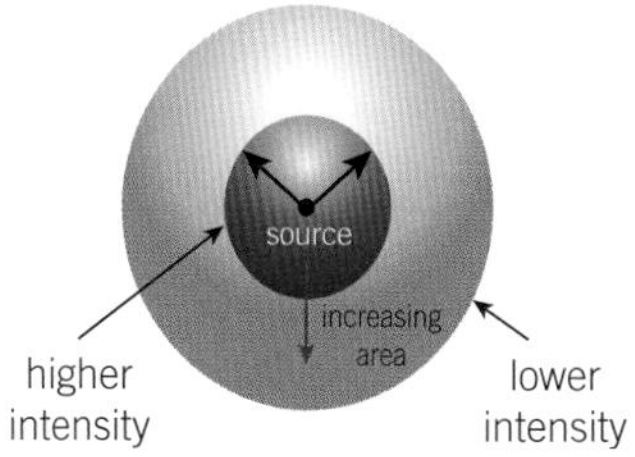

▲ **Figure 2** *The radiant power from a point source spreads out in a sphere, so the intensity of a wave depends on the area over which the power is spread out – the greater the area, the lower the intensity*

Worked example: Finding the power of the Sun

The average distance from the Earth to the Sun is 150 million km. The intensity of the radiation received by the upper atmosphere is 1400 W m^{-2}. Calculate the total power output of the Sun.

Step 1: Identify the correct equation to calculate the power from the Sun.

$$I = \frac{P}{A} = \frac{P}{4\pi r^2}$$

Step 2: Rearrange to make the power the subject.

$$P = I \times 4\pi r^2$$

Substitute in known values in SI units (including the distance in metres.

$P = 1400 \times 4\pi \times (150 \times 10^9)^2 = 4.0 \times 10^{26}$ W (2 s.f.)

We can see from the equation that the intensity has an inverse square relationship with the distance from the source ($I \propto 1/r^2$). If the distance doubles, the intensity decreases by a factor of 4 (2^2), and if the distance increases by a factor of 100 the intensity will be 100^2 times smaller (Figure 2).

Synoptic link

LDRs (light-dependent resistors) were introduced in Topic 9.9, The LDR.

Intensity and LDRs

We can use an LDR to investigate how the intensity varies with distance from a constant power source (like a simple filament lamp).

In order to determine the intensity a **calibration curve** is used. Each LDR has its own calibration curve that allows the user to convert the resistance of the LDR into intensity.

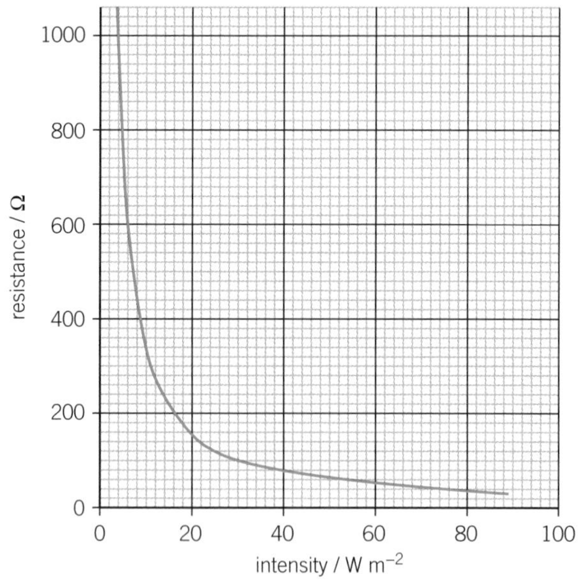

▲ **Figure 3** *Calibration curve for an LDR*

1 Use the data collected in Table 1 and the calibration curve to plot a graph of intensity against distance.

▼ **Table 1**

Distance from the source / cm	10	20	30	40	50	60
Resistance of the LDR / Ω	30	140	300	580	880	1400

2 Use the graph to test whether the intensity follows an inverse square relationship with distance from source.

Intensity and amplitude

When ripples travel out across the surface of a pond the intensity drops as the energy becomes more spread out. This causes a drop in amplitude. That is, the ripple height decreases the further the wave is from the source.

Decreased amplitude means a reduced average speed of the oscillating particles. Halving the amplitude results in particles oscillating with half the speed, and a quarter of the kinetic energy ($E_K = \frac{1}{2}mv^2$). So for any wave the intensity is directly proportional to the square of the amplitude. Double the amplitude of a wave and the intensity will quadruple.

$$\text{intensity} \propto (\text{amplitude})^2$$

Summary questions

1 State what happens to the intensity of a wave when the amplitude:
  a increases by a factor of 3; b decreases by a factor of 4. *(2 marks)*

2 Calculate the intensity when a power of 400 W is received over a cross-sectional area of 20 m^2. *(2 marks)*

3 Calculate the intensity 20 m from a source of light with a power of 60 W. *(3 marks)*

4 Figure 4 shows the cone of light created when light passes through a converging lens. Describe and explain how the intensity of light changes from A to B. *(4 marks)*

5 A satellite in orbit around the Earth uses two solar panels for power. The intensity of sunlight received at the height of the satellite is 1.4 kW m^{-2}. The surface area of each solar panel is 8.0 m^2. Calculate the total energy transferred to the panel in a period of 2.0 hours. *(4 marks)*

6 At a distance of 15 m from a point source the intensity of a sound wave is 1.0×10^{-4} W m^{-2}.
  a Show that the intensity 120 m from the source is approximately 1.6×10^{-6} W m^{-2}. *(3 marks)*
  b Discuss how the amplitude of the wave has changed. *(2 marks)*

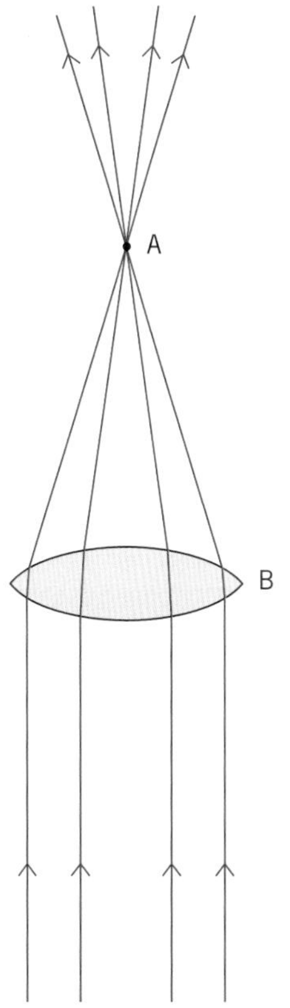

▲ **Figure 4**

Learning outcomes

Demonstrate knowledge, understanding, and application of:

- → the electromagnetic spectrum
- → properties of electromagnetic waves
- → the orders of magnitude of the wavelengths of the principal radiations from radio waves to gamma rays.

Travel through nothing

Electromagnetic waves (EM waves) do not need a medium. Unlike all other waves, they can travel through a vacuum. Without this ability to travel through space there could be no life on Earth, because this is how energy is transferred to our planet from the Sun.

What is an electromagnetic wave?

An EM wave is an example of a transverse wave, but it is a little more complex than a ripple on a pond. EM waves can be thought of as electric and magnetic fields oscillating at right angles to each other (Figure 2).

▲ **Figure 1** *These radio telescopes are designed to detect radio waves from objects deep in space, allowing physicists to learn more about our universe and ultimately about how life on Earth was possible*

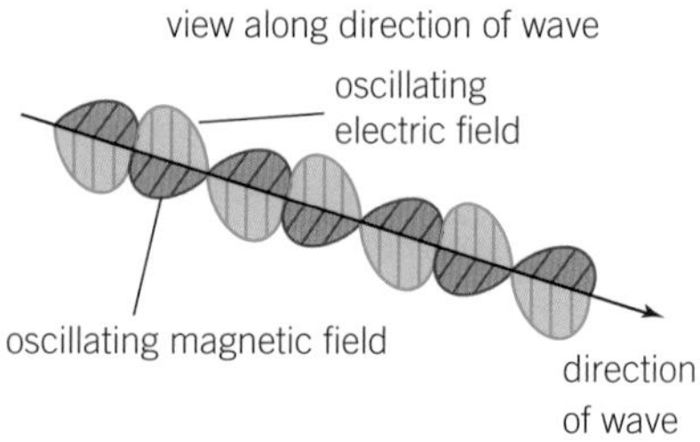

▲ **Figure 2** *An electromagnetic wave does not need a medium to be able to transfer energy*

The electromagnetic spectrum

The different types of EM wave are classified by wavelength. The full range of EM waves is called the **electromagnetic spectrum** and ranges from **radio waves**, with the longest wavelength, to **gamma rays**, with the shortest. Some radio waves have wavelengths longer than a million metres, whilst high-frequency gamma waves have wavelengths of just 10^{-16} m (less than the diameter of an atomic nucleus).

As you can see from Figure 3, the wavelength ranges of **X-rays** and gamma rays overlap. Unlike other parts of the spectrum, these EM

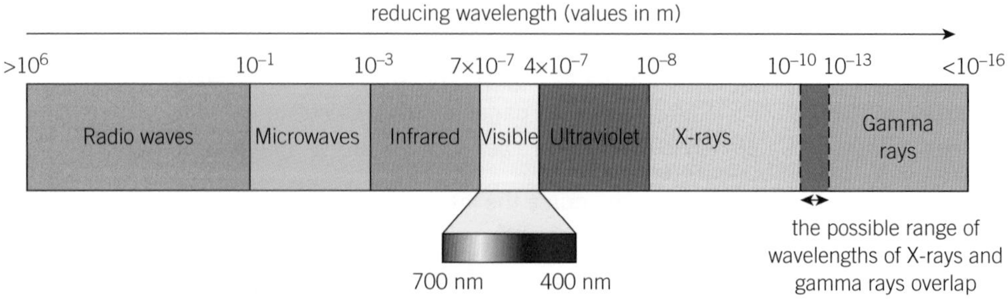

▲ **Figure 3** *The electromagnetic spectrum ranges from radio waves down to gamma rays*

waves are not classified by their wavelength, but by their origin. X-rays are emitted by fast-moving electrons, whereas gamma rays come from the unstable atomic nuclei.

Properties of EM waves

Like all waves, EM waves can be reflected, refracted, and diffracted. As EM waves are transverse waves they can also be plane polarised (see Topic 11.7, Polarisation of electromagnetic waves).

All EM waves travel at the same speed through a vacuum (c), $3.00 \times 10^8\,\mathrm{m\,s^{-1}}$. This is a very close approximation to their speed through air. Therefore, we can modify the wave equation for EM waves to

$$c = f\lambda$$

Using *c* to find wavelength

A radio station transmits at a frequency of 107.3 MHz. Calculate the wavelength of the radio wave to an appropriate number of significant figures.

Step 1: Identify the correct equation to calculate the speed of the wave.

$$c = f\lambda$$

Rearranging this equation for λ gives

$$\lambda = \frac{c}{f}$$

Step 2: Substitute in the known values in SI units and calculate the wavelength.

$$\lambda = \frac{3.00 \times 10^8}{107.3 \times 10^6} = 2.80\,\mathrm{m}\ (3\text{ s.f.})$$

Studying the universe

Only certain EM waves are transmitted through our atmosphere: some radio waves (wavelengths between 10 cm and 10 m), visible wavelengths, and longer wavelengths of UV make it to the surface, but most other frequencies are either reflected or absorbed by the atmosphere.

EM radiation from distant stars and galaxies is studied in great detail. Often this is only achievable by placing telescopes high on mountain tops or sending them into space.

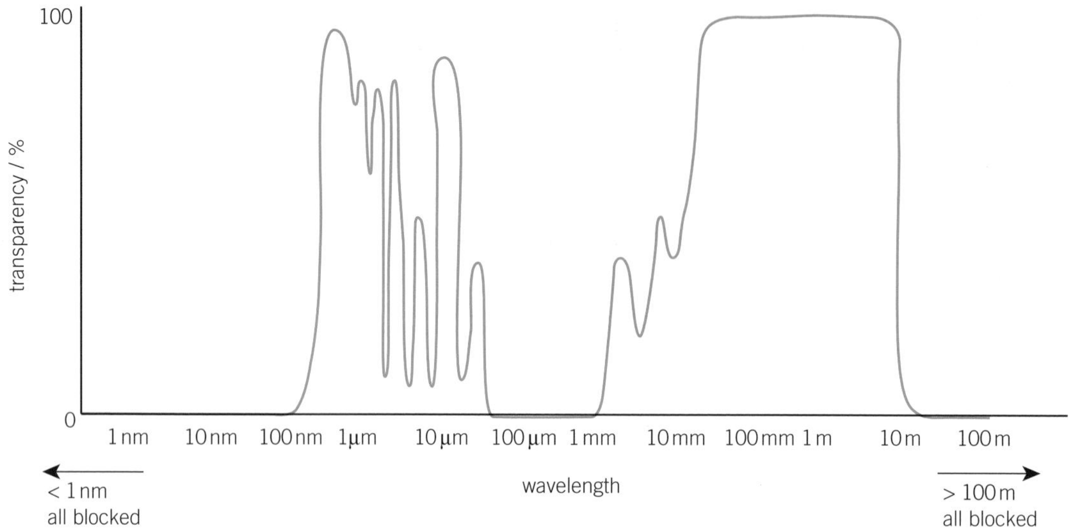

▲ **Figure 4** *Most frequencies of EM wave do not penetrate the Earth's atmosphere*

1 Give two reasons why telescopes are often built on mountains.
2 Outline the benefits and drawbacks of sending a telescope into space.
3 Look at Figure 4.
   a Determine the parts of the EM spectrum for which the radiation received at ground level is:
      i less than 10%; ii more than 80% of the radiation level above the atmosphere.
   b Calculate the highest and lowest frequencies of EM radiation that reach the Earth's surface.

Summary questions

1 List the EM spectrum in order of frequency from highest to lowest. *(2 marks)*

2 State the property of EM waves that confirms they are a type of transverse wave. *(1 mark)*

3 Calculate the wavelength of the following EM waves:
   a a radio wave with a frequency of 88.0 MHz; *(1 mark)*
   b microwaves with a frequency of 2.4 GHz; *(1 mark)*
   c X-rays with a frequency of 9.0×10^{16} Hz. *(1 mark)*

4 The human eye can detect EM waves with wavelengths from around 400 to 700 nm. Calculate the minimum and maximum frequencies of light that the eye can detect. *(2 marks)*

5 The Earth is on average 150 million km from the Sun. Calculate the time taken for light to travel from the Sun to the Earth. *(2 marks)*

6 A radar system uses microwave pulses that reflect off incoming aircraft. If the time delay from transmitting a pulse to receiving the reflection is 0.56 μs, calculate the distance to the aircraft. *(3 marks)*

11.7 Polarisation of electromagnetic waves

Specification reference: 4.4.1, 4.4.2

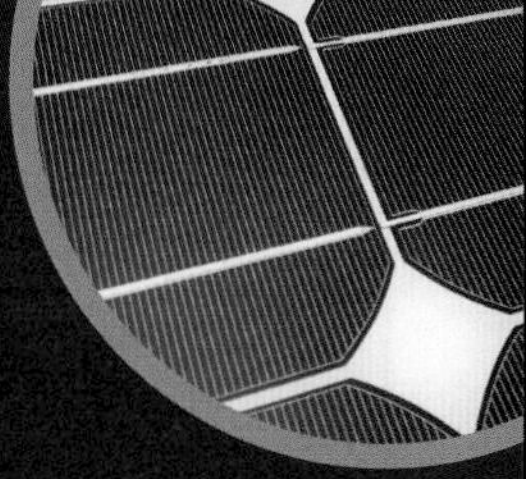

An echo from inflation?

We have seen how electromagnetic waves, like all transverse waves, can be plane polarised. In 2014 a pattern in the polarisation of the cosmic microwave background radiation was observed that may provide the first glimpse of evidence for gravitational waves and for inflation, a key part of the Big Bang theory (when the very early universe rapidly expanded from smaller than a proton to the size of a grapefruit).

Learning outcomes

Demonstrate knowledge, understanding, and application of:

→ plane polarised waves

→ the polarisation of electromagnetic waves.

Using polarising filters

Most naturally occurring electromagnetic waves are unpolarised. The electric field oscillates in random planes, all at 90° to the direction of energy transfer.

Unpolarised electromagnetic waves can be polarised using filters called polarisers. The nature of the polariser depends on the part of the electromagnetic spectrum to be polarised, but each filter only allows waves with a particular orientation through (Figure 1).

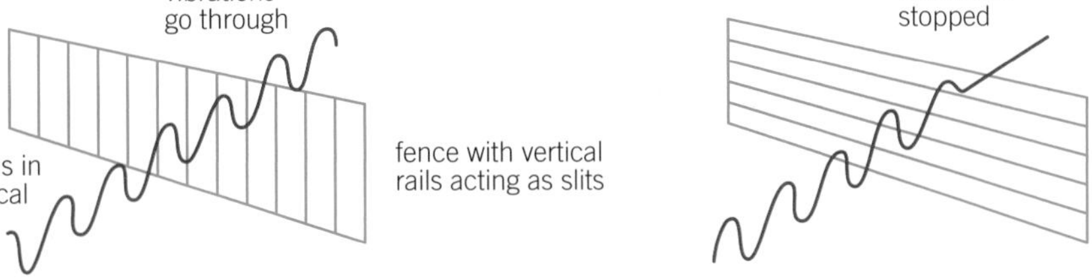

▲ **Figure 1** *A polarising filter acts like a slatted fence, only allowing electromagnetic waves polarised in the same direction as the filter to pass through*

Polarisation of light

Polaroid filters are plastic films that contain very long thin crystals and polarise light. They are used in sunglasses and over liquid crystal displays such as watches.

If you take two pieces of Polaroid filter, place them together, and rotate them, you can observe the effect of the plane polarisation of light passing through the filters (Figure 2).

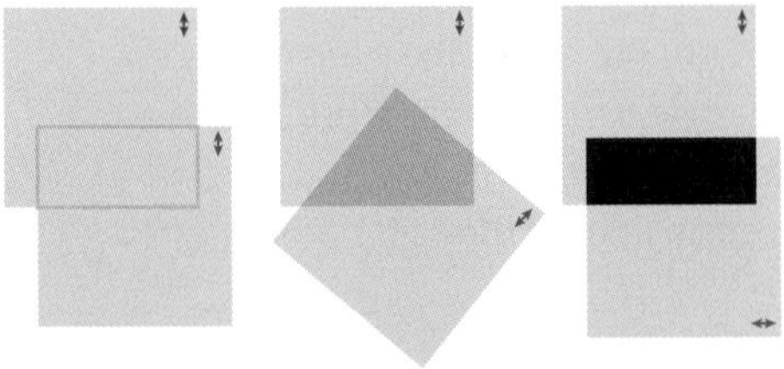

▲ **Figure 2** *As one filter is rotated with respect to the other, the intensity of the light passing through the filters varies*

Unpolarised light passing through the first filter becomes plane polarised. If the second filter (sometimes called the analyser) is in the same plane as the first, then the light passes through it unaffected. However, if the second Polaroid is slowly rotated, the intensity of the light transmitted through it drops. When the second filter has turned through 90°, no light is transmitted and the intensity falls to zero (Figure 3).

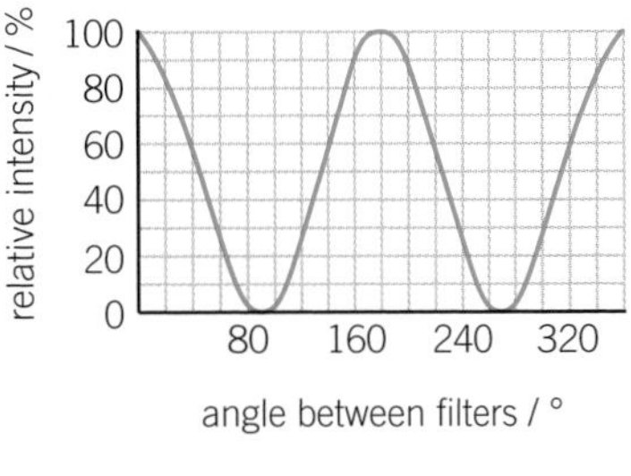

◀ **Figure 3** *Change in the intensity of the light transmitted through a pair of Polaroid filters as their relative orientation is rotated through 360°*

Polarisation of microwaves

Microwaves produced artificially tend to be plane polarised. Any unpolarised microwaves can be polarised like light, but in place of a Polaroid filter, a metal grille is used (Figure 4).

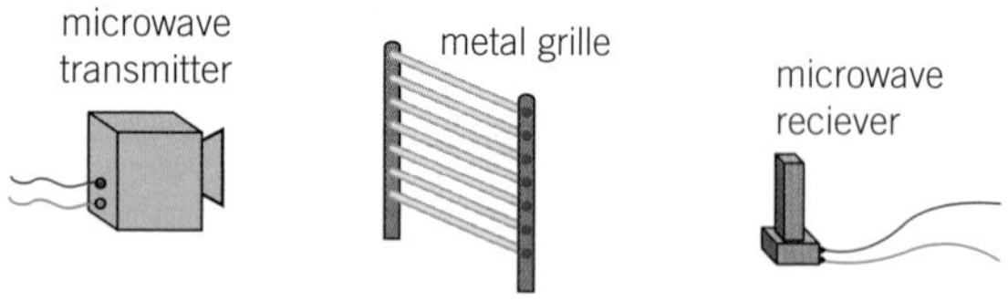

▲ **Figure 4** *A metal grille is able to polarise microwaves*

Inside the door of each microwave oven is a metal sheet with many holes in it. This allows light to pass through, enabling the user to see what is cooking, but at the same time preventing microwaves from escaping.

1. Outline why the intensity varies from a maximum value to zero.
2. A metal grille is placed between a source of plane polarised microwaves and a receiver. Describe and explain the effect of rotating the grille through 180° around the axis of the beam on the intensity recorded by the receiver.
3. Suggest why the metal sheet in the door of microwave ovens contains little holes, rather than a series of slits.

Aligning aerials

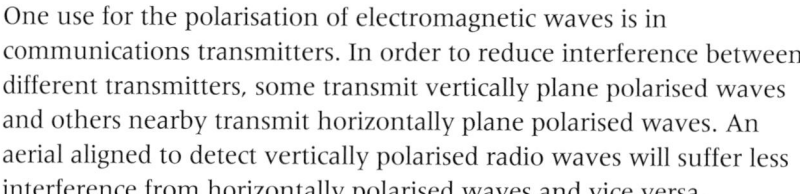

One use for the polarisation of electromagnetic waves is in communications transmitters. In order to reduce interference between different transmitters, some transmit vertically plane polarised waves and others nearby transmit horizontally plane polarised waves. An aerial aligned to detect vertically polarised radio waves will suffer less interference from horizontally polarised waves and vice versa.

Summary questions

1. State why the polarisation of light supports the view that light is a transverse wave. *(1 mark)*
2. Look at Figure 3. Explain why the maximum intensity occurs at 0°, 180°, and 360° and the minimum at 90° and 270°. *(2 marks)*
3. A student holds a polarising filter in front of a laptop screen and then rotates it. At a particular angle, the laptop screen appears to go dark.
   a. Suggest what you can deduce about the nature of light emitted from the laptop screen from the student's observation. *(1 mark)*
   b. Explain how the laptop screen can be viewed once again though the filter. *(3 marks)*
4. A beam of polarised light is directed normally at a polarising filter of cross-sectional area $9.0 \times 10^{-4}\,\mathrm{m^2}$. The polarising filter is slowly rotated in a plane at right angles to the beam. The transmitted intensity I plotted against the angle θ resembles Figure 3, with a maximum intensity of $20\,\mathrm{W\,m^{-2}}$.
   a. Calculate the power of light transmitted through the filter at $\theta = 0°$.
   b. Use the graph to calculate the ratio: $\dfrac{\text{amplitude of light at } 0°}{\text{amplitude of light at } 60°}$. *(2 marks)*

Seeing the light

Opticians use refractometers, which measure precisely how beams of light are refracted as they pass through the lens in a patient's eye. The angle at which the light refracts depends on a property of the material from which the lens is made called the **refractive index**. By accurately determining the refractive index of the lenses in the patient's eyes an appropriate prescription of glasses or contact lenses can be recommended.

Learning outcomes

Demonstrate knowledge, understanding, and application of:

→ refraction of light

→ refractive index

→ $n = \frac{c}{v}$; $n \sin\theta$ = constant.

Refractive index

Different materials refract light by different amounts. The angle at which the light is bent depends on the relative speeds of light through the two materials. Each material therefore has a refractive index, calculated using the equation

$$n = \frac{c}{v}$$

where n is the refractive index of the material (it has no units), c is the speed of light through a vacuum ($3.00 \times 10^8\,\text{m s}^{-1}$), and v is the speed of light through the material in m s^{-1}.

If $n = 1$ then the speed of light through the material is the same as the speed of light through a vacuum.

▼ **Table 1** *Refractive indices for some common substances*

Material	n
vacuum	1.00
air	1.00 (actually 1.000293)
water	1.33
olive oil	1.47
crown glass	1.52
diamond	2.42

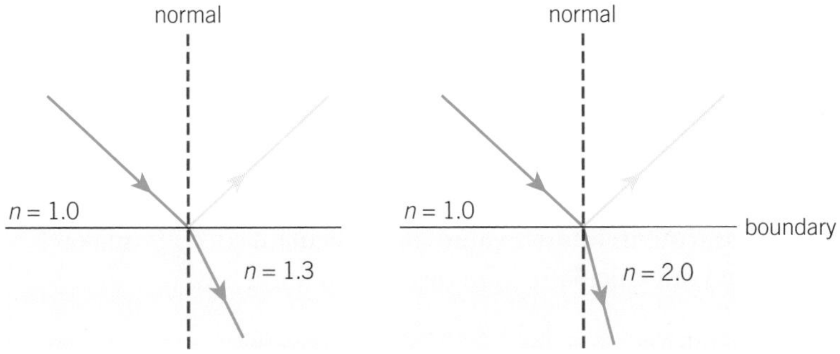

▲ **Figure 1** *The greater the refractive index, the more the light entering the material is refracted towards the normal*

Study tip

When drawing ray diagrams to show refraction, remember that angles are measured between the ray and the normal.

Worked example: The speed of light through olive oil

Use the data in Table 1 to determine the speed of light through olive oil.

Step 1: Identify the correct equation to use.

$$n = \frac{c}{v} \quad \text{which can be rearranged for } v \text{ to give}$$

$$v = \frac{c}{n}$$

Step 2: Substitute in known values in SI units and calculate the speed of light through the olive oil.

$$v = \frac{3.00 \times 10^8}{1.47} = 2.04 \times 10^8\,\text{m s}^{-1}$$

The speed of light through the material will always be less than the speed of light through a vacuum.

Summary questions

1. Describe the relationship between refractive index and the speed of light through a material. *(3 marks)*
2. The speed of light in ethanol is measured as $220 \times 10^6\ m\,s^{-1}$. Calculate the refractive index of ethanol. *(2 marks)*
3. Using the data in Table 1, calculate the speed of light through water. *(3 marks)*
4. Use the information in Figure 2 to determine the refractive index of material B. *(2 marks)*
5. A ray of light travels from olive oil to crown glass. It strikes the boundary between the media at an angle of 45°. Carefully draw a diagram showing the path of the refracted ray, labelling the angles with their correct values. *(4 marks)*
6. A ray of light travels from diamond into water. The light strikes the boundary between the diamond and the water at 20°. Calculate the angle of refraction. *(3 marks)*

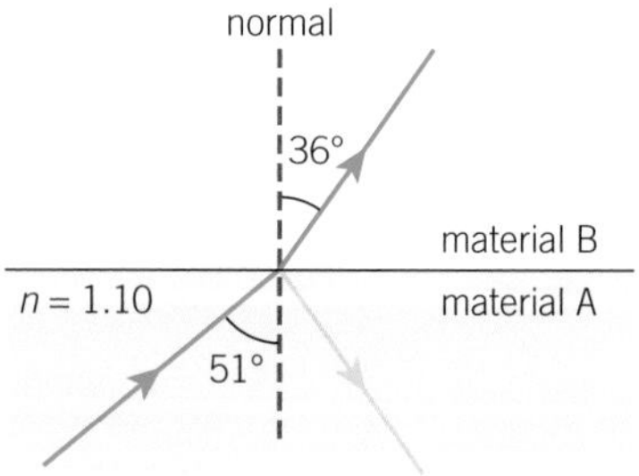

▲ **Figure 2**

Refraction law

The angles made by a ray of light at a boundary between two media was first investigated in 984 by the Persian physicist Ibn Sahl, then much later by Willebrord Snellius in 1621. Their findings can be simplified and expressed mathematically as

$$n \sin\theta = k$$

where n is the refractive index of the material, θ is the angle between the normal and the incident ray, and k is a constant.

We can apply this equation to describe what happens when light travels from one medium to another.

$$n_1 \sin\theta_1 = n_2 \sin\theta_2$$

Worked example: Calculating the angle of refraction

A ray of light travels from water to crown glass (a glass often used to make lenses). The light strikes the boundary between the two at an angle of 40.0° to the normal. Calculate the angle of refraction.

Step 1: Identify the correct equation to use.

$$n \sin\theta = k$$

Apply this equation to a ray of light travelling from water into glass.

$$n_{water} \sin\theta_{water} = n_{glass} \sin\theta_{glass}$$

Rearrange for $\sin\theta_{glass}$.

$$\sin\theta_{glass} = \frac{n_{water} \sin\theta_{water}}{n_{glass}}$$

Step 2: Substitute in known values and calculate $\sin\theta_{glass}$ (make sure your calculator is in degree mode).

$$\sin\theta_{glass} = \frac{1.33 \times \sin 40.0°}{1.52} = 0.562\ldots$$

$$\theta_{glass} = 34.2°$$

This angle is less than 40.0°, as the light has slowed down when it entered the glass from the water, bending towards the normal.

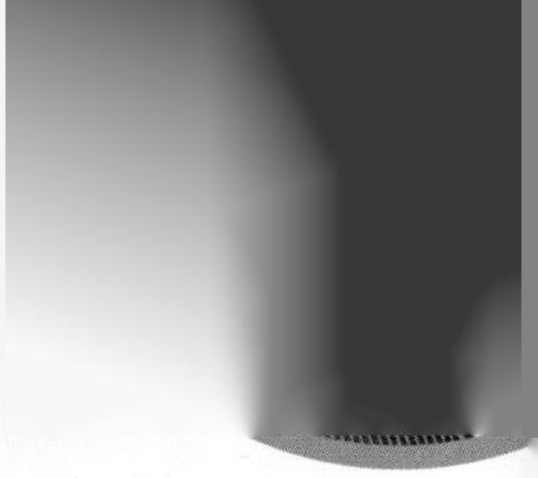

Giving diamonds their sparkle

Diamonds have a number of unusual physical properties. They are extremely hard (the term diamond comes from the ancient Greek word for unbreakable) and are superb thermal conductors. But it is perhaps their sparkle that makes diamonds so attractive.

At $n = 2.42$, the refractive index of diamond is one of the highest of all natural materials. Once light enters a diamond it is usually reflected off the inner surfaces several times before it leaves the diamond. A skilled diamond cutter exploits this property to give diamonds their characteristic sparkle.

Learning outcomes

Demonstrate knowledge, understanding, and application of:

→ critical angle and $\sin C = \frac{1}{n}$

→ total internal reflection for light.

▲ **Figure 1** *The high refractive index of diamond means light travels at a slow speed of 124 000 000 m s^{-1} through it*

Conditions for total internal reflection

The **total internal reflection** (TIR) of light occurs at the boundary between two different media. When the light strikes the boundary at a large angle to the normal, it is totally internally reflected. All the light is reflected back into the original medium. There is no light energy refracted out of the original medium.

Two conditions are required for TIR.

1. The light must be travelling through a medium with a higher refractive index as it strikes the boundary with a medium with a lower refractive index. For example, TIR is possible when light in glass meets air, but not the other way around.
2. The angle at which the light strikes the boundary must be above the **critical angle**. This angle depends on the refractive index of the medium.

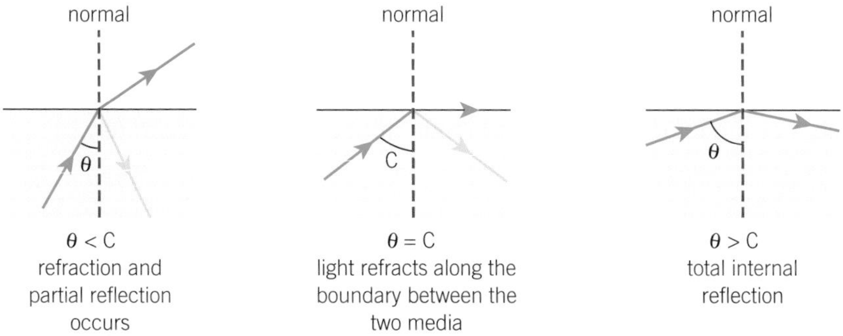

▲ **Figure 2** *Light meeting the boundary at exactly the critical angle travels along the boundary between the two media*

By using $n_1 \sin\theta_1 = n_2 \sin\theta_2$, you can determine the relationship between the refractive index of the medium and the critical angle when light travels from the medium into air.

At the critical angle C, θ_{air} is 90° (see Figure 2).

$$n \sin C = n_{air} \sin 90°$$

Both the refractive index of air and sin 90° are equal to 1, so this becomes

$$n \sin C = 1 \times 1$$

$$\sin C = \frac{1}{n}$$

From this we can see that the greater the refractive index the lower the critical angle.

Worked example: Crown glass

Crown glass has a refractive index of 1.52. Determine the critical angle between crown glass and air.

Step 1: Identify the correct equation to use.

$$\sin C = \frac{1}{n}$$

Rearranging for C gives

$$C = \sin^{-1}\left(\frac{1}{n}\right)$$

Step 2: Substitute in known values and calculate the critical angle (making sure your calculator is in degree mode).

$$C = \sin^{-1}\left(\frac{1}{1.52}\right) = 41.1°$$

If light strikes the internal surface of crown glass at above 41.1° then it will be totally internally reflected.

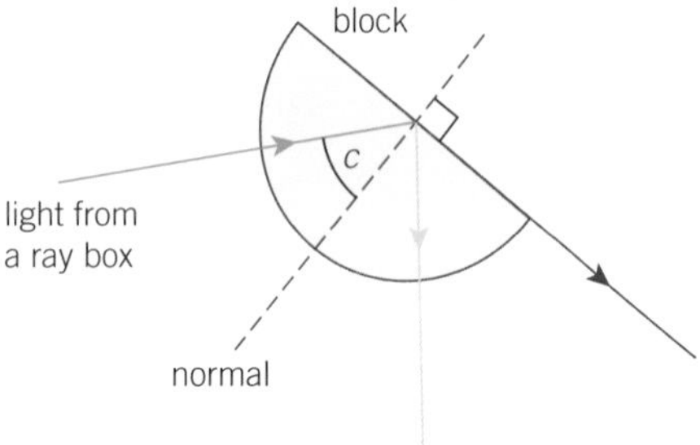

▲ **Figure 3** *Using a semi-circular block to measure the critical angle and calculate the refractive index of the block*

Determining refractive index from the critical angle

A simple experiment to determine the refractive index of a material can be carried out by carefully measuring the critical angle of a semi-circular block (Figure 3).

Directing the ray of light towards the centre of the semi-circular block ensures that light enters the block at 90° to the boundary and does not change direction, so the critical angle can be measured accurately.

▲ **Figure 4** *Optical fibres are usually made from flexible glass that has a high refractive index*

Optical fibres

Optical fibres are designed to totally internally reflect pulses of visible light (or occasionally infrared) travelling through them. They have many uses, including transmitting data for fast broadband connections and images from inside patients during keyhole surgery.

A simple optical fibre has a fine glass core surrounded by a glass cladding with a lower refractive index (Figure 5, left). Light travelling through the fibre is contained within the core because of total internal reflection at the core/cladding boundary.

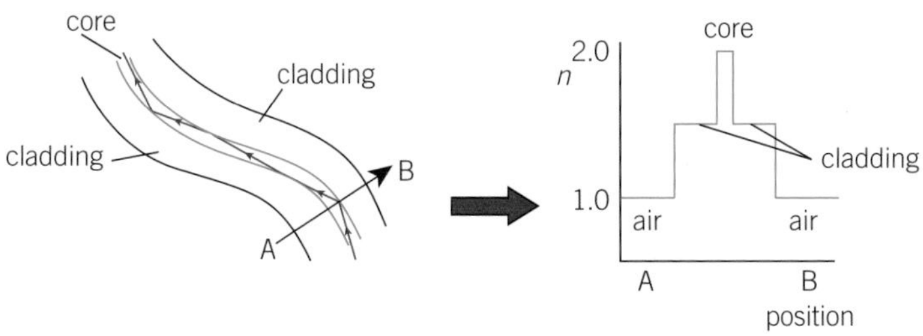

▲ **Figure 5** *Simple optical fibres are step indexed to ensure that the light remains inside the core of each fibre*

Simple optical fibres are step indexed. The refractive index changes suddenly between the core and cladding, and between the cladding and the air. When plotted on a graph of refractive index against distance across the fibre, the refractive index changes as a series of steps (Figure 5).

1 Explain why the cladding of an optical fibre must have a lower refractive index than the core.
2 Two pulses of light are sent along an optical fibre. One travels along the central axis of the fibre, the other undergoes many total internal reflections off the core–cladding boundary. Outline the differences in the received pulses at the end of the fibre.
3 In some optical fibres, called graded index fibres, the refractive index changes gradually across the fibre. It is lowest at the edge, increasing towards a maximum in the centre. Draw a diagram to show how a ray of light travels through a graded index optical fibre.

Summary questions

1 Describe what would happen to the critical angle if a semi-circular block with a lower refractive index was used in an investigation similar to the one described above. *(2 marks)*

2 Calculate the critical angle for diamond (refractive index for diamond is 2.42). *(2 marks)*

3 The critical angle for a Perspex (polyacrylate) block was measured as 42.8°. Calculate the refractive index of Perspex. *(2 marks)*

4 Use Figure 6 to calculate the refractive index of material A. *(3 marks)*

5 Crown glass has a critical angle of 41.1°. Draw diagrams to show the path of a ray of light that strikes the boundary between crown glass and air at:
a 41.1°; **b** 30.0°; **c** 65.0°. *(6 marks)*

6 The speed of light through flint glass is measured as 185 $\mathrm{Mm\,s^{-1}}$. Calculate the critical angle for flint glass with air. *(3 marks)*

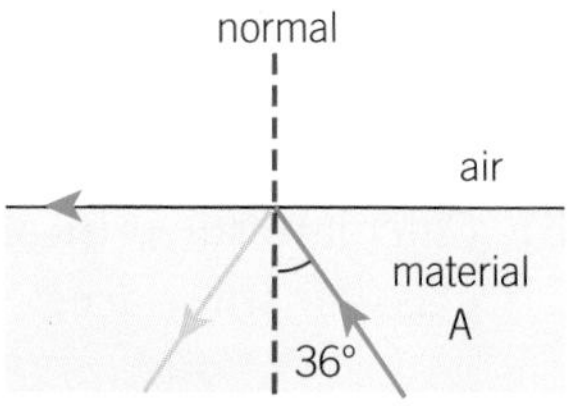

▲ **Figure 6**

Practice questions

1 **a** Explain what is meant by a *progressive wave*. *(2 marks)*

b Describe how a *transverse wave* differs from a *longitudinal wave*. *(2 marks)*

c (i) Explain what is meant by diffraction of a wave. *(1 mark)*

(ii) Describe how you could demonstrate that a sound wave of wavelength 0.10 m emitted from a loudspeaker can be diffracted. *(4 marks)*

Jan 2012 G482

2 **a** State two main properties of electromagnetic waves. *(2 marks)*

b A scientist is investigating a device emitting electromagnetic waves of frequency 15 GHz. Calculate the wavelength and identify the type of electromagnetic waves being investigated by the scientist. *(4 marks)*

3 **a** Define the refractive index of a transparent material such as glass. *(1 mark)*

b A ray of light is incident normally at the surface of a triangular shaped block of glass, see Figure 1.

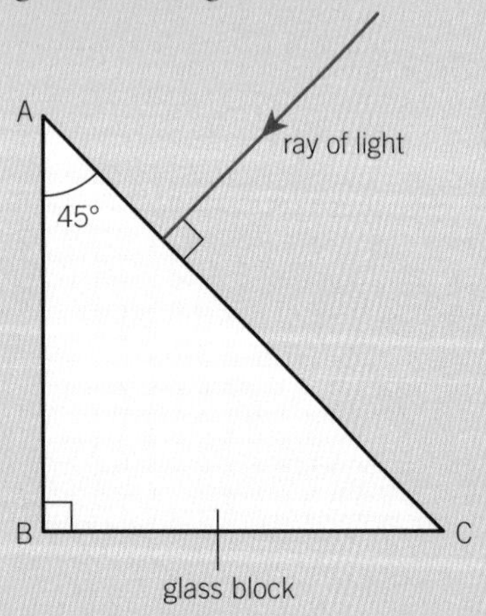

▲ **Figure 1**

The refractive index of the glass is 1.48.

(i) Calculate the speed of light in the glass block. *(2 marks)*

(ii) Calculate the critical angle for the glass-air interface. *(2 marks)*

(iii) Describe and explain the path of the ray of light **inside** the block. *(4 marks)*

4 **a** Define the following terms as applied to wave motion.

(i) *displacement and amplitude* *(2 marks)*

(ii) *frequency and phase difference* *(2 marks)*

b Figure 2 shows a transverse pulse on a *slinky*, a wound spring, at time $t = 0$. The pulse is travelling at a speed of $0.50\,\mathrm{m\,s^{-1}}$ from left to right. The front of the pulse is at point **X**, 0.25 m from the point **P**.

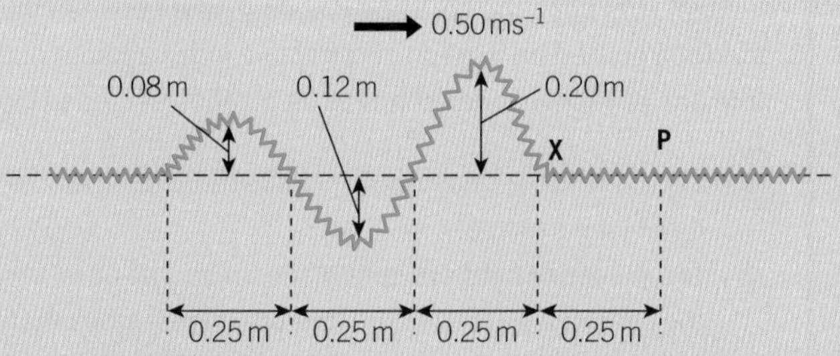

▲ **Figure 2**

On a copy of Figure 3, draw a displacement y against time t graph of the motion of point P on the slinky from $t = 0$ to $t = 2.5$ s.

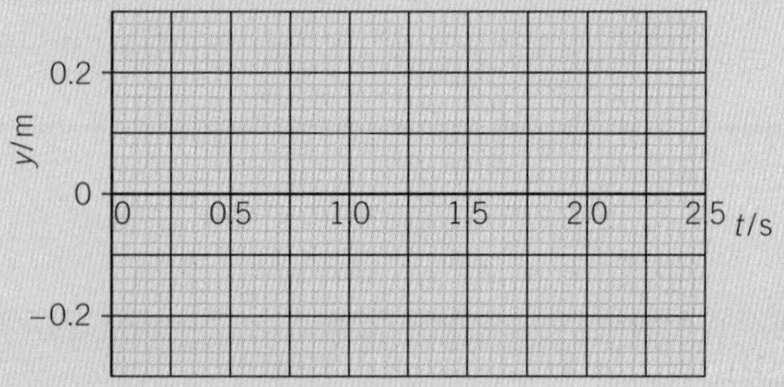

▲ **Figure 3**

(4 marks)

May 2012 G482

5 A transverse wave travels on a stretched cord from left to right. Figure 4 shows, at a given instant, the shape of the cord.

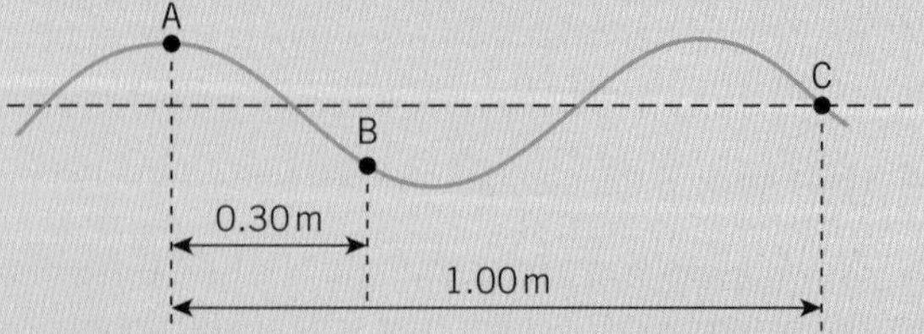

▲ **Figure 4**

The frequency of the wave is 2.4 Hz. **A**, **B** and **C** are particles on the rubber cord.

a Use Figure 4 to determine the speed of the wave. (*4 marks*)

b Calculate the phase difference, in degrees, between the points **A** and **B**. (*2 marks*)

c Describe and explain the direction in which **A** would move as the wave travels a very short distance. (*2 marks*)

June 2011 G482

6 a Describe a *plane polarised wave.* (*2 marks*)

b Light reflected from the surface of water is partially plane polarised in the horizontal direction. The reflected light is totally plane polarised when the angle of reflection is about 53°.

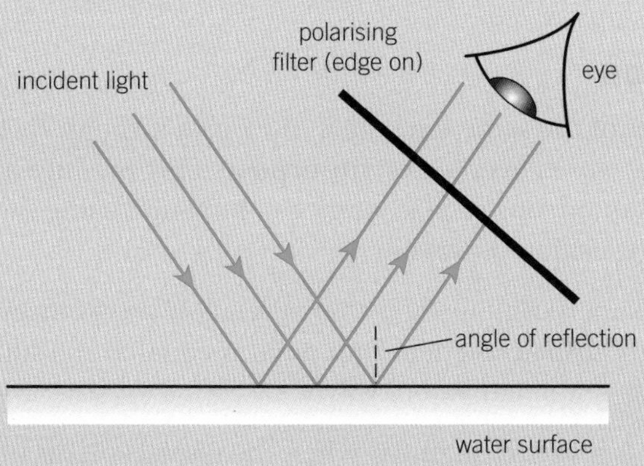

▲ **Figure 5**

Describe, referring to Figure 5, the experiment that you would perform using a polarising filter (a sheet of Polaroid) to determine whether the statement above is correct. Describe what you expect to observe. (*4 marks*)

June 2011 G482

7 Figure 6 shows the surface of water in a ripple tank at a given instant.

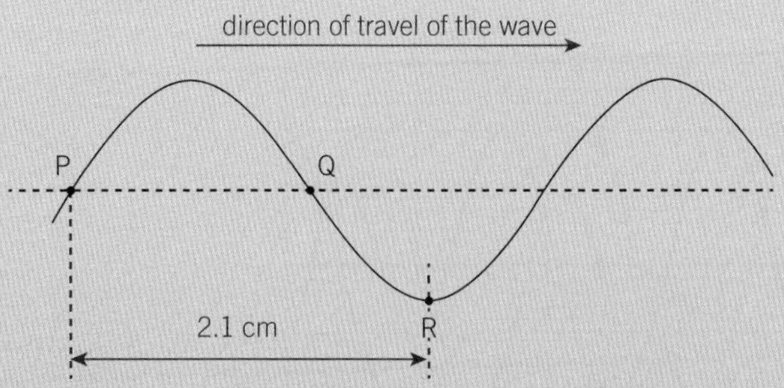

▲ **Figure 6**

a On a copy of Figure 6, show

(i) the wavelength of the wave (*1 mark*)

(ii) the amplitude of the wave. (*1 mark*)

b Compare and contrast the motion of particles at P and Q as the wave travels to the right. (*3 marks*)

c The distance between points P and R is 2.1 cm. The frequency of the wave is 20 Hz.

Calculate the speed of the wave. (*3 marks*)

d The speed of the waves in a ripple tank can be altered by changing the depth of the water.

State and explain the effect on the wavelength when the speed of the waves is doubled but the frequency is kept constant. (*2 marks*)

8 a A 1.2 mW laser emits light of wavelength 620 nm and a beam of diameter 0.82 mm.

Calculate

(i) the frequency of the light (*2 marks*)

(ii) the intensity of the beam of light. (*2 marks*)

b Figure 7 shows the path of a laser beam through a block of glass.

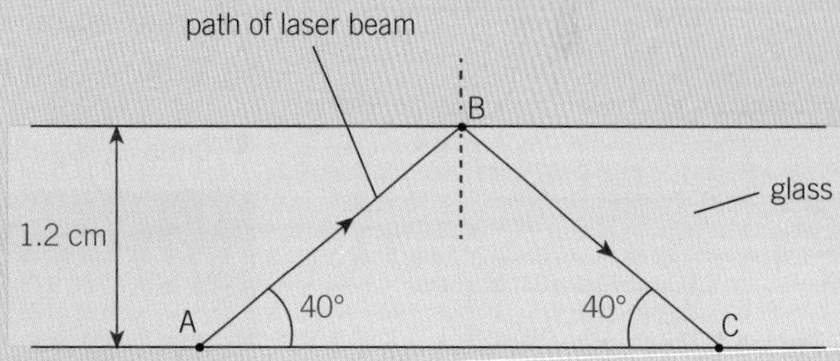

▲ **Figure 7**

The refractive index of glass is 1.50.

Calculate the time in ns it takes for the light to travel from A to B to C. (*4 marks*)

12 WAVES 2

12.1 Superposition of waves

Specification reference: 4.4.3

Learning outcomes

Demonstrate knowledge, understanding, and application of:

→ the principle of superposition of waves

→ graphical methods to illustrate the principle of superposition

→ interference.

Noise-cancelling headphones

Some headphones offer noise cancellation. This feature relies on the **principle of superposition of waves** to remove unwanted sounds from the listener's surroundings, allowing them to focus on the music. A microphone on the outside of the headphones detects the background noise. The speakers inside the headphones then produce waves that aim to perfectly cancel out all external sounds. Some people use these headphones without music to allow them to sleep in noisy environments like aircraft cabins.

Superposition

When two waves of the same type meet, they pass through each other. Where the waves overlap, or **superpose**, they produce a single wave whose instantaneous displacement can be found using the principle of superposition of waves.

The principle of superposition states that when two waves meet at a point the resultant displacement at that point is equal to the sum of the displacements of the individual waves.

As displacement is a vector quantity, when the displacements of two waves are added together the resultant can be greater or smaller than the individual displacements of each wave. Some examples are shown in Table 1 and Figure 2.

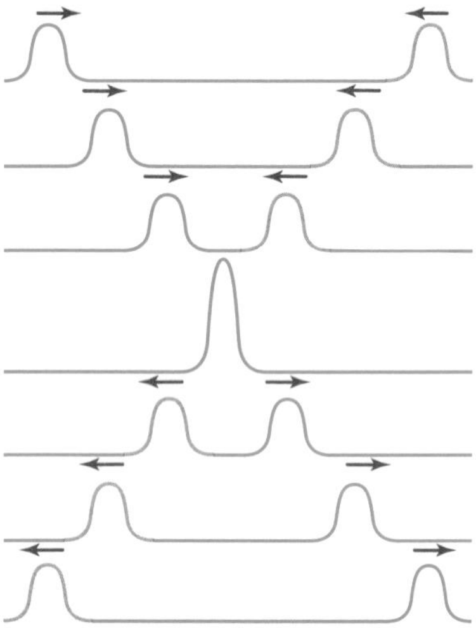

▲ **Figure 1** *Two pulses passing through each other superpose when they meet, in this case adding up to create a pulse with twice the amplitude of the individual pulses*

▼ **Table 1** *Superposition of wave 1 and wave 2*

	A	B	C	D
Time /s	0.000	0.040	0.138	0.260
Displacement of wave 1 /m	0.20	0.00	0.08	−0.18
Displacement of wave 2 /m	−0.14	−0.14	0.08	0.18
Resultant displacement /m	0.06	−0.14	0.16	0.00

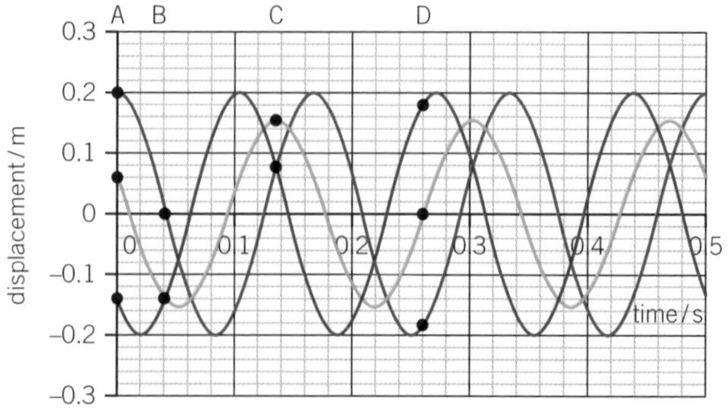

▲ **Figure 2** *More complex superposition of wave 1 (blue) and wave 2 (red) to produce the resultant wave (green)*

From superposition to interference

When two progressive waves continuously pass through each other they superpose and produce a resultant wave with a displacement equal to the sum of the individual displacements from the two waves. This effect is called **interference** and will be covered in more detail in Topic 12.2.

If the two waves are in phase then the maximum positive displacements (the peaks in a transverse wave) from each wave line up, creating a resultant displacement with increased amplitude. This is called **constructive interference** (Figure 3).

As intensity ∝ $(\text{amplitude})^2$, the increase in amplitude resulting from constructive interference increases the intensity: sound waves are louder, and light is brighter.

If two progressive waves are in antiphase, then the maximum positive displacement (the peak in a transverse wave) from one wave lines up with the maximum negative displacement (the trough) from the other, and the resultant displacement is smaller than for each individual wave. This is called **destructive interference**.

If the waves have the same amplitude the resultant wave will have zero amplitude – it is cancelled out completely (Figure 3).

The reduction in the displacement results in a drop in intensity at that point. Sounds are quieter, and light is dimmer. If the resultant wave has zero amplitude, the intensity falls to zero.

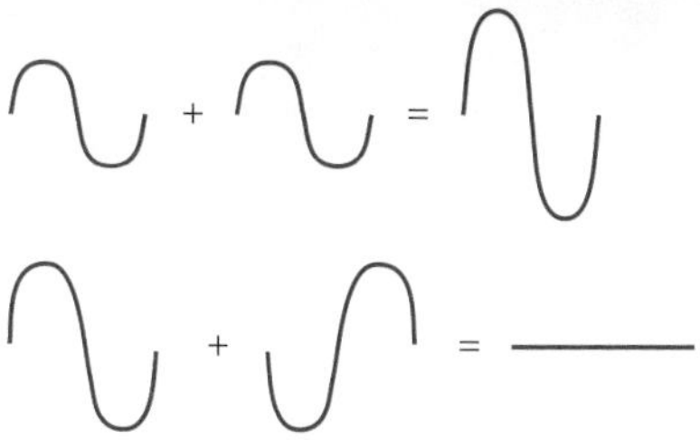

▲ **Figure 3** *Constructive (above) and destructive (below) interference*

Synoptic link

If you are unsure about the relationship between amplitude and intensity, look back at Topic 11.5, Intensity.

Summary questions

1. State and explain the type of interference used by noise-cancelling headphones. *(2 marks)*
2. Sketch two diagrams to illustrate the difference between constructive and destructive interference. *(2 marks)*
3. Determine the effect on the intensity when two waves of the same amplitude and perfectly in phase interfere. *(2 marks)*
4. Two sound waves are superposed. The first has a wavelength of 1.0 m and an amplitude of 5 mm. The second has a wavelength of 0.20 m and amplitude of 1 mm. Draw a displacement–distance graph for the resulting wave for distance in the range of 0 to 1.0 m. Assume the waves have a smooth sine-wave shape. *(6 marks)*
5. Two progressive waves travel towards and then pass through each other. Each wave travels at $1.0\,\text{m}\,\text{s}^{-1}$, and their starting positions are shown in Figure 4.

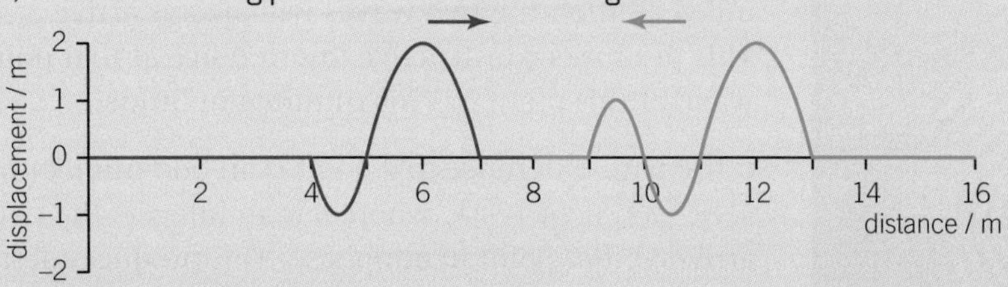

▲ **Figure 4**

Draw a series of six diagrams, each one second apart, to show how the resultant displacement changes as these two progressive waves superpose over 6 s. *(6 marks)*

Learning outcomes

Demonstrate knowledge, understanding, and application of:

- → interference, coherence, path difference, and phase difference
- → constructive interference and destructive interference in terms of path difference and phase difference
- → two-source interference using sound and microwaves.

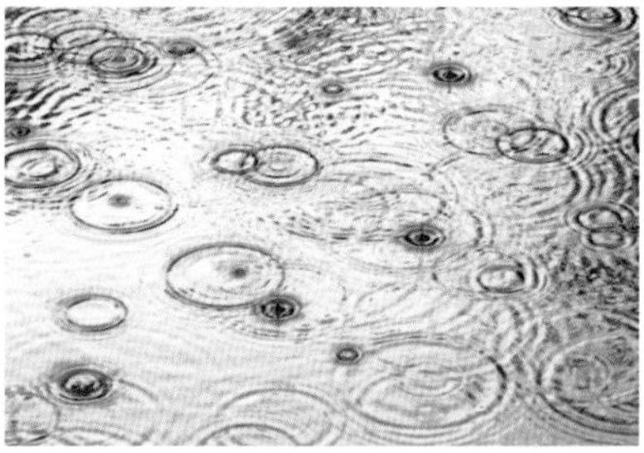

▲ **Figure 1** *As the ripples overlap they interfere, sometimes constructively, other times destructively, creating patterns*

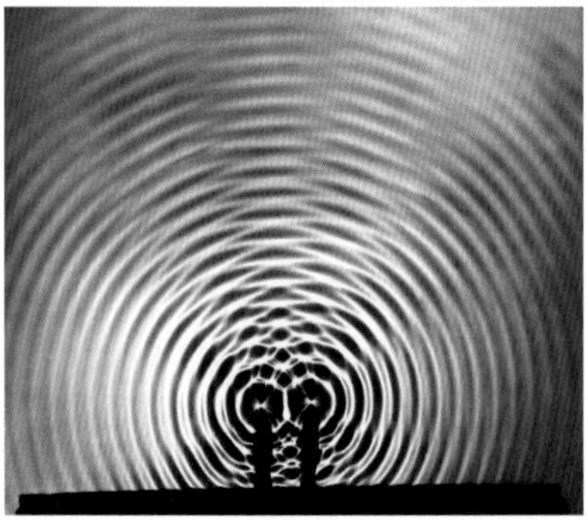

▲ **Figure 2** *A stable interference pattern formed by two dippers in a ripple tank producing coherent water waves*

Differences matter

If you watch the ripples on a pond when raindrops fall, you will see an **interference pattern** (Figure 1). As the wave caused by each raindrop travels outwards it overlaps with waves caused by other drops. At different points these superposed waves are in phase, interfering constructively, or out of phase, causing destructive interference. Random raindrops do not form a stable interference pattern but one that changes all the time. For a stable pattern, the waves must be **coherent** (Figure 2).

Forming stable interference patterns

Coherence refers to waves emitted from two sources having a *constant phase difference*. In order to be coherent the two waves must have the same frequency.

Filament lamps emit light of a range of different frequencies and ever-changing phase difference between different waves. In other words, they do not emit coherent light. Therefore, it is not possible to produce stable interference patterns using two filament lamps.

Path difference and phase difference

Interference patterns contain a series of **maxima** and **minima**. At a maximum the waves interfere constructively, at a minimum they interfere destructively. For example, a pair of loudspeakers emitting coherent sound waves produces an interference pattern with regions that are louder (maxima) and others that are quieter (minima) than the original waves. The same effect can be seen in Figure 2 with water waves. In places the water waves have increased amplitude (maxima) and in others the amplitude appears to be zero (minima).

In both cases these maxima and minima are a result of the two waves having travelled different distances from their sources. This difference in the distance travelled is called the **path difference**.

Figure 3 shows two sources emitting coherent waves. The wavelength of the progressive wave is λ. If the path difference to a point is zero or a whole number of wavelengths ($0, \lambda, 2\lambda, \ldots, n\lambda$, where n is an integer), then the two waves will always arrive at that point in phase. This produces constructive interference at that point. The resultant wave at this point has maximum amplitude.

If the path difference to a point is an odd number of half wavelengths ($\frac{1}{2}\lambda, \frac{3}{2}\lambda, \ldots, (n + \frac{1}{2})\lambda$, where n is an integer) the two waves will always arrive at that point in antiphase. This produces destructive interference. The resultant wave at this point has minimum amplitude.

At the central maxima, shown in Figure 4, the path difference is zero and so the phase difference is zero. At the first-order maxima the path difference is one whole wavelength, so the phase difference is 360° or

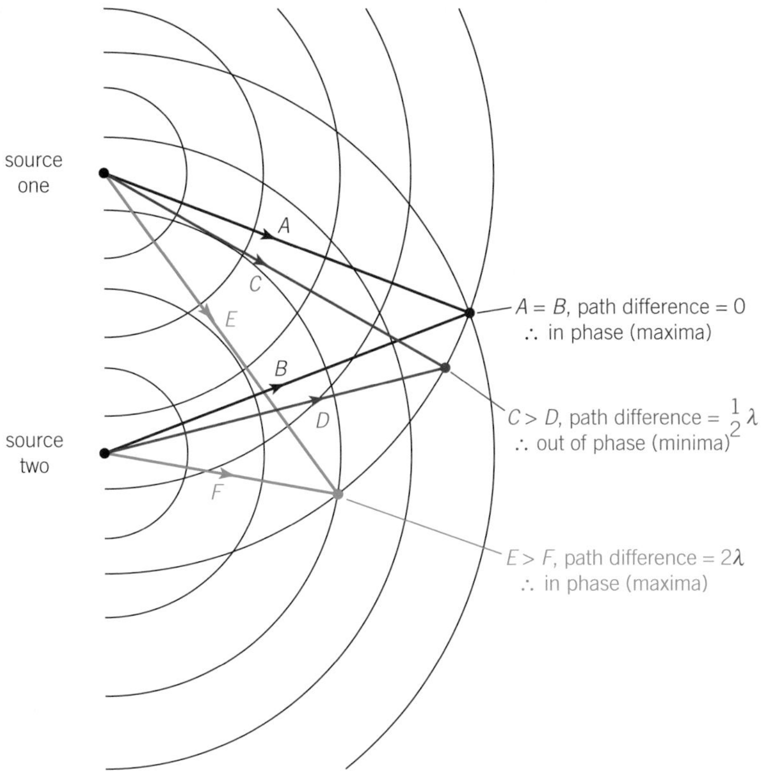

▲ **Figure 3** *The path difference determines the phase difference between the waves arriving at a given point*

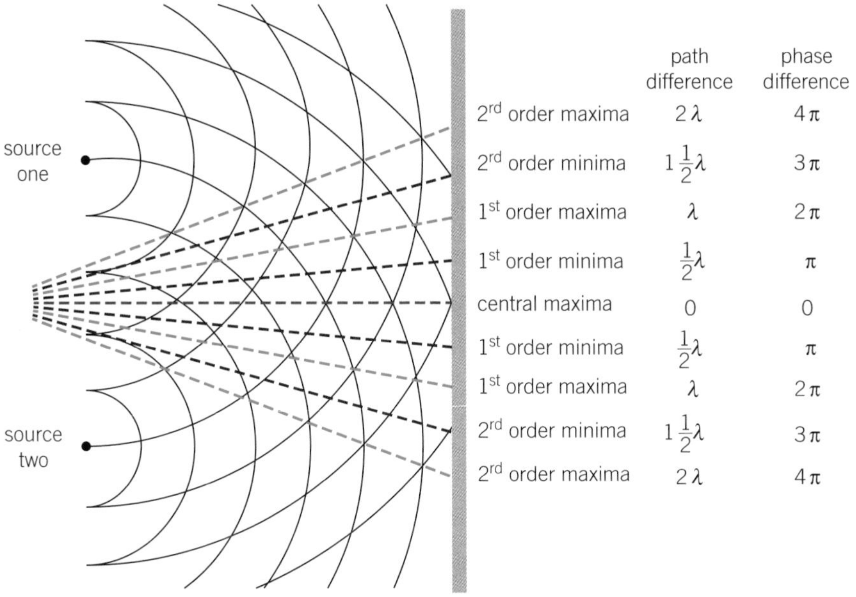

▲ **Figure 4** *Maxima and minima are caused by path differences resulting in phase difference between the waves*

2π radians. The peaks from the first waves perfectly line up with the peaks of the second waves and so constructive interference occurs.

At the first-order minima the path difference is half a wavelength, a phase difference of 180° or π radians. The peaks from the first waves line up with the troughs of the second waves, resulting in destructive interference.

Study tip

Path difference is a distance, measured in m, cm, or mm.
Do not confuse this with phase difference, which is measured in degrees or radians.

Interference in sound

Two loudspeakers connected to the same signal generator will emit coherent sound waves (Figure 5). The sound waves travel out from each loudspeaker and overlap, forming an interference pattern.

The interference pattern comprises a series of maxima (louder) and minima (softer). The positions of the maxima and minima can be detected with your ears or, more accurately, a microphone.

1. The speed of sound in air is approximately 340 m s^{-1}. The path difference at the first-order maxima is measured as 28 cm. Calculate the frequency of the sound.
2. Describe and explain what would happen to the interference pattern if the frequency of the sound from the loudspeakers were halved.

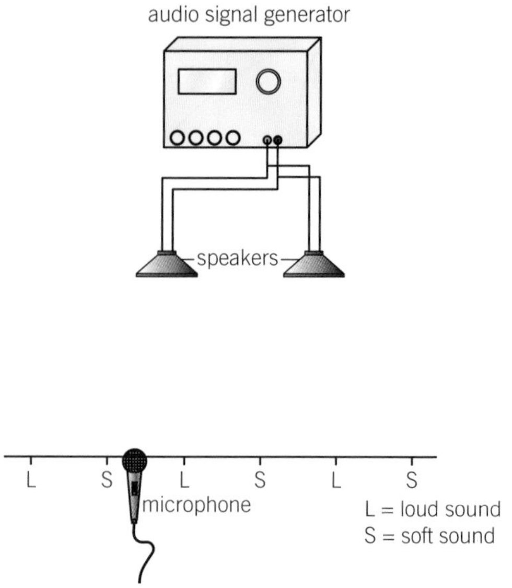

▲ **Figure 5** *The interference pattern from two loudspeakers forms alternating regions of louder and quieter sound*

Interference in microwaves

Producing coherent microwaves is more difficult than for sound. A single microwave source is used along with a pair of slits (a double slit), shown in Figure 6.

These diffracted microwaves overlap and form an interference pattern that can be detected with a microwave receiver connected to a voltmeter or an oscilloscope. Moving the receiver in an arc around the double slit detects the characteristic maxima and minima created as part of the interference pattern. The position of each can be carefully marked on paper.

1. Suggest a method to determine the wavelength of the microwaves used by measuring the path difference in an interference experiment. You should include details of any measurements taken and any subsequent calculations.
2. The frequency of the microwaves used is 24 GHz. With the help of a calculation, suggest a suitable width for each slit.
3. Describe how the equipment could be used to detect whether the microwaves are plane polarised.

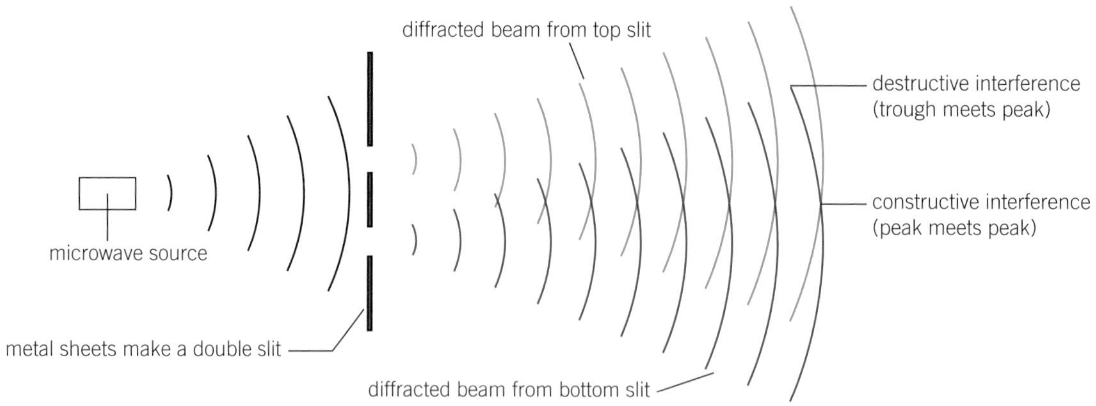

▲ **Figure 6** *The microwaves diffract at each slit, so each slit acts like a wave source and the diffracted waves from the two slits are coherent*

Thin film interference

The pattern of coloured light on thin oil films on water (Figure 7) is caused by interference. Light reflecting off the bottom surface of the oil interferes with the light reflected off the top surface. If the thickness of the oil results in a path difference that is a non-integer half number of wavelengths of light, the two sets of light waves are out of phase, destructive interference occurs, and the waves cancel out.

The colour results from the different wavelengths in white light and slight differences in the thickness of the oil layer. The distance the light travels through the oil before reflecting off the back surface differs. Different wavelengths of light are cancelled out by different thicknesses of oil. The wavelengths that are not cancelled out form the colours we observe.

1. Draw a diagram to show the path taken by two rays of light, one reflecting off the top surface of the oil, the other entering the oil before reflecting off the bottom surface.
2. Use your diagram to show that there is a path difference between the two waves.
3. Explain why the oil needs to be thicker in order to provide destructive interference for red light compared with blue light.

▲ **Figure 7** *The thin film of oil on the top of a puddle can cause beautiful, colourful patterns to form by interference*

Summary questions

1 State the type of interference formed at the following path differences:
- a 5λ; *(1 mark)*
- b 10λ; *(1 mark)*
- c 4.5λ. *(1 mark)*

2 Describe how the phase difference changes as you move away from the central maxima towards the third-order maxima. *(4 marks)*

3 Two coherent surface water waves are produced by a pair of dippers in a ripple tank. If the path difference to the second-order maxima is 9.0 cm, determine the wavelength of the water waves. *(2 marks)*

4 Two coherent sound waves of frequency 2.0 kHz form an interference pattern. The speed of sound in air is 340 m s^{-1}. Calculate:
- a the wavelength of the sound; *(1 mark)*
- b the path difference that causes a phase difference of 5π radians; *(2 marks)*
- c the path difference at the second-order minima. *(3 marks)*

12.3 The Young double-slit experiment

Specification reference: 4.4.3

Learning outcomes

Demonstrate knowledge, understanding, and application of:

- → Young double-slit experiment using visible light
- → $\lambda = \frac{ax}{D}$ for all waves where $a << D$.

Study tip

With visible light, the maxima and minima are known as bright and dark fringes respectively.

Was Newton ever wrong?

Sir Isaac Newton is perhaps the greatest physicist who ever lived. His contributions to mathematics and physics included the co-invention of calculus, the formulation of laws on motion and gravity, and the design of the first reflecting telescope. Newton demonstrated that white light is composed of a spectrum of colours. He rejected the idea that light was a wave. Instead he developed his own corpuscular (or particle) theory, in which he described light as a stream of tiny particles.

Newton died in 1726, but his ideas remained the accepted scientific theory for 75 more years until Thomas Young's wonderfully simple experiment demonstrated that light can form an interference pattern, conclusive proof that light must be acting as a wave.

Young double-slit experiment

Two coherent waves are needed to form an interference pattern. Young devised the method to achieve this that now bears his name. He used a **monochromatic** source of light (which can be achieved using a colour filter that allows only a specific frequency of light to pass) and a narrow single slit to diffract the light.

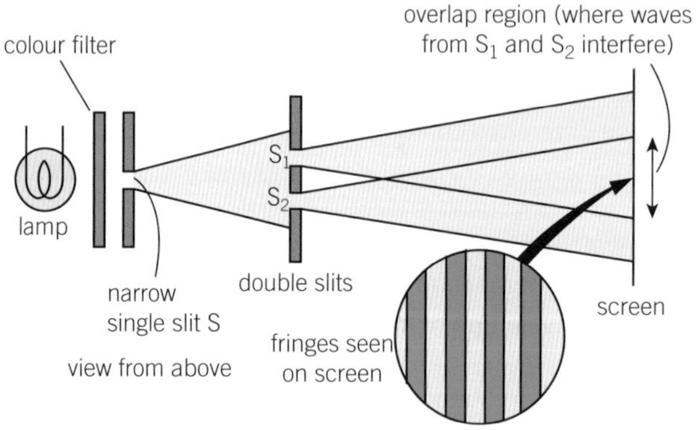

▲ **Figure 1** *The Young double-slit experiment made use of a single slit followed by a double slit to produce two sources of coherent light waves*

Light diffracting from the single slit arrives at the double slit in phase. It then diffracts again from the double slit. Each slit acts as a source of coherent waves, which spread from each slit, overlapping and forming an interference pattern that can be seen on a screen as alternating bright and dark regions called fringes.

The Young double-slit experiment successfully demonstrated the wave nature of light. Young used his experiment to determine the wavelength of various different colours of visible light.

Synoptic link

Was Newton really wrong? Quantum physics tell us that light can also be modelled as a particle: see Topic 13.1, The photon model.

A mathematical treatment

By considering the dimensions of different parts of a double-slit experiment we can derive a formula to calculate the wavelength λ of the light used to form the interference pattern.

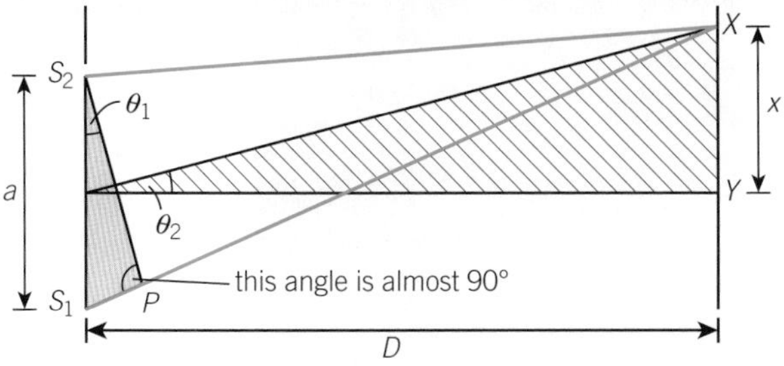

▲ **Figure 2** *The geometry of a typical double-slit experiment*

In Figure 2, the separation between the slits S_1 and S_2 is a. The interference pattern is observed on a distant screen a distance D from the slits, where $D >> a$. A bright fringe is seen at Y, and the next adjacent bright fringe is observed at X, where the separation between the fringes is x. The path difference S_1P must be equal to one whole wavelength λ, since $D >> a$. The two rays of light shown in blue are almost parallel to each other and the angles θ_1 and θ_2 are almost the same and very small. We can use the trigonometric approximation

$$\sin\theta_1 \approx \sin\theta_2 \approx \tan\theta_2$$

where $\sin\theta_1 = \lambda/a$ and $\tan\theta_2 = x/D$.

Therefore $\frac{\lambda}{a} \approx \frac{x}{D}$

This is often expressed as

$$\lambda = \frac{ax}{D}$$

This equation only applies if $a << D$ – the distance between the slits is much smaller than the distance from the screen to the double slits. Provided this is the case, the equation can be used to determine the wavelength of any wave producing an interference pattern from a double slit or two coherent sources, such as two loudspeakers connected to the same signal generator.

Study tip

The symbol ≈ means 'approximately equal to' and >> means 'very much greater than'.

Determining the wavelength of monochromatic light from a laser

A double slit can be used to determine the wavelength of light emitted from a laser (Figure 3). There is no need for a filter or single slit as the light from the laser is already monochromatic and in phase.

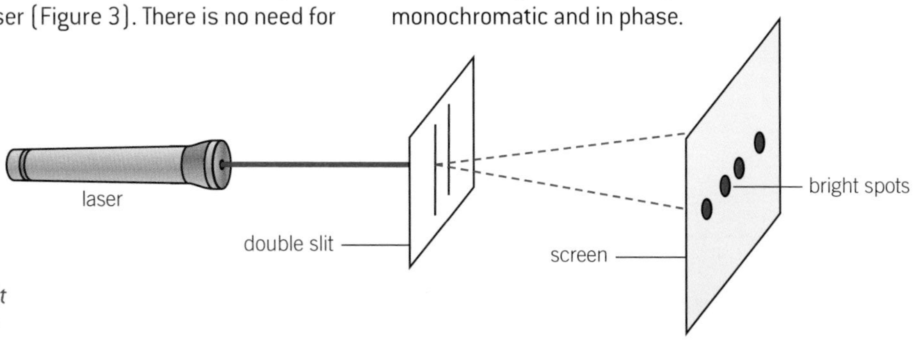

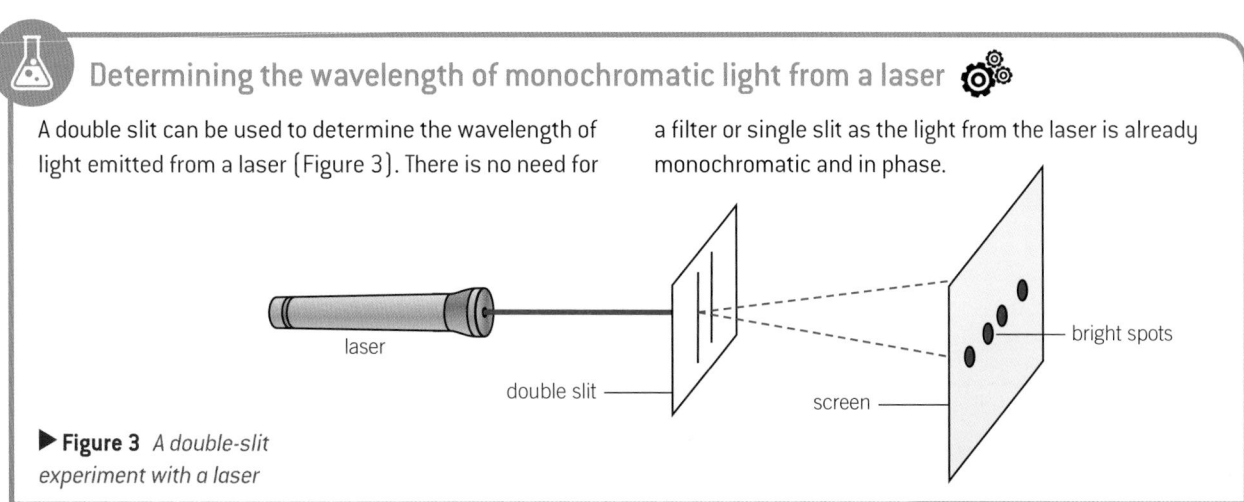

▶ **Figure 3** *A double-slit experiment with a laser*

By measuring the distance between several bright fringes, the separation x between adjacent fringes can be determined. If a is known (double slits are often labelled with this information) and D is measured directly, then λ can be calculated from:

$$\lambda = \frac{ax}{D}$$

1 Describe how the interference pattern would change if a laser emitting green light was used in place of the laser emitting red light.

2 Explain why it is better to measure the separation between multiple fringes to determine x, rather than between adjacent fringes.

3 A helium–neon laser is used to form an interference pattern on a screen 10 m from a double slit with a slit separation of 0.50 mm. Calculate the separation between adjacent fringes.

▲ **Figure 4** *The interference pattern formed from a helium–neon laser (λ = 632.8 nm)*

▲ **Figure 5**

Summary questions

1 Explain why it was necessary to use monochromatic light, along with a single slit and a double slit to produce a stable interference pattern for light. *(2 marks)*

2 The following measurements were taken during a double-slit experiment: separation between slits = 0.6 mm, separation between adjacent bright fringes = 1.4 mm, and distance from slits to screen = 1.6 m. Calculate the wavelength of light used. *(2 marks)*

3 The interference pattern in Figure 5 was obtained using a source of light of unknown wavelength. It is drawn to scale. The distance from the double slits with a slit separation of 1.0 mm to the screen was measured as 15 m. Determine the wavelength of the light. *(3 marks)*

4 The following measurements were taken during a double-slit experiment using light with a wavelength of 610 nm: separation between the slits = 0.40 mm and the separation between adjacent bright fringes = 1.8 mm. Calculate the distance from the slits to the screen. *(3 marks)*

5 Explain the effect on the interference pattern seen on the screen for a double-slit experiment when:

a a light source with a longer wavelength is used; *(1 mark)*

b the slit separation is doubled; *(2 marks)*

c the distance from slits to screen is increased by a factor of three; *(2 marks)*

d a light source with double the frequency is used. *(3 marks)*

Seeing UFOs

A lenticular cloud is a rare and beautiful natural phenomenon (Figure 1). They form over mountain ranges, but only if the conditions in the atmosphere are just right.

As the wind blows over the mountain peak a **stationary wave** (sometimes called **standing wave**) is formed. If the air contains enough moisture, the oscillations in the stationary wave make the water condense into the characteristic shape of this cloud.

Formation and properties of stationary waves

A stationary wave is not a single wave at all. It forms when two progressive waves with the same frequency (and ideally the same amplitude) travelling in opposite directions are superposed (Figure 2). As they have the same frequency, at certain points they are in antiphase. At these points their displacements cancel out. This forms a **node**, a point where the displacement is always zero, and therefore the amplitude and the intensity are zero.

At other points when the two waves are always in phase, an **antinode** is formed – the point of greatest amplitude and therefore intensity.

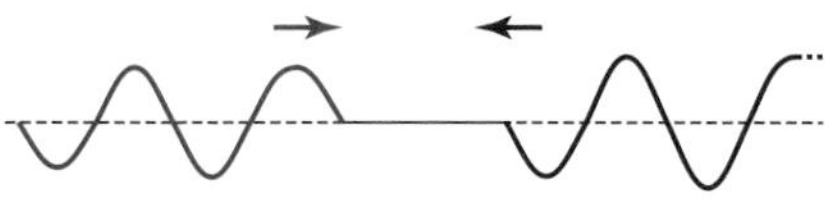

▲ **Figure 2** *Stationary waves are formed by the superposition of two waves of the same frequency travelling in opposite directions, for example, on a stretched string or by microwaves*

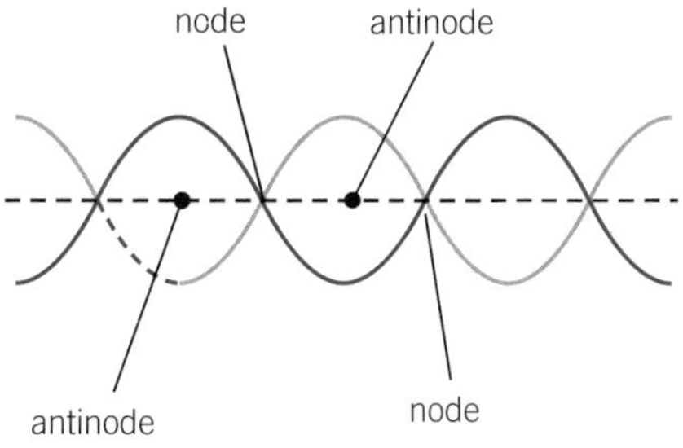

▲ **Figure 3** *A stationary wave is formed of a number of nodes and antinodes*

The separation between two adjacent nodes (or antinodes) is equal to half the wavelength of the original progressive wave (Figure 3), and the frequency is the same as that of the original waves. The wave profile for the stationary wave changes over time, creating the characteristic nodes and antinodes (Figure 4).

As the two progressive waves are travelling in opposite directions, there is no net energy transfer by a stationary wave, unlike a single progressive wave.

Learning outcomes

Demonstrate knowledge, understanding, and application of:

- → stationary (standing) waves: microwaves, stretched strings, and air columns
- → graphical representations of a stationary wave
- → similarities and differences between stationary and progressive waves
- → nodes and antinodes
- → the separation between adjacent nodes (or antinodes) is equal to $\frac{\lambda}{2}$.

▲ **Figure 1** *Because of their lens-like shape, if they form at night lenticular clouds are sometimes reported as UFOs*

Study tip

Unlike a progressive wave, a stationary waves have nodes and antinodes.

Study tip

Do not use terms like maxima and minima to describe antinodes and nodes. Maxima and minima are features of an interference pattern; antinodes and nodes are features of stationary waves.

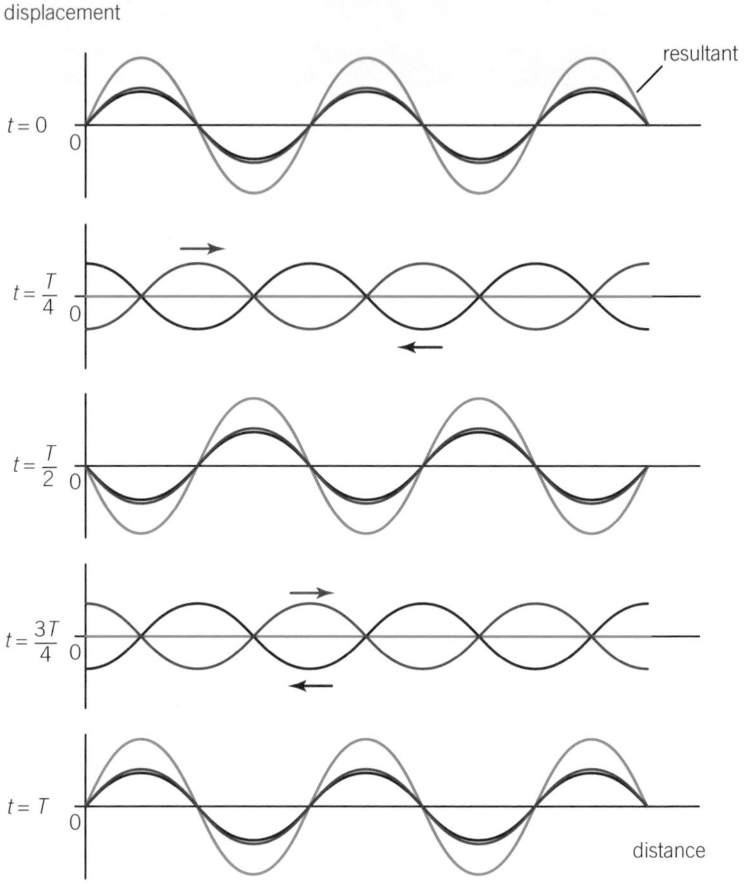

▲ **Figure 4** *A stationary wave (in green) is the resultant of two progressive waves travelling in opposite directions (red to the right, blue to the left)*

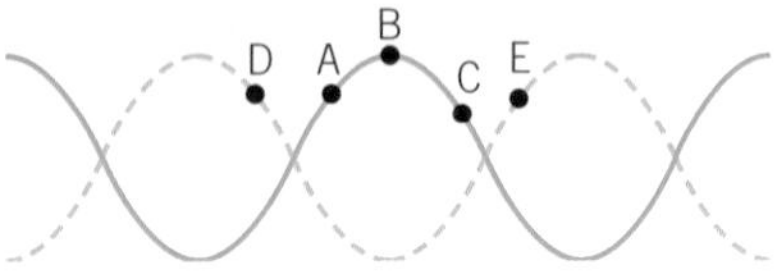

▲ **Figure 5** *Phase differences along a stationary wave are very different to phase differences along a progressive wave: particles A, B, and C are in phase with each other and in antiphase with particles D and E*

Phase differences along a stationary wave

In between adjacent nodes all the particles in a stationary wave are oscillating in phase with each other. They all reach their maximum positive displacement at the same time. However, their amplitudes differ, with the maximum amplitude at the antinode.

On different sides of a node the particles are in antiphase (they have a phase difference of π radians). The particles on one side of a node reach their maximum positive displacement at the same time as those on the other reach their maximum negative displacement.

Comparison of stationary and progressive waves

▼ **Table 1** *Summary of the properties of stationary and progressive waves*

	Progressive wave	Stationary wave
energy transfer	energy transferred in the direction of the wave	no net energy transfer
wavelength	minimum distance between two adjacent points oscillating in phase, for example, the distance between two peaks or two compressions	twice the distance between adjacent nodes (or antinodes) is equal to the wavelength of the progressive waves that created the stationary wave

▼ **Table 1** *(continued)*

	Progressive wave	Stationary wave
phase differences	the phase changes across one complete cycle of the wave	all parts of the wave between a pair of nodes are in phase, and on different sides of a node they are in antiphase
amplitude	all parts of the wave have the same amplitude (assuming no energy is lost to the surroundings)	maximum amplitude occurs at the antinode then drops to zero at the node

Forming a stationary wave using microwaves

A stationary wave can be formed by reflecting microwaves off a metal sheet so that two microwaves of the same frequency are travelling in opposite directions. A microwave receiver (Figure 6) will detect the changes in intensity between the nodes (lower/zero intensity) and antinodes (maximum intensity).

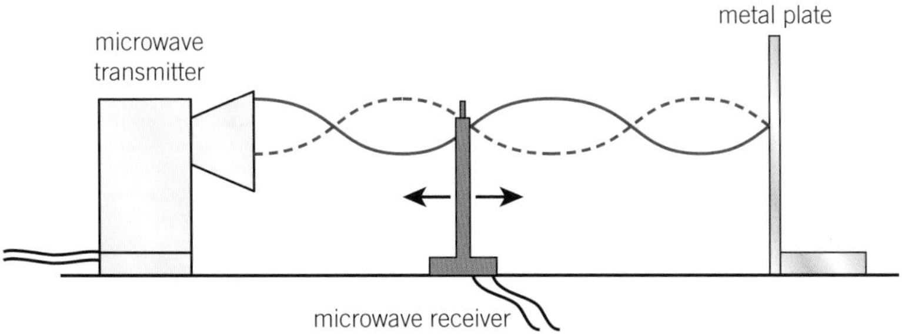

▲ **Figure 6** *Stationary microwaves*

The distance between the transmitter and metal sheet has to be adjusted until the receiver detects a series of nodes and antinodes. The distance between successive nodes or antinodes is equal to $\lambda/2$, where λ is the wavelength of the microwaves from the transmitter.

1 Describe how this experiment can be used to determine the wavelength of microwaves.
2 If the frequency of microwaves used in this experiment is 5.0 GHz, calculate the distance between adjacent nodes.
3 Explain why the node formed closest to the metal sheet provides the best cancellation whereas at other nodes the intensity does not fall to zero. (Hint: Think about the difference in distance travelled by the two waves forming the node, compared with the difference at the node closest to the transmitter.)

Summary questions

1 Describe two similarities between progressive and stationary waves. *(2 marks)*

2 Describe the phase difference between the two waves that form a stationary wave:
   a at a node; *(1 mark)*
   b at an antinode. *(1 mark)*

3 A standing wave has adjacent nodes 30 cm apart. Calculate the wavelength in metres of the progressive waves responsible for this standing wave. *(2 marks)*

4 Describe how the amplitude of a standing wave varies from one node to the next. *(3 marks)*

5 Figure 7 shows a standing wave formed on a stretched string.
   a State the phase differences between:
      i X and Y; ii X and Z. *(2 marks)*
   b If the period of the wave is 6.0 s sketch the string after:
      i 1.5 s; ii 3.0 s;
      iii 7.5 s. *(4 marks)*

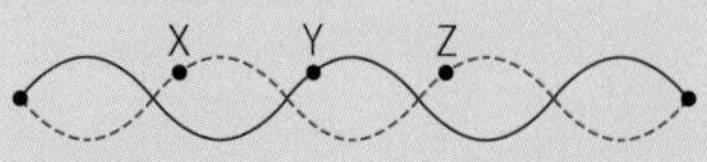

▲ **Figure 7**

Learning outcomes

Demonstrate knowledge, understanding, and application of:

- → the fundamental mode of vibration (first harmonic)
- → harmonics
- → stationary wave patterns for a stretched string.

▲ **Figure 1** *The strings of a violin are fixed at the ends so they are effectively nodes, resulting in the formation of a stationary wave when the bow rubs across a string*

Stringing out a note

Stringed instruments, like the violin, produce musical notes from stationary waves formed on their strings.

Each string has a **fundamental mode of vibration**. The frequency of this vibration is the **fundamental frequency**, and depends on the string's mass, tension, and length.

In addition, different **harmonics** are produced, with several different wavelengths. These wavelengths interfere with each other to produce the rich sound characteristic of stringed instruments. You can easily identify the type of stringed instrument – say guitar, violin, or piano – from its sound because of its unique combinations of harmonics.

Stationary waves on strings

If a string is stretched between two fixed points, these points act as nodes. When the string is plucked a progressive wave travels along the string and reflects off its ends. This creates two progressive waves travelling in opposite directions that then form a stationary wave.

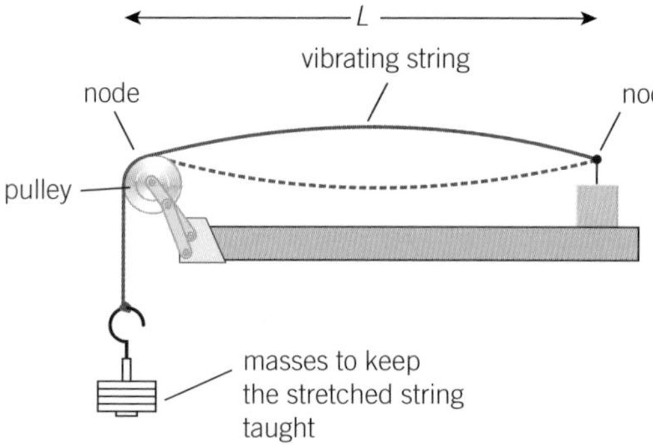

▲ **Figure 2** *A taut string vibrating in its fundamental mode of vibration, that is, at the fundamental frequency*

When the string is plucked, it vibrates in its fundamental mode of vibration (Figure 2), in which the wavelength of the progressive wave is double the length of the string ($2L$).

Harmonics and wavelength

The fundamental frequency f_0 is the minimum frequency of a stationary wave for a string. Along with this fundamental mode of vibration, the string can form other stationary waves called harmonics at higher frequencies (Table 1).

For a given string at a fixed tension, the speed of progressive waves along the string is constant. From the equation for progressive waves, $v = f\lambda$, you can see that as the frequency increases the wavelength must decrease in proportion. At a frequency of $2f_0$, the wavelength is half the wavelength at f_0.

▼ **Table 1** *The first five harmonics for a stationary wave between the fixed ends of a string – the frequency of each harmonic is an integer multiple of the fundamental frequency f_0 (in this case 20 Hz)*

Harmonic	Shape	Frequency / Hz	Frequency as a multiple of f_0	Wavelength of the progressive wave (where L is the length of the string)
1	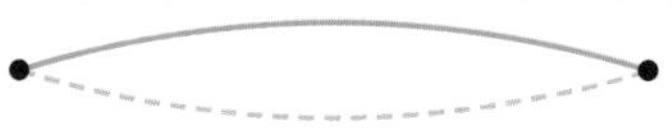	20	f_0	$2L$
2		40	$2f_0$	L
3		60	$3f_0$	$\frac{2}{3}L$
4		80	$4f_0$	$\frac{1}{2}L$
5		100	$5f_0$	$\frac{2}{5}L$

Investigating stationary waves on strings

Melde's experiment (Figure 3) is a simple way to investigate stationary waves on a string. A vibration generator can be used to change the frequency of the wave on the stretched string until a stable stationary wave is produced for a specific harmonic. The actual node on the right-hand side of the string is slightly to the right of the vibration generator (which cannot be a true node as it is vibrating).

In Figure 3, the string is vibrating at $2f_0$, approximately 214 Hz. If the generator is not set to an integer multiple of the fundamental frequency f_0 then no stationary wave will be formed. Instead the string will vibrate in non-distinct patterns.

With this apparatus the length of the string can be changed, or the fixed end can be replaced with a pulley and different masses attached to the string to vary its tension.

1. State the fundamental frequency (f_0) of the string in Figure 3.
2. The fundamental frequency of a string is 300 Hz. Sketch the shape of this string when the frequency of the vibration generator is set to:
   a 300 Hz; **b** 600 Hz; **c** 450 Hz.
3. The speed v of the progressive waves along a stretched string is given by $v = k\sqrt{T}$, where k is a constant and T is the tension in the string. Explain how increasing the tension affects the fundamental frequency.

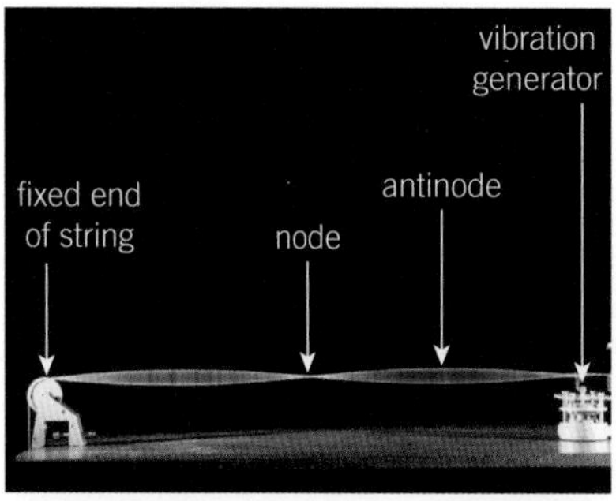

▲ **Figure 3** *Melde's string*

Summary questions

1 Describe how different harmonics on a string might be observed in a classroom. *(3 marks)*

2 Describe the effect on the fundamental frequency of a string if the length of the string doubles (assuming the tension remains constant). *(2 marks)*

3 The stationary wave in Figure 4 is produced at a frequency of 120 Hz and has a wavelength of 0.36 m.
   a State the fundamental frequency of the string. *(1 mark)*
   b Determine the wavelength of the original progressive wave. *(2 marks)*
   c Sketch the pattern produced if the frequency changes to:
      i 160 Hz; ii 100 Hz. *(2 marks)*

▲ **Figure 4**

4 The graph in Figure 5 was recorded using data collected from a violin playing a single note. It shows how the intensity varies at different frequencies (the frequency spectrum).
   a State the fundamental frequency for the vibrating string. *(1 mark)*
   b Explain how the graph shows the existence of harmonics on the string. *(2 marks)*

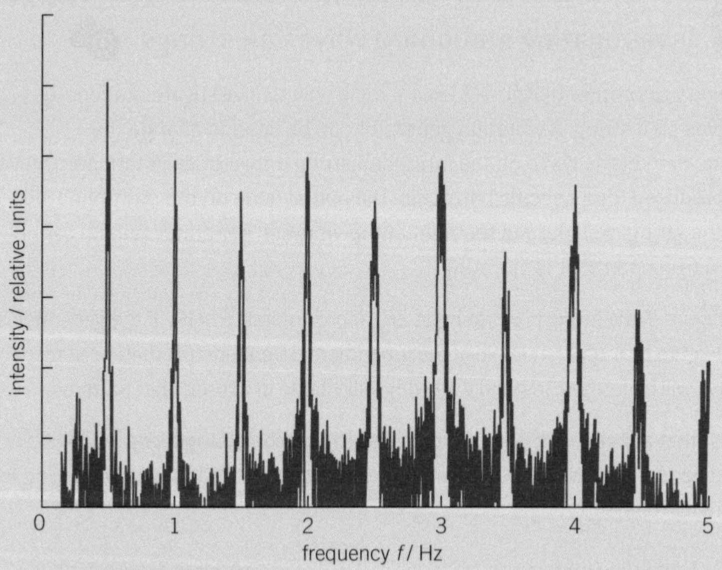

▲ **Figure 5**

5 A stationary wave is formed on a string of length 90 cm. At 3.6 kHz, six antinodes can be observed.
   a Determine the wavelength of the progressive waves on the string.
   b Calculate the speed of the progressive waves travelling along the string. *(4 marks)*

12.6 Stationary waves in air columns

Specification reference: 4.4.4

Playing the pipes

Stationary waves are formed not only by transverse waves like microwaves or waves on strings, but also by longitudinal waves, like sound. Most woodwind instruments, like these pan pipes (Figure 1), produce notes from stationary waves in air columns.

Gently blowing over the top of a tube creates a standing wave inside it, which produces a note at a particular frequency. The length of a tube determines the wavelength of the note it produces.

Learning outcomes

Demonstrate knowledge, understanding, and application of:

→ stationary wave patterns for air columns in closed and open tubes.

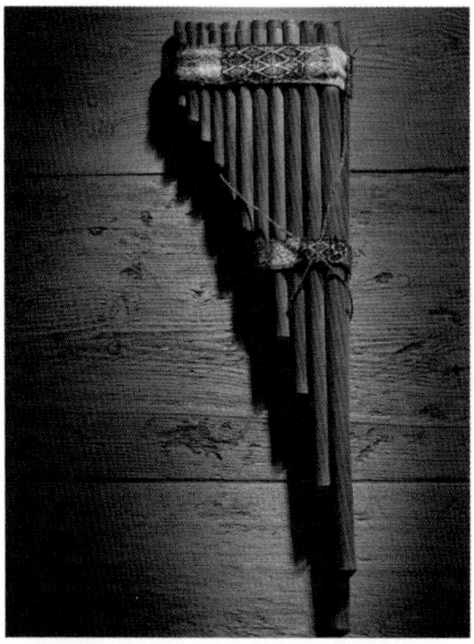

▲ **Figure 1** *Pan pipes are a traditional woodwind instrument played by Peruvians from the Andes*

Stationary waves with sound

Sound waves reflected off a surface can form a stationary wave. The original wave and the reflected wave travel in opposite directions and superpose (Figure 2).

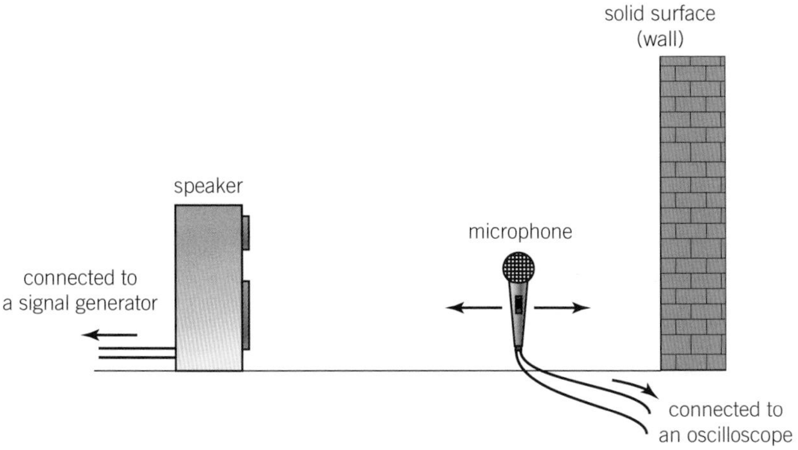

▲ **Figure 2** *Sound waves reflected off a solid surface can form a stationary wave in the same way as the stationary wave formed with microwaves described in Topic 12.4, Stationary waves*

Stationary sound waves can also be made in tubes by making the air column inside the tube vibrate at frequencies related to the length of the tube. The stationary wave formed depends on whether the ends of the tube are open or closed.

Stationary waves in a tube closed at one end

In order for a stationary wave to form in a tube closed at one end there must be an antinode at the open end and a node at the closed end. The air at the closed end cannot move, and so must form a node. At the open end, the oscillations of the air are at their greatest amplitude, so it must be an antinode.

The fundamental mode of vibration simply has a node at the base and an antinode at the open end. Harmonics are also possible (Figure 3).

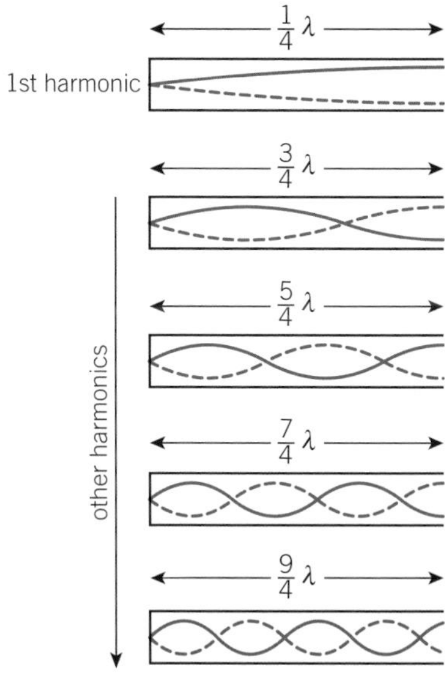

▲ **Figure 3** *Harmonics of a longitudinal wave in a tube closed at one end*

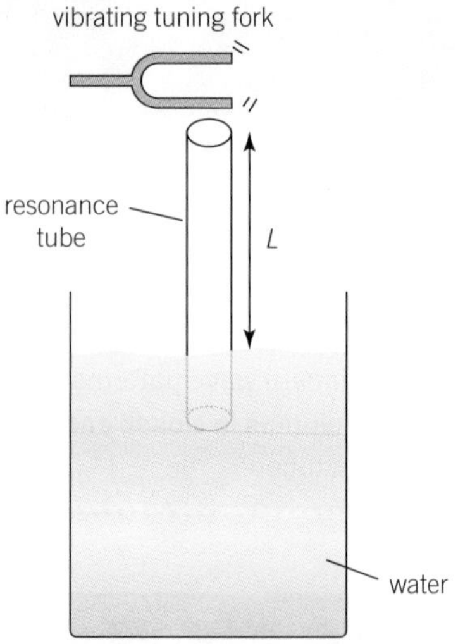

▲ **Figure 4** *The top end of the tube is open. The level of water determines the position of the closed end of this tube.*

▼ **Table 1** *Measurements of f and L*

Frequency of tuning fork / Hz	Length of air column /cm
200	38.5
220	34.1
280	26.4
340	21.5
440	15.0
520	12.3
600	9.8

Unlike stationary waves on stretched strings, in a tube closed at one end it is not possible to form a harmonic at $2f_0$ – there is no second (or fourth, sixth, ...) harmonic for a tube closed at one end. The frequencies of the harmonics in tubes closed at one end are always an odd multiple of the fundamental frequency ($3f_0$, $5f_0$, ...).

Speed of sound in a resonance tube

Holding a tuning fork above a tube closed at one end can form a stationary wave inside the tube. The air vibrates at the same frequency as the tuning fork. If the frequency of the tuning fork is at the fundamental frequency for the air column, the sound becomes loud as the air inside the tube resonates.

In the apparatus in Figure 4 the length of the tube can be changed by raising and lowering it in the water. When the frequency of the tuning fork matches f_0, the length of the tube above water L must be equal to $\frac{1}{4}\lambda$ (see Figure 3). The speed of sound in air can be calculated using $v = f\lambda = f \times 4L$, where f is the frequency of the tuning fork.

The length L corresponding to the fundamental mode of vibration is recorded for a number of different tuning forks. The results are shown in Table 1.

1 Copy the diagram of the experiment and draw in the stationary wave formed inside the tube. Label the position of any nodes and antinodes.

2 Plot a graph of L / m against $\frac{1}{f}$ / s.

a Show that the gradient of this graph is equal to $\frac{v}{4}$, where v is the speed of sound in air.

b Use your graph to find the speed of sound in air.

c Suggest a reason why the graph does not quite pass through the origin. (Hint: Look carefully at the diagram.)

Stationary waves in open tubes

A tube open at both ends must have an antinode at each end in order to form a stationary wave, as explained above (Figure 5). Unlike a tube closed at one end, harmonics at all integer multiples of the fundamental frequency (f_0, $2f_0$, $3f_0$, ...) are possible in an open tube.

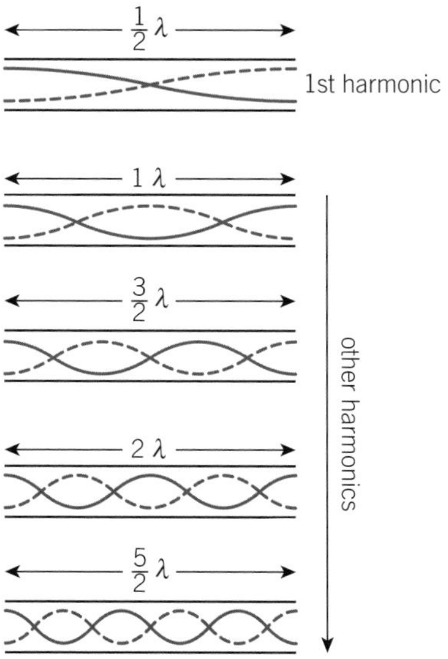

▲ **Figure 5** *Harmonics in an open tube*

Summary questions

1 Describe a simple method for the formation of a stationary sound wave using a speaker, a microphone, and a solid surface, and how to use it to determine the wavelength of sound produced from a signal generator. *(4 marks)*

2 The air column inside a tube closed at one end of length 1.2 m vibrates at its fundamental frequency. Calculate:

a the wavelength of the sound; *(2 marks)*

b the frequency of the sound produced (speed of sound through air = 340 m s^{-1}). *(1 mark)*

3 Repeat the calculations from question 2 for an open tube. *(3 marks)*

4 The air column inside an open tube is made to vibrate at $2f_0$, where f_0 is the fundamental frequency. Identify the nodes and antinodes in Figure 6 below. *(5 marks)*

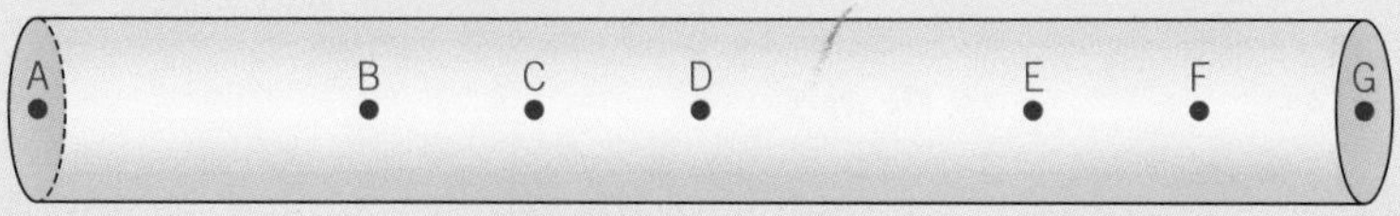

▲ **Figure 6**

5 Explain why it is not possible to produce a harmonic at $2f_0$, where f_0 is the fundamental frequency, in a tube closed at one end and open at the other. *(2 marks)*

Practice questions

1 This question is about the Young double slit experiment. See Figure 1. The fringe pattern seen on the screen is shown to the right.

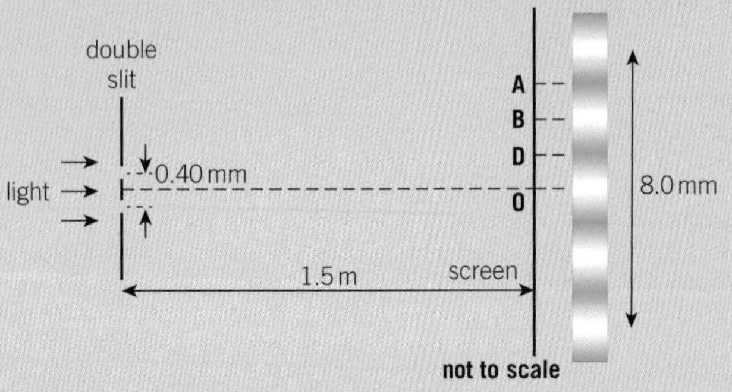

▲ **Figure 1**

Two parallel clear lines are scratched on a darkened glass slide 0.40 mm apart. When a beam of monochromatic visible light is shone through these slits, interference fringes are observed on a screen placed 1.5 m from the slide. The fringe at point **B** is bright and the fringe at point **D** is dark.

a Explain why this arrangement with two slits is used to produce visible fringes on the screen rather than two separate identical light sources. *(2 marks)*

b State the **phase difference** between the light waves from the two slits that meet on the screen in Figure 1 at points D and B. *(2 marks)*

c (i) Use Figure 1 to calculate the separation of adjacent bright fringes, the distance between **O** and **B**. *(1 mark)*

(ii) Show that the wavelength λ of the monochromatic light is about 5×10^{-7} m. *(3 marks)*

d Calculate the **path difference**, in nanometers, between the light waves from the two slits that meet on the screen in Figure 1 at point **A**. *(2 marks)*

June 2013 G482

2 Figure 2 shows two loudspeakers connected to a signal generator, set to a frequency of 1.2 kHz. A person walks in the direction **P** to **Q** at a distance of 3.0 m from the loudspeakers.

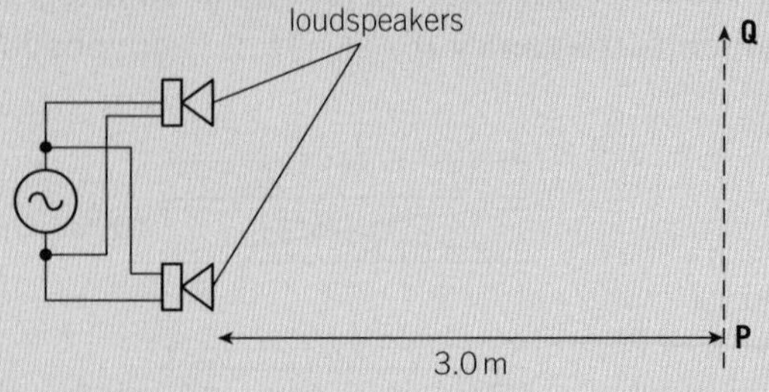

▲ **Figure 2**

(i) Calculate the wavelength λ of the sound waves omitted from the loudspeakers.

Speed of sound in air = $340\,\text{m s}^{-1}$ *(2 marks)*

(ii) Explain, either in terms of path difference or phase difference, why the intensity of the sound heard varies as the person moves along **PQ**. *(3 marks)*

(iii) The distance x between adjacent positions of maximum sound is 0.50 m. Calculate the separation a between the loudspeakers. Assume that the equation used for the interference of light also applies to sound. *(2 marks)*

(iv) The connections to one of the loudspeakers are reversed. Describe the similarities and differences in what the person hears. *(2 marks)*

Jan 2012 G482

3 Figure 3 shows a string stretched between two fixed points.

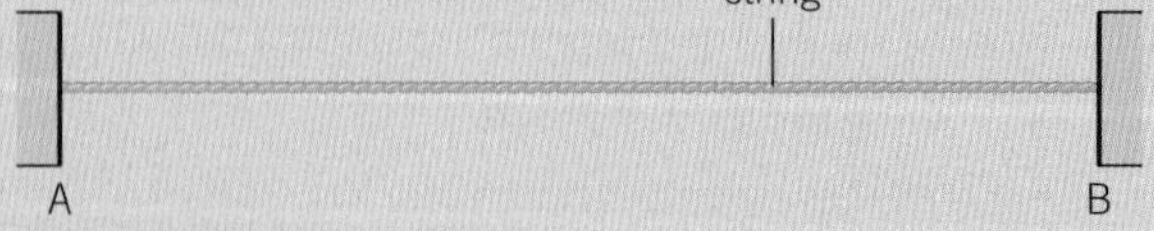

▲ **Figure 3**

The middle of the string is pulled up and then released. This creates a stationary wave on the string. The distance between the fixed supports is 32 cm.

a Explain how a stationary wave is produced on this stretched string. *(3 marks)*

b The string vibrates at its fundamental mode of vibration. Sketch the shape of the stationary wave produced. *(2 marks)*

c A stroboscope is used to determine the frequency of vibration of the string. The frequency is found to be 160 Hz. Calculate the speed of the transverse waves on the string. *(3 marks)*

4 a When used to describe stationary (standing) waves explain the terms

(i) node, *(1 mark)*

(ii) antinode. *(1 mark)*

b Figure 4 shows a string fixed at one end under tension. The frequency of the mechanical oscillator close to the fixed end is varied until a stationary wave is formed on the string.

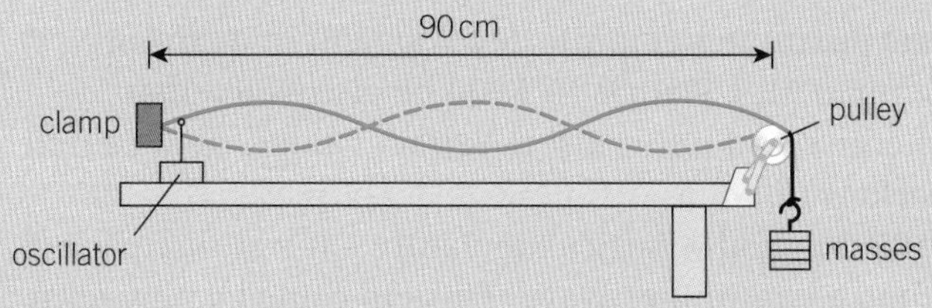

▲ **Figure 4**

(i) Explain with reference to a progressive wave on the string how the stationary wave is formed. *(3 marks)*

(ii) On a copy of Figure 4, label one node with the letter **N** and one antinode with the letter **A**. *(1 mark)*

(iii) State the number of antinodes on the string in Figure 4. *(1 mark)*

(iv) The frequency of the oscillator causing the stationary wave shown in Figure 4 is 120 Hz.

The length of the string between the fixed end and the pulley is 90 cm.

Calculate the speed of the progressive wave on the string. *(3 marks)*

c The speed v of a progressive wave on a stretched string is given by the formula

$$v = k\sqrt{W}$$

where k is a constant for that string. W is the tension in the string which is equal to the weight of the mass hanging from the end of the string.

In (**b**) the weight of the mass on the end of the string is 4.0 N. The oscillator continues to vibrate the string at 120 Hz. Explain whether or not you would expect to observe a stationary wave on the string when the weight of the suspended mass is changed to 9.0 N. *(3 marks)*

May 2009 G482

5 A vibrating tuning fork is held next to the open end of a horizontal tube. The tube is closed at the other end, see Figure 5. A stationary sound wave, of fundamental mode of vibration, is produced in the air column within the tube.

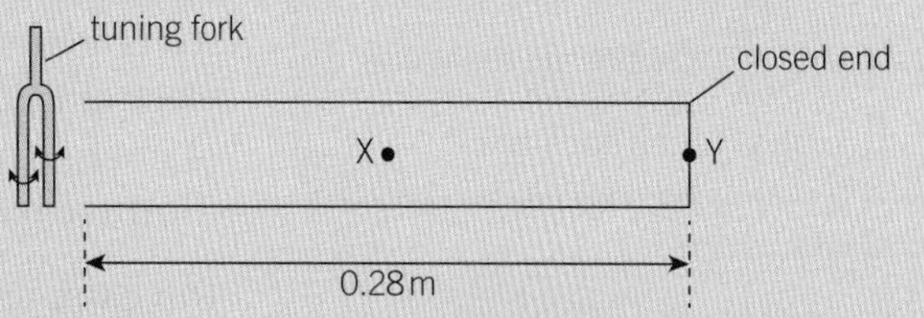

▲ **Figure 5**

The length of the air column in the tube is 0.60 m.

a On a copy of Figure 5, draw the stationary wave pattern produced in the air column within the tube.

Mark the positions of the node (**N**) and the antinode (**A**). *(2 marks)*

b State the oscillations of the air particles at points **X** and **Y**. *(2 marks)*

c State and explain the phase difference between air particles vibrating at the open end of the tube and at **X**. *(2 marks)*

d The stationary wave produced emits a note at its fundamental frequency f_0. The speed of sound in air is $340\,\text{m}\,\text{s}^{-1}$. Calculate the value of f_0. *(3 marks)*

13 QUANTUM PHYSICS
13.1 The photon model
Specification reference: 4.5.1

Learning outcomes
Demonstrate knowledge, understanding, and application of:
- → the particulate nature (photon model) of electromagnetic radiation
- → photons as quanta of energy of electromagnetic radiation
- → energy of a photon $E = hf$ and $E = \frac{hc}{\lambda}$
- → the electronvolt
- → using LEDs and the equation $eV = \frac{hc}{\lambda}$ to estimate the value of the Planck constant h.

Imaging the brain
A single-photon emission computed tomography (SPECT) scan, like the one in Figure 1, relies on the idea of electromagnetic radiation as photons. In the previous chapter we saw evidence for the wave nature of electromagnetic radiation. **Quantum physics** explores how, at the very small scale, we use different models to describe electromagnetic radiation and subatomic particles like electrons. Quantum physics lies behind much modern technology, from medical scans like this to broadband connections.

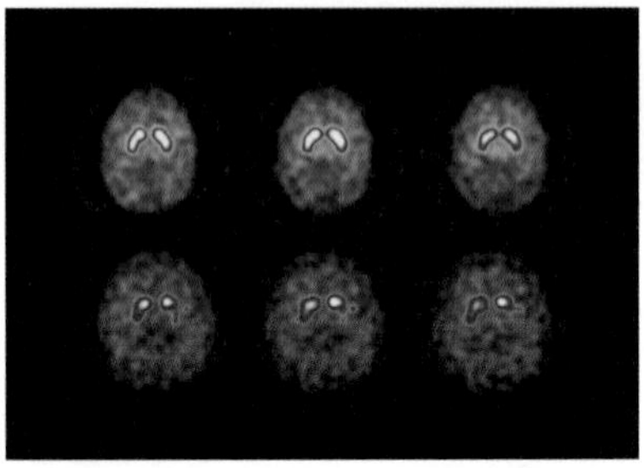

▲ **Figure 1** *This image was produced by recording single photons emitted as part of a brain scan – the top row shows normal brain activity, the bottom shows a patient with Parkinson's disease on one side*

Photons
One of the scientists instrumental in the development of quantum physics was the German Max Planck. In 1900 he discovered that electromagnetic energy could only exist in certain values – it appeared to come in little packets (quanta). His new model proposed that electromagnetic radiation had a particulate nature – it was tiny packets of energy, rather than a continuous wave. Einstein coined a new term for these packets, **photons**.

Developments in quantum physics have led to an understanding that we can use different models to describe electromagnetic radiation. For example, we use the photon model to explain how electromagnetic radiation interacts with matter, and the wave model to explain its propagation through space. These models are rigorously tested, and any model that does not match experimental data or makes incorrect predictions is modified or discarded, pushing our understanding forward.

Photon energy
The energy of each photon is directly proportional to its frequency. Specifically

$$E = hf$$

where E is the energy of the photon in J, f is the frequency of the electromagnetic radiation in Hz, and h is the **Planck constant**. It has an experimental value of 6.63×10^{-34} J s and will be used throughout the quantum physics topic.

We can combine this equation with the wave equation $c = f\lambda$ to express the energy of a photon in terms of its wavelength and c (the speed of light through a vacuum, $3.00 \times 10^{8}\,\text{m s}^{-1}$).

$$E = \frac{hc}{\lambda}$$

Synoptic link
The wave equation $c = f\lambda$ was introduced in Topic 11.2, Wave properties.

This is an intriguing equation. It has both wave elements (because of the λ) and particulate elements (photon, because of the E). From this equation we can also see the energy of a photon is inversely proportional to its wavelength ($E \propto \frac{1}{\lambda}$). Short-wavelength photons, like X-rays, have much more energy than long-wavelength radio waves. This energy partly explains why X-rays can damage the cells of our bodies.

Worked example: Photons of red light

A laser emits red light with a wavelength of 633 nm. Calculate the energy of each red photon emitted.

Step 1: Identify the correct equation to calculate the energy of a photon.

$$E = hf$$

We only know the wavelength of the photon, so we must use the wave equation to express frequency in terms of the wavelength and the speed of light.

$$c = f\lambda$$

$$f = \frac{c}{\lambda}$$

We can substitute this into our first equation, giving

$$E = \frac{hc}{\lambda}$$

Step 2: Substitute in the correct values, taking care with λ in nm.

$$E = \frac{6.63 \times 10^{-34} \times 3.00 \times 10^{8}}{633 \times 10^{-9}}$$

Calculate the energy of the photon in joules. $E = 3.14 \times 10^{-19}$ J

Study tip

A photon is a quantum of electromagnetic energy. The energy of a photon depends on frequency or wavelength of the electromagnetic radiation.

Questions of scale

At the subatomic scale of the quantum level, the SI unit of energy, the joule, is huge. Even the most energetic gamma photons only have energies of the order of tens of millijoules. A typical red photon has energy of around 3×10^{-19} J.

Just as we use the kWh as a unit of electrical energy when dealing with the billions of joules transferred to our homes, we often use another unit when measuring energies at the quantum scale, the **electronvolt** (eV).

The energy of 1 eV is defined as the energy transferred to or from an electron when it moves through a potential difference of 1 V.

We know that the work done on an electron moving through a p.d. is equal to the p.d. × the charge on the electron ($W = VQ = Ve$). Therefore, the work done on an electron as it moves through a p.d. of 1 V is given by

$$W = 1\,\text{V} \times 1.60 \times 10^{-19}\,\text{C} = 1.60 \times 10^{-19}\,\text{J}$$

Synoptic link

You can review the equation linking work and charge in Topic 9.2, Potential difference and electromotive force.

Study tip

When using the equation $E = hf$ make sure you don't muddle units. Energy should be in joules, not electronvolts.

Therefore, 1 eV is equivalent to 1.60×10^{-19} J, or 1 J is equivalent to 6.25×10^{18} eV. 1 eV is a tiny amount of energy. It is very common to see energies expressed as keV, MeV, or even GeV for high-energy particles or photons.

A typical infrared photon has an energy of 1.5 eV. You can imagine this energy as the equivalent of the kinetic energy gained by an electron travelling through a p.d. of 1.5 V (between the terminals of a typical AA cell).

▼ **Table 1** *Some photon energies expressed in both J and eV*

To convert from J to eV, divide by 1.60×10^{-19} →

Photon	Typical energy		Typical wavelength
	/ J	/ eV	/ m
radio	6.5×10^{-27}	4.0×10^{-8}	3.1×10^{1}
infrared	1.6×10^{-20}	0.1	1.2×10^{-5}
visible – red	3.0×10^{-19}	1.9	6.6×10^{-7}
visible – green	3.7×10^{-19}	2.3	5.4×10^{-7}
visible – blue	4.3×10^{-19}	2.7	4.6×10^{-7}
UV	9.6×10^{-19}	6.0	2.1×10^{-7}
X-ray	1.6×10^{-16}	1.0×10^{3}	1.2×10^{-9}
gamma	2.4×10^{-13}	1.5×10^{6}	8.3×10^{-13}

← To convert from eV to J, multiply by 1.6×10^{-19}

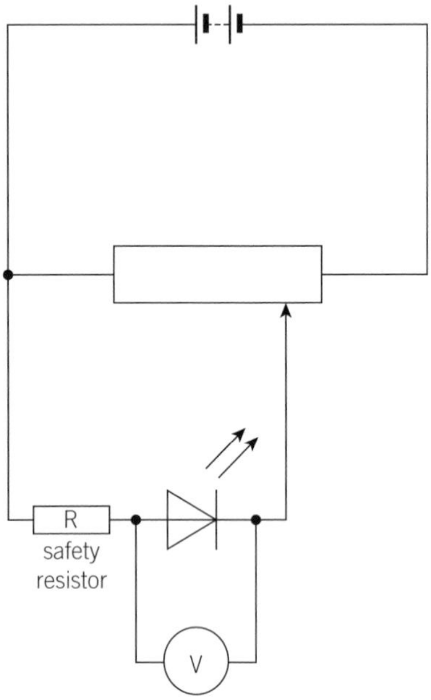

▲ **Figure 2** *A simple circuit containing an LED, a battery, a voltmeter, and a resistor (to protect the LED) can be used to give an approximate value for the Planck constant*

LEDs and the Planck constant

We can do a simple experiment with LEDs (Figure 2) to determine a value for the Planck constant by considering the energies of the photons they emit.

LEDs convert electrical energy into light energy. They emit visible photons when the p.d. across them is above a critical value (the threshold p.d.). Figure 3 shows the *I–V* characteristic for a typical LED.

When the p.d. reaches the threshold p.d. the LED lights up and starts emitting photons of a specific wavelength. At this p.d. the work done is given by $W = VQ$. This energy is about the same as the energy of the emitted photon. We can use the voltmeter to measure the minimum p.d. that is required to turn on the LED. A black tube placed over the LED helps to show exactly when the LED lights up. If we also know

Synoptic link

LEDs and threshold p.d. were covered in Topic 9.6, Diodes. You can review *I–V* characteristics in Topic 9.5, *I–V* characteristics.

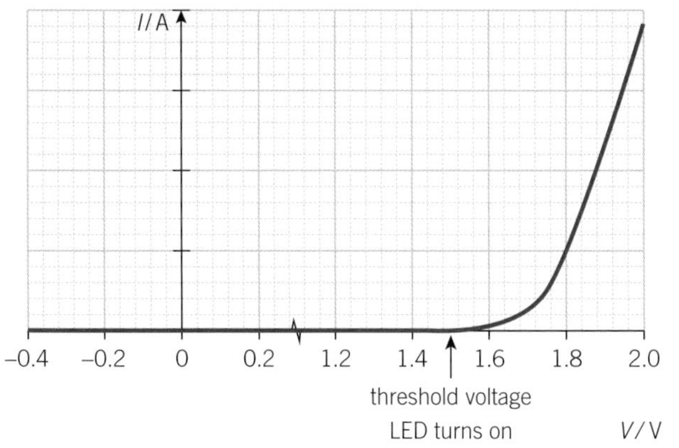

▲ **Figure 3** *The I–V characteristic and threshold p.d. for a typical LED*

the wavelength of the photons emitted by the LED we can determine the Planck constant.

At the threshold p.d., the energy transferred by an electron in the LED is approximately equal to the energy of the single photon it emits.

threshold p.d. × charge on electron ≈ energy of emitted photon

$$Ve = hf$$

Expressing this in terms of the wavelength of the emitted photon λ gives

$$eV = \frac{hc}{\lambda}$$

We can use this equation for a single LED and calculate h, but in order to obtain a more accurate value we should gather data using a variety of different-wavelength LEDs (the threshold p.d. dictates the colour of the LED). We can then plot a graph of V against $\frac{1}{\lambda}$ (Figure 4). The equation for a straight-line graph, $y = mx + c$, is equivalent to $V = \frac{hc}{\lambda e}$ here, so the Planck constant can be determined from the gradient of the graph, $\frac{hc}{e}$.

Synoptic link

For more information about the gradient of straight-line graphs, see Appendix A2, Recording results.

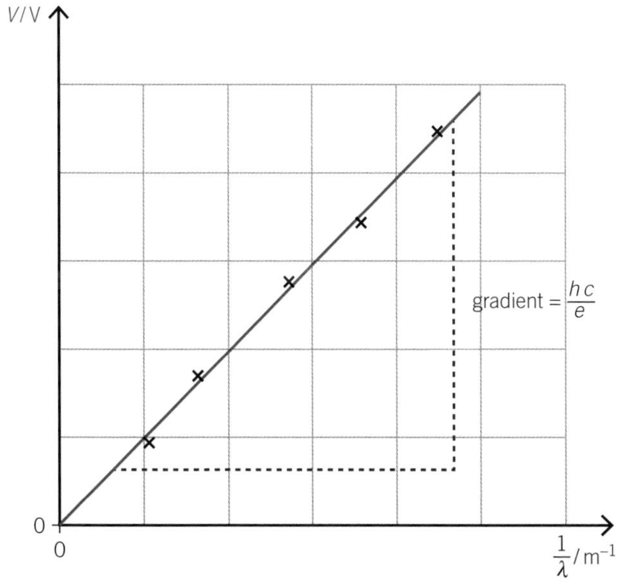

▲ **Figure 4** *A plot of V against* $\frac{1}{\lambda}$

Summary questions

1 Calculate the energy of each photon from the following frequencies:
  a infrared at 1.02×10^{14} Hz; *(1 mark)*
  b a radio wave at 97.0 MHz; *(1 mark)*
  c visible light at 6.00×10^{14} Hz. *(1 mark)*

2 State and explain which part of the visible spectrum has photons with the highest energy. *(3 marks)*

3 A photon has energy of 3.32×10^{-18} J. Calculate its wavelength. *(3 marks)*

4 Convert the following energies into eV:
  a 1.0 J; *(1 mark)*
  b 3.30×10^{-14} J; *(1 mark)*
  c 600 nJ. *(1 mark)*

5 Calculate the frequencies of the photons in Table 1. *(8 marks)*

6 Calculate the photon energies in eV of each of the following wavelengths:
  a X-rays at 4.50×10^{-10} m; *(2 marks)*
  b orange light at 600 nm. *(2 marks)*

7 a An LED emits orange photons with a wavelength of 620 nm. Calculate the threshold p.d. *(2 marks)*
  b Sketch the $I-V$ characteristics for an LED that emits red photons and one that emits blue photons. Use the graphs to explain the differences in the threshold p.d. *(6 marks)*

8 A laser emits blue light of wavelength 405 nm and radiant power 10 mW. Calculate the number of photons emitted per second. *(3 marks)*

9 A student is given an LED that emits monochromatic light of wavelength λ_R. Plan an experiment that uses this LED to determine a value for the Planck constant h. Suggest how the precision of the experiment can be improved. *(6 marks)*

Surface charging

Surface charging is a phenomenon experienced by all spacecraft. Outside the Earth's protective atmosphere, high-energy electromagnetic radiation causes electrons to be emitted from the metal parts of the spacecraft facing the Sun. This is called the **photoelectric effect**.

Surface charging results in some parts of the spacecraft carrying a positive charge, potentially leading to a damaging flow of charge through key electronic components inside the spacecraft. Engineers have to design solutions to this problem to ensure that charges cannot build up to potentially damaging levels.

Learning outcomes

Demonstrate knowledge, understanding, and application of:

→ the photoelectric effect and its demonstration

→ the interaction between one photon and one surface electron

→ work function and threshold frequency

→ the effects of the intensity of the incident radiation on the emission of photoelectrons.

The photoelectric effect

In 1887 Heinrich Hertz reported that when he shone UV radiation onto zinc, electrons were emitted from the surface of the metal.

This is the photoelectric effect. The emitted electrons are sometimes called **photoelectrons**. They are normal electrons, but their name describes their origin – emitted through the photoelectric effect (Figure 2).

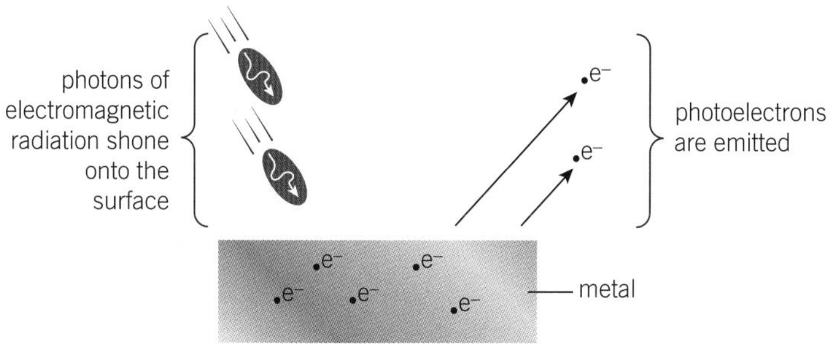

▲ **Figure 2** *The photoelectric effect occurs when electromagnetic radiation incident on the surface of a metal causes electrons to be emitted*

▲ **Figure 1** *The European Space Agency and its Japanese equivalent plan to launch the BepiColumbo probe, designed to minimise the hazards caused by surface charging, to Mercury in 2016*

The gold-leaf electroscope

A simple demonstration of the effect can be seen with a **gold-leaf electroscope**. These were originally designed to measure p.d. (an early voltmeter). However, we can use them to demonstrate how like electrical charges repel each other.

Briefly touching the top plate with the negative electrode from a high-voltage power supply will charge the electroscope. Excess electrons are deposited onto the plate and stem of the electroscope. Any charge developed on the plate at the top of the electroscope spreads to the stem and the gold leaf. As both the stem and gold leaf have the same charge, they repel each other, and the leaf lifts away from the stem (Figure 3). If a clean piece of zinc is placed on top of a negatively charged

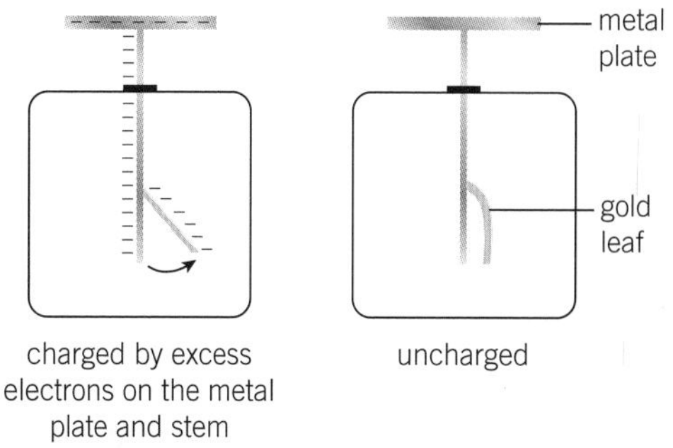

▲ **Figure 3** *Charged and uncharged electroscopes*

electroscope and UV radiation shines onto the zinc surface, then the gold leaf slowly falls back towards the stem. This shows that the electroscope has gradually lost its negative charge, because the incident radiation (in this case UV) has caused the free electrons to be emitted from the zinc. These electrons are known as photoelectrons.

Three key observations from the photoelectric effect

The electroscope experiment is simple, but it was revolutionary. When different frequencies of incident radiation were investigated in more detail, scientists at the time made three key observations.

1. Photoelectrons were emitted only if the incident radiation was above a certain frequency (called the **threshold frequency** f_0) for each metal. No matter how intense the incident radiation (how bright the light), not a single electron would be emitted if the frequency was less than the threshold frequency.
2. If the incident radiation was above the threshold frequency, emission of photoelectrons was instantaneous.
3. If the incident radiation was above the threshold frequency, increasing the intensity of the radiation did not increase the maximum kinetic energy of the photoelectrons. Instead more electrons were emitted. The only way to increase the maximum kinetic energy was to increase the frequency of the incident radiation.

These observations could not be explained using wave model of electromagnetic radiation. For example, if the threshold frequency for a particular metal is in the green part of the visible spectrum, bright red light does not cause emission, yet very dim blue light would. This does not fit with the wave model, in which the rate of energy transferred by the radiation is dependent on its intensity (brightness). The more intense the radiation, the more energy is transferred to the metal per second, and bright red light transfers more energy per second than dim blue light. Clearly a new model for electromagnetic radiation was needed to explain the observations.

Using photons to illuminate the photoelectric effect

The observations in the photoelectric effect can be explained if the wave model of light is replaced with the photon model. In 1905 Einstein published an explanation of the effect. Building on Planck's work, he proposed the idea of electromagnetic radiation as a stream of photons, rather than continuous waves.

He suggested that each electron in the surface of the metal must require a certain amount of energy in order to escape from the metal, and that each photon could transfer its exact energy to one surface electron in a one-to-one interaction.

As the energy of the photon is dependent on its frequency ($E = hf$), if the frequency of the photon is too low, the intensity of the light – that is, the number of photons per second – does not matter, as a single photon delivers its energy to a single surface electron in a one-to-one interaction. If a photon does not carry enough energy on its own to free an electron, the number of photons makes no difference. However, when the frequency of the light is above the threshold frequency f_0 for the metal, then each individual photon has enough energy to free a single surface electron and so photoelectrons are emitted (Figure 4).

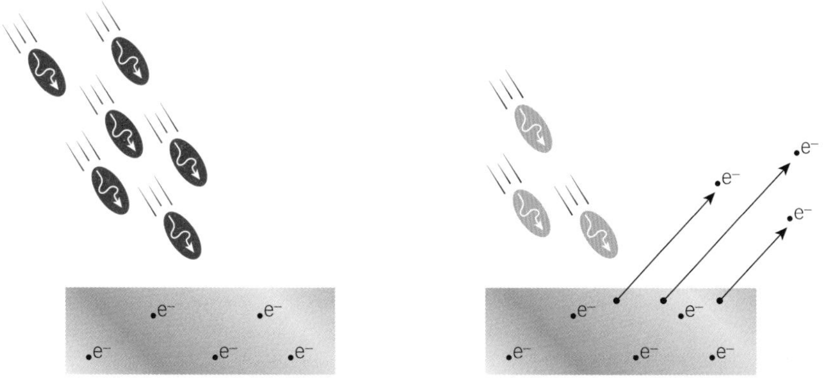

e⁻
if the photons have frequency below f_0 no photoelectrons are emitted.
if the photons have frequency above f_0 photoelectrons are emitted (even when the light is dimmer).

▲ **Figure 4** *Replacing the wave model with the photon model allowed Einstein to explain the photoelectric effect*

Study tip

The threshold frequency and work function are properties of the metal surface. They are not properties of electrons or photons.

This also explained why there was no time delay. As long as the incident radiation has frequency greater than, or equal to, the threshold frequency, as soon as photons hit the surface of the metal, photoelectrons are emitted. Electrons cannot accumulate energy from multiple photons. Only one-to-one interactions are possible between photons and electrons.

Einstein was also able to explain the third observation. Depending on their position relative to the positive ions in the metal, electrons would require different amounts of energy to free them. Einstein defined a constant for each metal which he called the **work function** ϕ. This is the *minimum* energy required to free an electron from the surface of the metal.

Increasing the intensity of the radiation means more photons per second hit the metal surface. As each photon interacts one-to-one with a single surface electron, as long as the radiation has frequency above the threshold frequency for the metal, more photons per second means a greater rate of photoelectrons emitted from the metal. The rate of emission of photoelectrons is directly proportional to the intensity of the incident radiation. Double the intensity and you double the number of photons per second, leading to a doubling in the number of electrons emitted from the metal per second.

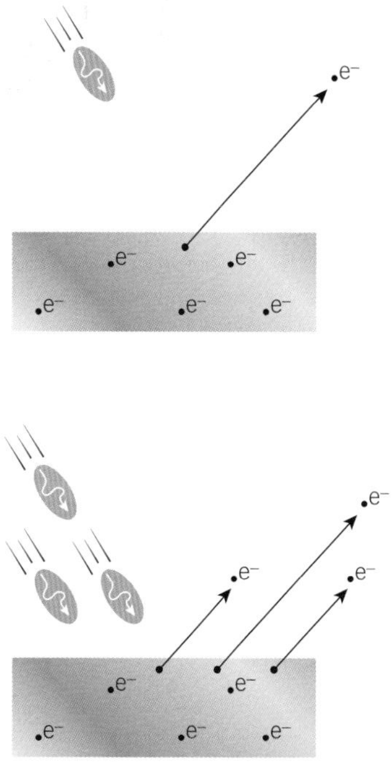

e⁻

▲ **Figure 5** *Intense radiation means a higher rate of photons landing and so a higher rate of electrons escaping*

Using the principle of conservation of energy, Einstein deduced that the kinetic energy of each photoelectron depends on how much energy was *left over* after the electron was freed from the metal (more

Study tip

Remember the three key ideas about the photoelectric effect: $E \propto f$ for photons, it is a one-to-one interaction (one photon causes the emission of one surface electron), and brighter light means a greater rate of photons incident on the metal.

on this in Topic 13.3, Einstein's photoelectric effect equation). At a given frequency all photons have the same amount of energy, and the metal a specific work function, so there is a maximum value of kinetic energy that any emitted photoelectrons can have. Increasing the intensity results in a greater rate of emission, but none of the emitted photoelectrons will move any faster.

The only way to increase the maximum kinetic energy of the emitted photoelectrons is to increase the frequency of the radiation. In this case each photon has more energy and so each electron has more kinetic energy after it has been freed from the metal.

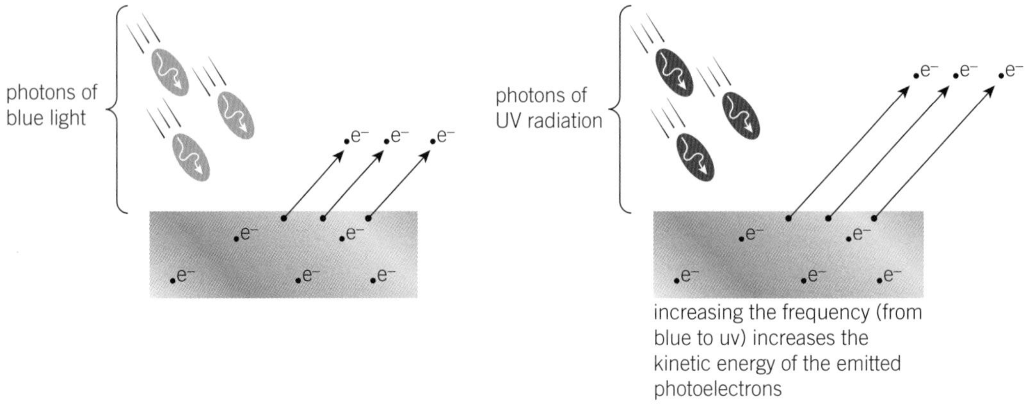

▲ **Figure 6** *Increasing the frequency results in the emission of photoelectrons with a higher kinetic energy (as there is more energy to spare after the electrons have been freed from the metal)*

Summary questions

1. Describe what would happen to an uncharged gold-leaf electroscope if its top surface were to come into contact with a positive electrode. *(2 marks)*
2. If a particular metal had a threshold frequency in the red part of the visible spectrum, explain what would happen to the metal if radiation was incident on its surface from:
   a. the infrared part of the spectrum; *(2 marks)*
   b. the blue part of the visible spectrum. *(1 mark)*
3. Explain why the maximum kinetic energy of photoelectrons emitted during the photoelectric effect depends on the frequency of the incident radiation. *(3 marks)*
4. The threshold wavelength λ_0 is the *longest* wavelength that will give rise to the photoelectric effect. Derive an expression for λ_0 in terms of the threshold frequency f_0. *(2 marks)*
5. State and explain the effect of quadrupling the intensity of incident radiation (keeping the frequency constant) on a metal surface emitting photoelectrons. *(3 marks)*

13.3 Einstein's photoelectric effect equation

Specification reference: 4.5.2

Seeing in the dark

In some kinds of night vision goggles the photoelectric effect is used to amplify light. In the construction of the goggles it is important to select materials with an appropriate work function. Often a semiconductor material such as gallium arsenide is used.

An image intensifier produces an image even in conditions that appear pitch black to the human eye. The emission of individual photoelectrons due to the photoelectric effect is used to build up a clear picture of the surroundings.

Learning outcomes

Demonstrate knowledge, understanding, and application of:

→ Einstein's photoelectric equation $hf = \phi + KE_{max}$

▲ **Figure 1** *As well as military applications, night vision technologies can be used in conservation to monitor animals in complete darkness, like these rhinos at a waterhole*

Conservation of energy and the photoelectric effect

In his model of the photoelectric effect Einstein applied the idea of conservation of energy to the photons and photoelectrons. By thinking about the energies involved he derived what is now simply known as **Einstein's photoelectric effect equation**. He realised that the energy of each individual photon must be conserved. This energy does two things:

- it frees a single electron from the surface of the metal in a one-to-one interaction
- any remainder is transferred into the kinetic energy of the photoelectron.

By using his idea of work function ϕ as the *minimum* energy to free the electron from a particular metal, he produced a general equation relating the energy of each photon, the work function of the metal, and the maximum kinetic energy of the emitted photoelectron.

According to the principle of conservation of energy, we have

energy of a single photon	=	minimum energy required to free a single electron from the metal surface	+	maximum kinetic energy of the emitted electron

$$hf = \phi + KE_{max}$$

All the terms in the equation, hf, ϕ, and KE_{max}, are energies, so all should be measured in joules, or consistently in electronvolts.

Worked example: Emitting photoelectrons from a metal surface

Photoelectrons with a maximum kinetic energy of 9.34×10^{-19} J are emitted from a metal with a work function of 2.40 eV. Calculate the frequency of the incident radiation.

Step 1: Identify the correct equation to calculate the frequency of the radiation.

$$hf = \phi + KE_{max}$$

Convert the work function into joules.

$$2.40\,\text{eV} \times 1.60 \times 10^{-19} = 3.84 \times 10^{-19}\,\text{J}$$

Step 2: Substitute the values into the equation to calculate the energy of a single photon.

$$hf = 3.84 \times 10^{-19} + 9.34 \times 10^{-19} = 1.32 \times 10^{-18}\,\text{J}$$

Using this value we can determine the frequency of the radiation.

$$f = \frac{1.32 \times 10^{-18}}{h} = \frac{1.32 \times 10^{-18}}{6.63 \times 10^{-34}} = 1.99 \times 10^{15}\,\text{Hz}$$

Study tip

$1\,\text{eV} = 1.6 \times 10^{-19}\,\text{J}$

Why *maximum* kinetic energy?

Some electrons in the surface of the metal are closer to the positive metal ions than others. Their relative positions affect how much energy is required to free them. The work function is the minimum energy required to free an electron from the metal – most electrons need a little more energy than the work function to free them.

An electron that requires the minimum amount of energy to free it (the work function of the metal) would have the most energy left over from the incident photon. Only a few of the emitted photoelectrons have this *maximum* kinetic energy – most have a little bit less, and so travel a little slower.

If a photon strikes the surface of the metal at the threshold frequency f_0 for the metal then it will only have enough energy to free a surface electron, with none left over to be transferred into kinetic energy of the electron.

In this case Einstein's photoelectric effect equation becomes

$$hf_0 = \phi + 0 \text{ or simply } hf_0 = \phi$$

A graph of KE_{max} against incident frequency

In Topic 13.2, The photoelectric effect, we learnt that the only way to increase the maximum kinetic energy of photoelectrons is to increase the frequency of the incident radiation. A graph of the maximum kinetic energy of the photoelectrons plotted against the frequency of the radiation on the surface can be seen in Figure 2.

Considering the general equation for a straight-line graph, $y = mx + c$, and rearranging $hf = \phi + KE_{max}$ to match, we get $KE_{max} = hf - \phi$. We can see that the gradient of this graph must equal the Planck constant h and the y-axis intercept is equal to $-\phi$, where ϕ is the work function of the metal. See Figure 2.

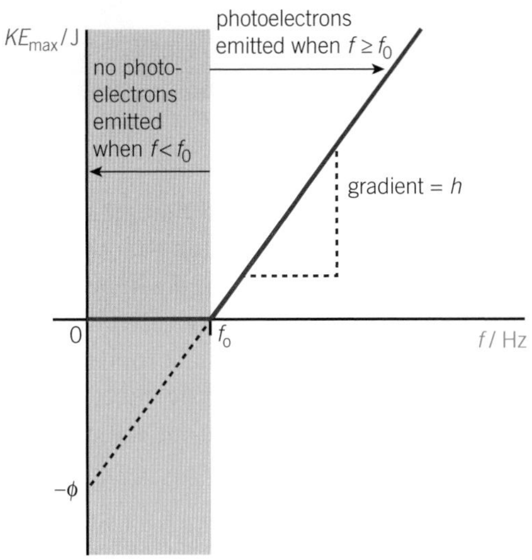

▲ **Figure 2** *A graph to show how the frequency f of the incident radiation on a metal surface affects the maximum kinetic energy KE of the emitted photoelectrons*

▼ **Table 1**

Metal	ϕ / eV
caesium	2.14
sodium	2.36
aluminium	4.08
zinc	4.30

Different metals

Every metal has a different work function, so the threshold frequency for each metal is different. Table 1 shows the work function for some metals.

In each case a graph of maximum kinetic energy against frequency will have the same gradient. This gradient is equal to h (Figure 3).

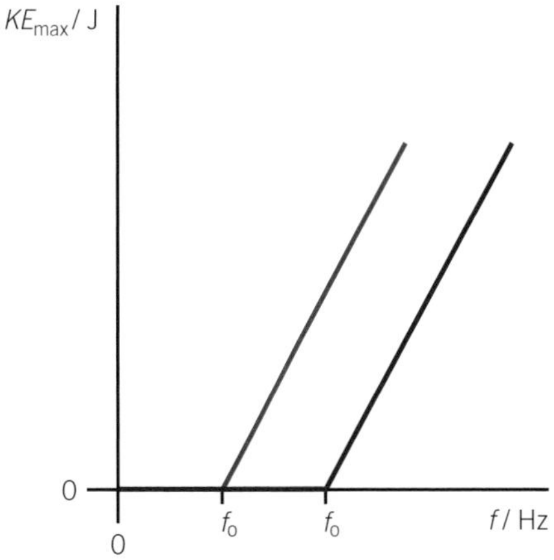

▲ **Figure 3** *Different metals have different work functions leading to different threshold frequencies*

Summary questions

1 Photoelectrons with a maximum kinetic energy of 2.68×10^{-19} J are emitted from a metal with a work function of 3.77×10^{-19} J. Calculate the energy of the incident photons. *(2 marks)*

2 Aluminium has a work function of 4.08 eV. Calculate the maximum kinetic energy in eV of photoelectrons emitted from aluminium when illuminated with photons with energy of:
a 5.20 eV; **b** 3.10 eV. *(3 marks)*

3 Using Table 1, calculate the threshold frequency for zinc and for sodium. *(4 marks)*

4 Explain why under monochromatic radiation, emitted photoelectrons have a range of different kinetic energies. *(3 marks)*

5 Calculate the maximum velocity of photoelectrons emitted from sodium when the incident radiation has a frequency of 1.48×10^{15} Hz. *(4 marks)*

6 The data in Table 2 was collected during an investigation into the photoelectric effect. Use this data to determine the threshold frequency and work function for the metal. *(4 marks)*

▼ **Table 2**

Frequency of incident radiation / Hz	Maximum kinetic energy of photoelectrons / eV
8.0×10^{14}	1.36
1.1×10^{15}	2.61

Learning outcomes

Demonstrate knowledge, understanding, and application of:

- → electron diffraction
- → diffraction of electrons by the atoms of graphite
- → the de Broglie equation $\lambda = \frac{h}{p}$.

Diffracting particles

In 1924 the French physicist Louis de Broglie proposed that all matter can have both wave and particle properties. He suggested that tiny subatomic particles, like electrons, have wavelengths and can be made to exhibit wave properties like diffraction. We have already seen in Topic 13.1, The photon model, that the particulate model of electromagnetic radiation can explain the photoelectric effect. So photons can be thought of as particles, yet electromagnetic radiation is also able to diffract, an example of wave behaviour.

This **wave–particle duality** is a model used to describe how all matter has both wave and particle properties. De Broglie, who was awarded the 1929 Nobel Prize in Physics for his insight, realised that all particles travel through space as waves. Anything with mass that is moving has wave-like properties. These waves are referred to as matter waves or de Broglie waves.

One of the largest objects to show its wave nature is the carbon-60 molecule (buckminsterfullerene, Figure 1). In 1999 it was used to form a diffraction pattern. The C_{60} had a wavelength of just 2.5×10^{-12} m, a hundred times smaller than the diameter of an atom.

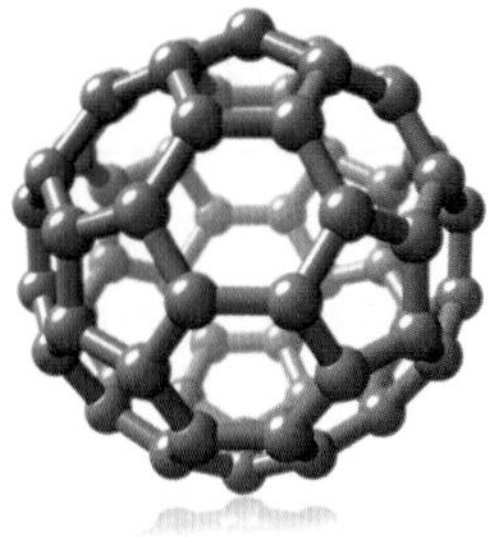

▲ **Figure 1** *Carbon-60 is one of the largest particles that has been successfully diffracted to date*

Electron diffraction

We would normally describe electrons as particles, as they have mass and charge. As a result they can be accelerated and deflected by electric and magnetic fields. This behaviour is only associated with particles. However, under certain conditions we can make electrons diffract. They spread out like waves as they pass through a tiny gap, and can even form diffraction patterns in the same way as light.

If an electron gun fires electrons at a thin piece of **polycrystalline graphite**, which has carbon atoms arranged in many different layers, the electrons pass between the individual carbon atoms in the graphite. The gap between the atoms is so small ($\sim 10^{-10}$ m) that it is similar to the wavelength of the electrons and so the electrons diffract, as waves, and form a diffraction pattern seen on the end of the tube (Figures 2 and 3).

We do not normally notice the wave nature of electrons because we need a tiny gap in order to observe electrons diffracting. For diffraction to occur the size of the gap through which the electrons pass must be similar to their wavelength.

This experiment beautifully demonstrates both the particle and wave nature of electrons. They are behaving as particles when they are accelerated by the high potential difference, they behave as waves when they diffract, and then they behave as particles again as they hit the screen with discrete impacts.

Synoptic link

Topic 9.3, The electron gun, explains how electron guns work.

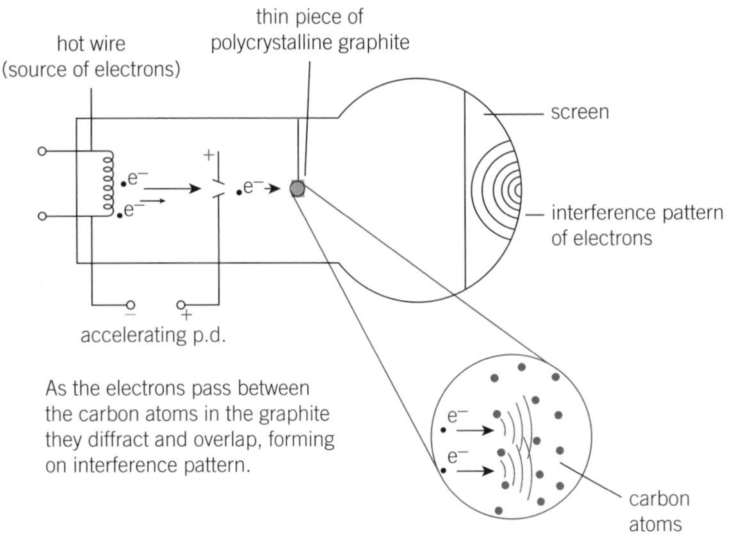

▲ **Figure 2** *When electrons pass through a piece of polycrystalline graphite they form a diffraction pattern*

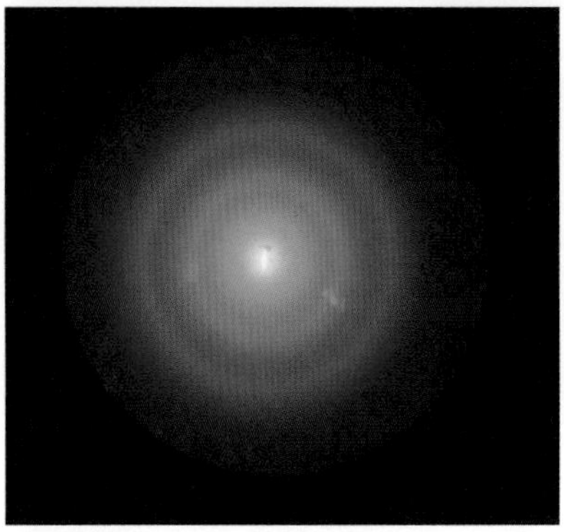

▲ **Figure 3** *These rings are the result of the diffraction of electrons*

The de Broglie equation

In developing wave–particle duality, de Broglie realised that the wavelength λ of a particle was inversely proportional to its momentum p. As the momentum of a particle increases by a certain factor its wavelength reduces by the same factor.

$$\lambda \propto \frac{1}{p}$$

Further investigation led to the development of what is now called the **de Broglie equation**.

$$\lambda = \frac{h}{p}$$

where λ is the wavelength in m, h is the Planck constant, and p is the momentum of the particle in $\mathrm{kg\,m\,s^{-1}}$.

Synoptic link

You have already learnt about the diffraction of waves in detail as part of Topic 11.4, Diffraction and polarisation. The same principles apply to the diffraction of particles like electrons.

Worked example: The wavelength of a fast-moving electron

Calculate the wavelength of an electron travelling at $2.00 \times 10^8\,\mathrm{m\,s^{-1}}$.

Step 1: Identify the correct equation to calculate the wavelength.

$$\lambda = \frac{h}{p}$$

As momentum is $p = mv$ we can substitute this into the previous equation giving

$$\lambda = \frac{h}{mv}$$

Study tip

It's important to be able to give examples of experiments which show the wave and particle natures of both light and electrons.

Synoptic link

Momentum is explored in Topic 7.2, Linear momentum.

Step 2: Substitute in known values in SI units and calculate the wavelength in metres.

$$\lambda = \frac{6.63 \times 10^{-34}}{9.11 \times 10^{-31} \times 2.00 \times 10^{8}} = 3.64 \times 10^{-12}\,\text{m}$$

Notice how small this wavelength is (less than the diameter of an atom). In order to diffract electrons tiny gaps are needed.

Accelerated particles

We can relate an object's kinetic energy E_k to its wavelength by considering both the de Broglie equation and our general equation for kinetic energy. From the worked example above

$$\lambda = \frac{h}{mv}$$

If we can find an expression for mv that includes E_k we can determine the relationship between a particle's kinetic energy and its wavelength.

$$E_k = \frac{1}{2}mv^2$$

$$2E_k = mv^2$$

$$2E_k m = m^2v^2$$

$$\sqrt{2E_k m} = mv$$

We can then substitute this expression into the de Broglie equation.

$$\lambda = \frac{h}{mv} = \frac{h}{\sqrt{2E_k m}}$$

As h and m are both constants, we can see that the wavelength of a particle is inversely proportional to the square root of its kinetic energy.

$$\lambda \propto \frac{1}{\sqrt{E_k}}$$

If the kinetic energy of a particle decreases by a factor of two, its wavelength increases by $\sqrt{2}$.

1 State the effect on the wavelength if the kinetic energy of a particle: **a** increases; **b** increases by a factor of three; **c** decreases by a factor of 10.
2 Calculate the wavelength of: **a** a proton that has a kinetic energy of 3.20×10^{-17} J; **b** an electron that has a kinetic energy of 600 eV.
3 An electron is accelerated through a p.d. of 100 V. Calculate its wavelength.

Crystallography

Understanding of the wave properties of electrons has led to many applications, including **crystallography**.

Crystallography is a method for determining the arrangement of atoms within a compound. The process often uses X-rays, which are fired at a crystal of the compound. The X-rays are diffracted as they pass through the gaps between the individual atoms. By studying the patterns formed by these diffracted beams, a crystallographer is able to produce a precise 3D model of the arrangement of the atoms within the substance.

Electrons are often used in a very similar way to X-rays. Using electrons instead of X-rays for crystallography offers advantages.

- Their wavelength can be tuned to be very similar in size to the gap between atoms, resulting in strong diffraction and a clear pattern for analysis.
- They can be used with very thin sheets or layers of material.

1 Describe how the wavelength of an electron can be altered using an electron gun.
2 Suggest why X-rays cannot be used to study thin sheets of material.
3 Calculate the wavelength of an electron with a kinetic energy of 40 eV and explain why such an electron would be useful for studying crystal structures.

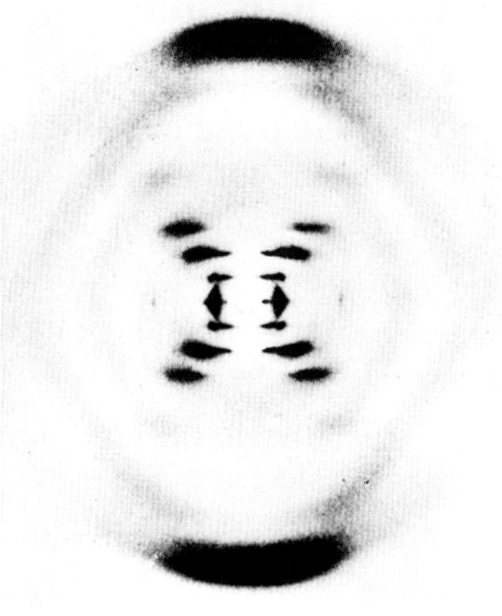

▲ **Figure 4** *Perhaps the most famous example of X-ray crystallography was Watson and Crick's use of Rosalind Franklin's image to determine the structure of DNA in 1953*

Beyond electrons

The de Broglie equation can be applied to all particles. Like electrons, protons and neutrons have been shown to have wave properties – they form diffraction patterns. However, as particles become larger their wave properties become harder to observe. The mass of individual protons is much greater than electrons, so at the same speed their momentum is significantly greater and therefore their wavelength is much smaller, and much harder to observe.

Summary questions

1 Calculate the wavelength of a proton that has a momentum of $1.67 \times 10^{-19}\,\text{kg m s}^{-1}$. *(2 marks)*

2 Explain why, when travelling at the same speed, a proton has a much smaller wavelength than an electron. *(3 marks)*

3 Explain why the wave properties of electrons are not evident in most experiments. *(3 marks)*

4 Calculate the speeds of electrons with the following wavelengths:
a $3.63 \times 10^{-10}\,\text{m}$; b 4.85 pm. *(4 marks)*

5 An electron is accelerated in an electron gun to a speed of $4.20 \times 10^{7}\,\text{m s}^{-1}$. Calculate:
a the wavelength of the electron; *(2 marks)*
b the speed of a neutron with the same wavelength as **a** (mass of neutron = 1.67×10^{-27} kg). *(2 marks)*

6 Calculate the wavelength of an electron travelling at a speed of $0.25\,c$ ($c = 3.00 \times 10^{8}\,\text{m s}^{-1}$). *(2 marks)*

Practice questions

1 a State what is meant by a photon. *(1 mark)*

b Describe and explain the photoelectric effect in terms of photons and surface electrons of a metal. *(4 marks)*

c A laser emits light of wavelength 6.3×10^{-7} m and of radiant power 2.0 mW. Calculate

(i) the energy of each photon in electronvolts (eV), *(3 marks)*

(ii) the number of photons emitted per second from the laser. *(2 marks)*

2 a Electromagnetic waves of wavelength 220 nm are incident on the surface of a metal. Electrons emitted from the surface of the metal have maximum speed 1.2×10^{6} m s^{-1}. Calculate

(i) the work function of the metal, *(4 marks)*

(ii) the threshold frequency of the metal. *(2 marks)*

b The intensity of the electromagnetic waves incident on the metal in **(a)** is increased. State and explain the effect this has on the maximum kinetic energy of the electrons emitted from the metal surface. *(2 marks)*

3 A negatively charged metal plate is exposed to electromagnetic waves of a range of frequencies. Figure 1 shows the variation of the maximum kinetic energy E_{max} of the emitted photoelectrons with frequency f of the incident electromagnetic waves.

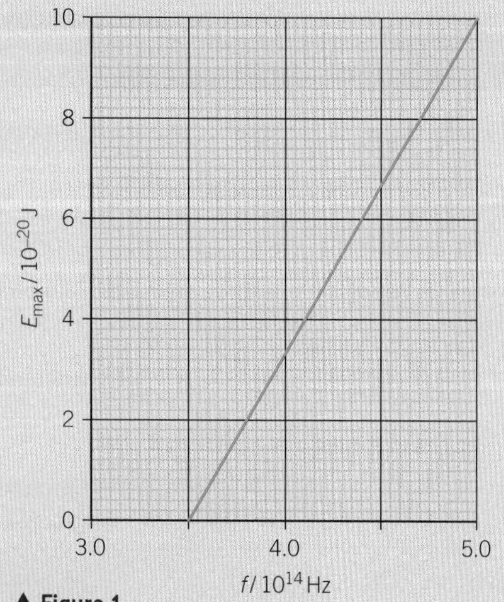

▲ **Figure 1**

a Explain why the straight line graph shown in Figure 1 does not pass through the origin. *(2 marks)*

b Determine the work function of the metal. *(2 marks)*

c The metal plate is replaced with one having a work function greater than the value calculated in (b). State and explain the change to the shape of the graph of E_{max} against f. *(4 marks)*

4 a Explain what is meant by the de Broglie wavelength of a particle. *(2 marks)*

b Electrons are accelerated through a potential difference of 200 V. Calculate the final de Broglie wavelength of these electrons. *(5 marks)*

c The electrons from (**b**) can be diffracted by the atoms in a solid. Suggest what you can deduce about the arrangement of the atoms in solids. *(1 mark)*

5 In a demonstration experiment of the photoelectric effect, light of wavelength 440 nm incident on a clean metal surface causes electrons to be emitted. No electrons are emitted from the surface when the wavelength of the incident light is greater than 550 nm.

a (i) Define the term *work function*. *(2 marks)*

(ii) Explain how the work function is related to the threshold frequency. *(2 marks)*

(iii) Calculate the value of the work function for this metal. *(2 marks)*

b (i) Show that the maximum speed of the emitted electrons in the experiment is about 4.5×10^{5} m s^{-1}. *(3 marks)*

(ii) Calculate the minimum de Broglie wavelength of an emitted electron. *(2 marks)*

6 a State two properties of a photon. *(2 marks)*

b Figure 2 shows a graph of E against $\frac{1}{\lambda}$, where E is the energy of a photon and λ is its wavelength.

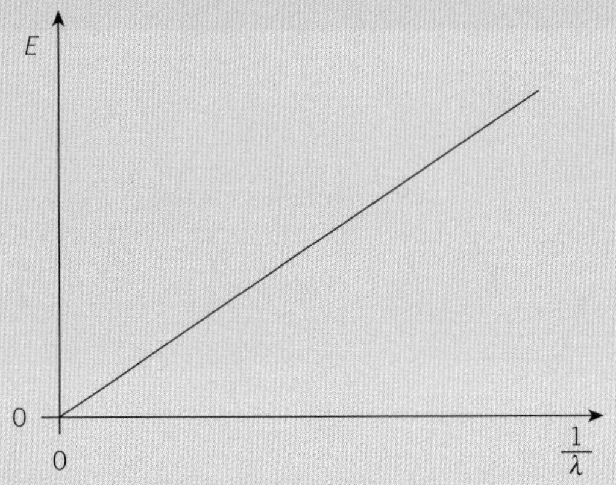

▲ Figure 2

(i) Explain why the graph of Figure 2 is a straight line. *(2 marks)*

(ii) State and explain what the gradient of the graph represents. *(2 marks)*

c The Sun emits electromagnetic radiation of average wavelength 5.5×10^{-7} m. The radiant power emitted from its surface is 6.3×10^{7} W m^{-2}. The radius of the Sun is 7.0×10^{8} m.

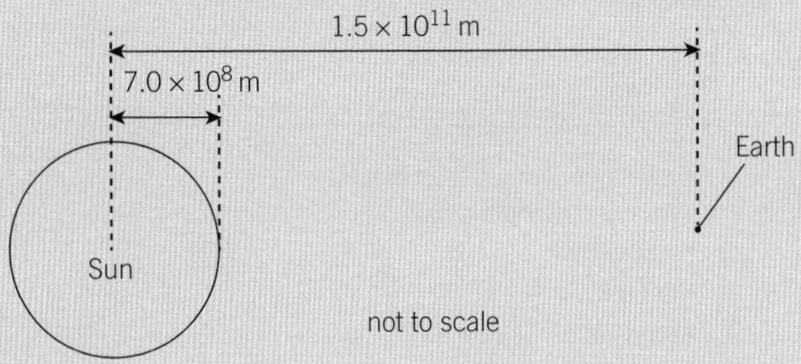

▲ Figure 3

(i) Estimate the total number of photons emitted per second from the surface of the Sun. *(4 marks)*

(ii) The Earth is about 1.5×10^{11} m from the Sun. Determine the average intensity of the radiation at the Earth from the Sun. *(3 marks)*

7 **a** State how the Planck constant h is used to model photons and the wave-like properties of particles. *(2 marks)*

b Explain what is meant by the photoelectric effect. *(1 mark)*

c Define threshold frequency of a metal. *(1 mark)*

d Electromagnetic radiation of wavelength 4.0×10^{-7} m is incident on the surface of a metal plate. The maximum energy of the emitted photoelectron is 1.6 eV.

Calculate the work function of the metal in eV. *(4 marks)*

8 A negatively charged metal plate is exposed to electromagnetic radiation of frequency f.

Figure 8 shows the variation of the maximum kinetic energy E_k of the photoelectrons emitted from the metal with frequency f.

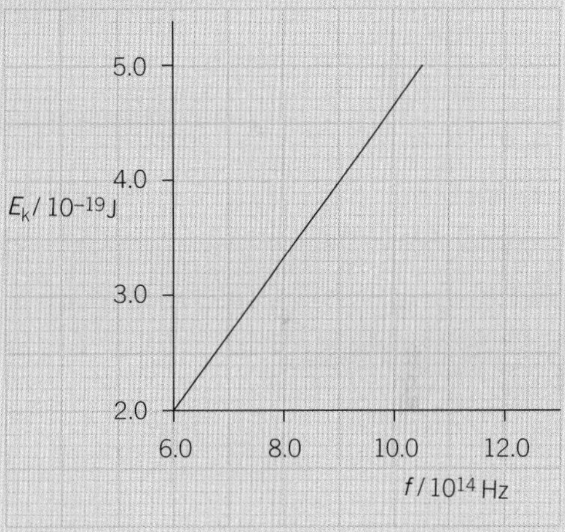

▲ Figure 4

a Use Figure 4 to determine the Planck constant. Explain your answer. *(3 marks)*

b State and explain the effect on the shape of the graph when the intensity of the incident radiation on the surface of the metal is increased. *(2 marks)*

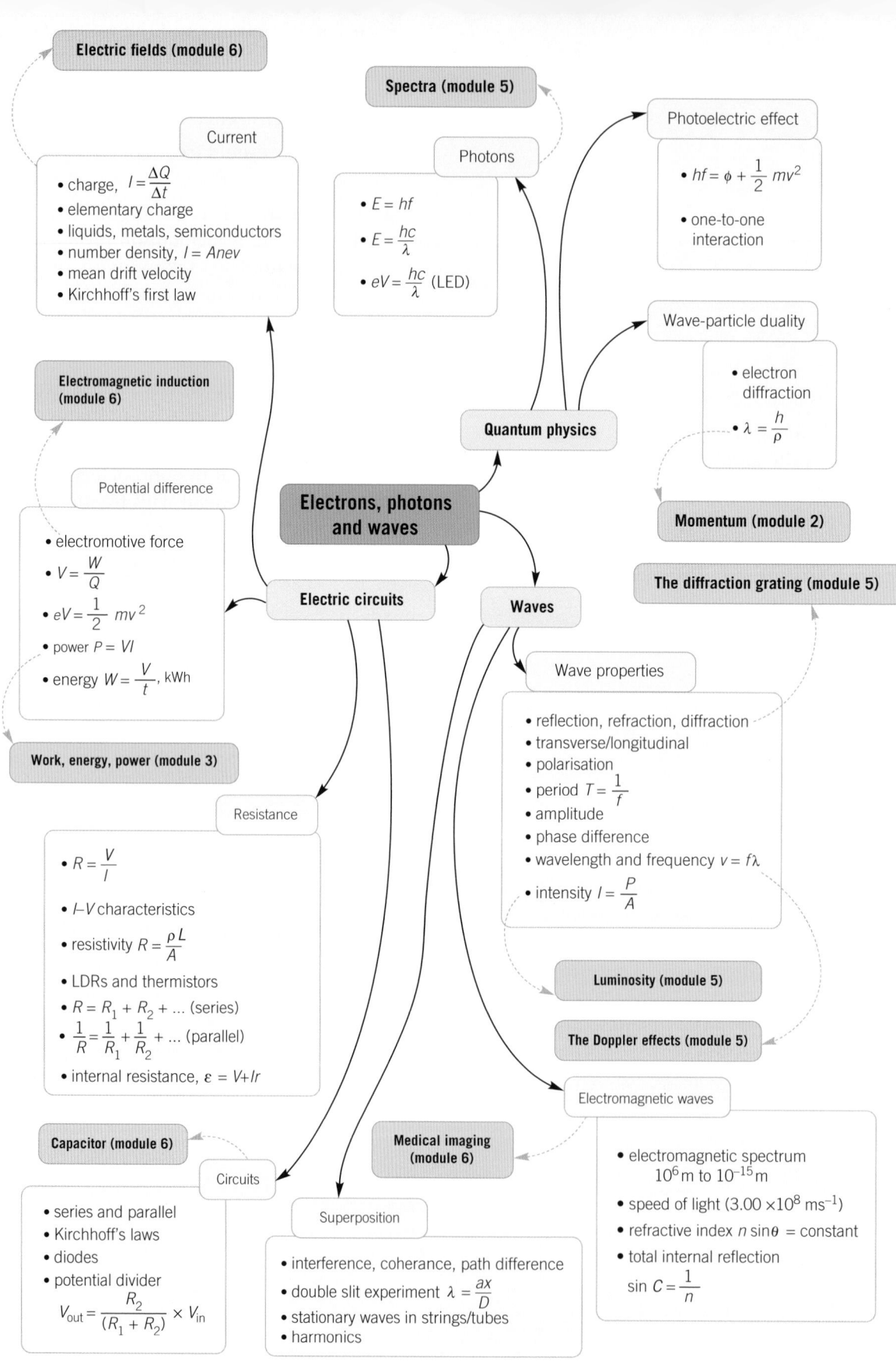

Electrons, photons and waves

Quantum physics

Photons
- $E = hf$
- $E = \frac{hc}{\lambda}$
- $eV = \frac{hc}{\lambda}$ (LED)

→ **Spectra (module 5)**

Photoelectric effect
- $hf = \phi + \frac{1}{2}mv^2$
- one-to-one interaction

Wave-particle duality
- electron diffraction
- $\lambda = \frac{h}{p}$

→ **Momentum (module 2)**

Electric circuits

Current
- charge, $I = \frac{\Delta Q}{\Delta t}$
- elementary charge
- liquids, metals, semiconductors
- number density, $I = Anev$
- mean drift velocity
- Kirchhoff's first law

→ **Electric fields (module 6)**

Potential difference
- electromotive force
- $V = \frac{W}{Q}$
- $eV = \frac{1}{2}mv^2$
- power $P = VI$
- energy $W = \frac{V}{t}$, kWh

→ **Electromagnetic induction (module 6)**

→ **Work, energy, power (module 3)**

Resistance
- $R = \frac{V}{I}$
- I–V characteristics
- resistivity $R = \frac{\rho L}{A}$
- LDRs and thermistors
- $R = R_1 + R_2 + \ldots$ (series)
- $\frac{1}{R} = \frac{1}{R_1} + \frac{1}{R_2} + \ldots$ (parallel)
- internal resistance, $\varepsilon = V + Ir$

Circuits
- series and parallel
- Kirchhoff's laws
- diodes
- potential divider

  $V_{out} = \frac{R_2}{(R_1 + R_2)} \times V_{in}$

→ **Capacitor (module 6)**

Waves

Wave properties
- reflection, refraction, diffraction
- transverse/longitudinal
- polarisation
- period $T = \frac{1}{f}$
- amplitude
- phase difference
- wavelength and frequency $v = f\lambda$
- intensity $I = \frac{P}{A}$

→ **The diffraction grating (module 5)**

→ **Luminosity (module 5)**

→ **The Doppler effects (module 5)**

Superposition
- interference, coherance, path difference
- double slit experiment $\lambda = \frac{ax}{D}$
- stationary waves in strings/tubes
- harmonics

Electromagnetic waves
- electromagnetic spectrum 10^6 m to 10^{-15} m
- speed of light (3.00×10^8 ms^{-1})
- refractive index $n \sin\theta$ = constant
- total internal reflection

  $\sin C = \frac{1}{n}$

→ **Medical imaging (module 6)**

Feeling the strain

How can we monitor extremely small changes in distance, such as small changes in the length of a load-bearing section of a bridge or tiny ground movement caused by seismic activity? A strain gauge can be used. It could be attached to the bridge or secured to the ground. The strain gauge consists of thin wire mounted on a non-conductive material (see Figure 1). Stretching the gauge decreases the cross-sectional area of the wire and increases its length. The resistance of the wire increases by a tiny amount. It would not be possible to measure these tiny changes in resistance using an ohmmeter. However, a circuit known as a Wheatstone bridge circuit can be used to monitor these tiny changes in resistance (see Figure 2). The variation in the voltage V can be monitored remotely using dataloggers and even using mobile phones.

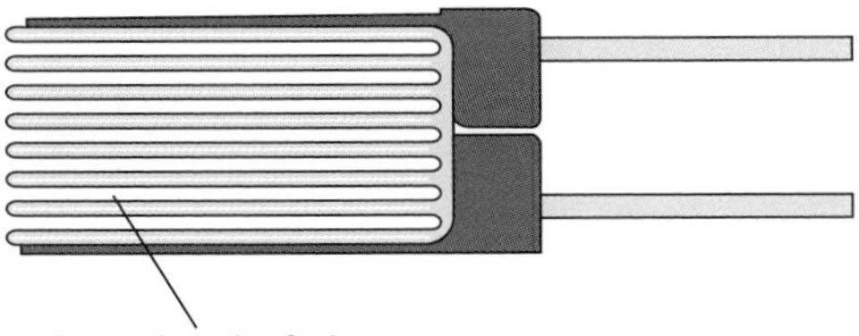

▲ **Figure 1** *Strain gauge*

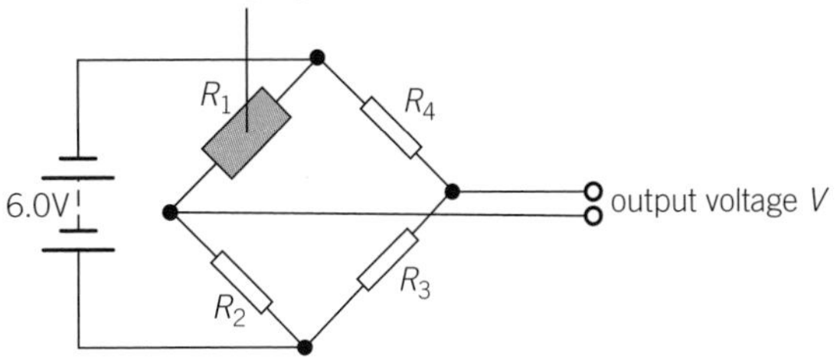

▲ **Figure 2** *Wheatstone circuit with a strain gauge*

The output voltage V is related to the values of the resistances R_1, R_2, R_3, and R_4 by the equation

$$V = \left(\frac{R_1}{R_1 + R_2} - \frac{R_4}{R_3 + R_4} \right) \times 6.0$$

1 List situations, other than the two mentioned here, where a strain gauge might be useful.
2 Show that the resistance of a strain gauge is directly proportional to length^2 of the wire.
3 Consider a Wheatstone bridge circuit where initially all the values of the resistances are the same and equal to 100 Ω. What is V equal to? Now imagine that there is 0.001% change in the resistance R_1 of the strain gauge. What is the output voltage now?

Energy and electrons

In Chapters 11, 12 and 13 you learnt about progressive waves, stationary waves and quantum physics. These are important topics in physics. Physicists have used these simple ideas to explain why the energy of electrons within an atom is quantised. You now have the knowledge to recreate the modelling process used in the development of quantum physics. This is an important aspect of How Science Works – theories and models are amalgamated to reveal something new about nature.

Here are some extension questions you can tackle to further improve your knowledge of atoms. These ideas will give you a head start when you study *Energy levels in atoms* in 6.4.

Imagine the hydrogen atom is one-dimensional, just like a string fixed at both ends. The size of the atom is about 10^{-10} m. You know that a moving electron has a de Broglie wavelength. The electron bound to the atom will produce a stationary wave.

1 What is the longest de Broglie wavelength of an electron that is bound to the atom?
2 Calculate the kinetic energy of this electron in joules and in eV.
3 Determine the other possible kinetic energies this electron can have within the atom.
4 Explain what is meant by the statement: *The energy of the electron within the atom is quantised.*

Section A

1 A student investigating an electrical experiment records the following measurements in the lab book.

- current in the LED = $120 \pm 8\,\text{mA}$
- potential difference across the LED = $1.8 \pm 0.2\,\text{V}$

What is the percentage uncertainty in the resistance of the LED?

A 4.4 %

B 6.7%

C 11%

D 18% *(1 mark)*

From the AS Paper 1 practice questions

2 Figure 1 shows a stationary wave pattern formed in an air column.

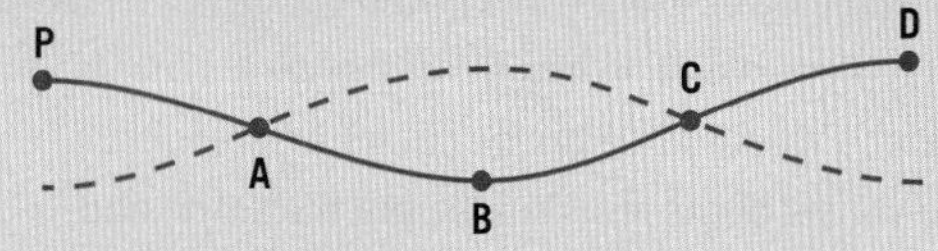

▲ **Figure 1**

Which point **A**, **B**, **C**, or **D** has a phase difference of 180° with reference to **P**?

(1 mark)

From the AS Paper 1 practice questions

3 A ray of monochromatic light is incident at a boundary between two transparent materials. The refractive index of the materials is 1.30 and 1.50. The angle of refraction for the emergent ray is 60°.

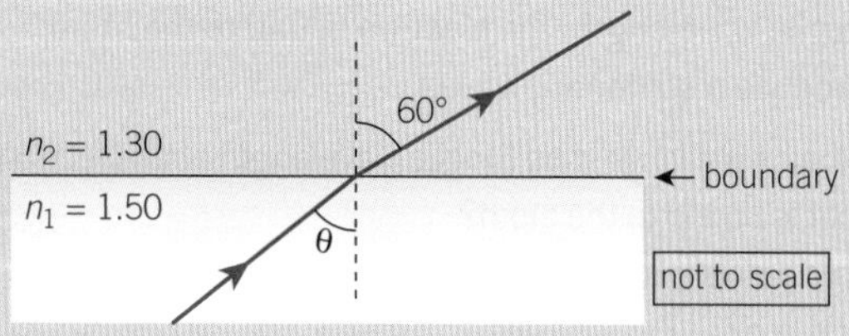

▲ **Figure 2**

What is the angle θ of incidence?

A 42°

B 49°

C 60°

D 88° *(1 mark)*

From the AS Paper 1 practice questions

4 Figure 3 shows the cross-section of a metal wire connected to a power supply. The charge carriers within the metal wire move from right to left.

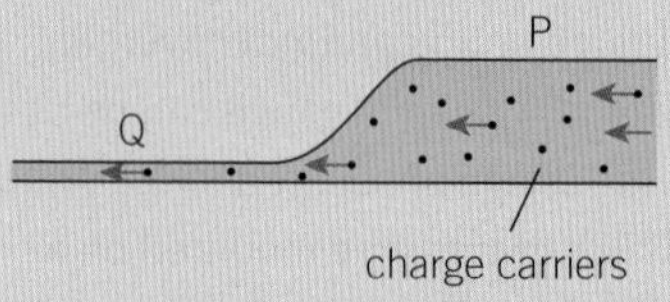

▲ **Figure 3**

The section Q of the wire is thinner than section P.

Which statement is correct?

A The direction of the conventional current is from right to left.

B The section Q of the wire has fewer charge carriers per unit volume.

C The current in both sections is the same.

D The charge carriers are negative ions.

(1 mark)

From the AS Paper 1 practice questions

5 A resistor **R** is connected in parallel with a resistor of resistance $10\,\Omega$. The total resistance of the combination is $6.0\,\Omega$. What is the resistance of resistor **R**?

A $0.067\,\Omega$

B $3.8\,\Omega$

C $4.0\,\Omega$

D $15\,\Omega$ *(1 mark)*

6 What is a reasonable estimate for the energy of a photon of visible light?

A $4 \times 10^{-19}\,\text{J}$

B $4 \times 10^{-18}\,\text{J}$

C $4 \times 10^{-16}\,\text{J}$

D $4 \times 10^{-11}\,\text{J}$ *(1 mark)*

7 The circuit in Figure 4 is constructed by a student in the laboratory.

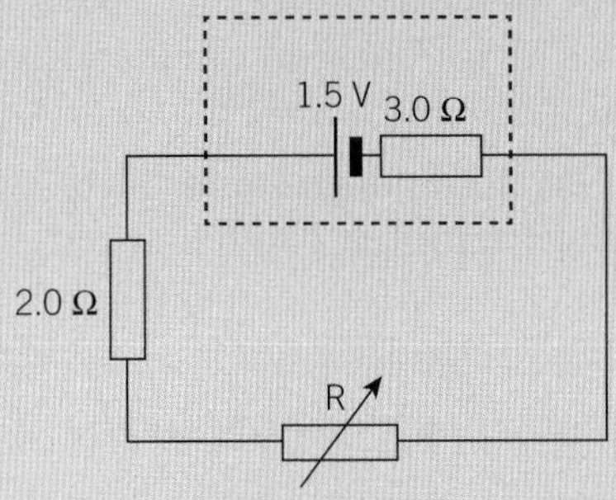

▲ **Figure 4**

The e.m.f. of the cell is 1.5 V and it has an internal resistance of 3.0 Ω. A resistor of resistance 2.0 Ω and a variable resistor R are connected in series to the terminals of the cell. The variable resistor is set to a resistance value of 7.0 Ω.

What is the value of the ratio

$$\frac{\text{power dissipated in R}}{\text{power supplied by the cell}}?$$

A 0.17

B 0.25

C 0.58

D 0.75 *(1 mark)*

From the AS Paper 1 practice questions

Section B

8 a Explain what is meant by coherent waves. *(1 mark)*

b State two ways in which a stationary waves differs from a progressive wave. *(2 marks)*

c Figure 5 shows a stationary pattern on a length of stretched string.

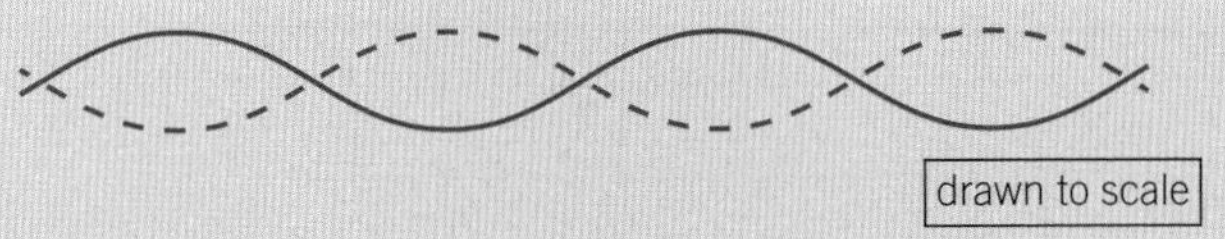

▲ **Figure 5**

The distances shown in Figure 5 are **drawn to scale**. The frequency of vibration of the string is 110 Hz.

(i) By taking measurements from Figure 5, determine the wavelength of the progressive waves on the string. *(2 marks)*

(ii) Calculate the speed of the progressive waves on the string. *(2 marks)*

From the AS Paper 1 practice questions

9 a Sketch a graph of energy E of a photon against frequency f of the electromagnetic radiation. *(1 mark)*

b Electromagnetic waves of frequency 8.93×10^{14} Hz are incident on the surface of metal. The work function of the metal is 3.20×10^{-19} J.

(i) Calculate the energy of the photons. *(2 marks)*

(ii) Calculate the maximum speed v_{max} of the photoelectrons emitted from the metal surface. *(3 marks)*

From the AS Paper 1 practice questions

10 a Define *refractive index* of a material. *(1 mark)*

b Figure 6 shows the path of a ray of light as it crosses the boundary between two materials A and B.

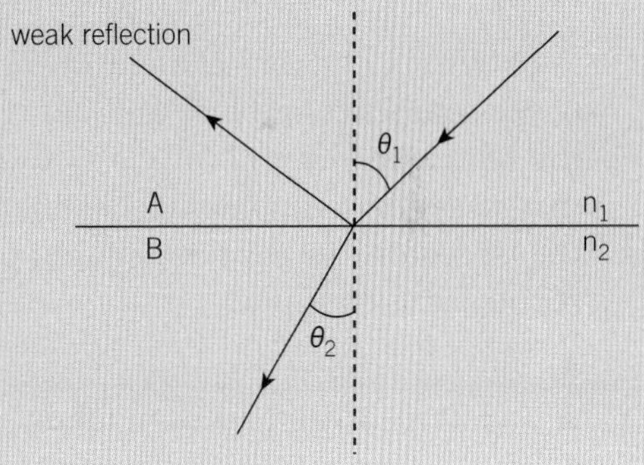

▲ **Figure 6**

The refractive index of material A is n_1 and the angle of incidence of the ray of light is θ_1. The angle of refraction in material B is θ_2 and the refractive index of material B is n_2. Write an equation that relates n_1, n_2, θ_1 and θ_2. *(1 mark)*

c A student is investigating the refraction of light by a transparent material by measuring the angles of incidence i and refraction r. Figure 7 show the results from the experiment.

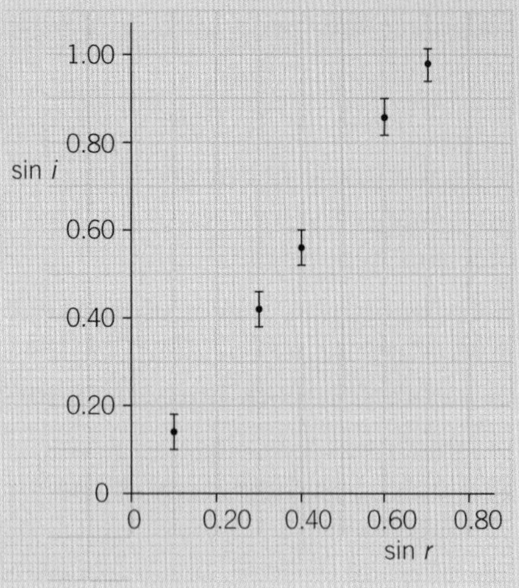

▲ **Figure 7**

Use Figure 7 to determine

(i) the refractive index of the material *(2 marks)*

(ii) the critical angle for this material. *(2 marks)*

d You are provided with a semi-circular glass block and a ray-box with a suitable supply. Design a laboratory experiment to determine the critical angle of the glass of the semi-circular block and hence the refractive index of the glass. You may use other equipment available in the laboratory. In your description pay particular attention to

- how the apparatus is used
- what measurements are taken
- how the data is analysed. *(4 marks)*

From the AS Paper 1 practice questions

11 a State *Ohm's law.* *(1 mark)*

b The *I-V* characteristic of a particular component is shown in Figure 8.

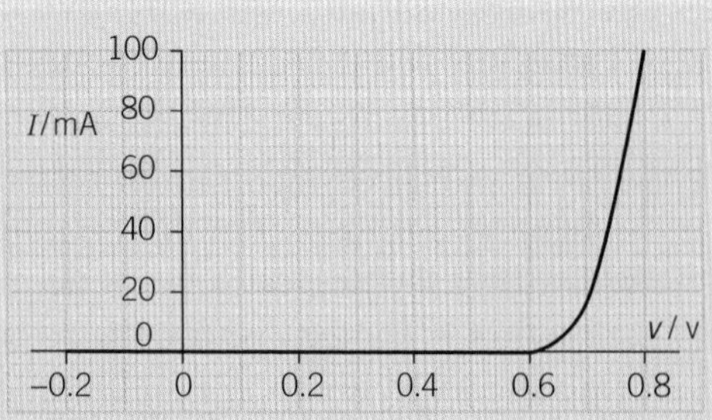

▲ **Figure 8**

(i) Use Figure 8 to describe how the resistance of this component depends on the potential difference (p.d.) across it. You may do calculations to support your answer. *(3 marks)*

(ii) Draw a circuit diagram for an arrangement that could be used to collect results to plot the graph shown in Figure 8. *(3 marks)*

c Figure 9 shows an electrical circuit.

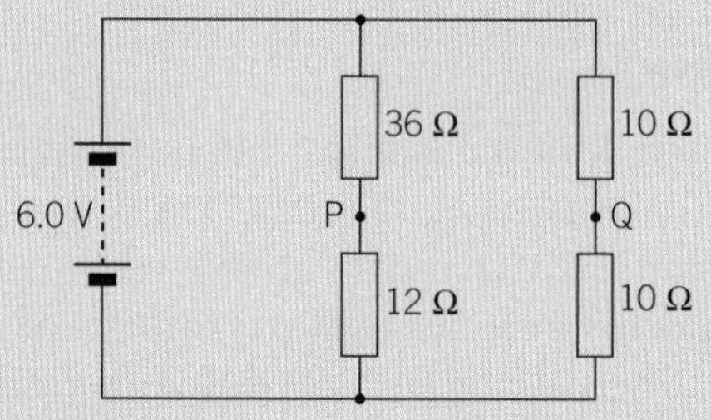

▲ **Figure 9**

The e.m.f. of the battery is 6.0 V and it has negligible internal resistance.

Calculate

(i) the current in the 36 Ω resistor *(2 marks)*

(ii) the potential difference across the 12 Ω resistor *(1 mark)*

(iii) the potential difference between points **P** and **Q**. *(2 marks)*

From the AS Paper 1 practice questions

12 Figure 2 shows the *I-V* characteristic of a blue light-emitting diode (LED).

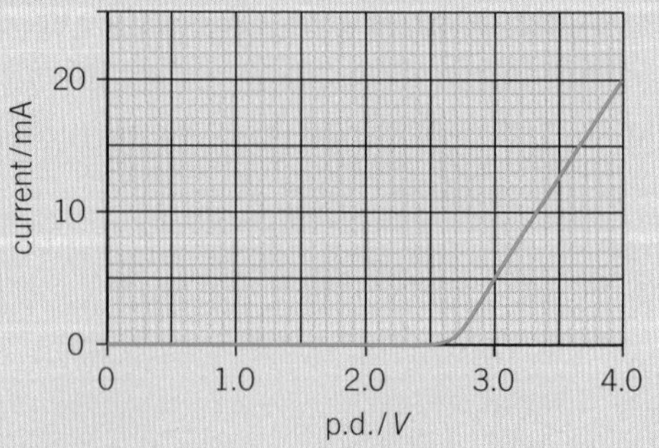

▲ **Figure 10a**

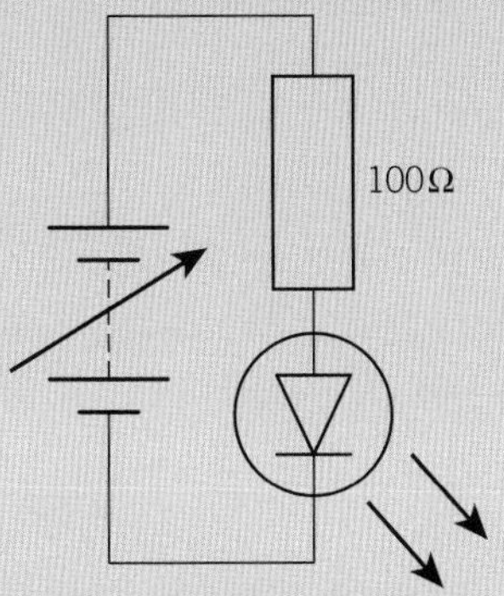

▲ **Figure 10b**

a (i) The data for plotting the *I-V* characteristic is collected using the components shown in Figure 10b. By drawing on a copy of Figure 10b complete the circuit showing how you would connect the two meters needed to collect these data. *(1 mark)*

(ii) When the current in the circuit of Figure 10b is 20 mA calculate the terminal potential difference across the supply. *(3 marks)*

b The energy of each photon emitted by the LED comes from an electron passing through the LED. The energy of each blue photon emitted by the LED is 4.1×10^{-19} J.

(i) Calculate the energy of a blue photon in electron volts. *(1 mark)*

(ii) Explain how your answer to (i) is related to the shape of the curve in Figure 10a. *(2 marks)*

c Calculate for a current of 20 mA

(i) the number *n* of electrons passing through the LED per second, *(2 marks)*

(ii) the total energy of the light emitted per second, *(2 marks)*

(iii) the efficiency of the LED in transforming electrical energy into light energy. *(2 marks)*

d The energy of a photon emitted by a red LED is 2.0 eV. The current in this LED is 20 mA when the p.d. across it is 3.4 V. Draw the *I-V* characteristic of this LED on a copy of Figure 10a. *(2 marks)*

Q4 G482 June 2014 paper
From the AS Paper 2 practice questions

13 a Show that the momentum *p* of a particle is given by the equation $p = \sqrt{2Em}$, where *m* is the mass of the particle and *E* is its kinetic energy. *(3 marks)*

b Slow-moving neutrons from a nuclear reactor are used to investigate the structure of complex molecules such as DNA. Neutrons can be diffracted by DNA. The mass of a neutron is 1.7×10^{-27} kg.

(i) Calculate the de Broglie wavelength of a neutron of kinetic energy 6.2×10^{-21} J. *(4 marks)*

(ii) Suggest why these slow-moving neutrons can be diffracted by DNA. *(1 mark)*

c Charged particles are accelerated in a laboratory by a group of scientists. The de Broglie wavelength of the particles is λ and their kinetic energy is *E*. Figure 11 shows a graph of λ^2 against $\frac{1}{E}$ for these accelerated particles.

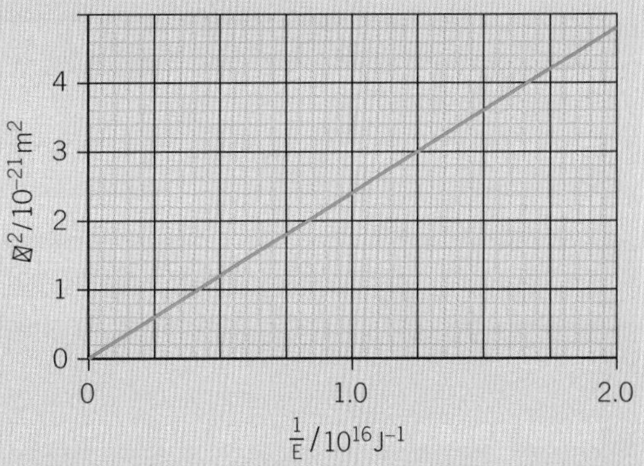

▲ **Figure 11**

Use Figure 11 to determine the mass *m* of the particles. *(4 marks)*

From the AS Paper 2 practice questions

MODULE 5

Newtonian world and astrophysics

Chapters in this module

14 Thermal physics
15 Ideal gases
16 Circular motion
17 Oscillations
18 Gravitational fields
19 Stars
20 Cosmology (the Big Bang)

Introduction

Newtonian mechanics has had an incredible impact across physics. In this module you will discover the wide range of this impact, from the vast orbits of stars and planets to the tiny interactions that cause pressure in gases. You will explore some of the most fundamental ideas in physics, from the concepts of heat and temperature and their relation to energy, to perhaps the most important question a physicist can ask — how did the Universe begin, and how might it end?

Thermal physics introduces ideas around temperature, matter, specific heat capacity and specific latent heat. You will learn about absolute zero and why sweating helps keep us cool.

Ideal gases explores how the microscopic motion of atoms can be modelled using Newton's laws and how this provides us with an understanding of pressure and temperature.

Circular motion builds on your understanding of motion and explores the mathematics of motion in circular paths of objects such as planets, artificial satellites, and rollercoasters.

Oscillations explores a new type of motion, seen in objects that vibrate back and forth. Examples include atoms vibrating in a solid and bridges swaying in the wind.

Gravitational fields develops ideas in circular motion, relating them to planetary motion and gravitational potential energy. You will learn how Newton's law of gravitation can be used to predict the motion of planets, stars, and distant galaxies.

Stars will cover the life cycle of stars, including our Sun, and explore some of the Universe's more fantastic objects like neutron stars and black holes. It also develops ideas on the analysis of electromagnetic radiation from space.

Cosmology (the Big Bang) explores ideas of the expansion of the Universe described by Hubble's law, the Big Bang theory, and the as yet unsolved mysteries of dark matter and dark energy.

Knowledge and understanding checklist

From your Key Stage 4 or first year A Level study you should be able to do the following questions. Work through each point, using your Key Stage 4/ first year A Level notes and the support available on Kerboodle.

- ☐ Describe how the internal energy and the motion of particles are different for different phases of matter (solids, liquids and gases).
- ☐ Explain how the motion of the molecules in a gas is related to both its temperature and its pressure, and so explain the relation between the temperature of a gas and its pressure at constant volume.
- ☐ Define the term specific heat capacity and distinguish between it and the term specific latent heat.
- ☐ Explain that motion in a circular orbit involves constant speed but changing velocity.
- ☐ Define weight, describe how it is measured, and describe the relationship between the weight of a body and the gravitational field strength.
- ☐ Use simple vector diagrams to illustrate forces, recall examples of ways in which objects interact by gravity, and describe how such examples involve interactions between pairs of objects that exert equal and opposite forces on each other.

Maths skills checklist

All physicists need to use maths in their studies. In this unit you will need to use many different maths skills, including the following examples. You can find support for these skills on Kerboodle and through MyMaths.

- ☐ **Use an appropriate number of significant figures.** You will need to do this throughout this module, including when calculating the time period of a planet based on its distance from the Sun, and the pressure exerted by an ideal gas.
- ☐ **Recognise and make use of appropriate units in calculations.** You will need to be able to do this when identifying the correct units when dealing with different astronomical distances.
- ☐ **Understand the relationship between degrees and radians and translate from one to the other.** You will need to do this when completing calculations involving circular motion.
- ☐ **Use calculators to handle sin *x*, cos *x*, and tan *x*, when *x* is expressed in degrees or radians.** You will need to do this when completing calculations involving circular motion and simple harmonic motion.
- ☐ **Interpret logarithmic plots.** You will need to this when working with the Hertzsprung–Russell diagram when learning about stars.

MyMaths.co.uk Bringing Maths Alive

14 THERMAL PHYSICS

14.1 Temperature

Specification reference: 5.1.1

Learning outcomes

Demonstrate knowledge, understanding, and application of:

→ thermal equilibrium

→ the absolute scale of temperature

→ temperature measurements in degrees Celsius and kelvin

→ $T(\mathrm{K}) \approx \theta(^\circ\mathrm{C}) + 273$.

▲ **Figure 1** *The three phases of water – solid, liquid, and gas – are in thermal equilibrium inside this triple-point cell: the ice does not melt, the water vapour does not condense*

The triple point

The **triple point** of a substance is one specific temperature and pressure where a strange thing happens. There, and nowhere else, the three **phases** of matter (solid, liquid, and gas) of that substance can exist in **thermal equilibrium**, that is, there is no net transfer of thermal energy between the phases.

For water, the triple point is at 0.01°C and 0.61 kPa, less than 1% of normal atmospheric pressure.

Temperature and thermal equilibrium

A simple way to think about temperature is as a measure of the hotness of an object on a chosen scale. The hotter an object is, the higher its temperature.

If one object is hotter than another there is a net flow of thermal energy from the hotter object into the colder one. This increases the temperature of the colder object and lowers the temperature of the hotter one. For example, when the outside air temperature is lower than your body temperature there is a net flow of energy from you to your surroundings.

When two objects are in thermal equilibrium there is no net flow of thermal energy between them. This means any objects in thermal equilibrium must be at the same temperature.

The zeroth law of thermodynamics

The zeroth (0th) law of thermodynamics was proposed after three laws were already recognised. (You do not need to know the other laws for this course, although you are already familiar with conservation of energy, one aspect of the first law.) It was deemed so fundamental to the study of thermal physics that it was named the zeroth law, coming before the others.

The zeroth law states that if two objects are each in thermal equilibrium with a third, then all three are in thermal equilibrium with each other. In other words, if both A and C are in thermal equilibrium with B, then A is in thermal equilibrium with C. This means that all three objects are at the same temperature.

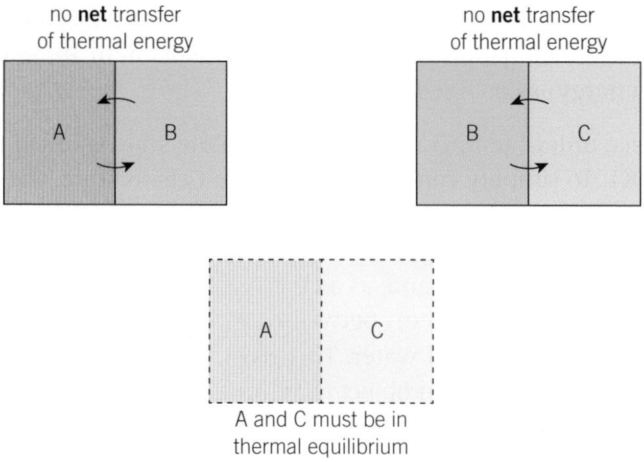

▲ **Figure 2** *The zeroth law introduces temperature as a physical property*

It may seem obvious, but the zeroth law means that objects have a measurable physical property that determines the direction of any transfer of thermal energy. This property is temperature.

The zeroth law forms the basis for a definition of temperature and thus for comparing temperatures, describes how thermometers work, and is important for the mathematical formulation of laws about the effect of changing temperature.

1. Describe the transfer of thermal energy from object A to object D if D is at a lower temperature than A.
2. In order to measure the temperature of an object accurately, simple liquid-in-glass thermometers must be at the same temperature as the object. Outline the reason.

Measuring temperature

In order to measure temperature a scale is needed that includes two fixed points at defined temperatures. The temperature of other objects can then be defined as a position on this scale.

Most of the world uses the **Celsius scale** proposed by the Swedish astronomer Anders Celsius in 1742. He suggested the freezing point and the boiling point of pure water (when the atmospheric pressure is 1.01×10^5 Pa) as the two fixed points, with 100 increments (or degrees) between 0°C and 100°C. By definition, any object at 100°C must be in thermal equilibrium with boiling water.

However, the Celsius scale is not perfect. Although its two fixed points seem simple to obtain, they vary significantly depending on the surrounding atmospheric pressure. For example, on top of a high mountain water boils at a lower temperature (as low as 70°C).

Synoptic link

The kelvin is one of the seven base units studied in Topic 2.1, Quantities and units.

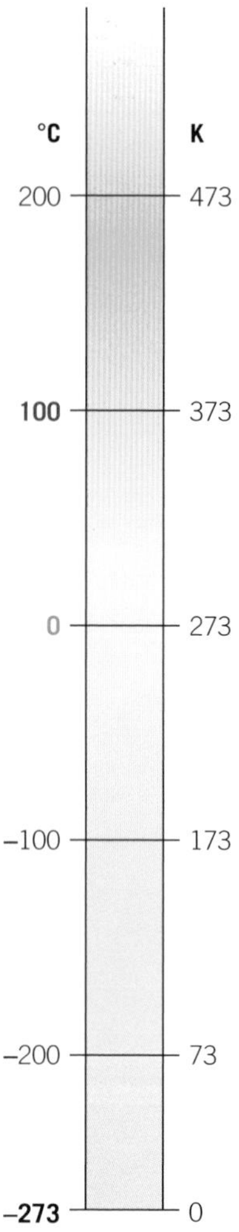

▲ **Figure 3** *To convert from temperature in °C to temperature in K, add 273*

Study tip

0 °C is equivalent to 273.15 K or approximately 273 K.

The **absolute temperature** scale (or **thermodynamic temperature scale**) uses the triple point of pure water and **absolute zero** (the lowest possible temperature, explored in more detail in Topic 14.3, Internal energy) as its fixed points.

The SI base unit of temperature on the absolute scale is called the **kelvin** (K). To simplify comparison with the Celsius scale, the scientific community agreed that the increments on the absolute scale would be the same size as those on the Celsius scale, so a change in temperature of 1 K is the same as a change of 1°C. As a result, there are exactly 273.16 increments between absolute zero (now defined as 0 K) and the triple point of water. This gives the following relationship between temperature of an object in °C and in K:

$$T(\text{K}) \approx \theta(°\text{C}) + 273$$

Temperatures in K are always positive, and the lowest temperature on the absolute scale is 0 K (see Figure 3).

Summary questions

1 Describe the net transfer of thermal energy between two objects A and B at the temperatures given in Table 1. *(3 marks)*

▼ **Table 1**

	Temperature of object A / °C	Temperature of object B / °C
a	100	0
b	50	50
c	−90	−40

2 Convert the following temperatures from °C into K:
a 0°C **b** 37.0°C **c** −120.5°C. *(3 marks)*

3 Convert the following temperatures from K into °C:
a 0 K **b** 200 K **c** 350 K. *(3 marks)*

4 Explain why is it not possible for an object to have a temperature of −50 K. *(1 mark)*

5 Describe the net transfer of thermal energy and any changes in temperature when a metal block at 300 K is placed in water at 15°C. *(3 marks)*

6 The typical core temperature of a star is about 10^7 K. When astronomers discuss the core temperatures of stars they often omit the unit °C or K. Suggest whether this is sensible. *(2 marks)*

7 Suggest a reason, in terms of thermal energy transfer, why a typical liquid-in-glass thermometer at room temperature placed into a cup of hot water does not give a truly accurate reading of the initial temperature of the water even when they reach thermal equilibrium. *(3 marks)*

14.2 Solids, liquids, and gases

Specification reference: 5.1.2

Floating and sinking

Liquid water is essential for life, and the simple fact that ice floats on water (Figure 1) has allowed complex life on Earth to survive the most extreme of ice ages. It means water freezes from the top downwards, so a small amount of liquid water can remain insulated underneath the ice except in the coldest of conditions.

Water is very unusual in this regard. It is one of only a few substances that is less dense in its solid phase than its liquid phase. To understand the reasons why, we need to look carefully at the nature of water molecules.

Learning outcomes

Demonstrate knowledge, understanding, and application of:

→ solids, liquids, and gases in terms of spacing, ordering, and motion of atoms or molecules

→ the simple kinetic model

→ Brownian motion.

The kinetic model

The **kinetic model** describes how all substances are made up of atoms or molecules, which are arranged differently depending on the phase of the substance.

In solids the atoms or molecules are regularly arranged and packed closely together, with strong electrostatic forces of attraction between them holding them in fixed positions, but they can vibrate and so have kinetic energy (Figure 2).

In liquids the atoms or molecules are still very close together, but they have more kinetic energy than in solids, and – unlike in solids – they can change position and flow past each other.

In gases, the atoms or molecules have more kinetic energy again than those in liquids, and they are much further apart. They are free to move past each other as there are negligible electrostatic forces between them, unless they collide with each other or the container walls. They move randomly with different speeds in different directions.

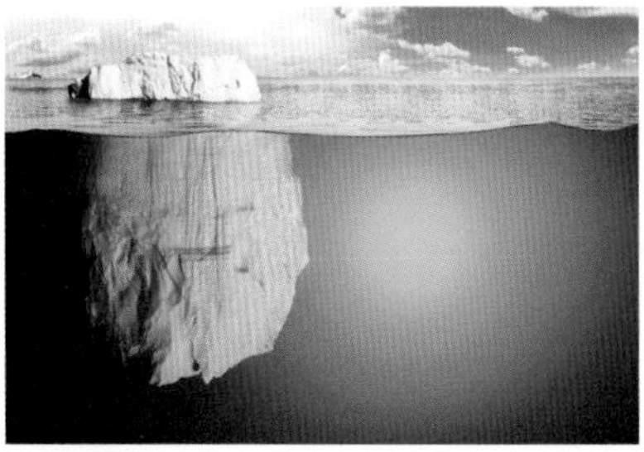

▲ **Figure 1** *An iceberg illustrates clearly how solid water is less dense than liquid water – before modern ship radar, icebergs were a significant hazard to shipping, perhaps most famously sinking the Titanic in 1912*

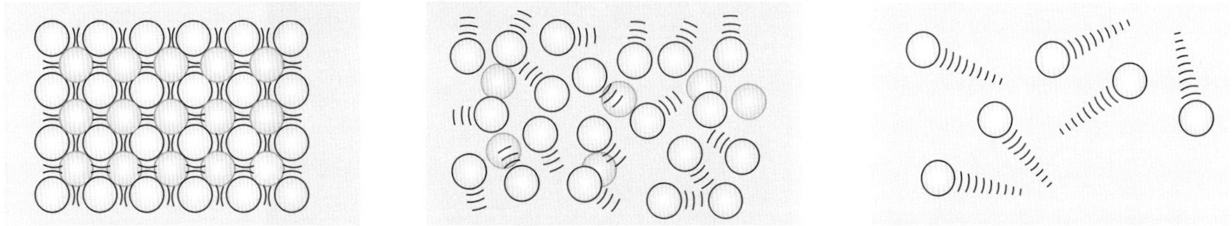

A solid is made up of particles (atoms or molecules) arranged in a regular 3-dimensional structure. There are strong forces of attraction between the particles. Although the particles can vibrate, they cannot move out of their positions in the structure.
When a solid is heated, the particles gain energy and vibrate more and more vigorously. Eventually they may break away from the solid structure and become free to move around. When this happens, the solid has turned into liquid: it has melted.

In a liquid the particles are free to move around. A liquid therefore flows easily and has no fixed shape. There are still forces of attraction between the particles.
When a liquid is heated, some of the particles gain enough energy to break away from the other particles. The particles which escape from the body of the liquid become a gas.

In a gas, the particles are far apart. There are almost no forces of attraction between them. The particles move about at high speed. Because the particles are so far apart, a gas occupies a very much larger volume than the same mass of liquid.
The molecules collide with the container. These collisions are responsible for the pressure which a gas exerts on its container.

▲ **Figure 2** *The kinetic model of three phases of matter and their differing energies*

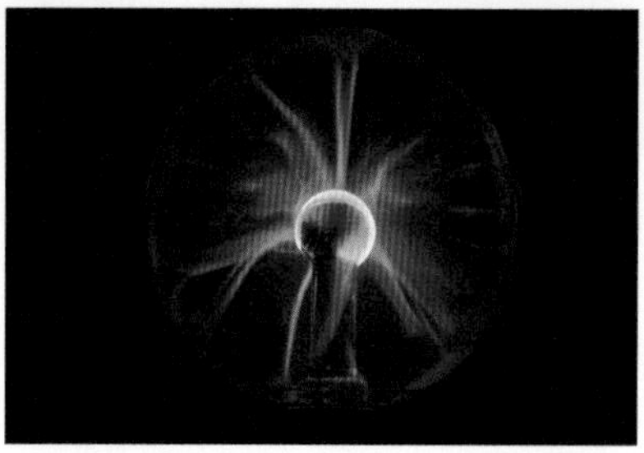

▲ **Figure 3** *Matter can exist in phases other than solid, liquid, and gas – in fact plasma, formed from gas so hot that its atoms are ionised, is the most common phase of matter in the universe, and can be made on Earth by applying a high potential difference across a gas at low pressure as in this plasma ball*

Observing Brownian motion

The idea that substances were made of particles (atoms or molecules) was discussed for centuries, but not confirmed until 1827 when Robert Brown looked through a microscope and recorded his observations of the random movements of fine pollen grains floating on water.

It was not until 1905 that Albert Einstein fully explained this **Brownian motion** in terms of collisions between the pollen grains and millions of tiny water molecules. He explained that these collisions were elastic and resulted in a transfer of momentum from the water molecules to the pollen grains, causing the grains to move in haphazard ways. This provided the first significant proof of the kinetic model – the idea that matter is made up of atoms and molecules and they have kinetic energy.

It is possible to observe Brownian motion in the laboratory using a smoke cell (Figure 4).

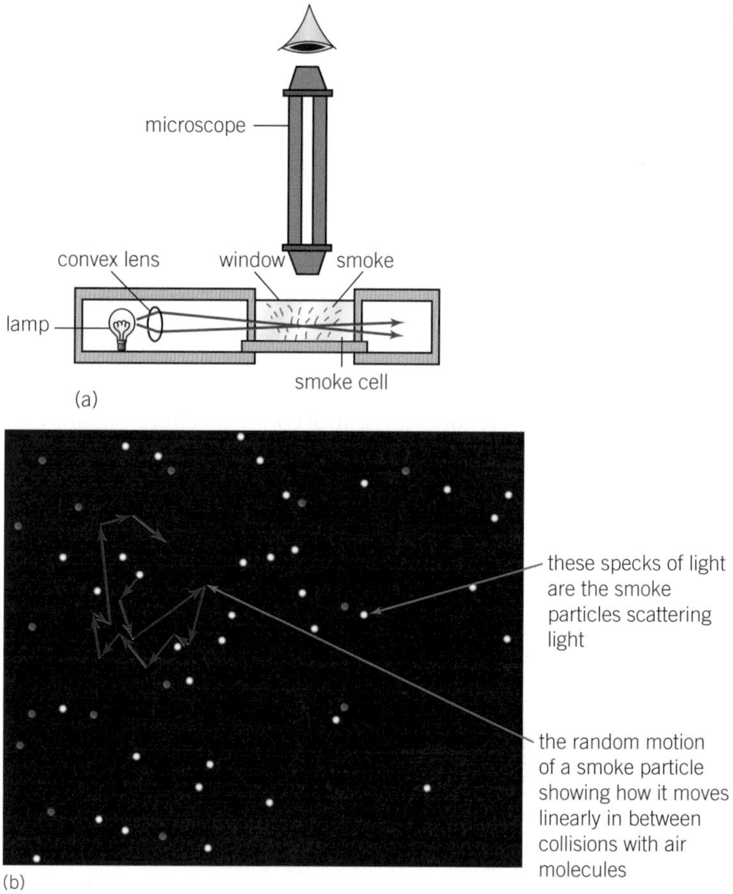

▲ **Figure 4** *Observing the random paths of smoke particles using a smoke cell*

Particles of smoke are large enough to be visible under a microscope. These particles move around in a random way. The random motion is caused by air molecules constantly striking the smoke particles. The air molecules themselves are in random motion. The mean kinetic energy of the smoke particles is the same as the mean kinetic energy of the air molecules. However, while the air molecules typically move with a speed around 500 m s^{-1}, the more massive smoke particles move much more slowly.

1 Sketch the path of a pollen grain being bombarded by water molecules.
2 Explain what happens to the motion of a smoke particle in air if the air temperature decreases.

Density

The spacing between the particles (atoms or molecules) in a substance in different phases affects the density of the substance. In general a substance is most dense in its solid phase and least dense in its gaseous phase. Unusually, solid water is less dense than liquid water. Water freezes into a regular crystalline pattern held together by strong electrostatic forces between the molecules. In this structure the molecules are held slightly further apart than in their random arrangement in liquid water, so ice is slightly less dense.

Synoptic link

Density was introduced in Topic 4.8, Density and pressure.

Summary questions

1 List the three main phases of a substance in order of the energy of the particles (atoms or molecules) in that substance. *(1 mark)*

2 Use diagrams of how atoms or molecules are arranged in solids, liquids, and gases to explain why gases have a much lower density than solids. *(1 mark)*

3 Water of mass 2.0 kg is gradually heated. Its volume is measured at each of the temperatures given in Table 1.
   a Use the data in Table 1 to determine the density of water at the of the temperatures shown. *(3 marks)*

▼ **Table 1** *Volume of 2.0 kg water at various temperatures*

Temperature / °C	5.0	20.0	40.0	60.0	90.0
Volume / 10^{-3} m^3	2.000	2.004	2.016	2.034	2.075

   b Explain why the volume of water increases as its temperature increases. *(2 marks)*
   c Suggest why, along with the melting of land ice, an increase in global temperature results in a rise in sea levels. *(1 mark)*

4 The mass of one water molecule is 3.0×10^{-26} kg. The density of ice is 920 kg m^{-3} and of water vapour (at boiling point) is 0.590 kg m^{-3}. Calculate the number of water molecules in:
   a 1.0 m^3 of ice b 1.0 m^3 of water vapour. *(5 marks)*

5 Use your values above to estimate the spacing between water molecules in ice and water vapour. *(5 marks)*

14.3 Internal energy

Specification reference: 5.1.2

Learning outcomes

Demonstrate knowledge, understanding, and application of:

→ internal energy as the sum of kinetic and potential energies in a system

→ absolute zero (0 K)

→ increase in internal energy with temperature

→ changes in internal energy during changes of phase

→ constancy of temperature during changes of phase.

The coldest place on Earth

The lowest natural temperature ever measured on Earth is –89.2°C (184 K), recorded in 1983 at the Russian Vostok research station in Antarctica (Figure 1). This is cold enough for carbon dioxide to solidify.

Lower temperatures have been achieved artificially in laboratories. The current record, set in 1999, is 100 pK, or 1.0×10^{-10} K (much colder than the deep space between galaxies, at 2.7 K). But it will never be possible to reach 0 K, and to understand why we need to understand what happens inside a substance as it changes temperature.

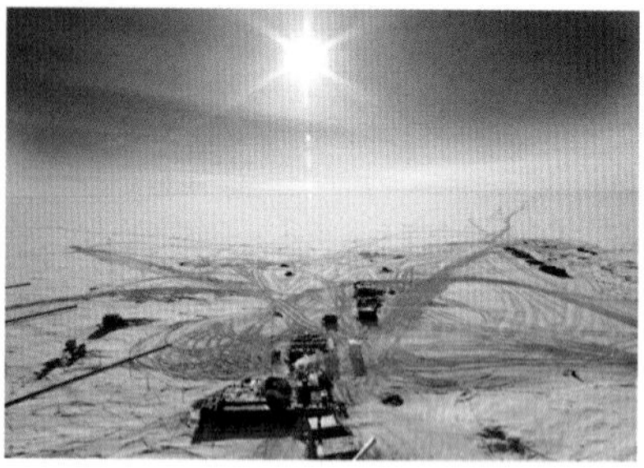

▲ **Figure 1** *The Vostok station still holds the official record for the coldest place on Earth – although satellite data in 2010 indicated a new low of –93.2°C (also in Antarctica). However, this was not confirmed by measurements on the ground*

Internal energy and absolute zero

The **internal energy** of a substance is defined as:

The sum of the randomly distributed kinetic and potential energies of atoms or molecules within the substance.

Consider a beaker of water at room temperature (Figure 2). The water contains a huge number of water molecules travelling at hundreds of meters per second. The internal energy of the water is the sum of all the individual kinetic energies of the water molecules in the glass and the sum of all the potential energies due to the electrostatic intermolecular forces between the molecules.

Now imagine cooling the beaker. The water will freeze and the water molecules move more slowly as the ice gets colder. Absolute zero is the lowest temperature possible. At this temperature the internal energy of a substance is a minimum. The kinetic energy of all the atoms or molecules is zero – they have stopped moving. However, the internal energy is not zero because the substance still has electrostatic potential energy stored between the particles. Even at 0 K, you cannot reduce the potential energy of the substance to zero.

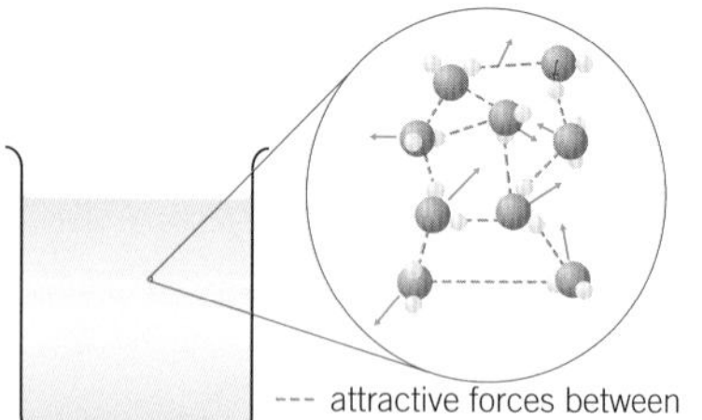
--- attractive forces between molecules determine the potential energy
→ movement determines the kinetic energy

▲ **Figure 2** *A beaker of water has an internal energy due to the kinetic and potential energies of the water molecules*

Increasing the internal energy of a body

Increasing the temperature of a body will increase its internal energy. As the temperature increases, the average kinetic energy of the atoms or molecules inside the body increases. In general, the hotter a substance, the faster the atoms or molecules that make up the substance move, and the greater the internal energy of the substance.

However, it is not only increasing the temperature of a body that increases its internal energy. When a substance changes phase, for example from solid to liquid, the temperature does not change, nor does the kinetic energy of the atoms or molecules. However, their electrostatic potential energy increases significantly.

If a solid substance is heated using a heater with a constant power output, a graph showing how the temperature increases with time can be recorded (Figure 3).

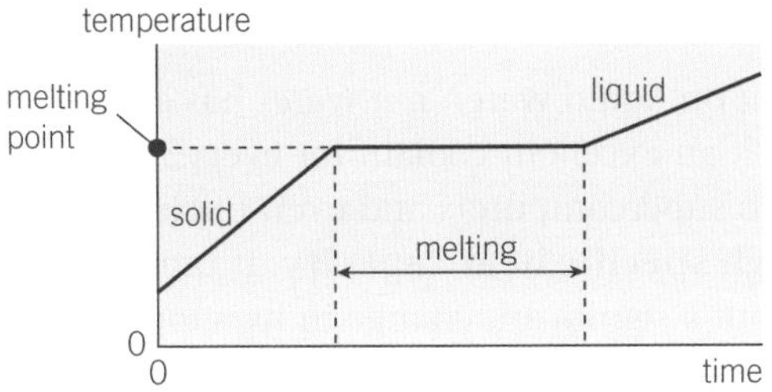

▲ **Figure 3** *The change from solid to liquid or from liquid to gas increases the internal energy of the substance, even though the temperature remains the same while the substance changes phase (the horizontal line on the graph)*

When a substance reaches its melting or boiling point, while it is changing phase the energy transferred to the substance does not increase its temperature. Instead the electrostatic potential energy of the substance increases as the electrical forces between the atoms or molecules change. Only once the phase change is complete does the kinetic energy of the atoms or molecules increase further, and so the temperature rises again.

In different phases the atoms or molecules of a substance have different electrostatic potential energies:

- Gas: The electrostatic potential energy is zero because there are negligible electrical forces between atoms or molecules.
- Liquid: The electrostatic forces between atoms or molecules give the electrostatic potential energy a negative value. The negative simply means that energy must be supplied to break the atomic or molecular bonds.
- Solid: The electrostatic forces between atoms or molecules are very large, so the electrostatic potential energy has a large negative value.

The electrostatic potential energy is lowest in solids, higher in liquids, and at its highest (0 J) in gases.

Synoptic link

You will learn more about average kinetic energy and temperature in Topic 15.4, the Boltzmann constant.

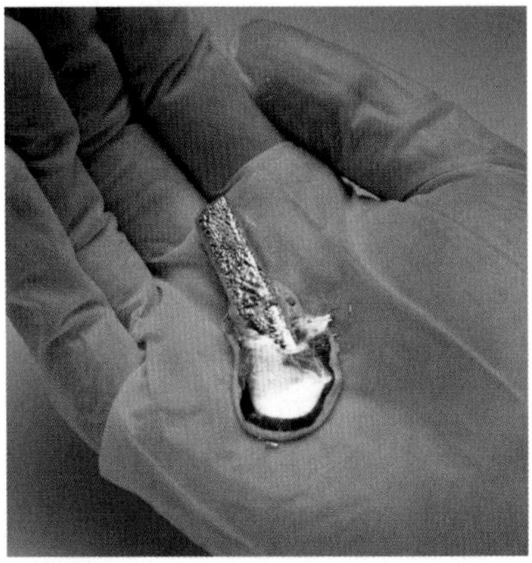

▲ **Figure 4** *The metal gallium has a melting point of 30°C, low enough to melt in a hand*

Summary questions

1. Explain why it is not possible to achieve a temperature lower than 0 K. *(1 mark)*
2. Describe what happens to the energy transferred to a substance being heated when it changes phase. *(2 marks)*
3. State two ways to increase the internal energy of a substance. *(2 marks)*
4. Explain why 1.0 kg of water at 0°C has more internal energy than 1.0 kg of ice at 0°C. *(2 marks)*
5. Explain, in terms of internal energy, why a window gets slightly warmer when water vapour condenses on its surface. *(2 marks)*

14.4 Specific heat capacity

Specification reference: 5.1.3

Learning outcomes

Demonstrate knowledge, understanding, and application of:

→ the specific heat capacity of a substance – $E = mc\Delta\theta$

→ an electrical experiment to determine the specific heat capacity of a metal or a liquid.

▲ **Figure 1** *Modern synthetic fluids are used to lubricate car engines and transfer force through hydraulic braking systems, yet ordinary water is used as the coolant to prevent the engine overheating*

Study tip

The term 'specific' in specific heat capacity refers to unit (1 kg) mass.

▼ **Table 1** *Some substances and their specific heat capacities*

Substance	c / J kg^{-1} K^{-1}
lead	129
silver	233
iron	449
aluminium	904
air*	1005
sodium	1230
paraffin wax	2200
water	4200
hydrogen*	14 300

**At constant pressure of 101 kPa*

The wonders of water

Not only does solid water float on liquid water, but water has another unusual property that makes it an excellent coolant for everything from car engines (Figure 1), to supercomputers and even nuclear reactors – its exceptionally high **specific heat capacity**. It can absorb a large amount of energy without a significant change in its temperature.

Specific heat capacity

The specific heat capacity of a substance is defined as the energy required per unit mass to change the temperature by 1 K (or 1°C), and has units of J kg^{-1} K^{-1}.

Water has a specific heat capacity of 4200 J kg^{-1} K^{-1}, that is, 4200 J are needed to increase the temperature of 1 kg of water by 1 K.

The specific heat capacity, c, of a substance is determined using the equation below:

$$c = \frac{E}{m \times \Delta\theta}$$

where E is the energy supplied to the substance in joules (J), m is the mass of the substance in kilograms (kg) and $\Delta\theta$ is the change in temperature of the substance. The change in temperature $\Delta\theta$ can be measured in K or °C, since both give the same numerical value for change.

This equation is normally written as

$$E = mc\Delta\theta$$

Different substances can have very different specific heat capacities. As you can see from Table 1, metals tend to have low values, and water has an exceptionally high value.

Worked example: Determining the mass of an aluminium tube

It takes 34.2 kJ to heat an aluminium tube from 20°C to 400°C. Assuming all the energy is transferred to the tube, calculate the mass of the tube.

Step 1: Select the equation and rearrange it to make mass the subject.

$E = mc\Delta\theta$ rearranged gives $m = \dfrac{E}{c\Delta\theta}$

Step 2: Substitute in known values in SI units, including a temperature change of 380°C, and calculate the mass.

$$m = \frac{3.42 \times 10^4}{904 \times 380} = 0.10\,\text{kg (2 s.f.)}$$

Determining specific heat capacity

A simple experiment using an electrical heater can be used to determine the specific heat capacity of a solid or liquid (Figures 2 and 3).

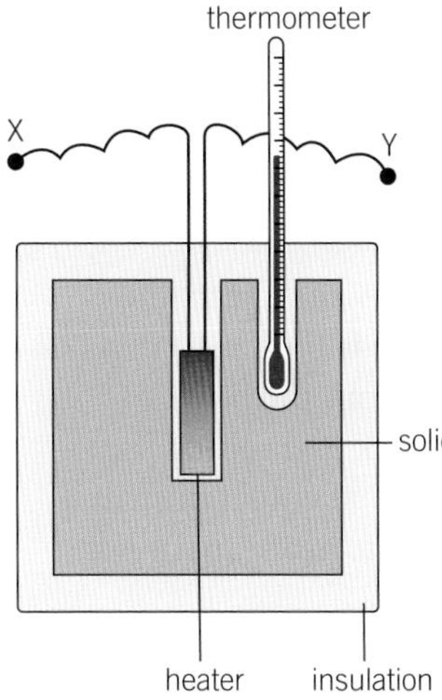

▲ **Figure 2** *When determining the specific heat capacity of a substance it is important to minimise the energy transferred from the substance to the surroundings by carefully insulating the substance*

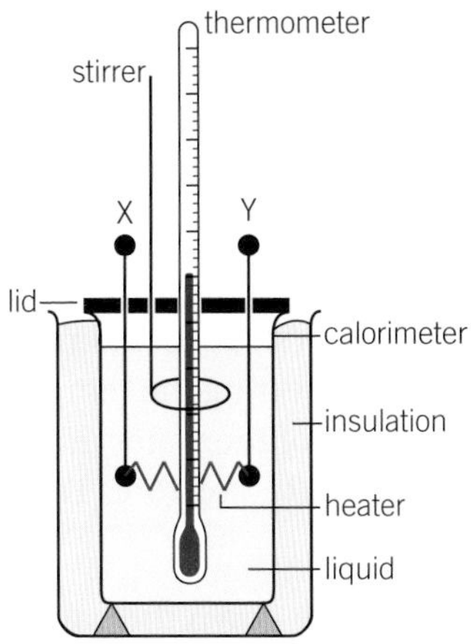

▲ **Figure 3** *When determining specific heat capacity of a liquid, the liquid must be carefully stirred to ensure it has uniform temperature throughout*

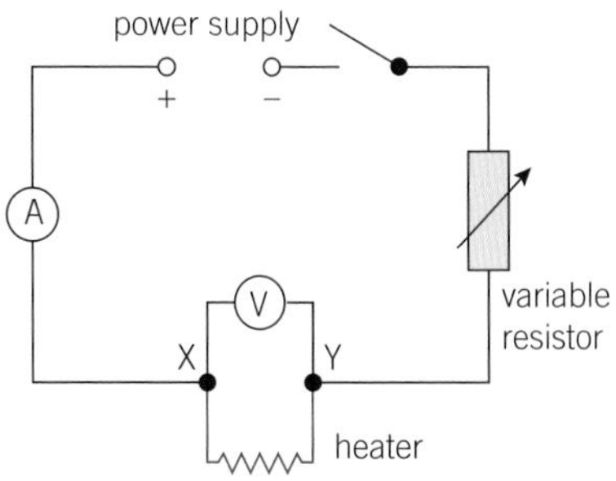

▲ **Figure 4** *Electrical circuit for the heater connected between terminals X and Y*

In both cases, the energy transferred from the heater to the substance is given by $E = IVt$, where I is the current in the heater, V is the potential difference across the heater, and t is the time taken to increase the temperature. Therefore the specific heat capacity of the substance can be determined using the equation

$$c = \frac{IVt}{m\Delta\theta}$$

Temperature–time graphs

Plotting a graph of temperature of the substance against time allows for a more accurate determination of the specific heat capacity (Figure 5).

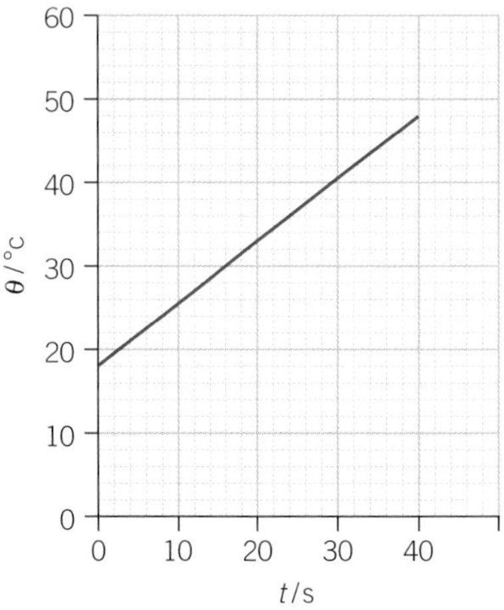

▲ **Figure 5** *A graph of temperature against time for a substance heated at a constant rate shows a linear relationship*

For a time Δt, the equation $E = mc\Delta\theta$ can be written as:

$$\frac{E}{\Delta t} = mc\frac{\Delta\theta}{\Delta t}$$

In Figure 5, $\frac{\Delta\theta}{\Delta t}$ is the gradient of the graph. $\frac{E}{\Delta t}$ is the constant power supplied, P, giving

$$P = mc\frac{\Delta\theta}{\Delta t} = mc \times \text{gradient}$$

Therefore the specific heat capacity of a substance can be determined using

$$c = \frac{P}{m \times \text{gradient}}$$

Method of mixtures

The method of mixtures is another way to determine specific heat capacity. Known masses of two substances at different temperatures are mixed together. Recording their final temperature at thermal equilibrium allows the specific heat capacity of one of the substances to be determined if the specific heat capacity of the other is known.

Worked example: Method of mixtures

A metal block of mass 100 g is heated in boiling water, reaching thermal equilibrium with the water at 100°C. It is then placed in 200 g of water at 20°C. Thermal energy flows from the block to the water, lowering the temperature of the block and raising the temperature of the water until they reach thermal equilibrium at a temperature θ_{final} of 26°C. Determine the specific heat capacity of the metal.

Step 1: Write down the values you know and select the equation you need to calculate the amount of energy transferred.

$m_{metal} = 0.100\,\text{kg}$, $\Delta\theta_{metal} = 74\,\text{K}$, $m_{water} = 0.200\,\text{kg}$, $\Delta\theta_{water} = 6\,\text{K}$

$E = mc\Delta\theta$

Step 2: Since thermal energy is transferred from the block to the water, energy transferred from metal block = energy transferred to water

$$m_{metal}c_{metal}\Delta\theta_{metal} = m_{water}c_{water}\Delta\theta_{water}$$

Step 3: Rearrange the equation and calculate c_{metal}.

$$c_{metal} = \frac{m_{water}c_{water}\Delta\theta_{water}}{m_{metal}\Delta\theta_{metal}} = \frac{0.200 \times 4200 \times 6}{0.100 \times 74} = 680\,\text{J kg}^{-1}\,\text{K}^{-1} \text{ (2 s.f.)}$$

Constant-volume-flow heating

Constant-volume-flow heating is a technique used to heat a fluid passing over a heated filament. It is used to heat water in some showers and dishwashers and to transfer energy away from heat sources like car engines or nuclear reactors.

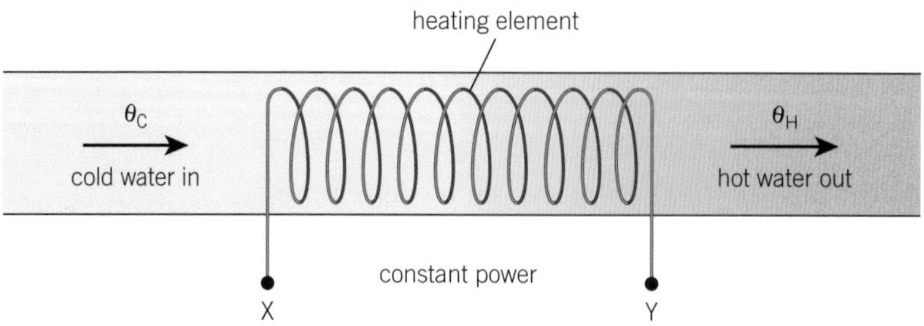

▲ **Figure 6** *The energy is supplied at a constant rate as the fluid passes over the heating element at a constant flow rate*

Because liquids are incompressible, a given volume of liquid in a pipe is equivalent to a given mass. The flow rate can therefore be regarded as the mass flowing through the pipe and passing over the heating element per unit time, in $kg\,s^{-1}$.

For constant-volume-flow heating, $E = mc\Delta\theta$ becomes

$$\frac{E}{\Delta t} = \frac{\Delta m}{\Delta t} c\Delta\theta$$

1 Calculate the power of an industrial heater with a flow rate of $1.20\,kg\,s^{-1}$ that heats water from 10°C to 80°C.

2 Calculate the energy supplied from a power shower with a flow rate of $0.050\,kg\,s^{-1}$, heating water from 20°C to 60°C, used for 15 minutes.

3 Energy is transferred from a small water-cooled nuclear reactor at a rate of 250 MW. Assuming all the energy is transferred to the water as thermal energy and the water increases in temperature by 80°C, calculate the diameter of the pipe needed to transfer water from the reactor with a maximum velocity of $3.0\,m\,s^{-1}$ (density of water = $1000\,kg\,m^{-3}$).

Summary questions

1 Calculate the energy required to raise the temperature of the following substances by 20°C.
a 1.0 kg of water **b** 600 g of aluminium **c** 4.2 μg of lead. *(4 marks)*

2 Describe an experiment that can be used to determine the specific heat capacity of a block of metal using an electrical heater, stating all the measurements that need to be taken. *(7 marks)*

3 This question is about a waterfall. Consider a 1 kg mass of water falling through a vertical drop of 450 m. Assuming all the energy is converted into thermal energy, calculate the difference in temperature between water at the top and bottom of the waterfall. *(3 marks)*

4 A 500 g mass of metal is heated using an electrical heater. The current in the heater and the potential difference across it are 2.0 A and 12 V. After 5.0 minutes the temperature of the metal has risen by 32°C. Calculate the specific heat capacity of the metal and identify the metal from Table 1. *(3 marks)*

5 A 60 W heater is used to heat a substance of mass 30 g. The graph in Figure 5 shows the change in temperature of the substance against time. Use the graph to determine the specific heat capacity of the substance. *(5 marks)*

6 A car of mass 1500 kg has two disc brakes of mass 8.0 kg. The material of the disc has a specific heat capacity of $500\,J\,kg^{-1}\,K^{-1}$. Assuming the kinetic energy of the car is transferred into thermal energy in the discs, calculate the increase in temperature of the brake discs when the car quickly decelerates from $20\,m\,s^{-1}$ to rest. *(3 marks)*

14.5 Specific latent heat

Specification reference: 5.1.3

Learning outcomes

Demonstrate knowledge, understanding, and application of:

→ specific latent heat of fusion and specific latent heat of vaporisation – $E = mL$

→ an electrical experiment to determine the specific latent heat of fusion and vaporisation.

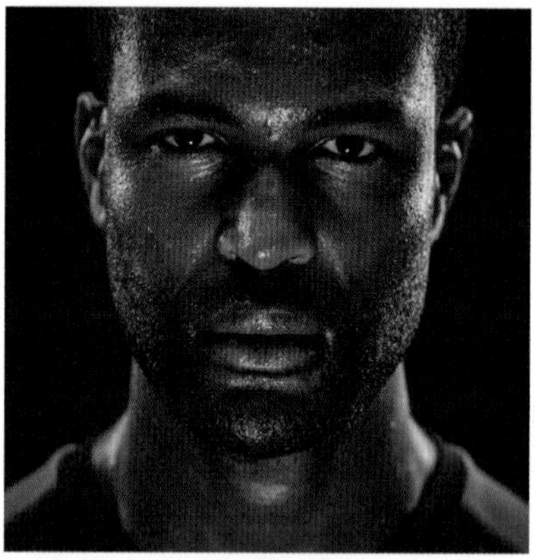

▲ **Figure 1** *Most mammals sweat, but only humans, other primates, and horses produce large volumes of sweat to keep cool (other mammals, like dogs, control their temperature by panting)*

Study tip

As with specific heat capacity, the term 'specific' refers to unit mass, 1 kg, and the term 'latent' comes from the Latin for hidden – although energy is being transferred, the temperature of the substance does not change. The energy is 'hidden' while it is changing phase.

Keeping cool

Humans sweat to keep cool. When sweat evaporates it requires energy to change from liquid to gas, so energy transfers from the skin to the sweat, cooling the skin and preventing us from getting dangerously hot.

The amount of energy needed to turn 1 kg of liquid into gas depends on a property of the liquid called the **specific latent heat**. This varies from substance to substance.

Specific latent heat

The specific latent heat of a substance, L, is defined as the energy required to change the phase per unit mass while at constant temperature. Therefore

$$L = \frac{E}{m}$$

where E is energy supplied to change the phase of mass m of the substance. There are two forms for the specific latent heat of a substance depending on the phase change.

- When the substance changes from solid to liquid phase we refer to the **specific latent heat of fusion**, L_f.
- When the substance changes from liquid to gas, we refer to the **specific latent heat of vaporisation**, L_v.

$$E = mL_f \qquad E = mL_v$$

Water has a specific latent heat of vaporisation of $2.26 \times 10^6\,\mathrm{J\,kg^{-1}}$, so it takes $2.26 \times 10^6\,\mathrm{J}$ to change 1 kg of liquid water into water vapour at constant temperature of 100 °C.

The specific latent heat of fusion L_f

When a substance is at its melting point it requires energy to change phase from solid to liquid. The energy transferred to the substance increases the internal energy of the substance without increasing its temperature.

To determine the specific latent heat of fusion, a heating circuit can be used, like the one used to determine the specific heat capacity of a substance in Topic 14.4 (Figure 4). A thermometer should be used to ensure the ice is at its melting point, not at a lower temperature, and the ice should be seen to be just starting to melt before the heater is switched on.

By measuring the potential difference V across the heater, the current I in the heater, and the time t during which the heater is used, the energy transferred to the ice can be determined using

$$E = IVt$$

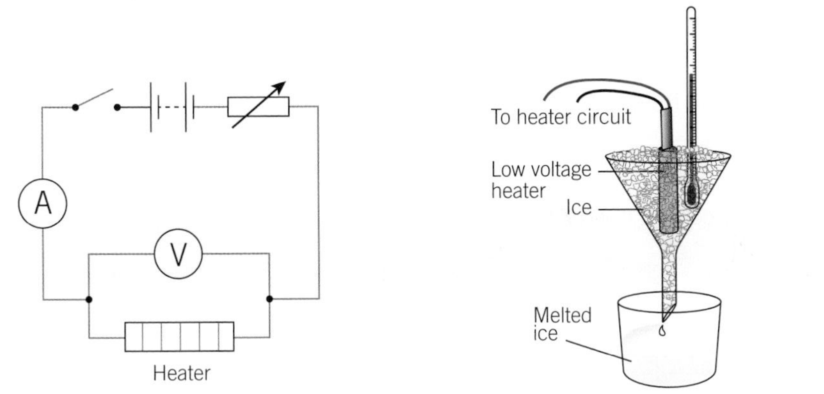

a *Heater circuit* *b* *Collecting melting ice*

▲ **Figure 2** *Determining the specific latent heat of fusion of water using an electrical heater*

It is important to accurately measure the mass m of the substance (the ice in this example) that changes phase from solid to liquid. The specific latent heat of fusion can then be determined using

$$L_f = \frac{IVt}{m}$$

The specific latent heat of vaporisation L_v

The energy required to change 1 kg of substance from its liquid phase to its gaseous phase at its boiling point is often considerably more than its specific latent heat of fusion, because there is a much larger difference between the internal energy of a gas and a liquid than between a liquid and solid. Consequently L_v is greater than L_f for most substances.

To determine L_v an electrical heater can be used with a condenser to collect and then measure the mass of liquid that changes phase (Figure 3).

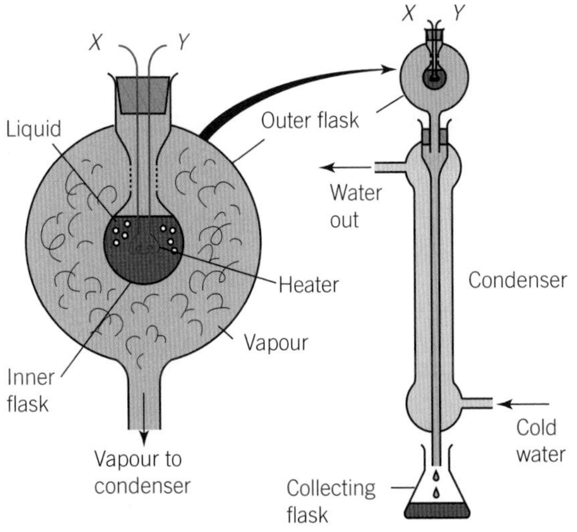

▲ **Figure 3** *Determining the specific latent heat of vaporisation requires a more complex arrangement to accurately measure the mass of liquid that changes phase*

As with the specific latent heat of fusion, the specific latent heat of vaporisation can be found using

$$L_v = \frac{IVt}{m}$$

where m is the mass of the substance that changed phase during heating.

Values for L_f and L_v

Table 1 lists values of L_f and L_v for various substances, along with their melting and boiling points at standard atmospheric pressure (101 kPa).

▼ **Table 1** *Some values for specific latent heat of fusion (L_f) and specific latent heat of vaporisation (L_v)*

	Melting point / °C	L_f / J kg^{-1}	Boiling point / °C	L_v / J kg^{-1}
water	0	3.30×10^5	100	2.26×10^6
lead	327	2.30×10^4	1750	8.71×10^5
aluminium	660	3.98×10^5	2450	1.14×10^7
silver	960	8.80×10^4	2190	2.33×10^6

In all cases it is important to remember that change of phase can occur the other way. When 1 kg of water freezes it transfers 330 000 J to its surroundings, as water in its solid phase has less internal energy than when in its liquid phase.

Combining specific latent heat and specific heat capacity

By considering the specific heat capacity and specific latent heat of a substance it is possible to determine the total energy required to heat and then change the phase of a substance. The graph in Figure 4 shows temperature plotted against time for a solid heated until it has completely turned into gas.

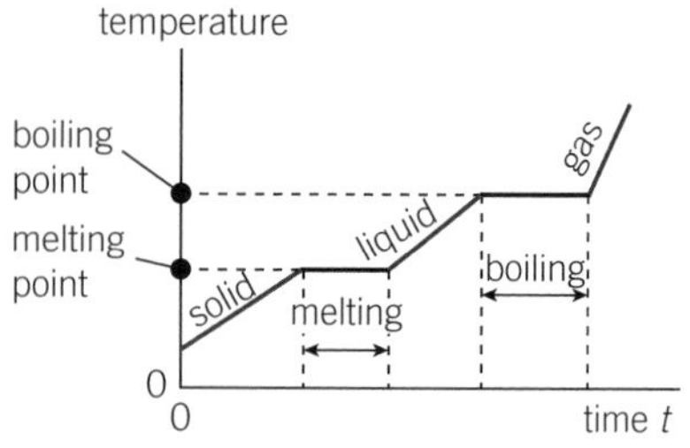

▲ **Figure 4** *A graph of temperature against time for a heated substance has several distinct sections*

This graph has four distinct sections before turning into a gas. The energy transferred to the substance in each section can be calculated using either the specific heat capacity or specific latent heat equation:

1 heating the solid to its melting point, $E = mc_{solid}\Delta\theta$
2 melting the solid at constant temperature, $E = mL_f$
3 heating the liquid to its boiling point, $E = mc_{liquid}\Delta\theta$
4 boiling the liquid at constant temperature, $E = mL_v$.

The total energy can then be determined by adding up the energy transferred in each section. This is illustrated in the worked example below.

Worked example: Turning ice into water vapour

Calculate the energy required to boil 2.0 kg ice initially at −40°C, and determine the percentage of this energy required to change the phase of the water from liquid to gas. The specific heat capacity of ice, c_{ice}, is 2.0×10^3 J kg^{-1} K^{-1}.

Step 1: By considering the energy transfers we can determine that the total energy.

total energy required = energy required to heat ice from −40°C to 0°C + energy required to melt ice + energy required to heat water from 0°C to 100°C + energy required to boil water

Step 2: Select the appropriate equations for each sections.

$E = m_{ice} c_{ice} \Delta\theta_{(-40\rightarrow 0)} + m_{ice} L_{f\,ice} + m_{water} c_{water} \Delta\theta_{(0\rightarrow 100)} + m_{water} L_{v\,water}$.

Substituting in known values and calculating the total energy required gives

$E = (2.0 \times 2.0 \times 10^3 \times 40) + (2.0 \times 3.30 \times 10^5) + (2.0 \times 4200 \times 100) + (2.0 \times 2.26 \times 10^6)$

$E = 6.18$ MJ

The energy required to completely boil the water $E = 2.0 \times 2.26 \times 10^6 = 4.52$ MJ.

Step 3: Convert the latent heat of vaporisation into a percentage of the total.

$\frac{4.52 \times 10^6}{6.18 \times 10^6} \times 100 = 73\%$ (2 s.f.)

About three-quarters of the energy was required to change the liquid into gas.

Summary questions

1 Calculate the energy required to change 2.5 kg of silver at its melting point from solid to liquid. *(2 marks)*

2 Describe why the specific latent heat of vaporisation is normally greater than the specific latent heat of fusion for a particular substance. *(1 mark)*

3 Calculate the energy transferred to the surroundings when 50 g of aluminium changes phase from liquid to solid. *(2 marks)*

4 A 24 W electrical heater is used to melt solid water already at its melting point. If the heater is left running for 20 minutes, calculate the mass of ice melted in that time. *(3 marks)*

5 The temperature–time graph in Figure 5 was obtained by heating a small piece of metal of mass 60 g. The specific heat capacity of the metal is 904 J kg^{-1} K^{-1} and the specific latent heat of fusion is 398 kJ kg^{-1}. Calculate:

a the rate of energy transfer to the metal from the heater *(3 marks)*

b the energy required to melt the metal. *(2 marks)*

6 A small lead bullet of mass 8.0 g travels at 400 m s^{-1}. The bullet strikes a concrete wall and melts on impact. Assuming the bullet is at 40°C on impact and that all the kinetic energy of the bullet is used to heat and then melt the bullet, calculate the temperature of the molten lead left on the wall. The specific heat capacity of lead is 129 J kg^{-1} K^{-1} (assume this is unchanged for molten lead). *(6 marks)*

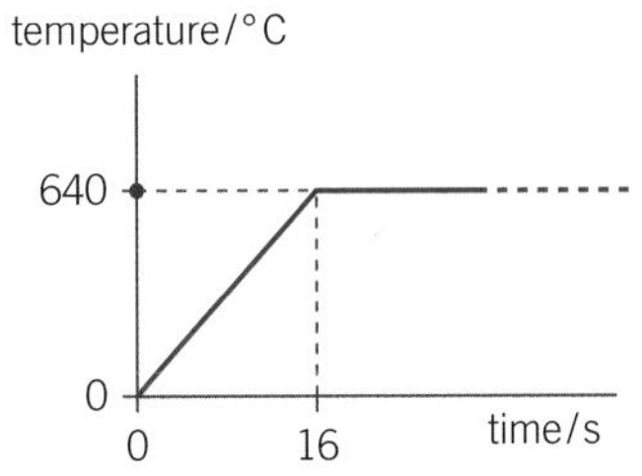

▲ **Figure 5** *A graph of temperature against time obtained when heating a small piece of metal*

Practice questions

1 a (i) Define *specific heat capacity.* (*1 mark*)

(ii) Describe the difference between the *latent heat of fusion* and the *latent heat of vaporisation.* (*1 mark*)

b The graph in Figure 1 shows the variation of temperature with time for a fixed mass of substance when heated by a constant power source. At **A** the substance is a solid; at **E** the substance is a vapour.

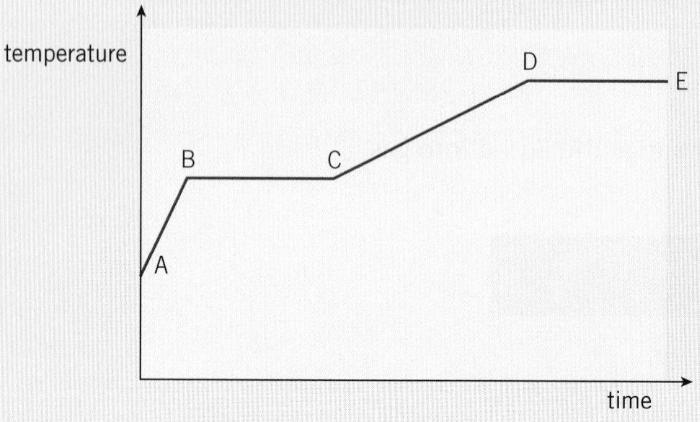

▲ Figure 1

(i) Describe the changes taking place in the kinetic energy and potential energy of the molecules for the following sections:

A to **B**

B to **C** (*2 marks*)

(ii) State and explain what you can conclude from Figure 1 about the specific heat capacity of the substance in the solid state compared with the specific heat capacity of the substance in the liquid state. (*2 marks*)

c The electric heating element of a bathroom shower has a power rating of 5.0 kW. An attempt is made to test the accuracy of this value by measuring the rate of flow of the water and the temperature of the water before and after passing the element.

The results of the test and other required data are as follows:

Temperature of water supply to the shower = 17.4 °C
Temperature of water after being heated by the element = 36.7 °C
Rate of flow of water = $3.60 \times 10^{-3}\,\mathrm{m^3\,min^{-1}}$
Density of water = $1000\,\mathrm{kg\,m^{-3}}$
Specific heat capacity of water = $4200\,\mathrm{J\,kg^{-1}\,K^{-1}}$

(i) Show that the power of the heating element is approximately 5 kW. (*4 marks*)

(ii) State and explain a possible source of uncertainty that might affect the reliability of the test. (*2 marks*)

Jun 2014 G484

2 A room measures 4.5 m × 4.0 m × 2.4 m. The air in the room is heated by a gas-powered heater from 12 °C to 21 °C. The density of the air, assumed to remain constant, is $1.3\,\mathrm{kg\,m^{-3}}$.

a Calculate the thermal energy required to raise the temperature of the air in the room. The specific heat capacity of air is $990\,\mathrm{J\,kg^{-1}\,K^{-1}}$. (*3 marks*)

b The heater has an output power of 2.3 kW. The heating gas has a density $0.72\,\mathrm{kg\,m^{-3}}$. Each cubic metre of heating gas provides 39 MJ of thermal energy.

Use your answer to (**a**) to calculate

(i) The time required to raise the temperature of the air from 12 °C to 21 °C, (*2 marks*)

(ii) The mass of heating gas used in this time. (*2 marks*)

c Suggest **two** reasons why the time required and the mass of heating gas will in practice be greater than the values calculated in (**b**). (*2 marks*)

Jun 2013 G484

3 a Show that the unit for specific heat capacity is $\mathrm{J\,kg^{-1}\,K^{-1}}$. (*1 mark*)

b A 5.0 kg meteorite lands on the surface of the Earth. Its initial temperature is 500 °C. It loses thermal energy at a constant rate of 120 W. The material of the meteorite has specific heat capacity of $600\,\mathrm{J\,kg^{-1}\,K^{-1}}$. Estimate its temperature after 1.0 hours. (*3 marks*)

c An electric kettle with 450 g of water is on at time $t = 0$. The initial temperature of the water is 15 °C. Figure 2 shows the variation of the temperature θ of the water with time t.

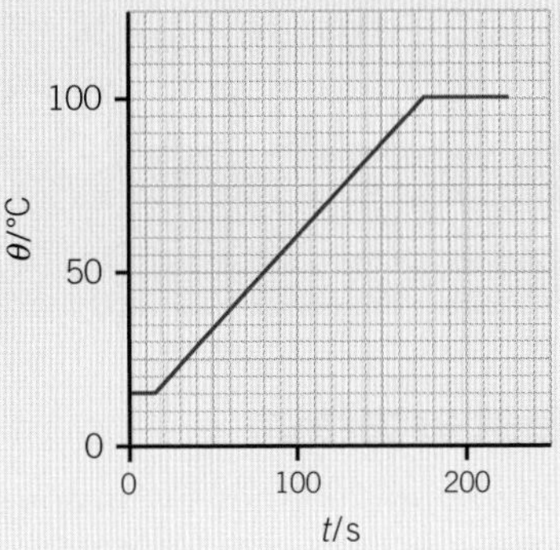

▲ Figure 2

(i) Explain the shape of the graph from $t = 0$ to 180 s. *(4 marks)*

(ii) The specific capacity of water is 4200 J kg^{-1} K^{-1}. Use Figure 3 to determine the power rating of the heating element of the kettle. *(3 marks)*

4 **a** Define *specific latent heat of vaporisation.* *(1 mark)*

b Derive the SI base unit for specific latent heat. *(3 marks)*

c A mass of 200 g of cold coffee is at temperature 18 °C.
Calculate the mass of steam at 100 °C that must be condensed into the coffee to raise its temperature to 70 °C.
specific latent heat of vaporisation of water = 2.3 MJ kg^{-1}
specific heat capacity of water or coffee = 4200 J kg^{-1} K^{-1} *(5 marks)*

5 **a** Explain what is meant by absolute zero. *(2 marks)*

b Define the internal energy of a substance. Explain how the internal energy of an ideal gas is different from that of a solid. *(3 marks)*

c Figure 3 shows a heater used to warm water travelling through an insulating pipe.

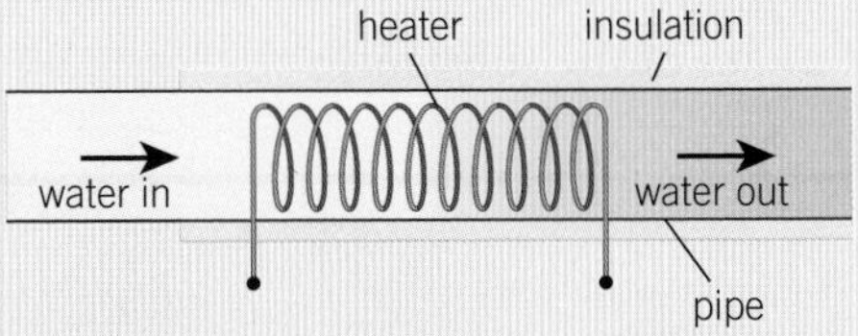

▲ Figure 3

The heater has a constant output power of 6.0 kW. The water is heated as it travels through the heater at 0.2 m s^{-1}. The internal cross-sectional area of the pipe is 4.9×10^{-4} m^2. The density of water is 1000 kg m^{-3}.

(i) Calculate the temperature of the water at which it leaves the heater. *(4 marks)*

(ii) State and explain two possible methods of increasing the temperature of the water leaving the pipe. *(2 marks)*

6 **a** Molten wax in a test tube is allowed to cool in a room. Figure 4 shows the variation of its temperature against time.

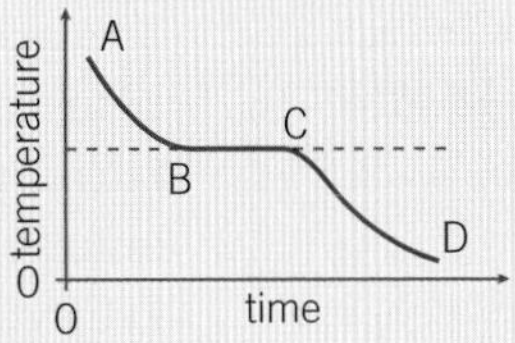

▲ Figure 4

Explain the shape of the graph shown in Figure 4 in terms of the energy of the molecules of wax. *(3 marks)*

b Figure 5 shows an arrangement used by a student to determine the specific latent heat of vaporisation of water.

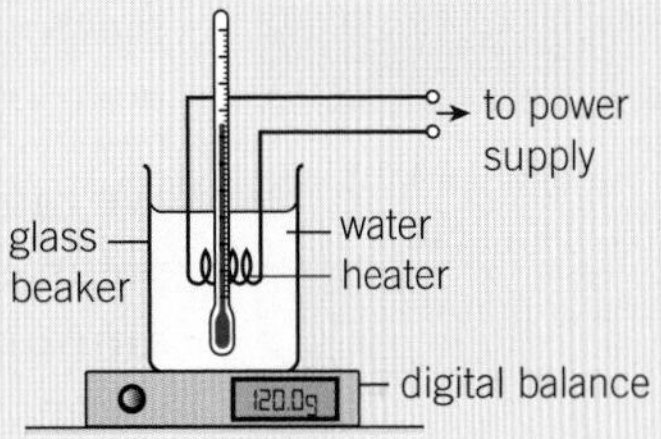

▲ Figure 5

The water in the beaker is heated by an electrical heater. The beaker is placed on a digital balance. The heater has a constant output power of 280 W. At time t = 0 the temperature of the water is 90 °C. The table below shows the results recorded by the student.

▼ Table 1

t/s	0	10.0	25.0	30.0	40.0	50.0	60.0	70.0
balance reading/g	120.0	120.0	119.5	119.0	117.7	116.8	115.8	114.8

(i) Explain why the balance reading remains constant for the first 10.0 s. *(1 mark)*

(ii) Use the table to determine the specific latent heat of vaporisation of water. *(3 marks)*

(iii) Explain why the value obtained in (ii) is larger than the data book value of 2.3×10^6 J kg^{-1}. *(1 mark)*

15 IDEAL GASES

15.1 The kinetic theory of gases

Specification reference: 5.1.4

Learning outcomes

Demonstrate knowledge, understanding, and application of:

- → amount of substance, measured in moles
- → the Avogadro constant N_A
- → the model of the kinetic theory of gases and its assumptions
- → pressure in terms of this model.

Synoptic link

You have met the kilogram along with the other SI base units in Topic 2.1, Quantities and units.

▲ **Figure 1** *An Avogadro Project sphere (in the centre of this measuring machine) is made of pure silicon and is the most spherical object ever made by humans – it is so perfectly spherical that if it were scaled up to the size of Earth, with a radius of 6370 km, its highest point would only be around 2 m above its lowest point*

Moving beyond the last artefact

The kilogram remains the only SI unit defined by means of an artefact – the international prototype kilogram, kept in a vault near Paris. Several alternative options for a universal definition are currently being explored. One of these, which is gaining favour amongst the scientific community, is led by the International Avogadro Project and aims to relate the kilogram to the mass of a particular atom (Figure 1). Using painstakingly manufactured silicon spheres, the project's workers hope to define the kilogram as the mass of a precise number of silicon atoms. This approach is already used to define another SI unit, the **mole**.

Particles and the mole

In order to understand how gases behave, not only must we study macroscopic (large-scale) properties like mass and temperature, but we must also understand what is going on at the particle level.

We can express the number of atoms or molecules in a given volume of gas using moles (mol), the SI unit of measurement for the **amount of substance**. This is a base quantity and is different from the mass of a substance. The amount of a substance indicates the number of elementary entities (normally atoms or molecules) within a given sample of substance.

One mole is defined as the amount of substance that contains as many elementary entities as there are atoms in 0.012 kg (12 g) of carbon-12. This number is called **the Avogadro constant**, N_A, and has been measured as 6.02×10^{23}.

By definition, 1 mol of any substance contains 6.02×10^{23} individual atoms or molecules. Therefore the total number of atoms or molecules in a substance, N, is given by the equation

$$N = n \times N_A$$

where n is the number of moles of the substance.

Molar mass

The **molar mass**, M, of a substance is the mass of one mole of the substance. Knowing the molar mass allows us to calculate the mass m of a sample of a substance if we know the number of moles, n, and vice versa:

$$m = n \times M$$

The molar mass of an element is simple to determine from the **nucleon number** (also called the mass number). Helium-4 has a nucleon number of 4. As a result the molar mass of helium-4 is 0.004 kg mol^{-1} (4 g mol^{-1}). Similarly, one mole of uranium-238 would have a mass of 0.238 kg (238 g).

It becomes a little more complex when dealing with molecules. Nitrogen forms N_2 molecules, that is, each molecule contains two nitrogen atoms, each with a molar mass of 0.014 kg mol^{-1}. The nucleon number of nitrogen is 14. Therefore the molar mass of nitrogen gas is 0.028 kg mol^{-1}.

A molecule of carbon dioxide (CO_2) contains one carbon atom (nucleon number 12) and two oxygen atoms (nucleon number 16). Therefore the molar mass of carbon dioxide is 0.044 kg mol^{-1} (= 0.012 + 0.016 + 0.016).

Table 1 gives the molar masses of these and some other common gases.

▼ **Table 1** *Molar masses of some common gases*

Substance	Elementary entities	Molar mass / kg mol^{-1}
hydrogen gas	H_2 molecules	0.002
helium gas	He atoms	0.004
oxygen gas	O_2 molecules	0.032
carbon dioxide gas	CO_2 molecules	0.044
neon gas	Ne atoms	0.020
argon gas	Ar atoms	0.040

1 Calculate the mass of 4.0 mol of helium gas.
2 Calculate the molar mass of methane (CH_4). The molar mass of carbon is 0.012 kg mol^{-1} and the molar mass of hydrogen is 0.001 kg mol^{-1}.
3 Calculate the number of molecules in 50 g of carbon dioxide.

The kinetic theory of gases

Studying how the atoms or molecules in a gas behave suggests basic laws relating the motion of these particles at the microscopic scale to macroscopic properties like the temperature and pressure of the gas.

The **kinetic theory of matter** is a model used to describe the behaviour of the atoms or molecules in an **ideal gas**. Real gases have complex behaviour, so in order to keep the model simple a number of assumptions are made about the atoms or molecules in an ideal gas.

Study tip

Mass represents the amount of matter in an object, measured in kg, whereas the amount of substance, measured in mol, indicates the number of elementary entities, such as atoms, ions, molecules, electrons, or other particles.

The assumptions made in the kinetic model for an ideal gas are as follows:

- The gas contains a very large number of atoms or molecules moving in random directions with random speeds.
- The atoms or molecules of the gas occupy a negligible volume compared with the volume of the gas.
- The collisions of atoms or molecules with each other and the container walls are perfectly elastic (no kinetic energy is lost).
- The time of collisions between the atoms or molecules is negligible compared to the time between the collisions.
- Electrostatic forces between atoms or molecules are negligible except during collisions.

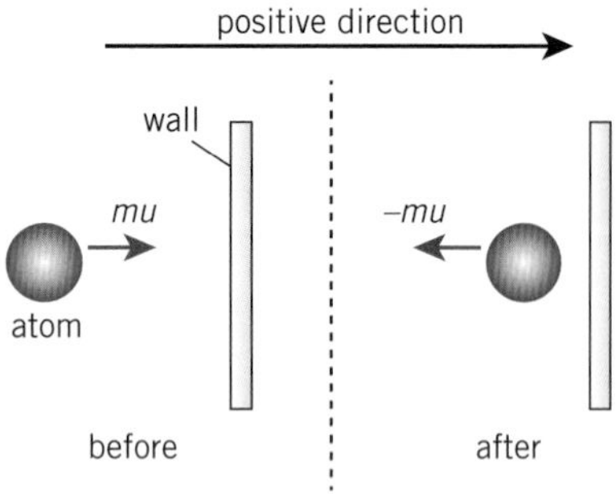

▲ **Figure 2** *The change in momentum of the atom is −2mu and not zero. Momentum is a vector quantity*

Using these assumptions and Newton's laws of motion, we can explain how the atoms or molecules in an ideal gas cause pressure.

The atoms or molecules in a gas are always moving, and when they collide with the walls of a container the container exerts a force on them, changing their momentum as they bounce off the wall.

When a single atom collides with the container wall elastically, its speed does not change, but its velocity changes from $+u\,\text{m}\,\text{s}^{-1}$ to $-u\,\text{m}\,\text{s}^{-1}$. The total change in momentum is $-2mu$ (see Figure 2).

The atom bounces between the container walls, making frequent collisions. According to Newton's second law, the force acting on the atom is $F_{\text{atom}} = \frac{\Delta p}{\Delta t}$, where $\Delta p = -2mu$ and Δt is the time between collisions with the wall. From Newton's third law, the atom also exerts an equal but opposite force on the wall.

A large number of atoms collide randomly with the walls of the container. If the total force they exert on the wall is F, then the pressure they exert on the wall is given by $p = \frac{F}{A}$, where A is the cross-sectional area of the wall.

Summary questions

1 Calculate the number of elementary entitles (atoms or molecules) in 3.0 mol of a substance. *(2 marks)*

2 Suggest why one mole of silicon has a different mass from one mole of aluminium. *(2 marks)*

3 A molecule of mass 5.3×10^{-26} kg travelling at $500\,\text{m}\,\text{s}^{-1}$ collides with a container wall. It collides at right angles to the wall. Calculate the change in the momentum of this molecule. *(2 marks)*

4 Calculate the number of moles there are in a substance containing:
  a 2.0×10^{24} molecules
  b 1.5×10^{17} atoms
  c 2.0×10^{24} molecules. *(3 marks)*

5 a The molar mass of copper is $64\,\text{g}\,\text{mol}^{-1}$ calculate the number of atoms in copper of mass 1.0 kg. *(2 marks)*
  b The molar mass of uranium is $235\,\text{g}\,\text{mol}^{-1}$. Calculate the mass of a single atom of uranium. *(2 marks)*

6 The density of lead is $11340\,\text{kg}\,\text{m}^{-3}$. Each lead atom has a mass of 3.46×10^{-25} kg. Calculate the number of moles of lead in a lead block with a volume of $0.20\,\text{m}^3$. *(4 marks)*

15.2 Gas laws

Specification reference: 5.1.4

On the rise

Weather balloons are launched into the upper atmosphere to measures changes in temperature and pressure, and air currents and atmospheric pollutants. As the balloon rises, the atmospheric pressure around it drops, causing it to expand.

The relationships between the temperature, pressure, and volume of an ideal gas can be described by a few simple **gas laws**.

Learning outcomes

Demonstrate knowledge, understanding, and application of:

- → the equation of state of an ideal gas $pV = nRT$, where n is the number of moles
- → techniques and procedures used to investigate pV = constant (Boyle's law) and $\frac{p}{T}$ = constant
- → an estimation of absolute zero using variation of gas temperature with pressure.

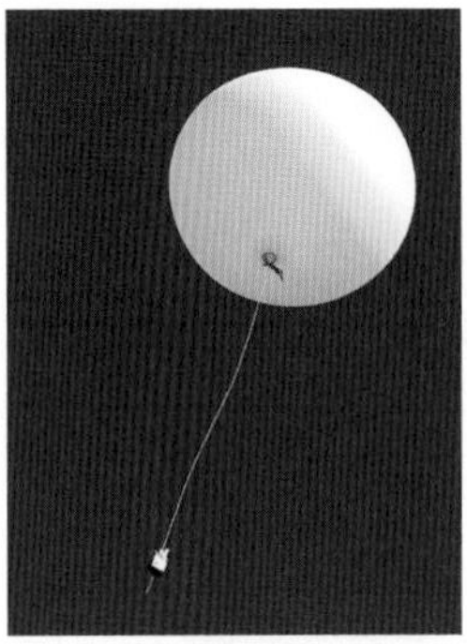

▲ **Figure 1** *Weather balloons expand as they rise, eventually bursting and parachuting back to Earth*

Pressure and Volume

If the temperature and mass of gas remain constant then the pressure p of an ideal gas is inversely proportional to its volume V. This can be expressed as

$$p \propto \frac{1}{V} \qquad \text{or} \qquad pV = \text{constant}$$

If a fixed mass of gas is kept in a sealed box, halving the volume of the box (slowly, to ensure the temperature remains constant) will compress the gas and double the pressure it exerts on the box.

Investigating Boyle's law

The relationship between the pressure of gas and its volume at a constant temperature was investigated by 1662 by Robert Boyle. He discovered the relationship $p \propto \frac{1}{V}$, which is now called **Boyle's Law**.

Boyle's experiments are simple to repeat in the classroom (Figure 2). If the pressure of a pressurised gas is slowly reduced, its volume increases. The gas must be in a sealed tube to ensure the amount of gas inside the tube remains fixed.

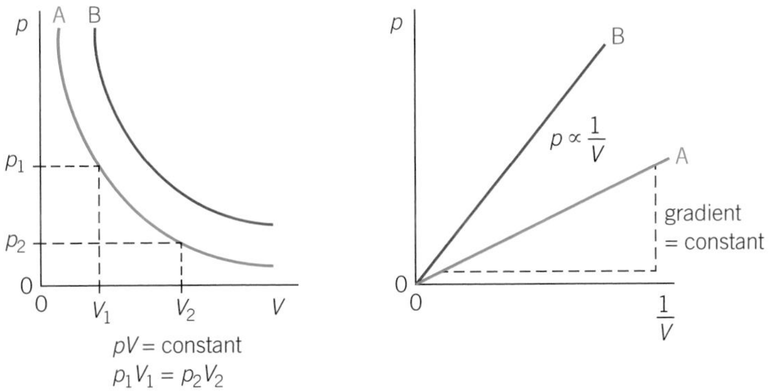

▲ **Figure 3** *Pressure–volume graphs for two gases at different temperatures – the straight line through the origin in the second graph shows that* $p \propto \frac{1}{V}$

Each line on the graph relates to a gas at a specific temperature. In this case B is at a higher temperature than A. The lines are called **isotherms** as they represent how the pressure and volume are related at one fixed temperature.

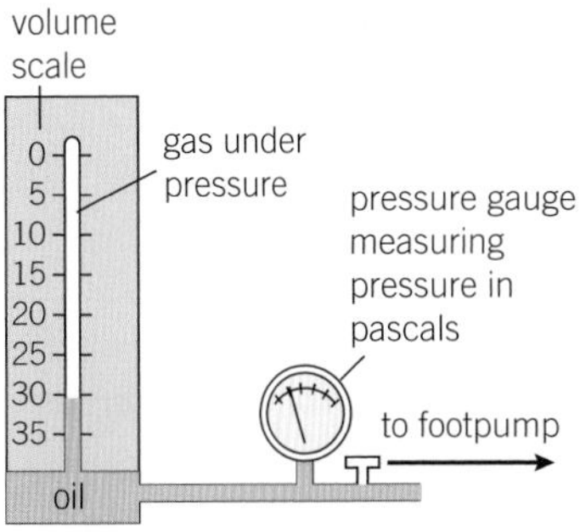

▲ **Figure 2** *Apparatus for investigating how changing the volume of a gas affects the pressure of the gas (Boyle's law)*

The graph in Figure 4 was produced in an investigation into Boyle's law.

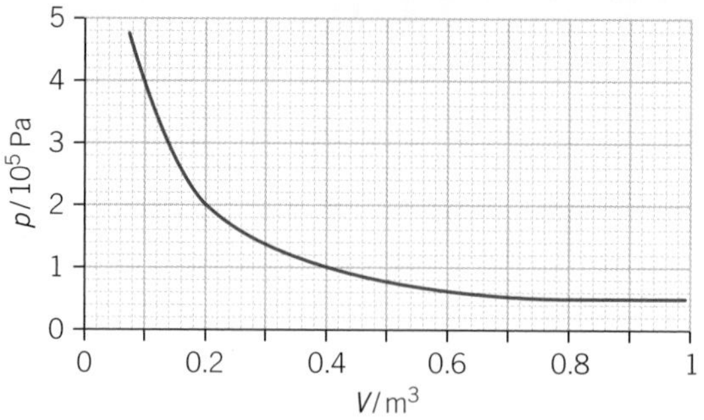

▲ **Figure 4** *A graph showing pressure against volume for a certain gas*

1 Explain why the pressure must be changed slowly.
2 Use the graph in Figure 4 to show that pV = constant.

Pressure and temperature

If the volume and mass of gas remain constant, the pressure p of an ideal gas is directly proportional to its absolute (thermodynamic) temperature T in kelvin. This relationship can be expressed as

$$p \propto T \qquad \text{or} \qquad \frac{p}{T} = \text{constant}$$

For a fixed mass of gas in a sealed container, doubling the temperature (say from 100 K to 200 K) will double the pressure the gas exerts on the container walls.

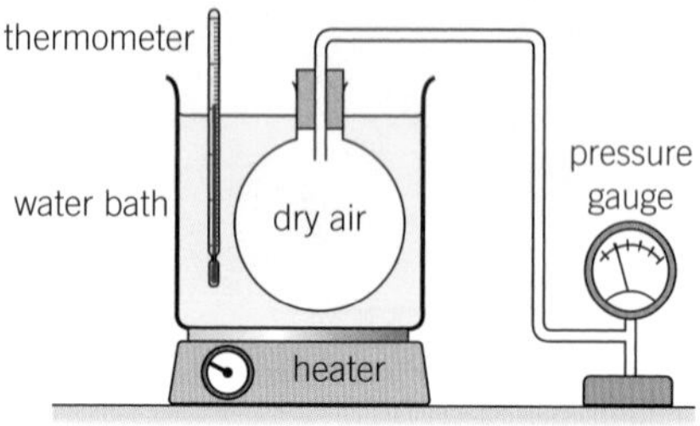

▲ **Figure 5** *Apparatus used to determine absolute zero through investigating how the temperature of a gas affects its pressure*

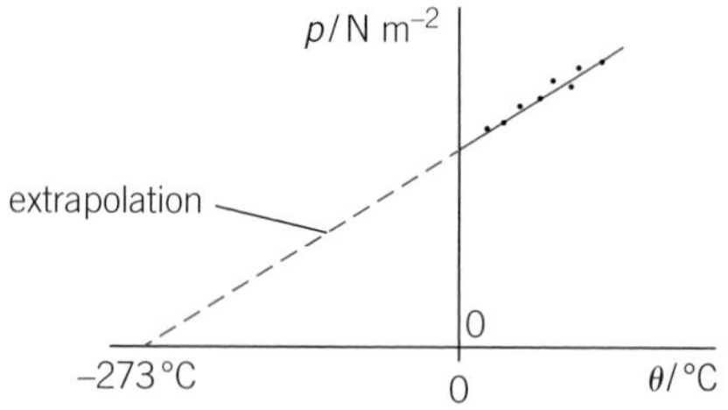

▲ **Figure 6** *A graph of pressure of gas against its temperature*

Estimating absolute zero

Because the expression above requires the absolute temperature T, an investigation into the relationship between the pressure of a fixed volume and mass of gas and its temperature can provide an approximate value for absolute zero.

With the set-up shown in Figure 5, the temperature of the water bath can be increased and the resulting increase in pressure of the gas inside the sealed vessel recorded.

At absolute zero the particles are not moving (the internal energy is at its minimum) so the pressure of the gas must be zero. Plotting a graph of pressure against temperature θ in Celsius from the experimental results gives a line that can be extrapolated back to a point where the pressure is zero (Figure 6).

1 Explain why the volume of the gas must remain fixed.

2 The data in Table 1 was collected in an investigation into the pressure of a fixed mass and volume of gas as temperature changed.

a Plot a graph of pressure p of the gas against temperature θ in °C (range of θ: −300°C to 100°C). Use your graph to determine a value for absolute zero.

b On your graph, sketch a second line to show the pattern you would expect if the experiment were repeated using the same mass of gas at a larger volume.

▼ **Table 1** *Table showing the variation of pressure with temperature for a fixed mass of gas at a constant volume*

θ / °C	p / 10^5 Pa
10	1.41
20	1.45
30	1.51
40	1.57
50	1.61
60	1.66
70	1.70

Combining the gas laws

By combing the two previously described gas laws we can show that for an ideal gas

$$\frac{pV}{T} = \text{constant}$$

If the conditions are changing from an initial state to a final state this can be written as

$$\frac{p_{\text{initial}} V_{\text{initial}}}{T_{\text{initial}}} = \frac{p_{\text{final}} V_{\text{final}}}{T_{\text{final}}} \quad \text{or simply} \quad \frac{p_1 V_1}{T_1} = \frac{p_2 V_2}{T_2}$$

Worked example: Volume of a weather balloon

A weather balloon with a volume of $2.0\,\text{m}^3$ is launched on a day when the atmospheric pressure is 101 kPa and the temperature is 20°C at ground level. It rises to a level where the air pressure is 20% of the pressure on the ground and the air temperature is −15°C. Calculate the volume of the balloon at this altitude.

Step 1: First convert the temperatures into kelvin.

20°C = 293 K and −15°C = 258 K

Step 2: Select the equation you need and rearrange it to find the final volume.

$$\frac{p_1 V_1}{T_1} = \frac{p_2 V_2}{T_2}$$

$$V_2 = \frac{p_1 V_1 T_2}{p_2 T_1}$$

Substituting in known values in SI units gives

$$V_2 = \frac{1.01 \times 10^5 \times 2.0 \times 258}{0.2 \times 1.01 \times 10^5 \times 293} = 8.8\,\text{m}^3 \text{ (2 s.f.)}$$

Study tip

Remember, temperatures must be stated in kelvin when using ideal gas equations.

The equation of state of an ideal gas

For one mole of an ideal gas, the constant in the combined relationship above is called the **molar gas constant**, *R*, and is equal to $8.31\,\text{J}\,\text{K}^{-1}\,\text{mol}^{-1}$. For n moles of gas the equation becomes

$$\frac{pV}{T} = nR \qquad \text{or} \qquad pV = nRT$$

This relationship is called the **equation of state of an ideal gas**. The molar gas constant is the same for all gases, as long as we can treat them as being ideal, so the equation above can be applied to trapped air in the laboratory or to helium in the atmosphere of distant stars.

Worked example: A pressurised container

A $3.50\,m^3$ pressurised container contains 425 moles of gas at 25.0°C. Calculate the pressure of the gas inside the container.

Step 1: Convert the temperature into kelvin.

25.0°C = 298 K

Step 2: Select the equation you need and rearrange it to make the pressure the subject.

$pV = nRT$, hence $p = \frac{nRT}{V}$

Substitute in known values and calculating the pressure of the gas inside the container.

$$p = \frac{425 \times 8.31 \times 298}{3.50} = 3.00 \times 10^5\,\text{Pa (3 s.f.)}$$

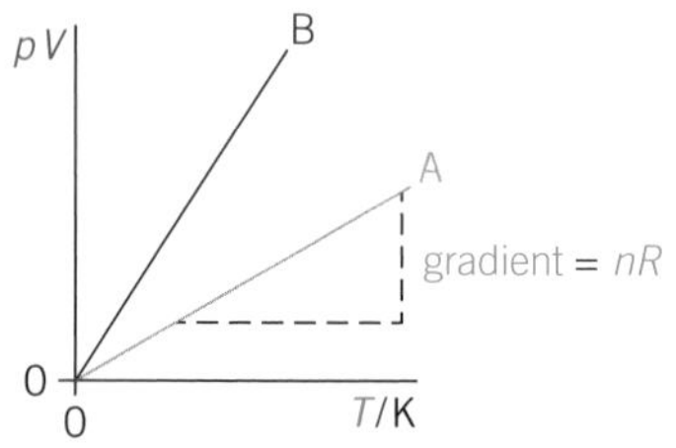

▲ **Figure 7** *A graph of pV against T for a fixed amount of gas produces a straight line through the origin. Gas B produces a steeper line than gas A, as gas B contains a greater number of moles than A.*

Graphical analysis

A graph of pV against T for a fixed amount of gas produces a straight line through the origin ($pV \propto T$). By considering the general equation of a straight line, $y = mx + c$, and the equation state of an ideal gas, $pV = nRT$, we can see the gradient of the graph is equal to nR. The greater the number of moles of gas, the steeper the line becomes.

Summary questions

1 A sealed container contains 60 moles of gas at temperature of 250 K and a pressure of 60 000 Pa. Calculate the volume of the container. *(2 marks)*

2 State the effect on the pressure of a fixed mass of gas at constant temperature if the volume of gas is:

a doubled;

b reduced to a third of its original value. *(2 marks)*

3 A fixed mass and volume of gas initially at a temperature of 20°C and pressure of 300 kPa is heated to 100°C. Calculate the change in pressure. *(4 marks)*

4 Using the values from Figure 4, plot a graph of p against $\frac{1}{V}$.
Use this graph to determine the number of moles of gas used in the experiment. The temperature of the gas during the experiment was a constant 20 °C. *(4 marks)*

5 Standard conditions for temperature and pressure (STP) are 0°C and 100 kPa. Calculate the volume occupied by 1 mol of air at STP. *(3 marks)*

6 Calculate the number of particles in a gas sample if, when the sample is in a sealed container of volume $0.25\,m^3$ at a temperature of 15°C, the pressure inside the container is 50 kPa. *(4 marks)*

7 Use the equation of state of an ideal gas to estimate the amount of air in your lungs. *(4 marks)*

15.3 Root mean square speed

Specification reference: 5.1.4

What happens when average velocity = 0 m s^{-1}?

We have already seen how the particles (atoms or molecules) in a gas move in random directions at different speeds. If we calculated the average velocity of the particles in a gas, because velocity is a vector the average would be 0 m s^{-1}. All the velocities of such a large number of particles would simply cancel out. So in order to describe the typical motion of particles inside the gas, we use a different measure, the **root mean square speed** (r.m.s. speed).

r.m.s. speed

In order to determine the r.m.s. speed, the velocity, c, of each atom or molecule in the gas is squared, c^2. Then the average of this squared velocity is found for all the gas particles, giving $\overline{c^2}$ – the bar is a symbol for 'mean'. This is the **mean square speed** of the gas particles. Finally the square root of this value is taken to give the r.m.s. speed, written as $\sqrt{\overline{c^2}}$ or $c_{r.m.s}$.

Worked example: Average speeds

A very small sample of gas contains just four molecules moving in one line. Their velocities in m s^{-1} are: −450, −50, 100, 400. Calculate the mean velocity, the mean speed $\bar{c}$, and the r.m.s. speed.

Step 1: For the mean velocity, you must take account of the signs of the velocities, because they are vectors.

$$\text{mean velocity} = \frac{(-450 - 50 + 100 + 400)}{4} = 0\,\text{m s}^{-1}$$

Step 2: Speed is a scalar, so mean speed $\bar{c}$ is calculated by ignoring the negative signs.

$$\bar{c} = \frac{(450 + 50 + 100 + 400)}{4} = 250\,\text{m s}^{-1}.$$

Step 3: To determine the r.m.s. speed, first square the speeds, then determine the mean.

$$\text{mean square speed} = \frac{(202\,500 + 2500 + 10\,000 + 160\,000)}{4} = 93\,750\,\text{m}^2\,\text{s}^{-2}$$

$$c_{r.m.s.} = \sqrt{93\,750} = 310\,\text{m s}^{-1}\ (2\ \text{s.f.}).$$

The average speed $\bar{c}$ is not the same as the r.m.s. speed.

Learning outcomes

Demonstrate knowledge, understanding, and application of:

- → the equation $pV = \frac{1}{3}Nm\overline{c^2}$ relating the number of particles and the mean square speed
- → root mean square speed and mean square speed.

Pressure at the microscopic level

The reason for our interest in r.m.s. speed is that it appears in the equation for the pressure and volume of a gas,

$$pV = \frac{1}{3}Nm\overline{c^2}$$

where p is the pressure exerted by the gas, V is the volume of the gas, N is the number of particles in the gas, m is the mass of each particle and $\overline{c^2}$ is the mean square speed of the particles.

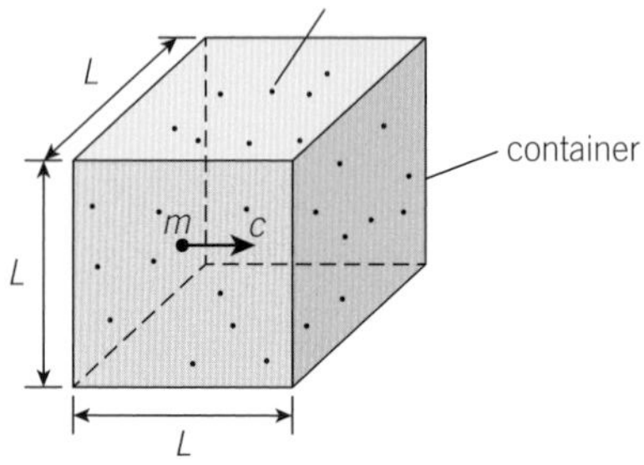

▲ **Figure 1** *Gas particles (atoms or molecules) in a container*

Derivation of $pV = \frac{1}{3}Nm\overline{c^2}$

The equation $pV = \frac{1}{3}Nm\overline{c^2}$ can be derived by considering how the movement of atoms or molecules of gas inside a sealed box gives rise to pressure.

Consider a single gas particle (atom or molecule) making repeated collisions with a container wall. The container is a cube with sides L. The gas particle has mass m and velocity c. It hits the surface of the wall at right angles.

The elastic collision results in a change in momentum of magnitude $2mc$ (see Topic 15.1, The kinetic theory of gases). The time t between collisions is the total distance covered by the particle divided by its speed. Therefore $t = \frac{2L}{c}$. According to Newton's second and third laws, the force exerted by the particle on the wall is:

$$\text{force} = \frac{\Delta p}{\Delta t} = 2mc \times \frac{c}{2L} = \frac{mc^2}{L}$$

If there are N particles in the container moving randomly, the average force exerted by each particle must be $\frac{m\overline{c^2}}{L}$, where $\overline{c^2}$ is the mean square speed of the particles.

On average, because of the random motion of the gas particles, about $\frac{1}{3}$ of the particles will be moving between two opposite faces of the container. Consequently the total force on one container wall of cross-sectional area L^2 due to collisions from all of the particles must be

$$\text{force} = \frac{m\overline{c^2}}{L} \times N \times \frac{1}{3} = \frac{Nm\overline{c^2}}{3L}$$

Finally, the pressure p exerted by the gas must equal to the total force exerted by all the particles divided by the cross-sectional area of the wall. Therefore

$$p = \frac{Nm\overline{c^2}}{3L} \times \frac{1}{L^2} = \frac{Nm\overline{c^2}}{3L^3} = \frac{Nm\overline{c^2}}{3V}$$

where V is the volume of the container. Therefore $pV = \frac{1}{3}Nm\overline{c^2}$.

1 Explain why, when considering the large number of particles in a sample of gas, it is a fair assumption that there must be about $\frac{1}{3}$ of the particles moving between two opposite faces of the container.
2 State the other ideal gas assumptions required for this derivation.

Distribution of particle speeds at different temperatures

The r.m.s. speed provides a useful way to describe the motion of the particles in a gas, but it is important to remember that it is an average. At any temperature, the random motion of the particles means that some are travelling very fast, whilst others are barely moving. The range of speeds of the particles in a gas at a given temperature is known as the **Maxwell–Boltzmann distribution**, shown in Figure 2.

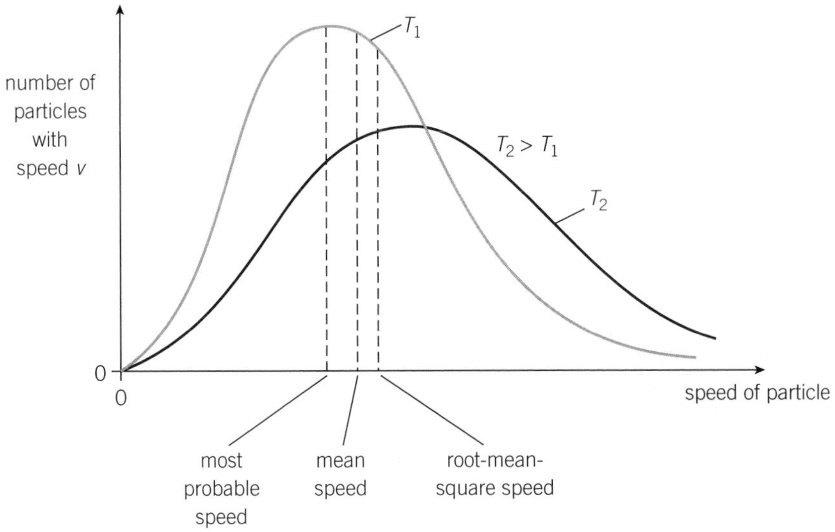

▲ **Figure 2** *The spread of speeds of particles in a gas is called the Maxwell–Boltzmann distribution, and is broader at the high temperature T_2 than at the low temperature T_1*

Changing the temperature of the gas changes the distribution. The hotter the gas becomes, the greater the range of speeds. The most common (modal) speed and the r.m.s. speed increase, and the distribution becomes more spread out.

Summary questions

1 Calculate the mean speed $\bar{c}$, mean squared speed $\overline{c^2}$, and r.m.s. speed $c_{r.m.s.}$ of a small group of atoms with the following velocities: $+100\,\mathrm{m\,s^{-1}}$, $-200\,\mathrm{m\,s^{-1}}$, $+150\,\mathrm{m\,s^{-1}}$, $-50\,\mathrm{m\,s^{-1}}$. *(3 marks)*

2 Describe how the speeds of the particles in a gas change as the temperature of the gas increases. *(2 marks)*

3 A gas cylinder contains nitrogen at a pressure of 800 kPa. The cylinder contains 4.0×10^{25} molecules and each molecule has a mass of 4.7×10^{-26} kg. The r.m.s. speed of the molecules is $450\,\mathrm{m\,s^{-1}}$. Calculate the volume of the cylinder. *(3 marks)*

4 Calculate the pressure inside the cylinder in question 3 if the r.m.s. speed of the molecules inside the cylinder increases to $600\,\mathrm{m\,s^{-1}}$. *(3 marks)*

5 One mole of oxygen has a mass of 0.032 kg. An oxygen cylinder has a volume of $0.020\,\mathrm{m^3}$ at a pressure of 140 kPa. It contains 2.0 moles of oxygen. Calculate:
 a the number of molecules inside the cylinder
 b the mass of each molecule
 c the r.m.s. speed of the oxygen molecules in the cylinder. *(6 marks)*

15.4 The Boltzmann constant

Specification reference: 5.1.4

Learning outcomes

Demonstrate knowledge, understanding, and application of:

→ the Boltzmann constant, $k = \frac{R}{N_A}$

→ $pV = NkT, \frac{1}{2}m\overline{c^2} = \frac{3}{2}kT$

→ internal energy of an ideal gas.

Where is all the helium?

Helium is the second most abundant element in the universe. It makes up about 24% of the known mass of the universe, yet on Earth it is exceptionally rare, making up just 0.0005% of our atmosphere. Where has it all gone?

At the temperatures experienced on Earth, and especially the high temperatures soon after Earth's formation, individual helium atoms can reach high enough speeds to escape the Earth's gravitational pull and fly into space. Luckily for life on Earth, nitrogen and oxygen molecules are not so fast. This topic explains how the average kinetic energy of the particles in a gas is related to its absolute temperature.

▲ **Figure 1** *Our atmosphere is approximately 78% nitrogen (N_2), 21% oxygen (O_2), and 1 % argon (Ar), with other gases making up significantly less than 1% (carbon dioxide, CO_2, is the next highest, making up 0.04% at current levels)*

The Boltzmann constant

Ludwig Boltzmann was an Austrian physicist, whose greatest achievement was arguably his work on statistical mechanics. He applied Newtonian mechanics to gas particles in order to model the behaviour of gases. He was able to explain how the microscopic properties of particles in substances relate to the macroscopic properties of the gas, including temperature and pressure.

The **Boltzmann constant**, k, is named is in his honour. As you will see later, it is used to relate the mean kinetic energy of the atoms or molecules in gas to the gas temperature. The Boltzmann constant is equal to the molar gas constant R divided by the Avogadro constant N_A:

$$k = \frac{R}{N_A} = 1.38 \times 10^{-23}\,\text{J}\,\text{K}^{-1}$$

A second equation of state of an ideal gas

We can use the Boltzmann constant to express the equation of state of an ideal gas in another way. You can substitute the definition of k into the ideal gas equation $pV = nRT$ to give $pV = nkN_A T$

The number of particles in the gas sample, N, is equal to $n \times N_A$. Therefore $pV = NkT$.

Synoptic link

You have me the equation $N = n \times N_A$ in Topic 15.1, The kinetic theory of gases.

Worked example: Moles in the classroom

A large school classroom has a volume of $600\,m^3$. On a typical day the atmospheric pressure in the classroom is 101 kPa and the temperature is 20°C. Calculate the number of particles of gas and the number of moles of gas inside the classroom.

Step 1: Select the appropriate equation and rearrange for N.

$$pV = NkT \quad \text{or} \quad N = \frac{pV}{kT}$$

$$N = \frac{1.01 \times 10^5 \times 600}{1.38 \times 10^{-23} \times 293} = 1.49... \times 10^{28} \text{ particles}$$

Step 2: Use $N = n \times N_A$ to calculate the number of moles.

$$n = \frac{N}{N_A} = 2.5 \times 10^4 \text{ mol (2 s.f.)}$$

Mean kinetic energy and temperature

By combining $pV = \frac{1}{3}Nm\overline{c^2}$ and $pV = NkT$, we can derive an expression which directly relates the mean kinetic energy of particles in a gas to the absolute temperature of the gas.

$$\frac{1}{3}Nm\overline{c^2} = NkT$$

The number of particles N is a constant and can be cancelled.

$$\frac{1}{3}m\overline{c^2} = kT$$

The left-hand side of the equation can be rewritten as $\frac{1}{3}m\overline{c^2} = \frac{2}{3} \times \frac{1}{2} \times m\overline{c^2}$, which gives

$$\frac{2}{3} \times \left(\frac{1}{2}m\overline{c^2}\right) = kT$$

Rearranging gives $\frac{1}{2}m\overline{c^2} = \frac{3}{2}kT$

The expression $\frac{1}{2}m\overline{c^2}$ is the mean average kinetic energy of the particles in the gas. Since all other values are constant,

$$E_k \propto T$$

This only applies if the temperature is measured in kelvin. Doubling the absolute temperature from 50 K to 100 K will double the average kinetic energy of the particles (atoms or molecules) in the gas.

Study tip

All gas atoms and molecules have the same mean kinetic energy E_k at a given temperature, with

$$E_k = \frac{3}{2}kT$$

or

$$E_k = (2.07 \times 10^{-23})T$$

Worked example: The speed of a helium atom

A helium atom has a mass of 6.64×10^{-27} kg. Calculate the r.m.s. speed of helium atoms in a gas at a temperature of 15.0°C.

Step 1: Convert the temperature into kelvin: 15.0°C = 288 K

Step 2: Rearrange the relationship $\left(\frac{1}{2}m\overline{c^2}\right) = \frac{3}{2}kT$ to make $\overline{c^2}$ the subject.

$$m\overline{c^2} = 3kT$$

$$\overline{c^2} = \frac{3kT}{m}$$

Take the square root to give $c_{\text{r.m.s}} = \sqrt{\frac{3kT}{m}}$

Finally, substitute in the values in SI units

$$c_{\text{r.m.s}} = \sqrt{\frac{3 \times 1.38 \times 10^{-23} \times 288}{6.64 \times 10^{-27}}} = 1.34 \times 10^{3}\,\text{m s}^{-1} \text{ (3 s.f.)}$$

The r.m.s speed of the helium atom is about $1.3\,\text{km s}^{-1}$.

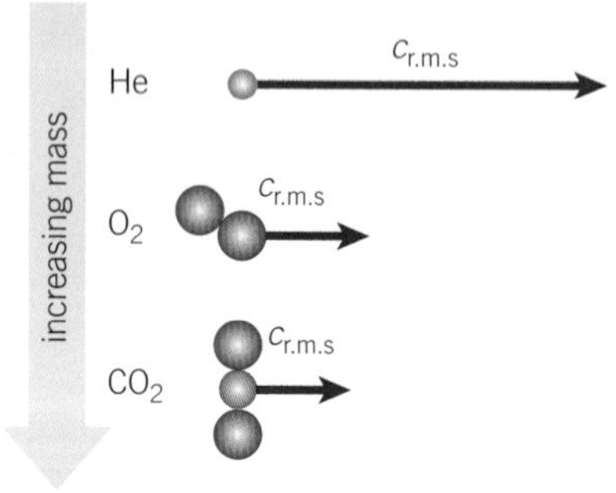

▲ **Figure 2** *The kinetic energy of the particles in different gases is the same at a given temperature, but their r.m.s. speeds vary, with lighter particles moving much faster*

Particle speeds at different temperatures

At a given temperature the atoms or molecules in different gases have the same average kinetic energy. The oxygen molecules and helium atoms around you, in spite of their different masses, have the same mean kinetic energy. However, as the particles have different masses their r.m.s. speeds will be different.

This explains why there is very little helium in the Earth's atmosphere. Helium atoms have a very small mass, which in turns means higher r.m.s. speeds. According to the Maxwell–Boltzmann distribution, some helium atoms have greater speeds than the r.m.s. speed. Over time, these faster-moving helium atoms have escaped from the Earth's atmosphere. The escape velocity for the Earth is about $11\,\text{km s}^{-1}$.

The internal energy of an ideal gas

The internal energy of a gas is the sum of the kinetic and potential energies of the particles inside the gas. One of the assumptions of an ideal gas (Topic 15.1) states that the electrostatic forces between particles in the gas are negligible except during collisions. This means that there is no electrical potential energy in an ideal gas. All the internal energy is in the form of the kinetic energy of the particles. Doubling the temperature of an ideal gas doubles the average kinetic energy of the particles inside the gas and therefore also doubles its internal energy.

Summary questions

1 Describe what happens to the absolute temperature of a gas if the r.m.s. speed of the particles in the gas:
  a increases
  b doubles
  c increases by a factor of 5. *(5 marks)*

2 Show that the Bolztmann constant k has a value of $1.38 \times 10^{-23}\,\mathrm{J\,K^{-1}}$. *(2 marks)*

3 A gas canister has a volume of $0.50\,\mathrm{m^3}$. The pressure inside the canister is 450 kPa and the temperature is 18°C. Calculate the number of particles of gas and the number of moles of gas inside the canister. *(4 marks)*

4 Explain why doubling the temperature of a real gas does not double the internal energy of the gas. *(2 marks)*

5 Show that the units of the Boltzmann constant are $\mathrm{J\,K^{-1}}$. *(2 marks)*

6 An oxygen molecule (O_2) has a mass of 5.3×10^{-26} kg. Calculate the r.m.s. speed of an oxygen molecule at room temperature (20°C). *(4 marks)*

7 Compare the kinetic energy and the r.m.s. speed at room temperature of a helium atom of mass 6.6×10^{-27} kg with your answer to question 6. *(4 marks)*

Practice questions

1 **a** Explain the term *absolute zero.* (*2 marks*)

b The temperature of an ideal gas is increased from 50 °C to 150 °C. Calculate and explain the percentage increase in the internal energy of the gas. (*3 marks*)

c A student conducts an experiment on a fixed mass of gas in a container. The temperature θ of the gas is changed and the pressure P inside the container measured. Figure 1 below shows the data points plotted by the student.

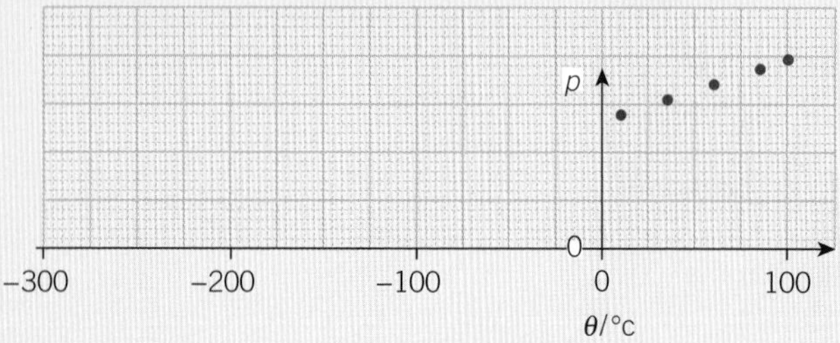

▲ Figure 1

(i) Use Figure 1 to determine a value for absolute zero on the Celcius temperature scale. Explain your answer. (*2 marks*)

(ii) Explain whether your answer in (**c**)(i) is accurate. (*1 mark*)

2 **a** One assumption required for the development of the kinetic model of a gas is that molecules undergo perfectly elastic collisions with the walls of their containing vessel and with each other.

(i) Explain what is meant by a *perfectly elastic collision.* (*1 mark*)

(ii) State **three** assumptions of the kinetic theory of gases. (*3 marks*)

b Figure 2 shows a cubical box of side length 0.20 m. The box contains one molecule of mass 4.8×10^{-26} kg moving with a constant speed of 500 m s^{-1}. The molecule collides elastically at right angles with the opposite faces **X** and **Y** of the box.

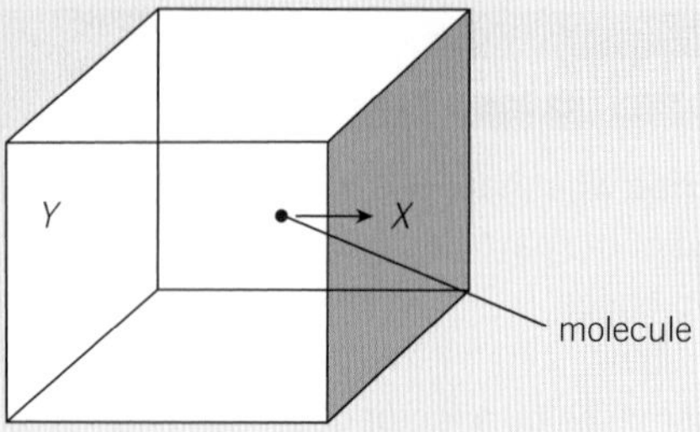

▲ Figure 2

(i) Calculate the change of momentum each time the molecule collides with face **X**. (*2 marks*)

(ii) Calculate the number of collisions made by the molecule with face **X** in 1.0 s. (*1 mark*)

(iii) Calculate the mean force exerted on the molecule by face **X**. (*2 marks*)

(iv) Hence state the force exerted on face **X** by the molecule. Justify your answer. (*1 mark*)

c The single molecule in the box in (**b**) is replaced by 3 moles of air at atmospheric pressure.

(i) Calculate the number of air molecules in the box. (*1 mark*)

(ii) Suggest why the pressure exerted by the air on each of the six faces of the box is the same. (*1 mark*)

(iii) The temperature of the air inside the box is increased. Explain in terms of the motion of the air molecules how the pressure exerted by the air will change. (*2 marks*)

Jun 2011 G484

3 **a** (i) The pressure p and volume V of a quantity of an ideal gas at absolute temperature T are related by the equations $pV = nRT$ and $pV = NkT$. In these equations identify the symbols n and N. (*1 mark*)

(ii) Choose one of the equations in (i) and show how Boyle's law follows from it. (*2 marks*)

(iii) Show that the product of pV has the same units as work done. (*1 mark*)

b The graph in Figure 3 shows the variation of pressure, p, with the reciprocal of volume, $\frac{1}{V}$, of 0.050 kg of oxygen behaving as an ideal gas.

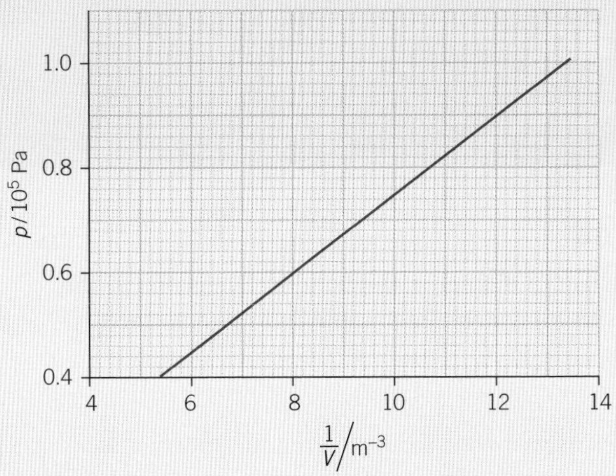

▲ Figure 3

(i) Use the graph to show that the variation of p with $\frac{1}{V}$ is taking place at constant temperature. *(2 marks)*

(ii) The molar mass of oxygen is $0.016\,\text{kg}\,\text{mol}^{-1}$. Calculate the temperature, in °C, of the oxygen in (**i**). *(3 marks)*

Jan 2013 G484

4 a Define the term *internal energy* of a material. *(2 marks)*

b A clay vase of mass 1.2 kg is heated in an oven at a temperature of 1100 °C. It is then removed and placed on a rack in a room to cool. The temperature of the room is 15 °C. The specific heat capacity of clay is $900\,\text{J}\,\text{kg}^{-1}\,\text{K}^{-1}$.

Calculate the total energy released from the vase as it cools from 1100 °C to 15 °C. *(2 marks)*

c (i) Calculate the root mean square (rms) speed of air molecules inside the oven at a temperature of 1100 °C. The molar mass of air is0.030 $\text{kg}\,\text{mol}^{-1}$. *(4 marks)*

(ii) The volume of the oven is $1.8\,\text{m}^3$. The pressure inside the oven remains constant at 100 kPa. Calculate the change in mass of the air in the oven as it cools from 1100 °C to 15 °C. *(4 marks)*

5 a Figure 4 shows the variation of pressure p with volume V for a gas in a piston.

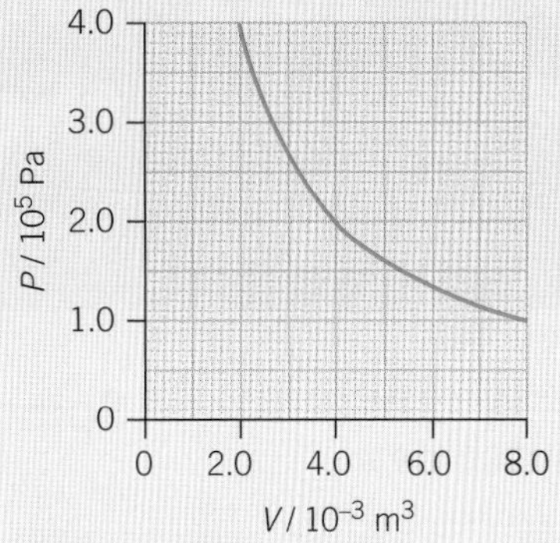

▲ Figure 4

(i) Use Figure 4 to show that the gas obeys Boyle's law. *(2 marks)*

(ii) The gas inside the cylinder is slowly compressed.
Use your knowledge of mechanics to explain how work is done on the gas. *(2 marks)*

b Figure 5 shows two cylinders X and Y linked together by a thin tube of negligible internal volume which is fitted with a tap. The tap is closed.

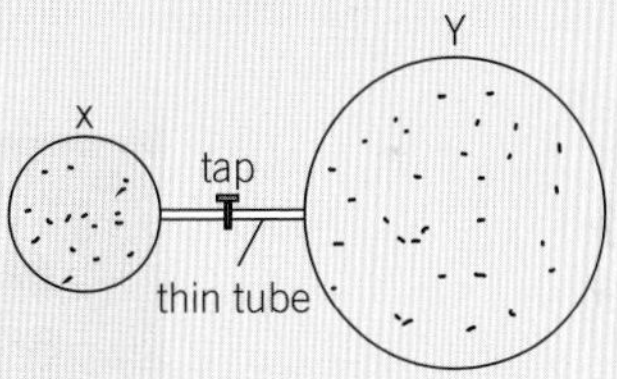

▲ Figure 5

The internal volume of X is $0.13\,\text{m}^3$ and it contains helium at a temperature of 27 °C and a pressure of $1.5 \times 10^5\,\text{Pa}$. The internal volume of Y is $0.50\,\text{m}^3$ and it too contains helium. The temperature of helium in Y is 27 °C and it exerts a pressure of $2.1 \times 10^5\,\text{Pa}$.

Helium may be assumed to be an ideal gas. The molar mass of helium is $4.0 \times 10^{-3}\,\text{kg}\,\text{mol}^{-1}$.

(i) Calculate the root mean square speed of the helium atoms. *(3 marks)*

(ii) Calculate the internal energy of the helium gas in X. *(3 marks)*

(iii) The tap is opened and the gases in X and Y are allowed to mix slowly. Calculate the final pressure exerted by the gases in the cylinders. *(4 marks)*

16 CIRCULAR MOTION

16.1 Angular velocity and the radian

Specification reference: 5.2.1

Learning outcomes

Demonstrate knowledge, understanding, and application of:

→ the radian as a measure of angle

→ the period and frequency of an object in circular motion

→ angular velocity ω, $\omega = \frac{2\pi}{T}$, or $\omega = 2\pi f$.

Spinning around

At 135 m tall and 120 m in diameter, the London Eye is one of the largest ferris wheels in the world. Each cabin follows a perfectly circular path, taking around 30 minutes to make one complete revolution at a speed of around $0.21\,\mathrm{m\,s^{-1}}$.

The description gives us two ways of measuring the speed of rotation. We could measure the speed of the cabins on the rim, which is different from the speed of points on the spokes nearer the centre, or we could measure the time taken to revolve through 360°, which is the same for any point on the wheel.

▲ **Figure 1** *Passengers on the London Eye have an average velocity of $0\,\mathrm{m\,s^{-1}}$, an average speed of $0.21\,\mathrm{m\,s^{-1}}$, and an average angular velocity of $3.5 \times 10^{-1}\,\mathrm{rad\,s^{-1}}$*

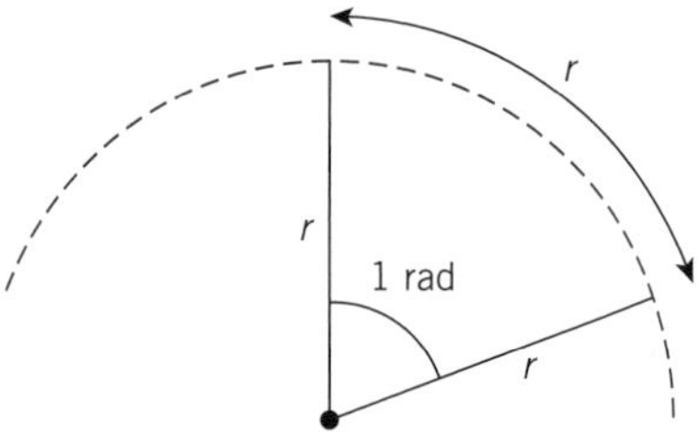

▲ **Figure 2** *An angle of 1 radian*

The radian

There is no scientific reason to have 360° in a circle. The most commonly accepted hypothesis is that it is a close approximation to the number of days in a year. Measuring angles in degrees is often convenient and easy, but the SI unit for angle is the **radian**. It has a precise definition. A radian is the angle subtended by a circular arc with a length equal to the radius of the circle. This is an angle of approximately 57.3° for any circle (Figure 2).

The angle in radians subtended by any arc is defined as follows:

$$\text{angle in radians} = \frac{\text{arc length}}{\text{radius}}$$

What is 360° in radians? For a complete circle, the arc length is equal to the circumference of the circle. Therefore

$$\text{angle in radians} = \frac{2\pi r}{r} = 2\pi \text{ radians}$$

Study tip

You can convert between degrees and radians by remembering that π radians is equal to 180°.

Therefore 360° is equal to 2π radians, or about 6.3 radians. To convert from degrees into radians, divide the angle in degrees by $\frac{180}{\pi}$.

Worked example: Expressing angular motion in degrees and radians

A spinning top completes 8.0 full revolutions before it comes to rest. Express the angle it moves through in degrees and radians.

Step 1: One full revolution or rotation is equal to 360°, therefore the angle moved through in 8.0 rotations is equal to:

$360° \times 8.0 = 2880°$

Step 2: One full revolution is equal to 2π rad, therefore the angle moved through in 8.0 rotations is equal to:

$2\pi \times 8.0 = 16\pi$ rad = 50 rad (2 s.f.)

Alternatively you can divide the angle in degrees by $\frac{180}{\pi}$, giving

$$\frac{2880}{\left(\frac{180}{\pi}\right)} = \frac{2280\pi}{180} = 16\pi \text{ rad} = 50 \text{ rad (2 s.f.)}$$

Angular velocity

To describe the motion of moving objects fully we need not only be able to describe their linear motion, but also how objects twist or rotate as they move (their circular motion). Any object moving in a circle or circular path moves through an angle θ in a certain time t (Figure 3). All points on the London Eye, for example, rotate through the same angle in the same period of time. This gives a method of describing movement in terms of angular motion– the wheel has an average **angular velocity** of $0.20° \text{ s}^{-1}$ (or $3.5 \times 10^{-3} \text{ rad s}^{-1}$).

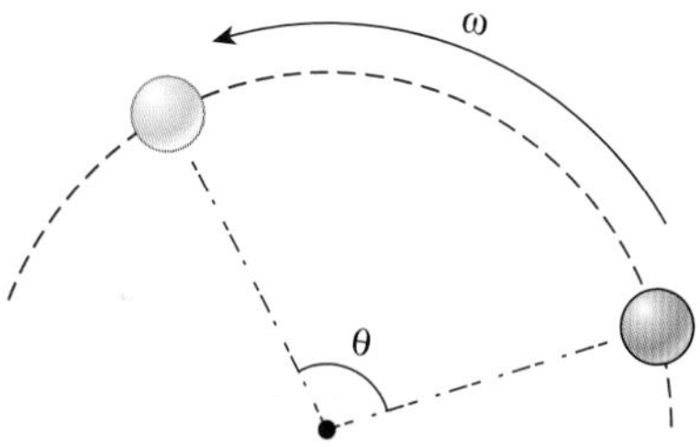

▲ **Figure 3** *The faster the object moves, the greater the angle moved through (subtended) in a given time*

The angular velocity ω of an object moving in a circular path is defined as the rate of change of angle. Therefore

$$\omega = \frac{\theta}{t}$$

In a time t equal to one period T, the object will move through an angle θ equal to 2π radians. Therefore

$$\omega = \frac{2\pi}{T}$$

The angular velocity is measured in radians per second.

As frequency f is the reciprocal of the period T, $f = \frac{1}{T}$, we can also express angular velocity ω as:

$$\omega = 2\pi f$$

Angular velocity can be expressed in several different units, including degrees per second ($° \text{ s}^{-1}$), revolutions per second (rev s^{-1}), and revolutions per minute (normally rpm). However, for this A Level Physics course, you should stick to radians per second (rad s^{-1}).

Worked example: The angular velocity of a hard disk

A spinning hard disk takes 8.33 ms to complete one full revolution. Calculate its angular velocity in rad s^{-1} and revolutions per second.

Step 1: Select the appropriate equation and substitute in known values in SI units.

$$\omega = \frac{2\pi}{T} = \frac{2\pi}{8.33 \times 10^{-3}} = 754 \, \text{rad}\, \text{s}^{-1} \text{ (3 s.f.)}$$

Step 2: Calculate the number of revolutions per second. This is the frequency of rotation.

$$f = \frac{1}{T} = \frac{1}{8.33 \times 10^{-3}} = 120 \, \text{rev}\, \text{s}^{-1} \text{ (2 s.f.)}$$

Alternatively, 1 rev s^{-1} is equal to 2π rad s^{-1}, so 1 rad s^{-1} is equal to $\frac{1}{2\pi}$ (or 0.159) rev s^{-1}. To express angular velocity in rev s^{-1} we need to multiply angular velocity in rad s^{-1} by $\frac{1}{2\pi}$.

$$\frac{754}{2\pi} = 120 \, \text{rev}\, \text{s}^{-1} \text{ (2 s.f.)}$$

Summary questions

1 Express the following angles in radians:
 a 180°
 b 45°. *(3 marks)*

2 Calculate the angular velocity of:
 a a roundabout which completes 1 revolution in 30 s
 b a spinning top with a time period of 0.10 s. *(2 marks)*

3 Calculate the angular velocity of the Earth as it moves around the Sun. (Hint: The orbital period of the Earth is 1 year). *(2 marks)*

4 A hard disk spins at 4500 rpm. Calculate the angular velocity in rad s^{-1} and the time taken to complete 50 revolutions. *(5 marks)*

5 A spinning disk has an angular velocity of 565 rad s^{-1}. Calculate the frequency of rotation and the time taken to complete 5400 revolutions. *(4 marks)*

6 An analogue clock has second, minute, and hour hands. Assuming the hands move round at a constant rate, calculate the angular velocity of each hand. *(6 marks)*

16.2 Centripetal acceleration

Specification reference: 5.2.2

Accelerating at constant speed

When you move in a circular path, for example on a loop rollercoaster, your direction is continuously changing. As a result your velocity is changing even if you travel at constant speed. This change in velocity means that objects following a circular path must be accelerating, and the greater the rate of change of velocity the greater the acceleration.

◀ **Figure 1** *People riding a loop-the-loop rollercoaster are always accelerating, even if their speed does not change*

The centre-seeking force

Any accelerating object requires a net (or resultant) force to be acting on it (Newton's first law). Any force that keeps a body moving with a uniform speed along a circular path is called a **centripetal force**. The term centripetal comes from the ancient Greek and means 'centre-seeking'. In the case of a rollercoaster, the source of the centripetal force comes from the way your weight and the normal contact force from your seat interact with each other. It is the changes in this interaction that make you feel weightless or pushed down into your seat, adding to the thrill of the ride.

A centripetal force is always perpendicular to the velocity of the object (Figure 2). This means that this force has no component in the direction of motion and so no work is done on the object. As a result its speed remains constant.

The centripetal force might be a gravitational attraction for a satellite in orbit around a planet, friction for a car going around a bend, or tension in the string when a yo-yo is swung around in a vertical circle. All three forces provide the centripetal force making an object move in a circular path.

Relating angular and linear velocity

At any point on a circular path, the linear velocity is always at a tangent to the circular path (at 90° to the radius – Figure 3).

For an object moving in a circle at constant speed, we can calculate its speed using the equation:

$$\text{speed} = \frac{\text{distance travelled}}{\text{time taken}}$$

Learning outcomes

Demonstrate knowledge, understanding, and application of:

- → a constant net force perpendicular to the velocity of an object, which causes it to travel in a circular path
- → constant speed in a circle, $v = r\omega$
- → centripetal acceleration, $a = \frac{v^2}{r}$ and $a = \omega^2 r$.

Synoptic link

Newton's first law, that acceleration requires the action of a resultant force, was covered in Topic 7.1, Newton's first and third laws of motion.

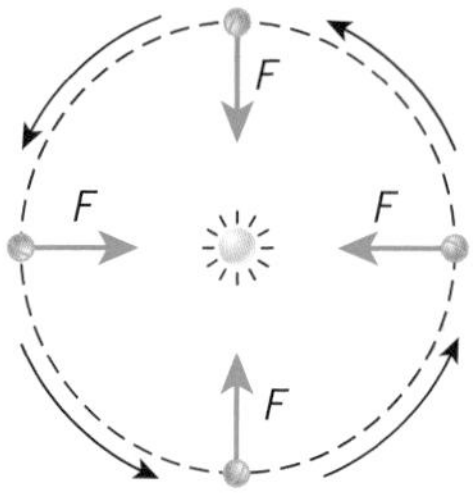

▲ **Figure 2** *The gravitational attraction between the Earth and Sun is the source of the centripetal force, always acting towards the centre, that makes the Earth follow an almost perfectly circular orbit*

Study tip

Remember that, in all cases of circular motion, there must be a net force towards the centre of the circle.

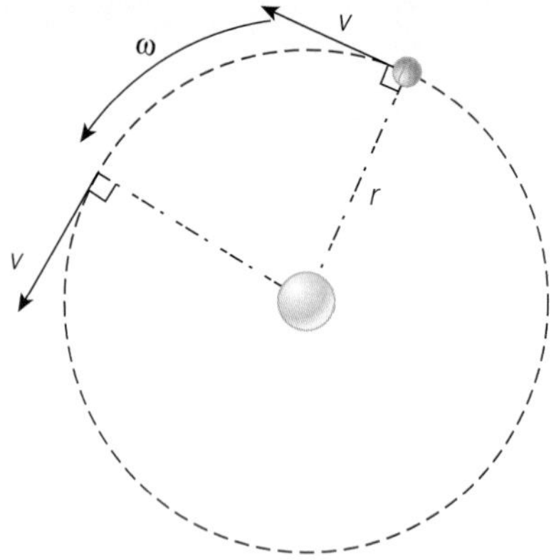

▲ **Figure 3** *The linear velocity v of an object moving in a circular path depends on the angular velocity and the radius of the path*

In one complete rotation, the distance travelled is the circumference of the circle, and the time is the period T. Therefore:

$$v = \frac{2\pi r}{T}$$

and since angular velocity $\omega = \frac{2\pi}{T}$ we can express the speed as

$$v = r\omega$$

For objects with the same angular velocity, the linear velocity at any instant is directly proportional to the radius. Double the radius and the linear velocity will also double. For example, consider three objects A, B, and C fixed on a spinning disk (Figure 4). All three have the same angular velocity because they all take the same time to complete one full revolution. However, A has the greatest linear velocity at any instant as it travels the greatest distance in the same time.

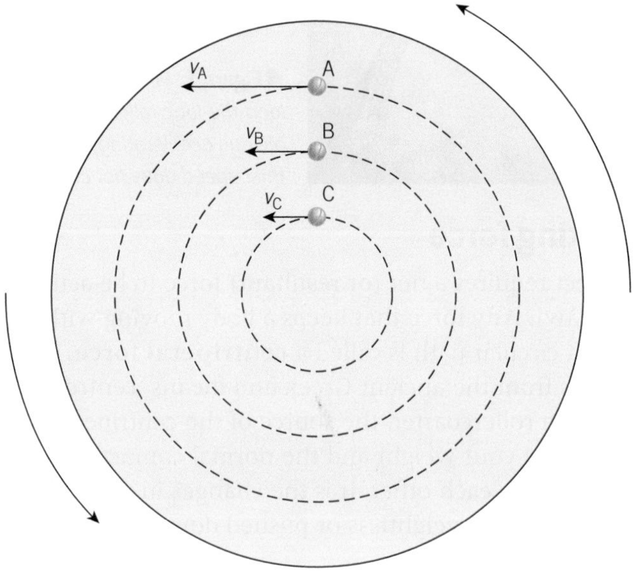

▲ **Figure 4** *As the radius of the path followed by the object increases, the linear velocity also increases*

Centripetal acceleration

The acceleration of any object travelling in a circular path at constant speed is called the **centripetal acceleration** and this, like the centripetal force, always acts towards the centre of the circle.

The centripetal acceleration a depends on the speed of the object, v, and the radius r of the circular path. Imagine sitting in a car as it goes around a roundabout. The faster the car travels the greater the acceleration we experience. The same is true if the radius of the path is smaller. In both cases the centripetal force required is greater as the acceleration is greater. The centripetal acceleration is given by the equation:

$$a = \frac{v^2}{r}$$

Combining this equation with $v = r\omega$ gives an alternative expression for centripetal acceleration in terms of angular velocity ω. That is

$$a = \omega^2 r$$

Synoptic link

Centripetal acceleration always acts towards the centre of the circular path followed by an object, from Newton's second law – see Topic 7.3, Newton's second law.

Worked example: Spinning on the equator

Calculate the centripetal acceleration of a person standing on the equator as the Earth rotates. The radius of the Earth = 6370 km. The period of the rotation is 24 hours = 86 400 seconds.

We can determine the acceleration using either $a = \frac{v^2}{r}$ or $a = \omega^2 r$.

Method 1: Using $a = \frac{v^2}{r}$

Firstly we can determine the linear speed using $v = \frac{2\pi r}{T}$

$$v = \frac{2\pi \times 6.370 \times 10^6}{86\,400} = 463\,\text{m s}^{-1}$$

Substituting this into $a = \frac{v^2}{r}$ gives

$$a = \frac{463^2}{6.370 \times 10^6} = 3.37 \times 10^{-2}\,\text{m s}^{-2}\ (3\text{ s.f.})$$

Method 2: Using $a = \omega^2 r$

Firstly we can determine the angular speed using $\omega = \frac{2\pi}{T}$

$$\omega = \frac{2\pi}{86\,400} = 7.27 \times 10^{-5}\,\text{rad s}^{-1}$$

Substituting this into $a = \omega^2 r$ gives:

$$a = (7.27 \times 10^{-5})^2 \times 6.370 \times 10^6 = 3.37 \times 10^{-2}\,\text{m s}^{-2}\ (3\text{ s.f.})$$

Either technique is valid, so it is just a matter of personal preference.

Deriving $a = \frac{v^2}{r}$

Consider an object moving at constant speed v along a circular path from A to B in a short time δt. In that time it subtends a small angle $\delta\theta$ (Figure 5).

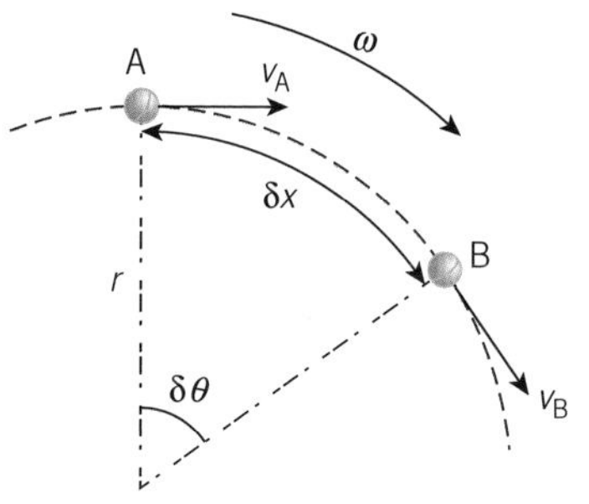

▲ **Figure 5** *An object moving at constant speed v along a circular path from A to B in a short time δt*

By definition the angle in radians is equal to the arc length δx divided by the radius r, giving

$$\delta\theta = \frac{\delta x}{r}$$

The velocity at A is v_A and the velocity at B is v_B. The two velocities are different as the direction has changed. The acceleration of the object can be determined using:

$$a = \frac{\delta v}{\delta t}$$

where δv is the change in velocity. If the distance between A and B is very small then the displacement from A to B, δx, is almost a straight line, and can be expressed as $\delta x = v\delta t$.

Substituting this into $\delta\theta = \frac{\delta x}{r}$ gives:

$$\delta\theta = \frac{v\delta t}{r}$$

$\delta\theta$ must also be the angle between the velocities v_A and v_B, so we have a second expression for $\delta\theta$ as long as the object is moving at constant speed and so the magnitude of v is constant.

$$\delta\theta = \frac{\delta v}{v}$$

Equating the two equations for $\delta\theta$ gives:

$$\delta\theta = \frac{v\delta t}{r} = \frac{\delta v}{v}$$

$$\frac{v^2}{r} = \frac{\delta v}{\delta t} = a$$

where a is the centripetal acceleration for an object moving at constant speed v in a circular path of radius r.

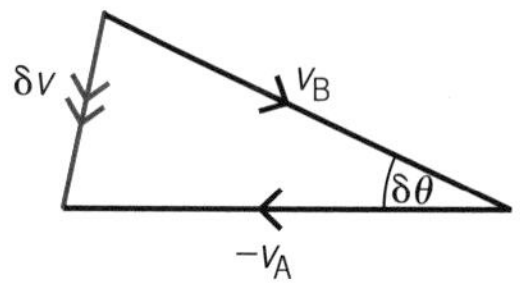

▲ **Figure 6** *δv is the difference of the vectors v_B and v_A*

1 By considering the angles involved, show that $\delta\theta = \frac{\delta v}{v}$.

2 Explain the significance of the direction of δv.

Summary questions

1 Name the actual force that provides the centripetal force in the following examples and state the direction of the force:
   a a satellite in orbit of the Earth
   b an electron going around a nucleus
   c a cyclist going around a bend. (*3 marks*)

2 Calculate the linear speed of a particle of dust 20 cm from the centre of a disc spinning at 6.0 rad s^{-1}. (*2 marks*)

3 Calculate the centripetal acceleration of the following objects:
   a a car moving around a bend of radius 60 m at a speed of 20 m s^{-1}
   b a yo-yo swung in a circle with a radius of 60 cm at 5.0 rad s^{-1}
   c an object moving in a circle of radius 1.5 m with a period of 750 ms. (*6 marks*)

4 An object moves in a circular path of radius 30 cm with a velocity of 1.40 m s^{-1}. Calculate the angular velocity and centripetal acceleration of the object. (*4 marks*)

5 A rollercoaster designer creates a loop where the acceleration at the bottom of the loop is five times greater than the acceleration of free fall. The radius of the loop is 12 m. Calculate the speed of the rollercoaster required to provide such an acceleration. (*3 marks*)

16.3 Exploring centripetal forces

Specification reference: 5.2.2

Hammer throwing

The hammer throw is one of the oldest Olympic sports. It has its origins in Highland games, with the first throw recorded at the end of the 18th century. Competitors spin round three or four times before releasing the hammer.

By studying the forces and techniques involved in throwing the hammer the greatest distance, the sport has evolved from a focus on brute strength to one of speed. Top hammer throwers rotate with an angular velocity of around $10\,\text{rad}\,\text{s}^{-1}$, resulting in the ball at the end of the hammer travelling at over $20\,\text{m}\,\text{s}^{-1}$ (nearly 50 mph) at launch.

Learning outcomes

Demonstrate knowledge, understanding, and application of:

→ centripetal force, $F = \frac{mv^2}{r}$ and $F = m\omega^2 r$

→ techniques and procedures used to investigate circular motion.

▲ **Figure 1** *The hammer world record currently stands at 86.74 m for men (7 kg ball), set in 1986 by Yuriy Sedykh, and 79.58 m for women (4 kg ball), set by Anita Włodarczyk in 2014*

Centripetal force

In Topic 16.2, Angular acceleration, you saw that any object traveling in a circular path must have a resultant force acting on it at right angles to its velocity (a centripetal force). The object's speed remains constant because the component of the force acting on the object in the direction of motion is zero.

Combining $F = ma$ with the equation for centripetal acceleration, $a = \frac{v^2}{r}$, we can determine an expression for centripetal force.

$$F = \frac{mv^2}{r}$$

For constant mass m and radius r, the centripetal force F is directly proportional to v^2. That is, $F \propto v^2$. Since $v = \omega r$, the centripetal force F can also be written in terms of the angular velocity ω as

$$F = \frac{m(\omega r)^2}{r} = \frac{m\omega^2 r^2}{r}$$

or

$$F = m\omega^2 r$$

This force F is always towards the centre of the circular path.

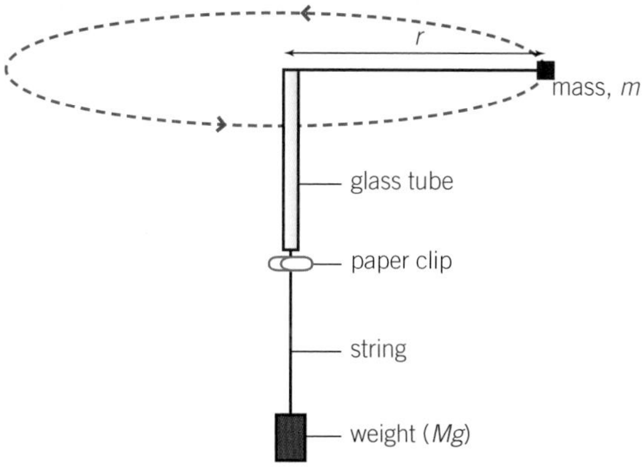

▲ **Figure 2** *A whirling bung on a string can be used as part of a simple investigation into centripetal force*

Investigating circular motion

You can investigate circular motion using the simple equipment in Figure 2. As the bung is swung in a horizontal circle the suspended weight remains stationary as long as the force it provides (Mg) is equal to the centripetal force required to make the bung travel in the circular path. If the centripetal force required is greater than the weight then the weight moves upwards. The paperclip acts as a marker to make this movement clearer. The weight and thus the centripetal force required for different masses, radii, and speeds (calculated from angular velocity and radius) can then be investigated.

Synoptic link

You first met the equation $F = ma$ in Topic 4.1, Force, mass, and weight.

Worked example: Taking a bend at top speed

A racing car of mass 1200 kg travels around a bend with a radius of 140 m. The maximum frictional force between the wheels and the track is 17 kN. Calculate the maximum speed at which the car can travel around the bend.

Step 1: Select the correct equation and rearrange to make r the subject.

$$F = \frac{mv^2}{r} \quad \text{so} \quad v = \sqrt{\frac{Fr}{m}}$$

Step 2: Substitute in the correct values using SI units, and calculate the maximum speed.

$$v = \sqrt{\frac{1.7 \times 10^4 \times 140}{1200}} = 45\,\text{m}\,\text{s}^{-1} \text{ (2 s.f.)}$$

Separating samples: The centrifuge

In 1864, Antonin Prandtl invented the first centrifuge in order to separate cream from milk. Centrifuges are now used widely in science, medicine, and industry, for example to separate the components of blood.

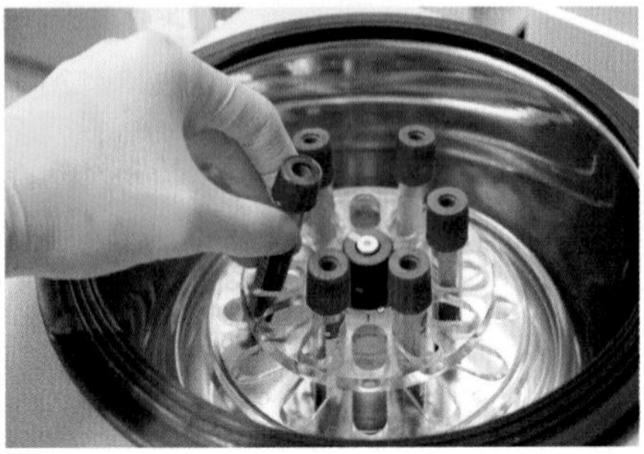

▲ **Figure 3** *Centrifuges are regularly used in chemistry and biology for separating particles suspended in liquids*

The centrifuge works by spinning liquids (or occasionally gases) at a very high speed. The particles in the liquid separate out as the tube holding them is spun. When spinning, the tubes swing out so they are horizontal. Particles with a greater mass – in the case of blood, the red blood cells – require greater centripetal force to follow the circular path, and so move outwards to the bottom of the tube. Lighter particles – the clear plasma in blood – end up near the top of the tube.

Gas centrifuges are used in the uranium enrichment required for nuclear weapons and most types of nuclear reactors. The heavier isotope of uranium (uranium-238) concentrates at the walls of the centrifuge as it spins, while the uranium-235 isotope stays close to the centre of the centrifuge. The urainium-235 is carefully extracted and concentrated, with tens of thousands of spins necessary to produce enough to make a nuclear weapon.

1. Draw a plan view of a spinning centrifuge containing test tubes to illustrate why heavier particles move towards the bottom of the tube.
2. Sketch a diagram to show the force(s) acting on a particle on the bottom of a test tube inside the fast-spinning centrifuge.
3. A centrifuge spins at 6000 rpm. Calculate the radius of the path followed by a particle of mass 2.0 mg if the maximum force on the particle is 63 mN.

More sources of centripetal force

In Topic 16.2, you saw several different sources of centripetal force, including friction, tension, and gravitational attraction (Figure 4).

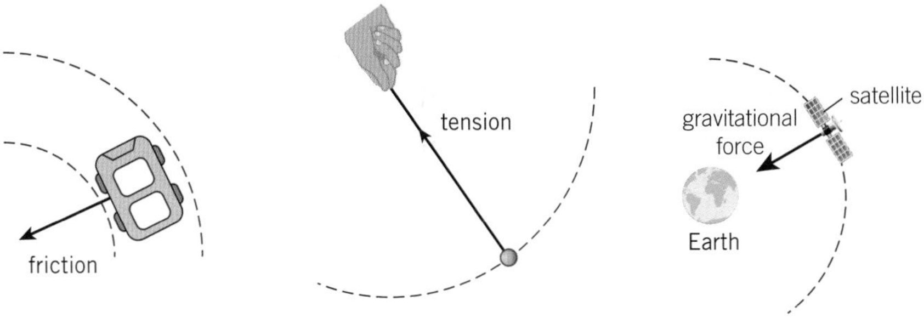

▲ **Figure 4** *Examples of sources of centripetal forces that make objects follow circular paths*

Banked surfaces

The greater the speed of an object following a circular path, the greater the centripetal force required to make it follow this path. A car approaching a bend must slow down in order to ensure the maximum frictional force between the tyres and the road is sufficient to provide the required centripetal force. If the car travels too fast it will follow a path of greater radius and leave the road.

For the same reason, the tracks in modern velodromes are banked up to angles of 45° so that track cyclists can travel at higher speeds. On the banked part of the track a horizontal component of the normal contact force, together with the frictional force from the tyres, provides the centripetal force required to follow the circular path at such high speed. The friction between the tyres and track would not be sufficient to allow the cyclist to travel at that speed on a flat surface.

A similar technique is used in some motor sports like NASCAR in the USA.

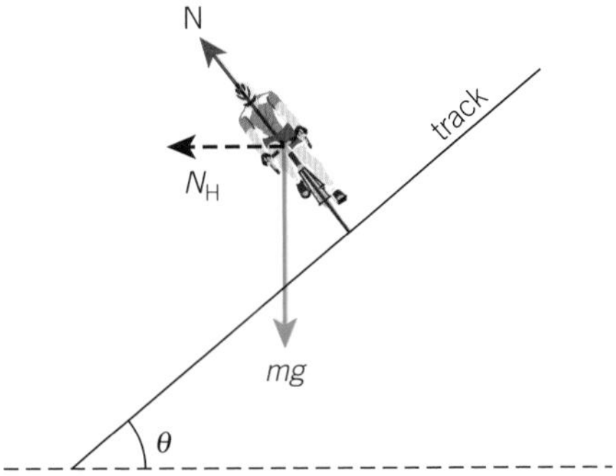

▲ **Figure 5** *A horizontal component of the normal contact force N acts towards the centre of the circle.* $N_H = N \sin \theta$.

▲ **Figure 6** *Using banked tracks, professional cyclists can reach much higher speeds than on a flat track*

Synoptic link

Circular motion is also very important when dealing with gravitational fields (Chapter 18, Gravitational fields) and the motion of charged particles in magnetic fields (Topic 23.3, Charged particles in magnetic fields).

At the fairground

A centripetal force can be due to changes in the normal contact force when an object is made to travel in a circular path. For example, in a ferris wheel, when the capsule is stationary the normal contact force N is equal to your weight (Figure 7a). However, when the wheel rotates a net force is required in order for you to travel in a circular path. At the top of the ride N reduces, resulting in a net force towards the centre of the circle. This also gives you a feeling of slightly reduced weight (Figure 7b). The opposite is true at the bottom of the ride, where N increases providing a net force upwards towards the centre of the circle.

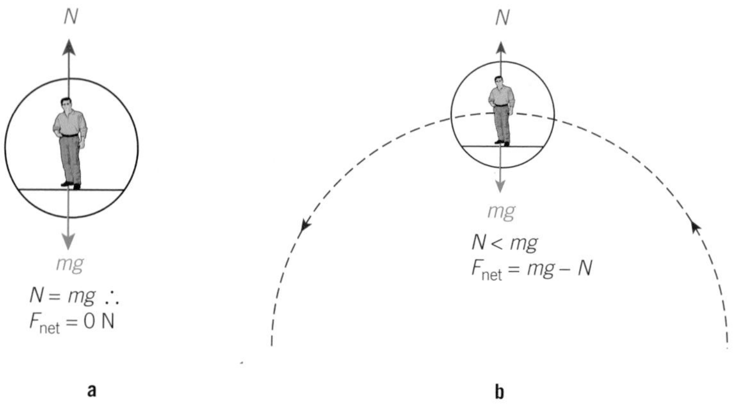

▲ **Figure 7** *Changes in the size of the normal contact force N provide the centripetal force required to follow a circular path*

The conical pendulum

A **conical pendulum** is a simple pendulum that, instead of swinging back and forth, rotates at constant speed, describing a horizontal circle. The time taken to complete each rotation depends only on the length of the pendulum string and the gravitational field strength.

Conical pendulums have been used when a smooth timing mechanism was required, for example to calibrate the tracking of sensitive telescopes that move slowly against the Earth's rotation to ensure they remain fixed on a certain point in the sky.

By applying Newton's laws of motion we can derive an expression relating the angle of the pendulum θ to its speed v and its radius r.

The horizontal component of the tension F_T in the string **N** provides the centripetal force F required for the circular motion of the pendulum.

$$F = ma = \frac{mv^2}{r}$$

$$F_T \sin\theta = \frac{mv^2}{r}$$

The vertical component of the tension in the string must be equal to the weight of the pendulum bob, because there is no acceleration in the vertical direction. Therefore:

$$F_T \cos\theta = mg$$

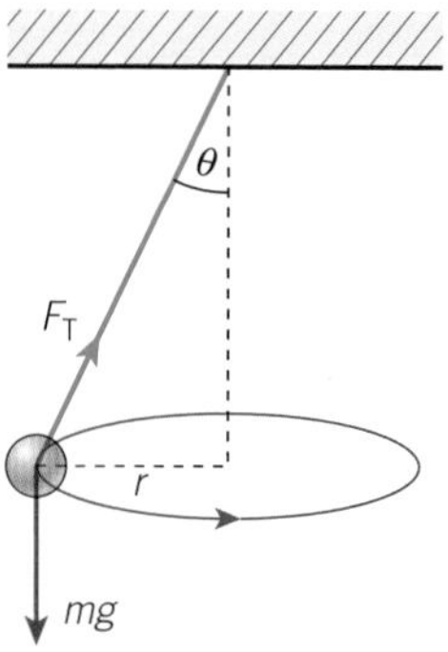

▲ **Figure 8** *A simple conical pendulum traces out a horizontal circle as it moves*

If we divide the first equation by the second we get

$$\tan\theta = \frac{v^2}{rg} \qquad \left(\text{Remember, } \frac{\sin\theta}{\cos\theta} = \tan\theta\right)$$

From this equation we can see that the angle of the pendulum is not affected by the mass of the pendulum bob.

1 Calculate the radius of a conical pendulum circling at 4.0 m s^{-1} and forming an angle of 30° to the vertical.

2 By considering the distance travelled by the pendulum bob in one complete revolution, show that the period of oscillation of a conical pendulum T is given by

$$T = 2\pi\sqrt{\frac{r}{g\tan\theta}}$$

Summary questions

1 If all other factors remain the same, state the effect on the centripetal force acting on an object when:

a the mass of the object doubles

b the speed of the object doubles

c the radius of the path followed by the object halves and the speed increases by a factor of 3. *(3 marks)*

2 A yo-yo is swung in a vertical circle. Explain using diagrams where the string is most likely to break. *(3 marks)*

3 Figure 9 shows a toy airplane travelling in a horizontal circle of radius 20 m. The mass of the airplane is 1.2 kg. The lift is equal to 16 N.

a Show that the centripetal force on the airplane is about 10 N. *(2 marks)*

b Show that the speed v of the airplane is 13 m s^{-1}. *(3 marks)*

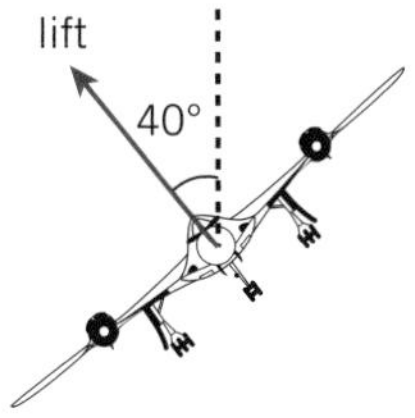

▲ **Figure 9** *A toy plane travelling in a horizontal circle*

4 The radius of path followed on a ferris wheel is 120 m. The wheel completes one revolution in 20 minutes. Calculate:

a the angular velocity

b the change in normal contact force from the top to the bottom. *(4 marks)*

5 The Earth has a mass of 6.0×10^{24} kg and is on average 150 million km from the Sun. Use this information to determine the centripetal force on the Earth as it moves around the Sun (Assume 1 year = 32×10^6 s). *(3 marks)*

6 The Earth has a radius of 6400 km. Scales on the Equator and at the North Pole give different readings for weight. If a person has a weight of 700 N, calculate the reading on the scale at:

a the North Pole

b the Equator. *(5 marks)*

Practice questions

1 Figure 1 shows apparatus used to investigate circular motion. The bung is attached by a continuous nylon thread to a weight carrier supporting a number of slotted masses which may be varied. The thread passes through a vertical glass tube. The bung can be made to move in a nearly horizontal circle at a steady high speed by a sustainable movement of the hand holding the glass tube. A constant radius r of rotation can be maintained by the use of a reference mark on the thread.

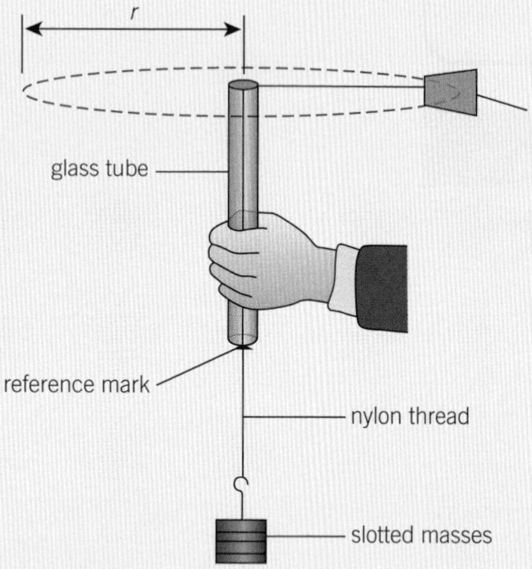

▲ Figure 1

a (i) Draw an arrow labelled **F** on Figure 1 to indicate the direction of the resultant force on the bung. *(1 mark)*

(ii) Explain how the speed of the bung remains constant even though there is a resultant force F acting on it. *(2 marks)*

b (i) Two students carry out an experiment using the apparatus in Figure 1 to investigate the relationship between the force F acting on the bung and its speed v for a constant radius. Describe how they obtain the values of F and v. *(5 marks)*

(ii) 1 Sketch, on a copy of Figure 2, the expected graph of F against v^2.

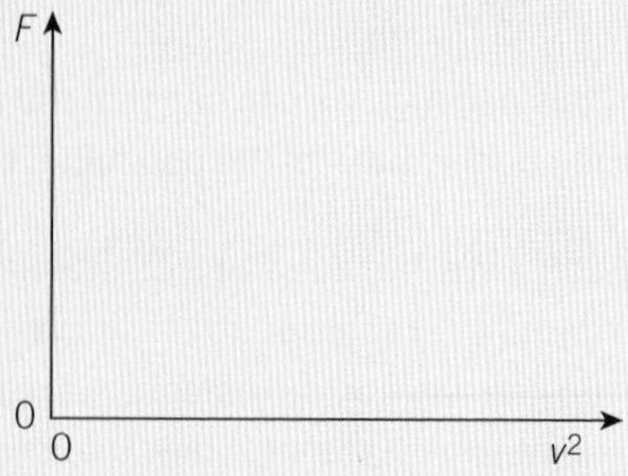

▲ Figure 2

(1 mark)

2 Explain how the graph can be used to determine the mass m of the bung. *(2 marks)*

Jun 2012 G484

2 a Figure 3 shows the London eye.

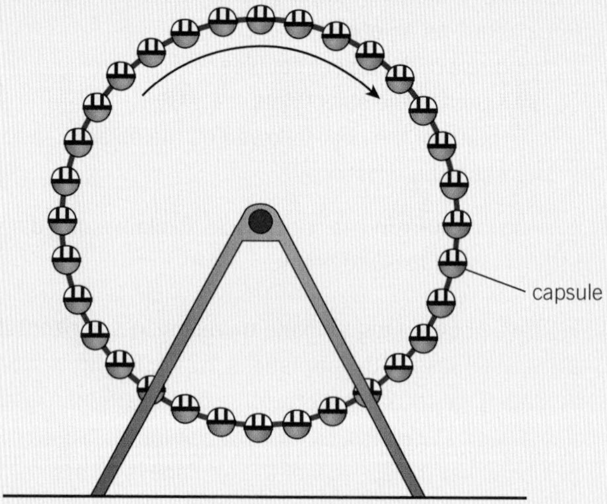

▲ Figure 3

(i) The wheel of the London eye has a radius of 60 m and each capsule travels at $0.26\,\text{m}\,\text{s}^{-1}$. Calculate the time taken for the wheel to make one complete rotation. *(1 mark)*

(ii) Each capsule has a mass of $9.7 \times 10^3\,\text{kg}$. Calculate the centripetal force which must act on the capsule to make it rotate with the wheel. *(2 marks)*

b Figure 4 shows the drum of a spin-dryer as it rotates. A dry sock **S** is shown on the inside surface of the side of the rotating drum.

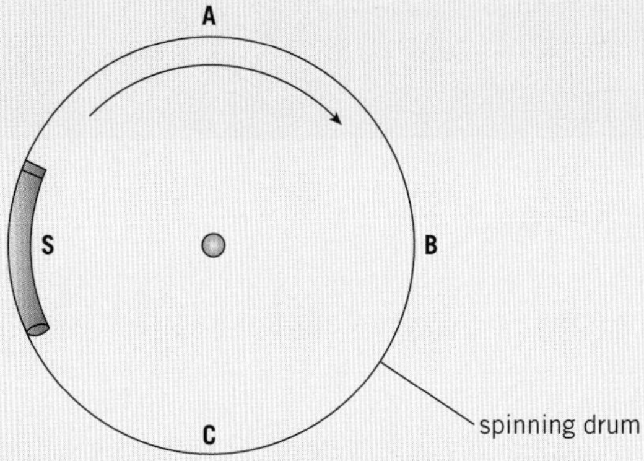

▲ Figure 4

(i) Draw arrows on a copy of Figure 4 to show the direction of the centripetal force acting on **S** when it is at points **A**, **B**, and **C**. *(1 mark)*

(ii) State and explain at which position, **A**, **B**, or **C** the normal contact force between the sock and the drum will be

1 the greatest *(2 marks)*

2 the least. *(1 mark)*

Jan 2010 G484

3 Figure 5a shows a car of mass 800 kg travelling at a constant speed of $15\,m\,s^{-1}$ around a bend on a level road following a curve of radius 30 m.

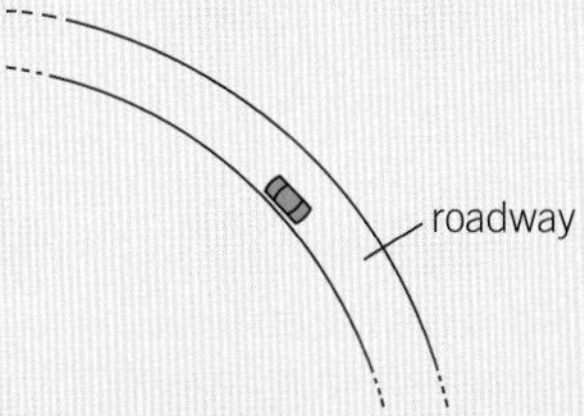

▲ Figure 5a

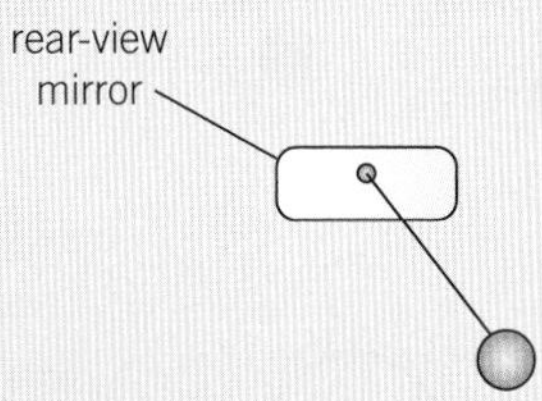

▲ Figure 5b

a Draw an arrow on a copy of Figure 5a to indicate the direction of the resultant horizontal force on the car at the position shown. *(1 mark)*

b Calculate the magnitude *F* of this force. *(3 marks)*

c A medallion hangs on a string from the shaft of the rear view mirror in the car. Figure 5b shows its position in the vertical plane, perpendicular to the direction of travel in Figure 5a.

(i) Draw and label arrows on Figure 5b to indicate the forces acting on the medallion. *(2 marks)*

(ii) On another occasion the car travels around the bend at $25\,m\,s^{-1}$. The angle of the string to the vertical is different. Explain how and why this is so. You may find it useful to sketch a vector diagram to aid your explanation. *(3 marks)*

Jan 2009 2824

4 a Define *angular velocity* of a rotating object. *(1 mark)*

b An astronomer observing a rotating dust cloud determines the speed *v* of the cloud at a distance *r* from its centre. Figure 6 shows the data points, together with the error bars, plotted for this rotating cloud.

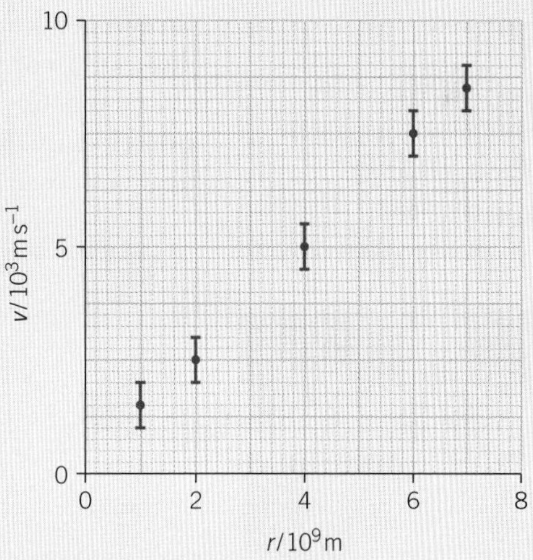

▲ Figure 6

(i) Explain how you can deduce from Figure 6 that the dust cloud has a constant angular velocity. *(2 marks)*

(ii) Calculate the period of the dust cloud. *(4 marks)*

(iii) Calculate the percentage uncertainty in your answer to **(b)(ii)**. *(3 marks)*

17 OSCILLATIONS

17.1 Oscillations and simple harmonic motion

Specification reference: 5.3.1

Learning outcomes

Demonstrate knowledge, understanding, and application of:

- → displacement, amplitude, period, frequency, angular frequency, and phase difference
- → angular frequency $\omega = 2\pi/T$ or $\omega = 2\pi f$
- → isochronous oscillators (the period of a simple harmonic oscillator is independent of its amplitude)
- → simple harmonic motion, $a = -\omega^2 x$
- → techniques and procedures used to determine the period and frequency of simple harmonic oscillations.

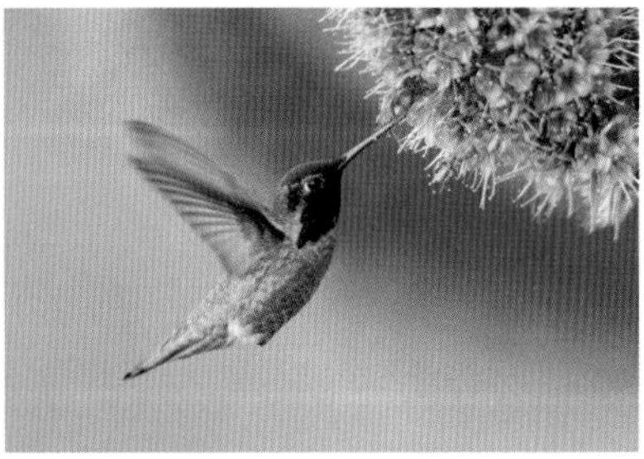

▲ **Figure 1** *The rapid oscillation of the wings of a hummingbird allow the bird to fly not only backwards, but – uniquely – also upside-down*

Fifty beats per second

We have seen how to analyse both linear and circular motion, but there are other, more complex types of motion too. Take the tip of a humming bird's wing. It moves rapidly in one direction, stops, and then heads in the opposite direction, over and over again. The wings typically complete 50 oscillations (or cycles) per second, producing an audible hum that gives the bird its name.

This chapter deals with different kinds of **oscillating motion**, most of which share many of the characteristics and mathematics of the oscillating particles we have seen in waves, along with aspects of circular motion.

Oscillating motion

There are many examples of oscillating motion, including a simple pendulum, the end of a ruler hanging over the edge of a desk, or a volume of water in a U-shaped tube. All share similar characteristics. In each case the object starts in an **equilibrium position**. A force is then applies to the object, displacing it, and it begins to oscillate.

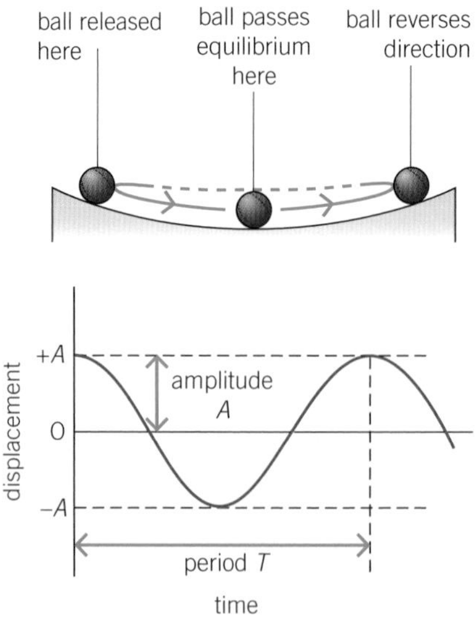

▲ **Figure 2** *A ball on a curved surface is an example of a simple oscillation – the displacement is zero at the equilibrium position*

Consider an object that is displaced from its equilibrium position and then released. It travels towards the equilibrium position at increasing speed. It then slows down once it has gone past the equilibrium position and eventually reaches maximum **displacement** (**amplitude**). It then returns towards its equilibrium position, speeding up, and once more slows down to a stop when it reaches maximum negative displacement. This motion is repeated over and over again.

Figure 2 shows how the displacement of an oscillating ball on a curved track varies with time. This graph is sinusoidal in shape and shows the amplitude A and period T of the oscillating ball. All examples of oscillation produce similar graphs.

Oscillating motion terminology

The key terms we use to describe oscillating motion are listed in Table 1.

▼ **Table 1** *Terminology for oscillating motion*

Quantity	Symbol and unit	Definition
displacement	x / m	the distance from the equilibrium position
amplitude	A / m	the maximum displacement from the equilibrium position
period	T / s	the time taken to complete one full oscillation
frequency	f / Hz	the number of complete oscillations per unit time

Synoptic link

Oscillating motion shares many ideas found in Chapter 11, Waves, and Chapter 16, Circular motion.

The terms are very similar to their wave counterparts, and just as with waves it is possible to compare the differences in displacement between two oscillating objects or the displacement of an oscillating object at different times. We use the term **phase difference** (denoted by the greek letter, ϕ) for this comparison. Two identical pendulums oscillating in step both reach their maximum positive displacement at the same time. Their phase difference is 0 rad. If they are in antiphase ($\phi = \pi$ rad) one pendulum will be at its maximum positive displacement when the other it at its maximum negative displacement.

Study tip

Phase difference is measured in radians or degrees.

Angular frequency

There is also a fundamental connection between oscillating motion and circular motion. One complete oscillation has many of the same characteristics as one complete revolution in circular motion.

Figure 3 shows an object **P** travelling at a constant angular velocity in a circle of radius A. If you consider only the motion of this object along the x-axis, the displacement x is given by $A\cos\theta$. The angle θ increases uniformly with time t, so the graph of displacement x against time t is very similar to that of the rolling ball in Figure 2.

Angular frequency is a term used to describe the motion of an oscillating object and is closely related to the angular velocity of an object in circular motion. The angular frequency of an oscillating object is given by the following equations.

$$\omega = \frac{2\pi}{T} \text{ or } \omega = 2\pi f$$

where T is the period of the oscillator and f is its frequency.

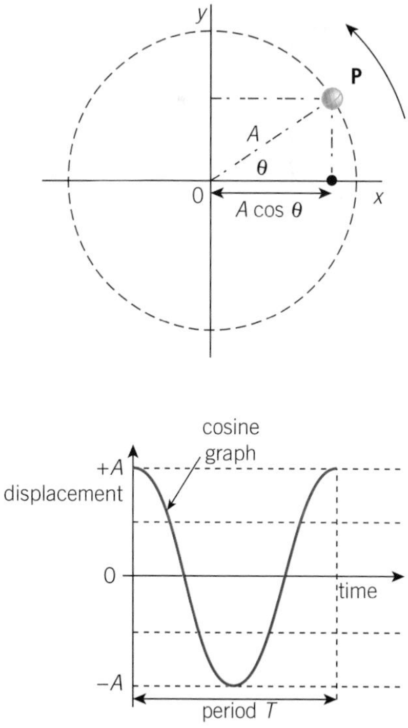

▲ **Figure 3** *Connection between circular motion and oscillatory motion*

Study tip

In simple harmonic motion, the acceleration is directly proportional to the displacement and is always directed towards a fixed point.

Simple harmonic motion

Simple harmonic motion (SHM) is a common kind of oscillating motion defined as oscillating motion for which the acceleration of the object is given by:

$$a = -\omega^2 x$$

where ω^2 is a constant for the object. From this equation we can see two key features of all objects moving with SHM:

- The acceleration a of the object is directly proportional to its displacement x, that is $a \propto x$.
- The minus sign means that the acceleration of the object acts in the direction opposite to the displacement (it returns the object to the equilibrium position).

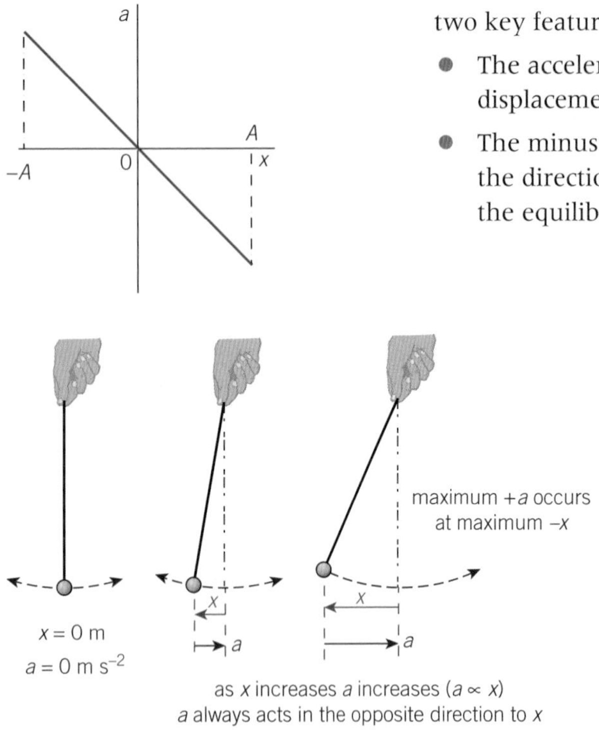

▲ **Figure 4** *A simple pendulum moves in simple harmonic motion, with $a \propto -x$. The acceleration is maximum when $x = \pm A$, where A is the amplitude.*

A graph of acceleration against displacement for any object moving in SHM shows these two important features (Figure 4). The gradient of the acceleration against displacement graph is equal to $-\omega^2$, where ω is the angular frequency of the oscillating object. Since the gradient for a particular oscillator is constant, this implies that the frequency (and the period) of the oscillator is also constant. An important aspect of SHM is that the period T of the oscillator is independent of the amplitude A of the oscillator. For example, the period of a simple pendulum moving in SHM does not depend on the amplitude of the swing. As the amplitude increases so does the average speed of the swing, so the period does not change. Such an oscillator is referred to as an **isochronous oscillator** (in Greek, 'iso' means 'the same' and 'chronos' means 'time').

Worked example: Frequency and maximum acceleration of a pendulum bob

A pendulum bob is in SHM with a period of 0.60 s and an amplitude of 5.0 cm. Calculate the frequency of the oscillations and the maximum acceleration of the bob.

Step 1: Calculate the frequency of the pendulum (not the angular frequency).

$$f = \frac{1}{T} = \frac{1}{0.60} = 1.7\ \text{Hz (2 s.f.)}$$

Step 2: To calculate the maximum acceleration, first determine the angular frequency.

$$\omega = \frac{2\pi}{T} = \frac{2\pi}{0.60} = 10.4\ldots\ \text{rad s}^{-1}$$

Step 3: Since the maximum acceleration occurs at the maximum amplitude

$$a = -\omega^2 x \text{ becomes } a_{max} = -\omega^2 A$$

Substituting in known values in SI units gives: $a_{max} = -10.4\ldots^2 \times 0.050 = -5.5\ \text{m s}^{-2}$ (2 s.f.)

The acceleration is a negative value as we have used a positive value for the displacement (0.050 m). The maximum positive acceleration would be 5.5 m s⁻², at the maximum negative displacement (−0.050 m).

Determining the period and frequency of objects moving with SHM

Figure 5 shows two simple harmonic oscillators – a simple pendulum and a mass attached to a spring. You can time a number of oscillations using a stopwatch and show that the period is independent of the amplitude.

The period of each oscillation can be measured using a stopwatch. Commonly, the time taken for several complete oscillations is measured, the process is repeated, and the period is then calculated.

A small pin or other marker can be placed at the equilibrium position. This **fiducial marker** provides a clear point from which to start and stop any timing measurements.

1 Suggest two reasons why a fiducial marker should be placed at the equilibrium position.
2 Explain why, when investigating the time period, it is sensible to record the time taken for several oscillations, rather than a single swing.

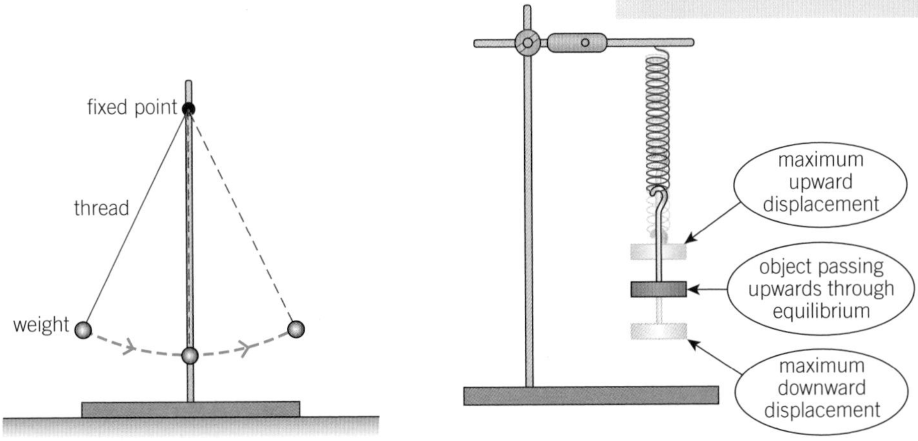

▲ **Figure 5** *You can time the oscillations of these objects and show they are both isochronous oscillators*

Summary questions

1 Sketch a graph of displacement against time for an oscillator consisting of a mass attached to the end of a spring. *(2 marks)*

2 Calculate the angular frequency of a pendulum that
 a has a period of 0.40 s
 b has a frequency of 0.75 Hz
 c completes 20 oscillations in 26 s. *(4 marks)*

3 Calculate the acceleration of an object moving in SHM with an angular frequency of 2.5 rad s^{-1} at each of the following displacements:
 a 0.12 cm b 0.00 cm. *(3 marks)*

4 State the phase difference between a pendulum at its equilibrium position and another identical pendulum which is:
 a at its maximum positive displacement
 b at its equilibrium position but moving in the opposite direction. *(2 marks)*

5 Two objects are moving in SHM. Their accelerations are given by the equations $a = -10x$ and $a = -40x$. Compare the motion of each object by calculating their periods. *(4 marks)*

6 Figure 6 shows a graph of acceleration a against displacement x for a vibrating metal strip. Use the graph to determine
 a the amplitude of the motion
 b the angular frequency ω of this oscillator. *(4 marks)*

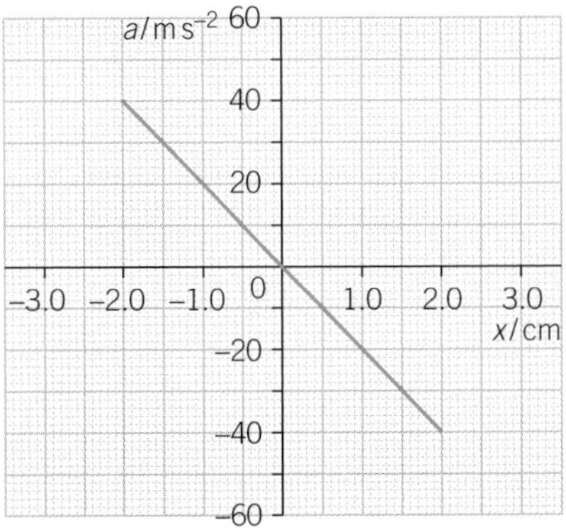

▲ **Figure 6**

17.2 Analysing simple harmonic motion

Specification reference: 5.3.1

Learning outcomes

Demonstrate knowledge, understanding, and application of:

→ solutions to the equation $a = -\omega^2 x$

→ velocity $v = \pm\omega\sqrt{A^2 - x^2}$, hence $v_{max} = \omega A$

→ graphical methods to relate the changes in displacement, velocity, and acceleration during simple harmonic motion.

Study tip

A sinusoidal graph is a sine or cosine shaped graph.

▲ **Figure 1** *The pendulum inside the tower swings back and forth with a period of precisely 2 s, which can be adjusted by adding and removing pennies*

Synoptic link

For more on velocity as the gradient of displacement–time graphs see Topic 3.2, Displacement and velocity.

Ticking clocks

The clock tower of Big Ben is one of the most famous examples of a pendulum clock. Like all pendulums, its pendulum is an example of an isochronous oscillator. The pendulum is 4.4 m long and has a mass of 310 kg. However, adding a single coin to the top of the pendulum has the effect of slightly lifting the pendulum's centre of gravity (effectively changing its length), reducing the duration of the swings by just 400 ms per day.

The pendulum inside the tower is a little more complex than a simple pendulum, but its swing, too, is another example of SHM. We can describe the motion of any object moving in SHM using just a few graphs and equations.

Using graphs to demonstrate SHM

We have seen how the graph of displacement against time for an oscillator moving in SHM has a sinusoidal shape. If no energy is transferred to the surroundings, then the amplitude A of each oscillation remains constant.

Taking a simple pendulum as an example, Figure 2(a) shows the variation of displacement x with time t. At zero displacement the pendulum is at, or moving through, its equilibrium position, and at the maximum displacements it is at the top of its swing. The pendulum is at its maximum positive displacement at time $t = 0$, giving the graph a cosine shape. It is also common to see a sine graph.

The gradient of a displacement–time graph is equal to the velocity v of the oscillator. At maximum displacements, the velocity is zero because the gradient of the graph is zero. The pendulum has momentarily stopped, before it returns towards its equilibrium position. The velocity and the gradient of the graph are at their maximum as the pendulum moves through its equilibrium position.

Figure 2(b) shows how the velocity v of the pendulum varies with time t, and Figure 2(c) shows how the acceleration a of the pendulum varies with time. The acceleration a can be determined from the gradient of the velocity–time graph. The acceleration–time graph is similar to the displacement–time graph, except 'inverted'. Therefore $a \propto -x$.

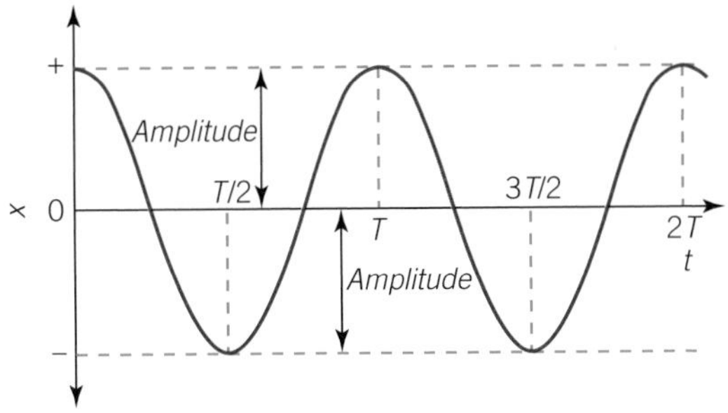

(a) *x-t* graph for a simple harmonic oscillator

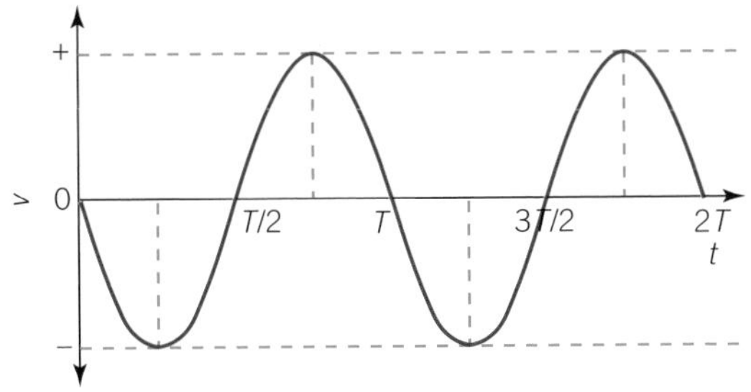

(b) *v-t* graph for a simple harmonic oscillator. You can get this graph by determining the gradient of the *x-t* graph.

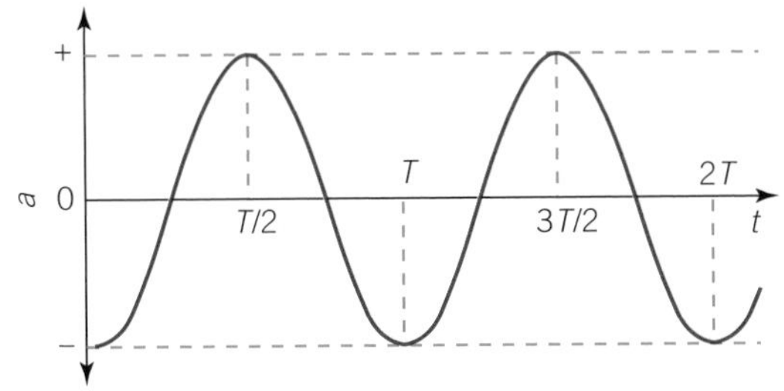

(c) *a-t* graph for the simple harmonic oscillator. You can get this graph by determining the gradient of the *v-t* graph.

▲ **Figure 2** *SHM graphs of (a) displacement, (b) velocity (the gradient of graph a), and (c) acceleration (the gradient of graph b) against time*

A mathematical treatment of SHM

Displacement

The defining equation for SHM is $a = -\omega^2 x$. As you have seen, the displacement x of a simple harmonic oscillator varies with time t in a sinusoidal manner. This means that the displacement–time graph can be described by either a sine or a cosine graph. The two commonly used equations for the displacement x are

$$x = A\cos\omega t \qquad x = A\sin\omega t$$

where ω is the angular frequency of the oscillator. These equations are solutions to the equation $a = -\omega^2 x$. You do not need to know how they are derived, but you must be able to apply them.

Which equation you use depends on *where* the oscillating object is at time $t = 0$. If the object begins oscillating from its amplitude (for example, a pendulum lifted from its equilibrium position and released), then at time $t = 0$ the object is at its positive amplitude, so use the cosine version.

If the object begins oscillating from its equilibrium position (for example, a pendulum flicked from its equilibrium position), then at time $t = 0$ the object is at its equilibrium position and its displacement is 0. In this case use the use the sine version.

Study tip

A sinusoidal graph doesn't necessarily mean a sine graph — it can be either a sine or a cosine graph.

Study tip

Calculators needs to be in radian mode (rad) before completing any SHM calculation involving sines or cosines.

Worked example: The displacement of a simple pendulum

A simple pendulum is released 30 cm from its equilibrium position and completes one oscillation every 1.4 s. Calculate its displacement 6.2 s after it was released.

Step 1: Calculate the angular frequency of the pendulum.

$$\omega = \frac{2\pi}{T} = \frac{2\pi}{1.4} \; (= 4.5\,\text{rad}\,\text{s}^{-1})$$

Step 2: Use this value to determine the displacement.

As the pendulum was released at its maximum displacement we use the cosine version of the equation. Make sure the calculator is in 'rad' mode.

$$x = A\cos\omega t = 0.30 \times \cos\left(\frac{2\pi}{1.4} \times 6.2\right) = -0.27\,\text{m (2 s.f.)}$$

As this is a negative value, the pendulum must be on the opposite side of the equilibrium position from the release point.

Study tip

For a simple harmonic oscillator, when velocity is a maximum the acceleration is zero, and when acceleration is maximum the velocity is zero.

Velocity

You know that in SHM velocity v varies with time. Have a closer look at Figures 2(a) and (b). Firstly, consider what happens to the shape of the velocity–time graph when the angular frequency ω of the oscillator is increased with no change in amplitude. The oscillator will be travelling the same distance in a shorter time interval. Therefore the maximum velocity v_{max} of the oscillator (when its displacement is $x = 0$) will increase, and thus the gradient of the displacement–time graph there will increase.

Next, consider the effect of increasing the amplitude A of the oscillator. Being an isochronous oscillator, it will travel a greater distance in the same time interval; hence the maximum velocity of the oscillator will also increase. You can calculate the velocity v of a simple harmonic oscillator at displacement x using the equation

$$v = \pm\omega\sqrt{A^2 - x^2}$$

The velocity at any particular displacement has a positive or a negative value, depending on the direction in which the oscillator is moving.

From this equation it follows that the velocity can vary between zero (at $x = A$) to its maximum values, $\pm v_{max}$, at the equilibrium position. At the equilibrium position, $x = 0$, and so the equation becomes

$$v_{max} = \omega A$$

Worked example: The velocity of an oscillating toy

A toy bird is suspended on a spring from the ceiling. It is pulled 6.0 cm downwards and, upon release, oscillates up and down with an angular frequency of 9.1 rad s^{-1}. Calculate the magnitude of the velocity of the toy after 3.4 s.

Step 1: First determine the displacement after 3.4 s

$$x = A\cos\omega t = 0.060 \times \cos(9.1 \times 3.4) = 0.0533...\,\text{m}$$

Step 2: Use the velocity equation to calculate the velocity at that displacement.

$v = \pm\omega\sqrt{A^2 - x^2} = \pm 9.1\sqrt{0.060^2 - 0.0533...^2} = \pm\ 0.25\,\text{m s}^{-1}$ (2 s.f.)

The magnitude of the velocity is therefore $0.25\,\text{m s}^{-1}$.

Summary questions

1 Sketch a graph of displacement against time for a pendulum moving in SHM through two complete oscillations. Label the points where the pendulum is stationary and the points where it is moving with maximum velocity. *(4 marks)*

2 Calculate the velocity of the pendulum in the worked example: "The displacement of a simple pendulum" on page 323 when the displacement is:

- **a** 0.20 m
- **b** 0.00 m
- **c** 0.30 m. *(5 marks)*

3 A simple pendulum completes 20 oscillations in 16 s. It is initially released from its amplitude of 0.16 m. Calculate its displacement after:

- **a** 0.40 s
- **b** 0.80 s
- **c** 19.30 s. *(6 marks)*

4 The graph in Figure 3 shows the motion of a mass–spring oscillator moving in SHM. Use the graph to determine:

- **a** the amplitude of the oscillation
- **b** the period
- **c** the angular frequency
- **d** the maximum velocity of the oscillator. *(5 marks)*

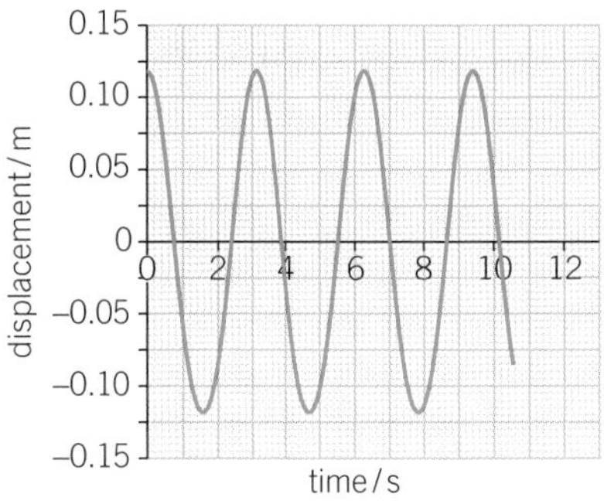

▲ **Figure 3**

5 The displacement of an object moving in SHM is given by $x = 0.12 \sin(3.5t)$. Calculate:

- **a** the amplitude of the oscillation
- **b** the angular frequency
- **c** the period
- **d** the displacement after:
  - **i** 0.0 s
  - **ii** 3.5 s
  - **iii** 14 s. *(7 marks)*

6 Sketch graphs of displacement against time, velocity against time, and acceleration against time for the object described in question 5 as it moves through two complete oscillations, starting from $t = 0$. *(6 marks)*

17.3 Simple harmonic motion and energy

Specification reference: 5.3.2

Learning outcomes

Demonstrate knowledge, understanding, and application of:

- → the interchange between kinetic and potential energy during simple harmonic motion
- → energy–displacement graphs for a simple harmonic oscillator.

▲ **Figure 1** *Motion on a swing can be modelled as simple harmonic motion, similar to that of a simple pendulum*

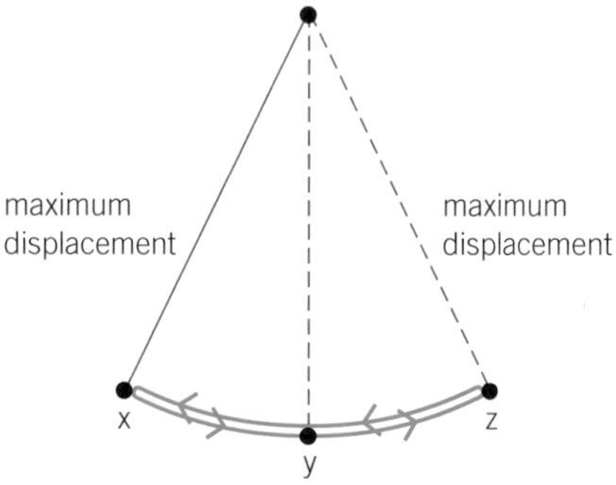

Position	E_p	E_k
x	E_{TOTAL}	0
y	0	E_{TOTAL}
z	E_{TOTAL}	0

▲ **Figure 2** *Transfer of gravitational potential energy to kinetic energy and back in a swinging pendulum*

▶ **Figure 3** *A graph of energy against displacement for an object moving in SHM with an amplitude A*

Legs forward, legs back...

Motion on a swing (Figure 1) can be modelled as another example of SHM. During each swing, energy is transferred from gravitational potential energy at the top of each swing to kinetic energy at the bottom, and vice versa. To swing higher, you need to move faster, which will transfer more kinetic energy into gravitational potential energy.

This same interchange between potential energy and kinetic energy occurs in all examples of simple harmonic motion. It can be described using a graph of energy against displacement.

Graphs of energy against displacement

For any object moving in SHM the total energy remains constant, as long as there are no losses due to frictional forces. Figure 2 shows the energy changes for a simple pendulum. At the amplitude the pendulum is briefly stationary and has zero kinetic energy. All its energy is in the form of potential energy (in this case, gravitational potential energy). As the pendulum falls it loses potential energy and gains kinetic energy. It has maximum velocity, and so maximum kinetic energy, as it moves through its equilibrium position. As the pendulum passes through the equilibrium position, it has no potential energy.

A similar interchange of potential energy and kinetic energy occurs for a mass–spring system. In this case, if the mass is oscillating vertically, the potential energy is in the form of gravitational potential energy (due to the position of the mass in the Earth's gravitational field) and elastic potential energy (stored in the spring). If the mass is oscillating horizontally, the potential energy is in the form of elastic potential energy only.

A graph of energy against displacement shows how the total energy of an oscillating system remains unchanged (the green line in Figure 3). There is a continuous interchange between potential energy E_p and kinetic energy E_k, but the sum at each displacement is always constant and equal to the total energy of the object.

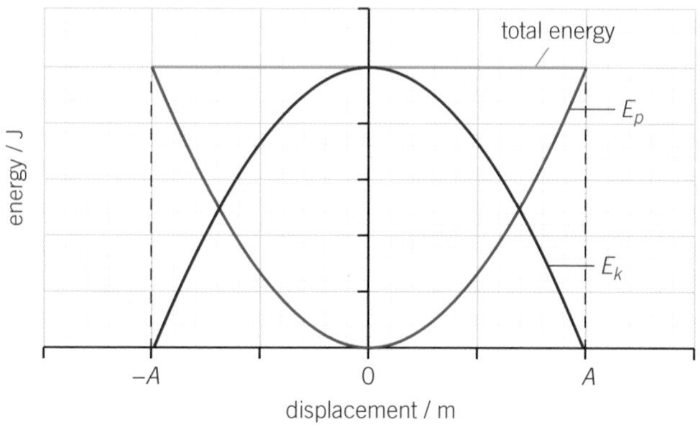

The red line on the graph represents the potential energy, and is zero at the equilibrium position. The purple line represents the kinetic energy and is zero at the amplitude.

Understanding the graphs

Figure 4 shows a simple spring–mass system of a glider on a horizontal track. When displaced, the glider will move in SHM. This system has two forms of energy – the elastic potential energy in the spring and the kinetic energy of the spring and glider. The displacement x of the glider is the same as the extension or compression of the spring.

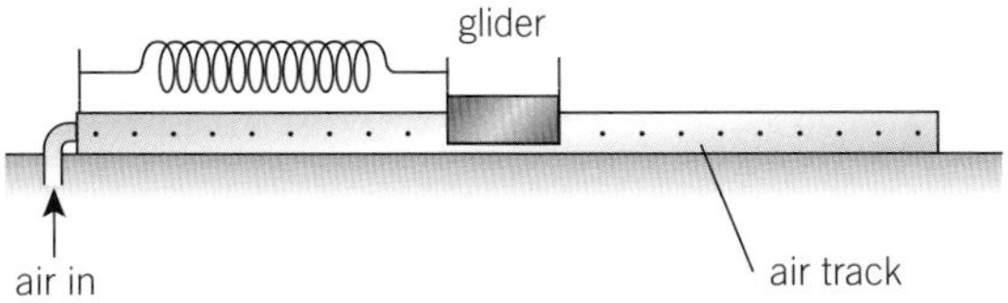

▲ **Figure 4** *A horizontal track carrying a glider attached to one end of a spring that is fixed at the other end and can be both compressed and extended*

Synoptic link

The equation for potential energy in a spring was introduced in Topic 6.2, Elastic potential energy.

- The elastic potential energy E_p is given by the equation $E_p = \frac{1}{2}kx^2$, where k is the force constant of the spring. A graph of E_p against x will therefore be a parabola (as E_p is proportional to x^2).
- The elastic potential energy is always positive and varies from $E_p = 0$ when $x = 0$ to $E_p = \frac{1}{2}kA^2$ when $x = A$ (amplitude).
- When $x = A$, the glider will be stationary for an instant. This means that it has no kinetic energy. The total energy of the oscillator must therefore be equal to $\frac{1}{2}kA^2$.
- The kinetic energy E_k of the glider at any instant must be the difference between the total energy and the elastic potential energy. Therefore

$$E_k = \frac{1}{2}kA^2 - \frac{1}{2}kx^2 = \frac{1}{2}k(A^2 - x^2)$$

- A graph of E_k against x will therefore be an inverted parabola (as seen in Figure 3).

Worked example: Calculating the kinetic energy of a mass–spring system

A mass of 400 g forms part of a mass–spring system. It is initially released from its amplitude of 10 cm and completes one oscillation every 0.80 s. Calculate its kinetic energy 2.1 s after release and determine this as a percentage of its maximum kinetic energy.

Step 1: Calculate the angular frequency of the mass–spring system.

$$\omega = \frac{2\pi}{T} = \frac{2\pi}{0.80} = 7.85\ldots \text{ rad s}^{-1}$$

Step 2: Use this value to determine the displacement after 2.1 s. Make sure the calculator is in 'rad' mode.

$$x = A\cos\omega t = 0.10 \times \cos(7.85... \times 2.1) = -0.0707...\,\text{m}$$

Step 3: Use the equation $v = \pm\omega\sqrt{A^2 - x^2}$ to determine the velocity at this displacement.

$$v = \pm 7.85... \sqrt{0.10^2 - 0.0707...^2} = 0.555...\,\text{m s}^{-1}$$

Step 4: Finally, determine the kinetic energy and the maximum kinetic energy.

$$E_k = \frac{1}{2}mv^2 = \frac{1}{2} \times 0.400 \times 0.555...^2 = 0.0616... = 0.061 \text{ (2 s.f.)}$$

Since the maximum velocity is given by $v_{max} = \omega A$, the maximum kinetic energy can be expressed as

$$E_{k\,max} = \frac{1}{2}m(\omega A)^2 = \frac{1}{2} \times 0.400 \times (7.85... \times 0.10)^2 = 0.123...\,\text{J}$$

Therefore, at 2.1 s the energy as a percentage of the maximum kinetic energy is given by

$$\frac{0.0616...}{0.123...} \times 100 = 50\% \text{ (2 s.f.)}$$

Summary questions

1 Sketch a graph showing how the kinetic energy and potential energy of a mass–spring system varies with displacement. State any assumptions made. Label your graph to show:

a the amplitude

b the maximum and minimum kinetic and potential energy. *(6 marks)*

2 The total energy of a mechanical oscillator is 1.6 J.

a State the maximum values of its kinetic energy and potential energy. *(2 marks)*

b The oscillator has mass 0.120 kg. Calculate its speed when the potential energy of the oscillator is 1.0 J. *(3 marks)*

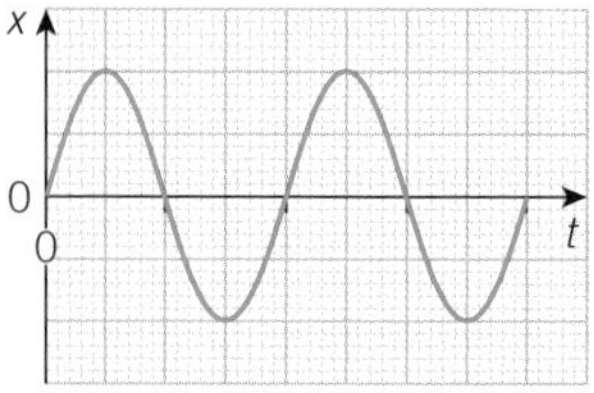

▲ Figure 5

3 Figure 5 shows the graph of displacement x against time t for an oscillator. Copy the graph and, aligned beneath it, sketch graphs to show the variation of:

a potential energy E_p with time t; *(2 marks)*

b kinetic energy E_k with time t. *(2 marks)*

4 A simple pendulum has a mass of 50 g. It is initially released from its amplitude of 0.050 m and has frequency 0.40 Hz. Calculate its kinetic energy 2.8 s after release. *(4 marks)*

5 Show that the maximum kinetic energy of an object moving in SHM is given by $E_{k\,max} = 2m\pi^2 f^2 A^2$. *(3 marks)*

17.4 Damping and driving

Specification reference: 5.3.3

Shocking

The suspension forks found on modern mountain bikes can be adjusted in several ways, affecting features like the travel, rebound, and, importantly, the amount of **damping**. Without damping to absorb the energy of the shock, after hitting a bump the front end of the bike would oscillate up and down like a mass–spring system.

Damping

An oscillation is damped when an external force that acts on the oscillator has the effect of reducing the amplitude of its oscillations. For example, a pendulum moving through air experiences air resistance, which damps the oscillations until eventually the pendulum comes to rest.

There are many forms of damping. When the damping forces are small, the amplitude of the oscillator gradually decreases with time, but the period of the oscillations is almost unchanged. This type of damping is referred to as **light damping**. This would be the case for a pendulum oscillating in air.

For larger damping forces, the amplitude decreases significantly, and the period of the oscillations also increases slightly. This type of **heavy damping** would occur for a pendulum oscillating in water. Now imagine an oscillator, such as a pendulum, moving through treacle or oil. In this example of very heavy damping, there would be no oscillatory motion. Instead the oscillator would slowly move towards its equilibrium position. Figure 2 shows the displacement–time graphs for light damping, heavy damping, and very heavy damping.

In all cases of damped motion, the kinetic energy of the oscillator is transferred to other forms (usually heat).

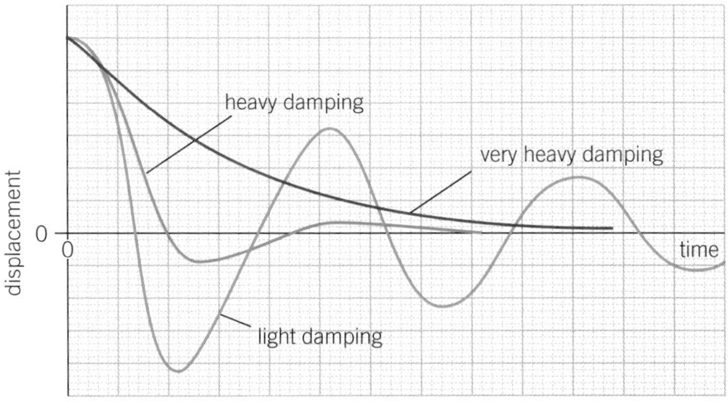

▲ **Figure 2** *The effects of different types of damping*

Learning outcomes

Demonstrate knowledge, understanding, and application of:

→ the effects of damping on an oscillatory system

→ free and forced oscillations

→ natural frequency and resonance

→ observing forced and damped oscillations for a range of systems.

▲ **Figure 1** *Damping is an essential feature of all suspension systems, absorbing the energy from bumps*

Synoptic link

There are many other examples of exponential decay in nature, including capacitance and radioactive decay – see Topic 21.4, Discharging capacitors, and Topic 25.3, Half-life and activity.

Free and forced oscillations

When a mechanical system is displaced from its equilibrium position and then allowed to oscillate without any external forces, its motion is referred to as **free oscillation**. The frequency of the free oscillations is known as the **natural frequency** of the oscillator.

A **forced oscillation** is one in which a periodic driver force is applied to an oscillator. In this case the object will vibrate at the frequency of the driving force (the **driving frequency**). For example, a mass hanging on a vertical spring can be forced to oscillate up and down at a given frequency if the top of the spring is held and the hand moves up and down. The hand is the driver and its motion provides a driver frequency that forces the mass–spring system to oscillate (Figure 3).

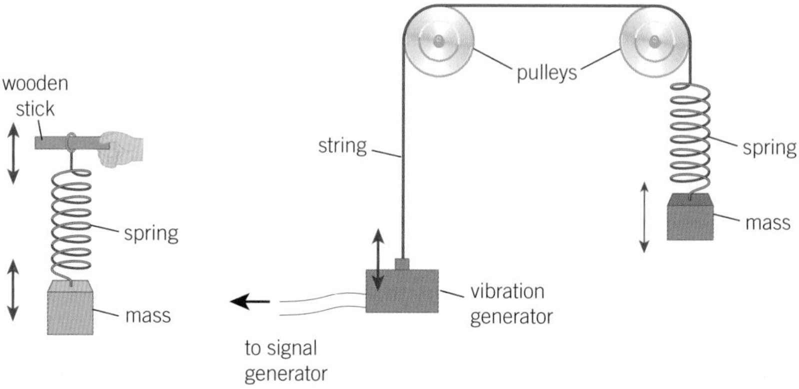

▲ **Figure 3** *Two examples of forced oscillation – to investigate damping, the object could be placed in water or oil*

If the driving frequency is equal to the natural frequency of an oscillating object, then the object will **resonate**. This will cause the amplitude of the oscillations to increase dramatically, and if not damped, the system may break. There is more on this in Topic 17.5.

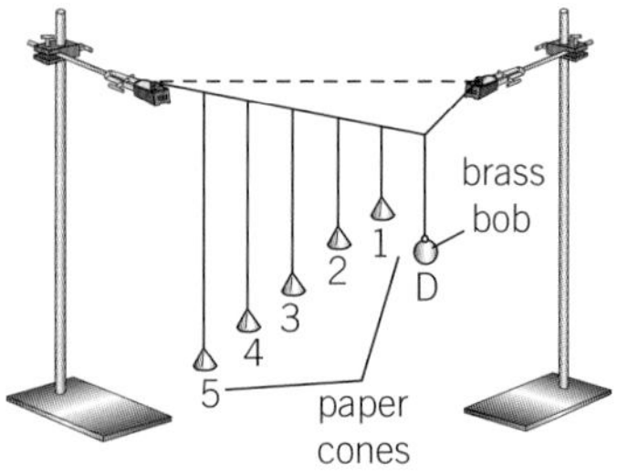

▲ **Figure 4** *Forced oscillation – Barton's pendulums*

Barton's pendulums provide another example of a forced oscillation (Figure 4). A number of paper cone pendulums of varying lengths are suspended from a string along with a heavy brass bob **D**. This heavy pendulum acts as the driver for the paper cone pendulums. The pendulum **D** oscillates at its natural frequency and forces all the other pendulums to oscillate at the same frequency. As pendulum 2 has the same length as pendulum **D** it has the same natural frequency. It will resonate and its amplitude will be greater than the other pendulums.

Damping and exponential decrease

In some examples of damping, the amplitude of a damped oscillating system decreases exponentially with respect to time. This is referred to as an **exponential decay**. A good example is a pendulum oscillating in air or a spring–mass system damped by air (Figure 5).

In any exponential decay the physical quantity (in this case amplitude) decreases by the same factor in equal time intervals. For example, for an amplitude A that decays exponentially, if it is measured every 4 seconds, then

$$\frac{A_4}{A_0} = \frac{A_8}{A_4} = \frac{A_{12}}{A_8} = \text{constant}$$

This constant-ratio property is the defining characteristic of an exponential decay.

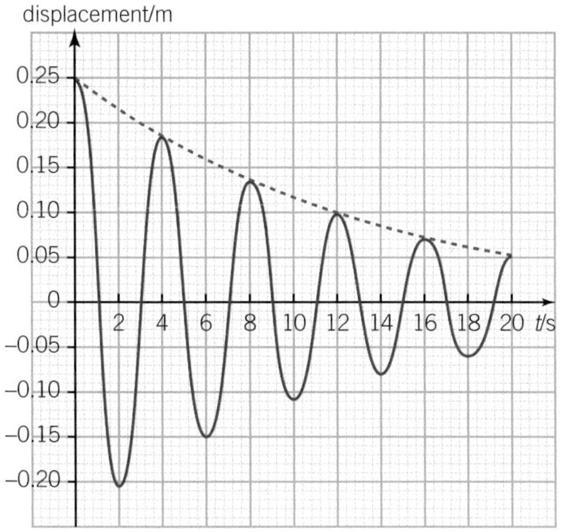

▲ **Figure 6** *Decay of the amplitude of a pendulum oscillating in air*

1 Determine the initial amplitude, the period, and the angular frequency of the pendulum in Figure 4.

2 Use the graph to determine if the amplitude decays exponentially.

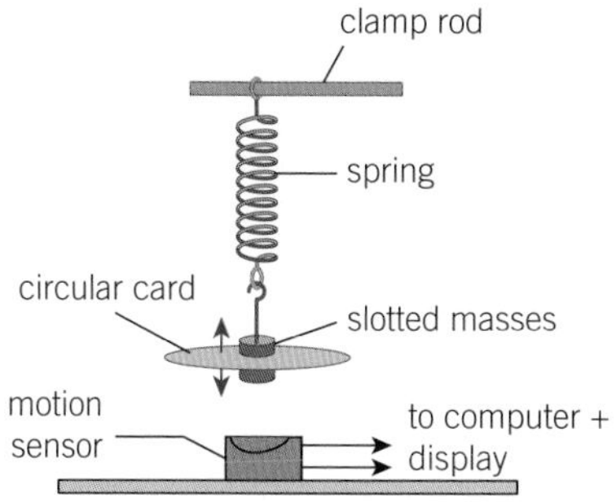

▲ **Figure 5** *An example of a damped spring–mass system*

Summary questions

1 Give one example of a free oscillation and one of forced oscillation. *(2 marks)*

2 Describe how damping affects the amplitude of an oscillating object. *(1 mark)*

3 Sketch a graph of displacement against time for a simple pendulum with

a no damping

b a small amount of damping. *(4 marks)*

4 A data-logger is used to monitor the oscillations of a lightly damped oscillator. The results are shown in Figure 7.

a State why the amplitude decreases with time. *(1 mark)*

b State one quantity that remains constant for the oscillations. *(1 mark)*

c Determine the natural frequency of the oscillator. *(2 marks)*

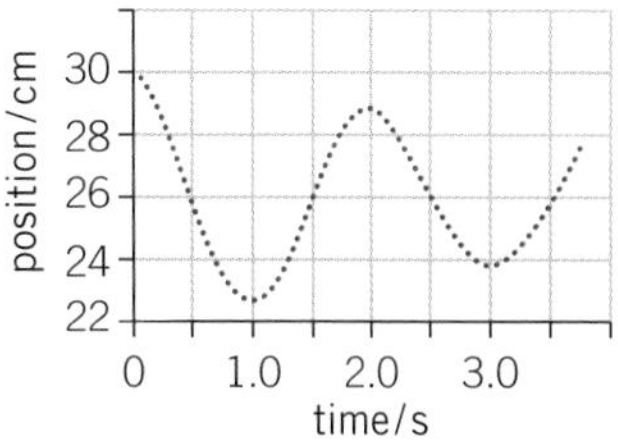

▲ **Figure 7**

5 The amplitude of a damped oscillator decreases exponentially with time. At time $t = 0$, the amplitude is 5.0 cm. The amplitude decreases to 90% after each period. The period of the oscillations is 1.0 s. Sketch the displacement–time graph for this damped oscillator up to a time of 5.0 s. *(4 marks)*

17.5 Resonance

Specification reference: 5.3.3

Learning outcomes

Demonstrate knowledge, understanding, and application of:

→ resonance and natural frequency

→ amplitude-driving frequency graphs for forced oscillators

→ practical examples of forced oscillations and resonance.

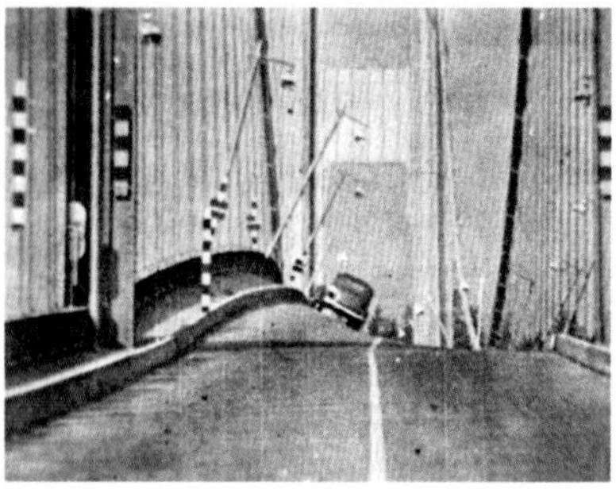

▲ **Figure 1** *The main span of the Tacoma Narrows Bridge in Washington state, USA, oscillated with an amplitude of several metres until eventually the bridge collapsed*

Oscillating to destruction

The original Tacoma Narrows Bridge was first opened in 1940. It was designed to withstand hurricane-force winds, yet a wind of just 40 mph famously brought it crashing down just four months later. The wind caused a type of resonance, causing the bridge to oscillate with increasing amplitude until eventually the structure failed and the bridge collapsed.

Resonance

Resonance is the effect that allows an opera singer to break a wine glass with just their voice. It occurs when the driving frequency of a forced oscillation is equal to the natural frequency of the oscillating object. In the case of the wine glass, resonance occurs when the frequency of the sound produced by the singer is equal to the natural frequency of the wine glass.

For a forced oscillator with negligible damping, at resonance

driving frequency = natural frequency of the forced oscillator

When an object resonates, the amplitude of the oscillation increases considerably. If the system is not damped, the amplitude will increase to the point at which the object fails – the glass breaks, or the bridge collapses. The greatest possible transfer of energy from the driver to the forced oscillator occurs at the resonant frequency. This is why the amplitude of the forced oscillator is maximum. In the case of the Tacoma Narrows Bridge, the kinetic energy from the wind was efficiently transferred to the bridge, leading to its ultimate collapse.

Examples of resonance

As well as causing a problem for engineers designing buildings and bridges, resonance can have useful effects:

- Many clocks keep time using the resonance of a pendulum or of a quartz crystal:
- Many musical instruments have bodies that resonate to produce louder notes.
- Some types of tuning circuits (for example in car radios) use resonance effects to select the correct frequency radio wave signal.
- Magnetic resonance imaging (MRI) enables diagnostic scans of the inside of our bodies to be obtained without surgery or the use of harmful X-rays.

Magnetic resonance imaging

Magnetic resonance imaging (MRI) relies on the resonance of hydrogen nuclei found in the water molecules within tissues inside the body.

Inside the scanner there is a strong magnetic field created by superconducting electromagnets. The hydrogen nuclei behave like tiny magnets and precess (a kind of rotation effect) in this magnetic field. The precession occurs at different natural frequencies depending on the type of molecule and thus occurs at different natural frequencies for different tissues in which the hydrogen nuclei are found. Radio waves from transmitting coils inside the scanner cause the nuclei to resonate and absorb energy.

When the radio waves from the transmitter are switched off, the hydrogen nuclei 'relax' and re-emit the energy gained as radio wave photons of specific wavelengths. These are detected by numerous receiving coils surrounding the scanner. The signals from these coils are processed by high-speed computers and the software helps to produce a three-dimensional image of the patient.

MRI scanning offers a number of advantages over some other forms of medical imaging. For example, unlike X-rays they do not expose the patent to ionising radiation and can produce clear images of soft tissue like the brain and heart (Figure 2).

1. Identify the driver and forced oscillators in an MRI scanner.
2. The natural frequency of the hydrogen nuclei depends on the magnetic field inside the MRI. Inside a certain MRI scanner, the natural frequency of the hydrogen nuclei is 128 MHz. Calculate the wavelength of the radio waves that would cause the nuclei to resonate.
3. Calculate the energy of the radio wave photons emitted from the relaxing hydrogen nuclei.

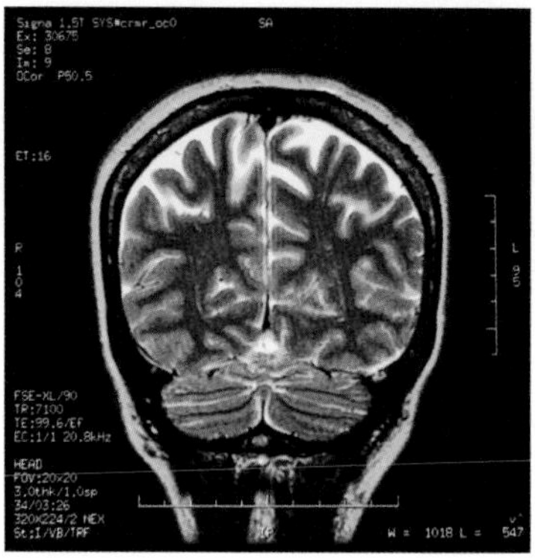

▲ **Figure 2** *An MRI scan of the brain*

Synoptic link

You can learn about other forms of medical imaging in Chapter 27, Medical imaging.

Resonance and damping

The Millennium Bridge in London was opened in June 2000 (Figure 3). It was quickly nicknamed the "wobbly bridge" after it was discovered to resonate when large numbers of people walked across it. As the bridge started to sway, pedestrians tended to match their step to the sway, providing a driving force that was very close to the natural frequency of the bridge. To prevent a possible collapse like that over the Tacoma Narrows, the bridge was closed for two years to allow engineers to install dampers.

Damping a forced oscillation has the effect of reducing the maximum amplitude at resonance. The degree of damping also has an effect on the frequency of the driver when maximum amplitude occurs. Figure 4 shows the effect of damping on the graph of amplitude against forcing frequency for an oscillator with a natural frequency f_0.

▲ **Figure 3** *The Millennium Bridge across the Thames, now with added dampers*

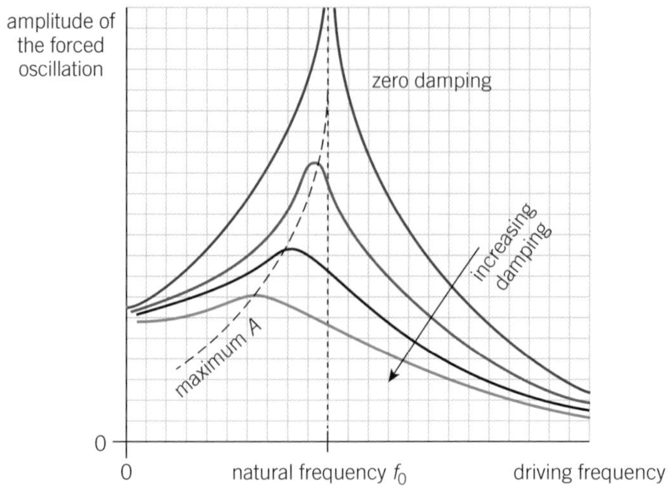

▲ **Figure 4** *Increasing damping can prevent the effects of resonance from becoming severe*

From Figure 4 you can see that:

- For light damping, the maximum amplitude occurs at the natural frequency f_0 of the forced oscillator.
- As the amount of damping increases:
  - the amplitude of vibration at any frequency decreases
  - the maximum amplitude occurs at a lower frequency than f_0
  - the peak on the graph becomes flatter and broader.

Summary questions

1 Sketch a graph of amplitude against driver frequency for an experiment on mass–spring forced oscillation carried out like the one in Figure 3 of Topic 17.4. Use this graph to explain the effect of resonance on an object. *(5 marks)*

2 Explain why a glass shatters when exposed to a sound at the natural frequency of the glass. *(3 marks)*

3 The side panel of a tumble dryer vibrates loudly when the dryer spins at a specific angular frequency. Explain why it only happens at a certain frequency and suggest a technique to reduce vibration. *(4 marks)*

4 An object is suspended from a spring. When displaced, the object executes 180 complete oscillations in a time of 1 minute. The upper end of the spring is then suspended from a mechanical oscillator. The spring–mass system is forced to oscillate. The frequency of the mechanical oscillator is gradually increased from zero. Sketch a graph to show the variation of amplitude of the object with the frequency of the mechanical oscillator. *(3 marks)*

5 An old van with an undamped suspension system drives over three speed bumps 10 m apart at a speed of $2.5\,\mathrm{m\,s^{-1}}$. The front end of the van begins to resonate. State the natural frequency of the suspension and explain why driving over the bumps at a different speed would reduce the amplitude of the oscillations. *(4 marks)*

Practice questions

1 a For a body undergoing simple harmonic motion describe the difference between

(i) displacement and amplitude *(2 marks)*

(ii) frequency and angular frequency. *(2 marks)*

b A harbour, represented in Figure 1, has vertical sides and a flat bottom. The surface of the water in the harbour is calm.

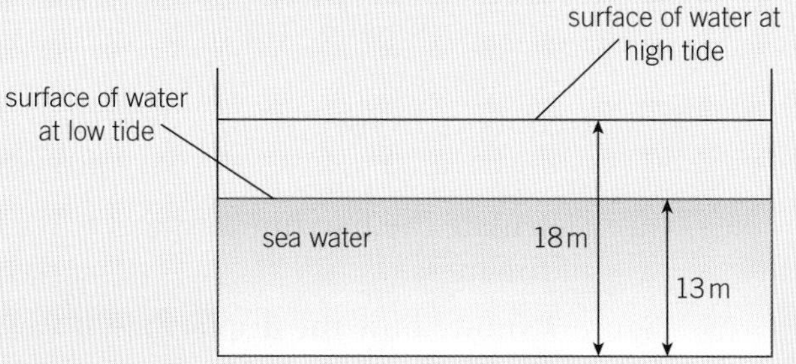

▲ Figure 1

The tide causes the surface of the water to perform simple harmonic motion with a period of 12.5 hours. The maximum depth of the water is 18 m and the minimum depth is 13 m.

(i) For the oscillation of the water surface, calculate

1 The amplitude *(1 mark)*

2 The frequency *(2 marks)*

(ii) Calculate the maximum vertical speed of the water surface. *(2 marks)*

(iii) Write an expression for the depth *d* in metres of water in the harbour in terms of time *t* in seconds. *(2 marks)*

Jan 2011 G484

2 Figure 2 shows a glider, tethered between two stretched springs, floating above a linear air track.

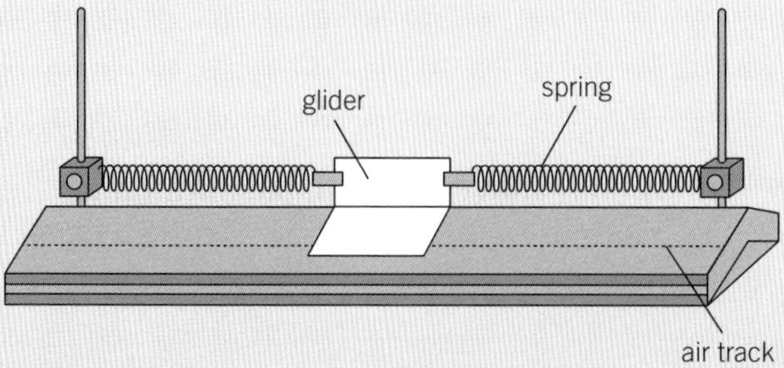

▲ Figure 2

The glider is pulled to one side and released. It oscillates in simple harmonic motion. The variation of the speed *v* of the glider with time *t* is shown in Figure 3.

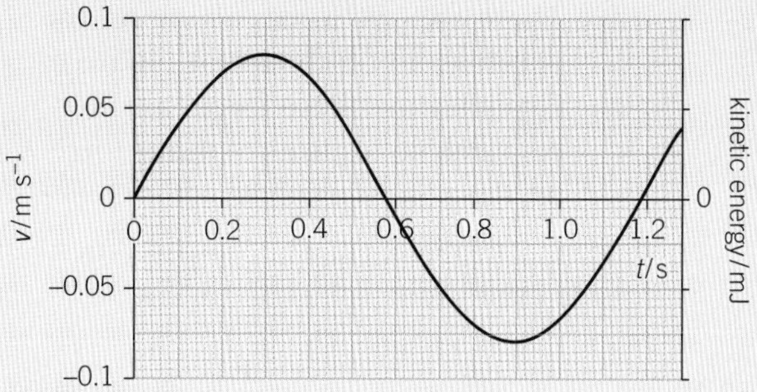

▲ Figure 3

a Calculate the frequency of the oscillation. *(2 marks)*

b Use Figure 3 to show the maximum acceleration of the glider is about 0.4 m s^{-2}. *(2 marks)*

c Show that the maximum displacement of the glider from equilibrium is about 15 mm. *(3 marks)*

d When the glider was initially pulled to one side, the increase in elastic potential energy stored in the springs was 1.2 mJ.

(i) On a copy of Figure 3 sketch a graph of the variation of the kinetic energy of the glider against time from the instant that it is released. Label the energy axis on the right hand side of the graph with a suitable scale. *(3 marks)*

(ii) Calculate the mass of the glider. *(2 marks)*

Jun 2008 2824

3 a With the help of a suitable sketch graph, explain what is meant by *simple harmonic motion*. *(3 marks)*

b Figure 4 shows a simple pendulum at the maximum amplitude of swing.

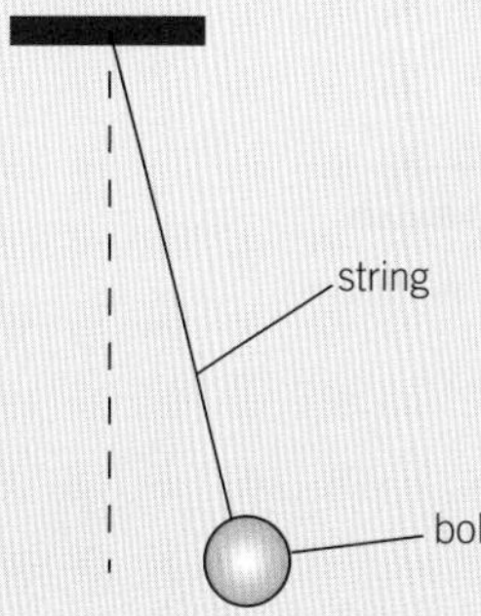

▲ Figure 4

(i) On Figure 4, draw an arrow to show the direction of the resultant force F acting on the pendulum bob. *(1 mark)*

(ii) There are two forces acting on the pendulum bob shown in Figure 4. Explain how these forces are responsible for the resultant force F on the pendulum bob. *(2 marks)*

c Figure 5 shows a graph of displacement x of the pendulum bob with time t.

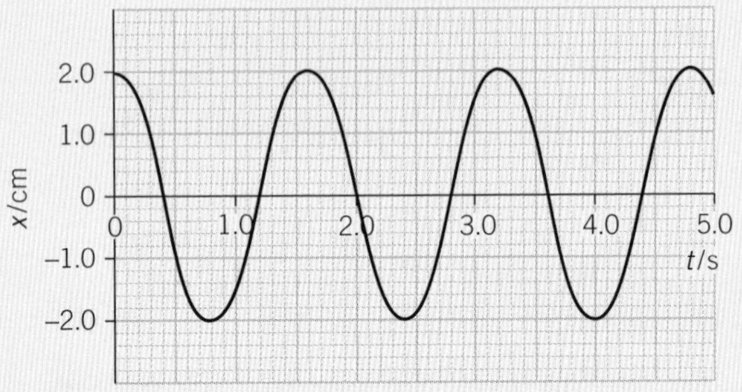

▲ Figure 5

(i) Use Figure 5 to determine the period T of the pendulum. *(2 marks)*

(ii) The mass of the pendulum bob is 50 g. Calculate the maximum kinetic energy of the pendulum bob. *(3 marks)*

(iii) State and explain the change, if any, to the shape of the graph shown in Figure 5 when the amplitude of the pendulum is halved. *(2 marks)*

4 Figure 6 shows a mass suspended from a spring.

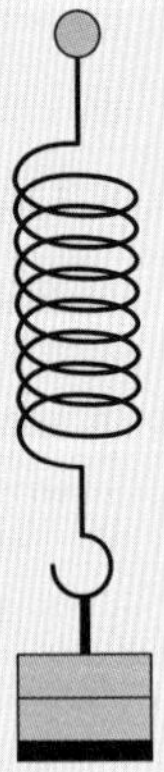

▲ Figure 6

a The mass is in equilibrium. By referring to the forces acting on the mass, explain what is meant by equilibrium. *(2 marks)*

b The mass in (**a**) is pulled down a vertical distance of 12 mm from its equilibrium position, it is then released and oscillates with simple harmonic motion.

(i) Explain what is meant by *simple harmonic motion*. *(2 marks)*

(ii) The displacement, x in mm, at a time t seconds after release is given by

$$x = 12\cos(7.85t).$$

Use this equation to show that the frequency of oscillation is 1.25 Hz. *(2 marks)*

(iii) Calculate the maximum speed v_{max} of the mass. *(2 marks)*

c Figure 7 shows how the displacement x of the mass varies with time t.

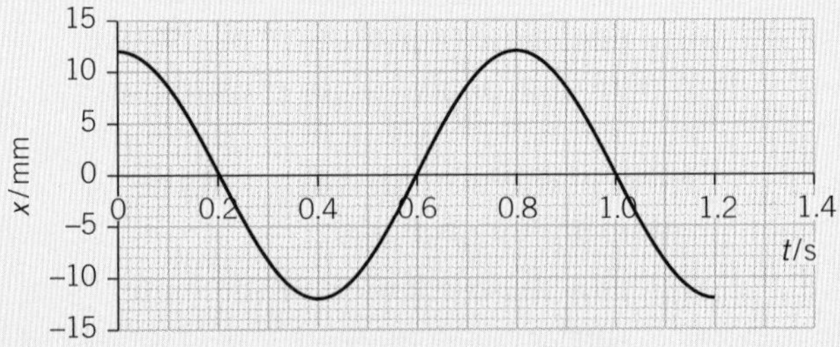

▲ Figure 7

Sketch on a copy of Figure 8 the graph of velocity against time for the oscillating mass.
Put a suitable scale on the velocity axis. *(3 marks)*

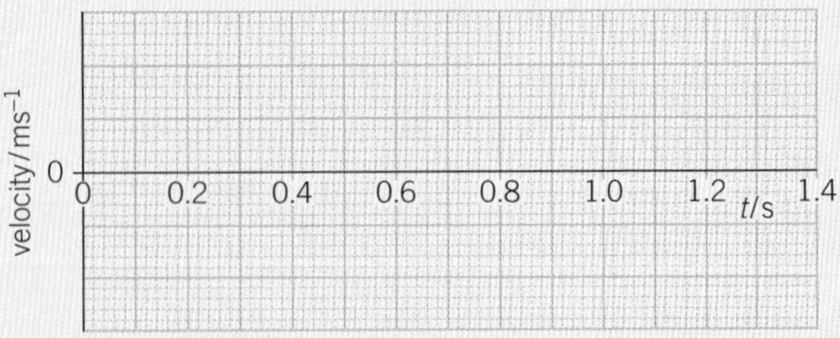

▲ Figure 8

Jan 2010 G484

5 a A body moves with simple harmonic motion. Define, in words, *simple harmonic motion*. *(2 marks)*

b A horizontal metal plate connected to a vibration generator is oscillating vertically with simple harmonic motion of period 0.090 s and amplitude 1.2 mm. There are dry grains of sand on the plate. Figure 9 shows the arrangement.

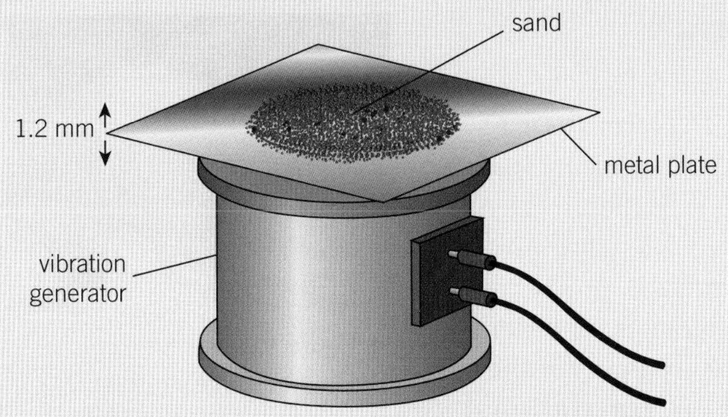

▲ Figure 9

(i) Calculate the maximum speed of the oscillating plate. *(2 marks)*

(ii) The frequency of the vibrating plate is kept constant and its amplitude is slowly increased from zero. The grains of sand start to lose contact with the plate when the amplitude is A_0. State and explain the necessary conditions when the grains of sand first lose contact with the plate. Hence calculate the value of A_0. *(4 marks)*

c The casing of a poorly designed washing machine vibrates violently when the drum rotates during the spin cycle. Figure 10 shows how the amplitude of vibration of the casing varies with the frequency of rotation of the drum.

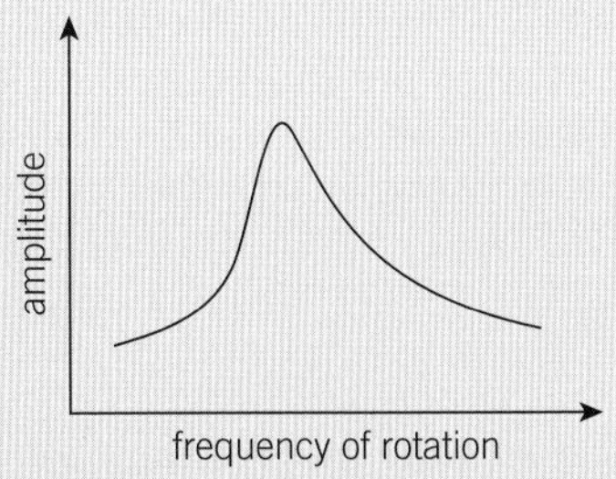

▲ Figure 10

(i) State the name of this effect and describe the conditions under which it occurs. *(2 marks)*

(ii) The design of the washing machine is improved to reduce the effect by adding a damping mechanism to the inside of the machine. Sketch on a copy of Figure 10 the new graph of amplitude against frequency of rotation expected for this improved design. *(2 marks)*

6 a Write an equation for *angular frequency* and state its SI unit. *(2 marks)*

b (i) Write an equation for the speed v of a mechanical oscillator in terms of its amplitude A and displacement x. *(1 mark)*

(ii) Show that the equation above may be written in the form

$$v^2 = a - bx^2$$

where a and b are constants for the oscillator. *(3 marks)*

(iii) Figure 11 shows an incomplete graph of v^2 against x^2 for a spring-mass oscillator being investigated by a student.

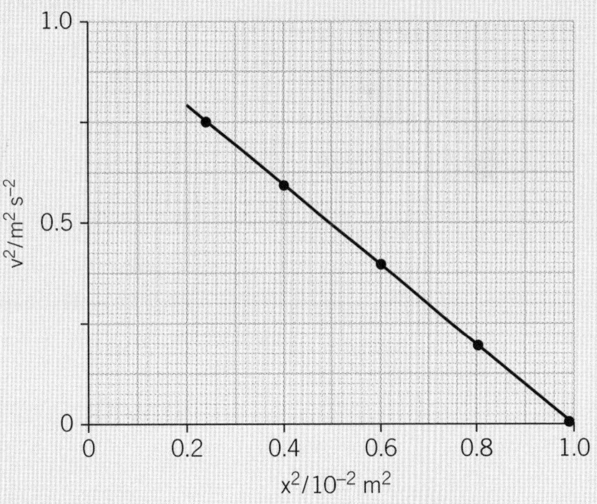

▲ Figure 11

Use Figure 11 to determine

1 the angular frequency of the oscillator *(3 marks)*

2 the maximum speed of the oscillator. *(3 marks)*

18 GRAVITATIONAL FIELDS

18.1 Gravitational fields

Specification reference: 5.4.1, 5.4.2

Learning outcomes

Demonstrate knowledge, understanding, and application of:

- → gravitational fields being due to mass
- → the mass of a spherical object modelled as a point mass at its centre
- → gravitational field lines to map gravitational fields
- → gravitational field strength; $g = \frac{F}{m}$
- → the uniformity of gravitational field strength close to the surface of the Earth and its numerical equivalence to the acceleration of free fall
- → the concept of gravitational fields as one of a number of forms of field giving rise to a force.

The space Ferrari

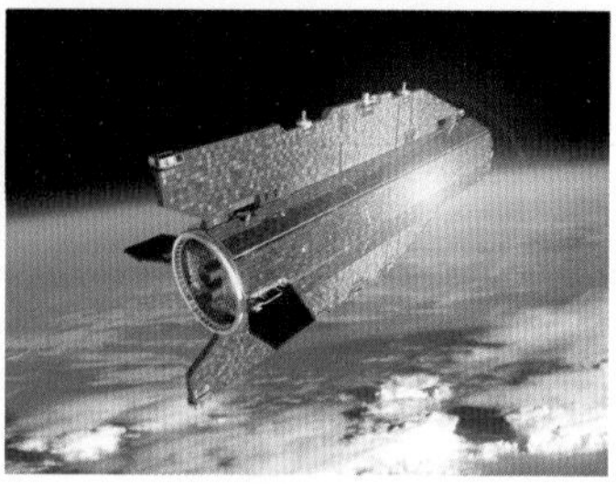

▲ **Figure 1** *Its aerodynamic shape, innovative ion drive, and stabilising fins gave GOCE the nickname 'the space Ferrari'*

The Gravity Field and Steady-State Ocean Circulation Explorer (GOCE) was launched by the European Space Agency in 2009. GOCE (Figure 1) was placed in an orbit just 260 km above the Earth's surface, much lower than other satellites, in order to obtain precise measurements of the strength of the Earth's **gravitational field**. The satellite contained sensitive accelerometers (similar to those that measure the orientation of smartphones) to record minute variations in the **gravitational field strength** produced by the Earth, giving scientists previously impossible insights into the structure of the Earth and the nature of ocean currents.

Gravitational fields

All objects with mass create a gravitational field around them. That includes you. This field extends all the way to infinity, but it gets weaker as the distance from the centre of mass of the object increases (Figure 2), becoming negligible at long distances.

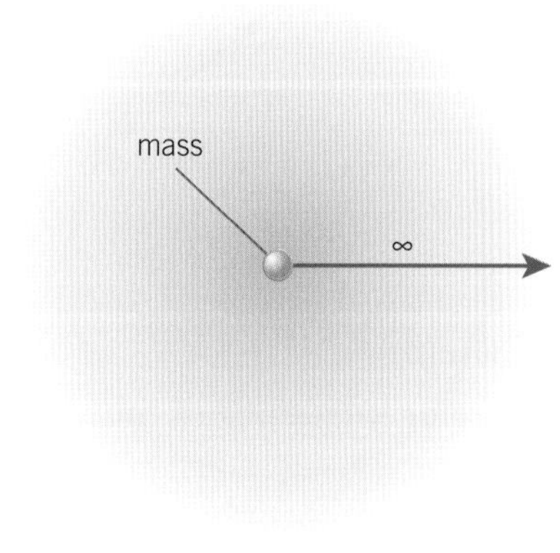

▲ **Figure 2** *A gravitational field is found around all objects with mass*

Any other object with mass placed in a gravitational field will experience an attractive force towards the centre of mass of the object creating the field. For objects on Earth, we call this gravitational attraction the object's *weight*.

Gravitational field strength

The gravitational field strength g at a point within a gravitational field is defined as the gravitational force exerted per unit mass on a small object placed at that point within the field.

This can be written as

$$g = \frac{F}{m}$$

where F is the gravitational force and m is the mass of the object in the gravitational field. Gravitational field strength has the unit $\mathrm{N\,kg^{-1}}$.

The unit $\mathrm{N\,kg^{-1}}$ is the same as $\mathrm{m\,s^{-2}}$, the SI unit for acceleration. You will recognise the equation $F = mg$ as the same as $F = ma$. In other words, gravitational field strength at a point is the same as the acceleration of free fall of an object at that point, $g = a$. On the surface of the Earth the gravitational field strength is approximately $9.81\,\mathrm{N\,kg^{-1}}$ at sea level near the equator, although it varies a little from place to place.

Gravitational field strength g is a vector quantity and always points to the centre of mass of the object creating the gravitational field (Figure 3).

Gravitational field patterns

We can map the gravitational field pattern around an object with **gravitational field lines** (also known as lines of force). These lines do not cross, and the arrows on the lines show the direction of the field, which is the direction of the force on a mass at that point in the field. Since gravitational force is always attractive, the direction of the gravitational field is always towards the centre of mass of the object producing the field. A stronger field is represented by field lines that are closer together. The field lines around a spherical mass, like a planet, form a **radial field**. The gravitational field strength decreases with distance from the centre of the mass, as shown by the field lines getting further apart (Figure 4).

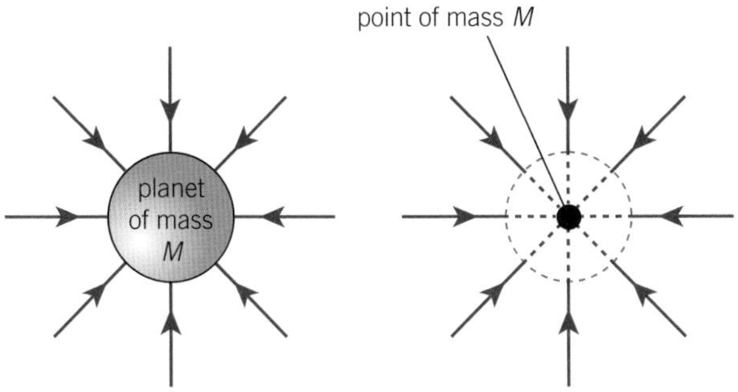

▲ **Figure 4** *The separation between the field lines indicates the magnitude of the gravitational field strength. You can model a spherical mass as a point mass*

The radial fields for a spherical mass and a single **point mass** are very similar. This means that we can model even a large planet or a star as a point mass, with field lines converging at the centre of mass of the object.

If the field lines are parallel and equidistant, the field is said to be a **uniform gravitational field**. In a uniform field, the gravitational field strength does not change. The gravitational field close to the surface of a planet is approximately uniform (Figure 5).

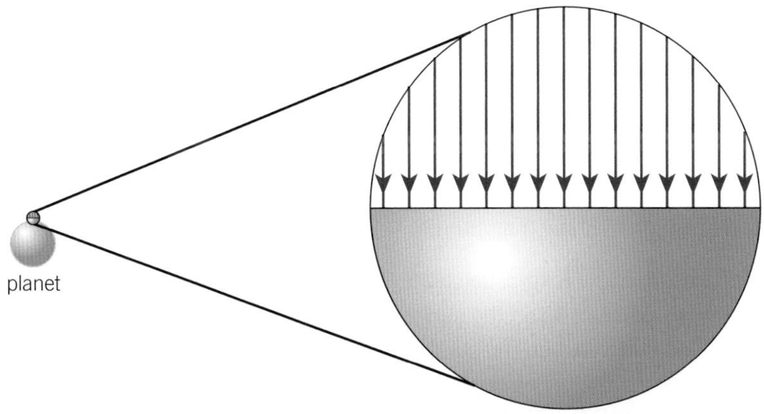

▲ **Figure 5** *Close to the surface of a planet the strength of the gravitational field does not change – it's a uniform field*

Synoptic link

Other types of fields exist, too. Electric fields are created by charged particles (see Topic 22.1, Electric fields) and moving charges create magnetic fields (see Topic 23.1, Magnetic fields).

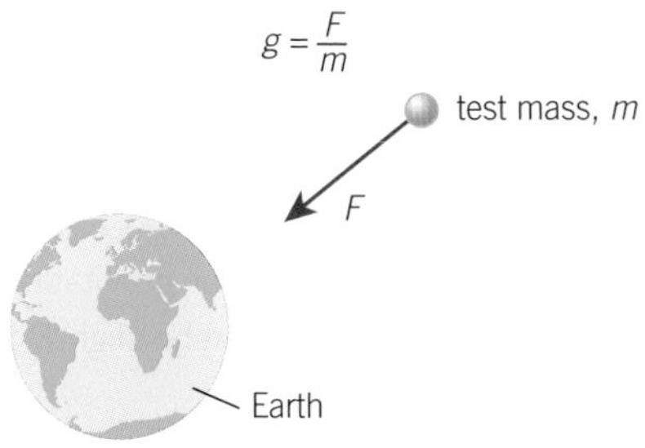

▲ **Figure 3** *Gravitational field strength is a vector quantity*

Synoptic link

From Topic 4.1, Force, mass, and weight, you know that the ratio of force to mass is equal to acceleration.

Study tip

Avoid using the term *gravity*. Instead be precise about your meaning: 'gravitational field strength', 'gravitational attraction', 'gravitational field'.

Gravimetry

Gravimetry is the precise measurement and study of a gravitational field. The Earth's gravitational field can be mapped, and minute variations can be detected. These local variations may be due to topography (e.g., mountains, craters from meteorite impacts) but also due to the composition of the Earth.

Gravimetry is used in mineral prospecting as denser rocks (such as those containing metal ores) cause higher than normal local gravitational fields on the Earth's surface. The same technique is used to search for oil. Since oil-bearing rocks have a lower density than the surrounding rock, the gravitational field strength above an oil deposit can be slightly lower than in the surrounding area.

The most accurate gravimeters are superconducting gravimeters, which use liquid helium to cool a superconducting sphere in a magnetic field. The weight of the sphere is balanced by the effects of the magnetic field. The electric current required to generate the magnetic field depends on the Earth's gravitational field strength at that point, and so *g* can be measured extremely precisely, to around 10^{-12} N kg^{-1}.

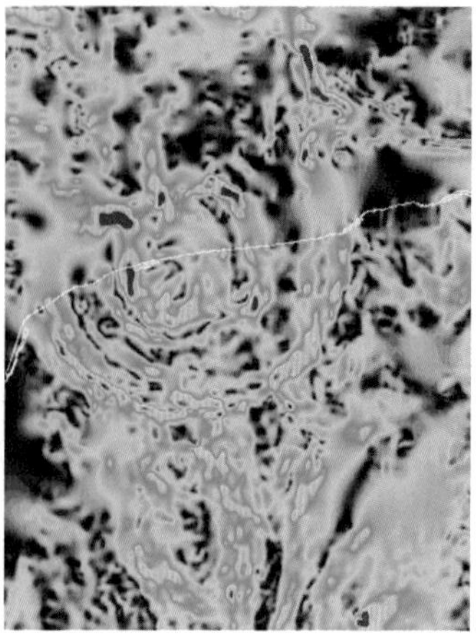

▲ **Figure 6** *Tiny variations in g reveal the crater from a huge meteorite impact on the Yucatan peninsula in Mexico – this impact is thought to have contributed to the extinction of the dinosaurs*

1 For every 1.0 g mass of the sphere, calculate the change in the gravitational force representing a variation of 10^{-12} N kg^{-1} in the field strength.

2 Sketch a diagram to show how the gravitational field lines over a gold deposit differ from those over its surroundings.

Summary questions

1 Suggest why *g* is a vector quantity. *(1 mark)*

2 Explain why the direction of the gravitational field strength at any point around a planet is always towards the centre of the planet. *(1 mark)*

3 Calculate the gravitational field strength required to produce the following forces on a 3.00 kg mass:
a 15.0 N; **b** 29.4 N; **c** 1.62 N. *(3 marks)*

4 Describe how you can use a newtonmeter and a known mass to determine the gravitational field strength on the top of a mountain. *(2 marks)*

5 Use the defining equation for gravitational field strength to show that alternative units of *g* are m s^{-2}. *(3 marks)*

6 The gravitational field strength *g* on the surface of Mars is 3.7 N kg^{-1}. Calculate the difference in the gravitational force experienced by an astronaut of mass of 75 kg on the surface of Mars compared with the gravitational force experienced by the same astronaut on the surface of the Earth. *(2 marks)*

7 Two balls of the same diameter and masses 1.0 kg and 5.0 kg are dropped from a tower. Determine the initial acceleration of each of the balls. Explain your answers. *(2 marks)*

18.2 Newton's law of gravitation

Specification reference: 5.4.2

The Schiehallion experiment

In 1774 the astronomer Nevil Maskelyne set out to determine the mean density of the Earth. The experiment involved measuring the minuscule deflection of a pendulum due to the gravitational attraction of the Scottish mountain of Schiehallion (Figure 1). Remarkably, the tiny gravitational attraction allowed Maskelyne and the mathematician Charles Hutton to determine the density (and subsequently the mass) of the Earth to within 20% of the true value. Using this result they were even able to determine the density of the Sun and the other planets known at the time. The size of the attractive gravitational force between all masses is described by **Newton's law of gravitation**.

Learning outcomes

Demonstrate knowledge, understanding, and application of:

→ Newton's law of gravitation

→ the equation $F = -\frac{GMm}{r^2}$.

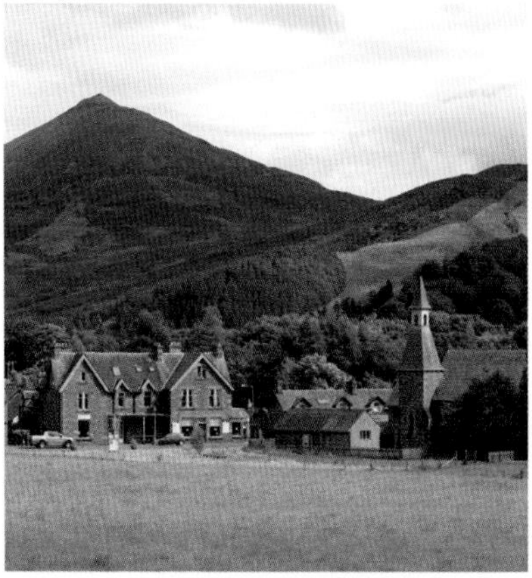

▲ **Figure 1** *The tiny gravitational attraction between the huge mountain and a small pendulum placed close to the mountain deflected the pendulum very slightly from the vertical*

Newton's law of gravitation

Newton's law of gravitation is sometimes described as a universal law of gravity, as it describes the forces between any objects that have mass. The fundamental law can be used to explain both the motion of the planets around the Sun, and why objects (e.g., apples) near the surface of the Earth fall towards the ground.

Consider two objects of masses M and m separated from each other by a distance r (Figure 2). Each object creates its own gravitational field, and the interaction of these fields gives rise to forces between the objects. According to Newton's third law of motion, the two objects must experience a force F of the same magnitude but in opposite directions.

Newton's law of gravitation states the force between two point masses is:

- directly proportional to the product of the masses, $F \propto Mm$
- inversely proportional to the square of their separation, $F \propto \frac{1}{r^2}$.

Therefore

$$F \propto \frac{Mm}{r^2}$$

We can write this as an equation using the **gravitational constant** G. A minus sign is also required to show that gravitational force is an attractive force. Therefore, an equation for Newton's law of gravitation is

$$F = -\frac{GMm}{r^2}$$

The value for G has been carefully determined from experiments as $6.67 \times 10^{-11}\,\mathrm{N\,m^2\,kg^{-2}}$.

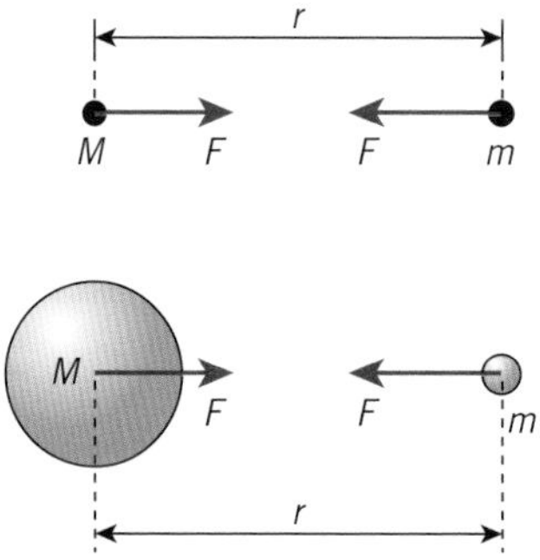

▲ **Figure 2** *Newton's law of gravitation describes the attractive force between point masses (above) and spherical masses (below)*

Synoptic link

You first met Newton's laws of motion in Chapter 7.

Study tip

As you saw in Topic 18.1, Gravitational fields, spherical objects can be modelled as point masses. You can therefore use the equation for Newton's law of gravitation for planets and stars. You just need to remember that the separation r is between the centres of the objects.

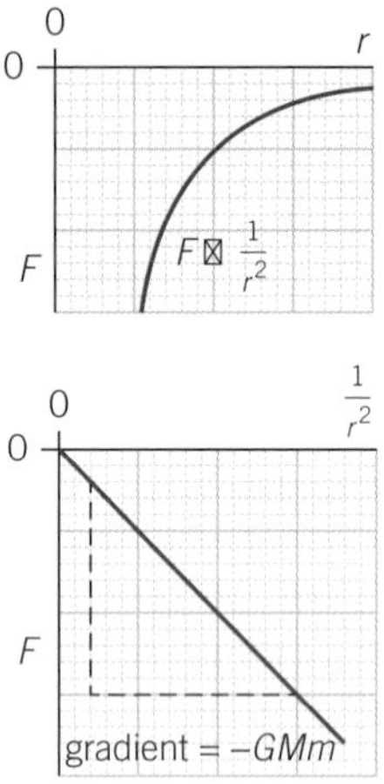

▲ **Figure 3** *A graph of F against r is a curve showing $F \propto \frac{1}{r^2}$ (inverse square law). A graph of F against $\frac{1}{r^2}$ is a straight line through the origin*

The attractive force F between objects decreases with distance in an inverse-square relationship ($F \propto \frac{1}{r^2}$). Double the distance and the force between objects will decrease by a factor of four (Figure 3).

Worked example: The gravitational force on an orbiting satellite

A satellite of mass 70.0 kg orbits the Earth at a height of 10 100 km above the surface. The mass of the Earth is 5.97×10^{24} kg and it has radius 6370 km. Calculate the magnitude of the gravitational force on the satellite due to the Earth.

Step 1: Determine the distance of the satellite from the centre of the Earth.

$$r = 6370 + 10\,100 = 16\,470\,\text{km}$$

Step 2: Use the equation for Newton's law of gravitation to calculate the size of the force on the satellite (ignore the minus sign).

$$F = \frac{GMm}{r^2} = \frac{6.67 \times 10^{-11} \times 5.97 \times 10^{24} \times 70.0}{(16\,470 \times 10^3)^2}$$

$$= 103\ \text{N (3 s.f.)}$$

Multiple objects

If several objects are involved, the resultant force can be determined by vector addition. If the interaction is in one dimension only (Figure 4), then the calculation uses addition and subtraction.

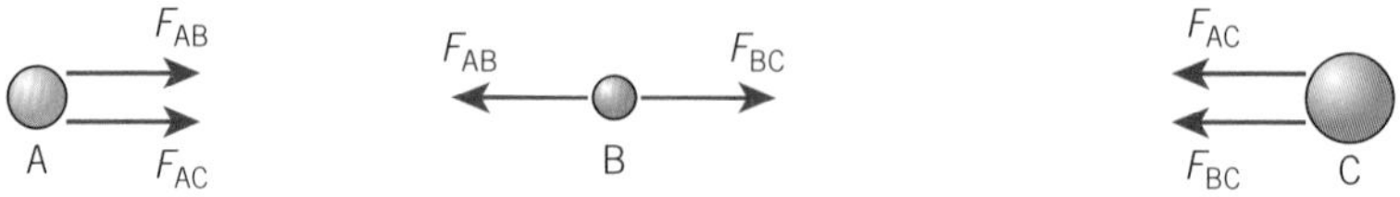

▲ **Figure 4** *Three large rocky asteroids A, B, and C in a line. The force on A is $F_{AB} + F_{AC}$, the force on B is $F_{BC} - F_{AB}$, and the force on C is $-F_{AC} - F_{BC}$.*

The same process applies in two dimensions, but in this case we need to use Pythagoras' theorem, or the sine or cosine rule if the vectors are not at 90°, to determine the resultant force.

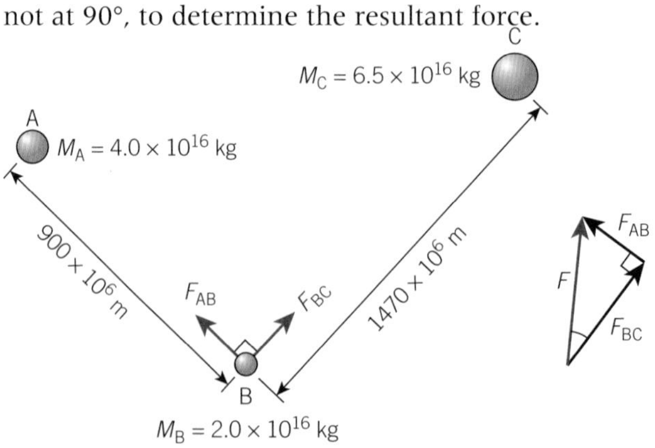

▲ **Figure 5** *The same three asteroids in different positions*

Worked example: Objects in two dimensions

Calculate the magnitude of the resultant force on asteroid B in Figure 5.

Step 1: Calculate the force on B from each asteroid.

$$F_{AB} = \frac{6.67 \times 10^{-11} \times 4.0 \times 10^{16} \times 2.0 \times 10^{16}}{(900 \times 10^6)^2}$$
$$= 66\,\text{kN}$$

$$F_{BC} = \frac{6.67 \times 10^{-11} \times 6.5 \times 10^{16} \times 2.0 \times 10^{16}}{(900 \times 10^6)^2}$$
$$= 40\,\text{kN}$$

$F_{AB} = 66\,\text{kN}$ towards A; $F_{BC} = 40\,\text{kN}$ towards C

Step 2: The forces are vectors acting on B. Using Pythagoras' theorem, the resultant force F is:

$$F = \sqrt{(F_{AB})^2 + (F_{BC})^2} = \sqrt{(66 \times 10^3)^2 + (40 \times 10^3)^2}$$
$$= 77\,\text{kN (2 s.f.)}$$

Synoptic link

The addition of vectors was introduced in Topic 2.4, Adding vectors.

Summary questions

1 Explain why the gradient of the graph of F against $\frac{1}{r^2}$ is equal to GMm. *(3 marks)*

2 Show that the gravitational force on a satellite of mass m_s at a height h above the surface of the Earth is given by

$$F = \frac{GM_E m_s}{(R_E + h)^2}$$

where M_E and R_E are the mass and radius of the Earth respectively. *(2 marks)*

3 Describe what happens to the gravitational force between two objects A and B when:
- **a** the mass of A doubles;
- **b** both masses A and B double;
- **c** the distance between A and B halves;
- **d** the mass of B doubles and the distance between A and B decreases by a factor of four. *(5 marks)*

4 Calculate the gravitational force between:
- **a** two protons of mass 1.67×10^{-27} kg separated by a distance of 1.0×10^{-14} m;
- **b** two students of mass 65 kg and 70 kg standing 1.5 m apart;
- **c** Saturn and the Sun (mass of Sun = 1.99×10^{30} kg, mass of Saturn = 5.68×10^{26} kg, average separation from the Sun to Saturn = 1400 million km). *(6 marks)*

5 The mean distance from the centre of the Earth to the centre of the Moon is about 380 000 km. The gravitational force between the Earth and the Moon is 2.03×10^{20} N. The mass of the Earth is 5.97×10^{24} kg. Calculate the mass of the Moon. *(3 marks)*

6 Use the information in question 5 to determine the resultant force on a probe of mass 120 kg when it is halfway between the Earth and the Moon. *(3 marks)*

18.3 Gravitational field strength for a point mass

Specification reference: 5.4.2

Learning outcomes

Demonstrate knowledge, understanding, and application of:

→ gravitational field strength $g = -\frac{GM}{r^2}$ for a point mass.

▲ **Figure 1** *In 1984 the American Bruce McCandless became the first ever human satellite, orbiting the Earth free from the tether of his nearby spacecraft*

'There's no gravity in space'. Is there?

It is common to hear the phrase 'zero gravity' used to describe the experience of astronauts in space, but it is misleading. Astronauts and their spacecraft are in a circular orbit around the Earth. The gravitational force between the Earth and the astronaut provides the centripetal force required for the circular motion. Even hundreds of kilometres above the Earth, they are definitely still inside the Earth's gravitational field, and although the gravitational field strength is smaller than on the surface it is far from being $0\,\text{N}\,\text{kg}^{-1}$.

Gravitational field strength in a radial field

You have seen how, in a radial field, the gravitational field strength decreases as the distance from the centre of mass of the object creating the field increases (Topic 18.1, Gravitational fields). Using the definition for gravitational field strength and the equation for Newton's law of gravitation we can derive an expression for the gravitational field strength g in a radial field.

Since $g = \frac{F}{m}$ and $F = -\frac{GMm}{r^2}$, we can substitute for force F to give $g = -\frac{GMm}{mr^2}$. So the gravitational field strength g at a distance r from the centre of an object of mass M is

$$g = -\frac{GM}{r^2}$$

The negative sign shows that the gravitational field strength at that point is in the opposite direction to the displacement r from the centre of mass — a gravitational field is an attractive field.

From this equation we can see that in a radial field the gravitational field strength at a point is:

- directly proportional to the mass of the object creating the gravitational field ($g \propto M$)
- inversely proportional to the square of the distance from the centre of mass of the object ($g \propto \frac{1}{r^2}$).

Worked example: g on the International Space Station

The radius of the Earth is 6370 km and it has mass 5.97×10^{24} kg. A 75 kg astronaut in the International Space Station (ISS) orbits at a height of 405 km above the surface of the Earth. Calculate the magnitude of the gravitational field strength at this altitude and the magnitude of the weight of the astronaut.

Step 1: Determine the distance of the ISS from the centre of mass of the Earth.

$$r = 6370 + 405 = 6775\,\text{km}$$

Step 2: Use the equation for gravitational field strength in a radial field to calculate g at this altitude. (The minus sign is not required for the magnitude.)

$$g = \frac{GM}{r^2} = \frac{6.67 \times 10^{-11} \times 5.97 \times 10^{24}}{(6.775 \times 10^6)^2} = 8.67\ldots \text{N kg}^{-1}$$

Step 3: The weight W of the astronaut is given by $W = mg$.

$$W = 75 \times 8.67\ldots = 650\,\text{N (2 s.f.)}$$

(For comparison the astronaut's weight on the surface of the Earth is $mg = 75 \times 9.81 = 740\,\text{N}$)

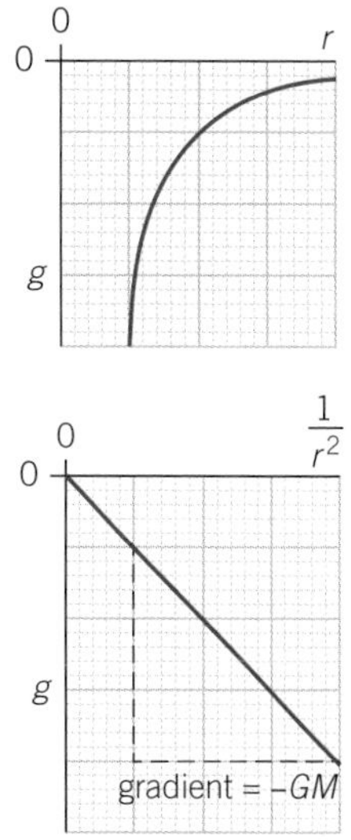

▲ **Figure 2** *A graph of g against r is a curve, showing $g \propto \frac{1}{r^2}$ (inverse square law). A graph of g against $\frac{1}{r^2}$ is a straight line through the origin*

Graphical work

A graph of gravitational field strength against distance from the centre of mass of the object creating a gravitational field is shown in Figure 2. The values of g are negative. As the distance from the centre of mass increases, g decreases until, at infinity, it reaches zero.

The graph of g against $\frac{1}{r^2}$ is a straight line through the origin with a gradient equal to $-GM$. This clearly shows that $g \propto -\frac{1}{r^2}$.

From the Earth to the Moon

When spacecraft travel from the Earth to the Moon, the gravitational field strength they experience varies throughout the journey.

On the surface of the Earth the magnitude of the gravitational field strength is 9.81 N kg^{-1}. The effect of the Moon's gravitational field is quite small. However, as the spacecraft travels further from the Earth the gravitational field strength of the Earth falls (Figure 3, section A). At the same time, the gravitational field strength from the Moon increases.

At B on the graph, the gravitational fields of the Earth and the Moon cancel out, meaning there is zero net gravitational field (ignoring the gravitational field of the Sun) at position Z, which is much closer to the Moon than the Earth.

As the spacecraft continues towards the Moon the gravitational field strength increases to a value of 1.60 N kg^{-1} on the surface of the Moon (C).

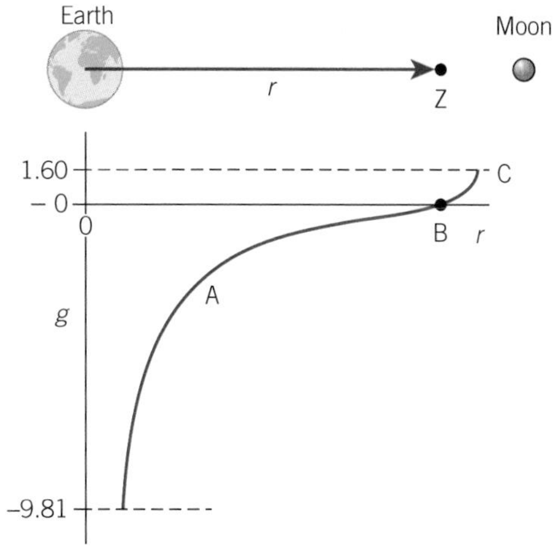

▲ **Figure 3** *A graph to show the resultant gravitational field strength on the journey from the Earth to the Moon*

1 Explain why the resultant gravitational field strength changes from a negative value near the Earth to a positive value near the Moon.

2 The distance between the centres of the Earth and the Moon is 3.8×10^8 m. Calculate the distance from the centre of mass of the Earth to position Z in Figure 3.

3 Explain why position Z is much closer to the Moon than the Earth and therefore explain why it requires much more energy to send a spacecraft to the Moon than for it to return from the Moon to the Earth.

Uniform gravitational fields

In a uniform gravitational field the gravitational field strength does not change. Close to the surface of the Earth g is fairly constant, and so the gravitational field can be considered approximately uniform. Even at the top of the tallest mountain, the increase in the distance from the centre of mass of the Earth is negligible.

Worked example: g in the Himalayas

It is suggested that objects are 'lighter' on the top of mountains as they are further from the centre of mass of the Earth. Calculate the percentage difference between the gravitational field strength on the surface of the Earth and on top of Mount Everest and comment on the effect.

radius of the Earth = 6370 km; mass of the Earth = 5.97×10^{24} kg; height of Mount Everest = 8840 m

Step 1: Use the equation for gravitational field strength in a radial field to calculate g at sea level and on top of Everest. (The minus sign is not required for the magnitude.)

$$g_{SL} = \frac{GM}{r^2} = \frac{6.67 \times 10^{-11} \times 5.97 \times 10^{24}}{(6.370 \times 10^6)^2} = 9.81\,\text{N kg}^{-1}\text{, as expected}$$

$$g_{Ev} = \frac{GM}{r^2} = \frac{6.67 \times 10^{-11} \times 5.97 \times 10^{24}}{(6.378840 \times 10^6)^2} = 9.79\,\text{N kg}^{-1}$$

Step 2: Calculate the percentage difference in g.

$$\frac{(9.81 - 9.79)}{9.81} \times 100 = 0.20\%$$

Objects will weigh slightly less, but the difference is negligible to a climber.

Summary questions

1 The Sun has a mass of 1.99×10^{30} kg and a diameter of 1.39 million km. Calculate the gravitational field strength at its surface. *(3 marks)*

2 Calculate the gravitational field strength 1.2×10^8 m from a point mass of 2.6×10^{23} kg. *(2 marks)*

3 State the effect on the gravitational field strength at a point in a radial field around a point mass when:
 a the mass of the point mass creating the field is halved;
 b the distance from the point mass increases by a factor of three;
 c the mass of the point mass decreases by a factor of four and the distance from the point mass halves. *(4 marks)*

4 Explain why, when moving an object from a height of 100 m to a height of 200 m above the surface of the Earth, the gravitational field strength does not decrease by a factor of four. *(2 marks)*

5 The Earth is not perfectly spherical. The radius at the equator is 6378 km and at the poles is 6371 km. Calculate the percentage change in the gravitational field strength between the equator and the poles. (Mass of the Earth = 5.97×10^{24} kg.) *(3 marks)*

6 Mars has a mass of 6.42×10^{23} kg and a surface gravitational field strength of 3.72 N kg^{-1}. Calculate the radius of Mars. *(3 marks)*

7 The surface gravitational field strength on Venus is 8.77 N kg^{-1} and it has a radius of 6.09 Mm. Calculate the mass of Venus. *(3 marks)*

18.4 Kepler's laws

Specification reference: 5.4.3

Back into darkness

Comet Lovejoy (Figure 1) is one of the most recently discovered great comets. Great comets appear only once a decade or so, and are so bright they are clearly visible at night, and sometimes even during the day.

The shape of the orbit of the comet, like that of all bodies orbiting the Sun, is governed by Kepler's laws of planetary motion. The laws explain why the comet travels much faster when it is closer to the Sun, giving us only a few months to enjoy its brilliance before it moves away, spending centuries in orbit beyond even the most distant planets.

Kepler's laws of planetary motion

Johannes Kepler was a brilliant German astronomer and mathematician. In the early years of the 17th century he published his three laws of planetary motion. These laws were based purely on observational data for the then known planets. Kepler had no knowledge of gravitational fields – his ideas helped Newton to formulate his law of gravitation.

Kepler's first law: The orbit of a planet is an **ellipse** with the Sun at one of the two foci (Figure 2).

An ellipse is a 'squashed' or elongated circle, with two foci. The orbits of all the planets are elliptical.

In most cases the orbits have a low **eccentricity** (a measure of how elongated the circle is), and so their orbits are modelled as circles. For example, at **aphelion** (the furthest point from the Sun) the Earth is 152 million km from the Sun, but at **perihelion** (the closest point to the Sun) the distance is 147 million km, a change of just 3%.

Kepler's second law: A line segment joining a planet and the Sun sweeps out equal areas during equal intervals of time.

As planets move on their elliptical orbit around the Sun, their speed is not constant. When a planet is closer to the Sun it moves faster. In Figure 3, between X and Y the planet moves faster than between P and Q. Kepler's second law states if the time interval from X to Y is the same as for P to Q (e.g., 1 month) the areas A and B must be the same.

This helps explain why we rarely see great comets. Their orbits are highly elliptical, and when they get close to the Sun, where we can see them, they move fast and so spend much less time on this part of their orbit than far away from the Sun. Comets spend most of their time too far from the Sun to be visible.

Kepler's third law: The square of the orbital period T of a planet is directly proportional to the cube of its average distance r from the Sun.

Learning outcomes

Demonstrate knowledge, understanding, and application of:

→ Kepler's three laws of planetary motion

→ the centripetal force on a planet from the gravitational force between it and the Sun

→ the equation $T^2 = \left(\frac{4\pi^2}{GM}\right)r^3$

→ the relationship for Kepler's third law $T^2 \propto r^3$ applied to systems other than our Solar System.

▲ **Figure 1** *Comet Lovejoy was a spectacular sight in 2011 and will next be visible in 2633 as it has an orbital period of 622 years*

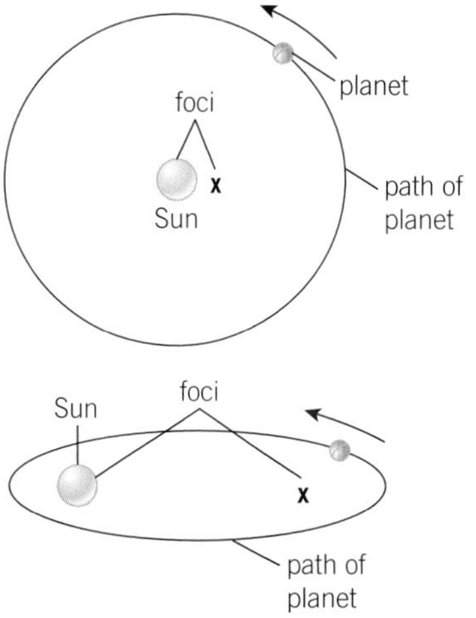

▲ **Figure 2** *The orbits of all planets are elliptical, but most of these ellipses are close to circles – here the left orbit is nearly circular whilst the right-hand orbit is highly elliptical*

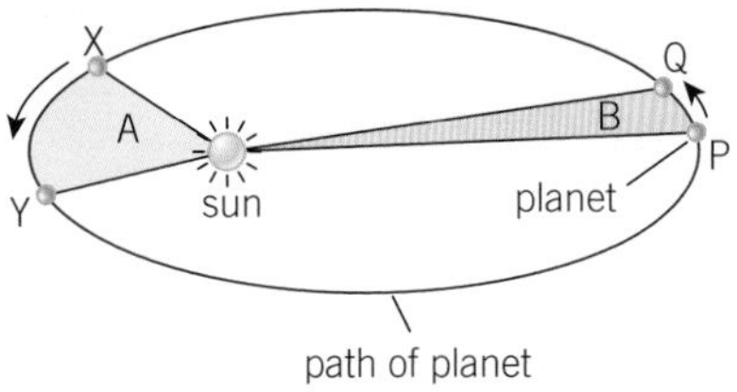

▲ **Figure 3** *Kepler's second law describes the motion of a planet in its elliptical path around the Sun*

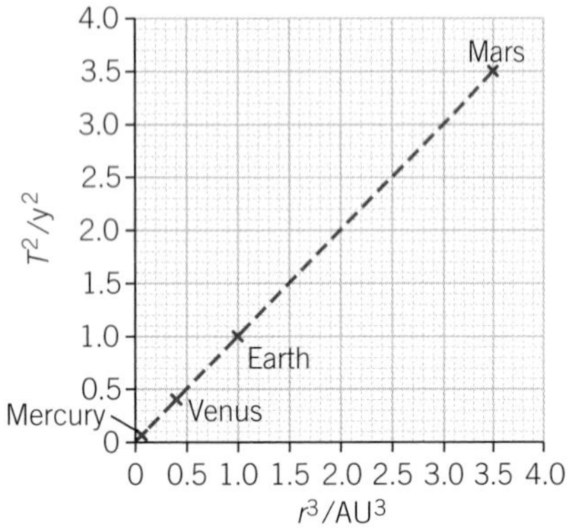

▲ **Figure 4** *A graph of T^2 against r^3 for the first four planets is a straight line through the origin — therefore $T^2 \propto r^3$*

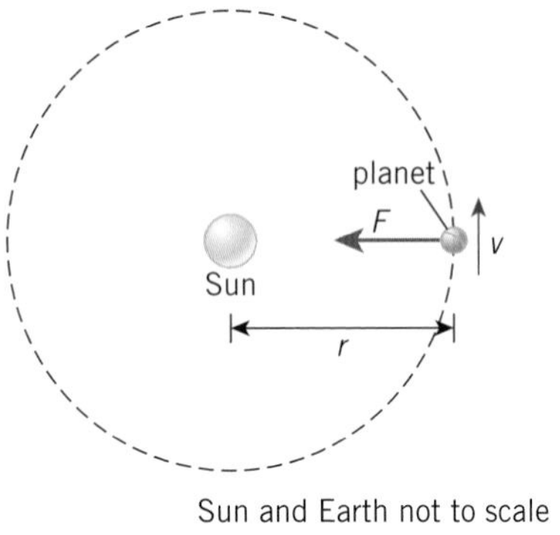

▲ **Figure 5** *Circular motion and Newton's law of gravitation are used in modelling orbits as circles*

Synoptic link

The mathematics of circular motion was studied in Chapter 16.

This can be written as a relationship as $T^2 \propto r^3$ or as

$$\frac{T^2}{r^3} = k$$

where k is a constant for the planets orbiting the Sun.

Table 1 shows data for the six inner planets in our Solar System. The values for T and r are relative to the values of T and r for Earth, 1.00 year and 1.50×10^{11} m, respectively. The mean distance between the Earth and the Sun is known as one **astronomical unit** (AU).

▼ **Table 1** *Data for the six inner planets*

Planet	T/y	T^2/y^2	r/AU	r^3/AU^3	$\frac{T^2}{r^3}$/y^2 AU^{-3}
Mercury	0.24	0.058	0.40	0.064	0.91
Venus	0.62	0.38	0.70	0.343	1.11
Earth	1.00	1.00	1.00	1.00	1.00
Mars	1.88	3.53	1.50	3.37	1.05
Jupiter	11.86	140.6	5.20	141	1.00
Saturn	29.46	867.9	9.50	857	1.01

In the units y^2 AU^{-3}, the ratio $\frac{T^2}{r^3}$ for the planets in the Solar System is approximately 1 — Kepler's third law is validated.

Modelling planetary orbits as circles

Most planets in the Solar System have almost circular orbits. We can therefore use the mathematics developed for circular motion along with Newton's law of gravitation to relate the orbital period T of a planet to its distance r from the Sun. In doing so, we can provide theoretical justification for Kepler's empirical third law.

Consider a planet of mass m orbiting the Sun at a distance r. The orbital speed of the planet is v and it has an orbital period T. The mass of the Sun is M. The centripetal force on the planet is provided by the gravitational force between it and the Sun. Therefore, the gravitational force F on the planet must be equal to the centripetal force. That is

$$\text{centripetal force on planet} = \text{gravitational force on planet}$$

$$\frac{mv^2}{r} = \frac{GMm}{r^2}$$

$$\text{or} \quad v^2 = \frac{GM}{r}$$

Since the planet is moving in a circle, the speed v of the planet can be determined by dividing the circumference of its orbit by its orbital period, $v = \frac{2\pi r}{T}$. Substituting this into the equation above gives

$$\frac{4\pi^2 r^2}{T^2} = \frac{GM}{r}$$

This can be rearranged to give

$$T^2 = \left(\frac{4\pi^2}{GM}\right) r^3$$

This equation is a mathematical version of Kepler's third law, because it shows that $T^2 \propto r^3$. It also means that:

- the ratio $\frac{T^2}{r^3}$ is a constant and equal to $\frac{4\pi^2}{GM}$
- the gradient of a graph of T^2 against r^3 must be equal to $\frac{4\pi^2}{GM}$.

This equation can be used to determine the orbital period of a planet or its distance from the Sun, as long as the other quantity is known. Similarly, it can be used to find the mass of an object from the period and radius of an object in orbit around it.

Worked example: Orbital period of Neptune

The planet Neptune has a mean distance of 4.50×10^{12} m from the Sun. Calculate the orbital period of Neptune in Earth years. The mass of the Sun is 1.99×10^{30} kg.

Step 1: Using $T^2 = \left(\frac{4\pi^2}{GM}\right)r^3$ gives $T = \sqrt{\left(\frac{4\pi^2}{GM}\right)r^3}$

Step 2: Substitute in the known values.

$$T = \sqrt{\left(\frac{4\pi^2}{6.67 \times 10^{-11} \times 1.99 \times 10^{30}}\right)(4.50 \times 10^{12})^3} = 5.206... \times 10^9 \text{ s}$$

Step 3: Change the period from seconds to years.

$$T = \frac{5.206... \times 10^9}{3.16 \times 10^7} = 165 \text{ Earth years}$$

Study tip

When solving problems involving orbits, a good starting point is the equation:

$$\frac{mv^2}{r} = \frac{GMm}{r^2}$$

Kepler's laws beyond planets

Kepler's laws apply not only to the planets in our Solar System but also to any smaller object in orbit around a larger one. This includes satellites and moons in orbits around planets.

The four innermost moons of Jupiter listed in Table 2 have elliptical orbits around their planet. The line joining each moon and Jupiter sweeps out segments of equal area in a given time and for each moon $T^2 \propto r^3$.

▼ **Table 2** *Data for the four largest moons of Jupiter*

Moon	r / 10^3 km	T / days
Io	420	1.8
Europa	670	3.6
Ganymede	1070	7.2
Callisto	1890	16.7

Summary questions

1 Sketch a single diagram showing the orbit of two planets around a star and use your diagram to describe Kepler's three laws of planetary motion. *(4 marks)*

2 Saturn is at a mean distance of 1400 million km from the Sun. Calculate the orbital period of Saturn in Earth years. The mass of the Sun is 1.99×10^{30} kg. *(3 marks)*

3 State the effect on the orbital period of a planet when the distance from a planet to the star:
a doubles; **b** increases by a factor three; **c** decreases by a factor of nine. *(6 marks)*

4 Use the data in Table 2 to confirm that the moons of Jupiter obey Kepler's third law. *(3 marks)*

5 Plot a graph of T^2 against r^3 for the moons shown in Table 2 and use your graph to show that the mass of Jupiter is approximately 1.9×10^{27} kg. *(5 marks)*

18.5 Satellites

Specification reference: 5.4.3

Learning outcomes

Demonstrate knowledge, understanding, and application of:

- → the relationship for Kepler's third law, $T^2 \propto r^3$ applied to systems other than our Solar System
- → geostationary orbit and the uses of geostationary satellites.

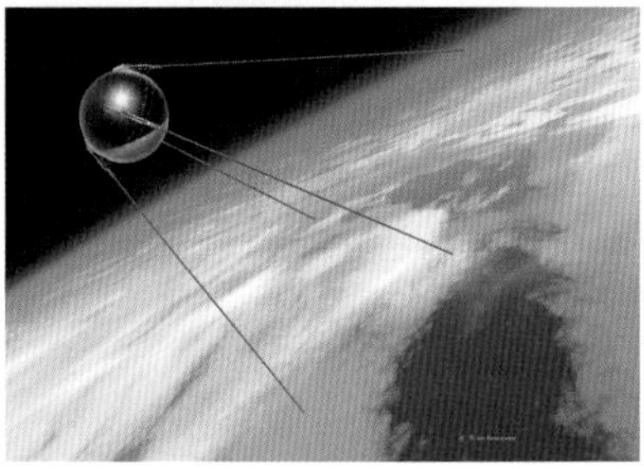

▲ **Figure 1** *Sputnik, the first satellite, sparked a revolution in communications (as first predicted in detail by the science fiction author Arthur C Clarke)*

The 50 cm sphere that started the space race

Sputnik 1 was the first artificial **satellite** of the Earth. Launched in 1957, it broadcast radio pulses back to the Earth below. These could easily be detected all over the globe as the satellite passed overhead.

The launch of Sputnik triggered the space race between the Soviet Union and the United States, which resulted in humans landing on the Moon just over a decade later. There are now hundreds of satellites in orbit, with a wide variety of uses.

Putting a satellite into orbit

Like planets in orbit around the Sun, satellites orbiting the Earth obey Kepler's laws of planetary motion. For simplicity we can model their orbit as a circle. The centripetal force on each satellite is provided by the gravitational force between it and the Earth. For any satellite in orbit, the gravitational force F is given by

$$F = \frac{mv^2}{r} = \frac{GMm}{r^2}$$

where m is the mass of the satellite, M is the mass of the Earth, v is the constant speed of the satellite, G is the gravitational constant, and r is the distance from the satellite to the centre of the Earth.

Since the only force acting on a satellite is the gravitational attraction between it and the Earth, it is always 'falling towards the Earth'. However, as it is travelling so fast, it travels such a great distance that as it falls the Earth curves away beneath it, keeping it at the same height above the surface.

All satellites must therefore be given exactly the right height and exactly the right speed for a stable orbit. Most satellites do not even have engines to adjust their speed or height once released. For any given satellite from the equation above we can see that the correct speed v for a stable orbit at distance r, from the centre of mass of the Earth is given by

$$v = \sqrt{\frac{GM}{r}}$$

The mass m of the satellite is not a factor in this equation. All satellites placed in a given orbit at a given height will be travelling at the same speed, even if their mass varies. Once launched they are normally above the atmosphere, so there is no air resistance to slow them down. As a result their speed remains constant.

Uses of satellites

Modern satellites have a wide variety of uses, including:

- communications: satellite phones (not mobile phones), TV, some types of satellite radio
- military uses: reconnaissance

- scientific research: both looking down onto the Earth to monitor crops, pollution, vegetation, and so on, and looking outwards to study the Universe (including several famous examples like the Hubble Space Telescope)
- weather and climate: predicting and monitoring the weather across the globe and monitoring long-term changes in climate
- global positioning (see later).

The desired use of a particular satellite affects the choice of orbit for the satellite.

Types of orbit

Satellites can be placed in one of a number of different orbits. For example, a polar orbit circles the poles. This type of orbit offers a complete view of the Earth over a given period as the Earth rotates beneath the path of the satellite, ensuring the satellite covers all parts of the globe after a number of orbits, so it is a useful orbit for mapping and reconnaissance.

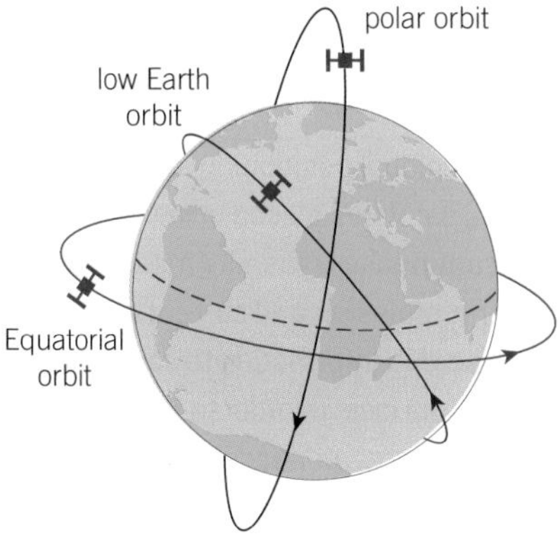

▲ **Figure 2** *Examples of different satellite orbits*

Satellites in orbit close to the Earth are often described as being in low Earth orbit. Since the relationship between the period of a satellite and the radius of its orbit is governed by Kepler's third law ($T^2 \propto r^3$), satellites in this kind of orbit take only a short time to orbit the Earth, less than 2 hours.

Some satellites are placed in orbit above the equator. These include **geostationary satellites**.

Geostationary satellites

As the height of the satellite increases so does its period, so it is possible to choose the satellite's period by selecting its height. A geostationary satellite is placed in a specific orbit (a geostationary orbit) so that it remains above the same point of the Earth whilst the Earth rotates. In order to be in a geostationary orbit, the satellite must:

- be in an orbit above the Earth's equator
- rotate in the same direction as the Earth's rotation
- have an orbital period of 24 hours.

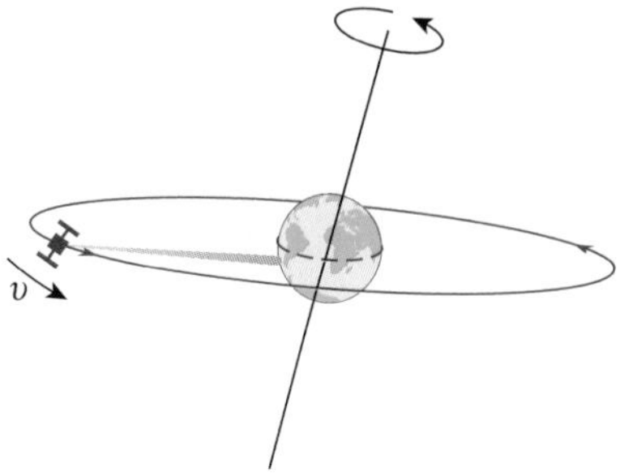

▲ **Figure 3** *Geostationary satellites must be in orbit above the equator at a specific height to give a period of 24 hours*

Worked example: Height of geostationary satellites

The radius of the Earth is 6370 km, and it has mass 5.97×10^{24} kg. Calculate the altitude (height above the ground) of geostationary satellites above the equator.

Step 1: Rearrange the equation for Kepler's third law to make r the subject.

$$T^2 = \left(\frac{4\pi^2}{GM}\right) r^3$$

$$r = \sqrt[3]{\frac{GMT^2}{4\pi^2}}$$

Step 2: The period of a geostationary satellite is 24 hours = 86 400 s. Substitute this and the other values in to calculate r.

$$r = \sqrt[3]{\frac{6.67\times10^{-11}\times5.97\times10^{24}\times86400^2}{4\pi^2}} = 4.22\times10^7\,\text{m}$$

Step 3: Subtract the radius of the Earth to obtain the altitude of the satellite.

altitude $= 4.22\times10^7 - 6.37\times10^6 = 36\times10^6$ m (2 s.f.)

Going beyond satnav

The global positioning system (GPS) is a series of 32 satellites placed in low Earth orbit. Each satellite continually transmits messages that include the time of transmission and current position. Receivers on the Earth use this information to calculate their precise position on the surface. In order to determine its position accurately, each receiver needs to receive signals from at least four different GPS satellites. Software calculates the position to within a few metres from the tiny time delay between the signals from each satellite.

GPS has revolutionised navigation, but has a number of uses besides the satellite navigation found in vehicles and smartphones:

- driverless vehicles: allowing vehicles to function safely without a human driver
- geofencing: sending and receiving information (e.g., alerts) when a vehicle, person, or even a pet enters or leaves a specific area
- mining: surveying to provide accurate positions (e.g., when drilling) both above and underground, with some systems precise to within centimetres
- tectonics: measuring the movement of seismic faults during even the smallest of earthquakes.

There are other systems similar to the United States' GPS. The Russian GLONASS system has 24 satellites in orbit, and the £5.5 bn Galileo project from the European Space Agency plans to have 30 satellites in orbit by 2020. Both systems can work alongside GPS to improve its accuracy and capacity.

1 Suggest why geostationary satellites are used for satellite television.

2 A geostationary satellite has a mass of 80 kg. Calculate its kinetic energy.

Summary questions

Data required for questions 3, 4, and 5: mass of the Earth = 5.97×10^{24} kg and radius of the Earth = 6370 km.

1 Use Kepler's third law to explain why satellites closer to the surface of the Earth take less time to orbit than those higher up. *(2 marks)*

2 Draw a labelled diagram to show the forces acting on a satellite in orbit. *(2 marks)*

3 A 180 kg satellite in a polar orbit travels at 6400 m s^{-1} circling the Earth nine times in one day. Calculate:
 a its orbital period;
 b the radius of its orbit;
 c the gravitational force acting on the satellite;
 d its centripetal acceleration. *(7 marks)*

4 Show that the lowest theoretical orbital period for a satellite around Earth is around 85 minutes. *(4 marks)*

5 Calculate the velocity of a satellite in orbit at a height of 5000 km above the surface of the Earth. *(4 marks)*

18.6 Gravitational potential

Specification reference: 5.4.4

Deep impact

Meteor Crater, in the Arizona desert, is just that. The crater was created around 50 000 years ago when a meteorite with a diameter of only 50 m crashed into the surface at a speed thought to be around 20 km s^{-1}. The meteorite gained kinetic energy as it accelerated towards the Earth, losing gravitational potential energy in the process. The enormous energies involved must have produced an explosion equivalent to 10 billion kilograms of TNT.

An understanding of why the meteorite gained energy on its approach to the Earth relies on an understanding of the **gravitational potential** around the Earth.

Learning outcomes

Demonstrate knowledge, understanding, and application of:

→ gravitational potential at a point as the work done in bringing unit mass from infinity to the point

→ gravitational potential at infinity

→ the expression for gravitational potential at a distance *r* from a point mass *M*

→ changes in gravitational potential.

▲ **Figure 1** *Meteor Crater in Arizona, USA, is over 150 m deep and has a diameter of over 1 km*

Gravitational potential

The gravitational potential V_g at a point in a gravitational field is defined as the work done per unit mass to move an object to that point from infinity.

This gives V_g the unit J kg^{-1}. Infinity refers to a distance so far from the object producing the gravitational field that the gravitational field strength is zero. Gravitational potential is a scalar quantity – it only has magnitude.

All masses attract each other. It takes energy, that is, external work must be done, to move objects apart. Gravitational potential is a maximum at infinity, where its value is taken to be 0 J kg^{-1}. This means that all values of gravitational potential are negative.

Study tip

'Unit mass' refers to an object with a mass of 1.0 kg.

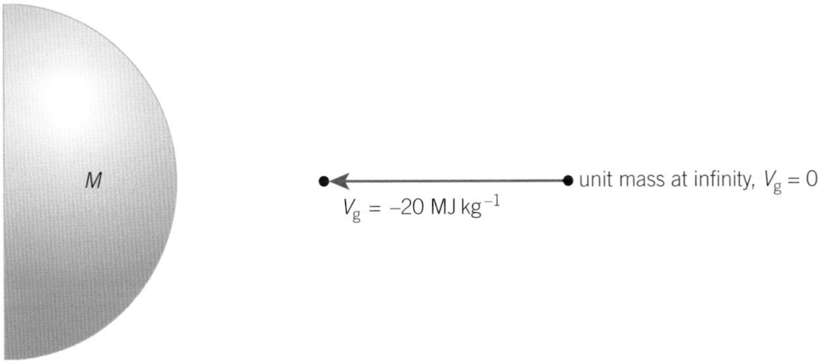

▲ **Figure 2** *All values of V_g are negative since gravitational force is attractive and $V_g = 0$ at infinity*

In Figure 2 the gravitational potential at the point X in the gravitational field of a planet of mass *M* is shown as −20 MJ kg^{-1}. In other words, it would require 20 000 000 J of work to move a unit mass from that point to infinity, 40 000 000 J to move 2.0 kg from that point to infinity, and so on.

Gravitational potential in a radial field

The gravitational potential at any point in a radial field around a point mass depends on two factors:

- the distance r from the point mass producing the gravitational field to that point
- the mass M of the point mass.

The gravitational potential $V_g \propto M$ and $\propto \frac{1}{r}$. You can use the idea of work done by a force and Newton's law of gravitation to show that V_g is given by the equation

$$V_g = -\frac{GM}{r}$$

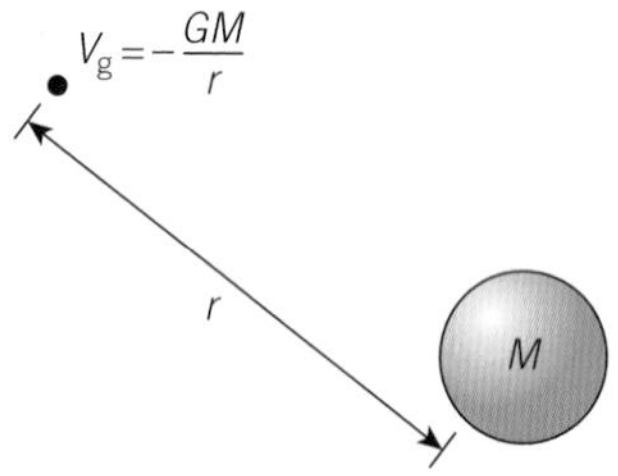

▲ **Figure 3** *At any point in a radial gravitational field, the gravitational potential V_g is given by $V_g = -\frac{GM}{r}$*

All values of V_g within the region of the gravitational field will be negative, and when $r = \infty$ then $V_g = 0$.

You have already seen how the gravitational field of a spherical object can be modelled as originating from a point mass (Topic 18.1, Gravitational fields). This means that we can apply the equation for gravitational potential above to a planet, as long as the distance r is measured from the centre of the planet, and is greater than or equal to the radius of the planet.

Worked example: The gravitational potential on the surface of the Earth

The radius of the Earth is 6370 km and it has mass 5.97×10^{24} kg. Calculate the gravitational potential on the surface of the Earth.

Step 1: The equation for gravitational potential can be used for the Earth because $r \geq$ radius of the Earth.

$$V_g = -\frac{GM}{r} = -\frac{6.67 \times 10^{-11} \times 5.97 \times 10^{24}}{6.370 \times 10^{6}} = -6.25... \times 10^{7}\,\text{J kg}^{-1}$$

This means it would take about 63 million joules to move a unit mass from the Earth to infinity.

Graph of V_g against r

A graph of the gravitational potential V_g around the Earth against the distance r from the centre of mass of the Earth is shown in Figure 4. Since $V_g \propto \frac{1}{r}$, if r doubles then V_g will halve. The potential will tend towards zero as r approaches infinity. Notice that the smallest value of r must be equal to the radius of the Earth. This gives the maximum magnitude of the potential, $63\,\text{MJ kg}^{-1}$.

A graph of V_g against $\frac{1}{r}$ will produce a straight line through the origin with a gradient equal to $-GM$.

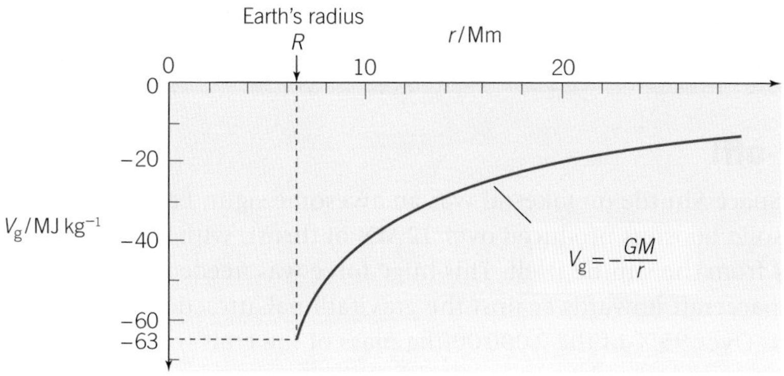

Figure 4 *The gravitational potential around the Earth varies depending on the distance from the centre of mass of the Earth*

Changes in gravitational potential

Moving from one point in a gravitational field to another results in a change in gravitational potential, ΔV_g.

- Moving towards a point mass results in a *decrease* in gravitational potential (e.g., from $-40\,\mathrm{MJ\,kg^{-1}}$ to $-60\,\mathrm{MJ\,kg^{-1}}$, giving $\Delta V_g = -20\,\mathrm{MJ\,kg^{-1}}$).
- Moving away from a point mass (towards infinity) results in an *increase* in gravitational potential (e.g., from $-30\,\mathrm{MJ\,kg^{-1}}$ to $-20\,\mathrm{MJ\,kg^{-1}}$, giving $\Delta V_g = +10\,\mathrm{MJ\,kg^{-1}}$).

Changes in gravitational potential are more complex if a number of masses are involved. Since gravitational potential is a scalar quantity, the total gravitational potential at any point is equal to the algebraic sum of the gravitational potentials from each mass at that point.

The graph in Figure 5 shows how the gravitational potential varies on a journey from the Earth to the Moon.

The gravitational potential rises as you move away from the surface of the Earth (A), before reaching a maximum value at B (note this value is not zero) and falling again as the distance to the Moon reduces (C).

Study tip

All values for gravitational potential are negative in a gravitational field. However, changes in gravitational potential can be negative or positive depending on the direction.

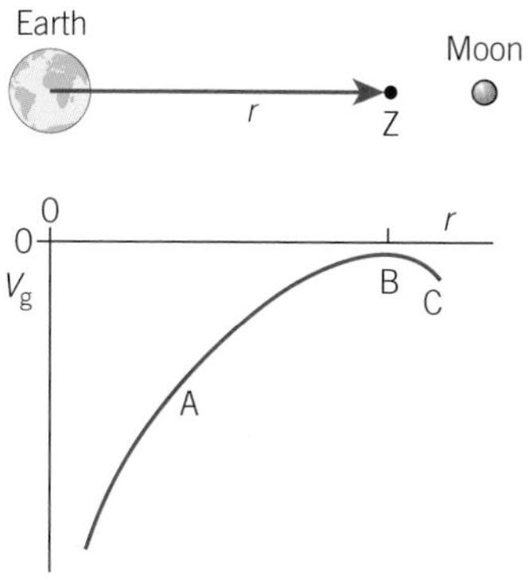

Figure 5 *The variation in gravitational potential between the Earth and the Moon*

Summary questions

1. Calculate the gravitational potential at a point 3.4×10^6 m from a 7.10×10^{21} kg point mass. *(2 marks)*
2. State the effect on the gravitational potential at a point in a radial field around a point mass if:
   a. the mass of the point mass creating the field doubles;
   b. the distance from the point mass decreases by a factor of four;
   c. the mass of the point mass increases by a factor of three and the distance from the point mass doubles. *(4 marks)*
3. Explain why a satellite in a circular orbit at a fixed height does not experience a change in gravitational potential. *(2 marks)*
4. The Sun has mass 1.99×10^{30} kg and diameter 1.39×10^9 m. Show that the gravitational potential on its surface is about $-1.9 \times 10^{11}\,\mathrm{J\,kg^{-1}}$. *(2 marks)*
5. Take six values from the graph in Figure 4 and plot a graph of V_g against $\frac{1}{r}$. Use your graph to determine the mass of the Earth. *(6 marks)*

18.7 Gravitational potential energy

Specification reference: 5.4.4

Learning outcomes

Demonstrate knowledge, understanding, and application of:

→ force–distance graph for a point or spherical mass; work done as area under graph

→ equation for gravitational potential energy at a distance *r* from a point mass *M*

→ escape velocity.

▲ **Figure 1** *The Shuttle was retired in 2011, but it remains one of the most iconic vehicles ever built*

Lift-off!

The Space Shuttle on take-off was an awesome sight. Each of the two solid boosters produced over 12 MN of thrust, with an additional 5 MN from the shuttle itself. This huge force was needed to accelerate the spacecraft upwards against the gravitational attraction of the Earth. Over 95% of the 2 000 000 kg mass of the craft was fuel. During launch the chemical energy in this vast amount of fuel was transferred into the kinetic and gravitational potential energy of the shuttle. Careful calculations were needed to ensure the shuttle ended up in a stable orbit.

Gravitational potential energy

The gravitational potential energy *E* of any object with mass *m* within a gravitational field is defined as the work done to move the mass from infinity to a point in a gravitational field. Therefore

$$E = mV_g$$

We often need to determine changes in gravitational potential energy. For an object of constant mass, the equation becomes

$$\Delta E = m\Delta V_g$$

Gravitational potential energy in a uniform gravitational field

In a uniform gravitational field, such as one close to the surface of a planet, in order to change the gravitational potential energy of an object, its height above the surface must be changed. This results in a change in gravitational potential, and so a change in gravitational potential energy.

On moving away from the surface of the Earth from A to B there is an increase in gravitational potential, so any object moving this way would gain gravitational potential energy (Figure 2). If an object moves from A to C there is no change in gravitational potential, so no change in gravitational potential energy. Moving from B to A results in a decrease in both gravitational potential and gravitational potential energy.

Study tip

The equation $E_p = mgh$ can only be used to determine the *changes* in gravitational potential energy in a uniform gravitational field.

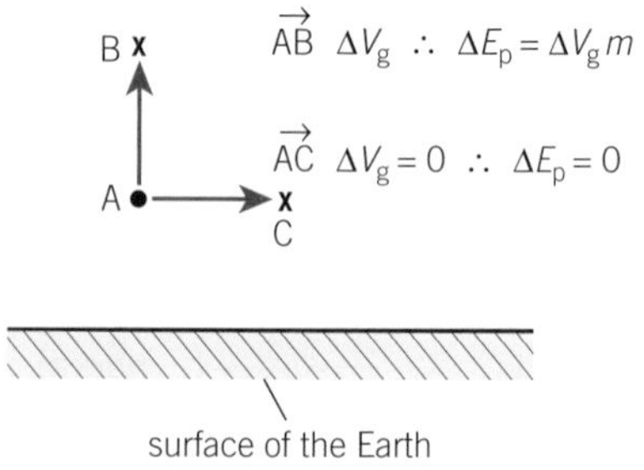

▲ **Figure 2** *Raising an object in a uniform gravitational field results in an increase in gravitational potential energy*

Gravitational potential energy in a radial field

In a radial gravitational field, since $V_g = -\frac{GM}{r}$, the gravitational potential energy can be written as

$$E = mV_g = -\frac{GMm}{r}$$

Any change in *r* results in a change in gravitational potential, and so a change in gravitational potential energy.

Moving an object from A to B results in a change in gravitational potential (Figure 3). As the gravitational potential decreases, so does the gravitational potential energy — it is usually transferred into kinetic energy as the object accelerates.

Worked example: Gravitational potential energy of a satellite

The mass of the Earth is 5.97×10^{24} kg and it has radius 6370 km. Calculate the gravitational potential energy of a 75.0 kg satellite at a height of 1200 km above the surface of the Earth.

Step 1: Determine the distance of the satellite from the centre of mass of the Earth.

$$6370 + 1200 = 7570\,\text{km}$$

Step 2: Calculate the gravitational potential energy.

$$E = -\frac{GMm}{r} = \frac{6.67 \times 10^{-11} \times 5.97 \times 10^{24} \times 75.0}{7.570 \times 10^{6}} = -3.95 \times 10^{9}\,\text{J (3 s.f.)}$$

Note that this is *not* the energy required to lift the satellite into orbit, since it already had a gravitational potential energy on the surface of the Earth. It will also need some kinetic energy in orbit.

Graphs of force against distance

In Topic 18.2, Newton's law of gravitation, you saw that the gravitational force between objects decreases with distance in an inverse-square relationship. Figure 4 shows how the **magnitude** of the gravitational force F varies with distance r from the centre of a spherical object.

The area under a force against distance graph is equal to the work done. The work done to move a mass 'up' from B to A (the object gains gravitational potential energy) is shown by the shaded part of the graph. The work done can be negative if the object moves (falls) from A to B (the object loses gravitational potential energy).

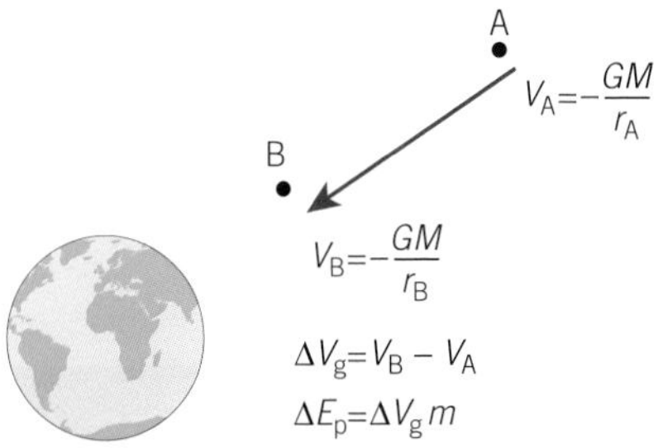

▲ **Figure 3** *Changing the distance from the centre of mass from an object like a planet results in a change in gravitational potential energy*

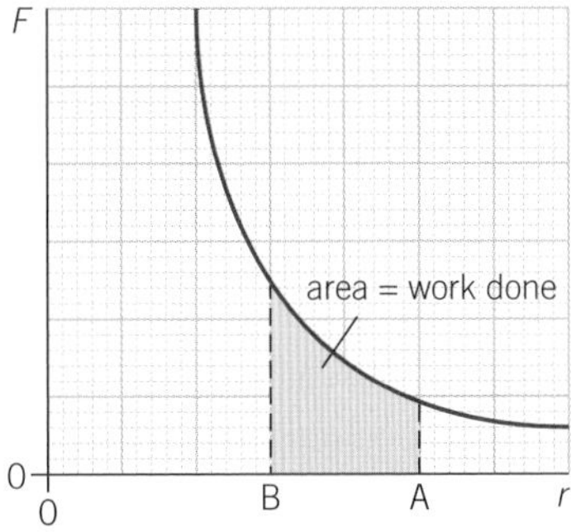

▲ **Figure 4** *The work done to move a mass from B to A inside a gravitational field is shown by the shaded part of the graph*

Synoptic link

The ideas about work here are similar to the ideas developed in Topic 6.2, Elastic potential energy, when calculating the work done to extend a spring.

Escape velocity

In order to escape the gravitational field of a mass like a planet, an object must be supplied with energy equal to the gain in gravitational potential energy needed to lift it out of the field.

Consider a projectile of mass m fired upwards. If we ignore air resistance, the kinetic energy of the projectile is transferred into gravitational potential energy as it rises. In order for the projectile to have just enough energy to leave the gravitational field, the loss of kinetic energy must equal the gain in gravitational potential energy.

$$\frac{1}{2}mv^2 = \frac{GMm}{r}$$

The minimum velocity v for this condition to be met is called the **escape velocity**. It is given by

$$v = \sqrt{\frac{2GM}{r}}$$

The escape velocity on a given planet is therefore the same for all objects regardless of their mass.

The escape velocity of gas atoms or molecules on different planets

The composition of the atmosphere on different planets depends on a number of factors including the average temperature of the planet and its surface gravitational field strength. If the temperature is high enough, atoms or molecules in certain gases will reach escape velocity and leave the gravitational field of the planet, drifting off into space.

In order to escape the gravitational field of a planet an individual atom or molecule from a gas must have a minimum speed equal to $\sqrt{\frac{2GM}{r}}$.

The average kinetic energy of a single gas atom or molecule is given by $\frac{1}{2}mc^2_{\text{r.m.s.}} = \frac{3}{2}kT$, where $c_{\text{r.m.s.}}$ is the r.m.s. speed of the molecules. At any given temperature, some molecules will be travelling faster than this r.m.s. speed. Molecules travelling with speed greater than $\sqrt{\frac{2GM}{r}}$ can escape.

1 Calculate the escape velocity for oxygen molecules from the surface of the Earth.

2 The mass of an oxygen molecule is 5.3×10^{-26} kg. Calculate the r.m.s. speed of the oxygen molecules in the Earth's atmosphere at 20°C. Therefore, explain why we still have oxygen molecules in the Earth's atmosphere.

Synoptic link

The average kinetic energy of a single gas atom or molecule and r.m.s. speed were studied in Chapter 15, Ideal gases.

Summary questions

1 Explain why in order for there to be a change in gravitational potential energy a mass needs to move vertically in a uniform field. *(2 marks)*

2 Calculate the gravitational potential energy of the following masses at a point in a gravitational field where the gravitational potential is -32 MJ kg^{-1}: **a** 40 kg; **b** 7.4 μg; **c** 1.67×10^{-27} kg (mass of a proton). *(3 marks)*

3 On the Apollo 14 mission to the Moon, astronaut Alan Shepard smuggled a golf club on board. He used it to strike a golf ball around 300 m on the surface. The Moon has mass 7.35×10^{22} kg and radius 1740 km. Calculate the velocity needed by the golf ball to escape from the surface of the Moon. *(3 marks)*

4 The Earth has mass 5.97×10^{24} kg and radius 6370 km. Calculate the change in gravitational potential energy required to lift 300 kg into an orbit 50 000 km above the surface of the Earth. *(4 marks)*

5 The Sun has mass 1.99×10^{30} kg and radius 6.96×10^{8} m. A very distant comet is 'caught' by the Sun's gravitational field and accelerates towards the centre of the Sun. Estimate the speed of the comet at the edge of the Sun. *(3 marks)*

Practice questions

1 a Figure 1 shows the Earth in space.

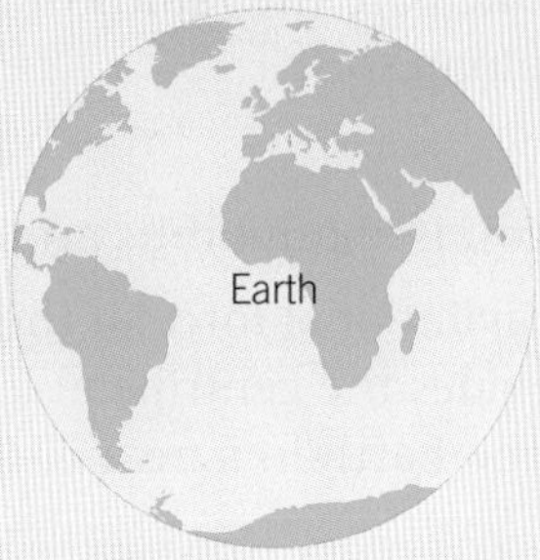

▲ Figure 1

(i) On a copy of Figure 1, draw lines to show the shape and direction of the gravitational field of the Earth. *(1 mark)*

(ii) The gravitational field strength, g, is uniform close to the Earth's surface. Describe the pattern of gravitational field lines close to the surface of the Earth. *(2 marks)*

b The planet Saturn has mass 5.7×10^{26} kg and radius 6.0×10^{7} m.

(i) Calculate the gravitational field strength g_s at Saturn's surface. *(2 marks)*

(ii) Saturn's second-largest moon, Rhea, has orbital radius 5.3×10^{8} m and mass 2.3×10^{21} kg.
Calculate for Rhea

1 its orbital speed v *(3 marks)*

2 its kinetic energy. *(1 mark)*

Jun 2013 G484

2 a Figure 2 shows an aeroplane flying in a horizontal circle at constant speed. The weight of the aeroplane is W and L is the lift force acting at right angles to the wings.

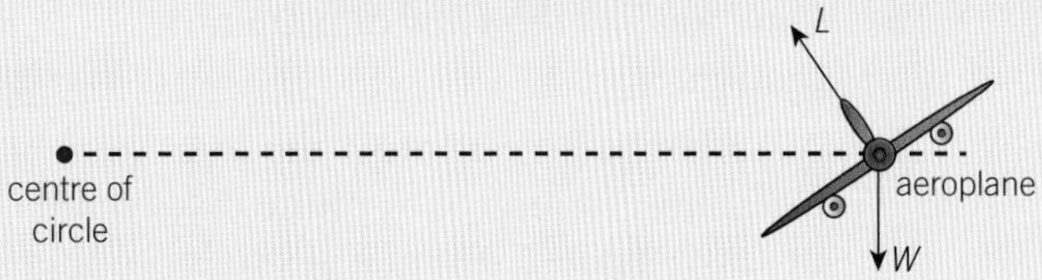

▲ Figure 2

(i) Explain how the lift force L maintains the aeroplane flying in a horizontal circle. *(2 marks)*

(ii) The aeroplane of mass 1.2×10^{5} kg is flying in a horizontal circle of radius 2.0 km. The centripetal force acting on the aeroplane is 1.8×10^{6} N. Calculate the speed of the aeroplane. *(2 marks)*

b Figure 3 shows a satellite orbiting the Earth at a constant speed v. The radius of the orbit is r.

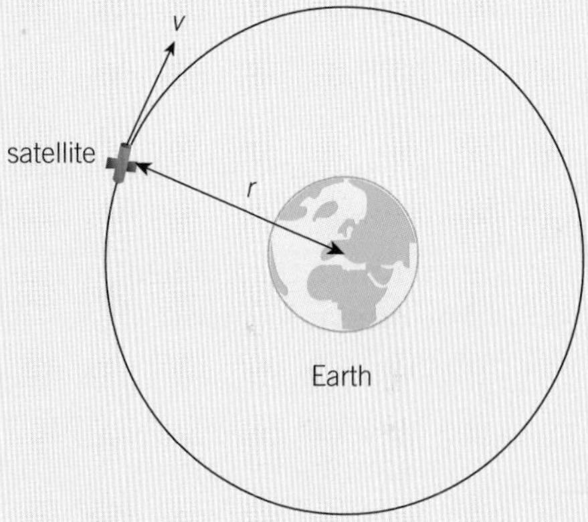

▲ Figure 3

Show that the orbital period T of the satellite is given by the equation

$$T^2 = \frac{4\pi^2 r^3}{GM}$$

Where M is the mass of the Earth and G is the gravitational constant. *(3 marks)*

c The satellites used in television communication systems are usually placed in geostationary orbits.

(i) State two features of geostationary orbits. *(2 marks)*

(ii) Calculate the radius of orbit of a geostationary satellite.

The mass of the Earth is 6.0×10^{24} kg. *(3 marks)*

3 a State *Kepler's third law* of planetary motion. *(1 mark)*

b A student is investigating the five innermost moons of Neptune discovered in 1989. Table 1 summarises some of the key data for these moons.

▼ Table 1

r/Mm	T/days	lg(r/Mm)	lg(T/days)
48.2	0.294	1.683	−0.532
50.1	0.311	1.700	−0.507
52.5	0.335		
62	0.429	1.792	−0.368
73.5	0.555	1.866	−0.256

The period of a moon is T and its distance from the centre of Neptune is r. The ideas of Kepler's laws can be applied to systems other than the Solar System.

(i) Complete the missing data in a copy of Table 1. *(1 mark)*

(ii) Plot a graph of lg(T/days) against lg(r/Mm) and draw a straight line of best fit. *(3 marks)*

(iii) Determine the gradient of the straight line drawn in **(b)(ii)**. *(2 marks)*

(iv) Discuss whether or not Kepler's third law is validated by the value determined in **(b)(iii)**. *(4 marks)*

4 a Define *gravitational field strength, g*. *(1 mark)*

b Explain why the acceleration due to gravity and the gravitational field strength at the Earth's surface have the same value. *(2 marks)*

c A space probe, with its engines shut down, orbits Mars at a constant distance of 3500 km above the centre of the planet in a time of 110 minutes.

(i) Calculate the speed of the space probe. *(2 marks)*

(ii) Show that the mass of Mars is about 6×10^{23} kg. *(2 marks)*

d (i) Write down an algebraic expression for g at the surface of a planet in terms of its mass M and radius R. *(1 mark)*

(ii) The acceleration due to gravity at the surface of Mars is 3.7 m s^{-2}. Calculate the radius of Mars, in kilometres. *(2 marks)*

Jun 2008 2824

5 This question is about orbits around the Sun.

a The gravitational force of the Sun, mass M, provides the centripetal force which holds the Earth in a near circular orbit of radius R.

By considering the Earth as an isolated planet moving in a circular orbit show that its speed v is given by the equation

$v = \sqrt{\frac{GM}{r}}$. *(3 marks)*

b A space observatory to monitor activity on the surface of the Sun has been placed in a circular orbit, which is 1% smaller than the orbit of the Earth, as shown in Figure 4.

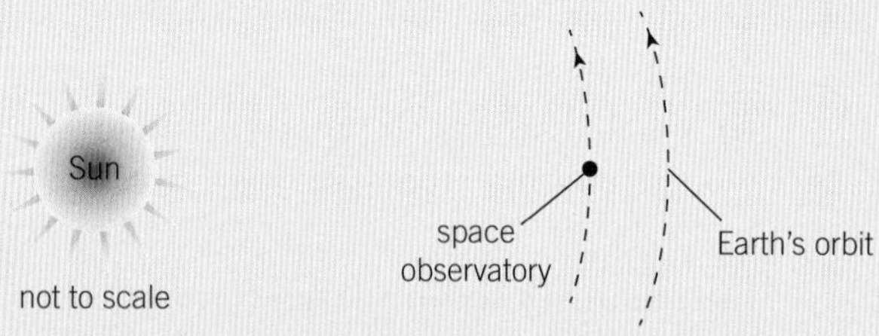

▲ Figure 4

Explain why the equation of part **(a)** predicts that the observatory should orbit the Sun in less than one year. *(2 marks)*

c Figure 5 shows the special case where the Earth and observatory are positioned so that both orbit the Sun in exactly one year.

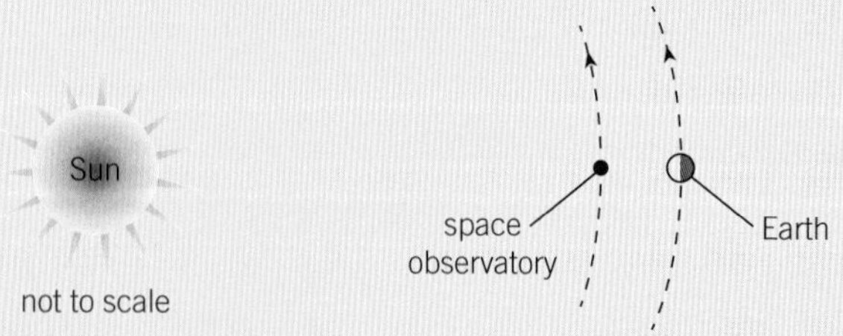

▲ Figure 5

(i) Explain why in this special case the speed of the observatory must be less than the speed of the Earth. (*1 mark*)

(ii) Draw labelled arrows on a copy of Figure 5 to show the directions of the gravitational forces acting on the observatory. Indicate, by length of arrow, which force is larger. (*1 mark*)

(iii) Explain how it is possible for the observatory to have an orbital period of one year. Suggest why this is convenient. (*3 marks*)

6 **a** Define *gravitational potential* at a point in space around a gravitating mass. (*1 mark*)

b The planet Mars has a radius of 3.38 Mm. The gravitational potential on its surface is $-12.7\,\text{MJ}\,\text{kg}^{-1}$.

(i) Explain the significance of the minus sign in the value of the potential. (*2 marks*)

(ii) Calculate

1 the mass of Mars (*3 marks*)

2 the mean density of Mars. (*2 marks*)

c Calculate the work done in raising an 80.0 kg space probe from the surface of Mars to a height of 1.25 Mars radii above its surface. (*3 marks*)

7 Figure 6 shows the variation of the magnitude of the Earth's gravitational field strength g with distance r from its centre.

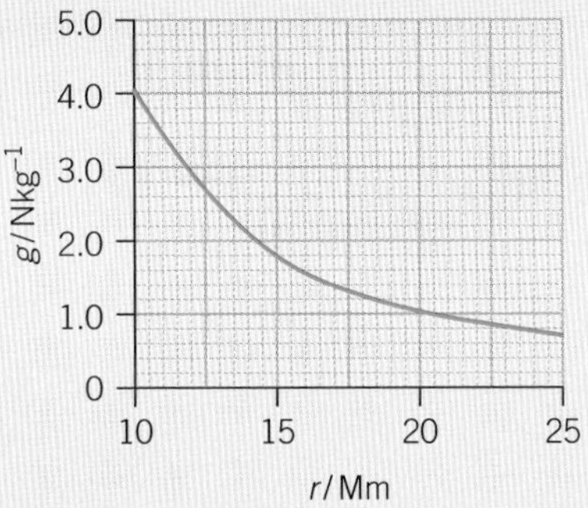

▲ Figure 6

a According to a student, g is inversely proportional to r.
Discuss this suggestion made by the student. (*2 marks*)

b Use Figure 6 to determine the mass M of the Earth. (*3 marks*)

c An artificial satellite of mass 310 kg is put into orbit at a distance $r = 15.0$ Mm. Calculate

(i) the speed of the satellite (*4 marks*)

(ii) the total energy of the satellite. (*3 marks*)

d Figure 7 shows a graph of lg (g) against lg (r) for the Earth

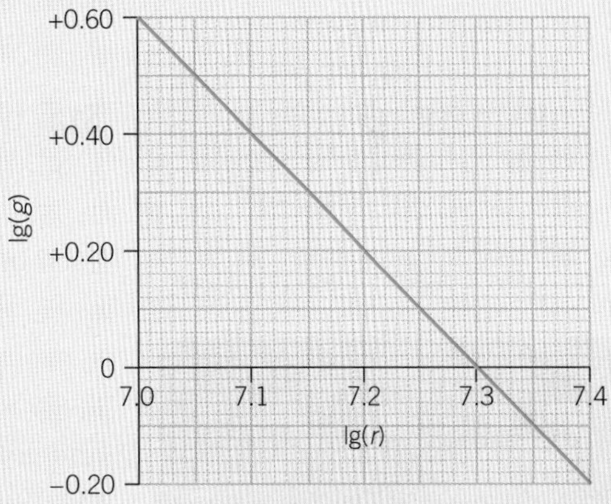

▲ Figure 7

(i) Explain why the graph shows a straight line. (*2 marks*)

(ii) Use Figure 7 to show that the gradient of the straight line is −2 and explain why this value will be the same for all the planets in the Solar System. (*2 marks*)

19 STARS

19.1 Objects in the Universe

Specification reference: 5.5.1

Learning outcomes

Demonstrate knowledge, understanding, and application of:

- → the terms planets, planetary satellites, comets, solar systems, galaxies, and the Universe
- → the formation of a star from interstellar dust and gas in terms of gravitational collapse, fusion of hydrogen into helium, radiation, and gas pressure.

▲ **Figure 1** *The Horsehead Nebula gets its name from one of the swirling clouds of dark dust and gases that resembles a horse's head*

Stellar nurseries

The **Universe** contains countless amazing objects, including the spectacular Horsehead Nebula in our own galaxy. **Nebulae** are gigantic clouds of dust and gas (mainly hydrogen), often many hundreds of times larger than our Solar System. The scale is difficult to comprehend.

Nebulae are often referred to as stellar nurseries, as they are the birthplace of all stars. Every star you see in the night sky is on its own journey from birth to eventual death. Our very own Sun was born in a nebula, and it too will eventually die.

Star birth

Nebulae are formed over millions of years as the tiny gravitational attraction between particles of dust and gas pulls the particles towards each other, eventually forming the vast clouds.

As the dust and gas get closer together this gravitational collapse accelerates. Due to tiny variations in the nebula, denser regions begin to form. These regions pull in more dust and gas, gaining mass and getting denser, and also getting hotter as gravitational energy is eventually transferred to thermal energy. In one part of the cloud a **protostar** forms – this is not yet a star but a very hot, very dense sphere of dust and gas.

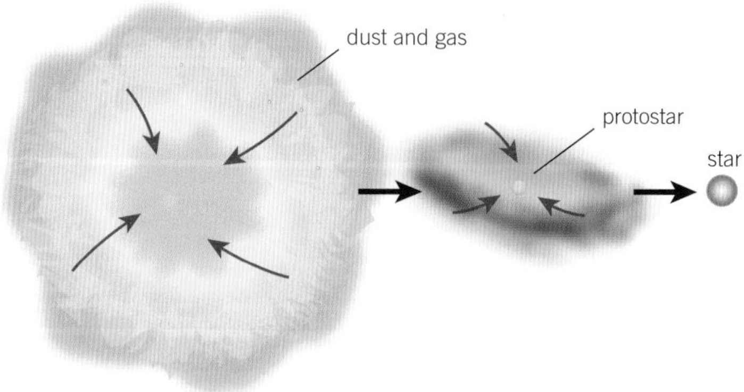

▲ **Figure 2** *A single protostar forming a star from the gravitational collapse of an interstellar cloud of dust and gas*

For a protostar to become a star, **nuclear fusion** needs to start in its core. Many protostars never reach this stage. Fusion reactions produce energy in the form of kinetic energy. Extremely high pressures and temperatures inside the core are needed in order to overcome the electrostatic repulsion between hydrogen nuclei in order to fuse them together to form helium nuclei. In some cases, as more and more mass is added to the protostar, it grows so large and the core becomes so hot that the kinetic energy of the hydrogen nuclei is large enough to overcome the electrostatic repulsion. Hydrogen nuclei are forced together to make helium nuclei as nuclear fusion begins. A star is born.

Star life

Once a star is formed, it remains in a stable equilibrium with almost a constant size. Gravitational forces act to compress the star, but the **radiation pressure** from the photons emitted during fusion and the **gas pressure** from the nuclei in the core push outwards. The force from this radiation and gas pressure balances the force from the gravitational attraction and maintains equilibrium.

Stars in this stable phase of their lives are described as being on their **main sequence**. How long a star remains stable depends on the size and mass of its core. The cores of large, massive supergiant stars are much hotter than those of small stars, releasing more power and converting the available hydrogen into helium in a much shorter time. Really massive stars are only stable for a few million years, whereas smaller stars like our Sun are stable for tens of billions of years.

What happens when a star runs low on hydrogen fuel in its core is discussed in Topic 19.2, The life cycle of stars.

Beyond stars

The Universe contains a variety of other objects beyond stars (Table 1).

▼ Table 1

Objects within the Universe	Description
Planets	A planet (named from the ancient Greek 'wanderer') is an object in orbit around a star with three important characteristics: • it has a mass large enough for its own gravity to give it a round shape (unlike the irregular shape of asteroids) • it has no fusion reactions (unlike a star) • it has cleared its orbit of most other objects (asteroids, etc.).
Dwarf planets*	A dwarf planet, like Pluto, has one important difference from a planet. Dwarf planets have not cleared their orbit of other objects. In Pluto's case there are many other bodies of comparable size close to its orbit.
Asteroids*	Asteroids are objects too small and uneven to be planets, usually in near-circular orbits round the Sun and without the ice present in comets.
Planetary satellites	A planetary satellite is a body in orbit around a planet. This includes moons and man-made satellites.

Synoptic link

Nuclear fusion and the formation of new elements in stars is covered in detail in Topic 26.4, Nuclear fusion.

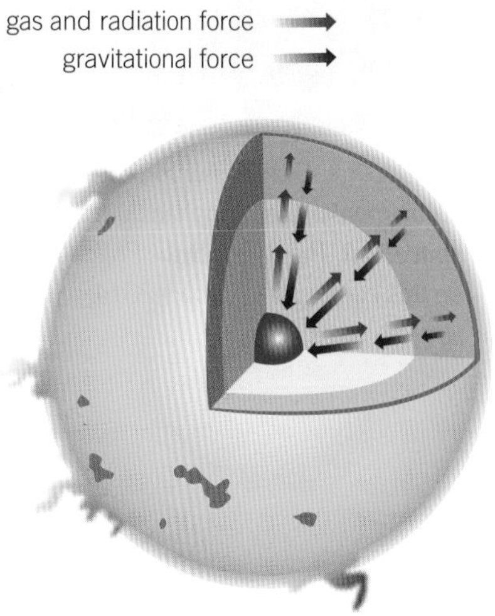

▲ **Figure 3** *The radiation and gas pressures in the core balance the pressure created by the gravitational attraction so that the shape of the star remains stable*

▲ **Figure 4** *Saturn is perhaps the most beautiful of all the planets orbiting the Sun, and has the largest number of planetary satellites and countless pieces of ice and rock that make up its rings*

▲ **Figure 5** *Comet Churyumov–Gerasimenko photographed by the Rosetta spacecraft. The space probe Philae landed on this comet in November 2014.*

▼ **Table 1** *(continued)*

Comets	Comets range from a few hundred metres to tens of kilometres across. They are small irregular bodies made up of ice, dust, and small pieces of rock. All comets orbit the Sun, many in highly eccentric elliptical orbits. As they approach the Sun, some comets develop spectacular tails.
Solar systems	Our Solar System contains the Sun and all objects that orbit it (planets, comets, etc.). It is one of many. In 2014 over 1100 other solar systems (sometimes called planetary systems) have been discovered.
Galaxies	A galaxy is a collection of stars, and interstellar dust and gas. On average a galaxy will contain 100 billion stars, a significant proportion of which have their own solar systems. Our galaxy is known as the Milky Way.

*Not examined in the A Level course

Defining what we mean by our Universe is a little more complex. Our Universe is quite literally everything! It is all electromagnetic radiation, energy and matter, all of space-time and everything that exists within it. This includes all the galaxies and all the contents of intergalactic space (including subatomic particles). This topic has given you some insight into the vastness of the Universe.

Summary questions

1 Sort the objects in Table 1 into a generalised list from smallest to largest. *(2 marks)*

2 Explain why nuclear fusion in the core of a star prevents further gravitational collapse. *(2 marks)*

3 Describe the similarities and differences between planets and comets. *(2 marks)*

4 Explain why larger stars tend to spend less time in their main-sequence phase. *(2 marks)*

5 The Sun has a radius of 7.0×10^5 km and an average density of 1410 kg m^{-3}. Calculate:

a the mass of the Sun;

b the ratio of the volume of Sun to the volume of the Earth ($m_{Earth} = 5.97 \times 10^{24}$ kg, $r_{Earth} = 6370$ km)

c the number of atoms within the Sun if the average spacing between each atom is around 10^{-10} m. *(5 marks)*

19.2 The life cycle of stars

Specification reference: 5.5.1

Birth, life, and death

Orion is one of the most instantly recognisable constellations. It is rather unusual, as it contains several bright stars in different stages of their life cycle. It even includes a nebula.

One of the brightest stars in Orion, Betelgeuse (top left), is a **red supergiant**. This huge star is in the last stages of its life and will soon 'explode' in an enormous **supernova** (in fact, it may have already happened, and the light from this blast could be on its 700-year journey towards us). Our Sun's ending will be far less spectacular than that of Betelgeuse. What happens to a star as it dies depends on the mass of the star.

Learning outcomes

Demonstrate knowledge, understanding, and application of:

- → evolution of a low-mass star like our Sun into a red giant and white dwarf
- → planetary nebula
- → characteristics of a white dwarf; electron degeneracy pressure; Chandrasekhar limit
- → evolution of a massive star into a red supergiant and then either a neutron star or black hole; supernova
- → characteristics of neutron stars and black holes.

Stars with low mass

Since the core of stars with low mass is cooler than that of more massive stars, they remain on their main sequence for much longer. However, eventually, often after billions of years, they run low on hydrogen fuel in their core. At this stage, they begin to move off the main sequence into the next phase of their lives.

Red giants

Stars between 0.5 $M_\odot$ and 10 $M_\odot$ will evolve into **red giants**. At the start of the red giant phase, the reduction in energy released by fusion in the core means that the gravitational force is now greater than the reduced force from radiation and gas pressure. The core of the star therefore begins to collapse. As the core shrinks, the pressure increases enough to start fusion in a shell around the core.

Red giant stars have inert cores. Fusion no longer takes place, since very little hydrogen remains and the temperature is not high enough for the helium nuclei to overcome the electrostatic repulsion between them. However, fusion of hydrogen into helium continues in the shell around the core. This causes the periphery of the star to expand as layers slowly move away from the core. As these layers expand, they cool, giving the star its characteristic red colour. In about 4 billion years from now, when our Sun expands into a red giant, it will engulf Mercury and Venus, stopping just short of the Earth.

Study tip

Solar mass $M_\odot$ is the mass of the Sun, about 1.99×10^{30} kg.

▲ **Figure 1** *The constellation of Orion the hunter contains many bright stars, three of which form its famous belt, and a nebula, just below the belt*

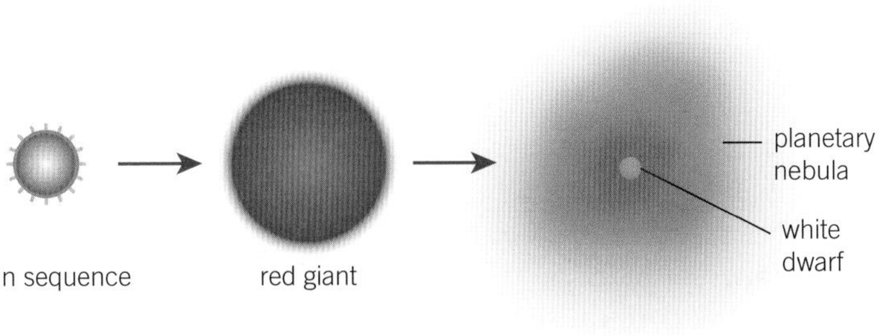

▲ **Figure 2** *The evolution of stars of lower mass, from main sequence to red giant and ending with white dwarf. The planetary nebula may collapse again to form another star, or even a solar system with its own planets (which explains its name).*

White dwarfs, electron degeneracy pressure, and the Chandrasekhar limit

Eventually most of the layers of the red giant around the core drift off into space as a **planetary nebula**, leaving behind the hot core as a **white dwarf**. The white dwarf is very dense, often with a mass around that of our Sun, but with the volume of the Earth. No fusion reactions take place inside a white dwarf. It emits energy only because it leaks photons created in its earlier evolution. The surface temperature of a white dwarf can be as much as 30 000 K.

▲ **Figure 3** *Subrahmanyan Chandrasekhar calculated the Chandrasekhar limit whilst still at university, and was awarded the 1983 Nobel Prize for Physics for his later work on black holes*

According to an important rule of quantum physics, the Pauli exclusion principle, two electrons cannot exist in the same energy state. When the core of a star begins to collapse under the force of gravity, the electrons are squeezed together, and this creates a pressure that prevents the core from further gravitational collapse. This pressure created by the electrons is known as **electron degeneracy pressure**.

But there is a limit. The electron degeneracy pressure is only sufficient to prevent gravitational collapse if the core has a mass less than $1.44 M_\odot$. This is called the **Chandrasekhar limit** – named after the astrophysicist Subrahmanyan Chandrasekhar (Figure 3), who improved the model used to describe the star when he was just 19 years old. This limit is the maximum mass of a stable white dwarf star. If the core is more massive than this, the star's life takes a more dramatic turn.

More massive stars

Stars with a mass greater than $10\ M_\odot$ live very different lives. Since their mass is much greater, their cores are much hotter. They consume the hydrogen in their core in much less time, some in only a few million years.

As with stars with smaller masses, when the hydrogen in the core runs low, the core begins to collapse under gravitational forces. However, as the cores of these more massive stars are much hotter, the helium nuclei formed from the fusion of hydrogen nuclei are moving fast enough to overcome electrostatic repulsion, so fusion of helium nuclei into heavier elements occurs.

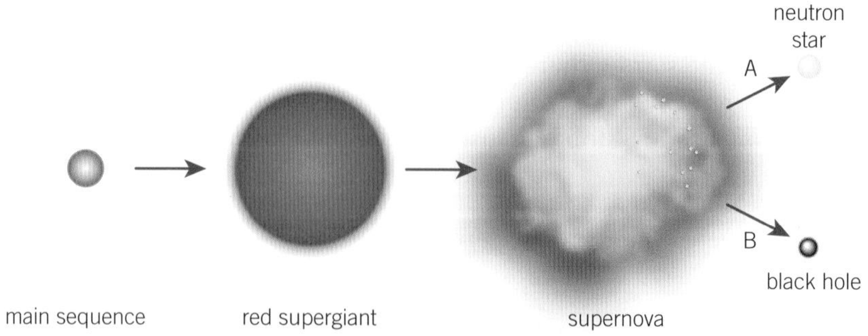

▲ **Figure 4** *The evolution of more massive stars*

Red supergiants

These changes in the core cause the star to expand, forming a red supergiant (sometimes called super red giant). Inside, the temperatures and pressures are high enough to fuse even massive nuclei together, forming a series of shells inside the star (Figure 5).

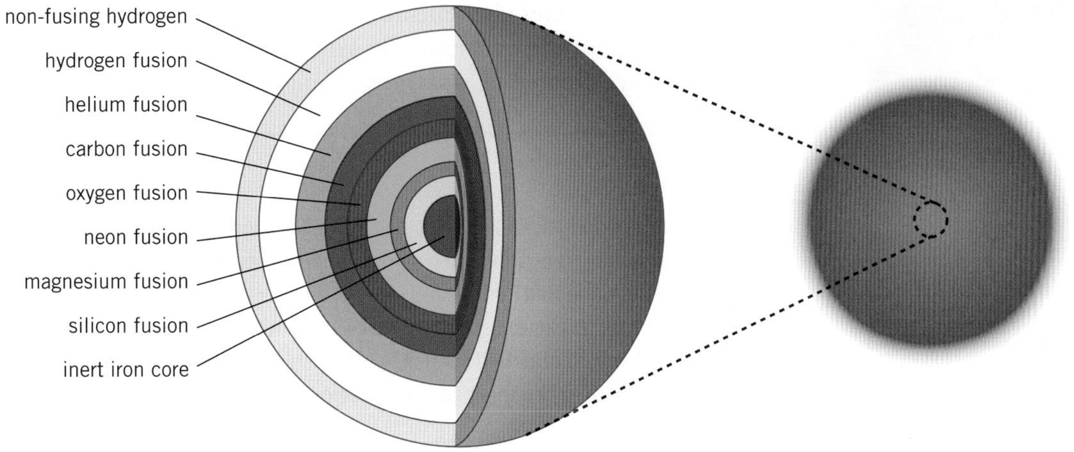

Figure 5 *Inside a red supergiant the core is made of onion-like layers in which different elements are created by fusion, with heavier elements deeper in, up to the central core, made of stable iron nuclei that cannot fuse any further*

This process continues until the star develops an iron core. Iron nuclei cannot fuse, because such reactions cannot produce any energy. This makes the star very unstable and leads to the death of the star in a catastrophic implosion of the layers that bounce off the solid core, leading to a shockwave that ejects all the core material into space. This 'explosion' is called a (type II) supernova.

Synoptic link

You will learn about the importance of the fusion of lighter elements up to iron in Topic 26.2, Binding energy.

Supernova and beyond

For more massive stars, at a critical point (depending on the mass of the star) the nuclear fusion taking place in the core suddenly becomes unable to withstand the crushing gravitational forces. The star collapses in on itself, leading to a supernova. Afterwards, the remnant core is compressed into one of two objects:

- **Neutron star**: If the mass of the core is greater than Chandrasekhar limit, the gravitational collapse continues, forming a neutron star. These strange stars are almost entirely made up of neutrons and can be very small – just 10 km in diameter. They have a typical mass of $2M_{\odot}$ and densities similar to that of an atomic nucleus ($\sim 10^{17}\,\text{kg}\,\text{m}^{-3}$).
- **Black hole**: If the core has a mass greater than about $3M_{\odot}$, the gravitational collapse continues to compress the core. The result is a gravitational field so strong that in order to escape it an object would need an escape velocity greater than the speed of light. Nothing, not even photons, can escape a black hole. Black holes vary in mass. Super-massive black holes with masses of several million $M_{\odot}$ are thought to be at the centre of most galaxies.

Supernovae are rare — the last recorded in our galaxy was back in 1604. But they are so luminous that we can see them in even the most distant of galaxies. Their output power is so great that sometimes they are brighter than the rest of their galaxy, radiating more energy in a few thousandths of a second than our Sun will in its entire lifetime.

Supernovae create all the heavy elements. Everything above iron in the Periodic Table was created in a supernova, and such events help

▲ **Figure 6** *The neutron star at the centre of the Crab Nebula sends out regular pulses, proving that the nebula must result from a historical supernova*

to distribute these heavier elements throughout the Universe. It is amazing to think that the copper in our wiring and the gold in our jewellery was once created by a star that went supernova.

LGM-1

In 1967 British astrophysicists Jocelyn Bell Burnell and Antony Hewish discovered a regularly repeating radio signal. It had a precise period of 1.337 s. Its regular nature made Bell Burnell and Hewish wonder whether the signal might come from an intelligent extraterrestrial civilization. They called it LGM-1, for 'Little Green Men'. Neither Bell Burnell nor Hewish really thought it was an alien signal, but nothing like this had been previously recorded in nature.

The signal turned out to be radio waves emitted from a rapidly spinning neutron star (dubbed a pulsar). As the star spins, the beam of radio waves sweeps across the Earth like the beam from a lighthouse. Many more pulsars have been discovered since, providing direct evidence for the existence of neutron stars left over after supernovae.

1 Explain why the Chandrasekhar limit must be exceeded in order for the core to form a neutron star.
2 The pulses received from LGM-1 came from a pulsar 2.18×10^{19} m from the Earth. Calculate the time taken for a pulse to travel from the star to Earth.
3 LGM-1 is estimated to have a radius 1.4 million times smaller than our Sun (solar radius 6.96×10^{8} m). Assuming LGM-1 has a density similar to that of a typical neutron star, calculate its mass.

The Schwarzschild radius

The Schwarzschild radius r_S of an object is the radius of an imaginary sphere sized so that, if all the mass of the object is compressed into the sphere, the escape velocity for the object would be greater than the speed of light. The radius of a black hole must be smaller than its Schwarzschild radius, so not even light can escape from its surface.

In 1916 the German astronomer Karl Schwarzschild used Einstein's theory of general relativity to calculate that for any object r_S is given by

$$r_S = \frac{2GM}{c^2}$$

where M is the mass of the object, c is the speed of light through a vacuum, and G is the gravitational constant.

1 Calculate the Schwarzschild radius for:
a the Earth (mass 5.97×10^{24} kg);
b the Sun (mass 1.99×10^{30} kg);
c an average human being.
2 Use the ideas of escape velocity (Topic 18.7, Gravitational potential energy) and kinetic and potential energy to derive the expression above for r_S.

Summary questions

1 Explain why heavier elements are not formed in the core of a red giant star. *(2 marks)*

2 Describe the process of the creation of heavier elements (nucleosynthesis) and how these elements come to be found distributed throughout the Universe. *(4 marks)*

3 Calculate the minimum and maximum values of the mass of a star that will form a red giant. *(2 marks)*

4 A neutron star has a density of 1.0×10^{17} kg m^{-3}. Calculate: **a** the mass of a 1.0 cm^3 piece of the star; **b** the volume of the star that would have the same mass as the Earth (5.97×10^{24} kg). *(4 marks)*

5 The red supergiant Betelgeuse is estimated to have a mass of between 8 and 20$M_\odot$ and a radius of between 950 and 1200 times the radius of our Sun. Use this information to determine the minimum and maximum values for the gravitational field strength on the surface of the star and its escape velocity. *(6 marks)*

19.3 The Hertzsprung–Russell diagram

Specification reference: 5.5.1

Nine million Suns

R136a1 is a star in the Tarantula Nebula. At 265 times the mass of our Sun it is the most massive star discovered to date. It is also the most luminous star currently known, with a **luminosity** 8 700 000 times that of our Sun.

There is an enormous variety of stars in our galaxy, with dramatic variations in mass, brightness, diameter, luminosity, surface temperature, and colour. Astrophysicists use many methods of classification to understand the variation. One of the most useful is the **Hertzsprung–Russell diagram**.

Learning outcomes

Demonstrate knowledge, understanding, and application of:

- → Hertzsprung–Russell (HR) diagram as luminosity–temperature plot
- → main sequence; red giants; red supergiants; white dwarfs.

The Hertzsprung–Russell diagram

The Hertzsprung–Russell (HR) diagram is a graph of stars in our galaxy showing the relationship between their luminosity on the y-axis and their average surface temperature on the x-axis. The temperature axis is a bit odd, with temperature increasing from right to left.

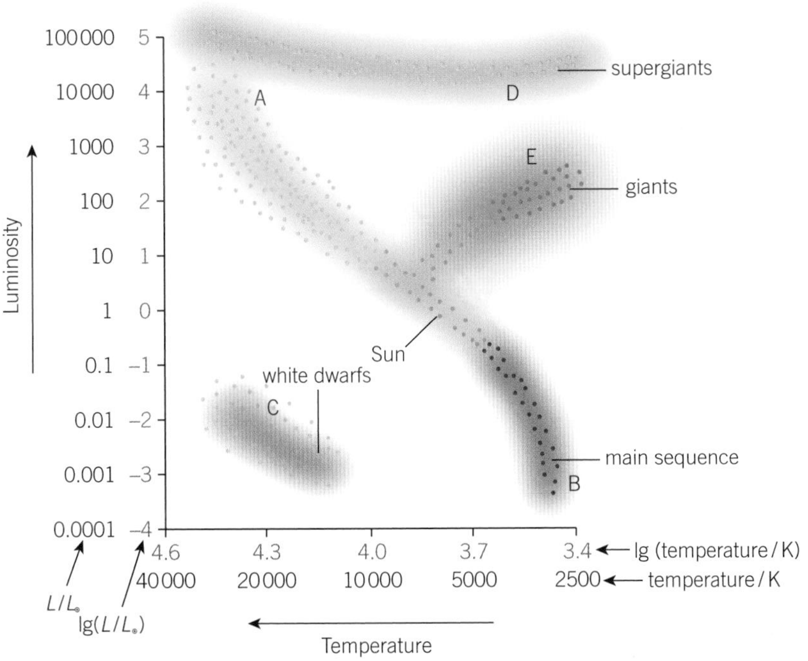

◀ **Figure 1** *The Hertzsprung–Russell diagram shows the positions of stars at various stages of their life cycle as a plot of luminosity against surface temperature (note the unusual axes)*

The luminosity of any star is the total radiant power output of the star. The luminosity of a star is related to its brightness – in general the greater the luminosity the brighter the star (more on luminosity in Topic 19.7, Stellar luminosity). Our Sun has a luminosity of 3.85×10^{26} W — it emits an incredible 385 000 000 000 000 000 000 000 000 J per second. In the HR diagram, luminosity is often plotted in units relative to

▲ **Figure 2** *The Tarantula Nebula is huge – 9.5×10^{18} m across – and contains millions of stars at different phases of their life cycle*

the Sun, where $1L_\odot = 3.85 \times 10^{26}$ W. Both luminosity and surface temperature of stars can vary widely – luminosity from less than $0.0001L_\odot$ to over $1\,000\,000L_\odot$, and surface temperature from 3000 K to 40 000 K. As a result, both scales in the HR diagram are normally logarithmic plots.

When stars are plotted on the HR diagram a pattern appears. The hottest, most luminous stars are in the top left at A, with the coolest, least luminous stars at B. Most stars on their main sequence form part of a curved line from A to B. Our Sun has a surface temperature of around 6000 K, and so sits near the middle of this line.

Very hot, dim stars like white dwarfs appear along a different line at C. Their surface temperatures can be many times greater than our Sun's. However, they are much smaller and less luminous.

Red supergiants are very luminous because of their enormous size, but they have a relatively low surface temperature. They are found around D. Smaller red giants are found in a line splitting from the main sequence at E.

Life cycle of stars from the HR diagram

The HR diagram is often used to show stellar evolution.

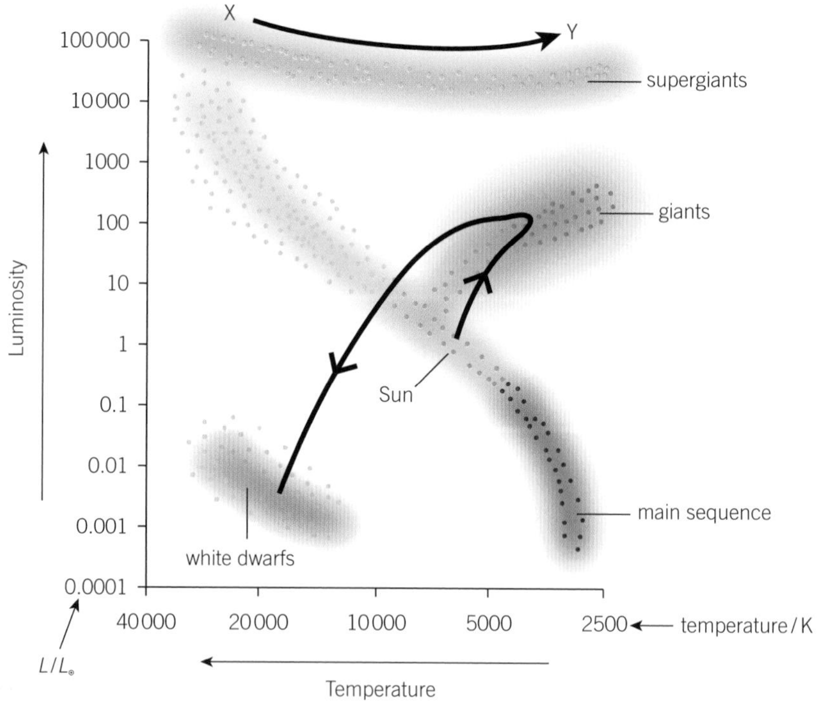

▲ **Figure 3** *Stellar evolution shown on a Hertzsprung–Russell diagram*

- Lower mass stars like our Sun evolve into red giants, moving away from the main sequence. They then gradually lose their cooler outer layers, and slowly move across the diagram, crossing the main sequence line to end up as white dwarfs.
- Higher mass stars start at X, before rapidly consuming their fuel and swelling into red supergiants at Y before they go supernova.

Summary questions

1. Sketch a Hertzsprung–Russell (HR) diagram and identify the positions of main-sequence stars, white dwarfs, and red giants. *(5 marks)*
2. Explain why when a main sequence star becomes a red giant it moves towards the upper right of an HR diagram. *(2 marks)*
3. Suggest where a black hole might appear on an HR diagram. *(2 marks)*
4. Calculate the maximum luminosity in W of the white dwarfs shown in Figure 2. *(2 marks)*
5. Use the HR diagram to determine the ratio of the temperature of the hottest stars to our Sun. *(2 marks)*

19.4 Energy levels in atoms

Specification reference: 5.5.2

Illuminating

From their first demonstration at the Paris Motor Show in 1910, neon tubes revolutionised lighting. Unlike the filament lamps of the day, neon and other tubes could produce signs in different colours, and they were much more efficient, not getting nearly as hot.

There is no hot filament inside the glass tube of a neon light, only neon or another gas at low pressure. When a large enough p.d. is applied across the tube, the gas glows, giving off a characteristic colour depending on the gas. The gas emits light because of the behaviour of electrons within its atoms.

Energy levels in gas atoms

When electrons are bound to their atoms in a gas they can only exist in one of a discrete set of energies, referred to as the **energy levels** (or energy states) of an electron. This seemed a very odd idea when physicists discovered it a hundred years ago.

- An electron cannot have a quantity of energy between two levels.
- The energy levels are negative because external energy is required to remove an electron from the atom. The negative values also indicate that the electrons are trapped within the atom or bound to the positive nuclei.
- An electron with zero energy is free from the atom.
- The energy level with the most negative value is known as the ground level or the **ground state**.

When an electron moves from a lower to a higher energy level within an atom in a gas, the atom is said to be **excited**. Raising an electron into higher energy levels requires external energy, for example, supplied by an electric field (as in the neon tube), through heating, or when photons of specific energy (and therefore frequency) are absorbed by the atoms (more on this in Topic 19.5, Spectra).

Figure 2 shows some electron energy levels within an atom. Each energy level has a specific negative value; in this example −6.8 eV, −3.0 eV, and −1.5 eV. An electron in the −3.0 eV energy level requires at least 3.0 eV to escape from the atom.

When an electron moves from a higher energy level to a lower one, it loses energy. Energy is conserved, so as the electron makes a transition between the levels, a photon is emitted from the atom. This transition between energy levels is sometimes called de-excitation.

In order for an electron to make a transition from −3.0 eV to −6.8 eV it must lose 3.8 eV. It emits this in the form of a photon with a specific energy of 3.8 eV. The energy of the photon $hf = 3.8$ eV. In general, the

Learning outcomes

Demonstrate knowledge, understanding, and application of:

→ energy levels of electrons in isolated gas atoms

→ the idea that energy levels have negative values

→ emission spectral lines from hot gases in terms of transition of electrons between discrete energy levels and emission of photons

→ the equations $hf = \Delta E$ and $\frac{hc}{\lambda} = \Delta E$.

▲ **Figure 1** *Neon signs are common in many cities, although LED displays are now replacing them*

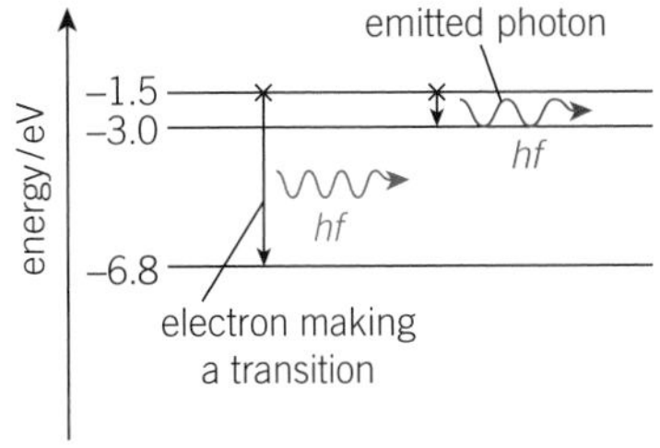

▲ **Figure 2** *Photons are emitted when electrons make transitions from higher to lower energy levels*

Synoptic link

You met photons in Topic 13.1, The photon model, and learned more about energy in discrete quantities in a different context in Topic 13.2, The photoelectric effect.

energy of any particular photon emitted in an electron transition from a higher to a lower energy level is given by

$$\Delta E = hf \qquad \text{and} \qquad \Delta E = \frac{hc}{\lambda}$$

where f is the frequency of the photon and ΔE is the difference in energy between the two energy levels.

Each element has its own unique set of energy levels, like fingerprints, so the energy levels of electrons in helium are different from those in hydrogen. Figure 3 shows the lowest five energy levels for hydrogen.

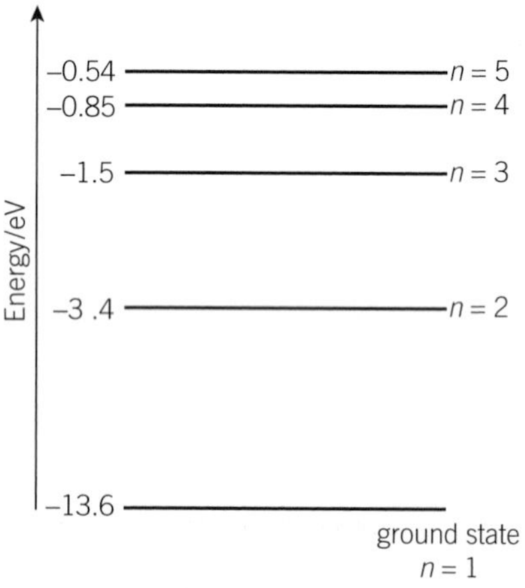

▲ **Figure 3** *The five lowest energy levels for electrons in hydrogen atoms, labelled with the principal quantum number, n*

Worked example: Emitting photons

Use the information in Figure 2 to determine the frequency of the two photons emitted.

Step 1: Calculate the change in energy in J for each electron as it drops from the higher energy level to the lower one.

First photon: $\Delta E = 6.8 - 1.5 = 5.3\,\text{eV} = 5.3 \times 1.60 \times 10^{-19}\,\text{J}$
$= 8.5 \times 10^{-19}\,\text{J}$

Second photon: $\Delta E = 3.0 - 1.5 = 1.5\,\text{eV} = 1.5 \times 1.60 \times 10^{-19}\,\text{J}$
$= 2.4 \times 10^{-19}\,\text{J}$

Step 2: Use the appropriate relationship to determine the frequency of the emitted photon.

$\Delta E = hf \qquad$ therefore $f = \frac{\Delta E}{h}$

First photon: $f = \frac{\Delta E}{h} = \frac{8.5 \times 10^{-19}}{6.63 \times 10^{-34}} = 1.3 \times 10^{15}\,\text{Hz}$ (2 s.f.)

Second photon: $f = \frac{\Delta E}{h} = \frac{2.4 \times 10^{-19}}{6.63 \times 10^{-34}} = 3.6 \times 10^{14}\,\text{Hz}$ (2 s.f.)

The first photon has higher frequency because the difference between the energy levels is greater ($f \propto \Delta E$).

Summary questions

1. An atom emits a photon of frequency 4.5×10^{15} Hz. Calculate the difference in energy between the two energy levels in the gas atom. *(2 marks)*
2. Explain why the wavelength of the emitted photon is shorter when an electron in an atom moves into the ground state from $n = 3$ than when it drops to the ground state from $n = 2$. *(3 marks)*
3. An electron moves from an energy level of −4.0 eV to −6.7 eV. Calculate the wavelength of the photon it emits and state in which part of the electromagnetic spectrum this photon belongs. *(4 marks)*
4. Use the energy levels for hydrogen in Figure 3 to calculate the possible frequencies of the photons emitted when an electron moves into $n = 1$, $n = 2$, and $n = 3$ from a higher energy level (nine possible photons in total). *(9 marks)*
5. A laser emits photons when electrons make transitions between energy levels. If a laser has a power output of 1.0 mW and emits 3.48×10^{15} photons per second, all of the same frequency, calculate the difference between the energy levels in eV. *(3 marks)*

19.5 Spectra

Specification reference: 5.5.2

The discovery of helium

Helium was discovered not from a sample on Earth, but by careful analysis of the light from the Sun. During a total eclipse in 1868, Pierre Janssen, a French astronomer, observed a bright yellow light with a wavelength of 587.49 nm in the spectrum of the gases surrounding the Sun. No element known at the time produced photons of this specific wavelength. The new element was named after Helios, the ancient Greek Sun god, by the English astronomer Sir Norman Lockyer later the same year.

The technique of analysing the light from stars proved so useful that the science of **spectroscopy** was born. Because different atoms have different **spectral lines**, the spectra from starlight can be used to identify the elements within stars, even those billions of miles away, without a direct, physical sample from the star.

Learning outcomes

Demonstrate knowledge, understanding, and application of:

- → different atoms have different spectral lines, which can be used to identify elements within stars
- → continuous spectra, emission line spectra, and absorption line spectra.

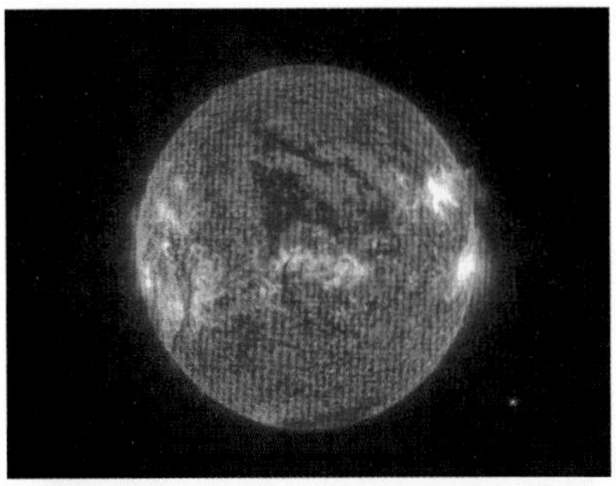

▲ **Figure 1** *The Sun is around one quarter helium*

Continuous, emission, and absorption spectra

There are three kinds of spectra.

- **Emission line spectra** — each element produces a unique emission line spectrum because of its unique set of energy levels.
- **Continuous spectra** — all visible frequencies or wavelengths are present. The atoms of a heated solid metal (e.g., a lamp filament) will produce this type of spectrum.
- **Absorption line spectra** — this type of spectrum has series of dark spectral lines against the background of a continuous spectrum. The dark lines have exactly the same wavelengths as the bright emission spectral lines for the same gas atoms.

If the atoms in a gas are excited (e.g., within the hot environment of stars), then when the electrons drop back into lower energy levels they emit photons with a set of discrete frequencies specific to that element. This produces a characteristic **emission line spectrum**. Each spectral line corresponds to photons with a specific wavelength. These spectra can be observed in a laboratory from heated gases. Each coloured line in Figure 2 represents a unique wavelength (or frequency) of photon emitted when an electron moves between two specific energy levels.

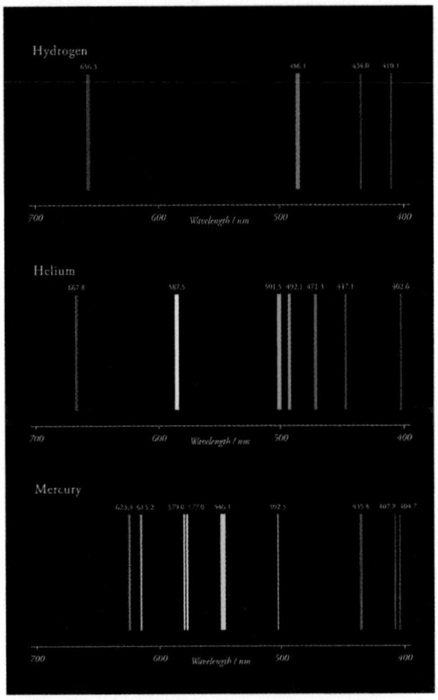

▲ **Figure 2** *Gas atoms of different elements produce different emission line spectra, which act like a unique fingerprint for each element*

Absorption line spectra

An absorption line spectrum is formed when light from a source that produces a continuous spectrum passes through a cooler gas. As the photons pass through the gas, some are absorbed by the gas atoms, raising electrons up into higher energy levels and so exciting the atoms. Only photons with energy exactly equal to the difference between the different energy levels are absorbed. This means that only specific wavelengths are absorbed, creating dark lines in the spectrum.

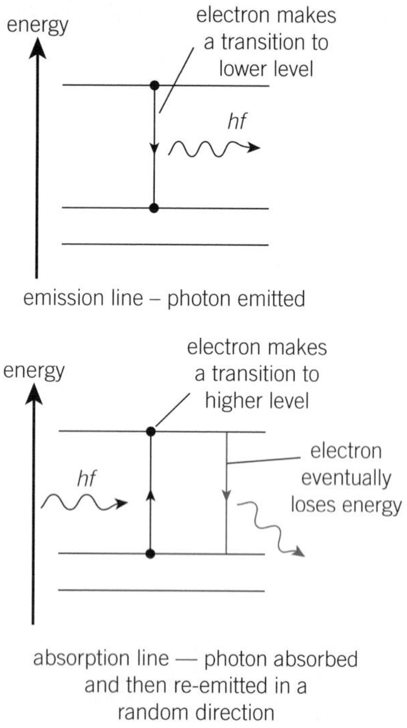

▲ **Figure 3** *A comparison of emission and absorption of photons*

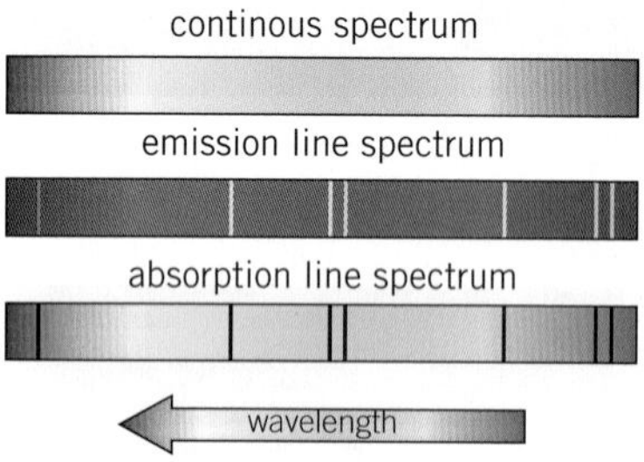

▲ **Figure 4** *The three main types of visible spectra — can you see a link between the wavelengths of the emission and absorption spectral lines?*

These lines show which photons have been absorbed by the gas atoms. Although the photons are re-emitted when the electron drops back down to a lower energy level atom, they are emitted in all possible directions, so the intensity in the original direction is greatly reduced (see Figure 3).

The absorption line spectrum for any gas is very nearly a negative of its emission line spectrum (Figure 4). In fact, a few lines from the emission line spectrum may not be visible in the absorption line spectrum, simply because in excited atoms electrons may return to their ground state in stages, releasing a photon each time, whereas absorption lines are mostly caused by electrons starting from their ground state.

Detecting elements within stars

When the light from a star is analysed, it is found to be an absorption line spectrum. Some wavelengths of light are missing – the photons have been absorbed by atoms of cooler gas in the outer layers of the star.

If we know the line spectrum of a particular element, we can check whether the element is present in the star, even for extremely distant stars. If a particular element is present then its characteristic pattern of spectral lines will appear as dark lines in the absorption line spectrum.

Summary questions

1 Describe the differences between a continuous spectrum and an emission line spectrum. *(2 marks)*

2 Explain why the wavelengths of the emission lines for gas atoms of a particular element have the same wavelengths as the dark absorption lines for the same atoms. *(2 marks)*

3 An absorption line at a wavelength of 682 nm is observed in the spectrum from a star. Determine the difference between the energy levels for the atoms in the gas responsible for this absorption line. *(3 marks)*

4 An absorption line is observed in the spectrum of a particular gas when electrons absorb photons and move between energy levels at −10.4 eV and −4.6 eV. Calculate the wavelength of these photons and so identify the part of the electromagnetic spectrum to which this absorption line belongs. *(4 marks)*

5 The value of the energy level in eV for the hydrogen atom is given by the equation $E_n = -\frac{13.6}{n^2}$

where n is an integer 1, 2, 3 etc.

a Draw an energy level diagram (to scale) for the hydrogen atom showing the five lowest energy levels. *(2 marks)*

b Determine the wavelength of the emitted photon when an electron makes a transition between levels $n = 3$ and $n = 2$. *(3 marks)*

19.6 Analysing starlight

Specification reference: 5.5.2

Colours on an optical disc

When white light shines onto an optical disc, it splits into beams of different colours (Figure 1) – try it by reflecting sunlight off the back of a DVD onto a wall. The disc has millions of equally spaced lines of microscopic pits on its surface that diffract the light to form an interference pattern.

A **diffraction grating** is an optical component with regularly spaced slits or lines that diffract and split light into beams of different colour travelling in different directions. These beams can be analysed to determine the wavelengths of spectral lines in the laboratory or from starlight.

Learning outcomes

Demonstrate knowledge, understanding, and application of:

→ use of a transmission diffraction grating to determine the wavelength of light

→ the condition for maxima $d\sin\theta = n\lambda$, where d is the grating spacing.

The diffraction grating

The fringes produced by passing light through a double slit are not very sharp, so it can be difficult to determine the position of the centre of each maximum. To overcome this limitation, a transmission diffraction grating can be used in place of the double slit. The grating consists of a large number of lines ruled on a glass or plastic slide. There can be as many as 1000 lines in a millimetre. Each line diffracts light like a slit. Using a large number of lines produces a clearer and brighter interference pattern.

▲ **Figure 1** *The small pits and grooves found on optical discs act like a diffraction grating, splitting light into beams of different colours*

When light passes through a diffraction grating it is split into a series of narrow beams. The direction of these beams depends on the spacing of the lines, or slits, of the grating and the wavelength of the light. Therefore, when white light is passed through a diffraction grating it splits into its component colours, making gratings especially useful in spectroscopy.

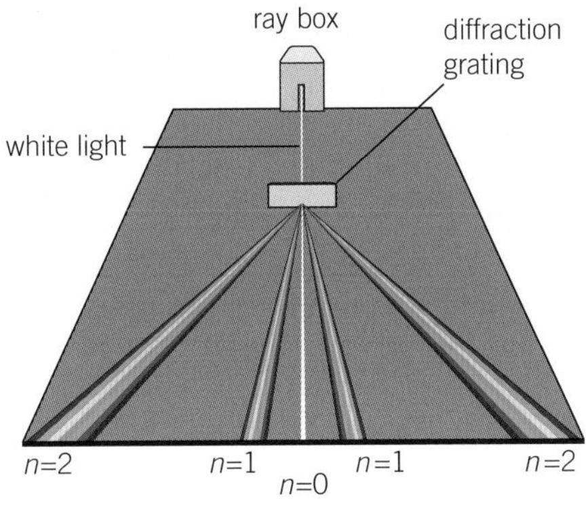

▲ **Figure 2** *Use of a diffraction grating*

Synoptic link

You saw in Topic 12.3, the Young double-slit experiment, that when light passes through a double slit it produces an interference pattern as a series of bright and dark fringes.

Forming maxima

Consider monochromatic light incident normally at a diffraction grating. The light is diffracted at each slit, and the interference pattern is the result of superposition of the diffracted waves in the space beyond the grating. Just as with the interference pattern created by a double slit, the formation of a maximum at a particular point depends on the path difference and the phase difference of the waves from all the slits.

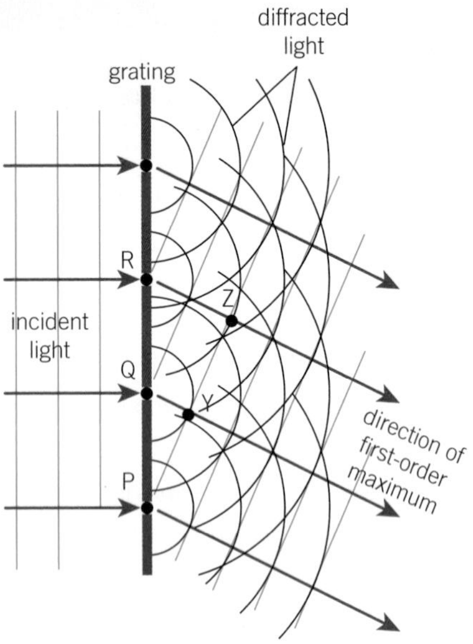

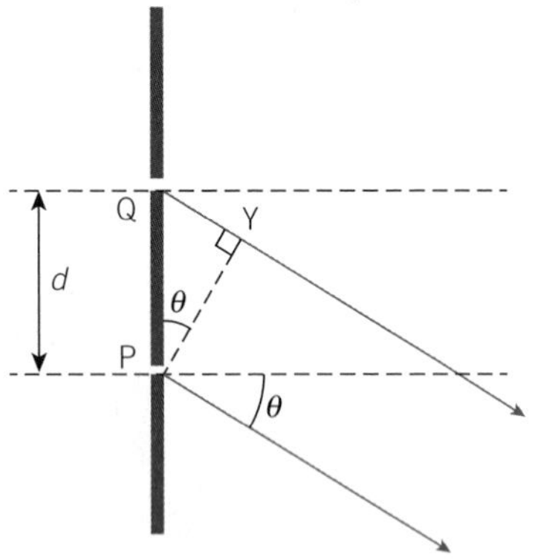

▲ **Figure 3** *Formation of the first-order maximum*

Study tip

The largest possible angle $\theta = 90°$. This makes $\sin\theta = 1$. The largest possible order number n that you can observe is therefore given by the equation $n_{max}\lambda = d$.

The zero-order maximum, $n = 0$, is formed when the path difference is zero, that is, at an angle $\theta = 0$. The angle θ is measured relative to the normal to the grating or to the direction of the incident light. Figure 3 shows the formation of one of the two first-order maxima, $n = 1$. The waves from adjacent slits P and Q have a path difference of exactly one whole wavelength λ. The same is true for waves from any two adjacent slits on the grating. Therefore, the distance QY is λ, distance RZ is 2λ and so on.

For the two nth-order maxima, the path difference QY at an angle θ will be equal to $n\lambda$ (in Figure 3, $n = 1$). The distance PQ is the separation between adjacent lines or slits on the grating. This distance is called the **grating spacing** d. From the triangle PQY, you can see that

$$\sin\theta = \frac{QY}{QP} = \frac{n\lambda}{d} \quad \text{or}$$

$$d\sin\theta = n\lambda$$

where n is an integer with values 0, 1, 2, etc. The equation is known as the **grating equation** and can be used to accurately determine the wavelength of monochromatic light.

Worked example: Finding the grating spacing

Monochromatic light from a laser of wavelength 532 nm is incident normally at a diffraction grating. The angle between the second-order maximum and the zero-order maximum is measured to be 32°. Calculate the grating spacing d.

Step 1: Rearrange $d\sin\theta = n\lambda$ for d.

$$d = \frac{n\lambda}{\sin\theta}$$

Step 2: Since the angle between the second-order maximum and the zero-order maximum is used, $n = 2$.

Therefore $d = \dfrac{2 \times 532 \times 10^{-9}}{\sin 32°} = 2.0 \times 10^{-6}\,\text{m}$ (2 s.f.)

Using a diffraction grating to determine the wavelength of light

Like the double-slit experiment, a diffraction grating can be used to determine the wavelength of monochromatic light. Measuring the angle between several maxima and the zero-order maximum and then plotting a graph of $\sin\theta$ against n will produce a straight line through the origin with a gradient of $\frac{\lambda}{d}$.

Many diffraction gratings are not labelled with the value of the grating spacing but with the numbers of lines (slits) per mm. Since grating spacing $= \left(\frac{1}{\text{lines per metre}}\right)$, a grating with 600 lines/mm has 600 000 lines/m and so a grating spacing of 1.67×10^{-6} m.

1 A laser emitting red light shines through a diffraction grating with 200 lines/mm. The angles between the zero-order maximum and the first six maxima are measured (Table 1).

▼ **Table 1** *Angle measurement for maxima in a diffraction pattern*

n	1	2	3	4	5	6
θ/°	7.2	14.7	22.2	30.6	39.0	49.2

a Plot a graph of $\sin\theta$ against n.

b Use your graph to determine the wavelength of the light emitted by the laser.

2 Calculate the largest number of orders that can be observed in this experiment.

3 On your graph sketch two other lines to show the relationship between n and θ if:

a a diffraction grating with twice as many lines/mm is used;

b a laser with a shorter wavelength is used.

Study tip

Remember the largest number of maxima you can see is rounded down. For example, if you calculate $n = 5.7$, maxima for $n = 0$ to $n = 5$ are visible – a total of 11 maxima.

Summary questions

1 Suggest why the maxima produced from a diffraction grating are brighter than those produced via the double-slit experiment. (*2 marks*)

2 Explain why the highest order maxima visible through a diffraction grating is given by $\frac{d}{\lambda}$. (*2 marks*)

3 A diffraction grating with grating spacing of 3.3×10^{-6} m is used to observe light from a star. The spectral line produces a first-order image at a diffraction angle of 8.6°. Calculate the wavelength of this spectral line. (*3 marks*)

4 Calculate the angle of the third-order maximum when light of wavelength 450 nm is incident on a diffraction grating with 350 lines/mm. (*4 marks*)

5 Calculate the maximum number of orders that can be observed with the arrangement in question 4. (*2 marks*)

6 A spectral line from a distant star is analysed using a diffraction grating with a grating spacing of 2.5×10^{-6} m. An absorption line in the first-order spectrum is observed at an angle of 13.4°. Calculate the energy in eV of the photons responsible for this spectral line. (*4 marks*)

▲ **Figure 4** *The diffraction pattern from a monochromatic light source obtained using a diffraction grating*

19.7 Stellar luminosity

Specification reference: 5.5.2

Learning outcomes

Demonstrate knowledge, understanding, and application of:

→ use of Wien's displacement law $\lambda_{max} \propto \frac{1}{T}$ to estimate the peak surface temperature of a star

→ luminosity L of a star; Stefan's law $L = 4\pi r^2 \sigma T^4$, where σ is the Stefan constant

→ use of Wien's displacement law and Stefan's law to estimate the radius of a star.

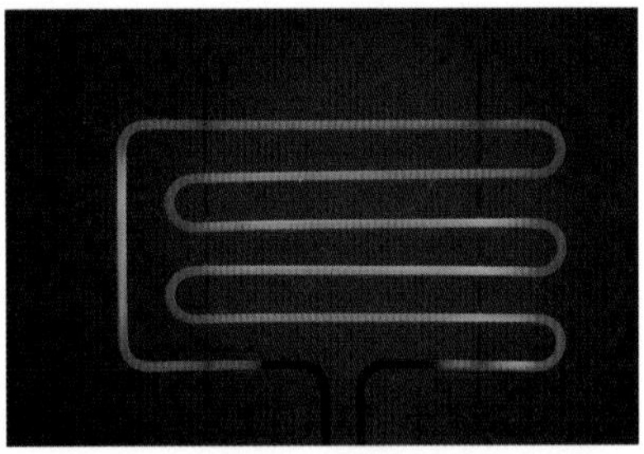

▲ **Figure 1** *A grill glows as it is heated, with a colour directly related to its temperature*

Red hot

You have already seen that the temperature of the surface of a star affects its colour, with the hottest stars glowing blue-white, and cooler stars a deeper shade of red. We can see the same effect when a metal is heated (Figure 1). At first the metal glows dull red, then reddish-orange as its temperature increases. If it does not melt, it will eventually glow white-hot if the temperature gets high enough.

Black-body radiation

At any given temperature above absolute zero, an object emits electromagnetic radiation of different wavelengths and different intensities. We can model a hot object as a **black body**. A black body is an idealised object that absorbs all the electromagnetic radiation that shines onto it and, when in thermal equilibrium, emits a characteristic distribution of wavelengths at a specific temperature. Figure 2 shows a graph of intensity against wavelength for electromagnetic radiation emitted by a black body at 6000 K.

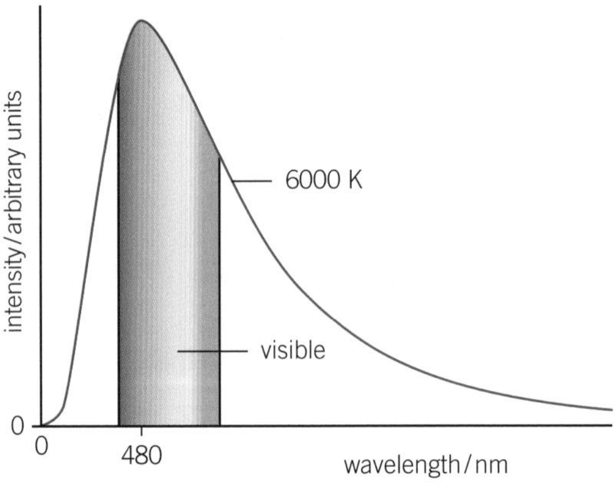

▲ **Figure 2** *The intensity–wavelength graph for a black body at a temperature of 6000 K*

Wien's displacement law

Wien's displacement law is a simple law that relates the absolute temperature T of a black body to the peak wavelength λ_{max} at which the intensity is a maximum. It can be applied to most objects, from stars to filament lamps, and even to mammals. Wien's displacement law states that λ_{max} is inversely proportional to T, that is

$$\lambda_{max} \propto \frac{1}{T}$$

It follows that for any black-body emitter $\lambda_{max}T$ = constant. The value of this constant is 2.90×10^{-3} m K and it is known as Wien's constant. (You do not need to memorise the value of this constant for this course).

Many objects, including stars, can be modelled as approximate black bodies (Table 1). This helps scientists to determine temperatures of objects simply by analysing the electromagnetic radiation they emit.

▼ **Table 1** *λ_{max} values for a number of different objects*

Object	λ_{max} / m	T / K
Healthy human	9.4×10^{-6}	310
Wood fire	1.9×10^{-6}	1500
Betelgeuse (red supergiant)	8.5×10^{-7}	3400
Sun	5.0×10^{-7}	5800
Sirius B (white dwarf)	1.2×10^{-7}	25 000

As the temperature of an object changes, so does the distribution of the emitted wavelengths. The peak wavelength reduces as the temperature increases, and the peak of the intensity–wavelength graph becomes sharper (Figure 3).

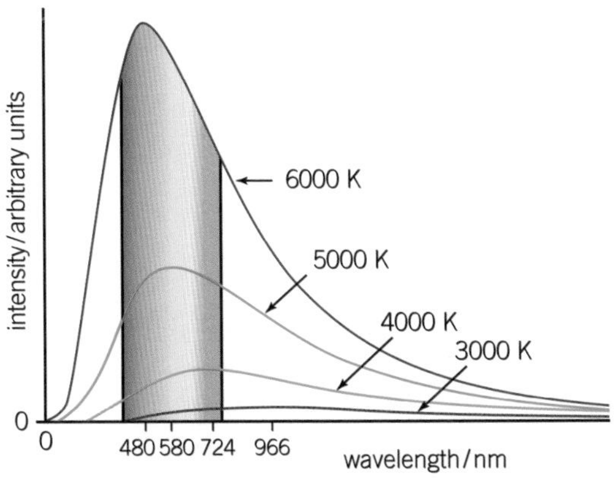

▲ **Figure 3** *The distribution of wavelengths changes as the temperature of the black-body emitter changes*

Stefan's law

Stefan's law, also known as the Stefan–Boltzmann law, states that the total power radiated per unit surface area of a black body is directly proportional to the fourth power of the absolute temperature of the black body. The total power radiated by a star is called luminosity (see Topic 19.3, The Hertzsprung–Russell diagram).

According to Stefan's law, the equation for the luminosity L in watts (W) of a star is given by the equation

$$L = 4\pi r^2 \sigma T^4$$

where r is the radius of the star in metres (m), T is the surface absolute temperature of the star in kelvin (K), and σ is the **Stefan constant**, $5.67 \times 10^{-8}\,\mathrm{W\,m^{-2}\,K^{-4}}$.

Stefan's law shows that the luminosity of a star is directly proportional:

- to its radius^2 ($L \propto r^2$)
- to its surface area ($L \propto 4\pi r^2$)
- to its surface absolute temperature^4 ($L \propto T^4$)

Synoptic link

You can refresh your memory by looking back to Topic 18.2, Newton's law of gravitation.

Wien's displacement law and Stefan's law can be used together to estimate the radius of a distant star. Once the radius is known, the mass and density of the star can be determined using Newton's law of gravitation. It is amazing what you can do just by analysing starlight.

Worked example: Radius of a star

The peak wavelength of radiation emitted by our Sun is about 500 nm, its surface temperature is 5800 K, and its luminosity is 3.85×10^{26} W. The peak wavelength emitted by a nearby star with a luminosity 10 times that of our Sun is 310 nm. Show that the radius of this star is approximately 840 000 km.

Step 1: To determine the surface temperature of the star use Wien's displacement law.

$$\lambda_{\text{max}} T = \text{constant}$$

$$\underbrace{500 \times 10^{-9} \times 5800}_{\text{Sun}} = \underbrace{310 \times 10^{-9} \times T_{\text{Star}}}_{\text{Star}}$$

Therefore

$$T_{\text{Star}} = \frac{500 \times 10^{-9} \times 5800}{310 \times 10^{-9}} = 9355\,\text{K}$$

Step 2: Use Stefan's law to determine the radius of the star.

$$L = 4\pi r^2 \sigma T^4, \text{ therefore } r = \sqrt{\frac{L}{4\pi \sigma T^4}}$$

The luminosity L of the star is $10 \times 3.85 \times 10^{26}$ W

$$r = \sqrt{\frac{10 \times 3.85 \times 10^{26}}{4\pi \times 5.67 \times 10^{-8} \times 9355^4}}$$

$$r = 8.399\ldots \times 10^{8}\,\text{m} = 840\,000\,\text{km (2 s.f.)}$$

Summary questions

1 State the SI unit for the luminosity of a star. (1 mark)

2 Use the data in Table 1 to show that $\lambda_{\text{max}} \propto \frac{1}{T}$. (3 marks)

3 The peak wavelength emitted by a red supergiant is 0.94 μm. Determine the surface temperature of the star. (3 marks)

4 Using Stefan's law, compare the luminosity of one star with another that has:

a double the surface temperature and the same radius;

b double the radius and half the surface temperature;

c half the mass, the same density, and three times the surface temperature. (6 marks)

5 The Sun has a radius of approximately 700 000 km and a surface temperature of 5800 K. Calculate the energy radiated by the Sun during one year. (3 marks)

6 The peak wavelength emitted by a distant star with a luminosity of 4.85×10^{31} W is measured as 305 nm using a diffraction grating. Calculate the radius of this star. (5 marks)

Practice questions

1 a (i) Describe the formation of a star such as our Sun and its most probable evolution. *(6 marks)*

(ii) Describe the probable evolution of a star that is much more massive than our Sun. *(2 marks)*

b The present mass of the Sun is 2.0×10^{30} kg. The Sun emits radiation at an average rate of 3.8×10^{26} J s^{-1}. Calculate the time in years for the mass of the Sun to decrease by one millionth of its present mass.
1 y = 3.2×10^{7} s *(3 marks)*

Jan 2011 G485

2 This question is about the light from low energy compact fluorescent lamps which are replacing filament lamps in the home.

a The light from a compact fluorescent lamp is analysed by passing it through a diffraction grating. Figure 1 shows the angular positions of the three major lines in the first order spectrum and the bright central beam.

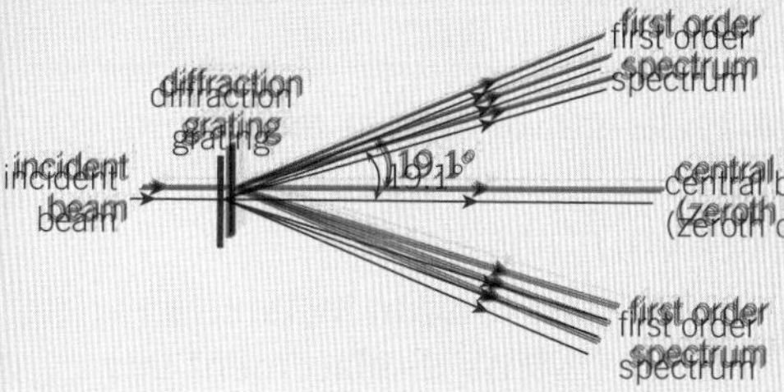

▲ Figure 1

(i) On a copy of Figure 1 label one set of the lines in the first order spectrum **R**, **G** and **V** to indicate which is red, green and violet. *(1 mark)*

(ii) Explain why the bright central beam appears white. *(1 mark)*

(iii) The line separation d on the grating is 1.67×10^{-6} m.

Calculate the wavelength λ of the light producing the first order line at an angle of 19.1° to the central bright beam. *(3 marks)*

b The wavelength of the violet light is 436 nm. Calculate the energy of a photon of this wavelength. *(3 marks)*

c The energy level diagram of Figure 2 is for the atoms emitting light in the lamp. The three electron transitions between the four levels **A**, **B**, **C**, and **D** shown produce the photons of red, green, and violet light. The energy E of an electron bound to an atom is negative. The ionisation level, not shown on the diagram, defines the zero of the vertical energy scale.

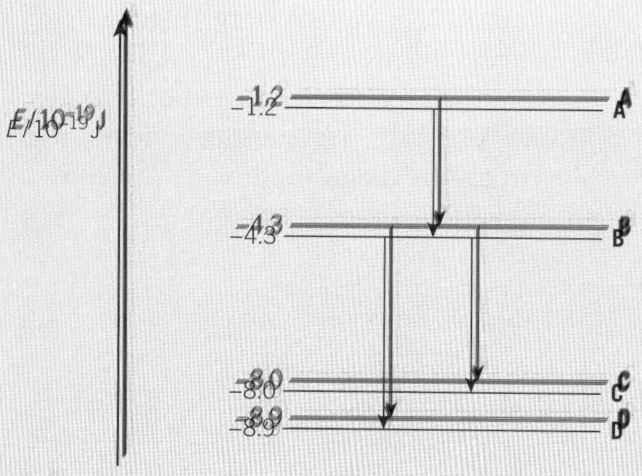

▲ Figure 2

Label the arrows on a copy of Figure 2 **R**, **G**, and **V** to indicate which results in the red, green, and violet photons. *(2 marks)*

Jan 2013 G482

3 a When a glowing gas discharge tube is viewed through a diffraction grating an emission line spectrum is observed.

(i) Explain what is meant by a *line spectrum*. *(2 marks)*

(ii) Describe how an absorption line spectrum differs from an emission line spectrum. *(1 mark)*

b A fluorescent tube used for commercial lighting contains excited mercury atoms. Two bright lines in the visible spectrum of mercury are at wavelength 436 nm and 546 nm.

$1\,\text{nm} = 10^{-9}\,\text{m}$

Calculate

(i) the energy of a photon of violet light of wavelength 436 nm *(3 marks)*

(ii) the energy of a photon of green light of wavelength 546 nm. *(1 mark)*

c Electron transitions between the three levels **A**, **B** and **C** in the energy level diagram for a mercury atom (Figure 3) produce photons at 436 nm and 546 nm. The energy E of an electron bound to an atom is negative. The ionisation level, not shown on the diagram, defines the zero of the vertical energy scale.

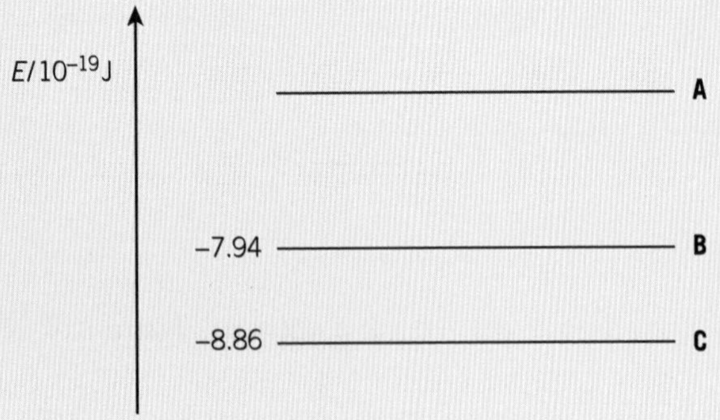

▲ Figure 3

(i) Draw two arrows on a copy of Figure 3 to represent the transitions which give rise to these photons. Label each arrow with its emitted photon wavelength. *(3 marks)*

(ii) Use your values for the energy of the photons from (b) to calculate the value of the energy level **A**. *(2 marks)*

d The light from a distant fluorescent tube is viewed through a diffraction grating aligned so that the tube and the lines on the grating are parallel. The light from the tube is incident as a parallel beam at right angles to the diffraction grating.

The line separation on the grating is $3.3 \times 10^{-6}\,\text{m}$.

Calculate the angle to the straight through direction of the first order green (546 nm) image of the tube seen through the grating. *(3 marks)*

Jun 2010 G482

4 a State *Wien's displacement law.* *(1 mark)*

b An astronomer is analysing light from stars in a particular cluster. Table 1 summarises some of the key data for five stars in this cluster.

▼ Table 1

λ_{max} / nm	T / K	
405	7200	
424	6800	
480	6000	
570	5100	
644	4500	

The wavelength of light at maximum intensity is λ_{max} and the surface temperature of the star is T.

(i) Use the last column in Table 1 to validate Wien's displacement law. *(3 marks)*

(ii) Hence determine the surface temperature of a white dwarf for which λ_{max} is 138 nm. *(3 marks)*

c The luminosity of the white dwarf in (b)(ii) is 1.0×10^{25} W. Determine its radius. *(3 marks)*

5 a Describe briefly the sequence of events which occur in the formation of a star, such as our Sun, from interstellar dust and gas clouds. *(4 marks)*

b Figure 4 shows the evolution of a star similar to our Sun on a graph of intensity of emitted radiation against temperature.

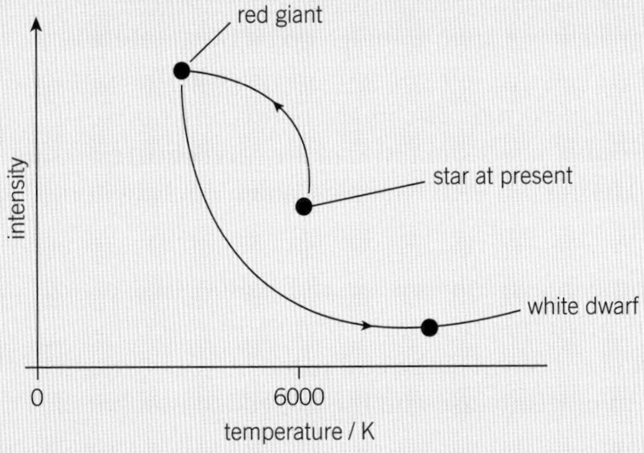

▲ **Figure 4**

(i) The final evolutionary stage of the star is a white dwarf. Describe some of the characteristics of a white dwarf. *(2 marks)*

(ii) Explain why, in its evolution, the star is brightest when at its coolest. *(2 marks)*

6 Figure 5 shows an incomplete Hertzsprung-Russell (HR) diagram. The approximate position of the Sun is labelled as **S**.

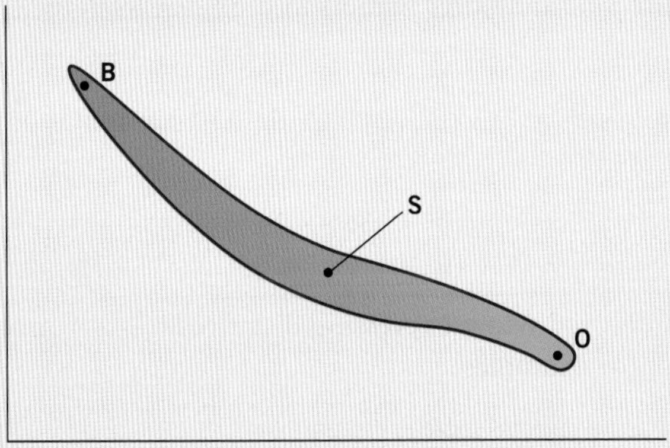

▲ **Figure 5**

a On a copy of Figure 5

(i) name the region of stars which is shaded *(1 mark)*

(ii) carefully label the axes *(2 marks)*

(iii) mark the regions occupied by red giants and white dwarfs. *(2 marks)*

b Describe the evolution of a star that is much more massive than our Sun. *(5 marks)*

c **B** and **O** show positions of two stars. Explain which star is likely to live longer. *(3 marks)*

7 a (i) Define the luminosity of a star. *(1 mark)*

(ii) An astronomer has made measurements on a distant star in our galaxy. The star has a surface temperature of (6000 ± 200 K and a radius of $(8.3 \pm 0.2) \times 10^7$ m. Calculate the luminosity of the star and the absolute uncertainty in this value. *(4 marks)*

b Show how the luminosity L of a star is related to its intensity at a distance r from the star. *(2 marks)*

c Wien's law related the peak wavelength λ of electromagnetic waves emitted from a star and its surface temperature T in kelvin. Figure 6 shows a graph of $\lg(\lambda / \text{m})$ against $\lg(T / K)$.

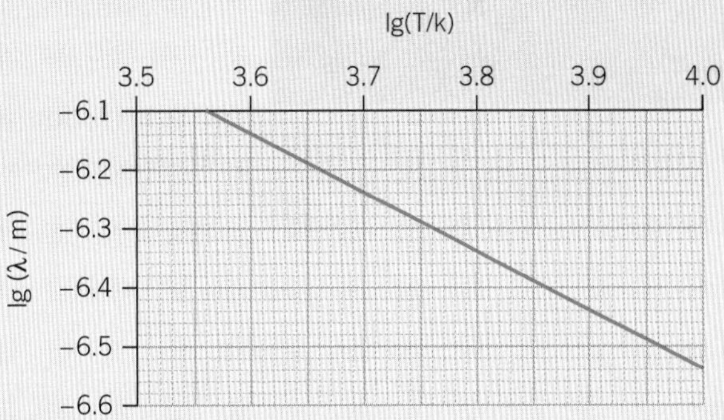

▲ **Figure 7**

(i) Explain why the graph has a gradient of −1. *(2 marks)*

(ii) Use Figure 6 to calculate the surface temperature of a star with λ = 480 nm. *(3 marks)*

20 COSMOLOGY (THE BIG BANG)

20.1 Astronomical distances

Specification reference: 5.5.3

Learning outcomes

Demonstrate knowledge, understanding, and application of:

→ distances measured in astronomical units, light-years, and parsecs

→ stellar parallax

→ the equation relating the parallax *p* in seconds of arc and the distance *d* in parsec.

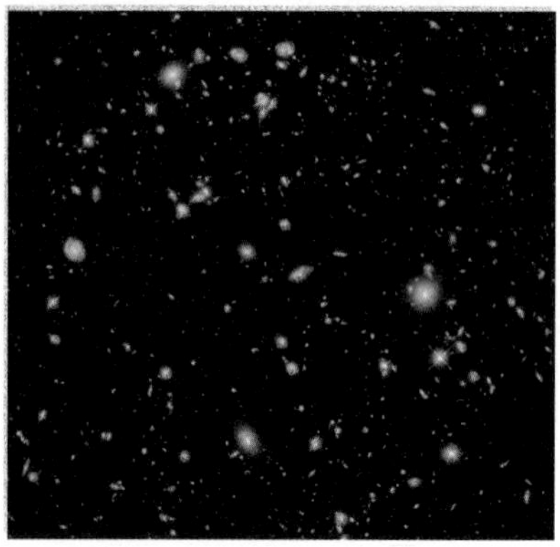

▲ **Figure 1** *The Hubble Ultra Deep Field shows some of the oldest galaxies in the Universe, some over 13 billion years old*

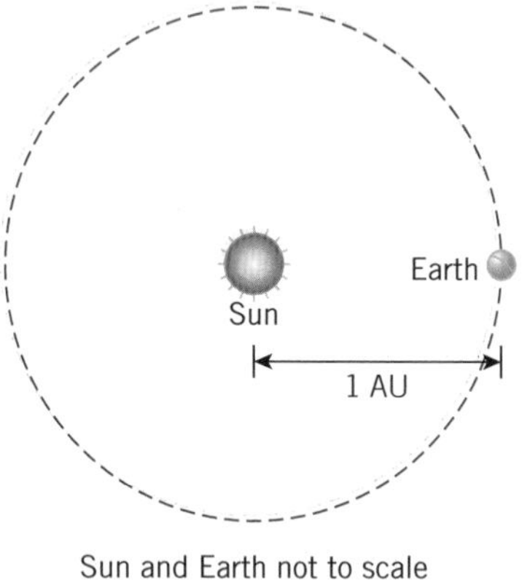

▲ **Figure 2** *As the Earth orbits the Sun in an ellipse, 1 AU is defined as the average distance from the Earth to the Sun*

Astronomical numbers

The Hubble Ultra Deep Field is a famous image taken by the Hubble Space Telescope in 2004. It shows a tiny region of space in the southern-hemisphere constellation of Fornax. The image of this tiny square of sky has an angular spread of just 2.4 minutes of arc (**arcminutes**) from edge to edge – about 0.04° or 7.0×10^{-4} radians, about equivalent to a patch of sky covered by a grain of sand held at arm's length.

Despite being so tiny, the image contains around 10 000 galaxies. We use the **cosmological principle** to assume that there is nothing special about this part of the sky. This typical patch of sky highlights the absolute vastness of our Universe and the incredible number of galaxies it must contain.

Units of distance

Astronomical distances can be expressed in metres using standard form. However, the distances are so vast that it is like giving the distance from Moscow to New York in millimetres. Instead, we use three main specialist units. In order of increasing length, they are the **astronomical unit**, the **light-year**, and the **parsec**.

The astronomical unit (AU)

The astronomical unit is the average distance from the Earth to the Sun, 150 million km, or 1.50×10^{11} m (Figure 2).

The astronomical unit is most often used to express the average distance between the Sun and other planets in the Solar System.

The light-year (ly)

The light-year is the distance travelled by light in a vacuum in a time of one year.

$$\text{distance} = \text{speed} \times \text{time} = 3.00 \times 10^8 \times (365 \times 24 \times 60 \times 60)$$
$$= 9.46 \times 10^{15}\,\text{m}$$

The light-year is often used when expressing distances to stars or other galaxies (see Table 1 for examples).

▼ **Table 1** *Distances to some objects visible in the sky*

Object	Distance / ly
Proxima Centauri (nearest star to the Sun)	4.24
Rigel (blue supergiant star in Orion's foot)	860
Diameter of the Milky Way	100 000
Andromeda galaxy (furthest object visible with the naked eye)	2 500 000

The parsec (pc)

Before defining the parsec, you need to be aware that professional astronomers prefer to measure angles not in degrees but arcminutes and arcseconds. There are 60 arcminutes in 1°, and 60 **arcseconds** in each arcminute. Therefore, 1 arcsecond = $\left(\frac{1}{3600}\right)^{\circ}$.

The parsec is defined as the distance at which a radius of one AU subtends an angle of one arcsecond.

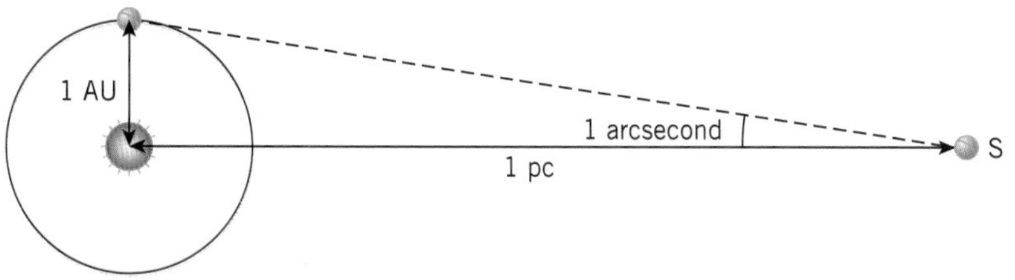

▲ **Figure 3** *The parsec is defined using the astronomical unit*

You can determine the value of 1 pc in metres by using the triangle in Figure 3, tan (1 arcsecond) = $\frac{1\,\text{AU}}{1\,\text{pc}}$.

Therefore

$$1\text{ pc} = \frac{1.50 \times 10^{11}}{\tan\left(\frac{1}{3600}\right)} = 3.1 \times 10^{16}\,\text{m (about 3.26 ly)}$$

It is worth looking more closely at Figure 3. Because the angle at point S is very small, the small-angle approximation ($\theta \approx \tan\theta$) can be used. If point S is at a distance of 2 pc, the angle subtended by the radius will be $\frac{1}{2}$ arcsecond, if 3 pc then $\frac{1}{3}$ arcsecond, and so on. If the point S is at a distance d parsec, then the angle subtended is simply $\frac{1}{d}$ arcsecond. This relationship will be useful in the next section.

Using stellar parallax to determine distances

Stellar parallax is a technique used to determine the distance to stars that are relatively close to the Earth, at distances less than 100 pc.

Parallax is the apparent shift in the position of a relatively close star against the backdrop of much more distant stars as the Earth orbits the Sun. You can mimic this effect by holding your thumb at arm's length in front of your face. First view the thumb with only your left eye, then the right. You will notice an apparent shift in the position of your thumb against the background (Figure 4). This illusion is exactly the same as the effect used for measuring stellar distances (Figure 5).

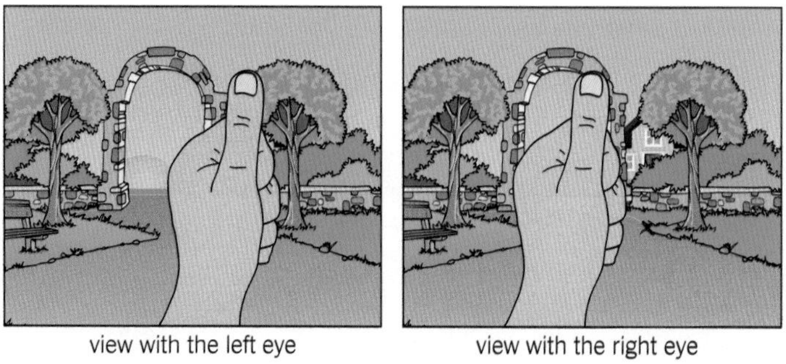

◀ **Figure 4** *Demonstrating parallax*

Synoptic link

You have already seen the astronomical unit used in Topic 18.4, Kepler's laws.

Study tip

Remember the light-year is not a unit of time, but one of distance.

Study tip

Seconds of arc and arcseconds are the same quantity.

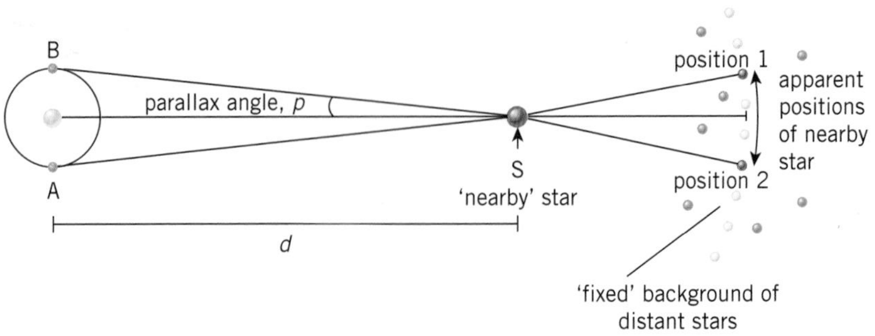

▲ **Figure 5** *Carefully recording the position of a nearby star in the sky against stars much further away allows the distance to the star to be determined using stellar parallax and simple trigonometry*

In Figure 5, when the Earth is in position A, the nearby star S appears in position 1. Six months later when the Earth is in position B, the star appears in position 2. Precise measurements can determine the **parallax angle** p. If p is measured in arcseconds, the distance to the nearby star in parsecs is given by

$$d = \frac{1}{p}$$

This is the equation you met in the previous section.

This technique is limited to stars less than 100 pc from the Earth, because as d increases the parallax angle decreases, eventually becoming too small to measure accurately, even with the most advanced astronomical techniques.

Study tip

When using $d = \frac{1}{p}$, make sure d is in parsecs and p is in arcseconds.

Summary questions

1 Explain what is meant by stellar parallax. *(2 marks)*

2 Calculate the distance from the Earth in parsecs to a star that makes a parallax angle of 0.018 arcseconds. *(2 marks)*

3 Using the data in Table 1, if needed, show that:
- a the Earth is approximately 8 light-minutes from the Sun; *(2 marks)*
- b Proxima Centauri is around 1.3 parsecs from the Earth. *(2 marks)*

4 Calculate the distance from the Earth in ly to a star that makes a parallax angle of 1.56×10^{-5}°. *(4 marks)*

5 The intensity of the light received from a star 16 ly away is measured as $2.3 \times 10^{-13}\,\mathrm{W\,m^{-2}}$. Calculate the luminosity of the star. *(4 marks)*

6 A tennis ball has a diameter of 6.75 cm. Calculate how far the ball would need to be from an observer to subtend an angle of 2.4 arcminutes. *(4 marks)*

20.2 The Doppler effect

Specification reference: 5.5.3

Much more than racing cars

The familiar 'neeeeeeeeaaawwwwwww' sound of a racing car moving past a stationary TV camera is perhaps the best known example of the **Doppler effect**, but there are many more. The Doppler effect is used to determine the speed of moving objects ranging from motorists and tennis balls (Figure 1) to rotating galaxies.

The Doppler shift

Whenever a **wave source** moves relative to an observer, the frequency and wavelength of the waves received by the observer change compared with what would be observed without relative motion.

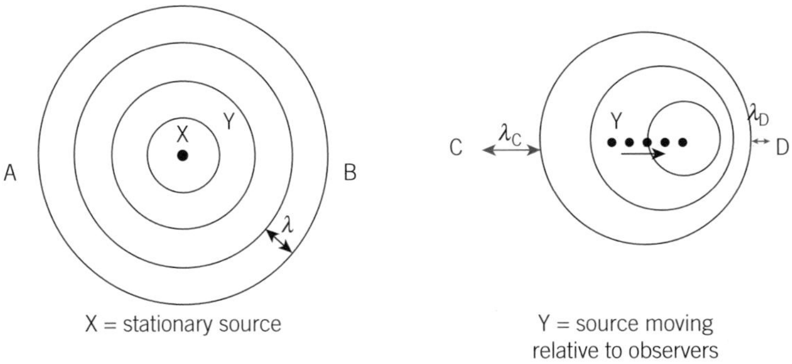

▲ **Figure 2** *Relative motion and the Doppler effect*

In the first diagram in Figure 2, the wave source is stationary relative to two observers at A and B. Both observers experience waves at the same frequency and wavelength λ as they were emitted from the source.

In the second diagram, the wave source Y is moving away from observer C towards observer D. In this example the waves received by D will be compressed. They have a shorter wavelength λ_D and so a higher frequency (this is the 'neeeeeee...' part as a racing car approaches a stationary observer). For observer C the waves are stretched out, the wavelength λ_C becomes longer and a lower frequency is observed (the '...aaawwwwwww' part as the racing car moves away). The faster the source moves, the shorter λ_D and the longer λ_C are.

In the case of the racing car, the Doppler effect applies to sound waves, but the effect happens with all types of waves. In the example of the radar gun determining the speed of a tennis ball, microwaves are reflected off the moving ball, so the ball acts like a source of microwaves.

Learning outcomes

Demonstrate knowledge, understanding, and application of:

- → the Doppler effect
- → Doppler shift of electromagnetic radiation
- → the Doppler equation for a source of electromagnetic radiation moving relative to an observer, $\frac{\Delta\lambda}{\lambda} \approx \frac{\Delta f}{f} \approx \frac{v}{c}$.

Synoptic link

You will learn about the use of the Doppler effect in medical ultrasonography to measure speed of blood flow in Topic 27.8, Doppler imaging.

▲ **Figure 1** *The speed of a tennis ball during a serve is measured from the Doppler shift of microwaves reflected off the ball*

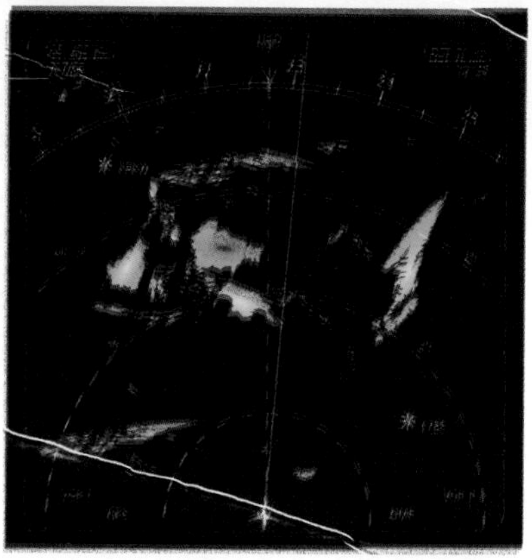

▲ Figure 3 *Pilots use information from their weather radar, like this, to navigate around potentially hazardous storms*

A storm on the way

Some types of weather radar use the Doppler effect to locate areas of precipitation (rain and snow). The radar transmits electromagnetic waves (usually microwaves), which are reflected off the precipitation back to a receiver. The wavelength of the reflected waves is Doppler-shifted, depending on the relative motion of the weather system and the receiver. Software is then used to plot the path of the weather system, and can even determine whether it is rain, snow, or hail. Modern weather radar, such as that found in commercial aircraft (Figure 3), is so sensitive that it can detect the motion of individual rain droplets and determine the intensity of the rain.

1 Explain how the differences in the reflected microwaves received by a weather radar reveal whether a rain storm is moving towards or away from the receiver.

2 Suggest how it might be possible for a weather radar to be able to distinguish between rain and hail using the intensity of the reflected microwaves.

Synoptic link

You first met electromagnetic waves in Topic 11.6, Electromagnetic waves. You learnt about the spectra of stars in Topic 19.5, Spectra. Remember that these spectra contain absorption lines that occur at specific wavelengths and are unique to the atoms of an element.

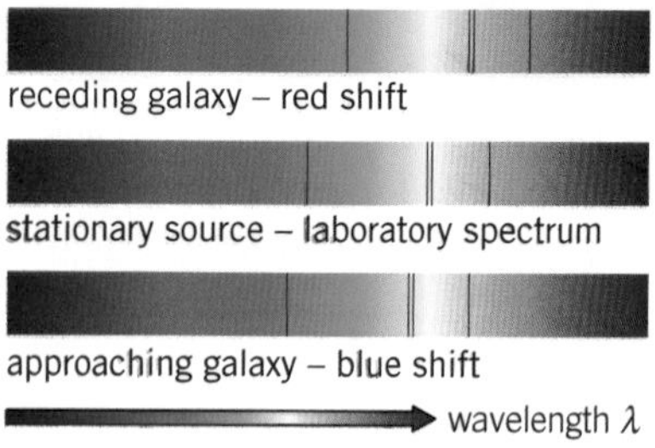

▲ Figure 4 *Blue and red shifts*

Doppler shifts in starlight

Light from stars can be analysed in many ways. One technique involves looking at the absorption lines in the spectra from stars.

The Doppler effect can be used to determine the relative velocity of a distant galaxy. First, the absorption spectrum of a specific element is determined in the laboratory. The same spectrum is observed in light from a distant galaxy. Any difference in the observed wavelengths of the absorption lines must be caused by the relative motion between the galaxy and the Earth.

- If the galaxy is moving towards the Earth the absorption lines will be **blue-shifted** – they move towards the blue end of the spectrum, because the wavelength appears shorter.
- If the galaxy is moving away from the Earth ('receding') the absorption lines will be **red-shifted** – they all move towards the red end of the spectrum, because the wavelength appears stretched.

This technique is very powerful. It can even be used to determine the speed of rotation of stars and galaxies.

A mathematical treatment

How fast the wave source moves relative to the observer affects the size of the observed shift in wavelength and frequency.

For electromagnetic waves, the **Doppler equation** below is very useful.

$$\frac{\Delta\lambda}{\lambda} \approx \frac{\Delta f}{f} \approx \frac{v}{c}$$

where λ is the source wavelength, $\Delta\lambda$ is the change in wavelength recorded by the observer, f is the source frequency, Δf is the change in frequency recorded by the observer, v is the magnitude of the relative velocity between the source and observer, and c is the speed of light through a vacuum ($3.00 \times 10^8\,\text{m s}^{-1}$). The Doppler equation can only be used for galaxies with speed far less than the speed of light.

The equation shows that the faster the source moves, the greater the observed change in wavelength and frequency. In Figure 4, the wavelength of each absorption line changes by the same percentage for a particular moving galaxy.

Worked example: Speed of a galaxy

In the laboratory an absorption line of hydrogen is observed at a wavelength of 656.4 nm. In a distant galaxy the same absorption line is observed at 658.1 nm. Calculate the speed of the galaxy and state whether it is moving towards or away from the Earth.

Step 1: Calculate the change in the wavelength

$$\Delta\lambda = 658.1 - 656.4 = 1.7\text{ nm}$$

The wavelength observed from the galaxy is longer than the wavelength in the laboratory, so the galaxy must be receding. The spectral line has been red-shifted.

Step 2: The Doppler equation $\frac{\Delta\lambda}{\lambda} \approx \frac{v}{c}$ can be rearranged to give $\frac{c\Delta\lambda}{\lambda} \approx v$.

Therefore

$$v \approx \frac{c\Delta\lambda}{\lambda} \approx \frac{3.00 \times 10^8 \times 1.7 \times 10^{-9}}{656.4 \times 10^{-9}} = 7.76... \times 10^5\text{ms}^{-1} = 780\text{kms}^{-1}\ (2\text{ s.f.})$$

Summary questions

1 A police siren is observed to change pitch as it passes a stationary observer. Use the Doppler effect to explain this observation. *(2 marks)*

2 Explain why the driver of a race car does not experience any Doppler shift in the sounds from the engine. *(2 marks)*

3 Light from a distant galaxy is red-shifted. Suggest how by measuring the red shift of different parts of the galaxy astronomers are able to determine the speed of rotation of the galaxy. *(4 marks)*

4 A particular absorption line is measured in a laboratory at a frequency of 5.12×10^{14} Hz. Calculate the frequency and wavelength associated with the same line in:

a a galaxy moving towards the Earth at 10.6 Mm s^{-1};

b a galaxy moving away from us at 25% of the speed of light. *(6 marks)*

5 Suggest why the light reflecting off a sprinter running towards a TV camera does not appear to be blue-shifted. *(2 marks)*

6 An absorption line is measured in the lab at a wavelength of 714.7 nm. In a distant galaxy the same absorption line is observed at 707.1 nm. Calculate the speed of the galaxy and state whether it is moving towards or away from the Earth. *(3 marks)*

20.3 Hubble's law

Specification reference: 5.5.3

Learning outcomes

Demonstrate knowledge, understanding, and application of:

→ Hubble's law $v \approx H_0 d$ for receding galaxies

→ galactic red shift and the model of an expanding Universe

→ Hubble constant H_0 in $km\,s^{-1}\,Mpc^{-1}$ and s^{-1}

→ the cosmological principle.

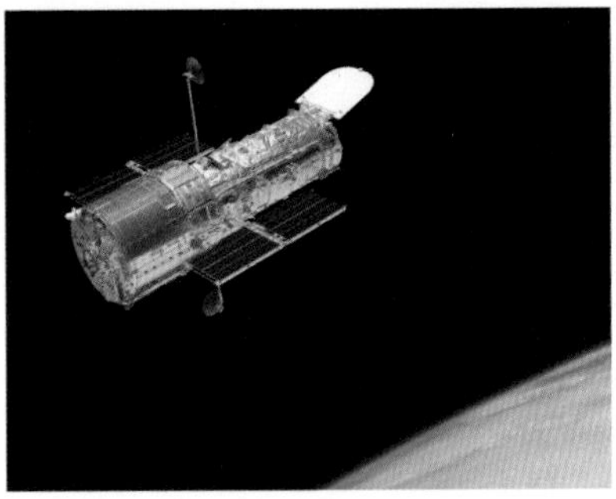

▲ **Figure 1** *The Hubble Space Telescope, named after Edwin Hubble, has allowed physicists to make huge leaps in our understanding of the Universe*

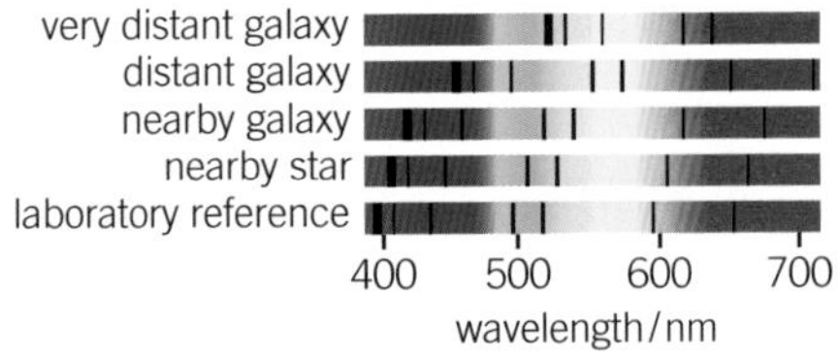

▲ **Figure 2** *The relative red shifts of near and distant objects revealed to Hubble that objects further away were moving faster relative to the Earth*

The work of Edwin Powell Hubble

Edwin Powell Hubble, an American astronomer, is generally thought of as one of the most important cosmologists of the 20th century. He showed there were many more galaxies in our Universe than people thought, and his work investigating the motion of distant galaxies led to the concept of the expanding Universe and the **Big Bang**.

Hubble's law

During the late 1920s Hubble analysed the Doppler shift in the absorption spectra of many distant galaxies. Using the data available then he made two key observations:

1 He confirmed earlier observations that the light from the vast majority of galaxies was red-shifted, that is, they had a relative velocity away from the Earth.

2 He found that in general the further away the galaxy was the greater the observed red shift and so the faster the galaxy was moving.

Using these observations he formulated what is now called **Hubble's law**: the recessional speed v of a galaxy is almost directly proportional to its distance d from the Earth, that is, $v \propto d$.

On a graph of recessional speed against distance for all galaxies, the plotted data should produce a straight line through the origin. As you can see from Figure 3, the spread in the original data points suggests that Hubble's law is valid, but there is a large uncertainty in the value for the gradient of the best-fit line.

The Hubble constant

The gradient of the graphs in Figure 3 is a constant of proportionality now called the **Hubble constant** H_0. From Hubble's law it follows that

$$v \approx H_0 d$$

The SI unit for the Hubble constant can be determined by dividing $m\,s^{-1}$ by m (the SI units). This gives the unit s^{-1}. However, as you see in Figure 3, cosmologists prefer to express speed in $km\,s^{-1}$ and distance in Mpc, giving an alternative unit of $km\,s^{-1}\,Mpc^{-1}$, which must be equivalent to s^{-1}. In 2013, the most reliable data gave $67.80 \pm 0.77\,km\,s^{-1}\,Mpc^{-1}$ for the Hubble constant, or about $2.2 \times 10^{-18}\,s^{-1}$.

Worked example: Hubble constant units

Convert $1\,km\,s^{-1}\,Mpc^{-1}$ into s^{-1}.

Step 1: Convert the speed into $m\,s^{-1}$ and then divide it by the distance of 1 Mpc in metres.

$$1\,km\,s^{-1}\,Mpc^{-1} = \frac{1.0 \times 10^{3}\,m\,s^{-1}}{10^{6} \times 3.1 \times 10^{16}\,m} = 3.2 \times 10^{-20}\,s^{-1}\ (2\ s.f.)$$

The expanding Universe

Hubble's law is the key evidence for the Big Bang theory (more on this in Topic 20.4) and the model of the **expanding Universe** following the Big Bang. This model is the most widely accepted explanation of the observation that the light from nearly all the galaxies we can see is red-shifted. It states that the fabric of space, and time, is expanding in all directions. It is not simply the galaxies moving away from each other, but the actual space itself expanding. As a result, any point, in any part of the Universe, is moving away from every other point in the Universe, and the further the points are apart the faster their relative motion away from each other. From our position on Earth this explains why light from more distant galaxies is more red-shifted, indicating that they are moving faster than those nearer to us.

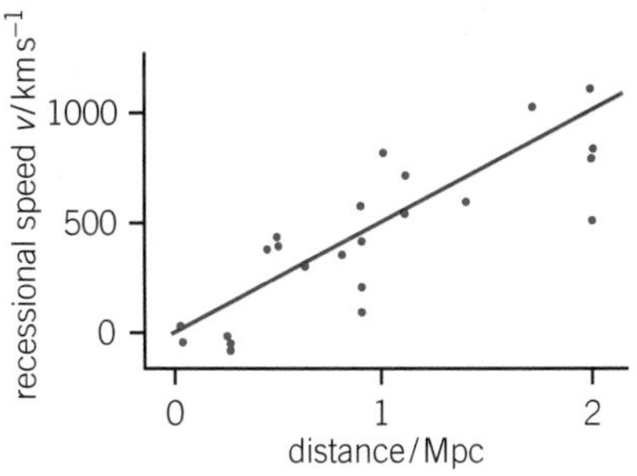

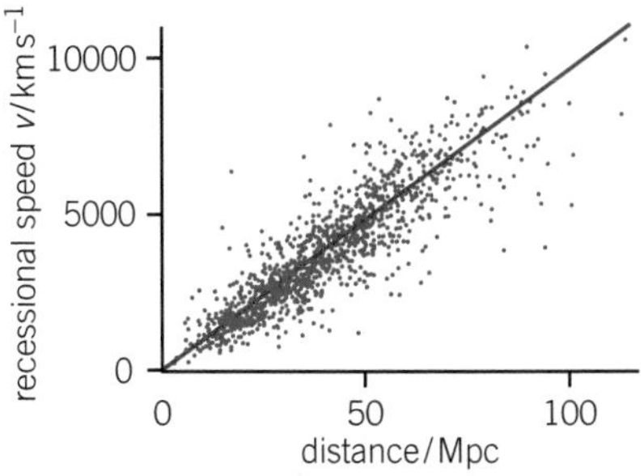

▲ **Figure 3** *Hubble's original 1929 data for the recessional speed v of galaxies against their distance d from the Earth (above) and (below) a plot with more recent data*

The cosmological principle

The **cosmological principle** is the assumption that, when viewed on a large enough scale, the Universe is **homogeneous** and **isotropic**, and the laws of physics are universal.

- The laws of physics can be applied across the Universe. This is a bold assumption. It means that the theories and models tested here on the Earth can be applied to everything within the Universe over all time and space.
- 'Homogeneous' means that matter is distributed uniformly across the Universe. For a very large volume, the density of the Universe is uniform. This means that the same type of structures (galaxies) are seen everywhere.
- 'Isotropic' means that the Universe looks the same in all directions to every observer. It follows that there is no centre or edge to the Universe.

In essence, the cosmological principle means that the Universe would look the same wherever you are.

Study tip

Remember, space itself is stretching, so think of galaxies being carried by the expanding Universe rather than simply moving through space.

Summary questions

1 Sketch a graph of the recessional speed of a galaxy against the distance of the galaxy from Earth, and use the graph to illustrate Hubble's law. *(2 marks)*

2 The value of the Hubble constant is about $2.2 \times 10^{-18}\,s^{-1}$. Calculate the distance from Earth of a galaxy with the following recessional speed:
a $160\,000\,m\,s^{-1}$; **b** $7.8 \times 10^{6}\,m\,s^{-1}$. *(4 marks)*

3 Use the lower graph in Figure 3 to confirm that the value for the Hubble constant is approximately $70\,km\,s^{-1}\,Mpc^{-1}$. *(3 marks)*

4 Calculate the recessional speed of a galaxy at the following distances from the Earth:
a $1.50 \times 10^{23}\,m$; **b** 25.0 Mpc; **c** 40 million ly. *(6 marks)*

5 In the laboratory an absorption line is observed at a wavelength of 638.9 nm. In a distant galaxy the same absorption line corresponds to a wavelength of 675.1 nm. Calculate the distance from the galaxy to the Earth in ly. *(4 marks)*

20.4 The Big Bang theory

Specification reference: 5.5.3

Learning outcomes

Demonstrate knowledge, understanding, and application of:

- → the Big Bang theory
- → experimental evidence for the Big Bang theory from microwave background radiation
- → the idea that the Big Bang gave rise to the expansion of space-time
- → estimation of the age of the Universe
- → $t \approx H_0^{-1}$.

▲ **Figure 1** *This horn antenna at Bell Laboratories in the USA unexpectedly revealed the microwave background radiation from the Big Bang*

It all started with the Big Bang

The Big Bang theory is one of the most important ideas in all of science. First proposed by the Belgian physicist and priest Georges Lemaître in 1931, it attempts to describe the origin and development of the early Universe. It suggests that at some moment in the past all the matter in the Universe was once contained in a single point, a singularity. This point is considered to be the beginning of the Universe – the beginning of space and time itself. This region was much hotter and denser than it is today. It then expanded outwards to become the dynamic Universe we see around us today.

In support of the Big Bang

In order to be accepted by the scientific community, any scientific theory needs to be supported by evidence. There are two key pieces of evidence for the Big Bang theory – Hubble's law and the **microwave background radiation**.

You have seen how Hubble's law shows that the Universe is expanding – the galaxies are receding from each other because the space itself is expanding in all dimensions. It follows that, if we could run time backwards, the Universe would be much smaller, denser, and hotter, and would eventually reach a single point. It is this single point that, according to the Big Bang theory, expanded out to form our present-day Universe.

However, the expanding Universe could be explained by other competing theories on the origin of the Universe. So by itself, it does not produce enough to establish the Big Bang. More evidence is needed.

Microwave background radiation

The second piece of evidence is the existence of microwave background radiation. In 1964 two American physicists, Robert Wilson and Arno Penzias, were attempting to detect signals from objects in space. They detected a uniform microwave signal they could not account for — not even trapping the pigeons whose droppings coated the antenna helped. Eventually they realised that they had accidentally discovered the microwave background radiation that could only be explained by the Big Bang and the expansion of space.

The Big Bang theory had earlier predicted the existence of this background microwave radiation. Its existence can be explained in two ways.

- When the Universe was young and extremely hot, space was saturated with high-energy gamma photons. The expansion of the Universe means that space itself was stretched over time. This expansion stretched the wavelength of these high-energy photons, so we now observe this primordial electromagnetic radiation as microwaves.

Study tip

Be careful not to fall into the "just a theory" mentality. A theory must have evidence to support it — Fred Hoyle's idea of the Steady State Universe, in which new matter is created as the Universe expands, had to be rejected by scientists as it failed to explain the microwave background.

- The Universe was extremely dense and hot when it was young. Expansion of space over billions of years has reduced that temperature to around 2.7 K. The Universe may be treated as a black-body radiator – at this temperature the peak wavelength would correspond to about 1 mm, in the microwave region of the spectrum.

None of the competing theories had predicted or could explain the origin of the microwave background radiation, so the Big Bang theory became the most widely accepted theory on the origin of the Universe. Penzias and Wilson received the 1978 Nobel Prize in Physics for their discovery.

The age of the Universe

We can estimate the age of the Universe by assuming that it has expanded uniformly over time since the Big Bang. Results from recent observations have shown that this is not the case – in fact, the expansion of the Universe is accelerating, so this assumption is poor, but it will give a crude indication of the age of the Universe.

Hubble's law shows that galaxies are receding from each other. This means that in the past they must have been closer together. If a galaxy at a distance d is moving away at a constant speed v, then a time $\frac{d}{v}$ must have elapsed since it was next to our galaxy. This time is therefore roughly the age of the Universe. The ratio $\frac{d}{v}$ is equal to $\frac{1}{H_0}$, so

$$\text{age of Universe } t \approx \frac{1}{H_0}$$

As discussed in Topic 20.3, an accurate determination of the Hubble constant is a considerable challenge for cosmologists. However, using $H_0 = 2.2 \times 10^{-18}\,\text{s}^{-1}$ gives the age as $4.5 \times 10^{17}\,\text{s}$ (~14 billion years).

The ESA Planck mission

In May 2009 the European Space Agency launched the Planck space observatory. Its mission was to measure precisely the tiny variations in the microwave background radiation. These fluctuations are another prediction of the Big Bang theory. It is these ripples which give rise to the present structure of galaxies.

The data collected by the Planck mission suggests that the value for the Hubble constant is $67.80 \pm 0.77\,\text{km}\,\text{s}^{-1}\,\text{Mpc}^{-1}$.

1. Using the average value of the Hubble constant determined by the Planck mission show that the Universe is approximately 14 billion years old.
2. Figure 2 shows temperature variations of about $\pm 10^{-4}$ K. Assuming Wien's law can be applied to the entire Universe, determine the percentage variation in the peak wavelength of the microwaves.

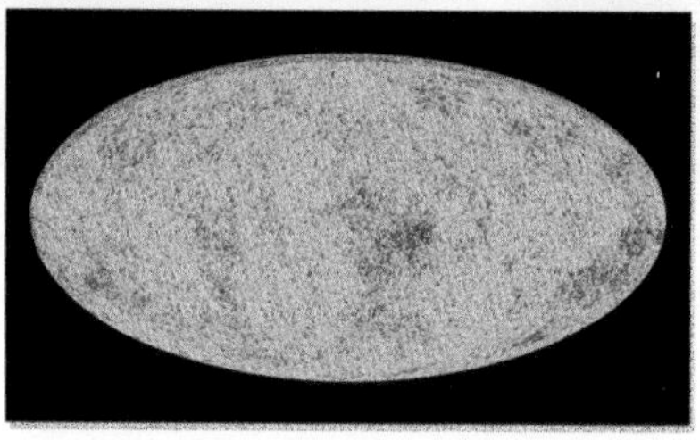

▲ Figure 2 *This image produced from Planck data shows the minuscule temperature variations from when our Universe was only 380 000 years old, with 'warmer' areas in red – most of the observations fit the predictions of the model well, but the uneven cold areas (dark blue) are unexpected*

Synoptic link

You met Wien's law in Topic 19.7, Stellar luminosity.

Summary questions

1. State two pieces of evidence in support of the Big Bang theory. *(2 marks)*
2. Explain the importance of the discovery of microwave background radiation. *(2 marks)*
3. If the Universe is between 11 billion and 15 billion years old, calculate the maximum and minimum possible values for the Hubble constant. *(4 marks)*
4. Use the information given in Topic 19.7, Stellar luminosity, on Wien's law to estimate:
   a. the dominant wavelength of the electromagnetic radiation in the Universe when its temperature was 10^{11} K *(2 marks)*
   b. the temperature of the Universe when it was full of visible light. *(2 marks)*

20.5 Evolution of the Universe

Specification reference: 5.5.3

Learning outcomes

Demonstrate knowledge, understanding, and application of:

- → the evolution of the Universe after the Big Bang to the present
- → current ideas about the composition of the Universe in terms of dark energy, dark matter, and a small percentage of ordinary matter.

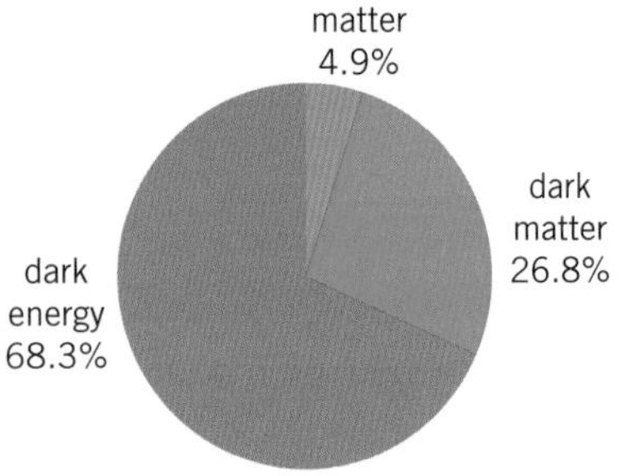

▲ **Figure 1** *Current theories indicate we do not understand 95% of what makes up our Universe*

We only understand 5% of our Universe!

In the late 1990s the world of physics was shaken by the discovery that the Universe appears to be expanding at an increasing rate. This called our understanding into question and led to the development of completely new ways to think about our Universe.

How this acceleration happens is not fully understood. The most widely accepted theory includes the concept of **dark energy**. It is suggested that this hypothetical form of energy fills all of space and tends to accelerate the expansion of the Universe. This, coupled with the discovery of **dark matter** a few years earlier, means that at our best estimate we currently only understand 5% of the stuff that makes up our Universe.

The evolution of the Universe

You have seen that the Universe is thought to be 13.7 billion years old. The evolution of the Universe is a story of expansion of space, cooling, and formation of matter, summarised in Table 1.

Our current ideas about the Universe

In the past few decades our understanding of the Universe has undergone dramatic changes. The two most significant are the discovery of dark energy and the discovery of dark matter.

▼ **Table 1** *The history of the Universe*

	Time after the Big Bang	Nature of the Universe
	The Big Bang	Time and space are created. The Universe is a singularity – it is infinitely dense and hot.
↓ decreasing temperature	10^{-35} s	The Universe expands rapidly, including a phase of incredible acceleration known as **inflation**. There is no matter in the Universe – instead it is full of electromagnetic radiation in the form of high-energy gamma photons. The temperature is about 10^{28} K.
	10^{-6} s	The first fundamental particles (quarks, leptons, etc.) gain mass through a mechanism that is not fully understood but involves the Higgs boson (discovered in 2013).
	10^{-3} s	The quarks combine to form the first hadrons, such as protons and neutrons. Most of the mass in the Universe was created within the first second through the process of pair production (high-energy photons transforming into particle–antiparticle pairs).
	1 s	The creation of matter stops after about 1 s, once the temperature has dropped to about 10^9 K.
	100 s	Protons and neutrons fuse together to form deuterium and helium nuclei, along with a small quantity of lithium and beryllium. The expansion of the Universe is so rapid that no heavier elements are created. During this stage, about 25% of the matter in the Universe is helium nuclei (known as primordial helium).

▼ Table 1 *(continued)*

	Time after the Big Bang	Nature of the Universe
↓ decreasing temperature	380 000 years	The Universe cools enough for the first atoms to form. The nuclei capture electrons. The electromagnetic radiation from this stage of the Universe is what can be detected as microwave background radiation.
	30 million years	The first stars appear. Through nuclear fusion in these stars the first heavy elements (beyond lithium) begin to form.
	200 million years	Our galaxy, the Milky Way, forms, as gravitational forces pull clouds of hydrogen and existing stars together.
	9 billion years	The Solar System forms from the nebula left by the supernova of a larger star. After the Sun forms the remaining material forms the Earth and other planets (around 1 billion years later). It is thought that around 1 billion years after the formation of the Earth (11 billion years after the Big Bang) primitive life on Earth begins.
	13.7 billion years (now)	Around 200 000 years ago the first modern humans evolve, and eventually study physics. The temperature of the Universe is 2.7 K.

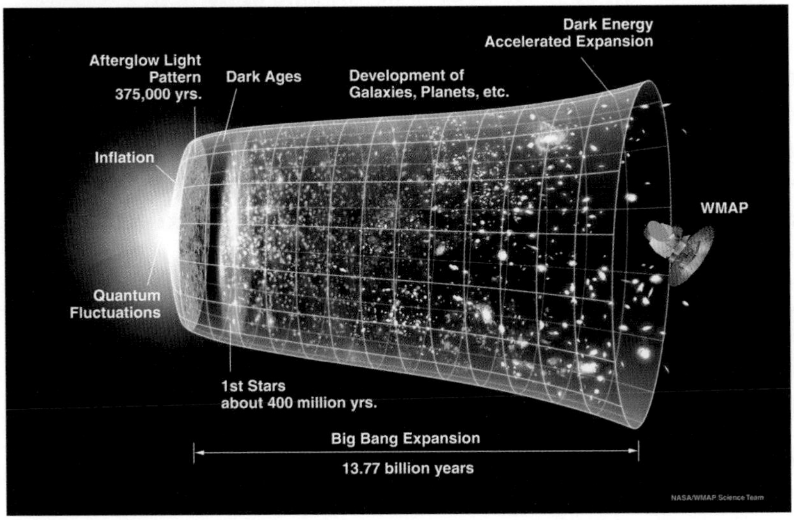

▲ **Figure 2** *Graphical model of the evolution of the Universe, from the earliest moment we can currently probe (to the left of the Figure). After several billion years of decelerating expansion as matter exerted gravitational force on itself, the expansion has more recently sped up again due to the dominating repulsive effects of dark energy.*

The accelerating Universe

The 2011 Nobel Prize in Physics was awarded for the discovery in 1999 that the expansion of the Universe is accelerating. Physicists Saul Perlmutter, Brian Schmidt, and Adam Riess were investigating the light from distant supernovae. They observed a particular type of supernova, a type Ia supernova, which produces a characteristic kind of light. On studying this light it was found to be less intense than predicted. The only possible conclusion was that the expansion of the Universe was accelerating. This acceleration needed a source of energy, one which had never been detected. They used the term 'dark energy' to describe

▲ Figure 3 *The Doppler effect is used to determine the speed of rotation of distant galaxies like the Andromeda galaxy*

a hypothetical form of energy that permeates all space. Dark energy is currently the best accepted hypothesis to explain the accelerating rate of expansion. It is estimated that dark energy, which remains as yet undetected, or even understood, makes up around 68% of our Universe.

Dark matter

In the late 1970s, astronomers studying the Doppler shift in light from galaxies found that the velocity of the stars in the galaxies did not behave as predicted. It was expected that their velocity would decrease as the distance from the centre of the galaxy increases. This is the effect observed in other gravitational systems where most of the mass is in the centre, including our Solar System and the moons of Jupiter.

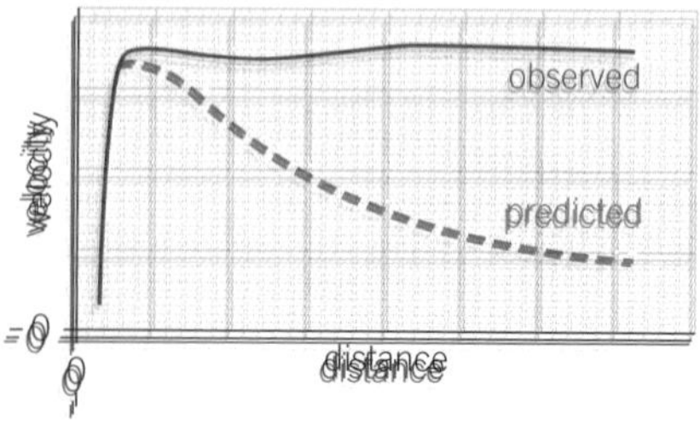

▲ Figure 4 *Observations of the velocities of stars in galaxies do not match predictions, leading to the idea of dark matter*

Figure 4 shows the differences between the predicted and observed velocities of the stars in a galaxy as you move out from the centre.

The observations can be explained if the mass of the galaxy is not concentrated in the centre. However, most of the matter we can see is in the centre. The current thinking is that there must be another type of matter which we cannot see. This dark matter is spread throughout the galaxy, explaining the observations. Calculations have shown that the Universe must be made of 27% of this kind of matter.

Not much is known about dark matter. We do know that it cannot be seen directly with telescopes and that it neither emits nor absorbs light. There are exotic speculations as to what it could be: black holes, gravitinos, weakly interacting massive particles (wimps), axions, Q-balls,... the list goes on. The truth is that dark energy and dark matter remain a mystery waiting to be solved by the next generation of physicists.

Summary questions

1 Describe the observations that led to the development of the idea of dark energy. *(2 marks)*

2 Explain why it was not possible for atoms to form until 380 000 years after the Big Bang. *(2 marks)*

3 Describe how the presence of dark matter accounts for the difference between the observation and prediction shown in Figure 4. *(3 marks)*

4 Sketch a timeline with a logarithmic scale to show the evolution of the Universe from the Big Bang until the present. *(4 marks)*

5 The Universe originated from a Big Bang. Table 2 below shows how the temperature T of the Universe has changed with time t since the Big Bang.

▼ Table 2

t / s	10^{-35}	10^{-12}	10^{-6}	10	4×10^{17}
T / K	10^{28}	10^{15}	10^{12}	10^{9}	2.7

By plotting a graph of $\lg t$ against $\lg T$, show that the temperature T and the time t are related by the equation $T^n t$ = constant, where n is an integer. Use your graph to determine the value for n. *(4 marks)*

Practice questions

1 a Calculate the distance of 1 light-year (ly) in metres. *(1 mark)*

b Figure 1 shows an incomplete diagram drawn by a student to show what is meant by a distance of 1 parsec (pc).

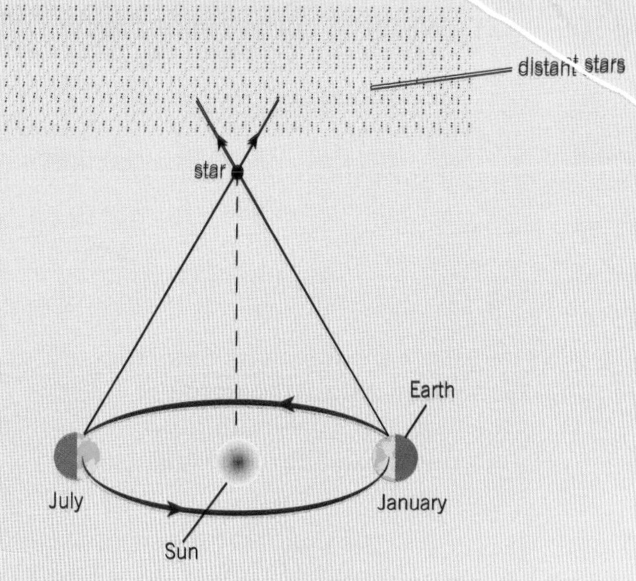

▲ Figure 1

Complete a copy of Figure 1 by showing the distances of 1 pc and 1 AU, and the parallax angle of 1 second of arc (1″). *(1 mark)*

c A recent supernova, SN2011fe, in the Pinwheel galaxy, M101, released 10^{44} J of energy. The supernova is 2.1×10^7 ly away.

(i) Calculate the distance of this supernova in pc.

$1\,\text{pc} = 3.1 \times 10^{16}\,\text{m}$ *(2 marks)*

(ii) Our Sun radiates energy at a rate of 4×10^{26} W. Estimate the time in years that it would take the Sun to release the same energy as the supernova SN2011fe. *(2 marks)*

d One of the possible remnants of a supernova event is a black hole. State **two** properties of a black hole. *(2 marks)*

Jun 2013 G485

2 a In the Universe there are about 10^{11} galaxies, each with about 10^{11} stars with each star having a mass of about 10^{30} kg. Estimate the attractive gravitational force between two galaxies separated by a distance of 4×10^{22} m. *(3 marks)*

b Explain why the galaxies do not collapse on each other. *(1 mark)*

c Describe qualitatively the evolution of the Universe immediately after the Big Bang to the present day. You are not expected to state the times for the various stages of the evolution. *(6 marks)*

d Figure 2 shows some absorption spectral lines of the spectrum of calcium as observed from a source on the Earth and from a distant galaxy.

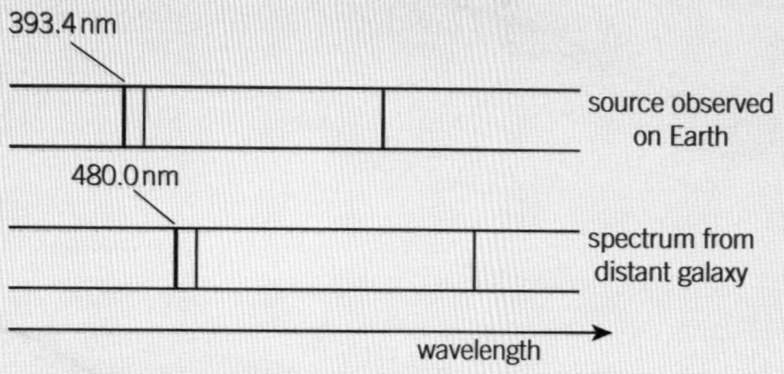

▲ Figure 2

(i) Describe an absorption spectrum. *(2 marks)*

(ii) Use Figure 1 to calculate the distance of the galaxy in Mpc. The Hubble constant has a value of $50\,\text{km}\,\text{s}^{-1}\,\text{Mpc}^{-1}$. *(3 marks)*

Jun 2012 G485

3 a Explain what is meant by the *Doppler effect*. *(2 marks)*

b A line in the spectrum of calcium has a wavelength of 397.0 nm when measured from a stationary laboratory source. The same spectral line is observed in five galaxies, resulting in the data shown in Table 1.

▼ Table 1

galaxy	distance / Mpc	wavelength / nm	velocity of recession / $\times 10^3$ km s^{-1}
A	50	400.3	2.5
B	300	416.9	15.0
C	620	438.0	31.0
D	980	461.8	49.0
E	1300	483.0	65.0

(i) State the equation for the change in wavelength produced by the Doppler effect and use it to explain why the light from galaxy E undergoes a much greater change in wavelength than the light received from galaxy A. *(2 marks)*

(ii) Plot a graph on a copy of Figure 3 of recession velocity against distance. *(2 marks)*

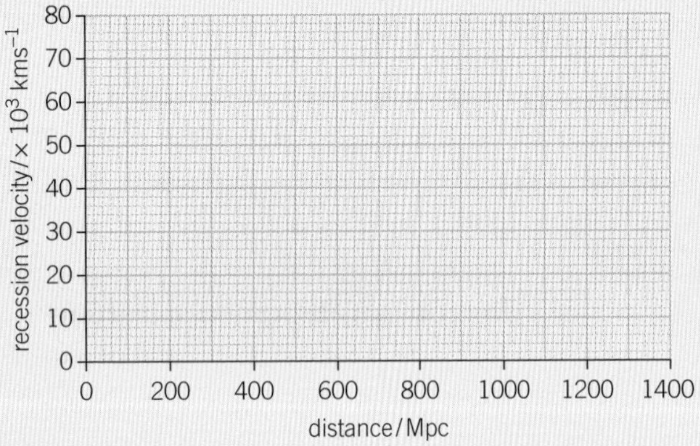

▲ Figure 3

(iii) Use the graph to find a value for the Hubble constant, giving a unit with your answer. *(3 marks)*

c Explain why Hubble's constant may not be a constant at all. *(2 marks)*

d The graph in Figure 3 has been used to support the Big Bang model of the Universe. Describe and explain one other piece of evidence which supports this model. *(3 marks)*

Jan 2010 2825/01

4 a A line in the hydrogen absorption spectrum has a wavelength of 656.3 nm when measured in the laboratory. Observation of a star shows the same absorption line to have a wavelength of 651.0 nm.

(i) Calculate the velocity of the star relative to Earth. *(3 marks)*

(ii) What else can be deduced about the star's motion from these measurements? Explain your answer. *(1 mark)*

(iii) How did Edwin Hubble use calculations of this type, together with other data, to develop our understanding of the Universe? *(5 marks)*

b What are the important properties of the cosmic microwave background radiation and how have these contributed to our understanding of the origin of the Universe? *(3 marks)*

Jun 2009 2825/01

5 a What is meant by the *Doppler Effect*? *(2 marks)*

b Figure 4 shows part of a *continuous spectrum* obtained from a light source in a laboratory. The spectrum is crossed by a single *absorption line* of wavelength 410 nm.

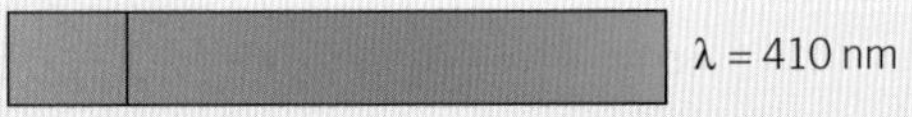

▲ Figure 4

(i) State what is meant by a *continuous* spectrum. *(1 mark)*

(ii) Explain how an *absorption line* occurs. *(2 marks)*

Figure 5 shows another four continuous spectra received from four different galaxies. The spectra are crossed by the same dark line as in Figure 4, but each one has become red shifted. The resulting wavelength is given beside the spectrum.

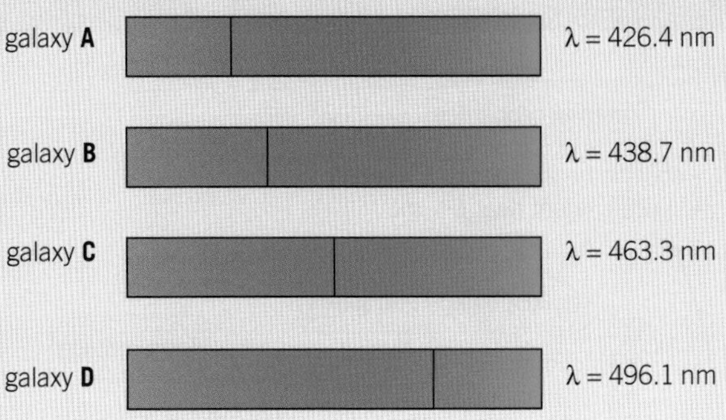

▲ Figure 5

(iii) What can be deduced about the galaxies from the fact that the lines are red shifted? *(1 mark)*

(iv) Calculate the change in wavelength $\Delta\lambda$ of the absorption line in galaxy **D**. Write your answer in the second column of a copy of Table 2. *(1 mark)*

▼ Table 2

galaxy	change in wavelength $\Delta\lambda$ / nm	velocity of galaxy / 10^7 m s^{-1}	Distance to galaxy / 10^{24} m
A	16.4	1.2	4.65
B	28.7	2.2	8.50
C	53.3	3.9	15.1
D			24.4

(v) Use the value of $\Delta\lambda$ to calculate the velocity of galaxy **D** relative to the observer. Write your answer in the third column of your copy of Table 2. *(2 marks)*

c Plot a graph of galaxy velocity against distance on a copy of Figure 6. Draw the best straight line through the points. *(2 marks)*

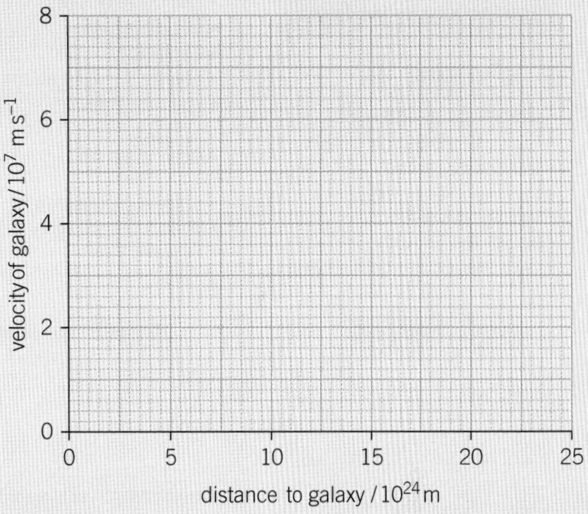

▲ Figure 6

d Use your graph to estimate the age of the Universe. Give a unit for your answer. *(3 marks)*

Jan 2009 2825/01

6 a Explain how the surface of a red giant can be cooler than the Sun but it can have a much greater luminosity. *(2 marks)*

b Antares is a bright star in the night sky. It has a parallax of 5.9 seconds of arc, mass of $12 M_\odot$ and radius $880 R_\odot$ ($M_\odot$ = mass of the Sun and $R_\odot$ = radius of the Sun). Calculate

(i) the distance in metres of Antares from the Earth *(3 marks)*

(ii) the surface gravitational field strength on Antares in terms of the Sun's surface gravitational field strength $g_\odot$. *(2 marks)*

Module 5 Summary

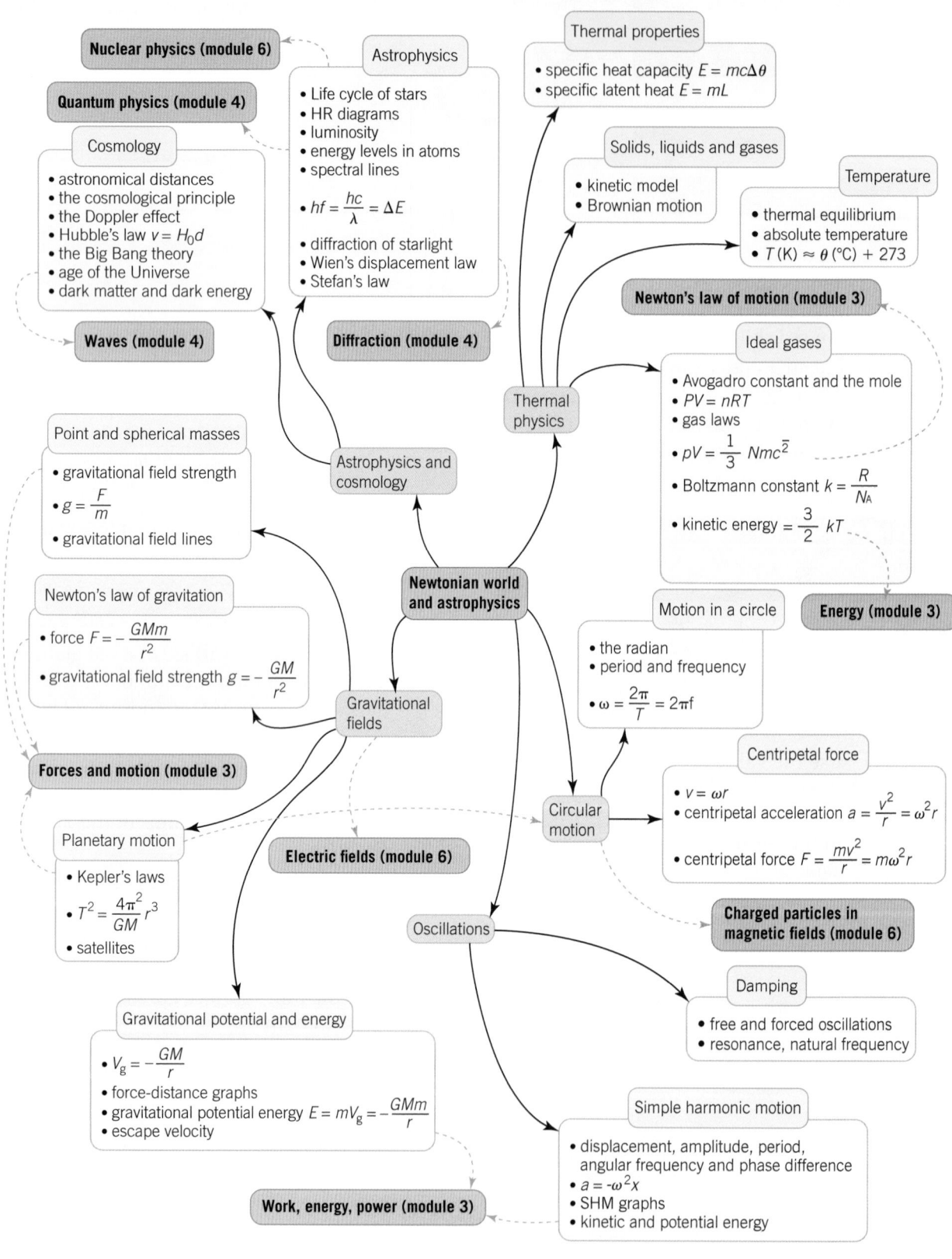

Newtonian world and astrophysics

Astrophysics and cosmology

Astrophysics
- Life cycle of stars
- HR diagrams
- luminosity
- energy levels in atoms
- spectral lines
- $hf = \frac{hc}{\lambda} = \Delta E$
- diffraction of starlight
- Wien's displacement law
- Stefan's law

Nuclear physics (module 6)

Quantum physics (module 4)

Diffraction (module 4)

Cosmology
- astronomical distances
- the cosmological principle
- the Doppler effect
- Hubble's law $v = H_0 d$
- the Big Bang theory
- age of the Universe
- dark matter and dark energy

Waves (module 4)

Thermal physics

Thermal properties
- specific heat capacity $E = mc\Delta\theta$
- specific latent heat $E = mL$

Solids, liquids and gases
- kinetic model
- Brownian motion

Temperature
- thermal equilibrium
- absolute temperature
- $T\,(\text{K}) \approx \theta\,(°\text{C}) + 273$

Newton's law of motion (module 3)

Ideal gases
- Avogadro constant and the mole
- $PV = nRT$
- gas laws
- $pV = \frac{1}{3} Nm\overline{c^2}$
- Boltzmann constant $k = \frac{R}{N_A}$
- kinetic energy $= \frac{3}{2} kT$

Energy (module 3)

Gravitational fields

Point and spherical masses
- gravitational field strength
- $g = \frac{F}{m}$
- gravitational field lines

Newton's law of gravitation
- force $F = -\frac{GMm}{r^2}$
- gravitational field strength $g = -\frac{GM}{r^2}$

Forces and motion (module 3)

Planetary motion
- Kepler's laws
- $T^2 = \frac{4\pi^2}{GM} r^3$
- satellites

Electric fields (module 6)

Gravitational potential and energy
- $V_g = -\frac{GM}{r}$
- force-distance graphs
- gravitational potential energy $E = mV_g = -\frac{GMm}{r}$
- escape velocity

Work, energy, power (module 3)

Circular motion

Motion in a circle
- the radian
- period and frequency
- $\omega = \frac{2\pi}{T} = 2\pi f$

Centripetal force
- $v = \omega r$
- centripetal acceleration $a = \frac{v^2}{r} = \omega^2 r$
- centripetal force $F = \frac{mv^2}{r} = m\omega^2 r$

Charged particles in magnetic fields (module 6)

Oscillations

Damping
- free and forced oscillations
- resonance, natural frequency

Simple harmonic motion
- displacement, amplitude, period, angular frequency and phase difference
- $a = -\omega^2 x$
- SHM graphs
- kinetic and potential energy

Fluorescence

Fluorescence is the emission of light by a substance that has absorbed light or other electromagnetic radiation. Usually, an atom within the substance is excited when it absorbs a high-energy ultraviolet photon. The atom then de-excites and emits lower-energy photons of visible light.

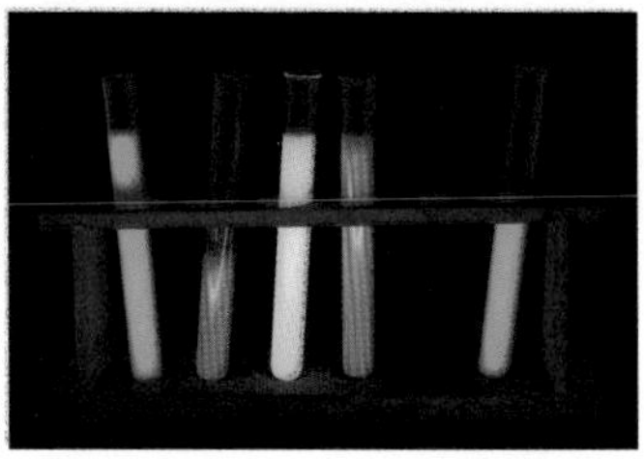

▲ **Figure 1** *Test tubes containing fluorescent solutions*

Fluorescent lamps, including the compact fluorescent lamp (CFL), emit visible light through fluorescence and are found in most homes. The lamp contains mercury vapour at low pressure, which emits ultraviolet light with wavelength ~250 nm when an electric current flows. The lining of the lamp absorbs the ultraviolet light, and emits visible light photons at wavelengths of 436 nm, 546 nm, and 579 nm.

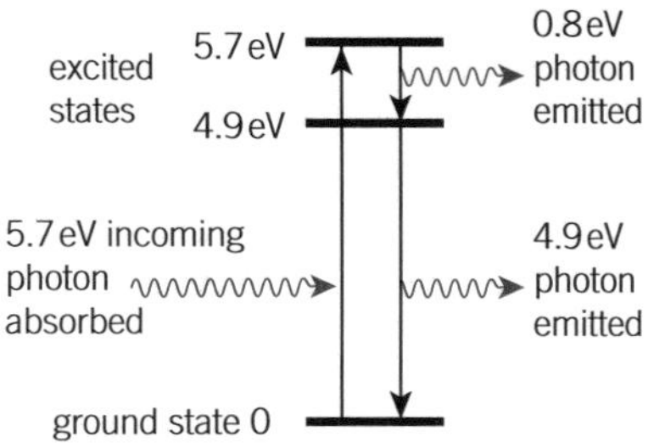

▲ **Figure 2** *Fluorescence can occur when a UV photon is absorbed by a substance that then emits several lower-energy visible photons.*

Many living things and organic fluids fluoresce, and so are visible to the human eye under ultraviolet light.

1 Draw a diagram to show the energy levels in the coating of a fluorescent lamp.
2 Calculate the wavelength of the photons in Figure 2 and determine which part of the electromagnetic spectrum they belong to.
3 Calculate the minimum velocity of an electron needed to excite the mercury vapour inside the fluorescent lamp (Figure 2), and suggest why the mercury needs to be at a low pressure.

Cepheid variables

A Cepheid variable is a type of star that pulsates, its luminosity varying over a set period of anything from 1 to 100 days. On Earth we can observe and measure this variation in brightness of the star.

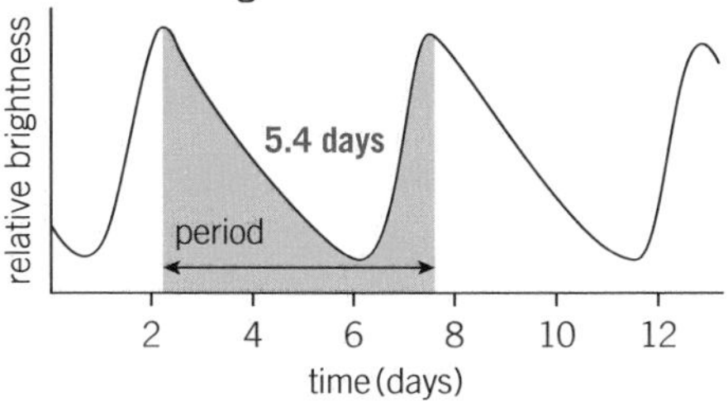

▲ **Figure 3** *The first Cepheid variable star studied had a period of 5.4 days*

Cepheid variables are typically very luminous stars, with luminosities 500 to 300 000 times greater than our Sun. The more luminous stars are typically cooler and larger in diameter.

Cepheid variables are one of a number of 'standard candles' used to accurately determine huge distances. By measuring the period and the intensity of the light detected on Earth, physicists are able to determine the distance to the star.

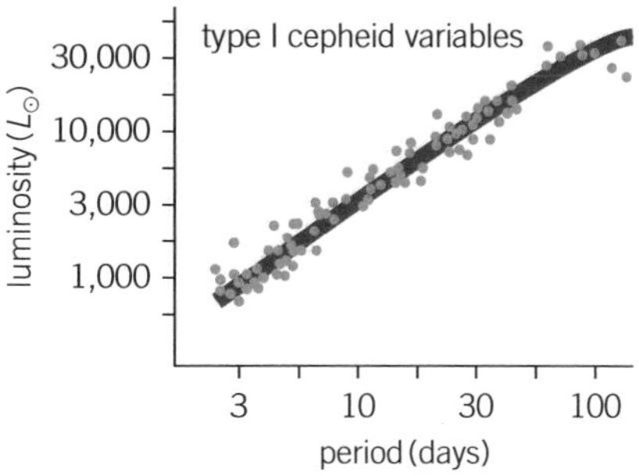

▲ **Figure 4** *The relationship between the luminosity of a Cepheid variable star and its time period*

1 Suggest why a star's temperature increases when it is compressed.
2 Cepheid variables form part of a so-called 'strip of instability'. Sketch an HR diagram and outline the position of this strip.
3 The intensity of the light from a Cepheid variable with a period of 30 days is measured on Earth to be 2.80×10^{-16} Wm^{-2}. Use the data in Figure 4 and the relationship between intensity, luminosity, and distance to determine the distance to the star in light years.

Module 5 practice questions

Section A

1 Which statement is correct for a substance at a temperature of 0 K?

A The substance has no potential energy.

B The substance has maximum kinetic energy.

C The substance has minimum internal energy.

D The substance has no thermal energy.

(1 mark)

2 A container has gas with a mixture of hydrogen and helium atoms. The temperature of the gas inside the container is kept constant.

Which statement is correct for these atoms?

A The hydrogen and the helium atoms have the same mean square speed.

B The hydrogen atoms have greater mean kinetic energy than the helium atoms.

C The hydrogen atoms have a greater mean square speed than the helium atoms.

D The hydrogen atoms have a smaller mean square speed than the helium atoms.

(1 mark)

3 The speed of a simple harmonic oscillator is $4.0\,\text{m s}^{-1}$ through the equilibrium position. What is the speed of the oscillator when its displacement is half the amplitude?

A $2.0\,\text{m s}^{-1}$ **B** $2.3\,\text{m s}^{-1}$

C $3.0\,\text{m s}^{-1}$ **D** $3.5\,\text{m s}^{-1}$

(1 mark)

4 The gravitational potential on the surface of a planet is V_g. The radius of the planet is R. What is the magnitude of the difference in the gravitational potential between a point on the surface of the planet and a point at a distance of $2R$ from the surface of the planet?

(1 mark)

A $0.50\,V_g$ **B** $0.67\,V_g$

C $0.75\,V_g$ **D** $0.89\,V_g$

5 Figure 1 shows some of the energy levels of an atom.

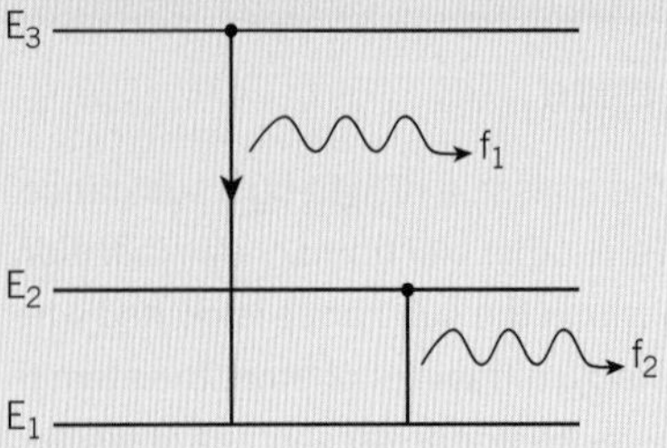

▲ **Figure 1**

An electron making a transition from energy level E_3 to E_1 emits a photon of frequency f_1. A transition from E_2 to E_1 produces a photon of frequency f_2.

What is the difference between the energy levels E_3 and E_2?

A hf_1 **B** hf_2

C $h(f_1 - f_2)$ **D** $h(f_1 + f_2)$

(1 mark)

6 A star of radius R and surface temperature T has luminosity L.

Which of the following stars has the same luminosity L?

	Radius of star	Surface temperature
A	$4R$	$2T$
B	$4R$	$\frac{T}{2}$
C	R	$2T$
D	$2R$	$\frac{T}{2}$

(1 mark)

7 The parallax of a star in the constellation of Orion is 0.25 arc seconds.

What is the distance of this star from the Earth?

A 0.25 ly **B** 0.25 pc

C 4.0 ly **D** 4.0 pc

(1 mark)

8 A galaxy is a distance d metres from the Earth and has a recessional speed of $v\,\text{m s}^{-1}$. What is the approximate age of the Universe in years?

A $\dfrac{(3.16 \times 10^7)d}{v}$ **B** $\dfrac{d}{(3.16 \times 10^7)v}$

C $\dfrac{(3.16 \times 10^7)v}{d}$ **D** $\dfrac{v}{(3.16 \times 10^7)d}$

(1 mark)

Section B

9 a Describe

(i) the motion of atoms in a solid at a temperature well below its melting point *(1 mark)*

(ii) the effect of a small increase in temperature on the motion of these atoms *(1 mark)*

(iii) the effect on the internal energy and temperature of the solid when it melts. *(2 marks)*

(b) Figure 2 shows the apparatus used to determine the specific heat capacity of a metal. A block made of the metal is heated by an electrical heater that produces a constant power of 48 W. In order to reduce heat loss from the sides, top, and bottom of the block, it is covered by a layer of insulating material.

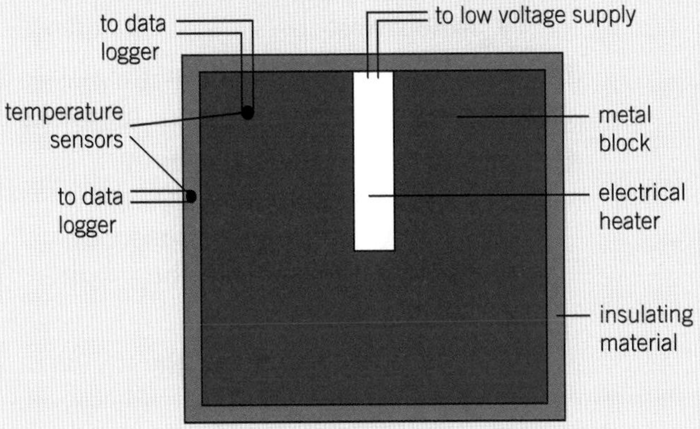

▲ Figure 2

Temperature sensors connected to a data logger show that the block and insulation are initially at the room temperature of 18 °C. The heater is switched on and after 720 seconds the sensors show that the temperature of the block is 54 °C and the average temperature of the insulating material is 38 °C.

(i) Use the information given above and the data shown below to determine the specific heat capacity of the metal block.
mass of metal block = 0.98 kg
power of heater = 48 W
specific heat capacity of the insulating material = 850 J kg^{-1} K^{-1}
mass of the insulating material = 0.027 kg *(4 marks)*

(ii) A second experiment is done without the insulating material and with the block again starting at 18 °C. Discuss whether the value of the specific heat capacity calculated from the second experiment is likely to be lower, the same or higher than the value calculated in (i). *(2 marks)*

OCR G484 January 2013

10 a (i) State what is meant by a *perfectly elastic collision*. *(1 mark)*

(ii) Explain, in terms of the behaviour of **molecules**, how a gas exerts a pressure on the walls of its container. *(4 marks)*

(iii) Explain, in terms of the behaviour of **molecules**, why the pressure of a gas in a container of constant volume increases when the temperature of the gas is increased. *(2 marks)*

(b) A weather balloon is filled with helium gas. Just before take-off the pressure inside the balloon is 105 kPa and its internal volume is 5.0×10^3 m^3. The temperature inside the balloon is 20 °C. The pressure, volume and temperature of the helium gas change as the balloon rises into the upper atmosphere.

(i) The balloon expands to a volume of 1.2×10^4 m^3 in the upper atmosphere where the temperature inside the balloon is –30 °C. Calculate the pressure inside the balloon. *(3 marks)*

(ii) Suggest why it is necessary to release helium from the balloon as it continues to rise. *(1 mark)*

OCR G484 January 2012

11 The apparatus shown in Figure 3 is used to determine an approximate value for absolute zero.

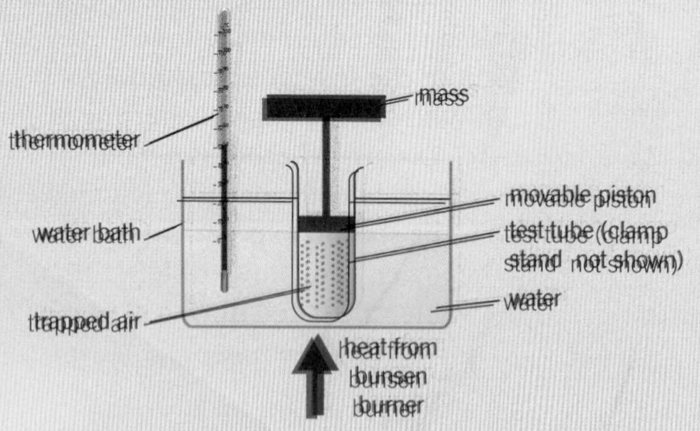

▲ Figure 3

A glass test tube of constant internal diameter is held in a vertical position by a clamp stand. Air is trapped between the closed end of the test tube and the air-tight piston. The piston can move up and down as the volume of the trapped air inside the test tube changes. A constant mass on top of the piston ensures a constant pressure is exerted on the trapped air inside the test tube. The test tube is placed in a water bath.

a Describe how the apparatus can be used to determine absolute zero. *(5 marks)*

b The maximum temperature of the trapped air using the apparatus shown in Figure 3 is 100 °C.

The molar mass of oxygen is $0.032\,\text{kg}\,\text{mol}^{-1}$.

Calculate the r.m.s. speed of oxygen molecules at this temperature. *(4 marks)*

12 a Define gravitational potential energy at a point. *(1 mark)*

b On Figure 4, sketch a graph to show the variation of the gravitational potential V_g for a planet with distance r from its centre. The value of V_g at the surface of the planet has been marked on the graph. *(2 marks)*

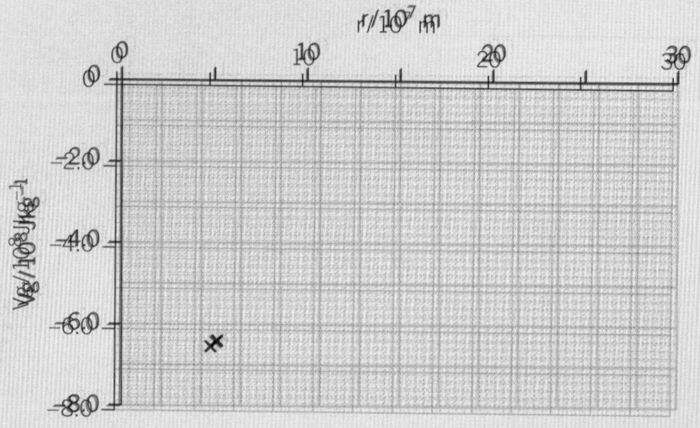

▲ Figure 4

c (i) Explain why gravitational potential V_g is negative. *(2 marks)*

(ii) Use Figure 4 to determine the mass M of the planet. *(2 marks)*

(iii) The planet in **(b)** has a moon in a circular orbit at a distance of 2.4×10^8 m from the centre of the planet. The mass of the moon is 1.1×10^{20} kg.

1 Show that speed v of the moon is given by the equation

$$v^2 = \frac{GM}{r}$$

where r is the orbital radius of the moon. *(2 marks)*

2 Calculate the **total** energy of the moon. *(4 marks)*

13 Figure 5 shows an arrangement used to investigate the tension F in the string and the speed v of a whirling bung.

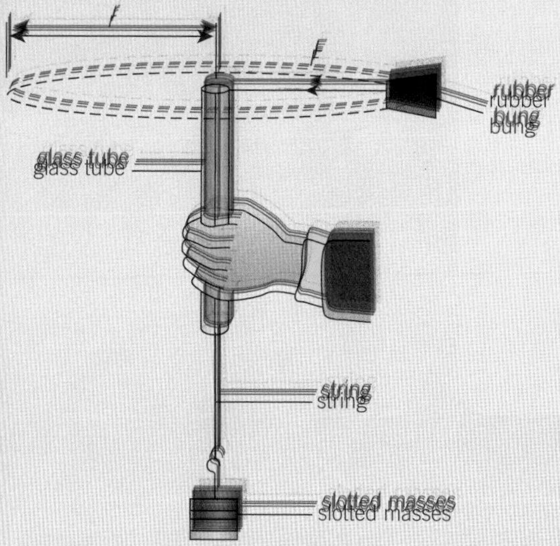

▲ Figure 5

The bung describes a horizontal circle of constant radius r. The tension in the string is altered by using different numbers of slotted masses. Figure 6 shows the data points and the associated error bars plotted by a student.

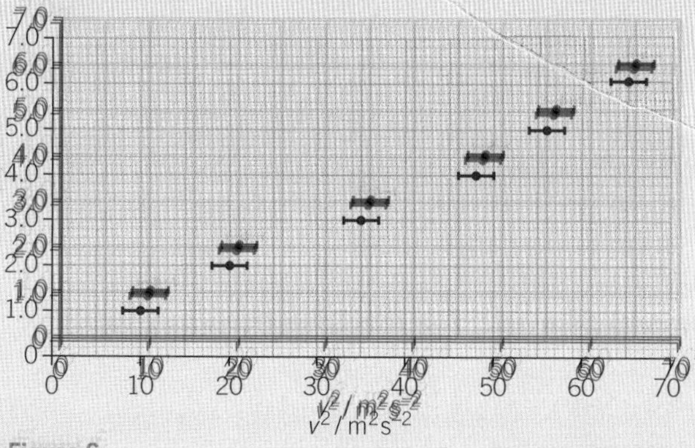

▲ Figure 6

a Plan an experiment that you can conduct to determine the speed of the whirling bung for a particular tension. Suggest one method of improving the precision of your experiment. *(4 marks)*

b (i) Show how the gradient G of the F against v^2 graph is related to the mass m of the bung and the radius r of the circle. *(2 marks)*

(ii) Determine a value for G and state its base units. *(2 marks)*

(iii) The radius of the circle is 32.0 cm. Use Figure 6 to determine the mass of the bung and the percentage uncertainty in the mass of the bung. *(3 marks)*

14 a One estimate of the age of the Universe is 13.7×10^9 years.

(i) Calculate the Hubble constant in km s^{-1} Mpc^{-1} using this age.

$1\,\text{pc} = 3.09 \times 10^{16}\,\text{m}$ *(3 marks)*

(ii) The wavelength of the hydrogen-alpha spectral line in the laboratory is 656 nm. Calculate the observed wavelength of this spectral line in the spectrum of the galaxy NGC 7469, which is 50.0 Mpc away from the Earth. *(4 marks)*

b State what is meant by the Big Bang. Describe how it explains the origin of the microwave background radiation. *(5 marks)*

MODULE 6

Particles and medical physics

Chapters in this module

Introduction

Physics is the study of all things great and small — this module will focus mostly on the smallest things imaginable, that is, particles. Topics covered include capacitors, electric fields, electromagnetism, nuclear physics, particle physics, and medical imaging.

Capacitors introduces the basic properties of capacitors and how they are used in electrical circuits. You will learn how they are used as an essential source of electrical energy in most modern electrical devices.

Electric fields develops the important concepts of Coulomb's law, uniform electric fields, electric potential, and energy. You will learn how electric fields relate to lightning strikes, smart windows, and even particle accelerators.

Magnetic fields explores magnetic fields, the motion of charged particles in magnetic fields, Lenz's law, and Faraday's law. You will learn how Faraday's law has had a dramatic and beneficial effect on society with important devices such as generators and transformers.

Particle physics develops ideas of the nature of the atom and its nucleus, as well as introducing a new world of fundamental particles. You will learn about how the nucleus was first discovered, and how we have since gone on to discover that even nucleons are made up of smaller particles.

Radioactivity explores the impact of unstable nuclei. You will learn that radioactivity is a truly random process, and yet still follows a predictable mathematical model.

Nuclear energy explores the meaning and consequences of Einstein's famous equation $E = mc^2$. You will learn about nuclear fission and its use in nuclear reactors, along with how nuclear fusion might one day provide cheap, clean energy.

Medical imaging introduces the variety of techniques used in modern diagnostic testing, including X-rays, CAT scans, PET scans and ultrasound scans. You will learn how physics has led to the development of a number of valuable non-invasive techniques used in hospitals today.

Knowledge and understanding checklist

From your Key Stage 4 or first year A Level study you should be able to do the following questions. Work through each point, using your Key Stage 4/ first year A Level notes and the support available on Kerboodle.

- ☐ Apply the equations relating p.d., current, quantity of charge, resistance, power, energy, and time, and solve problems for simple circuits.
- ☐ Apply Newton's first law to explain the motion of objects and apply Newton's second law in calculations relating forces, masses, and accelerations.
- ☐ Recall examples of ways in which objects interact by electrostatics and magnetism.
- ☐ Describe the attraction and repulsion between opposite poles and like poles for permanent magnets, and describe the characteristics of the magnetic field of a magnet.
- ☐ Describe how to show that a current can create a magnetic field, and how a magnet and a current-carrying conductor exert a force on one another. Show that Fleming's left-hand rule represents the relative orientations of the force, the conductor, and the magnetic field.
- ☐ Recall that some nuclei are unstable and may emit alpha particles, beta particles, or electromagnetic radiation as gamma rays.
- ☐ Use names and symbols of common nuclei and particles to write balanced equations that represent radioactive decay.
- ☐ Explain the concept of half-life, and how this is related to the random nature of radioactive decay.

Maths skills checklist

All physicists need to use maths in their studies. In this unit you will need to use many different maths skills, including the following examples. You can find support for these skills on Kerboodle and through MyMaths.

- ☐ **Apply the concepts underlying calculus.** You will need to do this as part of the capacitance, radioactive decay, and electromagnetic induction topics.
- ☐ **Estimate results.** You will need to do this when working on the relative sizes of atoms and the nucleus.
- ☐ **Understand simple probability.** You will need to do this as part of radioactive decay.
- ☐ **Use calculators to find and use power, exponential and logarithmic functions.** You will need to do this when solving problems in capacitor charge and discharge and radioactive decay.
- ☐ **Use ratios, fractions and percentages.** You will do this when studying transformers.

21 CAPACITANCE
21.1 Capacitors

Specification reference: 6.1.1

Learning outcomes

Demonstrate knowledge, understanding, and application of:

→ capacitance, $C = \frac{Q}{V}$

→ the unit farad

→ charging and discharging of capacitors in terms of the flow of electrons.

▲ **Figure 1** *Many mobile phones use touchscreen technology*

▲ **Figure 2** *Capacitors can look different, but they all store electrical charge*

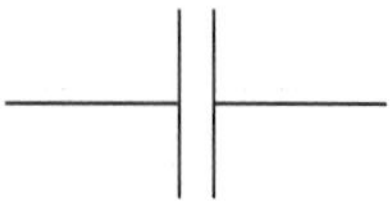

▲ **Figure 3** *The circuit symbol for a capacitor*

Response at your fingertips

Touchscreens – visual displays operated by touch (Figure 1) – are used in devices such as mobile phones, tablet computers, and ATMs. Capacitive touchscreens use a layer of a material that momentarily stores electrical charge at the exact point of contact. The next time you tap a touchscreen, remember that the screen is effectively made up of countless capacitors.

Capacitors

Capacitors are electrical components in which charge is separated. They come in a variety of shapes and sizes (Figure 2). However, their basic construction is the same. A capacitor consists of two metallic plates separated from each other by an insulator, often known as a dielectric, such as air, paper, ceramic, or mica – hence the circuit symbol for a capacitor shows two lines separated by a gap (Figure 3). You can easily make your own capacitor from two sheets of aluminium foil separated by newspaper.

Storing charge

Figure 4 shows a capacitor connected to a cell of e.m.f. ε. When the capacitor is connected to the cell, electrons flow from the cell for a very short time. They cannot travel between the plates because of the insulation. The very brief current means electrons are removed from plate **A** of the capacitor and at the same time electrons are deposited onto the other plate **B**. Plate **A** becomes deficient in electrons, that is, it acquires a net positive charge. Plate **B** gains electrons and hence acquires a negative charge. The current in the circuit must be the same at all points and charge must be conserved, so the two plates have an equal but opposite charge of magnitude Q. Therefore there is a potential difference (p.d.) across the plates. The current in the circuit falls to zero when the p.d. across the plates is equal to the e.m.f. ε of the cell. The capacitor is then fully charged. The net charge on the capacitor plates is zero.

The capacitor is therefore really a device that separates electrical charge into $-Q$ and $+Q$.

Capacitance

Commercial capacitors are usually marked with their capacitance value, which indicates the amount of charge Q that the capacitor can store for a given p.d. V.

The **capacitance** of a capacitor is defined as the charge stored per unit p.d. across it. That is

$$C = \frac{Q}{V}$$

where C is the capacitance in farads (F), Q is the charge stored and V is the potential difference across the capacitor. You can write this equation as:

$$Q = VC$$

For any capacitor, the greater the amount of positive and negative charge stored on the two plates, the greater the p.d. across them, so the charge on the capacitor is always proportional to the p.d. The unit of capacitance is the **farad** (F). You can see from the equation above that $1\,F = 1$ coulomb per volt ($C\,V^{-1}$).

There is room for confusion with the letter 'C' being used for both coulombs and capacitance. You just need to be vigilant and ensure that units and quantities are not mixed up in calculations.

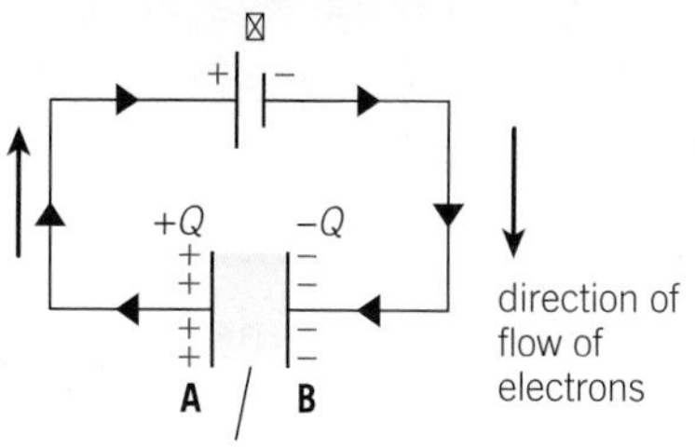

▲ **Figure 4** *A capacitor charges up because of flow of electrons*

Worked example: How many electrons?

A capacitor of capacitance 200 μF is connected to a 10 V supply. Calculate the number N of electrons removed from the positive plate of the capacitor.

Step 1: Calculate the charge Q stored by the capacitor.

$Q = VC = 10 \times 200 \times 10^{-6} = 2.0 \times 10^{-3}\,C$ $\quad (1\,\mu F = 10^{-6}\,F)$

Step 2: The number of electrons is equal to the charge on the plate divided by the charge e on an electron.

$$N = \frac{Q}{e} = \frac{2.0 \times 10^{-3}}{1.60 \times 10^{-19}} = 1.25 \times 10^{16} = 1.3 \times 10^{16} \text{ electrons (2 s.f.)}$$

This is 13 000 000 000 000 000 electrons!

Summary questions

1 A student inserts two bare copper wires into a lump of wet clay. The clay is dried out. The student suggests that he has made a capacitor. Explain whether he is correct. *(1 mark)*

2 The charge stored by a capacitor connected to a 12 V battery is 2.4×10^{-5} C. Calculate its capacitance in farads and microfarads. *(3 marks)*

3 A capacitor stores a charge of 30 pC when the p.d. across it is 3.2 V. Calculate the charge stored in pC when the p.d. is increased to 8.7 V. *(3 marks)*

4 A capacitor of capacitance 0.010 F is connected to a 1.5 V cell. Calculate the number of electrons delivered by the cell to the capacitor. *(3 marks)*

5 An uncharged capacitor of capacitance 1500 μF is connected in a circuit. The current in the circuit has a constant value of 80 μA. The capacitor is charged for a time of 60 s.

a Calculate the charge and the p.d. across the capacitor after a time of 60 s. *(4 marks)*

b Sketch a graph of p.d. V across the capacitor against time t over a period of 60 s. *(2 marks)*

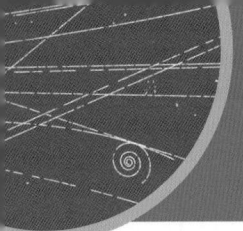

21.2 Capacitors in circuits

Specification reference: 6.1.1

Learning outcomes

Demonstrate knowledge, understanding, and application of:

- → total capacitance of capacitors in series, $\frac{1}{C} = \frac{1}{C_1} + \frac{1}{C_2} + \ldots$
- → total capacitance of capacitors in parallel, $C = C_1 + C_2 + \ldots$
- → analysis of circuits containing capacitors
- → investigation of circuits containing capacitors.

Connecting capacitors

Supercapacitors are compact specialist capacitors with capacitance values in the thousands of farads. They are used as alternatives to battery packs, memory backup devices, and emergency lighting. Unlike rechargeable batteries, which degrade over time, supercapacitors can be charged over and over again.

You are unlikely to have such capacitors in your laboratory. However, you can connect the capacitors available in order to get a desired value. You will need an enormous number of capacitors if you want to reach thousands of farads.

To predict the final capacitance of a combination, you first need to understand how capacitors behave in parallel and series combinations.

Capacitors in parallel

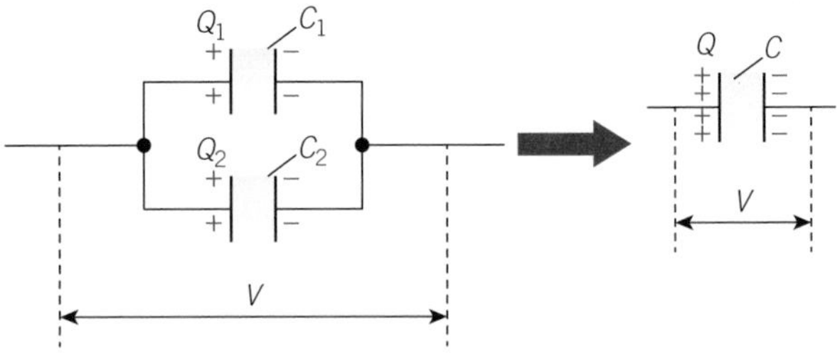

▲ **Figure 1** *Capacitors in parallel*

Figure 1 shows two capacitors of capacitances C_1 and C_2 connected in parallel. Together, their capacitance is greater than their individual capacitances, so the combination will store more charge for a given potential difference (p.d.).

For two or more capacitors in parallel:

- The p.d. V across each capacitor is the same.
- Electrical charge is conserved. Therefore, the total charge stored Q is equal to the sum of the individual charges stored by the capacitors, $Q = Q_1 + Q_2 + \ldots$.
- The total capacitance C is the sum of the individual capacitances of the capacitors, $C = C_1 + C_2 + \ldots$.

Capacitors in series

Figure 2 shows two capacitors of capacitances C_1 and C_2 connected in series. Together, their capacitance is less than their individual capacitances, so this combination will store less charge for a given p.d.

All the capacitors in series store the same charge. This is even true when they have different capacitances. The cell is connected to the left-hand plate of the capacitor of capacitance C_1 and to the right-hand

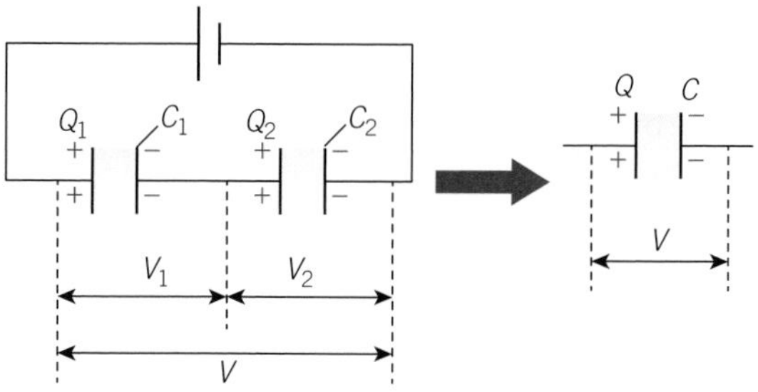

▲ **Figure 2** *Capacitors in series – each capacitor stores the same charge irrespective of the capacitors used*

plate of the capacitor of capacitance C_2. These plates acquire equal and opposite charges as electrons flow from and to these plates. The middle two plates are not connected to the cell because of the presence of the dielectric layers, but transfer of electrons between these plates ensures that they too acquire charge Q of the same magnitude. The overall charge of each capacitor is zero, but the magnitude of the charge on each plate is Q.

For two or more capacitors in series:

- According to Kirchhoff's second law, the total p.d. V across the combination is the sum of the individual p.d.s across the capacitors, $V = V_1 + V_2 + ...$
- The charge Q stored by each capacitor is the same
- The total capacitance C is given by the equation $\frac{1}{C} = \frac{1}{C_1} + \frac{1}{C_2} + ...$

To avoid mistakes when using the equation for total capacitance above, take care with the reciprocals. You can use the inverse button (x^{-1}) on your calculator. The equivalent equation is $C = (C_1^{-1} + C_2^{-1} + ...)^{-1}$.

Capacitor circuits

You can analyse complex circuits with the ideas developed above.

Worked example: Analysing a circuit

Calculate the total capacitance C of the circuit shown in Figure 3 and the p.d. measured by the digital voltmeter placed across the capacitor of capacitance 100 μF.

Step 1: Calculate the capacitance of the parallel combination.

$C = C_1 + C_2 = 500 + 200 = 700\,\mu F$

Step 2: Calculate the total capacitance of the whole circuit.

$C = (C_1^{-1} + C_2^{-1})^{-1} = (100^{-1} + 700^{-1})^{-1} = 87.5\,\mu F$

Step 3: Calculate the total charge stored.

$Q = VC = 3.0 \times 87.5 \times 10^{-6} = 2.63 \times 10^{-4}$ C

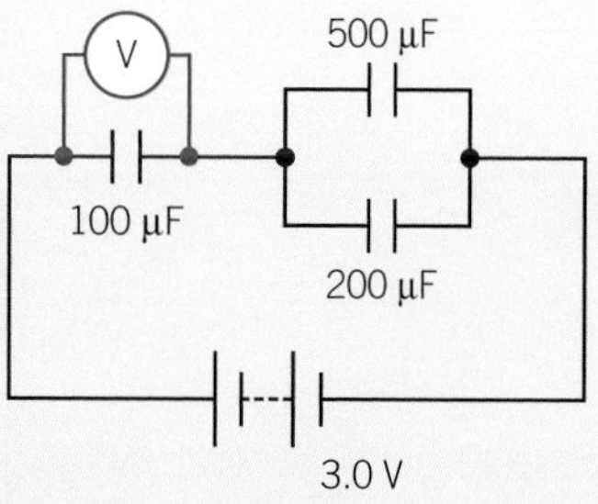

▲ **Figure 3** *What is the voltmeter reading?*

Step 4: The total charge is equal to the charge stored by the 100 μF capacitor, which is equal to the charge stored by the parallel capacitors together.

Apply the equation $Q = VC$ to the 100 μF capacitor to calculate the p.d. across it.

$$V = \frac{Q}{C} = \frac{2.63 \times 10^{-4}}{100 \times 10^{-6}} = 2.6\,\text{V (2 s.f.)}$$

The voltmeter reading will be 2.6 V.

Investigating circuits

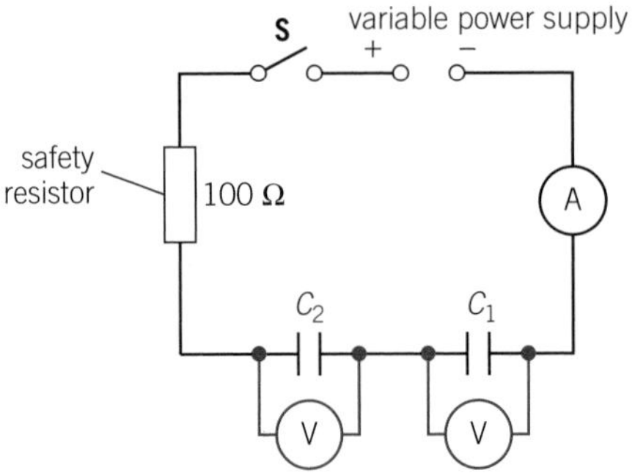

▲ **Figure 4** *A possible method for investigating capacitor circuits – the resistor protects the ammeter, and the final charges stored by the capacitors are independent of the resistor used*

A possible experimental layout for investigating capacitor circuits is shown in Figure 4. An ammeter, a resistor of 100 Ω, two capacitors, and a switch **S** are all connected in series to a power supply. When the switch **S** is closed, the ammeter briefly registers a current but very quickly settles down to a zero reading. This shows that electrons move in the circuit only until the capacitors are fully charged. The p.d.s V_1 and V_2 across each capacitor can be measured with the voltmeters.

Table 1 shows some typical results for two different settings of the power supply. The charges Q_1 and Q_2 stored by the capacitors are calculated. You can see that they are the same. Any difference is caused by the uncertainties in the voltmeter readings and the manufacturer's values for the capacitances.

▼ **Table 1** *Results from the circuit in Figure 4 for two settings of the power supply*

V_1 / V	V_2 / V	Charge stored by the 500 μF capacitor	Charge stored by the 1000 μF capacitor
4.00	2.00	$Q_1 = 4.00 \times 500$ $Q_1 = 2000\,\mu C$	$Q_2 = 2.00 \times 1000$ $Q_2 = 2000\,\mu C$
1.19	0.61	$Q_1 = 1.19 \times 500$ $Q_1 = 595\,\mu C$	$Q_2 = 0.61 \times 1000$ $Q_2 = 610\,\mu C$

Figure 5 shows a simple technique that may be used to confirm the capacitance rules for a series circuit. A multimeter set to 'capacitance' or 'farads' is used to determine the individual capacitance of each capacitor and then to determine the total capacitance by placing it across the combination of capacitors. This method does not require an external battery because the multimeter uses its own battery to show the readings. You can easily adapt this technique to measure the total capacitance of any simple or complex circuits.

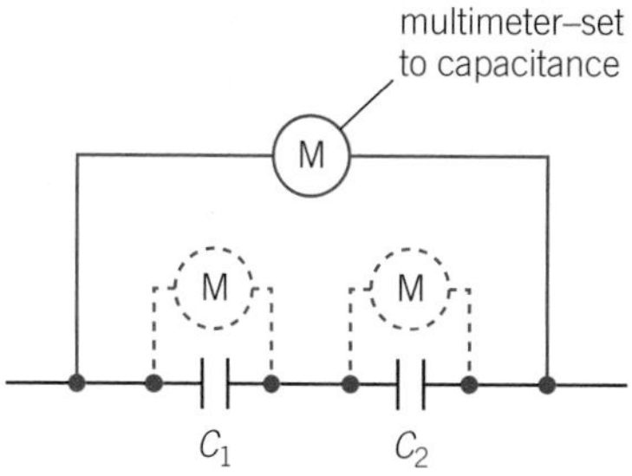

▲ **Figure 5** *Digital multimeters set on the capacitance or 'farads' setting can be used to measure the capacitance of individual capacitors or the entire circuit*

Proofs for the capacitance equations

You can use basic circuit rules to show the validity of the equations for capacitance.

Parallel (Figure 1)

The total charge stored Q is equal to the sum of the individual charges, that is

$Q = Q_1 + Q_2$

The p.d. V across each capacitor is the same because they are connected in parallel. You can use the equation $Q = VC$ for individual components or the entire circuit. Therefore

$VC = VC_1 + VC_2$

The p.d. V cancels out leaving the equation for total capacitance, $C = C_1 + C_2$.

Series (Figure 2)

According to Kirchhoff's second law

$V = V_1 + V_2$

The charge Q stored by each capacitor is the same. Once again, you can use the equation $Q = VC$ for individual components or the entire circuit. Therefore

$$\frac{Q}{C} = \frac{Q}{C_1} + \frac{Q}{C_2}$$

The charge Q cancels out, leaving the equation for total capacitance

$$\frac{1}{C} = \frac{1}{C_1} + \frac{1}{C_2}.$$

Summary questions

1. Calculate the total capacitance of two 100 pF capacitors connected in parallel and then in series. Comment on your answers. *(5 marks)*
2. Two 120 nF capacitors are connected in series to a 1.5 V cell. Calculate the total charge delivered to the capacitors by the cell. *(4 marks)*
3. A supercapacitor can have a capacitance of 4000 F. Calculate the number of identical 1000 μF capacitors you would need to make a capacitor with this capacitance. Explain your answer. *(3 marks)*
4. Calculate the total capacitance of the circuit shown in Figure 6. *(5 marks)*
5. Two capacitors of capacitances C and $2C$ are connected in series to a battery of electromotive force (e.m.f.) 6.0 V. Calculate the p.d. across each capacitor. Explain your answer. *(4 marks)*
6. An unmarked capacitor is connected in series with a capacitor of capacitance 20 nF. A multimeter shows the total capacitance to be 17 nF. Calculate the capacitance of the unmarked capacitor. *(3 marks)*

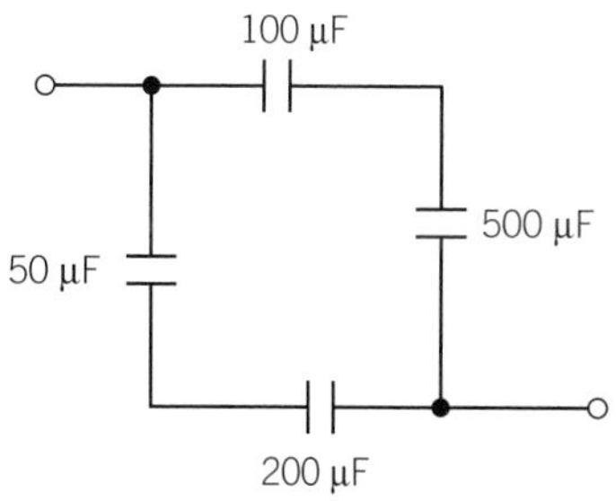

▲ **Figure 6**

21.3 Energy stored by capacitors

Specification reference: 6.1.2

Learning outcomes

Demonstrate knowledge, understanding, and application of:

- → p.d.–charge graphs for capacitors
- → energy stored by capacitors
- → $W = \frac{1}{2}QV = \frac{1}{2}\frac{Q^2}{C} = \frac{1}{2}V^2C$

▲ **Figure 1** *Where does the energy for all these flashes come from?*

Energy storage

Suppose you charge a capacitor of capacitance 1000 μF by connecting it to the terminals of a 3.0 V battery. If you then remove the capacitor from the circuit and connect it to a small filament lamp, you will see a brief flash of light – evidence that the capacitor stores energy. Mobile phone and camera flashes rely on capacitors to store and release the energy they need (Figure 1). The amount of energy stored in a capacitor depends on the value of the capacitance and the initial potential difference (p.d.) across it.

Pushing and removing electrons

Figure 2 shows an electron moving towards the negative plate of a capacitor that is being charged. This electron will experience a repulsive electrostatic force from all the electrons already on the plate. External work has to be done to push this electron onto the negative plate. Similarly, work is done to cause an electron to leave the positive plate of the capacitor. The external work is provided by the battery or power supply connected to the capacitor. In short, the energy stored in a capacitor comes from the energy of the battery or power supply.

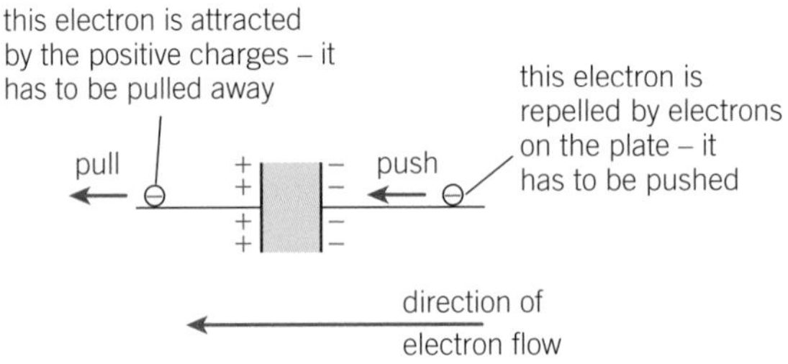

▲ **Figure 2** *External work has to be done on electrons when a capacitor is being charged*

Potential difference–charge graphs

How can you determine the energy stored in a capacitor? A good start would be the graph of p.d. against charge for a capacitor, like the one in Figure 3.

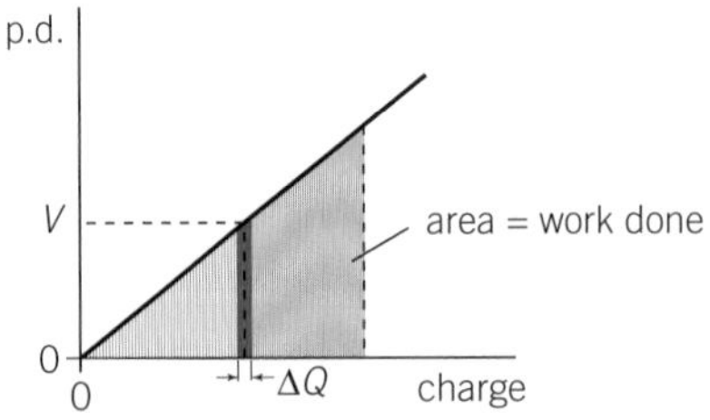

▲ **Figure 3** *Potential difference–charge graph for a capacitor*

You already know that work done = p.d. × charge. The small amount of work done, ΔW, to increase the charge stored in the capacitor by a small amount ΔQ is given by the equation

$$\Delta W \approx V \times \Delta Q$$

where the p.d. V does not change significantly. If you look at the graph in Figure 3, this is the area of the thin rectangular strip shown shaded under the graph. Hence ΔW is equal to the area of this strip. If you add all the similar strips together, then the area under the graph is the total work done on the charges for a charging capacitor.

- The area under a p.d.–charge graph = work done

Synoptic link

To review the relationship between work, p.d., and charge, see Topic 9.2, Potential difference and electromotive force.

Stored energy

The work done on the charges is the same as the energy stored in the capacitor. Just like springs, this energy is 'potential' or stored, and can be released. Since the area under a p.d.–charge graph is equal to the work done, you can derive an equation for the energy stored in a capacitor, W, from that area (Figure 4).

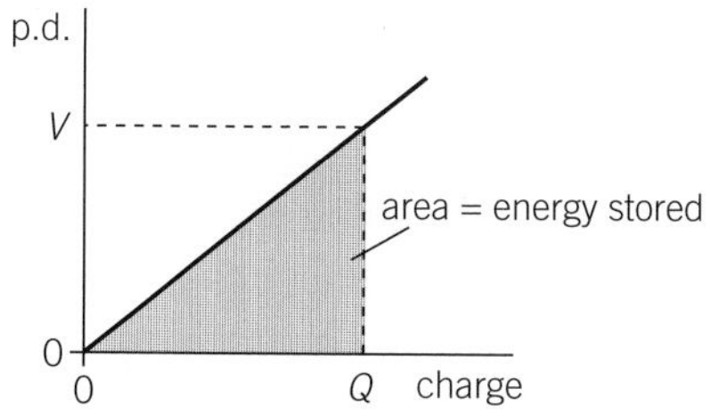

▲ **Figure 4** *The energy stored in the capacitor is the area of the shaded triangle*

W = area under graph = area of shaded triangle

$$W = \frac{1}{2}QV$$

where V is the p.d. when the charge is Q.

The charge stored by the capacitor is also given by the equation $Q = VC$, where C is the capacitance of the capacitor. Substituting this equation into $W = \frac{1}{2}QV$ gives us another very useful equation for the stored energy.

$$W = \frac{1}{2}QV = \frac{1}{2}V \times (VC)$$

or

$$W = \frac{1}{2}V^2C$$

For a given capacitor, W is directly proportional to V^2, so doubling the p.d. quadruples the energy stored.

Finally, if you substitute $V = \frac{Q}{C}$ into $W = \frac{1}{2}QV$, you get a third equation for the stored energy.

$$W = \frac{1}{2}\frac{Q^2}{C}$$

So you now have three equations for stored energy –

$$W = \frac{1}{2}QV \qquad W = \frac{1}{2}V^2C \qquad W = \frac{1}{2}\frac{Q^2}{C}$$

Worked example: Energy stored in a capacitor

A 0.10 F capacitor is connected to a power supply and then removed. The energy stored in the capacitor is 25 J. Calculate the initial p.d. across the capacitor.

Step 1: Write down the quantities you know and what you are trying to calculate.

$$W = 25\,\text{J},\ C = 0.10\,\text{F},\ V = ?$$

Step 2: Identify the correct equation to use and rearrange it to make V the subject.

$$W = \frac{1}{2}V^2C \quad \text{so} \quad V = \sqrt{\frac{2W}{C}}$$

Step 3: Substitute the values into this equation and calculate V.

$$V = \sqrt{\frac{2W}{C}} = \sqrt{\frac{2 \times 25}{0.10}} = 22\,\text{V (2 s.f.)}$$

The initial p.d. across the capacitor was 22 V.

Summary questions

1 A student plots a graph of potential difference V across a capacitor against the charge Q stored. State two quantities you can get from this graph. *(2 marks)*

2 Calculate the energy stored in a capacitor of capacitance 1000 μF when charged to a p.d. of:

a 2.0 V *(2 marks)*

b 6.0 V. *(2 marks)*

3 A 4.0 F capacitor is charged using a device that produces a constant current of 200 mA. Calculate the energy stored in the capacitor after a charging time of 10 minutes. *(3 marks)*

4 Two 300 μF capacitors are connected in series to a 24 V power supply. Calculate the total energy stored in the capacitors. *(3 marks)*

5 An inventor claims to have made a device that uses a 0.10 F capacitor and a 240 V supply to lift a 10 kg mass through a vertical height of about 29 m. With the help of calculations, explain whether or not this claim is plausible. *(4 marks)*

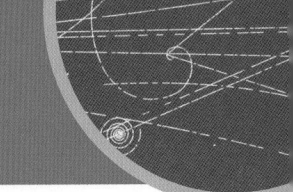

21.4 Discharging capacitors

Specification reference: 6.1.3

Exponential decay

Nature has many examples of physical quantities decreasing by the same factor in equal time intervals, such as the radioactivity of a sample of uranium salts, the height of a solution in a burette as it empties through the small tap, or the pressure of the atmosphere with increasing height (Figure 1). This constant-ratio pattern is called **exponential decay**. The p.d. across a capacitor discharging through a resistor decreases exponentially over time.

Exponential functions are governed by an important number called e, the base of natural logarithms, which has a value of 2.718... (you can see its value by using the e^x button on your calculator with $x = 1$). Like the constant π, e is an irrational number, first identified by two prominent mathematicians, Euler and Napier. It must not be confused with elementary charge, which has the same letter.

Learning outcomes

Demonstrate knowledge, understanding, and application of:

- → discharging a capacitor through a resistor
- → investigating the charge and the discharge of a capacitor
- → time constant CR of a capacitor–resistor circuit
- → $x = x_0 e^{-\frac{t}{CR}}$ and $x = x_0(1 - e^{-\frac{t}{CR}})$ for capacitor–resistor circuits
- → modelling of the equation $\frac{\Delta Q}{\Delta t} = \frac{-Q}{CR}$ for a discharging capacitor
- → exponential decay and the constant-ratio property of decay graphs.

Discharging a capacitor

Figure 2 shows a capacitor and a resistor connected in parallel to a battery. The capacitor has capacitance C and the resistance of the resistor is R. A digital voltmeter is placed across the resistor. The switch **S** in initially closed and the capacitor is fully charged. The p.d. across the capacitor and the resistor is equal to V_0. What happens when **S** is opened at time $t = 0$?

- The p.d. V across the capacitor or the resistor = V_0
- For the resistor, $V = IR$, therefore current I in the resistor = $\frac{V_0}{R}$.
- For the capacitor, $Q = VC$, therefore charge stored $Q = V_0C$.

The capacitor then discharges through the resistor. The charge stored by the capacitor decreases with time and hence the p.d. across it also decreases. The current in the resistor decreases with time as the p.d. across it decreases accordingly. Eventually, the p.d. V, the charge Q stored by the capacitor, and the current I in the resistor are all zero.

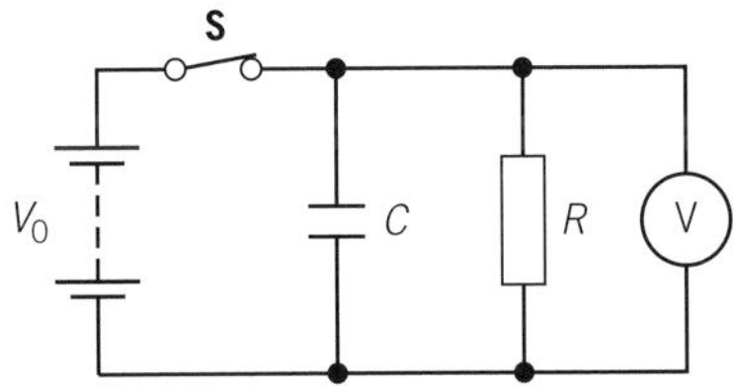

▲ **Figure 2** *A circuit to demonstrate the discharge of a capacitor through a resistor (you could use a datalogger instead of the voltmeter)*

▲ **Figure 1** *Atmospheric pressure decreases approximately exponentially with increasing height, at a rate of about 12% per kilometre above sea level*

What is the general relationship between p.d. V, charge Q, current I, and time t? Figure 3 shows the V–t, I–t, and Q–t graphs. They all show exponential decay over time after the switch is opened. As such, they all have the same shape. The equations for these quantities are as given by:

$$V = V_0 e^{-\frac{t}{CR}},\ I = I_0 e^{-\frac{t}{CR}},\ \text{and } Q = Q_0 e^{-\frac{t}{CR}}$$

where I_0 is the maximum current at $t = 0$, V_0 is the p.d. at $t = 0$, and Q_0 is the charge at $t = 0$.

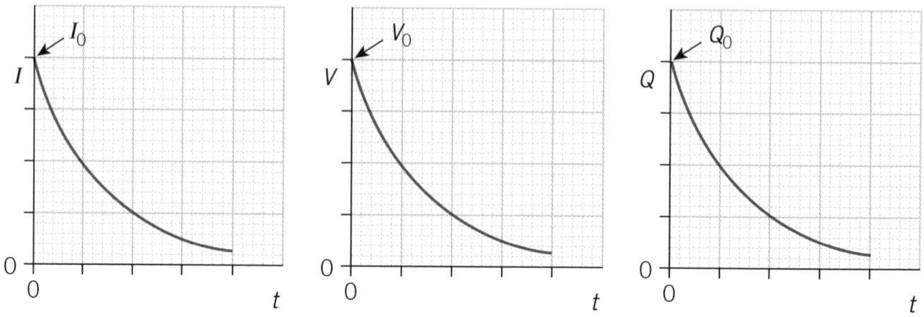

▲ **Figure 3** *The p.d. V across the resistor (or the capacitor), the current I in the resistor, and the charge Q stored by the capacitor all decrease exponentially with respect to time*

Worked example: Analysing a discharging capacitor

A 470 μF capacitor is fully charged to a p.d. of 10 V. It discharges through a resistor of resistance 100 kΩ. Calculate the initial current in the circuit and the p.d. across the capacitor after 30 s.

Step 1: Write down the quantities given in the question.

$C = 470 \times 10^{-6}$ F, $R = 100 \times 10^3\ \Omega$, $V_0 = 10$ V, $t = 30$ s

Step 2: Calculate the initial current I_0 using $V = IR$.

$$I_0 = \frac{V_0}{R} = \frac{10}{100 \times 10^3} = 1.0 \times 10^{-4}\ \text{A}$$

Step 3: Use the exponential decay equation to calculate the p.d. after $t = 30$ s. Before substituting values into the equation, you can calculate CR to make the substitution less daunting.

$CR = 470 \times 10^{-6} \times 100 \times 10^3 = 47$ FΩ

$$V = V_0 e^{-\frac{t}{CR}} = 10 \times e^{-\frac{30}{47}} = 5.3\ \text{V (2 s.f.)}$$

Constant-ratio property of exponential decay

Figure 4 shows the results from an experiment in which a charged capacitor of capacitance 470 μF was discharged through a resistor of resistance 100 kΩ. The graph of p.d. V across the capacitor against time t shows the characteristic shape of exponential decay with its constant-ratio property.

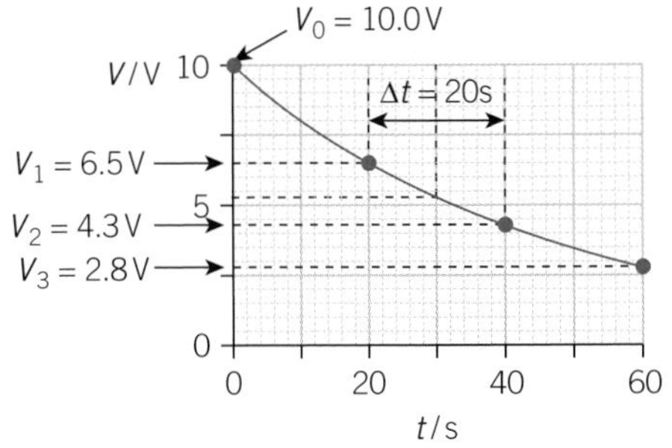

▲ **Figure 4** *Constant-ratio property of exponential decay graph. For a time interval of 20 s, we see that* $\frac{V_1}{V_0} \approx \frac{V_2}{V_1} \approx \frac{V_3}{V_2}$. *Try another interval* ***yourself*** *and you will see that the ratios of the p.d.s will still be the same*

Time constant

You will have noticed the product CR in the exponential decay equations. This product has the same unit as time, that is, you can easily show that a farad ohm is the same as a second (try it). When $t = CR$, the p.d. V across the resistor (or capacitor) is given by

$$V = V_0 e^{-\frac{t}{CR}} = V_0 e^{-\frac{CR}{CR}} = V_0 e^{-1} \approx 0.37\, V_0$$

The **time constant** of a capacitor–resistor circuit is equal to the product of capacitance and resistance (CR). It acts as a useful measure of how long the exponential decay will take in a particular capacitor–resistor circuit.

- The time constant τ for a discharging capacitor is equal to the time taken for the p.d. (or the current or the charge) to decrease to e^{-1} (about 37%) of its initial value.

Modelling exponential decay

The capacitor and the resistor in the circuit in Figure 2 are in parallel and hence have the same p.d. V. The charge Q stored by the capacitor is given by the equation $Q = VC$ and the current I in the circuit is given by the equation $I = \frac{V}{R}$. Since V is the same for both components, the current may be written as

$$I = \frac{V}{R} = \frac{Q}{CR}$$

For a capacitor, $I = -\frac{\Delta Q}{\Delta t}$. The negative sign is important because it shows that the charge on the capacitor decreases with time. Substituting for I in the equation above gives the following equation for a capacitor discharging through a resistor.

$$\frac{\Delta Q}{\Delta t} = -\frac{Q}{CR}$$

The exponential decay for charge, $Q = Q_0 e^{-\frac{t}{CR}}$, is a solution for the equation above. You do not need to be able to derive it for this course.

Synoptic link

Remember that the current in a circuit is the rate of flow of charge, $I = \frac{\Delta Q}{\Delta t}$. You studied this in Topic 8.1, Current and charge.

Study tip

Note from the equation that, as you would expect for an exponential decay, the rate of decay of charge $\frac{\Delta Q}{\Delta t}$ is directly proportional to the charge Q.

▼ **Table 1** *A spreadsheet showing how to predict the charge lost (ΔQ) and charge remaining (Q) on a capacitor over time*

t/s	Q/μC	ΔQ/μC
0.0	500.0	5.00
0.1	495.0	4.95
0.2	490.1	4.90
0.3	485.1	4.85
0.4	480.3	4.80
0.5	475.5	4.75
0.6	470.7	4.71
0.7	466.0	4.66
0.8	461.4	4.61
0.9	456.8	4.57
1.0	452.2	4.52
1.1	447.7	4.48
1.2	443.2	4.43
1.3	438.8	4.39
1.4	434.4	4.34

The equation $\frac{\Delta Q}{\Delta t} = -\frac{Q}{CR}$ can be used to model the decay of charge Q on the capacitor using a technique known as iterative modelling, as follows.

1 Start with a known value for the initial charge Q_0 and a known value for the time constant CR.

2 Choose a time interval Δt which is very small compared with the time constant.

3 Calculate the charge leaving the capacitor, ΔQ, in a time interval Δt using the equation

$$\Delta Q = \frac{\Delta t}{CR} \times Q$$

4 Calculate the charge Q left on the capacitor at the end of the period Δt by subtracting ΔQ from the previous charge.

5 Repeat the whole process for the subsequent multiples of the time interval Δt.

It would tedious to do this task by hand, but a spreadsheet makes it simple (Table 1). The results are shown for $Q_0 = 500.0\,\mu C$, $CR = 10.0\,s$, and $\Delta t = 0.1\,s$. The equation for modelling the charge **lost** in each time interval Δt is

$$\Delta Q = \frac{\Delta t}{CR} \times Q = \frac{0.1}{10.0} Q = 0.01Q$$

After each interval $\Delta t = 0.1\,s$, 1% of the previous charge has been lost from the capacitor. The constant-ratio property of exponential decay is clear.

Study tip

Remember that with natural logarithms (logs to the base e) –
$\ln V$ is the same as writing $\log_e V$
$\ln(AB) = \ln A + \ln B$
$\ln(e^{-x}) = -x$

Dealing with logarithms and experimental results

The p.d. V across a discharging capacitor is given by the equation $V = V_0 e^{-\frac{t}{CR}}$. Taking logs to base e of both sides gives

$$\ln V = \ln(V_0 e^{-\frac{t}{CR}}) = \ln V_0 + \ln e^{-\frac{t}{CR}} = \ln V_0 - \frac{t}{CR}$$

Compare this equation with the equation for a straight line, $y = mx + c$. You will notice that plotting $\ln V$ on the y-axis and t on the x-axis gives gradient $= -\frac{1}{CR}$ and y-intercept $= \ln V_0$.

Table 2 shows some experimental results for a discharging capacitor.

1 Copy the table and complete the values of $\ln V$ for the final column. Note that the values of V are quoted to 3 significant figures, so you must quote $\ln V$ to at least 3 decimal places.

2 Plot a graph of $\ln(V)$ against t and draw a straight best-fit line.

3 Determine the gradient of the line and hence determine the time constant CR of the circuit.

4 The resistance of the resistor used in the experiment was 33 kΩ. Determine the value of the capacitance C of the capacitor.

▼ **Table 2** *Change in p.d. across a capacitor with time*

t/s	V/V	ln(V/V)
0	9.00	
10	6.65	
20	4.91	
30	3.63	
40	2.68	
50	1.98	
60	1.46	

Summary questions

1 A 220 μF capacitor is charged to 2.0 V and then discharged through a 100 Ω resistor. Calculate the maximum current in the circuit and the maximum charge stored by the capacitor. *(2 marks)*

2 Calculate the time constant for these circuits.

a $C = 0.010$ F and $R = 1.0$ kΩ

b $C = 4700$ μF and $R = 1.5$ MΩ. *(2 marks)*

3 A charged capacitor discharges through a resistor. The initial p.d. across the capacitor is 9.0 V. Calculate the p.d. across the resistor after a time equal to 3 time constants. *(3 marks)*

4 Calculate the time constant of the circuit in Figure 5. *(3 marks)*

5 A 500 μF capacitor is discharged through a 200 kΩ resistor. Calculate the time taken for the current in the resistor to decrease to 25% of its initial value. *(4 marks)*

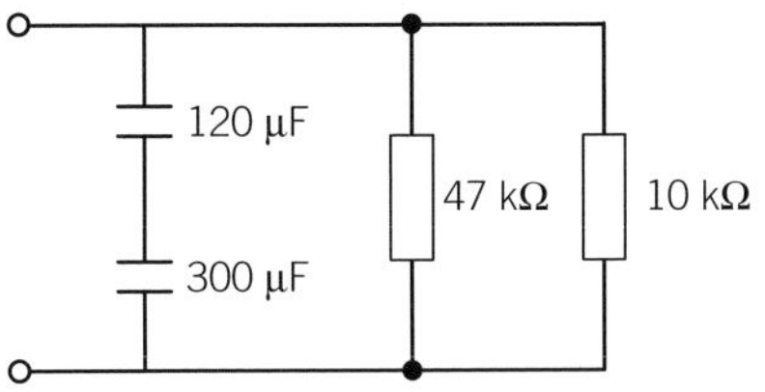

▲ Figure 5

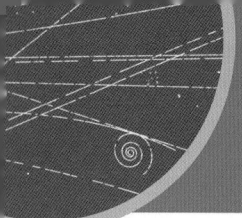

21.5 Charging capacitors

Specification reference: 6.1.3

Learning outcomes

Demonstrate knowledge, understanding, and application of:

→ charging a capacitor through a resistor

→ $x = x_0 e^{-\frac{t}{CR}}$ and

$x = x_0(1 - e^{-\frac{t}{CR}})$ for capacitor–resistor circuits.

▲ **Figure 1** *We rely on the capacitors in our phone chargers to change the alternating mains voltage into a direct voltage suitable for a mobile phone*

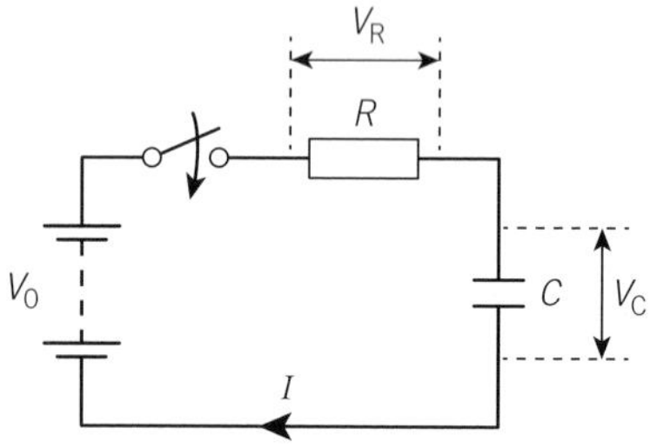

▲ **Figure 2** *Capacitor charging*

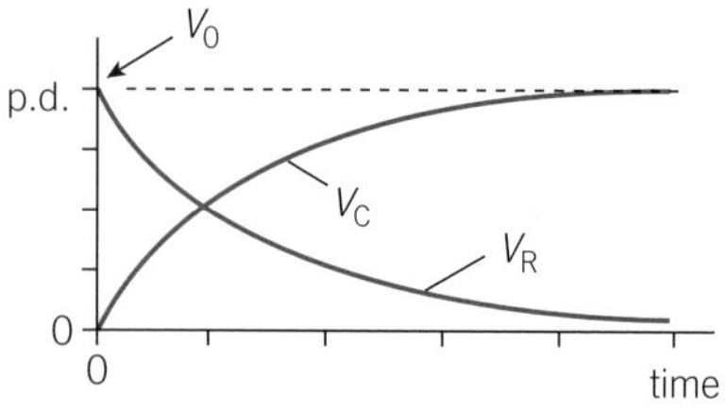

▲ **Figure 3** *Adding the values of V_C and V_R at any given time t gives V_0*

Smoothing

Mobile phones and other portable devices require regular charging. A phone charger takes in the fluctuating alternating voltage of the mains supply and converts it into a smooth direct voltage of about 5.0 V. This smoothing of voltage fluctuations is achieved with a network of components including capacitors.

Charging a capacitor

You have already seen how a capacitor discharges. In this topic you will analyse the charging of capacitors. Figure 2 shows a capacitor, a resistor, and a switch all connected in series to a battery. The battery provides a constant e.m.f. V_0. The capacitor has capacitance C and the resistance of the resistor is R. The capacitor is initially uncharged and the switch is open.

When the switch is closed, there is a maximum current in the circuit and the capacitor starts to charge up. The potential difference (p.d.) across the capacitor starts to increase from zero as it gathers charge. According to Kirchhoff's second law, the p.d. V_R across the resistor and the p.d. V_C across the capacitor must always add up to V_0. So V_R must decrease as V_C increases with time. After a long time, depending on the time constant CR of the circuit, the capacitor will be fully charged with a p.d. of V_0 and V_R will be zero. When this happens, the current I in the circuit is zero.

Important equations

The current I in the circuit decreases exponentially and is given by the equation

$$I = I_0 e^{-\frac{t}{CR}}$$

where I_0 is the maximum current at $t = 0$. Since $V = IR$, the p.d. across the resistor V_R also decreases exponentially with respect to time and hence is given by the equation

$$V_R = V_0 e^{-\frac{t}{CR}}$$

At any time t, $V_0 = V_R + V_C$, therefore

$$V_C = V_0 - V_0 e^{-\frac{t}{CR}} \text{ or } V_C = V_0(1 - e^{-\frac{t}{CR}}).$$

Figure 3 shows the typical p.d.–time graphs for the capacitor-charging circuit in Figure 2.

Here are some important rules which you can use to analyse circuits where a capacitor is charged though a resistor.

- The p.d. V, current I, and resistance R of the resistor are related by the equation $V = IR$.

- The p.d. V, charge Q, and capacitance C of the capacitor are related by the equation $Q = VC$.
- The current in the circuit is given by the equation $I = I_0 e^{-\frac{t}{CR}}$.
- The equation $x = x_0 (1 - e^{-\frac{t}{CR}})$ may be used for the capacitor, where x can be either charge Q on the capacitor or p.d. V across the capacitor.
- At any time t, the p.d. across the components adds up to V_0. In other words, $V_0 = V_R + V_C$.

Worked example: Analysing a charging capacitor

In the circuit shown in Figure 2, $C = 500\,\mu F$, $R = 100\,k\Omega$, and $V_0 = 6.0\,V$. Calculate the charge stored by the capacitor at time $t = 20\,s$.

Step 1: Write down the quantities given in the question.

$C = 500 \times 10^{-6}\,F$, $R = 100 \times 10^{3}\,\Omega$, $V_0 = 6.0\,V$, $t = 20\,s$

Step 2: Calculate the time constant of the circuit.

$CR = 500 \times 10^{-6} \times 100 \times 10^{3} = 50\,s$

Step 3: Calculate the p.d. across the capacitor after time $t = 20$ s.

$V_C = V_0 (1 - e^{-\frac{t}{CR}}) = 6.0 \times (1 - e^{-\frac{20}{50}}) = 1.97...\,V$

Step 4: Calculate the charge on the capacitor using $Q = VC$.

$Q = 1.97... \times 500 \times 10^{-6} = 9.9 \times 10^{-4}\,C$ (2 s.f.)

The charge stored by the capacitor at time $t = 20\,s$ is $9.9 \times 10^{-4}\,C$.

Summary questions

1 A capacitor is connected to a battery using copper wires. Explain why the capacitor can be fully charged after a very short time. *(1 mark)*

2 A capacitor is charged up through a fixed resistor using a battery. State two quantities that will decrease exponentially with respect to time. *(2 marks)*

3 Explain how a voltmeter connected across the resistor in the circuit shown in Figure 2 and a stopwatch may be used to determine the time constant CR of the circuit. *(2 marks)*

4 A 120 μF capacitor is charged through a 1.0 MΩ resistor using a power supply of constant output of 3.0 V. The initial p.d. across the capacitor is zero. After a time equal to one time constant, calculate the p.d. across the capacitor. *(2 marks)*

5 For the circuit in question 4, calculate the p.d. across the capacitor at time $t = 3.0$ minutes. *(3 marks)*

6 An uncharged capacitor of capacitance C is charged through a resistor of resistance R using a battery of e.m.f. V_0. Derive an equation for the time t taken for the p.d. across the capacitor and the resistor to be the same. *(4 marks)*

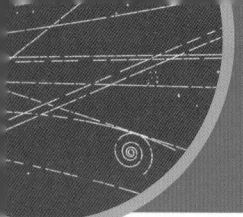

21.6 Uses of capacitors

Specification reference: 6.1.2

Learning outcomes

Demonstrate knowledge, understanding, and application of:

→ use of capacitors to store energy.

▲ **Figure 1** *Fermilab uses capacitors (blue and orange) capable of producing 13 megawatts of power*

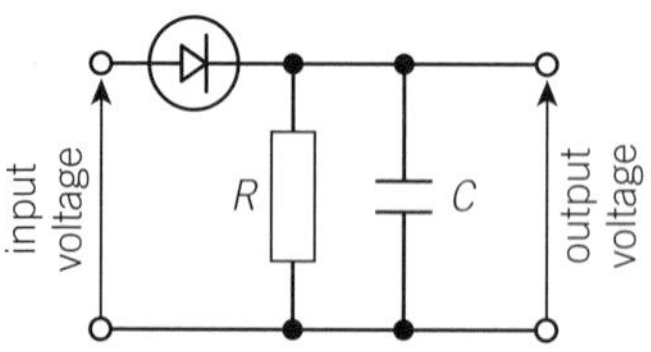

▲ **Figure 2** *(top) A rectifier circuit with a diode, resistor, and smoothing capacitor; (three graphs below) input and output voltages for the rectifier circuit*

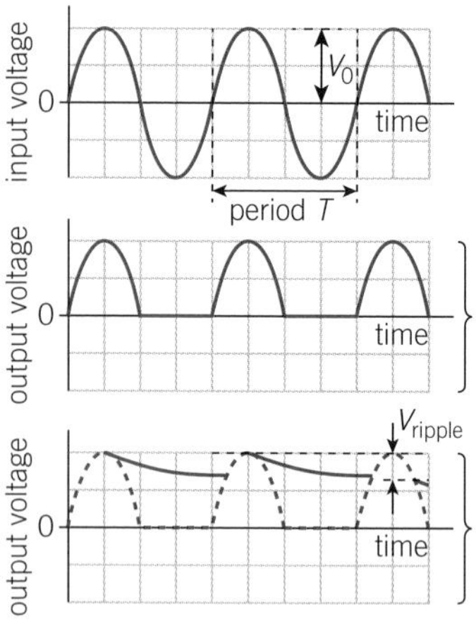

Power and capacitors

Capacitors are compact and can be charged easily to store energy. Unlike chemical cells, capacitors cannot store a great deal of energy in a small volume. However, capacitors can release the stored energy very quickly and thus generate high output power. This is exactly what happens in a camera flash. You can store about a joule of electrical energy in a normal capacitor, and when this is released in a time of 1 ms, you get a power output of about 1 kW.

Figure 1 shows specialist capacitors used at Fermilab in the USA to produce 13 MW of power for their particle accelerators. The accelerators at Fermilab have detected particles containing quarks.

Capacitors can be used to provide back-up power for computers and emergency lighting when the mains supply cuts out briefly.

Worked example: Flash

A cameral flash uses a 1.2 F capacitor and a cell of e.m.f. 1.5 V. Calculate the maximum output power from the flash in a discharge time of 1.1 ms.

Step 1: Write down the quantities given in the question.

$C = 1.2\,\text{F},\ V = 1.5\,\text{V},\ t = 1.1 \times 10^{-3}\,\text{s}$

Step 2: The maximum output power is the energy stored divided by the discharge time.

$$\text{power } P = \frac{\text{energy stored}}{\text{time}} = \frac{\frac{1}{2}V^2C}{t} = \frac{\frac{1}{2} \times 1.5^2 \times 1.2}{1.1 \times 10^{-3}} = 1.2... \times 10^3\,\text{W}$$

The power output is 1.2 kW (2 s.f.).

Smoothing capacitors

Domestic electricity is supplied as alternating current – the direction of the current changes rapidly and repeatedly as the supply voltage cycles from positive to negative and back. The simple rectifier circuit in Figure 2 changes an alternating input voltage to a smooth direct voltage. The diode allows current in one direction only. Without the capacitor, the output voltage from the circuit would consist of positive cycles only. With the capacitor, the output voltage is smoothed out and becomes almost direct voltage of constant value (Figure 2, right). The 'ripple' in the output voltage can be kept small by making the time constant of the circuit much greater than the period of the alternating voltage.

Ripples

For a conducting silicon diode with a threshold voltage of 0.7 V, the diode will allow the capacitor to charge up every time the alternating input voltage is greater than 0.7 V. As soon as the input voltage starts to decrease, the capacitor starts to discharge through the resistor. The rate of discharge depends on the time constant CR. The voltage V across the resistor is given by

$$V = V_0 e^{-\frac{t}{CR}}$$

As you can see from Figure 3, it is important that the time constant be much greater than the period of the input alternating voltage, otherwise the output voltage would not be very smooth.

The ripple voltage V_{ripple} is the difference between the maximum output voltage and the minimum output voltage. Therefore

$$V_{ripple} = V_0 - V_0 e^{-\frac{t}{CR}} = V_0 \left(1 - e^{-\frac{t}{CR}}\right)$$

When $t >> CR$, this equation approximates to

$$V_{ripple} = \frac{V_0 T}{CR}$$

where T is the period of the input alternating voltage.

1 Explain what would happen to output voltage from the rectifier circuit if the time constant were much smaller than the period of the input alternating voltage.
2 The period of the mains voltage is 20 ms and it has a peak voltage 340 V. Calculate the ripple voltage when a 1000 μF smoothing capacitor is used across a resistor of resistance 220 Ω.
3 Suggest suitable values for C and R for a ripple voltage that is 0.10% of the peak voltage for the mains voltage.

Summary questions

1 State two practical uses of capacitors. *(2 marks)*

2 The Z-machine at the Sandia National Laboratories in New Mexico, USA (Figure 3), uses high-voltage capacitors that are charged slowly and then discharged in 100 ns to produced 300 TW of power for their fusion research. Estimate the energy stored by the capacitors. *(2 marks)*

3 A 1000 μF capacitor is charged to 10 V. It is discharged through a filament lamp in 10 ms. Calculate the output power. *(3 marks)*

4 A 'square' voltage is applied to a rectifier circuit with a smoothing capacitor. The output is shown in Figure 4. Use Figure 4 to determine a value for the time constant of the resistor–capacitor network. *(4 marks)*

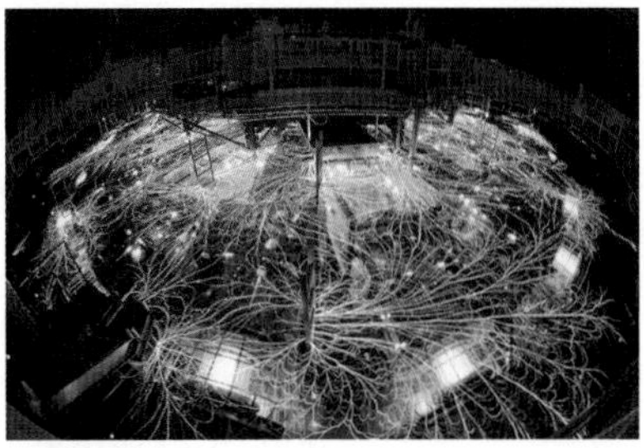

▲ **Figure 3** *The Z-machine*

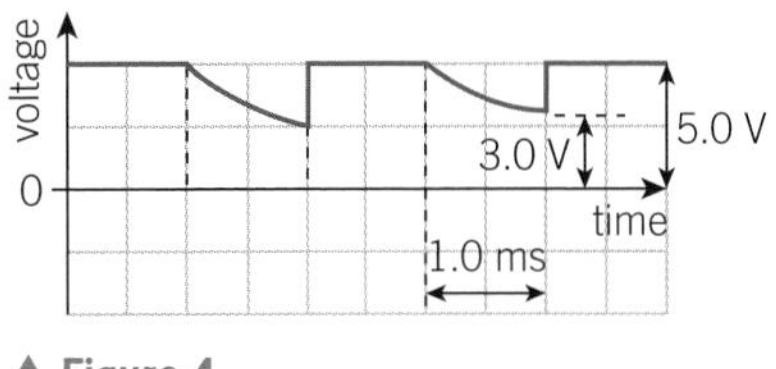

▲ **Figure 4**

Practice questions

1 a Figure 1 shows a circuit with a capacitor of capacitance C.

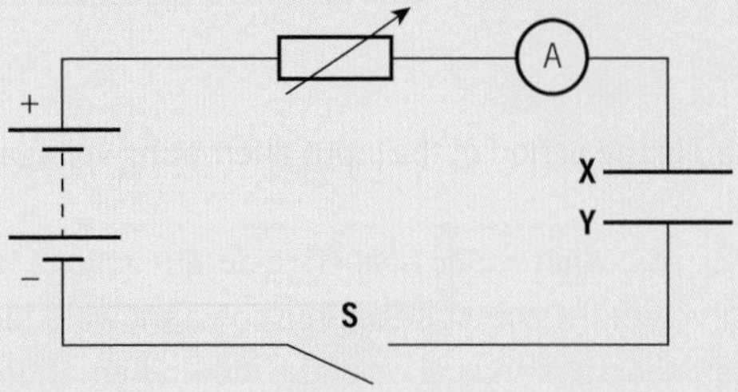

▲ **Figure 1**

The switch **S** is closed. The resistance of the variable resistor is manually adjusted so that the current in the circuit is kept **constant**.

(i) Explain in terms of movement of electrons how the capacitor plates **X** and **Y** acquire an equal but opposite charge. *(2 marks)*

(ii) The initial charge on the capacitor is zero. After 100 s, the potential difference across the capacitor is 1.6 V. The constant current in the circuit is 40 μA.

1 Calculate the capacitance C of the capacitor. *(3 marks)*

2 On a copy of Figure 2, sketch a graph to show the variation of potential difference V across the capacitor with time t. *(2 marks)*

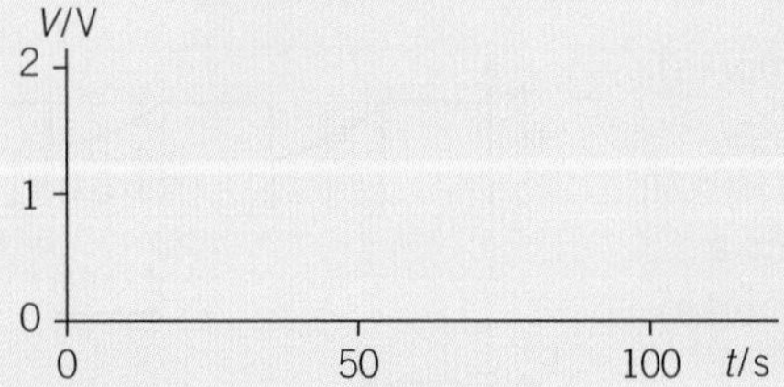

▲ **Figure 2**

b Figure 3 shows an arrangement used to determine the speed of a bullet.

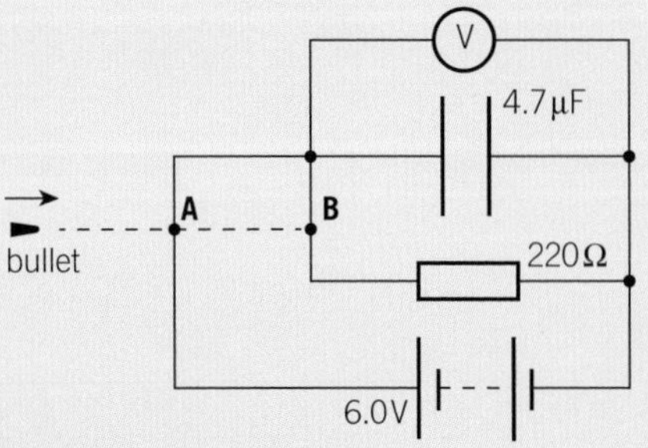

▲ **Figure 3**

The value of the resistance of the resistor and the value of the capacitance of the capacitor are shown in Figure 3. The voltmeter reading is initially 6.0 V. The bullet first breaks the circuit at **A**. The capacitor starts to discharge **exponentially** through the resistor. The capacitor stops discharging when the bullet breaks the circuit at **B**. The final voltmeter reading is 4.0 V.

(i) Calculate the time taken for the bullet to travel from **A** to **B**. *(3 marks)*

(ii) The separation between **A** and **B** is 0.10 m. Calculate the speed of the bullet. *(1 mark)*

Jan 2013 G485

2 Figure 4 shows a circuit being investigated by a student.

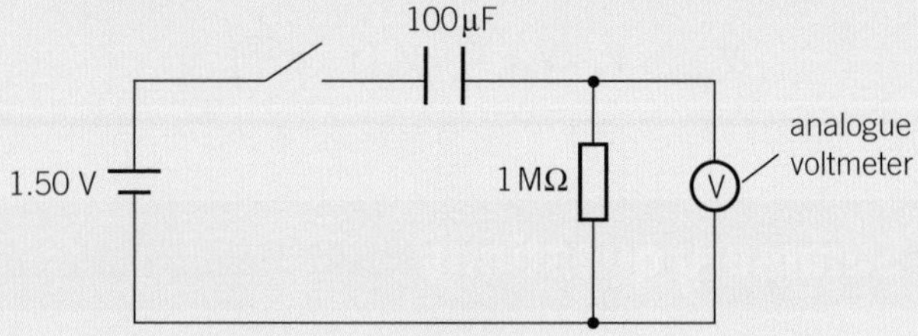

▲ **Figure 4**

The resistance and capacitance values of the components are shown in Figure 4. The voltmeter used by the student is an analogue one, that is, a moving-coil type. The cell has e.m.f. 1.50 V and has negligible internal resistance. The potential difference across the resistor is V_R at time t.

Table 1 shows data collected by the student.

▼ **Table 1**

t / s	V_R / V	ln (V_R / V)
0	1.49	
20	1.00	
40	0.65	
60	0.45	
80	0.32	
100	0.18	

a Use the first two columns of the table to validate that V_R decreases exponentially with respect to time. *(2 marks)*

b Complete the missing data in Table 1. *(1 mark)*

c Plot a graph of ln (V_R / V) against t and draw a straight line of best fit. *(3 marks)*

d Determine the gradient of the straight line drawn in **(c)** and hence determine the time constant of the circuit. *(4 marks)*

e Suggest why the time constant from **(d)** does not equal the product of the resistance of the resistor and the capacitance of the capacitor shown in Figure 4. *(2 marks)*

3 a Define *capacitance*. *(1 mark)*

b Figure 5 shows an arrangement of three identical capacitors connected to a 6.0 V battery.

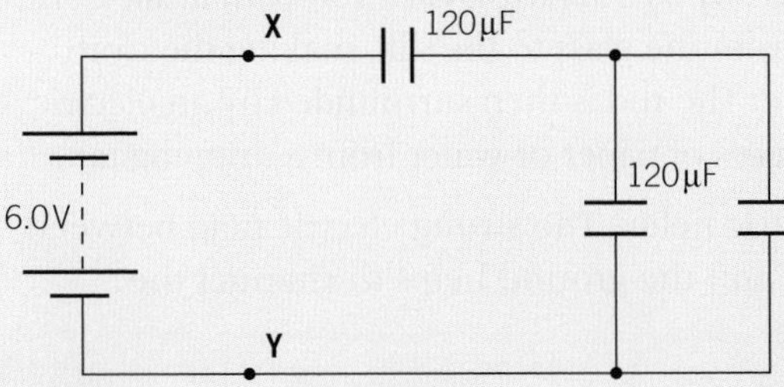

▲ Figure 5

Each capacitor has a capacitance of 120 μF.

(i) Show that the total capacitance of the circuit is 80 μF. *(2 marks)*

(ii) Calculate the total energy stored by the capacitors. *(2 marks)*

(iii) The battery is disconnected from the circuit shown in Figure 5. The p.d. between points **X** and **Y** remains at 6.0 V. A fixed resistor of resistance *R* is now connected between points **X** and **Y**. Figure 6 shows the variation of the p.d. across the resistor with time *t*.

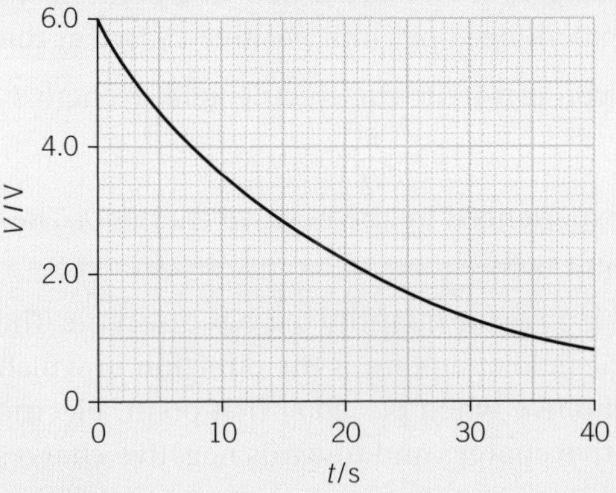

▲ Figure 6

1 Use Figure 6 to show that the circuit has a time constant of 20 s. *(1 mark)*

2 Hence, calculate the resistance *R* of the resistor. *(2 marks)*

Jan 2012 G485

4 a Define *capacitance*. *(1 mark)*

b Figure 7 shows a circuit consisting of a resistor and a capacitor of capacitance 4.5 μF.

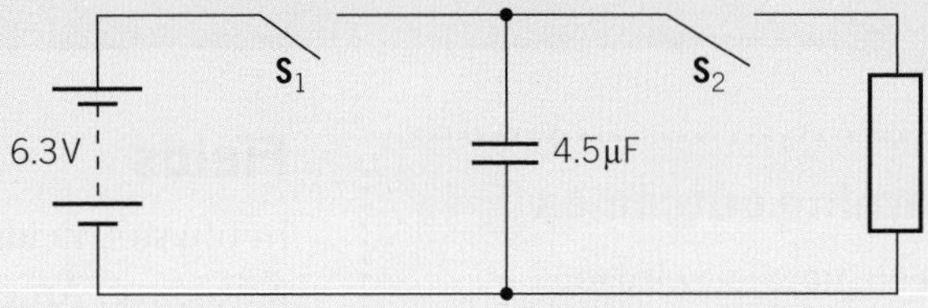

▲ Figure 7

Switch $\mathbf{S_1}$ is closed and switch $\mathbf{S_2}$ is left open. The potential difference across the capacitor is 6.3 V.

Calculate

(i) the charge stored by the capacitor *(1 mark)*

(ii) the energy stored by the capacitor. *(2 marks)*

c Switch $\mathbf{S_1}$ is opened and switch $\mathbf{S_2}$ is closed.

(i) Describe and explain in terms of the movement of electrons how the potential difference across the capacitor changes. *(3 marks)*

(ii) The energy stored in the capacitor decreases to zero. State where the initial energy stored in the capacitor is dissipated. *(1 mark)*

d Figure 8 shows the 4.5 μF capacitor now connected in parallel with a capacitor of capacitance 1.5 μF. Both switches are open and both capacitors are uncharged.

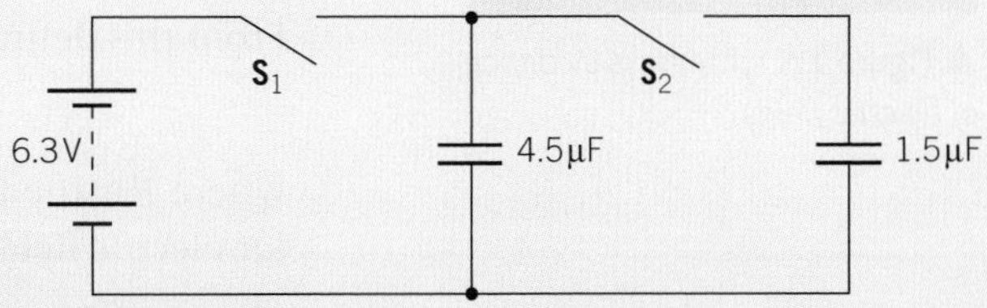

▲ Figure 8

Switch $\mathbf{S_1}$ is closed. The potential difference across the 4.5 μF capacitor is now 6.3 V. Switch $\mathbf{S_1}$ is opened and then switch $\mathbf{S_2}$ is closed.

(i) Calculate the total capacitance of the circuit when $\mathbf{S_2}$ is closed. *(1 mark)*

(ii) Calculate the final potential difference across the capacitors. *(2 marks)*

22 ELECTRIC FIELDS
22.1 Electric fields

Specification reference: 6.2.1

Learning outcomes

Demonstrate knowledge, understanding, and application of:

- → electric fields being due to charges
- → a uniformly charged sphere modelled as a point charge at its centre
- → electric field lines to map electric fields
- → electric field strength: $E = \frac{F}{Q}$.

▲ **Figure 1** *Lightning is evidence of electric fields*

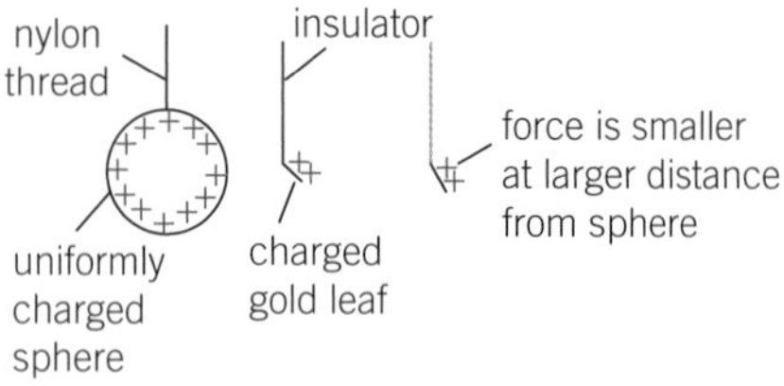

▲ **Figure 2** *The strength an electric field can be tested using a charged gold leaf – the angle it makes with the insulator indicates the force it experiences*

Fields

In physics, fields are regions in which an object will experience a force at a distance. An electric (or electrostatic) field is created by charged objects. Other charged objects in this electric field will experience force. You can create an electric field very easily by rubbing a glass rod with a silk cloth. Friction transfers electrons from the glass to the silk, making the cloth negative and the rod positive. The rod is then surrounded by an electric field – it will attract small pieces of paper or water from a dripping tap.

Nature creates its own electric fields. The strong electric field between the base of a thundercloud and the ground helps to channel the electric current of lightning.

Detecting electric fields

Figure 2 shows a positively charged metal ball. There must be an electric field around this ball. You can test the presence and the strength of the electric field using a thin strip of gold foil attached to the bottom of an insulator. The gold foil is given a constant positive charge by momentarily touching it to the charged ball. The charged foil experiences an electrostatic force when close to the ball. This force is smaller the further away the foil is from the charged ball, that is – the field created by the ball is stronger closer to the ball.

Electrons and protons are charged particles. Both create electric fields and so affect each other.

Electric field strength

The **electric field strength** of an electric field at a point in space is defined as the force experienced per unit positive charge at that point.

From this definition, you can write the electric field strength E as

$$E = \frac{F}{Q}$$

where F is the force experienced by the positive charge Q. The SI unit for electric field strength is NC^{-1}.

Electric field strength is a vector quantity – it has direction. The direction of the electric field at a point is the direction in which a positive charge would move when placed at that point. Electric fields point away from positive charges and towards negative charges.

Worked example: Calculating field strength

A negatively charged dust particle moves from left to right in a uniform electric field. It has a charge of magnitude 5.0×10^{-14} C and experiences a constant force of 8.6×10^{-10} N. Calculate the electric field strength and state the direction of the electric field. ➔

Step 1: The direction of the electric field is the direction in which a positive charge would move – that is, a negative charge will move in the opposite direction. Hence the direction of the electric field is from right to left.

Step 2: Write down the quantities given in the question.

$$F = 8.6 \times 10^{-10}\,\text{N},\ Q = 5.0 \times 10^{-14}\,\text{C},\ E = ?$$

Step 3: Use the equation for electric field to calculate E.

$$E = \frac{F}{Q} = \frac{8.6 \times 10^{-10}}{5.0 \times 10^{-14}} = 1.7 \times 10^{4}\,\text{N C}^{-1}\ \text{(2 s.f.)}$$

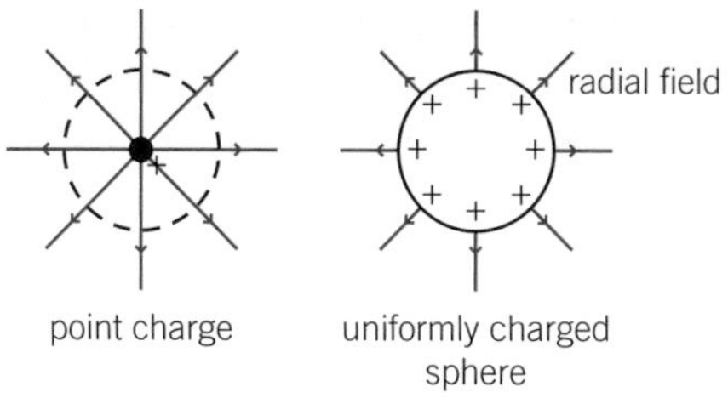

▲ **Figure 3** *A point charge and a uniformly charged sphere both produce radial fields – the field pattern for the charged sphere is the same as that for the point charged beyond the dotted sphere*

Electric field patterns

We use electric field lines (or lines of force) to map electric field patterns. These are, visual patterns that are very helpful in deducing the nature of the electric field created by individual charges and charges on conductors. Here are some important ideas:

- The arrow on an electric field line shows the direction of the field.
- Electric field lines are always at right angles to the surface of a conductor.
- Equally spaced, parallel electric field lines represent a uniform field – one in which the electric field strength is the same everywhere.
- Closer electric field lines represent greater electric field strength.

Figure 3 shows the electric field patterns for a point charge and a uniformly charged metal sphere. The electric field is radial in both cases. The field strength decreases with distance from the centre. You can model the uniformly charged sphere as a point charge at its centre. Figure 4 shows the electric field patterns for some charged conductors.

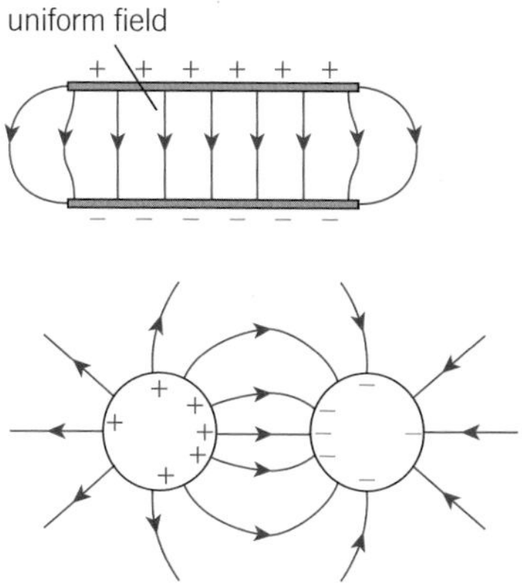

▲ **Figure 4** *The electric field patterns for two oppositely charged plates, which produce a uniform field between the plates, except close to the edges (left), and for two oppositely charged metal spheres*

Summary questions

1 The electric field strength is constant close to a conductor. State the field pattern produced by the electric field lines. *(1 mark)*

2 Figure 5 shows a field pattern correctly drawn by a student, but with the charges on the sphere and plate omitted. State the charges on the sphere and the plate. *(1 mark)*

3 Suggest how the field pattern would change if the sphere shown in Figure 3 had greater charge. *(2 marks)*

4 Calculate the force experienced by an electron in an electric field of field strength $4.0 \times 10^{5}\,\text{N C}^{-1}$. *(2 marks)*

5 A proton is accelerated from rest by a uniform electric field of field strength $2.0 \times 10^{5}\,\text{N C}^{-1}$. Assume mass of proton $= 1.7 \times 10^{-27}$ kg. Calculate the time it takes to travel a distance of 5.0 cm from its starting point. *(5 marks)*

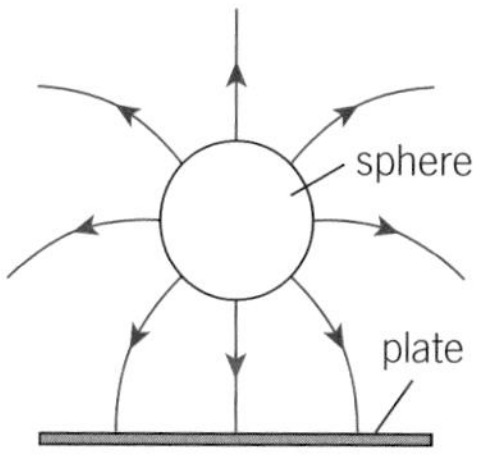

▲ **Figure 5**

22.2 Coulomb's law

Specification reference: 6.2.1, 6.2.2

Learning outcomes

Demonstrate knowledge, understanding, and application of:

- → a uniformly charged sphere modelled as a point charge at its centre
- → Coulomb's law, $F = \frac{Qq}{4\pi\varepsilon_0 r^2}$, for the force between two point charges
- → electric field strength $E = \frac{Q}{4\pi\varepsilon_0 r^2}$ for a point charge
- → similarities and differences between the gravitational field of a point mass and the electric field of a point charge.

Electrostatic forces

Charles Augustin Coulomb (1736–1806), a French physicist, formulated a law analogous to Newton's law of gravitation, but for charged particles. According to **Coulomb's law**, any two point charges exert an electrostatic (electrical) force on each other that is directly proportional to the product of their charges and inversely proportional to the square of the distance between them. This law applies to everything from large charged spheres down to microscopic atoms within all of us. Figure 1 is an image of palladium atoms deposited onto a graphite layer. The forces between the palladium atoms and the carbon atoms can be calculated using the equation for Coulomb's law.

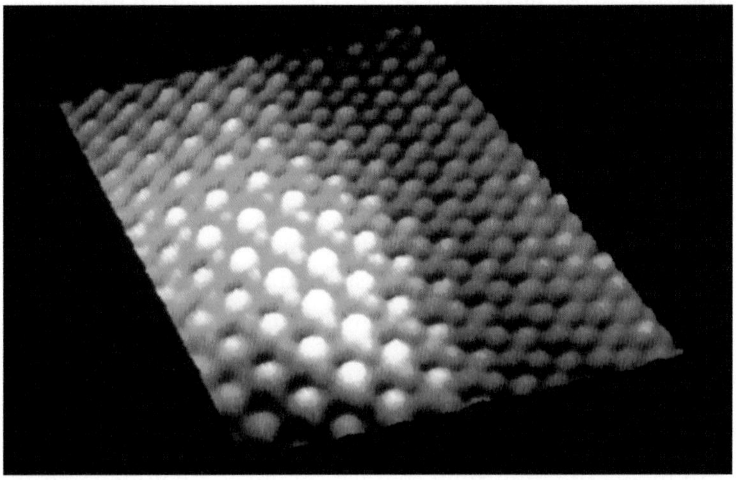

▲ **Figure 1** *Palladium atoms (white) spaced 0.4 nm apart on a graphite lattice (blue) in an image from a scanning tunnelling microscope – the forces between the atoms can be calculated using Coulomb's law*

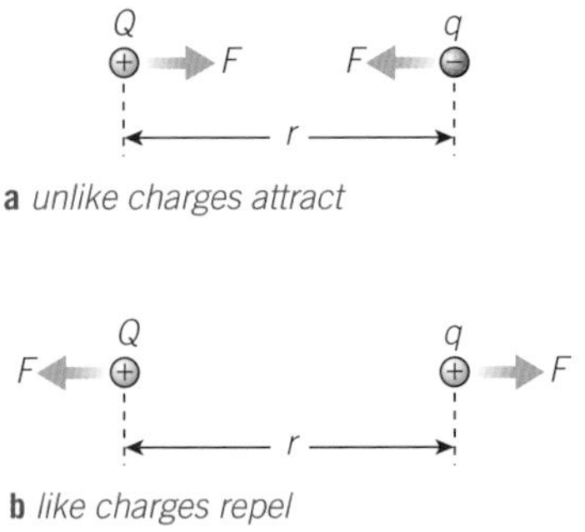

▲ **Figure 2** *Forces between point charges*

Synoptic link

You met Newton's third law in Topic 7.1, Newton's first and third laws of motion.

Coulomb's law

Figure 2 shows two like point charges and two unlike point charges with a separation r. The magnitudes of the charges are Q and q. The electrostatic force experienced by each point charge is F. The point charges interact, and according to Newton's third law will exert equal but opposite forces on each other. From the statement of Coulomb's law, we have

$$F \propto Qq$$

and

$$F \propto \frac{1}{r^2}$$

We can write this using the equation:

$$F = k\frac{Qq}{r^2}$$

where k is the constant of proportionality. This constant can be written in terms of the permittivity of free space ε_0 (pronounced epsilon-nought).

You will learn more about permittivity in Topic 22.3. ε_0 is equal to $8.85 \times 10^{-12}\,\mathrm{F\,m^{-1}}$.

$$k = \frac{1}{4\pi\varepsilon_0}$$

Thus the equation for Coulomb's law, which can be applied to any point charges, is written as

$$F = \frac{Qq}{4\pi\varepsilon_0 r^2}$$

Investigating Coulomb's law

As you saw in Topic 22.1, Electric fields, a uniformly charged sphere can be treated as if it is a point charge. You can therefore apply Coulomb's law to large uniformly charged spheres too. You just need to remember to take the separation r to be the separation between the centres of the two spheres.

Figure 3 shows a technique for investigating the forces between two charged metal spheres (in this case, table-tennis balls coated in conductive paint). The balls can be charged positively by touching each momentarily to the positive electrode of a high-tension supply. Lowering one of the charged spheres down towards the other will produce a larger reading on the balance.

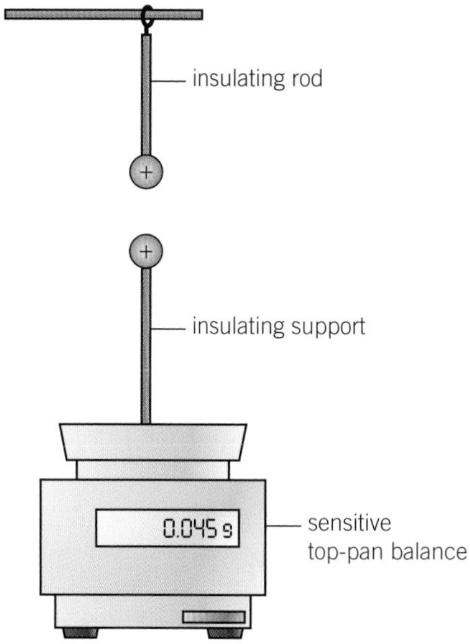

▲ **Figure 3** *Investigating forces between charged spheres*

Worked example: Forces between charged spheres

Two metal spheres of radii 1.8 cm and 2.5 cm are given charges of +4.0 nC and −5.2 nC respectively. Calculate the maximum force between these charged spheres.

Step 1: List the quantities you already know.

$r_1 = 1.8 \times 10^{-2}\,\mathrm{m}$, $r_2 = 2.5 \times 10^{-2}\,\mathrm{m}$,
$Q = 4.0 \times 10^{-9}\,\mathrm{C}$, $q = -5.2 \times 10^{-9}\,\mathrm{C}$

Step 2: The maximum force will be for the smallest separation possible, so when the separation r between the centres of the spheres is $(r_1 + r_2) = 4.3 \times 10^{-2}\,\mathrm{m}$, that is, almost touching but not quite (Figure 4).

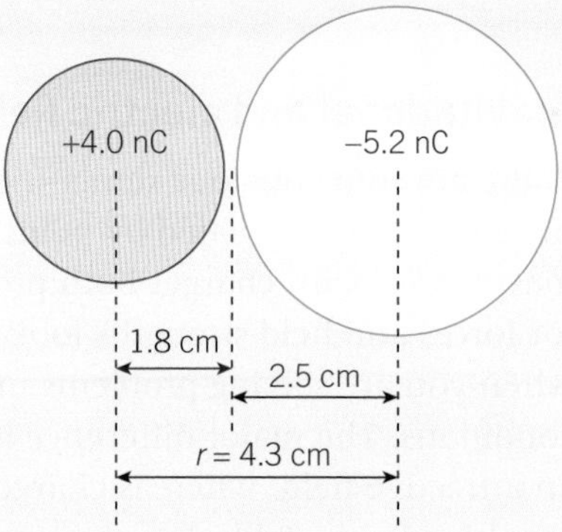

▲ **Figure 4**

Step 3: Use the equation for Coulomb's law to calculate the force.

$$F = \frac{Qq}{4\pi\varepsilon_0 r^2} = \frac{(4.0 \times 10^{-9}) \times (-5.2 \times 10^{-9})}{4\pi \times 8.85 \times 10^{-12} \times (4.3 \times 10^{-2})^2}$$

$F = -1.0 \times 10^{-4}\,\mathrm{N}$ (2 s.f.)

The negative sign simply means that the force between these oppositely charged spheres is attractive.

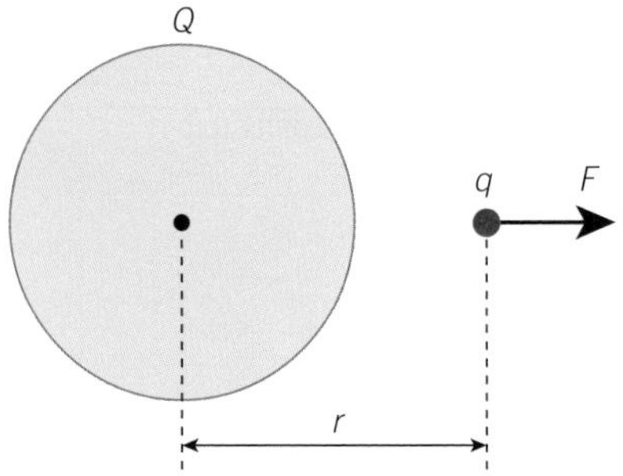

▲ **Figure 5** *What is the electric field strength at a distance r from the centre of the sphere?*

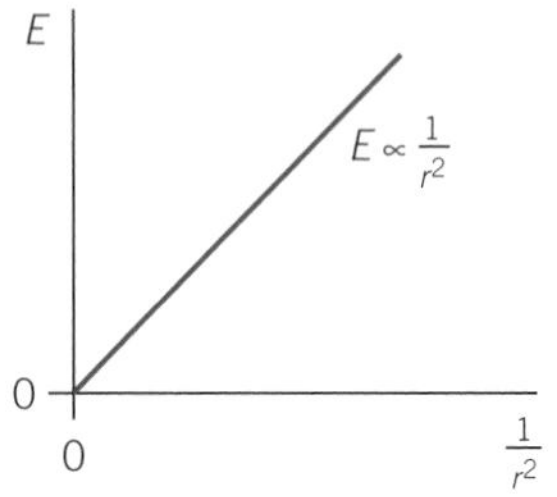

▲ **Figure 6** *A graph of E against $\frac{1}{r^2}$. What is the gradient of this straight line through the origin?*

Radial fields

Figure 5 shows a metal sphere with a charge $+Q$. You already know that the sphere produces a radial field, and that the separation between two adjacent electric field lines increases with the distance from the centre of the sphere. In other words the electric field strength decreases as you move further away from the centre of the sphere.

The electric field strength E at a distance r from the centre of the sphere is equal to the force F experienced by a positive 'test' charge divided by the charge q on the test charge. Therefore

$$E = \frac{F}{q} = \frac{Q\cancel{q}}{4\pi\varepsilon_0 r^2 \cancel{q}} = \frac{Q}{4\pi\varepsilon_0 r^2}$$

The electric field strength is thus directly proportional to the charge Q and obeys an inverse square law with distance r. A graph of E against $\frac{1}{r^2}$ will be a straight line through the origin (Figure 6).

Worked example: Intense electric field

The radius of a gold nucleus is about 7.0×10^{-15} m and it has 79 protons. Estimate the electric field strength at the surface of the nucleus.

Step 1: The charge on a single proton is 1.60×10^{-19} C. Use this to calculate the charge on the gold nucleus.

charge $Q = 79 \times 1.60 \times 10^{-19}\,\text{C} = 1.264 \times 10^{-17}\,\text{C}$

Step 2: Use the equation for the electric field strength to calculate E.

$r = 7.0 \times 10^{-15}\,\text{m}$, $Q = 1.264 \times 10^{-17}\,\text{C}$

$$E = \frac{Q}{4\pi\varepsilon_0 r^2} = \frac{1.264 \times 10^{-17}}{4\pi \times 8.85 \times 10^{-12} \times (7.0 \times 10^{-15})^2}$$

$$E = 2.3 \times 10^{21}\,\text{N}\,\text{C}^{-1} \text{ (2 s.f.)}$$

Gravitational and electric fields

There are some obvious similarities between gravitational fields and the electric fields created by point particles or spherical objects. Point masses and point charges both produce radial fields. The equations for forces and field strengths look similar, too. You need to be vigilant when you are solving problems under the pressure of examination conditions. The major difference is that masses always produce an attractive field, whereas charges can create both attractive and repulsive fields. Table 1 summarises the key facts about point (or spherical) masses and point (or spherical) charges.

Synoptic link

To review gravitational fields, look back at Chapter 18, Gravitational fields.

▼ **Table 1** *Comparison of gravitational and electric fields of particles (point masses and point charges)*

	Gravitational field	Electric field
property that creates the field	mass	charge
type of field produced	always attractive (direction of field always towards object)	positive point charges produce a repulsive field (direction of field away from object, repels a positive charge) negative point charges produce an attractive field (direction of field towards object, attracts a positive charge)
field strength	gravitational field strength is the force per unit mass $g = \frac{F}{m}$	electric field strength is the force per unit positive charge $E = \frac{F}{Q}$
force between particles	force ∝ product of masses $\text{force} \propto \frac{1}{\text{separation}^2}$	force ∝ product of charges $\text{force} \propto \frac{1}{\text{separation}^2}$
force and field strength equations	$F = -\frac{GMm}{r^2}$ $g = -\frac{GM}{r^2}$	$F = \frac{Qq}{4\pi\varepsilon_0 r^2}$ $E = \frac{Q}{4\pi\varepsilon_0 r^2}$
type of field	point masses produce a radial field	point charges produce a radial field

Summary questions

1 State one major difference between electric and gravitational fields produced by particles. *(1 mark)*

2 Calculate a value for $k = \frac{1}{4\pi\varepsilon_0}$ in SI units. *(2 marks)*

3 Two point charges of 2.0 nC are separated by a distance of 1.0 cm. Calculate the electrostatic force experienced by each charge. *(2 marks)*

4 Without detailed calculation, state and explain how your answer to question 3 would change when the same charges are separated by 3.0 cm. *(2 marks)*

5 The electric field at a distance of 5.0 cm from the centre of a charged dome (hemisphere) is $3.0 \times 10^4\,\text{N C}^{-1}$. Calculate the charge Q on the dome. *(3 marks)*

6 Calculate the force between a helium nucleus and a gold nucleus at a distance of 2.0×10^{-14} m. A gold nucleus has 79 protons and a helium nucleus has 2 protons. *(3 marks)*

7 Two electrons are 3.0×10^{-10} m apart.

Calculate the ratio: $\frac{\text{electrostatic force } F \text{ on electron}}{\text{weight of electron}}$ *(4 marks)*

22.3 Uniform electric fields and capacitance

Specification reference: 6.2.3

Learning outcomes

Demonstrate knowledge, understanding, and application of:

→ uniform electric field strength, $E = \frac{V}{d}$

→ parallel-plate capacitor; permittivity; $C = \frac{\varepsilon_0 A}{d}$, $C = \frac{\varepsilon A}{d}$, $\varepsilon = \varepsilon_r \varepsilon_0$

Smart windows

Some calculators and digital clocks use liquid crystal displays (LCDs). In these displays, the figures are formed by applying electric fields across the displays. Smart windows use a similar technology – see figure 1. When there is a current across the conductive layer, the liquid crystals in the conductive layer respond to the current's uniform electric field and align themselves parallel to the field. This allows light to pass through the gaps between the crystals. When the electric field is removed, the liquid crystals go back to their randomly orientated state and the window darkens.

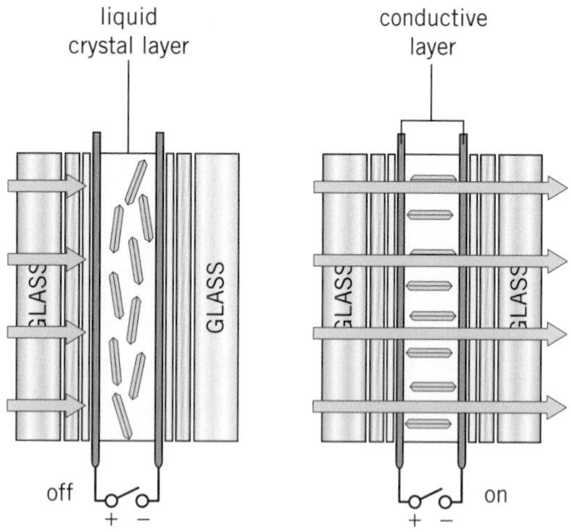

▲ **Figure 1** *The liquid crystal layer inside a smart window has a conductive layer on each side of it that acts like a parallel plate capacitor providing a uniform electric field*

Electric field between two parallel plates

You have already seen in Topic 22.1, Electric fields, that two oppositely charged parallel plates produce a uniform electric field in the region between the plates. The electric field strength E for this arrangement is uniform and is related to the p.d. V across the plates and their separation d. Figure 2 shows a small test charge Q between the plates. It experiences a constant force F given by the equation $F = EQ$ (we defined this equation in Topic 22.1, Electric fields). The charge will gain energy as it moves from the positive plate to the negative plate. According to the definition of p.d., V is equal to the work done per unit charge, and work done W on the charge is the product of the force F and the distance d. Therefore

$$W = Fd$$

or

$$VQ = (EQ)d$$

The charge Q cancels out and the equation for the magnitude of the electric field strength E is given by

$$E = \frac{V}{d}$$

This equation only works for parallel plates. It is useful in calculations, because you just need a voltmeter to measure V and a ruler to measure d in order to determine the electric field strength.

The unit for electric field strength is $\mathrm{N\,C^{-1}}$, but this equation shows that you can also use $\mathrm{V\,m^{-1}}$.

$$1\,\mathrm{N\,C^{-1}} = 1\,\mathrm{V\,m^{-1}}$$

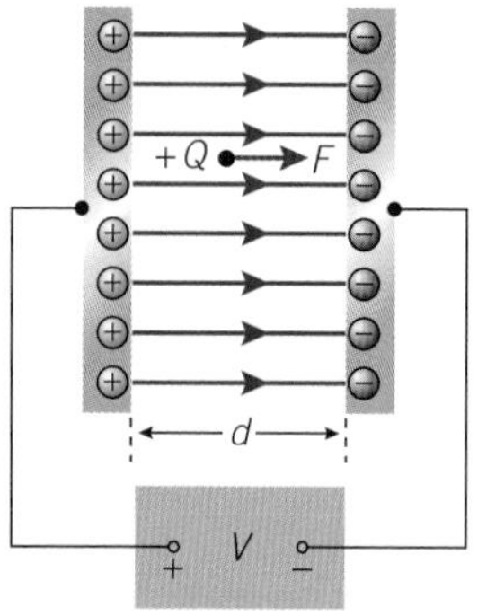

▲ **Figure 2** *Two parallel plates with opposite charges produce a uniform field*

Worked example: Accelerating particles

Figure 3 shows an arrangement used to accelerate electrons. The separation between the plates is 1.2 cm and the p.d. across the plates is 3.6 kV. Calculate the acceleration of an electron between the plates.

Step 1: Write down the quantities given in the question.

$$d = 1.2 \times 10^{-2}\,\text{m},\ V = 3.6 \times 10^{3}\,\text{V}$$

Step 2: Calculate the electric field strength between the plates.

$$E = \frac{V}{d} = \frac{3.6 \times 10^{3}}{1.2 \times 10^{-2}} = 3.0 \times 10^{5}\,\text{V m}^{-1}$$

Step 3: Calculate the force acting on the electron.

$$F = EQ = 3.0 \times 10^{5} \times 1.60 \times 10^{-19} = 4.8 \times 10^{-14}\,\text{N}$$

Step 4: Use $F = ma$ to calculate the acceleration.

$$a = \frac{F}{m} = \frac{4.8 \times 10^{-14}}{9.11 \times 10^{-31}} = 5.3 \times 10^{16}\,\text{m s}^{-2}\ (2\ \text{s.f.})$$

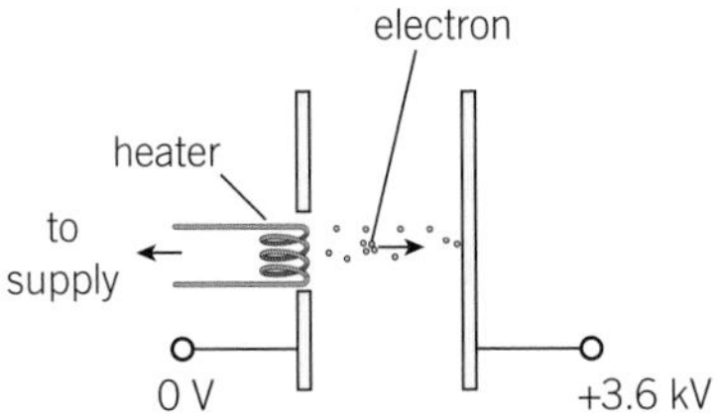

▲ **Figure 3** *Accelerating electrons*

Study tip

Electric field strength is defined by the equation $E = \frac{F}{Q}$ and not by $E = \frac{V}{d}$

Parallel plate capacitor

The capacitance of a parallel plate capacitor depends on the separation d between the plates, the area A of overlap between the plates, and the insulator (dielectric) used between the plates. For plates in a vacuum (or air), experiments show that the capacitance is proportional to the area ($C \propto A$) and inversely proportional to the separation between the plates ($C \propto \frac{1}{d}$). Therefore

$$C \propto \frac{A}{d}$$

The constant of proportionality in this relationship is the permittivity of free space ε_0, which you have already met in Topic 22.2, Coulomb's law. The equation for capacitance for a parallel plate capacitor is thus given by the equation

$$C = \frac{\varepsilon_0 A}{d}$$

When an insulator (or dielectric) other than a vacuum is used between the plates, the equation for capacitance uses ε, the permittivity for the insulator. This permittivity is always greater than ε_0, so sometimes we use the term relative permittivity ε_r (which has no units).

$$\varepsilon = \varepsilon_r \varepsilon_0$$

Therefore, the equation for capacitance may be written as

$$C = \frac{\varepsilon A}{d}$$

Table 1 lists the relative permittivities of some materials used in capacitors.

▼ **Table 1** *Relative permittivities ε_r for dielectrics used in capacitors*

Material	ε_r
vacuum	1 (by definition)
air	1.0006
perspex	3.3
paper	4.0
mica	7.0
barium titanate	1200

Worked example: A bin-bag capacitor

A capacitor is made from a plastic bin bag sandwiched between two sheets of aluminium foil 60 cm × 30 cm in size. The bin bag is 0.080 mm thick. Plastic has a relative permittivity of 4.0. Calculate the capacitance of this capacitor.

Step 1: Write down the quantities given in the question.

$$L = 0.60\,\text{m},\ x = 0.30\,\text{m},\ d = 0.080 \times 10^{-3}\,\text{m},\ \varepsilon_r = 4.0$$

Step 2: Substitute these values into the capacitance equation.

$$C = \frac{\varepsilon_0 \varepsilon_r A}{d} = \frac{8.85 \times 10^{-12} \times 4.0 \times (0.60 \times 0.30)}{0.080 \times 10^{-3}}$$

$$C = 7.965 \times 10^{-8}\,\text{F} = 8.0 \times 10^{-8}\,\text{F}\ (80\,\text{nF})\ (2\text{ s.f.})$$

Permittivity of free space ε_0

The charge stored on a capacitor made from large parallel plates can be measured directly by discharging it into a coulombmeter. Figure 4 shows an arrangement for determining the permittivity of free space. It can be adapted to determine the permittivity of any insulator placed between the plates.

Using the flying lead, the capacitor is charged to a p.d. of V. The charge Q stored by the capacitor is measured by tapping the flying lead to the plate of the coulombmeter.

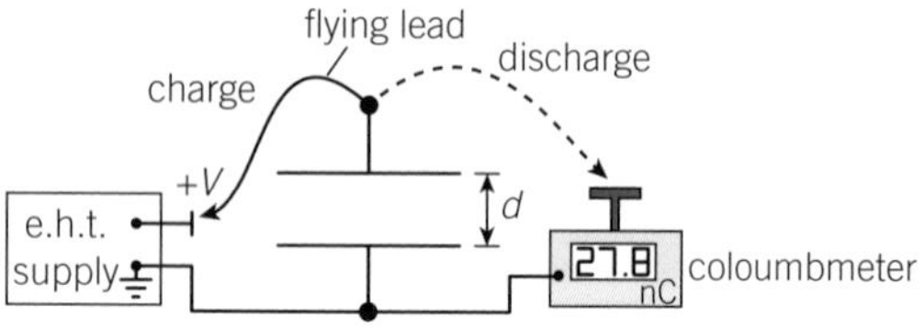

▲ **Figure 4** *Apparatus to determine permittivity*

Some typical results from one experiment are shown in Table 2.

diameter of each plate = 20.0 cm

separation between plates = 2.5 cm

▼ **Table 2** *Charge Q recorded by a coulombmeter for a capacitor charged to p.d. V*

V / V	Q / nC
500	5.6
1000	11.2
1500	16.7
2000	22.2
2500	27.8
3000	33.7

1. Plot a graph of Q against V.
2. Draw a straight best-fit line and determine the gradient of the line.
3. State what the gradient of the line represents.
4. Use the gradient to determine the permittivity of free space.

Millikan's experiment and quantisation of charge

Figure 5 shows an arrangement used by Robert Millikan to determine the elementary charge e. Tiny electrically charged oil droplets were observed through a microscope. By altering the electric field strength between the capacitor plates, he was able to hold individual droplets stationary. For a stationary droplet,

$$mg = EQ$$

where Q is the charge of the droplet, m is the mass of the droplet, g is the acceleration of free fall, and E is the electric field strength.

Millikan found that the charge Q on the droplets was quantised, that is, $Q = \pm ne$, where n is an integer. The droplets acquire charge through friction at the nozzle of the spray, by either losing or gaining electrons.

Millikan was awarded the Nobel prize in physics in 1923 for his work on determining the elementary charge and the photoelectric effect. You first encountered Millikan's oil-drop experiment in Topic 8.1, Current and charge.

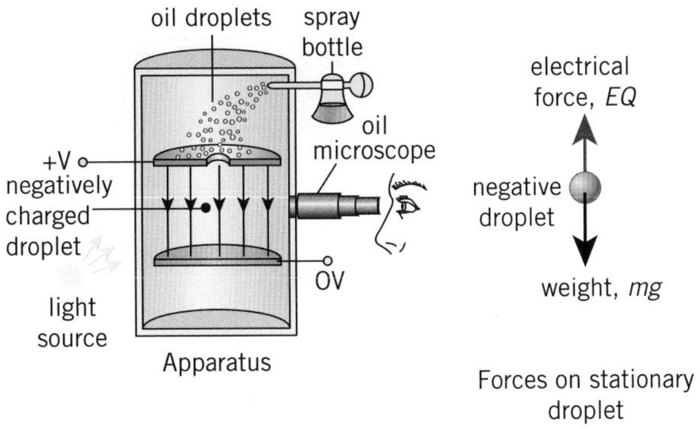

▲ **Figure 5** *The arrangement used by Millikan to investigate charges on oil droplets*

The radius r of the droplet is required in order to determine its mass, but it is impossible to measure r directly, because the droplets appear only as tiny specks of light in the microscope. Millikan's method for determining r was imaginative. The droplet falls vertically through the air at its terminal velocity when the electric field is switched off. The drag force F_d on the droplet is given by the equation

$$F_d = 6\pi\eta rv$$

where η is the viscosity of the air (about $1.8 \times 10^{-5}\,\text{N s m}^{-2}$).

At the terminal velocity, the drag force is equal to the weight of the droplet. Therefore

$$mg = 6\pi\eta rv.$$

The mass m of the droplet is equal to its volume multiplied by the density ρ of oil, hence

$$\frac{4\pi r^3 \rho g}{3} = 6\pi\eta rv$$

$$r^2 = \frac{9\eta v}{2\rho g}.$$

1 The density of oil is about $900\,\text{kg m}^{-3}$. Calculate the radius r of an oil droplet falling through a vertical distance of 5.0 mm in a time of 44 s.

2 Calculate the mass of this droplet.

3 The charged droplet was held stationary when the p.d. across the plates was 1.2 kV and the separation was 2.0 cm. Calculate the charge on the droplet and hence the number of electrons responsible for this charge.

Summary questions

1 State two SI units for electric field strength. *(1 mark)*

2 Calculate the electric field strength between two parallel plates separated by a distance of 1.0 cm and with a p.d. of 1.0 kV. *(2 marks)*

3 Calculate the acceleration of a proton between the plates in question 2.
Assume mass of proton = 1.7×10^{-27} kg *(3 marks)*

4 The capacitance of a parallel plate capacitor is 8.0 pF. State and explain its capacitance when:

a the separation between the plates is doubled *(2 marks)*

b the area of overlap between the plates is halved and their separation is also halved. *(2 marks)*

5 A capacitor is made from two circular plates of diameter 20 cm that are separated by an insulator. The insulator has relative permittivity of 4.0 and a thickness of 1.2 mm. The capacitor is connected to a 6.0 V battery Calculate the maximum charge stored by the capacitor. *(4 marks)*

6 An oil droplet of charge $2e$ is suspended between two charged parallel plates. The mass of the droplet is 2.5×10^{-15} kg. The separation between the plates is 1.2 cm. Calculate the p.d. across the plates. *(4 marks)*

22.4 Charged particles in uniform electric fields

Specification reference: 6.2.3

Learning outcomes

Demonstrate knowledge, understanding, and application of:

→ motion of charged particles in a uniform electric field.

Approaching light speed

Charged particles, such as electrons, can be accelerated by electric fields. A moderate electric field produced by a 1.5 V cell can accelerate electrons to speeds of about 700 km s^{-1}. Imagine what can be done with greater accelerating voltages – a linear particle accelerator uses electric fields to accelerate protons to speeds close to the speed of light, with kinetic energies of about 400 MeV (Figure 1).

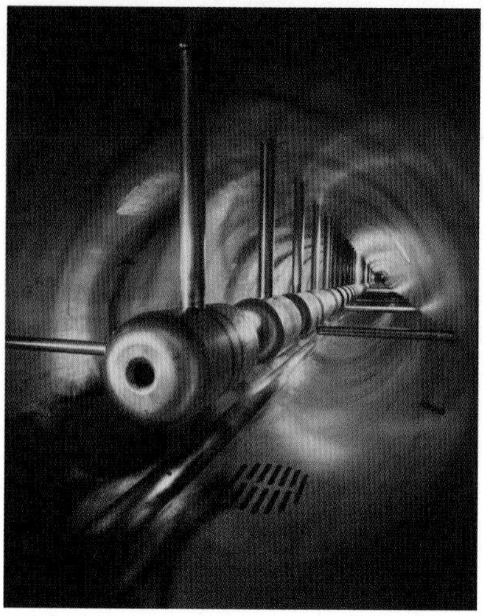

▲ **Figure 1** *Electric fields are used to accelerate protons to high speeds in this particle accelerator at the Fermi laboratory near Chicago in the USA*

Acceleration

Figure 2 shows two oppositely charged horizontal plates. Being negatively charged, an electron between the plates will travel away from the negative plate towards the positive plate, in the opposite direction of the electric field. The electron experiences a constant electrostatic force because of the uniform electric field between the plates, so it has a constant acceleration.

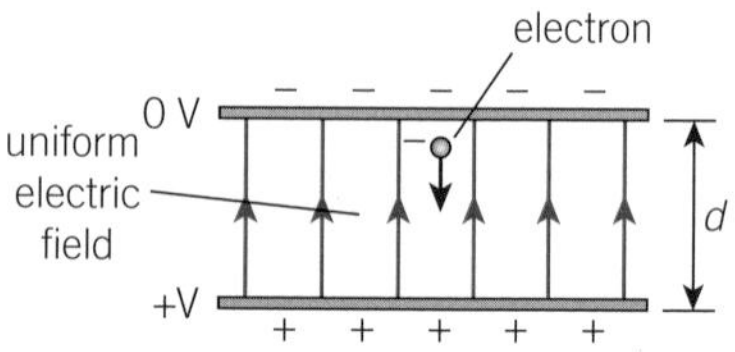

▲ **Figure 2** *The electron is accelerated by the electric field as it travels towards the positive plate*

You can use the following ideas to determine the motion of the electron (or any other charged particle) between the plates:

- electric field strength E between the plates is given by $E = \frac{V}{d}$, where V is the p.d. between the plates and d is the separation between the plates
- force F on the electron is given by $F = EQ = Ee$, where e is the elementary charge
- work done on the electron = $Vq = Ve$.

An electron travelling in the direction of the electric field, from positive to negative plate, will experience a deceleration. The motion of the electron is similar to that of a mass moving vertically upwards in the uniform gravitational field of the Earth.

Worked example: Slowing down and stopping

An electron is fired from a positive capacitor plate towards the negative plate along the direction of the electric field, with a velocity of $1.0 \times 10^7\,\mathrm{m\,s^{-1}}$. The p.d. across the plates is 600 V and their separation is 3.0 cm. Calculate the maximum distance the electron will travel.

Step 1: Write down the quantities given in the question.

$$V = 600\,\mathrm{V},\ d = 3.0 \times 10^{-2}\,\mathrm{m},\ u = 1.0 \times 10^7\,\mathrm{m\,s^{-1}},\ v = 0$$

Step 2: Calculate the force acting on the electron.

$$F = Ee = \frac{V}{d} \times \mathrm{e}$$

$$F = \frac{600 \times 1.6 \times 10^{-19}}{0.030} = 3.2 \times 10^{-15}\,\mathrm{N}$$

Step 3: Calculate the magnitude of the deceleration of the electron.

$$a = \frac{F}{m} = \frac{3.2 \times 10^{-15}}{9.11 \times 10^{-31}} = 3.513 \times 10^{15}\,\mathrm{m\,s^{-2}}$$

Step 4: Use an equation of motion to calculate how far it will travel before stopping and then falling back.

$$v^2 = u^2 + 2as$$

When the electron stops, $v = 0$.

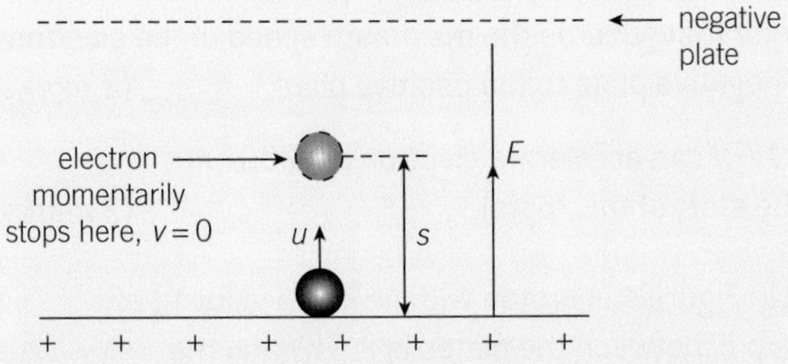

▲ **Figure 3** *The electron stops at a point between the plates where v = 0*

So

$$s = -\frac{u^2}{2a} = -\frac{(1.0 \times 10^7)^2}{2 \times (-3.513 \times 10^{15})}$$

$$s = 1.4 \times 10^{-2}\,\mathrm{m}\ (2\ \mathrm{s.f.})$$

The electron will travel about 1.4 cm from the positive plate before it turns back.

Synoptic link

You can review the equations of motion and the relationship between force, mass, and acceleration in Chapter 3, Motion, and Chapter 4, Forces in Action.

Charged particles moving at right angles to an electric field

An object thrown horizontally on the surface of the Earth describes a parabolic path – that is, its vertical motion is affected by the Earth's gravitational pull but its horizontal motion is unaffected. This is exactly what happens to a charged particle that enters a uniform electric field at a right angle. Figure 4 shows a particle of mass m, charge Q, and initial horizontal velocity v entering a uniform electric field at a right angle. The field strength E is provided by the two oppositely charged horizontal plates, and occupies a region of length L. You can predict the motion of this particle in the electric field by analysing its vertical and horizontal components of motion independently.

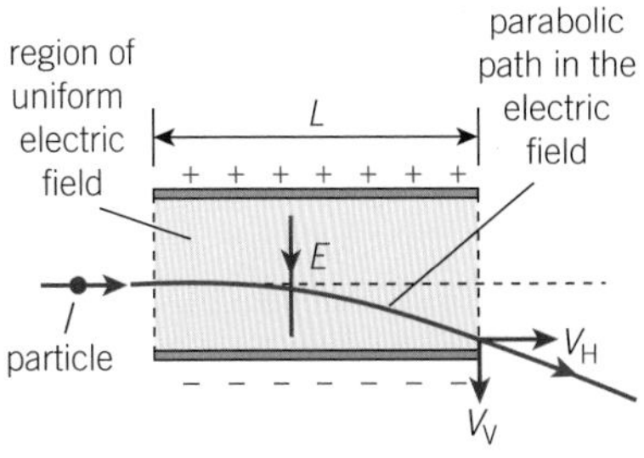

▲ **Figure 4** *A charged particle is deflected by an electric field*

Synoptic link

When the particle emerges from the field it will have a final velocity that can be calculated by applying Pythagoras's theorem to the two perpendicular components of its velocity v_V and v_H, and the particle's final direction can be obtained from $\tan\theta = \frac{v_V}{v_H}$, as described in Topic 2.4, Adding vectors.

For a charged particle moving in an electric field as in Figure 4, we see that for the **horizontal motion:**

- There is no acceleration, hence the horizontal velocity v_H of the particle remains constant, with velocity v.
- The time t spent in the field is given by the equation $t = \frac{L}{v}$.

And for the **vertical motion:**

- The vertical acceleration a of the particle is given by the equation $a = \frac{F}{m} = \frac{EQ}{m}$
- The initial vertical velocity $u = 0$.
- The final vertical component of the velocity v_V as the particle exits the field is given by the equation $v_V = u + at = 0 + \frac{EQ}{m} \times \frac{L}{v} = \frac{EQL}{mv}$

Summary questions

1 A proton is at rest between two parallel and uncharged plates. The plates are then connected to a power supply. Describe the motion of the proton. *(3 marks)*

2 Electrons are accelerated between two charged parallel plates. State and explain the only factor that governs the maximum speed of the electrons travelling from the negative plate to the positive plate. *(2 marks)*

3 Show that a p.d. of 1.5 V can accelerate electrons to 700 km s^{-1} (as mentioned at the start of this topic). *(3 marks)*

4 The particle shown in Figure 4 is proton with an initial velocity of 5.0×10^6 m s^{-1}. The p.d. between the plates is 2.5 kV and the separation is 2.0 cm. The length of each plate is 20 cm. Assume mass of proton = 1.7×10^{-27} kg Calculate:

a the final vertical velocity of the proton as it exits the electric field *(4 marks)*

b the vertical displacement of the proton as it exits the electric field. *(4 marks)*

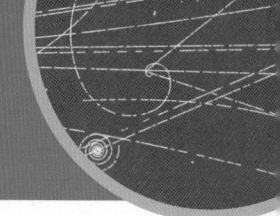

22.5 Electric potential and energy

Specification reference: 6.2.4

Electric potential energy

A stretched elastic band or a spring has stored energy. Charged particles can also store energy. Figure 1 shows an imaginary experiment involving positive charges being pushed closer together. The charges repel each other so you have to do work to decrease the separation between the charges, and you have to push harder as the separation decreases. All the work done is stored as electric potential energy. This stored energy is recoverable; – all you need to do is to let go!

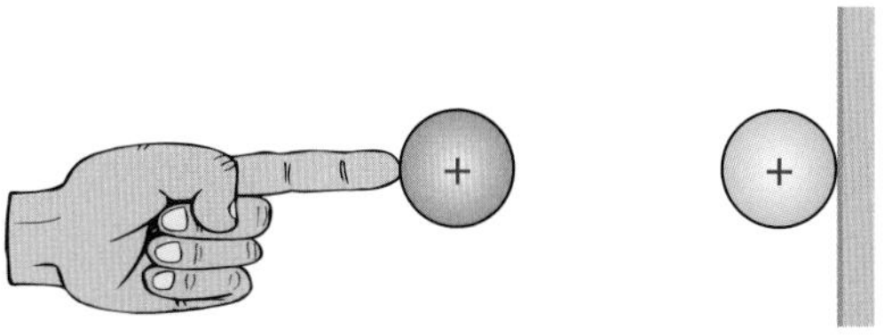

▲ **Figure 1** *Work must be done to bring these positive charges close together*

Learning outcomes

Demonstrate knowledge, understanding, and application of:

- → electric potential as the work done in bringing a unit charge from infinity to a point
- → electric potential $V = \frac{Q}{4\pi\varepsilon_0 r}$
- → capacitance $C = 4\pi\varepsilon_0 R$ for an isolated sphere
- → force–distance graphs for point or spherical charges
- → electric potential energy $E = Vq = \frac{Qq}{4\pi\varepsilon_0 r}$

Force–distance graphs for point and spherical charges

You already know that a uniformly charged sphere may be treated as if it were a point charge (see Topic 22.1, Electric fields) and that the force between the point charges is given by Coulomb's law, $F = \frac{Qq}{4\pi\varepsilon_0 r^2}$. The force F between two positive particles (or spheres) of charges Q and q varies with their separation as shown in Figure 2.

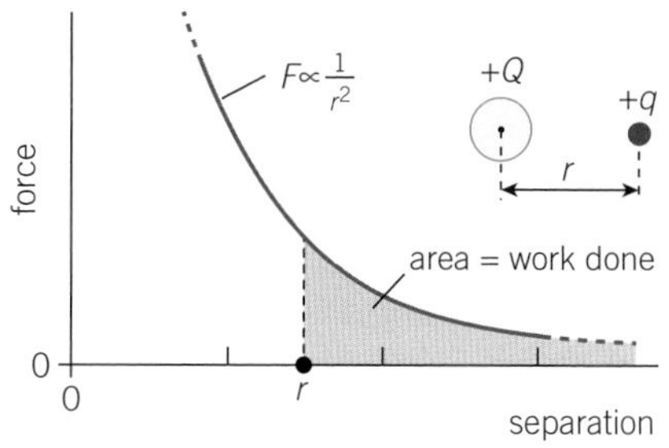

▲ **Figure 2** *Force against separation graph for two charged particles*

The area under a force–distance graph is equal to work done. The total work done to bring the particles from infinity to a separation r is the total area under the graph, shown by the shaded region in Figure 2. The total work done is the same as the electric potential energy E, which you can show is given by the equation

$$E = \frac{Qq}{4\pi\varepsilon_0 r}$$

Synoptic link

You learned that the area under a force–distance graph is equal to work done in Topic 18.7, Gravitational potential energy.

Study tip

The letter E is used for both electric field strength and electric potential energy, so you need to be vigilant when you see it.

If one of the particles has a negative charge, then the value for E will also be negative. The force between the particles will be attractive. The magnitude of E represents the external energy required to completely separate the charged particles to infinity. This is essentially what happens when atomic or molecular bonds are broken.

Worked example: Ionic solid

Common salt is made of sodium, Na^+, and chloride, Cl^-, ions (Figure 3). The separation between each pair of ions is 9.4×10^{-10} m. The magnitude of the charge on each ion is e, 1.60×10^{-19} C. Estimate the energy in electronvolts (eV) required to pull a pair of ions apart completely. You may ignore the effect of the other ions.

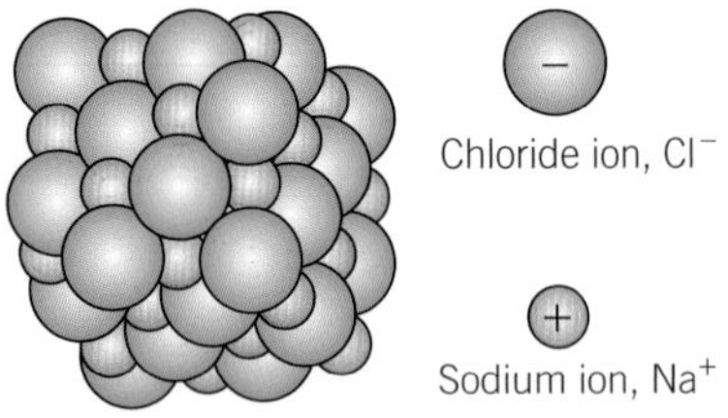

▲ **Figure 3** *The structure of sodium chloride*

Study tip

You met the electronvolt as a unit of energy in Topic 13.1, The photon model of light.

Step 1: Write down the quantities given in the question.

$Q = +1.60 \times 10^{-19}$ C (sodium), $q = -1.60 \times 10^{-19}$ C (chlorine), $r = 9.4 \times 10^{-10}$ m

Step 2: Use the equation for electric potential energy to calculate E.

$$E = \frac{Qq}{4\pi\varepsilon_0 r} = \frac{(1.60 \times 10^{-19})(-1.60 \times 10^{-19})}{4\pi \times 8.85 \times 10^{-12} \times 9.4 \times 10^{-10}}$$

$$= -2.44... \times 10^{-19}\,\text{J}$$

The negative sign means that external energy is required to pull these ions apart.

Step 3: Calculate the energy in electronvolts (remember $1\,\text{eV} = 1.60 \times 10^{-19}$ J).

$$\text{energy required} = \frac{-2.44... \times 10^{-19}}{1.60 \times 10^{-19}} = -1.5\,\text{eV (2 s.f.)}$$

Electric potential

The **electric potential** V at a point is defined as the work done per unit charge in bringing a positive charge from infinity to that point. If the test charge is q, the equation for V can be determined by dividing the electric potential energy E by q. Therefore

$$V = \frac{E}{q} = \frac{Q\not{q}}{4\pi\varepsilon_0 r\not{q}} = \frac{Q}{4\pi\varepsilon_0 r}$$

The unit for electrical potential is unit J C^{-1} or volts (V).

In your studies of circuits you came across the term potential difference (p.d.). You can use the same term here. **Electric potential difference** is the work done per unit charge between two points around the particle of charge Q (Figure 4).

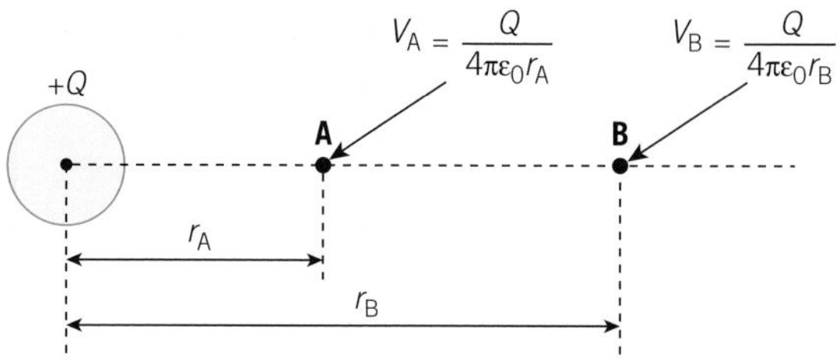

▲ **Figure 4** *Potential difference is the work done per unit charge, so for example the p.d. between points A and B has a magnitude* $V_A - V_B$

The electric p.d. between point **A** and **B** in Figure 4 is simply the difference in the potentials at these two points. Imagine placing a voltmeter between **A** and **B**. It would show the work done per unit charge. Sadly, a voltmeter that can measure electric potentials in the space around charged spheres and particles is not readily available.

Worked example: High electric potential

A football of diameter 22 cm is covered with aluminium foil and suspended from a nylon string. A high-tension supply is set to 5.0 kV. Its positive electrode is used to give a positive charge to the ball. Calculate the charge stored on the ball (treat it as a sphere).

Step 1: Write down the quantities given in the question.

$$V = 5000\,\text{V},\ d = 0.22\,\text{m},\ Q = ?$$

The potential on the surface of the sphere must be equal to 5000 V.

Step 2: Use the equation for electric potential to calculate the charge on the sphere.

$$V = \frac{Q}{4\pi\varepsilon_0 r}$$

$$Q = 4\pi\varepsilon_0 rV = 4\pi \times 8.85 \times 10^{-12} \times 0.11 \times 5000$$
$$= 6.1 \times 10^{-8}\,\text{C (2 s.f.)}$$

Back to capacitance

A capacitor is a device that stores charge. An isolated, charged sphere of radius R also stores charge. It too must be a capacitor, albeit a strange capacitor with a single 'plate'. The capacitance C of a charged sphere is the ratio of the charge it stores, Q, to the potential V at its surface. Therefore

$$C = \frac{Q}{V} = \frac{4\pi\varepsilon_0 RV}{V} = 4\pi\varepsilon_0 R$$

The equation above is the capacitance of an isolated sphere – one that is far away from other charged objects. You can imagine the Earth as being a huge capacitor floating in space. In spite of its radius of 6400 km, the capacitance of the Earth is only 7.1×10^{-4} F (710 μF).

Equipotentials

An equipotential line is a line, or a surface, along which the electric potential is the same – like a contour line for height on a map. Equipotential lines are very useful when interpreting the electric field strengths near charged conductors.

Like heights on a map, equipotential lines indicate the potential gradient, $\frac{\Delta V}{\Delta r}$.

If a test charge Q is moved through a uniform field, the change in electric potential energy is $-Q\Delta V$. This energy change is equal to the work done to move the charge, which is the force multiplied by the distance moved, $F\Delta r$.

The force F that acts on the charge in the uniform field depends on the electric field strength E and the charge Q through the equation

$$F = EQ$$

Thus $-Q\Delta V = EQ\Delta r$

So the electric field strength E is equal to $-1 \times$ the potential gradient, that is

$$E = -\frac{\Delta V}{\Delta r}$$

Figure 5 shows the equipotential lines and the electric field lines for two spheres with opposite charges.

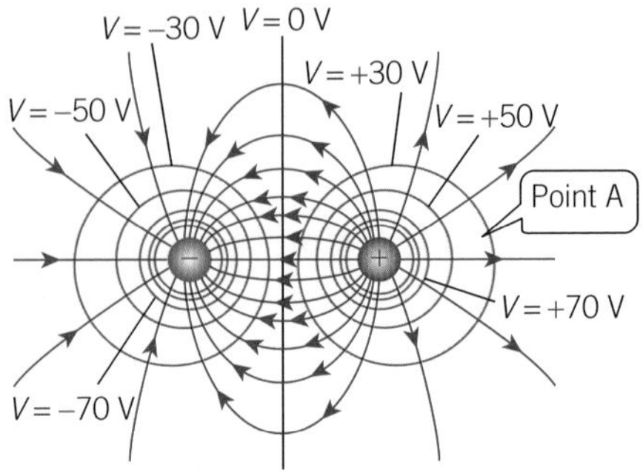

▲ **Figure 5** *Electric field and equipotential lines for two oppositely charged spheres*

1. State the link between electric field lines and an equipotential line.
2. The potential difference across the parallel plates of a capacitor is 400 V. Draw the electric field pattern and the equipotentials for this capacitor.
3. Estimate the magnitude of the electric field strength at point **A** in Figure 5.

Summary questions

1. The electric potential energy for two charged particles is 4×10^{-19} J. State and explain the value of the electric potential energy when the separation between the particles is doubled. *(2 marks)*
2. The electric potential at the surface of a sphere is 10 V. Explain what this means in terms of work done and unit charge. *(1 mark)*
3. Calculate the electric potential at the surface of a proton. The radius of a proton is about 8.8×10^{-16} m. *(2 marks)*
4. Calculate the electric potential energy of a proton and an electron separated by a distance of 1.0×10^{-10} m. Write your answers in joules (J) and in electronvolts (eV). *(3 marks)*
5. A sphere of radius 2.0 cm is charged to a potential of −6000 V. Calculate:
   a. the capacitance of the sphere *(2 marks)*
   b. the number of excess electrons on the surface of the sphere. *(3 marks)*
6. The electric potential energy of two protons is 1.0 MeV. Calculate their separation. *(4 marks)*

Practice questions

1 a Define *electric field strength* at a point in space. *(1 mark)*

b Figure 1 shows an evenly spaced grid.

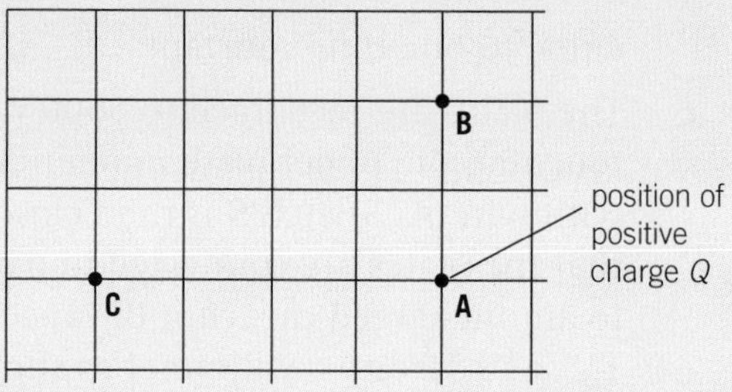

▲ Figure 1

A, **B**, and **C** are points on a grid. A positive charge Q is placed on the grid at point **A**. The magnitude of the electric field strength at point **B** due to the charge Q is $8.0 \times 10^5\,\text{NC}^{-1}$.

(i) Apart from the magnitude of the electric field strengths, state another difference between the electric field at points **B** and **C**. *(1 mark)*

(ii) Determine the magnitude of the electric field strength at point **C**. *(2 marks)*

c The simplest atom is that of hydrogen with one proton and one electron, see Figure 2.

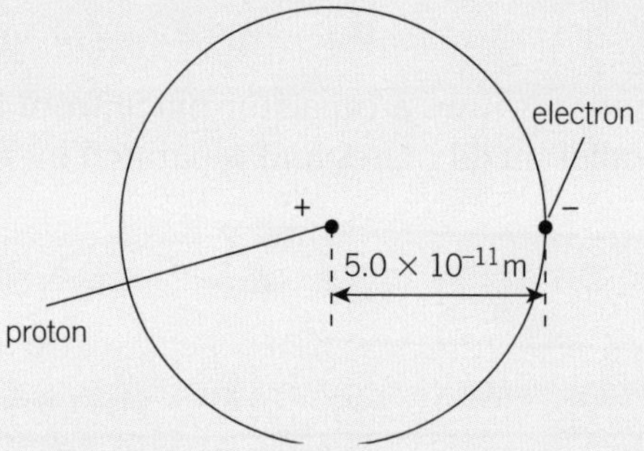

▲ Figure 2

The mean separation between the proton and the electron is shown in Figure 2.

(i) Calculate the magnitude of the electrical force F_E acting on the electron. *(3 marks)*

(ii) The gravitational force F_G acting on the electron due to the proton is very small compared with the electrical force F_E it experiences.

Calculate the ratio $\frac{F_E}{F_G}$. *(2 marks)*

(iii) A simplified model of the hydrogen atom suggests that the de Broglie wavelength of the electron is four times the mean separation between the proton and the electron shown in Figure 2.

Estimate

1 the momentum p of the electron *(3 marks)*

2 the kinetic energy E_k of the electron. *(3 marks)*

Jan 2013 G484

2 Figure 3 shows a close-up of the two electrodes of a spark plug.

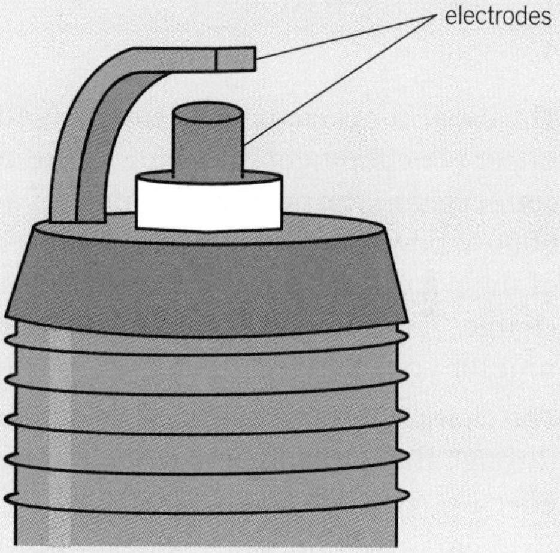

▲ Figure 3

The electrodes may be considered as two parallel plates. The electric field strength between the electrodes is almost uniform.

a Define electric field strength. *(1 mark)*

b The separation between the electrodes is 1.3 mm. An electric spark is produced when the electric field strength is $3.0 \times 10^6\,\text{V m}^{-1}$.

(i) Estimate the potential difference V between the electrodes when the spark is produced. *(2 marks)*

(ii) The electric spark lasts for 4.0×10^{-2} s and produces an average current of 2.7×10^{-9} A.

1 Calculate the charge transferred between the electrodes. *(2 marks)*

2 Calculate the number of electrons transferred between the electrodes. *(1 mark)*

(iii) Estimate the total energy transferred by the electrons in **(ii)**. *(2 marks)*

Jan 2012 G485

3 Figure 4 shows two parallel metal plates which act as a capacitor supported above a bench on an insulating rod which passes through the centre of each plate.

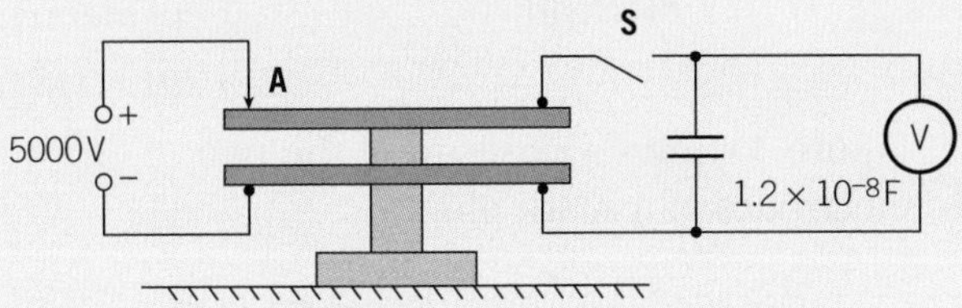

▲ Figure 4

a The capacitor is charged by touching the upper plate momentarily with a wire **A** connected to the positive terminal of a 5000 V power supply. The capacitance C of the plates is 1.2×10^{-11} F. Calculate the charge Q_0 on the plates. Give a suitable unit for your answer. *(3 marks)*

b The charge on the plates leaks away slowly through the insulating rod, which has an effective resistance R of $1.2 \times 10^{15}\,\Omega$.

(i) Show that the time constant for the plates to discharge through the rod is about 1.5×10^{4} s. *(1 mark)*

(ii) Show that the initial value of the leakage current is about 4×10^{-12} A. *(1 mark)*

(iii) Suppose that the plates continue to discharge at the constant rate calculated in **(ii)**. Show that the charge Q_0 would leak away in a time equal to the time constant. *(2 marks)*

(iv) Using the equation for the charge Q at time t

$$Q = Q_0 e^{-\frac{t}{RC}}$$

Show that, in practice, the plates only lose about $\frac{2}{3}$ of their charge in a time equal to one time constant. *(2 marks)*

c The plates are recharged to 5000 V by touching the upper plate momentarily with wire **A**. Switch **S** is then closed so that the plates are connected in parallel to an uncharged capacitor of capacitance 1.2×10^{-8} F and a voltmeter as shown in Figure 5.

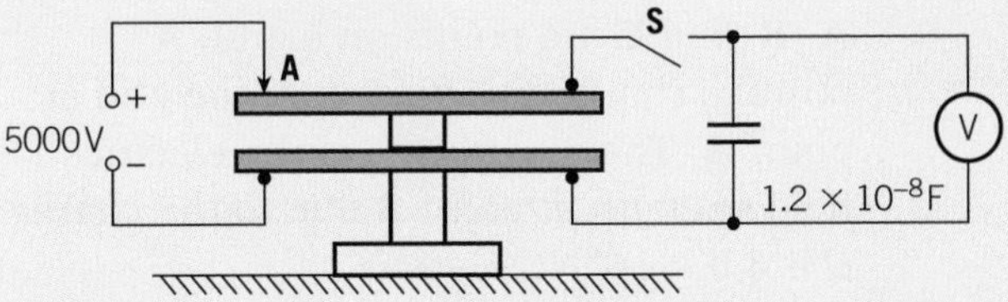

▲ Figure 5

(i) The charged and the uncharged capacitor act as two capacitors in parallel. The total charge Q_0 is shared instantly between the two capacitors. Explain why the charge left on the plates is $\frac{Q_0}{1000}$. *(3 marks)*

(ii) Hence or otherwise calculate the initial reading V on the voltmeter. *(2 marks)*

Jan 2010 2824

4 Figure 6 shows a capacitor made from two parallel metal plates and separated by air.

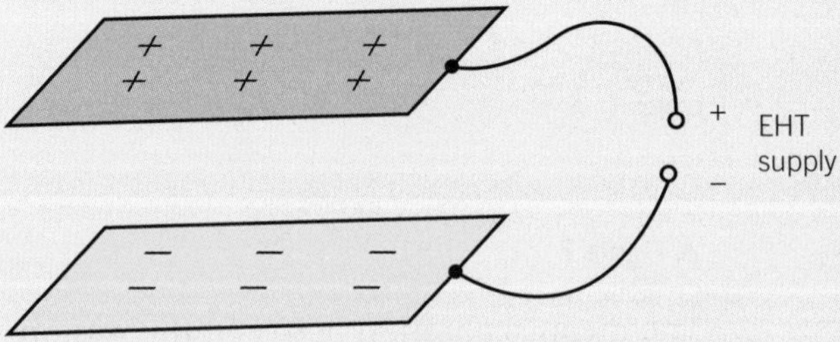

▲ Figure 6

a On a copy of Figure 6, draw electric field lines to show the electric field pattern between the plates. *(2 marks)*

b The separation between the capacitor plates is 2.0 cm and each plate has a surface area of 81 cm^2. The plates are connected to a 4.0 kV supply. A student disconnects the supply from and measures the charge stored on one of the plates using a coulombmeter. The reading on the coulombmeter is 14.0 ± 0.5 nC.

(i) Calculate the permittivity of free space ε_0 from these results. Include the absolute uncertainty in your answer. *(5 marks)*

(ii) Discuss whether or not the value of ε_0 in **(b)(i)** is accurate. *(2 marks)*

c Qualitatively explain the effect on the charge stored by the capacitor in **(b)** when a thick sheet of plastic is inserted in the space between the capacitor plates. *(2 marks)*

5 a Define *electric potential* at a point in space around a charged object. *(1 mark)*

b An isolated metal sphere is charged using a high-voltage supply. Discuss the factors that affect the charge stored on the surface of the sphere. *(2 marks)*

c A student is investigating the charge Q stored by a metal sphere of radius r. The sphere is always charged to a potential of 2.0 kV, see Figure 7. A coulombmeter is used to determine the charge stored on the sphere. Table 1 shows the results obtained by the student.

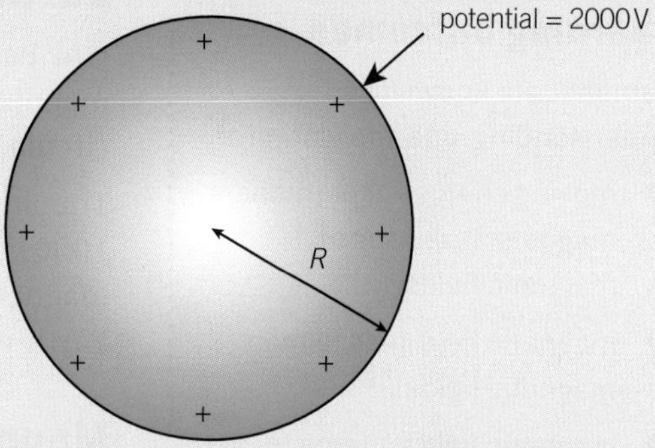

▲ Figure 7

▼ Table 1

r / cm	Q / 10^{-9} C
5.0	11.0
7.3	16.1
9.1	20.0
11.6	25.5
15.0	33.0

(i) Plot a graph of Q against r and draw a line of best fit. *(3 marks)*

(ii) Use your graph to determine the permittivity of free space. *(4 marks)*

23 MAGNETIC FIELDS

23.1 Magnetic fields

Specification reference: 6.3.1

Learning outcomes

Demonstrate knowledge, understanding, and application of:

→ moving charges or permanent magnets as causes of magnetic fields

→ magnetic field lines to map magnetic fields

→ magnetic field patterns for a long straight current-carrying conductor, a flat coil, and a long solenoid.

▲ **Figure 1** *The Earth's magnetic field protects us from the solar wind*

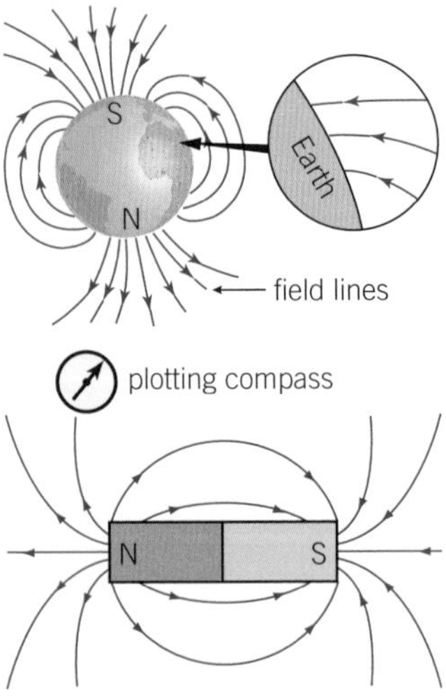

▲ **Figure 2** *The magnetic fields of the Earth and a bar magnet look similar – note the poles marked on the Earth*

Earth's magnetic field

Our Earth has a magnetic field, just like that of a bar magnet, with magnetic north and south poles. It is caused by the electrical currents circulating in the molten iron of the Earth's core.

The Sun emits streams of charged particles travelling at up to $1000\,\mathrm{km\,s^{-1}}$. Most of this solar wind is deflected by the Earth's magnetic field before it reaches the surface. Without this field, the Earth would be swept by ionising radiation and we could not live.

Magnetic fields

A magnetic field is a field surrounding a permanent magnet or a current-carrying conductor in which magnetic objects experience a force. You can detect the presence of a magnetic field with a small plotting compass. The needle will deflect in the presence of a magnetic field.

We use **magnetic field lines** (or lines of force) to map **magnetic field patterns** around magnets and current-carrying conductors. Magnetic field patterns are useful visual representations that help us to interpret the direction and the strength of the magnetic fields.

- The arrow on a magnetic field line is the direction in which a free north pole would move — the arrow points from north to south.
- Equally spaced and parallel magnetic field lines represent a uniform field, that is, the strength of the magnetic field does not vary.
- The magnetic field is stronger when the magnetic field lines are closer. For a bar magnet, the field is strongest at its north (N) and south (S) poles.
- Like poles (N–N or S–S) repel and unlike poles (S–N) attract.

Figure 2 shows the magnetic field patterns for a bar magnet and the Earth. You may already have seen how iron filings can reveal the magnetic field around a bar magnet. The field induces magnetism in the filings, which line up in the field. Figure 3 shows the magnetic field patterns between two unlike poles and two like poles.

Electromagnetism

When a wire carries a current, a magnetic field is created around the wire. The field is created by the electrons moving within the wire. Any charged particle that moves creates a magnetic field in the space around it. But how do we explain the magnetic field of a bar magnet? In fact, it is created by the electrons whizzing around the iron nuclei. You can visualise the iron atoms as tiny magnets, all aligned in the same direction.

Current-carrying conductors

For a current-carrying wire, the magnetic field lines are concentric circles centred on the wire and perpendicular to it. The direction of the magnetic field can be determined using the **right-hand grip rule**, shown in Figure 4. The thumb points in the direction of the conventional current, and the direction of the field is given by the direction in which the fingers curl around the wire.

The magnetic field patterns produced by a single coil and a solenoid are shown in Figure 5. Both the coil and the solenoid produce north and south poles at their opposite faces. The magnetic field pattern outside solenoid is similar to that for a bar magnet, and at the centre of the core of the solenoid it is uniform – you can tell this from parallel and equidistant magnetic field lines.

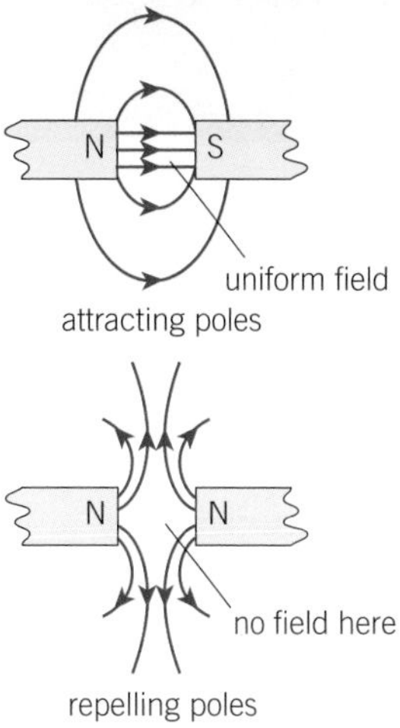

▲ **Figure 3** *Field patterns for attracting and repelling poles – which pair of poles produces a uniform magnetic field?*

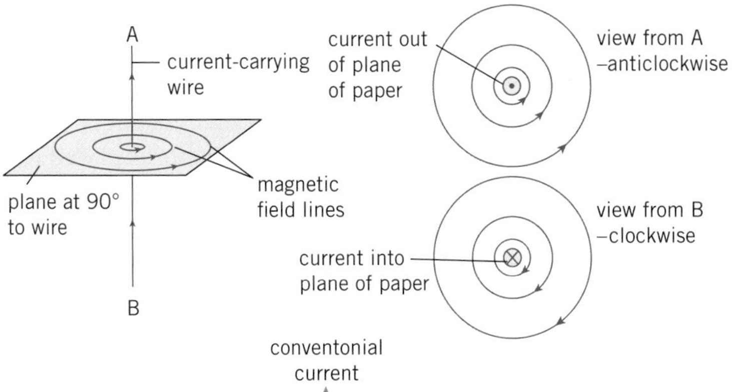

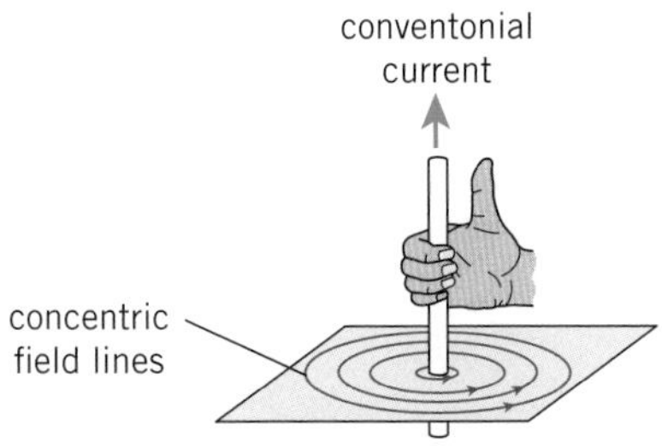

◀ **Figure 4** *The magnetic field around a long current-carrying wire and the right-hand grip rule (below)*

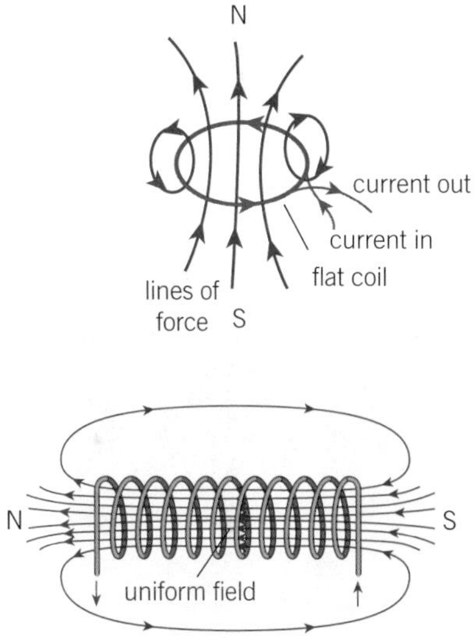

▲ **Figure 5** *Magnetic fields created by a current-carrying flat coil and a solenoid. You can use your right hand again to get the direction of the magnetic field for the solenoid. The fingers point in the direction of the conventional current and the thumb gives the direction of the magnetic field within the core of the solenoid.*

Summary questions

1 State and explain whether each of the following moving particles will produce a magnetic field.

a an electron; b a proton; c a neutron. *(3 marks)*

2 State two methods of producing a uniform magnetic field. *(2 marks)*

3 A horizontal current-carrying wire is shown in Figure 6. State the direction of the magnetic field at points A and B. *(2 marks)*

4 Suggest how the magnetic field pattern for a solenoid within its core (Figure 4) would change when the current is both reversed and increased. *(2 marks)*

5 Figure 7 shows the top view of two long and straight current-carrying wires placed very close to each other. The current in each wire is into the plane of the paper. Sketch the magnetic field pattern around these current-carrying wires. *(2 marks)*

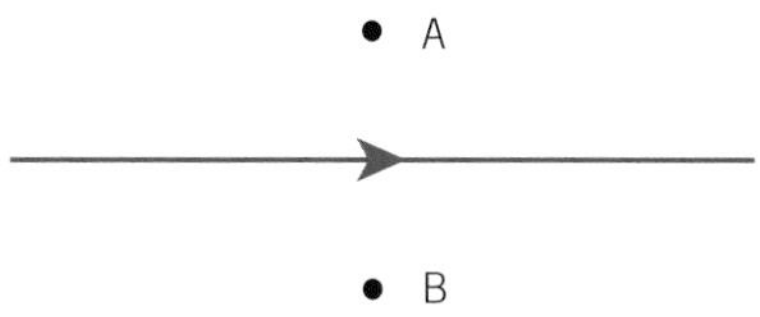

▲ **Figure 6**

▲ **Figure 7**

23.2 Understanding magnetic fields

Specification reference: 6.3.1

Learning outcomes

Demonstrate knowledge, understanding, and application of:

- → Fleming's left-hand rule
- → the force on a current-carrying conductor, $F = BIL\sin\theta$
- → the techniques and procedures used to determine the uniform magnetic flux density between the poles of a magnet using a current-carrying wire and digital balance
- → magnetic flux density and the unit tesla.

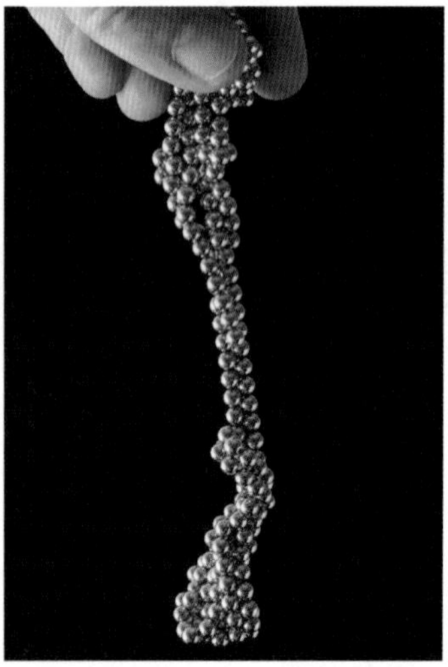

▲ **Figure 1** *Strong spherical magnets made from rare earth elements*

How strong?

Magnets made of alloys of neodymium, a rare earth element, are amongst the strongest available in the world. A coin-sized neodymium magnet can lift about 9 kg. Such magnets have enabled designers and engineers to reduce the size of devices such as the wafer-thin speakers used in mobile phones and electric motors for hybrid cars. The strength of magnets, and magnetic fields, is measured in **tesla** (T). The strength of a rare-earth magnet is about 1.3 T, compared with about 30 μT for the Earth's magnetic field at the equator.

Magnetic fields and forces

Fleming's left-hand rule

A current-carrying conductor is surrounded by its own magnetic field, as you learned in the previous topic. When the conductor is placed in an external magnetic field, the two fields interact just like the fields of two permanent magnets. The two magnets experience equal and opposite forces.

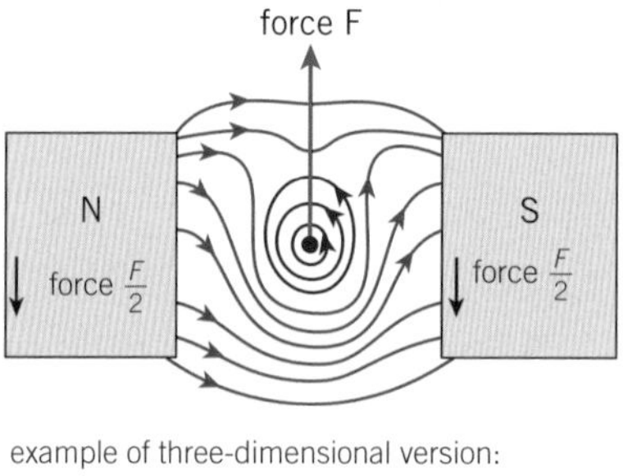

▲ **Figure 2** *The distorted magnetic field of the wire is responsible for catapulting it away from the poles*

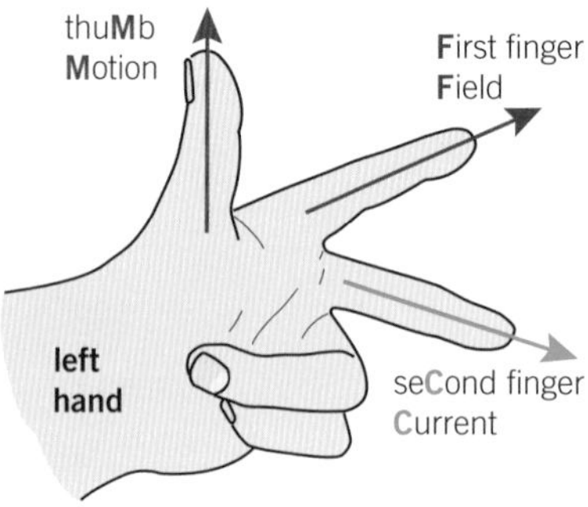

▲ **Figure 3** *Fleming's left-hand rule*

Figure 2 shows the resultant field pattern when a current-carrying wire is placed between the poles of a magnet. The direction of the force experienced by a current-carrying conductor placed perpendicular to the external magnetic field can be determined using **Fleming's left-hand rule**. Use your left hand as shown in Figure 3. The direction of your:

- first finger gives the direction of the external magnetic field
- second finger gives the direction of the conventional current
- thumb gives the direction of motion (force) of the wire

Magnetic flux density and the tesla

The magnitude of the force experienced by a wire in an external magnetic field depends on a number of factors. For example, the force is a maximum when the wire is perpendicular to the field and zero when it is parallel to the magnetic field. Experiments show that the magnitude of the force F experienced by the wire is directly proportional to the

- current I
- length L of the wire in the magnetic field
- $\sin\theta$, where θ is the angle between the magnetic field and the current direction
- the strength of the magnetic field.

Therefore

$$F = BIL\sin\theta$$

where B is the **magnetic flux density** – the strength of the field. It is analogous to electric field strength E for electric fields and gravitational field strength g for gravitational fields. The SI unit for magnetic flux density is the tesla (T). You can see from the equation above that $1\,T = 1\,N\,m^{-1}\,A^{-1}$.

The magnetic flux density is 1 T when a wire carrying a current of 1 A placed perpendicular to the magnetic field experiences a force of 1 N per metre of its length.

When the wire is perpendicular to the magnetic field, $\theta = 90°$ and $\sin\theta = 1$, therefore $F = BIL$. The direction of the force can be determined using Fleming's left-hand rule. You can therefore write the equation for magnetic flux density as

$$B = \frac{F}{IL}$$

Magnetic flux density is a vector quantity. It has both magnitude and a direction.

Worked example: Lifting up

A thin wire of weight $1.8 \times 10^{-3}\,N\,cm^{-1}$ is horizontal and perpendicular to a magnetic field of uniform flux density 0.15 T. The current in the wire is slowly increased from zero. The wire experiences a force vertically upwards. Calculate the size of the current in the wire such that the wire just starts to lift itself vertically.

Step 1: Write down the information given in the question.

$F = 1.8 \times 10^{-3}\,N$, $L = 0.01\,m$, $B = 0.15\,T$, $\theta = 0$

Step 2: For the wire to start moving, the force acting on it must be equal to its weight. Rearrange the equation $F = BIL$ with current I as the subject and then substitute all the values.

$$I = \frac{F}{BL} = \frac{1.8 \times 10^{-3}}{0.15 \times 0.01} = 1.2A$$

A current of 1.2 A in the wire will just start to lift the wire vertically.

Study tip

Remember that the equation $F = BIL$ only applies when B and I are perpendicular.

Determining magnetic flux density in the laboratory

Figure 4 shows apparatus for determining the magnetic flux density between two magnets. The magnets are placed on a top-pan balance. The magnetic field between them is almost uniform. A stiff copper wire is held perpendicular to the magnetic field between the two poles. The length L of the wire in the magnetic field is measured with a ruler. Using crocodile clips, a section of the wire is connected in series with an ammeter and a variable power supply. The balance is zeroed when there is no current in the wire. With a current I, the wire experiences a vertical upward force (predicted by Fleming's left-hand rule). According to Newton's third law of motion, the magnets experience an equal downward force, F, which can be calculated from the change in the mass reading, m, using $F = mg$, where g is the acceleration of free fall ($9.81\,\text{m}\,\text{s}^{-2}$). The magnetic flux density B between the magnets can then be determined from the equation $B = \frac{F}{IL}$.

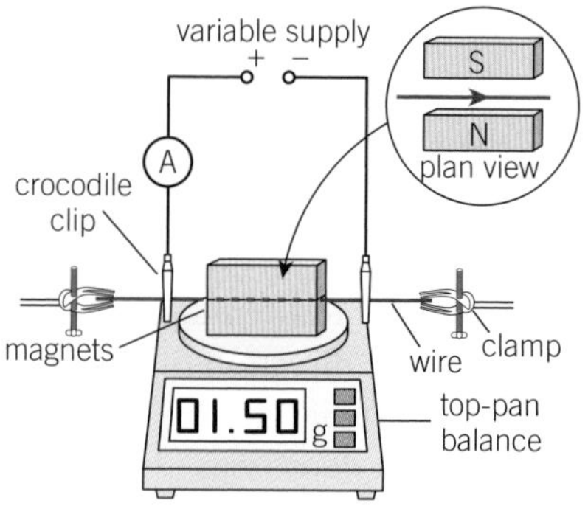

▲ **Figure 4** *An arrangement for determining B in the laboratory*

▼ **Table 1** *Results of an experiment using the apparatus in Figure 4*

Current I/A	Change in mass m/g	Force F/N
0.00	0	
1.00	0.31	
2.00	0.64	
3.00	0.89	
4.00	1.24	
5.00	1.50	
6.00	1.83	
7.00	2.14	

Analysing results

A student uses the arrangement in Figure 4 to determine the magnetic flux density B between two flat magnets held in a yoke. Table 1 shows the results.

1. Copy the table and complete the last column.
2. Plot a graph of force F against current I. Draw a straight best-fit line through the points.
3. Show that the gradient of the line graph is BL, where L is the length of the wire in the magnetic field.
4. The value of L is recorded as 5.0 ± 0.3 cm by the student. Determine the gradient and hence the value for B for the arrangement of the magnets. State the absolute uncertainty in your answer.

Summary questions

1 Explain why a current-carrying wire experiences a force when placed close to a magnet. *(1 mark)*

2 A current-carrying conductor is placed in a uniform magnetic field. In each case in Figure 5, use Fleming's left-hand rule to determine the direction of the force experienced by the conductor. *(3 marks)*

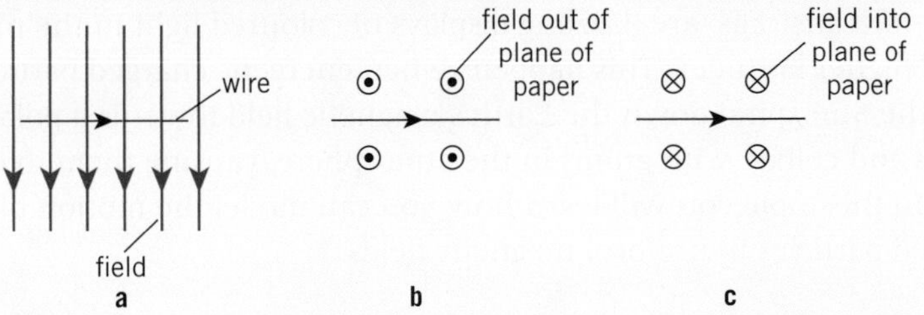

▲ **Figure 5**

3 Calculate the force per centimetre length on a straight wire placed perpendicular to a magnetic field of flux density 120 mT and carrying a current of 5.0 A. *(2 marks)*

4 A current-carrying wire placed perpendicular to a uniform magnetic field experiences a force of 5.0 mN. Determine the force on the wire when, separately:

a the current in the wire is quadrupled *(1 mark)*

b the magnetic flux density is doubled *(1 mark)*

c the length of wire in the wire is reduced to 30% of its original length. *(1 mark)*

5 A 2.8 cm length of copper wire carrying a current of 0.80 A is placed in a uniform magnetic field. The angle between the wire and the magnetic field is 38°. It experiences a force of 4.0 mN. Calculate the magnetic flux density of the field. *(3 marks)*

6 Figure 6 shows a square loop of wire of a simple electric motor placed in a uniform magnetic field of flux density B. The current in the loop is I.

a State and explain the initial direction of rotation of this coil. *(2 marks)*

b Show that torque of the couple acting on the loop is directly proportional to the cross-sectional area of the loop. *(3 marks)*

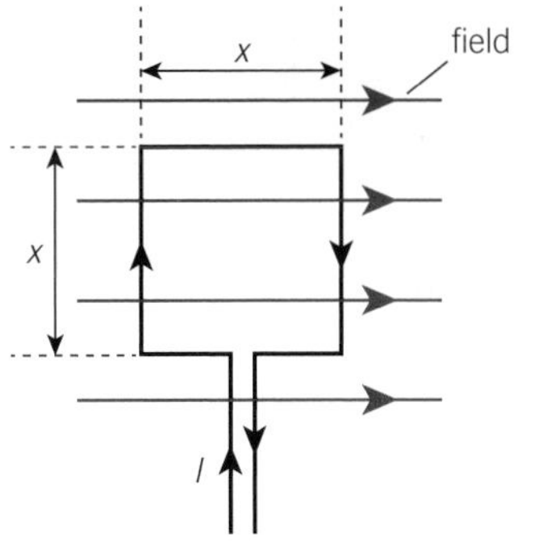

▲ **Figure 6**

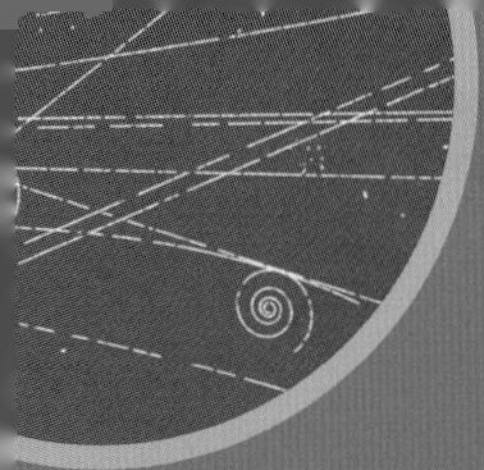

23.3 Charged particles in magnetic fields

Specification reference: 6.3.2

Learning outcomes

Demonstrate knowledge, understanding, and application of:

- → the force on a charged particle travelling at right angles to a uniform magnetic field, $F = BQv$
- → movement of charged particles in a uniform magnetic field
- → movement of charged particles moving in a region occupied by both electric and magnetic fields
- → velocity selector.

▲ **Figure 1** *The aurora borealis or northern lights*

The aurora

The aurora borealis, or northern lights, and its southern equivalent the aurora australis, are dazzling displays of coloured light in the night sky at higher latitudes. This happens when energetic charged particles from the Sun spiral down the Earth's magnetic field towards a polar region and collide with atoms in the atmosphere, causing them to emit light. In this topic you will learn how you can model the motion of charged particles in uniform magnetic fields.

Circular tracks

A charged particle moving in a magnetic field will experience a force. This effect can be demonstrated for a beam of electrons using an electron deflection tube (Figure 2). The force on the beam of electrons can be predicted using Fleming's left-hand rule. The beam of electrons is moving from left to right into a region of uniform magnetic field, shown in more detail in Figure 3. As the electrons enter the field, they experience a downward force. The electrons change direction, but the force F on each electron always remains perpendicular to its velocity. The speed of the electrons remains unchanged because the force has no component in the direction of motion. Once out of the field, the electrons keep moving in a straight line.

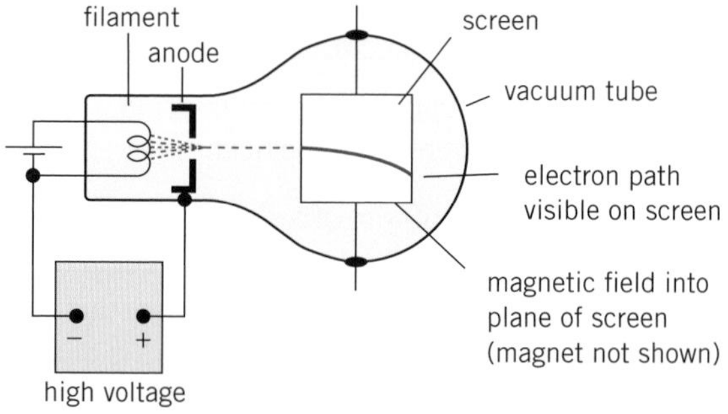

▲ **Figure 2** *An electron deflection tube*

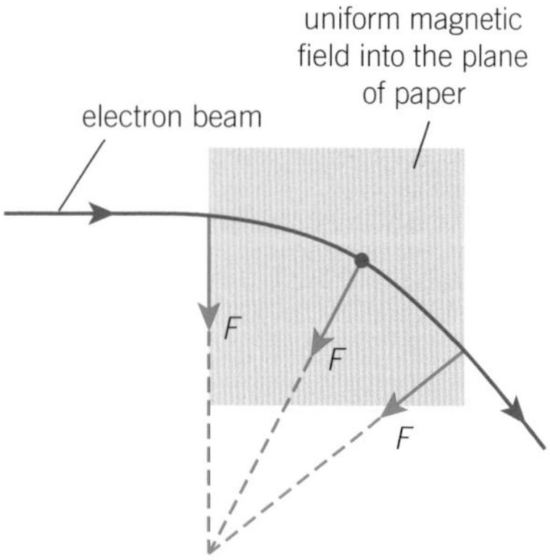

▲ **Figure 3** *The electrons travel in a circular path in the region of the uniform perpendicular magnetic field, and the force on the electrons is always at right angles to their motion*

A current-carrying wire in a uniform magnetic field experiences a force because each electron moving within the wire experiences a tiny force.

To find the force F acting on a charged particle of charge Q moving at a speed v at right angles to a uniform magnetic field of flux density B, consider a section of conductor, or a beam of charged particles (Figure 4). In a time t, all the charged particles contained within the shaded region go through section XY. The length L of the shaded region is vt, where v is the speed of the charged particle. The force F on the conductor is given by

$$F = BIL$$

Therefore

$$F = BI(vt)$$

However, current I is the rate of flow of charge. If there are N charged particles, each of charge Q, in the shaded region, the current is given by the equation

$$I = \frac{NQ}{t}$$

So the force acting on the conductor is given by

$$F = B \times \frac{NQ}{t} \times vt = NBQv$$

The force F on *each* charged particle must therefore be

$$F = \frac{NBQv}{N} = BQv$$

For an electron, or a proton, where $Q = e = 1.60 \times 10^{-19}$ C, this equation may be written as

$$F = Bev$$

Study tip

Check the direction in Figure 3 using Fleming's left-hand rule. Remember the conventional current is from right to left.

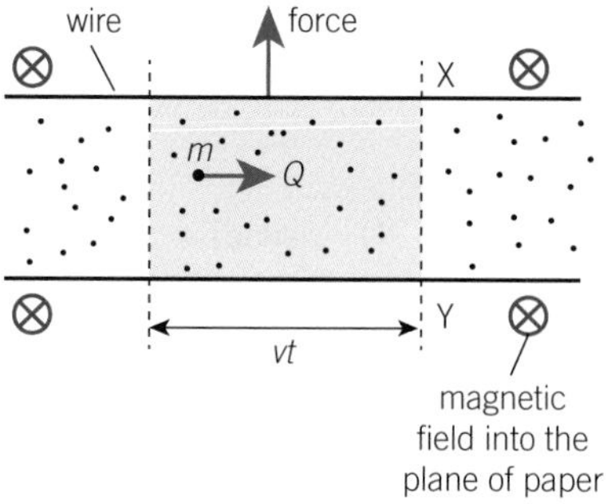

▲ **Figure 4** *Every charged particle experiences a tiny force within the conductor*

Worked example: Colossal acceleration

An electron travels perpendicular to a magnetic field of flux density 0.15 T. Calculate the acceleration of the electron given its speed is $5.0 \times 10^6\,\text{m}\,\text{s}^{-1}$.

Step 1: Write down the information that you have.

$B = 0.15\,\text{T}$, $v = 5.0 \times 10^6\,\text{m}\,\text{s}^{-1}$, $Q = e = 1.60 \times 10^{-19}\,\text{C}$

Step 2: Select the equations that you need.

The force acting on the electron is given by $F = BQv$ and the acceleration a can be calculated from $F = ma$. (The mass m of the electron is given in the Data Booklet.)

$F = BQv = ma$

$$a = \frac{BQv}{m} = \frac{0.15 \times 1.60 \times 10^{-19} \times 5.0 \times 10^6}{9.11 \times 10^{-31}} = 1.3 \times 10^{17}\,\text{m}\,\text{s}^{-2} \text{ (2 s.f.)}$$

Going round

Consider a charged particle of mass m and charge Q moving at right angles to a uniform magnetic field of flux density B. The particle will describe a circular path because the force acting on it is always perpendicular to its velocity. The centripetal force $\frac{mv^2}{r}$ on the particle is provided by the magnetic force BQv. Therefore

$$BQv = \frac{mv^2}{r} \quad \text{or} \quad r = \frac{mv}{BQ}$$

Synoptic link

In Topics 16.2, Angular acceleration, and 16.3, Centripetal force, you learnt about centripetal forces and accelerations. These ideas are very useful here too.

Study tip

The following equations are very useful when solving problems in which charged particles travel in circular paths:

- $F = BQv$
- $F = ma$
- $F = \frac{mv^2}{r}$
- $v = \frac{2\pi r}{T}$ (T = period)

The equation for the radius r shows that:

- faster-moving particles travel in bigger circles ($r \propto v$)
- more massive particles move in bigger circles ($r \propto m$)
- stronger magnetic fields make the particles move in smaller circles ($r \propto \frac{1}{B}$)
- particles with greater charge move in smaller circles ($r \propto \frac{1}{Q}$).

Worked example: Electrons in a magnetic field

A beam of electrons describes a circular path of radius 15 mm in a uniform magnetic field. The speed of the electrons is $8.0 \times 10^6\,\mathrm{m\,s^{-1}}$. Calculate the magnetic flux density B of the magnetic field.

Step 1: Write down the information given in the question.

$r = 1.5 \times 10^{-2}\,\mathrm{m}$, $v = 8.0 \times 10^6\,\mathrm{m\,s^{-1}}$, $Q = e = 1.60 \times 10^{-19}\,\mathrm{C}$

Step 2: Derive an equation for B from first principles and then substitute the values in.

$$BQv = \frac{mv^2}{r}$$

$$B = \frac{mv}{Qr} = \frac{9.11 \times 10^{-31} \times 8.0 \times 10^6}{1.60 \times 10^{-19} \times 1.5 \times 10^{-2}} = 3.0 \times 10^{-3}\,\mathrm{T}\ (2\ \mathrm{s.f.})$$

The magnetic flux density is about 3.0 mT.

Velocity selector

A **velocity selector** is a device that uses both electric and magnetic fields to select charged particles of specific velocity. It is a vital part of instruments such as mass spectrometers and some particle accelerators. It consists of two parallel horizontal plates connected to a power supply (Figure 5). They produce a uniform electric field of field strength E between the plates. A uniform magnetic field of flux density B is also applied perpendicular to the electric field. The charged particles travelling at different speeds to be sorted enter through a narrow slit Y. The electric and magnetic fields deflect them in opposite directions – only for particles with a specific speed v will these deflections cancel so that they travel in a straight line and emerge from the second narrow slit Z. For an undeflected particle

$$\text{electric force} = \text{magnetic force}$$

$$EQ = BQv$$

where Q is the charge on the particle. Thus the speed v depends only on E and B, that is

$$v = \frac{E}{B}$$

When E is $4.0 \times 10^5\,\mathrm{V\,m^{-1}}$ and $B = 0.10\,\mathrm{T}$, only particles with a speed of $4.0 \times 10^6\,\mathrm{m\,s^{-1}}$ will emerge from the slit Z.

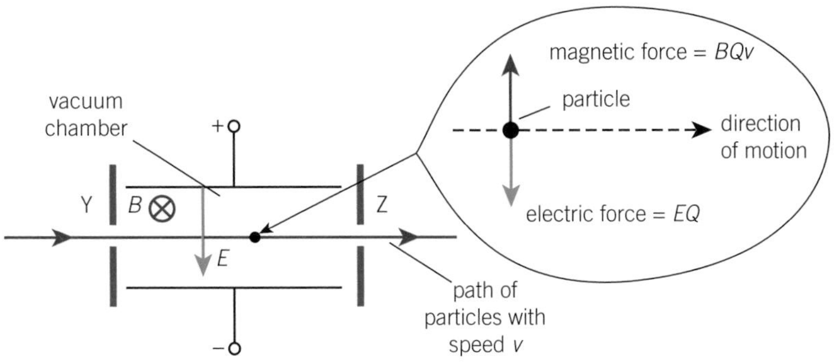

▲ **Figure 5** *A velocity selector*

Mass spectrometers

Mass spectrometers measure the masses and relative concentrations of atoms and molecules. They are used for all kinds of chemical analyses, from detecting the age of ancient rocks to examining pharmaceuticals. Figure 6 shows the basic structure of a mass spectrometer.

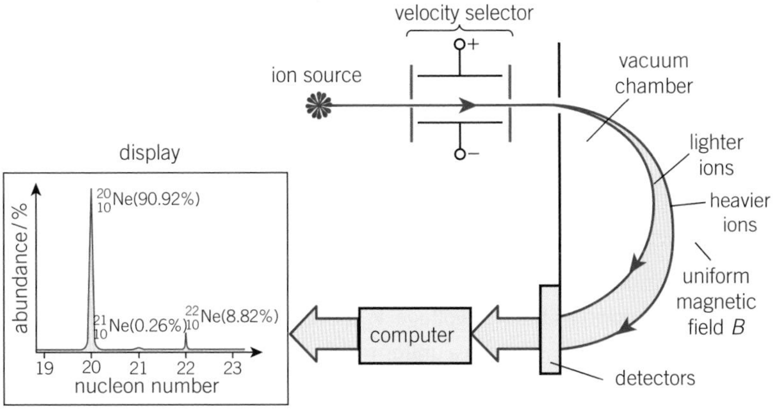

▲ **Figure 6** *A mass spectrometer*

Atoms from a sample are ionised and accelerated through a potential difference. They pass through a velocity selector and emerge with the same speed v before entering a uniform magnetic field of flux density B. The radius r of curvature of each ion is given by the equation $r = \frac{mv}{BQ}$, where Q is the charge on the ion and m is its mass . For a singly ionised atom, $Q = e$.

Since $r \propto m$, each different ion is deflected by a different amount onto the detector. The detector is connected to a computer programmed to show the relative abundance of each type of ion. Modern mass spectrometers are capable of identifying relative abundances as small as 1 in 10^{14} ions.

1 Suggest why it is important that all the ions have the same speed in a mass spectrometer.
2 Calculate the radius of curvature for a singly ionised carbon-13 ion travelling at a speed of $8.00 \times 10^4\,\mathrm{m\,s^{-1}}$ in a mass spectrometer with a field of flux density 0.750 T. The mass of the ion is 2.16×10^{-26} kg.
3 Estimate the radius of curvature of a singly ionised carbon-14 ion in the same mass spectrometer.

Synoptic link

You first met number density and the equation $I = Anev$ in Topic 8.4, Mean drift velocity.

Summary questions

1 Explain why a stationary charged particle in a magnetic field does not experience a magnetic force. *(1 mark)*

2 Calculate the maximum magnetic force experienced by an electron travelling at a speed of $6.0 \times 10^5\,\mathrm{m\,s^{-1}}$ in a uniform field of flux density 0.20 T. *(2 marks)*

3 A particle of charge +2e describes a circle of radius 2.5 cm in a uniform magnetic field of flux density 130 mT. Calculate its momentum. *(3 marks)*

4 The parallel plates of a velocity selector are connected to a 1.3 kV supply and have a separation of 2.5 cm. It selects particles of speed $4.0 \times 10^5\,\mathrm{m\,s^{-1}}$. Calculate the magnetic flux density of the magnetic field used in the velocity selector. *(3 marks)*

5 A proton travelling at $4.0 \times 10^6\,\mathrm{m\,s^{-1}}$ describes a circular path in a uniform magnetic field of flux density 800 mT. Calculate:

a the radius of the path; *(3 marks)*

b the period of the proton. *(2 marks)*

6 Show that the period of the proton in question 5 is independent of its speed. *(3 marks)*

The Hall probe

A Hall probe is a device used to measure magnetic flux density directly.

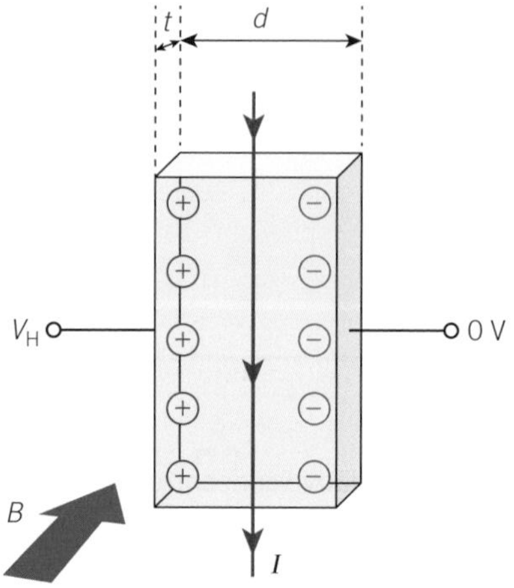

▲ **Figure 7** *Generation of a Hall voltage*

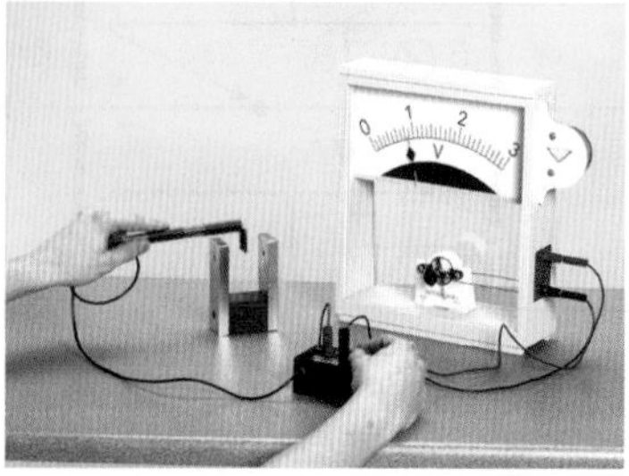

▲ **Figure 8** *Use of a Hall probe – the probe is being held between the poles of a U-shaped magnet, and a voltmeter. It is being used to detect the Hall voltage. The effect is strongest when the main current is aligned at right angles to the magnetic field, so rotating the probe causes the voltage to drop*

Figure 7 shows a thin slice of a semiconductor. It has thickness t and width d and carries a current I in the direction shown. An external magnetic field of flux density B is applied at right angles to the direction of the current. According to Fleming's left-hand rule, the electrons will be deflected towards the right-hand surface, where they accumulate, leaving the left-hand surface of the semiconductor with fewer electrons. As a result, a small potential difference, known as the Hall voltage, V_H, develops across the slice. The accumulated electrons create a uniform electric field of magnitude E, where

$$E = \frac{V_H}{d}$$

The Hall voltage V_H is given by the equation

$$V_H = \frac{BI}{nte}$$

where e is the elementary charge and n is the number density of the electrons within the semiconductor.

1 The internal electric field and the external magnetic field make the electrons travel undeflected through the semiconductor. The current is given by the equation $I = Anev$. Use this equation and the principles of a velocity selector to derive the equation

$$V_H = \frac{BI}{nte}$$

2 Suggest why semiconductors are preferable to metals in a Hall probe.

3 A flux density of 60 mT produces a Hall voltage of 14 mV. Calculate the Hall voltage when the flux density is 1.2 T.

23.4 Electromagnetic induction

Specification reference: 6.3.3

Turbines

You know that a current-carrying conductor produces magnetism, but can you produce electrical currents using magnetism? This question was tackled in the 1800s by the eminent scientist Michael Faraday, whose pioneering experiments revealed much about electromagnetic induction. Electromagnetic induction occurs in the generators in power stations, and in wind turbines. Figure 1 shows the inside of a large wind turbine. It generates electricity – induces an e.m.f. – by relative motion between a conductor and a magnetic field.

Learning outcomes

Demonstrate knowledge, understanding, and application of:

→ magnetic flux ϕ, the unit weber, $\phi = BA\cos\theta$

→ magnetic flux linkage.

▲ **Figure 1** *The generator inside a wind turbine produces electrical energy by electromagnetic induction*

Investigating electromagnetic induction

To induce an e.m.f. all you need is a coil and a magnet (Figure 2). A sensitive voltmeter attached to the coil shows no reading when the coil and the magnet are stationary. When the magnet is pushed towards the coil, an e.m.f. is induced across the ends of the coil, and when the magnet is pulled away a reverse e.m.f. is induced. Repeatedly pushing and pulling the magnet will induce an alternating current in the coil. The faster the magnet is moved, the larger is the induced e.m.f.

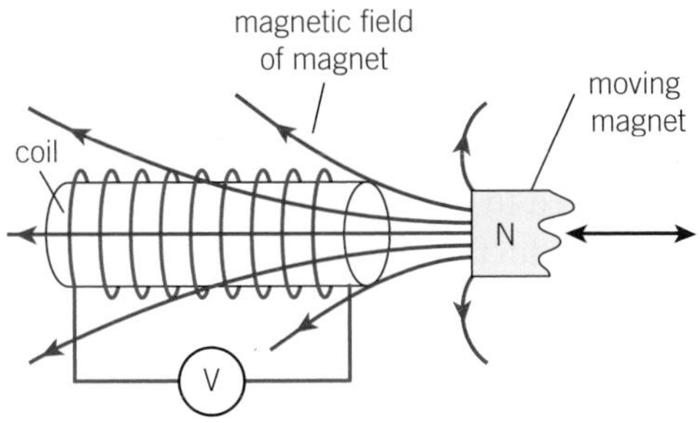

▲ **Figure 2** *Inducing an e.m.f. across the ends of a coil using a moving magnet*

There are other methods of inducing an e.m.f. in conductors. You can use a simple d.c. electric motor in reverse, for example using a falling mass to rotate the coil between the poles of the stationary magnet. The induced e.m.f. can be large enough to operate a lamp (Figure 3). An e.m.f. is induced in a loop of copper wire when it is moved perpendicular to the magnetic field lines of a magnet (Figure 4). The magnitude of the e.m.f. is bigger when the wire is pulled away faster from the magnetic field.

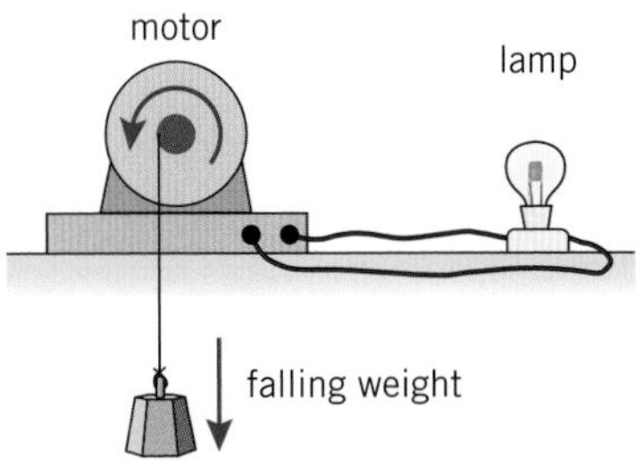

▲ **Figure 3** *Using a motor as a generator*

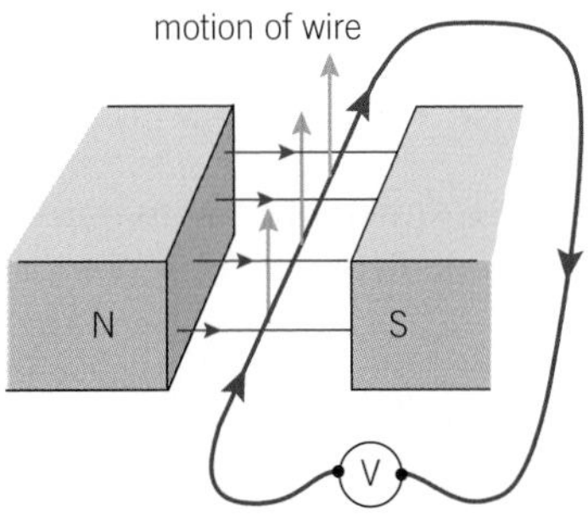

◀ **Figure 4** *Using a wire to produce an e.m.f.*

Explaining electromagnetic induction

Energy is always conserved – this principle cannot be violated. So where does the electrical energy produced in the coil shown in Figure 2 come from? Some of the work done to move the magnet is transferred into electrical energy. The motion of the coil (and the electrons in it) relative to the magnetic field makes the electrons move because they experience a magnetic force given by Bev, where B is the magnetic flux density, e is the elementary charge, and v is the relative speed between the coil and magnet. The moving electrons constitute an electrical current within the coil, so the process has produced electrical energy.

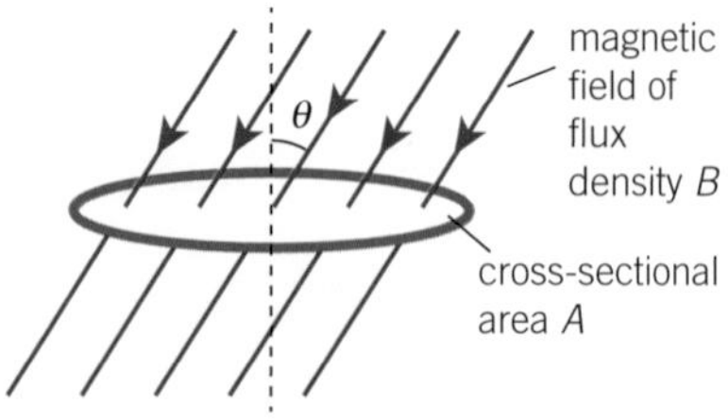

▲ **Figure 5** *Magnetic flux ϕ is the product of the component of the magnetic flux density perpendicular to the area and the cross-sectional area*

Magnetic flux and magnetic flux linkage

Every experiment demonstrating electromagnetic induction can be explained in terms of **magnetic flux**, ϕ.

Figure 5 shows a uniform magnetic field of flux density B passing through a region with a cross-sectional area A at an angle θ to the normal. The magnetic flux ϕ is defined as the product of the component of the magnetic flux density perpendicular to the area and the cross-sectional area, that is

$$\phi = (B\cos\theta) \times A \qquad \text{or} \qquad BA\cos\theta$$

When the field is normal to the area, $\cos 0° = 1$ and $\phi = BA$.

The SI unit for magnetic flux is the weber (Wb). From the equation above you can show that $1\,\text{Wb} = 1\,\text{T}\,\text{m}^2$.

Study tip

It is easy to confuse magnetic flux density B and magnetic flux ϕ, but they are very different. The units, T and Wb respectively, are the clue for identifying these two quantities.

Another quantity related to magnetic flux is called **magnetic flux linkage**. This is the product of the number of turns in the coil N and the magnetic flux, that is,

$$\text{magnetic flux linkage} = N\phi$$

The SI unit of magnetic flux linkage is also the **weber**, but sometimes weber-turns is also used to distinguish it from magnetic flux.

An e.m.f. is induced when…

An e.m.f. is induced in a circuit whenever there is a *change* in the magnetic flux linking the circuit. Since $\phi = BA\cos\theta$, you can induce an e.m.f. by changing B, A, or θ.

Worked example: Getting the terminology right

A coil has 200 turns and a core of cross-sectional area $1.0 \times 10^{-4}\,\text{m}^2$. The coil is placed at right angles to a magnetic field of flux density 0.30 T. Calculate the magnetic flux and magnetic flux linkage for the coil.

Step 1: Calculate the magnetic flux. At right angles, magnetic flux $\phi = BA = 0.30 \times 1.0 \times 10^{-4} = 3.0 \times 10^{-5}\,\text{Wb}$

Step 2: The magnetic flux linkage is $N\phi$. Therefore Magnetic flux linkage $= N\phi = 200 \times 3.0 \times 10^{-5} = 6.0 \times 10^{-3}\,\text{Wb}$

Summary questions

1 State the SI units for magnetic flux density, magnetic flux, and magnetic flux linkage. *(1 mark)*

2 Use the idea of magnetic flux to explain why an e.m.f. is induced in the coil shown in Figure 2. *(2 marks)*

3 A single loop of wire coil has a cross-sectional area $1.4 \times 10^{-4}\,m^2$. Calculate the maximum magnetic flux for this loop in a field of flux density 0.02 T. *(2 marks)*

4 Calculate the magnetic flux linkage for the coil shown in Figure 6. *(2 marks)*

5 The direction of the magnetic field is reversed for the coil shown in Figure 6. Calculate the change in the magnetic flux linkage. *(2 marks)*

6 In London, the Earth's magnetic field makes an angle of 66° with the horizontal and has flux density 4.9×10^{-5} T. Estimate the magnetic flux for a small coin lying on flat ground. *(3 marks)*

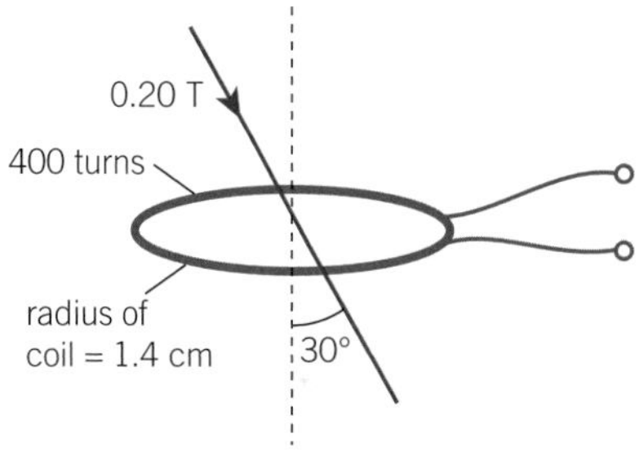

▲ **Figure 6**

23.5 Faraday's law and Lenz's law

Specification reference: 6.3.3

Learning outcomes

Demonstrate knowledge, understanding, and application of:

- → Faraday's law of electromagnetic induction
- → Lenz's law
- → e.m.f. = – rate of change of magnetic flux linkage, $\varepsilon = -\frac{\Delta(N\phi)}{\Delta t}$, techniques and procedures used to investigate magnetic flux using search coils
- → simple a.c. generator.

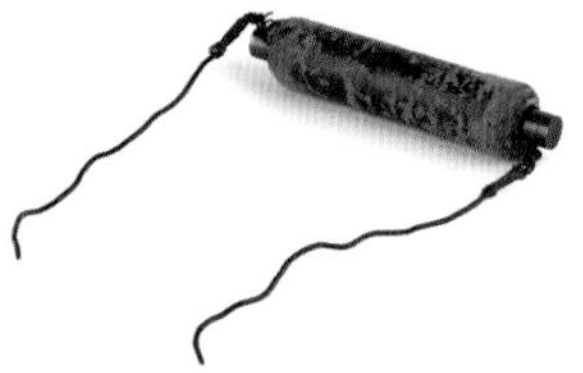

▲ **Figure 1** *The first generator in the world, made by Michael Faraday in 1831*

The first generator

Figure 1 shows the first ever electric generator – a coil of copper wound around a hollow core. Moving a magnetised iron rod through the coil induced an e.m.f. and hence a current in the coil. Faraday's imagination and inventiveness helped him to formulate a law for electromagnetic induction – a law that we now call **Faraday's law**.

Faraday's law of electromagnetic induction

In Topic 23.4, Electromagnetic induction, the idea of magnetic flux linkage was introduced. Faraday's law relates it to the magnitude of the induced e.m.f. in conductors.

Faraday's law: The magnitude of the induced e.m.f. is directly proportional to the rate of change of magnetic flux linkage.

We can write this mathematically as

$$\varepsilon \propto \frac{\Delta(N\phi)}{\Delta t}$$

where ε is the induced e.m.f and $\Delta(N\phi)$ is the change in magnetic flux linkage in a time interval Δt.

This relationship can be written as an equation where the constant of proportionality is equal to –1. The reasons for the negative sign will be given later when we examine **Lenz's law**.

$$\varepsilon = -\frac{\Delta(N\phi)}{\Delta t}$$

The equation above is as simple and elegant as Newton's second law in mechanics, and like all fundamental laws, it can explain a variety of phenomena.

Worked example: Search coil

A search coil (used to measure variations in magnetic flux) is made of thin copper wire with 2000 turns and a mean cross-sectional area of 1.4 cm^2. It is placed between the poles of a strong magnet at right angles to the magnetic field of flux density 0.30 T and then quickly removed from the field in a time of 80 ms. The ends of the search coil are connected to an oscilloscope (Figure 2). Calculate the magnitude of the average e.m.f. induced across the ends of the search coil.

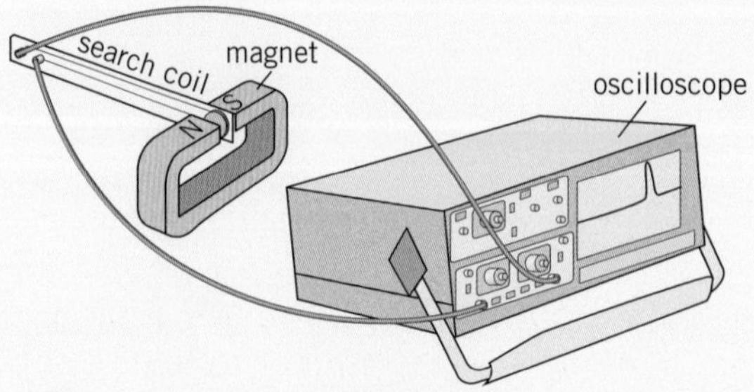

▲ **Figure 2** *A search coil in use*

Step 1: To find ε you first need to calculate $\Delta(N\phi)$. The final flux linkage for the coil is zero. The initial flux linkage can be calculated using $N\phi = NBA\cos\theta$ (where $\theta = 90°$). It is important to convert the cross-sectional area of the coil into m^2 when calculating the change in the flux linkage.

$(N\phi)$ = final flux linkage – initial flux linkage

$$\Delta(N\phi) = 0 - 2000 \times (0.30 \times 1.4 \times 10^{-4})$$
$$= -8.40 \times 10^{-2}\,\text{Wb}\ (1\,\text{cm}^2 = 10^{-4}\,\text{m}^2)$$

Step 2: Calculate the induced e.m.f. ε using Faraday's law.

$$\varepsilon = -\frac{\Delta(N\phi)}{\Delta t} = \frac{8.40 \times 10^{-2}}{0.08} = 1.1\,\text{V (2 s.f.)}$$

Lenz's law

Figure 3 shows the coil and magnet arrangement that you have already met in Topic 23.4. The only difference here is that there is no voltmeter – instead the wires are connected together so that any induced currents in the coils are large enough to create their own strong magnetic fields. The direction of the induced e.m.f., and hence the current, changes direction when the magnet is pulled away from coil instead of being pushed into the coil. Why does this happen?

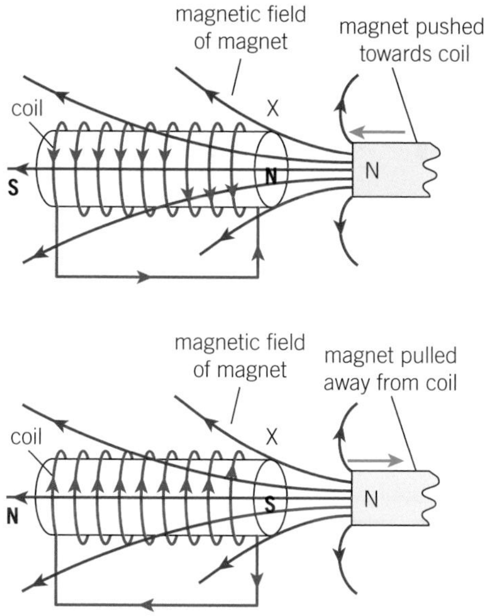

▲ **Figure 3** *The coil and the magnet repel (above) or attract (below) each other.*

Figure 3 shows what happens when the magnet and the end X of the coil are brought closer together. In the upper image, the induced current is such that the end X of the coil has a north polarity. You have to do work to push the magnet towards the coil. The work done on the magnet is equal to the electrical energy produced in the coil. The end X cannot be a south pole. If it could be, then the principle of conservation of energy would be violated, with attraction between the coil and the magnet creating electrical energy from nowhere.

When the magnet is pulled away from the coil, the motion of the magnet must once again be opposed so that you must do work. The end X therefore has a south polarity and the induced e.m.f. and current are reversed (lower part of Figure 3). **Lenz's law** is an expression of conservation of energy.

Lenz's law: The direction of the induced e.m.f. or current is always such as to oppose the change producing it.

The negative sign in the equation for Faraday's law is mathematical way of expressing Lenz's law. In most calculations, you can ignore this minus sign. However, it is a reminder that energy cannot be created from nothing.

The alternating current generator

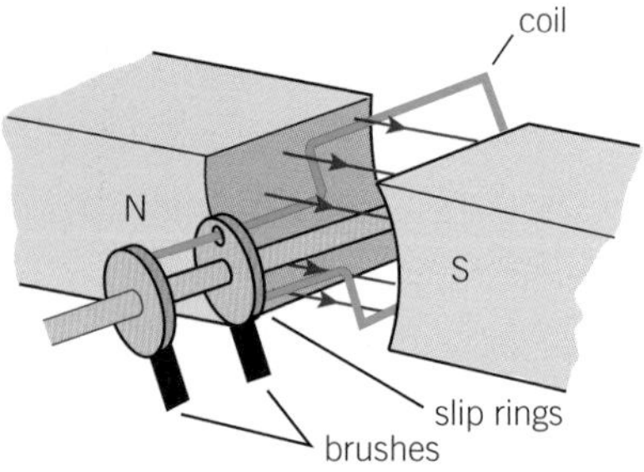

▲ **Figure 4** *An a.c. generator*

Our lives would be completely different without the mains electricity from generators spinning away to producing an alternating e.m.f. of frequency 50 Hz. We can explain the principles of an alternating current (a.c.) generator using Faraday's law.

The simple a.c. generator in Figure 4 consists of a rectangular coil of cross-sectional area A and N turns of coil rotating in a uniform magnetic field of flux density B. The flux linkage for the coil is

$$\text{flux linkage} = N\phi = N(BA\cos\theta) = BAN\cos\theta$$

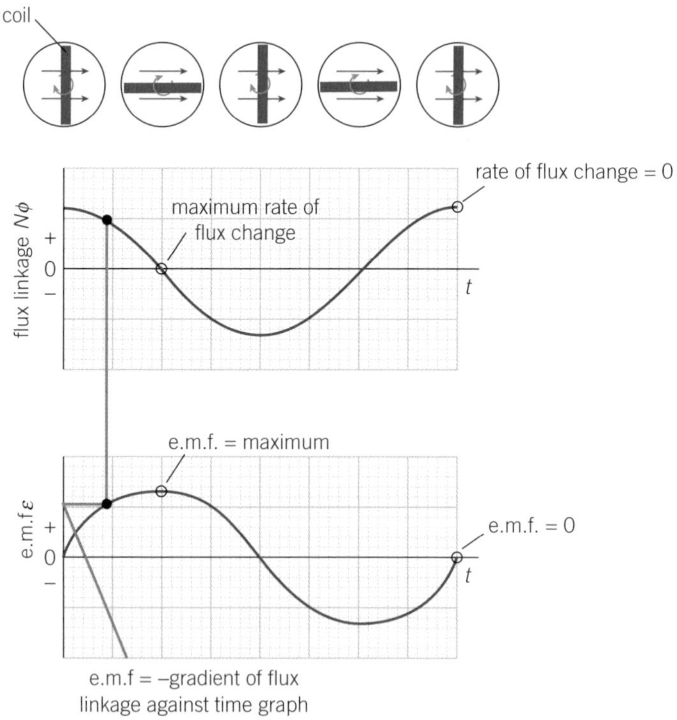

▲ **Figure 5** *The variation of flux linkage with time (above) and of the induced e.m.f. with time (below)*

As the coil rotates at a steady frequency, the flux linkage changes with time t as shown in the first graph in Figure 5. This variation is referred to as sinusoidal and is caused by the changing $\cos\theta$ factor.

According to Faraday's law, the induced e.m.f. $\varepsilon = -\frac{\Delta(BAN\cos\theta)}{\Delta t}$.

- The magnitude of the gradient from the magnetic flux linkage against time graph is equal to the induced e.m.f. ε.
- For a given generator, B, A, and N are all constant, therefore $\varepsilon \propto -\frac{\Delta(\cos\theta)}{\Delta t}$.

The lower graph in Figure 5 shows the variation of e.m.f. ε with time t. The maximum induced e.m.f. is directly proportional to:

- the magnetic flux density B
- the cross-sectional area A of the coil
- the number of turns N
- the frequency f of the rotating coil.

Study tip

For a generator, the induced e.m.f. ε is at its maximum when the flux linkage is zero and ε is zero when the flux linkage is at its maximum.

Summary questions

1 State what the minus sign represents in the equation for Faraday's law. *(1 mark)*

2 Figure 6 shows the variation of flux linkage with time for three coils. State and explain the e.m.f. induced in the coil in each case. *(3 marks)*

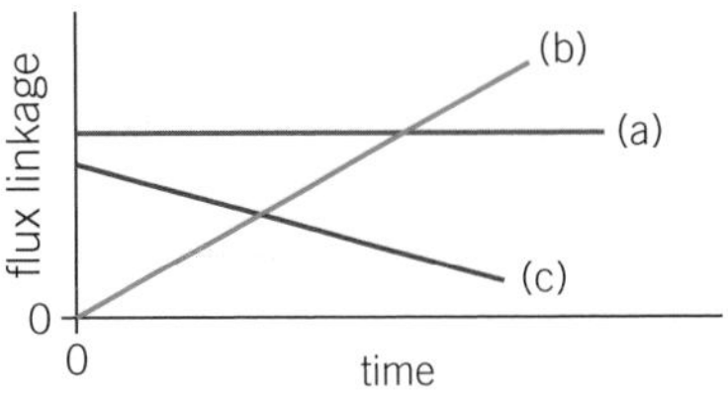

▲ **Figure 6**

3 A coil connected to a voltmeter is placed next to one end of a long current-carrying solenoid. The voltmeter reads zero. When the current in the solenoid is switched off, the voltmeter shows a reading for a very short interval of time and then goes back to zero. Explain these observations. *(3 marks)*

4 The north pole of a bar magnet is placed on top of a square coil of cross-sectional area $3.0 \times 10^{-4}\,\text{m}^2$. The coil has 800 turns. The magnet is quickly removed from the coil in a time of 0.12 s. The average induced e.m.f. in the coil is 32 mV. Calculate the magnetic flux density at the pole of the magnet. *(4 marks)*

5 Explain why a large current-carrying coil can produce dangerously high 'back' e.m.f. when the current is suddenly switched off. *(3 marks)*

6 A horizontal copper wire of length L forms part of a circuit. It is moved with a constant speed v in a region of vertical magnetic field of flux density B. Use Faraday's law to show that the induced e.m.f. ε across the ends of the wire is given by the expression $\varepsilon = BvL$. *(3 marks)*

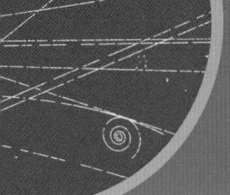

23.6 Transformers

Specification reference: 6.3.3

Learning outcomes

Demonstrate knowledge, understanding, and application of:

- → the simple laminated, iron-cored transformer
- → $\frac{n_s}{n_p} = \frac{V_s}{V_p} = \frac{I_p}{I_s}$ for an ideal transformer
- → techniques and procedures used to investigate transformers.

▲ **Figure 1** *A transformer – note the fins for air cooling*

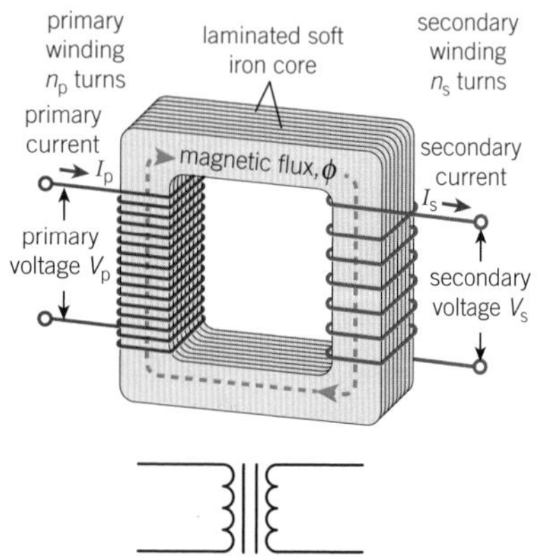

▲ **Figure 2** *The structure of an iron-core transformer and its circuit symbol*

Study tip

Transformers will not work with steady direct current because there is no changing magnetic flux.

Transformers change voltages

One important use of electromagnetic induction is in transformers, which change alternating voltages to higher or lower values. Power stations use transformers to convert the supply from 25 kV up to 400 kV. Mobile phone chargers have transformers that change the mains voltage of 230 V down to lower values such as 5 V. In this topic you will learn about iron-core transformers.

Step-up and step-down transformers

A simple transformer (Figure 2) consists of a laminated iron core, a primary (input) coil, and a secondary (output) coil. An alternating current is supplied to the primary coil. This produces a varying magnetic flux in the soft iron core. The secondary coil, which is wound round the same core, is linked by this changing flux. The iron core ensures that all the magnetic flux created by the primary coil links the secondary coil and none is lost. According to Faraday's law of electromagnetic induction, a varying e.m.f. is produced across the ends of the secondary coil.

The input voltage V_p and the output voltage V_s are related to the number n_p of turns on the primary coil and number n_s of turns on the secondary coil by the **turn-ratio equation**

$$\frac{n_s}{n_p} = \frac{V_s}{V_p} \text{ for an ideal transformer}$$

- A **step-up transformer** has more turns on the secondary than on the primary coil, and $V_s > V_p$.
- A **step-down transformer** has fewer turns on the secondary than on the primary coil, and $V_s < V_p$.

Worked example: Step-down transformer

A step-down transformer changes 230 V mains voltage to 5.0 V. The transformer has 920 turns on its primary coil. Calculate the number of turns on its secondary coil.

Step 1: Rearrange the turn-ratio equation.

$$\frac{n_s}{n_p} = \frac{V_s}{V_p}$$

$$n_s = \frac{V_s n_p}{V_p}$$

Step 2: Calculate the number of turns on the secondary coil.

$$n_s = \frac{V_s n_p}{V_p} = \frac{5.0 \times 920}{230} = 20 \text{ turns}$$

Experimenting with transformers

Figure 3 shows an arrangement that you can use in the laboratory to investigate transformers. A multimeter set to 'alternating voltage' can be used to measure the input V_p and output V_s voltages, or you can use an oscilloscope instead. Thin insulated copper wires are used to make primary and secondary coils. You can change the number of turns on one or both coils to see what happens to V_s for a fixed value of V_p and vice versa.

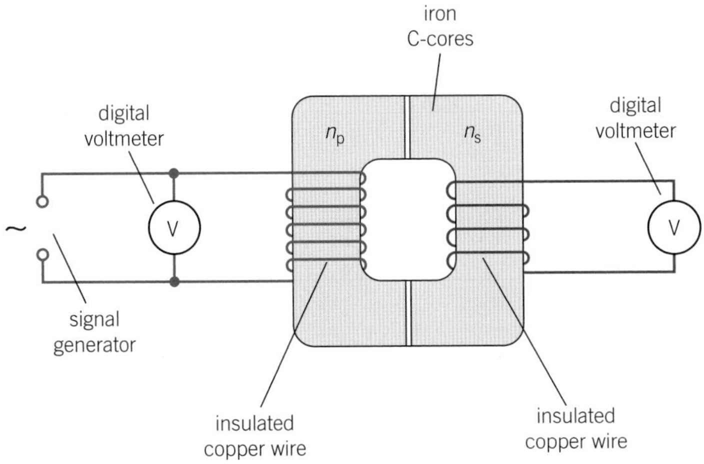

▲ **Figure 3** *Apparatus for investigating transformers*

Efficient transformers

For a 100% efficient transformer, the output power from the secondary coil is equal to the input power into its primary coil. Since power is the product of voltage and current, we have

$$V_s I_s = V_p I_p$$

or

$$\frac{I_p}{I_s} = \frac{V_s}{V_p}$$

Thus, in a step-up transformer, the voltage is stepped up but the current is stepped down. Increasing the voltage by a factor of 100 will decrease the output current by a factor of 100. Similarly, in a step-down transformer, the voltage is stepped down and the current is stepped up.

Transformers can be made efficient by using low-resistance windings to reduce power losses due to the heating effect of the current. Making a laminated core with layers of iron separated by an insulator helps to minimise currents induced in the core itself (eddy currents), so this too minimises loses due to heating. The core is made of soft iron, which is very easy to magnetise and demagnetise, and this also helps to improve the overall efficiency of the transformer.

The National Grid

In the UK, electrical power is transported across the country by the National Grid. This network consists of transformers and cables on pylons and underground. All a.c. generators in large power stations produce an alternating voltage of about 25 kV at a precise frequency of 50 Hz. Figure 4 shows how a system of cables and transformers distributes electrical power across the country.

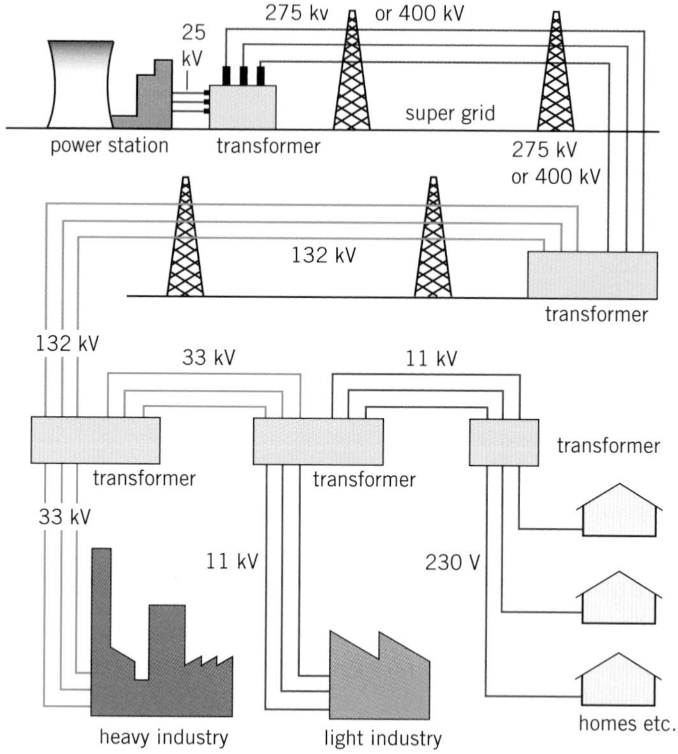

▲ **Figure 4** *The National Grid system*

Electrical power is transmitted at high voltage so as to minimise heat losses in the transmission cables. To deliver a power P_0 at a voltage V, the current I required is given by the equation $I = \frac{P_0}{V}$. For transmission cables of resistance R, the power loss P_L due to heating in the cables is given by the equation $P_L = I^2R = \frac{P_0^{\,2}R}{V^2}$. The higher the transmission voltage V, the smaller are the power losses through heating $(P_L \propto \frac{1}{V^2})$.

A small power station produces 1 MW. Calculate:

1 the current in the transmission cables operating at 400 kV
2 the power losses when the resistance of the cables is 500 Ω
3 the percentage of power lost when power is transmitted at 400 kV
4 the percentage of power lost when the same power is transmitted along the same cables at 40 kV. Comment on your answers to questions 3 and 4.

Summary questions

1 Explain the purpose of the iron core in a transformer. *(1 mark)*

2 Design a step-up transformer that will increase the output voltage by a factor of 20. *(1 mark)*

3 State two reasons why a transformer may not have 100% efficiency. *(2 marks)*

4 An old mobile phone charger has an inbuilt transformer that produces an output voltage of 5.2 V. The input voltage is 230 V and the primary coil has 500 turns. Calculate the number of turns on the secondary coil. *(2 marks)*

5 An electronic device uses a transformer with turns of ratio 20 : 1 to step down the mains voltage from 230 V. Calculate the output voltage from the transformer. *(2 marks)*

6 A transformer is used to step down 230 V mains voltage to 12 V. A 60 W lamp connected to the secondary coil is lit normally. The primary coil has 1000 turns. Calculate:
   a the number of turns on the secondary coil; *(2 marks)*
   b the current in the primary coil. *(2 marks)*

Practice questions

1 **a** Define *magnetic flux density.* (*1 mark*)

b Figure 1 shows an arrangement used by a student to determine the magnetic flux density between the poles of a magnet.

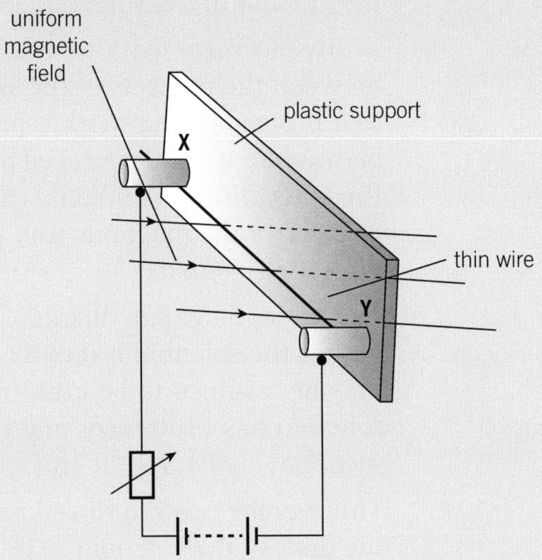

▲ Figure 1

A thin copper wire is placed horizontally on the electrical contacts **X** and **Y**. The separation between **X** and **Y** is 5.0 cm. The magnetic field between the poles of the magnet is at right angles to the wire. The current in the circuit is slowly increased from zero until the wire momentarily lifts off the contacts. The current *I* when this happens is recorded by the student, together with the mass *m* of the wire. The student repeats the experiment with copper wire of different thicknesses.

Table 1 shows data collected by the student.

▼ Table 1

I / A	m / g	F / 10^{-3} N
0.36	0.21	2.1
0.59	0.28	2.7
0.95	0.39	
1.34	0.51	
1.70	0.62	

(i) Name the equipment used to measure the mass and determine the maximum percentage uncertainty in the measurement of mass. (*4 marks*)

(ii) The force acting on the wire when it just lifts off the contacts **X** and **Y** is *F*. Complete the last column on a copy of Table 1. (*1 mark*)

(iii) Plot a graph of *F* against *I* and draw a straight line of best fit. (*3 marks*)

(iv) The straight line of best fit does not pass through the origin. Suggest a likely systematic error in this experiment. (*1 mark*)

(v) Use your graph to determine the magnetic flux density of the magnetic field between the poles of the magnet. (*4 marks*)

2 Figure 2 shows a section through a mass spectrometer.

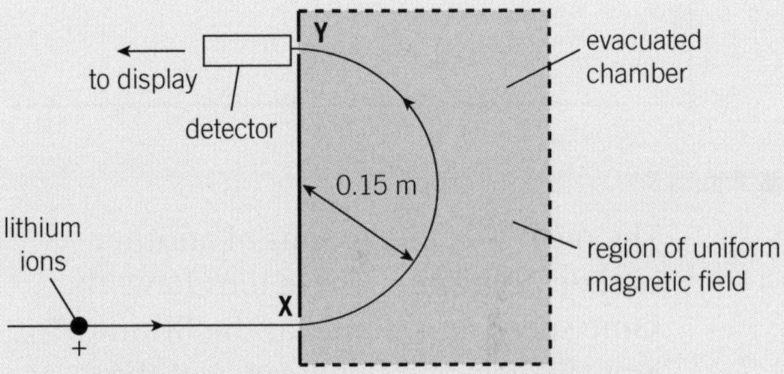

▲ Figure 2

A beam of positive lithium ions enter the evacuated chamber through the hole at **X**. the ions travel through a region of uniform magnetic field. The magnetic field is directed vertically into the plan of the diagram. The ions exit and are detected at **Y**.

a Name the rule that may be used to determine the direction of the force acting on the ions. (*1 mark*)

b Explain why the speed of the ions travelling from **X** to **Y** in the magnetic field does not change despite the force acting on the ions. (*1 mark*)

c The lithium-7 ions are detected at **Y**. All the ions have the same speed $4.0 \times 10^5\,\mathrm{m\,s^{-1}}$ and charge, $+1.6 \times 10^{-19}\,\mathrm{C}$. The radius of the semi-circular path of the ions in the magnetic field is 0.15 m. The mass of a lithium-7 ion is 1.2×10^{-26}.

(i) Calculate the force acting on a lithium ion as it moves in the semi-circle. *(2 marks)*

(ii) Calculate the magnitude of the magnetic flux density *B*. *(2 marks)*

(iii) The current recorded by the detector at **Y** is $4.8 \times 10^{-9}\,\mathrm{A}$. Calculate the number of lithium-7 ions reaching the detector per second. *(2 marks)*

d Figure 3 shows the variation of current *I* in the detector with magnetic flux density *B*.

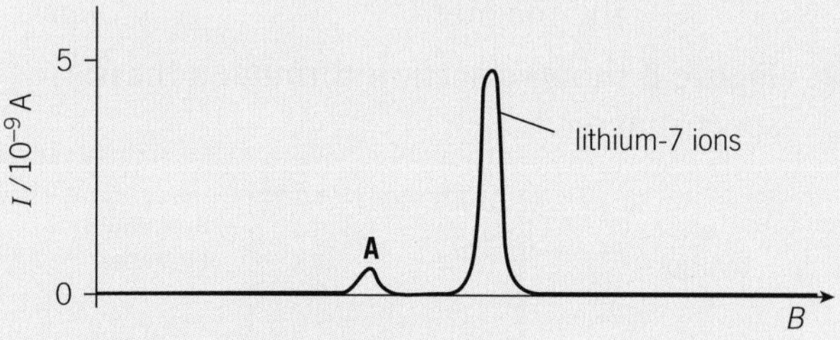

▲ **Figure 3**

The peak **A** is due to ions of another isotope of lithium. These ions have the same speed and charge as the lithium-7 ions. Explain the significance of the 'height' and position of peak **A**. *(2 marks)*

Jan 2013 G485

3 a Define *magnetic flux*. *(1 mark)*

b Figure 4 shows a solenoid connected to a battery and the magnetic field through it when the switch **S** is closed.

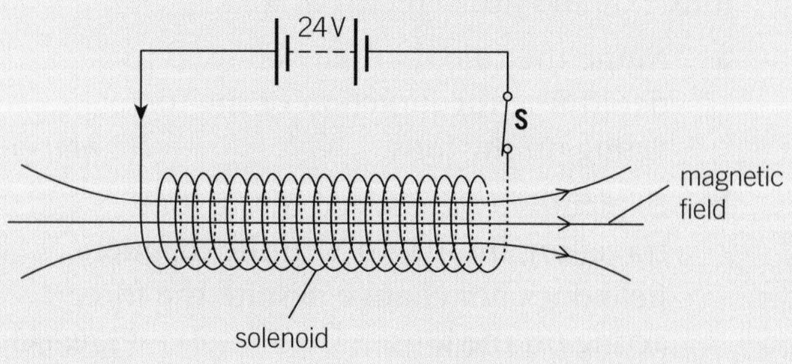

▲ **Figure 4**

(i) The battery has an e.m.f. of 24 V and negligible internal resistance. The solenoid is made from copper wire. The wire has radius $4.6 \times 10^{-4}\,\mathrm{m}$ and total length 130 m. The resistivity of copper is $1.7 \times 10^{-8}\,\Omega\mathrm{m}$. Calculate the current in the solenoid. *(3 marks)*

(ii) A tiny electrical spark is created between the contacts of the switch **S** as it is opened. The spark is produced because an e.m.f. is induced across the ends of the solenoid by the collapse of the magnetic flux linked with the solenoid.

The initial magnetic flux density within the solenoid is 0.090 T and may be assumed to be uniform. The solenoid has 1100 turns and cross-sectional area $1.3 \times 10^{-3}\,\mathrm{m^2}$.

The average e.m.f. induced across the ends of the solenoid is 150 V. Estimate the time taken for the magnetic flux to collapse to zero. *(3 marks)*

Jun 2013 G486

4 a Define *magnetic flux*. *(1 mark)*

b Figure 5 shows a generator coil of 500 turns and cross-sectional area $2.5 \times 10^{-3}\,\mathrm{m^2}$ placed in a magnetic field of magnetic flux density 0.035 T. The plane of the coil is perpendicular to the magnetic field.

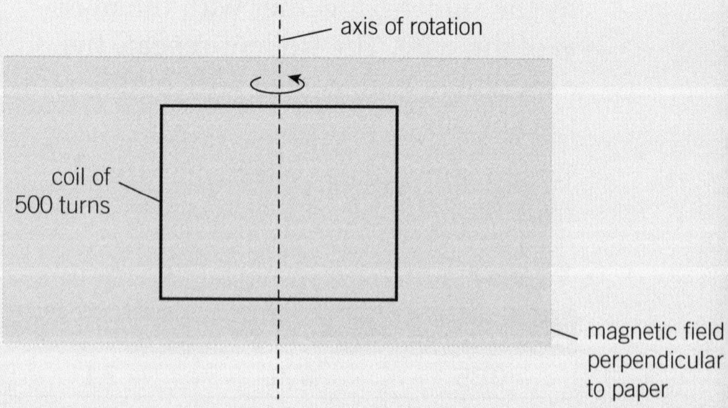

▲ **Figure 5**

Calculate the magnetic flux linkage for the coil in this position. Give a unit for your answer. *(3 marks)*

c The coil is rotated about the axis in the direction shown in Figure 5.

Figure 6 shows the variation of the magnetic flux ϕ against time t as the coil is rotated.

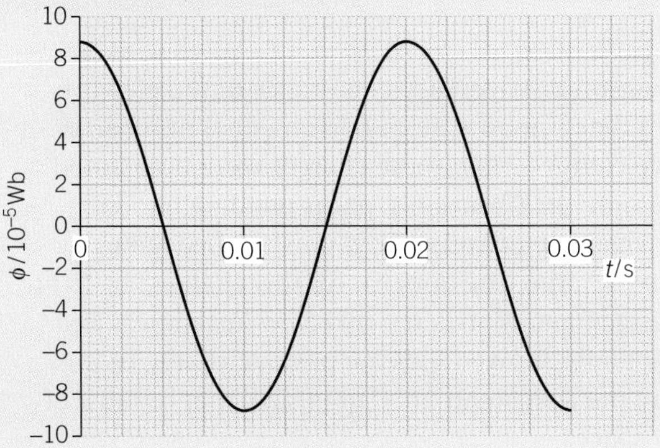

▲ **Figure 6**

(i) Explain why the magnitude of the magnetic flux through the coil varies as the coil varies. *(2 marks)*

(ii) State Faraday's law of electromagnetic induction. *(1 mark)*

(iii) Use Figure 6 to describe and explain the variation with time of the induced e.m.f. across the ends of the coil. *(3 marks)*

(iv) Use Figure 6 to determine the magnitude of the average induced e.m.f. for the coil between the times 0 s and 0.005 s. *(2 marks)*

(v) State and explain the effect on the magnitude of the maximum induced e.m.f. across the ends of the coil when the coil is rotated at twice the frequency. *(2 marks)*

Jun 2010 G485

5 Figure 7 shows part of an accelerator used to produce high-speed protons. The protons pass through an evacuated tube that is shown in the plane of the paper.

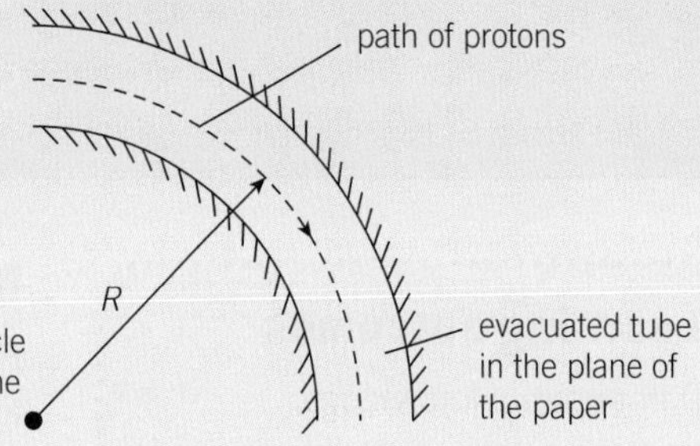

▲ **Figure 7**

The protons are made to travel in a circle of radius R by a magnetic field of flux density B.

a State clearly the direction of the magnetic flux density B that produces the circular motion of the protons. *(1 mark)*

b Show that the relationship between the velocity v of the protons and the radius R is given by $v = \frac{BQR}{m}$ where Q and m are the charge and mass of a proton respectively. *(1 mark)*

c Calculate the magnetic flux density B of the magnetic field needed to keep protons in a circular orbit of radius 0.18 m. The time for one complete orbit is 2.0×10^{-8} m. *(3 marks)*

d Explain why the magnetic field does not change the speed of the protons. *(2 marks)*

Jan 2011 G485

24 PARTICLE PHYSICS
24.1 Alpha-particle scattering experiment

Specification reference: 6.4.1

Learning outcomes

Demonstrate knowledge, understanding, and application of:

- the alpha-particle scattering experiment
- relative sizes of the atom and the nucleus.

Nuclear model

Englishman J. J. Thomson discovered the existence of the electron in 1897. He proposed that a neutral atom had an equal number of electrons and positive charges. How these charges were distributed was unknown at the time. In Thomson's 'plum-pudding' model, an atom contained negative electrons (the plums) embedded in a uniform sea of positive charge (the dough). All this changed in 1911 when Rutherford, Geiger, and Marsden – two New Zealanders and a German – experimentally showed that the positive charge of the atom existed in a tiny nucleus about 10^{-14} m in size, that is, most of the atom was empty space.

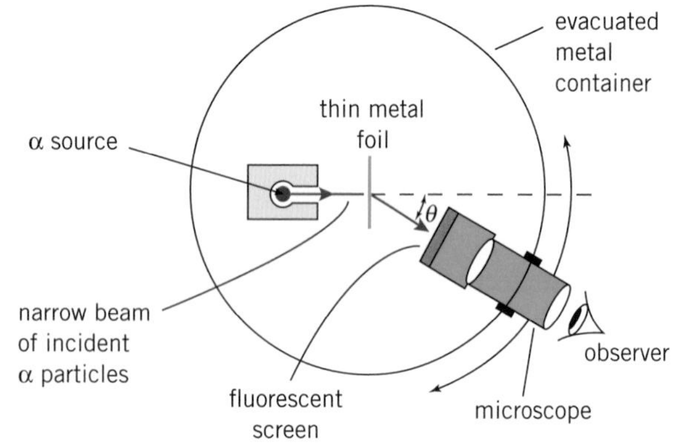

▲ **Figure 1** *Apparatus for the alpha-scattering experiment of Rutherford, Geiger, and Marsden*

Rutherford's alpha-scattering experiment

Figure 1 shows a simplified version of the arrangement used in Rutherford's experiments. A narrow beam of alpha particles, all of the same kinetic energy, from a radioactive source were targeted at a thin piece of gold foil which was only a few atomic layers thick. The alpha particles were scattered by the foil and detected on a zinc sulfide screen mounted in front of a microscope. Each alpha particle hitting this fluorescent screen produced a tiny speck of light. The microscope was moved around in order to count the number of alpha particles scattered through different values of the angle θ per minute, for θ from zero to almost 180°.

Observations and conclusions

The scattering experiment led to the following two significant observations, which could not support Thomson's plum-pudding model of the atom.

- Most of the alpha particles passed straight through the thin gold foil with very little scattering. About 1 in every 2000 alpha particles was scattered.
- Very few of the alpha particles – about 1 in every 10 000 – were deflected through angles of more than 90°.

These significant observations can be explained in terms of a new model of the atom – the nuclear model. The first observation meant that most of the atom was empty space with most of the mass concentrated in a small region – the **nucleus**. The second observation led to the conclusion that the nucleus has a positive charge, because it repelled the few positive alpha particles that came near it. In fact, the charge on the nucleus is quantised and given by $+Ze$, where Z is the atomic number of the element (the proton number for the nucleus – see Topic 24.2, The nucleus) and e is the elementary charge 1.60×10^{-19} C.

Microscopic interactions

The scattering of the alpha particles from the gold nuclei can be modelled from Coulomb's law with the alpha particle having a charge $+2e$ and the gold nucleus having a charge $+Ze$.

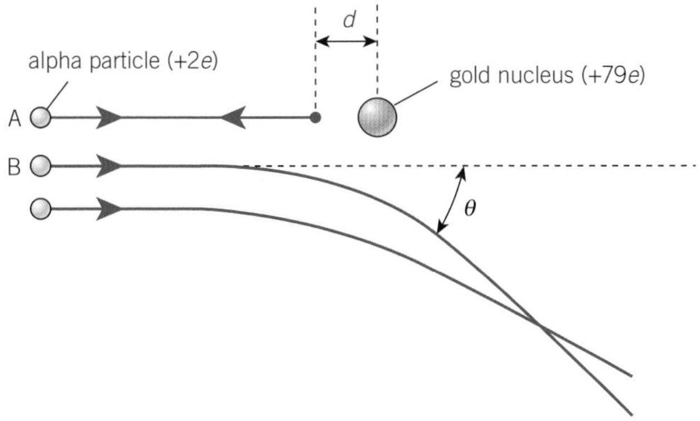

▲ **Figure 2** *Scattering of alpha particles by the positive gold nucleus*

Figure 2 shows the paths of some alpha particles as they pass close to the heavy gold nucleus. The alpha particle A makes a head-on collision with the nucleus and rebounds back with a scattering angle 180°. The minimum distance between the alpha particle and the gold nucleus is d. The probability of such a collision is very small because of the tiny diameter of the nucleus. The alpha particle B makes an oblique collision with the nucleus and is scattered through an angle θ.

Sizes of the atom and the nucleus

Rutherford predicted the fraction of alpha particles that would be scattered through an angle θ. He found that departures from his predictions started to occur for more energetic alpha particles that managed to get much closer to the nucleus. From his experiments, Rutherford concluded that the nucleus had a radius of about 10^{-14} m.

In one of the experiments, Rutherford used alpha particles of kinetic energy 1.2×10^{-12} J (about 7.7 MeV). The distance d of closest approach between an alpha particle and the gold nucleus can be calculated using the idea of conservation of energy. At this distance, the alpha particle momentarily stops. Therefore

initial kinetic energy of alpha particle = electrical potential energy at distance d

$$1.2 \times 10^{-12} = \frac{Qq}{4\pi\varepsilon_0 d} \qquad (Q = Ze = 79e \text{ and } q = 2e)$$

$$1.2 \times 10^{-12} = \frac{79 \times 2 \times (1.60 \times 10^{-19})^2}{4\pi \times 8.85 \times 10^{-12} \times d}$$

$$d = 3.0 \times 10^{-14} \approx 10^{-14}\,\text{m}$$

This calculation gives an *upper limit* for the radius of the gold nucleus. More energetic alpha particles might get closer. In Topic 24.2, The nucleus, you will see that the order of magnitude value for the radius of a nucleus is about 10^{-15} m. The radius of most atoms is about 10^{-10} m. So the nucleus is about 10^5 times smaller than the atom. If a nucleus is represented by a dot of diameter 1 mm, then the outermost electron of the atom would be 100 m away!

Rutherford's modelling

One of Rutherford's predictions, based on electrostatic repulsion between the alpha particle and the gold nucleus, was that the number *N* of alpha particles scattered through an angle θ was inversely proportional to $\sin^4\left(\frac{\theta}{2}\right)$ (You do not need to recall this for this course). Table 1 shows some of the actual results collected by Geiger and Marsden, working under Rutherford's direction.

▼ **Table 1** *Number N of alpha particles scattered through angle θ*

θ/°	15	30	45	60	120	150
N	132 000	7800	1435	477	52	33

1. Suggest why only a small number of alpha particles were scattered through large angles.
2. Calculate the force experienced by an alpha particle at a distance of 10^{-14} m from the centre of the gold nucleus.
3. Use the table to show that *N* is inversely proportional to $\sin^4\left(\frac{\theta}{2}\right)$.

Summary questions

1. In Rutherford's alpha-scattering experiment, most of the alpha particles were not scattered. What can you conclude about the nature of atoms? *(1 mark)*
2. State the approximate radii of the atom and the nucleus. *(1 mark)*
3. In a visual model of the atom, the nucleus is represented by an apple of diameter 8 cm. Estimate the diameter of the atom in this model. *(2 marks)*
4. Figure 3 shows the path of an alpha particle close to the nucleus of lead. Draw arrows to represent the force on the alpha particle when at points A, B, and C. *(2 marks)*

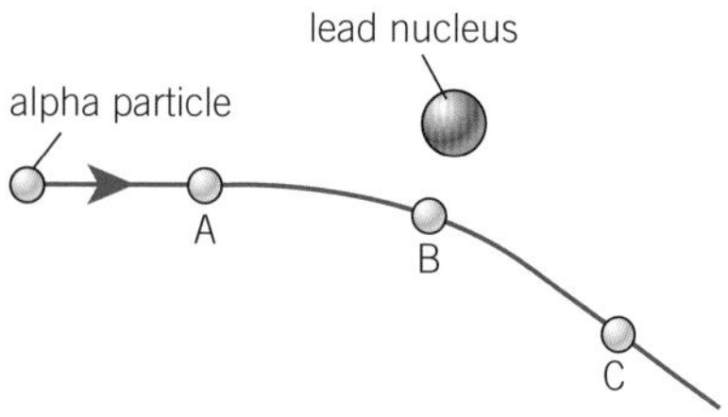

▲ **Figure 3**

5. Alpha particles of kinetic energy 8.8 MeV are fired at lead atoms. The charge on the nucleus of lead is 82*e*. Calculate:
   a. the minimum distance the alpha particles approach to the nucleus of lead *(4 marks)*
   b. the maximum electrostatic force experienced by the alpha particle. *(3 marks)*
6. A tiny droplet of oil diameter 1.0 mm is placed on water. The oil spreads out as a circular disc of thickness approximately one atom thick. Estimate the radius of this oil disc. *(3 marks)*

Specification reference: 6.4.1

Learning outcomes

Demonstrate knowledge, understanding, and application of:

- → the simple nuclear model of the atom
- → protons, neutrons, and electrons
- → proton number, nucleon number, isotopes, notation for the representation of nuclei
- → the strong nuclear force and its short-range nature
- → radius of nuclei – $R = r_0 A^{\frac{1}{3}}$
- → mean densities of atoms and nuclei.

Neutrons

In 1930, Bothe and Becker in Germany bombarded a beryllium target with alpha particles. They noticed that a very penetrating, non-ionising radiation was emitted from beryllium. They incorrectly assumed that they were observing gamma rays. In 1932 in Cambridge, Chadwick showed that the alpha particles hitting the beryllium nuclei were knocking **neutrons** from its nuclei. Chadwick was awarded the 1935 Nobel Prize for Physics for his discovery of the neutron. Neutrons carry no charge and exist in all nuclei except hydrogen.

The nuclear model of the atom

The nucleus of an atom contains positive protons and uncharged neutrons. Figure 1 shows a helium nucleus. The proton and the neutron have approximately the same mass. The term **nucleon** is used to refer to either a proton or a neutron. The proton has a charge of $+e$, where e is the elementary charge. A neutral atom has the same number of electrons and protons.

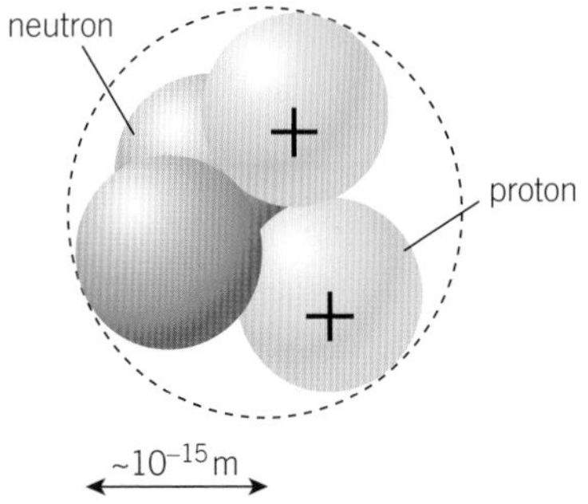

▲ **Figure 1** *A helium nucleus (alpha particle) with two protons and two neutrons*

Isotopes

The nucleus of an atom for a particular element is represented as

$$^{A}_{Z}\mathrm{X}$$

where X is the chemical symbol for the element, A is the **nucleon number** (the total number of protons and neutrons), and Z is the **proton number** (also known as **atomic number**). The number of neutrons N in the nucleus is thus $N = (A - Z)$.

Isotopes are nuclei of the same element that have the same number of protons but different numbers of neutrons. All isotopes of an element undergo the same chemical reactions.

The grid in Figure 2 shows the isotopes of hydrogen (H), helium (He), and lithium (Li).

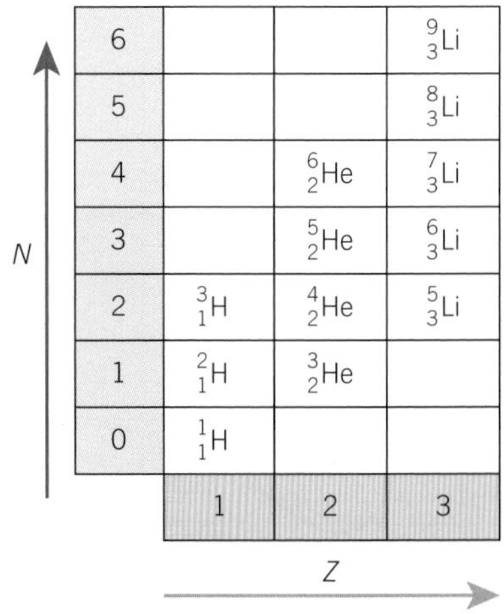

▲ **Figure 2** *Some isotopes shown by proton and neutron number on an N–Z grid*

Atomic mass units

The masses of atoms and nuclear particles are often expressed in **atomic mass units** (u). One atomic mass unit (1 u) is one-twelfth the mass of a neutral carbon-12 atom (Table 1). The experimental value of 1 u is about 1.661×10^{-27} kg.

▼ **Table 1** *The masses of some particles in atomic mass units (u), where 1 u = 1.661×10^{-27} kg.*

Particle	electron	proton	neutron	helium-4 nucleus	carbon-12 nucleus	iron-56 nucleus	uranium-235 nucleus
Mass / u	0.00055	1.00728	1.00867	4.00151	11.99671	55.79066	234.99343

Study tip

The approximate mass of a particle is its nucleon number in atomic mass units. For example, the mass of a carbon-12 nucleus ≈ 12 u and the mass of a uranium-235 nucleus ≈ 235 u.

Nuclear size and density

The radius of the nucleus depends on the nucleon number A of the nucleus. Fast-moving electrons have a de Broglie wavelength of about 10^{-15} m. Diffraction of such electrons has been used to determine the radii of isotopes. Experiments have shown that the radius R of a nucleus is given by the equation

$$R = r_0 A^{\frac{1}{3}}$$

where r_0 has an approximate value of 1.2 fm (1 fm = 10^{-15} m). The simplest nucleus is that of hydrogen – ${}^1_1H$, with $A = 1$. You can therefore think of r_0 as roughly the radius of a proton.

The nucleus of an atom is very small, massive, and hence extremely dense. All nuclei have a density of about 10^{17} kg m^{-3} – about 10^{14} times denser than water. A spoonful of nuclear material would have a mass of about a thousand million tonnes. Ordinary matter, made of atoms and not just nuclei, has a density of around 10^3 kg m^{-3}.

Synoptic link

You first met the de Broglie equation in Topic 13.4, Wave–particle duality.

Worked example: Density of a helium nucleus

Calculate the approximate density of a helium-4 nucleus and of a helium atom.

Step 1: Calculate the volume of the helium-4 nucleus.

$$\text{volume of nucleus} = \frac{4}{3}\pi R^3 = \frac{4}{3}\pi (r_0 A^{\frac{1}{3}})^3 = \frac{4}{3}\pi r_0^{\,3} A$$

$$\text{volume of nucleus} = \frac{4}{3}\pi \times (1.2 \times 10^{-15})^3 \times 4 = 2.895... \times 10^{-44}\,\text{m}^3$$

Step 2: The approximate mass of the helium-4 nucleus is 4 u, and density = $\frac{\text{mass}}{\text{volume}}$.

$$\text{density of nucleus} = \frac{4 \times 1.661 \times 10^{-27}}{2.895... \times 10^{-44}} = 2.3 \times 10^{17}\,\text{kg m}^{-3}\ (2\ \text{s.f.})$$

Step 3: The mass of the electrons is negligible, so the mass of the helium atom is about 4 u. It has a radius of about 10^{-10} m.

$$\text{density of atom} = \frac{4 \times 1.661 \times 10^{-27}}{\frac{4}{3}\pi \times (10^{-10})^3} = 1600\,\text{kg m}^{-3}\ (2\ \text{s.f.})$$

Synoptic link

You can review this law in Topic 22.2, Coulomb's law.

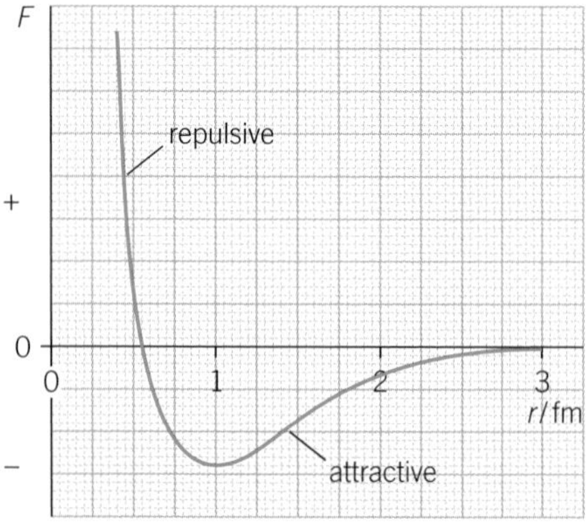

▲ **Figure 3** *A graph showing how the nuclear force F varies with separation r for two nucleons*

Nature of the strong nuclear force

In a helium-4 nucleus, the two protons are separated by a distance of about 10^{-15} m and exert a large repulsive electrostatic force on each other. According to Coulomb's law, the repulsive electrostatic force F is given by

$$F = \frac{Qq}{4\pi\varepsilon_0 r^2}$$

$$= \frac{(1.60 \times 10^{-19})^2}{4\pi \times 8.85 \times 10^{-12} \times (10^{-15})^2} \approx 230\,\text{N}$$

This is an extremely large repulsive force, so why do the protons not fly apart? The attractive gravitational force between the protons is far too small (about 10^{-34} N) to keep them together, so there must be another, much stronger force acting on the protons. This force is the **strong nuclear force**.

The strong nuclear force acts between all nucleons. It is a very short range force, effective over just a few femtometres. Figure 3 shows the variation of the strong nuclear force F between two nucleons with separation r. The force is attractive to about 3 fm and repulsive below about 0.5 fm.

Nuclear radii

High-speed electrons have a de Broglie wavelength small enough to be diffracted by individual nuclei. The de Broglie wavelength λ of such electrons is given by the equation

$$\lambda = \frac{hc}{E}$$

where h is the Planck constant, c is the speed of light, and E is the kinetic energy of the electron.

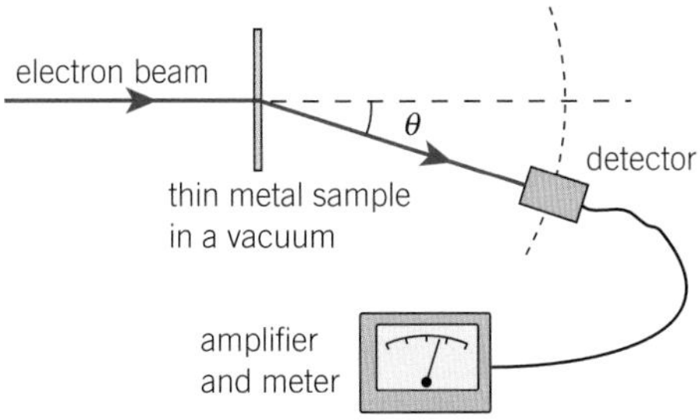

a *outline of experiment*

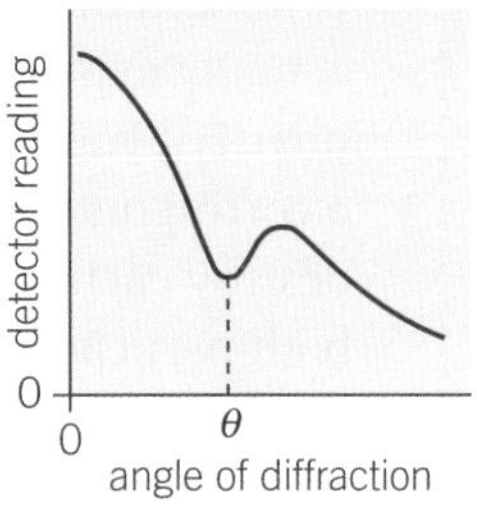

b *typical results*

◀**Figure 4** *A high-speed electron diffraction experiment and a typical result*

Figure 4 shows the arrangement used to carry out the experiment, and a typical result. The first diffraction minimum occurs at an angle θ, which is related to the radius R of the nucleus by the equation

$$\sin\theta = \frac{0.61\lambda}{R}$$

1 Show that the radius R of a nucleus is given by the equation

$$R = \frac{0.61hc}{E\sin\theta}$$

2 Electrons of energy 420 MeV give a diffraction minimum angle of 44° for oxygen-16 nuclei. Calculate the radius R of the oxygen nucleus.

3 Compare your answer for R in the question above with the radius obtained using $R = r_0 A^{\frac{1}{3}}$.

Summary questions

1 State how many protons and neutrons there are in a helium-4 nucleus. *(1 mark)*

2 State how many protons, neutrons, and electrons there are in the atoms of the following isotopes:
a ${}^{6}_{2}\text{He}$; **b** ${}^{9}_{3}\text{Li}$; **c** ${}^{56}_{26}\text{Fe}$; **d** ${}^{235}_{92}\text{U}$. *(4 marks)*

3 Calculate the nuclear radii in fm of all the isotopes shown in question 2. *(4 marks)*

4 Calculate the approximate density of the uranium-235 nucleus. How does it compare with the value for helium-4 in the worked example? *(5 marks)*

5 A neutron star of mass 4.0×10^{30} kg has a radius of about 12 km. Calculate its mean density. Comment on the answer. *(3 marks)*

6 For two protons separated in the nucleus of an atom by a distance of about 10^{-15} m, calculate the ratio gravitational force on proton/electrostatic force on proton. *(4 marks)*

7 According to a student, the mean density of a nucleus is independent of its nucleon number A. Deduce whether or not this assumption is correct. *(3 marks)*

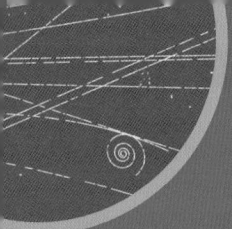

24.3 Antiparticles, hadrons, and leptons

Specification reference: 6.4.2

Learning outcomes

Demonstrate knowledge, understanding, and application of:

- → particles and antiparticles: electron–positron, proton–antiproton, neutron–antineutron, and neutrino–antineutrino
- → relative masses and charges of particles and their corresponding antiparticles
- → classification, examples, and behaviour of hadrons
- → classification, examples, and behaviour of leptons.

Antimatter

Antimatter is not just a useful device for science-fiction writers – it actually exists in nature. Antimatter was first predicted by the theoretical physicist Paul Dirac in 1928. His theory predicted that every particle has a corresponding **antiparticle**, and that if the two meet they completely destroy each other in a process called annihilation, where the masses of both particle and antiparticle are converted into a high-energy pair of photons. An antiparticle has the opposite charge to the particle (if the particle has charge) and exactly the same rest mass as the particle.

The antiparticle of the electron is the **positron**. A positron has mass 9.11×10^{-31} kg – like an electron, and charge $+1.60 \times 10^{-19}$ C – the opposite of the charge on an electron. The antiproton, antineutron, and antineutrino are the antiparticles of the proton, neutron, and neutrino respectively. Most antiparticles are symbolised by a bar over the letter for the particle. For example, the symbol for a neutrino is the Greek letter ν and the symbol for an antineutrino is $\bar{\nu}$.

Fundamental forces

In order to study subatomic particles, you need to be aware of the four fundamental forces in nature that can explain all known interactions. Table 1 shows these four forces and some of their characteristics. You have already met all but one of these fundamental forces – the **weak nuclear force** is responsible for inducing beta-decay within unstable nuclei. You will study two types of beta decay in Topic 24.5, Beta decay.

▼ **Table 1** *The four fundamental forces or interactions*

Fundamental force	Effect	Relative strength	Range
strong nuclear	experienced by nucleons	1	~10^{-15} m
electromagnetic	experienced by static and moving charged particles	10^{-3}	infinite
weak nuclear	responsible for beta-decay	10^{-6}	~10^{-18} m
gravitational	experienced by all particles with mass	10^{-40}	infinite

Fundamental particles?

In some particle accelerators, protons are accelerated to enormous speeds and then smashed together. Some of the kinetic energy of the protons is transformed into mass in the form of an incredible array of particles like baryons, mesons, kaons, and pions. The important question is – are these particles fundamental?

When physicists talk about a **fundamental particle**, they mean a particle that has no internal structure and hence cannot be divided into smaller bits. The four types of particles mentioned above are not fundamental particles, and nor are protons and neutrons, because they are all composed of quarks, as you will see in Topic 24.4. Quarks *are* considered to be fundamental particles, as are electrons and neutrinos.

Hadrons and leptons

Subatomic particles are classified into two families – **hadrons** and **leptons**.

- Hadrons are particles and antiparticles that are affected by the strong nuclear force. Examples include protons, neutrons, and mesons. Hadrons, if charged, also experience the electromagnetic force. Hadrons decay by the weak nuclear force.
- Leptons are particles and antiparticles that are not affected by the strong nuclear force. Examples include electrons, neutrinos, and muons. Leptons, if charged, also experience the electromagnetic force.

At the Large Hadron Collider (LHC) at CERN in Geneva (Figure 1), hadrons are used to probe fundamental particles. In 2013, CERN announced the discovery of the Higgs boson, the existence of which had been predicted in order to explain why all particles have the property of mass.

Figure 2 shows a small fraction of the hundreds of hadrons discovered in the last fifty or so years. They are all composed of quarks, and as such, are not fundamental particles.

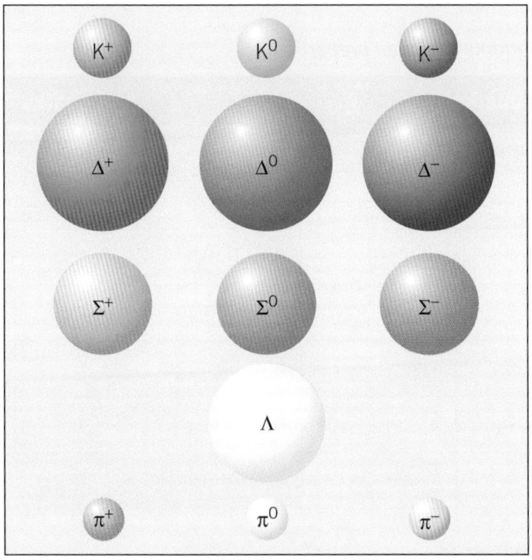

▲ **Figure 2** *Some of the many hadrons known to exist*

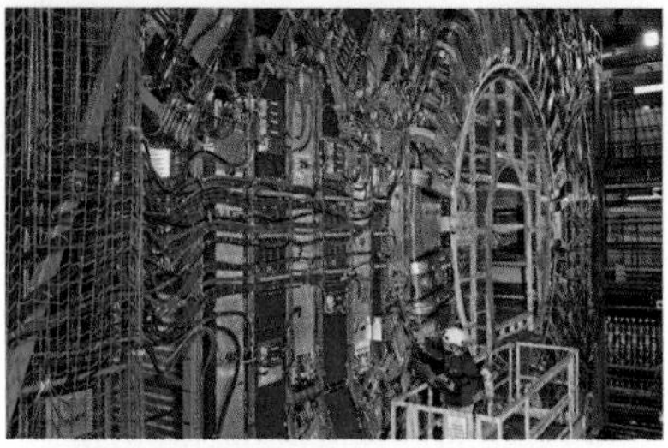

▲ **Figure 1** *The CMS detector in the LHC, shown here, was one of two experiments that gave physicists clear evidence for the existence of the Higgs boson*

Study tip

The names 'hadron' and 'lepton' come from Greek – they mean 'thick' and 'lightweight', respectively.

Summary questions

1. State two fundamental forces with an infinite range. (*1 mark*)
2. State the forces that will affect all protons. (*1 mark*)
3. What are hadrons? Give one example of a particle that is a hadron. (*2 marks*)
4. The mass of a proton p is 1.7×10^{-27} kg and it has a charge $+e$. Write a symbol for the antiproton and state its mass and charge. (*2 marks*)
5. The muon μ^- is a particle that is not affected by the strong nuclear force. It has a mass of 1.9×10^{-28} kg. Calculate the mass of the antimuon μ^+ as a multiple of electron masses and state whether this antiparticle is a hadron or a lepton. (*2 marks*)
6. Use the data sheet on page 565 to determine the properties of an antineutron. (*2 marks*)

24.4 Quarks

Specification reference: 6.4.2

Learning outcomes

Demonstrate knowledge, understanding, and application of:

- → the simple quark model of hadrons in terms of up, down, and strange quarks and their anti-quarks
- → the quark model of the proton and the neutron
- → the charges of the up, down, strange, anti-up, anti-down, and anti-strange quarks as fractions of the elementary charge *e*.

James Joyce and particle physics

In Topic 24.3, Antiparticles, hadrons, and leptons, it was mentioned that all hadrons are made of **quarks**. You will look in details at quarks in this topic. The unusual name 'quark' was coined by the American physicist Murray Gell-Mann, one of the people who first postulated their existence in the 1960s. The name comes from a single line in James Joyce's novel *Finnegans Wake*, 'three quarks for Muster Mark'.

Hadrons and quarks

Quarks, together with leptons, are the building blocks of all matter. They are considered to be fundamental particles. Any particle that contains quarks is called a hadron. Amazingly, it only takes a small number of quarks and anti-quarks to make up the hundreds of hadrons discovered in collisions in particle accelerators.

The **standard model** of elementary particles requires six quarks and their six anti-quarks. The six types of quarks are up, down, charm, strange, top, and bottom. They are denoted by the symbols u, d, c, s, t, and b. Their corresponding anti-quarks are anti-up, anti-down, anti-charm, anti-strange, anti-top, and anti-bottom ($\bar{u}$, $\bar{d}$, $\bar{c}$, $\bar{s}$, $\bar{t}$, and $\bar{b}$). All quarks have a charge Q that is a fraction of the elementary charge e. For example, the up quark has a charge $+\frac{2}{3}e$, often written as just $+\frac{2}{3}$ for simplicity.

All the quarks are listed in Table 1, but for this course you need to know only about the up, down, and strange quarks and their anti-quarks.

▼ **Table 1** *The quarks and their properties*

Quarks			Anti-quarks		
Name	Symbol	Charge Q/e	Name	Symbol	Charge Q/e
up	u	$+\frac{2}{3}$	anti-up	$\bar{u}$	$-\frac{2}{3}$
down	d	$-\frac{1}{3}$	anti-down	$\bar{d}$	$+\frac{1}{3}$
charm	c	$+\frac{2}{3}$	anti-charm	$\bar{c}$	$-\frac{2}{3}$
strange	s	$-\frac{1}{3}$	anti-strange	$\bar{s}$	$+\frac{1}{3}$
top	t	$+\frac{2}{3}$	anti-top	$\bar{t}$	$-\frac{2}{3}$
bottom	b	$-\frac{1}{3}$	anti-bottom	$\bar{b}$	$+\frac{1}{3}$

Protons and neutrons

All hadrons experience the strong nuclear force. In fact, it is the individual quarks that are bound together within the particle by the attractive strong nuclear force. The force is so strong that it may not be possible to separate the individual quarks.

A proton consists of three quarks – up, up, and down, or simply u u d. The total charge of the proton is the sum of the individual charges of the quarks. Even at this subatomic level, the principle of conservation of charge is upheld. Therefore

$$\text{proton charge } Q = (+\tfrac{2}{3})e + (+\tfrac{2}{3})e + (-\tfrac{1}{3})e = +1e$$

A neutron also consists of three quarks, but this time they are up, down, down, or u d d. You can show that the total charge of the neutron is zero. You can determine the charge of any hadron using Table 1, as illustrated in the worked example below.

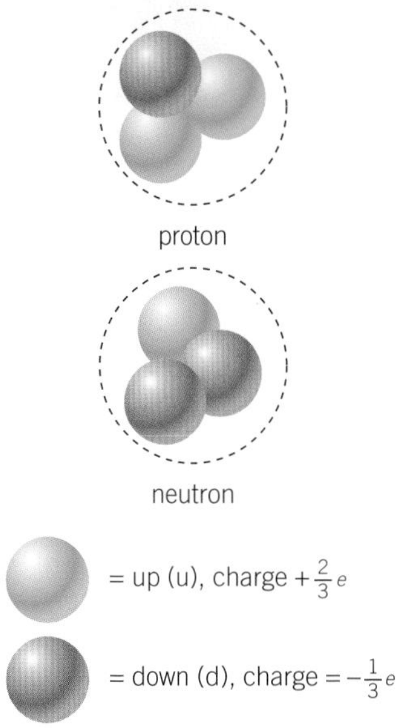

▲ **Figure 1** *The quark combinations of the proton and the neutron*

Worked example: What is the charge?

One hadron formed in particle collisions is Λ, which has the composition u d s. What is the charge on the Λ particle?

Step 1: Look up the charges of each quark and add them together.
charge Q = charge of the up quark + charge of the down quark + charge of the strange quark

$Q = (+\frac{2}{3})e + (-\frac{1}{3})e + (-\frac{1}{3})e = 0$

The Λ particle has no charge.

Mesons and baryons

Baryons are any hadrons made with a combination of three quarks. Protons and neutrons are baryons, as are antiprotons because they have the combination $\bar{u}\,\bar{u}\,\bar{d}$. **Mesons** are the hadrons made with a combination of a quark and an anti-quark. Figure 2 lists all the quark combinations for the mesons. As you can see, the properties of all hadrons can be explained in terms of combinations of quarks.

Study tip

The names 'baryon' and 'meson' for the two types of hadron come from Greek. They mean 'heavy' and 'medium', respectively.

Summary questions

1 List all the positive quarks. *(1 mark)*

2 State the quark combinations for the proton and the neutron. *(2 marks)*

3 Compare baryons and mesons. *(2 marks)*

4 Write the anti-quark combination of the antineutron. *(1 mark)*

5 Determine the charge Q of these hadrons:
  a K^+ meson $u\,\bar{s}$
  b π^0 meson $u\,\bar{u}$. *(4 marks)*

6 A hadron named Z(4430) was discovered at the LHC in Geneva in 2014. It has the quark combination $c\,\bar{c}\,d\,\bar{u}$. Determine the charge of this particle. *(2 marks)*

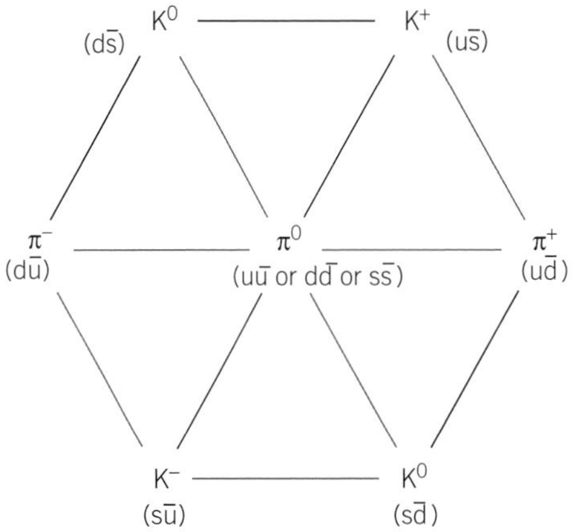

▲ **Figure 2** *Quark combinations for the mesons using u, d, s, and their anti-quarks. The blue lines join particles with the same charge, the red lines join particles with the same strangeness (another property of quarks – you don't need to know about it right now)*

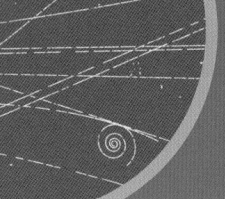

24.5 Beta decay

Specification reference: 6.4.2

Learning outcomes

Demonstrate knowledge, understanding, and application of:

- → beta-minus (β^-) and beta-plus (β^+) decay, and the quark models for these decays
- → quark transformation equations balanced in terms of charge
- → decay of particles in terms of the quark model.

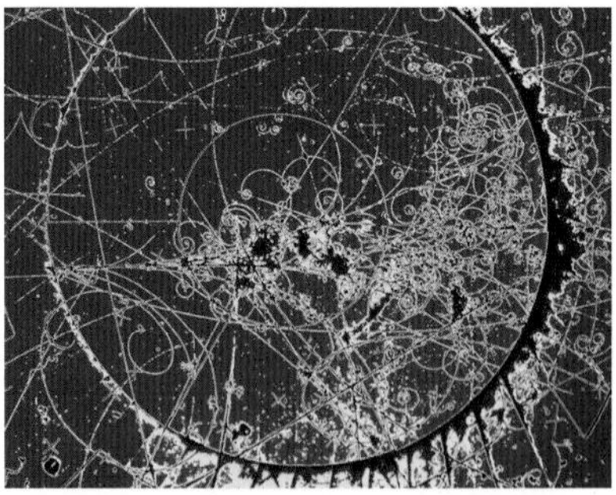

▲ **Figure 1** *Evidence for the elusive neutrino – a muon neutrino enters the detector (a bubble chamber) on the left, leaving no track, and interacts with a neutron at the edge of the large circle, producing a spray of other particles with paths in the applied magnetic field that depend on their charge and mass*

Synoptic link

You will learn more about radioactive decay and how it can occur in Chapter 25, Radioactivity.

The enigmatic neutrino

Neutrinos are quite mysterious fundamental particles that carry no charge and may have a tiny mass – less than a millionth the mass of an electron. These leptons exist in abundance in our universe. There can be as many as 100 million neutrinos per cubic metre around you, but these elusive particles do not interact much with matter because of their tiny mass and no charge – most of them pass straight through you, into the Earth and out the other side.

The existence of the neutrino was predicted by the Austrian–Swiss theoretical physicist Wolfgang Pauli in 1930 in order to explain **beta decay** in terms of conservation laws. There are three types of neutrinos: the electron neutrino ν_e, muon neutrino ν_μ, and tau neutrino ν_τ. Each neutrino also has its antiparticle. In this course, we are only interested in the electron neutrino and the electron antineutrino $\bar{\nu}_e$.

Beta decay

You should remember from your GCSE course that unstable nuclei emit radiation in various forms. Alpha (α) radiation is the emission of helium nuclei, beta (β) radiation is the emission of either electrons (β^-) or positrons (β^+), and gamma radiation (γ) is the emission of high-energy gamma photons. These emissions must be something to do with changes taking place within the nuclei of the atoms. In the case of beta decay, the changes occur to the neutrons or the protons. The force responsible for beta decay is the weak nuclear force.

In β^- decay, a neutron (${}^1_0\text{n}$) in an unstable nucleus decays into a proton (${}^1_1\text{p}$), an electron (${}^{\ 0}_{-1}\text{e}$), and an electron antineutrino ($\bar{\nu}_e$) in a process represented by the decay equation shown below.

$$\text{beta-minus } (\beta^-) \text{ decay} - \quad {}^1_0\text{n} \rightarrow {}^1_1\text{p} + {}^{\ 0}_{-1}\text{e} + \bar{\nu}_e$$

Notice that the nucleon number *A* and proton (atomic) number *Z* are conserved, as is the total charge.

In β^+ decay, a proton (${}^1_1\text{p}$) decays into a neutron (${}^1_0\text{n}$), a positron (${}^{\ 0}_{+1}\text{e}$), and an electron neutrino (ν_e) in the following way.

$$\text{beta-plus } (\beta^+) \text{ decay} - \quad {}^1_1\text{p} \rightarrow {}^1_0\text{n} + {}^{\ 0}_{+1}\text{e} + \nu_e$$

Again, nucleon and proton numbers are conserved, and so is total charge.

Quark transformation

You can zoom in even closer to see what happens within a nucleus during beta decay. As you already know from Topic 24.4, Quarks, neutrons, and protons are composed of quarks. Each type of beta decay is associated with the decay of a specific quark within the proton or the neutron. Figure 2 illustrates what happens in β^- and β^+ decays.

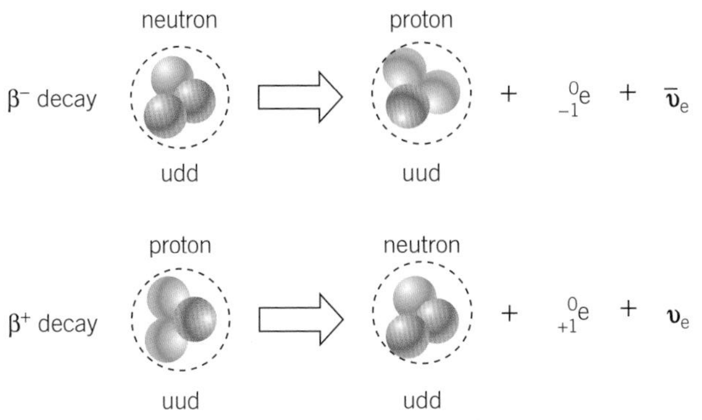

▲ **Figure 2** *Quark transformations in beta decays*

As you can see from Figure 2, in β^- decay one of the down quarks becomes an up quark, and in the process an electron and an electron antineutrino are emitted.

$$d \rightarrow u + {}^{0}_{-1}e + \bar{\nu}_e$$

The charge on the left-hand side is $-\frac{1}{3}e$ and the total charge on the right-hand side is $\frac{2}{3}e + (-1)e = -\frac{1}{3}e$. The decay equation is balanced in terms of charge.

Similarly, in β^+ decay, one of the up quarks becomes a down quark, and in the process a positron and an electron neutrino are emitted.

$$u \rightarrow d + {}^{0}_{+1}e + \nu_e$$

The total charge on both sides of the equation is $+\frac{2}{3}e$. As expected, charge has been conserved in this decay too.

Charge is a quantity that must be conserved in any reaction or decay involving charged particles. To date, no violation of this important principle has ever been observed.

Summary questions

1 State the charge on a neutrino. *(1 mark)*

2 Name the force responsible for the two types of beta decay. *(1 mark)*

3 **a** Write an equation for beta-plus decay in terms of a neutron and a proton. *(1 mark)*
b State all the quantities conserved in this decay. *(1 mark)*

4 Complete this reaction for beta-minus decay –

udd $\rightarrow$ u ? d + $^{?}_{-1}e$ + $\bar{?}$ *(1 mark)*

5 Use the information given at the start of this topic to estimate the number of electron neutrinos that will make a total mass of 1.0 kg. *(2 marks)*

6 The equation below shows a likely decay of a hadron. A pi⁻ meson (see Figure 2 in Topic 24.4) is $d\bar{u}$.

$$X \rightarrow p + \pi^-$$

Predict the likely quark composition of the particle X. *(2 marks)*

Practice questions

1 Table 1 shows some of the isotopes of phosphorus and, where they are unstable, the type of decay.

▼ Table 1

Isotope	$^{29}_{15}P$	$^{30}_{15}P$	$^{31}_{15}P$	$^{32}_{15}P$	$^{33}_{15}P$
Type of decay	β^+	β^+	stable	β^-	β^-

a State the difference between each of the isotopes shown in the table. *(1 mark)*

b Describe the structure of the proton in terms of up (u) and down (d) quarks. *(1 mark)*

c Describe what happens in a beta-plus (β^+) decay using a quark model. *(2 marks)*

d State **two** quantities conserved in a beta decay. *(1 mark)*

e Examine the isotopes in Table 1 and suggest what determines whether an isotope emits β^+ or β^-. *(1 mark)*

Jun 2013 G485

2 a Describe the nature of the strong nuclear force. *(3 marks)*

b The mass *m* of a neutron is approximately equal to the mass of a proton, with *m* about 1.7×10^{-27} kg. The radius *R* of a nucleus in fm is given by the equation $R = 1.2A^{\frac{1}{3}}$, where *A* is the nucleon number of the nucleus. (1 fm = 10^{-15} m)

(i) Calculate the average density of a $^{63}_{29}Cu$ nucleus and a $^{235}_{92}U$ nucleus. Comment on your answers. *(5 marks)*

(ii) Figure 1 shows a sketch graph of lg (*R*) against lg (*A*).

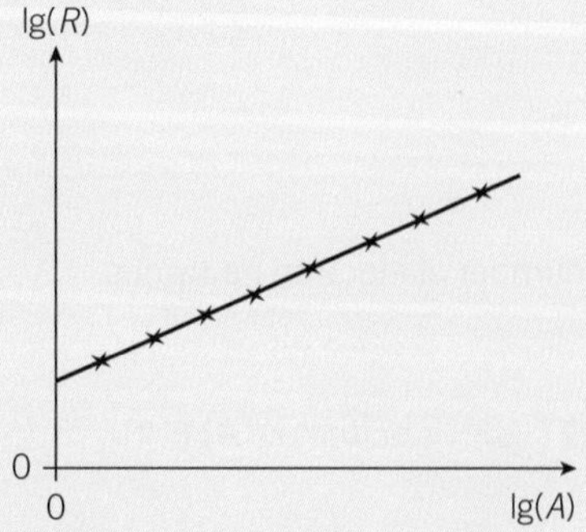

▲ Figure 1

1 Explain why the graph shows a straight line. *(2 marks)*

2 What is the expected value for the gradient? *(1 mark)*

3 a In experiments carried out to determine the nature of atoms, alpha particles were fired at thin metal foils. Describe how the alpha-particle scattering experiments provide evidence for the existence, charge, and size of the nucleus. *(5 marks)*

b Describe the nature and range of the **three** forces acting on the protons and neutrons in the nucleus. *(5 marks)*

c The radius of a $^{235}_{92}U$ nucleus is 8.8×10^{-15} m. The average mass of a nucleon is 1.7×10^{-27} kg.

(i) Estimate the average density of this nucleus. *(3 marks)*

(ii) State one assumption made in your calculation. *(1 mark)*

Jun 2011 G485

4 a Copy and complete the following decay equations for the carbon and phosphorus isotopes.

(i) carbon decay

$^{15}_{6}C \rightarrow {}^{\cdots}_{\cdots}e + {}^{\cdots}_{\cdots}N + \ldots\ldots$

(ii) phosphorus decay

$^{30}_{15}P \rightarrow {}^{\cdots}_{\cdots}e + {}^{\cdots}_{\cdots}Si + \ldots\ldots$ *(2 marks)*

b State the two beta decays in terms of a quark model of the nucleons.

(i) beta-plus decay *(1 mark)*

(ii) beta-minus decay *(1 mark)*

c Name the force responsible for beta decay. *(1 mark)*

Jun 2010 G485

5 a Explain what is meant by the term *hadron*, and give one example of a hadron. *(2 marks)*

b Figure 2 shows the quark composition of a proton and its approximate size.

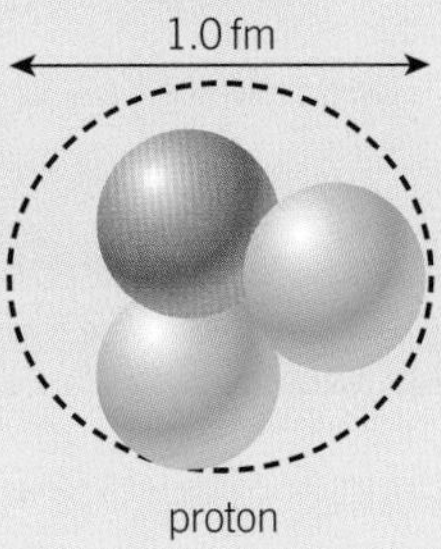

▲ Figure 2

(i) State the quark composition of a proton. *(1 mark)*

(ii) State the quark composition of an anti-proton. *(1 mark)*

(iii) Estimate the electrostatic force between the positive quarks of the proton. *(4 marks)*

(iv) Explain why it is impossible to dislodge the quarks from within the proton. *(1 mark)*

6 This question is about the nuclei of carbon–10 and uranium–237. Both nuclei are beta-minus emitters. The proton number of the carbon nucleus is 6 and the proton number of the uranium nucleus is 92.

a Determine the number of neutrons in each nucleus. *(2 marks)*

b Write a nuclear decay equation showing the changes taking place to a nucleon within each nucleus following a beta-minus decay. *(2 marks)*

c The radius r of a nucleus in fm is related to the nucleon number A by the equation

$$R = 1.2\,A^{1/3}$$

The mass of a nucleon is about 1.7×10^{-27} kg.

(i) Calculate the mean density of the carbon–10 nucleus. *(3 marks)*

(ii) Without any further calculations, explain why the density of uranium–237 nucleus is about the same as that of the carbon nucleus. *(2 marks)*

7 **a** Figure 3 shows the path of an alpha particle travelling past a stationary nucleus of gold.

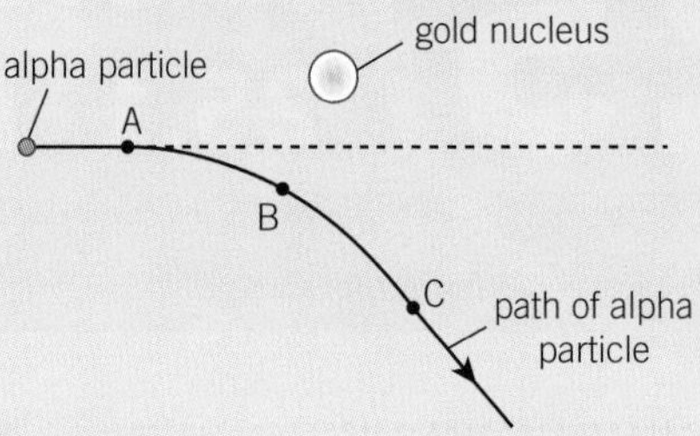

▲ Figure 3

(i) Describe how the electrostatic force F acting on the alpha particle changes from A to C. *(2 marks)*

(ii) The separation between the gold nucleus and the alpha particle at position B is 5.0×10^{-14} m. The proton number for the gold nucleus is 79. Calculate the electrostatic force acting on the alpha particle. *(3 marks)*

b Alpha particles are fired at gold foil by a group of scientists. The group are investigating how the distance r of closest approach of alpha particles to gold nuclei depends on the initial kinetic energy E of the alpha particles.
Figure 4 shows some of the results collected by the scientists.

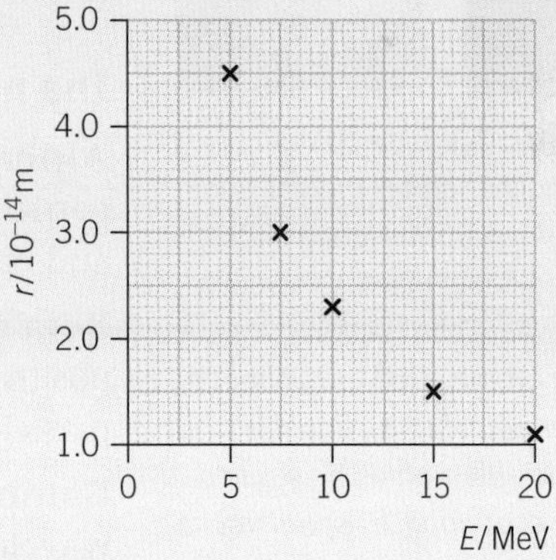

▲ Figure 4

The alpha particles are repelled by the gold nuclei.

(i) Use Figure 4 to show that $r \propto \frac{1}{E}$. *(2 marks)*

(ii) Explain why $r \propto \frac{1}{E}$. *(2 marks)*

(iii) Suggest why the $r \propto \frac{1}{E}$ relationship is no longer valid for r less than about 3 fm. Estimate the **minimum** kinetic energy in MeV of the alpha particles when this is the case. *(3 marks)*

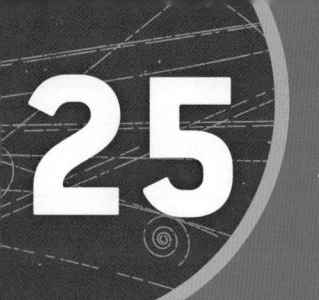

25 RADIOACTIVITY

25.1 Radioactivity

Specification reference: 6.4.3

Learning outcomes

Demonstrate knowledge, understanding, and application of:

- → radioactive decay
- → α-particles, β-particles, and γ-rays
- → nature, penetration, and range of these radiations, and techniques used to investigate their absorption.

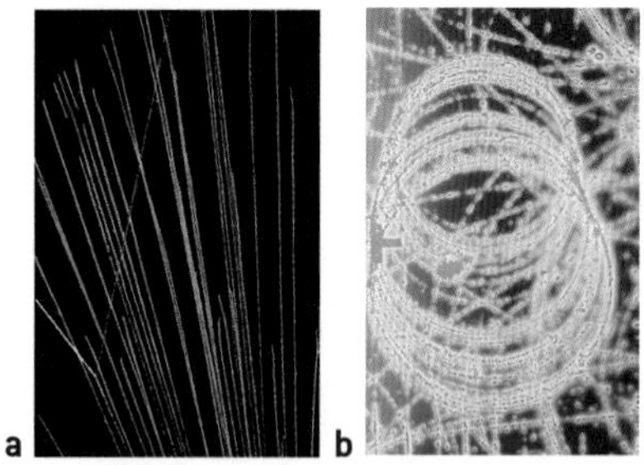

▲ **Figure 1** *a) In this false-colour version of a cloud-chamber picture from the 1920s, alpha particles leave tracks (green) as they shoot upwards through the chamber, where one (yellow) collides with a proton (red) – this image was one of several taken by English physicist Patrick Blackett as he studied alpha-particle scattering; b) spiral tracks left in a cloud chamber by beta particles*

Study tip

Remember that the source is radioactive and not the radiation.

Synoptic link

The changes within nuclei that result in the emission of radiation will be explored in Chapter 26, Nuclear physics.

Types of radiation

Radioactivity was accidentally discovered in 1896 by the French physicist Henri Becquerel. He thought that uranium salts might produce X-rays when exposed to sunlight, but after postponing an experiment in which he intended to use photographic plates to record these rays he found that even in the dark the uranium salts had emitted invisible radiation that fogged plates wrapped in lightproof paper.

Investigations carried out by Becquerel, Ernest Rutherford, Marie and Pierre Curie, and Frederick Soddy at the turn of the 20th century showed that radioactive substances emitted different types of radiation – alpha (α), beta (β), and gamma (γ). All three are described as **ionising radiations** because they can ionise atoms by removing some of their electrons, leaving positive ions.

A **cloud chamber** can be used to detect the presence of these types of radiation. It contains air saturated with vapour at a very low temperature. When air molecules are ionised, liquid condenses onto the ions to leave tracks of droplets marking the path of the radiation (Figure 1).

The nature of alpha, beta, and gamma radiations

Alpha radiation consists of positively charged particles. Each alpha particle comprises two protons and two neutrons (a helium nucleus), and has charge $+2e$, where e is the elementary charge.

Beta radiation consists of fast-moving electrons (β^-) or fast-moving positrons (β^+). A beta-minus particle has charge $-e$ and a beta-plus particle has charge $+e$.

Gamma radiation (or rays) consists of high-energy photons with wavelengths less than about 10^{-13} m. They travel at the speed of light and carry no charge.

All are emitted from the nuclei of atoms as a result of changes within unstable nuclei.

The effect of electric and magnetic fields

Figure 2 shows how a uniform electric field provided by two oppositely charged parallel plates can distinguish between the different types of radiation. The negative beta-minus particles (electrons) are deflected towards the positive plate, whilst the positive alpha and beta-plus (positron) particles are deflected towards the negative plate. Alpha particles are deflected less than beta particles because of their greater mass. The paths of the beta-minus and beta-plus particles are mirror images. Gamma rays are not deflected, because they are uncharged.

For a uniform magnetic field (Figure 2), the direction of the force on each particle can be determined using Fleming's left-hand rule. Again, the uncharged gamma rays are not deflected.

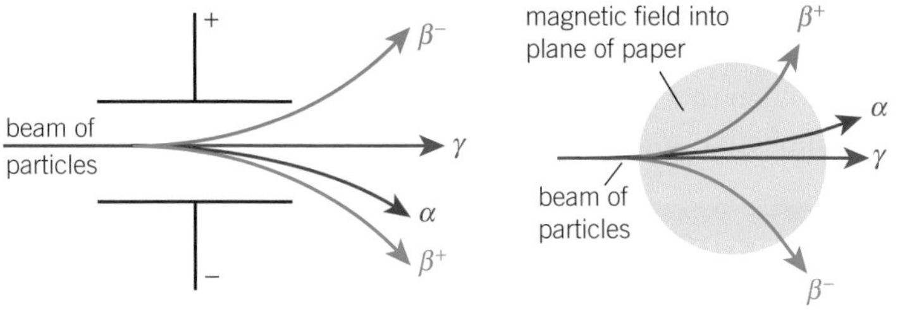

▲ **Figure 2** *The effect of a uniform electric field and a uniform magnetic field on the paths of different types of radiation*

Synoptic link

You already know from Topic 22.4, Charged particles in electric fields, and Topic 23.3, Charged particles in magnetic fields, how these fields affect charged particles.

Absorption experiments

Alpha particles, beta particles, and gamma rays all cause ionisation, which affects how they can penetrate different materials.

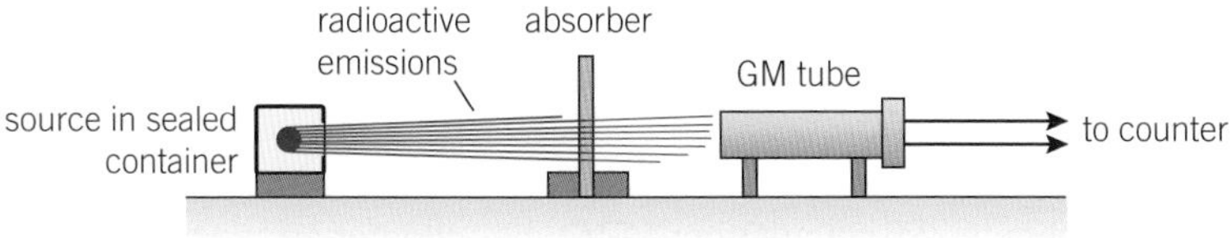

▲ **Figure 3** *Experimenting with radiation and different absorbers*

Figure 3 shows how a Geiger–Müller (GM) tube and a counter may be used to investigate the absorption of α, β^-, and γ radiation by different materials. The GM tube is kept at a fixed distance from the radioactive source. Each ionising particle, or photon, detected by the GM tube produces a single count or click.

Everything around us (including your own body) produces a small amount of radiation. This **background radiation** must be measured before you conduct any absorption experiments. The background count rate is the count rate without the radioactive source present, and depends on where in the country you are – it is typically around 0.4 counts per second, about 20 counts per minute.

The count rate for a particular absorber is then determined. You can use different thicknesses of the same absorber to investigate how this affects the count rate. The recorded count rate also includes the background count rate. In order to get the true or **corrected count rate**, you subtract the background count rate from the measured count rate.

1 Why is it not possible to carry out a similar absorption experiment in a school or college laboratory with positrons?

Absorption of alpha, beta, and gamma radiations

The large mass and charge of alpha particles mean they interact with surrounding particles to produce strong ionisation, and therefore they have a very short range in air. It takes only a few centimetres of air to absorb most alpha particles. A thin sheet of paper completely absorbs them.

The small mass and charge of beta particles make them less ionising than alpha particles. This means that they have a much longer range in air, about a metre. It takes about 1–3 mm of aluminium to stop most beta particles.

Gamma rays have no charge, and this makes them even less ionising than beta particles. You can show that for gamma rays the count rate decays exponentially with the thickness of a lead absorber. You need a few centimetres of lead to absorb a significant proportion of gamma rays.

The dangers of radioactivity

All types of radiation cause ionisation, which means that they can damage living cells. This is why radioactive sources are stored in lead-lined storage containers. When transferring radioactive sources, for your own protection you must use a pair of tongs with long handles in order to keep the source as far from your body as possible. Never handle radioactive sources with bare hands.

Study tip

Background count is the number of counts recorded without the radioactive source present.

Summary questions

1 Complete the missing words in Table 1 below. *(3 marks)*

▼ **Table 1**

Alpha radiation	Beta radiation	Gamma radiation
________ nucleus 2 protons + 2 neutrons charge ______	β^- is an ________ charge $-e$ β^+ is a ________ charge $+e$	photon of gamma radiation of ________ less than about 10^{-13} m ________ charge

2 List the different radiations in order of decreasing ionisation effect. *(1 mark)*

3 A student gets 250 counts from a GM tube–counter arrangement for a radioactive source in 2.0 minutes. The background count is 48 counts in this time. Calculate the corrected count rate in counts per minute and counts per second. *(2 marks)*

4 A single alpha particle can produce about 10^4 ions per mm in air. The typical range of an alpha particle in air is about 2.5 cm. It takes about 10 eV of energy to produce a single ion. Estimate the initial kinetic energy in MeV of an alpha particle. *(3 marks)*

5 Use your answer to question 4 to estimate the speed of the alpha particle. (The mass of an alpha particle is 6.6×10^{-27} kg.) *(3 marks)*

6 Table 2 below shows the results from an absorption experiment using gamma rays from a cobalt-60 source and an absorber made of lead.

▼ **Table 2**

Thickness of lead x / mm	1.8	3.1	6.7	9.8	15.7	19.6
Corrected count rate C / counts s^{-1}	3.8	3.5	2.7	2.3	1.6	1.4

a Plot a graph of ln$C$ against x. *(3 marks)*

b Use this graph to estimate the half-thickness of lead – this is the thickness of lead that will absorb half of the gamma ray photons. Explain your answer. *(5 marks)*

25.2 Nuclear decay equations

Specification reference: 6.4.3

Transmutation

In the early 1900s, the hands and numbers of clocks were painted with a mixture of zinc sulfide and radioactive radium. The energetic particles emitted from radium's radioactive decay made the zinc sulfide glow in the dark. Many painters developed cancers, and the use of radium in paints was stopped when the risks of radioactivity were better understood.

The nuclei of radium atoms emit alpha particles, and in doing so, they change (transmute) into the new nuclei of radon atoms. The unstable radon nuclei in turn decay into nuclei of another element. This process does stop eventually when stable nuclei are formed. The nucleus before the decay is known as the **parent nucleus**, and the new nucleus after the decay is called the **daughter nucleus**.

Basic characteristics and conservation rules

Table 1 provides a reminder of the basic characteristics of types of radiation.

▼ **Table 1** *The different types of radiation*

Radiation	Symbol	Charge	Mass / u	Typical speed / m s^{-1}
alpha	${}^{4}_{2}\text{He}$ or α	+2e	4.00151	~ 10^6
beta-minus	${}^{0}_{-1}\text{e}$ or β^- or e^-	−e	0.00055	~ 10^8
beta-plus	${}^{0}_{+1}\text{e}$ or β^+ or e^+	+e	0.00055	~ 10^8
gamma	γ (also ${}^{0}_{0}\gamma$)	0	0	speed of light, 3.00×10^8

You already know that conservation laws are important in physics. These conservation ideas can also be applied when nuclei decay. In all nuclear reactions, the nucleon number *A* and proton (atomic) number *Z* must be conserved. However, as Albert Einstein showed, mass and energy are interchangeable — the energy released in nuclear reactions is produced from mass.

Alpha decay

The nuclear transformation equation below shows a parent nucleus X decaying into a daughter nucleus Y when it emits an alpha particle.

$$ {}^{A}_{Z}\text{X} \rightarrow {}^{A-4}_{Z-2}\text{Y} + {}^{4}_{2}\text{He} $$

parent nucleus daughter nucleus

Loss of an alpha particle removes two protons and two neutrons from a parent nucleus, so the nucleon number drops by four. The daughter has a different proton number so is a different element. The equation is balanced, with the total nucleon and proton numbers before and after being the same. Energy is also released in the decay.

Learning outcomes

Demonstrate knowledge, understanding, and application of:

→ nuclear decay equations for alpha, beta-minus, and beta-plus decays

→ balancing nuclear transformation equations.

▲ **Figure 1** *The glow from the hands and numbers of this old clock are caused by a radioactive substance*

Synoptic link

In Newtonian physics, you came across the principles of conservation of energy and momentum (Topic 5.2, Conservation of energy, and Topic 7.5, Collisions in two dimensions). Kirchhoff's second law is related to the idea that charge too is conserved (Topic 10.1, Kirchhoff's laws and circuits). In fact, the conservation of mass and energy is a little more complicated – as you will learn in Topic 26.1, Einstein's mass – energy equation, mass and energy are interchangeable.

Synoptic link

In Topic 24.5, Beta decay, you learnt that there are two types of beta decay: β^- and β^+.

Worked example: Radium

A radium-226 nucleus ($^{226}_{88}\text{Ra}$) decays by alpha emission to become a nucleus of radon (Rn). Predict the isotope of radon produced in this decay.

Step 1: Determine the final nucleon and proton numbers for the radon nucleus.

$$A = 226 - 4 = 222 \text{ and } Z = 88 - 2 = 86$$

Step 2: Represent the daughter nucleus using the correct chemical symbol and A and Z numbers.

daughter nucleus: $^{222}_{86}\text{Rn}$

Beta decay

Beta decay is caused by the weak nuclear force. Radioactive nuclei that emit beta-minus radiation are characterised as having too many neutrons for stability. The weak nuclear force is responsible for one of the neutrons decaying into a proton. In the process an electron ($^{0}_{-1}\text{e}$) is emitted, together with an electron anti-neutrino ($\overline{\nu}_e$). The nucleon and proton numbers must balance, as shown in the general nuclear transformation equation for beta-minus decay below, together with a couple of examples.

$$^{A}_{Z}\text{X} \rightarrow {}^{A}_{Z+1}\text{Y} + {}^{0}_{-1}\text{e} + \overline{\nu}_e$$

parent nucleus   daughter nucleus

- strontium-90: $^{90}_{38}\text{Sr} \rightarrow {}^{90}_{39}\text{Y} + {}^{0}_{-1}\text{e} + \overline{\nu}_e$
- helium-6: $^{6}_{2}\text{He} \rightarrow {}^{6}_{3}\text{Li} + {}^{0}_{-1}\text{e} + \overline{\nu}_e$

Radioactive nuclei that emit beta-plus radiation often have too many protons for stability. Once again, the weak nuclear force initiates changes within the parent nucleus by transforming one of the protons into a neutron. In the process a positron ($^{0}_{+1}\text{e}$) is emitted together with an electron neutrino (ν_e). Again, the nucleon and proton numbers balance in the general nuclear transformation equation:

$$^{A}_{Z}\text{X} \rightarrow {}^{A}_{Z-1}\text{Y} + {}^{0}_{+1}\text{e} + \nu_e$$

parent nucleus   daughter nucleus

- potassium-37: $^{37}_{19}\text{K} \rightarrow {}^{37}_{18}\text{Ar} + {}^{0}_{+1}\text{e} + \nu_e$
- fluorine-17: $^{17}_{9}\text{F} \rightarrow {}^{17}_{8}\text{O} + {}^{0}_{+1}\text{e} + \nu_e$

Study tip

Potassium-37 is found in most foods, including bananas.

Gamma decay

Gamma photons are emitted if a nucleus has surplus energy following an alpha or beta emission. The composition of the nucleus remains the same. The nuclear decay equation when a gamma photon is emitted is shown below.

$$^{A}_{Z}X \rightarrow {}^{A}_{Z}X + \gamma$$

Decay chains – a complete story

The radioactive decay of nuclei is complex, because the daughter nuclei can themselves be radioactive. An ancient rock containing uranium will therefore also contain its daughters, their daughters, and so on. All of them will emit their own characteristic radiation.

Figure 2 shows the decay chain for a parent radium-226 nucleus. After a very long time, following many transformations, the chain ends in a stable isotope of lead-206. The half-life of each isotope is also shown in Figure 2. Half-life will be covered in greater depth in Topic 25.3, Half-life and activity.

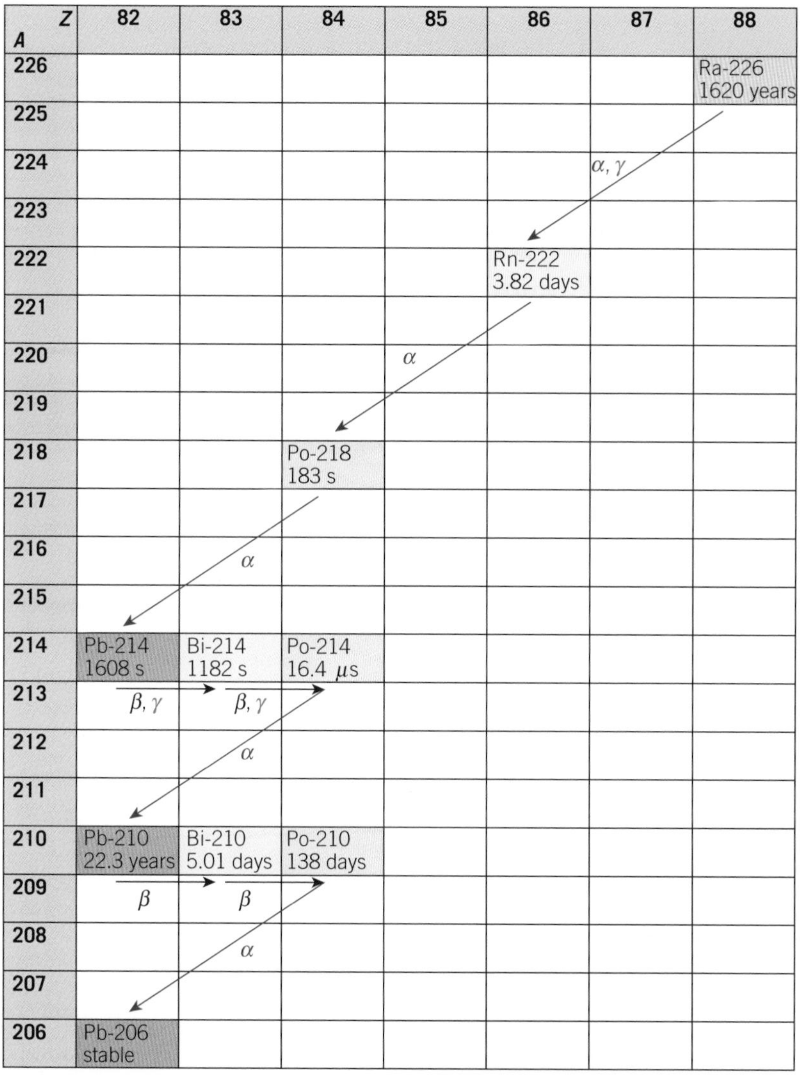
Z
A
82
83
84
85
86
87
88
226
225
224
223
222
221
220
219
218
217
216
215
214
213
212
211
210
209
208
207
206
Ra-226
1620 years
α, γ
Rn-222
3.82 days
α
Po-218
183 s
α
Pb-214
1608 s
Bi-214
1182 s
Po-214
16.4 μs
β, γ
β, γ
α
Pb-210
22.3 years
Bi-210
5.01 days
Po-210
138 days
β
β
α
Pb-206
stable

▲ **Figure 2** *The decay chain for radium-226 – a sample initially made of pure radium-226 can end up as 10 different isotopes*

Patterns for stability

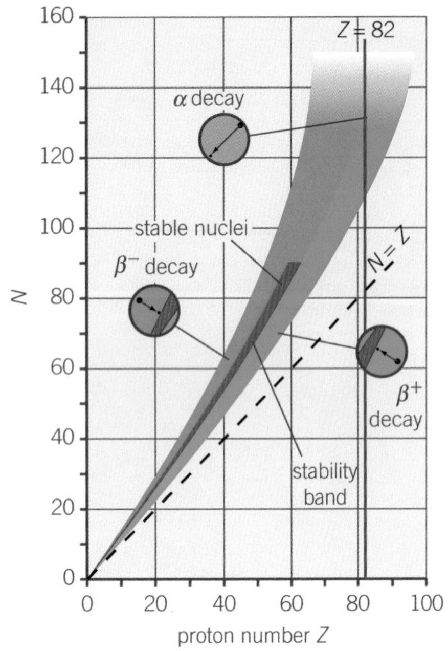

▲ **Figure 3** *N–Z plot of nuclei*

Figure 3 shows a graph of number of neutrons N against proton number Z. All stable nuclei lie on a very narrow band known as the stability band (brown). The ratio of neutrons to protons in stable nuclei gradually increases as the number of protons in the nuclei increases. Only nuclei with proton numbers less than about 20 are stable with an equal number of protons and neutrons. Most nuclei have more neutrons than protons.

The stability band is surrounded by possible unstable nuclei. You can determine the likely decay of an unstable nucleus from its position relative to the stability band.

- Nuclei with more than 82 protons are likely to decay by emitting alpha particles.
- Nuclei to the right of the band have too many protons (proton-rich) and will likely decay by beta-plus.
- Nuclei to the left of the band have too many neutrons (neutron-rich) and will likely undergo beta-minus decay.

1 The only stable isotope of aluminium is $^{27}_{13}Al$. State and explain whether the isotope $^{29}_{13}Al$ is proton-rich or neutron-rich.

2 The only stable isotope of phosphorus is $^{31}_{15}P$. There are six other phosphorus isotopes with nucleon numbers ranging from 28 to 34. List all six of these isotopes and identify whether they are likely to be β^+ or β^- emitters.

Summary questions

1 State two numbers conserved in the nuclear transformation shown below.

$^{14}_{6}C \rightarrow ^{14}_{7}N + ^{0}_{-1}e + \overline{\nu}_e$ *(1 mark)*

2 For the reaction shown in question 1, state:

a the force responsible for the decay; *(1 mark)*

b the type of decay; *(1 mark)*

c the nucleon number of the daughter nucleus. *(1 mark)*

3 Complete the following nuclear transformation equations for alpha decay.

a $^{238}_{92}U \rightarrow ^{4}_{2}He + ^{?}_{?}Th$ *(1 mark)*

b $^{222}_{?}Rn \rightarrow ^{4}_{2}He + ^{?}_{84}Po$ *(2 marks)*

4 Complete the following nuclear transformation equation for beta-plus decay.

$^{?}_{?}N \rightarrow ^{13}_{6}C + ? + \nu_e$ *(2 marks)*

5 A nucleus of uranium-234 ($^{234}_{92}U$) transforms into an isotope of lead (Pb) after emitting five alpha particles. Predict the lead isotope formed. *(3 marks)*

6 In Figure 4, the final nucleus X is an isotope of lead (Pb). Use Figure 4 to identify the isotope X. *(4 marks)*

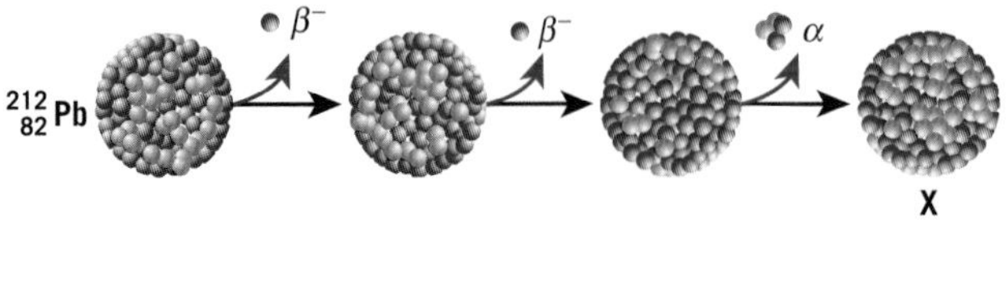

◀ **Figure 4**

25.3 Half-life and activity

Specification reference: 6.4.3

Random and spontaneous

When doing experiments with a radioactive source, you would have noticed that the counts, or clicks, from a GM tube do not show a regular pattern. The clicks are random. This suggests that the radioactive nuclei themselves must also decay in a random manner. In fact, radioactive decay is described as a random and a spontaneous event.

It is *random* because:

- we cannot predict when a particular nucleus in a sample will decay or which one will decay next
- each nucleus within a sample has the same chance of decaying per unit time.

It is *spontaneous* because the decay of nuclei is not affected by:

- the presence of other nuclei in the sample
- external factors such as pressure.

You can simulate the random behaviour of unstable nuclei by flipping coins or rolling a large number of dice. You can even use popcorn cooking in a microwave oven. The kernels represent the undecayed nuclei and a single pop represents a single decay. At the start, there are many unpopped kernels and the popping rate is high. As the amount of unpopped corn decreases, so does the popping rate.

Learning outcomes

Demonstrate knowledge, understanding, and application of:

→ the spontaneous and random nature of decay

→ activity of a source

→ decay constant λ of an isotope; $A = \lambda N$

→ simulation of radioactive decay.

▲ **Figure 1** *What does making popcorn have in common with radioactive nuclei?*

Half-life

A large number of six-sided dice can be used to simulate the decay of naturally decaying radioactive nuclei. Imagine starting off with 216 dice. Each die represents a single undecayed nucleus in a sample. Each throw represents a small interval of time. Assume the number '1' appearing on the top face represents a decay. The probability of decay for each die is $\frac{1}{6}$. This means that with each throw about 1 in 6 dice will 'decay' and about 5 in 6 will remain undecayed. The actual number decaying and remaining will be determined by chance. The decay is random because you cannot predict which dice will decay – you can only state their probability of decay.

After the first throw, you would expect about $\frac{216}{6} = 36$ dice to 'decay' and 180 to remain. After the second throw, you would expect about another 30 to decay, with about 150 left. After n throws, you would expect $\left(\frac{5}{6}\right)^n \times 216$ dice to be left. This constant-ratio property means that the number of dice left decays exponentially with the number of throws. You can show that it takes about 3.8 throws to halve the number of dice each time.

A radioactive sample behaves similarly. The two main differences are the number of nuclei, which could be many trillions, and the probability of decay, which depends on the isotope. The probability of decay governs how quickly the nuclei decay and therefore the **half-life** of the isotope.

The half-life of an isotope is the average time it takes for half the number of active nuclei in the sample to decay.

This means that after a time equal to one half-life $t_{\frac{1}{2}}$, the number N of undecayed nuclei in a sample will have halved. The number N must therefore decay exponentially with time (Figure 2).

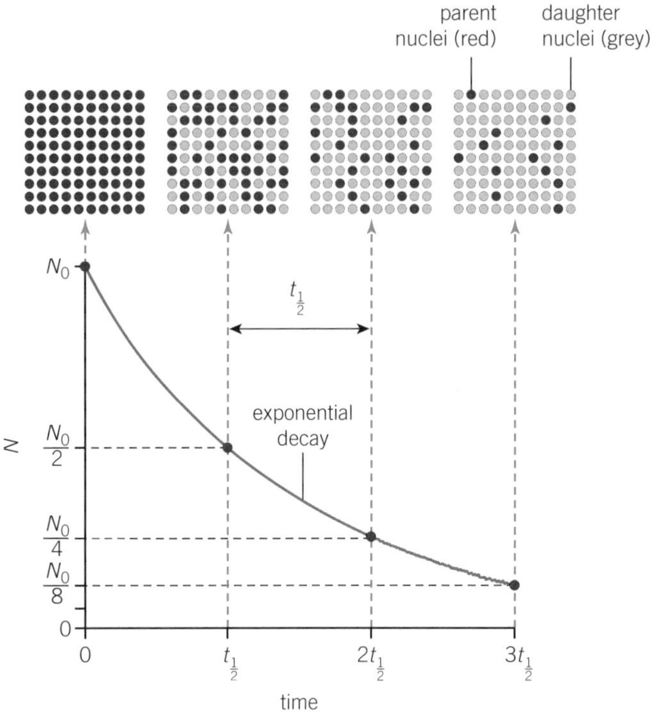

▲ **Figure 2** *Exponential decay in the number of undecayed nuclei from $N = N_0$ at time $t = 0$ – notice how the graph starts to show more statistical variations as the number of nuclei becomes smaller*

Unstable isotopes have half-lives ranging from fractions of a second to billions of years. For example, beryllium-8 has a half-life of about 8×10^{-17} s, whereas thorium-232 has a half-life of 14 billion years. Lead-204 is extraordinary – it has a half-life 10 million times longer than the age of our Universe.

▲ **Figure 3** *Some spacecraft, such as NASA's Curiosity rover (shown), are powered using radioactive isotopes. The radioactive source is housed in a thermally insulated container, which absorbs all the radiation energy emitted. Thermocouples transfer this heat to electrical energy*

Activity *A*

The count rate measured using a GM tube and a counter is only a fraction of the rate at which particles, or photons are emitted by a radioactive source. The actual **activity** of the source will be much higher than the count rate.

The activity A of a source is the rate at which nuclei decay or disintegrate.

You can also think of the activity as the number of alpha, beta, or gamma photons emitted from the source per unit time. Activity is measured in decays per second. An activity of one decay per second is one **becquerel** (1 Bq). An activity of 2000 Bq from an alpha source means 2000 nuclei decay per second or 2000 alpha particles are emitted per second. The activity depends on the number of undecayed nuclei present in the source and on the half-life of the isotope.

Worked example: Power from a radioactive source

The activity of an alpha-emitting source is 5.0×10^{12} Bq. The kinetic energy of each alpha particle is 4.0 MeV. Calculate the power emitted by this source.

Step 1: Calculate the energy of each alpha particle.

$$1\,\text{eV} = 1.60 \times 10^{-19}\,\text{J}$$

$$\therefore \text{energy of an } \alpha\text{-particle} = 4.0 \times 10^{6} \times 1.60 \times 10^{-19} = 6.4 \times 10^{-13}\,\text{J}$$

Step 2: The activity means that there are 5.0×10^{12} α-particles emitted per second.

Therefore, energy emitted per second $= 5.0 \times 10^{12} \times 6.4 \times 10^{-13}$
$= 3.2\,\text{J s}^{-1}$ (2 s.f.)

Step 3: Power is the rate of energy emitted.

Therefore, power = 3.2 W

Decay constant λ

Consider a source with a very large number of nuclei, with N undecayed nuclei at time $t = 0$ that decay into stable daughter nuclei. As you have learnt, the decay is both random and spontaneous. In a small interval of time Δt, it would be reasonable to assume that the number of nuclei disintegrating would be directly proportional to both N and Δt, that is

$$\Delta N \propto N\Delta t$$

Therefore

$$\frac{\Delta N}{\Delta t} \propto -N$$

The minus sign is included to show that the number of nuclei is decreasing with time. $\frac{\Delta N}{\Delta t}$ is the rate of decay of the nuclei, that is, the activity A of the source. For a source containing a known isotope, the relationship may be written with a constant λ, the **decay constant** of the isotope. Therefore

$$A = \lambda N$$

The minus sign has now been omitted because we just need to know the value of the activity. The decay constant has the SI unit s^{-1} (or h^{-1} or even y^{-1}, but not Bq).

The decay constant can be defined as the probability of decay of an individual nucleus per unit time.

Summary questions

1 Define activity and state its SI unit. *(2 marks)*

2 A beta-emitting source has an activity of 4.0 kBq. Calculate the number of beta particles emitted in a period of 1.0 minute. State any assumption made. *(3 marks)*

3 At time $t = 0$, there are 5000 undecayed nuclei in a source. The half-life of the isotope is 20 s.

- **a** Predict the number of undecayed nuclei after 100 s. *(2 marks)*
- **b** State and explain what will happen to the activity after 100 s. *(2 marks)*

4 A GM tube detects 2.5% of the activity of a source and measures 200 counts per second. Estimate the activity of the source in Bq. *(2 marks)*

5 An alpha-emitting source has an activity of 8.6×10^{6} Bq. The decay constant of the isotope is $2.0 \times 10^{-6}\,\text{s}^{-1}$. Calculate the number of nuclei in the source. *(2 marks)*

6 Calculate the power emitted from an alpha-emitting source with an activity of 1.0 MBq and each alpha particle having kinetic energy 4.6 MeV. *(3 marks)*

7 A strontium-90 source has mass 3.0 μg. The decay constant of strontium-90 isotope is $1.1 \times 10^{-9}\,\text{s}^{-1}$. Calculate the activity of the source in MBq. The molar mass of strontium is $0.090\,\text{kg mol}^{-1}$. *(4 marks)*

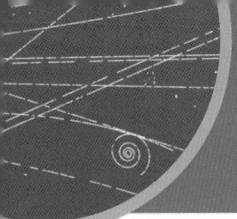

25.4 Radioactive decay calculations

Specification reference: 6.4.3

Learning outcomes

Demonstrate knowledge, understanding, and application of:

→ the half-life of an isotope; $\lambda t_{\frac{1}{2}} = \ln(2)$

→ techniques used to determine the half-life of an isotope

→ the equations $A = A_0 e^{-\lambda t}$ and $N = N_0 e^{-\lambda t}$.

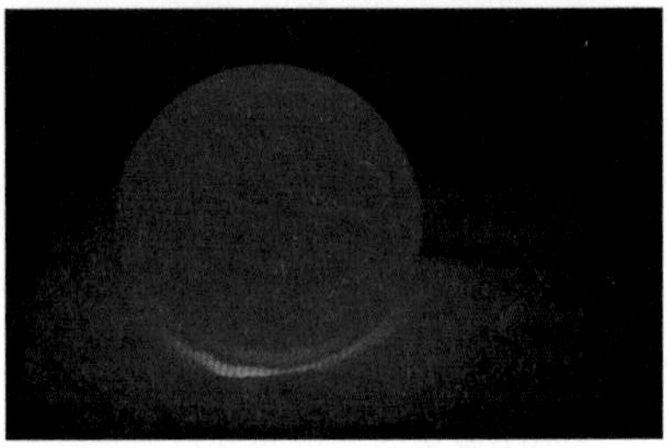

▲ **Figure 1** *A pellet of plutonium, illuminated by the glow of its own radioactivity*

Study tip

You can also calculate the number of nuclei N left in a sample if you know the number of half-lives elapsed, n, using the equation: $N = (0.5)^n N_0$, where $n = \frac{t}{t_{\frac{1}{2}}}$.

Synoptic link

You have met other types of exponential decay already, such as the damping of simple harmonic motion in Topic 17.4, Damping and driving, and the discharge of capacitors in Topic 21.4, Discharging capacitors.

Determining half-life

Plutonium is a highly toxic and carcinogenic substance and is very dangerous even in tiny amounts. The isotope plutonium-239 has a half-life of 24 000 years. Do you have to record results for decades before you can determine its half-life? In fact, to determine the half-life of any isotope, all you need to know is how many nuclei are present in the source and its activity. The activity can easily be determined using radiation detectors, and the number of nuclei can either be measured directly with a mass spectrometer or calculated from its mass.

Exponential decay

The mathematical solution to the decay equation $\frac{\Delta N}{\Delta t} = -\lambda N$ is $N = N_0 e^{-\lambda t}$, where N_0 is the number of undecayed nuclei at time $t = 0$, N is the number of undecayed nuclei in the sample at time t, and e is the base of natural logarithms, 2.718. You do not need to be able to derive this equation, but you are expected to apply it to solve problems.

The number of undecayed nuclei decreases exponentially with time. The activity A of the source is directly proportional to N. Therefore, the activity also decreases exponentially with time and is given by the equation $A = A_0 e^{-\lambda t}$, where A_0 is the activity at time t.

Decay constant and half-life

The decay constant λ of an isotope is related to its half-life $t_{\frac{1}{2}}$. You can use your knowledge of natural logarithms and the decay equation $N = N_0 e^{-\lambda t}$ to determine this link.

After a time $t = t_{1/2}$, $N = \frac{N_0}{2}$.

Therefore

$$\frac{N_0}{2} = N_0 e^{-\lambda t_{1/2}} \quad \text{or} \quad \frac{1}{2} = e^{-\lambda t_{1/2}}$$

This equation can also be written as

$$e^{\lambda t_{1/2}} = 2$$

By taking natural logarithms (ln) of both sides, we end up with

$$\ln(e^{\lambda t_{1/2}}) = \ln(2) \quad \text{or} \quad \lambda t_{1/2} = \ln(2)$$

The value of ln (2) is about 0.693. The decay constant and half-life are inversely proportional to each other. The decay constant of uranium-237 isotope, with a half-life of 6.8 days, is going to be much smaller than that of nitrogen-16, which has a half-life of 7.4 s.

Worked example: Thorium-227

A freshly prepared sample of thorium-227 has 4.0×10^{12} nuclei. The isotope of thorium-227 has a half-life of 18 days. Calculate its activity after 22 days.

Step 1: You can work in days, but it is best to convert the half-life into seconds when calculating the decay constant λ.

$$\lambda t_{1/2} = \ln(2)$$

$$\lambda = \frac{\ln(2)}{t_{1/2}} = \frac{\ln(2)}{18 \times 24 \times 3600} = 4.457 \times 10^{-7}\,\text{s}^{-1}$$

To avoid rounding errors, you must leave more significant figures in this intermediate answer.

Step 2: Calculate the initial activity of the source using $A = \lambda N$.

Initial activity $A_0 = 4.457 \times 10^{-7} \times 4.0 \times 10^{12} = 1.783 \times 10^{6}\,\text{Bq}$

Step 3: Use $A = A_0 e^{-\lambda t}$ to calculate the activity after 22 days. Once again, you need to convert the time t into seconds.

$$A = A_0 e^{-\lambda t} = 1.783 \times 10^{6} \times e^{-(4.457 \times 10^{-7} \times 22 \times 24 \times 3600)} = 7.6 \times 10^{5}\,\text{Bq (2 s.f.)}$$

You can do this calculation all at once on a calculator, but it is best to double-check that you have entered the data correctly.

Note: You could have also used the following method for the last step. The number of half-lives $n = \frac{22}{18} = 1.222...$

$$A = A_0 e^{-\lambda t} = 1.783 \times 10^{6} \times (0.5)^{1.222...} = 7.6 \times 10^{5}\,\text{Bq (2 s.f.)}$$

Measuring half-life

Protactinium-234 is a suitable isotope to use in an experiment to measure half-life, because its half-life is short. The protactinium-234 isotope is produced from the decay of thorium-234, which is itself produced from the decay of uranium-238. A sealed plastic bottle containing an organic solvent and a solution of uranyl(VI) nitrate in water is used to separate the protactinium from thorium. This works because the compound of the protactinium daughter isotope is soluble in the organic solvent, whereas the parent thorium compound is not.

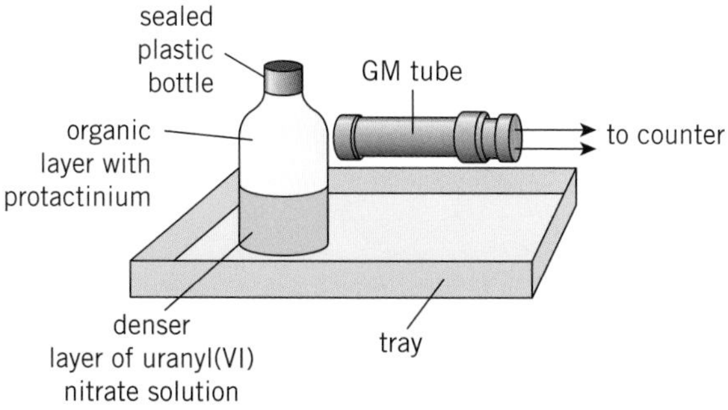

▲ **Figure 2** *A practical arrangement for determining the half-life of protactinium-234*

The background count rate is firstly determined in the absence of the source. The plastic bottle is shaken for about 15 s to dissolve the protactinium in the organic solvent, which floats to the top. The end-window of the GM tube is placed opposite the organic layer (Figure 2). In order to avoid contamination, the GM tube must not touch the bottle. The counts from the decaying protactinium can be recorded by taking a 10 s count every half-minute. The corrected count rate is directly proportional to the activity of the source. Therefore, the half-life of protactinium-234 can be determined by plotting a graph of corrected count rate against time.

Analysing data

Unlike the half-life of protactinium-234, that of radon-222 cannot safely be measured in a school or college laboratory. Radon is a gas, and because of possible leakage problems, it is safest to leave the collection of data to specialists.

Table 1 shows data collected by a university researcher. The corrected count rate at time t is C. The absolute uncertainty in each value of C is also provided.

▼ Table 1

t / s	C / counts s^{-1}
0	50.0 ± 2.2
30	32.1 ± 1.8
60	22.9 ± 1.5
90	15.2 ± 1.3
120	9.5 ± 1.0
150	6.8 ± 0.8

1 Use the table to plot a graph of ln (C) against t. Include the error bars for the ln (C) values. Draw a best-fit straight line through the error bars.
2 Explain why the plot produces a straight line.
3 Determine the gradient of the best-fit line and therefore the half-life of radon-222.
4 Describe how you can determine the absolute uncertainty in your value for the half-life.

Summary questions

1 Calculate the decay constant in s^{-1} of the following isotopes:
  a lithium-8: half-life = 0.84 s; *(2 marks)*
  b sodium-24: half-life = 15 h. *(2 marks)*

2 The decay constant of uranium-238 is 4.9×10^{-18} s^{-1}. Calculate its half-life in seconds and in years. *(3 marks)*

3 The isotope of polonium-210 has a half-life of 140 days. A radioactive source has 8.0×10^{10} nuclei of this isotope. Calculate the initial activity of the source. *(4 marks)*

4 Use the information given in this topic to determine the ratio

$$\frac{\text{decay constant of uranium-237}}{\text{decay constant of nitrogen-16}}$$ *(3 marks)*

5 Americium-241 is used in domestic smoke detectors. The half-life of this isotope is 430 y. The activity of a particular americium-241 source is 4.8 kBq. Calculate:
  a the number of americium-241 nuclei present in the source; *(4 marks)*
  b the activity of the source after 25 y. *(2 marks)*

6 Estimates show that since the creation of the Earth, the amount of uranium-235 on the Earth has dropped to 1.2% of its initial amount. The half-life of the isotope of uranium-235 is 710 million years. Estimate the age of the Earth in years. *(4 marks)*

▲ **Figure 3** *How old is the Earth?*

25.5 Modelling radioactive decay

Specification reference: 6.4.3

Using spreadsheets

The technique of iterative modelling that you have already seen used for discharging capacitors can be used to predict the number of undecayed nuclei in a sample after a certain time. The starting point of this process is the decay equation $\frac{\Delta N}{\Delta t} = -\lambda N$.

Learning outcomes

Demonstrate knowledge, understanding, and application of:

→ graphical methods and spreadsheet modelling of the equation $\frac{\Delta N}{\Delta t} = -\lambda N$ for radioactive decay.

Modelling exponential decay

In Topic 25.4, Radioactive decay calculations, you learnt that the decay equation $\frac{\Delta N}{\Delta t} = -\lambda N$ has the exact solution $N = N_0 e^{-\lambda t}$, where N_0 is the number of undecayed nuclei in the sample at time $t = 0$, N is the number of undecayed nuclei at time t, and e is the base of natural logarithms. In this section we will use a different approach to predict the number of undecayed nuclei in a source.

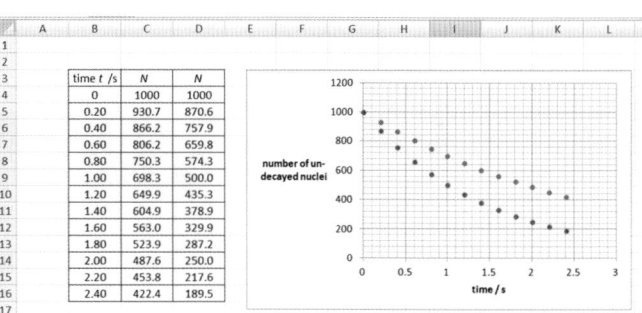

time t /s	N	N
0	1000	1000
0.20	930.7	870.6
0.40	866.2	757.9
0.60	806.2	659.8
0.80	750.3	574.3
1.00	698.3	500.0
1.20	649.9	435.3
1.40	604.9	378.9
1.60	563.0	329.9
1.80	523.9	287.2
2.00	487.6	250.0
2.20	453.8	217.6
2.40	422.4	189.5

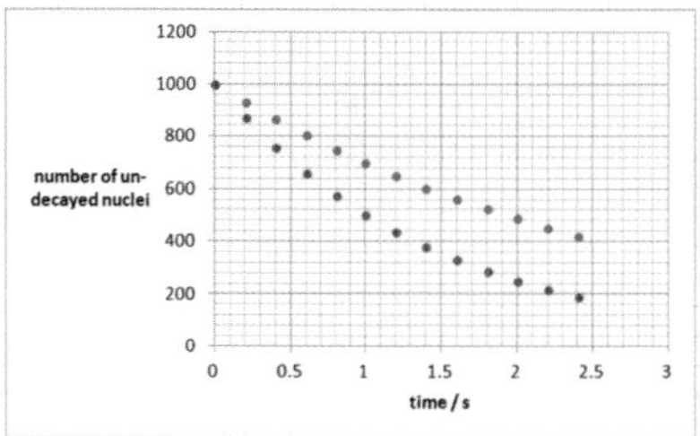

▲ **Figure 1** *You can use a spreadsheet to model the decay of nuclei*

Procedure

1 Start with a given number N_0 of undecayed nuclei in the sample.

2 Choose a very small interval of time Δt. The value of Δt must be very small compared with the half-life $t_{\frac{1}{2}}$ of the isotope, so that you can assume that the activity of the source does not change significantly in this small interval.

3 Calculate the number of nuclei decaying, ΔN, within the source during the time interval Δt using the equation

$$\frac{\Delta N}{\Delta t} = \lambda N \quad \therefore \quad \Delta N = (\lambda \Delta t) \times N$$

(The minus sign can be ignored here, because ΔN has already been defined as the number of nuclei *decreasing* in a time Δt.)

4 Calculate the number N of undecayed nuclei in the source at the end of the period Δt by subtracting ΔN from the previous value for N.

5 Repeat step 4 for subsequent multiples of the time interval Δt.

As with capacitors, it is easier to carry out the steps above using a spreadsheet. To illustrate how this is done, consider the following example.

- $N_0 = 1000$
- half-life $t_{\frac{1}{2}} = 1.00\,\text{s}$
- decay constant $\lambda = 0.693\,\text{s}^{-1}$
- $\Delta t = 0.10\,\text{s}$

Synoptic link

In Topic 21.4, Discharging capacitors, you saw how a spreadsheet can be used to predict the charge left on a capacitor.

Summary questions

1. Show that a half-life of 1.00 s gives a decay constant λ of 0.693 s^{-1}. *(1 mark)*
2. State one way in which you could improve the iterative modelling method shown in this spread so that there is better agreement between the values shown in columns 2 and 3 in Table 1. *(1 mark)*
3. Explain why it would not be sensible to use a time interval Δt of 0.25 s in an iterative modelling method for an isotope with a half-life of 1.00 s. *(1 mark)*
4. Suggest a suitable time interval Δt for an iterative modelling method for an isotope with a decay constant of 3.0×10^{-2} s^{-1}. *(2 marks)*
5. Use a spreadsheet to carry out the iterative modelling process with the following values and for time t up to 60 s: $N_0 = 1000$; half-life $t_{\frac{1}{2}} = 50$ s; $\Delta t = 1.0$ s.
   a. Determine your value for the number of undecayed nuclei after a time of 30 s. *(3 marks)*
   b. Calculate the value for the number of undecayed nuclei after a time of 30 s using the equation $N = N_0 e^{-\lambda t}$. Discuss how this value compares with your answer in (a). *(3 marks)*

The equation for modelling the number of nuclei decaying in each time interval $\Delta t = 0.10$ s is

$$\Delta N = (0.693 \times 0.10) \times N = 0.0693N$$

This means that the number of nuclei decaying in 0.10 s will be 6.93% of the initial number of nuclei, so after each period of 0.10 s, the number of undecayed nuclei in the source must be 93.07% of the previous number of nuclei. The number of nuclei therefore decrease exponentially with time.

Table 1 shows the results from a spreadsheet. The second column shows the number of undecayed nuclei using the iterative modelling method and the third column is the actual number of undecayed nuclei calculated using the equation $N = N_0 e^{-\lambda t}$. As you can see, the match is extremely good, and it can be made even better with a smaller interval of time, like 0.01 s.

▼ **Table 1** *Spreadsheet calculation modelling radioactive decay*

	Iterative modelling method	Using $N = N_0 e^{-\lambda t}$
time t / s	N	N
0.00	1000	1000
0.10	930.7	933.0
0.20	866.2	870.6
0.30	806.2	812.3
0.40	750.3	757.9
0.50	698.3	707.1
0.60	649.9	659.8
0.70	604.9	615.6
0.80	563.0	574.3
0.90	523.9	535.9
1.00	487.6	500.0
1.10	453.8	466.5
1.20	422.4	435.3

25.6 Radioactive dating

Specification reference: 6.4.3

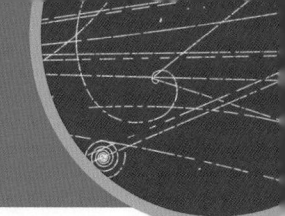

Carbon-dating

All living things on the Earth contain carbon atoms. Through photosynthesis plants absorb carbon dioxide from the atmosphere, with all the isotopes of carbon, and incorporate these isotopes into their tissues – animals eat the plants, or eat other animals that have eaten the plants, and therefore take in the carbon.

Atmospheric carbon is mainly the stable isotope, carbon-12, but also a tiny amount of the radioactive isotope carbon-14. Carbon-14 has a half-life of about 5700 years and is produced continuously in the upper atmosphere by cosmic rays. The ratio of carbon-14 to carbon-12 nuclei in atmospheric carbon is almost constant at 1.3×10^{-12}. The ratio is the same in all living things. Once an organism dies, it stops taking in carbon, whilst the total amount of carbon-14 it contains continues to decay, so this ratio decreases over time. The activity from carbon-14 in a sample of organic material is proportional to the number of undecayed carbon-14 nuclei. The time since the organism died can therefore be determined by comparing the activities, or the ratios of carbon-14 to carbon-12 nuclei, of the dead material and similar living material. **Carbon-dating** of organic materials as old as 50 000 years is possible with samples as small as nanograms using mass spectrometry.

Learning outcomes

Demonstrate knowledge, understanding, and application of:

→ radioactive dating, such as carbon-dating.

▲ **Figure 1** *Carbon-dating can be used on the wrappings of this Egyptian mummy to find its age*

Atmospheric carbon-14

High-speed protons in cosmic rays from space colliding with atoms in the upper atmosphere produce neutrons. These neutrons in turn collide with nitrogen-14 nuclei in the atmosphere to form carbon-14 nuclei. The carbon-14 nuclei eventually emit beta-minus particles (electrons) and become nitrogen-14 again, so the amount of nitrogen-14 in the atmosphere is replenished.

$$^{1}_{0}\text{n} + ^{14}_{7}\text{N} \rightarrow ^{14}_{6}\text{C} + ^{1}_{1}\text{p}$$

$$^{14}_{6}\text{C} \xrightarrow[\text{half-life} = 5700\text{ y}]{} ^{14}_{7}\text{N} + ^{0}_{-1}\text{e} + \bar{\nu}_{e}$$

Worked example: Dead wood

A wooden axe found in an Egyptian tomb is found to have an activity of 0.38 Bq. The activity of an identical mass of wood cut from a living tree is 0.65 Bq. Calculate the age of the wood used to make the axe.

Step 1: Calculate the decay constant λ of the isotope of carbon-12. Remember to change the half-life into seconds.

$$\lambda = \frac{\ln(2)}{t_{\frac{1}{2}}} = \frac{\ln(2)}{5700 \times 3.16 \times 10^{7}} = 3.848 \times 10^{-12}\,\text{s}^{-1}$$

Step 2: Use the equation $A = A_0 e^{-\lambda t}$ for activity to determine the age t of the wood.

$$0.38 = 0.65\,e^{-(3.848 \times 10^{-12})t} \quad \text{or} \quad \frac{0.38}{0.65} = e^{-(3.848 \times 10^{-12})t}$$

Take natural logarithms (ln) of both sides.

$$\ln\left(\frac{0.38}{0.65}\right) = -3.848 \times 10^{-12} \times t \qquad \text{(remember, } \ln(e^{-x}) = -x\text{)}$$

$$t = \frac{\ln\left(\frac{0.38}{0.65}\right)}{-3.848 \times 10^{-12}} = 1.395... \times 10^{11}\,\text{s} \quad \text{so age} = 4400\,\text{years (2 s.f.)}$$

Limitations to carbon-dating

There are several limitations to the technique of carbon-dating. It assumes that the ratio of carbon-14 atoms to carbon-12 atoms has remained constant over time. Increased emission of carbon dioxide due to burning fossil fuels may have reduced this ratio, as would natural events such as volcanic eruptions. The ratio may also be affected by solar flares from the Sun and by the testing of nuclear bombs. The tiny amounts of carbon-14 present in organisms also means that the activities are extremely small, about 15 counts per minute for 1 g of carbon – comparable to the background count rate.

▲ **Figure 2** *These are some of the oldest rocks on Earth, dated at 3.7–3.8 billion years old*

Dating rocks

You cannot use carbon-14 to date rocks on the Earth or meteors formed during the creation of the Solar System, because its half-life is not long enough for these ages. Instead, geologists use the decay of rubidium-87 to date ancient rocks. Nuclei of rubidium-87 emit beta-minus particles and transform into stable nuclei of strontium-87. The half-life of the isotope rubidium-87 is about 49 billion years, so it is a good candidate for dating ancient rocks – Earth has been dated to about 4.5 billion years old and the Universe is about 13.7 billion years old.

Summary questions

1 State why carbon-14 is found in all living organisms. *(1 mark)*

2 State how atmospheric carbon-14 is produced in the Earth's atmosphere. *(2 marks)*

3 All living organisms contain the isotope carbon-14.

  a Use the information provided in this topic to estimate the activity of 1.0 kg of a living tree due to the decay of carbon-14. *(2 marks)*

  b Explain why dating an ancient wooden axe by measuring its activity may be problematic. *(1 mark)*

4 A living wood is found to have an activity of 1.5 Bq. Calculate the activity of a dead wood that is 2000 years old and has the same mass as the living sample. *(3 marks)*

5 A tool made of wood and found in a cave is analysed. The concentration of carbon-14 atoms in this wood is determined using a mass spectrometer. The amount of carbon-14 in the wood is 69% of that in the same mass of wood from a living tree. Estimate the age in years of the dead wood in the tool. *(4 marks)*

6 In some rocks from Scotland, 0.56% of the rubidium-87 originally present is found to have decayed since the rocks were formed. Calculate the age in years of these rocks. *(4 marks)*

Practice questions

1 The radioactive nucleus of plutonium ($^{238}_{94}Pu$) decays by emitting an alpha particle ($^{4}_{2}He$) of kinetic energy of 5.6 MeV with a half-life of 88 years. The plutonium nucleus decays into an isotope of uranium.

a State the number of neutrons in the **uranium** isotope. *(1 mark)*

b The mass of an alpha particle is 6.65×10^{-27} kg.

(i) Show that the kinetic energy of the alpha particle is about 9×10^{-13} J. *(1 mark)*

(ii) Calculate the speed of the alpha particle. *(2 marks)*

c In a space probe, a source containing plutonium-238 nuclei is used to generate 62 W for the onboard electronics.

(i) Use your answer to **(b)(i)** to show that the initial activity of the sample of plutonium-238 is about 7×10^{13} Bq. *(1 mark)*

(ii) Calculate the decay constant of the plutonium-238 nucleus.

1 year = 3.16×10^{7} s *(2 marks)*

(iii) The molar mass of plutonium-238 is 0.24 kg. Calculate

1 the number of plutonium-238 nuclei in the source *(2 marks)*

2 the mass of plutonium in the source. *(1 mark)*

Jun 2012 G485

2 Radon-220 ($^{220}_{86}Rn$) nuclei decay by alpha emission and transform into polonium (Po) nuclei. Figure 1 shows a graph of ln (A) against t for a pure sample of radon-220, where A is the activity in Bq and t is the time in seconds.

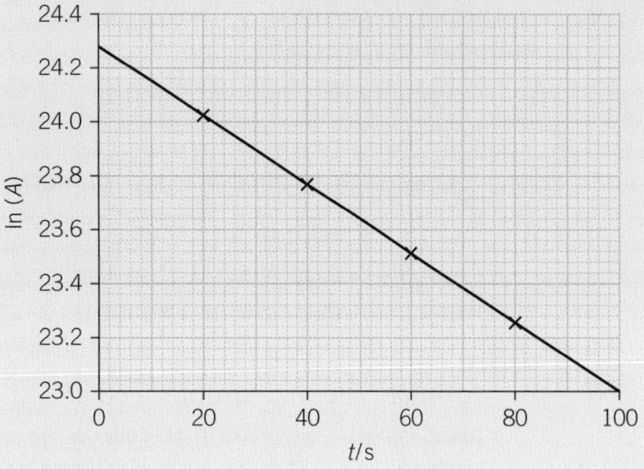

▲ **Figure 1**

a Write a nuclear transformation equation for the decay of a single nucleus of $^{220}_{86}Rn$. *(2 marks)*

b Use Figure 1 to determine

(i) the half-life of radon-220 *(4 marks)*

(ii) the initial mass of radon-220.
molar mass of radon-220 = 0.220 kg mol^{-1} *(4 marks)*

c State and explain how the shape of the graph will change when the initial mass of radon-220 is doubled. *(2 marks)*

3 The isotopes of carbon-14 ($^{14}_{6}C$) and carbon-15 ($^{15}_{6}C$) are beta-minus emitters. Table 1 shows the maximum kinetic energy of each electron emitted and the half-life of the isotope.

▼ **Table 1**

isotope	maximum kinetic energy / MeV	half-life
$^{14}_{6}C$	0.16	5560 years
$^{15}_{6}C$	9.8	2.3 s

a State one property common to all isotopes of an element. *(1 mark)*

b The neutrons and protons inside each isotope experience fundamental forces. Name the two fundamental forces experienced by both neutrons and protons. *(2 marks)*

c An isotope of carbon-15 decays into an isotope of nitrogen (N).

(i) Complete the nuclear reaction below.

$^{15}_{6}C \rightarrow ^{\cdots\cdots}_{\cdots\cdots}N + ^{0}_{-1}e + \bar{\upsilon}$ *(1 mark)*

(ii) Use the quark model to state the changes taking place within the nucleus of the carbon-15 atom. *(1 mark)*

d (i) Estimate the maximum speed of an electron from the nucleus of carbon-14. *(2 marks)*

(ii) Suggest why the actual speed of the electron is much less than your answer in (i). *(1 mark)*

e (i) Calculate the decay constant λ in s^{-1} of carbon-14. *(2 marks)*

(ii) The molar mass of carbon-14 is $14\,g\,mol^{-1}$. Show that 1.0 mg of carbon-14 has 4.3×10^{19} nuclei. *(1 mark)*

(iii) Calculate the activity of the 1.0 mg mass of carbon-14. *(2 marks)*

f The isotope of carbon-14 is very useful in determining the age of a relic (e.g. ancient wooden axe) using a technique known as carbon-dating. Describe carbon-dating and explain one of its major limitations. *(4 marks)*

Jan 2012 G485

4 a A sample of a radioactive isotope contains 4.5×10^{23} active undecayed nuclei. The half-life of the isotope is 12 hours. Calculate

(i) the initial activity of the sample *(2 marks)*

(ii) the number of active nuclei of the isotope remaining after 36 hours *(1 mark)*

(iii) the number of active nuclei of the isotope remaining after 50 hours. *(2 marks)*

b Explain why the activity of a radioactive material is a major factor when considering the safety precautions in the disposal of nuclear waste. *(2 marks)*

Jun 2010 G485

5 a Describe what is meant by the spontaneous and random nature of radioactive decay of unstable nuclei. *(2 marks)*

b Define the *decay constant.* *(2 marks)*

c Explain the technique of radioactive carbon-dating. *(4 marks)*

d The activity of a sample of living wood was measured over a period of time and averaged to give 0.249 Bq. The same mass of a sample of dead wood was measured in the same way and the activity was 0.194 Bq. The half-life of carbon-14 is 5570 years.

(i) Calculate

1 the decay constant in y^{-1} for the carbon-14 isotope *(1 mark)*

2 the age of the sample of dead wood in years *(2 marks)*

(ii) Suggest why the activity was measured over a long time period and then averaged. *(1 mark)*

(iii) Explain why the method of carbon-dating is not appropriate for samples that are greater than 10^5 years old. *(1 mark)*

Jan 2011 G485

6 This question is about the radioisotope americium-241 used in smoke detectors. Figure 2 shows a cross-section through a simplified smoke detector mounted on the ceiling.

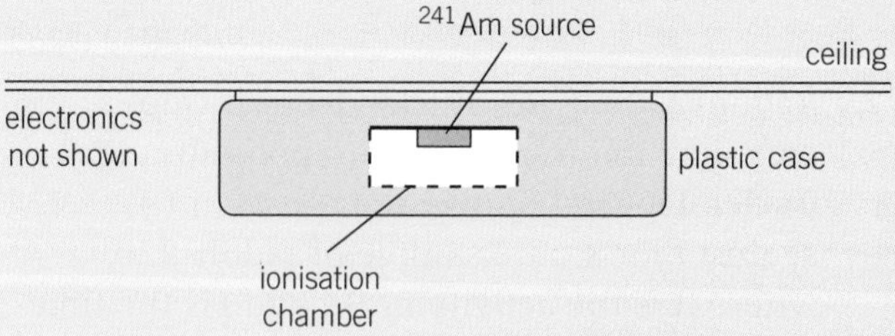

▲ **Figure 2**

The alpha particles emitted by the americium ionise the air inside the ionisation chamber maintaining a small current in a circuit including the ionisation chamber in series. When smoke enters the chamber the ions are absorbed and the current falls, causing the alarm to sound.

a Americium-241 occurs naturally from the decay of plutonium-241 by beta minus emission, or is made artificially by the bombardment of plutonium-240 inside a nuclear reactor. The nuclear equations for each of these processes are shown below with letters substituted for some of the symbols.

$$^{241}_{Z}\text{Pu} \rightarrow {}^{241}_{95}\text{Am} + \beta^-$$

$$^{240}_{Z}\text{Pu} + \text{X} \rightarrow {}^{241}_{95}\text{Am} + \beta^-$$

Write down

(i) the numerical value of the letter Z *(1 mark)*

(ii) what Z represents *(1 mark)*

(iii) the correct name of particle X. *(1 mark)*

b A typical smoke detector contains 2.5×10^{-10} kg of americium-241.

(i) Show that the source contains about 6×10^{14} nuclei of americium-241. *(2 marks)*

(ii) The half-life of americium-241 is 480 years. Show that its decay constant is about $4.6 \times 10^{-11}\,\text{s}^{-1}$.

1 year = 3.15×10^7 s *(1 mark)*

(iii) Calculate the activity of the americium-241 in the smoke detector. Give a suitable unit with your answer. *(3 marks)*

(iv) Estimate the time it takes for the activity to fall by one percent. *(3 marks)*

c Nuclei of americium-241 decay by alpha particle emission. Suggest

(i) why the americium is not a hazard when it is inside the detector *(1 mark)*

(ii) how a small speck of the source could be hazardous if it came out of the plastic case. *(2 marks)*

Jun 2009 2824

7 This question is about the activity of a small sample of vanadium-52. A researcher measures the activity A of a pure sample of vanadium-52 from time $t = 0.0$ to $t = 7.0$ minutes. The results are shown in Table 2.

▼ Table 2

t/mins	A/Bq	ln (A/Bq)
0.0	3740	
1.0	3180	
2.0	2680	
3.0	2200	
4.0	1700	
5.0	1400	
6.0	1200	
7.0	1040	

a Use the table to determine an approximate value for the half-life of the isotope vanadium-52 in minutes. Explain your reasoning. *(2 marks)*

b Copy Table 2 and complete the ln (A/Bq) column. *(1 mark)*

c The researcher plots a graph of ln (A/Bq) against t in minutes.

(i) Explain why the magnitude of the gradient of the graph is equal to the decay constant λ of the isotope. *(2 marks)*

(ii) Plot a graph of ln A against t and determine λ in min^{-1}. *(4 marks)*

(iii) Determine the half-life of the isotope in minutes. *(2 marks)*

(iv) Explain why there is a scatter of the data points in your graph in (ii). *(1 mark)*

26 NUCLEAR PHYSICS
26.1 Einstein's mass–energy equation
Specification reference: 6.4.4

Learning outcomes
Demonstrate knowledge, understanding, and application of:
- → Einstein's mass–energy equation, $\Delta E = \Delta mc^2$
- → energy released or absorbed in simple nuclear reactions
- → creation and annihilation of particle–antiparticle pairs.

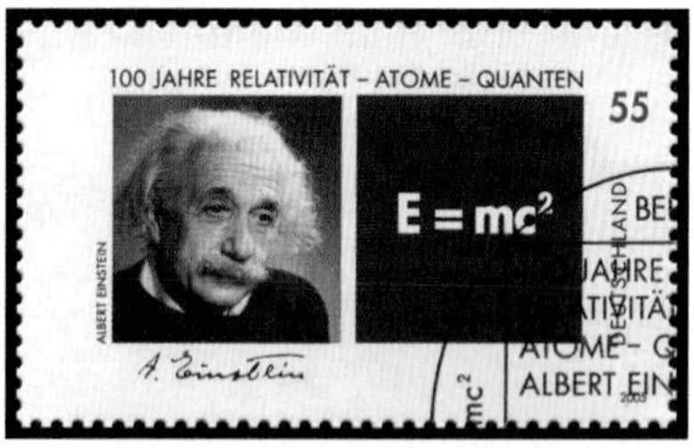

▲ **Figure 1** *Albert Einstein, shown here on a German stamp, suggested that mass and energy are equivalent*

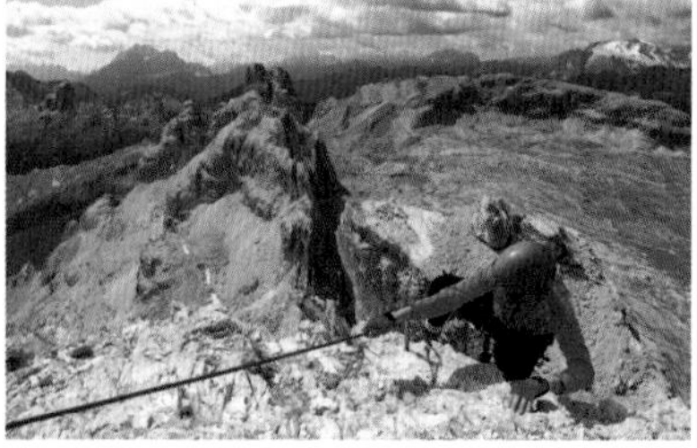

▲ **Figure 2** *The mass of this climber is more than her rest mass because of her gain in gravitational potential energy*

$E = mc^2$
The idea that mass and energy are equivalent was proposed by Albert Einstein in 1905 with his famous equation $E = mc^2$, where E is energy, m is mass, and c is the speed of light in a vacuum. This equation has two interpretations.

The first is that *mass is a form of energy*. The interaction of an electron–positron pair illustrates this idea well – the particles completely destroy each other (**annihilation**) and the entire mass of the particles is transformed into two gamma photons.

The second interpretation is that *energy has mass*. The change in mass Δm of an object, or a system, is related to the change in its energy ΔE by the equation $\Delta E = \Delta mc^2$. A moving ball has kinetic energy, implying that its mass is greater than its **rest mass**. The same happens to electrons in particle accelerators. However, because they can have speeds close to the speed of light, their mass could be a hundred times greater than their rest mass.

Similarly, a decrease in the energy of a system means the mass of the system must also decrease. For example, the mass of a mug of hot tea decreases as it cools and loses thermal energy. However, the change in mass is negligibly small (the conversion factor c^2 is enormous).

Everyday situations
Consider a person with rest mass 70 kg sitting in a stationary car. Now imagine the car travelling at a steady speed of $15\,\text{m s}^{-1}$ ($\approx 55\,\text{km h}^{-1}$). The person has gained kinetic energy, an increase in energy ΔE. The person will therefore have increased mass. The change in mass Δm can be calculated using the mass–energy equation $\Delta E = \Delta mc^2$.

$$\Delta E = \Delta mc^2$$

$$\Delta m = \frac{\Delta E}{c^2} = \frac{\frac{1}{2}mv^2}{c^2} = \frac{\frac{1}{2} \times 70 \times 15^2}{(3.00 \times 10^8)^2} = 8.8 \times 10^{-14} \approx 10^{-13}\,\text{kg}$$

This is a minuscule change in mass and is not noticeable.

Natural radioactive decay
Unstable nuclei decay by emitting either particles or photons. In alpha decay, the parent nucleus emits an alpha particle, creating a daughter nucleus, which recoils in the opposite direction. The alpha particle and the daughter nucleus have kinetic energy. You cannot simply use the principle of conservation of energy to explain this event. It makes more sense to discuss it, and other nuclear reactions, in terms of conservation of mass–energy.

The total amount of mass and energy in a system is conserved. Since energy is released in radioactive decay, there must be an accompanying decrease in mass. In simple terms, this means that the total mass of the alpha particle and the daughter nucleus in the example above must be less than the mass of the parent nucleus. This decrease in mass Δm is equivalent to the energy released ΔE.

Similarly, beta decay is accompanied by a decrease in mass.

Synoptic link

You can review the radioactive decay of unstable nuclei in Topic 25.2, Nuclear decay equations.

Worked example: Decay of carbon-14

The decay of a carbon-14 nucleus is represented by the decay equation

$$^{14}_{6}\text{C} \rightarrow {}^{14}_{7}\text{N} + {}^{0}_{-1}\text{e} + \overline{\nu}_e$$

A carbon-14 nucleus is initially at rest. Use Table 1 to calculate the total kinetic energy released by the decay of a single carbon-14 nucleus.

Step 1: Determine the change in mass Δm in this reaction.

initial mass = $2.3253914 \times 10^{-26}$ kg

final mass = $(2.3252723 + 0.0000911) \times 10^{-26}$ kg

$\Delta m = [(2.3252723 + 0.0000911) - 2.3253914] \times 10^{-26}$

$= -2.800 \times 10^{-31}$ kg

The minus sign shows that the mass decreases, therefore energy must be released.

Step 2: Use Einstein's mass–energy equation to calculate the energy released.

kinetic energy released = ΔE

$\Delta E = \Delta mc^2 = 2.800 \times 10^{-31} \times (3.00 \times 10^8)^2 = 2.52 \times 10^{-14}$ J

▼ **Table 1** *Rest masses of some particles*

Particle	Mass / 10^{-26} kg
$^{14}_{6}$C nucleus	2.3253914
$^{14}_{7}$N nucleus	2.3252723
$^{0}_{-1}$e (electron)	0.0000911
$\overline{\nu}_e$ (electron antineutrino)	negligible

Annihilation and creation

Positrons are the antiparticles of electrons. When they meet, they annihilate each other, and their entire mass is transformed into energy in the form of two identical gamma photons. This is not science fiction. It does happen, and medical physicists have exploited this phenomenon in positron emission tomography (PET). A PET scanner is used to examine the function of organs, including the brain.

Consider an electron–positron pair annihilating each other.

- change in mass $\Delta m = 2m_e$ (m_e = mass of electron or positron $= 9.11 \times 10^{-31}$ kg)
- energy released $\Delta E = \Delta mc^2 = 2m_ec^2$
- minimum energy of two gamma photons $= 2m_ec^2$
- minimum energy of each gamma photon $= m_ec^2$

Therefore, the minimum energy of each photon is 8.2×10^{-14} J or about 0.51 MeV. If the interacting particles also have kinetic energy, then the energy of each photon would be even greater.

Synoptic link

You will learn more about PET scanners in Topic 27.5, PET scans.

▲ **Figure 3** *A photon creates an electron and a positron in a bubble chamber – the electron and the positron curve in opposite directions in a magnetic field*

▲ **Figure 4** *The LHC can create two opposing proton beams that smash into each other, with each individual proton having up to 4 TeV (0.64 µJ) of kinetic energy!*

In **pair production**, a single photon vanishes and its energy creates a particle and a corresponding antiparticle. Figure 3 shows such an event. The pair produced here is an electron–positron pair. Since an electron is equivalent to a minimum energy of 0.51 MeV, the minimum energy of the photon creating the electron–positron pair must be 2 × 0.51 = 1.02 MeV.

Nuclear reactions

In a particle accelerator, like the LHC at CERN in Geneva, very energetic protons are smashed together. Their kinetic energy is transformed into matter. Under the right conditions, energy in whatever form can be transformed into matter, just as a gamma photon, which is electromagnetic energy, can change into an electron–positron pair.

Soon after the Big Bang and the creation of the Universe, the temperatures were so high that particle–antiparticle pairs of all sorts were being created and destroyed in interactions. Particle accelerators provide a means of recreating the conditions in the very early Universe.

Consider the nuclear reaction below. Two protons, travelling at speeds close to that of light, collide and produce a proton, a neutron, and a hadron called a π^+ meson.

$$^1_1\text{p} + {}^1_1\text{p} \rightarrow {}^1_1\text{p} + {}^1_0\text{n} + \pi^+$$

The total rest mass of the particles after the collision is greater than that before. The increase Δm multiplied by c^2 must be equal to the minimum kinetic energy of the colliding protons.

Summary questions

1 Use Einstein's idea about mass and energy to state and explain whether there is an increase or a decrease in the mass of the following systems:
- **a** a person running; *(1 mark)*
- **b** wood burning; *(1 mark)*
- **c** electrons decelerating. *(1 mark)*

2 Calculate the equivalent energy for the masses below:
- **a** mass of a proton 1.7×10^{-27} kg; *(2 marks)*
- **b** 1 kg mass. *(2 marks)*

3 There is a decrease in mass of 9.6×10^{-30} kg when a single nucleus of polonium-210 emits an alpha particle. Calculate the energy released in a single decay of polonium-210. *(2 marks)*

4 Calculate the increase in the mass of an electron with kinetic energy 1.0 keV. *(3 marks)*

5 Compare the increase in the mass of an electron accelerated through a potential difference of 1.0 MV with its rest mass. *(4 marks)*

6 The nuclear transformation equation below shows the decay of a single thorium-228 nucleus.

$$^{228}_{90}\text{Th} \rightarrow {}^{224}_{88}\text{Ra} + {}^4_2\text{He}$$

Use the information given below to calculate the energy released in the single decay of thorium-228. *(4 marks)*

Mass of thorium-228 nucleus = 3.7853×10^{-25} kg; mass of radium-224 nucleus = 3.7187×10^{-25} kg; mass of helium-4 (alpha particle) = 6.625×10^{-27} kg.

26.2 Binding energy

Specification reference: 6.4.4

Deuterium nucleus

Deuterium is an isotope of hydrogen. A nucleus of deuterium consists of one proton and one neutron. Now imagine separating these two nucleons. All nucleons are bound together by the strong nuclear force, so they can only be separated by doing work to overcome that force. External energy has to be supplied to make this happen. According to Einstein's mass–energy equation, energy and mass are equivalent, therefore the total mass of the separated nucleons must be greater than the mass the deuterium nucleus.

Is this really true? We can use a mass spectrometer to determine the mass of particles accurately. In terms of unified atomic mass units u (1.661×10^{-27} kg), a deuterium nucleus has mass 2.013553 u, a proton has mass 1.007276 u, and a neutron has mass 1.008665 u. The total mass of the separated proton and neutron is indeed more than the mass of the deuterium nucleus. The difference is 0.002388 u, which is equivalent to an energy of about 3.5×10^{-13} J or 2.2 MeV. In simple terms, this means that a minimum energy of 2.2 MeV is needed to completely separate the nucleons of a deuterium nucleus.

Suppose we could reverse the process and construct a deuterium nucleus from a proton and a neutron. This time, an energy of 2.2 MeV would be released – most likely in the form of a photon.

Mass defect and binding energy

In the example of the deuterium nucleus above, the difference in mass of 0.002388 u is known as the **mass defect** of the deuterium nucleus.

The mass defect of a nucleus is defined as the difference between the mass of the completely separated nucleons and the mass of the nucleus.

The energy difference of 2.2 MeV for the deuterium nucleus is known as its **binding energy**.

The binding energy of a nucleus is defined as the minimum energy required to completely separate a nucleus into its constituent protons and neutrons.

To calculate the binding energy of a nucleus, you can use Einstein's mass–energy equation.

$$\text{binding energy of nucleus} = \text{mass defect of nucleus} \times c^2$$

The binding energy is not the same for all nuclei. A uranium-235 has 92 protons and 143 neutrons, and you would expect the external energy required to split this nucleus into its constituent protons and neutrons to be much greater than 2.2 MeV – there are many more strong nuclear bonds to be broken.

Learning outcomes

Demonstrate knowledge, understanding, and application of:

- → mass defect; binding energy; binding energy per nucleon
- → binding energy per nucleon against nucleon number curve; energy changes in reactions
- → binding energy of nuclei using $\Delta E = \Delta mc^2$ and masses of nuclei.

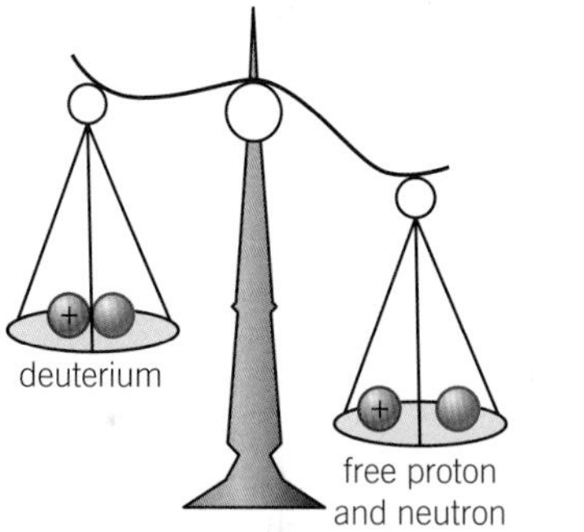

▲ **Figure 1** *A deuterium nucleus has less mass than its separated nucleons*

Study tip

Be careful with your use of the terms 'atoms' and 'nuclei'. Binding energy holds the nucleus together and not the atom.

Worked example: Binding energy of uranium nucleus

The mass of a uranium-235 ($^{235}_{92}U$) nucleus is 235.004393 u. Calculate its binding energy in MeV.

Step 1: Calculate the total mass of the constituent nucleons.

The uranium-235 nucleus has 92 protons and (235 − 92) = 143 neutrons

Therefore, mass of nucleons = (92 × 1.007276) + (143 × 1.008665)
= 236.908487 u

Step 2: Calculate the mass defect for the uranium-235 nucleus.

mass defect = 235.004393 − 236.908487 = (−)1.904094 u

Step 3: Change the mass defect from u to kg ($1\,u = 1.661 \times 10^{-27}$ kg).

mass defect = $1.904094 \times 1.661 \times 10^{-27} = 3.162... \times 10^{-27}$ kg

Step 4: Calculate the binding energy and convert it from J to eV using the conversion factor $1\,eV = 1.60 \times 10^{-19}$ J

binding energy = mass defect × c^2

binding energy = $3.162... \times 10^{-27} \times (3.00 \times 10^8)^2 = 2.84... \times 10^{-10}$ J

binding energy = $\frac{2.84... \times 10^{-10}}{1.60 \times 10^{-19}} = 1.779... \times 10^9$ eV = 1780 MeV (3 s.f.)

Binding energy per nucleon

You would expect a uranium-235 nucleus to have a greater binding energy than a deuterium nucleus because uranium-235 has more nucleons than a deuterium nucleus. To compare how easy it is to break up nuclei, it would be sensible to determine the average **binding energy per nucleon** of nuclei. The greater the binding energy (BE) per nucleon, the more tightly bound are the nucleons within the nucleus, or in other words a nucleus is more stable if it has a greater BE per nucleon.

The binding energy per nucleon for uranium-235 is 1780 MeV/235 ≈ 7.6 MeV and for deuterium is 2.2 MeV/2 ≈ 1.1 MeV per nucleon.

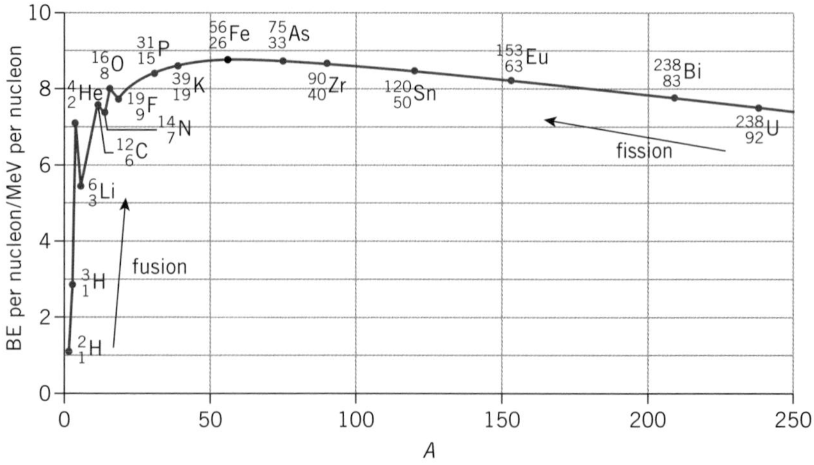

▲ **Figure 2** *Graph of binding energy per nucleon against nucleon number A for nuclei*

Figure 2 is a graph of BE per nucleon against nucleon number A. The shape of the graph helps us to understand processes such as natural radioactive decay, **fission**, and **fusion**. The last two processes are covered in greater depth in the next two topics. From the graph you can see that:

- For nuclei with $A < 56$, the BE per nucleon increases as A increases.
- For nuclei with $A > 56$, the BE per nucleon decreases as A increases.
- The nucleus of iron-56 ($^{56}_{26}Fe$) has the greatest BE per nucleon – it is the most stable isotope in nature.
- The helium-4 nucleus (alpha particle), with its two protons and two neutrons, has an abnormally greater BE per nucleon than its immediate neighbours. The same goes for carbon-12 and oxygen-16 nuclei.
- Energy is released in natural radioactive decay. Figure 2 can be used to show that in cases of spontaneous decay the total binding energy of the parent nucleus is less than the binding energy of the daughter nucleus and the alpha particle. The difference is the energy released in the decay as kinetic energy.
- In a fusion process, two low A number nuclei join together to produce a higher A number nucleus. The newly formed nucleus has much greater binding energy than the initial nuclei and therefore energy is released. Fusion is the process by which the Sun and other stars produce their energy. Thanks to fusion, we have life on Earth.
- In a fission process, a high A number nucleus splits into two lower A number nuclei. Energy is released because the two nuclei produced have higher binding energy than the parent nucleus. All fission reactors use this process to produce energy.

Summary questions

1 State the SI units of mass defect and binding energy. *(1 mark)*

2 State the link between binding energy and mass defect. *(1 mark)*

3 Show that a mass defect of 0.002368 u is equivalent to a binding energy of about 3.5×10^{-13} J. *(3 marks)*

4 The binding energy of the nucleus of iron-56 is 7.8×10^{-11} J. Calculate its BE per nucleon in joules per nucleon and in MeV per nucleon. *(3 marks)*

5 Use Figure 2 to estimate the binding energy in MeV of:
 a a helium-4 nucleus; *(2 marks)*
 b an oxygen-16 nucleus; *(2 marks)*
 c a uranium-238 nucleus. *(2 marks)*

6 The mass of the beryllium-8 nucleus ($^{8}_{4}Be$) is 1.33×10^{-26} kg. The mass of a proton or a neutron is about 1.67×10^{-27} kg. Use this information to calculate the binding energy per nucleon of the beryllium-8 nucleus in both J per nucleon and MeV per nucleon. *(4 marks)*

26.3 Nuclear fission

Specification reference: 6.4.4

Learning outcomes

Demonstrate knowledge, understanding, and application of:

- → induced nuclear fission; chain reaction
- → basic structure of a fission reactor; components – fuel rods, control rods and moderator
- → environmental impact of nuclear waste
- → balancing nuclear transformation equations.

▲ **Figure 1** *Lise Meitner with Otto Hahn. They worked together in Berlin for many years before Meitner was forced to flee the Nazi regime*

Induced fission

In 1938, two German physicists, Otto Hahn and Fritz Strassmann, discovered traces of lighter elements in a sample of uranium that was being irradiated by slow neutrons. This was explained by Lise Meitner and Otto Frisch in terms of **induced fission**. The uranium-235 nuclei were absorbing the slow neutrons, becoming unstable, and splitting up into two approximately equal halves plus fast neutrons. Energy is released in each fission reaction, as Einstein's mass–energy equation explains. The energy released per fission event can be as much as 200 MeV.

The first ever nuclear reactor was built secretly in 1942 in a squash court at the University of Chicago, USA, by a team led by the Italian Enrico Fermi.

In the UK, about 20% of our electrical energy comes from power stations using nuclear fuel. A kilogram of uranium-235 can produce millions of times more energy than a kilogram of coal. The biggest drawback of nuclear power is the production of radioactive and hazardous waste.

The process of induced fission

Uranium is the most common fuel used in nuclear power stations. Uranium obtained from mined ore consists of about 99.3% uranium-238 isotope and 0.7% uranium-235. The uranium-235 isotope easily undergoes fission on absorbing a slow neutron. These slow neutrons are also known as **thermal neutrons** because their mean kinetic energy is similar to the thermal energy of particles in the reactor core. Uranium-238 nuclei are more likely to capture (that is, absorb) the neutrons than to undergo fission.

Both uranium-235 and uranium-238 nuclei can split spontaneously without absorbing neutrons, but this is very rare. However, uranium-236 nuclei have a much greater chance of splitting spontaneously. A typical induced fission reaction of uranium-235 by a thermal neutron is shown below.

$$ {}^{1}_{0}\mathrm{n} + {}^{235}_{92}\mathrm{U} \longrightarrow {}^{236}_{92}\mathrm{U} \longrightarrow {}^{141}_{56}\mathrm{Ba} + {}^{92}_{36}\mathrm{Kr} + 3\,{}^{1}_{0}\mathrm{n} $$

thermal neutron and uranium nucleus	unstable uranium-236 nucleus	daughter nuclei and fast neutrons

The uranium-235 nucleus captures a thermal neutron and becomes a highly unstable nucleus of uranium-236. In less than a microsecond, the uranium-236 nucleus splits. The daughter nuclei produced in this example are barium-141 and krypton-92, but there are many other possible variants too. Three fast neutrons are also produced. The nuclear equation is balanced, with the total number of protons (92) and nucleons (236) being conserved.

Fission energy

The total mass of the particles after the fission reaction is always less than the total mass of the particles before the reaction. The difference in mass Δm corresponds to the energy ΔE released in the reaction. Put another way, the total binding energy of the particles after fission is greater than the total binding energy before it. The difference in the binding energies is equal to the energy released.

The energy released in a single fission reaction is a combination of kinetic energy of the particles produced and the energy of photons and neutrinos emitted (Figure 2).

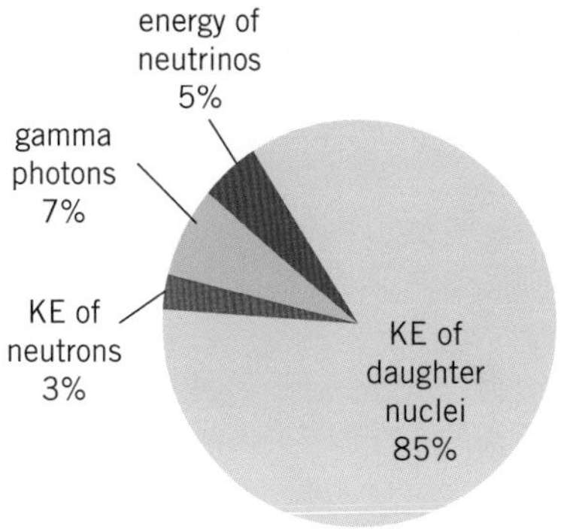

▲ **Figure 2** *The energy released from fission*

Worked example: Energy from fission

The difference in mass Δm in the induced fission of uranium-235 above is about 0.185 u. Calculate the total energy released in this reaction in MeV.

Step 1: Change Δm from u to kg ($1\,u = 1.661 \times 10^{-27}\,kg$).

$$\Delta m = 0.185 \times 1.661 \times 10^{-27} = 3.07 \times 10^{-28}\,kg$$

Step 2: Use Einstein's mass–energy equation to calculate the energy released.

energy released $\Delta E = \Delta mc^2$

$$\Delta E = 3.07 \times 10^{-28} \times (3.00 \times 10^{8})^2 = 2.76 \times 10^{-11}\,J$$

Step 3: Change the energy from joules to electronvolts ($1\,eV = 1.60 \times 10^{-19}\,J$).

$$\Delta E = \frac{2.76 \times 10^{-11}}{1.60 \times 10^{-19}} = 172.5 \times 10^{6}\,eV = 170\,MeV \text{ (2 s.f.)}$$

Study tip

You can check your calculations using $1\,u \approx 930\,MeV$.

Chain reaction

The fission of a uranium-235 nucleus is more likely with slow neutrons than fast neutrons. Consider what might happen if the three fast neutrons produced in a fission reaction can be slowed down, so that they too can instigate further fission reactions in other uranium-235 nuclei. A **chain reaction** becomes possible, with these three neutrons starting three more reactions, which each produce another three neutrons, and so on (Figure 3). After n generations of fission events, the number of neutrons would be 3^n – the growth in neutron numbers will be exponential. The rate of energy release will also grow exponentially with time. This is the last thing we want inside a nuclear reactor. The steady production of power from a nuclear reactor is controlled by ensuring that, on average, one slow neutron survives between successive fission reactions. How this is achieved is discussed later.

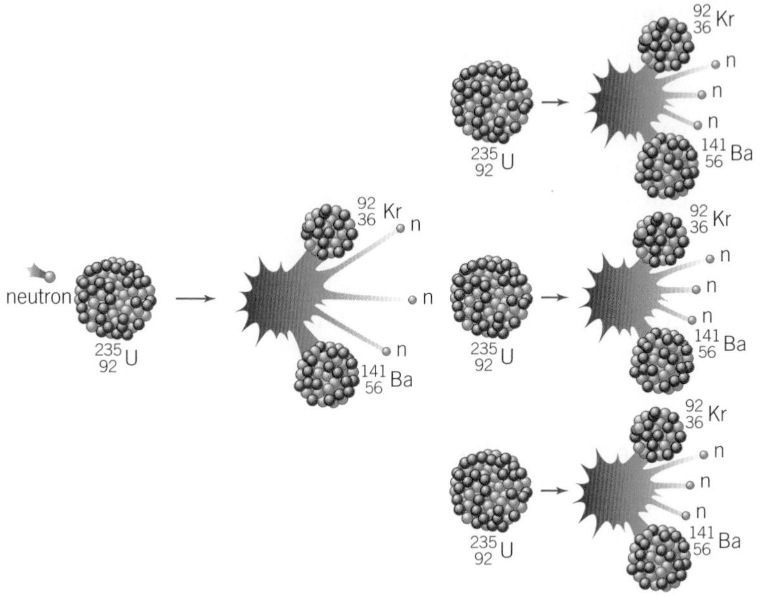

▲ **Figure 3** *An uncontrolled chain fission reaction*

Inside a fission reactor

There are many different designs of fission reactors, but the key components are the same. Fuel rods are spaced evenly within a steel–concrete vessel known as the reactor core. A **coolant** is used to remove the thermal energy produced from the fission reactions within the fissile fuel. The fuel rods are surrounded by the **moderator**, and **control rods** can be moved in and out of the core (Figure 4).

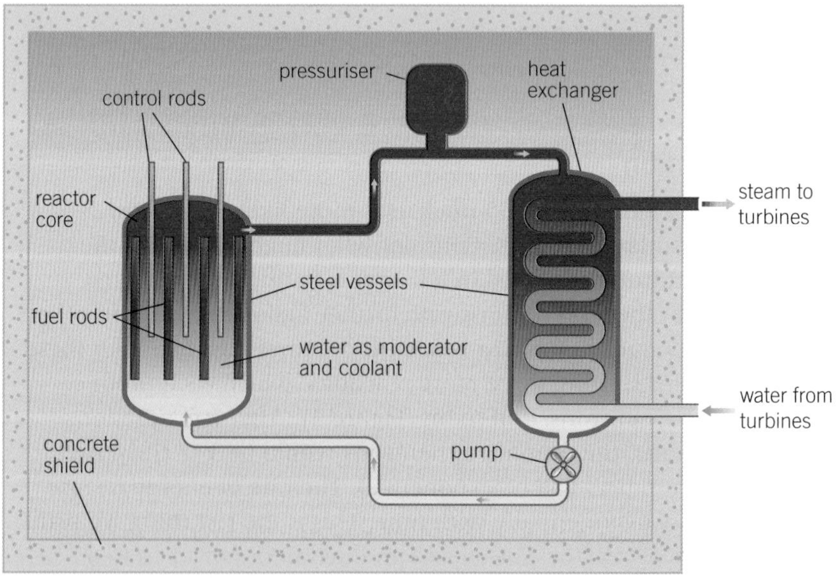

▲ **Figure 4** *The main components of a water-cooled reactor*

Fuel rods

Fuel rods contain enriched uranium, which consists mainly of uranium-238 with 2–3% uranium-235.

Moderator

The role of the moderator is to slow down the fast neutrons produced in fission reactions. The material for a moderator must be cheap and readily available, and must not absorb the neutrons in the reactor.

Induced fission of uranium-235 produces fast neutrons with kinetic energies up to 1 MeV. The chance of these fast neutrons being absorbed by uranium-235 nuclei is quite small, whereas thermal neutrons have a greater chance of producing induced fission. The mean kinetic energy of thermal neutrons is about $\frac{3}{2}kT$, where k is the Boltzmann constant and T is the thermodynamic temperature of the reactor core. The root mean square speed of thermal neutrons is a few km s^{-1}.

Fast-moving neutrons just bounce off the massive uranium nuclei with negligible loss of kinetic energy. However, when they collide elastically with protons (or deuterium) in water or with carbon nuclei, they transfer significant kinetic energy and slow down. Water and carbon are therefore good candidates for a moderator.

In many reactors, the moderator is also the coolant. In a pressurised water reactor (PWR), the water acts both as a moderator and a coolant. The output electrical power from a PWR is about 700 MW.

> **Synoptic link**
>
> Boltzmann's constant was covered in Topic 15.4, The Boltzmann constant.

> **Synoptic link**
>
> Review Topic 15.3, Root mean square speed, if you need to remind yourself about r.m.s. speeds.

Control rods

The control rods are made of a material whose nuclei readily absorb neutrons, most commonly boron or cadmium. The position of the control rods is automatically adjusted to ensure that exactly one slow neutron survives per fission reaction. To slow down or completely stop the fission, the rods are pushed further into the reactor core.

Environmental impact

Neutrons of intermediate kinetic energies are readily absorbed by uranium-238 nuclei within the fuel rods. These nuclei of uranium-238 quickly decay into nuclei of plutonium-239.

$$^{238}_{92}\text{U} + {}^{1}_{0}\text{n} \rightarrow {}^{239}_{92}\text{U} \xrightarrow[t_{\frac{1}{2}} = 24\,\text{min}]{\beta^- \text{ decay}} {}^{239}_{93}\text{Np} \xrightarrow[t_{\frac{1}{2}} = 2.4\,\text{days}]{\beta^- \text{ decay}} {}^{239}_{94}\text{Pu}$$

Plutonium-239 is one of the most hazardous materials produced in nuclear reactors. It is extremely toxic as well as radioactive, and it has a half-life of 24 thousand years. The daughter nuclei produced from its numerous fission reactions are also radioactive.

Radioactive waste from nuclear reactors cannot be disposed of as normal waste. Decisions about its disposal affect not only us, but also future generations. The storage of radioactive waste presents us with both practical and ethical issues. High-level radioactive waste, which includes spent fuel rods, has to be buried deep underground for many centuries because isotopes with long half-lives must not enter our water and food supplies. The burial locations need to be geologically stable, secure from attack, and designed for safety.

Governments and popular campaigns around the world are now pushing for cleaner, renewable energy resources such as wind and solar power.

Neutrons in nuclear reactors

The interaction of neutrons with nuclei inside a nuclear reactor depends on their speed. The idea of 'cross-section' is used to indicate the chance of a neutron being absorbed by a specific nucleus. You can think of the nucleus as a disc that will absorb the neutron if it hits this disc. The cross-sectional area is measured in m^2, or in barns, where 1 barn = 10^{-28} m^2.

Table 1 shows the cross-sections for capture not leading to fission and for fission, for different nuclei with thermal neutrons and fast neutrons. Thermal neutrons have mean kinetic energy of about 0.1 eV, and fast neutrons have mean KE greater than about 100 eV.

▼ **Table 1** *Neutron capture cross-sections*

Component	Nuclei	Cross-section / barns			
		Thermal neutrons		Fast neutrons	
		Fission	Capture	Fission	Capture
Fuel	uranium-235	590	99	1.9	0.56
	uranium-238	1.2×10^{-5}	2.7	0.043	0.33
	plutonium-239	750	270	1.8	0.50
Control rod	boron-10	0	3800	0	2.7
	cadmium-48	0	2500	0	0.27
Moderator	hydrogen-1 (proton)	0	0.67		5.1×10^{-4}
	hydrogen-2 (deuterium)	0	1.3×10^{-3}	0	1.1×10^{-4}
	carbon-12	0	2.0×10^{-3}	0	1.0×10^{-5}

1. State and explain the advantages of using boron-10 rather than cadmium-48 in control rods.
2. Heavy water molecules contain deuterium nuclei rather than the protons found in ordinary water. Explain why heavy water is more suitable than ordinary water as a moderator.
3. Use the table to explain why fast neutrons must be slowed down in a nuclear reactor.
4. The mass of a neutron is about 1.7×10^{-27} kg. Determine the root mean square speed of thermal neutrons.

Summary questions

1. State the roles of the moderator and the control rods in a nuclear reactor. *(2 marks)*
2. Use the worked example and Figure 2 to estimate the kinetic energies (in MeV) of all the neutrons and of the daughter nuclei produced following fission of a uranium-235 nucleus. *(2 marks)*
3. In fission reactions the binding energy per nucleon increases from about 7.5 MeV to about 8.5 MeV (see Figure 2 in Topic 26.2). Estimate the energy released (in MeV) from the fission of plutonium-239. *(3 marks)*
4. The average energy produced in a fission reaction of a single uranium-235 nucleus is about 170 MeV. The molar mass of uranium-235 is 0.235 kg mol^{-1}. Calculate the total energy (in joules) released by the fission of 1.0 kg of uranium-235 (the Avogadro constant $N_A = 6.02 \times 10^{23}$ mol^{-1}). *(4 marks)*
5. A neutron loses about 28% of its kinetic energy when it elastically collides with a carbon nucleus. Estimate the number of collisions with carbon that a neutron must make to reduce its kinetic energy from 1 MeV (fast neutron) to about 0.1 eV (thermal neutron). *(4 marks)*

26.4 Nuclear fusion

Specification reference: 6.4.4

All thanks to fusion

The very existence of almost all life on the Earth depends on the light coming from our Sun. The Sun generates its energy by **fusion**, a process in which small nuclei are combined to make larger nuclei. The fusing of small nuclei produces enormous energy – typically several MeV per fusion reaction. The energy released in fusion reactions can be explained in terms of small changes in the mass of nuclei and Einstein's mass–energy equation (see Figure 2 in Topic 26.2, Binding energy). Our Sun converts more than a billion kilograms of matter into energy every second.

Learning outcomes

Demonstrate knowledge, understanding, and application of:

→ nuclear fusion; fusion reactions and temperature

→ balancing nuclear transformation equations.

Fusion reactions

The only way to make nuclei fuse is to bring them close together, to within a few 10^{-15} m, so that the short-range strong nuclear force can attract them into a larger nucleus. All nuclei have a positive charge, so they will repel each other. The repulsive electrostatic force between nuclei is enormous at small separations. At low temperatures, the nuclei cannot get close enough to trigger fusion. However, at higher temperatures, they move faster and can get close enough to absorb each other through the strong nuclear force.

▲ **Figure 1** *The energy produced in the Sun comes from fusion and enables life on Earth*

The conditions for fusion are just right in the core of our stars like our Sun. The temperature is close to 1.4×10^{7} K and the density is $1.5 \times 10^{5}\,\mathrm{kg\,m^{-3}}$. The enormous density ensures a high number of fusion reactions per second.

Examples of fusion

There are many different types of fusion reactions that can take place within stars – they all release energy. Fusion reactions often occur in cycles or sequences. One such cycle is shown here.

- Two protons fuse together to produce a deuterium nucleus ($^{2}_{1}\mathrm{H}$), a positron, and a neutrino. This reaction produces about 2.2 MeV of energy.

$$^{1}_{1}\mathrm{p} + {}^{1}_{1}\mathrm{p} \rightarrow {}^{2}_{1}\mathrm{H} + {}^{\;\,0}_{+1}\mathrm{e} + \nu$$

- You can easily explain this energy using the graph of binding energy per nucleon against nucleon number (Figure 2 in Topic 26.2). The two single protons have zero binding energy and the deuterium nucleus has a binding energy of $1.1 \times 2 = 2.2$ MeV. The difference in the binding energies, of 2.2 MeV, is the energy released in this fusion reaction.
- The deuterium nucleus from the first reaction fuses with a proton. A helium-3 nucleus is formed and 5.5 MeV of energy is released.

$$^{2}_{1}\mathrm{H} + {}^{1}_{1}\mathrm{p} \rightarrow {}^{3}_{2}\mathrm{He}$$

Study tip

In all nuclear reactions, including fusion, the nucleon number A and proton (atomic) number Z are conserved.

- The helium-3 from the second reaction combines with another helium-3 nucleus. A helium-4 nucleus is formed, together with two protons and 12.9 MeV of energy.

$$^{3}_{2}\text{He} + ^{3}_{2}\text{He} \rightarrow ^{4}_{2}\text{He} + 2^{1}_{1}\text{p}$$

The whole cycle is repeated again with the two protons. This cycle is known as the proton–proton cycle or the hydrogen-burning cycle, and is one of the main production routes for helium in stars. The proton–proton cycle occurs around 9×10^{37} times each second inside the Sun.

Fusion on the Earth

There are no power stations using fusion yet. The main problems centre on maintaining high temperatures for long enough to sustain fusion and on confining the extremely hot fuel within a reactor. At present, all experimental fusion reactors produce energy for a very short period of time and in much smaller quantities than must be supplied to start the reaction. In some experiments, powerful lasers have been used to heat and compress a small pellet containing deuterium and tritium.

ITER

In Europe, hopes for fusion reactors rest with the International Thermonuclear Experimental Reactor (ITER), which will carry out important tests in 2027 at temperatures ten times higher than the interior of the Sun. ITER is designed to produce more power than it uses – from 50 MW of input power to 500 MW of output power.

ITER will use a mixture of deuterium and tritium as fuel, which will be heated to temperatures greater than 1.5×10^{8} K. At such temperatures, the electrons are stripped off the deuterium and tritium atoms leaving positive nuclei, so the fuel becomes a plasma. In the ITER reactor (Figure 2), the plasma will be compressed into a doughnut-shaped ring. It would lose its thermal energy if it were to touch the sides of the reactor, so it will be kept away from the walls by strong magnetic fields produced by superconductors and by electrical currents passed through the plasma.

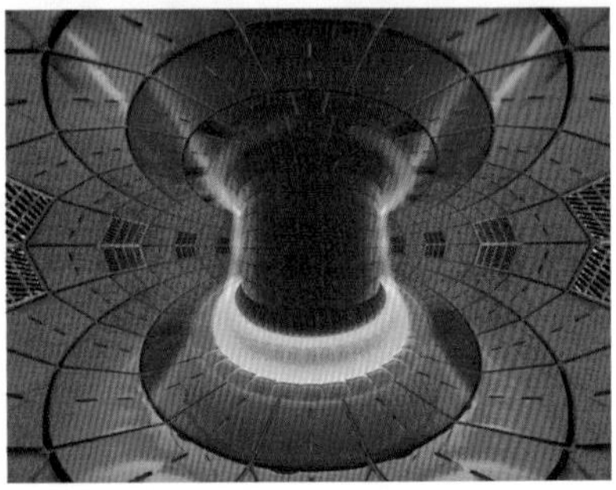

▲ **Figure 2** *The planned ITER will produce energy from fusion*

The fusion of a tritium nucleus ($^{3}_{1}$H) and a deuterium nucleus ($^{2}_{1}$H) produces a helium nucleus, a neutron, and 17.6 MeV of energy.

$$^{3}_{1}\text{H} + ^{2}_{1}\text{H} \rightarrow ^{4}_{2}\text{He} + ^{1}_{0}\text{n}$$

About 80% of the fusion energy is carried by the neutron. The neutrons will be absorbed by a lithium blanket around the reactor. Apart from heating the blanket, the interaction of the neutrons with the lithium nuclei will produce more tritium, which will be recycled for fusion.

1. Plasma can be modelled as an ideal gas. Calculate the mean kinetic energy of the tritium and deuterium nuclei at a temperature of 1.5×10^{8} K.
2. Explain why the total mean energy required for fusion of a tritium nucleus and a deuterium nucleus is twice your answer to question 1.

Summary questions

1 Show that the nuclear transformation equation for fusion below is balanced, with all nucleons and protons being conserved. *(2 marks)*

$${}^{1}_{1}\text{p} + {}^{1}_{1}\text{p} \rightarrow {}^{2}_{1}\text{H} + {}^{0}_{+1}\text{e} + \nu$$

2 Explain why fusion cannot occur at low temperatures. *(2 marks)*

3 The Sun converts more than 10^9 kg of matter into energy per second. Estimate the rate of energy production by the Sun. *(2 marks)*

4 In the fusion reaction shown below, 5.5 MeV energy is released.

$${}^{2}_{1}\text{H} + {}^{1}_{1}\text{p} \rightarrow {}^{3}_{2}\text{He}$$

Calculate the decrease in mass in this single reaction. *(3 marks)*

5 Two deuterium nuclei can fuse together to form a helium nucleus.

a Write a nuclear transformation equation for this fusion reaction. *(2 marks)*

b Use Figure 2 in Topic 26.2 to show that the energy released in this reaction is about 4×10^{-12} J. *(4 marks)*

c Use your answer to (b) to calculate the maximum energy that can be produced by the fusion of 1.0 kg of deuterium. The molar mass of deuterium is 0.002 kg mol^{-1}. *(4 marks)*

d Estimate how long 1.0 kg of deuterium fuel would last in a proposed 500 MW output fusion reactor that will have an efficiency of 50% (that is, convert 50% of the energy released into useful energy). *(1 mark)*

Practice questions

1 The nuclear reaction represented by the equation

$^{235}_{92}U + ^{1}_{0}n \rightarrow ^{94}_{39}Y + ^{139}_{53}I + 3^{1}_{0}n$

Takes place in the core of a nuclear reactor at a power station.

a Describe how this reaction can lead to a chain reaction. *(1 mark)*

b Explain the role of fuel rods, control rods, and moderator in a nuclear reactor. *(5 marks)*

c In the nuclear reactor of a power station, each fission reaction of uranium produces 3.2×10^{-11} J of energy. The electrical power output of the power station is 3.0 GW. The efficiency of the system that transforms nuclear energy into electrical energy is 22%. Calculate

(i) the total power output of the reactor core *(1 mark)*

(ii) the total energy output of the reactor core in one day

1 day = 8.64×10^{4} s *(1 mark)*

(iii) the mass of uranium-235 converted in one day. The mass of a uranium-235 nucleus is 3.9×10^{-25} kg. *(2 marks)*

d Discuss the physical properties of nuclear waste that make it dangerous. *(2 marks)*

Jun 2012 G485

2 a In the core of a nuclear reactor, one of the many fission reactions of the uranium-235 nucleus is shown below.

$^{235}_{92}U + ^{1}_{0}n \rightarrow ^{140}_{54}Xe + ^{94}_{38}Sr + 2^{1}_{0}n$

(i) State **one** quantity that is conserved in this fission reaction. *(1 mark)*

(ii) Figure 1 illustrates this fission reaction.

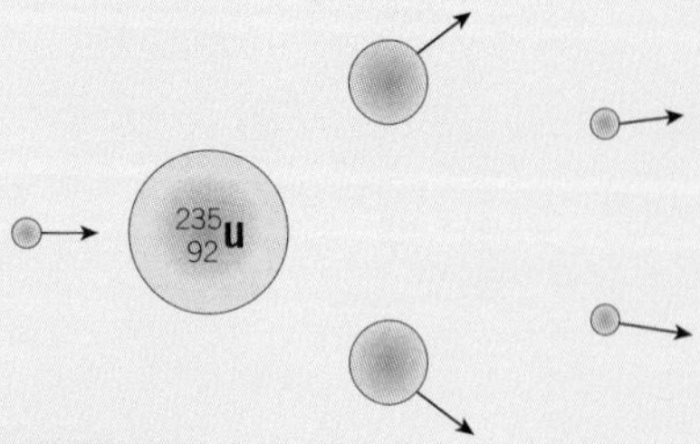

▲ Figure 1

Label all the particles in a copy of Figure 1 and extend the diagram to show how a chain reaction might develop. *(2 marks)*

b Fusion of hydrogen nuclei is the source of energy in most stars. A typical reaction is shown below.

$^{2}_{1}H + ^{2}_{1}H \rightarrow ^{3}_{2}He + ^{1}_{0}n$

The $^{2}_{1}H$ nuclei repel each other. Fusion requires the $^{2}_{1}H$ nuclei to get very close and this usually occurs at very high temperatures, typically 10^{9} K.

(i) Use the data below to calculate the energy released in the fusion reaction above.

Mass of $^{2}_{1}H$ nucleus = 3.343×10^{-27} kg

Mass of $^{3}_{2}He$ nucleus = 5.006×10^{-27} kg

Mass of $^{1}_{0}n$ = 1.675×10^{-27} kg *(3 marks)*

(ii) State in what form the energy in **(b)(i)** is released. *(1 mark)*

(iii) The $^{2}_{1}H$ nuclei in stars can be modelled as an ideal gas. Calculate the mean kinetic energy of the $^{2}_{1}H$ nuclei at 10^{9} K. *(2 marks)*

(iv) Suggest why some fusion can occur at a temperature as low as 10^{7} K. *(1 mark)*

Jan 2013 G485

3 a The following nuclear reaction occurs when a slow-moving neutron is absorbed by an isotope of uranium-235.

$^{1}_{0}n + ^{235}_{92}U \rightarrow ^{141}_{56}Ba + ^{92}_{36}Kr + 3^{1}_{0}n$

(i) Explain how this reaction is able to produce energy. *(2 marks)*

(ii) State in what form the energy is released in such a reaction. *(1 mark)*

b The binding energy per nucleon of each isotope in **(a)** is given in Table 1.

▼ Table 1

isotope	binding energy per nucleon / MeV
$^{235}_{92}U$	7.6
$^{141}_{56}Ba$	8.3
$^{92}_{36}Kr$	8.7

(i) Explain why the neutron $^{1}_{0}n$ does not appear in the table above. *(1 mark)*

(ii) Calculate the energy released in the reaction shown in **(a)**. *(2 marks)*

Jun 2010 G485

4 a Describe the process of induced nuclear fission. *(2 marks)*

b Explain how nuclear fission can provide energy *(2 marks)*

c Suggest a suitable material which can be used as a moderator in a fission reactor and explain its role. *(3 marks)*

Jan 2011 G485

5 In a particular fission reaction a uranium-235 nucleus absorbs a neutron and undergoes fission to a barium-141 nucleus and a krypton-92 nucleus. The reaction is as follows:

$$^{235}_{92}U + ^{1}_{0}n \rightarrow ^{141}_{56}Ba + ^{92}_{36}Kr + 3^{1}_{0}n$$

Data: binding energies per nucleon for these nuclei are:

$^{235}_{92}U$ 7.6 MeV; $^{141}_{56}Ba$ 8.4 MeV; $^{92}_{36}Kr$ 8.6 MeV

a Show that the energy released when one $^{235}_{92}U$ nucleus undergoes fission in this way is about 200 MeV. *(3 marks)*

b Calculate how much energy is released when 1.00 kg of uranium-235 undergoes fission. Assume that every fission generates the same amount of energy as the reaction stated above. *(3 marks)*

Jan 2009 2825/04

6 This question is about particles and their antiparticles.

a State the mass and charge of an *antiproton*. *(2 marks)*

b State where an antiproton might be found. *(1 mark)*

c When a proton and an antiproton meet, γ-photons are produced.

(i) Describe these photons as fully as you can for a slow-moving proton-antiproton collision. No calculation is required. *(3 marks)*

(ii) A proton and an antiproton are moving with almost the same high speed and in the same direction. Each possesses 8.00×10^{-11} J of kinetic energy. The two particles meet. Calculate the frequency of the γ-photons produced. *(4 marks)*

7 a Describe the processes of fission and fusion of nuclei. Distinguish clearly between them by highlighting **one** similarity and **one** difference between the two processes. State the conditions required for each process to occur in a sustained manner. *(7 marks)*

b The fission of a uranium-235 nucleus releases about 200 MeV of energy, whereas the fusion of four hydrogen-1 nuclei releases about 28 MeV. However the energy released in the fission of one kilogramme of uranium-235 is less than the average released in the fusion of one kilogramme of hydrogen-1. Explain this by considering the number of particles in one kilogramme of each. *(4 marks)*

Jan 2007 2824

27 MEDICAL IMAGING
27.1 X-rays

Specification reference: 6.5.1

Learning outcomes

Demonstrate knowledge, understanding, and application of:

→ the basic structure of an X-ray tube; components – heater (cathode), anode, target metal and high voltage supply

→ production of X-ray photons from an X-ray tube.

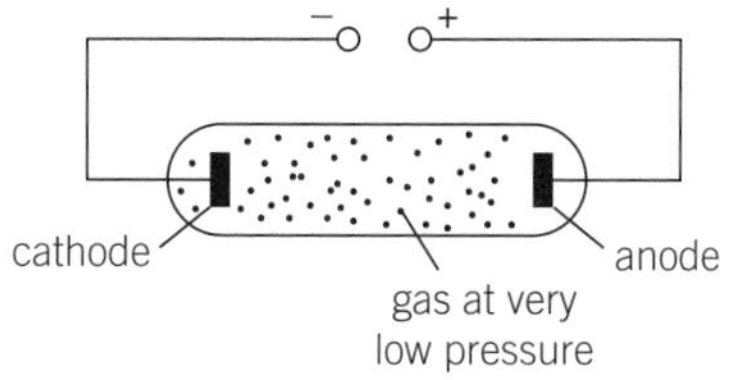

▲ **Figure 1** *A discharge tube containing gas at low pressure through which charge can flow*

Synoptic link

X-rays are part of the electromagnetic spectrum, which you studied in Topic 11.6, Electromagnetic waves.

The first X-ray picture

Wilhelm Röntgen discovered **X-rays** in 1895. He was investigating the light emitted by gases in a discharge tube when a p.d. is applied between its two electrodes (Figure 1). When the gas in the tube was at extremely low pressure, the tube went dark, but he noticed that a fluorescent plate near his apparatus glowed. When he placed his hand between the tube and the plate, he could see shadows of the bones in his hand. The unknown rays from the tube were passing through soft tissue but were stopped by bone. We now call these rays X-rays.

Röntgen took the world's first X-ray picture (Figure 2). He did not know that intense X-rays are very harmful. Modern medical X-ray imaging uses low-intensity X-rays for very short exposure times, so is relatively safe, yet produces amazing images of structures within the body (Figure 3).

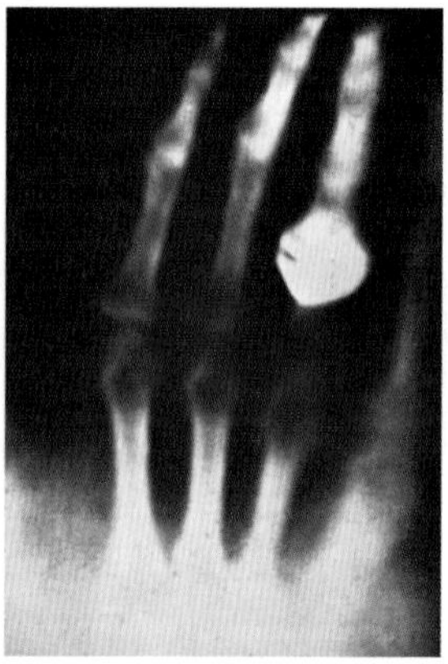

▲ **Figure 2** *The first recorded X-ray image shows the hand of Anna Bertha Röntgen, Wilhelm Röntgen's wife – note the ring on her finger*

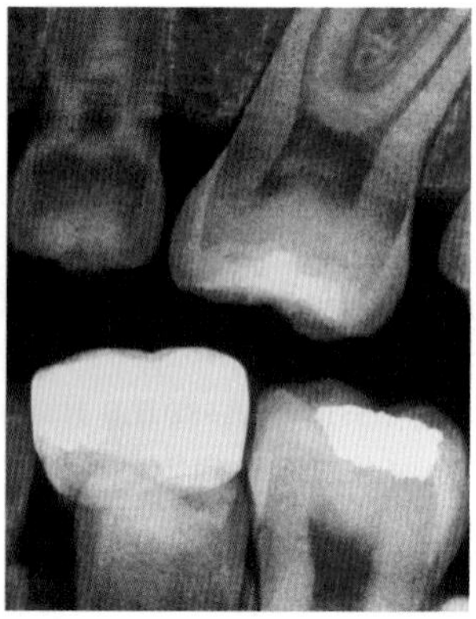

▲ **Figure 3** *Modern X-ray images are sophisticated and have good contrast – this colour image shows a child's teeth with fillings*

The nature of X-rays

Experiments performed on the newly discovered X-rays showed that they could be polarised, were diffracted by atoms in crystals, and had extremely short wavelengths (range 10^{-8} to 10^{-13} m). They are electromagnetic waves and therefore travel through a vacuum at the speed of light.

X-ray photons have 10–10 000 times more energy than a photon of visible light, depending on their wavelength. X-rays are harmful to living cells and can kill them. It is this property of X-rays that is used in the treatment of cancer.

Production of X-rays

X-ray photons are produced when fast-moving electrons are decelerated by interaction with atoms of a metal such as tungsten. The kinetic energy of the electrons is transformed into X-ray photons.

Figure 4 shows a patient having a radiograph (X-ray image) taken. The X-ray machine is above the patient. It contains an **X-ray tube** that produces X-ray photons that pass through the patient to the detection plate below. Digital detection plates have replaced photographic plates, because the images can be stored and shared on computers and can be enhanced to detect subtle changes in tissues and bones.

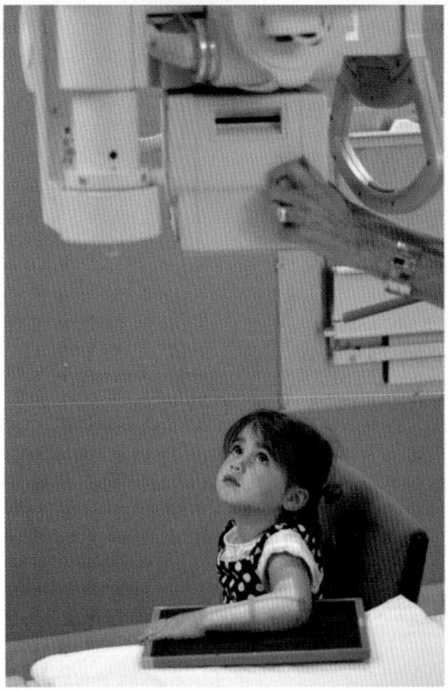

▲ **Figure 4** *The X-ray tube is housed inside the machine above the young patient*

An X-ray tube (Figure 5) consists of an evacuated tube containing two electrodes. The tube is evacuated so that electrons pass through the tube without interacting with gas atoms. An external power supply is used to create a large p.d. (typically 30–100 kV) between these electrodes. The cathode (negative) is a heater, which produces electrons by **thermionic emission**. These electrons are accelerated towards the anode (positive). The anode is made from a metal, known as the **target metal**, such as tungsten, that has a high melting point.

X-ray photons are produced when the electrons are decelerated by hitting the anode. The energy output of X-rays is less than 1% of the kinetic energy of the incident electrons. The remainder of the energy is transformed into thermal energy of the anode. In many X-ray tubes, oil is circulated to cool the anode, or the anode is rotated to spread the heat over a large surface area.

The anode is shaped so that the X-rays are emitted in the desired direction through a window. The X-ray tube is lined with lead to shield the radiographer from any X-rays emitted in other directions.

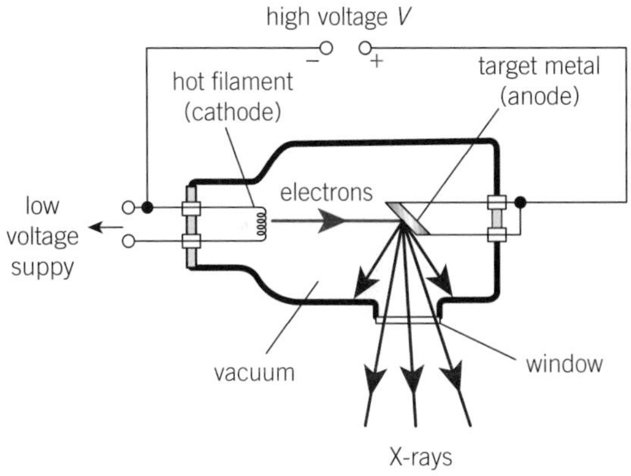

▲ **Figure 5** *An X-ray tube*

The shortest wavelength

An electron accelerated through a potential difference V gains kinetic energy eV, where e is the elementary charge. Since one electron releases one X-ray photon, from the principle of conservation of energy, the maximum energy of a photon from an X-ray tube must equal the maximum kinetic energy of a single electron.

maximum energy of X-ray photon = maximum kinetic energy of electron

The energy of a photon is equal to the Planck constant h × frequency f, and maximum frequency of the emitted X-rays f is the speed c divided by the minimum wavelength λ, so

$$hf = eV$$

$$\frac{hc}{\lambda} = eV \qquad \text{therefore} \qquad \lambda = \frac{hc}{eV}$$

The wavelength from an X-ray tube is inversely proportional to the accelerating potential difference. Increasing the tube current just increases the intensity of the X-rays.

Synoptic link

One electron is responsible for producing one photon – this one-to-one mechanism is similar to that described in Topic 13.2, The photoelectric effect.

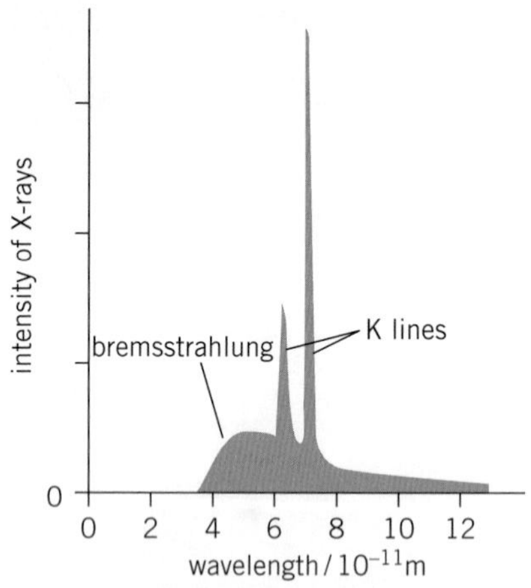

▲ **Figure 6** *A typical X-ray spectrum for molybdenum*

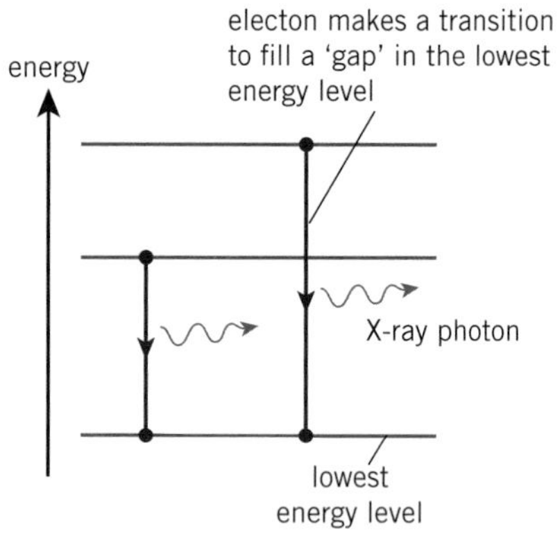

▲ **Figure 7** *How K-lines are produced by transitions between electron energy levels*

Characteristic spectrum

Figure 6 shows a typical X-ray spectrum, a graph of the intensity of the X-rays from an X-ray tube against wavelength for a particular supply voltage. The target metal used is molybdenum.

The range of decelerations of the electrons inside the X-ray tube produces the broad background of bremsstrahlung. 'Bremsstrahlung' means 'braking radiation' in German. The narrow, intense lines are referred to as the K-lines, and are characteristic of the target metal. The bombarding electrons can remove electrons in the metal atoms close to the nuclei. So the gaps created in the lower energy levels of the metal atoms are quickly filled by electrons dropping from higher energy levels. These transitions release photons of specific energies and therefore wavelengths (Figure 7).

1 Use Figure 6 to estimate the accelerating p.d. for the X-ray tube.
2 Estimate the difference between the two energy levels responsible for the most intense K-line in the X-ray spectrum of molybdenum.
3 Suggest how the shape of the graph would change when the accelerating p.d. is increased.

Summary questions

1 State a typical value for the wavelength of X-rays. *(1 mark)*

2 Use the wavelength from question 1 to calculate:
a the frequency of the X-rays; *(2 marks)*
b the energy of a single X-ray photon. *(2 marks)*

3 An X-ray tube is connected to a 65 kV supply. Calculate:
a the kinetic energy of an electron at the anode; *(2 marks)*
b the maximum energy of an X-ray photon. *(1 mark)*

4 The tube current in an X-ray tube is 21 mA. Calculate the number of electrons hitting the anode per second. *(2 marks)*

5 The X-ray tube in question 4 has an efficiency of 0.60%. Estimate the number of X-ray photons emitted from the tube per second. *(2 marks)*

6 Calculate the shortest wavelength of X-rays from an X-ray tube operating at 100 kV. Explain your answer. *(4 marks)*

27.2 Interaction of X-rays with matter

Specification reference: 6.5.1

Absorption of X-rays

Figure 1 shows a digital X-ray image of a patient's leg. Clearly, bones absorb more X-ray photons than do soft tissues and muscles. X-ray photons interact with the atoms of the material they pass through. The photons can be scattered or absorbed by the atoms, and this reduces the intensity of the X-rays. The term **attenuation** is used to describe the decrease in the **intensity** of an electromagnetic radiation as it passes through matter. So you can say that bone attenuates X-rays more than soft tissues.

Attenuation mechanisms

The intensity of a parallel (collimated) beam of X-rays will decrease as it passes through matter. There are four attenuation mechanisms by which X-ray photons interact with atoms (Figure 2). Each mechanism reduces the intensity of the collimated beam in the original direction of travel.

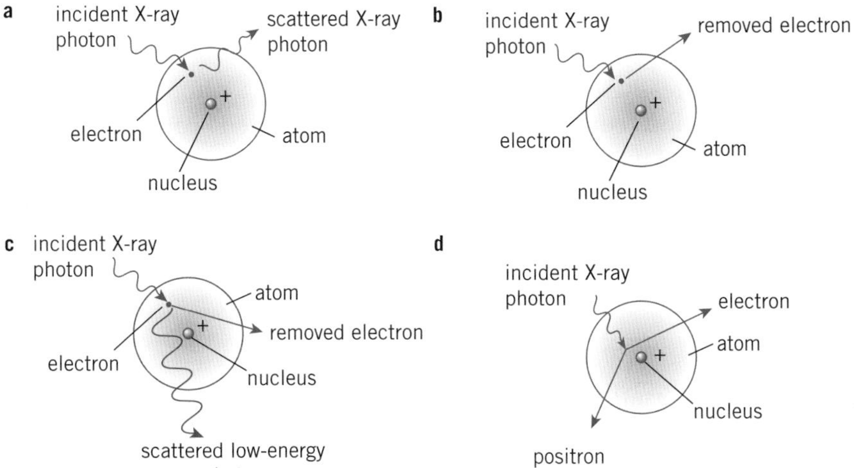

▲ **Figure 2** *Attenuation mechanisms: (a) simple scatter – the X-ray photon is scattered elastically by an electron; (b) photoelectric effect – the X-ray photon disappears and removes an electron from the atom; (c) Compton scattering – the X-ray photon is scattered by an electron, its energy is reduced, and the electron is ejected from the atom; (d) pair production – the X-ray photon disappears to produce an electron–positron pair*

Simple scatter

This mechanism is important for X-ray photons with energy in the range 1–20 keV. The X-ray photon interacts with an electron in the atom, but has less energy than the energy required to remove the electron, so the X-ray photon simply bounces off (is scattered) without any change to its energy. The X-ray machines used in hospitals use p.d.s greater than 20 kV, so this type of mechanism is insignificant for hospital radiography.

Learning outcomes

Demonstrate knowledge, understanding, and application of:

- → attenuation of X-rays
- → X-ray attenuation mechanisms: simple scatter, photoelectric effect, Compton effect, and pair production
- → $I = I_0 e^{-\mu x}$
- → X-ray imaging with contrast media.

Synoptic link

Intensity is defined as the radiant power per unit cross-sectional area and has SI unit W m^{-2}. See Topic 11.5, Intensity.

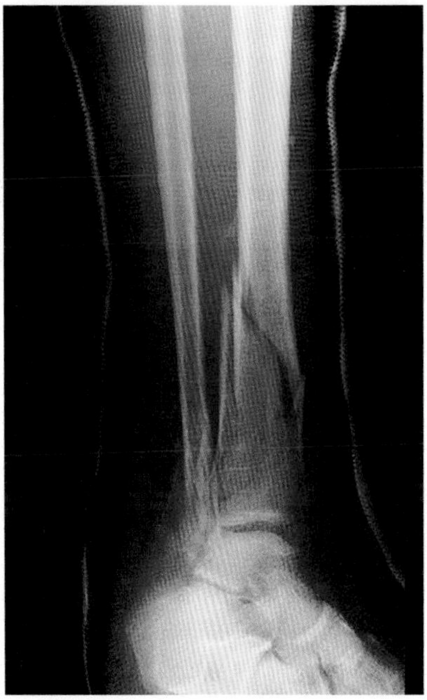

▲ **Figure 1** *You can easily identify the outline of the fibreglass cast around the leg and of course the broken bone in this X-ray image*

Synoptic links

You will recognise the value of 1.02 MeV from Topic 26.1, Einstein's mass–energy equation. It is the minimum energy required for a photon to create an electron–positron pair.

X-rays interact with matter as photons. You have already seen this in Topic 13.2, The photoelectric effect.

Photoelectric effect

This mechanism is significant for X-ray photons with energy less than 100 keV. The X-ray photon is absorbed by one of the electrons in the atom. The electron uses this energy to escape from the atom. Attenuation of X-rays by this type of mechanism is dominant when an X-ray image is taken, because hospital X-ray machines typically use 30–100 kV supplies.

Compton scattering

This mechanism is significant for X-ray photons with energy in the range 0.5–5.0 MeV. The incoming X-ray photon interacts with an electron within the atom. The electron is ejected from the atom, but the X-ray photon does not disappear completely – instead it is scattered with reduced energy. In the interaction, both energy and momentum are conserved. (Yes, photons do have momentum, but this concept is not covered at A Level.)

Pair production

This mechanism only occurs when X-ray photons have energy equal to or greater than 1.02 MeV. An X-ray photon interacts with the nucleus of the atom. It disappears and the electromagnetic energy of the photon is used to create an electron and its antiparticle, a positron.

Attenuation coefficients

You have already seen that X-ray photons interact with matter and this interaction reduces the intensity of a collimated beam of X-rays in the original direction of travel. The transmitted intensity of X-rays depends on the energy of the photons and on the thickness and type of the substance. For a given substance and energy of photons, the intensity falls exponentially with thickness of substance. The transmitted intensity I is given by the equation

$$I = I_0 e^{-\mu x}$$

where I_0 is the initial intensity before any absorption, x is the thickness of the substance, and μ is the **attenuation coefficient** or the **absorption coefficient** of the substance. Bone is a better absorber of X-rays than muscle, so bone has a larger value of μ than muscle. The SI unit of the attenuation coefficient is m^{-1}, but you can use cm^{-1} and mm^{-1}.

Worked example: Absorption by bone

A collimated beam of X-rays from a 100 kV supply is incident on bone. The initial intensity of the beam is $18\,W\,m^{-2}$. The attenuation coefficient of bone is $0.60\,cm^{-1}$. Calculate the intensity of the beam after it has passed through 7.0 mm of bone.

Step 1: Write down all the quantities given. It is important to have the values of μ and x in consistent units (here cm^{-1} and cm, respectively).

$$I_0 = 18\,W\,m^{-2},\ \mu = 0.60\,cm^{-1},\ x = 0.70\,cm$$

Step 2: Substitute the values into the exponential decay equation and calculate the transmitted intensity I.

$$I = I_0 e^{-\mu x} = 18 \times e^{-(0.60 \times 0.70)} = 12\,W \text{ (2 s.f.)}$$

Contrast medium

Soft tissues have low absorption coefficients, so a contrast medium is used to improve the visibility of their internal structures in X-ray images. The two most common are iodine and barium compounds, both of which are relatively harmless to humans.

Barium and iodine are elements with large atomic number Z. For X-ray imaging, the predominant interaction mechanism is the photoelectric effect, for which the attenuation coefficient is proportional to the cube of the atomic number ($\mu \propto Z^3$). The average atomic number for soft tissues is about 7. This means that iodine ($Z = 53$) and barium ($Z = 56$) are about 430 times and 510 times more absorbent than soft tissues, respectively.

Iodine is used as a contrast medium in liquids, for example, to view blood flow. An organic compound of iodine is injected into blood vessels so that doctors can diagnose blockages in the blood vessels and the structure of organs such as the heart from the X-ray image (Figure 3).

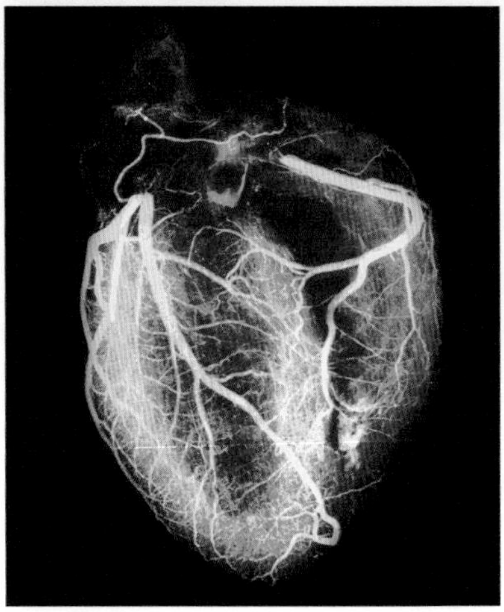

▲ **Figure 3** *An X-ray image (angiogram) of a healthy heart, obtained by injecting iodine into the circulatory system so that the blood vessels show up clearly*

Barium sulfate is often used to image digestive systems. It is given to a patient in the form of a white liquid mixture (a 'barium meal'), which the patient swallows before an X-ray image is taken. Figure 4 shows an X-ray image of the intestine of a patient who has had a barium meal. The pale regions are where the barium has accumulated.

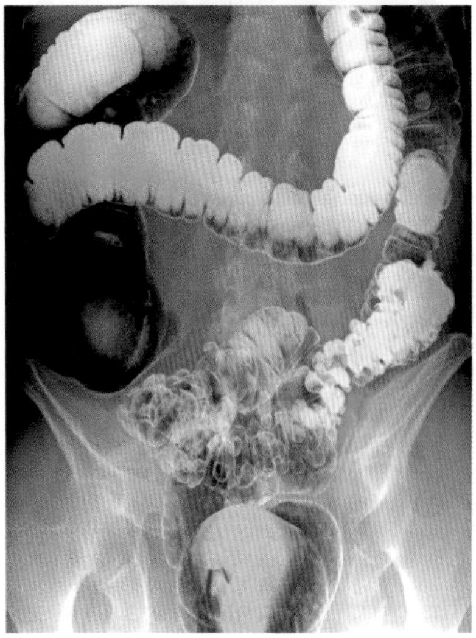

▲ **Figure 4** *A coloured X-ray image of a patient's intestine after a barium meal – notice how the outline of the intestine is easy to identify against the surrounding soft tissues*

Therapeutic use

X-rays are also used for therapy rather than imaging. Specialised X-ray machines, called linacs (linear accelerators), are used to create high-energy X-ray photons. These photons are used to kill off cancerous cells. They do so by the mechanisms of Compton scattering and pair production.

Summary questions

1. Name the attenuation mechanisms in which an electron inside an atom is involved. *(1 mark)*
2. Explain why simple scattering is not an important mechanism when X-ray images are taken in a hospital. *(2 marks)*
3. The attenuation coefficient of muscle is $0.21\,\text{cm}^{-1}$ for X-ray photons of energy 100 keV. Convert this attenuation coefficient into m^{-1}. *(1 mark)*
4. Use the information in question 3 to calculate the percentage of intensity of X-rays transmitted for muscle of thickness 0.80 cm. *(3 marks)*
5. Calculate the maximum wavelength of an X-ray photon responsible for pair production. *(3 marks)*
6. Use the information given in question 3 to calculate the thickness of muscle that will reduce the transmitted intensity of X-rays by half. *(3 marks)*

27.3 CAT scans

Specification reference: 6.5.1

Learning outcomes

Demonstrate knowledge, understanding, and application of:

- → computerised axial tomography (CAT) scanning and the necessary components
- → the advantages of a CAT scan over an X-ray image.

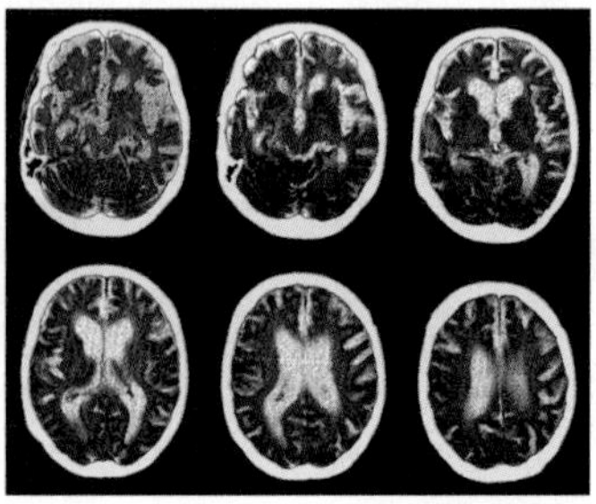

▲ **Figure 1** *A CAT scan yielded these virtual slices through the head of a patient with Alzheimer's disease – you can see the growing cavities (white) in the brain (brown)*

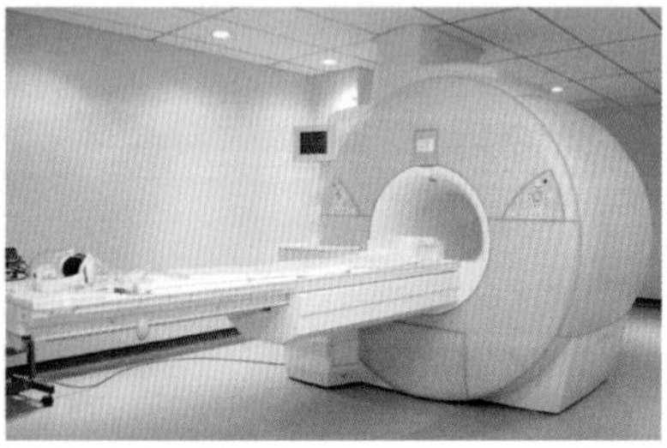

▲ **Figure 2** *A modern CAT scanner*

Three-dimensional imaging

A conventional X-ray image provides a quick and cheap way to examine patients' internal structures. X-rays pass through the patient, and the intensity of the transmitted X-rays is recorded as a two-dimensional image on an electronic plate. Overlapping bones and tissues cannot be differentiated, and without the use of a contrast medium, different soft tissues are difficult to distinguish.

Figure 1 shows cross-sectional images of a head from a computerised axial tomography (CAT) scanner. A CAT scanner records a large number of X-ray images from different angles and assembles them into a three-dimensional image with the help of sophisticated software.

The scanning process and the analysis of electrical signals from detectors is controlled by a computer (and so the term 'computerised'). The term 'axial' refers to the images taken in the axial plane, cross-sections through the patient. Finally, 'tomography' is made up of two Greek words, 'tomos' meaning slice and 'graphein' meaning to record.

Computerised axial tomography

In a modern CAT scanner (Figure 2), the patient lies on their back on a horizontal examination table that can slide in and out of a large vertical ring or gantry. The gantry houses an X-ray tube on one side and an array of electronic X-ray detectors on the opposite side. The X-ray tube and the detectors opposite it rotate around within the gantry.

The X-ray tube produces a fan-shaped beam of X-rays that is typically only 1–10 mm thick. The thin beam irradiates a thin slice of the patient, and the X-rays are attenuated by different amounts by different tissues. The intensity of the transmitted X-rays is recorded by the detectors, which send electrical signals to a computer (Figure 3).

Each time the X-ray tube and detectors make a 360° rotation, a two-dimensional image or 'slice' is acquired. By the time the X-ray tube has made one complete revolution, the table has moved about 1 cm through the ring. In the next revolution, the X-ray beam irradiates the next slice through the patient's body. So the X-ray beam follows a spiral path during the 10–30 minute scan.

The radiographer can view each two-dimensional slice through the patient. In addition, the slices can be manipulated by sophisticated software to produce a three-dimensional image of the patient. This three-dimensional image can be rotated and zoomed on a display.

The technology of CAT scanners is still developing. The CAT scanners described above have X-ray detectors that rotate with the X-ray tube, but there are CAT scanners with a complete stationary ring of X-ray detectors but still with a rotating X-ray tube.

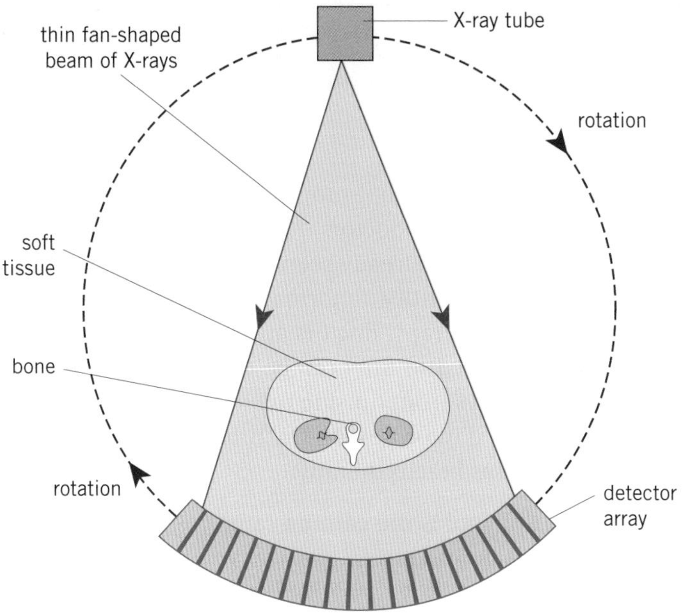

▲ **Figure 3** *The X-ray tube and the detectors rotate around the patient*

Advantages and disadvantages

A single traditional X-ray scan is quicker and cheaper than a CAT scan. However, CAT scans can be used to create a three-dimensional image of the patient that helps doctors to assess the shape, size, and position of disorders such as tumours. CAT scans can distinguish between soft tissues of similar attenuation coefficients.

X-rays are ionising radiation and as such are harmful. Some CAT scans can be quite prolonged and so expose the patients to a radiation dose equivalent to several years of background radiation, much more than a simple X-ray.

Patients have to remain very still during the scanning process, because any movement blurs the slice. Remaining still can be quite tricky for some patients, especially for the very young.

Summary questions

1 Name the main components of a CAT scanner. *(2 marks)*

2 State one advantage and one disadvantage of a CAT scan over an X-ray image. *(2 marks)*

3 Suggest why a thin beam of X-rays is necessary in a CAT scanner. *(1 mark)*

4 Explain what is meant by a 'slice' in CAT scanning. *(1 mark)*

5 Suggest how the CAT image of the blood flow in the head shown in Figure 4 may have been obtained. *(2 marks)*

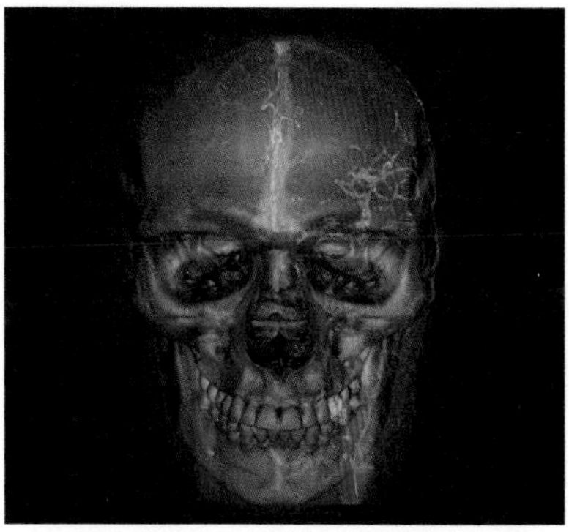

▲ **Figure 4** *Blood supply to the head of a 38-year-old man – the left carotid artery (on the right of this image) is highlighted*

27.4 The gamma camera

Specification reference: 6.5.2

Learning outcomes

Demonstrate knowledge, understanding, and application of:

- → medical tracers technetium-99m and fluorine-18
- → the gamma camera and its components, formation of gamma camera images
- → diagnosis using the gamma camera.

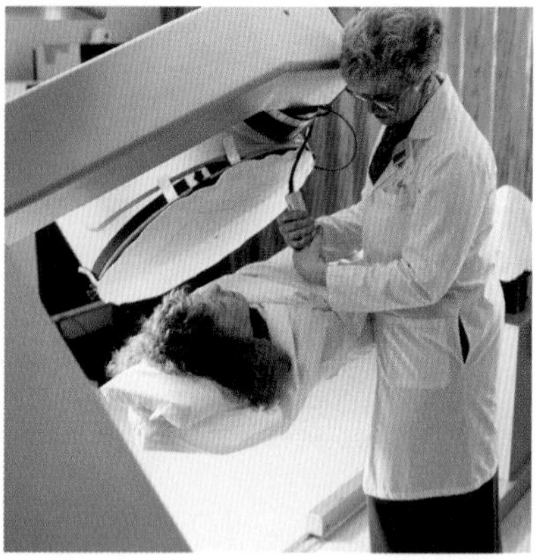

▲ **Figure 1** *A modern gamma camera in use on a patient*

Diagnosis and therapy

In medicine, radioactive isotopes (radioisotopes) are used in both diagnosis and therapy. In diagnosis, doctors try to find out what is wrong with the patient. You have already seen how a CAT scanner can be used as a diagnostic tool to identify abnormalities inside a patient without surgery. In radiation therapy, doctors attempt to cure the patient using ionising radiation. Tumours can be targeted by gamma radiation or high-energy X-rays from outside the patient or by using a radioactive source implanted in or next to the tumour inside the patient, a technique known as brachytherapy.

Another valuable diagnostic tool is the gamma camera. This is a detector of gamma photons emitted from radioactive nuclei injected into the patient.

Choosing the right radioisotopes heading and weight

Radioisotopes used for medical imaging have to be placed inside the patient and their radiation detected from the outside. This makes gamma-emitting sources ideal because gamma photons are the least ionising and can penetrate through the patient to be detected externally. Alpha and beta sources cause more ionisation. They are dangerous and are not used for imaging techniques.

Radioisotopes chosen for medical imaging must have a short half-life to ensure high activity from the source so that only a small amount is required to form the image. The other benefit is the patient is not subjected to a high dosage of radiation that continues long after the procedure.

Many of the radioisotopes used in medicine are produced artificially. Fluorine-18, for example, which is used in PET scans, has a short half-life and has to be produced on-site at the hospital using a particle accelerator. You will look in more detail at PET scanners in Topic 27.5, PET scans. Technetium-99m (Tc-99m) is an extremely versatile radioisotope that can be used to monitor the function of major organs such as the heart, liver, lungs, kidneys, and brain. The isotope is produced from the natural radioactive decay of molybdenum-99.

$$ {}^{99}_{42}\text{Mo} \xrightarrow{67\,\text{h}} {}^{99\text{m}}_{43}\text{Tc} + {}^{0}_{-1}\text{e} + \bar{\nu}_\text{e} $$

$$ {}^{99\text{m}}_{43}\text{Tc} \xrightarrow{6.0\ \text{h}} {}^{99}_{43}\text{Tc} + \gamma $$

$$ {}^{99}_{43}\text{Tc} \xrightarrow{210\,000\ \text{years}} {}^{99}_{44}\text{Ru} + {}^{0}_{-1}\text{e} + \bar{\nu}_\text{e} $$

The Mo-99 isotope decays by beta-minus emission with a half-life of 67 hours. Tc-99m is a daughter nucleus in this decay, and it too is unstable (the 'm' means 'metastable', and refers to a nucleus that stays in a high-energy state, with more energy than the stable nucleus, for a longer period than expected). The Tc-99m isotope loses energy by emitting a gamma photon with energy of exactly 140 keV, with a half-life of about 6.0 hours. Stable Tc-99 remains, which has an extremely long half-life of 210 000 years.

Medical tracers used in diagnosis

In order to ensure that the radioisotope reaches the correct organ or tumour, the radioisotope has to be chemically combined with elements that will target the desired tissues to make a **radiopharmaceutical**, also known as a **medical tracer**. For example, technetium-99m can be chemically combined with sodium and oxygen to make the inorganic chemical compound $NaTcO_4$. This compound, once injected into the patient, will target the cells in the brain. The Tc-99m in the compound travels through the patient's body. Its progress through the body can be traced using a gamma camera as the Tc-99m emits gamma photons. The concentration of the radiopharmaceutical can be used to identify irregularities in the function of the body.

Use of the gamma camera

A gamma camera (Figure 2) detects the gamma photons emitted from the medical tracer (usually based on technetium-99m) injected into the patient, and an image is constructed indicating the concentration of the tracer within the patient's body.

The gamma photons travel towards the **collimator**, a honeycomb of long, thin tubes made from lead. Any photons arriving at an angle to the axis of the tubes are absorbed by the tubes, so only those travelling along the axis of the tubes reach the **scintillator**.

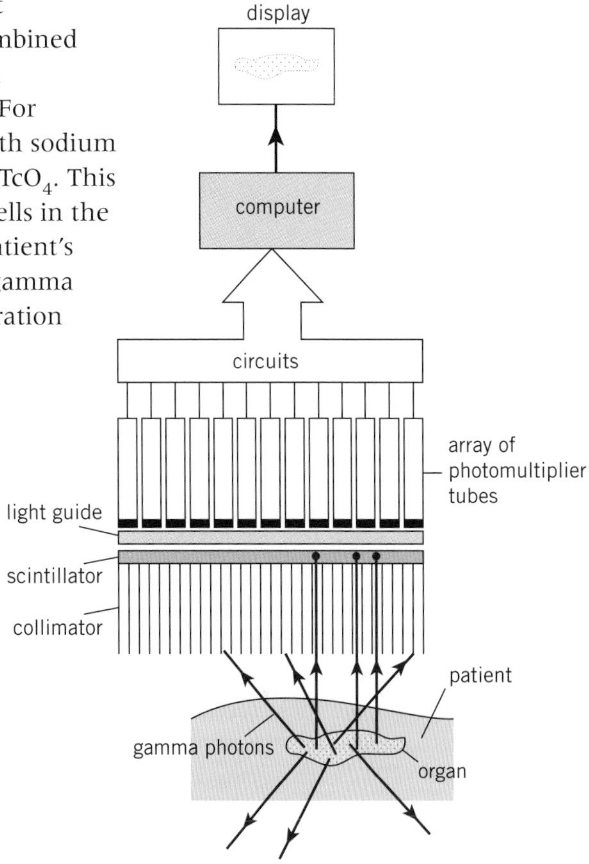

▲ **Figure 2** *The components of a gamma camera*

The scintillator material is often sodium iodide. A single gamma photon striking the scintillator produces thousands of photons of visible light. Not all the gamma photons produce these tiny flashes, because the chance of a gamma photon interacting with the scintillator is about 1 in 10.

The photons of visible light travel through the light guide into the **photomultiplier tubes**. These tubes are arranged in a hexagonal pattern. A single photon of light entering a photomultiplier tube is converted into an electrical pulse (voltage). The outputs of all the photomultiplier tubes are connected to a computer. With the help of sophisticated software, the electrical signals from the tubes can be processed very quickly to locate the impacts of the gamma photons on the scintillator. These impact positions are used to construct a high-quality image that shows the concentrations of the medical tracer within the patient's body. The final image is displayed on a screen (Figure 3).

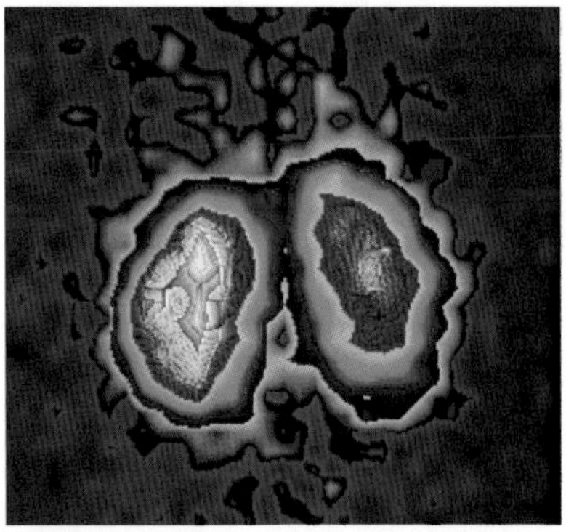

▲ **Figure 3** *A gamma camera image of a patient's kidneys, seen from the back – the kidney on the right is infected and less active than the normal one on the left, so it has taken up less Tc-99m*

Study tip

Use the term 'photons' to describe the operation of a gamma camera and not 'gamma rays' or 'visible light'.

A gamma camera differs from an X-ray imaging technique in one very important respect – it produces an image that shows the function and processes of the body rather than its anatomy.

Photomultipliers

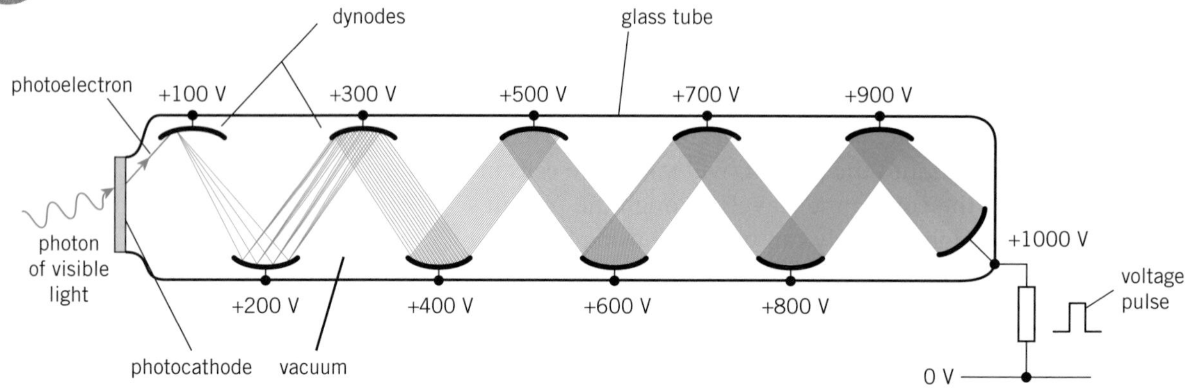

▲ **Figure 4** *Photomultiplier tube*

Figure 4 shows the details of a simple photomultiplier tube. A single photon of visible light hitting the photocathode produces a photoelectron. This electron is accelerated to the first electrode (dynode), which is held at a potential of +100 V. The high-speed impact of this electron at the dynode produces an average of four secondary electrons. These secondary electrons are then accelerated towards the second dynode at a higher potential of +200 V. Each electron creates four secondary electrons on average as this process is repeated at successive dynodes, and the number of electrons grows exponentially. With ten dynodes, the number of electrons arriving at the anode from one photon can be as many as a million. The electrons collected at the anode pass through a resistor and produce a tiny voltage pulse.

1 State the function of a photocathode in a photomultiplier tube.

2 Show that a photomultiplier with ten dynodes produces about 10^6 electrons for every incident photon of visible light.

3 Calculate the total charge represented by 10^6 electrons.

Synoptic link

You first met photoelectrons in Topic 13.2, The photoelectric effect.

Summary questions

1 Name two radioisotopes used as medical tracers. *(1 mark)*

2 Suggest why a Tc-99m-based medical tracer has to be produced on-site in a hospital. *(1 mark)*

3 State the function of a photomultiplier tube. *(1 mark)*

4 State one advantage of a gamma scan over an X-ray scan. *(1 mark)*

5 Explain why the decay of Tc-99 within the patient is not a major concern during a gamma scan. *(2 marks)*

6 The typical initial activity of a Tc-99m-based medical tracer is about 500 MBq. Use the half-life from the text to calculate the initial number of Tc-99m nuclei. *(3 marks)*

27.5 PET scans

Specification reference: 6.5.2

Fluorine-18

Fluorine-18 is a versatile radiopharmaceutical (medical tracer) used in positron emission tomography (PET). The isotope is a positron emitter with a half-life of about 110 minutes. A nucleus of fluorine-18 decays into a nucleus of oxygen-18, a positron, a neutrino, and a gamma photon.

$$^{18}_{9}\text{F} \rightarrow\ ^{18}_{8}\text{O} +\ ^{0}_{+1}\text{e} + \nu_e + \gamma$$

Fluorine-18 has to be made either on-site or in a specialist laboratory near the hospital with a particle accelerator. In one method, high-speed protons collide with oxygen-18 nuclei and produce fluorine-18 nuclei and neutrons. Non-radioactive oxygen-18 is easy to find – about 20% of natural oxygen is this isotope. A single collision is shown by the nuclear transformation equation

$$^{1}_{1}\text{p} +\ ^{18}_{8}\text{O} \rightarrow\ ^{18}_{9}\text{F} +\ ^{1}_{0}\text{n}$$

Diagnosis using PET scans

Just as in CAT scans, a PET scan produces slices through the body that can be used to construct a detailed three-dimensional image, but gamma radiation is used instead of X-rays.

Most PET scanners use a medical tracer called fluorodeoxyglucose (FDG), which is similar to naturally occurring glucose but is tagged with a radioactive fluorine-18 atom in place of one oxygen atom. The advantage of using FDG is that our bodies treat it like normal glucose. When FDG is injected into the patient it accumulates in tissues with a high rate of respiration. The activity from the FDG in the body is monitored using gamma detectors.

Another medical tracer used for PET scanning is carbon monoxide made using the carbon-11 isotope. This isotope emits a positron and has a half-life of about 20 minutes. Carbon monoxide is very good at clinging onto haemoglobin molecules in the red blood cells, so it can be transported through the body and the concentrations of carbon monoxide can be monitored in a PET scan.

The PET scanner

Figure 2 shows the principles of a PET scanner. The patient lies on a horizontal table and is surrounded by a ring of gamma detectors. Each detector consists of a photomultiplier tube and a sodium iodide scintillator, and produces a voltage pulse or signal for every gamma photon incident at its scintillator. The detectors are all connected to a high-speed computer.

Learning outcomes

Demonstrate knowledge, understanding, and application of:

→ medical tracers: fluorine–18

→ positron emission tomography (PET)

→ diagnosis using PET scanning.

▲ **Figure 1** *A particle accelerator facility in Russia where medical tracers for PET scanners are produced*

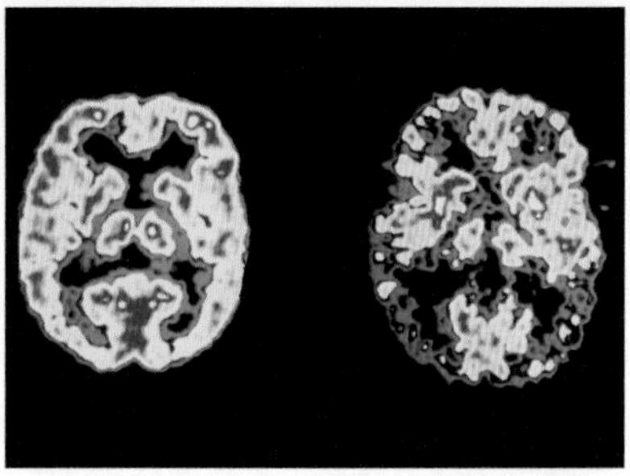

▲ **Figure 3** *PET scans can diagnose abnormal activity in the brain, such as in this comparison between the activity (red and yellow) in a normal brain (the scan on the left) and the brain of a person with Alzheimer's disease (the scan on the right)*

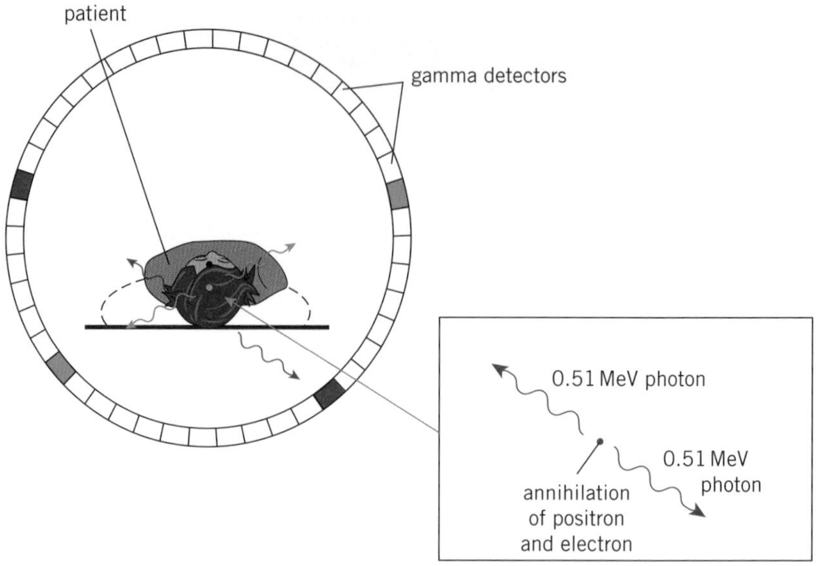

▲ **Figure 2** *A patient surrounded by a ring of gamma detectors*

The patient is injected with FDG. The PET scanner detects the gamma photons emitted when the positrons from decaying fluorine-18 nuclei annihilate with electrons inside the patient. Note that the gamma photons detected for the PET scan come from the annihilation of the positrons, not the gamma photons emitted by the decaying fluorine-18 nuclei. On average, a positron travels about 1 mm from its emission point before it annihilates an electron.

The annihilation of a positron and an electron produces two gamma photons travelling in opposite directions, so momentum is conserved (as mentioned in Topic 27.2, Interaction of X-rays with matter). The computer can determine the point of annihilation from the difference in the arrival times of these photons at the two diametrically opposite detectors and the speed of photons c ($3.00 \times 10^8\,\mathrm{m\,s^{-1}}$). The voltage signals from all the detectors are fed into the computer, which analyses and manipulates these signals to generate an image (scan) on a display screen in which different concentrations of the tracer show up as areas of different colours and brightness.

Advantages and disadvantages of PET

PET is a non-invasive technique (the patient is not subjected to the risks of surgery). PET scans are used to help diagnose different types of cancers, to help plan complex heart surgery, and to observe the function of the brain. It can help doctors identify the onset of certain disorders of the brain, such as Alzheimer's disease (Figure 3). PET scans are also being used to assess the effect of new medicines and drugs on organs.

One major disadvantage of PET is that the technique is very expensive because of the facilities required to produce the medical tracers. PET scanners are found only at larger hospitals, and only patients with complex health problems are recommended for PET scans.

Summary questions

1 Name the medical tracer (radiopharmaceutical) that contains fluorine-18 nuclei. *(1 mark)*

2 Describe one method used to produce fluorine-18 nuclei. *(1 mark)*

3 Describe the construction and function of a gamma detector used in PET scanners. *(2 marks)*

4 Annihilation of a positron and an electron produces two gamma photons. Calculate the time difference between the arrival times of these photons if one of them travels 5.00 cm further than the other. Comment on your answer. *(3 marks)*

5 A patient is injected with FDG. A typical PET scan takes about 20 minutes. Calculate the percentage drop in the original activity of FDG by the time the scan finishes. *(3 marks)*

27.6 Ultrasound

Specification reference: 6.5.3

Ultrasound scans

We can hear sound with frequencies in the range from about 20 Hz to 20 kHz. Ultrasound is simply longitudinal sound waves with frequency greater than 20 kHz, beyond the range of human hearing. Although ultrasound is inaudible to us, some animals such as bats and dolphins use ultrasound to communicate and hunt prey.

The benefits of using ultrasound to form images of the internal structures of the body are obvious. It is non-ionising and therefore harmless, it is non-invasive (no surgery is necessary, so no risk of infection), and it is quick.

Ultrasound used for medical imaging has frequencies in the range of 1–15 MHz. Like audible sound, ultrasound can be refracted as it travels between substances, reflected at the boundary between two substances, and diffracted by small structures or apertures. The wavelength of ultrasound in the human body is less than 1 mm, so ultrasound can be used to identify features as small as a few millimetres.

An **ultrasound transducer** is a device used both to generate and to receive ultrasound. It changes electrical energy into sound and sound into electrical energy, by means of the **piezoelectric effect**.

The piezoelectric effect

Some crystals, such as quartz, produce an electromotive force (e.m.f.) when they are compressed, stretched, twisted, or distorted. This piezoelectric effect is a reversible process. In order words, when an external p.d. is applied across the opposite faces of the crystal, the electric field can either compress or stretch the crystal (Figure 2). The strain experienced by the crystal is no more than about 0.1%.

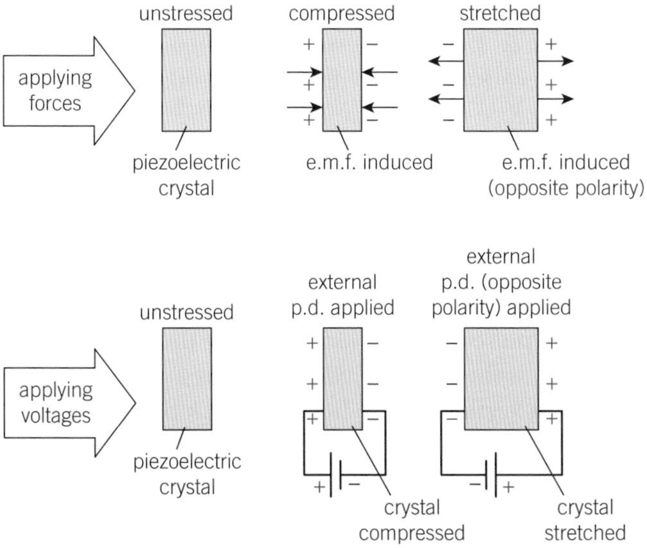

▲ **Figure 2** *Piezoelectric effect*

Learning outcomes

Demonstrate knowledge, understanding, and application of:

→ ultrasound frequency
→ the piezoelectric effect
→ ultrasound transducers
→ ultrasound A-scans and B-scans.

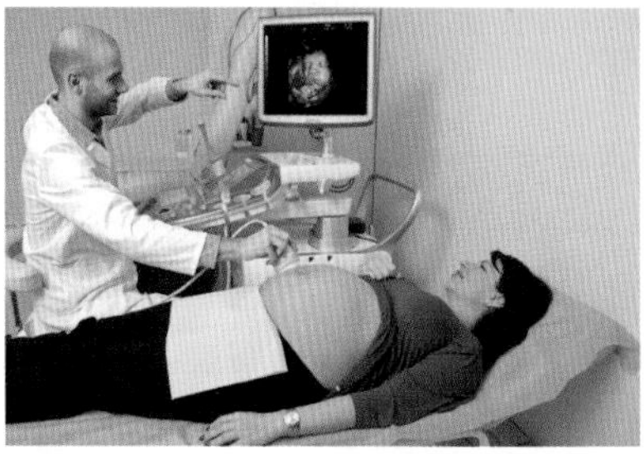

▲ **Figure 1** *An ultrasound transducer being used for a fetal scan (a B-scan – see later)*

Ultrasound transducer

To generate ultrasound, a high-frequency (e.g. 5 MHz) alternating p.d. is applied across opposite faces of a crystal. This repeatedly compresses and expands the crystal. The frequency is chosen to be the same as the natural frequency of oscillation of the crystal. The result is that the crystal resonates, and produces an intense ultrasound signal.

An ultrasound transducer emits *pulses* of ultrasound, typically 5000 pulses every second (a frequency of 5 kHz pulses of ultrasound of frequency 5 MHz).

The same transducer is also used to detect ultrasound. Any ultrasound incident on the crystal will make it vibrate, so the crystal is compressed and expanded by tiny amounts. This vibration generates an alternating e.m.f. across the ends of the crystal, which can be detected by electronic circuits.

Modern ultrasound transducers use either lead zirconate titanate (a ceramic) or polyvinylidene fluoride (a polymer) instead of quartz. Figure 3 shows the basic construction of an ultrasound transducer used in hospitals.

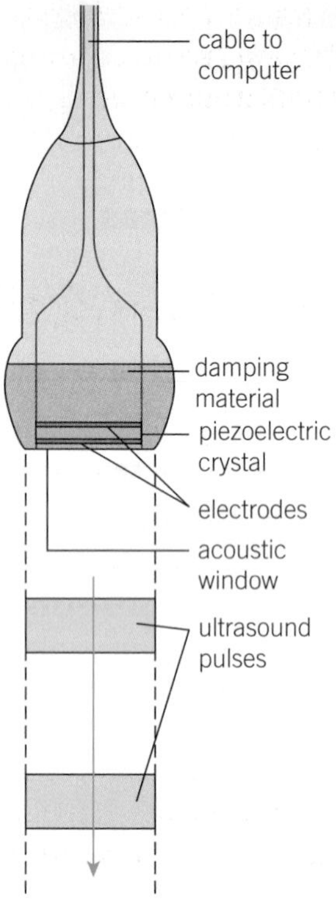

▲ **Figure 3** *An ultrasound transducer*

A-scans

The simplest type of ultrasound scan is called an A-scan. A single transducer is used to record along a straight line through the patient. An A-scan can be used to determine the thickness of bone or the distance between the lens and retina in the eye. This technique is being superseded by more elaborate techniques such as the B-scan (see later), but it provides a useful insight into the principles of using ultrasound to scan internal structures.

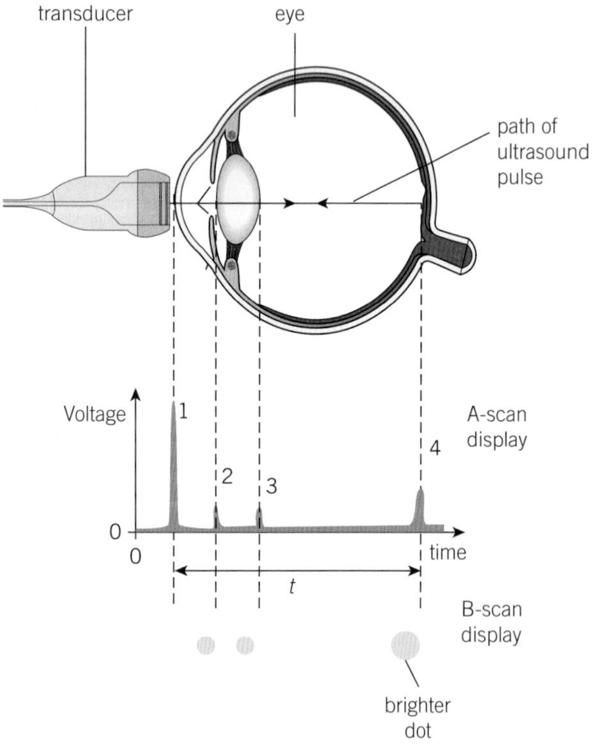

▲ **Figure 4** *An A-scan display from an ultrasound measurement of the eye*

Consider a transducer sending ultrasound pulses into the body of a patient. Each pulse of ultrasound will be partly reflected and partly transmitted at the boundary between any two different tissues. The reflected or 'echo' pulse will be received at the transducer. It will have less energy than the original pulse because of energy losses within the body and also because some of the energy of the original pulse is transmitted though the boundary.

The pulsed voltage at the ultrasound transducer is displayed on an oscilloscope screen or computer screen as a voltage against time plot. Figure 4 shows an idealised scan of the eye. The voltage pulse 1 is the voltage pulse responsible for sending the ultrasound pulse into the eye. The voltage pulses 2 and 3 are due to the reflections at the front and back of the eye lens. The voltage pulse 4 is due to the reflection at the back of the eye (retina). The amplitudes of the voltage signals are attenuated, as already explained.

The time interval t is the time taken for the ultrasound pulse to travel from the front of the transducer to the retina and then back to the transducer. The total distance travelled by the ultrasound pulse is $2L$, where L is the distance between the transducer and the retina. The value of L can be calculated if the average speed v of the ultrasound in the eye is known.

Study tip

The A in A-scan stands for amplitude.

Worked example: Eyeballing

The average speed of ultrasound in the eye is $1550\,\text{m}\,\text{s}^{-1}$. The time interval t in the A-scan shown in Figure 4 is $27\,\mu\text{s}$. Determine the approximate length L of the eyeball.

Step 1: Calculate the total distance travelled by the ultrasound in the time interval t.

$$\text{distance} = vt = 1550 \times 27 \times 10^{-6} = 0.04185\,\text{m}$$

Step 2: The distance 0.042 m is equal to twice the distance L (the ultrasound has to travel to the retina and then back to the transducer).

Therefore, $L = \dfrac{0.04185}{2} = 0.021\,\text{m}$ (2 s.f.)

The length L of the eyeball is 2.1 cm.

B-scans

When you see images of an ultrasound scan, they are most likely to be a B-scan, which provides a two-dimensional image on a screen. In a B-scan (also known as a 2D scan), the transducer is moved over the patient's skin. The output of the transducer is connected to a high-speed computer. For each position of the transducer, the computer produces a row of dots on the digital screen – each dot corresponds to the boundary between two tissues. The brightness of the dot is proportional to the intensity of the reflected ultrasound pulse. The collection of dots produced correspond to the different positions of the transducer over the patient, making a two-dimensional image of a section though the patient (Figure 5).

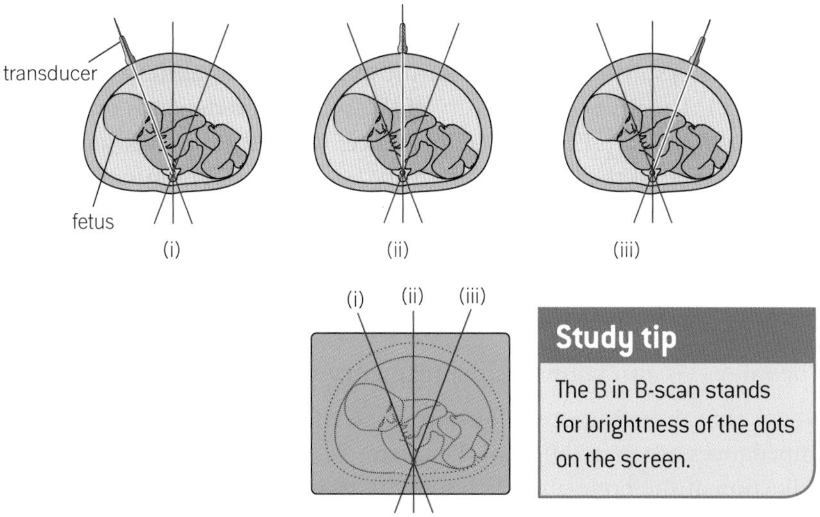

▲ **Figure 5** *A B-scan is effectively a multiple of A-scans*

Study tip

The B in B-scan stands for brightness of the dots on the screen.

Summary questions

1. State the nature of ultrasound. *(1 mark)*
2. State the typical frequency of ultrasound used for ultrasound scanning. *(1 mark)*
3. The speed of ultrasound in air is about $340\,\text{m}\,\text{s}^{-1}$.
   - **a** Calculate the wavelength of ultrasound in air from a transducer working at 10 MHz. *(2 marks)*
   - **b** Ultrasound travels faster in the body than in air. State and explain how this will affect the wavelength of sound from the same transducer in the body. *(2 marks)*
4. State the major difference between an A-scan and a B-scan. *(1 mark)*
5. The ultrasound pulses from the transducer are emitted at a rate of about 5 kHz. The speed of ultrasound in the body is about $1600\,\text{m}\,\text{s}^{-1}$. Explain why such a pulse rate would be suitable for ultrasound scanning of a patient. *(4 marks)*

27.7 Acoustic impedance

Specification reference: 6.5.3

Learning outcomes

Demonstrate knowledge, understanding, and application of:

→ acoustic impedance of a medium; $Z = \rho c$

→ reflection of ultrasound at a boundary

→ $\frac{I_r}{I_0} = \frac{(Z_2 - Z_1)^2}{(Z_2 + Z_1)^2}$

→ impedance (acoustic) matching

→ the use of gel in ultrasound scanning.

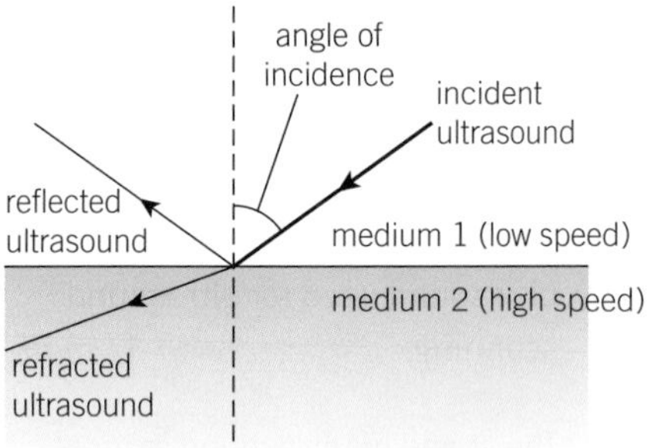

▲ **Figure 1** *Ultrasound will be both reflected and refracted at a boundary between two media*

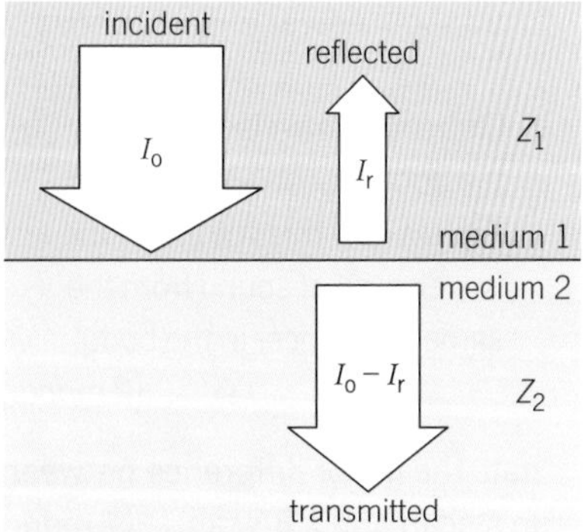

▲ **Figure 2** *Reflection and transmission of normally incident ultrasound at a boundary between media*

Study tip

Be careful: *c* is the speed of the ultrasound in the substance and not the speed of light.

What happens at a boundary?

When a uniform beam of ultrasound is incident at a boundary between two substances (media), a proportion of its intensity will be reflected and the remainder will be refracted (Figure 1). The fraction of the ultrasound intensity reflected at the boundary depends on the **acoustic impedance** of both media.

Acoustic impedance Z

The acoustic impedance Z of a substance is defined as the product of the density ρ of the substance and the speed c of ultrasound in that substance, that is

$$Z = \rho c$$

The SI unit of acoustic impedance is $\mathrm{kg\,m^{-2}\,s^{-1}}$.

Table 1 lists data for some important substances in ultrasound scanning.

▼ **Table 1** *Data for some substances encountered in ultrasound scans*

Substance	ρ / kg m^{-3}	c / km s^{-1}	Z / 10^6 kg m^{-2} s^{-1}
air	1.3	0.340	0.000 442
fat	950	1450	1.38
soft tissue (average)	1060	1540	1.63
muscle	1070	1580	1.69
skin	1070	1590	1.70
bone (average)	1900	4000	7.60

Reflected intensity

Consider a collimated beam of ultrasound incident at a boundary between two substances with acoustic impedances Z_1 and Z_2 (Figure 2). The reflected intensity of the ultrasound depends on the values of Z_1 and Z_2. For normal incidence, when the angle of incidence is 0°, the ratio of the reflected intensity I_r to the incident intensity I_0 is given by the equation

$$\frac{I_r}{I_0} = \frac{(Z_2 - Z_1)^2}{(Z_2 + Z_1)^2} \quad \text{or} \quad \frac{I_r}{I_0} = \left(\frac{Z_2 - Z_1}{Z_2 + Z_1}\right)^2$$

The ratio $\frac{I_r}{I_0}$ is also known as the **intensity reflection coefficient**. Notice that there is more reflection when the values of the acoustic impedances are very different. For example, there will be greater reflection at the bone–muscle boundary than at the blood–muscle boundary. With the exception of bone, the acoustic impedances of most substances that make up the human body are quite similar, so bones are easier to distinguish in an ultrasound scan than different types of soft tissues (Figure 3).

Worked example: Reflected intensities

A beam of ultrasound is incident normally at the boundary between muscle and bone. Calculate the percentage of the incident intensity reflected at this boundary.

Step 1: Use Table 1 to find the values of the acoustic impedances. It does not matter which is Z_1 and which is Z_2 – the answer will be the same because of the squaring.

$Z_{1\,(\text{muscle})} = 1.69 \times 10^6\,\text{kg}\,\text{m}^{-2}\,\text{s}^{-1}$ and $Z_{2\,(\text{bone})} = 7.60 \times 10^6\,\text{kg}\,\text{m}^{-2}\,\text{s}^{-1}$

Step 2: Substitute the values carefully and solve the equation. You do not have to include the 10^6 factors, because they cancel each other out.

$$\frac{I_r}{I_0} = \left(\frac{Z_2 - Z_1}{Z_2 + Z_1}\right)^2 = \left(\frac{1.69 - 7.60}{1.69 + 7.60}\right)^2 = 0.40 \text{ (2 s.f.)}$$

The percentage of the incident intensity reflected at the boundary is 40%. The remainder is transmitted through the boundary.

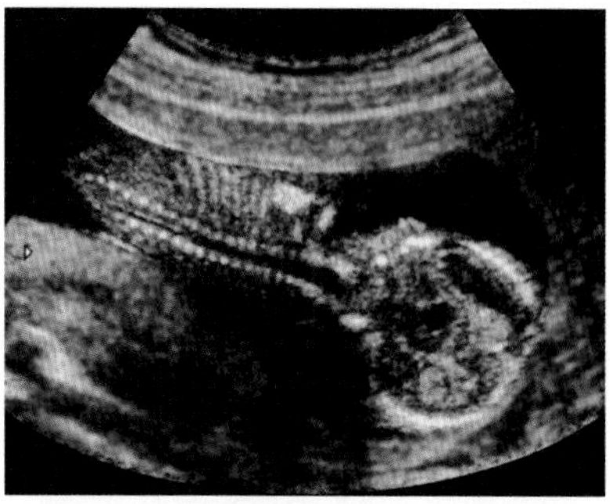

▲ **Figure 3** *An ultrasound scan of a thirteen-week-old fetus in the womb – the head is at the bottom right, and the bones of the spine and ribs in the centre are much easier to see than the soft organs*

Acoustic matching – coupling gel

When an ultrasound transducer is placed on the skin of a patient air pockets will always be trapped between the transducer and the skin. The air–skin boundary means that about 99.9% of the incident ultrasound will be reflected before it even enters the patient. To overcome this problem, a special gel, called a **coupling gel**, with acoustic impedance similar to that of skin is smeared onto the skin and the transducer. The gel fills air gaps between the transducer and the skin and ensures that almost all the ultrasound enters the patient's body. The terms **impedance matching** or **acoustic matching** are used when two substances (e.g. coupling gel and skin) have similar values of acoustic impedance. In this case negligible reflection occurs at the boundary between the two substances.

Summary questions

1 Define acoustic impedance of a substance. *(1 mark)*

2 State what causes a large fraction of reflection of ultrasound at the boundary between substances. *(1 mark)*

3 Lead zirconate titanate is used in the construction of modern ultrasound transducers. It has acoustic impedance $2.9 \times 10^7\,\text{kg}\,\text{m}^{-2}\,\text{s}^{-1}$ and density $5600\,\text{kg}\,\text{m}^{-3}$. Calculate the speed of ultrasound in this material. *(2 marks)*

4 Calculate the percentage of the incident intensity reflected at the fat–muscle boundary. *(2 marks)*

5 The coupling gel used in ultrasound scans has acoustic impedance of $1.65 \times 10^6\,\text{kg}\,\text{m}^{-2}\,\text{s}^{-1}$.
   a Show that the percentage of the incident intensity reflected at the air–skin boundary is 99.9%. *(2 marks)*
   b Calculate the percentage of the incident intensity reflected at the gel–skin boundary. *(2 marks)*

27.8 Doppler imaging

Specification reference: 6.5.3

Learning outcomes

Demonstrate knowledge, understanding, and application of:

→ the Doppler effect in ultrasound

→ the speed of blood v in the body: $\frac{\Delta f}{f} = \frac{2v\cos\theta}{c}$.

Synoptic link

You have already met the Doppler effect with electromagnetic waves in Topic 20.2, The Doppler effect.

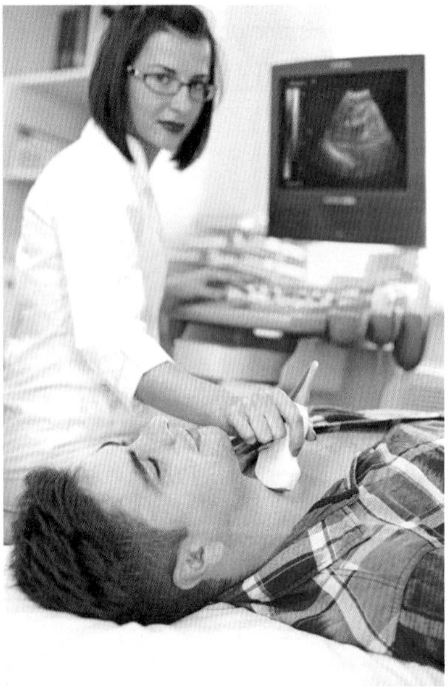

▲ **Figure 1** *A patient undergoing a Doppler ultrasound investigation of the thyroid gland*

Doppler ultrasound

The frequency of ultrasound changes when it is reflected off a moving object – the Doppler effect. Doppler ultrasound, a non-invasive technique, uses the reflection of ultrasound from iron-rich blood cells to help doctors to evaluate blood flow through major arteries and veins, such as those in the arms, legs, neck, and even the heart. The technique can be used to reveal blood clots (thrombosis), identify the narrowing of the walls caused by accumulation of fatty deposits (atheroma), and evaluate the amount of blood flow to a transplanted kidney or liver.

Colour Doppler scans

During Doppler ultrasound, the ultrasound transducer is pressed lightly over the skin above the blood vessel. The transducer sends pulses of ultrasound and receives the reflected pulses from inside the patient. Ultrasound reflected off tissues will return with the same frequency and wavelength, but that reflected off the many moving blood cells will have a changed frequency. The frequency is increased when the blood is moving towards the transducer and decreased when the blood is receding from the transducer. The frequency shift or change in frequency, Δf, is directly proportional to the speed v (of approach or recession) of the blood (see later). The transducer is connected to a computer that produces a colour-coded image to show the direction and speed of the blood flow on a screen (Figure 2).

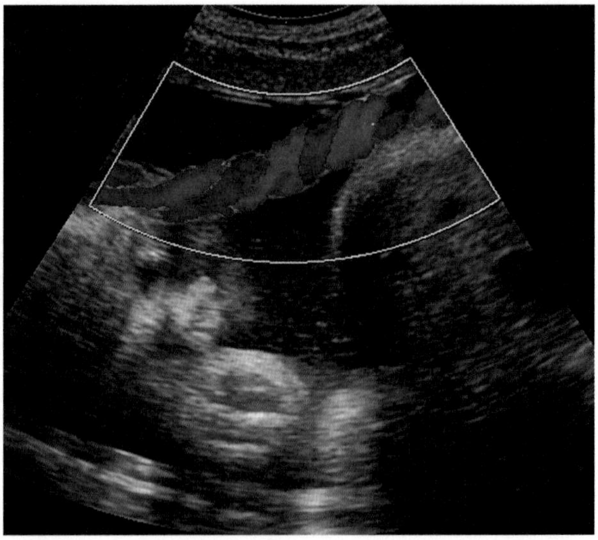

▲ **Figure 2** *Coloured Doppler ultrasound scan showing umbilical blood flow – the fetus is lying across the bottom left, and oxygenated (arterial) blood, which is flowing from mother to fetus, is red, whilst deoxygenated (venous) blood, which is flowing from fetus to mother, is blue*

Determining the speed of blood

Ultrasound in scans has a frequency in the range 5 to 15 MHz, and in blood flow analysis this can give a Doppler shift up to 3 kHz.

Figure 3 shows an ultrasound transducer placed over a blood vessel. The axis of the probe is held at an angle θ to the blood vessel. The change in the observed ultrasound frequency Δf is given by the equation

$$\Delta f = \frac{2fv\cos\theta}{c}$$

where f is the original ultrasound frequency, v is the speed of the moving blood cells, and c is the speed of the ultrasound in blood. Note that the Doppler shift in frequency is directly proportional to the speed of the blood flow. The probe has to be held at an angle to the skin – holding it at right angles would give no observed change in frequency because cos90° = 0. The typical angle used is about 60°.

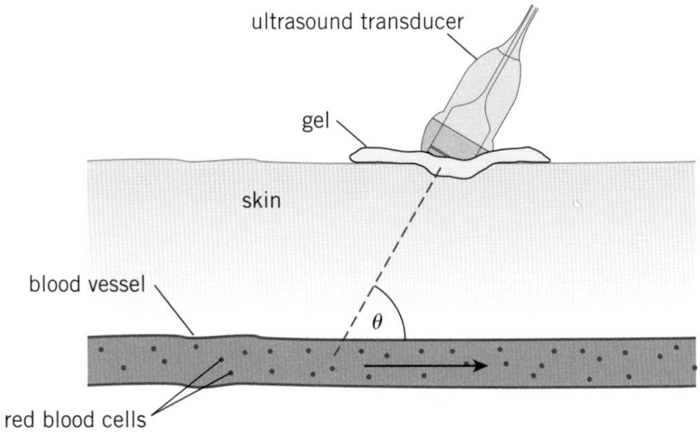

▲ **Figure 3** *Ultrasound transducer used to determine the speed of blood flow*

Worked example: The speed of blood

Doppler ultrasound technique is used on a patient's blood vessel. The transducer is held at an angle of 60° to the blood vessel and emits ultrasound of frequency 10 MHz. The observed Doppler shift is 1.5 kHz. The speed of ultrasound in blood is 1600 m s^{-1}. Calculate the speed of blood flow.

Step 1: List all the quantities given.

$\Delta f = 1500\,\text{Hz}$, $f = 10 \times 10^6\,\text{Hz}$, $\theta = 60°$, $c = 1600\,\text{m s}^{-1}$ (2 s.f.)

Step 2: Rearrange the equation and then substitute the values to calculate the speed v of the blood flow.

$$v = \frac{c\Delta f}{2f\cos\theta} = \frac{1600 \times 1500}{2 \times 10 \times 10^6 \times \cos 60°} = 0.24\,\text{m s}^{-1} \text{ (2 s.f.)}$$

The speed of the blood flow is about 24 cm s^{-1}.

Summary questions

1 In the technique of Doppler ultrasound, what is responsible for producing the change in frequency of the ultrasound? *(1 mark)*

2 Explain why the transducer is not placed at right angles to the surface of the patient's skin. *(2 marks)*

3 The Doppler shift in frequency for blood travelling at a speed of 12 cm s^{-1} is 500 Hz. Calculate the speed of blood for a Doppler shift of 700 Hz. Explain your answer. *(3 marks)*

4 Ultrasound of frequency 7.0 MHz is directed at an angle of 60° to the blood vessel of a patient. The diameter of the blood vessel is about 1.5 mm and the Doppler shift in frequency is 900 Hz. The speed of ultrasound in the blood is 1600 m s^{-1}. Calculate the volume of blood flowing thorough the patient's blood vessel per second. *(4 marks)*

Practice questions

1 a State two main properties of X-ray photons. *(2 marks)*

b Figure 1 shows an X-ray photon interacting with an atom to produce an electron-positron pair in a process known as pair production.

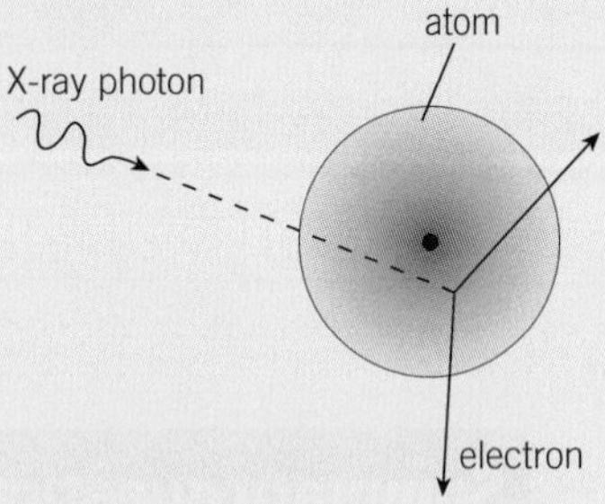

▲ **Figure 1**

Calculate the maximum wavelength of X-rays that can produce an electron-positron pair. *(3 marks)*

c Name an element used as a contrast material in X-ray imaging. Explain why contrast materials are used in the diagnosis of stomach problems. *(3 marks)*

Jun 2013 G485

2 a State two main properties of ultrasound. *(2 marks)*

b Describe how the piezoelectric effect is used in an ultrasound tansducer both to emit and receive ultrasound. *(2 marks)*

c Explain why a gel is used between the ultrasound transducer and the patient's skin during a scan. *(2 marks)*

d Explain a method using ultrasound to determine the speed of blood in an artery in the arm. *(4 marks)*

Jun 2013 G485

3 a State the equation that relates the transmitted intensity I of a beam of X-rays after passing through a medium of thickness x to the incident intensity I_0. Give the meaning of any terms used. *(2 marks)*

b (i) Table 1 contains data of the transmitted intensity I through a medium for varying thickness x of the medium. Copy and complete Table 1.

▼ **Table 1**

intensity I after passing through thickness x / W m^{-2}	thickness x of medium / m	ln (I / W m^{-2})
1.32×10^8	0.5×10^{-2}	
4.14×10^6	1.0×10^{-2}	
1.29×10^5	1.5×10^{-2}	
4.06×10^3	2.0×10^{-2}	
1.27×10^2	2.5×10^{-2}	

(1 mark)

(ii) On a copy of Figure 2 plot ln (I) against thickness x and draw the line of best fit. *(3 marks)*

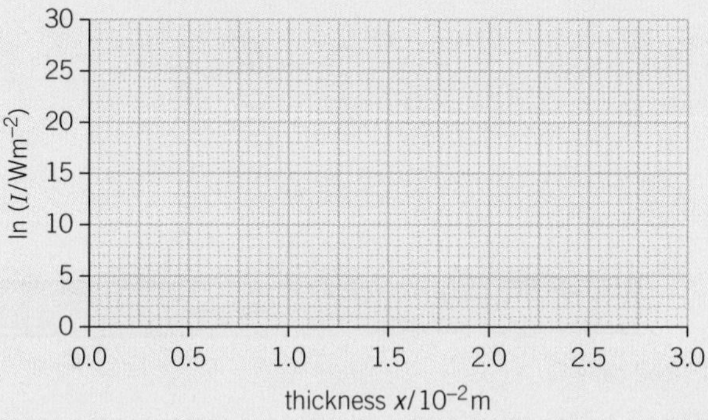

▲ **Figure 2**

(iii) Determine the gradient *(2 marks)*

(iv) Use your graph to find the incident intensity I_0. *(2 marks)*

(v) Calculate the thickness x of the medium that would reduce the transmitted intensity to $\frac{1}{4}$ of its original intensity. *(3 marks)*

Jun 2008 2825/02

4 a Describe briefly how X-rays are produced in an X-ray tube. *(2 marks)*

b Describe the Compton effect in terms of an X-ray photon. *(2 marks)*

c A beam of X-rays of intensity 3.0×10^9 W m^{-2} is used to target a tumour in a patient. The tumour is situated at a depth of 1.7 cm in soft tissue. The attenuation (absorption) coefficient μ of soft-tissues is 6.5 cm^{-1}.

(i) Show that the intensity of the X-rays at the tumour is about $5 \times 10^4\,W\,m^{-2}$. *(2 marks)*

(ii) The cross-sectional area of the X-ray beam at the tumour is $5\,mm^2$. The energy required to destroy the malignant cells of the tumour is 200 J. The tumour absorbs 10% of the energy from the X-rays. Calculate the total exposure time required to destroy the tumour. *(3 marks)*

Jan 2013 G485

5 **a** Describe the *piezoelectric effect*. *(1 mark)*

b Describe how ultrasound scanning is used to obtain diagnostic information about internal structures of a body. In your description include the differences between an A-scan and a B-scan. *(4 marks)*

c Table 2 shows the speed of ultrasound, density and acoustic impedance for muscle and bone.

▼ **Table 2**

material	speed of ultrasound / $m\,s^{-1}$	density / $kg\,m^{-3}$	acoustic impedance / $10^6\,kg\,m^{-2}\,s^{-1}$
muscle	1590	1080	1.72
bone	4080	1750	7.14

(i) Show that the unit for acoustic impedance is $kg\,m^{-2}\,s^{-1}$. *(1 mark)*

(ii) An ultrasound pulse is incident at right angles to the boundary between bone and muscle. Calculate the fraction of reflected intensity of the ultrasound. *(2 marks)*

Jan 2011 G485

6 The reflections of ultrasound pulses from interfaces within the body may be displayed as an A-scan on an oscilloscope. Figure 3a illustrates an A-scan display of the intensity of the reflected pulses against time for the interfaces **A**, **B**, **C** and **D** of Figure 3b.

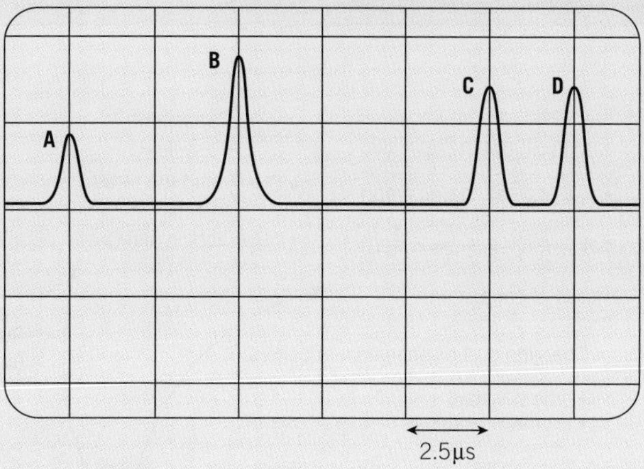

▲ **Figure 3a**

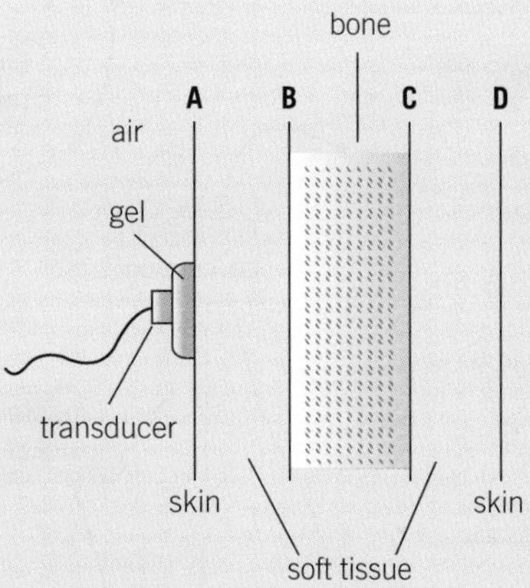

▲ **Figure 3b**

Data:

speed of ultrasound in bone = $4000\,m\,s^{-1}$

speed of ultrasound in soft tissue = $1500\,m\,s^{-1}$

oscilloscope time-base setting = $2.5\,\mu s$ per division

a Calculate

(i) the time interval between the observed reflections from the front edge **B** and the rear edge **C** of the bone *(2 marks)*

(ii) the thickness of the bone *(3 marks)*

(iii) the distance of the front edge **B** of the bone from the skin **A**. *(2 marks)*

b State and explain how the trace in Figure 3a might change if an acoustic coupling medium such as gel is **not** used between the transducer and the skin. *(2 marks)*

Jan 2010 2825/02

Module 6 Summary

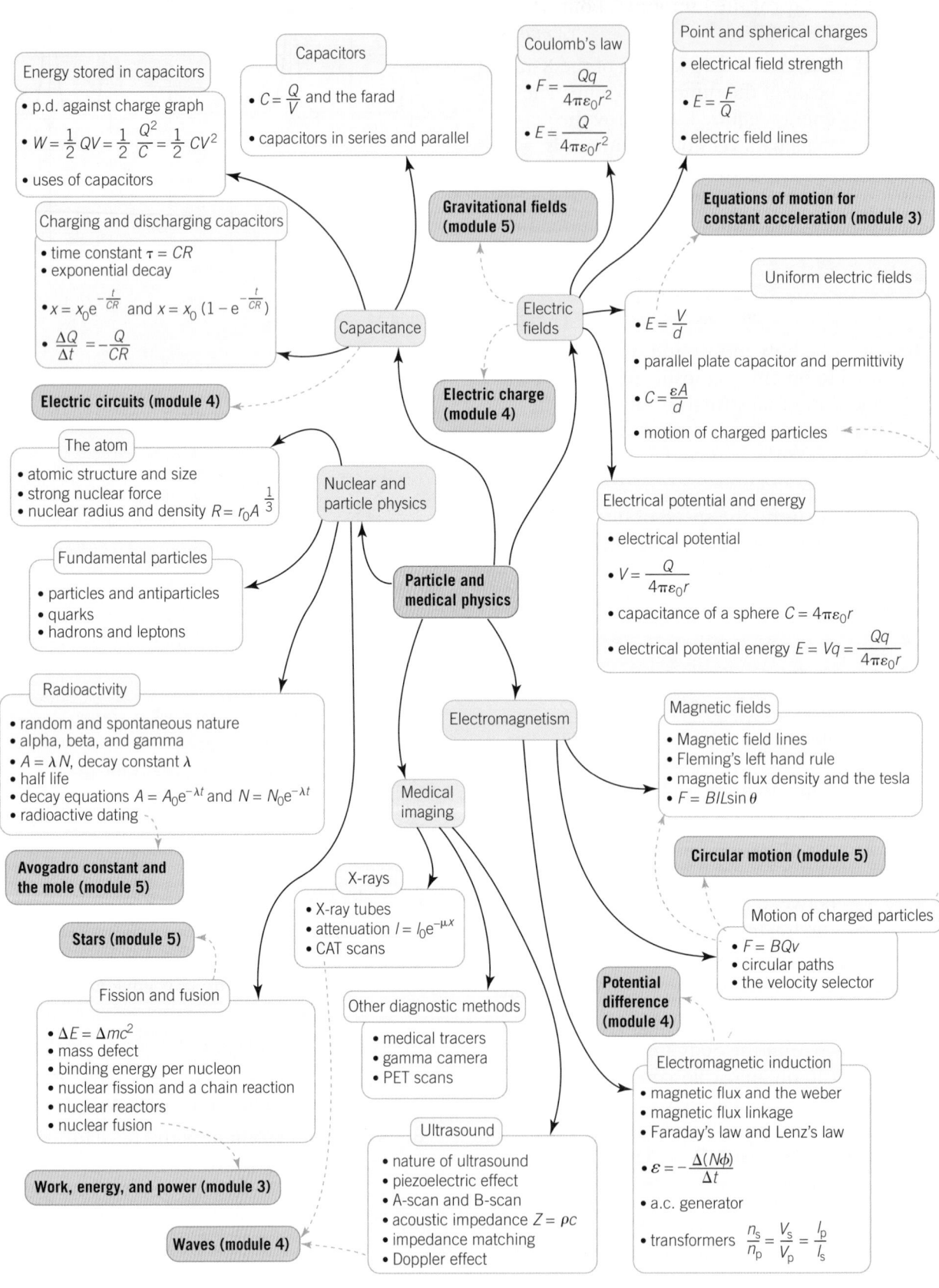

Particle and medical physics

Capacitance

Energy stored in capacitors
- p.d. against charge graph
- $W = \frac{1}{2}QV = \frac{1}{2}\frac{Q^2}{C} = \frac{1}{2}CV^2$
- uses of capacitors

Capacitors
- $C = \frac{Q}{V}$ and the farad
- capacitors in series and parallel

Charging and discharging capacitors
- time constant $\tau = CR$
- exponential decay
- $x = x_0 e^{-\frac{t}{CR}}$ and $x = x_0(1 - e^{-\frac{t}{CR}})$
- $\frac{\Delta Q}{\Delta t} = -\frac{Q}{CR}$

Electric circuits (module 4)

Electric fields

Coulomb's law
- $F = \frac{Qq}{4\pi\varepsilon_0 r^2}$
- $E = \frac{Q}{4\pi\varepsilon_0 r^2}$

Point and spherical charges
- electrical field strength
- $E = \frac{F}{Q}$
- electric field lines

Gravitational fields (module 5)

Equations of motion for constant acceleration (module 3)

Electric charge (module 4)

Uniform electric fields
- $E = \frac{V}{d}$
- parallel plate capacitor and permittivity
- $C = \frac{\varepsilon A}{d}$
- motion of charged particles

Electrical potential and energy
- electrical potential
- $V = \frac{Q}{4\pi\varepsilon_0 r}$
- capacitance of a sphere $C = 4\pi\varepsilon_0 r$
- electrical potential energy $E = Vq = \frac{Qq}{4\pi\varepsilon_0 r}$

Nuclear and particle physics

The atom
- atomic structure and size
- strong nuclear force
- nuclear radius and density $R = r_0 A^{\frac{1}{3}}$

Fundamental particles
- particles and antiparticles
- quarks
- hadrons and leptons

Radioactivity
- random and spontaneous nature
- alpha, beta, and gamma
- $A = \lambda N$, decay constant λ
- half life
- decay equations $A = A_0 e^{-\lambda t}$ and $N = N_0 e^{-\lambda t}$
- radioactive dating

Avogadro constant and the mole (module 5)

Stars (module 5)

Fission and fusion
- $\Delta E = \Delta mc^2$
- mass defect
- binding energy per nucleon
- nuclear fission and a chain reaction
- nuclear reactors
- nuclear fusion

Work, energy, and power (module 3)

Medical imaging

X-rays
- X-ray tubes
- attenuation $I = I_0 e^{-\mu x}$
- CAT scans

Other diagnostic methods
- medical tracers
- gamma camera
- PET scans

Ultrasound
- nature of ultrasound
- piezoelectric effect
- A-scan and B-scan
- acoustic impedance $Z = \rho c$
- impedance matching
- Doppler effect

Waves (module 4)

Electromagnetism

Magnetic fields
- Magnetic field lines
- Fleming's left hand rule
- magnetic flux density and the tesla
- $F = BIL\sin\theta$

Circular motion (module 5)

Motion of charged particles
- $F = BQv$
- circular paths
- the velocity selector

Potential difference (module 4)

Electromagnetic induction
- magnetic flux and the weber
- magnetic flux linkage
- Faraday's law and Lenz's law
- $\varepsilon = -\frac{\Delta(N\phi)}{\Delta t}$
- a.c. generator
- transformers $\frac{n_s}{n_p} = \frac{V_s}{V_p} = \frac{I_p}{I_s}$

Nuclear weaponry

Inside a nuclear warhead is a small amount of fissile material. When the weapon detonates a chain reaction is initiated, releasing a huge amount of energy in a short time. Most modern warheads go further than the simple fission devices developed during WWII. They are designed to harness the energy released during fission and use it to superheat material surrounding the warhead, until it is hot enough to initiate nuclear fusion. These are known as thermonuclear weapons, or hydrogen bombs.

▲ **Figure 1** *An array of nuclear warheads*

The energy released by nuclear weapons is so huge that it tends to be measured in the equivalent of tonnes of TNT, where each tonne is equal to 4.2 GJ of energy. The first warhead tested was equivalent to around 20 kT of TNT, despite having a mass of around 1 T. Modern day thermonuclear weapons can have yields in the megatons (MT).

1 Explain why fissile material, rather than regular explosives, is used to initiate the fusion reaction.
2 The most powerful weapon tested was the Russian 'Tsar' bomb. It had a yield of around 50 MT. Calculate the energy released in joules and the change in mass needed to release this energy.
3 Explain why, after a nuclear explosion, the surrounding area is dangerous to humans.

Inkjet printers

You saw in Topic 22.4 how charged particles in uniform electric fields can follow parabolic paths. This effect is used in some types of printer — inside each print cartridge is a piezoelectric crystal (Topic 27.6) that changes shape when a potential difference is applied across it. This pushes a droplet of ink out of the cartridge.

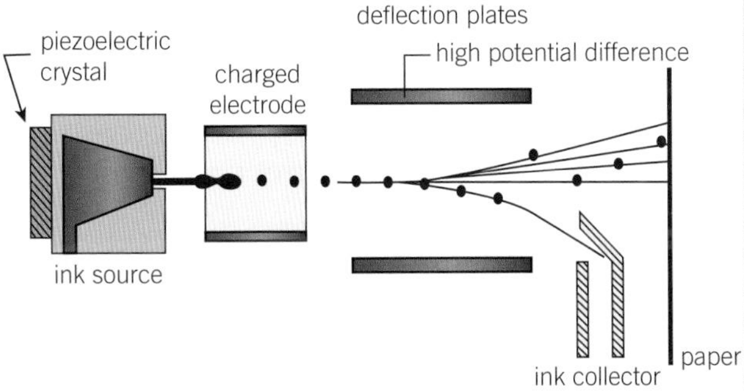

▲ **Figure 2** *The workings of an inkjet printer*

The tiny drops of ink are fired towards the page, becoming charged as they pass through an electrode. The charged droplets pass between parallel plates and are deflected by the electric field by the amount required to place them precisely on the page.

Unwanted droplets are deflected towards an ink collector (or 'gutter') and collected for reuse. Only a small fraction of fired droplets are initially used to print — the majority are recycled.

1 Explain how varying the charge on a droplet affects the size and direction of its deflection.
2 A droplet travelling at $25\,\text{m s}^{-1}$ has a charge of $-1.2\,\text{G}e$ and passes between a pair of parallel plates 0.5 mm long and 1.0 mm apart. The potential difference across the plates is 800 V.
   a Sketch a diagram of this arrangement.
   b Calculate the velocity (magnitude and direction) of the droplet as it leaves the plate.
   c If the gutter is at the same height as the bottom plate, determine the distance from the plate to the gutter required for the droplet to be collected.
3 Each ink droplet has a mass of around $5 \times 10^{-10}\,\text{kg}$ and each page is normally around 3 mm from the cartridge. Discuss, with aid of suitable calculations, whether the manufacturers of print cartridges should be concerned about the droplets falling under gravity.

Module 6 practice questions

Section A

1 Figure 1 shows a capacitor of capacitance 100 μF connected in a circuit.

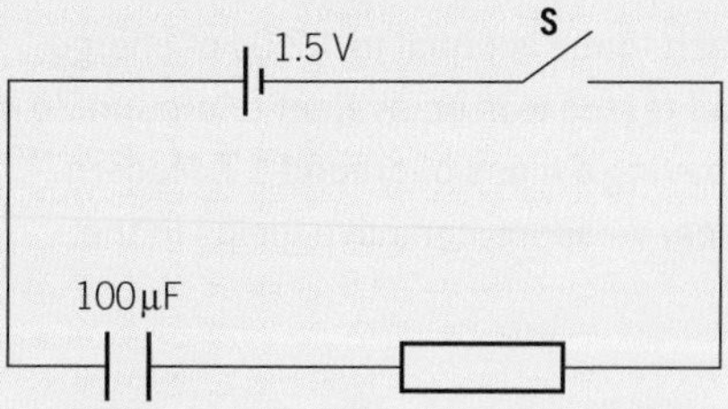

▲ Figure 1

The cell has e.m.f. of 1.5 V and has negligible internal resistance. The switch **S** is closed. After some time the potential difference across the resistor is 0.4 V. What is the charge stored by the capacitor?

A 40 μC B 100 μC

C 110 μC D 150 μC

(1 mark)

2 Which is the correct statement for electric field strength?

A It can be measured in tesla.

B Its unit is equivalent to $N\,kg^{-1}$.

C It is the force experienced per unit positive charge.

D It has a constant value between two positively charged spheres.

(1 mark)

3 The capacitance of a parallel plates capacitor is 8.0 pF. The area of overlap between the plates is halved and the separation between the plates is quadrupled.

What is the capacitance of the capacitor in this new arrangement?

A 1.0 pF B 2.0 pF

C 4.0 pF D 8.0 pF

(1 mark)

4 An isolated positively charged metal sphere has radius R. The electric potential at a distance of $2R$ from the centre of the sphere is +100 V.

What is the electric potential on the surface of the sphere?

A + 25 V B + 50 V

C + 200 V D + 400 V

(1 mark)

5 A current-carrying wire placed at right angles to a uniform magnetic field experiences a force of 100 μN. The wire rotates in the magnetic field until the angle between the magnetic field and the wire is 40°. What is the **change** in the force experienced by the current-carrying wire?

A 23 μN B 36 μN

C 64 μN D 77 μN

(1 mark)

6 The change in mass of the nuclei in a nuclear reaction is –0.10 u.

What is the amount of energy in joules released in the reaction?

A $0.1 \times (3.0 \times 10^8)^2$

B $0.1 \times 1.66 \times 10^{-27} \times (3.00 \times 10^8)^2$

C $0.1 \times 1.60 \times 10^{-19} \times (3.00 \times 10^8)^2$

D $0.1 \times 6.02 \times 10^{23} \times (3.0 \times 10^8)^2$

(1 mark)

7 A radioactive sample has one type of isotope of calcium.

The isotope decays into stable daughter nuclei.

What is the percentage of isotope decayed in the sample after 2.5 half-lives?

A 18% B 40%

C 60% D 82%

(1 mark)

8 Which statement is **correct** about the Compton effect?

A The X-ray photon does not eject an electron.

B The X-ray photon is scattered with reduced energy.

C The X-ray photon produces an electron-positron pair.

D The X-ray photon is completely absorbed by an electron. *(1 mark)*

Section B

9 a Define *capacitance*. (*1 mark*)

b Figure 2 shows an electrical circuit.

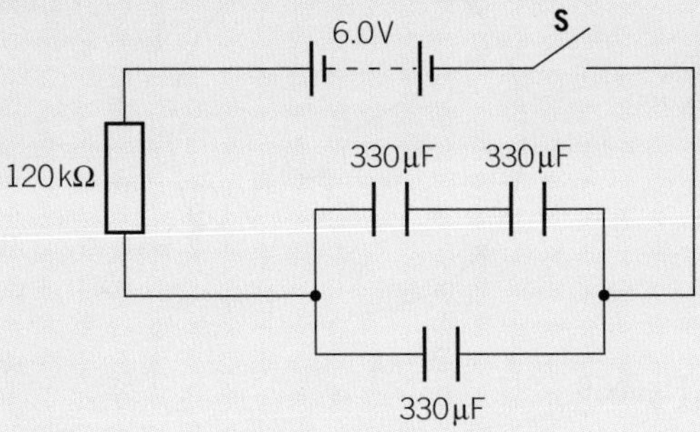

▲ **Figure 2**

All three capacitors are identical. The capacitance of each capacitor is 330 µF. The resistor has resistance 120 kΩ. The battery has e.m.f. 6.0 V and negligible internal resistance.

Calculate

(i) the time constant of the circuit shown in Figure 2 (*3 marks*)

(ii) the energy stored by the capacitors when they are fully charged. (*2 marks*)

c A student is provided with an unmarked capacitor and an unmarked resistor. He connects up the circuit shown in Figure 3.

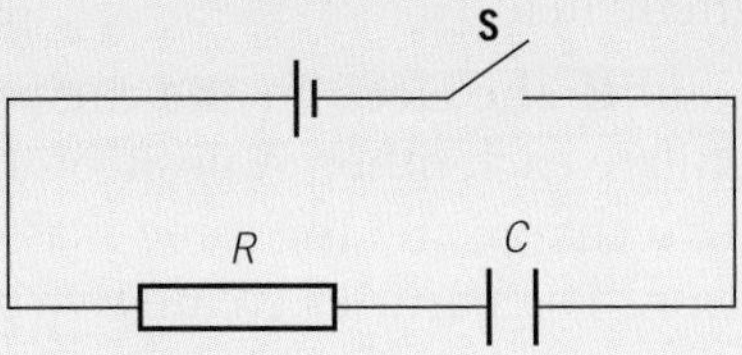

▲ **Figure 3**

(i) Plan an experiment based on the circuit shown in Figure 2 to determine *CR*, where *C* is the capacitance of the capacitor and *R* is the resistance of the resistor. (*4 marks*)

(ii) Figure 4 shows a graph plotted by another student using the circuit shown in Figure 3.

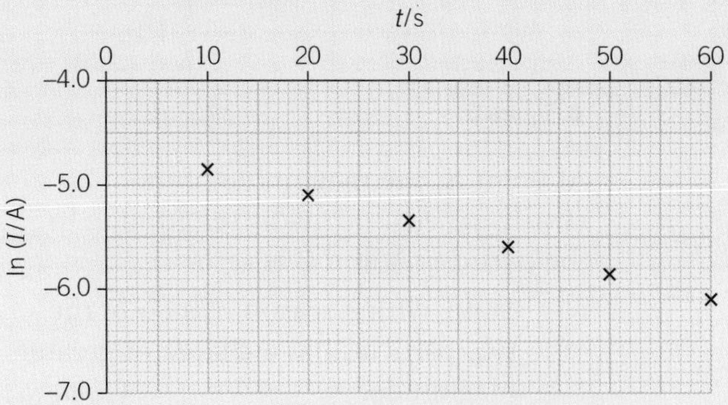

▲ **Figure 4**

The current in the circuit is *I* at a time after the switch **S** is closed.

1 Show how the gradient *G* of the ln (*I*) against *t* graph is related to *CR*. (*2 marks*)

2 Determine a value for *CR*. (*3 marks*)

3 Calculate the maximum current in the circuit. (*2 marks*)

10 a Figure 5 shows a negatively charged metal sphere close to a positively charged metal plate.

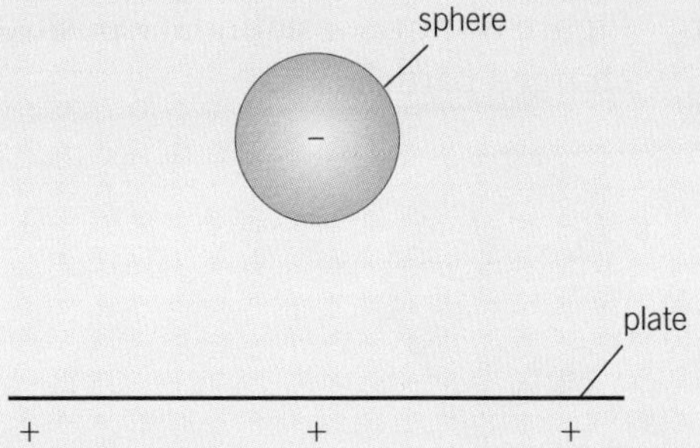

▲ **Figure 5**

On a copy of Figure 5, draw a minimum of five field lines to show the electric field pattern between the plate and the sphere. (*2 marks*)

b Figure 6 shows two positively charged particles **A** and **B**.

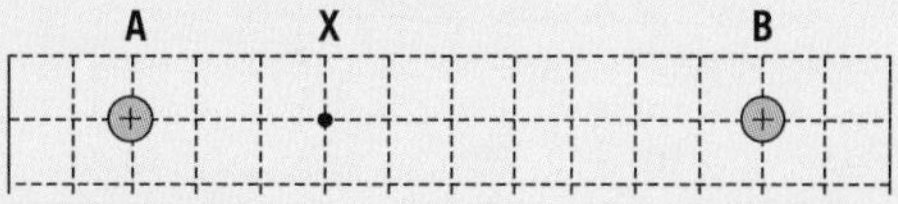

▲ Figure 6

At point **X**, the magnitude of the **resultant** electric field strength due to the particles **A** and **B** is zero.

(i) State, with a reason, which of the two particles has a charge of greater magnitude. *(1 marks)*

(ii) On a copy of Figure 7 sketch the variation of the resultant electric field strength E with distance d from the particle **A**. *(3 marks)*

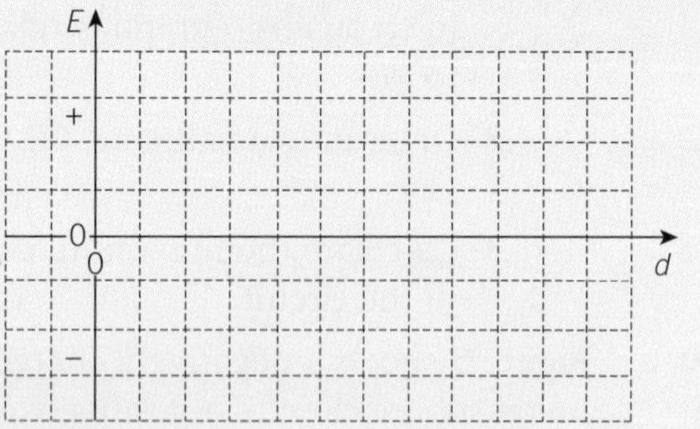

▲ Figure 7

c Figure 8 shows a stationary positively charged particle.

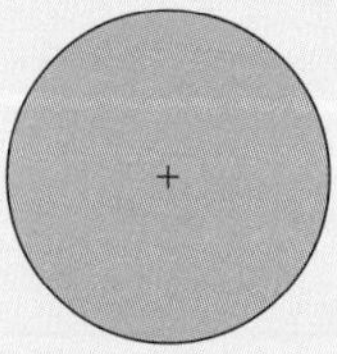

▲ Figure 8

This particle creates both electric and gravitational fields in the space around it. Explain why the **ratio** of the electric field strength E to the gravitational field strength g at any point around this charge is independent of its distance from the particle. *(1 mark)*

Jun 2014 G485

11 Figure 9 shows the circular track of an electron moving in a uniform magnetic field.

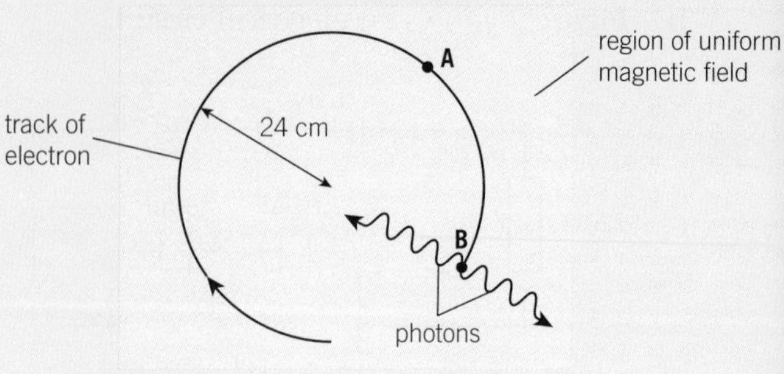

▲ Figure 9

The magnetic field is perpendicular to the plane of Figure 9. The speed of the electron is $6.0 \times 10^7\,\mathrm{m\,s^{-1}}$ and the radius of the track is 24 cm. At point **B** the electron interacts with a stationary positron.

a (i) On a copy of Figure 9, draw an arrow to show the force acting on the electron when at point **A**. Label this arrow **F**. *(1 mark)*

(ii) Explain why this force does not change the speed of the electron. *(1 mark)*

b Calculate the magnitude of the force F acting on the electron due to the magnetic field when it is at **A**. *(2 marks)*

c Calculate the magnetic flux density of the magnetic field. *(2 marks)*

d At point **B**, the electron and the positron annihilate each other. A positron has a positive charge and the same mass as the electron. The particles create two gamma ray photons. Calculate the wavelength of the gamma rays assuming the kinetic energy of the electron is negligible. *(3 marks)*

Jun 2012 G485

12 Figure 10 shows the variation of the magnetic flux **linkage** with time *t* for a small generator.

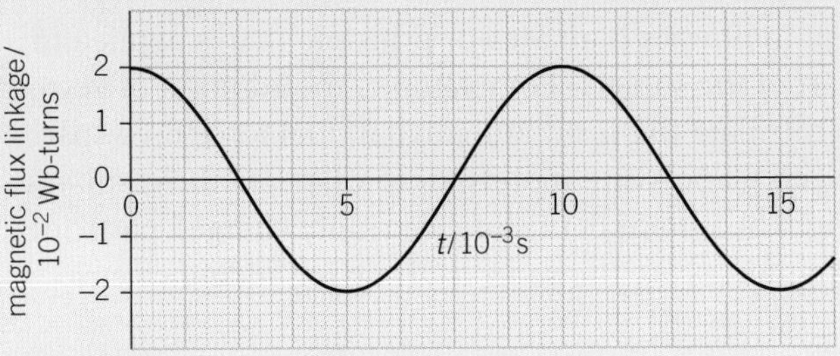

▲ **Figure 10**

The generator has a flat coil of negligible resistance that is rotated at a steady frequency in a uniform magnetic field. The coil has 400 turns and cross-sectional area $1.6 \times 10^{-3}\,m^2$. The output from the generator is connected to a resistor of resistance 150 Ω.

a Use Figure 10 to

(i) calculate the frequency of rotation of the coil *(1 mark)*

(ii) calculate the magnetic flux density *B* of the magnetic field *(3 marks)*

(iii) show that the **maximum** electromotive force (e.m.f.) induced in the coil is about 12 V. *(3 marks)*

b Hence calculate the **maximum** power dissipated in the resistor. *(2 marks)*

Jun 2012 G485

13 a State two properties of X-rays. *(2 marks)*

b Explain what is meant by the *Compton effect.* *(2 marks)*

c The intensity *I* of a collimated beam of X-rays decreases exponentiatlly with thickness *x* of the material through which the beam passes according to the equation $I = I_0 e^{-\mu x}$. The attenuation (absorption) coefficient μ depends on the material.

(i) State what I_0 represents in this equation. *(1 mark)*

(ii) Bone has an attenuation coefficient of $3.3\,cm^{-1}$. Calculate the thickness in cm of bone that will reduce the X-ray intensity by half. *(3 marks)*

d Explain the purpose of using a contrast medium such as barium when taking X-ray images of the body. *(2 marks)*

Jun 2012 G485/01

14 a State **two** main properties of ultrasound. *(2 marks)*

b Describe how the piezoelectric effect is used in an ultrasound transducer both to emit and receive ultrasound. *(2 marks)*

c Explain why a gel is used between the ultrasound transducer and the patient's skin during a scan. *(2 marks)*

d Explain a method using ultrasound to determine the speed of blood in an artery in the arm. *(4 marks)*

Jun 2013 G485/01

UNIFYING CONCEPTS

▲ **Figure 1** *Illustration of an Apollo spacecraft over the moon. The resultant gravitational force on the spacecraft at any point in its flight can be calculated using vectors.*

A unifying concept in physics is an idea or principle that is applied to more than one part of the subject. For example, the addition of vectors can be applied to calculate the resultant gravitational force on a space probe flying between planets and moons, or to calculate the resultant electric field strength at a point between electrically charged particles. The rules for adding forces are the same as those for adding electric field strengths.

To be a good physicist, it is important that you can identify connections between different topics and apply your knowledge and understanding of each topic to answer novel questions.

In the Unified Physics examination paper, question styles will include short answer questions, practical questions, problem-solving questions, and extended response questions. These questions cover different combinations of topics from different modules, though it is impossible to cover every single one in any given set of questions.

Table 1 shows how just one learning outcome links to many other topics and modules in A Level Physics.

▼ **Table 1**

Learning outcome	Links with other modules		What you might be asked about...
4.4.1(d) – the wave equation $v = f\lambda$	5.5.2(a)	Energy levels	Wavelength of electromagnetic waves emitted from hot objects.
	5.5.2(g)	Diffraction grating	Determining energy levels of atoms.
	5.5.2(i)	Wien's displacement law	Predicting the temperature of a star from an observed frequency spectrum.
	5.5.3(e)	Doppler effect	Determining changes in frequency and wavelength of stellar radiation.
	6.4.4(c)	Creation and annihilation of particle-antiparticle pairs	Determining the energy from the frequency of the photons.
	6.5.1(b)	Production of X-ray photons	Determining the shortest wavelength of X-rays for a given accelerating voltage.
	6.5.2(d)	PET scanner	Determining the frequency of the emitted photons.
	6.5.3(a)	Ultrasound	Determining the wavelength of ultrasound in soft tissues.
	6.5.3(g)	Doppler effect in ultrasound	Determining the change in wavelength of ultrasound due to Doppler shift.

Analysing and answering a synoptic question

Let's look at a synoptic question and the knowledge and understanding you would need to answer it. Links with different topics are shown, with each number in square brackets representing a different topic.

Diffraction gratings[1] were covered in Topic 19.6, Analysing starlight. See the Worked Example: Finding the grating spacing.

Emission spectra[2] were covered in Topic 19.5, Spectra.

1 A diffraction grating is used to observe the emission spectrum from hydrogen in the laboratory.

Figure 1 shows part of the observed emission spectrum.

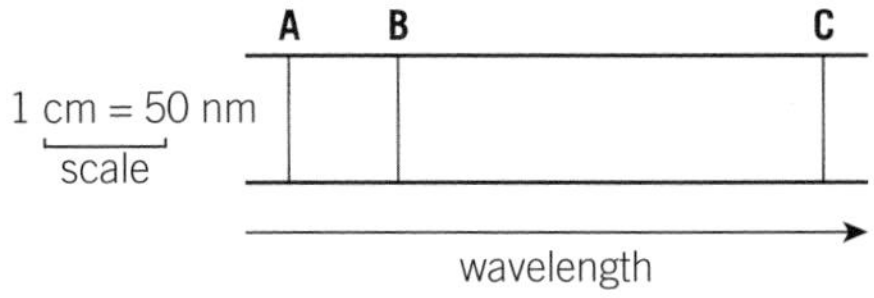

▲ **Figure 1**

Three emission lines **A**, **B**, and **C** are shown in Figure 1. The spectral line *C* corresponds to electromagnetic waves of wavelength 660 nm and the first order image is produced at an angle of 30° to the direction of the incident light to the diffraction grating.

Prefixes[3] were covered in 2.1, Quantities and units. The prefix nano, n, represents a factor of 10^{-9}.

a The relative positions of the spectral lines in Figure 1 are drawn to scale, with a separation of 1.0 cm representing a change in wavelength of 50 nm. Show that spectral lines **A** and **B** have wavelengths of 430 nm and 490 nm respectively. *(2 marks)*

Just measure the distance between the **spectral lines**[2], ensuring you reduce the chance of measurement errors (see **Appendix A3**[4]).

Make sure you use the correct **prefix**[3] for each unit.

Answer to part **(a)**

The distance between spectral lines **A** and **C** is 4.4 cm.

This is equivalent to a change in wavelength 4.4 × 50 = 220 nm.

So the wavelength λ_C of spectral line **C** is:

$\lambda_C = 660 - 220 = 440$ nm (which matches 430 nm within experimental uncertainty) ✓

Similarly, the wavelength λ_B of spectral line **B** is:

$\lambda_B = 660 - (3.5 \times 50) = 485$ nm (which matches 490 nm within experimental uncertainty[4]) ✓

Check **Appendix A3**[4] for a reminder on uncertainty.

b For the spectral line **A**, calculate

(i) the first order image produced from the diffraction grating[1] *(2 marks)*

(ii) the maximum number of orders[1] that can be observed. *(2 marks)*

Answer to part **(b)**

Part (b)(i)

$d \sin\theta = n\lambda$, with $n = 1$ for the first order

Therefore $\dfrac{\sin\theta}{\lambda} = \text{constant}$

$$\frac{\sin\theta}{430 \times 10^{-9}} = \frac{\sin 30}{660 \times 10^{-9}} \quad ✓$$

$\sin\theta = 0.3257...$

$\theta = 19°$ (2 s.f.) ✓

This formula was covered in **Topic 19.6, Analysing starlight**[1].

Manipulation of formulae with trigonometric functions was covered in the **Wave chapters, 11 and 12**[5].

It is good practice to use the values given in the question rather than your values from part (a).

Part (b)(ii)

For $n = 1$ and $\lambda = 660\,\text{nm}$, the angle $\theta = 30°$.

$$\therefore d = \frac{1 \times 660 \times 10^{-9}}{\sin 30} = 132 \times 10^{-6}\,\text{m} \quad ✓$$

$d \sin\theta = n\lambda$ and the maximum value for n[1] can be calculated using $\theta = 90°$.

$$n_{max} = \frac{d \sin 90}{\lambda} = \frac{1.32 \times 10^{-6} \times 1}{430 \times 10^{-9}} = 3.06...$$

$\therefore n_{max} = 3$[1] ✓

Remember that this has to be an integer, as described in **Topic 19.6, Analysing starlight**[1].

Photons[6] were covered in Topic 13.1, The photon model. The energy of a photon is given by the equations $E = hf$ or $E = \dfrac{hc}{\lambda}$.

Energy level diagrams[7] were covered in Topic 19.4, Energy levels in atoms.

c The emission spectrum shown in Figure 1 is due to electrons making transitions to the **same** final energy level.

(i) Calculate the energy in eV of the photon[6] of wavelength 656 nm. *(2 marks)*

(ii) Sketch an energy level diagram[7] responsible for the three spectral lines and draw lines between the energy levels to show the electron transitions which give rise to the these spectral lines. *(4 marks)*

Answer to part **(c)**

Part (c)(i)

$$E = \frac{hc}{\lambda}$$

$$E = \frac{6.63 \times 10^{-34} \times 3.00 \times 10^{8}}{656 \times 10^{-9}} = 3.032... \times 10^{-19}\,\text{J} \quad ✓$$

$$E = \frac{3.032... \times 10^{-19}}{160 \times 10^{-19}} = 1.895...\,\text{eV} = 1.90\,\text{eV (to 2 s.f.)} \quad ✓$$

1 eV = 1.60 × 10^{-19} J[6], see Topic 13.1, The photon model.

Part (c)(ii)

Energy of photon[6] of wavelength 434 nm: $E = 1.895 \times \frac{656}{434} = 2.86\,\text{eV}$ ✓

Energy of photon[6] of wavelength 486 nm: $E = 1.895 \times \frac{656}{434} = 2.56\,\text{eV}$ ✓

The energy level diagram[7] showing the transitions:

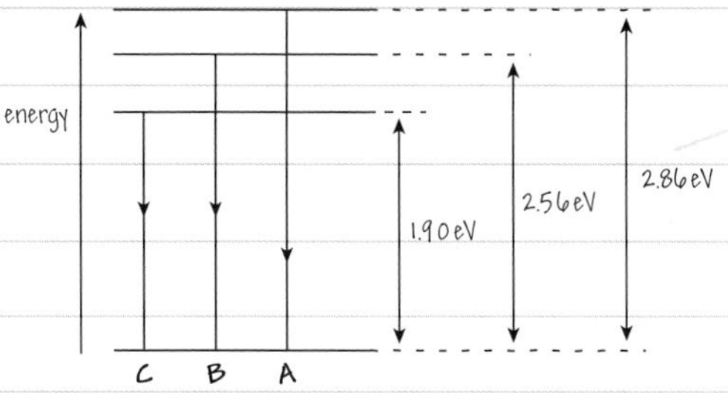

✓ ✓

You need to calculate the energy of the emitted photons before you can proceed any further.

The energy of a photon is the difference between two **energy levels[7]**.

d The same three spectral lines are observed from a galaxy receding[8] from the Earth at a speed of 0.30*c*, where *c* is the speed of light in a vacuum.

Describe and explain how the observed spectrum[2] from this galaxy will differ from Figure 1. *(4 marks)*

Receding galaxies and Doppler Effect[8] were covered in Topic 20.2, The Doppler effect. The fractional shift in the observed wavelength is covered by the Doppler equation $\frac{\Delta\lambda}{\lambda} \approx \frac{v}{c}$. Have a look at the Worked Example: Speed of a galaxy.

Answer to part **(d)**

According to the Doppler Effect[8], each spectral line[2] is shifted towards the longer wavelength end of the spectrum. ✓

The fractional increase in the wavelength is given by the Doppler equation $\frac{\Delta\lambda}{\lambda} \approx \frac{v}{c} = 0.30$ ✓

The observed wavelengths are:

$656 \times 1.30 = 853\,\text{nm}$

$486 \times 1.30 = 632\,\text{nm}$

and $434 \times 1.30 = 564\,\text{nm}$ ✓

The spectral lines corresponding to 632 nm and 564 nm lie in the visible region of the electromagnetic spectrum, but the 853 nm is in the infrared region of the electromagnetic spectrum[9]. ✓

The **Electromagnetic spectrum[9]** was covered in Topic 11.6, Electromagnetic waves. Figure 3 in Topic 11.6 shows the wavelengths of the different regions.

1 a Figure 2 shows an electrical circuit set up by a student.

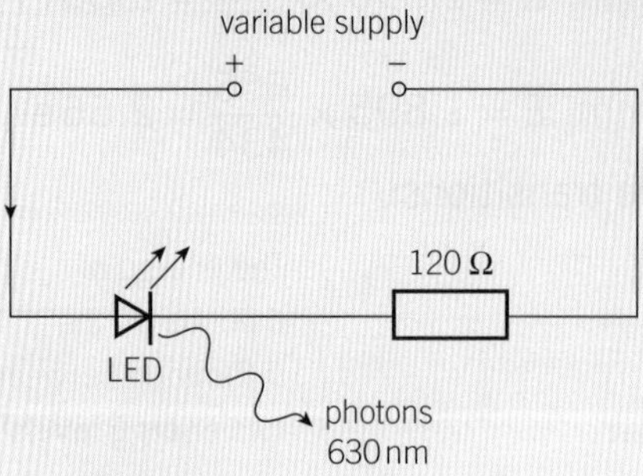

▲ **Figure 2**

The variable supply has negligible internal resistance. In one experiment, the potential difference (p.d.) across the terminals of the supply is 2.50 V and the p.d. across the light-emitting diode (LED) is 2.02 V. The resistance of the resistor in the circuit is 120 Ω.

(i) Calculate the number of electrons passing through the LED per second. *(3 marks)*

(ii) Calculate the total power dissipated in the circuit. *(1 mark)*

(iii) The photons emitted from the LED have wavelength 630 nm. Assume a single electron passing through the LED is responsible for one photon emitted from the LED. Estimate the radiant power emitted from the LED. *(3 marks)*

b The circuit shown in Figure 2 is adapted by a group of students to investigate the relationship between the minimum p.d. V across an LED that emits light and the wavelength λ of the emitted light. Several coloured LEDs are used. Table 2 shows the results collected and processed by the group.

▼ **Table 2**

V / V	λ / nm	$\frac{1}{\lambda}$ / 10^6 m^{-1}
1.94	630 ± 10	1.59 ± 0.03
2.00	615 ± 10	1.63 ± 0.03
2.30	535 ± 10	1.90 ± 0.03
2.50	505 ± 10	
2.62	475 ± 10	2.11 ± 0.04

The absolute uncertainty in the wavelengths is also shown in the table.

(i) The students decide to plot a graph of V against $\frac{1}{\lambda}$.
Explain why this is a sensible decision. *(2 marks)*

(ii) Show that the missing value of $\frac{1}{\lambda}$ in the table is $(1.98 \pm 0.04) \times 10^6\,m^{-1}$. *(2 marks)*

(iii) Use your answer from (ii) to complete a copy of the graph in Figure 3. Four of the points have already been plotted for you. *(1 mark)*

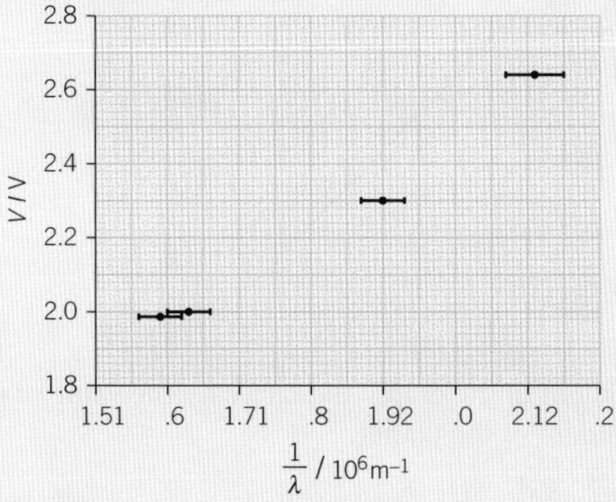

▲ **Figure 3**

(iv) Use the graph to determine the gradient and the absolute uncertainty in the value of the gradient. *(3 marks)*

(v) Explain why the theoretical value of the gradient is about $1.2 \times 10^{-6}\,V\,m$. *(2 marks)*

2 Information in physics can be represented in graphical form.

a Figure 4 shows a velocity-time graph for a car.

Use Figure 4 to sketch a graph of total distance travelled over time from time $t = 0$ to $t = 30\,s$. *(4 marks)*

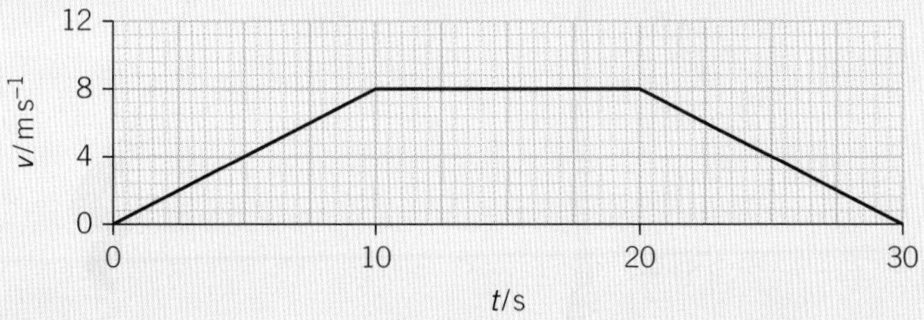

▲ **Figure 4**

b Figure 5 shows a graph of lg (g / N kg^{-1}) against lg (r / m) for a planet, where g is the gravitational field strength at a distance r from the centre of the planet.

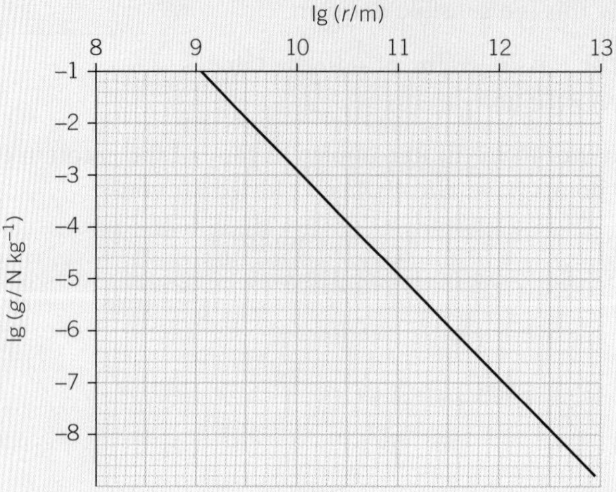

▲ **Figure 5**

(i) Determine the gradient of the graph. Explain why the value of the gradient is independent of the planet.

(3 marks)

(ii) Use Figure 5 to determine the mass of the planet.

(3 marks)

3 At a theme park there is a roundabout which consists of a central post supporting a bar, 4.0 m long, which rotates freely in a horizontal plane. At the ends of the bar, chains are attached. The other end of each chain has a flat seat on which a child can sit. The arrangement is shown in Figure 6. Throughout the question air resistance should be neglected.

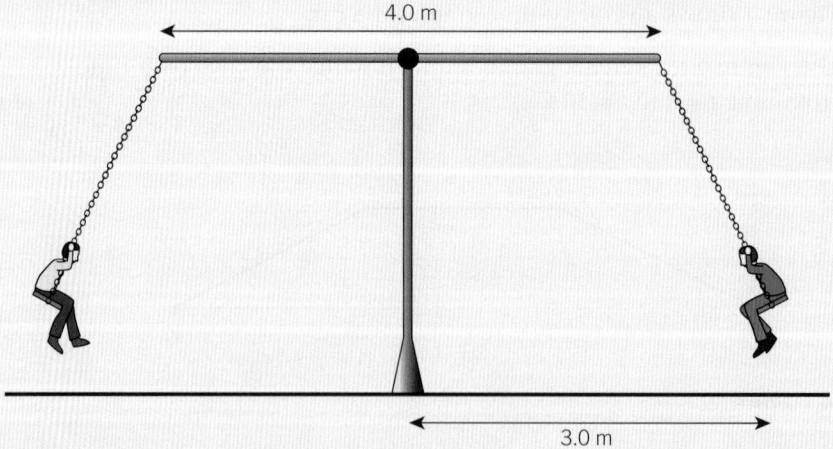

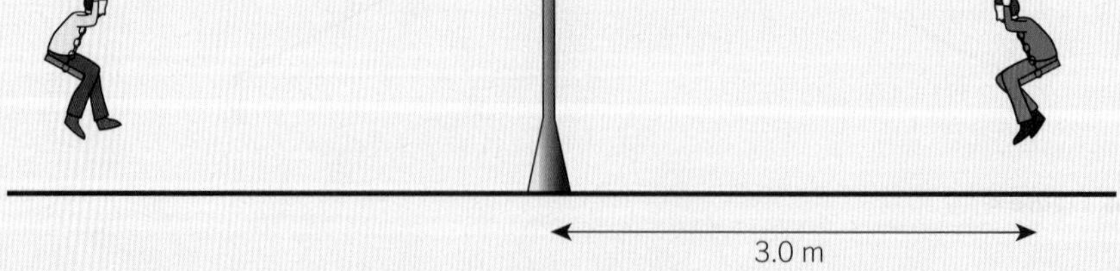

▲ **Figure 6**

The figure shows two children, each of weight 220 N, rotating in a horizontal circle of radius 3.0 m. Each revolution takes 4.7 s.

a On a copy of Figure 6, draw arrows to represent the **two** forces acting on one of the children. On each arrow name the object that provides the force. *(3 marks)*

b Calculate

(i) the mass of each child *(1 mark)*

(ii) the speed of each child *(1 mark)*

(iii) the kinetic energy of each child *(2 marks)*

(iv) the magnitude and direction of the resultant force on each child *(3 marks)*

(v) the angle of the chain to the vertical. *(2 marks)*

c For a child on this roundabout a force diagram that is sometimes drawn is shown in Figure 7a. A triangle of forces for these three forces is given in Figure 7b.

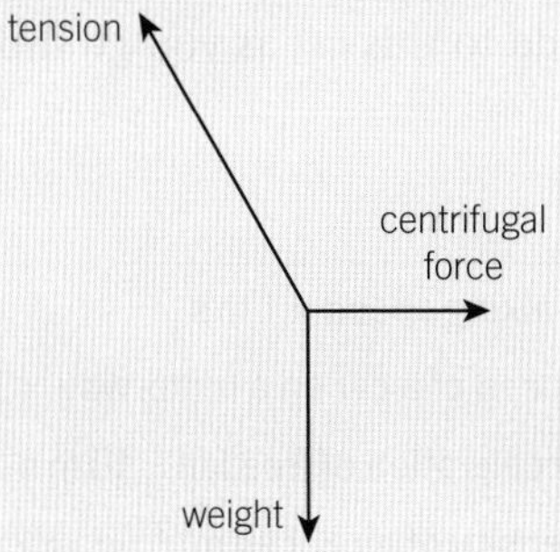

▲ Figure 7a

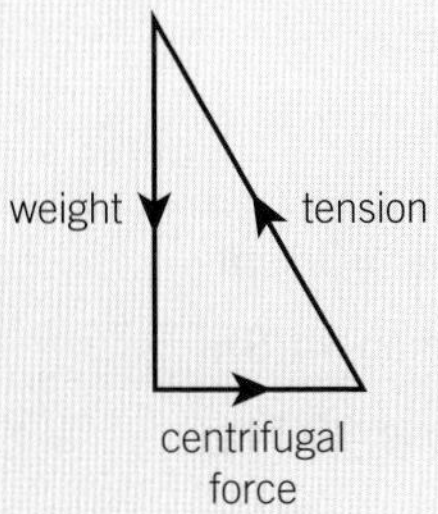

▲ Figure 7b

Explain why these diagrams are **incorrect** even though they can be used to obtain the correct value for the magnitude of the resultant force on the child. *(3 marks)*

Jun 2010 2826/01

4 Conservation laws are used in physics in many different situations. In this question you are asked to state three conservation laws and then to explain how some other laws can be regarded as conservation laws.

a State **three** conservation laws. *(3 marks)*

b Each of the following examples follows from a conservation law. In each case state what is being conserved and explain how each demonstrates conservation.

(i) Kirchhoff's first law states that — the sum of the electric currents into any point in an electrical circuit equals the sum of the currents out of the point. *(3 marks)*

(ii) Kirchhoff's second law states that — the sum of the electromotive forces (e.m.f.s) around any loop in a circuit is equal to the sum of the potential differences (p.d.s) around the loop. *(3 marks)*

(iii) When a cannon fires a cannonball the cannon recoils. *(3 marks)*

(iv) The Earth continually receives electromagnetic radiation from the Sun but its mean temperature stays almost constant over a few years. *(4 marks)*

A1 – Physical quantities and units

Being sensible

You already know that all physical quantities have a numerical value and a unit. For example, speed is a physical quantity and its SI base unit is m s^{-1}.

Synoptic link

You first met physical quantities in Topic 2.1, Quantities and units.

When carrying out calculations or experiments, it is always sensible to have a close look at the answer to see if it is reasonable. Experience of doing many calculations in physics will help you gauge whether or not an answer is reasonable. The list below provides helpful benchmarks in mechanics for you to make well-reasoned estimates of other physical quantities. Try to add to this list as you go through your physics course.

▼ **Table 1**

Length of a ruler = 30 cm	walking speed = 1 m s^{-1}
height of a person = 1.5 m	speed of a car on a motorway = 30 m s^{-1}
mass of an apple = 0.1 kg	acceleration of free fall = 10 m s^{-2}
mass of a person = 70 kg	density of air = 1 kg m^{-3}
mass of a car = 1000 kg	density of water = 1000 kg m^{-3}

Making estimates

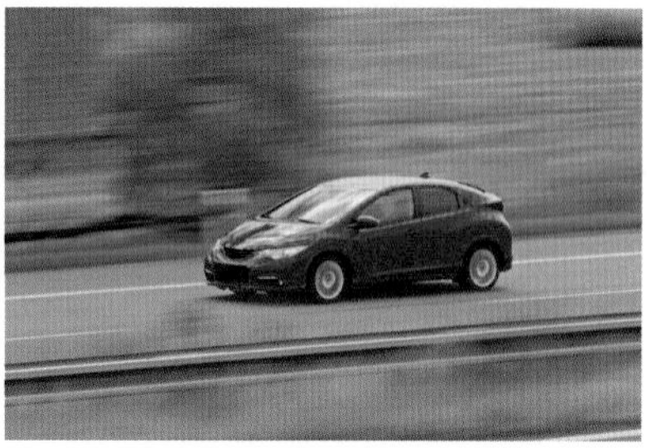

▲ **Figure 1** *Estimating the kinetic energy of a car*

What is the kinetic energy of the car shown in Figure 1?

You can make an estimate of the energy using the equation for kinetic energy and some of the values listed in Table 1. Estimates use known facts and sensible assumptions to arrive at an answer, so they are not just wild guesses. You should not be plucking random values from your head when making estimates.

The kinetic energy E_k of a car is given by the equation

$$E_k = \frac{1}{2} \text{mass} \times \text{speed}^2$$

$$E_k = \frac{1}{2} \times 1000 \times 30^2 = 4.5 \times 10^5\,\text{J} = 5 \times 10^5\,\text{J (1 s.f.)}$$

Assumptions are needed for estimates. In the estimation above, we assumed that the car is travelling at a constant speed.

Working out units

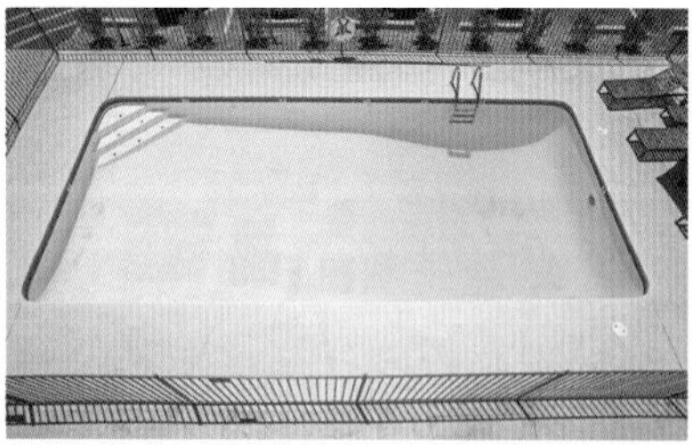

▲ **Figure 2** *Estimating the mass of water in a swimming pool*

Most physical quantities can be expressed by combinations of six base units – kg, m, s, A, K, and mol.

The area of a rectangular room can be found by multiplying its length and width together. This means that the derived unit for area is m × m = m^2.

All derived units can be worked out using an appropriate equation and then multiplying and/or dividing the base units. For example, speed is distance divided by time. This means that the derived unit for speed is m ÷ s = m s^{-1}.

Checking if an equation is correct in terms of units

An equation describing a relationship between physical quantities can only be correct if both sides have the same SI units. If the left-hand side of an equation has the units $\text{kg}\,\text{m}^{-2}\,\text{s}^{-1}$, then the right-hand side of the equation must also have the same units, $\text{kg}\,\text{m}^{-2}\,\text{s}^{-1}$. The equation is **homogeneous** in terms of units, where homogeneous means identical. Checking the homogeneity of physical equations using SI units is a powerful method of assessing whether or not an equation is correct.

All balanced out

For an object that started from rest, the final velocity is given by the equation $v^2 = 2as$, where a is the acceleration and s is the displacement of the object. You can show that the equation is homogeneous by determining the units on the left- and right-hand sides of the equal sign and showing them to be identical.

Since numbers have no units, you will not need to worry about the 2 in this equation.

▼ **Table 2**

quantity	v	a	s	'left hand side' v^2	'right hand side' $2as$
base unit	m s^{-1}	m s^{-2}	m	$(\text{m s}^{-1})^2 = \text{m}^2\,\text{s}^{-2}$	$(\text{m s}^{-2}) \times (\text{m}) = \text{m}^2\,\text{s}^{-2}$

Using an equation to find SI units

The drag force F acting on an object falling through the air at a speed v is given by the equation $F = kAv^2$, where A is the area of the object. To determine the units for k you must rearrange the equation with k as the subject, then substitute all the known SI units for the quantities.

$$k = \frac{F}{Av^2}$$

The quantity k has the SI units kg m^{-3}.

▼ **Table 3**

quantity	F	A	v	$k = \frac{F}{Av^2}$
base unit	kg m s^{-2}	m^2	m s^{-1}	$\frac{\text{kg m s}^{-2}}{\text{m}^2 \times (\text{m s}^{-1})^2} = \frac{\text{kg m s}^{-2}}{\text{m}^2 \times \text{m}^2\,\text{s}^{-2}} = \text{kg m}^{-3}$

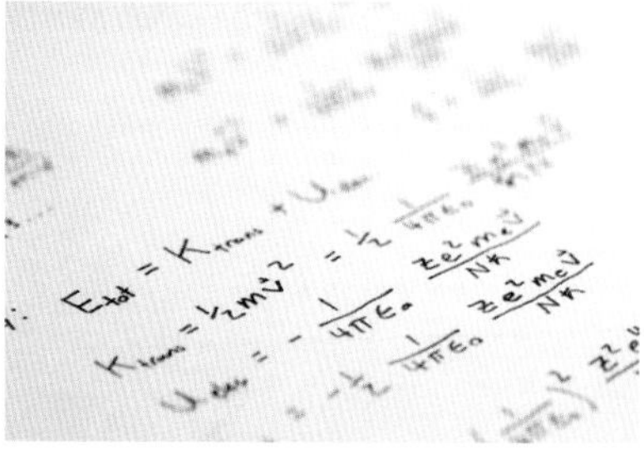

▲ **Figure 3** *Are these equations homogeneous?*

A2 – Recording results and straight lines

Table of results

Tables are useful for displaying a lot of information at once. This is usually numerical data, for example, readings taken from measuring devices during experiments. You can clearly see the values of each quantity in a table, although it may be difficult to spot patterns or trends in the data.

▼ **Table 1**

V/V	I/A	R/Ω
0.15	1.01	0.15
0.32	2.12	0.15
0.38	2.42	0.16

Table 1 shows a section of a typical results table from an investigation of an electrical component. The measured quantities are potential difference V (or p.d.) and the current I. The units are also included. The first two columns have headings V/ V and I/ A. The slash is used to separate the quantity from its unit. The final column is for resistance R of the component, which is calculated by dividing V by I. The unit of electrical resistance, Ω, for this processed data is also included.

You have to think about the exactness to which the quantities are measured. All the values for V and I are measured to 2 decimal places. Since the values in the final column are calculated from V and I, values of resistance must be written to the correct number of significant figure. The V values are written to 2 significant figures and the I values are all to 3 significant figures. Therefore the values of R can only be quoted to 2 significant figures – equal to the *lowest* significant figures used in the calculation.

Plotting graphs

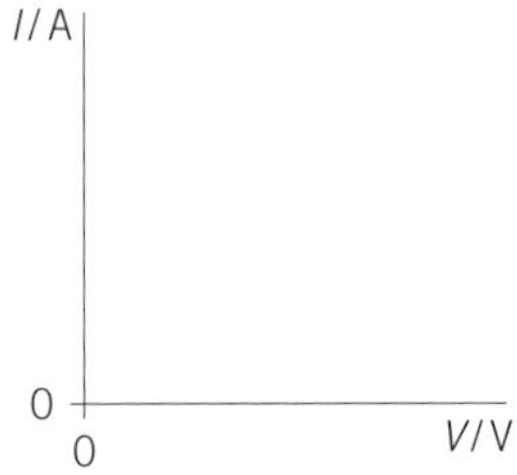

▲ **Figure 1** *The axes of the graph must have the correct labels*

In physics, you often plot a graph of your results in order to identify any trends or patterns. What do you plot on the vertical axis (y-axis) and on the horizontal axis (x-axis)? As a general rule, you plot the **independent variable**, the one you deliberately change in an experiment, on the x-axis, and the **dependent variable**, the variable which changes as a result, on the y–axis.

You must label both axes correctly with the quantities and their units. The labels are simply those you would have used in your column headings (see Table 1).

Here are a few useful tips when plotting graphs:

- Never choose scales which are multiples of 3, 7, 11, or 13.
- Ensure that you are using most of the graph paper to plot your results.
- Plot your points using small crosses.
- For a straight line graph, draw the straight **best fit line** using a long ruler.
- When calculating the gradient (see below), clearly show the triangle used – the triangle should be large, using at least half of the line drawn.

Gradients

You can determine the **gradient** of a straight line graph using a large triangle. For a curve, you have to draw a tangent first and then determine its gradient using a large triangle. The gradient is worked out using the following equation:

$$\text{gradient} = \frac{\text{change in } y}{\text{change in } x} = \frac{\Delta y}{\Delta x}$$

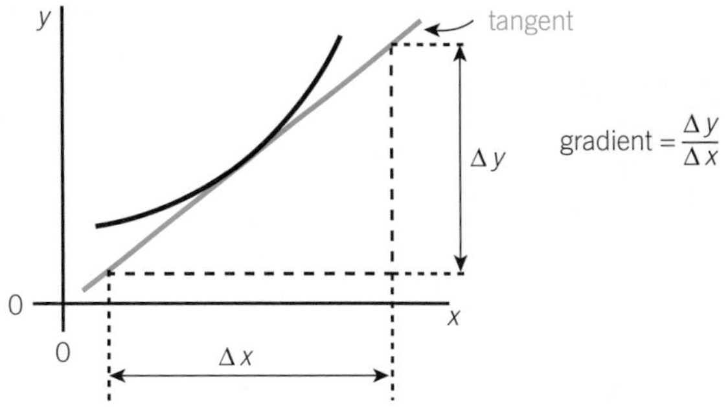

▲ **Figure 2** *Use a large triangle when determining the gradient*

Straight lines

The general equation for a straight line is written in the form

$y = mx + c$

where m is the gradient and c is the y-intercept (Figure 3).

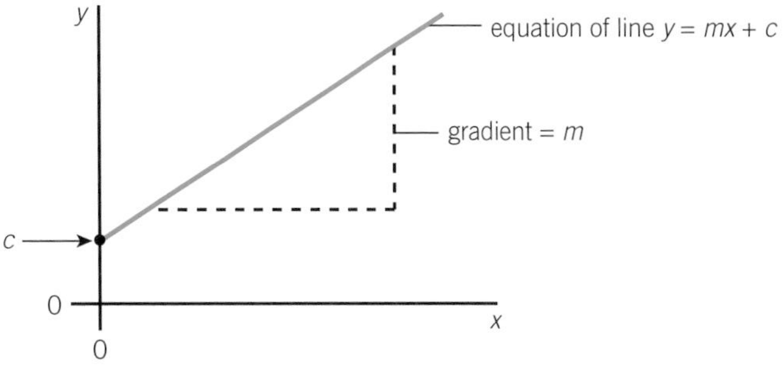

▲ **Figure 3** $y = mx + c$

In your physics work, you will come across equations that can easily be presented in the form of a straight line equation. For example, $v = u + at$. This is the equation for an object accelerating in a straight line. The initial velocity of the object is u, the acceleration is a, and v is the final velocity after a time t. If you plot t on the x-axis and v on the y-axis, then the gradient of the line must be acceleration a, and the y-intercept must be u (Figure 4).

$v = at + u$

$y = mx + c$

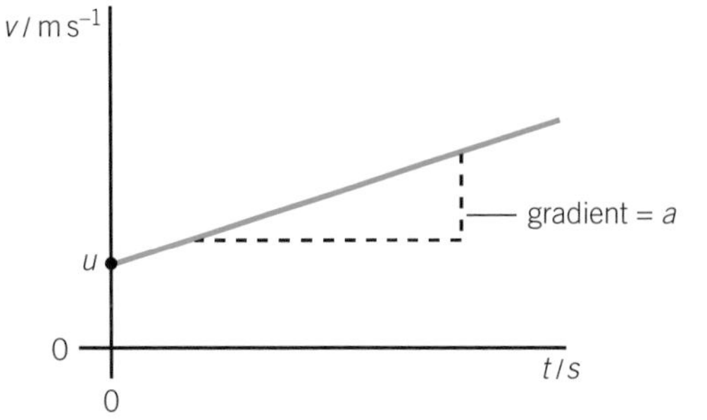

▲ **Figure 4** *The y-intercept is* u *and the gradient is* a

Curves to straight lines

Not all the graphs you plot in physics produce straight lines. Consider the equation $s = \frac{1}{2}at^2$ for an object accelerating from rest. The constant acceleration of the object is a and its displacement after a time t is s. If you plot s against t you will get a curve – a parabola. It is difficult to extract useful information from such a curve. However, you can turn this equation into a straight line equation as shown below.

$s = \frac{1}{2}at^2 + 0$

$y = mx + c$

You will get a straight line graph by plotting t^2 on the x-axis and s on the y-axis. The gradient of the line is $\frac{1}{2}a$ (Figure 5). By plotting the data in this way, you can determine the acceleration a of the object by multiplying the gradient by 2.

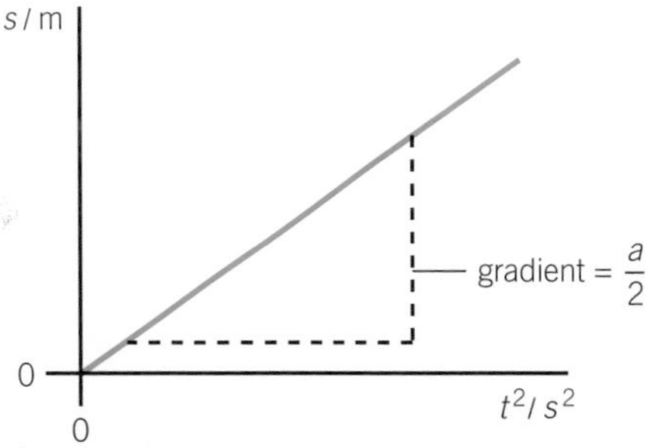

▲ **Figure 5** *The gradient of this straight line is* $\frac{a}{2}$ *and the line passes through the origin*

A3 – Measurements and uncertainties

▲ **Figure 1** *An error made when you use a calculator is a mistake that can be corrected but measurement errors are more difficult to deal with*

Errors are not mistakes

Making measurements and using instruments is a key part of scientific activity. No measurement can ever be perfect. When you think about errors you may think about mistakes. If you make a mistake in an experiment, you may be may be able to do the experiment again without the mistake. However, an experiment will still contain errors. An error in a scientific sense is the difference between the result you get and the correct result. Errors are usually caused by measuring devices, even if they are used correctly, or by the design of the experiment itself.

Measurement errors

A **true value** is the value that would be obtained in an ideal measurement. True values may be values found in a data book, or the values you expect to get in your experiments. A **measurement error** is the difference between a measured value and the true value for the quantity being measured. Remember that mistakes are not counted as errors here.

Random errors

Random errors can happen when any measurement is being made. They are measurement errors in which measurements vary unpredictably. There can be many reasons for this, including

- factors that are not controlled in the experiment
- difficulty in deciding on the reading given by a measuring device.

Random errors cannot be corrected. All you can hope to do is reduce their effect by making more measurements and reporting the mean value.

Systematic errors

Systematic errors are measurement errors in which the measurements differ from the true values by a consistent amount each time a measurement is made. Reasons for this include

- the way in which measurements are taken
- faulty measuring devices.

For example, poor contact between a thermometer and the object whose temperature is being measured will cause systematic errors. A faulty measuring device may give readings that are consistently too high or too low. This may be because it has not been calibrated correctly. A faulty device may give a **zero error**, in which the reading is not zero when the quantity being measured is zero (Figure 2).

Unlike random errors, it may be possible to correct for systematic errors.

▲ **Figure 2** *Zero error on an ammeter*

Precision and accuracy

In everyday life, people often use the words *precise* and *accurate* to mean the same thing – this is not the case in physics.

- **Accuracy** is to do with how close a measurement result is to the true value – the closer it is, the more **accurate** it is.
- **Precision** is to do with how close repeated measurements are to each other – the closer they are to each other, the more **precise** the measurement is.

Figure 3 is a visual way of appreciating the terms accuracy and precision using a dartboard as an example. You are aiming for bullseye – it represents the true value.

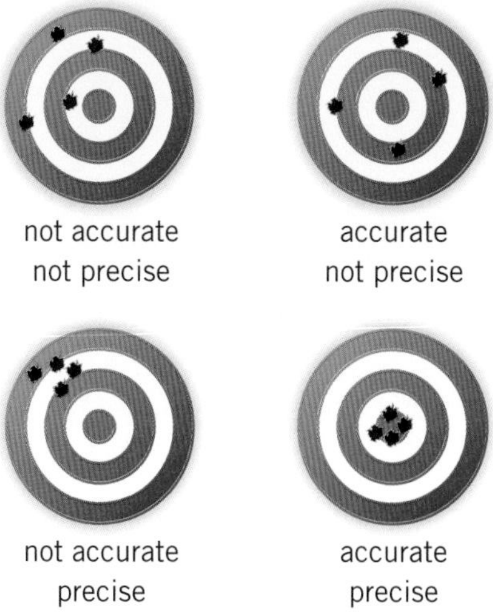

▲ **Figure 3** *Accuracy and precision*

Uncertainty

Random and systematic errors mean that you rarely obtain the same value for a particular measurement. Consider using a micrometer to measure the diameter of a supposedly uniform copper wire. The readings below show the readings along the length of the copper wire:

$$0.53\,\text{mm},\ 0.49\,\text{mm},\ 0.52\,\text{mm},\ 0.51\,\text{mm}$$

The smallest scale division of this instrument is 0.01 mm but the readings above show a spread much greater than this.

The **mean** value for the diameter can be calculated by adding together the values for each repeat reading, then dividing by the number of readings. The mean diameter of the copper wire is

$$\frac{0.53\,\text{mm} + 0.49\,\text{mm} + 0.52\,\text{mm} + 0.51\,\text{mm}}{4} = 0.51\,\text{mm}.$$

The **range** of the measurements is 0.04 mm. This is the difference between the smallest and largest readings (0.53 mm – 0.49 mm).

The **uncertainty** in the measurement is an interval within which the true value can be expected to lie. The **absolute uncertainty** in the mean value of a measurement can be approximated as half the range. This is often expressed as ± value. In this example, you can write the diameter as its mean value ± absolute uncertainty

$$\text{diameter} = 0.51 \pm 0.02\,\text{mm}$$

The **percentage uncertainty** in the diameter can be calculated from its absolute uncertainty and mean value as follows:

$$\%\ \text{uncertainty in diameter} = \frac{\text{absolute uncertainty}}{\text{mean value}} \times 100$$

$$= \frac{0.02}{0.51} \times 100 = 3.9\%$$

Finally, what do you do when repeat measurements give identical values or you have just taken a single measurement? In this situation you can approximate the absolute uncertainty to be equal to the **resolution** of the measuring instrument. This is the smallest change in the measured quantity that the instrument can show. The micrometer readings for the copper wire are written to 2 decimal places, with ±0.01 mm being the smallest change it could show.

Therefore, if all the readings were 0.53 mm, or you just had a single reading of 0.53 mm, the diameter of the copper wire may be written as

$$\text{diameter} = 0.53 \pm 0.01\,\text{mm}$$

Analysing uncertainties

Uncertainties can help you identify where the greatest errors in an experiment are, giving you the chance to improve it using a different method or measuring instrument.

The final uncertainty in an answer depends on how quantities are combined. Here are three important rules about the way uncertainties propagate.

1 **Adding or subtracting quantities**

When you add or subtract quantities in an equation, you add the absolute uncertainties for each value.

What is the extension?

The original length of a spring is 2.5 ± 0.1 cm and the final length is 15.0 ± 0.2 cm. Calculate the extension of the spring and the absolute uncertainty.

Step 1: Calculate the extension by subtracting the lengths.

$$\text{extension} = 15.0 - 2.5 = 12.5\,\text{cm}$$

Step 2: Add the absolute uncertainties.

$$\text{absolute uncertainty} = 0.1 + 0.2 = 0.3$$

Step 3: Write the answer in the normal convention.

$$\text{extension} = 12.5 \pm 0.3\,\text{mm}$$

2 **Multiplying or dividing quantities**

When you multiply or divide quantities, you add the percentage uncertainties for each value.

What is the resistance?

The current I in a resistor is 1.60 ± 0.02 A and the potential difference V across the resistor is 6.00 ± 0.20 V. Calculate the resistance and the absolute uncertainty.

Step 1: Calculate the resistance R of the resistor.

$$R = \frac{V}{I} = \frac{6.00}{1.60} = 3.75\,\Omega$$

Step 2: Calculate the percentage uncertainty in each measurement.

$$\%\text{ uncertainty in } I = \frac{0.02}{1.60} \times 100 = 1.25\%$$

$$\%\text{ uncertainty in } V = \frac{0.20}{6.00} \times 100 = 3.33\%$$

Step 3: Add the percentage uncertainties.

% uncertainty in $R = 1.25 + 3.33 = 4.58\%$

Step 4: Calculate the absolute uncertainty in R.

absolute uncertainty in $R = 0.0458 \times 3.75 = 0.17\,\Omega$

Step 5: The values of V and I are quoted to 3 significant figures, therefore the final answer for the resistance must also be written to 3 significant figures.

$R = 3.75 \pm 0.17\,\Omega$ (3 s.f.)

3 **Raising a quantity to a power**

When a measurement in a calculation is raised to a power n, your percentage uncertainty is increased n times. The power n can be an integer or a fraction.

Cross-sectional area of a wire

The diameter of a wire is recorded as $0.51 \pm 0.02\,\text{mm}$. Calculate the cross-sectional area of the wire and the absolute uncertainty.

Step 1: Calculate the cross-sectional area A of the wire.

$$A = \frac{\pi d^2}{4} = \frac{\pi \times (0.51 \times 10^{-3})^2}{4} = 2.04 \times 10^{-7}\,\text{m}^2$$

Step 2: The percentage uncertainty in A is equal to 2 times the percentage uncertainty in d.
(The π and the 4 are numbers and therefore have no uncertainty associated with them.)

$$\%\text{ uncertainty in } A = 2 \times \left(\frac{0.02}{0.51} \times 100\right) = 7.84\%$$

Step 3: Calculate the absolute uncertainty in A.

absolute uncertainty in $A = 0.0784 \times 2.04 \times 10^{-7} = 0.16 \times 10^{-7}\,\text{m}^2$

Step 4: The diameter of the wire is quoted to 2 significant figures, therefore the final answer for the cross-sectional must also be written to 2 significant figures.

$A = (2.0 \pm 0.2) \times 10^{-7}\text{m}^2$ (2 s.f.)

Graphs

Straight line graphs are important in physics because you can use them to formulate relationships between physical quantities. As indicated in Appendix A2, Recording results, you plot points using small crosses. If the points appear to lie on a straight line, then you can draw your straight line of best fit using a long ruler. You must ignore a point that is much further than any other point from the best fit line. This point is referred to as being **anomalous**.

The uncertainty in a measurement can be used to give a small range or **error bar** for each measurement. Instead of plotting just the points on a graph, you can plot an error bar for all of your measurements.

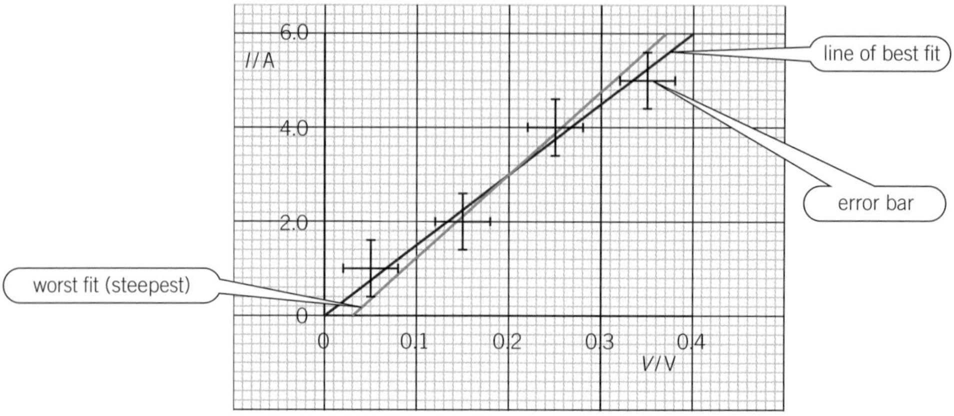

▲ **Figure 4** *Error bars are useful when you draw the line of best fit and the worst line for your measurements*

Your straight best fit line must pass through all the error bars (Figure 4). You would use this line to determine the value of the gradient. How can you determine an approximate value for the uncertainty in the gradient? You would draw the **line of worst fit** – the least acceptable straight line through the data points – this can either be the steepest or the shallowest line.

The absolute uncertainty in the gradient is the positive difference between the gradient of the line of best fit and the gradient of the line of worst fit.

The percentage uncertainty in the gradient can be calculated as

$$\% \text{ uncertainty in gradient} = \frac{\text{absolute uncertainty}}{\text{gradient best fit line}} \times 100\%$$

5d. Physics A Data Sheet

Data, Formulae and Relationships

The data, formulae and relationships in this datasheet will be printed for distribution with the examination papers.

Data

Values are given to three significant figures, except where more – or fewer – are useful.

Physical constants

acceleration of free fall	g	$9.81\ \mathrm{m\,s^{-2}}$
elementary charge	e	$1.60 \times 10^{-19}\ \mathrm{C}$
speed of light in a vacuum	c	$3.00 \times 10^{8}\ \mathrm{m\,s^{-1}}$
Planck constant	h	$6.63 \times 10^{-34}\ \mathrm{J\,s}$
Avogadro constant	N_A	$6.02 \times 10^{23}\ \mathrm{mol^{-1}}$
molar gas constant	R	$8.31\ \mathrm{J\,mol^{-1}\,K^{-1}}$
Boltzmann constant	k	$1.38 \times 10^{-23}\ \mathrm{J\,K^{-1}}$
gravitational constant	G	$6.67 \times 10^{-11}\ \mathrm{N\,m^{2}\,kg^{-2}}$
permittivity of free space	ε_0	$8.85 \times 10^{-12}\ \mathrm{C^{2}\,N^{-1}\,m^{-2}\,(F\,m^{-1})}$
electron rest mass	m_e	$9.11 \times 10^{-31}\ \mathrm{kg}$
proton rest mass	m_p	$1.673 \times 10^{-27}\ \mathrm{kg}$
neutron rest mass	m_n	$1.675 \times 10^{-27}\ \mathrm{kg}$
alpha particle rest mass	m_α	$6.646 \times 10^{-27}\ \mathrm{kg}$
Stefan constant	σ	$5.67 \times 10^{-8}\ \mathrm{W\,m^{-2}\,K^{-4}}$

Quarks

up quark	charge $= +\frac{2}{3}e$
down quark	charge $= -\frac{1}{3}e$
strange quark	charge $= -\frac{1}{3}e$

Conversion factors

unified atomic mass unit	$1\ \mathrm{u} = 1.661 \times 10^{-27}\ \mathrm{kg}$
electronvolt	$1\ \mathrm{eV} = 1.60 \times 10^{-19}\ \mathrm{J}$
day	$1\ \mathrm{day} = 8.64 \times 10^{4}\ \mathrm{s}$
year	$1\ \mathrm{year} \approx 3.16 \times 10^{7}\ \mathrm{s}$
light year	$1\ \mathrm{light\ year} \approx 9.5 \times 10^{15}\ \mathrm{m}$
parsec	$1\ \mathrm{parsec} \approx 3.1 \times 10^{16}\ \mathrm{m}$

Mathematical equations

arc length = $r\theta$

circumference of circle = $2\pi r$

area of circle = πr^2

curved surface area of cylinder = $2\pi rh$

surface area of sphere = $4\pi r^2$

area of trapezium = $\frac{1}{2}(a + b)h$

volume of cylinder = $\pi r^2 h$

volume of sphere = $\frac{4}{3}\pi r^3$

Pythagoras' theorem: $a^2 = b^2 + c^2$

cosine rule: $a^2 = b^2 + c^2 - 2bc\cos A$

sine rule: $\frac{a}{\sin A} = \frac{b}{\sin B} = \frac{c}{\sin C}$

$\sin\theta \approx \tan\theta \approx \theta$ and $\cos\theta \approx 1$ for small angles

$\log(AB) = \log(A) + \log(B)$

(Note: $\lg = \log_{10}$ and $\ln = \log_e$)

$\log\left(\frac{A}{B}\right) = \log(A) - \log(B)$

$\log(x^n) = n\log(x)$

$\ln(e^{kx}) = kx$

Formulae and relationships

Module 2 – Foundations of physics

vectors

$F_x = F\cos\theta$

$F_y = F\sin\theta$

Module 3 – Forces and motion

uniformly accelerated motion

$v = u + at$

$s = \frac{1}{2}(u + v)t$

$s = ut + \frac{1}{2}at^2$

$v^2 = u^2 + 2as$

force

$F = \frac{\Delta p}{\Delta t}$

$p = mv$

turning effects

$moment = Fx$

$torque = Fd$

density

$\rho = \frac{m}{V}$

pressure

$p = \frac{F}{A}$

$p = h\rho g$

work, energy and power

$W = Fx\cos\theta$

$\text{efficiency} = \frac{\text{useful energy output}}{\text{total energy input}} \times 100\%$

$P = \frac{W}{t}$

$P = Fv$

springs and materials

$F = kx$

$E = \frac{1}{2}Fx;\ E = \frac{1}{2}kx^2$

$\sigma = \frac{F}{A}$

$\varepsilon = \frac{x}{L}$

$E = \frac{\sigma}{\varepsilon}$

Module 4 – Electrons, waves and photons

charge $\Delta Q = I\Delta t$

current $I = Anev$

work done $W = VQ;\ W = \varepsilon Q;\ W = VIt$

resistance and resistors

$R = \frac{\rho L}{A}$

$R = R_1 + R_2 + \ldots$

$\frac{1}{R} = \frac{1}{R_1} + \frac{1}{R_2} + \ldots$

power $P = VI$, $P = I^2R$ and $P = \frac{V^2}{R}$

internal resistance $\varepsilon = I(R + r);\ \varepsilon = V + Ir$

potential divider $V_{out} = \frac{R_2}{R_1 + R_2} \times V_{in}$

$\frac{V_1}{V_2} = \frac{R_1}{R_2}$

waves $v = f\lambda$

$f = \frac{1}{T}$

$I = \frac{P}{A}$

$\lambda = \frac{ax}{D}$

refraction $n = \frac{c}{v}$

$n\sin\theta = \text{constant}$

$\sin C = \frac{1}{n}$

quantum physics $E = hf$

$E = \frac{hc}{\lambda}$

$hf = \phi + KE_{max}$

$\lambda = \frac{h}{p}$

Module 5 – Newtonian world and astrophysics

thermal physics $E = mc\Delta\theta$

$E = mL$

ideal gases $pV = NkT; \; pV = nRT$

$pV = \frac{1}{3}Nm\overline{c^2}$

$\frac{1}{2}m\overline{c^2} = \frac{3}{2}kT$

$E = \frac{3}{2}kT$

circular motion

$\omega = \frac{2\pi}{T}; \; \omega = 2\pi f$

$v = \omega r$

$a = \frac{v^2}{r}; a = \omega^2 r$

$F = \frac{mv^2}{r}; F = m\omega^2 r$

oscillations

$\omega = \frac{2\pi}{T}; \; \omega = 2\pi f$

$a = -\omega^2 x$

$x = A\cos\omega t; \; x = A\sin\omega t$

$v = \pm\,\omega\sqrt{A^2 - x^2}$

gravitational field

$g = \frac{F}{m}$

$F = -\frac{GMm}{r^2}$

$g = -\frac{GM}{r^2}$

$T^2 = \left(\frac{4\pi^2}{GM}\right)r^3$

$V_g = -\frac{GM}{r}$

energy $= -\frac{GMm}{r}$

astrophysics

$hf = \Delta E; \frac{hc}{\lambda} = \Delta E$

$d\sin\theta = n\lambda$

$\lambda_{\text{max}} \propto \frac{1}{T}$

$L = 4\pi r^2\sigma T^4$

cosmology

$\frac{\Delta\lambda}{\lambda} \approx \frac{\Delta f}{f} \approx \frac{v}{c}$

$p = \frac{1}{d}$

$v = H_0 d$

$t = H_0^{-1}$

Module 6 – Particles and medical physics

capacitance and capacitors

$C = \frac{Q}{V}$

$C = \frac{\varepsilon_0 A}{d}$

$C = 4\pi\varepsilon_0 R$

$C = C_1 + C_2 + \ldots$

$\frac{1}{C} = \frac{1}{C_1} + \frac{1}{C_2} + \ldots$

$W = \frac{1}{2}QV; \; W = \frac{1}{2}\frac{Q^2}{C}; \; W = \frac{1}{2}V^2C$

$\tau = CR$

$x = x_0 e^{\frac{-t}{CR}}$

$x = x_0(1 - e^{\frac{-t}{CR}})$

electric field $E = \frac{F}{Q}$

$F = \frac{Qq}{4\pi\varepsilon_0 r^2}$

$E = \frac{Q}{4\pi\varepsilon_0 r^2}$

$E = \frac{V}{d}$

$V = \frac{Q}{4\pi\varepsilon_0 r}$

energy $= \frac{Qq}{4\pi\varepsilon_0 r}$

magnetic field $F = BIL\sin\theta$

$F = BQv$

electromagnetism

$\phi = BA\cos\theta$

$\varepsilon = -\frac{\Delta(N\phi)}{\Delta t}$

$\frac{n_s}{n_p} = \frac{V_s}{V_p} = \frac{I_p}{I_s}$

radius of nucleus $R = r_0 A^{\frac{1}{3}}$

radioactivity $A = \lambda N; \; \frac{\Delta N}{\Delta t} = -\lambda N$

$\lambda t_{\frac{1}{2}} = \ln(2)$

$A = A_0 e^{-\lambda t}$

$N = N_0 e^{-\lambda t}$

Einstein's mass-energy equation $\Delta E = \Delta mc^2$

attenuation of X-rays $I = I_0 e^{-\mu x}$

ultrasound $Z = \rho c$

$\frac{I_r}{I_0} = \frac{(Z_2 - Z_1)^2}{(Z_2 + Z_1)^2}$

$\frac{\Delta f}{f} = \frac{2v\cos\theta}{c}$

Glossary

absolute scale of temperature A scale for measuring temperature based on absolute zero and the triple point of pure water, with gradations equal in size to those of the Celsius scale; unit kelvin (K)

absolute zero The lowest possible temperature, the temperature at which substances have minimum internal energy

absorption coefficient A measure of the absorption of X-ray photons by a substance, also known as attenuation coefficient – SI unit m^{-1}

absorption line spectrum A set of specific frequencies of electromagnetic radiation, visible as dark lines in an otherwise continuous spectrum on spectroscopy. They are absorbed by atoms as their electrons are excited between energy states by absorbing the corresponding amount of energy in the form of photons – every element has a characteristic line spectrum

acceleration The rate of change of velocity, a vector quantity

acceleration of free fall The rate of change of velocity of an object falling in a gravitational field, symbol g

acoustic impedance The product of the density ρ of a substance and the speed c of ultrasound in that substance – symbol Z, SI unit $kg\,m^{-2}\,s^{-1}$

acoustic matching (or impedance matching) The use of two substances with similar acoustic impedance to minimise reflection of ultrasound at the boundary between them

activity The rate at which nuclei decay or disintegrate in a radioactive source, measured in becquerels (Bq) or decays per second

alpha radiation Ionising radiation consisting of particles comprising two protons and two neutrons (a helium nucleus), with a charge of $+2e$

air resistance The drag or resistive force experienced by objects moving through air

ammeter A device used to measure electric current — it must be placed in series and ideally have zero resistance

amount of substance A measure of the amount of matter in moles

ampère The base SI unit of electric current, symbol A, defined as the current flowing in two parallel wires in a vacuum 1 m apart such that there is an attractive force of 2.0×10^{-7} N per metre length of wire between them

amplitude *(waves)* The maximum displacement from the equilibrium position (can be positive or negative)

angle of incidence The angle between the direction of travel of an incident wave and the normal at a boundary between two media

angle of reflection The angle between the direction of travel of a reflected wave and the normal at a boundary between two media

angular frequency A quantity used in oscillatory motion – equal to the product of frequency f and 2π

angular velocity The rate of change of angle for an object moving in a circular path – symbol ω

anion A negatively charged ion, one which is attracted to an anode

annihilation The complete destruction of a particle and its antiparticle in an interaction that releases energy in the form of identical photons

anode A positively charged electrode

antiparallel *(vectors)* In the same line but opposite directions

antiparticle The antimatter counterpart of a particle, with the opposite charge to the particle (if the particle has charge) and exactly the same rest mass as the particle

antiphase Particles oscillating completely out of step with each other (one reaches its maximum positive displacement as the other reaches its maximum negative displacement) are in antiphase

aphelion The furthest point from the Sun in an orbit

Archimedes' principle The upthrust on an object in a fluid is equal to the weight of fluid it displaces

arcminute A minute of arc; 1° = 60 arcminutes

arcsecond A second of arc; 1 arcminute = 60 arcseconds

astronomical unit The mean distance from the Earth to the Sun, i.e. 150 million km or 1.50×10^{11} m

atomic mass unit One atomic mass unit (1 u) is one-twelfth the mass of a neutral carbon-12 atom

atomic number The number of protons in a nucleus – symbol Z

attenuation The decrease in the intensity of electromagnetic radiation as it passes through matter and/or space

attenuation coefficient A measure of the absorption of X-ray photons by a substance, also known as absorption coefficient – SI unit m^{-1}

average speed The rate of change in distance calculated over a complete journey

average velocity The change in displacement Δs for a journey divided by the time taken Δt; $\Delta s/\Delta t$

Avogadro constant 6.02×10^{23}, the number of atoms in 0.012 kg (12 g) of carbon-12; symbol N_A

background radiation The radiation emitted by the surroundings, which must be measured before radiation produced in an experiment can usefully be measured

baryon Any hadron made with a combination of three quarks

base unit One of seven units that form the building blocks of the SI measurement system

battery A collection of cells that transfers chemical energy into electrical energy

becquerel A unit of activity – one becquerel is an activity of one decay per second

beta decay A neutron in an unstable nucleus decays into a proton, an electron, and an electron antineutrino (β^- decay), or a proton into a neutron, a positron, and an electron neutrino (β^+ decay)

beta radiation Ionising radiation consisting of fast-moving electrons (β^-) or (β^+) emitted from unstable nuclei, with a charge of $-e$ or $+e$, respectively

Big Bang The theory that at a moment in the past all the matter in the Universe was contained in a singularity (a single point), the beginning of space and time, that expanded rapidly outwards

binding energy The minimum energy required to completely separate a nucleus into its constituent protons and neutrons

binding energy per nucleon The binding energy divided by the number of protons and neutrons in the nucleus; the greater the binding energy per nucleon, the more tightly bound are the nucleons within the nucleus

black body An idealised object that absorbs all the electromagnetic radiation incident on it and, when in thermal equilibrium, emits a characteristic distribution of wavelengths at a specific temperature

black hole The remnant core of a massive star after it has gone supernova and the core has collapsed so far that in order to escape it an object would need an escape velocity greater than the speed of light, and therefore nothing, not even photons, can escape

blue shift The shortening of observed wavelength that occurs when a wave source is moving towards the observer – in astronomy, if a galaxy is moving towards the Earth, the absorption lines in its spectrum will be blue-shifted, that is, moved towards the blue end of the spectrum

Boltzmann constant The molar gas constant R divided by the Avogadro constant N_A, a constant that relates the mean kinetic energy of the atoms or molecules in a gas to the gas temperature – symbol k

Boyle's Law The pressure of an ideal gas is inversely proportional to its volume, provided that the mass of gas and the temperature do not vary

braking distance Distance travelled by a vehicle from the time the brakes are applied until the vehicle stops

breaking strength The stress value at the point of fracture, calculated by dividing the breaking force by the cross-sectional area

brittle Property of a material that does not show plastic deformation and deforms very little (if at all) under high stress

Brownian motion The continuous random motion of small particles suspended in a fluid, visible under a microscope

capacitance The charge stored per unit potential difference across a capacitor

capacitor A component that stores charge, consisting of two plates separated by an insulator (dielectric)

carbon dating A method for determining the age of organic material, by comparing the activities, or the ratios, of carbon-14 to carbon-12 nuclei of the dead material of interest and similar living material

cathode A negatively charged electrode

cation A positively charged ion, one which is attracted to a cathode

cell A device that transfers chemical energy into electrical energy

Celsius scale A temperature scale with 100 degrees between the freezing point and the boiling point of pure water (at atmospheric pressure 1.01×10^3 Pa), 0°C and 100°C

centre of gravity An imaginary point at which the entire weight of an object appears to act

centre of mass A point through which any externally applied force produces straight-line motion but no rotation

centripetal acceleration The acceleration of any object travelling in a circular path at constant speed, which always acts towards the centre of the circle

centripetal force A force that keeps a body moving with a constant speed in a circular path

chain reaction A reaction in which the neutrons from an earlier fission stage are responsible for further fission reactions leading to an exponential growth in the rate of the reactions

Chandrasekhar limit The mass of a star's core beneath which the electron degeneracy pressure is sufficient to prevent gravitational collapse, 1.44 solar masses

charge carrier A particle with charge that moves through a material to form an electric current — for example, an electron in a metal wire

closed system An isolated system that has no interaction with its surroundings

cloud chamber A detector of ionising radiation consisting of a chamber filled with air saturated with vapour at a very low temperature so that droplets of liquid condense around ionised particles left along the path of radiation

coherence Two waves sources, or waves, that are coherent have a constant phase difference

collimator Part of a gamma camera, a honeycomb of long, thin tubes made from lead that absorbs any photons arriving at an angle to the axis of the tubes so that a clear picture is obtained

comet A small, irregular body made of ice, dust, and small pieces of rock in an (often highly eccentric elliptical) orbit around the Sun – as they approach the Sun, some comets develop spectacular tails

component One of the two perpendicular vectors obtained by resolving a vector

compression The decrease in length of an object when a compressive force is exerted on it

compression *(waves)* A moving region in which the medium is denser or has higher pressure than the surrounding medium

compressive deformation A change in the shape of an object due to compressive forces

compressive force Two or more forces together that reduce the length or volume of an object

conical pendulum A simple pendulum that, instead of swinging back and forth, rotates in a horizontal circle at constant speed

conservation of charge A conservation law which states that electric charge can neither be created nor destroyed — the total charge in any interaction must be the same before and after the interaction

constant speed Motion in which the distance travelled per unit time stays the same

constant velocity Motion in which the change in displacement per unit time stays the same

constructive interference Superposition of two waves in phase so that the resultant wave has greater amplitude than the original waves

continuous spectrum A spectrum in which all visible frequencies or wavelengths are present (a heated solid metal such as a lamp filament will produce this type of spectrum)

control rods Rods made of a material whose nuclei readily absorb neutrons (commonly boron or cadmium), which can be moved into or out of a reactor core to ensure that exactly one slow neutron survives per fission reaction or to completely stop the fission reaction

conventional current A model used to describe electric current in a circuit — conventional current travels from positive to negative — it is the direction in which positive charges would travel

coolant A substance that removes the thermal energy produced from reactions within a fission reactor

corrected count rate The radiation count rate measured in an experiment minus the background count rate

cosmological principle The assumption that, when viewed on a large enough scale, the Universe is homogeneous and isotropic, and the laws of physics are universal

coulomb The derived SI unit of electrical charge, symbol C; 1 coulomb of electric charge passes a point in one second when there is an electric current of one ampère, $1\,C = 1\,A\,s$

Coulomb's law Any two point charges exert an electrostatic (electrical) force on each other that is directly proportional to the product of their charges and inversely proportional to the square of their separation

couple A pair of equal and opposite forces acting on a body but not in the same straight line

coupling gel A gel with acoustic impedance similar to that of skin smeared onto the transducer and the patient's skin before an ultrasound scan in order to fill air gaps and ensure that almost all the ultrasound enters the patient's body

critical angle The angle of incidence at the boundary between two media that will produce an angle of refraction of $90°$

crystallography A method for determining the structure of a substance by studying the interference patterns produced by waves passing through a crystal of the substance

damping An oscillation is damped when an external force that acts on the oscillator has the effect of reducing the amplitude of its oscillations

dark energy A hypothetical form of energy that fills all of space and would explain the accelerating expansion of the Universe

dark matter A hypothetical form of matter spread throughout the galaxy that neither emits nor absorbs light – it could explain the differences between the predicted and observed velocities of stars in galaxies

daughter nucleus A new nucleus formed following a radioactive decay

de Broglie equation An equation relating the wavelength and the momentum of a particle:

$$\lambda = \frac{h}{p}$$

decay constant The probability of decay of an individual nucleus per unit time

density The mass per unit volume of a substance

derived quantity A quantity that comes from a combination of base units

derived unit A unit used to represent a derived quantity, such as N for force

destructive interference Superposition of two waves in antiphase so that the waves cancel each other out and the resultant wave has smaller amplitude than the original waves

diffraction The phenomenon in which waves passing through a gap or around an obstacle spread out

diffraction grating A glass or plastic slide on which as many as 1000 lines in a millimetre are ruled, at a spacing that diffracts visible wavelengths of light

diode A semiconductor component that allows current only in one particular direction

displacement The distance travelled in a particular direction — it is a vector with magnitude and a direction

displacement (*waves*) The distance from the equilibrium position in a particular direction — displacement is a vector, so it has a positive or negative value

drag force The resistive force exerted by a fluid on an object moving through it

Doppler effect The change in the frequency and wavelength of waves received from an object moving relative to an observer compared with what would be observed without relative motion

Doppler equation $\frac{\Delta\lambda}{\lambda} \approx \frac{\Delta f}{f} \approx \frac{v}{c}$, where λ is the source wavelength, $\Delta\lambda$ is the change in wavelength recorded by the observer, f is the source frequency, Δf is the change in frequency recorded by the observer, v is the magnitude of the relative velocity between the source and observer, and c is the speed of light through a vacuum ($3.00 \times 10^8\,\text{m s}^{-1}$)

driving frequency The frequency with which the periodic driver force is applied to a system in forced oscillation

ductile Property of a material that has a large plastic region in a stress–strain graph, so can be drawn into wires

eccentricity A measure of the elongation of an ellipse

efficiency The ratio of useful output energy to total input energy, often expressed as a percentage

elastic deformation A reversible change in the shape of an object due to a compressive or tensile force — removal of stress or force will return the object to its original shape and size (no permanent strain)

elastic limit The value of stress or force beyond which elastic deformation becomes plastic deformation, and the material or object will no longer return to its original shape and size when the stress or force is removed

elastic potential energy The energy stored in an object because of its deformation

electric charge A physical property, symbol q or Q, either positive or negative, measured in coulombs, C, or as a relative charge

electric current The rate of flow of charge, symbol I, measured in ampères, A; normally a flow of electrons in metals or a flow of ions in electrolytes

electric field strength The force experienced per unit positive charge at that point

electric potential The work done by an external force per unit positive charge to bring a charge from infinity to a point in an electric field – unit volt or J C^{-1}

electric potential difference The work done by an external force per unit positive charge to move a charge between two points in an electric field

electricity meter A device that measures the electrical energy supplied in kWh to a house from the grid

electrolyte A liquid containing ions that are free to move and so to conduct electricity

electromagnetic spectrum The full range of frequencies of electromagnetic waves, from gamma rays to radio waves

electromagnetic wave Transverse waves with oscillating electric and magnetic field components, such as light and X-rays, that do not need a medium to propagate — they travel at a speed of $3.0 \times 10^8\,\text{m s}^{-1}$ in a vacuum

electromotive force (e.m.f.) Defined as the energy transferred from chemical to electrical energy per unit charge

electron degeneracy pressure A quantum-mechanical pressure created by the electrons in the core of a collapsing star due to the Pauli exclusion principle

electron gun A device that uses a large accelerating potential difference to produce a narrow beam of electrons

electronvolt A derived unit of energy used for subatomic particles and photons, defined as the energy transferred to or from an electron when it passes through a potential difference of 1 volt; 1 eV is equivalent to $1.60 \times 10^{-19}\,\text{J}$

elementary charge The electric charge equivalent to the charge on a proton, $1.60 \times 10^{-19}\,\text{C}$; symbol e

elementary particle A fundamental particle

ellipse An elongated 'circle' with two foci

emission line spectrum A set of specific frequencies of electromagnetic radiation, visible as bright lines in spectroscopy, emitted by excited atoms as their electrons make transitions between higher and lower energy states, losing the corresponding amount of energy in the form of photons as they do so – every element has a characteristic line spectrum

energy The capacity for doing work, measured in joules, J

energy level A discrete (quantised) amount of energy that an electron within an atom is permitted to possess

equation of state of an ideal gas $pV = nRT$, where n is the number of moles of gas

equilibrium A body is in equilibrium when the net force and net moment acting on it are zero

equilibrium position (*waves*) The resting position of waves or particles in an oscillation

escape velocity The minimum velocity at which an object has just enough energy to leave a specified gravitational field

excited (an atom) Containing an electron or electrons that have absorbed energy and been boosted into a higher energy level

expanding Universe The idea that the fabric of space and time is expanding in all directions and that as a result any point, in any part of the Universe, is moving away from every other point in the Universe, and the further the points are apart the faster their relative motion away from each other

exponential decay A constant-ratio process in which a quantity decreases by the same factor in equal time intervals

extension The increase in length of an object when a tensile force is exerted on it

Faraday's law The magnitude of the induced e.m.f. is directly proportional to the rate of change of magnetic flux linkage

fiducial marker A marker for a point used as a fixed basis for measurement or comparison

filament lamp An electrical component containing a narrow filament of wire that transfers electrical energy into heat and light

fission A process in which a large nucleus splits into two smaller nuclei after absorbing a neutron

Fleming's left-hand rule A mnemonic for the direction of the force experienced by a current-carrying wire placed perpendicular to the external magnetic field: on the left hand, the first finger gives the direction of the external magnetic field, the second finger gives the direction of the conventional current, and the thumb gives the direction of motion (force) of the wire

fluid A substance that can flow, including liquids and gases

force A push or pull on an object, measured in newtons, N

force constant A quantity determined by dividing force by extension (or compression) for an object obeying Hooke's law — called constant of proportionality k in Hooke's law, measured in $\mathrm{N\,m^{-1}}$

force–extension graph A graph of force against extension (or compression), with the area under the graph equal to the work done on the material

force–time graph A graph of net force against time, with the area under the graph equal to the impulse

forced oscillation An oscillation in which a periodic driver force is applied to an oscillator

free oscillation The motion of a mechanical system displaced from its equilibrium position and then allowed to oscillate without any external forces

free electron An electron in a metal that is not bound to an atom and is free to move — sometimes called a delocalised electron

free fall The motion of an object accelerating under gravity with no other force acting on it

free-body diagram A diagram that represents the forces acting on a single object

frequency (*oscillations*) The number of complete oscillations per unit time – unit Hertz (Hz)

frequency *(waves)* The number of wavelengths passing a given point per unit time

fundamental frequency The lowest frequency at which an object (e.g., an air column in a pipe or a string fixed at both ends) can vibrate

fundamental mode of vibration A vibration at the fundamental frequency

fundamental particle A particle that has no internal structure and hence cannot be split into smaller particles

fusion A process in which two smaller nuclei join together to form one larger nucleus

galaxy A collection of stars and interstellar dust and gas bound together by their mutual gravitational force

gamma radiation Ionising radiation consisting of high-energy photons, with wavelengths less than about 10^{-13} m, which travel at the speed of light

gamma rays Short-wavelength electromagnetic waves, with wavelengths from 10^{-10} m to 10^{-16} m

gas laws The laws governing the behaviour of ideal gases, like Boyle's law

gas pressure In stars, the pressure of the nuclei in the star's core pushing outwards and counteracting the gravitational force pulling the matter in the star inwards

geostationary satellite A satellite that remains in the same position relative to a spot on the Earth's surface, by orbiting in the direction of the Earth's rotation over the equator with a period of 24 hours

gold-leaf electroscope A device with a metallic stem and a gold leaf that can be used to identify and measure electric charge — a device that was historically used as a voltmeter for measuring large voltages

gradient In a graph, the change in the vertical axis quantity divided by the corresponding change in the horizontal axis quantity

grating equation An equation that can be used to determine accurately the wavelength of monochromatic light sent through a diffraction grating, $d\sin\theta = n\lambda$

grating spacing The separation between adjacent lines or slits in a diffraction grating

gravitational constant, *G* The constant in Newton's law of gravitation $F = -\frac{GMm}{r^2}$, with a value determined from experiment of $6.67 \times 10^{-11}\,\text{N kg}^{-2}\,\text{m}^2$

gravitational field A field created around any object with mass, extending all the way to infinity, but diminishing as the distance from the centre of mass of the object increases

gravitational field lines Lines of force used to map the gravitational field pattern around an object having mass

gravitational field strength, *g* The gravitational force exerted per unit mass at a point within a gravitational field

gravitational potential The work done per unit mass to bring an object from infinity to a point in the gravitational field – unit J kg^{-1}

gravitational potential energy The capacity for doing work as a result of an object's position in a gravitational field

ground state The energy level with the most negative value possible for an electron within an atom – the most stable energy state of an electron

hadron A particle or antiparticle that is affected by the strong nuclear force, and, if charged, by the electromagnetic force – for example, a proton

half-life The average time it takes for half the number of active nuclei in a sample of an isotope to decay

harmonic A whole-number multiple of the fundamental frequency

heavy damping Damping that occurs when the damping forces are large and the period of the oscillations increases slightly with the rapid decrease in amplitude

Hertzsprung–Russell diagram A graph showing the relationship between the luminosity of stars in our galaxy (on the *y*-axis) and their average surface temperature (on the *x*-axis, with temperature increasing from right to left)

homogeneous Uniform in terms of the distribution of matter across the Universe when viewed on a sufficiently large scale

Hooke's law The force applied is directly proportional to the extension of the spring unless the limit of proportionality is exceeded

Hubble constant The gradient of a best-fit line for a plot of recessional speed against distance from Earth of other galaxies

Hubble's law The recessional speed *v* of a galaxy is almost directly proportional to its distance *d* from the Earth

hysteresis loop A loop-shaped plot obtained when, for example, loading and unloading a material produce different deformations

ideal gas A model of a gas including assumptions that simplify the behaviour of real gases

impedance matching (or acoustic matching) The use of two substances with similar acoustic impedance to minimise reflection of ultrasound at the boundary between them

impulse The area under a force–time graph — the product of force and the time for which the force acts

in phase Particles oscillating perfectly in time with each other (reaching their maximum positive displacement at the same time) are in phase

inelastic collision A collision in which kinetic energy is transferred to other forms, e.g. heat

induced fission Nuclear fission occurring when a nucleus becomes unstable on absorbing another particle (such as a neutron)

inflation A phase of astonishing acceleration of the expansion of the Universe thought to have occurred $10^{-35}\,\text{s}$ after the Big Bang

infrared waves Electromagnetic waves, with wavelengths from $10^{-3}\,\text{m}$ to $7 \times 10^{-7}\,\text{m}$

instantaneous speed The speed at the moment it is measured — speed over an infinitesimal interval of time

intensity *(waves)* The radiant power passing through a surface per unit area — unit W m^{-2}

intensity reflection coefficient The ratio of reflected intensity over incident intensity for ultrasound incident at a boundary

interference Superposition of two progressive waves from coherent sources to produce a resultant wave with a displacement equal to the sum of the individual displacements from the two waves

interference pattern A pattern of constructive and destructive interference formed as waves overlap

internal energy The sum of the randomly distributed kinetic and potential energies of the atoms, ions, or molecules within the substance

internal resistance The resistance of a source of e.m.f. (e.g a cell) due to its construction, which causes a loss in energy/voltage as the charge passes through the source, symbol *r*, SI unit ohm, Ω

ion An atom that has either lost or gained electrons and so has a net charge

ionic solution An ionic compound dissolved in a liquid to form an electrolyte

ionising radiation Any form of radiation that can ionise atoms by removing an electron to leave a positive ion

isochronous oscillator An oscillator that has the same period regardless of amplitude

isotherm A line on a pressure–volume graph that connects points at the same temperature

isotopes Nuclei of the same element that have the same atomic number (number of protons) but different nucleon numbers (numbers of neutrons)

isotropic The same in all directions (for example the Universe, appearing the same to any observer regardless of position)

***I–V* characteristic** A description of the relationship between the electric current in a component and the potential difference across it — in most cases this is usually in the form of a simple graph of I against V

kelvin The SI base unit of the absolute (thermodynamic) scale of temperature

Kepler's first law of planetary motion The orbit of a planet is an ellipse with the Sun at one of the two foci

Kepler's second law of planetary motion A line segment connecting a planet to the Sun sweeps out equal areas during equal intervals of time

Kepler's third law of planetary motion The square of the orbital period T of a planet is directly proportional to the cube of its average distance r from the Sun

kilowatt-hour A derived unit of energy, most often associated with paying for electrical energy, symbol kWh (1 kWh = 3.6 MJ). Energy in kWh can be calculated by multiplying the power in kW by the time in hours

kinetic energy The energy associated with an object as a result of its motion

kinetic model A model that describes all substances as made of atoms, ions, or molecules, arranged differently depending on the phase of the substance

kinetic theory of matter *see* kinetic model

Kirchhoff's first law At any point in an electrical circuit, the sum of currents into that point is equal to the sum of currents out of that point, electrical charge is conserved

Kirchhoff's second law In a closed loop of an electrical circuit, the sum of the e.m.f.s is equal the sum of the p.d.s

law of reflection The angle of incidence is equal to the angle of reflection

Lenz's law The direction of the induced e.m.f. or current is always such as to oppose the change producing it

lepton A fundamental particle or antiparticle that is not affected by the strong nuclear force – for example, an electron

light-dependent resistor An electrical component with a resistance that decreases as the light intensity incident on it increases

light damping Damping that occurs when the damping forces are small and the period of the oscillations is almost unchanged

light-emitting diode A type of diode that emits light when it conducts electricity

light-year The distance travelled by light in a vacuum in a time of one year (9.46×10^{15} m)

limit of proportionality The value of stress or force beyond which stress is no longer directly proportional to strain

linear momentum The product of the mass and velocity of a particle, measured in kg m s^{-1} or N s

loading *(electrical circuits)* Connecting a component or a device across the teminals of a source of e.m.f. or across another component

loading curve A force–extension graph

longitudinal wave A wave in which the medium is displaced in the same line as the direction of energy transfer — oscillations of the medium particles are parallel to the direction of the wave travel

lost volts The potential difference across the internal resistor of a source of e.m.f.

luminosity The total radiant power output of a star – symbol L, unit W

magnetic field lines Lines of force drawn to represent a magnetic field pattern

magnetic field patterns Visual representations used in interpreting the direction and strength of magnetic fields

magnetic flux The product of the component of the magnetic flux density perpendicular to a given area and that cross-sectional area: $\phi = BA\cos\theta$

magnetic flux density The strength of a magnetic field – defined by the equation F/IL, where F is the force acting on current-carrying conductor placed at right angles to a magnetic field, I is the current in the conductor and L is the length of the conductor in the magnetic field – symbol B, unit tesla (T)

magnetic flux linkage The product of the number of turns in a coil N and the magnetic flux ϕ

main sequence The main period on an H–R diagram in a star's life, during which it is stable

mass Amount of matter, a base quantity measured in kilograms, kg

mass defect The difference between the mass of a nucleus and the mass of its completely separated constituent nucleons

maximum *(waves)* The point of greatest amplitude in an interference pattern, produced by constructive interference

Maxwell–Boltzmann distribution The distribution of the speeds of particles in a gas

mean drift velocity The average velocity of electrons as they move through a wire, symbol v, unit ms^{-1}

mean square speed The mean of the squared velocities (of all the particles in a gas)
medical tracer A radiopharmaceutical, that is, a compound labelled with a radioisotope that can be traced inside the body using a gamma camera
meson Any hadron comprising a combination of a quark and an anti-quark
microwave background radiation The microwave signal of uniform intensity detected from all directions of the sky, which fits the profile for a black body at a temperature of 2.7 K
microwaves Long-wavelength electromagnetic waves, with wavelengths from 10^{-1} m to 10^{-3} m
minimum *(waves)* The point of least amplitude in an interference pattern, produced by destructive interference
moderator A substance used to slow down the fast neutrons produced in fission reactions so that they can propagate the fission reaction
molar gas constant The constant in the equation of state of an ideal gas – symbol R, 8.31 J K^{-1} mol^{-1}
molar mass The mass of one mole of a substance
mole The amount of substance that contains as many elementary entities as there are atoms in 0.012 kg (12 g) of carbon-12
moment The product of force and perpendicular distance from a pivot or stated point
monochromatic light Light of a single frequency
natural frequency The frequency of a free oscillation
nebula (plural nebulae) A cloud of dust and gas (mainly hydrogen), often many hundreds of times larger than our Solar System
negative *(charge)* One type of electric charge; negatively charged objects attract positively charged ones, and repel other negative charges
negative temperature coefficient (NTC) A relationship in which a variable decreases as temperature increases, for example the resistance of NTC thermistors
neutrino A lepton (a fundamental particle) that carries no charge and may have a tiny mass, less than a millionth the mass of an electron
neutron An electrically neutral particle, a hadron, found in the nucleus of atoms
neutron star The remnant core of a massive star after the star has gone supernova and (if the mass of the core is greater than the Chandrasekhar limit) the core has collapsed under gravity to an extremely high density (similar to that of an atomic nucleus, ~10^{17} kg m^{-3}), as it is almost entirely made up of neutrons
Newton's first law of motion A body will remain at rest or continue to move with constant velocity unless acted upon by a resultant force
Newton's second law of motion The rate of change of momentum of an object is directly proportional to the resultant force and takes place in the direction of the force
Newton's third law of motion When two objects interact, each exerts an equal but opposite force on the other during the interaction
Newton's law of gravitation The force between two point masses is directly proportional to the product of the masses and inversely proportional to the square of the separation between them;

$$F = -\frac{GMm}{r^2}$$

node For a stationary wave, a point where the amplitude is always zero
non-ohmic component A component that does not obey Ohm's law, e.g filamant lamp and diode
normal An imaginary line perpendicular to a surface such as the boundary between one medium and another (e.g., air and glass)
normal contact force The force exerted by a surface on an object, which acts perpendicularly to the surface
nuclear fusion *see* fusion
nucleon A particle in the nucleus of an atom, either a proton or a neutron
nucleon number The total number of protons and neutrons in a nucleus (also called the mass number); symbol A
nucleus The small, positively charged region at the centre of an atom where most of the mass of the atom is concentrated
number density The number of free electrons per cubic metre of a material, symbol n, unit m^{-3}
ohm The derived SI unit of resistance, symbol Ω — defined as the resistance of a component that has a potential difference of 1 V per unit ampere
Ohm's law The potential difference across a conductor is directly proportional to the current in the component as long as its temperaure remains constant
ohmic conductor A conductor that obeys Ohm's law
optical fibre A fibre made of glass designed with a varying refractive index in order to totally internally reflect pulses of visible or infrared light travelling through it
oscillating motion Repetitive motion of an object around its equilibrium position
oscilloscope An instrument that displays an electrical signal as a voltage against time trace on a screen
out of phase Particles that are neither in phase, nor in antiphase, are out of phase

pair production The replacement of a single photon with a particle and a corresponding antiparticle of the same total energy

parallax angle The angle of the apparent shift in the position of a relatively close star against the backdrop of much more distant stars as the Earth makes a qaurter an orbit around the Sun

parallel *(vectors)* In the same line and direction

parallel circuit A type of branching electrical circuit in which there is more than one path for the current — components in parallel have the same potential difference

parent nucleus A nucleus before the occurence of radioactive decay

partially polarised Description of a transverse wave in which there are more oscillations in one particular plane, but the wave is not completely plane polarised — occurs when transverse waves reflect off a surface

parsec The distance at which a radius of one AU subtends an angle of one arcsecond

path difference The difference in the distance travelled by two waves from their source to a specific point

peak The maximum positive amplitude of a transverse wave

perihelion The closest point to the Sun in an orbit

perfectly elastic collision A collision in which no kinetic energy is transferred

period *(oscillations)* The time taken to complete one oscillation

period *(waves)* The time taken for one complete wavelength to pass a given point

phase A phase of matter is its state (solid, liquid, or gas)

phase difference The difference between the displacements of particles along a wave, or the difference between the displacements of particles on different waves, measured in degrees or radians, with each complete cycle or a difference of one wavelength representing 360° or 2π radians

phase difference *(for oscillating motion)* The difference in displacement between two oscillating objects or the displacement of an oscillating object at different times – symbol ϕ

photoelectric effect The emission of photoelectrons from a metal surface when electromagnetic radiation above a threshold frequency is incident on the metal

photoelectric effect equation Einstein's equation relating the energy of a photon, the work function of a metal, and the maximum kinetic energy of any emitted photoelectrons: $hf = \phi + KE_{max}$

photoelectrons Electrons emitted from the surface of a metal by the photoelectric effect

photomultiplier tube An apparatus that converts a photon of visible light into an electrical pulse, for example as part of a gamma camera

photon A quantum of electromagnetic energy — photon energy E is given by $E = hf$, where h is the Planck constant and f is the frequency of the electromagnetic radiation

piezoelectric effect The production of an electromotive force (e.m.f.) by some crystals, such as quartz, when they are compressed, stretched, twisted, or distorted

pivot A point about which a body can rotate

Planck constant Symbol h, an important constant in quantum mechanics, 6.63×10^{-34} J s

planet An object in orbit around a star with a mass large enough for its own gravity to give it a round shape, that undergoes no fusion reactions, and that has cleared its orbit of most other objects

planetary nebula The outer layers of a red giant that have drifted off into space, leaving the hot core behind at the centre as a white dwarf

planetary satellite A body in orbit around a planet – it may be natural (a moon) or artificial

plane polarised Description of a transverse wave in which the oscillations are limited to only one plane

plastic deformation An irreversible change in the shape of an object due to a compressive or tensile force — removal of the stress or force produces permanent deformation

plumb-line A string with a weight used to provide a vertical reference line

point mass A mass with negligible volume

polarisation The phenomenon in which oscillations of a transverse wave are limited to only one plane

polarity The type of charge (positive or negative) or the orientation of a cell relative to a component

polycrystalline graphite Thin layers of graphite with regularly arranged carbon atoms in different orientations

polymeric Description of a material comprising of long-chain molecules, such as rubber, which may show large strains

positive *(charge)* One type of electric charge — positively charged objects attract negatively charged ones, and repel other positive charges

positron The antiparticle of the electron

potential difference (pd) Defined as the energy transferred from electrical energy to other forms (heat, light, etc.) per unit charge.

potential divider An electrical circuit designed to divide the potential difference across two or more components (often two resistors) in order to produce a specific output

potential divider equation An equation relating the output potential difference from a simple potential divider containing a pair of resistors:
$V_{out} = \frac{R_2}{(R_1 + R_2)} \times V_{in}$

potentiometer An electrical component with three terminals and some form of sliding contact that can be adjusted to vary the potential difference between two of the terminals

power The rate of work done, measured in watts, W

prefix A word or letter placed before another one, for example, 5.0 km is 5.0×10^3 m

pressure The force exerted per unit cross-sectional area, measured in pascals, Pa

principle of conservation of energy The total energy of a closed system remains constant — energy cannot be created nor can it be destroyed

principle of conservation of momentum Total momentum of a system remains the same before and after a collision

principle of moments For a body in rotational equilibrium, the sum of the anticlockwise moments about a point is equal to the sum of the clockwise moments about the same point

principle of superposition of waves When two similar types of waves meet at a point the resultant displacement at that point is equal to the sum of the displacements of the individual waves

progressive wave A wave in which the peaks and troughs, or compressions and rarefactions, move through the medium as energy is transferred

projectile An object that is thrown or propelled on the surface of the Earth

proton A positively charged particle, a hadron, found in the nucleus of atoms

proton number The atomic number, that is, the number of protons in a nucleus – symbol Z

protostar A very hot, very dense sphere of condensing dust and gas that is on the way to becoming a star

P-waves Primary waves — longitudinal waves that travel through the Earth from an earthquake

Pythagoras' theorem The square of the length of the hypotenuse of a right-angled triangle equals the sum of the squares of the lengths of the other two sides

quark An elementary particle that can exist in six forms (plus their antiparticles) and joins with other quarks to make up hadrons

quantisation The availability of some quantities, such as energy or charge, only in certain discrete values

quantity A property of an object, substance, or phenomenon that can be measured

quantum mechanics The branch of physics dealing with phenomena on the very small scale, often less than the size of an atom

radial field A symmetrical field that diminishes with distance^2 from its centre, such as the gravitational field around a spherical mass or the electrical field around a spherical charged object

radian The angle subtended by a circular arc with a length equal to the radius of the circle (approximately 57.3°)

radiation pressure Pressure from the photons in the core of a star, which acts outwards to counteract the pressure from the gravitational force pulling the matter in the star inwards

radioactivity The process by which unstable nuclei split, or decay, emitting ionising radiation (alpha particles, beta particles, and gamma rays)

radiopharmaceutical A radioisotope chemically combined with elements that will target particular tissues in order to ensure that the radioisotope reaches the correct organ or tumour for diagnosis or treatment

radio waves Long-wavelength electromagnetic waves, with wavelengths greater than 10^{-1} m

rarefaction *(waves)* A moving region in which the medium is less dense or has less pressure than the surrounding medium

ray A line representing the direction of energy transfer of a wave, perpendicular to the wavefronts

red giant An expanding star at the end of its life, with an inert core in which fusion no longer takes place, but in which fusion of lighter elements continues in the shell around the core

red shift The lengthening of observed wavelength that occurs when a wave source is moving away from the observer – in astronomy, if a galaxy is moving away from the Earth (receding), the absorption lines in its spectrum will be red-shifted

red supergiant A huge star in the last stages of its life before it 'explodes' in a supernova

reflection The change in direction of a wave at a boundary between two different media, so that the wave remains in the original medium

refraction The change in direction of a wave as it changes speed when it passes from one medium to another

refractive index The refractive index of a material $n = \frac{c}{v}$, where c is the speed of light through a vacuum and v is the speed of light through the material

relative charge A simplified measurement of the electric charge of a particle or object, measured as multiples of the elementary charge

resistance A property of a component calculated by dividing the potential difference across it by the current in it, symbol R, unit ohm, Ω

resistivity A property of a material, measured in Ω m, defined as the product of the resistance of a component made of the material and its cross-sectional area divided by its length

resistor An electrical component that obeys Ohm's law, transferring electrical energy to thermal energy

resistor circuit Two or more resistors arranged to provide a specific resistance

resolving a vector Splitting a vector into two component vectors perpendicular to each other

resonance The increase in amplitude of a forced oscillation when the driving frequency matches the natural frequency of the oscillating system

rest mass The mass of an object, such as a particle, when it is stationary

restoring force A force that tries to return a system to its equilibrium position

resultant vector A single vector that has the same effect as two or more vectors added together

right-hand grip rule For a current-carrying wire, the thumb points in the direction of the conventional current, and the direction of the field is given by the direction in which the fingers of the right hand would curl around the wire

root mean square speed The square root of the mean square speed (of all the particles in a gas)

satellite A body orbiting around planet

scalar quantity A quantity with magnitude (size) but no direction

scintillator Part of a gamma camera, often made of sodium iodide, which produces thousands of photons of visible light when struck by a single gamma photon

semiconductor A material with a lower number density than a typical conductor, for example silicon

series An arrangement of electrical components connected end-to-end that means that the current is the same in each component

series circuit A type of electrical circuit where the components are connected end-to-end

SI Système International d'Unités (International System of Units)

simple harmonic motion Oscillating motion for which the acceleration of the object is directly proportional to its displacement and is directed towards some fixed point – characterised by the equation $a = -\omega^2 x$

solar system A planetary system consisting of a star and at least one planet in orbit around it – our own Solar System contains the Sun and all the objects that orbit it

specific heat capacity The energy required per unit mass to change the temperature by 1 K (or 1°C); unit $\mathrm{J\,kg^{-1}\,K^{-1}}$

specific latent heat The energy required to change the phase per unit mass while at constant temperature – symbol L

specific latent heat of fusion The energy required to change unit mass of a substance from solid to liquid while at constant temperature – symbol L_f

specific latent heat of vaporisation The energy required to change unit mass of a substance from liquid to gas while at constant temperature – symbol L_v

spectral line A line in an emission line spectrum or absorption line spectrum at a specific wavelength

spectroscopy A technique in physics in which spectral lines are identified and measured in order to identify elements present within stars

standard form Mathematical notation in which a number is shown with the decimal point placed after the first digit, followed by ×10 raised to an appropriate power

standard model The current theory of particle physics that deals with elementary particles (quarks, electrons, etc.) and their interactions

standing wave A wave that remains in a constant position with no net transfer of energy and is characterised by its nodes and antinodes — also called a stationary wave

stationary wave A wave that remains in a constant position with no net transfer of energy and is characterised by its nodes and antinodes — also called a standing wave

Stefan constant The constant σ in Stefan's law, $L = 4\pi r^2 \sigma T^4$, relating the luminosity L of a star to its surface area $4\pi r^2$ and its absolute surface temperature T: $\sigma = 5.67 \times 10^{-8}\,\mathrm{W\,m^{-2}\,K^{-4}}$

stellar parallax A technique used to determine the distance to stars that are relatively close to the Earth (less than 100 pc) by comparing their apparent positions against distant stars at times 6 months apart

step-down transformer A transformer with fewer turns on the secondary than on the primary coil, and a lower output voltage than input voltage

step-up transformer A transformer with more turns on the secondary than on the primary coil, and a higher output voltage than input voltage

stiffness The ability of an object to resist deformation

stopping distance The total distance travelled from the time when a driver first sees a reason to stop to the time when the vehicle stops, the sum of the thinking distance and the braking distance

strain see 'tensile strain'

stress see 'tensile stress'

strong material A material with a large value for the ultimate tensile strength

strong nuclear force One of the four fundamental forces in nature, acting on hadrons and holding nuclei together

superconductivity A phenomenon in which the resistivity of a material falls to almost zero when the material is cooled below a certain temperature

supernova The implosion of a red supergiant at the end of its life, which leads to subsequent ejection of stellar matter into space, leaving an inert remnant core

superposition *(waves)* Overlap of two waves at a point in space

S-waves Secondary waves: transverse waves that travel through the Earth from an earthquake

target metal A metal with a high melting point used for the anode in an X-ray tube, for example tungsten

tensile deformation A change in the shape of an object due to tensile forces

tensile force Equal and opposite forces acting on a material to stretch it

tensile strain The extension per unit length, a dimensionless quantity

tensile stress The force per unit cross-sectional area, measured in Pa

tension The pulling force exerted by a string, cable, or chain on an object

terminal p.d. The potential difference across an electrical power source — when there is no current this is equal to the e.m.f. of the source, but if there is a current in the source this is equal to the e.m.f. minus the lost volts

terminal velocity The constant speed reached by an object when the drag force (and upthrust) is equal and opposite to the weight of the object

thermal equilibrium A state in which there is no net flow of thermal energy between the objects involved, that is, objects in thermal equilibrium must be at the same temperature

thermal neutron A neutron in a fission reactor with mean kinetic energy similar to the thermal energy of particles in the reactor core – also known as a slow neutron

thermionic emission The emission of electrons (or other charge carriers) from the surface of a heated piece of metal

thermionic emission The emission of electrons from the surface of a hot metal wire

thermistor An electrical component that has a resistance that decreases as the temperature increases (a negative temperature coefficient)

thermodynamic scale of temperature *see* absolute scale of temperature

thinking distance The distance travelled by a vehicle from when the driver first perceives a need to stop to when the brakes are applied

threshold frequency The minimum frequency of the electromagnetic radiation that will cause the emission of an electron from the surface of a particular metal — symbol f_0, measured in Hz

threshold voltage The minimum potential difference at which a diode begins to conduct

time constant The product of capacitance and resistance, *CR*, for a capacitor–resistor circuit – equal to the time taken for the p.d. (or the current or the charge) to decrease to e^{-1} (about 37%) of its initial value when the capacitor discharges through a resistor – symbol τ

time of flight The time taken for an object to complete its motion

timebase The time interval represented by one horizontal square on an oscilloscope screen

torque (of a couple) The product of one of the forces of a couple and the perpendicular distance between the forces

total internal reflection The reflection of all light hitting a boundary between two media back into the original medium when the light is travelling through the medium with the higher refractive index and the incidence angle at the boundary is greater than the critical angle

transverse wave A wave in which the medium is displaced perpendicular to the direction of energy transfer — the oscillations of medium particles are perpendicular to the direction of travel of the wave

triangle of forces Three forces acting at a point in equilbrium, represented by the sides of a triangle

triple point For a given substance, one specific temperature and pressure at which all three phases of that substance can exist in thermodynamic equilibrium

trough The maximum negative amplitude of a transverse wave

turn-ratio equation Equation for a transformer: $\frac{V_s}{V_p} = \frac{n_s}{n_p}$, where output voltage is V_s, input voltage is V_p, n_s is the number of turns on the secondary coil and n_p is the number of turns on the primary coil

ultimate tensile strength The maximum stress that a material can withstand before it breaks

ultraviolet Electromagnetic waves, with wavelengths from 4×10^{-7} m to 10^{-8} m

ultrasound transducer A device used both to generate and to receive ultrasound, which changes electrical energy into sound and sound into electrical energy

uniform gravitational field A gravitational field in which the field lines are parallel and the value for g remains constant

Universe Everything that exists within space and time

unpolarised Description of a transverse wave in which the oscillations occur in many planes

upthrust The upward buoyant force exerted on a body immersed in a fluid

vector quantity A quantity with magnitude (size) and direction

vector triangle A triangle constructed to scale to determine the resultant of two vectors

velocity A vector quantity equal to the rate of change of displacement

velocity selector A device that uses both electric and magnetic fields to select charged particles of specific velocity

visible light Electromagnetic waves, with wavelengths from 4×10^{-7} m to 7×10^{-7} m

volt The derived SI unit of potential difference and electromotive force, symbol V, defined as the energy transferred per unit charge, whether energy is either transferred to or from the charges — 1 V is the p.d. across a component when 1 J of energy is transferred per 1 C passing through the component

voltage See 'potential difference'

voltmeter A device used to measure potential difference — it must be placed in parallel across components and ideally have an infinite resistance

wave equation An equation that relates the frequency f in hertz, the wavelength λ in metres, and the wave speed v in m s^{-1}: $v = f\lambda$

wave profile A graph showing the displacement of the particles in the wave against the distance along the wave

wave source A source of waves, such as light or sound – the object moving relative to an observer of the Doppler effect

wave speed The distance travelled by the wave per unit time

wavefront A line of points in phase with each other in a wave, perpendicular to the direction of energy transfer

wavelength The minimum distance between two points oscillating in phase, for example the distance from one peak to the next or from one compression to the next

wave–particle duality A theory that states that matter has both particle and wave properties and also electromagnetic radiation has wave and particulate (photon) nature

weak nuclear force One of the four fundamental forces in nature, responsible for inducing beta-decay within unstable nuclei

weight The gravitational force on an object, measured in newtons, N

white dwarf A very dense star formed from the core of a red giant, in which no fusion occurs

Wien's displacement law The peak wavelength λ_{max} at which the intensity of radiation from a black body is a maximum is inversely proportional to the absolute temperature T of the black body

work The product of force and the distance moved in the direction of the force, measured in J

work function The minimum energy needed to remove a single electron from the surface of a particular metal; symbol φ, measured in J

X-rays Short-wavelength electromagnetic waves, with wavelengths from 10^{-8} m to 10^{-13} m, which can be used in medical imaging

X-ray tube A piece of equipment that produces X-ray photons by firing electrons from a heated cathode across a large p.d. in an evacuated tube – X-ray photons are produced when the electrons are decelerated by hitting the target metal of the anode

yield point A point on a stress–strain graph beyond which the deformation is no longer entirely elastic

Young modulus The ratio of tensile stress to tensile strain when these quantities are directly proportional to each other, measured in Pa

Answers

2.1

4.5×10^{-8} s

1 a Distance and time, respectively. [2]
b 0.60 m and 0.040 s (or 4.0×10^{-2} s) [2]

2 a 0.000 000 000 1 s (or 1.0×10^{-10} s) [1]
b 0.000 000 000 15 m (or 1.5×10^{-10} m) [1]
c 16 000 000 K (or 1.6×10^{7} K) [1]

3 a 2.0×10^{-10} m [1]
b 4.0×10^{5} m [1]
c 3.5×10^{-5} s [1]
d 2.5×10^{-4} A [1]
e 7.56×10^{-7} s [1]

4 a 534 km [1]
b 12.74 Mm [1]
c 75 µm [1]
d 140 nA [1]

2.2

25°C

−40°F = −40°C

1 base unit for force: base unit for mass × base unit for acceleration
base unit: kg × m s^{-2} [1]
base unit: kg m s^{-2} [1]

2 a base unit for force constant : base unit for force × base unit for extension
base unit: kg m s^{-2} ÷ m [1]
base unit: kg s^{-2} [1]
b base unit for work done: base unit for force × base unit for distance
base unit: kg m s^{-2} × m [1]
base unit: kg $m^2 s^{-2}$ [1]
c base unit for pressure: base unit for force × base unit for area
base unit: kg m s^{-2} ÷ m^2 [1]
base unit: kg $m^{-1} s^{-2}$ [1]

3 In the last three units, the first letter is a prefix for the factor – nano (10^{-9}), milli (10^{-3}), and mega M (10^3). [1]
There is a space in the first unit showing that it is a derived unit, newton metres. [1]
The units are: 1 newton metre (energy or work), 1 nanometre (length), one millinewton (force), and one meganewton (force). [1]

4 base unit for number density: inverse of base unit for volume [1]
base unit: m^{-3} [1]

2.3

1 Both masses should either be in grams or in kilograms.
The answer should be either 0.650 kg or 650 g. [1]

2 Distance and displacement have the same unit, metres (m). [1]
However, distance is a scalar (magnitude only) and displacement is a vector (magnitude and direction). [1]

3 Both time and energy are scalars. [1]
Dividing these two scalar quantities produces a scalar quantity (power). [1]

4 a distance = 12.0 cm [1]
b displacement = 6.0 cm [1]
c average speed = $\frac{\text{distance}}{\text{time}} = \frac{0.120}{20}$ [1]
average speed = 6.0×10^{-2} m s^{-1} [1]

5 Displacement is the direct distance between two points. Therefore, it must always be either equal to or less than the distance. [1]

2.4

1 a resultant velocity = 0.5 m s^{-1} [1]
b resultant velocity = 0.5 + 2.0 = 2.5 m s^{-1} [1]
c resultant velocity = 0.5 − 1.0 = −0.5 m s^{-1} [1]

2 a A correct triangle drawn, with the sides correctly labelled. [1]
The arrows all 'follow' and are in a clockwise direction. [1]
$F^2 = 2.0^2 + 3.0^2$ [1]
$F = \sqrt{13} = 3.6$ N [1]
$\theta = \tan^{-1}\left(\frac{2.0}{3.0}\right) = 33.7...° = 34°$ (2 s.f.) [1]
b A correct triangle drawn, with the sides correctly labelled. [1]
The arrows all 'follow' and are in a clockwise direction. [1]
$F^2 = 13.5^2 + 6.0^2$ [1]
$F = \sqrt{218.25} = 14.77... = 14.8$ N (2 s.f.) [1]
$\theta = \tan^{-1}\left(\frac{6.0}{13.5}\right) = 24°$ (2 s.f.) [1]

3 a $v = 0.90 + 0.30 = 1.2$ m s^{-1} [1]
Direction: due north [1]
b $v = 0.90 - 0.30 = 0.60$ m s^{-1} [1]
Direction: due north [1]
c $v^2 = 0.30^2 + 0.90^2$ [1]
$v = \sqrt{0.90} = 0.949... = 0.95$ m s^{-1} (2 s.f.) [1]
$\theta = \tan^{-1}\left(\frac{0.30}{0.90}\right) = 18°$ (2 s.f.)
Direction: bearing of 18° from the north (18° east of north) [1]

2.5

1 $\theta = 0°$: $F_x = F\cos\theta = 10\cos 0° = 10\,N$ [1]

$\theta = 45°$: $F_x = F\cos\theta = 10\cos 45° = 7.1\,N$ [1]

$\theta = 90°$: $F_x = F\cos\theta = 10\cos 90° = 0\,N$ [1]

The horizontal component of the force decreases as the angle θ is increased. [1]

2 Horizontal: $F_x = F\cos\theta = 1650\cos 35° = 1352\,N$ [1]

Vertical: $F_y = F\sin\theta = 1650\sin 35° = 946\,N$ [1]

3 a Northwards: $F_x = F\cos\theta = 350\cos 40° = 268\,N$ [1]

b Eastwards: $F_y = F\sin\theta = 350\sin 40° = 224.98... = 220\,N$ (2 s.f.) [1]

(Note: this is the same as $350\cos 50°$)

4 Vertical force: $F_x = F\cos\theta = 6.5\cos 20° = 6.1\,kN$ [1]

Horizontal force: $F_y = F\sin\theta = 6.5\sin 20° = 2.2\,kN$ [1]

2.6

1 The forces are equivalent to 1.0 N to the right and 1.0 N downward. [1]

$F^2 = 1.0^2 + 1.0^2$ [1]

$F = 1.4\,N$ [1]

Direction: $\theta = \tan^{-1}\left(\frac{1.0}{1.0}\right) = 45°$ [1]

2 Total force in x direction = $8.0\cos 20° + 8.0\cos 20° = 15.0\,kN$ [1]

Total force in y direction = $8.0\sin 20° - 8.0\sin 20° = 0$ [1]

Resultant force $F = 15.0\,kN$ [1]

The resultant force is in the x direction (from left to right). [1]

3 Total force in x direction = $12.5 + 11.8\cos 25° + 8.7\cos 35° = 30.3\,kN$ [3]

Total force in y direction = $11.8\sin 25° + 12.5\sin 0° - 8.7\sin 35° \approx 0\,kN$ [1]

Resultant force $F = 30.3\,kN$ [1]

The resultant force is in the x direction (from left to right). [1]

4 $F^2 = 4.0^2 + 3.0^2 - 2 \times 4.0 \times 3.0 \times \cos 40°$ (cosine rule) [1]

$F = 2.57...$ [1]

$\frac{3.0}{\sin\theta} = \frac{2.57...}{\sin 40}$ (sine rule) [1]

$\theta = 48.6... = 49°$ (2 s.f.) [1]

3.1

1 a $v = \frac{x}{t} = \frac{180}{9.0}$ [1]

$v = 20\,m\,s^{-1}$ [1]

b $v = \frac{x}{t} = \frac{2000}{6.5 \times 60}$ [1]

$v = 5.12... = 5.1\,m\,s^{-1}$ (2 s.f.) [1]

2 $v = \frac{x}{t} = \frac{19.2}{24 \times 3600}$ [1]

$v = 2.22... \times 10^{-4} = 2.2 \times 10^{-4}\,m\,s^{-1}$ (2 s.f.) [1]

3 $x = vt = 31 \times 19$ [1]

$x = 589 = 590\,m$ (2 s.f.) [1]

4 $t = \frac{x}{v} = \frac{12000 \times 10^3}{240}$ [1]

$t = 5.0 \times 10^4\,s$ [1]

$t = \frac{5.0 \times 10^4}{3600} = 13.8... = 14\,h$ (2 s.f.) [1]

5 a $x = (25 \times 2.0 \times 60) + 800$ [1]

$x = 3800\,m$ [1]

b $v = \frac{x}{t} = \frac{3800}{120 + 50}$ [1]

$v = 22.35... = 22\,m\,s^{-1}$ (2 s.f.) [1]

6 v = gradient [1]

$v = \frac{\Delta x}{\Delta t} = \frac{1600 - 400}{90 - 60}$ [1]

$v = 40\,m\,s^{-1}$ [1]

3.2

1 From $t = 0$ to $t = 4.0\,s$: constant velocity (moving away). [1]

From $t = 4.0\,s$ to $t = 6.0\,s$: constant velocity (coming back). [1]

The magnitude of the velocity between $t = 4.0\,s$ to $t = 6.0\,s$ is greater than the magnitude of the velocity between $t = 0$ to $t = 4.0\,s$. [1]

2 v = gradient [1]

velocity v at $t = 2.0\,s$: $v = \frac{\Delta s}{\Delta t} = \frac{0.40}{4.0}$ [1]

$v = 0.10\,m\,s^{-1}$ [1]

velocity v at $t = 5.0\,s$: $v = \frac{\Delta s}{\Delta t} = \frac{0 - 0.40}{6.0 - 4.0}$ [1]

$v = -0.20\,m\,s^{-1}$ (in opposite direction from earlier velocity) [1]

3 total distance x travelled = $0.40 \times 2 = 0.80\,m$ and total time $t = 6.0\,s$ [1]

$v = \frac{x}{t} = \frac{0.80}{6.0}$ [1]

$v = 0.133... \approx 0.13\,m\,s^{-1}$ (2 s.f.) [1]

4 a $v = \frac{x}{t} = \frac{2\pi r}{t}$ [1]

$v = \frac{2\pi \times 0.80}{8.0}$ [1]

$v = 0.63\,m\,s^{-1}$ (2 s.f.) [1]

b change in displacement = $\sqrt{0.80^2 + 0.80^2} = 1.131...\,m$ [1]

$v = \frac{\Delta s}{\Delta t} = \frac{1.131...}{\frac{1}{4} \times 8.0}$ [1]

$v = 0.57\,m\,s^{-1}$ (2 s.f.) [1]

3.3

1 $a = \frac{\Delta v}{\Delta t} = \frac{8.0 - 0}{12}$ [1]

$a = 0.67\,m\,s^{-2}$ (2 s.f.) [1]

2 $a = \frac{\Delta v}{\Delta t} = \frac{10 - 40}{60}$ [1]

$a = -0.50\,m\,s^{-2}$ (magnitude of $0.50\,m\,s^{-2}$) [1]

3 From $t = 0$ to $t = 1.0$ s: constant acceleration. [1]

From $t = 1.0$ s to $t = 3.0$ s: constant acceleration of smaller magnitude. [1]

4 a Maximum acceleration is when gradient of v–t is maximum; this occurs at $t < 1.0$ s. [1]

$a = \frac{v}{t} = \frac{2.0-0}{1.0}$ [1]

$a = 2.0\,\text{m s}^{-2}$ [1]

b Graph showing constant acceleration between $t = 0$ and $t = 1.0$ s and a different constant acceleration between $t = 1.0$ s and $t = 3.0$ s. [1]

Constant values of accelerations $2.0\,\text{m s}^{-2}$ and $0.5\,\text{m s}^{-2}$ marked on the graph. [1]

5 a $a = \frac{v}{t} = \frac{7.5-0}{0.75}$ [1]

$a = 10\,\text{m s}^{-2}$ [1]

b a = gradient [1]

$a = \frac{v}{t} = -\frac{0-10}{1.50-0.25}$ [1]

$a = -8.0\,\text{m s}^{-2}$ (The negative sign implies deceleration.) [1]

3.4

1 From $t = 0$ to $t = 75$ s: constant acceleration. [1]

From $t > 75$ s: constant velocity (of $15\,\text{m s}^{-1}$). [1]

2 a (displacement or distance = area under the graph)

distance = $\left(\frac{1}{2} \times 75 \times 15\right) + (15 \times 25)$ (area of triangle and rectangle) [2]

distance = 937.5 = 940 m (2 s.f.) [1]

b $v = \frac{s}{t} = \frac{937.5}{100}$ [1]

$v = 9.4\,\text{m s}^{-1}$ (2 s.f.) [1]

3 a (displacement or distance = area of triangle + area of rectangle + area of trapezium)

distance = $\left(\frac{1}{2} \times 10 \times 4.0\right) + (10 \times 5.0) + \left[\frac{1}{2}(10 + 5.0) \times 4.0\right]$ [3]

distance = 100 m [1]

b $v = \frac{s}{t} = \frac{100}{13.0}$ [1]

$v = 7.69... = 7.7\,\text{m s}^{-1}$ (2 s.f.) [1]

4 a Constant deceleration (allow 'acceleration'). [1]

Stops momentarily at $t = 4.0$ s. [1]

After 4.0 s, it starts to return. [1]

b total distance = area enclosed by line and t-axis [1]

total distance = $\left(\frac{1}{2} \times 4.0 \times 4.0\right) \times 2 = 16$ m [1]

c average speed = $\frac{\text{distance}}{\text{time}} = \frac{16}{8.0}$ [1]

average speed = $2.0\,\text{m s}^{-1}$ [1]

average velocity = $\frac{\text{change in displacement}}{\text{time}}$; change in displacement = 0 [1]

Therefore, average velocity = 0 [1]

3.5

Look at the derivation of $s = ut + \frac{1}{2}at^2$ and follow a similar process.

1 $s = ?$, $u = 13.4\,\text{m s}^{-1}$, $v = 22.3\,\text{m s}^{-1}$, $t = 8.7$ s;

$s = \frac{1}{2}(u + v)t$ [1]

$s = \frac{1}{2} \times (13.4 + 22.3) \times 8.7$ [1]

$s = 155$ m (3 s.f.) [1]

2 $a = ?$, $s = 200$ m, $v = 4.2\,\text{m s}^{-1}$, $u = 3.2\,\text{m s}^{-1}$;

$v^2 = u^2 + 2as$ [1]

$a = \frac{v^2 - u^2}{2s} = \frac{4.2^2 - 3.2^2}{2 \times 200} = 0.019\,\text{m s}^{-2}$ (2 s.f.) [1]

3 $s = 100$ m, $a = ?$, $t = 9.58$ s; [1]

$a = \frac{2s}{t^2} = \frac{2 \times 100}{9.58^2}$ [1]

$a = 2.17... = 2.2...\,\text{m s}^{-2}$ (2 s.f.) [1]

4 a $s = 400$ m, $u = 0$, $a = ?$, $t = 4.6$ s; [1]

$a = \frac{2s}{t^2} = \frac{2 \times 400}{4.6^2}$ [1]

$a = 37.8... = 38\,\text{m s}^{-2}$ (2 s.f.) [1]

b $u = 0$, $a = 37.8...\,\text{m s}^{-2}$, $t = 4.6$ s; $v = u + at$ [1]

$v = 0 + (37.8... \times 4.6)$ [1]

$v = 173.9... = 170\,\text{m s}^{-1}$ (2 s.f.) [1]

5 $s = ?$, $u = 0$, $a = 9.81\,\text{m s}^{-2}$, $t_1 = 3.0$ s and $t_2 = 5.0$ s [1]

distance = $\left[\frac{1}{2} \times 9.81 \times 5.0^2\right] - \left[\frac{1}{2} \times 9.81 \times 3.0^2\right]$ [1]

distance = 78.48 = 78 m (2 s.f.) [1]

6 $s = 30$ m, $u = 28\,\text{m s}^{-1}$, $v = 0$, $a = ?$; $v^2 = u^2 + 2as$ [1]

$a = \frac{0 - 28^2}{2 \times 30}$ [1]

$a = -13.06... = -13\,\text{m s}^{-2}$ (magnitude is $13\,\text{m s}^{-2}$ to 2.s.f) [1]

3.6

1 stopping distance = thinking distance + braking distance [1]

stopping distance = (22 × 1.5) + 38 [1]

stopping distance = 71 m [1]

2 thinking distance = speed of car × reaction time [1]

Assuming the reaction time is constant, thinking distance ∝ speed of car. [1]

3 a braking distance $s = 75$ m (from Table 1) [1]

$s = 75$ m, $u = 31.1\,\text{m s}^{-1}$, $v = 0$, $a = ?$; $v^2 = u^2 + 2as$ [1]

$a = \frac{0 - 31.1^2}{2 \times 75}$ [1]

$a = -6.448... = -6.4\,\text{m s}^{-2}$ (magnitude is $6.4\,\text{m s}^{-2}$ to 2 s.f.) [1]

b $u = 31.1\,\text{m s}^{-1}$, $v = 0$, $a = -6.448\,\text{m s}^{-2}$, $t = ?$;

$v = u + at$ [1]

$t = \dfrac{0-31.1}{-6.448}$ [1]

$t = 4.8\,\text{s}$ (2 s.f.) [1]

4 thinking distance = area under graph up to $0.50\,\text{s} = 20 \times 0.50 = 10\,\text{m}$ [1]

braking distance = area under graph from

$t = 0.50\,\text{s}$ to $3.5\,\text{s} = \dfrac{1}{2} \times 3.0 \times 20 = 30\,\text{m}$ [1]

stopping distance = 10 + 30 = 40 m [1]

5 $v^2 = u^2 + 2as$ [1]

$v = 0$, therefore, $0 = u^2 + 2as$ or $s = \dfrac{-u^2}{2a}$ [1]

For a constant deceleration of magnitude a, $s \propto u^2$. [1]

3.7

1 Gradient calculated using a large triangle.
gradient = $4.89\,\text{m s}^{-2}$ for the results;
allow $\pm 0.03\,\text{m s}^{-2}$

2 $g = 2 \times \text{gradient} = 2 \times 4.89$
$g = 9.78\,\text{m s}^{-2}$

3 % difference = $\dfrac{9.78-9.81}{9.81} \times 100$

% difference = (−)0.31%

1 The acceleration of free fall is the same for both and equal to g ($9.81\,\text{m s}^{-2}$). [1]

Assumption: There is negligible air resistance acting on the falling objects. [1]

2 $s = H = ?$, $u = 0$, $a = 9.81\,\text{m s}^{-2}$, $t = 2.3$ s; $s = ut + \dfrac{1}{2}at^2$ [1]

$H = \dfrac{1}{2} \times 9.81 \times 2.3^2$ [1]

$H = 25.9\ldots = 26\,\text{m}$ (2 s.f.) [1]

3 Object dropped from a given height. [1]

Time of fall measured and measurements repeated. [1]

Explanation of how g is determined using $s = \dfrac{1}{2}gt^2$ [1]

Correct account of how experiment is made precise, e.g: Drop a heavy object from a larger distance, so percentage uncertainty in the distance is smaller. [1]

4 a $s = 9.5\,\text{m}$, $u = 0\,\text{m s}^{-1}$, $t = 1.5\,\text{s}$, $a = ?$ [1]

$a = \dfrac{2s}{t^2} = \dfrac{2\times 9.5}{1.5^2}$ [1]

$a = g = 8.44\ldots \approx 8.4\,\text{m s}^{-2}$ (2 s.f.) [1]

b The presence of drag (air resistance) would give a lower experimental value for the acceleration of free fall than the true value. [1]

5 a $s = 0.125\,\text{m}$ (± 0.001 m), $u = 0\,\text{m s}^{-1}$,

$t = 4 \times 0.04 = 0.16\,\text{s}$, $a = ?$ [1]

$a = \dfrac{2s}{t^2} = \dfrac{2\times 0.125}{0.16^2}$ [1]

$a = g = 9.77\ldots = 9.8\,\text{m s}^{-2}$ (2 s.f.) [1]

b $s = 0.071\,\text{m}$ (± 0.001 m), $u = 0\,\text{m s}^{-1}$,

$t = 3 \times 0.04 = 0.12$ s, $a = ?$ [1]

$a = \dfrac{2s}{t^2} = \dfrac{2\times 0.071}{0.12^2}$ [1]

$a = g = 9.86\ldots\,\text{m s}^{-2}$ [1]

average acceleration = $\dfrac{(9.77 + 9.86\ldots)}{2}$ = $9.8\,\text{m s}^{-2}$ (2 s.f.) [1]

3.8

1 $v = \sqrt{7.0^2 + 9.0^2}$ [1]

$v = 11\,\text{m s}^{-1}$ (2 s.f.) [1]

2 a Vertically:

$s = 29\,\text{m}$, $u = 0$, $a = 9.81\,\text{m s}^{-2}$, $t = ?$; [1]

$t = \sqrt{\dfrac{2s}{a}} = \sqrt{\dfrac{2 \times 29}{9.81}}$ [1]

$t = 2.4\,\text{s}$ (2 s.f.) [1]

b Horizontally: [1]

distance = $vt = 320 \times 2.4315$ [1]

distance = 780 m (2 s.f.) [1]

c vertical velocity = $9.81 \times 2.4315 = 23.8\ldots\,\text{m s}^{-1}$ [1]

$v^2 = 320^2 + 23.8\ldots^2$ [1]

$v = 321\,\text{m s}^{-1}$ (3 s.f.) [1]

3 a Vertically:

$s = ?$, $u = 22.0\sin 35°$, $v = 0$, $a = -9.81\,\text{m s}^{-2}$;

$v^2 = u^2 + 2as$ [2]

$s = \dfrac{0-(22\sin 35)^2}{-2\times 9.81}$ [1]

$s = 8.1\,\text{m}$ (2 s.f.) [1]

b Vertically:

$v = -22.0\sin 35°$, $u = 22.0\sin 35°$, $a = -9.81\,\text{m s}^{-2}$, $t = ?$; $v = u + at$ [1]

$t = \dfrac{-22.0\sin 35 - 22.0\sin 35}{-9.81}$ [1]

$t = 2.57\ldots\,\text{s}$ [1]

Horizontally:

distance = $vt = 22\cos 35° \times 2.57\ldots$ [1]

distance = 46 (2 s.f.) [1]

4 v–t graph shows straight line of negative gradient. [1]

The initial vertical velocity is $+13\,\text{m s}^{-1}$ ($12.6\,\text{m s}^{-1}$) at $t = 0$. [1]

The final velocity is $-13\,\text{m s}^{-1}$ at about 2.6 s. [1]

4.1

1 Graph showing $m = m_0$ when v is much smaller than c. The curve is asymptotic to $v = c$ line.

2 a $m = \dfrac{m_0}{\sqrt{1-(v/c)^2}} = \dfrac{9.1\times10^{-31}}{\sqrt{1-0.10^2}}$

$m = 9.146 \times 10^{-31} \approx 9.1 \times 10^{-31}$ kg

b $m = \dfrac{m_0}{\sqrt{1-(v/c)^2}} = \dfrac{9.1\times10^{-31}}{\sqrt{1-0.999^2}}$

$m = 2.04 \times 10^{-29}$ kg (an increase in the mass by a factor of about 22)

3 The mass would be infinite when $v = c$. It would require an infinite force or an infinite amount of energy to move the electron.

1 a $F = ma;\ a = \dfrac{F}{m} = \dfrac{500}{1200}$ [1]

$a = 0.42\,\text{m s}^{-2}$ (2 s.f.) [1]

b $v = u + at = 0 + (0.417 \times 6.0)$ [1]

$v = 2.5\,\text{m s}^{-1}$ (2 s.f.) [1]

2 $W = mg;\ m = \dfrac{W}{g} = \dfrac{1.1}{9.81}$ [1]

$m = 0.112...$ kg, therefore $m = 110$ g (2 s.f.) [1]

3 $F = ma;\ a = \dfrac{F}{m} = \dfrac{5800}{0.046}$ [1]

$a = 1.3 \times 10^5\,\text{m s}^{-2}$ (2 s.f.) [1]

4 $a = \dfrac{(v-u)}{t} = \dfrac{28}{9.6}$ [1]

$a = 2.92\,\text{m s}^{-2}$ (2 s.f.) [1]

5 net force $= 10^{-16}\sqrt{2.0^2 + 1.5^2}$ [1]

net force $= 2.5 \times 10^{-16}$ N [1]

$a = \dfrac{F}{m} = \dfrac{2.5 \times 10^{-16}}{1.7 \times 10^{-27}}$ [1]

$a = 1.5 \times 10^{11}\,\text{m s}^{-2}$ (2 s.f.) [1]

direction: $\theta = \tan^{-1}\left(\dfrac{2.0}{1.5}\right) = 53°$ to the 1.5×10^{-16} N force. [2]

6 $v^2 = u^2 + 2as;\ a = \dfrac{(0 - 420^2)}{2 \times 0.098}$ [1]

$a = 9.0 \times 10^5\,\text{m s}^{-2}$ [1]

$F = ma = 0.0080 \times 9.0 \times 10^5$, $F = 7.2 \times 10^3$ N [1]

4.2

1 The centre of mass will be at the 50 cm mark (the middle) only if the ruler is uniform in both shape and density of the material. This may not have been the case [1]

2 a [1]

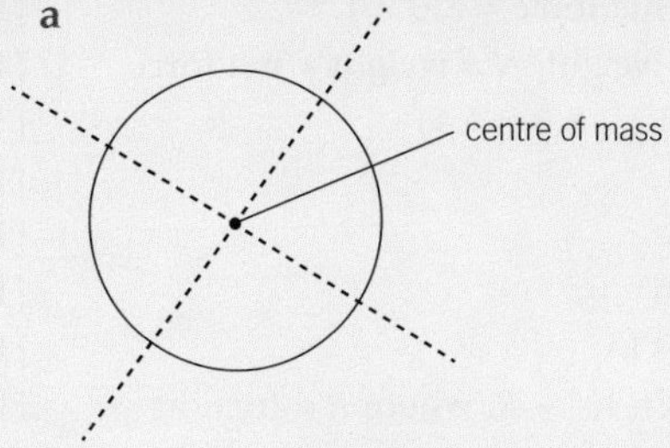

b [1]

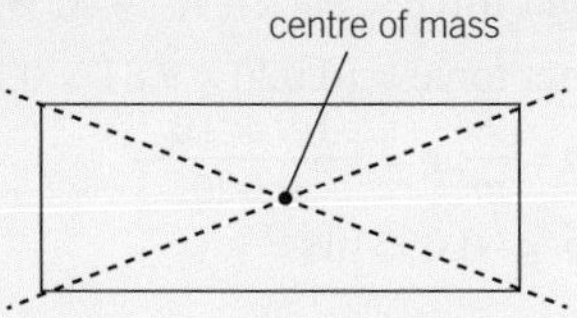

c [1]

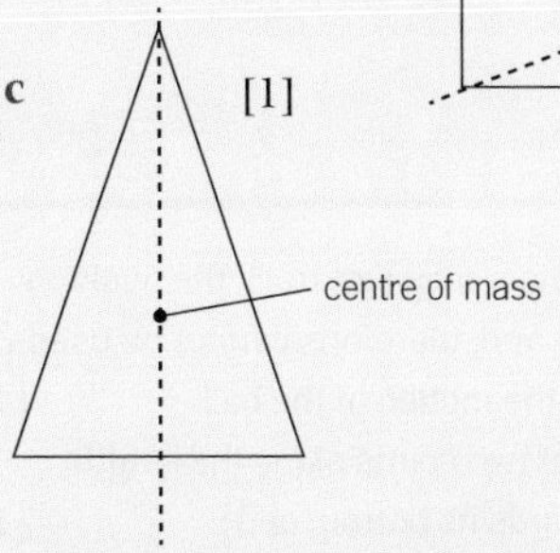

3 By trial and error, horizontally balance the card on the edge of the ruler. [1]

Mark points on the card and then draw a straight line to show the line along which the card was balanced. The centre of mass will lie on this line. [1]

Repeat the procedure above for a different orientation of the card and draw another straight line. [1]

The centre of mass is located at the point of intersection of these two lines. [1]

4 The weight of the ball is due to the plastic wall. [1]

The ball is a symmetrical object so the centre of mass will be in the centre. [1]

5 Push the object with the point of a pencil (or equivalent) and by trial and error determine the point on the object where the object does not rotate when pushed. [1]

The centre of mass will lie along the line of action of the force applied by the pencil when there is no rotation. [1]

Repeat this procedure from a different orientation. [1]

The centre of mass is located at the point of intersection of these two (imaginary) lines of action. [1]

4.3

1 a [1]

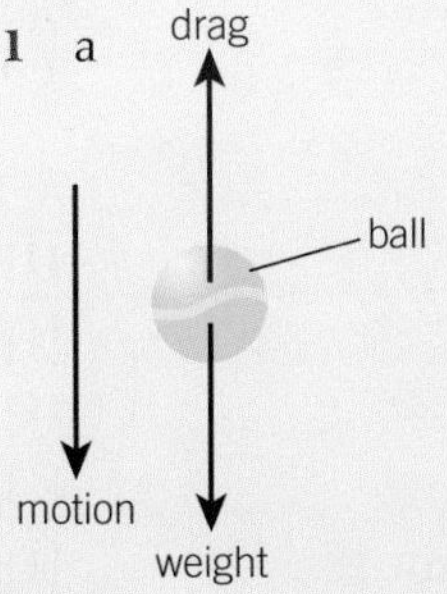

b [1]

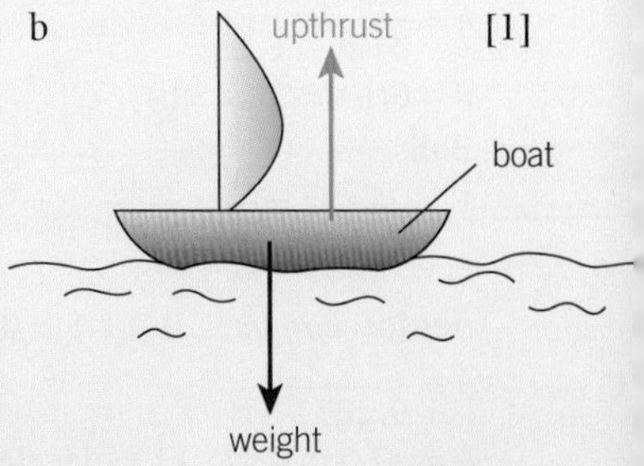

2 net force = ma; net force = 8.0×1.5 [1]
net force = N – weight; N = weight + net force [1]
$N = (8.0 \times 9.81) + (8.0 \times 1.5)$ [1]
$N = 90\,\text{N}$ (2 s.f.) [1]

3 $a = g\sin 30°$ [1]
$a = 9.81 \times \sin 30°$ [1]
$a = 4.9\,\text{m}\,\text{s}^{-2}$ (2 s.f.) [1]

4 net force = $mg \sin 30° - R$, where R = friction [1]
net force = $(0.020 \times 9.81 \times 0.5) - 0.10 = -1.9 \times 10^{-3}$ [1]
$a = \frac{F}{m}$; $a = \frac{-1.9 \times 10^{-3}}{0.020}$ [1]
$a = -0.095\,\text{m}\,\text{s}^{-2}$ [1]

4.4

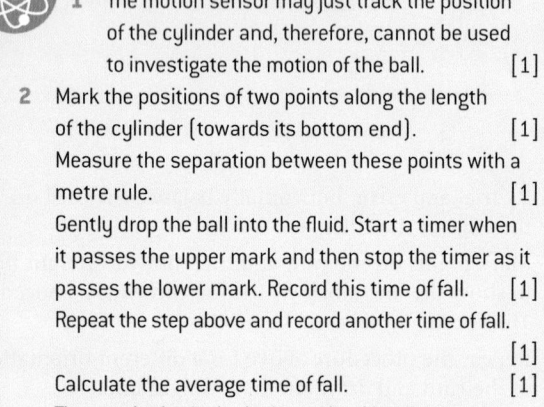

1 The motion sensor may just track the position of the cylinder and, therefore, cannot be used to investigate the motion of the ball. [1]

2 Mark the positions of two points along the length of the cylinder (towards its bottom end). [1]
Measure the separation between these points with a metre rule. [1]
Gently drop the ball into the fluid. Start a timer when it passes the upper mark and then stop the timer as it passes the lower mark. Record this time of fall. [1]
Repeat the step above and record another time of fall. [1]
Calculate the average time of fall. [1]
The terminal velocity is determined by dividing the distance between the two marks by the average time of fall. [1]

1 Change her area by changing shape (e.g., extending arms or opening a parachute). [1]

2 The weight is equal to the drag force at terminal velocity. [1]
drag = $mg = 0.120 \times 9.81$ [1]
drag = $1.2\,\text{N}$ (2 s.f.) [1]

3 drag $\propto$ speed^2 [1]
The drag will increase by a factor of $3^2 = 9$. [1]
drag = $1.0 \times 9 = 9.0\,\text{kN}$ [1]

4 Skydiver:
$W = mg$; $m = \frac{W}{g} = \frac{800}{9.81}$ [1]
$m = 81.5\ldots\,\text{kg}$ [1]
net force = ma; $800 - 300 = 81.5\ldots a$ [1]
$a = \frac{500}{81.5\ldots}$ [1]
$a = 6.1\,\text{m}\,\text{s}^{-2}$ (2 s.f.)
Ball:
$W = mg$; $m = \frac{W}{g} = \frac{2.0}{9.81}$ [1]
$m = 0.203\ldots\,\text{kg}$ [1]
net force = ma; $2.0 - 2.5 = 0.203\ldots a$ [1]
$a = \frac{-0.5}{0.203\ldots}$ [1]
$a = -2.5\,\text{m}\,\text{s}^{-2}$ (2 s.f.) (deceleration) [1]

5 a $D = 0.20 \times 1.5^2 = 0.45\,\text{N}$ [1]
$W = mg = 0.30 \times 9.81 = 2.943\,\text{N}$ [1]
net force = ma; $2.943 - 0.45 = 0.30a$ [1]
$a = \frac{2.493}{0.30}$ [1]
$a = 8.3\,\text{m}\,\text{s}^{-2}$ (2 s.f.) [1]

b At terminal velocity, drag is equal to weight. Therefore, $D = 2.94\,\text{N}$. [1]
$D = 2.94 = 0.20v^2$ [1]
$v = \sqrt{\frac{2.94}{0.20}}$ [1]
$v = 3.8\,\text{m}\,\text{s}^{-1}$ (2 s.f.) [1]

4.5

1 2.0 N force: moment = $Fx = 2.0 \times 0.10 = 0.2\,\text{N}\,\text{m}$ (anticlockwise) [1]
4.0 N force: moment = $Fx = 4.0 \times 0 = 0$ [1]
6.0 N force: moment = $Fx = 6.0 \times 0.18 = 1.08\,\text{N}\,\text{m}$ (clockwise) [1]

2 a clockwise moment = $(18 \times 0.150) + (35 \times 0.360)$ [2]
clockwise moment = $15\,\text{N}\,\text{m}$ (2 s.f.) [1]

b sum of clockwise moments = sum of anticlockwise moments [1]
$15.3 = 0.032 \times F$ [1]
$F = 480\,\text{N}$ (2 s.f.) [1]

3 The line of action of the weight must fall beyond the base. [1]
$\tan\theta = \frac{1.5}{5.0}$ [1]
$\theta = 17°$ (2 s.f.) [1]

4 sum of clockwise moments = sum of anticlockwise moments [1]
$(18 \times 0.40) + (20 \times 0.60) = F\sin 60° \times 0.75$ [3]
$F = \frac{19.2}{(0.75 \sin 60°)}$ [1]
$F = 30\,\text{N}$ (2 s.f.) [1]

4.6

1 It will move forward. [1]
It will also spin about its centre (of mass) as it moves. [1]

2 force ≈ 2 N (allow a force in the range 0.5–10 N) [1]
torque = $Fd = 2 \times 0.04$ [1]
torque = $0.08\,\text{N}\,\text{m}$ [1]

3 a It will move to the right. [1]
It will also spin in a clockwise direction. [1]

b It will spin in an anticlockwise direction. [1]
This is because there is a net couple in the anti-clockwise direction. [1]

4 a moment = $(F \times [d + x]) - F \times x$ [1]
moment = $Fd + Fx - Fx = Fd$ [1]

b The moment about **A** is the same as the torque of the couple. [1]

4.7

1 The net force is zero – the resultant of the three forces is zero. [1]

2 a A correct triangle drawn, with the sides correctly labelled. [1]

The arrows all 'follow' and are in a clockwise direction. [1]

b $T^2 = 5.0^2 + 12^2$ [1]

$T = \sqrt{169} = 13\,\text{N}$ [1]

c The resultant in any direction must be zero. [1]

The resultant of the 5.0 N and 12 N forces is equal to T but in the opposite direction. [1]

3 $\sin 60° = \dfrac{T_1}{3.8 \times 10^3}$ [1]

$T_1 = 3.8 \times 10^3 \times \sin 60° = 3.3 \times 10^3\,\text{N}$ (2 s.f.) [1]

$\sin 30° = \dfrac{T_2}{3.8 \times 10^3}$ [1]

$T_2 = 3.8 \times 10^3 \times \sin 30° = 1.9 \times 10^3\,\text{N}$ (2 s.f.) [1]

4 a $\dfrac{T_1}{\sin 40} = \dfrac{0.500 \times 9.81}{\sin 110}$ (sine rule) [1]

$T_1 = 3.4\,\text{N}$ (2 s.f.) [1]

$\dfrac{T_2}{\sin 30} = \dfrac{0.500 \times 9.81}{\sin 110}$ (sine rule) [1]

$T_2 = 2.6\,\text{N}$ (2 s.f.) [1]

b If the strings are horizontal, with the angles zero, there can be no upward vertical component of forces from either T_1 or T_2 to balance the downward weight mg. [1]

It would be impossible for the slotted masses to be in equilibrium, therefore, the strings cannot be horizontal. [1]

4.8

1 $m = \rho V = 1.3 \times 140$ [1]

$m = 180\,\text{kg}$ (2 s.f.) [1]

2 $p = \dfrac{F}{A} = \dfrac{8.0}{\pi \times (3.75 \times 10^{-3})^2}$ [1]

$p = 2.4 \times 10^5$ Pa (2 s.f.) [1]

3 $\rho = \dfrac{m}{V} = \dfrac{0.080}{(85-70)\times 10^{-6}}$ ($1\,\text{cm}^3 = 10^{-6}\,\text{m}^2$) [1]

$\rho = 5.3 \times 10^3\,\text{kg}\,\text{m}^{-3}$ (2 s.f.) [1]

4 area $A = (0.15 \times 0.075)\,\text{m}^2$ [1]

$F = pA = 1.0 \times 10^5 \times (0.15 \times 0.075)$ [1]

$F = 1.1 \times 10^3\,\text{N}$ (2 s.f.) [1]

5 volume $V = \frac{4}{3}\pi r^3 = \frac{4}{3} \times \pi \times (12 \times 10^3)^3$ [1]

$\rho = \dfrac{m}{V} = \dfrac{3.0 \times 10^{24}}{\frac{4}{3} \times \pi \times (12 \times 10^3)^3}$ [1]

$\rho = 4.1 \times 10^{11}\,\text{kg}\,\text{m}^{-3}$ (2 s.f.) [1]

6 mass of gold $= 0.582 \times 3.34 \times 10^{-6} \times 1.93 \times 10^4 = 3.75\ldots \times 10^{-2}\,\text{kg}$ [1]

mass of copper $= 0.418 \times 3.34 \times 10^{-6} \times 8.96 \times 10^3 = 1.25\ldots \times 10^{-2}\,\text{kg}$ [1]

$\rho = \dfrac{(3.75\ldots + 1.25\ldots) \times 10^{-2}}{3.34 \times 10^{-6}}$ [1]

$\rho = 1.50 \times 10^4\,\text{kg}\,\text{m}^{-3}$ (3 s.f.) [1]

Assumption: gold and copper do not react together and their atoms do not become more or less densely packed when they are mixed. [1]

4.9

+

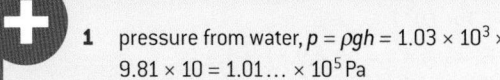

1 pressure from water, $p = \rho g h = 1.03 \times 10^3 \times 9.81 \times 10 = 1.01\ldots \times 10^5$ Pa

atmospheric pressure $= 1.01\ldots \times 10^5$ Pa

Therefore, total pressure is 2.02×10^5 Pa (2 s.f.)

2 force $F = pA = (\rho g h + \text{atmospheric pressure})A$

$F = [(1.03 \times 10^3 \times 9.81 \times 330) + 1.01\ldots \times 10^5] \times 1.2\ldots \times 10^{-2}$

$F = 4.12 \times 10^4$ (2 s.f.)

3 Pressure in the lungs at a depth of 10 m is twice the (atmospheric) pressure at the water surface. If the diver does not exhale, this air will expand without a way to escape because of the pressure difference of about 10^5 Pa between the air in the lungs of the diver and the surrounding air.

1 $p = \rho g h = 1.35 \times 10^4 \times 9.81 \times 0.765$ [1]

$p = 1.01 \times 10^5$ Pa (2 s.f.) (the same as atmospheric pressure on Earth's surface) [1]

2 $p = \rho g h = 1.0 \times 10^3 \times 9.81 \times 610$ [1]

$p = 6 \times 10^6$ Pa (2 s.f.) [1]

3 Immediately after release, the ball experiences two forces – upthrust and weight. [1]

Upthrust is greater than the weight of the ball, so the ball accelerates upwards. [1]

As it travels vertically upwards it also experiences drag due to water. [1]

As it pops out of the water, the only force is weight, so it decelerates and falls back onto the surface of the water. [1]

Eventually it remains still on the surface of the water, with upthrust equal to weight. [1]

4 a upthrust $= 1.54 - 1.34 = 0.20\,\text{N}$ [1]

b upthrust = weight of water displaced; mass of water $= \dfrac{0.20}{9.81} = 2.03\ldots \times 10^{-2}\,\text{kg}$ [1]

(volume of water displaced = volume V of bar)

$V = \dfrac{2.03\ldots \times 10^{-2}}{1.0 \times 10^3} = 2.03\ldots \times 10^{-5}\,\text{m}^3$ [1]

$\rho = \dfrac{m}{V} = \dfrac{(1.54 / 9.81)}{2.03\ldots \times 10^{-5}}$ [1]

$\rho = 7.70 \times 10^3\,\text{kg}\,\text{m}^{-3}$ (3 s.f.) [1]

5 weight of block $= (\rho_s \times x^3)g$, where x = length of each side of cube [1]

weight of water displaced $= (\rho \times fx^3)g$, where f is the fraction of volume submerged [1]

weight of block = weight of water displaced [1]

$(\rho_s \times x^3)g = (\rho \times fx^3)g$

Therefore, $f = \frac{\rho_s}{\rho}$, as required. [1]

5.1

1 $W = Fx = 24 \times 0.50$ [1]

$W = 12\,J$ [1]

2 $W = Fx = 430 \times 1000$ [1]

$W = 4.3 \times 10^5\,J$ [1]

3 $W = Fx = mg \times x = 60 \times 9.81 \times 5.8$ [1]

$W = 3.4 \times 10^3\,J$ (2 s.f.) [1]

4 $x = 3.1 \times \sin 10 = 0.538\ldots\,m$ [1]

$W = Fx = mg \times x = 38 \times 9.81 \times 0.538\ldots$ [1]

$W = 200\,J$ (2 s.f.) [1]

5 $W = Fx\cos\theta = 65 \times 5.0 \times \cos 52$ [1]

$W = 200\,J$ [1]

All the work done is transformed to thermal energy. [1]

6 work done = change in KE [1]

$F \times 0.030 = 1.4 \times 10^3$ [1]

$F = \frac{1.4 \times 10^3}{0.030}$ [1]

$F = 4.7 \times 10^4\,N$ (2 s.f.) [1]

5.2

1 a Potential means 'stored'. [1]

b Kinetic energy to thermal energy. [1]

2 Heat [1]

$20 - 5 = 15\,J$ [1]

3 a Electrical energy to light. [1]

b Electrical energy to sound. [1]

4 a The car is travelling at constant velocity, so KE of car does not change. [1]

b Thermal energy = 100 – 25 = 75% [1]

5.3

1 $E_k = \frac{1}{2}mv^2 = \frac{1}{2} \times 1500 \times 10^2$ [1]

$E_k = 7.5 \times 10^4\,J$ [1]

2 $E_p = mgh = 9.4 \times 10^4 \times 9.81 \times 1500$ [1]

$E_p = 1.4 \times 10^9\,J$ (2 s.f.) [1]

3 loss in energy = change in GPE [1]

loss in energy $= mg\Delta h = 0.120 \times 9.81 \times (0.90 - 0.70)$ [1]

loss in energy = 0.24 J (2 s.f.) [1]

4 a $E_p = mgh = 400 \times 9.81 \times 55$ [1]

$E_p = 2.15\ldots \times 10^5\,J = 2.2 \times 10^5\,J$ (2 s.f.) [1]

b Energy is conserved, therefore kinetic energy $= 2.2 \times 10^5\,J$ (2 s.f.) [1]

c $E_k = \frac{1}{2}mv^2$; $v = \sqrt{\frac{2E_k}{m}}$ [1]

$v = \sqrt{\frac{2 \times 2.15\ldots \times 10^5}{400}}$ [1]

$v = 33\,m\,s^{-1}$ (2 s.f.) [1]

5 $v = \sqrt{2gh} = \sqrt{2 \times 9.81 \times 110}$ [1]

$v = 46\,m\,s^{-1}$ (2 s.f.) [1]

Assumption: All GPE transferred to KE – no losses. [1]

6 $E_p = mgh = 0.80 \times 9.81 \times 1.2 = 9.4176\,J$ [1]

$E_k = \frac{1}{2}mv^2 = \frac{1}{2} \times 0.80 \times 3.2^2 = 4.096\,J$ [1]

work done against drag = 9.4176 – 4.096 = 5.3 J (2 s.f.) [1]

7 $E_k = \frac{1}{2}mv^2 = \frac{1}{2} \times 0.030 \times 240^2 = 864\,J$ [1]

work done = change in KE; $864 = F \times 0.085$ [1]

$F = \frac{864}{0.085} = 1.0 \times 10^4\,N$ (2 s.f.) [1]

5.4

1 $P = \frac{W}{t} = \frac{240}{30}$ [1]

$P = 8.0\,W$ [1]

2 energy $= Pt = 2000 \times 60$ [1]

energy $= 1.2 \times 10^5\,J$ [1]

3 electrical energy $= Pt = 60 \times 3600$ [1]

electrical energy $= 2.16 \times 10^5\,J$ [1]

light energy $= 0.05 \times 2.16 \times 10^5 = 1.1 \times 10^4\,J$ (2 s.f.) [1]

4 $E_k = \frac{1}{2}mv^2 = \frac{1}{2} \times 1200 \times 18^2$ [1]

$E_k = 1.944 \times 10^5\,J$ [1]

rate of work done $= \frac{1.944 \times 10^5}{20} = 9.7 \times 10^3\,J\,s^{-1}$ or W (2 s.f.) [1]

5 $P = Fv = (4 \times 210 \times 10^3) \times 250$ [1]

$P = 2.1 \times 10^8\,W$ [1]

6 work done $= Fx = 15 \times 1.4 = 21\,J$ [1]

input energy $= 3.5 \times 30 = 105\,J$ [1]

efficiency $= \frac{21}{105} \times 100 = 20\%$ [1]

7 rate of loss of GPE $= \frac{600 \times 10^6}{0.40} = 1.5 \times 10^9\,J\,s^{-1}$ or W [1]

(mass per second) $\times g \times h = 1.5 \times 10^9$ [1]

mass per second $= \frac{1.5 \times 10^9}{9.81 \times 50} = 3.05\ldots \times 10^6\,kg\,s^{-1}$ [1]

volume per second $= \frac{3.05\ldots \times 10^6}{\rho} = \frac{3.05\ldots \times 10^6}{1000}$ [1]

volume per second $= 3.1 \times 10^3\,m^3\,s^{-1}$ (2 s.f.) [1]

6.1

1 Extension is proportional to load, so new extension is $\frac{6.0 \times 12}{4.0} = 18\,\text{mm}$

Assumption: the spring obeys Hooke's law.

2 $k = \frac{F}{x} = \frac{4.0 \times 9.81}{0.012} = 3270\,\text{N m}^{-1}$

1 Both obey Hooke's law and elastic behaviour. [1]

A has a greater value of force constant because of the larger gradient. [1]

2 Straight-line graph up to 5.0 N. [1]

Correct value of extension (8 mm) shown on the graph. [1]

Curved section beyond the elastic limit. [1]

3 a $k = \frac{F}{x} = \frac{4.0}{5 \times 10^{-3}}$ [1]

$k = 800\,\text{N m}^{-1}$ [1]

b $F = \frac{32}{5} \times 4.0$ [1]

$F = 26\,\text{N}$ (2 s.f.) [1]

Assumption: Hooke's law is obeyed. [1]

4 a [1]

F/N
0.98
1.96
2.94
3.92
4.91
5.89
6.87

Correct values of F in the final column. [1]

b Correct labelling of axes, including the units. [1]

Correct plotting of all points. [1]

Correct straight line through first 5 points followed by a smooth curve. [1]

c The graph shows a linear relationship between force and length. [1]

The graph crosses the L-axis at the original length of the spring; therefore, the graph shows Hooke's law is obeyed. [1]

Force at the elastic limit is about 4.9 N. [1]

d force constant = gradient of the linear section of the graph [1]

force constant ≈ $23\,\text{N m}^{-1}$ [1]

5 a force = mg = (0.280 × 9.81) N and $x = 0.294 - 0.200 = 0.094\,\text{m}$ [1]

$k = \frac{F}{x} = \frac{0.280 \times 9.81}{0.094}$ [1]

$k = 29.2\,\text{N m}^{-1}$ (3 s.f.) [1]

b (i) Force is shared equally by each spring. The extension of each spring is halved. [1]

Since $k = \frac{F}{x}$, this implies that $k \propto \frac{1}{x}$. [1]

The combined force constant is doubled; force constant = 29.2 × 2 = $58.4\,\text{N m}^{-1}$ [1]

b (ii) Force is the same in each spring. The extension of combined springs is doubled. [1]

Since $k = \frac{F}{x}$, this implies that $k \propto \frac{1}{x}$. [1]

The combined force constant is halved; force constant = $\frac{29.2}{2} = 14.6\,\text{N m}^{-1}$ [1]

6.2

1 $E = \frac{1}{2}Fx = \frac{1}{2} \times 18 \times 0.20$ [1]

$E = 1.8\,\text{J}$ [1]

2 $E = \frac{1}{2}kx^2$, therefore: $1.5 = \frac{1}{2}k \times (2.0 \times 10^{-3})^2$ [1]

$k = \frac{1.5 \times 2}{(2.0 \times 10^{-3})^2}$ [1]

$k = 7.5 \times 10^5\,\text{N m}^{-1}$ [1]

3 work done = area under graph (area of trapezium) [1]

work done = $\frac{1}{2} \times 0.10 \times (5 + 15)$ [1]

work done = 1.0 J [1]

4 elastic potential energy = $E = \frac{1}{2}kx^2 = \frac{1}{2} \times 120 \times 0.04^2$ [1]

elastic potential energy = $9.6 \times 10^{-2}\,\text{J}$ [1]

Assume all elastic potential energy is transferred to gravitational potential energy. [1]

$mgh = 9.6 \times 10^{-2}\,\text{J}$ [1]

$h = \frac{9.6 \times 10^{-2}}{0.008 \times 9.81} = 1.2\,\text{m}$ (2 s.f.) [1]

6.3

1 The molecular chains are easier to untangle for smaller forces, hence the F–x graph has a smaller gradient.

The molecular chains are difficult to extend further when they are fully extended, hence the $F - x$ graph has a steeper gradient.

2 Rubber returns to its original length when the force is completely removed, and so shows elastic behaviour. The hysteresis loop shows that all the energy stored is not returned – some energy is transferred to thermal energy. So rubber is poor at storing energy.

3 a Not all the energy stored in the material is returned – so the landings are less bumpy.

b The work done on the material is greater than the energy returned. The area of the hysteresis loop is the energy transferred to thermal energy, so the tyres warm up.

1 Rubber will return to its original length when the forces are removed. [1]

2 The force–extension graph for rubber is not a straight line through the origin. [1]

It does not obey Hooke's law: force is not proportional to extension. Therefore, a rubber band cannot have a force constant. [1]

3 The metal wire shows elastic behaviour up to its elastic limit and polythene does not show any elastic behaviour. [1]

Polythene shows plastic deformation, as does the metal wire beyond its elastic limit. [1]

4 thermal energy = area of the hysteresis loop [1]

area of loop ≈ 4.0 squares [1]

Therefore, thermal energy ≈ 4.0 × (0.1 × 10) = 4 J [1]

6.4

1 area under graph $= \frac{1}{2} \times \sigma \times \varepsilon$

area under graph $= \frac{1}{2} \times \frac{F}{\rho} \times \frac{x}{L}$

area under graph $= \frac{\frac{1}{2}Fx}{V}$, $V = AL$ – the volume of the material.

elastic potential energy $= \frac{1}{2}Fx$

Therefore, the area under the graph is energy stored per unit volume.

2 Energy is equivalent to work done, and work done = force × distance.

Force is given by the equation $F = ma$. It has base unit kg m s^{-2}.

(Volume has base units m^3.)

Therefore, energy per unit volume has base units:

$$\frac{\text{kg m s}^{-2} \times \text{m}}{\text{m}^3} = \text{kg m}^{-1}\,\text{s}^{-2}$$

stress = force/area

(Area has base units m^2.)

Therefore, stress has base units: $\frac{\text{kg m s}^{-2}}{\text{m}^2} = \text{kg m}^{-1}\,\text{s}^{-2}$

Both base units are the same.

3 energy per unit volume $= \frac{1}{2} \cdot$ stress $\cdot$ strain

$120 \times 10^6 = \frac{1}{2} \times$ stress $\times 0.40$

stress $= 6.0 \times 10^8$ Pa

Assumption: the silk obeys Hooke's law.

1 It is the maximum stress that can be applied to a material before it breaks. [1]

2 Cast iron is a brittle material.

3 $\sigma = \frac{6.3}{\pi \times (0.10 \times 10^{-3})^2} = 2.00... \times 10^8$ Pa [1]

$\varepsilon = \frac{1.048 - 1.035}{1.035} = 1.25... \times 10^{-2}$ [1]

Young modulus $= \frac{2.00... \times 10^8}{1.25... \times 10^{-2}} = 1.6 \times 10^{10}$ Pa (2 s.f.) [1]

4 $\left(\text{ultimate tensile strength} = \frac{F}{A}\right)$ [1]

$220 \times 10^6 = \frac{F}{\pi \times (0.60 \times 10^{-3})^2}$ [1]

$F = 220 \times 10^6 \times \pi \times (0.60 \times 10^{-3})^2$ [1]

$F = 250$ N (2 s.f.) [1]

5 a $E = \sigma \div \varepsilon = \frac{F}{A} \div \frac{x}{L}$ [1]

$E = \frac{F}{A} \times \frac{L}{x} = \frac{FL}{Ax}$ [1]

b $E = \frac{FL}{Ax} = \frac{300 \times 2.500}{\pi \times (0.42 \times 10^{-3})^2 \times 1.4 \times 10^{-2}}$ [1]

$E = 9.7 \times 10^{10}$ Pa (2 s.f.) [1]

c The elastic limit of the material is not exceeded, so stress ∝ strain. [1]

7.1

1 The resultant force is zero. [1]

2 Force on the Earth = 600 N [1]

The force is the same because the person and the Earth interact with each other. According to Newton's third law, the magnitude of the force experienced by each is the same. [1]

3 The force experienced by each is the same but in opposite directions. [1]

Hence, the resultant force = 0. [1]

4 The runner exerts a backward force on the Earth. [1]

According to Newton's third law, the Earth exerts an equal forward force on the person. [1]

5 a The acceleration of free fall would be 9.81 m s^{-2}, assuming there is no drag. [1]

b The force on the Earth is the same as the weight of the bird (Newton's third law). [1]

$F = ma$

$a = \frac{F}{m} = \frac{150}{6.0 \times 10^{24}}$ [1]

$a = 2.5 \times 10^{-23}$ m s^{-1} [1]

7.2

1 Total energy and linear momentum are both conserved. [1]

2 The loss of momentum is numerically equal to the gain in momentum (principle of conservation of momentum); therefore, $\Delta p = 120$ kg m s^{-1}. [1]

$\Delta v = \frac{\Delta p}{m} = \frac{120}{2.0}$ [1]

$\Delta v = 60$ m s^{-1} [1]

3 The initial momentum is zero; therefore, the final momentum must also be zero.

momentum of cannon = momentum of shell [1]

$1200 \times v = 20 \times 300$ [1]

$v = 5.0$ m s^{-1} [1]

4 a total initial momentum = total final momentum

$(300 \times 2.5) + (400 \times -4.0) = 300v + 0$ [1]

$v = \frac{-850}{300}$ [1]

$v = -2.8\,\text{m s}^{-1}$ (2 s.f.) [1]

b initial KE $= \frac{1}{2} \times 300 \times 2.5^2 + \frac{1}{2} \times 400 \times 4.0^2$

$= 4.13... \times 10^3\,\text{J}$ [1]

final KE $= \frac{1}{2} \times 300 \times 2.8^2 = 1.1176 \times 10^3\,\text{J}$ (2 s.f.) [1]

loss in KE $= 4.13... \times 10^3 - 1.1176 \times 10^3$

$= 3.0 \times 10^3\,\text{J}$ (2 s.f.) [1]

c The kinetic energy is not conserved. [1]

Therefore, the collision is inelastic. [1]

7.3

1 $F = \frac{\Delta p}{\Delta t} = \frac{1.2 \times 10^4}{5.0}$ [1]

$F = 2.4 \times 10^3\,\text{N}$ [1]

2 $\Delta p = F \times \Delta t = 150 \times 0.025$ [1]

$\Delta p = 3.75 = 3.8\,\text{kg m s}^{-1}$ (2 s.f.) [1]

3 $F = \frac{\Delta p}{\Delta t} = \frac{1.0 \times 10^6 \times 3.4 \times 10^3 - 2.3 \times 10^6 \times 1.2 \times 10^3}{200}$ [1]

$F = 3.2 \times 10^6\,\text{N}$ [1]

4 $\Delta p = (0.150 \times -15) - (0.150 \times 15) = -4.5\,\text{kg m s}^{-1}$ [1]

$F = \frac{\Delta p}{\Delta t} = \frac{4.5}{0.025}$ (magnitude only) [1]

$F = 180\,\text{N}$ [1]

5 change in momentum *per second* $= 2.5 \times 4.0$

$= 10\,\text{kg m s}^{-2}$ [1]

Therefore, force $= 10\,\text{N}$ [1]

7.4

1 Since $p = \frac{h}{\lambda}$, momentum p is inversely proportional to the wavelength λ.

2 $p = \frac{h}{\lambda} = \frac{6.63 \times 10^{-34}}{500 \times 10^{-9}}$

$p = 1.3 \times 10^{-27}\,\text{kg m s}^{-1}$ (2 s.f.)

3 area $= \frac{1}{9.1 \times 10^{-6}}$

area $= 1.1 \times 10^5\,\text{m}^2$ (2 s.f.)

1 N s and kg m s^{-1} [2]

2 impulse $= F \times \Delta t = 200 \times 5.0$ [1]

impulse $= 1000\,\text{N s}$ [1]

3 Δp = impulse $= 1.1\,\text{N s}$

$\Delta v = \frac{\Delta p}{m} = \frac{1.1}{0.050}$ [1]

$\Delta v = 22\,\text{m s}^{-1}$ [1]

4 a change in momentum = area under graph [1]

$\Delta p = \frac{1}{2} \times 1.5 \times 10^{-17} \times 6.0 \times 10^{-6}$

$= 4.5 \times 10^{-23}\,\text{kg m s}^{-1}$ [1]

b $4.5 \times 10^{-23} = 1.7 \times 10^{-27}v - (1.7 \times 10^{-27} \times 5.0 \times 10^4)$ [1]

$v = \frac{1.3 \times 10^{-22}}{1.7 \times 10^{-27}}$ [1]

$v = 7.6 \times 10^4\,\text{m s}^{-1}$ (2 s.f.) [1]

7.5

1 The angle would be 90°. [1]

2 The diagram is incorrect because:

After the collision, the total momentum in the original direction of travel of X is $5\,\text{kg m s}^{-1}$. This is not the same as the initial momentum in this direction. [1]

Also, there can be no momentum at right angles to the original direction of travel of X because the initial momentum in this direction was zero. [1]

3 The initial momentum must equal the final momentum. The final momentum is the vector sum of the two momentums. [1]

Therefore

initial momentum $= \sqrt{3.0^2 + 2.0^2}$ [1]

initial momentum $= 3.6\,\text{kg m s}^{-1}$ (2 s.f.) [1]

4 final momentum = total initial momentum [1]

final momentum $= \sqrt{(0.250 \times 3.0)^2 \times (0.150 \times 4.0)^2}$ [1]

final momentum $= 0.96...\,\text{kg m s}^{-1}$ [1]

final velocity $v = \frac{0.96...}{0.400} = 2.4\,\text{m s}^{-1}$ (2 s.f.) [1]

8.1

1 The charge was due to an excess of electrons, each with charge e. The total charge on the drop must be ne, where n is the number of excess of electrons.

2 Measure diameter to determine the radius, r, and then use the equations for volume and density to determine the mass.

$\rho = \frac{m}{V}$; therefore, $m = \rho V$ and $V = \frac{4}{3}\pi r^3$;

therefore, $m = \rho \frac{4}{3}\pi r^3$

Weight is given by $W = mg$; therefore, the weight of the drop is given by $W = \rho \frac{4}{3}\pi r^3 g$

3 To allow the uncertainty and so the accuracy to be determined. To ensure the data does not appear to be more accurate than it actually is.

1 a $2.0 \times 1.60 \times 10^{-19}\,\text{C} = 3.2 \times 10^{-19}\,\text{C}$ [1]

b $-5.0 \times 1.60 \times 10^{-19}\,\text{C} = -8.0 \times 10^{-19}\,\text{C}$ [1]

c $-12 \times 1.60 \times 10^{-19}\,\text{C} = -1.9 \times 10^{-18}\,\text{C}$ (2 s.f.) [1]

d $41 \times 1.60 \times 10^{-19}\,\text{C} = 6.6 \times 10^{-18}\,\text{C}$ (2 s.f.) [1]

2 **a** −10 *e* must be 10 electrons, each with charge *e*. [1]

b $e = 1.60 \times 10^{-19}$ C number of electrons

$= \dfrac{\text{total charge}}{\text{charge on each electron}}$ [1]

$\dfrac{15}{1.60 \times 10^{-19}} = \text{number of electrons}$ [1]

$\dfrac{15}{1.60 \times 10^{-19}} = 9.4 \times 10^{19}$ electrons (2 s.f.) [1]

3 $I = \dfrac{\Delta Q}{\Delta t}$ therefore $\Delta Q = I \Delta t$ [1]

$\Delta t = 4.0$ hours = 14400 seconds [1]

$\Delta Q = 500 \times 10^{-3} \times 14400 = 7200$ C [1]

[1 mark for 2.0 C]

4 $Q = ne$ therefore

$Q = 5.0 \times 10^{14} \times 1.60 \times 10^{-19} = 8.0 \times 10^{-5}$ C [1]

$I = \dfrac{\Delta Q}{\Delta t} = \dfrac{8.0 \times 10^{-5}}{1.0} = 8.0 \times 10^{-5}$ A [1]

5 number of electrons

$= \dfrac{\text{total charge}}{\text{charge on each electron}}$

$= \dfrac{9000}{1.60 \times 10^{-19}} = 5.625 \times 10^{22}$ electrons [1]

$\Delta t = 2.0$ hours = 7200 s [1]

average number of electrons per second

$= \dfrac{\text{number of electrons}}{\text{time taken}} = \dfrac{5.625 \times 10^{22}}{7200}$

$= 7.8 \times 10^{18}$ electrons per second (2 s.f.) [1]

6 Two weeks = 1.2×10^6 s.

$\Delta Q = I\Delta t = 6.2 \times 1.2 \times 10^6 = 7.44 \times 10^6$ C [1]

number of electrons $= \dfrac{\text{total charge}}{\text{charge on each electron}}$

$= \dfrac{7.44 \times 10^6}{1.60 \times 10^{-19}}$

$= 4.7 \times 10^{25}$ electrons (2 s.f.) [1]

7 1 A s is equivalent to 1 C [1]

1.0 mA h = 1.0×10^{-3} A h = 60×10^{-3} A min = 3.6 A s; therefore, 1.0 mA h = 3.6 C [1]

5000 mA h = 5000 × 3.6 = 18000 C [1]

8.2

1 Conventional current is from the positive terminal to the negative terminal. [1]

Electrons flow from the negative terminal to the positive terminal. [1]

2 Any valid comparisons, examples include:

Similarities:

Both are examples of flows of charge. [1]

Differences:

In metals the charge carriers are electrons. In ionic solutions the charge carriers are ions. [1]

In metals the charge carriers are negative. In ionic solutions the charge carriers can be positive or negative. [1]

In metals electrons flow in the opposite direction to the conventional current. In ionic solutions the charge carriers can either flow in the same direction or in the opposite direction to the conventional current. [1]

Must contain at least one similarity and one difference for 4 marks.

3 As Figure 5 with labelled electrodes [1]

Cu^{2+} moving towards negative cathode [1]

SO_4^{2-} moving towards positive anode [1]

4 Cations move from positive to negative (towards the cathode). The same direction as the conventional current. [1]

Anions move from negative to positive (towards the anode). The opposite direction to the conventional current. [1]

5 $I = \dfrac{\Delta Q}{\Delta t}$ and $\Delta t = 3.0$ minutes = 180 s [1]

Each cation has a relative charge of +2e.

$\Delta Q = 2 \times 1.60 \times 10^{-19} \times 6.0 \times 10^{14} = 1.9 \times 10^{-4} C$ [1]

$I = \dfrac{1.9 \times 10^{-4}}{180} = 1.1 \times 10^{-6}\,\text{A} = 1.1\,\mu\text{A}$ (2 s.f.) [1]

8.3

1 The total/net charge in any interaction must be the same before and after the interaction. [2]

[A simple, 'The charge in any interaction must be the same before and after the interaction' gains 1 mark]

2 As Figure 2 [1]

The sum of the current into a point must equal the sum of the current out of the point. [1]

3 **a** **i** 7 A towards the 2 A

ii 5 A away from the junction [both required for 1]

b **iii** 4 A towards the junction [1]

c **iv** 2 A to the left [1]

v 5 A to the left [1]

vi 7 A towards the junction [1]

4 Current in wire A =

$I = \dfrac{1.9 \times 10^{21} \times 1.60 \times 10^{-19}}{60} = 5.06...$ A [2]

Current in wire B = 15 A − 5.06... A = 9.9 A (2 s.f.) [1]

5 Discussion should include:

Charge must be conserved

Charge is due to electrons/ions

Therefore, the total number of electrons/ions must be conserved

Current is a flow of charge

Rate of flow of charge into a point must be equal to the rate of flow of charge from that point

[1 mark for each valid point, with up to three total marks]

6 Two protons have a net charge of $+2e$ (3.20×10^{-19} C) [1]

Any particles created in the collision must give rise to the same net charge. For example, if the positive charges are measured after the collision and found to be $+5e$, this suggests a particle (or several particles) with a charge of $-3e$ must have been created, ensuring the net charge remains at $+2e$. [1]

8.4

1 Show that the equation is homogenous with respect to base units.

$I = A\ n\ e\ v$

I [A], A [m^2], n [m^{-3}], e [C], v [m s^{-1}]

Therefore

$[A] = [m^2] \times [m^{-3}] \times [C] \times [m\,s^{-1}]$

$[A] = [C\,s^{-1}]$

$[A] = [A]$

2 a mean drift velocity increases

b mean drift velocity increases

c mean drift velocity halves

1 Conductors have the greatest number density [1]
followed by semiconductors, and then insulators [1]

2 $I = A\ n\ e\ v$ and n for copper $= 8.5 \times 10^{28}\ \text{m}^{-3}$ [1]

$I = 5.50 \times 10^{-8} \times 8.5 \times 10^{28} \times 1.60 \times 10^{-19} \times 2.0 \times 10^{-3}$ [1]

$I = 1.5\ \text{A}$ (2 s.f.) [1]

3 $I = A\ n\ e\ v$ therefore $v = \dfrac{I}{A\ n\ e}$ [1]

$v = \dfrac{500 \times 10^{-3}}{7.10 \times 10^{-6} \times 5.86 \times 10^{28} \times 1.60 \times 10^{-19}}$ [1]

$v = 7.51 \times 10^{-6}\ \text{m s}^{-1}$ (3 s.f.) [1]

4 a From $v = \dfrac{I}{A\ n\ e}$ it follows that $v \propto \dfrac{1}{A}$ [1]

Therefore, if the cross-sectional area increases, the mean drift velocity decreases. [1]

b From $v = \dfrac{I}{A\ n\ e}$ it follows that $v \propto \dfrac{1}{n}$. [1]

As copper has a higher n than zinc, n increases, so the mean drift velocity decreases. [1]

c From $v = \dfrac{I}{A\ n\ e}$ it follows that $v \propto \dfrac{1}{A}$. [1].

As $A = \pi r^2$, if r decreases by a factor of 3, A will decrease by a factor of 9 (3^2). [1]

As A decreases by a factor of 9, the mean drift velocity must increase by a factor of 9. [1]

5 $I = A\ n\ e\ v$; therefore, $v = \dfrac{I}{A\ n\ e}$ [1]

$A = \pi r^2$; therefore, $A = \pi \times \left(\dfrac{1.0 \times 10^{-3}}{2}\right)^2$ [1]

$= 7.85... \times 10^{-7}\ \text{m}^2$

$v = \dfrac{I}{A\ n\ e} = \dfrac{3.0 \times 10^{-3}}{7.85... \times 10^{-7} \times 6.6 \times 10^{28} \times 1.60 \times 10^{-19}}$

$v = 3.6 \times 10^{-7}\ \text{ms}^{-1}$ (2 s.f.) [1]

6 $I = A\ n\ e\ v$; therefore, $n = \dfrac{I}{A\ v\ e}$ [1]

$n = \dfrac{I}{A\ v\ e} = \dfrac{12 \times 10^{-3}}{8.2 \times 10^{-6} \times 72 \times 1.60 \times 10^{-19}}$ [1]

$n = 1.27... \times 10^{20}$ [1]

Giving a ratio of $\dfrac{1.27... \times 10^{20}}{8.5 \times 10^{28}}$ or 1.5×10^{-9}: 1 (2 s.f.) [1]

9.1

1 a Diode [1]

b Thermistor [1]

c Capacitor [1]

2 a Cell and lamp [1]
Correctly drawn in series [1]

b Battery [1]
Resistor and ammeter [1]
Correctly drawn in series [1]

c Power supply and two resistors [1]
Correctly drawn in series [1]

3 Cell, open switch, lamp in series [1]
Voltmeter correctly drawn in parallel with lamp [1]

4 Any two from [2]
Circuit not complete
Negative terminal labelled incorrectly
Positive terminal labelled incorrectly

5 Power supply, ammeter and lamp [1]
Correctly drawn in series [1]
Voltmeter in parallel across lamp [1]

9.2

1 Power supply/cell/battery, voltmeter and lamp [1]
Voltmeter in parallel across lamp [1]

2 Potential difference is used when work is done by the charge carriers. A transfer of energy from the charge carriers to the component, transferring electrical energy into other forms. [1]

Electromotive force is used when work is done on the charge carriers. A transfer of energy to the charge carriers from the cell/battery/power supply, transferring other forms of energy (chemical, light, etc.) into electrical energy. [1]

3 $V = \frac{W}{Q}$; therefore, $W = VQ$ [1]

$W = 80 \times 4.0 = 320\,J$ [1]

4 Calculate the p.d. across a filament lamp when 168 J of energy is transferred to the lamp by 14 C of charge. [2]

$V = \frac{W}{Q}$ [1]

$V = \frac{168}{14} = 12\,V$ [1]

5 One volt is the potential difference across a component when 1 J of energy is transferred per unit charge passing through the component. 1 V is equal to 1 J of energy transferred per coulomb of charge. [1]

$V = \frac{W}{Q}$, V [V], W [J], Q [C]

From $\Delta Q = I\Delta t$ [C] = [A s]

Therefore

[V] = [J] × [A^{-1}] × [s^{-1}] [1]

From $W = Fx$ [J] = [N m]

Therefore:

[V] = [N] × [m] × [A^{-1}] × [s^{-1}]

From $F = ma$ [N] = [kg m s^{-2}] [1]

Therefore:

[V] = [kg] × [m] × [s^{-2}] × [m] × [A^{-1}] × [s^{-1}]

Which becomes:

[V] = [kg] × [m^2] × [s^{-3}] × [A^{-1}]

Therefore, in base units the volt is equal to

$kg\,m^2\,s^{-3}\,A^{-1}$ [1]

6. Time = 6.0 × 60 × 60 = 21600 s [1]

$\Delta Q = I\Delta t = 500 \times 10^{-3} \times 21600 = 10800\,C$ [1]

$V = \frac{W}{Q}$; therefore, $W = VQ$ [1]

$W = 1000 \times 10800 = 11\,MJ$ (2 s.f.) [1]

9.3

1 Total $V = 50 \times 100 = 5000\,V$

$eV = \frac{1}{2}mv^2$ therefore $v = \sqrt{\frac{2eV}{m}}$

$v = \sqrt{\frac{2 \times 1.60 \times 10^{-19} \times 5000}{9.11 \times 10^{-31}}}$

$v = 4.2 \times 10^7\,ms^{-1}$ (2 s.f.)

2 Electrons must spend the same time in each electrode as they move along the accelerator (in order to ensure they leave the electrode at the same time as the potential difference changes and accelerates them towards the next one). The electrons are moving faster as they move along the accelerator. Therefore, electrodes must increase in length.

1 Electrons are emitted from the hot wire/filament at the rear of the electron gun. [1]

There is a large p.d. between the filament and an anode. [1]

Electrons are accelerated towards the anode. [1]

They pass through a hole/gap in the anode. [1]

2 $eV = \frac{1}{2}mv^2$ therefore kinetic energy = eV [1]

kinetic energy = $1.60 \times 10^{-19} \times 12000$ [1]

kinetic energy = $1.9 \times 10^{-15}\,J$ (2 s.f.) [1]

3 kinetic energy = $\frac{1}{2}mv^2$ therefore [1]

$v = \sqrt{\frac{2 \times 1.8 \times 10^{-15}}{9.11 \times 10^{-31}}} = 6.0 \times 10^7\,ms^{-1}$ (2 s.f.) [1]

4 $v = 0.09 \times 3.00 \times 10^8 = 2.7 \times 10^7\,ms^{-1}$ [1]

$eV = \frac{1}{2}mv^2$ therefore $V = \frac{\frac{1}{2}mv^2}{e}$ [1]

$V = \frac{\frac{1}{2} \times 9.11 \times 10^{-31} \times (2.7 \times 10^7)^2}{1.60 \times 10^{-19}}$ [1]

$V = 2100\,V$ (2 s.f.) [1]

5 Velocity of electron will be greater than the proton. [1]

The kinetic energy of each will be the same (as they have the same magnitude charge). [1]

Mass of proton is greater, so travels more slowly at the same kinetic energy. [1]

9.4

1 Graph of I against V [1] with a straight line through the origin. [1]

$V \propto I$ [1] for a conductor at constant temperature [1]

2 $R = \frac{V}{I} = \frac{5.2}{0.50} = 10\,\Omega$ (2 s.f.) [1]

3 $R = \frac{V}{I} = \frac{2.4}{80 \times 10^{-3}} = 30\,\Omega$ [1]

4 One ohm is the resistance of a component when a p.d. of 1 V is produced per ampere of current. [1]

(See 9.2 question 3 to get the base units of the volt.)

$R = \frac{V}{I}$

R [Ω], V [kg $m^2\,s^{-3}\,A^{-1}$], I [A] [1]

[Ω] = [kg $m^2\,s^{-3}\,A^{-1}$] × [A^{-1}]

Which becomes:

[Ω] = [kg $m^2\,s^{-3}\,A^{-2}$]

Therefore, in base units the ohm is equal to $kg\,m^2\,s^{-3}\,A^{-2}$ [1]

5 $I = \frac{\Delta Q}{\Delta t} = \frac{54}{180} = 0.30\,A$ [1]

$R = \frac{V}{I}$ therefore $V = IR = 0.30 \times 1200 = 360\,V$ [1]

6 $I = \frac{\Delta Q}{\Delta t}$ and $V = \frac{W}{Q}$ therefore $V = \frac{W}{I \Delta t}$ [1]

$V = \frac{500}{1.5 \times 60} = 5.55\ldots \text{V}$ [1]

$R = \frac{V}{I} = \frac{5.55\ldots}{1.5} = 3.7\,\Omega$ (2 s.f.) [1]

9.5

1 Power supply, ammeter, voltmeter, variable resistor/potentiometer

2 Increasing the resistance of the variable resistor will reduce the current in the circuit.

3 Set up either circuit as shown in Figure 2.
Record values for *I* and *V*.
Adjust variable resistor/potential divider to produce a range of values.
Include negative values.
Repeat and average.
Plot a graph of *I* against *V*.

1 a A [1] $V \propto I$ [1]

b Component A: $R = \frac{V}{I} = \frac{4.0}{2.0} = 2.0\,\Omega$ [1]

Component B: $R = \frac{V}{I} = \frac{4.0}{1.6} = 2.5\,\Omega$ [1]

2 Graph of *I* against *V* with axes labelled (including units) [1]
Points plotted correctly [1]
Line of best fit drawn [1]

3 a $1.8\,\Omega \pm 0.2\,\Omega$ [1]
b $2.2\,\Omega \pm 0.2\,\Omega$ [1]
c $4.5\,\Omega \pm 0.3\,\Omega$ [1]

4 Filament lamp [1]
Any three from: [3]
As the current increases more electrons/charge carriers pass through the lamp per unit time/second.
The rate of collisions between electrons and positive ions increases.
Each collision transfers energy to the ions/ions gain more energy.
Ions vibrate more.
Temperature of the wire increases.
Resistance increases.

5 Graph of *I* against *V*.
Two straight lines through the origin with different gradients [1]
Steeper line labelled room temperature / shallower line labelled higher temperature. [1]
Explanation.
Hotter wire has a greater resistance. [1]
This results in a lower current at the same p.d. / shallower line. [1]

9.6

1 Any two from: [2]
Voltmeter connected in series
Voltmeter in the wrong place (i.e., not across diode)
Missing ammeter
Diode wrong way round

2 A and B [1]
Diode in series with B is the correct orientation to allow current in bulb B. [1]
Diode in series with C will not allow current in bulb C. [1]

3 Graph of I against V with axes labelled [1]
Axes include correct units [1]
Points plotted correctly [1]
Line of best fit drawn [1]

4 a $\infty\,\Omega$ / very large [1]
b $650\,\Omega$ +/– $1000\,\Omega$ [1]
c $12\,\Omega$ +/– $5\,\Omega$ [1]

5 a 0 values for the current on the negative side as p.d. increase. [1]
Current remains 0 until a small increase in positive p.d. [1]
Current then increases linearly. [1]

b On the negative side the $-I$ is directly proportional to $-V$ (as the current is in the resistor only). [1]
On the positive side the graph is the same as a standard diode (as the current passes through the diode when its resistance drops low enough). [1]

9.7

1 Resistance of a normal wire drops as it gets cooler.
Resistance of a superconductor drops as it gets cooler, but at a critical temperature falls to $0\,\Omega$.

2 No energy transferred to heat when there is a current in a component.
Allows very high currents.

1 Resistance applies to a particular component. [1]
Resistivity is a property of the material. [1]

2 $R = \frac{\rho L}{A}$ [1]

$R = \frac{1.7 \times 10^{-8} \times 1.0}{3.32 \times 10^{-6}}$ [1]

$R = 5.1\,\text{m}\Omega$ (2 s.f.) [1]

3 Increasing the temperature of the metal:
Positive ions in the metal have more energy. [1]
The positive ions vibrate more. [1]
Increasing the resistivity. [1]

4 $R = \frac{\rho L}{A}$ therefore $\rho = \frac{RA}{L}$ [1]

$A = \pi r^2$ [1]

$A = \pi \times \left(3.0 \times 10^{-3}\right)^2 = 2.82... \times 10^{-5}\,\text{m}^2$ [1]

$\rho = \frac{170 \times 2.82... \times 10^{-5}}{12}$ [1]

$\rho = 4.0 \times 10^{-4}\,\Omega\text{m}$ (2 s.f.) [1]

5 a $R = \frac{\rho L}{A}$ therefore $R \propto L$. As L doubles R doubles. [1]

b $R = \frac{\rho L}{A}$ therefore $R \propto \frac{1}{A}$ [1]. As A doubles R halves. [1]

c As $A = \pi r^2$ if r halves A decreases by a factor of 4 (2^2) [1]

$R = \frac{\rho L}{A}$ therefore $R \propto \frac{1}{A}$. A decreases by a factor of 4, R increases by a factor of 4. [1]

d As $V = L \times A$, if V remains constant, if L doubles then A halves. [1]

From $R \propto L$ as L doubles R doubles and from

$R \propto \frac{1}{A}$ as A halves R doubles. [1]

Therefore R increases by a factor of 4. [1]

6 a Reduce the impact of random errors [1]

To ensure the wire is of uniform diameter [1]

b i Graph of R against L with axes labelled [1]

Axes include correct units [1]

Points plotted correctly [1]

Line of best fit drawn [1]

ii gradient $= \frac{R}{L}$ [1]

From $R = \frac{\rho L}{A}$, $\frac{R}{L} = \frac{\rho}{A}$ therefore gradient $= \frac{\rho}{A}$

and $\rho = \text{gradient} \times A$ [1]

$A = \pi \times \left(\frac{0.46 \times 10^{-3}}{2}\right)^2 = 1.66... \times 10^{-7}\,\text{m}^2$ [1]

$\rho = \text{gradient} \times A$

$\rho = -6.12 \times 1.66... \times 10^{-7}$

$\rho = 1.0 \times 10^{-6}\,\Omega\text{m}$ (2 s.f.) $\pm 0.2 \times 10^{-6}\,\Omega\text{m}$ [1]

c Adjusting the variable resistor keeps the current at 0.50 A / Reducing the resistance of the variable resistor as the length of the wire increases keeps the current at 0.50 A [1]

Changing the current will change the temperature of the wire. [1]

Changing the temperature will affect the resistance of the wire. [1]

7 Any four from: [4]

Set up circuit as Figure 1.

Record values of V and I for different thicknesses of wire of the same length (L).

Adjust variable resistor (if present) to ensure I remains constant.

Repeat readings and average.

Calculate R for different thicknesses using $R = \frac{V}{I}$

Calculate A of wire using $A = \pi r^2$

Plot a graph of R against $\frac{1}{A}$

Measure gradient.

gradient $= \rho L$

Find resistivity of the wire from $\rho = \frac{\text{gradient}}{L}$

9.8

1 a As the temperature drops the resistance increases. [1]

b Use the thermistor as a sensor. At a given temperature the thermistor will have a specific resistance. As the resistance changes this can be used to adjust the temperature of the lorry. If the resistance drops, the temperature should be increased, and so on, keeping the temperature at a set level. [2]

2 Graph of R against T with axes labelled (include correct units), smooth curve of decreasing gradient [1]

With significant change in R over the range of 100–300 °C [1]

Large change in R over the range of temperatures found in ovens [1]

3 Calculate R using $R = \frac{V}{I}$ [1]

Graph of R against T with axes labelled [1]

Axes include correct units [1]

Points plotted correctly [1]

Line of best fit drawn [1]

4 a $64\,\Omega \pm 5\,\Omega$ [1]

b $15\,\Omega \pm 3\,\Omega$ [1]

5 $R = \frac{V}{I} = \frac{1.03}{45.2 \times 10^{-3}} = 22.8\,\Omega$ [2]

Temperature = 17 °C ± 2 °C [1]

9.9

1 To reduce the effect of the atmosphere (IR absorbed by water vapour) / to reduce the impact of cloud cover (i.e. it is more likely there is a cloud-free day).

2 When the planet passes in front of the star there is a small but detectable drop in intensity (as some of the light is blocked by the planet). This drop in intensity might be detectable by an LDR, resulting in a slight increase in resistance.

1 LDR resistance increases as it gets darker. [1]
LDR resistance decreases as it gets lighter. [1]

2 LED should be LDR [1]
Table missing units [1]

3 Any valid comparisons, examples include: [4]
Resistance of resistor and thermistor affected by temperature (unaffected by light intensity)
Resistance of thermistor increases as it gets colder.
Resistance of resistor decreases as it gets colder.
Resistance of LDR changes as the light level changes (unaffected by temperature).
Thermistor and LDR are made from semiconductors.

4 a 20–40 (arbitrary units) [1]
b 0.06–0.08 (arbitrary units) [1]

9.10

1 Circuit diagram with power supply/cell/battery, filament lamp in series with an ammeter. Voltmeter connected in parallel across lamp. [1]
Measure V and I and calculate power using $P = IV$. [1]

2 $P = IV$ [1]
$P = IV = 5.0 \times 8.0 = 40\,W$ [1]

3 a $P = IV$ therefore $I = \frac{P}{V}$ [1]
$I = \frac{1200}{20} = 60\ A$ [1]
b $P = \frac{W}{t}$ therefore $W = Pt$ [1]
$W = 1200 \times 3600 = 4.3\,\text{MJ}$ (2 s.f.) [1]

4 a $P = I^2R$ therefore at constant current $P \propto R$ [1]
If R doubles P doubles. [1]
b $P = I^2R$ therefore if the resistance remains unchanged $P \propto I^2$ [1]
If I doubles P increases by a factor of 4 (2^2). [1]

5 Find the watt in base units.
From $P = \frac{W}{t} \rightarrow [\text{W}] = [\text{J s}^{-1}]$
From $W = Fx \rightarrow [\text{J}] = [\text{N}] \times [\text{m}]$
Therefore $[\text{W}] = [\text{N m s}^{-1}]$
From $F = ma \rightarrow [\text{N}] = [\text{kg m s}^{-2}]$
Therefore the watt in base units is $\text{kg m}^2\text{s}^{-3}$. [2]
Express V A in base units.
From $V = \frac{W}{Q} \rightarrow [\text{V}] = [\text{J C}^{-1}]$
From $W = Fx \rightarrow [\text{J}] = [\text{N}] \times [\text{m}]$
Therefore $[\text{V}] = [\text{N m C}^{-1}]$
From $F = ma \rightarrow [\text{N}] = [\text{kg m s}^{-2}]$
Therefore $[\text{V}] = [\text{kg m}^2\text{s}^{-2}\,\text{C}^{-1}]$
From $\Delta Q = I\Delta t \rightarrow [\text{C}] = [\text{A s}]$
Therefore $[\text{V}] = [\text{kg m}^2\text{s}^{-3}\,\text{A}^{-1}]$
Therefore $[\text{V A}] = [\text{kg m}^2\text{s}^{-3}\,\text{A}^{-1}\,\text{A}] = [\text{kg m}^2\text{s}^{-3}]$ [2]

9.11

1 $P = \frac{W}{t}$ therefore $W = Pt$ [1]
$W = 60 \times 7200 = 430\,kJ$ [1]

2 number of units = 0035387 – 0034512 = 875 units [1]
Cost = 875 × 0.12 = £105 [1]

3 From definition 1 kW h is the energy transferred by a 1 kW device in 1 hour. [1]
$P = \frac{W}{t}$ therefore $W = Pt$ [1]
$W = 1000 \times 3600 = 3.6\,\text{MJ}$ [1]

4 a $P = \frac{W}{t}$ therefore $W = Pt$
In SI units, $P = 9000\,\text{W}$ and $t = 900$ s [1]
$W = 9000 \times 900 = 8.1\,\text{MJ}$ [1]
b $P = \frac{W}{t}$ therefore $W = Pt$
In kW h units, $P = 9.0\,\text{kW}$ and $t = 0.25$ hours [1]
$W = 9 \times 0.25 = 2.3\,\text{kW h}$ [1]

5 $P = \frac{W}{t}$ therefore $W = Pt$
In kW h units, $P = 0.060\,\text{kW}$ and $t = 840$ hours [1]
$W = 0.060 \times 840 = 50.4$ kW h [1]
Cost = 50 × 0.112 = £5.60 [1]

6 Filament lamps: in kW h units, $P = 18 \times 0.100 = 1.8\,\text{kW}$ and LEDs: in kW h units, $P = 18 \times 0.015 = 0.27\,\text{kW}$ [1]
$t = 60 \times 60 \times 2 \times 365 = 2.62 \times 10^6\,\text{s}$ [1]
Energy transferred by lamps = $W = 1.8 \times 10^3 \times 2.62 \times 10^6 = 4.716 \times 10^9$ J [1]
Energy transferred by LEDs = $W = 0.27 \times 10^3 \times 2.62 \times 10^6 = 707.4 \times 10^6$ J [1]
Energy saved = $4.716 \times 10^9 - 707.4 \times 10^6 = 4.0$ GJ [1]

10.1

1 A: 2.0 A [1]
B: 2.0 A [1]
C: 0.6 A [1]
D: 3.2 A [1]
E: 1.2 A [1]
F: 1.8 A [1]

2 A: 4.0 V [1]
B: 12 V [1]
C: 9.0 V [1]
D: 8.0 V [1]
E: 4.0 V [1]
F: 4.0 V [1]
G: 3.0 V [1]

3 a Two resistors in series with power supply. [1]
5.0 V across each resistor. [1]
b Two resistors in parallel with power supply. [1]
10 V across each resistor. [1]

4 a Two resistors in series with power supply. 3.3 V across one resistor (the one with lower resistance) [1] and 6.7 V across the other resistor (the one with higher resistance). [1]

b Two resistors in parallel with power supply. 10 V across each resistor. [2]

5 Current in the added branch. [1]

Current in the previous branches remains unaffected. [1]

More current drawn from supply. [1]

6 a Correctly drawn circuit with two lamps on the first branch [1] and one lamp on the second. [1]

b First branch has twice the resistance as the second. [1]

Therefore, the lamps on this branch have half the current of the second branch through them. [1]

First branch 2.0 A through each lamp [1]

Second branch 4.0 A [1]

c $P = IV$

First branch: Current in each bulb = 2.0 A and p.d. = 6.0 V [1] therefore $P = IV = 2.0 \times 6.0 = 12$ W [1]

Second branch: Current in bulb = 4.0 A and p.d. = 12 V [1] therefore $P = IV = 4.0 \times 2 = 48$ W [1]

d As before, the first branch has twice the resistance as the second therefore the lamps on this branch have half the current of the second branch through them. [1]

First branch $\left(1\frac{2}{3}\right)$ 1.7 A through each lamp [1]

Second branch has twice the current. $\left(3\frac{1}{3}\right)$ 3.3 A [1]

10.2

1 a resistance increases. [1]

b resistance decreases. [1]

2 a $R = 5.0 + 9.0 = 14\,\Omega$ [1]

b $\frac{1}{R} = \frac{1}{6.0} + \frac{1}{4.0} = \frac{5}{12}$ [1]

therefore $R = \frac{12}{5} = 2.4\,\Omega$ [1]

c $\frac{1}{R} = \frac{1}{10} + \frac{1}{15} + \frac{1}{20} = \frac{13}{60}$ [1]

therefore $R = \frac{60}{13} = 4.6\,\Omega$ [1]

3 a $\frac{1}{R} = \frac{1}{(150+100)} + \frac{1}{200} = \frac{9}{1000}$ [1]

therefore $R = \frac{1000}{9} = 111\,\Omega$ (3 s.f.) [1]

b For the parallel part of the circuit

$\frac{1}{R} = \frac{1}{(20+10)} + \frac{1}{20} = \frac{1}{12}$ [1]

therefore $R = \frac{12}{1} + 60 = 72\,\Omega$ [1]

c For each branch $\frac{1}{R} = \frac{1}{10} + \frac{1}{10} + \frac{1}{10} = \frac{3}{10}$ [1]

therefore $R = \frac{10}{3} \times 3 = 10\,\Omega$ [1]

4 Single 50 Ω

Single 100 Ω

Single 200 Ω [1 mark in total for all three singles correct]

50 Ω in series 100 Ω [1]

50 Ω in series 200 Ω [1]

200 Ω in series 100 Ω [1]

All three in series [1]

50 Ω in parallel 100 Ω [1]

50 Ω in parallel 200 Ω [1]

200 Ω in parallel 100 Ω [1]

All three in parallel [1]

(50 Ω in parallel 100 Ω) in series with 200 Ω [1]

(50 Ω in parallel 200 Ω) in series with 100 Ω [1]

(200 Ω in parallel 100 Ω) in series with 50 Ω [1]

(50 Ω in series 100 Ω) in parallel with 200 Ω [1]

(50 Ω in series 200 Ω) in parallel with 100 Ω [1]

(200 Ω in series 100 Ω) in parallel with 50 Ω [1]

5 $\frac{1}{R} = \frac{1}{R_1} + \frac{1}{R_2} + \frac{1}{R_3}$ therefore $\frac{1}{R} - \left(\frac{1}{R_1} + \frac{1}{R_2}\right) = \frac{1}{R_3}$ [1]

$\frac{1}{1030} - \left(\frac{1}{2200} + \frac{1}{4700}\right) = \frac{1}{R_3} = 303... \times 10^{-6}\,\Omega^{-1}$ [1]

$R_3 = \frac{1}{303... \times 10^{-6}} = 3290\,\Omega$ (3 s.f.) [1]

10.3

1 Four marks for all 7, three marks for 6 correct, two marks for 5 correct, one mark for 4 correct, zero marks for less than 4 correct.

I: current

V: potential difference

W: work done/energy transferred

P: power

R: resistance

Q and ΔQ: charge

t and Δt: time

2 a $\Delta Q = I\Delta t$ [1]

$\Delta Q = 0.30 \times 45 = 14$ C (2 s.f.) [1]

b $W = Pt$ [1]

$W = 0.20 \times 120 = 24$ J [1]

3 Current through resistor B = 0.50 – 0.40 = 0.10 A [1]

b Resistor A: $V = IR$ [1]

therefore $V = 0.50 \times 5.0 = 2.5$ V [1]

Resistor D: = IV, $V = \frac{P}{I}$ [1]

therefore $V = \frac{0.50}{0.50} = 1.0$ V [1]

c $R = \frac{V}{I} = \frac{1.0}{0.50} = 2.0\,\Omega$ [2]

4 Resistor A: Using $P = I^2R$, $P = 0.50^2 \times 5.0$ $= 1.3\,\text{W}$ (2 s.f.) [1]

Resistor B: Using $P = I^2R$, $P = 0.10^2 \times 8.0 = 80\,\text{mW}$ [1]

Resistor C: The p.d across B,
$V = IR = 0.10 \times 8.0 = 0.80\,\text{V}$ = p.d across C [1]
$P = 0.40 \times 0.80 = 320\,\text{mW}$ [1]

Resistor D: 0.50 W
Total power = 1.3 + 0.080 + 0.32 + 0.50 = 2.2 W [1]

therefore $V = \frac{P}{I} = \frac{2.2}{0.50} = 4.4\,\text{V}$ [1]

10.4

1 a $\varepsilon = V + Ir$ therefore $V = \varepsilon - Ir$

Total e.m.f = 1.5 + 1.5 = 3.0 V

Total internal resistance = 0.75 + 0.75 = 1.5 Ω

$V = \varepsilon - Ir = 3.0 - (0.80 \times 1.5) = 1.8\,V$

b Find the current in the cell. $\varepsilon = I(R+r)$ therefore $\frac{\varepsilon}{(R+r)} = I = \frac{3.0}{(11.5)} = 0.26...\,A$

$V = \varepsilon - Ir = 3.0 - (0.26... \times 1.5) = 2.6\,V$ (2 s.f.)

2 a Total e.m.f = 1.5 V

Total internal resistance = $\frac{1}{R} = \frac{1}{R_1} + \frac{1}{R_2}$,
$R = 0.38\,\Omega$

$V = \varepsilon - Ir = 1.5 - (0.80 \times 0.38) = 1.2\,\text{V}$

b Find the current in the cell. $\varepsilon = I(R+r)$ therefore $\frac{\varepsilon}{(R+r)} = I = \frac{1.5}{(10.38)} = 0.14\,\text{A}$

$V = \varepsilon - Ir = 1.5 - (0.14 \times 0.38) = 1.4\,\text{V}$

1 Graph of terminal p.d. against *I* with axes labelled
Axes include correct units
Points plotted correctly
Line of best fit drawn

2 Gradient is constant.

3 Intercept = e.m.f = 1.6 V
Gradient = internal resistance = 0.50 Ω

4 To ensure it does not heat up, changing its internal resistance.

1 A low internal resistance is needed to provide a large current [1], for example, in a car battery.
A high internal resistance is needed to ensure a high current cannot be produced (for safety reasons) [1]
For example in a school high voltage power supply. [1]

2 As the current increases the terminal p.d decreases. [2]

3 a Lost volts = $Ir = 1.5 \times 2.0 = 3.0\,\text{V}$ [1]

Terminal p.d. = e.m.f – lost volts = 9.0 – 3.0 = 6.0 V [1]

Using $P = I^2R$, $P = 1.5^2 \times 2.0 = 4.5\,\text{W}$ [1]

Using $V = IR$, $R = \frac{6.0}{1.5} = 4.0\,\Omega$ [1]

4 a For the parallel part of the circuit

$\frac{1}{R} = \frac{1}{90} + \frac{1}{45} = \frac{1}{30}$ therefore $R = \frac{30}{1} + 50 = 80\,\Omega$ [2]

b Terminal p.d. = $IR = 0.10 \times 80 = 8.0\,\text{V}$ [1]
Lost volts = 12 – 8.0 = 4.0 V [1]

c Lost volts = Ir therefore
$r = \frac{\text{lost volts}}{I} = \frac{4.0}{0.1} = 40\,\Omega$ [1]

5 Graph of terminal p.d against *I* with axes labelled (include correct units), showing a straight line with a negative gradient [1]
Y-axis intercept labelled e.m.f [1]
Gradient labelled -*r* [1]
Second line with double the e.m.f and double the gradient [1]

10.5

1 The resistance of this part of the potential divider drops. Lowering V_{out}.

2 a $V_{out} = \frac{R_2}{(R_1 + R_2)} \times V_{in}$

$= \frac{4700}{(2200 + 4700)} \times 12 = 8.2\,V$ (2 s.f.)

b Resistance of the loaded part of the circuit
$= \frac{1}{R} = \frac{1}{4700} + \frac{1}{10000} = \frac{147}{470000}$ therefore

$R = \frac{470000}{147} = 3197.2...\,\Omega$

$V_{out} = \frac{R_2}{(R_1 + R_2)} \times V_{in}$

$= \frac{3197.2...}{(2200 + 3200)} \times 12 = 7.1\,\text{V}$ (2 s.f.)

c Resistance of the loaded part of the circuit =
$\frac{1}{R} = \frac{1}{4700} + \frac{1}{100} = \frac{12}{1175}$ therefore

$R = \frac{1175}{12} = 97.9...\,\Omega$

$V_{out} = \frac{R_2}{(R_1 + R_2)} \times V_{in}$

$= \frac{97.9...}{(2200 + 98)} \times 12 = 0.51\,\text{V}$ (2 s.f.)

1 Potential divider circuit drawn with two resistors in series. [1]
Diagram labelled with 20 V V_{in} and V_{out} across one of the resistors. [1]
Both resistors must have the same resistance. [1]
Therefore, the p.d. will be shared equally between them, each one receiving 10 V. [1]

2 a $V_{out} = \frac{R_2}{(R_1 + R_2)} \times V_{in} = \frac{90}{(270 + 90)} \times 12 = 3.0\,\text{V}$ [2]

$V_{out} = \frac{R_2}{(R_1 + R_2)} \times V_{in} = \frac{120}{(30 + 120)} \times 60 = 48\,\text{V}$ [2]

3 a Potential divider with two resistors drawn correctly. V_{out} connected across 30 Ω resistor [1]

b Potential divider with two resistors drawn correctly. V_{out} connected across 90 Ω resistor [1]

4 $\frac{V_1}{V_2} = \frac{R_1}{R_2}$ therefore $R_2 = \frac{R_1 \times V_2}{V_1}$ [1]

$V_1 = 360 - 3.0 = 357\,V$ [1]

$R_2 = \frac{110 \times 3.0}{357} = 0.92\ \Omega$ (2 s.f.) [1]

10.6

1 Potential divider containing a resistor and thermistor in series.

V_{out} connected across the resistor.

2 Maximum:

$V_{out} = \frac{R_2}{(R_1 + R_2)} \times V_{in} = \frac{50 \cdot 10^6}{(1000 + 50 \cdot 10^6)} \times 9.0$

$= 9.0\ V\ \ (8.9998\ \ V)$

Minimum: $V_{out} = \frac{R_2}{(R_1 + R_2)} \times V_{in}$

$= \frac{500}{(1000 + 500)} \times 9.0 = 3.0\ V$

1 Any valid two: [2]

Can be made very compact

Uses fewer components

Can easily be made into a rotary dial.

Allows the full range of output potential difference from 0 V to V_{in}.

2 Potential divider containing a resistor and an LDR in series. [2]

V_{out} connected across the LDR. [1]

3 a $V_{2200} = 12 - 6.0 = 6.0\,V$ [1]

$R_2 = \frac{2200 \times 6.0}{6.0} = 2200\ \Omega$ [1]

(As 6.0 V is half of V_{in}, the resistance of the thermistor must equal the resistance of the resistor)

b $V_{2200} = 12 - 10 = 2.0\,V$ [1]

$R_2 = \frac{2200 \times 10}{2.0} = 11000\ \Omega$ [1]

c $V_{2200} = 12 - 1.0 = 11\,V$ [1]

$R_2 = \frac{2200 \times 1.0}{11} = 200\ \Omega$ [1]

4 Increasing the temperature would increase the value of V_{out}. [2]

5 a Graph of R against T with axis labelled (include correct units) and a smooth curve of decreasing gradient [1]

Values at 0°C and at 100°C drawn correctly. [1]

b Find V_1 when $V_{out} = 4.0$ V. $V_1 = 12 - 4.0 = 8.0\,V$ [1]

$\frac{V_1}{V_2} = \frac{R_1}{R_2}$ therefore $R_2 = \frac{R_1 \times V_2}{V_1}$.

$R_2 = \frac{220 \times 4.0}{8.0} = 110\ \Omega$ [1]

Read value of temperature at 110 Ω. [1]
(around 30 °C – depending on the sketch graph)

6 Allows V_{out} to be varied [1] at a specific temperature. [1]

11.1

1 Any valid three, for example (answers must contain at least one similarity and one difference) [3]

Similarities:

Progressive waves

Transfer energy

Differences:

Transverse wave – oscillations are perpendicular to the direction of the wave's movement.

Longitudinal wave – oscillations are parallel to the direction of the wave's movement.

Transverse wave – contains peaks and troughs.

Longitudinal wave – contains compressions and rarefactions.

2 Transverse wave – Fix one end of the slinky, hold the other end, move this end perpendicular to the body of the slinky. [1]

Longitudinal wave – Fix one end of the slinky, hold the other end, move this end parallel to the body of the slinky. [1]

3 A: Vertically downwards [1]

B: Vertically upwards [1]

C: Vertically downwards [1]

4 Particles are closer together [1]

Stronger restoring force/vibrations are passed more rapidly from one particle to the next. [1]

5 Diagram should include: Compressions [1], Rarefactions [1], particles vibrating parallel to the direction of energy transfer [1]

11.2

1 a $\varphi = \frac{x}{\lambda} \times 360^\circ = \frac{20}{40} \times 360^\circ = 180^\circ$

b $\varphi = \frac{x}{\lambda} \times 360^\circ = \frac{40}{40} \times 360^\circ = 360^\circ$

c $\varphi = \frac{x}{\lambda} \times 360^\circ = \frac{80}{40} \times 360^\circ = 720^\circ$

2 a $x = \frac{\varphi \times \lambda}{360^\circ} = \frac{90^\circ \times 1.60}{360^\circ} = 40$ cm

b $x = \frac{\varphi \times \lambda}{360^\circ} = \frac{540^\circ \times 1.60}{360^\circ} = 2.4$ m

c 5π rad $= 900^\circ$ $x = \frac{\varphi \times \lambda}{360^\circ} = \frac{900^\circ \times 1.60}{360^\circ}$
$= 4.0$ m

1 Increasing the timebase results in a smaller time period for each square on the screen.
The wave trace will appear more compressed.
2 Period of oscillation = 0.02 s, therefore one complete cycle will be completed in each square.

1 $v = f\lambda$ [1]

$v = 2.0 \times 0.50 = 1.0 \text{ m s}^{-1}$ [1]

2 A: down [1]

D: up [1]

3 Connect a microphone to an oscilloscope. [1]
Blow the whistle and record the number of divisions n between successive peaks of the signal displayed on the oscilloscope. [1]

Find the period T by multiplying n by the time base. [1]

Frequency f of the sound calculated using $f = 1/T$. [1]

4 $f = \frac{1}{T}$ [1]

$f = \frac{1}{2.0 \times 10^{-3}} = 500 \text{ Hz}$ [1]

$v = f\lambda$ therefore $\lambda = \frac{v}{f}$ [1]

$\lambda = \frac{340}{500} = 0.68 \text{ m}$ [1]

5 a i Same shaped wave profile as in question. [1]

Wave profile shifted a quarter-cycle to the right. [1]

ii Same shaped wave profile as in question. [1]

Wave profile shifted a half-cycle to the right. [1]

b i 0.3 m [1]

ii 0.0 m [1]

iii 0.3 m [1]

6 a 90° or $\frac{\pi}{2}$ rad [1]

b 180° or π rad [1]

c 270° or $\frac{3}{2}$ π rad [1]

11.3

1 Any valid three, for example (answers must contain at least one similarity and one difference) [3]

Similarities:

Property of all waves

Frequency does not change

Differences:

Reflection – speed and wavelength do not change

Refraction – speed and wavelength change

Reflection – wave does not change medium

Refraction – wave changes from one medium to another

2 Normal drawn in each example (at 90° to the surface). [1]

In each case the angle of incidence = angle of reflection. [3]

3 Normal drawn correctly and partial reflection shown in diagram. [1]

Ray of light bends away from the normal. [1]

4 $v = f\lambda$ therefore if f is constant [1]
$v \propto \lambda$ [1]

5 General shape, including a different wavelength at each end of the pool [1]

λ increases moving towards the deep end [1]

λ reduces moving towards the shallow end [1]

11.4

1 Only transverse waves can be plane polarised. [1]

Sound is a longitudinal wave. [1]

Therefore, sound waves cannot be plane polarised

2 Any two examples of transverse waves: [2]

e.g.

(Visible) Light; Microwaves; Radio waves; Infrared waves; Ultraviolet; S waves

3 Diffraction effects are most significant when the wavelength is a similar size to the gap/obstacle. [1]

Wavelength of light is much smaller than most gaps / sound waves have a larger wavelength. [1]

4 3.0 m wave diffracts more. [1]

3.0 cm wave does not really diffract as the wavelength is much smaller than the size of the gap. [1]

3.0 m wave diffracts significantly as the wavelength is the same size of the gap. [1]

5 Higher frequency means a smaller wavelength. [1]

Radio waves have a longer wavelength; therefore, they diffract over the hill reaching the bottom of the valley. [1]

The TV signal has a shorter wavelength; therefore, does not diffract as significantly, failing to reach the bottom of the valley. [1]

6 a $\lambda = \frac{v}{f} = \frac{340}{1200} = 0.28 \text{ m}$ (2 s.f.) [1]

Very similar to the gap therefore significant diffraction. [1]

b $\lambda = \frac{v}{f} = \frac{340}{1.0 \times 10^6} = 340 \,\mu\text{m}$ [1]

Much smaller than the gap therefore no diffraction. [1]

11.5

1 Graph of intensity against distance with axis labelled (including units), points plotted correctly and line of best fit drawn.

2 Use $I = \frac{k}{distance^2}$ to check if the data follows an inverse square relationship.

1 a Intensity increases by a factor of 9 (3^2). [1]

b Intensity decreases by a factor of 16 (4^2). [1]

2 $I = \frac{P}{A}$ [1]

$I = \frac{P}{A} = \frac{400}{20} = 20\,\text{W m}^{-2}$ [1]

3 $I = \frac{P}{4\pi r^2}$ [1]

$I = \frac{60}{4 \times \pi \times (20)^2}$ [1]

$I = 12\ \text{mW m}^{-2}$ (2 s.f.) [1]

4 From $I = \frac{P}{A}$ if the power is constant [1]

then: $I \propto \frac{1}{A}$ [1]

As the area reduces [1]

Intensity increases. [1]

5 Intensity = $1.4\,\text{kW m}^{-2}$ therefore power received by each $8.0\,\text{m}^2$ panel is: $1400 \times 8.0 = 11\,200\,\text{W}$ [1]

Total power received = $11\,200 \times 2 = 22\,400\,\text{W}$ [1]

$P = \frac{W}{t}$ therefore $W = Pt$ [1]

$W = 22400 \times 7200 = 160\ \text{MJ}$ [1]

6 a $I = \frac{P}{4\pi r^2}$ therefore $P = I \times 4\pi r^2$ [1]

$P = 1.0 \times 10^{-4} \times 4 \times \pi \times (15)^2 = 0.28\ \text{W}$ (2 s.f.) [1]

$I = \frac{P}{4\pi r^2}$ therefore the intensity at 120 m

$= I = \frac{0.28}{4 \times \pi \times 120^2} = 1.6 \times 10^{-6}\,\text{W m}^{-2}$ (2 s.f.) [1]

b Intensity has fallen by a factor of 65. [1]

As intensity $\propto$ (amplitude)^2 the amplitude will have decreased by a factor of 8.1 ($\sqrt{65}$) (2 s.f.) [1]

1 Lower chance of cloud cover, intensity of electromagnetic waves is greater (less energy absorbed by the atmosphere)

2 All EM waves can be detected, no atmospheric distortion, no weather/cloud cover

Drawbacks: Cost, difficult to repair

3 a i Gamma rays, X-rays, Ultraviolet, longer λ of IR, radio waves longer than around 10 m

ii most wavelengths of visible light, microwaves, radio waves up to 10 m

b $v = f\lambda$ therefore if $f = \frac{v}{\lambda}$

Highest frequency corresponds to a λ of around 200 nm:

$f = \frac{v}{\lambda} = \frac{3.00 \times 10^8}{200 \times 10^{-9}} = 1.5 \times 10^{15}\ \text{Hz}$

Lowest frequency corresponds to a λ of around

10 m: $f = \frac{v}{\lambda} = \frac{3.00 \times 10^8}{10} = 30\,\text{MHz}$

11.6

1 Gamma rays, X-rays, ultraviolet, visible light, infrared, microwaves, radio waves [2]

(1 mark if one incorrect, 0 marks if more than one incorrect)

2 Polarisation [1]

3 a $\lambda = \frac{v}{f} = \frac{3.00 \times 10^8}{88 \times 10^6} = 3.4\,\text{m}$ (2 s.f.) [1]

b $\lambda = \frac{v}{f} = \frac{3.00 \times 10^8}{2.4 \times 10^9} = 0.13\,\text{m}$ (2 s.f.) [1]

c $\lambda = \frac{v}{f} = \frac{3.00 \times 10^8}{9.0 \times 10^{16}} = 3.3\,\text{nm}$ (2 s.f.) [1]

4 $v = f\lambda$ therefore if $f = \frac{v}{\lambda}$

Highest frequency:

$f = \frac{v}{\lambda} = \frac{3.00 \times 10^8}{400 \times 10^{-9}} = 7.5 \times 10^{14}\ \text{Hz}$ [1]

Lowest frequency:

$f = \frac{v}{\lambda} = \frac{3.00 \times 10^8}{700 \times 10^{-9}} = 4.3 \times 10^{14}\ \text{Hz}$ (2 s.f.) [1]

5 $t = \frac{\textit{distance travelled}}{\textit{speed}} = \frac{150 \times 10^9}{3.00 \times 10^8}$ [1]

$t = 500\,\text{s}$ [1]

6 distance travelled = speed $\times$ time taken [1]

Time taken for pulse to reach aircraft = 0.28 µs [1]

distance travelled = $3.00 \times 10^8 \times 0.28 \times 10^{-6}$ = 84 m [1]

1 When the Polaroids are aligned the intensity is at a maximum value.

As one is rotated the intensity is reduced.

When the second Polaroid is aligned at 90° to the first, the intensity is zero.

2 As the grille is rotated the intensity falls.

It falls to zero when the gaps in the grille are in the opposite plane to the microwaves (after 90°).

As the grille is rotated further the intensity increases again.

It reaches the maximum value again when the gaps in the grille are in the same plane as the microwaves (after 180°).

3 Holes will not allow any orientation of plane polarised microwaves through.

11.7

1 Only transverse waves can be plane polarised. [1]

2 At 0°, 180°, and 360° the Polaroids are aligned, so the maximum intensity is received. [1]

At 90° and 270°the Polaroids are aligned in opposite planes, so the intensity falls to zero. [1]

3 a It must be plane polarised. [1]

b Rotate the Polaroid further [1] until it is at 90° from the minimum intensity. [1]

In this orientation the Polaroid is aligned in the same plane as the light emitted from the screen. [1]

4 a $20 \times 9.0 \times 10^{-4} = 1.8 \times 10^{-3}$ W [1]

b $\frac{100}{28} = 3.6$ (2 s.f.) [1]

11.8

1 As the speed of light through the material decreases the refractive index increases. [1]

From $n = \frac{c}{v}$ [1] the refractive index is inversely proportional to the speed of light through the material $n \propto \frac{1}{v}$ [1]

2 $n = \frac{c}{v}$ [1]

$n = \frac{3.00 \times 10^8}{220 \times 10^6} = 1.4$ (2 s.f.) [1]

3 $n = \frac{c}{v}$ therefore $v = \frac{c}{n}$ [1]

$v = \frac{3.00 \times 10^8}{1.33}$ [1]

$v = 2.3 \times 10^8\ \mathrm{m\,s^{-1}}$ (2 s.f.) [1]

4 $n_1 \sin\theta_1 = n_2 \sin\theta_2$ therefore $n_2 = \frac{n_1 \sin\theta_1}{\sin\theta_2}$ [1]

$n_2 = \frac{1.10 \times \sin 51}{\sin 36} = 1.5$ (2 s.f.) [1]

5 Normal and partial reflection shown [1]

Angle θ_2 is less than θ_1 [1]

Angle $\theta_2 = 43^o$ [2]

(from workings below)

$n_1 \sin\theta_1 = n_2 \sin\theta_2$ therefore $\sin\theta_2 = \frac{n_1 \sin\theta_1}{n_2}$

$= \frac{1.47 \times \sin 45}{1.52} = 0.638\ldots$

Angle $\theta_2 = \sin^{-1} 0.638.. = 43^o$

6 $n_1 \sin\theta_1 = n_2 \sin\theta_2$ therefore $\sin\theta_2 = \frac{n_1 \sin\theta_1}{n_2}$ [1]

$\sin\theta_2 = \frac{2.42 \times \sin 20}{1.33} = 0.622..$ [1]

$\theta_2 = \sin^{-1} 0.622\ldots = 38^\circ$ (2 s.f.) [1]

11.9

1 In order for TIR the light must be travelling from a material of higher refractive index to one of a lower refractive index.

2 The pulse reflected multiple times arrives after the pulse transmitted through the centre of the fibre. The pulse reflected multiple times has a lower amplitude when it leaves the fibre than the pulse transmitted through the centre of the fibre.

3 Light follows a curved path (sinusoidal).

1 From $\sin C = \frac{1}{n}$ [1]

If the refractive index decreases the critical angle increases. [1]

2 $\sin C = \frac{1}{n}$ therefore $C = \sin^{-1}\left(\frac{1}{n}\right)$ [1]

$C = \sin^{-1}\left(\frac{1}{2.42}\right) = 24^\circ$ (2 s.f.) [1]

3 $\sin C = \frac{1}{n}$ therefore $n = \frac{1}{\sin C}$ [1]

$n = \frac{1}{\sin 42.8} = 1.47$ (3 s.f.) [1]

4 C = 36° [1]

$\sin C = \frac{1}{n}$ therefore $n = \frac{1}{\sin C}$

$n = \frac{1}{\sin 36}$ [1]

$n = 1.7$ (2 s.f.) [1]

5 a At the critical angle therefore light travels along boundary. [2]

b Below the critical angle therefore light is partially reflected and refracted. [2]

c Above the critical angle therefore light is totally internally reflected. [2]

6 $n = \frac{c}{v} = \frac{3.00 \times 10^8}{185 \times 10^6} = 1.62\ldots$ [1]

$\sin C = \frac{1}{n}$ therefore $C = \sin^{-1}\left(\frac{1}{n}\right)$ [1]

$C = \sin^{-1}\left(\frac{1}{1.62\ldots}\right) = 38.0^\circ$ (3 s.f.) [1]

12.1

1 Destructive interference [1]

In order to cancel out the sound waves from the surrounding environment [1]

2 As figure 3. [2]

3 intensity $\propto$ (amplitude)^2 [1]

As the amplitude doubles the intensity will increase by a factor of 4 (2^2) [1]

[1 mark only if mentioned only increasing]

4 Two clearly distinct waves with the correct wavelength [1] and amplitude. [1]

Waves have a clear sinusoidal shape. [1]

Constructive interference where the waves both have positive displacement [1] and where the waves both have negative displacement. [1]

Destructive interference where one wave has positive displacement and the other has negative displacement. [1]

5 See Figure 1. [6]

12.2

1 Path difference at the 1st order maxima = 1λ.

$f = \frac{v}{\lambda} = \frac{340}{0.28} = 1200\,\text{Hz}$ (2 s.f.)

2 If frequency is halved, the wavelength is doubled. Maxima and minima would be further apart (the path difference at the 1st order maxima would be doubled).

1 Use a ruler to carefully measure the path from the centre of the first slit to the 1st order maxima.
Use a ruler to carefully measure the path from the centre of the second slit to the 1st order maxima.
Calculate the difference.
Path difference at the 1st order maxima = 1λ.
Repeat for different maxima and minima (being careful to relate the path difference to the wavelength (e.g., path difference at the 2nd order maxima = 2λ, etc)).
Average the values for the wavelength.

2 Most significant diffraction when the gap size = wavelength.

$\lambda = \frac{v}{f} = \frac{3.00 \times 10^8}{24 \times 10^9} = 1.3\,\text{cm}$ (2 s.f.)

3 Use the metal sheets to create a single gap. Place in front of the microwave source and rotate the source through 180°. If the intensity received by the receiver drops then rises again the source is plane polarised.

1 Diagram showing light reflecting off the top surface (obeying the law of reflection) and light reflecting off the bottom surface after travelling through the oil (refracting on entry and exit).

2 The light travelling through the oil travels a greater distance (approx. 2 x the thickness of the oil).

3 Wavelength of red light is greater than blue light.
The path difference must be greater in order to produce destructive interference; therefore, the oil needs to be thicker.

1 a Constructive [1]
b Constructive [1]
c Destructive [1]

2 Increases [1]
Zero in the centre (at the central maxima) [1]
Moving through 180° or π rad at the first-order minima, 360° or 2π rad at the first-order maxima, etc [1]
Reaching 1080° or 6π rad at the third-order maxima. [1]

3 At the second-order maxima the path difference = 2 λ [1]
Therefore, the wavelength = 4.5 cm [1]

4 a the wavelength of the sound;
$\lambda = \frac{v}{f} = \frac{340}{2000} = 0.17\,\text{m}$ [1]
b At a phase difference of 5π radians the path difference = 2.5 λ [1]
Path difference = 2.5 × 0.17 = 0.43 m [1]
c At the second-order minima the phase difference = 3π radians [1]
At a phase difference of 3π radians the path difference = 1.5λ [1]
Path difference = 1.5 × 0.17 = 0.26 m [1]

12.3

1 There would be green light in place of red. The separation between fringes would be smaller due to the shorter wavelength.

2 This would reduce percentage uncertainty in measurements as the distance will be greater.

3 $\lambda = \frac{ax}{D}$ therefore $x = \frac{\lambda D}{a}$

$x = \frac{632.8 \times 10^{-9} \times 10}{0.50 \times 10^{-3}} = 13\text{ mm}$

1 Waves must be coherent to form a stable interference pattern [1]
Using a monochromatic source, a single slit and double slit results in two sources of coherent light [1]

2 $\lambda = \frac{ax}{D}$ [1]

$\lambda = \frac{0.6 \times 10^{-3} \times 1.4 \times 10^{-3}}{1.6} = 530\text{ nm}$ [1]

3 $x = 8.3$ mm [1]
x measured across several fringes [1]

$\lambda = \frac{1.0 \times 10^{-3} \times 8.3 \times 10^{-3}}{15} = 550\text{ nm} \pm 50\text{ nm}$ [1]

4 $\lambda = \frac{ax}{D}$ therefore $D = \frac{ax}{\lambda}$ [1]

$D = \frac{0.40 \times 10^{-3} \times 1.8 \times 10^{-3}}{610 \times 10^{-9}}$ [1]

$D = 1.2$ m (2 s.f.) [1]

5 a As λ increases x increases. [1]
b $x = \frac{\lambda D}{a}$ therefore if other factors remain constant $x \propto \frac{1}{a}$ [1]
As a doubles x halves. [1]
c $x = \frac{\lambda D}{a}$ therefore if other factors remain constant $x \propto D$ [1]
As D increases by a factor of 3 x increases by a factor of 3. [1]
d Double the frequency and the wavelength halves. [1]
$x = \frac{\lambda D}{a}$ therefore if other factors remain constant $x \propto \lambda$ [1]
As λ halves x halves. [1]

12.4

1 Measure the distance between adjacent nodes (or antinodes) – this is equal to $\frac{\lambda}{2}$.
Multiply the average distance by 2.

2 $\lambda = \frac{v}{f} = \frac{3.00 \times 10^8}{5.0 \times 10^9} = 0.060\,\text{m}$

Therefore the distance between nodes = 0.030 m

3 Close to the metal sheet both waves which form the stationary wave have similar amplitudes (they have travelled similar distances from the source).
At other nodes there is a greater difference between the amplitudes of the two waves (as they have travelled different distances). Resulting in non-perfect cancellation.

1 Comprised of oscillations/vibrations. [1]
Particles in the waves have frequency, period of oscillation and amplitude. [1]

2 a They are in antiphase [1]
b They are in phase [1]

3 The distance between adjacent nodes is equal to $\frac{\lambda}{2}$. [1]
The wavelength of the parent waves = 0.30 × 2 = 0.60 m [1]

4 At the node the amplitude = 0 [1]
Moving away from the node the amplitude increases (reaching a maximum at the antinode). [1]
Moving past the antinode, the amplitude reduces back to zero at the node. [1]

5 a (i) 180° or π rad. [1]
(ii) 360° or 2π rad. [1]
b (i) Flat line [1]
(ii) Simple sinusoidal wave (two waves on top of each other) [2]
(iii) As first diagram [1]

12.5

1 107 Hz
2 See Table 1
a At fundamental frequency
b At 2nd harmonic
c No pattern forms (not an integer multiple of the fundamental frequency)

3 As T increases v increases. From $v = f\lambda$ as λ is constant (as the length of the string is constant), f increases. As T increases by a factor of 2, f increases by a factor of $\sqrt{2}$.

1 A stretched string fixed at one end [1] with a vibration generator (connected to a signal generator) at the other end. [1]
Adjust the frequency of the signal generator. [1]

2 If L increases f_o decreases. [1]
If L doubles f_o halves. [1]

3 a 3rd harmonic, therefore the fundamental frequency must be $\frac{120}{3} = 40$ Hz. [1]
b Length of string = 0.36 m, therefore the distance between nodes = 0.12 cm [1]
Wavelength = 2 × 0.12 = 0.24 cm [1]
(Using $\lambda = \frac{2}{3}$ L results in both marks)
c (i) 4th harmonic, see Table 1 [1]
(ii) no pattern formed (not an integer multiple of f_o) [1]

4 a 0.5 Hz +/- 0.05 Hz [1]
b Intensity increases at integer multiple of f_o. [1]
Due to the string also vibrating at these harmonics. [1]

5 a 6th harmonic, node to node distance = $\frac{0.90}{6}$ = 0.15 m [1]
Wavelength = 2 × 0.15 = 0.30 cm [1]
b $v = f\lambda$ [1]
$v = 3600 \times 0.30 = 1080\ \text{m s}^{-1}$ [1]

12.6

1 See figure 3 at the fundamental frequency
2 a gradient = Lf
The wavelength of the progressive waves = $4L$.
$v = f\lambda$, becomes $v = f \times 4L$ therefore $\frac{v}{4} = fL$.
b Gradient = 85 m s^{-1} ± 5 m s^{-1}
Therefore speed of sound = 85 x 4 = 340 m s^{-1}
c Tuning fork is slightly above the tube.
Measurements of L contain slight systematic error and are shorter than $\frac{\lambda}{4}$.

1 Connect speaker to signal generator. [1]
Position speaker in front of solid surface, with microphone in between. [1]
Adjust frequency of sound until a number of nodes and antinodes are detected using the microphone. [1]
Measure the distance between nodes = $\frac{\lambda}{2}$. Therefore the wavelength of the sound waves = node-to-node distance × 2 [1]

2 a $L = \frac{\lambda}{4}$ therefore $\lambda = 4L$ [1]
$\lambda = 4 \times 1.2 = 4.8\,\text{m}$ [1]
b $f = \frac{v}{\lambda} = \frac{340}{4.8} = 71\,\text{Hz}$ [1]

3 $L = \frac{\lambda}{2}$ therefore $\lambda = 2L$ [1]

$\lambda = 2 \times 1.2 = 2.4$ m [1]

$f = \frac{v}{\lambda} = \frac{340}{2.4} = 142$ Hz (3 s.f.) [1]

4 Nodes: B and E [2]

Antinodes: A, D and G [3]

(1 mark for each correctly identified. Deduct 1 mark for each incorrect letter, minimum mark = 0)

5 At $2f_0$ there would need to be an antinode at each end of the tube. [1]

In a closed tube there must be a node at the closed end. [1]

13.1

1 a $E = 6.63 \times 10^{-34} \times 1.02 \times 10^{14}$
$= 6.76 \times 10^{-20}$ J (3 s.f.) [1]

b $E = 6.63 \times 10^{-34} \times 97.0 \times 10^{6}$
$= 6.43 \times 10^{-26}$ J (3 s.f.) [1]

c $E = 6.63 \times 10^{-34} \times 6.00 \times 10^{14}$
$= 3.98 \times 10^{-19}$ J (3 s.f.) [1]

2 Violet [1] Highest frequency. [1]

From $E = hf$, the higher the frequency the greater the energy. [1]

3 $E = \frac{hc}{\lambda}$ therefore $\lambda = \frac{hc}{E}$ [1]

$\lambda = \frac{6.63 \times 10^{-34} \times 3.00 \times 10^{8}}{3.32 \times 10^{-18}}$ [1]

$\lambda = 59.9$ nm (3 s.f.) [1]

4 a 6.3×10^{18} eV (2 s.f.) [1]

b 206 keV (3 s.f.) [1]

c 3.8×10^{12} eV (2 s.f.) [1]

5 Note, the exact values in this question may vary depending on the method used to calculate the frequency.

Radio: $f = 9.7$ MHz [1]

Infrared: $f = 2.5 \times 10^{13}$ Hz [1]

Visible – red: $f = 4.5 \times 10^{14}$ Hz [1]

Visible – green: $f = 5.6 \times 10^{14}$ Hz [1]

Visible – blue: $f = 6.5 \times 10^{14}$ Hz [1]

UV: $f = 1.4 \times 10^{15}$ Hz [1]

X-ray: $f = 2.5 \times 10^{17}$ Hz [1]

gamma: $f = 3.6 \times 10^{20}$ Hz [1]

6 a $E = \frac{6.63 \times 10^{-34} \times 3.00 \times 10^{8}}{4.50 \times 10^{-10}} = 4.42 \times 10^{-16}$ J [1]

$= 2760$ *eV* [1]

b $E = \frac{6.63 \times 10^{-34} \times 3.00 \times 10^{8}}{600 \times 10^{-9}} = 3.3 \times 10^{-19}$ J [1]

$= 2.1$ *eV* [1]

7 a $eV = \frac{hc}{\lambda}$ [1]

$V = \frac{hc}{e\lambda} = \frac{6.63 \times 10^{-34} \times 3.00 \times 10^{8}}{1.60 \times 10^{-19} \times 620 \times 10^{-9}}$

$= 2.0$ V (2 s.f.) [1]

b General shape of I–V characteristic for diode (see topic 9.6) [3]

Threshold p.d. lower for the red LED [1]

Red photons have a lower frequency [1], therefore less energy (lower threshold p.d) is required [1]

8 Energy of each photon,

$E = \frac{hc}{\lambda} = \frac{6.63 \times 10^{-34} \times 3.00 \times 10^{8}}{405 \times 10^{-9}} = 4.91... \times 10^{-19}$ J [1]

10 mW = 10×10^{-3} J s^{-1} [1]

Number of photons = $\frac{10 \times 10^{-3}}{4.91... \times 10^{-19}}$

$= 2.0 \times 10^{-16}$ photons per second (2 s.f.) [1]

9 Connect the LED to a variable supply. [1]
A safety resistor is also connected in series with the LED. [1]
Connect a voltmeter across the LED. [1]
Slowly increase the p.d. across the LED until it just emits light. Record the p.d. V across the LED. [1]
The Planck constant is calculated using $eV = hc/\lambda$ or $h = eV\lambda/c$, where e is the alimentary charge and c is the speed of light in a vacuum. [1]
Improvement: Carry out the experiment in a dark room or place a black tube over the LED to judge when the LED just starts to emit light. [1]

13.2

1 Become (positively) charged [1]
Gold leaf would rise/move away from the stem [1]

2 a No emission [1] Infrared photons are below threshold frequency (have insufficient energy) [1]

b Emission of photoelectrons as blue photons are above threshold frequency [1]

3 Energy transferred to each electron comes from a single photon in a one-to-one interaction. [1]

Energy of each photon depends on its frequency ($E = hf$). [1]

Greater the frequency, the higher the energy of the photon and so the greater the maximum kinetic energy of the electron. [1]

4 Maximum [1] wavelength that would cause photoelectric emission from the surface of a metal. [1]

5 Increased emission [1]

Number of emitted electrons per second would quadruple. [1]

Quadrupling the intensity results in four times the number of photons; therefore, four times the number of electrons emitted per second. [1]

13.3

1 $hf = \phi + KE_{MAX} = 3.77 \times 10^{-19} + 2.68 \times 10^{-19}$ [1]

$hf = 6.45 \times 10^{-19}\ J$ [1]

2 a $KE_{MAX} = hf - \phi = 5.20 - 4.08 = 1.12$ eV [1]

b $hf < \phi$ [1] therefore no electrons emitted [1]

3 $hf_o = \phi$ therefore $f_o = \dfrac{\phi}{h}$

Zinc: $\phi = 4.30$ eV $= 6.88 \times 10^{-19}$ J [1]

Sodium: $\phi = 2.36$ eV $= 3.776 \times 10^{-19}$ J [1]

Zinc: $f_o = \dfrac{6.88 \times 10^{-19}}{6.63 \times 10^{-34}} = 1.04 \times 10^{15}$ Hz (3 s.f.) [1]

Sodium: $f_o = \dfrac{3.776 \times 10^{-19}}{6.63 \times 10^{-34}} = 5.70 \times 10^{14}$ Hz (3 s.f.) [1]

4 With monochromatic radiation all photons have the same frequency and therefore the same energy. [1]

Each electron requires a specific energy to free it from the surface of the metal (the lowest energy required to free an electron is equal to the work function of the metal). [1]

The kinetic energy of the emitted electron is a result of the remaining energy after the electron has been freed. [1]

5 $hf = \phi + KE_{MAX}$ therefore $KE_{MAX} = hf - \phi$

Sodium: $\phi = 2.36$ eV $= 3.78 \times 10^{-19}$ J

$KE_{MAX} = 6.63 \times 10^{-34} \times 1.48 \times 10^{15} - 3.78 \times 10^{-19}$

$= 6.03 \times 10^{-19}$ J (3 s.f.) [1]

$KE_{MAX} = \dfrac{1}{2}mv^2$ therefore $v = \sqrt{\dfrac{2 \times KE_{MAX}}{m}}$ [1]

$v = \sqrt{\dfrac{2 \times 6.03 \times 10^{-19}}{9.11 \times 10^{-31}}}$ [1]

$v = 1.15$ Mm s^{-1} (3 s.f.) [1]

6 $hf = \phi + KE_{MAX}$ therefore $\phi = hf - KE_{MAX} = (6.63 \times 10^{-34}) \times (8.0 \times 10^{14}) - (1.36 \times 1.60 \times 10^{-19})$ [1]

$\phi = 3.128 \times 10^{-19} = 3.1 \times 10^{-19}$ J (2 s.f.) [1]

$hf_0 = \phi$ therefore $f_0 = \dfrac{\phi}{h} = \dfrac{3.128 \times 10^{-19}}{6.63 \times 10^{-34}}$ [1]

$f_0 = 4.7 \times 10^{14}$ Hz (2 s.f.) [1]

13.4

1 a It decreases.

b It decreases by a factor of 1 over $\sqrt{3}$.

c It increases by a factor of $\sqrt{10}$.

2 a 2.03×10^{-12} m

b 5.01×10^{-11} m

3 1.23×10^{-10} m

1 Adjust the accelerating p.d. This changes the velocity/momentum of the electron and therefore its wavelength.

2 They are too penetrating. They may damage the material.

3 $E_K = 40$ eV $= 6.4 \times 10^{-18}$ J

$\mathcal{E}_K = \dfrac{1}{2}mv^2$ therefore

$v = \sqrt{\dfrac{2 \times \mathcal{E}_K}{m}} = \sqrt{\dfrac{2 \times 6.4 \times 10^{-18}}{9.11 \times 10^{-31}}} = 3.7 \times 10^6$ m s^{-1}

$\lambda = \dfrac{h}{p}$ as momentum, $p = mv$ we can say $\lambda = \dfrac{h}{mv}$

$\lambda = \dfrac{6.63 \times 10^{-34}}{9.11 \times 10^{-31} \times 3.7 \times 10^6} = 2.0 \times 10^{-10}$ m (2 s.f.)

Similar size to the space between the atoms in the crystal, resulting in significant diffraction.

1 $\lambda = \dfrac{h}{p}$ [1]

$\lambda = \dfrac{6.63 \times 10^{-34}}{1.67 \times 10^{-19}} = 3.97 \times 10^{-15}$ m (3 s.f.) [1]

2 The proton has a greater mass [1]

Therefore, at the same velocity the momentum of the proton is greater than the electron [1]

As $\lambda = \dfrac{h}{p}$ the wavelength of the proton must be smaller [1]

3 Wavelength of the electron is very small [1]

In order to observe diffraction the electron must pass through a gap a similar size to its wavelength [1]

This does not happen in the course of most experiments using electrons [1]

4 a $v = \dfrac{h}{\lambda m} = \dfrac{6.63 \times 10^{-34}}{3.63 \times 10^{-10} \times 9.11 \times 10^{-31}}$ [1]

$v = 2.00 \times 10^6$ m s^{-1} [1]

b $v = \dfrac{h}{\lambda m} = \dfrac{6.63 \times 10^{-34}}{4.85 \times 10^{-12} \times 9.11 \times 10^{-31}}$ [1]

$v = 150 \times 10^6$ m s^{-1} [1]

5 a $\lambda = \dfrac{h}{p}$ as momentum, $p = mv$ therefore $\lambda = \dfrac{h}{mv}$

$\lambda = \dfrac{6.63 \times 10^{-34}}{9.11 \times 10^{-31} \times 4.20 \times 10^7}$ [1]

$\lambda = 1.73 \times 10^{-11}$ m [1]

b $v = \dfrac{h}{\lambda m} = \dfrac{6.63 \times 10^{-34}}{1.73 \times 10^{-11} \times 1.67 \times 10^{-27}}$ [1]

$v = 22.9$ km s^{-1} (3 s.f.) [1]

6 $0.25\,c = 0.25 \times 3.00 \times 10^8 = 75 \times 10^6$ m s^{-1} [1]

$\lambda = \dfrac{h}{mv} = \dfrac{6.63 \times 10^{-34}}{9.11 \times 10^{-31} \times 75 \times 10^6}$

$= 9.7 \times 10^{-12}$ m (2 s.f.)

[1]

14.1

1 There is a net flow of thermal energy from object A to object D.

2 If the thermometer is not at the same temperature as the object there would be a net flow of thermal energy from the object to the thermometer. This would change the temperature of the object (depending on the relative temperatures of the thermometer and the object).

1 a Net flow of thermal energy from A to B. [1]

b No net flow of thermal energy. [1]

c Net flow of thermal energy from B to A. [1]

2 a 273 K [1]

b 310 K [1]

c 152.5 K [1]

3 a −273 °C [1]

b −73 °C [1]

c 77 °C [1]

4 0 K is the lowest possible temperature.

5 15 °C is 288 K

temperature of metal > temperature of water [1]

There is a net flow of thermal energy from the metal to the water. [1]

This increases the temperature of the water and reduces the temperature of the metal block. [1]

6 Yes it is sensible. [1]

The difference in temperatures expressed in K and °C are negligible at such large temperatures (1000000273 K ~ 1000000000 °C) [1]

7 When the thermometer is placed in the hot water, there is a net flow of thermal energy from the water to the thermometer. [1]

This reduces the temperature of the water (and increases the temperature of the thermometer). [1]

This net flow of energy stops when the thermometer and the water are at the same temperature (which is lower than the initial temperature of the water) [1]

14.2

1 Several straight lines of different lengths showing random motion. See Figure 4 in the main content pages.

2 At lower temperatures particles move more slowly. Motion remains random but becomes less haphazard.

1 Highest Energy
Gas
Liquid
Solid
Lowest Energy [1]

2 Particles are much further apart in gases than in solids. [1]

3 a Use of density = $\frac{\text{mass}}{\text{volume}}$ [1]

Temperature / °C	5.000	20.000	40.000	60.000	90.000
Density / kg m^{-3}	1000	998.0	992.1	983.3	963.9

Two marks for all five correct, one mark for three or more correct.

b At higher temperatures the average speed of the particles increases. [1]

This increases the rate of collision between the particles, causing the liquid to expand. [1]

c As the temperature increases the volume of water increases – leading a rise in sea level. [1]

4 a Use of density = $\frac{\text{mass}}{\text{volume}}$ rearranged to

mass = density × volume [1]

Mass of 1.0 m^3 of ice = 920 × 1.0 = 920 kg [1]

Number of particles = $\frac{920}{3.0 \times 10^{-26}}$

= 3.1×10^{28} particles [1]

b Mass of 1.0 m^3 of water vapour = 0.590 × 1.0 = 0.590 kg [1]

Number of particles = $\frac{0.590}{3.0 \times 10^{-26}}$

= 2.0×10^{25} particles [1]

5 Ice: 3.1×10^{28} particles in 1.0 m^3

Cube roots: $\sqrt[3]{3.1 \times 10^{28}} = 3.14 \times 10^9$ particles along each face. [1]

As each face is 1.0 m long, spacing of each particle $= \frac{1.0}{3.14 \times 10^9} = 3.2 \times 10^{-10}$ m [1]

Water vapour: 2.0×10^{25} particles in 1.0 m^3

(Assuming a cube of gas): [1]

Cube root: $\sqrt[3]{2.0 \times 10^{25}} = 2.7 \times 10^8$ particles along each face. [1]

Spacing of each particle = $\frac{1.0}{2.7 \times 10^8} = 3.7 \times 10^{-9}$ m [1]

14.3

1 0 K is the lowest temperature as at this temperature the internal energy of a substance is at its minimum value. The kinetic energy of all the atoms or molecules is zero - they have stopped moving. [1]

2 Increases the internal energy of the substance [1] as the electrostatic potential increases when a substance changes phase (from solid to liquid, or from liquid to gas). [1]

3 Increase the temperature of the substance. [1]

Change the phase of the substance from solid to liquid, or from liquid to gas. [1]

4 The average kinetic energy of the atoms or molecules in 1.0 kg of water at 0 °C is the same as the average kinetic energy of the atoms or molecules in 1.0 kg of ice at 0 °C.

However, the atoms or molecules in 1.0 kg of water have a higher electrostatic potential energy. [1]

As the internal energy is the sum of the kinetic and potential energies of the atoms or molecules within the substance. The water has a higher internal energy. [1]

5 When the water vapour condenses there is a decrease in the electrostatic potential energy of the particles in the water. [1]

This energy is transferred from the water to the window. [1]

14.4

1 $\frac{E}{t} = \frac{\Delta m}{t} c\Delta\theta$

$\frac{E}{t} = 1.20 \times 4200 \times (80 - 10) = 350\,\text{kW}$

2 $\frac{E}{t} = \frac{\Delta m}{t} c\Delta\theta$

$\frac{E}{t} = 0.050 \times 4200 \times (60 - 20) = 8400\,\text{W}$

$E = Pt = 8400 \times (15 \times 60) = 7.6\,\text{MJ}$

3 $\frac{E}{t} = P = \frac{\Delta m}{t} c\Delta\theta$ rearranging for flow rate $\frac{\Delta m}{t} = \frac{P}{c\Delta\theta}$

$\frac{\Delta m}{t} = \frac{250 \times 10^6}{4200 \times 80} = 740\,\text{kg s}^{-1}$

From density $= \frac{\text{mass}}{\text{volume}}$

volume per second $= \frac{\text{mass per second (flow rate)}}{\text{density}}$

$= \frac{740}{1000} = 0.74\,\text{m}^3\,\text{s}^{-1}$

In 1 second the water moves 3.0 m. Therefore the cross sectional area of the water (and therefore the pipe) is given by volume = length × cross sectional area

cross sectional area $= \frac{\text{volume}}{\text{length}} = \frac{0.74}{3} = 0.25\,\text{m}^3$

From cross sectional area $= \pi r^2$ the radius = 0.28 m therefore the diameter = 0.56 m

1 a Use of $E = mc\Delta\theta$ [1]

Water: $E = 1.0 \times 4200 \times 20 = 84000\,\text{J}$ [1]

b Aluminium: $E = 0.600 \times 904 \times 20 = 10800\,\text{J}$ [1]

c Lead: $E = 4.2 \times 10^{-6} \times 129 \times 20 = 10.8\,\text{mJ}$ [1]

2 Appropriate diagram [2]

Measurements: Current [1]

Potential difference [1]

Initial temperature [1]

Final temperature [1]

Time [1]

3 Change in GPE is converted into thermal energy. [1]

Loss in GPE $= mg\Delta h = 1.0 \times 9.81 \times 450 = 4400\,\text{J}$ [1]

$E = mc\Delta\theta$ Therefore $\Delta\theta = \frac{E}{mc} = \frac{4400}{1.0 \times 4200} = 1.0\,°\text{C}$ [1]

4 $c = \frac{IVt}{m\Delta\theta}$ [1]

$c = \frac{2.0 \times 12 \times (5.0 \times 60)}{0.500 \times 32}$ [1]

$c = 450\,\text{J kg}^{-1}\,\text{K}^{-1}$ this corresponds to iron in Table 1 [1]

5 From $E = mc\Delta\theta$: $P = mc\frac{\Delta\theta}{\Delta t}$ [1]

From the graph $\frac{\Delta\theta}{\Delta t}$ = gradient [1]

Gradient $= 0.75\,°\text{C s}^{-1} \pm 0.04\,°\text{C s}^{-1}$ [1]

$P = mc\frac{\Delta\theta}{\Delta t} = mc$ gradient and therefore

$c = \frac{P}{m \times \text{gradient}}$ [1]

$c = \frac{60}{0.030 \times 0.75} = 2700\,\text{J kg}^{-1}\,\text{K}^{-1}$ +/− $200\,\text{J kg}^{-1}\,\text{K}^{-1}$ [1]

6 Drop in kinetic energy of car

$= \frac{1}{2}mv^2 = \frac{1}{2} \times 1500 \times 20^2 = 300\,\text{kJ}$ [1]

As there are two discs the energy dissipated by each disc = 150 kJ [1]

$E = mc\Delta\theta$ therefore $\Delta\theta = \frac{E}{mc} = \frac{150000}{8.0 \times 500} = 38\,°\text{C}$ [1]

14.5

1 $E = mL_f$ [1]

$E = 2.5 \times 88000 = 220\,\text{kJ}$ [1]

2 There is a greater change in internal energy changing phase from liquid to gas than from solid to liquid. [1]

3 $E = mL_f$ [1]

$E = 0.050 \times 398000 = 20\,\text{kJ}$ [1]

4 Energy transferred to the water $= Pt = 24 \times (60 \times 20)$ $= 28800\,\text{J}$ [1]

$E = mL_f$, $m = \frac{E}{L_f}$ [1]

$m = \frac{28800}{3.30 \times 10^4} = 0.87\,\text{kg}$ [1]

5 a $\frac{E}{\Delta t} = mc\frac{\Delta\theta}{\Delta t}$ [1]

$\frac{E}{t} = 0.060 \times 904 \times \frac{640}{16}$ [1]

$\frac{E}{\Delta t} = 2200\,\text{W}$ [1]

b $E = mL_f$ [1]

$E = 0.060 \times 398000 = 24000\,\text{J}$ [1]

6 Kinetic energy of bullet $= \frac{1}{2}mv^2 = \frac{1}{2} \times 0.008 \times 400^2$ $= 640\,\text{J}$ [1]

Energy required to heat the bullet to its melting point (327 °C):

$E = mc\Delta\theta = 0.008 \times 129 \times (327 - 40) = 296\,\text{J}$ [1]

Energy required to melt the lead = $E = mL_f$
$= 0.008 \times 23000 = 184\,\text{J}$ [1]

Energy remaining = 640 – (296 + 184) = 160 J [1]

$E = mc\Delta\theta$ Therefore $\Delta\theta = \frac{E}{mc} = \frac{160}{0.008 \times 129} = 155\,°\text{C}$ [1]

Therefore final temperature of lead = 155 + 327
= 480 °C (2 s.f.) [1]

15.1

1 mass of gas $= n \times M = 4.0 \times 0.004 = 0.016\,\text{kg}$

2 M of $CH_4 = 0.012 + (4 \times 0.001) = 0.016\,\text{kg mol}^{-1}$

3 M of $CO_2 = 0.012 + (2 \times 0.016) = 0.044\,\text{kg mol}^{-1}$

mass of gas $= n \times M$ therefore $n = \frac{\text{mass of gas}}{M} = \frac{0.050}{0.044}$

$= 1.1\,\text{mol}$

$N = n \times N_A = 1.1 \times 6.02 \times 10^{23} = 6.8 \times 10^{23}$ molecules

1 $N = n \times N_A = 3.0 \times 6.02 \times 10^{23}$ [1]
$N = 1.8 \times 10^{24}$ atoms or molecules [1]

2 The number of atoms in 1 mol of silicon is the same as the number of atoms in 1 mol of aluminium. [1]
However, the atoms have a different mass (silicon atoms have a greater mass than aluminium atoms). [1]

3 Initial momentum = mu and final momentum = $-mu$ [1]
Therefore change in momentum, $\Delta p = 2\,mu$ [1]

4 a $N = n \times N_A$ therefore $n = \frac{N}{N_A}$
$n = \frac{N}{N_A} = \frac{2.0 \times 10^{24}}{6.02 \times 10^{23}} = 3.3\,\text{mol}$ [1]

b $n = \frac{N}{N_A} = \frac{1.5 \times 10^{17}}{6.02 \times 10^{23}} = 2.5 \times 10^{-7}\,\text{mol}$ [1]

c $n = \frac{N}{N_A} = \frac{2.0 \times 10^{24}}{6.02 \times 10^{23}} = 3.3\,\text{mol}$ [1]

5 a $m = n \times M = \frac{N}{N_A} \times M$ therefore $N = \frac{m \times N_A}{M}$ [1]
$= \frac{1.0 \times 6.02 \times 10^{23}}{64 \times 10^{-3}} = 9.4 \times 10^{24}$ [1]

b $m = n \times M = \frac{N}{N_A} \times M$
Find m when $N = 1$ [1]
$M = \frac{1}{6.02 \times 10^{23}} \times 235 \times 10^{-3} = 3.9 \times 10^{-25}\,\text{kg}$ [1]

6 Mass of lead, density $= \frac{\text{mass}}{\text{volume}}$ therefore
mass = density × volume [1]
mass $= 11340 \times 0.20 = 2300\,\text{kg}$ [1]
Number of atoms $= \frac{2300}{3.46 \times 10^{-25}} = 6.6 \times 10^{27}$ atoms [1]
$n = \frac{N}{N_A} = \frac{6.6 \times 10^{27}}{6.02 \times 10^{23}} = 11 \times 10^3\,\text{mol}$ [1]

15.2

1 To ensure the temperature of the gas remains constant.

2 Select values of p and V.

p / Nm^{-2}	V / m^3	pV
440 000	0.11	48 000
200 000	0.24	48 000
60 000	0.80	48 000

1 Changing the volume would also affect temperature and/or pressure.

2 a Graph of p against θ with axis labelled (including units).
Points plotted correctly.
Line of best fit drawn.

2 b Same x-axis intercept.
Shallower gradient.

1 $pV = nRT$ therefore $V = \frac{nRT}{p}$ [1]
$V = \frac{60 \times 8.31 \times 250}{60000} = 2.1\,\text{m}^3$ [1]

2 a $p \propto \frac{1}{V}$ therefore if V is doubled, p halves. [1]

b $p \propto \frac{1}{V}$ therefore if V reduces by a factor of 3,
p increases by a factor of 3. [1]

3 $\frac{p}{T}$ = constant [1]
Initially: $\frac{300000}{293} = 1023.9\,\text{Pa K}^{-1}$ [1]
$p = \text{constant} \times T$
After the change $p = 1023.9 \times 373 = 382\,000$ [1]
Therefore the change = 82 000 Pa [1]

4 Graph of p against $\frac{1}{V}$ with axis labelled (including units)
Points plotted correctly [1]
Line of best fit drawn [1]
Determination of gradient = 48 000 +/– 2000 [1]
Gradient = nRT therefore $n = \frac{\text{gradient}}{RT}$
$n = \frac{48000}{8.31 \times 293} = 20\,\text{mol}$ [1]

5 $pV = nRT$ therefore $V = \frac{nRT}{p}$ [1]
$V = \frac{1 \times 8.31 \times 273}{100000}$ [1]
$V = 0.023\,m^3$ [1]

6 $pV = nRT$ therefore $n = \frac{pV}{RT}$ [1]
$n = \frac{50000 \times 0.25}{8.31 \times 288}$ [1]
$n = 5.2\,mol$ [1]
$N = n \times N_A = 5.2 \times 6.02 \times 10^{23}$
$= 3.1 \times 10^{24}$ particles (atoms or molecules) [1]

7 $pV = nRT$ therefore $n = \dfrac{pV}{RT}$ [1]

Temperature of air inside the lungs ~ 300 K

Volume of the lungs ~ 5800 ml → $0.0058\,m^3$ [1]

Pressure in the lungs ~ atmospheric pressure = 100000 Pa

$n \approx \dfrac{100000 \times 0.0058}{8.31 \times 300}$ [1] – mark awarded for using your estimates

$n \approx 0.23\,mol$ [1]

15.3

1 If there are large numbers of particles, approximately $\frac{1}{3}$ will be moving (or have components of their velocity) in each of the 3 dimensions (x, y and z).

2 Elastic collisions

Collisions only occur with the side of the container

Volume of gas/container is much larger than the volume of the particles

1 Mean speed $= \dfrac{100+200+150+50}{4} = 125\,m\,s^{-1}$ [1]

Mean square speed $= \dfrac{100^2+200^2+150^2+50^2}{4}$

$= 19000\,m^2\,s^{-2}$ [1]

Root mean square speed $= \sqrt{19000} = 140\,m\,s^{-1}$ [1]

2 Particles gain kinetic energy as the temperature increases [1]

Therefore the speeds of the particles increases [1]

3 $pV = \frac{1}{3}Nm\overline{c^2}$ therefore $V = \dfrac{\frac{1}{3}Nm\overline{c^2}}{p}$ [1]

$V = \dfrac{\frac{1}{3} \times 4.0 \times 10^{25} \times 4.7 \times 10^{-26} \times 450^2}{800000}$ [1]

$V = 0.16\,m^3$ [1]

4 $pV = \frac{1}{3}Nm\overline{c^2}$ therefore $p = \dfrac{\frac{1}{3}Nm\overline{c^2}}{V}$ [1]

$p = \dfrac{\frac{1}{3} \times 4.0 \times 10^{25} \times 4.7 \times 10^{-26} \times 600^2}{0.16}$ [1]

$p = 1.4\,MPa$ [1]

5 a $N = n \times N_A = 2.0 \times 6.02 \times 10^{23}$

$= 1.2 \times 10^{24}$ molecules [1]

b mass of molecule $= \dfrac{M}{N_A} = \dfrac{0.032}{6.02 \times 10^{23}}$

$= 5.3 \times 10^{-26}\,kg$ [1]

c $pV = \frac{1}{3}Nm\overline{c^2}$ therefore $\overline{c^2} = \dfrac{pV}{\frac{1}{3}Nm}$ [1]

$\overline{c^2} = \dfrac{140000 \times 0.020}{\frac{1}{3} \times 1.2 \times 10^{24} \times 5.3 \times 10^{-26}}$ [1]

$\overline{c^2} = 132000\,m^2\,s^{-2}$ [1]

$c_{r.m.s.} = \sqrt{132000} = 360\,m\,s^{-1}$ [1]

15.4

1 a The temperature increases. [1]

b If the speed doubles the kinetic energy will quadruple [1] as a result the temperature will quadruple. [1]

c If the speed increases by a factor of 5 the kinetic energy will increase by a factor of 25 (5^2) [1] as a result the temperature will increase by a factor of 25. [1]

2 $k = \dfrac{R}{N_A}$ [1]

$k = \dfrac{8.31}{6.02 \times 10^{23}} = 1.38 \times 10^{-23}\,J\,K^{-1}$ [1]

3 $pV = NkT$ rearranged to give $N = \dfrac{pV}{kT}$

18 °C = 291 K [1]

$N = \dfrac{450000 \times 0.50}{1.38 \times 10^{-23} \times 291}$ [1]

$N = 5.60 \times 10^{25}$ particles (atoms or molecules) [1]

$N = n \times N_A$ therefore $n = \dfrac{N}{N_A}$

$= \dfrac{5.60 \times 10^{25}}{6.02 \times 10^{23}} = 93.1\,mol$ [1]

4 Doubling the temperature of a real gas doubles the average kinetic energy of the atoms or molecules in the gas. [1]

However, unlike an ideal gas, the atoms or molecules in a real gas also have potential energies and so in a real gas the internal energy is equal to the sum of random distribution of kinetic and potential energies of the atoms or molecules within the gas. Doubling the kinetic energy does not double the internal energy. [1]

5 Using $\frac{1}{2}m\overline{c^2} = \frac{3}{2}kT$ [1]

[J] = k [K]

k = [J] [K^{-1}] [1]

6 $\frac{1}{2}m\overline{c^2} = \frac{3}{2}kT$ [1]

Therefore: $\overline{c^2} = \dfrac{\frac{3}{2}kT}{\frac{1}{2}m} = \dfrac{3kT}{m}$ and $c_{r.m.s.} = \sqrt{\dfrac{3kT}{m}}$ [1]

$c_{r.m.s.} = \sqrt{\dfrac{3 \times 1.38 \times 10^{-23} \times 293}{5.3 \times 10^{-26}}}$ [1]

$c_{r.m.s.} = 480\,m\,s^{-1}$ [1]

7 Kinetic energy of the helium atom is the same at the same temperature. [1]

Speed of the helium atom is greater [1]

The oxygen molecule has 8 times the mass of the helium atom [1]

Therefore to have the same kinetic energy the helium atom must be travelling 2.8 times ($\sqrt{8}$) times faster. [1]

16.1

1 a To convert from degree to radians: divide by $\frac{180}{\pi}$ [1]

$\frac{180}{\frac{180}{\pi}} = \pi \text{ rad} = 3.14\,\text{rad}$ [1]

b $\frac{45}{\frac{180}{\pi}} = \frac{1}{4}\pi \text{ rad} = 0.79\,\text{rad}$ [1]

2 a $\omega = \frac{2\pi}{T} = \frac{2\pi}{30} = 0.21\,\text{rad s}^{-1}$ [1]

b $\omega = \frac{2\pi}{T} = \frac{2\pi}{0.10} = 63\,\text{rad s}^{-1}$ [1]

3 $T = 365 \times 24 \times 60 \times 60 = 32 \times 10^6\,\text{s}$ in one year [1]

$\omega = \frac{2\pi}{T} = \frac{2\pi}{32 \times 10^6} = 2.0 \times 10^{-7}\,\text{rad s}^{-1}$ [1]

4 $4500\,\text{rpm} = \frac{4500}{60} = 75$ revolutions per second. [1]

$\omega = 2\pi f = 2\pi \times 75 = 470\,\text{rad s}^{-1}$ [1]

Time taken to complete 50 revolutions = 50 × period [1]

Time taken to complete 50 revolutions = $50 \times \frac{1}{75}$

Time taken to complete 50 revolutions = 0.67 s [1]

5 $\omega = 2\pi f$ therefore $f = \frac{\omega}{2\pi}$ [1]

$f = \frac{565}{2\pi} = 90\,\text{Hz}$ [1]

Time taken to complete 5400 revolutions = 5400 × period

Time taken to complete 5400 revolutions = $5400 \times \frac{1}{90}$ [1]

Time taken to complete 5400 revolutions = 60 seconds [1]

6 Period of each hand:

Second hand = 60 s [1]

Minute hand = 1 hour = 3600 s [1]

Hour hand = 12 hours = 43 200 s [1]

Second hand: $\omega = \frac{2\pi}{T} = \frac{2\pi}{60} = 0.10\,\text{rad s}^{-1}$ [1]

Minute hand: $\omega = \frac{2\pi}{T} = \frac{2\pi}{3600} = 1.7 \times 10^{-3}\,\text{rad s}^{-1}$ [1]

Second hand: $\omega = \frac{2\pi}{T} = \frac{2\pi}{43200} = 1.5 \times 10^{-4}\,\text{rad s}^{-1}$ [1]

16.2

1 a Gravitational attraction (gravity) [1]

b Electrostatic attraction [1]

c Friction (between the tyre and road) [1]

2 $v = r\omega$ [1]

$v = 0.20 \times 6.0 = 1.2\,\text{m s}^{-1}$ [1]

3 a $a = \frac{v^2}{r} = \frac{20^2}{60}$ [1] $a = 6.7\,\text{m s}^{-2}$ [1]

b $a = \omega^2 r = 5.0^2 \times 0.60$ [1] $a = 15\,\text{m s}^{-2}$ [1]

c $\omega = \frac{2\pi}{T} = \frac{2\pi}{750 \times 10^{-3}} = 8.4\,\text{rad s}^{-1}$ [1]

$a = \omega^2 r = 8.4^2 \times 1.5 = 110\,\text{m s}^{-2}$ [1]

4 $v = r\omega$ therefore $\omega = \frac{v}{r}$ [1]

$\omega = \frac{v}{r} = \frac{1.40}{0.30} = 4.7\,\text{rad s}^{-1}$ [1]

$a = \frac{v^2}{r} = \frac{1.40^2}{0.30}$ [1]

$a = 6.5\,\text{m s}^{-2}$ [1]

5 $a = 5 \times 9.81 = 49.1\,\text{m s}^{-2}$ [1]

$a = \frac{v^2}{r}$ therefore $v = \sqrt{ar}$ [1]

$v = \sqrt{ar} = \sqrt{49.1 \times 12} = 24\,\text{m s}^{-1}$ [1]

16.3

1 Diagram to show at the same force and speed the heavier particles follow a path of greater radius. This has the effect of the particles moving towards the bottom of the tube as it spins.

2 Diagram to show particle on the inside of the centrifuge. One force acting on the particle, a normal contact force from the wall of the centrifuge towards the centre of the centrifuge.

3 6000 rpm = 100 revolutions per second = $628.4\,\text{rad s}^{-1}$

$F = m\omega^2 r$

$r = \frac{63 \times 10^{-3}}{(2.0 \times 10^{-6} \times 628.4^2)} = 0.079\,\text{m}$

1 $\tan\theta = \frac{v^2}{rg}$ therefore $r = \frac{v^2}{\tan\theta g} = \frac{4.0^2}{\tan 30 \times 9.81}$ $= 2.8\,\text{m}$

2 $\tan\theta = \frac{v^2}{rg}$ therefore $v^2 = \tan\theta rg$

Since $v = \frac{2\pi r}{t}$ then $v^2 = \frac{4\pi^2 r^2}{t^2}$

$\frac{4\pi^2 r^2}{t^2} = \tan\theta rg$ therefore $4\pi^2 r = t^2 \tan\theta g$

$t^2 = \frac{4\pi^2 r}{g\tan\theta}$ therefore $t = \sqrt{\frac{4\pi^2 r}{g\tan\theta}} = 2\pi\sqrt{\frac{r}{g\tan\theta}}$

1 a Use: $F = \frac{mv^2}{r}$

Since $F \propto m$, if the mass doubles the force required doubles. [1]

b Since $F \propto v^2$, if the speed doubles the force required quadruples (increases by 2^2). [1]

c Since $F \propto v^2$, if the speed increases the force required increases by a factor of 9 (2^2), and since $F \propto \frac{1}{r}$, if the radius halves the force required doubles. Therefore the force needed increases by a factor of 18. [1]

2 Most likely to break at the bottom [1]

This is where the tension in the string is greatest [1] as the tension at the bottom, T, is given by:

$T = mg + \frac{mv^2}{r}$ [1]

3 a $F = 16 \times \sin 40°$ [1]

$= 10\,N$ (2 s.f.) [1]

b $F = \frac{mv^2}{r}$ [1]

$v = \sqrt{\frac{Fr}{m}} = \sqrt{\frac{10 \times 20}{1.2}}$ [1]

$v = 13\,m\,s^{-1}$ (2 s.f.) [1]

4 a $T = 20$ minutes $= 1200\,s$

$\omega = \frac{2\pi}{T} = \frac{2\pi}{1200} = 5.2 \times 10^{-3}\,rad\,s^{-1}$ [1]

b At the top the normal contact force $= N_{top} = mg - m\omega^2 r$ [1]

At the bottom the normal contact force $= N_{bottom} = m\omega^2 r + mg$ [1]

Therefore the change in the normal contact force is $2m\omega^2 r$ [1]

5 $t = 1$ year $= 32 \times 10^6\,s$

$v = \frac{2\pi r}{t} = \frac{2\pi \times 150 \times 10^9}{32 \times 10^6}$ [1]

$v = 29 \times 10^3\,m\,s^{-1}$ [1]

$F = \frac{mv^2}{r} = \frac{6.0 \times 10^{24} \times (29 \times 10^3)^2}{150 \times 10^9} = 3.3 \times 10^{22}\,N$ [1]

6 a At the North the no centripetal force required, therefore the scale reading would be 700 N [1]

b $t = 1$ day $= 86400\,s$

$v = \frac{2\pi r}{t} = \frac{2\pi \times 6400 \times 10^3}{86400}$ [1] $v = 470\,m\,s^{-1}$ [1]

The centripetal force acting on the person is given by:

$F = \frac{mv^2}{r} = \frac{\left(\frac{700}{9.81}\right) \times (470)^2}{6400 \times 10^3}$ [1]

$F = 2.5\,N$ [1]

Therefore the reading on the scale 700 N – 2.5 N = 697.5 N [1]

17.1

1 At this point the object is moving at its highest speed; reducing the uncertainty in the timing measurements.
The object will continue to move through the equilibrium position, even if the motion is damped.

2 Reduce the effect of random errors.
Reduce the uncertainty in the timing measured due to human errors in the timings (reaction time, etc.).

1 Sinusoidal in shape (either sine or cosine) [1]

Constant period and amplitude [1]

2 a $\omega = \frac{2\pi}{T} = \frac{2\pi}{0.40} = 16\,rad\,s^{-1}$ [1]

b $\omega = 2\pi f = 2\pi \times 0.75 = 4.7\,rad\,s^{-1}$ [1]

c Period $= \frac{26}{20} = 1.3\,s$ [1]

$\omega = \frac{2\pi}{T} = \frac{2\pi}{1.3} = 4.8\,rad\,s^{-1}$ [1]

3 a $a = -\omega^2 x$ [1]; $a = 2.5^2 \times 0.12 \times 10^{-3} = 7.5 \times 10^{-4}$ [1]

b When $x = 0$, $a = 0\,m\,s^{-2}$ [1]

4 a $\frac{\pi}{2}$ rad = 1.6 rad [1]

b π rad = 3.1 rad [1]

5 $\omega = \frac{2\pi}{T}$ therefore $T = \frac{2\pi}{\omega}$ [1]

First object. $\omega^2 = 10$ therefore $\omega = 3.1\,rad\,s^{-1}$ / Second object. $\omega^2 = 40$ therefore $\omega = 6.3\,rad\,s^{-1}$ [1]

First object. $T = \frac{2\pi}{\omega} = \frac{2\pi}{3.1} = 2.0\,s$ /

Second object. $T = \frac{2\pi}{\omega} = \frac{2\pi}{6.3} = 1.0\,s$ [1]

The time period the second object is half that of the first object. [1]

6 a $A = 2.0\,cm$ [1]

b Since $= -\omega^2 x$, $-\omega^2 = \frac{a}{x} =$ gradient [1]

Gradient $= \frac{\text{rise}}{\text{step}} = \frac{-80}{0.040} = -2000\,s^{-2}$ [1]

$-\omega^2 = -2000$ therefore $\omega = \sqrt{2000} = 45\,rad\,s^{-1}$ [1]

17.2

1 Sinusoidal shape (either sine or cosine) [1]

Two complete oscillations with the same period and constant amplitude [1]

Pendulum stationary at the maximum displacements (positive and negative) [1]

Pendulum moving fastest as it passes through the equilibrium position (the x-axis) [1]

2 a $\omega = \frac{2\pi}{T} = \frac{2\pi}{1.4} = 4.5\,rad\,s^{-1}$

$v = \pm\omega\sqrt{A^2 - x^2}$ [1]

$v = \pm 4.5 \times \sqrt{0.30^2 - 0.20^2} = \pm 1.0\,m\,s^{-1}$ [1]

b At $x = 0.00\,m$, $v = \pm\omega\sqrt{A^2 - x^2}$ becomes

$v = \pm\omega A$ [1]

$v = \pm 4.5 \times 0.30 = \pm 1.35\,m\,s^{-1}$ [1]

c $A = 0.30\,m$ therefore if $x = A$ $v = 0\,m\,s^{-1}$ [1]

3 a Period of pendulum $= \frac{16}{20} = 0.80\,s$ [1]

Angular frequency of pendulum

$\omega = \frac{2\pi}{T} = \frac{2\pi}{0.80}\,rad\,s^{-1}$ [1]

Since it was released from its amplitude,

$x = A\cos\omega t$ [1]

$x = A\cos\omega t = 0.16\cos\left(\frac{2\pi}{0.80} \times 0.40\right) = -0.16\,m$ [1]

b $x = A\cos\omega t = 0.16\cos(\frac{2\pi}{0.80}\times 0.80) = 0.16\,\text{m}$ [1]

c $x = A\cos\omega t = 0.16\cos(\frac{2\pi}{0.80}\times 19.30) = 0.11\,\text{m}$ [1]

4 a $A = 0.12\,\text{m}$ [1]

b Period = 3.1 s ± 0.2 s [1]

c $\omega = \frac{2\pi}{T} = \frac{2\pi}{3.1} = 2.0\,\text{rad s}^{-1}$ [1]

d $v_{max} = \omega A$ [1]

$v_{max} = 2.0 \times 0.12 = 0.24\,\text{m s}^{-1}$ [1]

5 a 0.12 m [1]

b From $x = 0.12\sin(3.5t)$, $\omega = 3.5\,\text{rad s}^{-1}$ [1]

c $\omega = \frac{2\pi}{T}$ therefore $T = \frac{2\pi}{\omega}$ [1]

$T = \frac{2\pi}{3.5} = 1.8\,\text{s}$ [1]

d (i) $x = 0\,\text{m}$ (sine function therefore at $t = 0$ $x = 0$) [1]

(ii) $x = 0.12\sin(3.5 \times 3.5) = -0.037\,\text{m}$ [1]

(iii) $x = 0.12\sin(3.5 \times 14) = -0.11\,\text{m}$ [1]

6 Displacement against time:

Sine graph [1]

Amplitude = 0.12 m

Period = 1.8 s [1]

Velocity against time:

Cosine graph [1]

Period = 1.8 s [1]

Acceleration against time:

Negative sine graph [1]

Period = 1.8 s [1]

17.3

1 a See Figure 3 in the main content pages.

Assumption: no frictional or other energy losses. [1]

The amplitude is the maximum displacement (maximum values on the x-axis) – both positive and negative. [1]

b Kinetic energy:

Maximum – When displacement = 0 (y-axis intercept). [1]

Minimum – At amplitude [1]

Potential energy:

Maximum – At amplitude [1]

Minimum – When displacement = 0 (y-axis intercept). [1]

2 a Maximum kinetic energy = 1.6 J (when potential energy = 0 J) [1]

Maximum potential energy = 1.6 J (when kinetic energy = 0 J) [1]

b Kinetic energy = Total energy – potential energy = 1.6 – 1.0 = 0.60 J [1]

$E_k = \frac{1}{2}mv^2$ therefore $v = \sqrt{\frac{2E_k}{m}}$ [1]

$v = \sqrt{\frac{2E_k}{m}} = \sqrt{\frac{2\times 0.60}{0.120}} = 3.2\,\text{m s}^{-1}$ [1]

3 a Starts at zero, sine shape [1]

At maximum displacement (both positive and negative) maximum potential energy – graph does not go negative – 'two humps' per oscillation. [1]

b Starts at maximum value, cosine shape [1]

At maximum displacement (both positive and negative) zero kinetic energy – graph does not go negative – 'two humps' per oscillation. [1]

4 $\omega = 2\pi f = 2\pi \times 0.40 = 2.5\,\text{rad s}^{-1}$ [1]

$x = A\cos\omega t = 0.050\cos(2.5\times 2.8) = 0.038\,\text{m}$ [1]

$v = \pm\omega\sqrt{A^2 - x^2} = \pm 2.5\times\sqrt{0.050^2 - 0.038^2}$

$= \pm 0.081\,\text{m s}^{-1}$ [1]

$E_k = \frac{1}{2}mv^2 = \frac{1}{2}\times 0.050\times 0.081^2 = 170\times 10^{-6}\,\text{J}$ [1]

5 $E_{k_{max}} = \frac{1}{2}mv_{max}^{\ 2}$

$v_{max} = \omega A$ and $\omega = 2\pi f$ therefore $v_{max} = 2\pi fA$ [1]

$v_{max}^{\ 2} = 4\pi^2 f^2 A^2$ [1]

$E_{k_{max}} = \frac{1}{2}mv_{max}^{\ 2} = 2m\pi^2 f^2 A^2$ [1]

17.4

1 Initial amplitude: 0.25 m

Period: 4.0 s

Angular frequency: $\omega = \frac{2\pi}{T} = \frac{2\pi}{4.0} = 1.6\,\text{rads}^{-1}$

2 Using a time interval of 2.0 s

$\frac{A_{2.0}}{A_0} = \frac{0.21}{0.25} = 0.84$

$\frac{A_{4.0}}{A_{2.0}} = \frac{0.18}{0.21} = 0.86$

$\frac{A_{6.0}}{A_{4.0}} = \frac{0.16}{0.18} = 0.89$

$0.84 \approx 0.86 \approx 0.89$ therefore likely to be an exponential decay.

1 Any valid example (one of each) [2] for example:

Free:

Mass-spring system

Pendulum

Ruler over the edge of a desk

Forced:

Mass-spring system attached to a vibration generator (or person)

Barton's pendulums

Person on a swing

2 Damping reduces the amplitude over time. [1]

3 a Sinusoidal, with constant period [1]

Constant amplitude [1]

b See light damping in Figure 2 in the main content pages.

Sinusoidal, with constant period [1]

Amplitude reduces with each oscillation [1]

4 a Oscillations are (lightly) damped [1]

b Period [1] (allow position of equilibrium position – 26 cm)

c Period = 2.0 s [1]

$f = \frac{1}{T} = \frac{1}{2.0} = 0.50\,\text{Hz}$ [1]

5 Cosine graph [1]

Constant period of 1.0 s and initial displacement (amplitude) = 5.0 cm [1]

Amplitude decreases after each oscillation [1]

Graph showing an exponential decrease in amplitude: First cycle = 5.0 cm, second ~ 4.5 cm, third ~ 4.1 cm, fourth ~ 3.6 cm, and fifth ~ 3.3 cm [1]

17.5

1 Driver: Radio wave transmitting coils (inside the scanner).

Forced oscillator: hydrogen nuclei.

2 $c = f\lambda$ therefore $\lambda = \frac{c}{f} = \frac{3.00 \times 10^8}{128 \times 10^6} = 2.3\,\text{m}$

3 $E = hf = 6.63 \times 10^{-34} \times 128 \times 10^6 = 8.49 \times 10^{-26}\,\text{J}$

1 See Figure 4 in the main content pages.

Graph of amplitude against driver frequency [1]

Natural frequency labelled on x-axis [1]

As the driver frequency approaches the natural frequency the amplitude increases [1]

Reaching a maximum at the natural frequency [1]

Above the natural frequency, as the driver increases the amplitude drops [1]

2 Glass resonates [1]

Amplitude of the oscillations increases dramatically [1]

Eventually the amplitude of the oscillations is so large the glass breaks [1]

3 The spinning drum creates a driving force on the panel. [1]

The driver frequency is related to the angular frequency of the spinning drum. [1]

At a specific angular frequency, the driver frequency is equal to the natural frequency of the panel, making it resonate. [1]

To reduce the amplitude of the panel it should be damped; any valid example installing rubber/foam pads, etc [1]

4 See Figure 4 in the main content pages.

Graph of amplitude against driver frequency [1]

Natural frequency = $\frac{180}{60} = 3.0\,\text{Hz}$, natural frequency labelled on x-axis [1]

As the driver frequency approaches the natural frequency the amplitude increases, reaching a maximum at the natural frequency. Above the natural frequency, as the driver increases the amplitude drops [1]

5 Time between bumps = period of oscillation

$= \frac{1}{\text{natural frequency of van}}$ [1]

Time between bumps = $\frac{10}{2.5} = 4.0\,\text{s}$ [1]

Natural frequency = $f = \frac{1}{T} = \frac{1}{4.0} = 0.25\,\text{Hz}$ [1]

At a different speed the time between the bumps will be different, therefore the driver frequency will not equal the natural frequency of the van. [1]

18.1

1 $10^{-15}\,\text{N}$

2 The field lines are closer together above the gold deposit.

1 It has a magnitude and direction (represented by an arrow). [1]

2 Gravitational field strength is always attractive and points towards the centre of mass of the object causing the gravitational field. [1]

3 a $g = \frac{F}{m} = \frac{15.0}{3.00} = 5.00\,\text{N kg}^{-1}$ [1]

b $g = \frac{F}{m} = \frac{29.4}{3.00} = 9.80\,\text{N kg}^{-1}$ [1]

c $g = \frac{F}{m} = \frac{1.62}{3.00} = 0.540\,\text{N kg}^{-1}$ [1]

4 Hold the newtonmeter vertically in a fixed position and suspend the known mass from it. [1]

Measure the value of force shown in the newtonmeter and use this, along with the known mass m, to find g using $g = \frac{F}{m}$. [1]

5 $g = \frac{F}{m}$ [1]

$F = ma$ therefore in base units $F = \text{kg m s}^{-2}$ [1]

m in base units: kg

g therefore: $\text{kg m s}^{-2}\,\text{kg}^{-1} = \text{m s}^{-2}$ [1]

6 On Earth: $F = mg = 75 \times 9.81 = 740\,\text{N}$ and On Mars $F = mg = 75 \times 3.7 = 280\,\text{N}$ [1]

Change = 740 – 280 = 460 N [1]

7 $a = \frac{F}{m}$

$F = mg$

$a = \frac{mg}{m} = g$ [1]

Both balls would initially accelerate at $9.81\,\text{m}\,\text{s}^{-2}$. [1]

18.2

1 Since $F = \frac{GMm}{r^2}$ and $y = mx + c$ [1]

F on the y-axis and $\frac{1}{r^2}$ on the x-axis [1]

Gradient $= GMm$ ($F = GMm\frac{1}{r^2} \rightarrow y = mx$) [1]

2 $F = \frac{GMm}{r^2}$

$M = M_E$ and $m = ms$ [1]

and $r = R_E + h$. [1]

Therefore $F = \frac{GM_E m_s}{(R_E + h)^2}$

3 a Since $F \propto M$ if M doubles F doubles. [1]

b Since $F \propto Mm$ if M doubles and m doubles, F quadruples. [1]

c Since $F \propto \frac{1}{r^2}$ if r halves F quadruples. [1]

d Since $F \propto M$ if M doubles F doubles and since $F \propto \frac{1}{r^2}$ if r decreases by a factor of four F increases by a factor of 16. [1]

Therefore in total F increases by a factor of 32. [1]

4 a $F = \frac{GMm}{r^2}$

$= \frac{6.67 \times 10^{-11} \times 1.67 \times 10^{-27} \times 1.67 \times 10^{-27}}{(1.0 \times 10^{-14})^2}$ [1]

$F = 1.9 \times 10^{-36}\,\text{N}$ [1]

b $F = \frac{GMm}{r^2} = \frac{6.67 \times 10^{-11} \times 65 \times 70}{(1.5)^2}$ [1]

$F = 1.3 \times 10^{-7}\,\text{N}$ [1]

c $F = \frac{GMm}{r^2}$

$= \frac{6.67 \times 10^{-11} \times 1.99 \times 10^{30} \times 5.68 \times 10^{26}}{(1400 \times 10^9)^2}$ [1]

$F = 3.8 \times 10^{22}\,\text{N}$ [1]

5 $F = \frac{GMm}{r^2}$ therefore $m = \frac{Fr^2}{GM}$ [1]

$m = \frac{2.03 \times 10^{20} \times (380 \times 10^6)^2}{6.67 \times 10^{-11} \times 5.97 \times 10^{24}}$ [1]

$m = 7.36 \times 10^{22}\,\text{kg}$ [1]

6 Net force on probe $F = \frac{GM_E m_{probe}}{r_{E\text{ to probe}}^2} - \frac{GM_M m_{probe}}{r_{M\text{ to probe}}^2}$ [1]

$F = \frac{6.67 \times 10^{-11} \times 5.97 \times 10^{24} \times 120}{(190 \times 10^6)^2}$

$- \frac{6.67 \times 10^{-11} \times 7.36 \times 10^{22} \times 120}{(190 \times 10^6)^2}$ [1]

$F = 1.31\,\text{N}$ (towards the Earth) [1]

18.3

1 Gravitational field strength is a vector quantity. Since gravitational fields are always attractive, the gravitational fields of the Earth and Moon are in opposite directions.

2 At position Z the net gravitational field strength is zero, therefore:

$-\frac{GM_{Earth}}{r_{Earth\text{ to }Z}^2} = -\frac{GM_{Moon}}{r_{Moon\text{ to }Z}^2}$

$\sqrt{\frac{M_{Earth}}{M_{Moon}}} = \frac{r_{Earth\text{ to }Z}}{r_{Moon\text{ to }Z}}$

$\sqrt{\frac{5.97 \times 10^{24}}{7.35 \times 10^{22}}} = 9.01 = \frac{r_{Earth\text{ to }Z}}{r_{Moon\text{ to }Z}}$

Therefore the distance must be 9 times greater from the Earth than to the Moon.

$\frac{380000 \times 9}{10} = 342\,000$ km from the Earth (38 000 km from Moon).

3 The mass of the Moon is much smaller than that of the Earth, therefore its gravitational field strength is much weaker at the same distance from its centre of mass.

In order to send a spacecraft from the Earth to the Moon, work must be done up to the point where the Moon's gravitational field becomes greater than the Earth's (and so attracts the spacecraft). This point is much closer to the Moon than the Earth.

1 Since diameter = 1.39 million km, radius = 695×10^6 m [1]

$g = -\frac{GM}{r^2} = -\frac{6.67 \times 10^{-11} \times 1.99 \times 10^{30}}{(695 \times 10^6)^2}$ [1]

$g = -275\,\text{N}\,\text{kg}^{-1}$ [1]

2 $g = -\frac{GM}{r^2} = -\frac{6.67 \times 10^{-11} \times 2.60 \times 10^{23}}{(1.2 \times 10^8)^2}$ [1]

$g = -1.2 \times 10^{-3}\,\text{N}\,\text{kg}^{-1}$ [1]

3 a Since $g \propto M$, if M halves, g halves. [1]

b Since $g \propto \frac{1}{r^2}$, if r increases by a factor of three, g decreases by a factor of 9 (3^2). [1]

c Since $g \propto M$, if M decreases by a factor of four, g decreases by a factor of four, and since $g \propto \frac{1}{r^2}$, if r decreases by a factor of two, g increases by a factor of four. [1]. Therefore overall there is no change in g. [1]

4 $g \propto \frac{1}{r^2}$ however r is measured from the centre of mass. [1]

Moving from 100 m above to surface to 200 m above the surface does not double r (6400.1 km to 6400.2 km). [1]

5 At the poles: $g = -\frac{GM}{r^2} = -\frac{6.67\times10^{-11}\times5.97\times10^{24}}{(6371\times10^3)^2}$

$= 9.81\,\text{N kg}^{-1}$ [1]

At the equator: $g = -\frac{GM}{r^2} = -\frac{6.67\times10^{-11}\times5.97\times10^{24}}{(6378\times10^3)^2}$

$= 9.79\,\text{N kg}^{-1}$ [1]

% change = 0.2 % [1]

6 $g = -\frac{GM}{r^2}$ therefore $r = \sqrt{\frac{GM}{g}}$ [1]

$r = \sqrt{\frac{GM}{g}} = \sqrt{\frac{6.67\times10^{-11}\times6.42\times10^{23}}{3.72}}$ [1]

$r = 3.4\times10^6\,\text{m}$ [1]

7 $g = \frac{GM}{r^2}$ therefore $M = \frac{gr^2}{G}$ [1]

$M = \frac{8.77\times(6.09\times10^6)^2}{6.67\times10^{-11}}$ [1]

$= 4.88\times10^{24}\,\text{kg}$ [1]

18.4

1 See Figures 2 and 3 in the main content pages.

Diagram should include two planets at different distances from the star. [1]

Diagram showing two elliptical orbits with the Sun at a focus [1] Kepler's first law.

Diagram should also show that a line segment joining a planet and the Sun sweeps out equal areas during equal intervals of time (for one or both orbits). [1] Kepler's second law.

The diagram should also show (in words) $T^2 \alpha r^3$ [1]

2 $T^2 = \left(\frac{4\pi^2}{GM}\right)r^3$ gives $T = \sqrt{\left(\frac{4\pi^2}{GM}\right)r^3}$ [1]

$T = \sqrt{\left(\frac{4\pi^2}{6.67\times10^{-11}\times1.99\times10^{30}}\right)(1400\times10^9)^3}$ [1]

$T = 900\times10^6\,\text{s} = 28.6$ years [1]

3 a Since $T^2 \alpha r^3$ if r doubles T^2 increases by a factor of 8 (2^3)[1], therefore T increase by a factor of $\sqrt{8}$ (2.8) [1]

b Since $T^2 \alpha r^3$ if r increases by a factor of three T^2 increases by a factor of 27 (3^3)[1], therefore T increases by a factor of $\sqrt{27}$ (5.2) [1]

c Since $T^2 \alpha r^3$ if r decreases by a factor of nine T^2 decreases by a factor of 729 (9^3)[1], therefore T decreases by a factor of $\sqrt{729}$ (27) [1]

4 Kepler's third law: $\frac{T^2}{r^3} = k$ [1]

Calculation of k for each moon [1] (any units e.g.)

Moon	$r/\times10^3$ km	T/ days	$k\times10^{-17}$ days^2 km^{-3}
Io	420	1.8	4.37
Europa	670	3.6	4.31
Ganymede	1070	7.2	4.23
Callisto	1890	16.7	4.13

k is approximately constant. (<6% variation). [1]

5 Graph of T^2 against r^3, axis labelled correctly with quantities and units, points plotted correctly and line of best fit drawn. [3]

Gradient $= \frac{4\pi^2}{GM}$ [1]

Therefore $M = \frac{4\pi^2}{G\times\text{gradient}} \approx 1.9\times10^{27}\,\text{kg}$ [1]

18.5

1 They remain above the same point on the surface of the Earth. Providing continuous coverage for a certain area.

2 Speed of the satellite: $v = \frac{2\pi r}{T} = \frac{2\times\pi\times42\times10^6}{86400}$

$= 3.1\times10^3\,\text{ms}^{-1}$

Kinetic energy of the satellite: $E_k = \frac{1}{2}mv^2$

$= \frac{1}{2}\times80\times(3.1\times10^3)^2 = 380\,\text{MJ}$

1 Kepler's third law states $T^2 \alpha r^3$ [1], therefore if r reduces, T also reduces. [1]

2 Only one force [1]

Gravitational attraction towards the centre of mass of the Earth [1]

3 a Nine times in one day, therefore the period $= \frac{24}{9} = 2.7\,\text{hours} = 9600\,\text{s}$ [1]

b $T^2 = \left(\frac{4\pi^2}{GM}\right)r^3$ therefore: $r = \sqrt[3]{\frac{T^2}{\left(\frac{4\pi^2}{GM}\right)}}$ [1],

$r = \sqrt[3]{\frac{9600^2}{\left(\frac{4\pi^2}{6.67\times10^{-11}\times5.97\times10^{24}}\right)}} = 9.8\times10^6\,\text{m}$ [1]

c $F = \frac{mv^2}{r}$ [1] $F = \frac{180\times6400^2}{9.8\times10^6} = 750\,\text{N}$ [1]

d $a = \frac{v^2}{r}$ [1] $a = \frac{v^2}{r} = \frac{6400^2}{9.8\times10^6} = 4.2\,\text{m s}^{-2}$ [1]

4 The smallest value for $r = 6370\,\text{km}$ (radius of the Earth). [1]

$T^2 = \left(\frac{4\pi^2}{GM}\right)r^3$ [1]

$T = \sqrt{\left(\frac{4\pi^2}{6.67\times10^{-11}\times5.97\times10^{24}}\right)\times(6370\times10^3)^3}$ [1]

$T = 5060\,\text{s} = 85\,\text{mins}$ [1]

5 $T^2 = \left(\frac{4\pi^2}{GM}\right)r^3$ and $r = 6370 + 5000\,\text{km} = 11\,370\,\text{km}$ [1]

$T = \sqrt{\left(\frac{4\pi^2}{6.67\times10^{-11}\times5.97\times10^{24}}\right)\times(11370\times10^3)^3}$ [1]

$T = 12\,100\,\text{s}$ [1]

$v = \frac{2\pi r}{T} = \frac{2\times\pi\times11370\times10^3}{12100} = 5900\,\text{m s}^{-1}$ [1]

18.6

1 $V_g = -\frac{GM}{r}$ [1]

$V_g = -\frac{6.67\times10^{-11}\times7.10\times10^{21}}{3.4\times10^{6}} = -140\,\text{kJ kg}^{-1}$ [1]

2 a Since $V_g \propto M$, if M doubles, V_g doubles. [1]

b Since $V_g \propto \frac{1}{r}$, if r decreases by a factor of four, V_g increases by a factor of 4. [1]

c Since $V_g \propto M$, if M increases by a factor of three, V_g increases by a factor of three and since $V_g \propto \frac{1}{r}$, if r doubles, V_g decreases by a factor of two. [1]

Resulting in a total increase by a factor of 1.5. [1]

3 Since $V_g = -\frac{GM}{r}$, as M is constant [1] and r is constant (at a fixed height) [1] V_g must be constant.

4 $V_g = -\frac{GM}{r}$ and radius = 695 Mm [1]

$V_g = -\frac{6.67\times10^{-11}\times1.99\times10^{30}}{695\times10^{6}}$

$= -1.91\times10^{11}\,\text{J kg}^{-1} \approx -1.9\times10^{11}\,\text{J kg}^{-1}$ [1]

5 Graph of V_g against $\frac{1}{r}$, axis labelled correctly with quantities and units, points plotted correctly and line of best fit drawn. [4]

Gradient = $-GM$ [1]

Therefore $M = -\frac{\text{Gradient}}{G} \approx 5.94\times10^{24}\,\text{kg}$ [1]

18.7

1 $v=\sqrt{\frac{2GM}{r}}=\sqrt{\frac{2\times6.67\times10^{-11}\times5.97\times10^{24}}{6.37\times10^{8}}}$

$=11000\,\text{ms}^{-1}$

2 $E_k=\frac{3}{2}kT=\frac{3}{2}\times1.38\times10^{-23}\times293=6.1\times10^{-21}\,\text{J}$

$E_k=\frac{1}{2}mv^2$ therefore $v=\sqrt{\frac{2E_k}{m}}=\sqrt{\frac{2\times6.1\times10^{-21}}{5.3\times10^{-26}}}$

$=480\,\text{ms}^{-1}$

Much less than the escape velocity.

1 For there to be a change in gravitational potential energy a fixed mass must experience a change in gravitational potential. [1]

At a fixed height in a uniform gravitational field the gravitational potential is constant. Only a change in vertical height will results in a change in gravitational potential and so a change in gravitational potential energy. [1]

2 a $E = mV_g = 40 \times -32\times10^{6} = -1.3\times10^{9}\,\text{J}$ [1]

b $E = mV_g = 7.4\times10^{-9} \times -32\times10^{6} = 0.24\,\text{J}$ [1]

c $E = mV_g = 1.67\times10^{-27} \times -32\times10^{6}$

$= -5.3\times10^{-20}\,\text{J}$ [1]

3 $v = \sqrt{\frac{2GM}{r}}$ [1]

$v = \sqrt{\frac{2\times6.67\times10^{-11}\times7.35\times10^{22}}{1740\times10^{3}}}$ [1]

$v = 2400\,\text{m s}^{-1}$ [1]

4 On the surface: $V_g = -\frac{GM}{r} = -\frac{6.67\times10^{-11}\times5.97\times10^{24}}{6370\times10^{3}}$

$= -62.5\,\text{MJ kg}^{-1}$ [1]

At a height of 50 000 km the surface: $V_g = -\frac{GM}{r}$

$= -\frac{6.67\times10^{-11}\times5.97\times10^{24}}{56370\times10^{3}} = -7.06\,\text{MJ kg}^{-1}$ [1]

$\Delta V_g = 55.4\,\text{MJ kg}^{-1}$ [1]

$\Delta E = m\Delta V_g = 300\times55.4\times10^{6} = 1.66\times10^{10}$ [1]

5 As the comet accelerates towards the Sun it loses gravitational potential energy and gains kinetic energy. Therefore: $\frac{1}{2}mv^2 = \frac{GMm}{r}$ [1]

On impact $v = \sqrt{\frac{2GM}{r}}$ [1]

$v = \sqrt{\frac{2\times6.67\times10^{-11}\times1.99\times10^{30}}{696\times10^{6}}} = 620\,\text{km s}^{-1}$ [1]

19.1

1 Comets

Planetary Satellites or Dwarf Planets

Planets

Solar Systems

Galaxies [2]

2 Fusions produces gas and radiation pressure [1]

This pushes outwards, against the gravitational collapse [1]

3 Similarities:

In orbit of star

Elliptical orbit / Obey Kepler's laws of planetary motion

Differences:

Planets are spherical / comets are irregular

Comets are mainly ice / small pieces of rock

The orbits of comets are much more eccentric (elliptical)

Comets are not large enough to have cleared their orbit of most other objects

At least one similarity [1] and one difference [1]

4 Core is hotter / Greater rate of fusion in the core [1]

Star depletes its hydrogen in the core in a shorter time [1]

5 a $m = \rho V$ and $V = \frac{4}{3}\pi r^3$ therefore $m = \rho\frac{4}{3}\pi r^3$

$m = 1410 \times\frac{4}{3}\times\pi\times700000\times10^{33} = 2.0\times10^{30}\,\text{kg}$ [1]

b Volume of Earth: $V = \frac{4}{3}\pi r^3 = \frac{4}{3} \times \pi \times (6370 \times 10^3)^3 = 1.1 \times 10^{21}\ \text{m}^3$ and

Volume of Sun: $V = \frac{4}{3}\pi r^3 = \frac{4}{3} \times \pi \times (700\,000 \times 10^3)^3 = 1.4 \times 10^{27}\ \text{m}^3$ [1]

Ratio $= 1.4 \times 10^{27} : 1.1 \times 10^{21}$ or 1.3 million:1 [1]

5 c Accept alternative techniques: Volume of occupied by each atom:

$V = \frac{4}{3}\pi r^3 = \frac{4}{3} \times \pi \times (1.0 \times 10^{-10})^3 = 4.2 \times 10^{-30}\ \text{m}^3$ [1]

Number of atoms $= \frac{1.4 \times 10^{27}}{4.2 \times 10^{-30}} = 3.3 \times 10^{56}$ [1]

19.2

1 In order for the mass to be large enough to provide a gravitational attraction strong enough to overcome the electron degeneracy pressure.

2 $t = \frac{d}{s} = \frac{2.18 \times 10^{19}}{3.00 \times 10^8} = 7.27 \times 10^{10}\ \text{s}$ (around 2300 years)

3 $m = \rho V$ and $V = \frac{4}{3}\pi r^3$ therefore $m = \rho \frac{4}{3}\pi r^3$

Radius of star $= \frac{700000000}{1400000} = 500\ \text{m}$

$m = 1.0 \times 10^{17} \times \frac{4}{3} \times \pi \times 500^3 = 5.2 \times 10^{25}\ \text{kg}$

1 a $r_s = \frac{2GM}{c^2} = \frac{2 \times 6.67 \times 10^{-11} \times 5.97 \times 10^{24}}{(3.00 \times 10^8)^2} = 8.8\ \text{mm}$

b $r_s = \frac{2GM}{c^2} = \frac{2 \times 6.67 \times 10^{-11} \times 1.99 \times 10^{30}}{(3.00 \times 10^8)^2} = 2.9\ \text{km}$

c Assuming a mass of 75 kg: $r_s = \frac{2GM}{c^2}$

$= \frac{2 \times 6.67 \times 10^{-11} \times 75}{(3.00 \times 10^8)^2} = 1.1 \times 10^{-25}\ \text{m}$

2 In order to reach escape velocity, the kinetic energy lost must equal the gravitational potential energy gained.

$\frac{1}{2}mv^2 = \frac{GMm}{r}$. To be a black hole, $v = c$. Therefore:

$\frac{1}{2}mc^2 = \frac{GMm}{r}$

$\frac{1}{2}c^2 = \frac{GM}{r}$

$r = \frac{2GM}{c^2}$

1 Temperature is not high enough, [1] therefore the kinetic energy of larger nuclei is not large enough to overcome the electrostatic repulsion and fuse together. [1]

2 Any four below:

Elements formed in red supergiants

Core is large enough, therefore hot enough to fuse larger nuclei together

Fusion takes place in shells around the core

Eventually when the core becomes iron the star will explode in a supernova

Elements above iron created in the supernova

The supernova distributes heavier elements throughout the universe.

3 Minimum $= 0.5 \times 1.99 \times 10^{30} = 9.95 \times 10^{29}\ \text{kg}$ [1]

Maximum $= 10 \times 1.99 \times 10^{30} = 1.99 \times 10^{31}\ \text{kg}$ [1]

4 a $1.0\ \text{cm}^3 = 1.0 \times 10^{-6}\ \text{m}^3$ [1]

$m = \rho V = 1.0 \times 10^{17} \times 1.0 \times 10^{-6} = 1.0 \times 10^{11}\ \text{kg}$ [1]

b $V = \frac{m}{\rho} = \frac{5.97 \times 10^{24}}{1.0 \times 10^{17}}$ [1]

$V = 59.7 \times 10^6\ \text{m}^3$ [1]

5 Minimum value for g and v occur at the minimum mass and maximum radius [1]

Mass $= 8 \times 1.99 \times 10^{30} = 1.59 \times 10^{31}\ \text{kg}$

Radius $= 1200 \times 700 \times 10^6 = 8.40 \times 10^{11}\ \text{m}$

$g = -\frac{6.67 \times 10^{-11} \times 1.59 \times 10^{31}}{(8.40 \times 10^{11})^2} = -1.50 \times 10^{-3}\ \text{N kg}^{-1}$ [1]

$v = \sqrt{\frac{2GM}{r}} = \sqrt{\frac{2 \times 6.67 \times 10^{-11} \times 1.59 \times 10^{31}}{8.40 \times 10^{11}}}$

$= 50.3\ \text{km s}^{-1}$ [1]

Maximum value for g and v occur at the maximum mass and minimum radius [1]

Mass $= 20 \times 1.99 \times 10^{30} = 3.98 \times 10^{31}\ \text{kg}$

Radius $= 950 \times 700 \times 10^6 = 6.65 \times 10^{11}\ \text{m}$

$g = -\frac{6.67 \times 10^{-11} \times 3.98 \times 10^{31}}{(6.65 \times 10^{11})^2} = -6.00 \times 10^{-3}\ \text{N kg}^{-1}$ [1]

$v = \sqrt{\frac{2GM}{r}} = \sqrt{\frac{2 \times 6.67 \times 10^{-11} \times 3.98 \times 10^{31}}{6.65 \times 10^{11}}}$

$= 89.4\ \text{km s}^{-1}$ [1]

19.3

1 See Figure 2 in the main content pages.

Y axis luminosity with correct scale [1]

X axis temperature with correct scale [1]

Main sequence correct position [1]

White dwarves correct position [1]

Red giants correct position [1]

2 Expands so its luminosity increases (moving it up) [1]

Surface cools (moving it to the right) [1]

3 Luminosity is zero but temperature is unknown [1]

Would not appear on the diagram [1]

4 Valid value for relative luminosity 0.01–0.04 [1]

For example: $0.02 \times 3.85 \times 10^{26} = 7.7 \times 10^{24}\ \text{W}$ [1]

5 Hottest stars approximately 30 000 K (25 000–35 000) [1]

Temperature of the Sun = 6000 K

30 000:6000 therefore 5:1 [1]

19.4

1 $\Delta E = hf$ [1]

$\Delta E = 6.63 \times 10^{-34} \times 4.5 \times 10^{15} = 3.0 \times 10^{-18}\,\text{J}$ [1]

2 $n = 3$ to the ground state is a greater change in energy than $n = 2$ to the ground state. [1]

Therefore the energy of the photon from $n = 3 \rightarrow n = 1$ is greater [1]

Therefore the frequency is greater and the wavelength is shorter ($hf = \Delta E$ and $\frac{hc}{\lambda} = \Delta E$) [1]

3 $\Delta E = \frac{hc}{\lambda}$ [1]

Change in energy = 2.7 eV = $4.3 \times 10^{-19}\,\text{J}$ [1]

$\lambda = \frac{hc}{\Delta E} = \frac{6.63 \times 10^{-34} \times 3.00 \times 10^{8}}{4.3 \times 10^{-19}} = 460\,\text{nm}$ [1]

Visible (blue) [1]

4

From	To	ΔE /eV	ΔE /J	f / Hz
5	1	13.06	2.09×10^{-18}	3.15×10^{15}
4	1	12.75	2.04×10^{-18}	3.08×10^{15}
3	1	12.10	1.94×10^{-18}	2.92×10^{15}
2	1	10.20	1.63×10^{-18}	2.46×10^{15}
5	2	2.86	4.58×10^{-19}	6.90×10^{14}
4	2	2.55	4.08×10^{-19}	6.15×10^{14}
3	2	1.90	3.04×10^{-19}	4.59×10^{14}
5	3	0.96	1.5×10^{-19}	2.3×10^{14}
4	3	0.65	1.0×10^{-19}	1.6×10^{14}

One mark for each correct frequency [9]

5 Energy of each photon

$= \frac{\text{energy radiated per second}}{\text{number of photons emitted per second}}$ [1]

Energy of each photon $= \frac{1.0 \times 10^{-3}}{3.48 \times 10^{15}} = 2.9 \times 10^{-19}\,\text{J}$ [1]

$2.9 \times 10^{-23}\,\text{J} = 1.8\,\text{eV}$ [1]

19.5

1 Continuous spectrum: All frequencies and wavelengths present. [1]

Emission spectrum: Only certain/specific frequencies and wavelengths present. [1]

2 The energy levels involved are the same (for a specific wavelength). [1]

The energy of the photon emitted or absorbed is the same, implying the same wavelength. [1]

3 $\Delta E = \frac{hc}{\lambda}$ [1]

$\Delta E = \frac{6.63 \times 10^{-34} \times 3.00 \times 10^{8}}{682 \times 10^{-9}}$ [1]

$\Delta E = 2.92 \times 10^{-19}\,\text{J}$ [1]

4 $\Delta E = \frac{hc}{\lambda}$ [1]

Change in energy = 5.8 eV = $9.3 \times 10^{-19}\,\text{J}$ [1]

$\lambda = \frac{hc}{\Delta E} = \frac{6.63 \times 10^{-34} \times 3.00 \times 10^{8}}{9.3 \times 10^{-19}} = 210\,\text{nm}$ [1]

UV [1]

5 a The energy levels calculated correctly as −13.6 eV, −3.40 eV, −1.51 eV, −0.84 eV and 0.54 eV. [1]

The levels are drawn to scale. [1]

b energy of a photon $= 13.6\left(\frac{1}{2^3} - \frac{1}{3^2}\right) = 1.89\,\text{eV}$ [1]

energy in joules $= 1.89 \times 1.60 \times 10^{-19}\,\text{J}$

$\lambda = \frac{hc}{\Delta E} = \frac{6.63 \times 10^{-34} \times 3.00 \times 10^{8}}{1.89 \times 1.60 \times 10^{-19}}$ [1]

$\lambda = 6.58 \times 10^{-7}\,\text{m}$ [1]

19.6

1 a Graph of $\sin\theta$ against n

Labels and units

Points plotted correctly

Line of best fit

b Gradient $= \frac{\lambda}{d}$

gradient ≈ 0.126 and $d = 5.00 \times 10^{-6}\,\text{m}$

$\lambda \approx 5.0 \times 10^{-6} \times 0.126 \approx 630\,\text{nm}$

2 $\frac{d}{\lambda} = \frac{5.00 \times 10^{-6}}{630 \times 10^{-9}} = 7.9$ therefore the 7th order maxima will be the highest order observed.

3 a d reduces by a factor of two therefore gradient increases by a factor of two.

b λ decreases therefore gradient also decreases.

1 More light passing through grating [1]

therefore maxima are brighter [1]

2 $d\sin\theta = n\lambda$ therefore $n = \frac{d\sin\theta}{\lambda}$ [1]

Maximum value of $\theta = 90°$, therefore $\sin\theta = 1$,

giving $n = \frac{d \times 1}{\lambda} = \frac{d}{\lambda}$ [1]

3 $d\sin\theta = n\lambda$ therefore $\lambda = \frac{d\sin\theta}{n}$ [1]

$\lambda = \frac{3.3 \times 10^{-6} \times \sin 8.6}{1}$ [1]

$\lambda = 490\,\text{nm}$ [1]

4 $d\sin\theta = n\lambda$ therefore $\theta = \sin^{-1}\left(\frac{n\lambda}{d}\right)$ [1]

$d = \frac{1}{350000} = 2.86 \times 10^{-6}\,\text{m}$ [1]

$$\theta = \sin^{-1}\left(\frac{3 \times 450 \times 10^{-9}}{2.86 \times 10^{-6}}\right)$$ [1]

$\theta = 28°$ [1]

5 $\frac{d}{\lambda} = \frac{2.86 \times 10^{-6}}{450 \times 10^{-9}} = 6.4$ [1]

Therefore the 6th order maxima will be the highest order observed. [1]

6 $d \sin\theta = n\lambda$ therefore $\lambda = \frac{d \sin\theta}{n}$

$\lambda = \frac{2.5 \times 10^{-6} \times \sin 13.4}{1}$ [1]

$\lambda = 580\,\text{nm}$ [1]

$\Delta E = \frac{hc}{\lambda} = \frac{6.63 \times 10^{-34} \times 3.00 \times 10^{8}}{580 \times 10^{-9}}$ [1]

$\Delta E = 3.4 \times 10^{-19}\,\text{J} = 2.1\,\text{eV}$ [1]

19.7

1 Units are J s^{-1}, therefore W. [1]

2 Use of $\lambda_{max} T = \text{constant}$ [1]

Correct calculations for the constant: [1]

Object	λ_{max}/m	T/K	$\lambda_{max}T$/m K
Sun	5×10^{-7}	5800	2.9×10^{-3}
Healthy human	9.4×10^{-6}	310	2.9×10^{-3}
Wood fire	1.9×10^{-6}	1500	2.9×10^{-3}

$\lambda_{max} T$ is a constant, therefore $\lambda_{max} \propto \frac{1}{T}$ [1]

3 $\lambda_{max} T = \text{constant}$ [1]

Constant = 3.0×10^{-3} m K [1]

$T = \frac{\text{constant}}{\lambda_{max}} = \frac{3.0 \times 10^{-3}}{0.94 \times 10^{-6}} = 3200\,\text{K}$ [1]

4 a $L = 4\pi r^2 \sigma T^4$

Since $L \propto T^4$, if T doubles L increases by a factor of 16 (2^4). [1]

b Since $L \propto r^2$, if r doubles L increases by a factor of 4 (2^2) and since $L \propto T^4$, if T halves L decreases by a factor of 16 (2^4). [1]

Therefore the luminosity decreases by a factor of 4. [1]

c Since $L \propto T^4$, if T increases by a factor of three L increases by a factor of 81 (3^4). [1]

Since $V = \frac{m}{\rho}$, if the density is the same, half the mass will give half the volume, and since $V = \frac{4}{3}\pi r^3$, if the volume halves the radius will decrease by a factor of $\sqrt[3]{2}$.

Since $L \propto r^2$, if r decreases by a factor of $\sqrt[3]{2}$, L decreases by of $\sqrt[3]{2^2}$. [1]

Therefore the luminosity increases by a factor of 51. [1]

5 $L = 4\pi r^2 \sigma T^4$ [1]

$L = 4 \times \pi \times (700 \times 10^{6})^2 \times 5.67 \times 10^{-8} \times 5800^4$
$= 3.95 \times 10^{26}\,\text{W}$ [1]

Total energy = luminosity × time
$= 3.95 \times 10^{26} \times (365 \times 24 \times 60 \times 60) = 1.25 \times 10^{34}\,\text{J}$ [1]

6 $\lambda_{max} T = \text{constant}$ [1]

Constant = 3.0×10^{-3} m K

$T = \frac{\text{constant}}{\lambda_{max}} = \frac{3.0 \times 10^{-3}}{305 \times 10^{-9}} = 9800\,\text{K}$ [1]

$L = 4\pi r^2 \sigma T^4$ therefore $r = \sqrt{\frac{L}{4\pi\sigma T^4}}$ [1]

$r = \sqrt{\frac{4.85 \times 10^{31}}{4\pi \times 5.67 \times 10^{-8} \times 9800^4}}$ [1]

$r = 85.9\,\text{Gm}$ [1]

20.1

1 The technique used to determine the distance to stars less than 100 pc from the Earth. [1]

It relies on the apparent motion of nearby stars against the fixed background of distant stars. [1]

2 $d = \frac{1}{p} = \frac{1}{0.018}$ [1]

$d = 56\,\text{pc}$ [1]

3 a Distance from Earth to the Sun = 1.50×10^{11} [1]

Time taken for light to travel this distance

$= \frac{1.50 \times 10^{11}}{3.00 \times 10^{8}} = 500\,\text{s} = 8\,\text{min}\,20\,\text{seconds}$ [1]

b Proxima Centauri is 4.24 ly from Earth and 1 ly $= 9.46 \times 10^{15}\,\text{m}$ [1]

Therefore distance $= 9.46 \times 10^{15} \times 4.24 = 4.01 \times 10^{16}\,\text{m}$ [1]

4 $p = 1.56 \times 10^{-5}° = 0.0562$ arcseconds [1]

$d = \frac{1}{p} = \frac{1}{0.0562} = 17.8\,\text{pc}$ [1]

Distance in ly = 17.8 × 3.26 [1]

Distance = 58 ly [1]

5 $I = \frac{P}{4\pi r^2}$ therefore $P = I4\pi r^2$ [1]

$r = 16\,\text{ly} = 16 \times 9.46 \times 10^{15} = 1.51 \times 10^{17}\,\text{m}$ [1]

$P = 2.3 \times 10^{-13} \times 4 \times \pi \times (1.51 \times 10^{17})^2$ [1]

$P = 6.6 \times 10^{22}\,\text{W}$ [1]

6 2.4 arcminutes is equal to 0.040°. [1]

Consider a right angle triangle, distance d is from observer to the ball (adjacent side). The opposite side must be $\frac{6.75}{2} = 3.38\,\text{cm}$ [1]

$\tan\theta = \frac{opp}{adj} = \frac{\text{radius of ball}}{d}$ therefore

$d = \frac{3.38 \times 10^{-2}}{\tan 0.040}$ [1]

$d = 48\,\text{m}$ [1]

20.2

1 If the rain is moving towards the receiver the received wavelength will be lower than the transmitted wavelength.

2 Different forms of precipitation may reflect more or less energy (e.g. hail may reflect more microwaves than water droplets). Differences in the reflected microwaves affects the intensity received by the receiver.

1 When the car is moving towards the receiver the sound waves are compressed (shorter wavelength), therefore sound higher pitch. [1]

When the car is moving away from the receiver the sound waves are stretched (longer wavelength), therefore sounds lower pitch. [1]

2 No relative motion between driver and wave source. [1]

Therefore the waves received by the driver have the same wavelength/frequency as the source. [1]

3 Each side of the rotating galaxy will have a different relative velocity and so a different red shift. [1]

The side rotating away from us will have a greater relative velocity (and so a greater red shift) than the side rotating towards us. [1]

Using $\frac{\Delta\lambda}{\lambda} \approx \frac{\Delta f}{f} \approx \frac{v}{c}$ the relative velocity of each side can be calculated. [1]

The differences in these relative velocities can be used to determine the speed of rotation. [1]

4 a $\frac{\Delta f}{f} \approx \frac{v}{c}$ therefore $\Delta f \approx \frac{vf}{c}$

$\approx \frac{10.6\times10^{6}\times5.12\times10^{14}}{3.00\times10^{8}}$ [1]

$\Delta f \approx 0.18 = 10^{14}$ Hz therefore in the galaxy the line is observed at: $5.12\times10^{14} + 0.18\times10^{14}$ $= 5.30\times10^{14}$ Hz [1]

$\lambda = \frac{c}{f} = \frac{3.00\times10^{8}}{5.30\times10^{14}} = 566\,\text{nm}$ [1]

b $b\frac{\Delta f}{f} \approx \frac{v}{c}$ therefore $\Delta f \approx \frac{vf}{c}$

$\approx \frac{0.25\times3.00\times10^{8}\times5.12\times10^{14}}{3.00\times10^{8}}$ [1]

$\Delta f \approx 1.280\times10^{14}$ Hz therefore in the galaxy the line is observed at: $5.118\times10^{14} - 1.28\times10^{14}$ $= 3.84\times10^{14}$ Hz [1]

$\lambda = \frac{c}{f} = \frac{3.00\times10^{8}}{3.84\times10^{14}} = 781\,\text{nm}$ [1]

5 Speed of runner is small (compared with c) [1]

Therefore the change in wavelength of the reflected waves are too small for the human eye to observe [1]

6 $\Delta\lambda = 7.6\,\text{nm}$ [1]

$\frac{\Delta\lambda}{\lambda} \approx \frac{v}{c}$ therefore $\frac{\Delta\lambda c}{\lambda} \approx v \approx \frac{7.6\times10^{-9}\times3.00\times10^{8}}{714.7\times10^{-9}}$

$= 3.19\times10^{6}\,\text{m s}^{-1}$ [1]

As the observe wavelength is shorter the galaxy must be moving towards the Earth. [1]

20.3

1 See Figure 3 in the main content pages.

Graph of recessional velocity of a galaxy against the distance of the galaxy from Earth [1]

Straight line through the origin showing the velocity of receding galaxies (the recessional velocity) is approximately proportional to their distance from the Earth. [1]

2 a $v \approx H_0 d$ therefore $d \approx \frac{v}{H_0} = \frac{160000}{2.2\times10^{-18}}$ [1]

$d \approx 7.3\times10^{22}$ m [1]

b $v \approx H_0 d$ therefore $d \approx \frac{v}{H_0} = \frac{7.8\times10^{6}}{2.2\times10^{-18}}$ [1]

$d \approx 3.5\times10^{24}$ m [1]

3 Hubble constant = gradient [1]

Determination of gradient [1]

Gradient $\approx 70\,\text{km s}^{-1}\,\text{Mpc}^{-1}$ [1]

4 a $1.50\times10^{23}\,\text{m} = 4.84$ Mpc [1]

$v \approx H_0 d = 70\times4.84 = 340\,\text{km s}^{-1}$ [1]

b $v \approx H_0 d \approx 70\times25.0 = 1800$ [1]

c $40\times10^{6}\,\text{ly}\times3.0\times10^{8}\,\text{m s}^{-1}\times(365\times24\times60\times60)$ $= 3.784\times10^{23}$ m [1]

$= 12.2$ Mpc [1]

$v \approx H_0 d = 70\times12.2 = 850\,\text{m s}^{-1}$ [1]

5 $\frac{\Delta\lambda}{\lambda} = \frac{v}{c}$ so $v = \frac{c\Delta\lambda}{\lambda}$ [1]

$= 1.70\times10^{4}\,\text{m s}^{-1}$ [1]

$d \approx \frac{v}{H_0}$ [1]

$d \approx 243\,\text{Mpc} = 7.9\times10^{8}$ ly [1]

20.4

1 $67.80 \pm 0.77\,\text{km s}^{-1}\,\text{Mpc}^{-1} = 2.19\times10^{-18}\,\text{s}^{-1}$

$t \approx H_0^{-1}$ therefore $t \approx \frac{1}{2.19\times10^{-18}} = 4.57\times10^{17}$ s

$= 14.5$ billion years

2 $\lambda_{max}T = \text{constant}$

constant $= 3.0\times10^{-3}$ m K

$\lambda_{max} = \frac{\text{constant}}{T} = \frac{3.0\times10^{-3}}{2.7001} = 1.11\,106\,996\times10^{-3}$ m

$\lambda_{max} = \frac{\text{constant}}{T} = \frac{3.0\times10^{-3}}{2.6999} = 1.111\,152\,265\times10^{-3}$ m

$\frac{1.111152265\times10^{-3}}{1.11106996\times10^{-3}} = 1.000\,07$ or 0.007%

1 Expanding universe (from red shift – Hubble's law) [1]
Microwave background radiation [1]

2 Predicted by the big bang theory [1]
No other theory could account for/ explain the origin of this radiation, therefore the big bang theory was widely accepted. [1]

3 $t \approx H_0^{-1}$ therefore $H_0 \approx \frac{1}{t}$ [1]
Minimum value for $t = 3.46 \times 10^{17}$ s and maximum value for $t = 4.73 \times 10^{17}$ s [1]
Maximum value: $H_0 \approx \frac{1}{3.46 \times 10^{17}} = 2.89 \times 10^{-18}\,\text{s}^{-1}$ [1]
Minimum value: $H_0 \approx \frac{1}{4.73 \times 10^{17}} = 2.11 \times 10^{-18}\,\text{s}^{-1}$ [1]

4 a $\lambda_{max} T$ = constant
Constant = 3.0×10^{-3} m K
$\lambda_{max} = \frac{\text{constant}}{T} = \frac{3.0 \times 10^{-3}}{1 \times 10^{11}}$ [1]
$= 3.0 \times 10^{-14}$ m [1]

b $\lambda_{max} T$ = constant
Constant = 3.0×10^{-3} m K
Visible light around 500 nm [1]
$T = \frac{\text{constant}}{\lambda max} = \frac{3.0 \times 10^{-3}}{500 \times 10^{-9}} = 6000\,\text{K}$ [1]

20.5

1 Distant galaxies are accelerating (universe appears to be expanding at an increasing rate). [1]
For this acceleration to happen a source of energy is needed; this is Dark Energy. [1]

2 Universe was too hot. [1]
Atoms were ionised; eventually the universe cooled enough for nuclei to hold on to their electrons and so form atoms. [1]

3 If the matter were just concentrated in the middle of the galaxy the rate of rotation would drop away / this is not observed, instead the rate of rotation remains fairly constant moving outwards from the centre of the galaxy. [1]
Additional mass (matter) is needed to account for the faster rotation at the edges. [1]
This mass is in the form of Dark Matter. [1]

4 See the data in Table 1 in the main content pages. [4]

5 All correct values of lg t and lg T. [1]
Correct graph plotted with line of best fit. [1]
lg t = constant – n lg T [1]
$n = 2$ [1]

21.1

1 Yes it is a capacitor because it consists of two metallic electrodes which are separated by insulation. [1]

2 $C = QV = \frac{2.4 \times 10^{-5}}{12}$ [1]
$C = 2.0 \times 10^{-6}$ F [1]
$C = 2.0\,\mu$F [1]

3 $Q \propto V$ [1]
$Q = \frac{8.7}{3.2} \times 30$ [1]
$Q = 81.6\,\text{pC} \approx 82\,\text{pC}$ [1]

4 $Q = VC = 1.5 \times 0.010 = 1.5 \times 10^{-2}$ C [1]
number of electrons $= \frac{1.5 \times 10^{-2}}{1.6 \times 10^{-19}}$ [1]
number of electrons $= 9.4 \times 10^{16}$ [1]

5 a charge Q = current × time $= 80 \times 10^{-6} \times 60$ [1]
$Q = 4.8 \times 10^{-3}$ C [1]
$V = \frac{Q}{C} = \frac{4.8 \times 10^{-3}}{1500 \times 10^{-6}}$ [1]
$V = 3.2$ V [1]

b A straight line graph passing through the origin. [1]
Graph shows $V = 3.2$ V at $t = 60$ s. [1]

21.2

1 Parallel: $C = C_1 + C_2 = 100 + 100 = 200$ pF [1]
Series: $C = (C_1^{-1} + C_2^{-1})^{-1} = (100^{-1} + 100^{-1})^{-1}$ [1]
$C = 50$ pF [1]
The total capacitance for the parallel circuit is **twice** the capacitance of a single capacitor [1] and the total capacitance for the series circuit is **half** the capacitance of a single capacitor. [1]

2 $C = (C_1^{-1} + C_2^{-1})^{-1} = (120^{-1} + 120^{-1})^{-1}$ [1]
$C = 60$ nF [1]
$Q = VC = 60 \times 10^{-9} \times 1.5$ [1]
$Q = 9.0 \times 10^{-8}$ C [1]

3 Total capacitance of N identical 1000 μF capacitors in parallel = $N \times 1000\,\mu$F [1]
Therefore, $N \times 1000 \times 10^{-6} = 4000$ [1]
$N = 4 \times 10^6$ (4 million) in parallel [1]

4 $C_1 = (100^{-1} + 500^{-1})^{-1} = 83.3\,\mu$F [2]
$C_2 = (50^{-1} + 200^{-1})^{-1} = 40\,\mu$F [2]
Total capacitance = $C_1 + C_2 = 83.3 + 40 \approx 123\,\mu$F [1]

5 The charge stored by each is the same and the p.d. across the combination is 6.0 V. [1]
$V = \frac{Q}{C} \propto \frac{1}{C}$, hence the p.d. across the capacitor with capacitor $2C$ will be half the p.d. across the capacitor with capacitance C. [1]
Therefore, p.d. across $C = 4.0$ V [1] and the p.d. across $2C = 2.0$ V. [1]

6 $\frac{1}{17} = \frac{1}{C} + \frac{1}{20}$ [1]
$\frac{1}{C} = \frac{1}{17} - \frac{1}{20}$ [1]
$C = 113\,\text{nF} \approx 110\,\text{nF}$ [1]

21.3

1 The area under the graph is equal to work done (or energy stored). [1]
The gradient of the graph is C^{-1}. [1]

2 a $W = \frac{1}{2}V^2C = \frac{1}{2} \times 2.0^2 \times 1000 \times 10^{-6}$ [1]
$W = 2.0 \times 10^{-3}$ J [1]

b $W = \frac{1}{2}V^2C = \frac{1}{2} \times 6.0^2 \times 1000 \times 10^{-6}$ [1]

$W = 1.8 \times 10^{-2}$ J [1]

3 charge Q = current × time = 0.200 × 600 = 120 C [1]

$W = \frac{1}{2}\frac{Q^2}{C} = \frac{1}{2} \times \frac{120^2}{4.0}$ [1]

$W = 1.8 \times 10^3$ J [1]

4 $C = \left(C_1^{-1} + C_2^{-1}\right)^{-1} = \left(300^{-1} + 300^{-1}\right)^{-1} = 150\,\mu\text{F}$ [1]

$W = \frac{1}{2}V^2C = \frac{1}{2} \times 24^2 \times 150 \times 10^{-6}$ [1]

$W = 4.32 \times 10^{-2}\,\text{J} \approx 4.3 \times 10^{-2}\,\text{J}$ [1]

5 $W = \frac{1}{2}V^2C = \frac{1}{2} \times 240^2 \times 150 \times 0.10$ (W = energy stored in capacitor) [1]

$W = 2.88 \times 10^3$ J [1]

Gain in GPE = mgh = 10 × 9.81 × 29 = 2.84 × 10^3 J [1]

The values are compared, followed by a valid comment, e.g. more work can be done than energy gained so the claim is possible if little energy is lost. [1]

21.4

1 All values of ln (V/V) are correct.

t / s	V / V	ln (V / V)
0	9.00	2.197
10	6.65	1.895
20	4.91	1.591
30	3.63	1.289
40	2.68	0.986
50	1.98	0.683
60	1.46	0.378

2 Correct labelling of axes and use of correct scale. Correct plotting of all point.

3 The gradient determined using a large triangle. The value of the gradient is (−) 3.0 × 10^{-2} s^{-1} (allow ± 5 %)

$|\text{gradient}| = \frac{1}{CR}$, therefore $CR = \frac{1}{3.0 \times 10^{-2}}$

$CR = 33$ s

4 $C = \frac{\text{time constant}}{R} = \frac{33}{33 \times 10^3}$

$C = 1.0 \times 10^{-3}$ F (1000 μF)

1 Maximum current $= \frac{V_0}{R} = \frac{2.0}{100} = 2.0$ A [1]

Maximum charge $= V_0C = 2.0 \times 10^{-2} \times 220 \times 10^{-6} = 4.4 \times 10^{-4}$ C [1]

2 a $CR = 0.01 \times 1000 = 10$ s [1]

b $CR = 4700 \times 10^{-6} \times 1.5 \times 10^6 = 7.05 \times 10^3\,\text{s} \approx 7.1 \times 10^3$ s (about 2 hours) [1]

3 $t = 3CR$ and $V = V_0 e^{-\frac{t}{CR}}$ [1]

$V = 9.0 \times e^{-\frac{3CRt}{CR}}$ [1]

$V = 0.45$ V [1]

4 $C = \left(120^{-1} + 300^{-1}\right)^{-1} = 85.7\,\mu\text{F}$ [1]

$R = \left(47^{-1} + 10^{-1}\right)^{-1} = 8.25\,\text{k}\Omega$ [1]

$CR = 85.7 \times 10^{-6} \times 8.25 \times 10^6 = 0.71$ s [1]

5 $CR = 500 \times 10^{-6} \times 200 \times 10^3 = 100$ s [1]

$I = I_0 e^{-\frac{t}{CR}} = I_0 e^{-\frac{t}{100}}$ and $0.25 = e^{-\frac{t}{100}}$ [1]

$\ln(0.25) = -\frac{t}{100}$ [1]

$t = -\ln(0.25) \times 100 \approx 140$ s [1]

21.5

1 The time constant of the circuit is very small because of the small resistance of the copper wires. Therefore the capacitor charges up in a very short time. [1]

2 The current in the circuit and the p.d. across the resistor will both decrease exponentially with respect to time.

3 Start the stopwatch when the switch is closed. Stop the stopwatch when the voltmeter reading has dropped to 37% of its initial reading of V_0. [1]

The time recorded on the stopwatch is equal to the time constant. [1]

4 $V_c = V_0\left(1 - e^{-\frac{t}{CR}}\right) = 3.0 \times \left(1 - e^{-\frac{CR}{CR}}\right) = 3.0 \times 0.63$ [1]

$V_c = 1.896\,\text{V} \approx 1.9$ V [1]

5 $CR = 120 \times 10^{-6} \times 1.0 \times 10^6 = 120$ s [1]

$V_c = V_0\left(1 - e^{-\frac{t}{CR}}\right) = 3.0 \times \left(1 - e^{-\frac{180}{120}}\right) = 3.0 \times 0.78$ [1]

$V_c = 2.33\,\text{V} \approx 2.3$ V [1]

6 $V_C = V_R$, therefore $V_0\left(1 - e^{-\frac{t}{CR}}\right) = V_0 e^{-\frac{t}{CR}}$ [1]

$\frac{1}{2} = e^{-\frac{t}{CR}}$ [1]

$\ln(0.5) = -\frac{t}{CR}$ [1]

$t = 0.69CR$ [1]

21.6

1 There is very little smoothing of the output voltage.

2 $V_{ripple} = \frac{V_0T}{CR} = \frac{340 \times 0.020}{1000 \times 10^{-6} \times 220}$

$V_{ripple} = 31$ V

3 $0.001 = \frac{0.020}{CR}$

$CR = 20$ s

Any suitable values, e.g. $C = 20\,\mu$F and $R = 1.0$ MΩ

1 Any two sensible suggestions, e.g. camera flashes, smoothing capacitors, etc. [2]

2 energy = power × time $\approx 300 \times 10^{12} \times 100 \times 10^{-9}$ [1]

energy $\approx 3.0 \times 10^{7}$ J [1]

3 power $= \frac{1}{2}\frac{V^2C}{t} = \frac{1}{2} \times \frac{10^2 \times 1000 \times 10^{-6}}{10 \times 10^{-3}}$ [2]

power = 5.0 W [1]

4 $V = V_0 e^{-\frac{t}{CR}}$

$3.0 = 5.0e^{-\frac{0.01}{CR}}$ [1]

$\ln(0.60) = -\frac{0.01}{CR}$ [1]

$CR = -\frac{0.01}{\ln(0.60)}$ [1]

$CR \approx 2.0 \times 10^{-2}$ s [1]

22.1

1 The electric field lines are parallel and spaced equally. [1]

2 The sphere has positive charge and the plate has negative charge. [1]

3 The field lines will still be radial. [1]

There would be more field lines at the sphere indicating greater field strength. [1]

4 $F = EQ = 4.0 \times 10^5 \times 1.6 \times 10^{-19}$ [1]

$F = 6.4 \times 10^{-14}$ N [1]

5 $F = EQ = 2.0 \times 10^5 \times 1.6 \times 10^{-19}$ [1]

$F = 3.2 \times 10^{-14}$ N [1]

$a = \frac{F}{m} = \frac{3.2 \times 10^{-14}}{1.7 \times 10^{-27}} = 1.88 \times 10^{13}\ \text{m s}^{-2}$ [1]

$s = \frac{1}{2}at^2$

$0.050 = \frac{1}{2} \times 1.88 \times 10^{13} \times t^2$ [1]

$t = 7.3 \times 10^{-8}$ s [1]

22.2

1 Gravitational field is always attractive but electric fields can be both attractive and repulsive. [1]

2 $k = \frac{1}{4\pi \times 8.85 \times 10^{-12}} \approx 9.0 \times 10^9\ \text{mF}^{-1}$ [2]

3 $F = \frac{Qq}{4\pi\varepsilon_0 r^2} = \frac{(2.0 \times 10^{-9})^2}{4\pi \times 8.85 \times 10^{-12} \times 0.010^2}$ [1]

$F = 3.6 \times 10^{-4}$ N [1]

4 The distance increases by a factor of 3, therefore the force will decrease by a factor of $3^2 = 9$. [1]

force $= \frac{3.6 \times 10^{-4}}{9} = 4.0 \times 10^{-5}$ N [1]

5 $E = \frac{Q}{4\pi\varepsilon_0 r^2}$

$3.0 \times 10^4 = \frac{Q}{4\pi\varepsilon_0 0.050^2}$ [1]

$Q = 3.0 \times 10^4 \times 4\pi \times 8.85 \times 10^{-12} \times 0.050^2$ [1]

$Q = 8.3 \times 10^{-9}$ C [1]

6 $Q = 79e$ and $q = 2e$ [1]

$F = \frac{Qq}{4\pi\varepsilon_0 r^2} = \frac{79 \times 2 \times (1.6 \times 10^{-19})^2}{4\pi \times 8.85 \times 10^{-12} \times (2.0 \times 10^{-14})^2}$ [1]

$F = 91$ N [1]

7 $F = \frac{Qq}{4\pi\varepsilon_0 r^2} = \frac{(1.6 \times 10^{-19})^2}{4\pi \times 8.85 \times 10^{-12} \times (3.0 \times 10^{-10})^2}$ [1]

$F = 2.557 \times 10^{-9}$ N [1]

weight $= mg = 9.11 \times 10^{-31} \times 9.81 = 8.937 \times 10^{-30}$ N [1]

ratio $= \frac{2.557 \times 10^{-9}}{8.937 \times 10^{-30}} \approx 2.9 \times 10^{20}$ [1]

22.3

1 Graph drawn with correctly-labelled axes, including units.

2 Gradient determined using a "large triangle" on a straight line of best fit: $1.12 \times 10^{-11}\ \text{C V}^{-1}$ (allow $\pm 0.03 \times 10^{-11}\ \text{C V}^{-1}$)

3 The gradient is equal to the capacitance of the capacitor made from the parallel plates.

4 $C = \frac{\varepsilon_0 A}{d}$

$\varepsilon_0 = \frac{1.12 \times 10^{-11} \times 0.025}{\pi \times 0.10^2}$

$\varepsilon_0 = 8.9 \times 10^{-12}\ \text{F m}^{-1}$

1 terminal velocity $= \frac{5.0 \times 10^{-3}}{44}$

$r^2 = \frac{9\eta v}{2\rho g} = \frac{9 \times 1.8 \times 10^{-5} \times \left(\frac{5.0 \times 10^{-3}}{44}\right)}{2 \times 900 \times 9.81}$

$r = 1.02 \times 10^{-6}\ \text{m} \approx 1.0 \times 10^{-6}$ m

2 mass = volume × density $= \frac{4}{3}\pi \times (1.02 \times 10^{-6})^3 \times 900$

mass $= 4.0 \times 10^{-15}$ kg

3 $mg = EQ$

$mg = \frac{VQ}{d}$

$Q = \frac{mgd}{V} = \frac{4.0 \times 10^{-15} \times 9.81 \times 2.0 \times 10^{-2}}{1200}$

$Q = 6.56 \times 10^{-19}$ C

number of electrons $= \frac{Q}{e} = \frac{6.56 \times 10^{-19}}{1.6 \times 10^{-19}} = 4.1 \approx 4$

1 N C^{-1} and V m^{-1} [1]

2 $E = \frac{V}{d} = \frac{1.0 \times 10^3}{1.0 \times 10^{-2}}$ [1]

$E = 1.0 \times 10^5\ \text{V m}^{-1}$ [1]

3 $F = EQ = 1.0 \times 10^5 \times 1.6 \times 10^{-19} = 1.6 \times 10^{-14}$ N [1]

$a = \frac{F}{m} = \frac{1.6 \times 10^{-14}}{1.7 \times 10^{-27}}$ [1]

$a = 9.4 \times 10^{12}\ \text{m s}^{-1}$ [1]

4 a $C \propto \frac{1}{d}$

$C = \frac{8.0}{2}$ [1]

$C = 4.0\,\text{pF}$ [1]

b $C \propto \frac{A}{d}$

Both A and d change by the same factor, so the ratio is the same. [1]

$C = 8.0\,\text{pF}$ [1]

5 $C = \frac{\varepsilon_0 \varepsilon_r A}{d} = \frac{8.85 \times 10^{-12} \times 4.0 \times \pi \times 0.10^2}{1.2 \times 10^{-3}}$ [1]

$C = 9.268 \times 10^{-10}\,\text{F}$ [1]

$Q = VC = 6 \times 9.268 \times 10^{-10}$ [1]

$Q = 5.56 \times 10^{-9}\,\text{C} \approx 5.6\,\text{nC}$ [1]

6 $mg = EQ = \frac{VQ}{d}$ [1]

$V = \frac{mgd}{Q} = \frac{2.5 \times 10^{-15} \times 9.81 \times 1.2 \times 10^{-2}}{2 \times 1.6 \times 10^{-19}}$ [2]

$V = 920\,\text{V}$ [1]

22.4

1 The proton will be attracted towards the negative plate (or away from the positive plate).

The proton moves in the direction of the electric field.

It experiences a constant force and hence will have a constant acceleration between the plates.

2 The maximum kinetic energy of an electron = Ve.

Hence the only factor that affects the maximum speed of the electron is the p.d. V between the plates.

3 $KE = Ve = \frac{1}{2}mv^2$ [1]

$v = \sqrt{\frac{2Ve}{m}} = \sqrt{\frac{2 \times 1.5 \times 1.6 \times 10^{-19}}{9.11 \times 10^{-31}}}$ [1]

$v = 7.26 \times 10^5\,\text{m s}^{-1} \approx 700\,\text{km s}^{-1}$ [1]

4 a $E = \frac{V}{d} = \frac{2.5 \times 10^3}{0.020} = 1.25 \times 10^5\,\text{V m}^{-1}$ [1]

$v_v = \frac{EQL}{mv} = \frac{1.25 \times 10^5 \times 1.6 \times 10^{-19} \times 0.20}{1.7 \times 10^{-27} \times 5.0 \times 10^6}$ [2]

$v_v = 4.71 \times 10^5\,\text{m s}^{-1} \approx 4.7 \times 10^5\,\text{m s}^{-1}$ [1]

b $a = \frac{F}{m} = \frac{EQ}{m} = \frac{1.25 \times 10^5 \times 1.6 \times 10^{-19}}{1.7 \times 10^{-27}}$ [1]

time spent in field = $\frac{0.20}{5.0 \times 10^6}$ [1]

$s = \frac{1}{2}at^2 = \frac{1}{2} \times \frac{1.25 \times 10^5 \times 1.6 \times 10^{-19}}{1.7 \times 10^{-27}} \left(\frac{0.20}{5.0 \times 10^6}\right)^2$ [1]

$s = 9.4 \times 10^{-3}\,\text{m}\ (9.4\,\text{mm})$ [1]

22.5

1 The electric field lines are always at right angles to an equipotential line.

2 Correct lines for the potentials, and labelled with values.

The field lines are correctly drawn.

The field lines at right angles to the equipotential lines and the field direction is correct.

See diagram below.

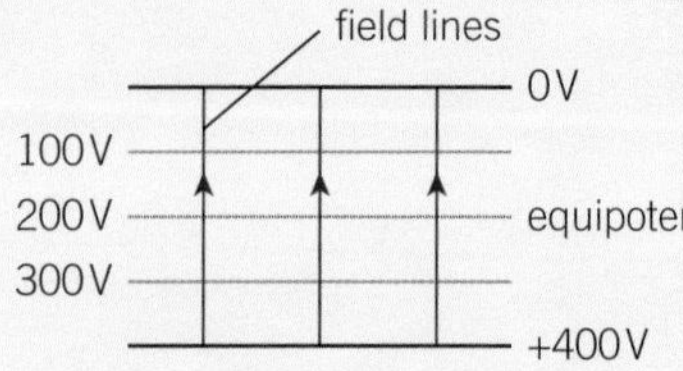

3 $\Delta V = 70 - 50 = 20\,\text{V}$ and $\Delta r \approx 5\,\text{mm}$

$E = \frac{\Delta V}{\Delta r} = \frac{20}{0.005} = 4.0 \times 10^3\,\text{V m}^{-1}$ (magnitude only)

1 $E \propto \frac{1}{r}$, therefore as r is doubled E will halve. [1]

$E = 2 \times 10^{-19}\,\text{J}$ [1]

2 10 J of work is done in bringing a unit charge from infinity to the surface of the sphere. [1]

3 $V = \frac{Q}{4\pi\varepsilon_0 r} = \frac{1.6 \times 10^{-19}}{4\pi \times 8.85 \times 10^{-12} \times 8.8 \times 10^{-16}}$ [1]

$V = 1.64 \times 10^6\,\text{V} \approx 1.6 \times 10^6\,\text{V}$ [1]

4 $V = \frac{Qq}{4\pi\varepsilon_0 r} = \frac{(1.6 \times 10^{-19})^2}{4\pi \times 8.85 \times 10^{-12} \times 1.0 \times 10^{-10}}$ [1]

$E = 2.3 \times 10^{-18}\,\text{J}$ [1]

$E = \frac{2.3 \times 10^{-18}}{1.6 \times 10^{-19}} \approx 14\,\text{eV}$ [1]

5 a $C = 4\pi\varepsilon_0 r = 4\pi \times 8.85 \times 10^{-12} \times 0.02$ [1]

$C = 2.23 \times 10^{-12}\,\text{F} \approx 2.2 \times 10^{-12}\,\text{F}$ [1]

b $Q = VC = 6000 \times 2.23 \times 10^{-12}$ [1]

$Q = 1.34 \times 10^{-8}\,\text{C}$ (magnitude only) [1]

number of electrons $= \frac{1.34 \times 10^{-8}}{1.6 \times 10^{-19}}$

$= 8.4 \times 10^{10}$ electrons [1]

6 energy $= 1.0 \times 10^6 \times 1.6 \times 10^{-19}$ [1]

$1.0 \times 10^6 \times 1.6 \times 10^{-19} = \frac{Qq}{4\pi\varepsilon_0 r}$

$= \frac{(1.6 \times 10^{-19})^2}{4\pi \times 8.85 \times 10^{-12} \times r}$ [2]

$r = 1.44 \times 10^{-15}\,\text{m} \approx 1.4 \times 10^{-15}\,\text{m}$ [1]

23.1

1 a Yes, because it is a charged particle that is moving. [1]

b Yes, because it is a charged particle that is moving. [1]

c No, because a neutron has no charge and whether it is moving or not is irrelevant. [1]

2 Use two opposite magnetic poles. The magnetic field is uniform in the space between two opposite poles of a magnet (or magnets). [1]

Use a current-carrying solenoid. The magnetic field is uniform at the centre of a solenoid carrying a current. See Figure 4 in the main content pages. [1]

3 At A, the field direction is out of the plane of the paper. [1]

At B, the field direction is into the plane of the paper. [1]

4 The field direction would be reversed and there would be more and closer magnetic field lines. [2]

5 The field pattern is correct. [1]

The direction of the field is correct. [1]

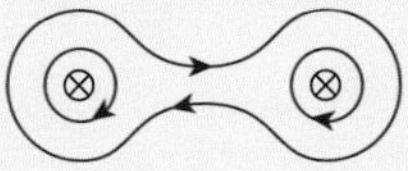

23.2

1 Correct values for F in the last column.
(Note: $F = mg$)

F / 10^{-3} N
0
3.04
6.28
8.73
12.2
14.7
18.0
21.0

2 Graph drawn with correctly-labelled axes, correctly-plotted points and a straight line through the data points.

3 $F = BIL$

Comparing the equation $F = (BL)I$ with $y = mx$, we have: gradient = BL.

4 gradient = 3.0×10^{-3} N A^{-1} (Allow $\pm 0.1 \times 10^{-3}$ N A^{-1})

$BL = 3.0 \times 10^{-3}$ or $B \times 0.05 = 3.0 \times 10^{-3}$

$B = 6.0 \times 10^{-2}$ T

Uncertainty in $B = \frac{0.3}{5.0} \times 6.0 \times 10^{-2} = 0.4 \times 10^{-2}$ T (1 s.f.)

$B = (6.0 \pm 0.4) \times 10^{-2}$ T

1 A current-carrying is surrounded by its own magnetic field and the interaction of this field and field of the magnet produces a force on the wire. [1]

2 a Into the plane of the paper. [1]

b Up the page. [1]

c Down the page. [1]

3 $F = BIL\sin\theta = 0.120 \times 5.0 \times 0.01 \times \sin 90°$ [1]

$F = 6.0 \times 10^{-3}$ N [1]

4 a $F = BIL\sin\theta$, $F \propto I$ therefore the force is $5.0 \times 4 = 20$ mN [1]

b $F = BIL\sin\theta$, $F \propto B$ therefore the force is $5.0 \times 2 = 10$ mN [1]

c $F = BIL\sin\theta$, $F \propto L$ therefore the force is $0.30 \times 5.0 = 1.5$ mN [1]

5 $F = BIL\sin\theta$, $4.0 \times 10^{-3} = B \times 0.80 \times 0.028 \times \sin 38°$ [1]

$$B = \frac{4.0 \times 10^{-3}}{0.80 \times 0.028 \times \sin 38}$$ [1]

$B = 0.29$ T [1]

6 a The length of the loop on the left-hand side and perpendicular to the field experiences a force into the plane of the paper and the opposite side experiences a force out of the plane of the paper. [1]

Hence the loop will rotate clockwise if observed from the top. [1]

b torque = $Fd = (BIx) \times x$ [1]

torque = $(BI) \times x^2$ [1]

The product BI is constant; therefore torque $\propto$ cross-sectional area x^2. [1]

23.3

1 The radius of the path is given by the equation $r = \frac{mv}{BQ}$, so it is important v to be the same so that $r \propto m$.

2 $r = \frac{mv}{BQ} = \frac{2.16 \times 10^{-26} \times 8.00 \times 10^{4}}{0.750 \times 1.6 \times 10^{-19}}$

$r = 1.44 \times 10^{-2}$ m

3 radius $\approx \frac{14}{13} \times 1.44 \times 10^{-2}$

radius $\approx 1.55 \times 10^{-2}$ m

1 magnetic force = electric force; $BQv = EQ$, therefore $v = \frac{E}{B}$

$V_H = Ed = Bvd$

$v = \frac{I}{Ane}$, therefore $V_H = \frac{BId}{Ane}$

$A = td$, therefore $V_H = \frac{BI}{nte}$

2 The number density n of electrons is smaller for semiconductors, hence the Hall voltage is larger $\left(V_H \propto \frac{1}{n}\right)$.

3 $V_H = \frac{1.2}{0.060} \times 0.014$

$V_H = 0.28$ V

1 The force on the particle is given by $F = BQv$ and if the speed v is zero, then the force will also be zero. [1]

2 $F = BQv = 0.20 \times 1.6 \times 10^{-19} \times 6.0 \times 10^5$ [1]

$F = 1.92 \times 10^{-14}\,\text{N} \approx 1.9 \times 10^{-14}\,\text{N}$ [1]

3 $BQv = \frac{mv^2}{r}$ therefore $mv = BQr$ ($p = mv$) [1]

$p = 0.130 \times (2 \times 1.6 \times 10^{-19}) \times 0.025$ [1]

$p = 1.04 \times 10^{-21}\,\text{kg m s}^{-1}$ [1]

4 $E = \frac{1300}{0.025} = 5.2 \times 10^4\,\text{V m}^{-1}$ [1]

$v = \frac{E}{B}$

$B = \frac{E}{v} = \frac{5.2 \times 10^4}{4.0 \times 10^5}$ [1]

$B = 0.13\,\text{T}$ [1]

5 a $BQv = \frac{mv^2}{r}$ [1]

$r = \frac{mv}{BQ} = \frac{1.7 \times 10^{-27} \times 4.0 \times 10^6}{0.800 \times 1.6 \times 10^{-19}}$ [1]

$r = 0.053 = 5.3 \times 10^{-2}\,\text{m}$ [1]

b $T = \frac{2\pi r}{v} = \frac{2\pi \times 0.053}{4.0 \times 10^6}$ [1]

$T = 8.3 \times 10^{-8}\,\text{s}$ [1]

6 period $T = \frac{2\pi r}{v}$ [1]

$r = \frac{mv}{BQ}$ [1]

Therefore $T = \frac{2\pi m}{BQ}$ which is independent of the radius r. [1]

23.4

1 magnetic flux density → T; magnetic flux → Wb and magnetic flux linkage → Wb (or Wb-turns) [1]

2 $\phi = BA$. The cross-sectional area A is constant. [1]

The magnetic flux ϕ linking the coil changes because the magnetic flux density B changes when the magnet is moved relative to the coil. [1]

3 $\phi = BA\cos\theta = 0.02 \times 1.4 \times 10^{-4} \times \cos 0°$ [1]

$\phi = 2.8 \times 10^{-6} \approx 3 \times 10^{-6}\,\text{Wb}$ [1]

4 $\phi = BA\cos\theta = 0.20 \times [\pi \times 0.014^2] \times \cos 30°$ [1]

$\phi = 1.07 \times 10^{-4}\,\text{Wb}$

$N\phi = 400 \times 1.07 \times 10^{-4} \approx 4.3 \times 10^{-2}\,\text{Wb turns}$ [1]

5 B is a vector quantity, hence the change in the flux density is 0.40 T. [1]

change in flux $= 2 \times 1.07 \times 10^{-4}\,\text{Wb}$

change in flux linkage $= 400 \times 2 \times 1.07 \times 10^{-4} \approx 8.6 \times 10^{-2}\,\text{Wb turns}$ [1]

6 radius of coin ≈ 1.0 cm; $A = \pi \times 0.01^2$ (allow ± 30%) [1]

$\phi = BA\cos\theta = 4.9 \times 10^{-5} \times [\pi \times 0.01^2] \times \cos 24°$ [1]

$\phi = 1.41 \times 10^{-8}\,\text{Wb} \approx 1.4 \times 10^{-8}\,\text{Wb}$ [1]

23.5

1 The minus sign is a consequence of conservation of energy. [1]

2 a No change in the flux linkage, hence no induced e.m.f. [1]

b Constant rate of change of flux linkage, hence a constant induced e.m.f. [1]

c Constant rate of change of flux linkage, hence a constant induced e.m.f., but in opposite direction to that in (b). [1]

3 The initial reading of the voltmeter is zero, because there is no change in the flux linkage. [1]

When the current is switched off, the magnetic field collapses in a very short time, hence a large e.m.f. is induced. [1]

Explanation justified in terms of $\varepsilon = \frac{\Delta(N\phi)}{\Delta t}$ and Δt very small. [1]

4 $\varepsilon = \frac{\Delta(N\phi)}{\Delta t} = \frac{800 \times B \times 3.0 \times 10^{-4}}{0.12} = 0.032$ [2]

$B = \frac{0.032 \times 0.12}{800 \times 3.0 \times 10^{-4}}$ [1]

$B = 1.6 \times 10^{-2}\,\text{T}$ [1]

5 A large coil is linked by its own magnetic field. [1]

When the current is switched off, this magnetic field collapses in a very short time and induces a very large e.m.f. in the opposite direction. [1]

Explanation justified in terms of $\varepsilon = \frac{\Delta(N\phi)}{\Delta t}$ and Δt very small. [1]

6 $\varepsilon = \frac{\Delta(N\phi)}{\Delta t}$ and $N = 1$ [1]

In a time interval t, the area 'swept' by the wire $= L \times vt$ [1]

$\varepsilon = \frac{\Delta(N\phi)}{\Delta t} = B\frac{\Delta A}{\Delta t} = \frac{BLvt}{t} = BLv$ [1]

23.6

1 $I = \frac{P}{V}$

$\text{current} = \frac{1.0 \times 10^6}{400 \times 10^3} = 2.5\,\text{A}$

2 $PL = I^2R = 2.5^2 \times 500$

power loss ≈ 3.1 kW

3 $\%\text{ of power lost} = \frac{3.1 \times 10^3}{1.0 \times 10^6} \times 100 = 0.31\%$

4 $I = \frac{P}{V}$

$\text{current} = \frac{1.0 \times 10^6}{40 \times 10^3} = 25\,\text{A}$

$P = I^2R = 25^2 \times 500 = 3.1 \times 10^2\,\text{kW}$

$\%\text{ of power lost} = \frac{3.1 \times 10^5}{1.0 \times 10^6} \times 100 = 31\%$ this is a 100 fold increase.

1 The iron core ensures that all the magnetic flux created by the primary coil links the secondary coil. [1]

2 Any transformer where the turn ratio is 20:1. For example, $n_p = 2000$ and $n_s = 100$. [1]

3 Heat losses in the windings (cables) due to the current in them. [1]

Heat losses in the core due to eddy currents. [1]

4 $\frac{n_s}{n_p} = \frac{V_s}{V_p}$

$\frac{n_s}{500} = \frac{5.2}{230}$ [1]

$n_s = 11.3 \approx 11$ turns [1]

5 $\frac{n_s}{n_p} = \frac{V_s}{V_p}$

$20^{-1} = \frac{V_s}{230}$ [1]

$V_s = 11.5\,V \approx 12\,V$ [1]

6 a $\frac{n_s}{n_p} = \frac{V_s}{V_p}$

$\frac{n_s}{1000} = \frac{12}{230}$ [1]

$n_s = 52$ turns [1]

b For a 100 % efficient transfer, the input and output powers are the same. [1]

$I_p = \frac{P}{V} = \frac{60}{230} = 0.26\,A$ [1]

24.1

1 The chance of getting close to the tiny nuclei of atoms is very small. Hence, fewer alpha particles are deflected at large angles.

2 $F = \frac{Qq}{4\pi\varepsilon_0 r^2} = \frac{79 \times 2 \times (1.6 \times 10^{-19})^2}{4\pi \times 8.85 \times 10^{-12} \times (10^{-14})^2}$

$F \approx 360\,N$

3 If true, then $N\sin^4\left(\frac{\theta}{2}\right) =$ constant.

Two pairs of values used from the table to show that $N\sin^4\left(\frac{\theta}{2}\right) \approx$ constant

1 Most of the atom is empty space (vacuum). [1]

2 atom ~ 10^{-10} m and nucleus ~ 10^{-15} m (allow 10^{-14} m). [1]

3 diameter ≈ 8.0 cm × 10^5 [1]

diameter ≈ 8000 m [1]

4 Correct directions of the forces (lines joining centre of particles). [1]

Magnitude of the force is greatest at B – shown by a longer force arrow. [1]

5 a $8.8 \times 10^6 \times 1.6 \times 10^{-19} = \frac{Qq}{4\pi\varepsilon_0 d}$ $(Q = Ze = 82e$ and $q = 2e)$ [2]

$1.408 \times 10^{-12} = \frac{82 \times 2 \times (1.6 \times 10^{-19})^2}{4\pi \times 8.85 \times 10^{-12} \times d}$ [1]

$d = 2.68 \times 10^{-14}\,m \approx 2.7 \times 10^{-14}\,m$ [1]

b $F = \frac{Qq}{4\pi\varepsilon_0 r^2} = \frac{82 \times 2 \times (1.6 \times 10^{-19})^2}{4\pi \times 8.85 \times 10^{-12} \times (2.68 \times 10^{-14})^2}$ [2]

$F = 53\,N$ [1]

6 initial volume = final volume [1]

$\frac{4}{3}\pi \times (0.5 \times 10^{-3})^3 = 10^{-10} \times (\pi \times r^2)$ [1]

$r = 1.29\,m \approx 1.3\,m$ [1]

24.2

1 $R = \frac{0.61\lambda}{\sin\theta}$

$R = \frac{0.61}{\sin\theta} \times \frac{hc}{E} = \frac{0.61hc}{E\sin\theta}$

2 $R = \frac{0.61hc}{E\sin\theta} = \frac{0.61 \times 6.63 \times 10^{-34} \times 3.0 \times 10^8}{(420 \times 10^6 \times 1.6 \times 10^{-19})\sin 44}$

$R = 2.6 \times 10^{-15}\,m$

3 $R = r_0 A^{1/3} = 1.2 \times 10^{-15} \times 16^{1/3}$

$R = 3.0 \times 10^{-15}\,m$

This value and the value of 2.6 fm from Q2 are close enough.

1 A helium nucleus has 2 protons and 2 neutrons. [1]

2 a 2 protons, 4 neutrons, and 2 electrons. [1]

b 3 protons, 6 neutrons, and 3 electrons. [1]

c 26 protons, 30 neutrons, and 26 electrons. [1]

d 92 protons, 143 neutrons, and 92 electrons. [1]

3 a $R = r_0 A^{1/3} = 1.2 \times 6^{1/3} = 2.18 \approx 2.2$ fm

b $R = r_0 A^{1/3} = 1.2 \times 9^{1/3} = 2.496 \approx 2.5$ fm

c $R = r_0 A^{1/3} = 1.2 \times 56^{1/3} = 4.59 \approx 4.6$ fm

d $R = r_0 A^{1/3} = 1.2 \times 235^{1/3} = 7.405 \approx 7.4$ fm

4 volume $= \frac{4}{3}\pi \times (7.405 \times 10^{-15})^3 = 1.70 \times 10^{-42}\,m^{-3}$ [1]

mass ≈ 235u = $235 \times 1.66 \times 10^{-27} = 3.90 \times 10^{-25}$ kg [1]

density $= \frac{\text{mass}}{\text{volume}} = \frac{3.90 \times 10^{-25}}{1.70 \times 10^{-42}}$ [1]

density $= 2.3 \times 10^{17}\,kg\,m^{-3}$ [1]

The density of the uranium nucleus is the same as that of the helium nucleus. [1]

5 density $= \frac{\text{mass}}{\text{volume}} = \frac{4.0 \times 10^{30}}{\frac{4}{3}\pi \times 12000^3}$ [1]

density $= 5.5 \times 10^{17}\,kg\,m^{-3}$ [1]

This density is similar to the density of nuclei. [1]

6 ratio $= \frac{Gm^2}{r^2} \div \frac{e^2}{4\pi\varepsilon_0 r^2} = \frac{Gm^2 4\pi\varepsilon_0}{e^2}$ [1]

ratio $= \{6.67 \times 10^{-11} \times [1.7 \times 10^{-27}]^2 \times 4\pi\varepsilon_0\} \div \{(1.6 \times 10^{-19})^2\}$ [2]

ratio $= 8.4 \times 10^{-37}$ [1]

7 radius = $r_0 A^{1/3}$

volume = $\frac{4}{3}\pi r^3 \propto A$ [1]

mass ≈ $Au \propto A$ [1]

density = $\frac{\text{mass}}{\text{volume}}$; therefore density does not depend on A (assumption correct). [1]

24.3

1 Gravitational force and electromagnetic force have an infinite range. [1]

2 Electromagnetic and strong nuclear force. [1]

3 Hadrons are particles (including antiparticles) that experience the strong nuclear force. [1]

A suitable example given, e.g proton, neutron or meson. [1]

4 symbol: $\overline{\text{p}}$ [1]

mass = 1.7×10^{-27} kg and charge = $-e$. [1]

5 It is a lepton. [1]

mass = $\frac{1.9 \times 10^{-28}}{9.11 \times 10^{-31}} = 210$; mass = $210\,m_e$ [1]

6 charge = 0 and mass ≈ 1.7×10^{-27} kg [1]

It is a hadron. [1]

24.4

1 Positive quarks are: up, charm and top. [1]

2 proton: uud and neutron: udd [2]

3 Baryons and mesons are hadrons. [1]

Baryons have 3 quarks and mesons have one quark and one anti-quark. [1]

4 $\overline{\text{udd}}$ [1]

5 a $Q = +\frac{2}{3} + \frac{1}{3} = +1$ [2]

b $Q = +\frac{2}{3} - \frac{2}{3} = 0$ [2]

6 $Q = +\frac{2}{3} - \frac{2}{3} - \frac{1}{3} - \frac{2}{3} = -1$ [2]

24.5

1 Zero [1]

2 Weak nuclear force [1]

3 a ${}^{1}_{1}\text{p} \rightarrow {}^{1}_{0}\text{n} + {}^{0}_{+1}\text{e} + \nu_e$ [1]

b Nucleon and proton (atomic) numbers are conserved, as is charge. [1]

4 udd → uud + ${}^{0}_{-1}\text{e} + \overline{\nu_e}$ [1]

5 number = $\frac{1}{10^{-6} \times 9.11 \times 10^{-31}}$ [1]

number ≈ 1.1×10^{36} [1]

6 uud + d$\overline{\text{u}}$ → (u$\overline{\text{u}}$)udd [1]

Therefore x is likely to have the composition udd. [1]

25.1

1 Positrons would immediately be annihilated by the large quantity of electrons available in all matter.

1 Alpha radiation: Helium nucleus / charge $+2e$ [1]

Beta radiation: Beta-minus is an electron / beta-plus is a positron [1]

Gamma radiation: wavelength / zero charge [1]

2 Alpha particles, beta particles, and gamma rays. [1]

3 Corrected counts in 2.0 minutes = 250 – 48 = 202 counts

count rate = $\frac{202}{2}$ = 101 counts min^{-1} [1]

count rate = $\frac{202}{2 \times 60} \approx 1.7$ counts s^{-1} [1]

4 $KE = 10^4 \times 25 \times 10$ [2]

$KE = 2.5 \times 10^6$ eV = 2.5 MeV [1]

5 $\frac{1}{2}mv^2 = 2.5 \times 10^6 \times 1.6 \times 10^{-19}$ [1]

$\frac{1}{2} \times 6.6 \times 10^{-27} \times v^2 = 4.0 \times 10^{-13}$ [1]

$v = 1.1 \times 10^7$ m s^{-1} [1]

6 a Correct values of ln(C) calculated [1]

Correct plot of ln(C) against x plotted and a best fit line drawn. [2]

b If exponential decay then $C = C_0 e^{-kx}$

$\ln(C) = \ln(C_0) - kx$ [1]

A graph of ln(C) against x will be a straight line with gradient k. [1]

half-thickness = $\frac{\ln(2)}{-k}$ [1]

$k \approx (-)\ 0.057$ mm^{-1} (allow ± 5%) [1]

half-thickness = $\frac{\ln(2)}{-0.057} \approx 12$ mm [1]

25.2

1 ${}^{29}_{13}$Al has 2 more neutrons than ${}^{27}_{13}$Al.

It is therefore neutron-rich.

2 The beta-plus emitters are ${}^{28}_{15}$P, ${}^{29}_{15}$P and ${}^{30}_{15}$P because they are proton-rich.

The beta-minus emitters are ${}^{32}_{15}$P, ${}^{33}_{15}$P and ${}^{34}_{15}$P because they are neutron-rich.

1 Nucleon and proton (atomic) numbers are both conserved. [1]

2 a Weak nuclear force. [1]

b Beta-minus decay. [1]

c The nitrogen-14 is the daughter; therefore nucleon number = 14. [1]

3 a ${}^{238}_{92}\text{U} \rightarrow {}^{4}_{2}\text{He} + {}^{234}_{90}\text{Th}$ [1]

b ${}^{222}_{86}\text{Rn} \rightarrow {}^{4}_{2}\text{He} + {}^{218}_{84}\text{Po}$ [2]

4 ${}^{13}_{7}\text{N} \rightarrow {}^{13}_{6}\text{C} + {}^{0}_{+1}\text{e} + \nu_e$ [2]

5 Nucleon number will decrease by 4 × 5 = 20 [1]

Proton number will decrease by 2 × 5 = 10 [1]

Therefore lead isotope is ${}^{234-20}_{92-10}\text{Pb} = {}^{214}_{82}\text{Pb}$ [1]

6 After the first beta-minus decay: ${}^{212}_{83}\mathrm{Y}_1$ [1]

After the second beta-minus decay: ${}^{212}_{84}\mathrm{Y}_2$ [1]

After the alpha decay: ${}^{212-4}_{84-2}\mathrm{X}$ [1]

Therefore the lead isotope is ${}^{208}_{82}\mathrm{X}$ [1]

25.3

1 The activity A of a source is the rate at which nuclei decay or disintegrate. [1]

The SI unit of activity is the Bq. [1]

2 number decaying = 4000 × 60 [1]

number decaying = 2.4×10^5 [1]

Assumption: the activity remains constant over the 1.0 minute period. [1]

3 a The time of 100 s if 5 half-lives. [1]

number of nuclei left = $\left(\frac{1}{2}\right)^5 \times 5000 = 156 \approx 160$ [1]

b $A \propto N$ [1]

Therefore the activity will drop to $\frac{1}{32}$ of its initial value. [1]

4 activity = $\frac{100}{2.5} \times 200$ [1]

activity = 8×10^3 Bq [1]

5 $A = \lambda N$

$8.6 \times 10^6 = 2.0 \times 10^{-6} \times N$ [1]

$N = 4.3 \times 10^{12}$ nuclei [1]

6 power = activity × energy of each alpha particle [1]

power = $1.0 \times 10^6 \times (4.6 \times 10^6 \times 1.6 \times 10^{-19})$ [1]

power = 7.36×10^{-7} W ≈ 7.4×10^{-7} W [1]

7 $N = \frac{3.0 \times 10^{-6} \times 10^{-3}}{0.090} \times 6.02 \times 10^{23} = 2.01 \times 10^{16}$ [2]

$A = \lambda N$

$A = 1.1 \times 10^{-9} \times 2.01 \times 10^{16}$ [1]

$A = 2.2 \times 10^7$ Bq = 22 MBq [1]

25.4

1 Points plotted correctly, including the error bars.
Axes labelled and a sensible scale used.
Correct line of best fit drawn.

2 The activity decays exponentially with time and so does the count rate C, therefore $C = C_0 e^{-\lambda t}$.

$\ln(C) = \ln(C_0) - \lambda t$

Therefore a $\ln(C) - t$ graph will be a straight line with a gradient of $-\lambda$.

3 The gradient of the line is – 0.0134. (Allow ± 5.0%)

Therefore $\lambda = 0.0134\ \mathrm{s}^{-1}$

$t_{1/2} = \frac{\ln 2}{\lambda} = \frac{\ln 2}{0.0134} = 52\ \mathrm{s}$

4 Draw a worst fit line through the error bars.
Determine the gradient of this line and hence the extreme value for the half-life.
The absolute uncertainty is the difference between this extreme value and the value obtained in 3.

1 a $\lambda = \frac{\ln 2}{t_{1/2}} = \frac{\ln 2}{0.84}$ [1]

$\lambda = 0.83\ \mathrm{s}^{-1}$ [1]

b $\lambda = \frac{\ln 2}{t_{1/2}} = \frac{\ln 2}{15 \times 3600}$ [1]

$\lambda = 1.28 \times 10^{-5}\ \mathrm{s}^{-1} \approx 1.3 \times 10^{-5}\ \mathrm{s}^{-1}$ [1]

2 $t_{1/2} = \frac{\ln 2}{\lambda} = \frac{\ln 2}{4.9 \times 10^{-18}}$ [1]

$t_{1/2} = 1.41 \times 10^{17}$ s ≈ 1.4×10^{17} s [1]

$t_{1/2} = \frac{1.41 \times 10^{17}}{3.16 \times 10^7} \approx 4.5 \times 10^9$ y [1]

3 $\lambda = \frac{\ln 2}{t_{1/2}} = \frac{\ln 2}{140 \times 24 \times 3600}$ [1]

$\lambda = 5.73 \times 10^{-8}\ \mathrm{s}^{-1}$ [1]

$A = \lambda N = 5.73 \times 10^{-8} \times 8.0 \times 10^{10}$ [1]

$A = 4.58 \times 10^3$ Bq ≈ 4.6×10^3 Bq [1]

4 ratio = $\frac{\text{half-life of nitrogen-16}}{\text{half-life of uranium-237}}$ [1]

ratio = $\frac{7.4}{6.8 \times 24 \times 3600}$ [1]

ratio = $1.26 \times 10^{-5} \approx 1.3 \times 10^{-5}$ [1]

5 a $\lambda = \frac{\ln 2}{t_{1/2}} = \frac{\ln 2}{430 \times 3.16 \times 10^7}$ [1]

$\lambda = 5.11 \times 10^{-11}\ \mathrm{s}^{-1}$ [1]

$A = \lambda N$

$4.8 \times 10^3 = 5.11 \times 10^{-11} \times N$ [1]

$N = 9.4 \times 10^{13}$ [1]

b $A = A_0 e^{-\lambda t} = 4.8 \times 10^3 (0.5)^{\frac{25}{430}}$ [1]

$A = 4.6 \times 10^3$ Bq [1]

6 $N = N_0 e^{-\lambda t}$ or $N = N_0 \times (0.5)^{\frac{t}{t_{1/2}}}$ [1]

Therefore, $0.012 = (0.5)^{\frac{t}{710 \times 10^6}}$ [1]

$\ln(0.012) = \ln(0.5) \times \frac{t}{710 \times 10^6}$ [1]

$t = 4.53 \times 10^9$ y ≈ 4.5 billion years [1]

25.5

1 $\lambda = \frac{\ln 2}{t_{1/2}} = \frac{\ln 2}{1.00} = 0.693\ \mathrm{s}^{-1}$ [1]

2 Use a time period Δt smaller than 0.10 s. [1]

3 In a period of $\Delta t = 0.25$ s the activity cannot be assumed to be constant, so the agreement between the modelling process and the actual values for N would be poor. [1]

4 $t_{1/2} = \frac{\ln 2}{\lambda} = \frac{\ln 2}{3 \times 10^{-2}} = 23.1\ \mathrm{s}$ [1]

Therefore Δt must be much smaller than 23.1 s, e.g. 1.0 s or even 0.01 s. [1]

5 a $t_{1/2} = \frac{\ln 2}{\lambda} = \frac{\ln 2}{50} = 0.013863....\ \mathrm{s}^{-1}$ [1]

$\Delta N = (0.013863.... \times 1.0) \times N = 0.013863...N$ [1]

Modelling used to predict N = 657.8 or 658 [1]

b $N = N_0e^{-\lambda t}$ or $N = N_0 \times (0.5)^{\frac{t}{t_{1/2}}}$ [1]

$N = 1000\ (0.5)^{(30/50)} = 659.8$ or 660 [1]

The % difference between this value and the value in (a) is about 0.3 %. [1]

25.6

1 All living things take in carbon from atmospheric carbon dioxide. [1]

2 Carbon-14 nuclei are formed in the upper atmosphere of the Earth when neutrons (produced from collisions of cosmic particles) collide with nitrogen-14 nuclei. [1]

See reaction: ${}^{1}_{0}n + {}^{14}_{7}N \rightarrow {}^{14}_{6}C + {}^{1}_{1}p$ [1]

3 a 1 g of carbon produces 15 counts per minute or $\frac{15}{60} = 0.25\,Bq$ [1]

Therefore, 1 kg will give an activity of $0.25 \times 1000 = 250\,Bq$ [1]

b The count rate from a small sample is comparable to the background count rate. [1]

4 $A = A_0e^{-\lambda t}$ or $A = A_0(0.5)^{\frac{t}{t_{1/2}}}$ [1]

$A = 1.5 \times (0.5)^{2000/5700}$ [1]

$A = 1.176\,Bq \approx 1.2\,Bq$ [1]

5 $N = N_0e^{-\lambda t}$ or $N = N_0 \times (0.5)^{\frac{t}{t_{1/2}}}$ [1]

Therefore, $0.69 = (0.5)^{\frac{t}{5700}}$ [1]

$\ln(0.69) = \ln(0.5) \times \frac{t}{5700}$ [1]

$t = 3.051 \times 10^3\,y \approx 3.1 \times 10^3$ years [1]

6 $N = N_0e^{-\lambda t}$ or $N = N_0 \times (0.5)^{\frac{t}{t_{1/2}}}$ [1]

Therefore, $0.9944 = (0.5)^{\frac{t}{49 \times 10^9}}$ (% of rubidium left = 99.44 %) [1]

$\ln(0.9944) = \ln(0.5) \times \frac{t}{49 \times 10^9}$ [1]

$t \approx 4.0 \times 10^8$ years (400 million years) [1]

26.1

1 a The person has greater energy because of increased *KE*, so the mass of the person would be greater. [1]

b The burning wood loses chemical energy so its mass will decrease. [1]

c The *KE* of the electron will decrease; hence its mass will also decrease. [1]

2 a $\Delta E = \Delta mc^2 = 1.7 \times 10^{-27} \times (3.0 \times 10^8)^2$ [1]

$\Delta E = 1.53 \times 10^{-10}\,J \approx 1.5 \times 10^{-10}\,J$ [1]

b $\Delta E = \Delta mc^2 = 1.0 \times (3.0 \times 10^8)^2$ [1]

$\Delta E = 9.0 \times 10^{16}\,J$ [1]

3 $\Delta E = \Delta mc^2 = 9.6 \times 10^{-30} \times (3.0 \times 10^8)^2$ [1]

$\Delta E = 8.64 \times 10^{-13}\,J \approx 8.6 \times 10^{-13}\,J$ [1]

4 $\Delta E = 1.0 \times 10^3 \times 1.6 \times 10^{-19} = 1.6 \times 10^{-16}\,J$ ($1\,eV = 1.6 \times 10^{-19}\,J$) [1]

$\Delta m = \frac{\Delta E}{c^2} = \frac{1.6 \times 10^{-16}}{(3.0 \times 10^8)^2}$ [1]

$\Delta m = 1.78 \times 10^{-33}\,kg \approx 1.8 \times 10^{-33}\,kg$ [1]

5 $\Delta E = 1.0\,MeV = 1.0 \times 10^6 \times 1.6 \times 10^{-19} = 1.6 \times 10^{-13}\,J$ [1]

$\Delta m = \frac{\Delta E}{c^2} = \frac{1.6 \times 10^{-13}}{(3.0 \times 10^8)^2}$ [1]

$\Delta m = 1.78 \times 10^{-30}\,kg$ [1]

$\text{increase} = \frac{1.78 \times 10^{-30}}{9.1 \times 10^{-31}} = 1.956$

The increase in the mass is about twice the rest mass, so the electron has a mass three times greater than at rest. [1]

6 $\Delta m = (3.7187 \times 10^{-25} + 6.625 \times 10^{-27}) - 3.7853 \times 10^{-25} = -3.50 \times 10^{-29}\,kg$ [2]

$\Delta E = \Delta mc^2 = 3.50 \times 10^{-29} \times (3.0 \times 10^8)^2$ [1]

$\Delta E = 3.15 \times 10^{-12}\,J \approx 3.2 \times 10^{-12}\,J$ [1]

26.2

1 mass defect is measured in kg and binding energy in J. [1]

2 binding energy of nucleus = mass defect of nucleus × c^2 [1]

3 mass = $0.002368 \times 1.66 \times 10^{-27}\,kg$ [1]

binding energy = mass defect × $c^2 = 0.002368 \times 1.66 \times 10^{-27} \times (3.00 \times 10^8)^2$ [1]

binding energy = $3.54 \times 10^{-13}\,J \approx 3.5 \times 10^{-13}\,J$ [1]

4 BE per nucleon = $\frac{7.8 \times 10^{-11}}{56} = 1.393 \times 10^{-12}\,J$ per nucleon [1]

BE per nucleon = $\frac{1.393 \times 10^{-12}}{1.60 \times 10^{-19}}$ [1]

BE per nucleon = 8.7 MeV per nucleon [1]

5 a binding energy = 7.1×4 [1]

binding energy ≈ 28 MeV [1]

b binding energy = 8.0×16 [1]

binding energy ≈ 130 MeV [1]

c binding energy = 7.5×238 [1]

binding energy ≈ 1800 MeV [1]

6 mass defect = $(8 \times 1.67 \times 10^{-27}) - 1.33 \times 10^{-26} = 6.00 \times 10^{-29}\,kg$ [1]

binding energy per nucleon = $\frac{6.00 \times 10^{-29} \times (3.0 \times 10^8)^2}{8}$ [1]

binding energy per nucleon = $6.75 \times 10^{-13} \approx 6.8 \times 10^{-13}\,J$ per nucleon [1]

binding energy per nucleon = $\frac{6.75 \times 10^{-13}}{1.60 \times 10^{-19}} \approx$ 4.2 MeV per nucleon [1]

26.3

1 Boron-10 does not lead to fission reactions and it has a high probability of just capturing (absorbing) neutrons.

Boron-10 has a larger cross section for capture than cadmium-48, which makes it suitable for use in control rods.

[Note, too, that cadmium is very toxic, unlike boron.]

2 The protons in ordinary water are $\frac{0.67}{1.3\times10^{-3}} \approx 520$ times more likely to capture the thermal neutrons within the reactor than deuterium nuclei. These neutrons are wanted for the reaction; the moderator is intended to slow down fast neutrons.

3 Thermal neutrons are $\frac{590}{1.9} \approx 310$ times more likely to cause fission of uranium-235 than fast neutrons.

4 $0.1 \times 1.6 \times 10^{-19} = \frac{1}{2} \times 1.7 \times 10^{-27} \times v^2$ (v = rms speed)

$v = 4.3\ \text{km s}^{-1}$

1 A moderator slows down fast neutrons produced in fission reactions. [1]

Nuclei within control rods absorb neutrons inside a reactor. [1]

2 Neutrons: kinetic energy = 170 × 0.03 = 5.1 MeV [1]

Daughter nuclei: kinetic energy = 170 × 0.85 = 145 MeV [1]

3 energy released = difference in the BE [1]

energy released = (8.5 – 7.5) × 239 [1]

energy released = 239 MeV ≈ 240 MeV [1]

4 number of uranium-235 nuclei = $\frac{1.0}{0.235} \times 6.02 \times 10^{23}$ [1]

energy released in joules = $170 \times 10^6 \times 1.6 \times 10^{-19}$ [1]

total energy released = $\frac{1.0}{0.235} \times 6.02 \times 10^{23} \times 170 \times 10^6 \times 1.6 \times 10^{-19}$ [1]

total energy released = 7.0×10^{13} J [1]

5 $0.1 = 1.0 \times 10^6 \times (0.72)^n$ [1]

$1.0 \times 10^7 = (0.72)^n$ [1]

$n = \frac{\log(10^7)}{\log(0.72)}$ [1]

n ≈ 49 collisions [1]

26.4

1 mean $E_k = \frac{3}{2}kT = \frac{3}{2} \times 1.38 \times 10^{-23} \times 1.5 \times 10^8$

mean $E_k = 3.11 \times 10^{-15} \approx 3.1 \times 10^{-15}$ J

2 It takes two nuclei, each with mean E_k of 3.1×10^{-15} J, to produce a single fusion reaction. Therefore mean energy required for fusion is about 6.2×10^{-15} J.

1 nucleon number before = 1 + 1 = 2; nucleon number after = 2 [1]

proton number before = 1 + 1 = 2; proton number after = 1 + 1 = 2 [1]

2 At low temperatures the nuclei will be travelling too slowly to be able to get close enough for the strong nuclear force to bring about fusion. [2]

3 $\Delta E = \Delta mc^2 = 10^9 \times (3.0 \times 10^8)^2$ (in one second) [1]

rate of energy production = 9.0×10^{25} W [1]

4 energy from each reaction = 5.5 MeV = $5.5 \times 10^6 \times 1.6 \times 10^{-19}$ J [1]

$\Delta E = \Delta mc^2$

$\Delta m = \frac{5.5\times10^6 \times 1.6\times10^{-19}}{(3.0\times10^8)^2}$ [1]

$\Delta m = 9.8 \times 10^{-30}$ kg [1]

5 a $^2_1H + ^2_1H \rightarrow ^4_2He$ [2]

b binding energy per nucleon of $^2_1H \approx 1.1$ MeV and binding energy per nucleon of $^4_2He \approx 7.1$ MeV (read from graph) [1]

change in BE = energy released = (4 × 7.1) – (4 × 1.1) = 24 MeV [1]

convert to joules: energy released = $24 \times 10^6 \times 1.6 \times 10^{-19}$ [1]

energy released = 3.8×10^{-12} J ($\approx 4 \times 10^{-12}$ J) [1]

c Total number of nuclei = $\frac{1.0}{0.002} \times 6.02 \times 10^{23} = 3.0 \times 10^{26}$ [1]

Total number of helium-4 pairs = 1.5×10^{26} [1]

energy = $1.5 \times 10^{26} \times 3.8 \times 10^{-12}$ [1]

energy = 5.7×10^{14} J [1]

d time = $\frac{5.7\times10^{14}}{500\times10^6\times2} = 5.7 \times 10^5$ s (≈ 6.6 days) [1]

27.1

1 $\frac{hc}{\lambda} = eV$; $V = \frac{hc}{\lambda e}$

The minimum wavelength = 3.5×10^{-11} m from Figure 6 in the main content pages. (Allow ± 5%)

$V = \frac{hc}{e\lambda} = \frac{6.63\times10^{-34} \times 3.0\times10^8}{1.6\times10^{-19} \times 3.5\times10^{-11}}$

$V \approx 36$ kV

2 The wavelength of the K-line = 7.5×10^{-11} m from Figure 6 in the main content pages. (Allow ± 5%)

$\Delta E = \frac{hc}{\lambda} = \frac{6.63\times10^{-34} \times 3.0\times10^8}{7.5\times10^{-11}}$

$\Delta E = 2.652 \times 10^{-15}$ J $\approx 2.7 \times 10^{-15}$ J

3 The kinetic energy of the electron at the anode will increase.

The maximum energy of the X-ray photon will also increase.

Therefore the shortest wavelength will decrease.

The wavelengths of the K-lines will be unaffected because they depend on the target metal (and not the p.d. used)

1 Any value in the range 10^{-8} m to 10^{-13} m. [1]

2 a $f = \frac{c}{\lambda} = \frac{3.0 \times 10^8}{\lambda}$ [1]

f in the range 3×10^{16} Hz to 3.0×10^{21} Hz [1]

[or same calculation for other acceptable value from q1]

b $E = \frac{hc}{\lambda} = \frac{6.63 \times 10^{-34} \times 3.0 \times 10^8}{\lambda}$ [1]

E in the range 2.0×10^{-17} J to 2.0×10^{-12} J [1]

3 a energy = $eV = 65 \times 10^3 \times 1.6 \times 10^{-19}$ [1]

energy = 1.04×10^{-14} J $\approx 1.0 \times 10^{-14}$ J [1]

b energy of photon = energy of electron [1]

energy = 1.04×10^{-14} J $\approx 1.0 \times 10^{-14}$ J [1]

4 number of electrons per second = $\frac{21 \times 10^{-3}}{1.6 \times 10^{-19}}$ [1]

number of electrons per second = $1.31 \times 10^{17}\,\text{s}^{-1}$ $\approx 1.3 \times 10^{17}\,\text{s}^{-1}$ [1]

5 number of photons per second = $0.60 \times 10^{-2} \times 1.31 \times 10^{17}$ [1]

number of photons per second = $7.9 \times 10^{14}\,\text{s}^{-1}$ [1]

6 maximum energy of X-ray photon = maximum kinetic energy of an electron [1]

$\frac{hc}{\lambda} = eV; V = \frac{hc}{\lambda e}$ [1]

$\lambda = \frac{6.63 \times 10^{-34} \times 3.0 \times 10^8}{1.6 \times 10^{-19} \times 100 \times 10^3}$ [1]

$\lambda = 1.2 \times 10^{-11}$ m [1]

27.2

1 Simple scatter, photoelectric effect, and Compton scattering. [1]

2 X-ray machines use supply voltages 30 – 100 kV. This means that X-ray photons have energy greater than 30 keV. [1]

Simple scatter is dominant for photons with energy in the range 1 to 20 keV. [1]

3 $21\,\text{m}^{-1}$ [1]

4 $\frac{I}{I_0} = e^{-\mu x} = e^{-(0.21 \times 0.80)} = 0.8454$ [2]

% transmitted intensity = 85 % [1]

5 energy of X-ray photon = $1.02 \times 10^6 \times 1.6 \times 10^{-19}$ J [1]

$E = \frac{hc}{\lambda}$

wavelength $\lambda = \frac{6.63 \times 10^{-34} \times 3.0 \times 10^8}{1.02 \times 10^6 \times 10^{-19}}$ [1]

$\lambda = 1.2 \times 10^{-12}$ m [1]

6 $\frac{I}{I_0} = e^{-\mu x} = e^{-0.21x} = 0.5$ [1]

$\ln(0.5) = -0.21x$ [1]

$x = \frac{\ln(0.5)}{-0.21} \approx 3.3$ cm [1]

27.3

1 Gantry which has a rotating X-ray tube and X-ray detectors, movable table, computer and display. [2]

2 Advantage: Produces three-dimensional image or can image soft tissues. [1]

Disadvantage: Ionising or can take longer. [1]

3 A thin beam is used so that a thin section of the patient's body can be scanned. [1]

4 A slice means a two-dimensional cross-sectional image through the patient. [1]

5 A contrast material would have been injected into the blood vessel of the patient before the CAT scan. [1]

The contrast medium used would be iodine based. [1]

27.4

1 To produce a single photoelectron when a single photon of visible light hits the photocathode.

2 Each dynode produces 4 secondary electrons, therefore total number of electrons = 4^{10}.

Total number of electrons = $1.049 \times 10^6 \approx 10^6$

3 charge = $10^6 \times 1.6 \times 10^{-19}$

charge $\approx 1.6 \times 10^{-13}$ C

1 Technetium-99m and fluorine-18. [1]

2 It has a short half-life (of 6.0 h) and hence cannot be stored. [1]

3 To change a single photon of visible light into an electrical pulse. [1]

4 Gamma scans can be used to diagnose the function of the body rather than just imaging flesh and bone. [1]

5 Technetium-99 has a very long half-life (of 210 000 y). [1]

Its activity would be extremely small compared with the time taken for the scan. [1]

6 $\lambda = \frac{\ln 2}{t_{1/2}} = \frac{\ln 2}{6.0 \times 3600} = 3.209 \times 10^{-5}\,\text{s}^{-1}$ [1]

$A = \lambda N$;

$500 \times 10^6 = 3.209 \times 10^{-5} \times N$ [1]

$N = 1.56 \times 10^{13} \approx 1.6 \times 10^{13}$ [1]

27.5

1 FDG or fluorodeoxyglucose. [1]

2 High-speed protons collide with oxygen-18 nuclei to produce fluorine-18 and neutrons. [1]

3 A gamma detector consists of a photomultiplier tube and its own scintillator. [1]

A single gamma photon incident at the detector will produce a voltage pulse. [1]

4 time difference = $\frac{0.0500}{3.0 \times 10^8}$ [1]

time difference = 1.7×10^{-10} s [1]

This is a very small time difference, so the computer must have the capability to analyse and manipulate the signals from the detectors in much shorter times. [1]

5 [Half-life = 110 minutes (text)]

$A = A_0(0.5)^n$; $n = \frac{20}{110} = 0.1818...$ [1]

fraction of activity left = $(0.5)^{0.1818...} = 0.88$ [1]

Therefore percentage drop in activity = 100 – 88 = 12 % [1]

27.6

1 Ultrasound is sound with frequency greater than 20 kHz. [1]

2 Any frequency in the range 1 to 15 MHz. [1]

3 a $v = f\lambda$

$340 = 10 \times 10^6 \times \lambda$ [1]

$\lambda = 3.4 \times 10^{-5}$ m [1]

b wavelength ∝ speed [1]

The wavelength of the ultrasound will be greater (than 3.4×10^{-5} m). [1]

4 An A-scan is a one-dimensional scan and produces no image. A B-scan is a collection of A-scans and produces a two-image on a screen. [1]

5 The time between successive pulses is $\frac{1}{5000} = 2.0 \times 10^{-4}$ s. [1]

In this interval of time, total distance travelled by pulse = $2.0 \times 10^{-4} \times 1600 = 0.32$ m, so the 'depth' that can be imaged is 0.16 m. This is reasonable for most scans. [2]

At this frequency, the transducer can detect reflections before the next pulse is sent into the body. [1]

27.7

1 acoustic impedance = density of substance × speed of ultrasound in substance [1]

2 The reflection at a boundary is large when the acoustic impedances of the substances are very different. [1]

3 $Z = \rho c$

$2.9 \times 10^7 = 5600 \times c$ [1]

$c = 5200\,\text{m s}^{-1}$ [1]

4 $\frac{I_r}{I_0} = \left(\frac{Z_2 - Z_1}{Z_2 + Z_1}\right)^2 = \left(\frac{1.38 - 1.69}{1.38 + 1.69}\right)^2$ [1]

ratio = 0.010 or 1 % [1]

5 a $\frac{I_r}{I_0} = \left(\frac{Z_2 - Z_1}{Z_2 + Z_1}\right)^2 = \left(\frac{0.000442 - 1.70}{0.000442 + 1.70}\right)^2$ [1]

ratio = 0.99896 or 99.9 % [1]

b $\frac{I_r}{I_0} = \left(\frac{Z_2 - Z_1}{Z_2 + Z_1}\right)^2 = \left(\frac{1.65 - 1.70}{1.65 + 1.70}\right)^2$ [1]

ratio = 0.00022 or 0.022 % [1]

27.8

1 The reflected of the ultrasound from the iron-rich blood cells is responsible for the change in the ultrasound frequency. [1]

2 The Doppler shift in frequency depends on $\cos\theta$, and $\theta = 90$ would give no change in frequency. [2]

3 Explanation: $\Delta f \propto v$ [1]

Therefore $v = \frac{700}{500} \times 12 = 16.8\,\text{cm s}^{-1} \approx 17\,\text{cm s}^{-1}$ [2]

4 $v = \frac{c\Delta f}{2f\cos\theta} = \frac{1600 \times 900}{2 \times 7.0 \times 10^6 \times \cos 60}$ [1]

$v = 0.2057....\,\text{m s}^{-1}$ [1]

volume per second = $\pi \times (0.75 \times 10^{-3})^2 \times 0.2057$ [1]

volume per second = $3.6 \times 10^{-7}\,\text{m}^3\,\text{s}^{-1}$ [1]

Index

Acknowledgements

COVER: Bizroug / Shutterstock; **p2-3**: Sarahbean/Shutterstock; **p6-7**: Martyn F. Chillmaid/Science Photo Library; **p8**: Nasa/Vrs/Science Photo Library; **p10**: Tr3gin/Shutterstock; **p12**: Aleksandar Todorovic/ Shutterstock; **p14**: Piotr Wawrzyniuk/Shutterstock; **p16**: Roger Bamber/ Alamy; **p18**: US Coast Guard Photo/Alamy; **p20-21**: OlegDoroshin/ Shutterstock; **p22**: Paul White - Transport Infrastructures/Alamy; **p24**: Bikeriderlondon/Shutterstock; **p27** (T): Villiers Steyn/Shutterstock; **p27** (B): Christoff/Shutterstock; **p29**: Pavel Polkovnikov/Shutterstock; **p31**: Denis Tabler/Shutterstock; **p32**: Peteri/Shutterstock; **p35** (T): Bob Mawby/Shutterstock; **p35** (B): SP-Photo/Shutterstock; **p37**: Ria Novosti/ Science Photo Library; **p38**: Kenneth Eward/Biografx/Science Photo Library; **p40**: Dick Kenny/Shutterstock; **p46**: Andrew Brookes/National Physical Laboratory/Science Photo Library; **p47**: NASA; **p49** (T): Heidi Coppock-Beard/Getty Images; **p49** (B): Don Hammond/Design Pics/Alamy; **p51**: Greg Epperson/Shutterstock; **p52**: Greg Epperson/Shutterstock; **p54** (T): Robert L Kothenbeutel/Shutterstock; **p54** (B): Justin Kase z12z/ Alamy; **p57**: Awe Inspiring Images/Shutterstock; **p60**: Lassedesignen/ Shutterstock; **p62**: Monkey Business Images/Shutterstock; **p65**: Lawrence Livermore National Laboratory/ University Of California/ Science Photo Library; **p66**: Rob Hyrons/Shutterstock; **p67**: Alexis Rosenfeld/Science Photo Library; **p69** (T): Mevans/iStockphoto; **p69** (B): GARY DOAK/Alamy; **p72**: Baloncici/iStockphoto; **p74** (T): Mar.K/ Shutterstock; **p74** (B): Jesus Keller/Shutterstock; **p75**: John Hanley/ Shutterstock; **p76**: Tom Hirtreiter/Shutterstock; **p77**: Germanskydiver/ Shutterstock; **p79**: Pearl Bucknall/Alamy; **p80**: Bikeriderlondon/ Shutterstock; **p82**: David Parker/Science Photo Library; **p84** (T): Danlsaunders/iStockphoto; **p84** (B): Charistoone-images/Alamy; **p86**: Peter Marshall/Alamy; **p88**: Irina Papoyan/Shutterstock; **p90**: Melis/ Shutterstock; **p92**: Hellen Grig/Shutterstock; **p96**: Nitinut380/ Shutterstock; **p100**: NASA/Science Photo Library; **p101**: Iliuta Goean/ Shutterstock; **p102**: NASA/Science Photo Library; **p103**: Ken Schulze/ Shutterstock; **p104**: RobertCrum/iStockphoto; **p105**: fStop Images - Caspar Benson/Getty Images; **p107**: Loren Winters, Visuals Unlimited/ Science Photo Library; **p109**: Science Photo Library; **p115** (T): Weber/ Shutterstock; **p115** (B): Sunny Forest/Shutterstock; **p120-121**: Matteis/ Look at Sciences/Science Photo Library; **p122**: Jhaz Photography/ Shutterstock; **p126**: Sergey Nivens/Shutterstock; **p127**: Mihai Simonia/ Shutterstock; **p128**: Jon Le-Bon/Shutterstock; **p129**: CERN/Science Photo Library; **p131** (T): Nobeastsofierce/Shutterstock; **p131** (B): Oktay Ortakcioglu/iStockphoto; **p134**: Iceninephoto/iStockphoto; **p138**: Science Photo Library; **p139**: Trevor Clifford Photography/Science Photo Library; **p140**: Peter Menzel/Science Photo Library; **p142** (T): Dino Osmic/Shutterstock; **p142** (B): Martyn F. Chillmaid/Science Photo Library; **p143**: Gavran333/Shutterstock; **p145**: Francisco Javier Gil/ Shutterstock; **p148**: Power and Syred/Science Photo Library; **p151** (T): Alexandru Nika/Shutterstock; **p151** (B): Science Source/Science Photo Library; **p153**: Idea for life/Shutterstock; **p156**: David Parker/Science Photo Library; **p157** (T): Spfotocz/Shutterstock; **p157** (B): Martyn F. Chillmaid/Science Photo Library; **p160**: Martyn F. Chillmaid/Science Photo Library; **p161**: NASA/JPL-Caltech/Science Photo Library; **p162**: Prill/Shutterstock; **p163**: StockPhotosArt/Shutterstock; **p165** (T): Pi-Lens/ Shutterstock; **p165** (B): Martyn F. Chillmaid/Science Photo Library; **p166**: Martyn F. Chillmaid/Science Photo Library; **p170**: Jim Corwin/ Science Photo Library; **p177**: S Corvaja/European Space Agency/Science Photo Library; **p181**: Public Health England/Science Photo Library; **p186**: Paulthepunk/iStockphoto; **p190**: Trevor Clifford Photography/ Science Photo Library; **p194** (T): Xieyouding/iStockphoto; **p194** (B): Jag_cz/Shutterstock; **p197**: Furtseff/Shutterstock; **p205**: Philippe Plailly/ Science Photo Library; **p206**: Adrian Davies/Alamy; **p207**: M-gucci/ iStockphoto; **p210**: American Spirit/Shutterstock; **p217**: 123dartist/ Shutterstock; **p218**: Asharkyu/Shutterstock; **p224** (T): Mr Twister/ Shutterstock; **p224** (B): Berenice Abbott/Science Photo Library; **p227**: TFoxFoto/Shutterstock; **p230**: Edward Kinsman/Science Photo Library; **p231**: Russell Shively/Shutterstock; **p234**: Ad_doward/iStockphoto; **p235**: Andrew Lambert Photography/Science Photo Library; **p237**: Aastock/Shutterstock; **p242**: PR Michel Zanca/ISM/Science Photo Library; **p247**: European Space Agency/P. Carril/Science Photo Library; **p251**: PobladuraFCG/iStockphoto; **p254**: Evan Oto/Science Photo Library; **p255**: Andrew Lambert Photography/Science Photo Library; **p257**: Science Photo Library; **p261**: General-fmv/Shutterstock; **p266-267**: BeholdingEye/iStockphoto; **p268**: Photo by Mark McLinden; **p271**: Niyazz/Shutterstock; **p272**: Magnetix/Shutterstock; **p274**: Ria Novosti/ Science Photo Library; **p275**: Charles D. Winters/Science Photo Library; **p276**: Scottchan/Shutterstock; **p280**: Cameron Whitman/Shutterstock; **p286**: Andrew Brookes, National Physical Laboratory/Science Photo Library; **p289**: Armin Rose/Shutterstock; **p296**: Image of taken from weather balloon. Dr Warren Houghton and Sixth Form Students, Exeter School; **p302**: QQ7/Shutterstock; **p305**: Gregdx/Shutterstock; **p309**: Muzsy/Shutterstock; **p310**: Suthep/Shutterstock; **p311**: Dmitry Yashkin/ Shutterstock; **p316**: Keneva Photography/Shutterstock; **p320**: S.Borisov/ Shutterstock; **p324**: Tom Wang/Shutterstock; **p327**: Kochneva Tetyana/ Shutterstock; **p330**: Library of Congress/Science Photo Library; **p331** (T): Allison Herreid/Shutterstock; **p331** (B): Philip Bird LRPS CPAGB/ Shutterstock; **p336**: European Space Agency,P. Carril/Science Photo Library; **p338**: Geological Survey of Canada/Science Photo Library; **p339**: Cristapper/Shutterstock; **p342**: NASA/Science Photo Library; **p345**: Alex Cherney, Terrastro.Com/Science Photo Library; **p348**: Detlev Van Ravenswaay/Science Photo Library; **p351**: Action Sports Photography/ Shutterstock; **p354**: Jason and Bonnie Grower/Shutterstock; **p382**: NASA/ESA/STSCI/R. Ellis (Caltech), and The UDF 2012 Team/Science Photo Library; **p385**: Michael Krinke/iStockphoto; **p386**: Ivsanmas/ Shutterstock; **p388**: MarcelClemens/Shutterstock; **p390**: NASA/Science Photo Library; **p391**: European Space Agency, the Planck Collaboration/ Science Photo Library; **p393**: WMAP Science Team/NASA; **p394**: PlanilAstro/Shutterstock; **p360**: Reinhold Wittich/Shutterstock; **p362**: European Space Agency/Rosetta/Osiris Team/Science Photo Library; **p361**: MarcelClemens/Shutterstock; **p363**: Yuriy Kulik/Shutterstock; **p364**: Physics Today Collection/American Institute of Physics/Science Photo Library; **p366**: Creativemarc/Shutterstock; **p367**: Robert Gendler/ Science Photo Library; **p369**: Sam Aronov/Shutterstock; **p371** (T): Xfox01/Shutterstock; **p373**: Anton Gvozdikov/Shutterstock; **p375**: GIPhotoStock/Science Photo Library; **p376**: Stocksnapper/Shutterstock; **p399**: Samuel Burt/iStockphoto; **p404-405**: Zmeel/iStockphoto; **p406** (T): MikhailSh/Shutterstock; **p406** (B): Yurazaga/Shutterstock; **p412**: OJO_Images/iStockphoto; **p415**: MarcelClemens/Shutterstock; **p420**: NanD_Phanuwat/Shutterstock; **p422**: David Hay Jones/Science Photo Library; **p423**: Randy Montoya/Sandia National Laboratories/Science Photo Library; **p426**: Ross Ellet/Shutterstock; **p428**: Philippe Plailly/ Science Photo Library; **p436**: Fermilab/Science Photo Library; **p446**: Koya979/Shutterstock; **p448**: Underworld/Shutterstock; **p452**: Pi-Lens/ Shutterstock; **p456**: Martyn F. Chillmaid/Science Photo Library; **p457**: Teun van den Dries/Shutterstock; **p460**: Royal Institution of Great Britain / Science Photo Library; **p464**: Vichie81/Shutterstock; **p477**: James King-Holmes/Science Photo Library; **p480**: CERN/Science Photo Library; **p484** (L): Science Photo Library; **p484** (R): Lawrence Berkeley Laboratory/Science Photo Library; **p487**: Ted Kinsman/Science Photo Library; **p491**: Joop Hoek/Shutterstock; **p492**: NASA/Science Photo Library; **p494**: U.S. Dept. Of Energy/Science Photo Library; **p496**: MarcelClemens/Shutterstock; **p497**: Author; **p499**: Mikhail Zahranichny/ Shutterstock; **p500**: Natural History Museum, London/Science Photo Library; **p504** (T): Chrisdorney/Shutterstock; **p504** (B): Florin Stana/ Shutterstock; **p506** (T): Goronwy Tudor Jones, University Of Birmingham/Science Photo Library; **p506** (B): James King-Holmes/Science Photo Library; **p510**: Emilio Segre Visual Archives/American Institute Of Physics/Science Photo Library; **p515**: Gajus/Shutterstock; **p516**: David Parker/Science Photo Library; **p520** (L): Science Photo Library; **p520** (R): James King-Holmes/Science Photo Library; **p521**: ChameleonsEye/ Shutterstock; **p523**: Living Art Enterprises/Science Photo Library; **p525** (T): Science Photo Library; **p525** (B): DU Cane Medical Imaging Ltd/ Science Photo Library; **p526** (T): Zephyr/Science Photo Library; **p526** (B): Ezz Mika Elya/Shutterstock; **p527**: Zephyr/Science Photo Library; **p528**: Stevie Grand/Science Photo Library; **p529**: CNRI/Science Photo Library; **p531**: Ria Novosti/Science Photo Library; **p532**: Dr Robert Friedland/ Science Photo Library; **p533**: Bart78/Shutterstock; **p538**: Sovereign, ISM/ Science Photo Library; **p539** (R): Olesia Bilkei/Shutterstock; **p539** (L): Dr Najeeb Layyous/Science Photo Library; **p545**: Scanrail/iStockphoto; **p550**: Deltev Van Ravenswaay/Science Photo Library; **p556** (T): Mino Surkala/Shutterstock; **p556** (B): Tim Roberts Photography/Shutterstock; **p557**: Gecko753/iStockphoto; **p560** (T): Dotshock/Shutterstock; **p560** (B): Charistoone-stock/Alamy

Artwork by Q2A media

With thanks to Dr Houghton, Mr Schramm, and Sixth Form pupils from Exeter School for the photo on page 296, taken from a weather balloon.